W9-BGQ-860

Studies in
Natural Products Chemistry

Volume 23
Bioactive Natural Products (Part D)

Studies in Natural Products Chemistry
edited by Atta-ur-Rahman

Vol. 1 Stereoselective Synthesis (Part A)
Vol. 2 Structure Elucidation (Part A)
Vol. 3 Stereoselective Synthesis (Part B)
Vol. 4 Stereoselective Synthesis (Part C)
Vol. 5 Structure Elucidation (Part B)
Vol. 6 Stereoselective Synthesis (Part D)
Vol. 7 Structure and Chemistry (Part A)
Vol. 8 Stereoselective Synthesis (Part E)
Vol. 9 Structure and Chemistry (Part B)
Vol. 10 Stereoselective Synthesis (Part F)
Vol. 11 Stereoselective Synthesis (Part G)
Vol. 12 Stereoselective Synthesis (Part H)
Vol. 13 Bioactive Natural Products (Part A)
Vol. 14 Stereoselective Synthesis (Part I)
Vol. 15 Structure and Chemistry (Part C)
Vol. 16 Stereoselective Synthesis (Part J)
Vol. 17 Structure and Chemistry (Part D)
Vol. 18 Stereoselective Synthesis (Part K)
Vol. 19 Structure and Chemistry (Part E)
Vol. 20 Structure and Chemistry (Part F)
Vol. 21 Bioactive Natural Products (Part B)
Vol. 22 Bioactive Natural Products (Part C)
Vol. 23 Bioactive Natural Products (Part D)

Foreword

Volume 23 of Studies in Natural Products Chemistry represents the fourth volume of this series devoted to the field of bioactive natural products. This volume contains a large number of interesting articles on natural product isolated from terrestrial and marine sources with comprehensive discussions on bioactivity aspects. The articles, which are written by eminent authorities in their respective fields, should trigger further research in developing new pharmacophores against various diseases.

I would like to express my thanks to Mr. Muhammad Yaqoob for assistance in the preparation of the index. I am also grateful to Mr. Waseem Ahmed for typing and to Mr. Mahmood Alam for secretarial assistance.

It is hoped that this volume will be of wide interest to phytochemists, medicinal chemists and pharmacologists working in universities and industry.

Atta-ur-Rahman
Ph.D. (Cantab), Sc.D. (Cantab)

February, 2000

Preface

The 23rd volume of *Studies in Natural Products Chemistry*, edited by Atta-ur-Rahman, adds another attractive example to the traditional series, not to be missed in any library of a scientist interested in aspects of purification, structure-activity relationships, biosynthesis and pharmaceutical application of natural products. The first volume of *Studies in Natural Products Chemistry* was published in 1988, and since then, in steady continuation, two volumes per year were offered to the scientific community due to Atta-ur-Rahman's efficient and talented editorship. In the past, topics like stereoselective synthesis, structure elucidation and structure and synthesis were reviewed. The last three volumes, including the present one, are devoted to bioactive natural products, following the actual trend in the field. The availability of batteries of specific *in vitro* assays allow the screening of numerous bioactivities of natural compounds, efficiently uncovering new pharmacophores, structure-activity relationships, new leads for rational drug designs or new natural product drugs.

The bioactive secondary metabolites from several marine sources, sponges and tunicates (Kobayashi/Ishibashi), sea plumes (Katz/Adamczeski) and ascidians (Bowdem) cover a broad range of different chemical structures, terpenoids, unsaturated fatty acids, steroids, alkaloids, cyclic peptides and polysaccharides.
In three chapters, bioactive constituents are collected based on their isolation from different plant families: Teuricum (Ulubelen/Topçu/Sönmez), Dipterocarpaceae (Seo/Kinghorn) and Zingiberaceae (Pancharoen/Prawat/Tuntiwachwuttikul). The genus Teuricum is a rich source of neo-clerodane diterpenoids. Besides terpenoids, cumarins, flavonoids, phenolics and quinones, resveratrol oligomers (oligostilbenoids) are the most characteristic compound class identified from Dipterocarpaceae and have been associated with antibacterial, antiviral and cytotoxic activities. The members of the Zingiberaceae family produce a wide range of different chemical structures: terpenoids, diarylheptanoids, arylalkanoids, phenylpropanoids and -butanoids, cyclohexane oxides and flavonoids. Ethnomedical uses and characteristic bioactive compounds of Celatraceae species are surveyed in the chapter of González/Bazzocchi/Moujir/Jiménez: Polyester sesquinterpenes and pyridine-sesquiterpene alkaloids with insect antifeedant, sesquiterpene alkaloids with immunosupressive and antitumoral or diterpene triepoxides with potent antileukemic and immunosuppresive activities.
The antifungal properties of sampangines (natural as well as synthetic compounds) and cumarins, widely distributed in the plant kingdom, are surveyed by Orabi and Sardari/Nishibe/Daneshtalab, respectively. Remarkable are the common antitumor, antiviral and antiparasitic activities of chemically very different plant constituents like ansamycin antibiotics (Funayama/Cordell), glucans (Kogan), quassinoids (Okana/Fukamiya/Lee), ellagitannins (Yoshida/Hatano/Ito/Okuda) and organosulfur compounds of garlic (Kamel/Saleh). However, the precise mechanism of action of the different classes of compounds often has still to be elucidated. As the title "Chemistry of Some Natural Products of Biological Interest" suggests, Mahato presents a potpourri of structure, chemistry, NMR, biosynthetic and some bioactivity data on a whole series of plant constituents, diterpenes (cleomeolide, verticillol), triterpenes (mimusopic and mimusopsicacids), saponins (acaciasides, bacosides, bacopasaponins, corchorusins), macrocyclic spermidine alkaloids, flavonoid- and phenylpropanoid glycosides. Most recent aspects of the biochemical mechanisms involved in the teratogenic effects caused by Veratrum alkaloids (e.g. jervine or cyclopamine), are reported by Gaffield and one of the last chapters gives a survey on antioxidant phenolic metabolites from fruit and vegetables (Tomás-Barberán/Ferreres/Gil) which might be beneficial agents for the prevention of cardiovascular diseases and cancer.

The present volume of *Studies in Natural Products Chemistry* is, like its predecessors, a fascinating collection reflecting the current trend in the field. The editor, Prof. Atta-ur-Rahman, is to be congratulated on once more having attracted prominent contributions offering the reader comprehensive information on well-seleced topics covering exciting views on the bioactivity of natural products. The scientific community is eagerly looking foreward to future volumes of this traditional and fascinating series.

Wolfgang Voelter

CONTENTS

CONTRIBUTORS

Madeline Adamczeski	Department of Chemistry, College of Arts and Sciences, American University, 4400 Massachusetts Avenue, NW Washington, DC 20016-8014, USA
F.A. Tomas-Barberan	Department of Food Science and Technology, CEBAS (CSIC), P.O. Box 4195, Murcia 30080, Spain
I.L. Bazzochi	Institute Universitario de Bio-Organica Antonio Gonzalez, Astrofisico F. Sanchez, 2, 38206, La Laguna Tenerife, Canary Islands, Spain
Bruce F. Bowden	School of Biomedical and Molecular Sciences, James Cook University, Townsville 4811, Qld., Australia
Geoffrey A. Cordell	Professor and Head, Department of Medicinal Chemistry and Pharmacognosy (M/C 781), College of Pharmacy, 833 S. Wood St., Chicago, Illinois, 60612-7231, USA
M. Daneshtlab	Faculty of Pharmacy and Pharmaceutical Sciences, University Alberta, Edmonton, AB, Canada T6G 2N8
F. Ferreres	Department of Food Science and Technology, CEBAS (CSIC), P.O. Box 4195, Murcia 30080, Spain
N. Fukamiya	Faculty of Integrated Arts and Sciences, Hiroshima University, 1-7-1 Kaagamiyama, Higashi-Hiroshima 739-8521, Japan
Shinji Funayama	Associate Professor of Natural Product Chemistry, Department of Bioscience and Biotechnology, Faculty of Engineering, Aomori University, 2-3-1 Kohbata, Aomori 030-0943, Japan
William Gaffield	Western Regional Research Centre, U.S. Department of Agriculture, Albany, CA 94710, U.S.A.
M.I. Gil	Department of Food Science and Technology, CEBAS (CSIC), P.O. Box 4195, Murcia 30080, Spain
Antonio González González	Institute Universitario de Bio-Organica "Antonio Gonzalez", Universidad de La Laguna Avda. Astrofisico F^{co}. Sanchez, 2, 38206, La Laguna, Tenerife, Islas Canarias, Spain

Tsutomu Hatano — Professor of Pharmacognosy, Faculty of Pharmaceutical Sciences, Okayama University, Tsushima, Okayama 700-8530, Japan

Masami Ishibashi — Faculty of Pharmaceutical Scienes, Chiba University, Chiba 263-8522, Japan

Hideyuki Ito — Professor of Pharmacognosy, Faculty of Pharmaceutical Sciences, Okayama University, Tsushima, Okayama 700-8530, Japan

I.A. Jimenez — Institute Universitario de Bio-Organica Antonio Gonzalez, Astrofisico F. Sanchez, 2, 38206, La Laguna, Tenerife, Canary Island, Spain

Alae Kamel — Department of Chemistry, Texas Southern University, 3100 Cleburne Avenue Houston, Texas 77004, U.S.A.

E.Katz — Department of Chemistry, College of Arts and Sciences, American University, 4400 Massachusetts Avenue, NW Washington, DC 20016-8014, USA

A. Douglas Kinghorn — Professor of Pharmacognosy, Assistant Department Head, Senior University Scholar, Department of Medicinal Chemistry and Pharmacognosy (MC 781), College of Pharmacy, 833 South Wood Street, Chicago, Illinois 60612-7231, USA

Jun'ichi Kobayashi — Faculty of Pharmaceutical Sciences, Hokkaido University, Sapporo 060-0812, Japan

Grigorij Kogan — Institute of Chemistry, Slovak Academy of Sciences, 84238, Bratislava, Slovakia

Eun-Kyoung Seo — Seo Chemistry and Life Sciences Group, Research Triangle Institute, P.O. Box 12194, Research Triangle Park, North Carolina 27709, U.S.A.

K.H. Lee — School of Pharmacy, The University of North Carolina, Chapel Hill, NC 27599-7360, U.S.A.

S.B. Mahato — Indian Institute of Chemical Biology, Calcutta-700032, India

L.Moujir — Departamento de Microbiología y Biología Celular, Universidad de La Laguna, 38206, La Laguna, Spain

S. Nishibe — Faculty of Pharmaceutical Sciences, Health Sciences, University of Hokkaido, Ishikari-Tobetsu, Hokkaido, 061-02, Japan

Masayoshi Okano   Faculty of Integrated Arts and Sciences, Hiroshima University, 1-7-1 Kaagamiyama, Higashi-Hiroshima 739-8521, Japan

Takuo Okuda   Professor of Pharmacognosy, Faculty of Pharmaceutical Sciences, Okayama University, Tsushima, Okayama 700-8530, Japan

Khaled Y. Orabi   Department of Pharmacognosy, College of Pharmacy, King Saud University, P.O. Box 2457, Riyadh 11451, Saudi Arabia

Orasa Pancharoen   Department of Chemistry, Silpakorn University, Nakorn Pathom 73000, Thailand

Uma Prawat   Department of Chemistry, Rajabhat Institute Phuket, Phuket 83000, Thailand

M. Saleh   Department of Chemistry, Texas Southern University, 3100 Cleburne Avenue Houston, Texas 77004, U.S.A.

Soroush Sardari   Faculty of Pharmacy and Pharmaceutical Sciences, University Alberta, Edmonton, AB, Canada T6G 2N8

Ufuk Sönmez   Faculty of Pharmacy, University of Istanbul, 34452, Istanbul, Turkey

Gulaçti Topçu   TUBITAK, Marmara Research Center, P.O. Box 21, 41470, Gebze, Kocadi, Turkey

Pittaya Tuntiwachwuttikul   Department of Chemistry, Silpakorn University, Nakorn Pathom 73000, Thailand

Ayhan Ulubelen   Faculty of Pharmacy, University of Istanbul, 34452, Istanbul, Turkey

Takashi Yoshida   Professor of Pharmacognosy, Faculty of Pharmaceutical Sciences, Okayama University, Tsushima, Okayama 700-8530, Japan

Bioactive Natural Products

Atta-ur-Rahman (Ed.) *Studies in Natural Products Chemistry, Vol. 23*

MICROBIAL MODELS OF MAMMALIAN METABOLISM. SAMPANGINES

KHALED Y. ORABI

Department of Pharmacognosy, College of Pharmacy, King Saud University, P.O. Box 2457, Riyadh 11451, Saudi Arabia

ABSTRACT: Sampangine is an antifungal drug candidate with a remarkable *in vitro* antifungal activity, however, it lacks *in vivo* activity. In pursuit of overcoming this obstacle, the metabolic profile of sampangine and two analogs, benzosampangine and 3-methoxysampangine were evaluated using microorganisms and mammals. On the other hand, to facilitate the mammalian metabolism study, *in vitro* microbial models were developed to predict potential mammalian metabolites of sampangine.

Microbial metabolism studies on sampangine have resulted in the isolation and characterization of two metabolites, sampangine-4'-O-methyl-β-glucopyranoside (SAMM1) and sampangine-β-glucopyranoside (SAMM2).

Mammalian metabolism studies on sampangine gave two water-soluble conjugates. One of those conjugates, SAM MM1, has been identified as sampangine-β-glucuronic acid.

On the other hand, Benzosampangine has been metabolized by microorganisms into benzosampangine-β-glucopyranoside, in addition to two unidentified metabolites.

Screening of microorganisms for their ability to metabolize 3-methoxysampangine resulted in the production of more polar conjugates, 3-MeOSAMM1 and 3-MeOSAMM2. 3-MeOSAMM1 is thought to be the demethylated derivative, 3-hydroxysampangine, since it was produced by a known demethylator (*Streptomyces spectabilis*), and gave blue color with ferric chloride spray reagent. No other data are available for the other metabolite.

In vitro antifungal activities evaluation for sampangine metabolites demonstrated comparable activities to sampangine. However, an *in vivo* efficacy study revealed that SAMM2 is inactive. The antifungal activity profile of sampangine metabolites led us to conclude that metabolism *per se* is not the cause for the *in vivo* inactivity of sampangine, hence, further studies on the other pharmacokinetic parameters (bioavailability, distribution, clearance, etc.) will be necessary.

INTRODUCTION

By the end of 1997, the World Health Organization announced that about thirty million cases of acquired immunodeficiency syndrome (AIDS) have been reported worldwide with approximately an equal number of unreported cases. These numbers are escalating fast. Due to high mortality and morbidity rates of AIDS, special attention has been and still being paid towards the discovery and development of therapeutic agents to either cure the cause or alleviate the symptoms of this syndrome.

The characteristic breakdown in the immune system that is associated with AIDS is often manifested in the form of serious opportunistic infections (OI). Since the course of immunosuppression in AIDS patients cannot be currently halted or reversed, the only recourse is treatment of the OI in these individuals, even though the infection is a reflection of a more complicated underlying immune disorder.

Unfortunately, treatment of opportunistic infections in AIDS patients poses many problems including; long duration of therapy, more frequent and severe drug toxicities, the relative ineffectiveness of the available systemic antifungal agents, and the development of resistance to the existing agents. These significant shortcomings have intensified the search for new, more effective, and less toxic prototype antifungal agents.

SAMPANGINES

The above-mentioned serious drawbacks have intensified the search for new, more effective and less toxic prototype antifungal agents. Natural products have, in the past, provided such prototype bioactive compounds and it is logical that new antifungal agents could be developed from this source.

3-Methoxysampangine

Sampangines entered the antifungal candidate arena in 1990, when a group of researchers at the University of Mississippi [1] reported the isolation, structural elucidation and *in vitro* antifungal activity evaluation of 3-methoxysampangine from the root bark of the west African tree, *Cleistopholis patens* (Benth) Engl. and Diels (Annonaceae), where they isolated 3-methoxysampangine together with other three known alkaloids using bioassay-directed fractionation. In this study, they reported the minimum inhibitory concentrations (MIC) of 3-methoxysampangine, 3.12 µg/ml for *C. albicans* and *Aspergillus fumigatus* (amphotericin B; 0.78 µg/ml and 1.56 µg/ml, receptively) and 0.2 µg/ml for *C. neoformans* (amphotericin B; 3.12 µg/ml). Due to the scarcity of 3-methoxysampangine (0.000156% yield), many attempts were made to synthesize it. In 1992, Peterson *et al.*, successfully reported [2] the total synthesis, structural elucidation and antimycotic/antimycobacterial activity evaluation of 3-methoxysampangine, and other derivatives. The above procedure, patterned after Bracher's reported synthesis of sampangine [3], afforded 3-methoxysampangine in only 0.17% yield. Later in 1993, Zjawiony *et al.*, used another procedure to give 40% yield [4].

Sampangine

In 1986 Rao *et al.*, reported [5] the isolation and structural elucidation of a new copyrine alkaloid, sampangine (0.001% yield), from the chloroform extract of the stem bark of *Cananga odorata* Hook, F. and Thomas. (Annonaceae). However, they did not report biological activity of this alkaloid. In 1989, Franz Bracher reported the total synthesis of cleistopholine and sampangine [3]. Sampangine was produced in a 26% yield, and this was a significant breakthrough in sampangine chemistry. Chemically, sampangine is 7H-Naphtho[1, 2, 3-*ij*][2, 7]naphthyridin-7-one. Following its re-synthesis by Peterson *et al.* in 1992 [2], sampangine was also shown to possess antifungal activity against both *C. albicans* (MIC= 1.56 μg/ml) and *C. neoformans* (MIC= 0.78 μg/ml) (amphotericin B; 0.78 μg/ml and 3.12 μg/ml, respectively). However, subsequent studies to evaluate its *in vivo* efficacy, sampangine was found to be virtually inactive *in vivo* in a mouse model of cryptococcosis. Although sampangine shows some reduction of organism burden in the candidiasis recovery model, such reductions are not statistically significant[1]. In light of its remarkable *in vitro* activity, one explanation for this observation could be poor pharmacokinetics.

Benzosampangine

A synthetic analog of sampangine (it has not been reported to occur in nature), benzosampangine was synthesized by Peterson *et al.* [2] in 1992.

Benzosampangine was shown to possess good *in vitro* antifungal activity against *C. albicans* and *Aspergillus fumigatus* with MIC=0.39 μg/ml (amphotericin B; 0.78 μg/ml and 0.39 μg/ml, respectively), and *C. neoformans* with MIC= 1.56 μg/ml (amphotericin B; 0.39 μg/ml). *In vivo* evaluation of benzosampangine in a murine cryptococcosis model, showed it to be capable of significantly reducing the brain tissue burden of *C. neoformans* and significantly improving survival times, as compared to untreated infected control animals. On the other hand, benzosampangine is inactive *in vivo* in the candidiasis model. It is worthwhile to note that the level of activity of benzosampangine certainly does not compare to that of amphotericin B, however, given that no information regarding the pharmacokinetics of sampangines is known, the ability to demonstrate any level of *in vivo* efficacy was considered encouraging.

---

[1] Clark, A.M.; Hufford, C.D.; Peterson, J.R.; Zjawiony, J.K.; Liu, S.; Walker, L.A. In Abstract, *The 31st Interscience Conference on Antimicrobial Agents and Chemotherapy*, Chicago, IL, Sep. 29-Oct. 2, 1991.

Other Derivatives

More than thirty sampangine derivatives have now been synthesized by different groups of researchers at The University of Mississippi. The derivatives can be categorized according to the position of substitution. The chemical structures of these analogs are presented in Fig. (1).

A-ring analogs - include 3-hydroxysampangine, 3-methoxysampangine, 3-methylsampangine, 3-bromosampangine, 3-fluorosampangine, and 3-cyanosampangine.

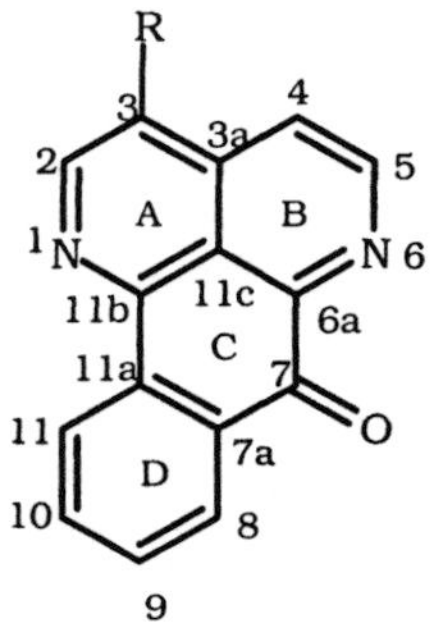

R = H; Samapangine
R = OCH$_3$; 3-Methoxysampangine
R = CH$_3$; 3-Methylsampangine
R = Br; 3-Bromosampangine
R = OH; 3-Hydroxysampangine
R = F; 3-Fluorosampangine
R = CN; 3-Cyanosampangine

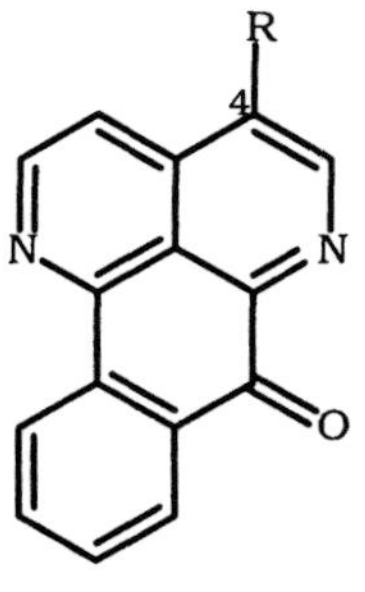

R = OCH$_3$; 4-Methoxysampangine
R = CH$_3$; 4-Methylsampangine
R = Br; 4-Bromosampangine
R = OH; 4-Hydroxysampangine
R = F; 4-Fluorosampangine
R = CN; 4-Cyanosampangine
R = N$_3$; 4-Azidosampangine
R = NH$_2$; 4-Aminosampangine
R = Cl; 4-Chlorosampangine
R = NHPh; 4-Anilinosampangine
R = CH$_2$Br; 4-Bromomethylsampangine

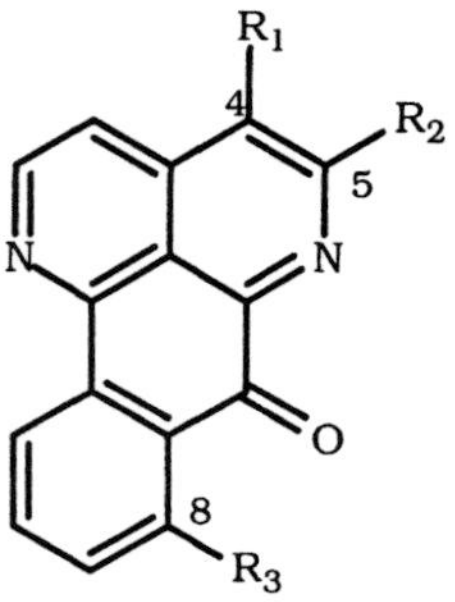

R$_1$ = CH$_3$, R$_2$ = H, R$_3$ = OCH$_3$; 4-Methyl-8-methoxysampangine
R$_1$ = Br, R$_2$ = OCH$_2$CH$_3$, R$_3$ = H; 4-Bromo-5-ethoxysampangine
R$_1$ = R$_2$ = H, R$_3$ = OCH$_3$, 8-Methoxysampangine

Fig. (1). Sampangine and its analogs.

Fig. (1). contd.....

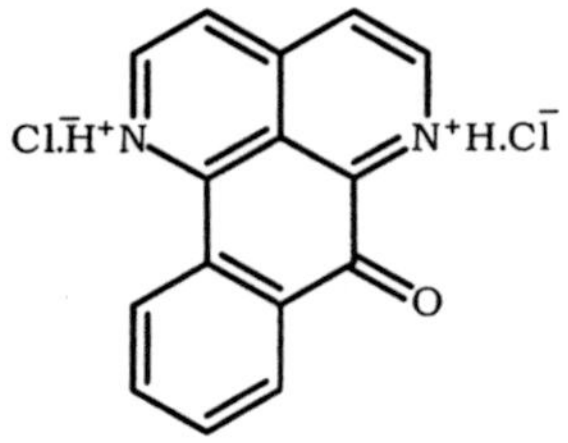

Sampangine-1,6-dihydroxhloride salt

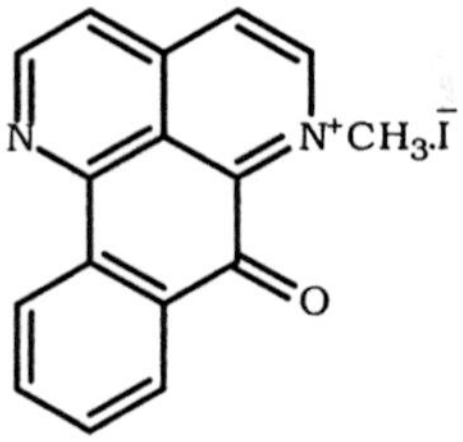

Sampangine-6-methiodide

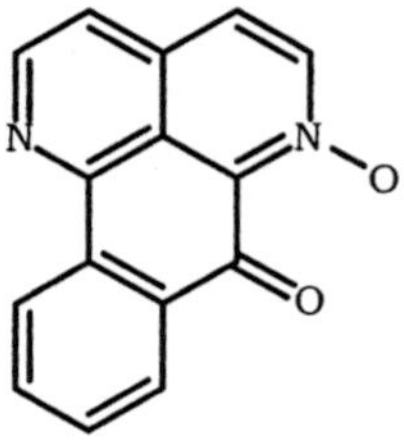

Sampangine-6-oxide

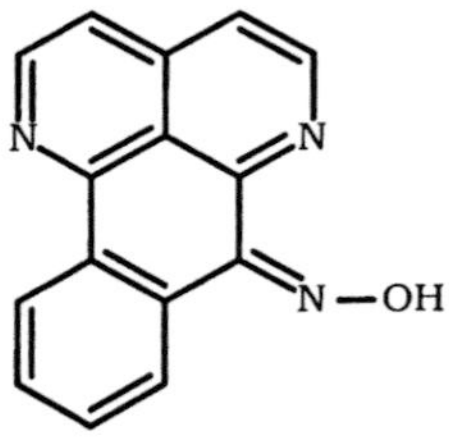

Samapngine-6-oxime

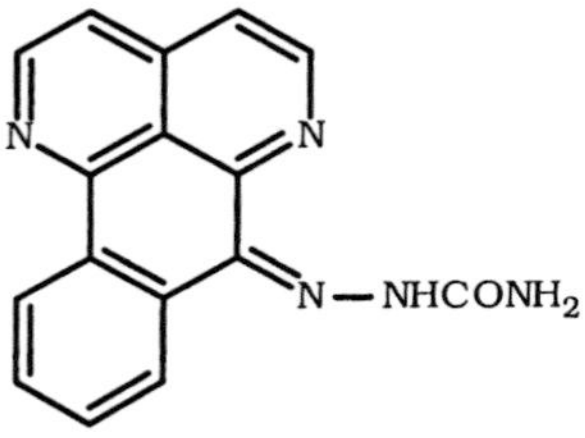

Sampangine-7-semicarbazone

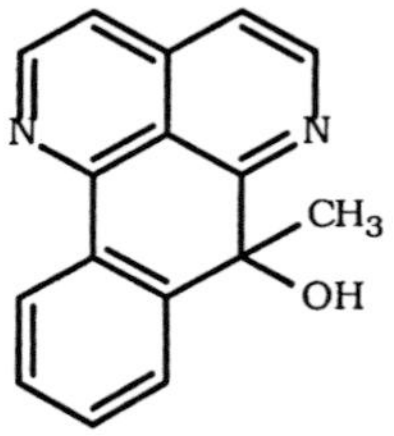

7-Hydroxy-7-methyl-7-deoxosampangine

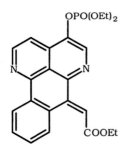

Wittig reaction product

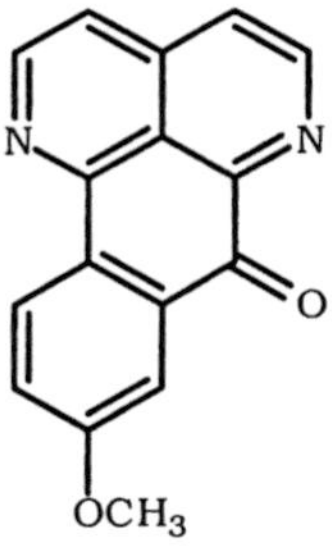

9-Methoxysampangine

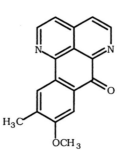

9,10-Dimethylsampangine

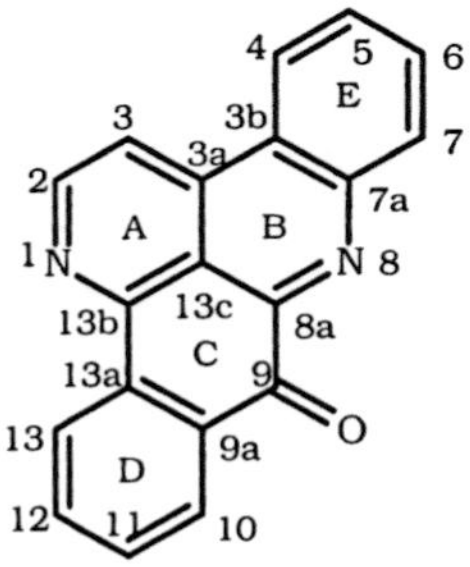

Benzosampangine

B-ring analogs - include 4-bromosapmpangine, 4-methoxysampagine, 4-azidosampangine, 4-methylsampangine, 4-chlorosampangine, 4-anilinosampangine, 4-aminosampangine, 4-bromomethylsampangine, 4-bromo-5-ethoxysampangine, sampangine-6-oxide, 4-methyl-8-methoxysampangine, sampangine-6-methiodide, 4-hydroxysampangine, 4-fluorosampangine, 4-cyanosampangine, sampangine-1,6-dihydrochloride salt

C-ring analogs - include sampangine-7-oxime, sampangine-7-semicarbazone, 7-hydroxy-7-methyl-7-deoxosampangine, and a Wittig reaction product.

D-ring analogs - include 8-methoxysampangine, 9-methoxysampangine, 8-methoxy-4-methylsampangine, 9,10-dimethylsampangine.

Table 1 summarizes the antifungal/antimycobacterial activities of sampangine analogs.

Evaluation of *in vitro* antifungal activities of these analogs revealed the following:

1)   Activity against *Cryptococcus neoformans and Candida albicans* was significantly affected by the position and type of substituent, with MIC values ranging from 0.1 µg/ml to 100 µg/ml.

2)   It seems apparent that alteration of the carbonyl group generally results in loss of activity. The only C-Ring modified analogs that retain good *in vitro* activity are sampangine-7-oxime, and 7-semicarbazone (as will be seen later, the microbial and mammalian metabolites of sampangine are also active) in which the sp^2 character of position 7 is retained.

3)   Sampangine (and presumably the sampangine class of antibiotics) acts by a mechanism of action different from that of amphotericin B, based on the fact that no difference in activity is observed in the presence or absence of ergosterol (amphotericin B will have a reduced activity in the presence of ergosterol) [1].

4)   Substitution at position 4 will reduce the anticandidal activity, while some 4-substituted analogs retain activity against *C. neoformans*.

The planar geometries of sampangines suggested a possible mode of activity may be through an interaction with DNA via intercalation of the planar chromophore between adjacent DNA base pairs, exhibiting a DNA binding affinity of 3×10^3 M^{-1} (which is relatively weak) and site exclusion size of one drug bound per ten base pairs at saturation[2]. Interestingly, 3-methylsampangine demonstrates an association constant approximately four times that of the parent sampangine. Since the methyl group is not

[1]Clark, A.M.; Zhu, X.; Walker, L.A.; Zjawiony, J.K.; Mukherjee, A.; Hufford, C.D. In Abstract, *The 33rd Interscience Conference on Antimicrobial Agents and Chemotherapy*, New Orleans, LA, Oct. 17-20, 1993.

[2]Peterson, J.R.; Zjawiony, J.K.; Clark, A.M.; Hufford, C.D.; Graves, D.E.; Walker, L.A. In Abstract, *The 31st Interscience Conference on Antimicrobial Agents and Chemotherapy*, Chicago, IL, Sep. 29-Oct. 2, 1991.

involved in a direct interaction with DNA, it apparently facilitates a favorable structural geometry to the drug-DNA complex. Based on the relatively low DNA binding constant and potent biological activity, the mechanism(s) by which biological activity is achieved is probably through a protein-mediated mechanism.

Table 1. *In vitro* Antifungal and Antimycobacterial Activities of Sampangines

Compound	Organism [a]						Ref.
	Ca[b]		Cn[c]		Mi[d]		
	MIC	MFC	MIC	MFC	MIC	MBC	
3-methoxysampangine	1.56	25	0.2	3.12	1.56	25	2, 4
3-methylsampangine	0.39	50	0.2	6.25	0.39	6.25	2
3-bromosampangine	12.5	50	0.2	12.5	0.78	0.78	6
sampangine	6.25	6.25	0.2	0.2	0.2	0.2	3
3-fluorosampangine	12.5	NT	0.2	NT	1.56	NT	4
3-cyanosampangine	50	NT	12.5	NT	25	NT	4
4- bromosampangine	12.5	50	0.2	0.2	3.12	NT	2
4- methoxysampangine	3.12	100	12.5	100	3.12	NT	2
4-azidosampangine	25	100	6.25	100	NT	NT	2
4-aminosampangine	NT	NT	6.25	NT	NT	NT	2
4-chlorosampangine	50	50	6.25	50	3.12	NT	2
4-anilinosampangine	NT	NT	12.5	NT	NT	NT	6
sampangine-1,6-dihydrochloride salt	NT	NT	0.39	NT	NT	NT	6
4-methylsampangine	12.5	100	0.2	0.2	3.12	NT	6
4-bromomethylsampangine	NT	NT	0.39	3.12	3.12	NT	6
sampangine-6-methiodide	25	50	0.2	0.2	12.5	NT	6
sampangine-6-oxide	25	50	0.2	0.2	12.5	NT	6
4-methyl-8-methoxysampangine	12.5	25	0.78	3.12	0.78	NT	6
4-hydroxysampangine	50	NT	50	NT	50	NT	4
4-fluorosampangine	50	NT	0.78	NT	6.25	NT	4
4-cyanosampangine	50	NT	6.25	NT	6.25	NT	4
benzosampangine	0.39	25	0.2	100	0.78	NT	2
sampangine-7-oxime	50	100	3.12	3.12	12.5	50	6
sampangine-7-semicarbazone	50	100	3.12	6.25	NT	NT	6
7-hydroxy-7-methyl-7-deoxosampangine	NT	NT	1.56	50	NT	NT	6
8-methoxysampangine	25	50	3.12	6.25	25	25	6

[a]Activity expressed as minimum inhibitory concentration (MIC), minimum fungicidal concentration (MFC), and minimum bacteriocidal concentration (MBC) (μg/ml) using two-fold serial broth macrodilution technique.
[b]Ca = *Candida albicans* NIH B311 in yeast nitrogen broth.
[c]Cn = *Cryptococcus neoformans* ATCC 52675 in Mycophil broth.
[d]Mi = *Mycobacterium intracellulare* ATCC 23068 in Middlebrook broth.
NT = not tested.

In general, the sampangines are very active against the yeasts *Cryptococcus neoformans* and *Candida albicans*. Additionally, a number of the sampangines also exhibit strong *in vitro* activity against the filamentous fungus *Aspergillus*. It is also noteworthy that virtually all of the sampangines are also quite active *in vitro* against *Mycobacterium*, but are only marginally to moderately active against the gram-positive bacterium, *S. aureus*, and generally inactive against the gram-negative bacteria, *E. coli* and *P. aeruginosa*.

3-methylsampangine appears to be the most active of the analogs. 4-substituted sampangines, in general, appear to be less active relative to 3-substituted sampangines (Table 1).

The *in vivo* activities of selected sampangines in murine models of candidiasis and crytococcosis have also been investigated. In addition to benzosampangine, 4-methoxysampangine appears to be effective in reducing the organism burden in the candidiasis recovery models. Sampangine, while reducing somewhat the tissue burden, this reduction is not reproducibly statistically significant.

As explained previously, sampangine is inactive *in vivo* and in light of its remarkable *in vitro* activity, one explanation for this observation could be poor pharmacokinetics; thus the problem might be solved by developing good understanding of its pharmacokinetics. Metabolism is only one, but extremely important, parameter of pharmacokinetics. Therefore, by careful study of the metabolism of sampangine it may be possible to shed light on the failure of sampangine to be effective *in vivo*.

METABOLISM STUDIES

Although xenobiotic metabolism has evolved in the last decades to become a mature science, the history of xenobiotic metabolism is much older than just the past few decades. Xenobiotic metabolism was born in 1841 with the discovery of the biotransformation of benzoic acid into hippuric acid, and the major pathways were discovered in the 19th century [7]. The first half of the 20th century saw an accumulation of data on many xenobiotics, recognized the role of enzymes, and bathed in the belief that metabolism means detoxification. The state of knowledge in the middle of the 20th century was summarized in a most comprehensive and authoritative manner by R.T. Williams [8]. The second half of the 20th century has seen the maturation of xenobiotic metabolism. From the many examples available, it became possible to deduce generalizations and rules and to use them as guides for future advances [9, 10]. In recent years, the contributions of molecular biology and computational chemistry have been particularly impressive.

An explanation for the current and still growing interest in xenobiotics metabolism resides in the central position it has acquired in the investigation of drugs and xenobiotics for both fundamental and applied research.

In drug research and development, metabolism is of pivotal importance due to the interconnectedness between pharmacokinetic and pharmacodynamic processes. Very early in the testing of a newly isolated or synthesized promising compound metabolic studies must be initiated to identify the metabolites, the pathways by which they are formed and the possible intermediates. Based on these findings, the metabolites can be synthesized and tested for their own pharmacological and toxicological effects.

In preclinical and early clinical studies, much pharmacokinetic data must be obtained and relevant criteria must be satisfied before a drug candidate can enter large-scale clinical trials [11]. As a result of these demands, the metabolic and pharmacokinetic registration files of a new drug may well be larger than the pharmacological and toxicological registration files. Moreover, in xenobiotic (drug) research and development, the major aim is the marketing of safe and useful products, with metabolic studies being central in the safety evaluation of new chemicals [12].

The phenomenon of metabolism originates from the fact that mammals usually treat any exogenous drug as non-self and work hard to weaken the effect and/or the concentration of this "foreign" substance. The biological defenses against this chemical invasion are the physiological pathways of molecular elimination via the liver and kidneys. Mammals have developed a first-line defensive arsenal, consisting of an extensive ensemble of enzymes which oxidize or reduce, hydrolyze or conjugate these foreign xenobiotic molecules. The essential design principle of these enzymes is non-selectivity: they recognize common, general, chemical features of organic molecules, not the molecules themselves. This system of xenobiotic metabolism, which is adapted to molecular invasion, is as vital to mammalian survival as the immunological apparatus which has been adapted to deal with invasion by organisms or antigenic molecules. At the molecular level, they share the same design principles of non-selectivity.

Built-in non-discriminatory activity has, however, got a down side: the immunological system sometimes sees self as foreign and becomes self destructive and the xenobiotic metabolic system sometimes generates products more reactive and biologically destructive than the original substrate.

Events involved in the metabolism of xenobiotics can include the phenomena of absorption, distribution, and excretion [13]. The study of the disposition - or fate - of xenobiotics in living systems includes the consideration of their absorption into the organism, how and where they are distributed and stored, the chemical and biochemical transformations they may undergo, and how and by what route(s) they are finally excreted and returned to the environment. For purposes of this article, "metabolism" will be used synonymously with the more strictly defined term "biotransformation".

Thus, the behavior of xenobiotics may be characterized as "What the compound does to the body" and "What the body does to the compound". In pharmacology one speaks of "pharmacodynamic effects" to indicate what a drug does to the body, and "pharmacokinetic effects" to indicate what the body does to the drug.

Pharmacokinetic effects will have a decisive influence on the intensity and duration of pharmacodynamic effects, and metabolism will generate new chemical entities (metabolites) which may have distinct pharmacodynamic properties of their own. Conversely, by its own pharmacodynamic effects, a compound may affect the state of the organism (e.g., hemodynamic changes, enzyme activities, etc.) and hence its capacity to handle xenobiotics.

To understand the concept of the metabolism, an elaboration of the "elimination" process (i.e., biotransformation and excretion) is necessary. Molecules circulating in the blood may undergo one of three distinct fates, which will contribute to the elimination of a given xenobiotic compound: excretion unchanged, non-enzymatic chemical transformation, and enzymatic metabolism.

In quantitative terms, enzymatic metabolism greatly predominates over non-enzymatic transformation while the fraction excreted unchanged decreases with increasing lipophilicity of the xenobiotics. The metabolism of drugs and other xenobiotics is typically a biphasic process in which the compound first undergoes functionalization reaction (phase I reaction) of oxidation, reduction or hydrolysis, which introduces or unveils within its structure a functional group (e.g., -OH, -NH$_2$) suitable for linkage with an endogenous moiety in the second step of conjugation (phase II reaction). Although biotransformations occur in several sites, the liver is the main site of biotransformation of drugs by either phase I and/or phase II reactions. In some cases, compounds possessing a suitable functionality do undergo conjugation directly, while phase I metabolites may be excreted prior to conjugation. Also well documented are the reactions of functionalization which a number of conjugates undergo prior to their excretion.

Model Systems for Drug Metabolism Studies

Traditionally, drug metabolism studies have relied on the use of model systems to predict metabolic pathways in human. For this purpose, either *in vivo* whole animal system (utilizing small laboratory animal models) or *in vitro* enzyme systems (microsomal preparations, tissue cultures or perfused organ systems) have been widely employed [14, 15]. In biomedical research, animal systems continue to serve an important role as models for biological responses in man.

The Hopkins Center for Alternatives for Animal Testing estimates that approximately 17 to 22 million animals are used annually in U.S. laboratories and that testing of a new chemical will cost approximately $ 1.5 million [16]. Other statistical data have estimated that at least 70 million animals are used annually in the U.S. alone for toxicity testing as well as biomedical and behavioral research [17]. Worldwide, this figure may be as high as 250 million animals [18].

When whole animal systems are to be used to conduct the *in vivo* studies, the biological fluids and tissues of these animals are then examined for the presence of the parent drug and its metabolites. Such metabolites are usually present in microquantities and, as a result, are often very difficult to isolate and identify chemically. Although direct structural information is often obtainable (though mass spectral techniques) on the mere microgram quantities of metabolites usually available from mammalian metabolic systems, there is no substitute for having large quantities for complete structure elucidation and biological evaluation. Organic synthetic methods do provide an obvious means for providing metabolites of relatively simple drugs, but the preparation of metabolites of more complex molecules may require vast expenditures in time and effort.

Due to the strong political involvement of animal welfare groups, the enormous expense of animal studies, and the recognized difficulties of dealing with such complex systems experimentally, interest has turned to the development of alternative systems for decreasing the use of animals in laboratory studies [19-21].

The introduction of *in vitro* techniques (microsomal preparations, tissue cultures, perfused organ systems, or purified enzymes) was meant to overcome many of those problems unique to whole animals systems. Because liver is the major site of chemical and drug biotransformation, it is the tissue of choice for metabolism and toxicity studies, the majority of which are conducted using microsomal preparations [16]. While there is considerable impetus to perform metabolism studies using this technique, they may be complicated by several commonly encountered practical problems, including [13]:

1) microsomal fractions are often influenced by the manner of preparation. (the reliability and reproducibility of such factors as the method of homogenization and the choice of media can be very important [9]);
2) although mammalian tissue preparations occasionally may be utilized in synthesizing preparative quantities of metabolites, there are difficulties related to the stability of such systems [22];
3) biological specimens obtained from mammalian systems usually yield mere microgram quantities of metabolites, which are not

sufficient for complete structure elucidation and biological evaluation;

4)   the need for sensitive analytical methods suitable for the determination of minute quantities of drugs and their metabolites in biological media;

5)   qualitative and quantitative metabolic differences among various species of animals, this being especially critical when attempting to compare lower animals to man;

6)   procurement of sufficient quantities of metabolites for complete structure elucidation and biological testing; and

7)   difficulty in detecting the metabolites due to the presence of an multitude of naturally occurring substances that might interfere with identification of metabolites.

It is occasionally suggested that mammalian tissue (e.g., liver microsomes) and organ perfusion preparations may be useful in synthesizing preparative quantities of metabolites. However, these systems are generally not useful in providing gram quantities of metabolites, they are usually unstable and complex to operate, and they are far from simple to employ on an routine basis [22-24].

Tissue culture techniques generally require very exacting media with numerous components and are often plagued by bacterial contamination. A recent interest in the use of cultured hepatocytes as models for metabolism offers some advantages, but still requires a tedious procedure of perfusion of a section of liver and seeding of isolated hepatocytes in a complex nutrient medium, followed by daily maintenance of the cultures [25].

Microbial Models for Mammalian Metabolism

The use of microbial systems as *in vitro* models for drug metabolism in humans overcome most, if not all, of these obstacles [14]. The concept of "microbial models of mammalian metabolism" can be defined as the use of microorganisms (bacteria, yeasts, and fungi) to facilitate the study of xenobiotic biotransformation in mammals, including man [26]. This concept was first formalized by Rosazza and Smith [13] as a result of a systematic study of microbiological hydroxylations on a number of aromatic substrates. Direct parallels were observed with patterns of hydroxylations described earlier for *in vitro* and *in vivo* mammalian systems. The concept that fungi (being eukaryotes) can mimic the biotransformations performed by intact mammals or organ preparations due to similarities in the enzyme systems between fungi and mammals was formalized in an initial review [27]. Subsequent reviews [14, 28-34] followed and provided an overview of: a) the methodologies involved, b) parallels in the types of reactions observed, c) the enzymatic (biochemical) basis for observed parallels, and d) numerous specific examples supporting the basis and utility of microbial models concept.

Based on this concept, Smith and Rosazza proposed that microorganisms could serve as convenient and reliable models of mammalian metabolism [27]. This observation was also noted by Ferris and coworkers [35]. As a means of facilitating drug metabolism studies, Smith and Rosazza [27, 36] suggested that it might be possible to define microbial transformation systems which could mimic many of the kinds of biotransformation observed in mammals. The methodology used in this technique was initially reviewed by Smith and Rosazza and later more extensively reviewed by Goodhue [37]. Usually a two-stage fermentation procedure as outlined by Smith and Rosazza is used for microbial metabolism studies; good success using a soy bean meal-glucose medium was reported by Smith and Rosazza [27]. In other cases the use of peptone-glucose medium has been very successful and has the added advantage of being a completely solubilized medium [38-40].

The literature abounds with reports of microbial transformations which mimic many of the types of transformations observed in mammals [13, 27, 36, 41-44].

Microbial metabolism methodology involves the following steps:

Selection: Generally, a large number of microorganisms will be selected for preliminary or initial screening for their ability to metabolize a drug substrate. As with other studies, in microbial transformations, this choice may involve a rather empirical approach suing a broad "random screen" in which many cultures, chosen more or less at random from the culture collection, are screened for their ability to biotransform the drug in any fashion. More directed approaches based on a number of factors including literature precedent, experience and intuition typically yield more predictable results with considerably fewer cultures which are chosen based on their reported ability to catalyze specific phase I biotransformation, e.g., ketone reduction, aromatic hydroxylation, etc. In this type of study, the ultimate "model" will present a "composite" of several cultures, each exhibiting the individual routes of metabolism. Alternatively, one may choose (again, based on literature precedent) cultures known to exhibit the broad range of reactions observed in mammalian systems (fungi of the genus *Cunninghamella* fall into this category).

Screening: The preliminary evaluation of the metabolic capabilities of the selected cultures, using uniform microbiological culturing techniques is the "screening" aspect. Fig. (2) illustrates the usual two-stage screening/fermentation procedure that has been employed. During the screening stage, the "screen/confirm" rule is applied i.e., the production of general (common) metabolites by the "screened" microorganisms is confirmed and then examined in more detail to detect any minor metabolite, as a prelude to preparative scale production of metabolites.

This rule can be easily accomplished by comparing the colors and R_f values of metabolites on thin-layer chromatogram (TLC) produced by each microorganism which was tested. Optimization of the biotransformations (if necessary) involves the systematic alteration of fermentation parameters (medium composition, aeration, pH, the influence of growth on metabolism, etc.) and other parameters (time-course studies to determine optimal time of substrate addition and harvesting, mode of substrate addition, etc.) to enhance yields.

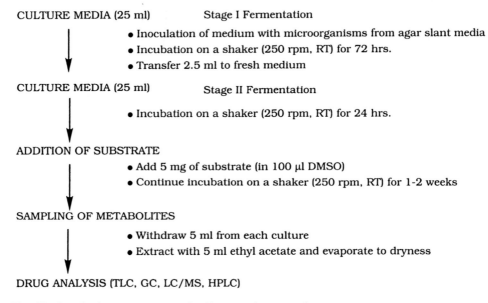

CULTURE MEDIA (25 ml)   Stage I Fermentation

- Inoculation of medium with microorganisms from agar slant media
- Incubation on a shaker (250 rpm, RT) for 72 hrs.
- Transfer 2.5 ml to fresh medium

CULTURE MEDIA (25 ml)   Stage II Fermentation

- Incubation on a shaker (250 rpm, RT) for 24 hrs.

ADDITION OF SUBSTRATE

- Add 5 mg of substrate (in 100 µl DMSO)
- Continue incubation on a shaker (250 rpm, RT) for 1-2 weeks

SAMPLING OF METABOLITES

- Withdraw 5 ml from each culture
- Extract with 5 ml ethyl acetate and evaporate to dryness

DRUG ANALYSIS (TLC, GC, LC/MS, HPLC)

Fig. (2). Standard two-stage screening/fermentation procedure.

Controls: Once "hits" have been confirmed in the screening stage, control studies are conducted. The purpose of the control studies is to ensure that the metabolites produced by microorganisms are "real" and not artifacts of the fermentation or isolation procedure.

Three control studies are routinely utilized for this purpose:

Substrate Control: The Substrate (drug) is incubated in the suitable culture medium under the identical conditions of the screen, but without inoculation of microorganisms. This control study is usually conducted prior to the screening experiment to establish the stability of the substrate in the medium.

Culture Control: The organism is grown in the culture medium under the identical conditions of the screen, but with no substrate present. TLC comparison of the culture control extract with that of the test (with substrate) will reveal the presence of any entity usually produced by the microorganism independent of the substrate. Products extracted from the media will also be detected in this study. The importance of this control is to avoid isolation of compounds that are not metabolites related to the substrate, but are products of secondary metabolism of the organism.

Autoclaved Substrate Control: The culture is grown for 1-2 weeks and then sterilized by autoclaving to ensure that the organism is dead and that all enzymes are denatured, after which the substrate is added and the "autoclaved culture" is incubated for another 2 weeks. TLC analysis of the extract of the autoclaved substrate control will reveal any compound that is produced as a result of nonenzymatic alteration of the substrate (under the influence of certain compounds, e.g., acids, that are being excreted by the microorganism during its normal life cycle and not due to a certain enzyme action on the substrate). Such compounds are not true metabolites.

Preparative-scale Fermentation: When the control studies have proven that a microorganism is performing a desired metabolism, preparative-scale fermentations are then conducted for the isolation, purification, and structural elucidation of greater quantities of the metabolite(s) produced.

Development of Microbial Models

Several approaches are utilized in the development of microbial models of mammalian metabolism. The model could be either retrospective, prospective, or parallel in relation to the mammalian studies [26].

The Retrospective Approach: Microbial metabolism studies are conducted after the mammalian routes of metabolism have been elucidated. Such studies can be used to confirm tentative assignments of metabolites in mammals, and support the premise that the same routes of metabolism are observed in both systems.

The Prospective Approach: In this approach the routes of metabolism are first elucidated in microbial systems and then used to predict (and the metabolites facilitate the confirmation of) routes of metabolism in mammals. Also it could be used to explore unanswered questions in mammalian metabolism. For example, microbial studies may reveal a previously undetected mammalian metabolite, elucidate the previously unexplored stereochemical disposition of the xenobiotic in mammalian systems, or be used to study mechanistic aspects concerning the formation of the metabolites.

The Parallel Approach: This approach involves the simultaneous study of biotransformations in mammalian and microbial species. Most studies to date have been retrospective, but the trend is to make such studies more prospective. Occasionally, it is difficult to find a clear cut distinction between those approaches. In a "retrospective" study (i.e., the mammalian routes of metabolism have already been explored), the microbial model, isolated metabolites, or analytical systems developed may be used in a prospective fashion to confirm tentative metabolites or explore unanswered questions in mammalian metabolism, such as the stereochemical course or mechanism of biotransformation.

Microbial systems show distinct advantages in certain areas which make them attractive to drug metabolism studies [13, 14, 26]:

*Ease of Experimental Design:*It is easy to manipulate the experimental parameters. For example, a reductive reaction (such as ketone reduction to the corresponding alcohol) can be enhanced by rendering the culture more reductive by lowering aeration or by the addition of glucose.

Preparative-scale Bioconversions: The concentrations of the substrate used are usually much higher than what could be obtained form *in vivo* animal or *in vitro* enzyme or tissue culture systems. The high substrate concentration coupled with the "cleaner" extracts from microbial cultures lead to easier detection, isolation, and identification of metabolites. This facet is particularly important if the same metabolite is suspected to occur in mammals, but the structure is unknown or equivocal because metabolite levels are limiting. Further, the metabollite can be evaluated for biological activity, including toxicity, yielding information which is exceedingly important in the early phases of drug development. In addition, the metabolite may serve as a standard reference material for conducting mammalian metabolism studies to determine if that metabolite is found in higher organisms.

Single or Multiple Pathways Modeling: A single microbial model (one culture) may exhibit the entire complex of divergent routes of mammalian metabolism seen with a particular substrate in numerous mammalian species. For example, *Cunninghamella bainieri* was shown to produce several known oxidative mammalian metabolites of N-n-propylamphetamine. Alternatively, one culture may exhibit a single route of metabolism in lieu of all others. This is particularly useful if the pathway is minor and/or quantitatively masked by other routes of metabolism in mammals, and thus hard to study. In this case, the ultimate "microbial model" may be a composite of cultures, each exhibiting the individual reactions.

Cost: The cost of maintaining a culture collection or that involved in the methodologies employed is only a fraction of that of maintaining laboratory animals or even tissue culture systems. Most microorganisms are maintained on stock agar slants at 4°C and transferred only every 4-6 months.

Low Animal Demand: The use of microbial systems in metabolism studies should lessen the demand of animals which would be beneficial from both humanitarian and economic standpoints.

While such advantages may be cited, it would not be suggested that microbial models will replace all animal metabolism studies, anymore than the animal models have "replaced" human metabolism studies. But such systems should be particularly useful in predicting routes of metabolism resulting in bioactivation, detoxification, or generation of toxic metabolites, and to uncover aspects of mammalian metabolism difficult to approach by other methods.

It is of extreme importance to study metabolism of new drug candidates. As new antifungal drug candidates, the sampangines should be studied thoroughly, particularly their metabolism. As explained previously, sampangine is inactive *in vivo* and this could perhaps be due to its metabolism. This problem might be solved by developing a good understanding of its pharmacokinetics. Metabolism is one important parameter of pharmacokinetics, so by careful study of the metabolism of sampangine we might be able to solve this *in vivo* inactivity obstacle. Another advantage that could be obtained from this study is the possibility of developing "superior" metabolites, i.e., metabolites with better pharmacokinetics, toxicokinetics, pharamacodynamics, and/or pharmaceutical properties than sampangines.

This chapter is an account of the utilization of microbial systems to predict the mammalian metabolites of the promising antifungal agent candidate, sampangine, and two of its analogs, namely benzosampangine and 3-methoxysampangine.

RESULTS AND DISCUSSION

Sampangine

Introduction to the Problem

The basic idea to study the microbial biotransformation of sampangines was initiated when a researcher in the Research Institute of Pharmaceutical Sciences, School of Pharmacy, The University of Mississippi, MS 38677, USA, observed the production of red urine after intraperitoneal administration of sampangine to rats. Initially, this red color was thought to be due to red blood cells in the urine (a phenomenon could be explained due to sampangine toxicity). Microscopical examination of this urine revealed the absence of blood, suggesting the presence of a red metabolite(s) in the urine. Moreover, the *in vivo* studies showed that sampangine is inactive and this could perhaps be due to its metabolism, a problem which might be solved by developing a good understanding of its pharmacokinetics, particularly its metabolism. Thus a microbial model of sampangine metabolism was investigated. Strikingly, some of the screened cultures acquired a red color only a few days after the incubation of sampangine. TLC analysis of these cultures revealed their capability to metabolize sampangine into a red metabolite(s). Preparative-scale fermentations, isolation of microbial metabolites, and the use of these metabolites to predict and confirm the identity of those mammalian metabolites formed the basis of this project.

Acquisition of the Starting Material, Sampangine

Due to the scarcity of sampangine in nature (0.001% yield), total synthesis was necessary in order to provide adequate supplies. Using the elegant method of Bracher [3] sampangine was obtained amply (26% yield). In fact, all the sampangine used in this project was obtained through total synthesis.

Stability Evaluation of the Starting Material

Before microorganisms are screened for their ability to metabolize sampangine, it would be necessary to establish the starting material's stability in the medium under the conditions to be used in the screening. Also, starting material and metabolite recovery studies from the medium should be conducted. A substrate control experiment was conducted in which sampangine was incubated in a complex medium (medium α) which consists of (per liter of distilled water): glucose, 20 g; NaCl, 5 g; K_2HPO_4, 5 g;yeast extract, 5 g; peptone, 5 g. Sampling of the medium after 7 days revealed that sampangine could be extracted with ethylacetate from medium α, at pH 7. TLC analysis indicated that sampangine is stable in the medium.

Screening of Microorganisms for their Ability to Metabolize Sampangine

Screening of fungi for their ability to metabolize the antifungal sampangine implies that sampangine should be added in subminimal inhibitory concentrations to the screened cultures. Also, it is worthy to note that sampangine exhibited no antifungal activity against cultures that have the ability to metabolize it. With this in mind a total of seventy microorganisms (Table 2) were evaluated for their ability to metabolize sampangine using the standard two-stage screening/fermentation procedure. TLC analysis of the culture extracts was used to identify the ability of these organisms to metabolize sampangine. Substrate control and culture control studies were utilized to ensure that these metabolites were a result of enzymatic activity and not a consequence of degradation or other non-metabolic changes and not related to the medium composition. Based on TLC analyses and control studies, some of the following microorganisms were shown to be capable of complete conversion of sampangine to two polar compounds, and *Beauvaria bassiana* ATCC 7159 and *Rhizopus arrhizus* ATCC 11145 were chosen for preparative scale fermentation to produce sufficient quantities of those two compounds for structural elucidation and other biological studies. It is noteworthy that all cultures that demonstrated the ability to metabolize sampangine turned bright red within a few days after sampangine was added.

Table 2.   Microorganisms Used in the Screening of SAM, BZSAM and 3-MeOSAM

Microorganisms [a]	SAM	BZSAM	3-MeOSAM
	Metabolite Production [b]		
Absidia glauca ATCC 22752 [c]	SAMM2	BZSAMM1	(-)
Achlya racemosa ATCC 11392	(-)	(-)	NT
Acrasis rosea ATCC 26823	(-)	(-)	NT
Acrodictys erecta ATCC 24083	(-)	(-)	NT
Acrothecium capsici ATCC 10714	(-)	(-)	NT
Agaricus campestris ATTCC 26815	(-)	NT	NT
Allomyces arbuscula ATCC 10983	(-)	(-)	NT
Aspergillus alliaceus NRRL 315 [d]	(-)	(-)	NT
Aspergillus flavipes ATCC 1030	(-)	(-)	NT
Aspergillus flavus NRRL 501	NT	NT	(-)
Aspergillus flavus NRRL 626	(-)	(-)	(-)
Aspergillus fumigatus ATCC 26934	(-)	NT	NT
Aspergillus niger ATCC 10549	(-)	NT	(-)
Aspergillus ochraceus ATCC 18500	(-)	(-)	(-)
Aspergillus tamarii NRRL 8101	(-)	(-)	(-)
Aureobasidium pullulans ATCC 9348	(-)	(-)	NT
Beauvaria bassiana* ATCC 7159	SAMM1	(-)	(-)
Botrytis allii ATCC 9435	(-)	NT	NT
Bullera alba ATCC 18568	(-)	(-)	NT
Calonectria decora ATCC 14767	(-)	NT	(-)
Candida tropicalis ATCC 20021	NT	NT	(-)
Cantharellus cibarius ATCC 13228	(-)	(-)	NT
Ceratobasidium cornigerum ATCC 38315	(-)	NT	NT
Chaetomium cochliodes NRRL 2320	(-)	NT	NT
Chlorencoelia versiformis ATCC 24031	(-)	NT	NT

(Table 2). contd.....

Microorganisms [a]	SAM	BZSAM	3-MeOSAM
	Metabolite Production [b]		
Cladosporium resinae ATCC 22712	(-)	(-)	NT
Cokeromyces recurvatus ATCC 13568	(-)	NT	NT
Coniophora puteana ATCC 12675	(-)	NT	NT
Coprinus species ATCC 16789	(-)	NT	(-)
Cordyceps militaris ATCC 34164	(-)	NT	NT
Coriolus antarcticus ATCC 34581	(-)	NT	NT
Cryptococcus macerans Ziffer	(-)	NT	NT
Cunninghamella blakesleeana ATCC 8688a	(-)	**BZSAMM1**	(-)
Cunninghamella echinulata NRRL 3655	(-)	NT	NT
Cunninghamella echinulata ATCC 9244	(-)	(-)	(-)
Cunninghamella elegans ATCC 9245	SAMM2	NT	NT
Cunninghamella species NRRL 5695	SAMM2	BZSAMM1	(-)
Cylindrocarpon radicicola 11011	(-)	NT	(-)
Cylindrocephalum aureum ATCC 12720	(-)	(-)	NT
Dacrymyces deliquescens var. *deliquescens* 13292	(-)	(-)	(-)
Dactylaria haptotyla ATCC 28924	(-)	(-)	NT
Debaryomyces polymorphus ATCC 20280	(-)	NT	NT
Dictyostelium mucoroides ATCC 2682	(-)	(-)	NT
Doratomyces microsporus ATCC 16225	SAMM1	(-)	(-)
Eupenicillium javanicum ATCC 26879	(-)	(-)	NT
Filobasidiella neoformans ATCC 10226	SAMM1	(-)	(-)
Filosporella annelidica ATCC 32834	(-)	(-)	(-)
Fomes pinicola ATCC 15341	(-)	NT	NT
Fusarium oxysporum ATCC 7601	(-)	(-)	NT
Fusarium oxysporum f. cepae ATCC 11711	(-)	NT	NT
Fusarium solani ATCC 12823	(-)	(-)	NT

(Table 2). contd.....

Microorganisms [a]	SAM	BZSAM	3-MeOSAM
	Metabolite Production [b]		
Fusarium solani f. sp. *cucurbitae* CSih # c-5 [e]	(-)	BZSAMM1	(-)
Geotrichum amycelicum ATCC 24658	NT	(-)	NT
Gliocladium deliquescens ATCC 10097	(-)	NT	NT
Gluconobacter oxydans ATCC 621	(-)	NT	NT
Gongronella butleri ATCC 22822	(-)	(-)	NT
Hansenula anomala ATCC 20170	(-)	NT	NT
Helicodendron triglitziense ATCC 16770	(-)	(-)	NT
Helicostylum piriforme QM 6945 [f]	(-)	NT	NT
Helminthosporium species NRRL 4671	(-)	(-)	NT
Lipomyces lipofer ATCC 107242	(-)	NT	NT
Melanospora ornata ATCC 26180	(-)	(-)	(-)
Mortierella zonata ATCC 13309	(-)	(-)	NT
Mucor griseo-cyanus ATCC 1207a	(-)	NT	NT
***Mucor remannianus* 1839 CSih [e]**	(-)	BZSAMM2	(-)
Penicillium chrysogenum ATCC 9480	(-)	NT	NT
Polyporus brumalis ATCC 34487	NT	NT	(-)
Rhizopogon species ATCC 36060	(-)	BZSAMM1	(-)
***Rhizopus arrhizus* ATCC 11145**	SAMM2	(-)	(-)
Rhizopus stolonifer ATCC 24795	NT	(-)	(-)
Rhodococcus species ATCC 21146	(-)	NT	NT
Saccharomyces cerevisiae ATCC 2366	(-)	NT	NT
Saccharomyces cerevisiae NRRLY-2034	(-)	NT	NT
Saccharomyces lipolytica ATCC 16617	NT	(-)	NT
Schizosaccharomyces pombe ATCC 20130	(-)	(-)	NT
Sepedonium chrysospermum ATCC 13378	NT	(-)	NT
Septomyxa affinis ATCC 6737	NT	NT	(-)

(Table 2). contd.....

Microorganisms [a]	SAM	BZSAM	3-MeOSAM
	Metabolite Production [b]		
Streptomyces griseus ATTC 13968 (medium α)	NT	NT	(-)
Streptomyces griseus ATTC 13968 (medium θ)	NT	NT	(-)
Streptomyces lavendulae L-105	(-)	(-)	NT
Streptomyces platensis NRRL 2364	NT	(-)	NT
Streptomyces punipalus UI-3529 [g]	NT	(-)	NT
Streptomyces rimosus ATCC 23955	NT	BZSAMM3	(-)
Streptomyces rimosus NRRL 2234	NT	(-)	(-)
Streptomyces roseochromogenus ATCC 13400	NT	(-)	(-)
Streptomyces spectabilis ATCC 27465 (medium α)	NT	NT	(-)
Streptomyces spectabilis ATCC 27465 (medium θ)	NT	NT	3-MeOSAMM1 3-MeOSAMM2
Stysanus microsporus ATCC 2833	NT	(-)	NT
Talaromyces ucrainicus ATCC 18352	NT	(-)	NT
Thamnidium elegans ATCC 18191	NT	(-)	NT
Trametes zonata ATCC 38279	NT	(-)	NT
Trichophyton mentagrophytes ATCC 9972	NT	(-)	(-)

[a]All microorganisms were grown only in medium α, while *Streptomyces griseus* ATCC 13968 and *Streptomyces spectabilis* ATCC 27465 were grown in both medium α and medium θ

[b]Metabolites are given a designation, SAMM (SAMpangine Metabolite), BZSAMM (BenZoSAMpangine Metabolite) or 3-MeOSAMM (3-MethOxySAMpangine Metabolite) followed by a number according to the chronological order of their isolation. Metabolite production denoted (-) indicates no metabolite production, as shown by TLC analysis. Cultures that were chosen for preparative scale fermentation are denoted in bold type.

[c]American Type Culture Collection, Rockville, Maryland, USA.

[d]Northern Regional Research Laboratories, Peoria, Illinois, USA.

[e]Cultures obtained from Dr. Charles J. Sih, Department of Pharmaceutical Biochemistry, University of Wisconsin, Madison, Wisconsin, USA.

[f]Quartermaster Culture Collection, Quartermaster Research and Engineering Command, United States Army Natick Laboratories, Natick, Massachusetts 01760, USA.

[g]Culture obtained from Dr. John P. Rosazza, Division of Medicinal Chemistry and Natural Products, University of Iowa, Iowa City, Iowa.

NT: Not Tested.

Preparative-scale Biotransformation of Sampangine by *Rhizopus arrhizus*

A preparative-scale biotransformation of sampangine was performed with *Rhizopus arrhizus* ATCC 11145. Using the standard two-stage fermentation procedure, the microorganism was incubated with 640 mg sampangine in 4L of medium α. TLC analysis of the water-saturated butanol extract of an aliquot from the culture broth detected the complete conversion of sampangine into one major metabolite after 5 days of incubation. This metabolite (designated **SAMM2**) was more polar than sampangine. By this time the culture turned into a bright red color and no deposited sampangine was seen on the wall of the fermentation flasks. The fermentation process was terminated and the incubation mixtures were combined, filtered to remove the mycelia, and the filtrate was extracted with water-saturated butanol (4Lx4). The combined extracts were evaporated to dryness *in vacuo* at 40ºC to afford a red residue (11.54g). This residue was purified by column chromatography over a silica gel column, using chloroform-methanol-ammonia (87; 12.5: 0.5) mixture as

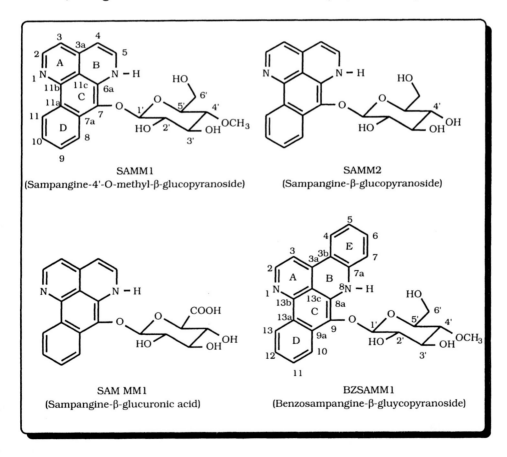

SAMM1
(Sampangine-4'-O-methyl-β-glucopyranoside)

SAMM2
(Sampangine-β-glucopyranoside)

SAM MM1
(Sampangine-β-glucuronic acid)

BZSAMM1
(Benzosampangine-β-gluycopyranoside)

eluent. Fractions were collected and combined based on TLC analysis. This afforded a pure red amorphous powder (SAMM2, 913 mg, 83.5% yield) Fig. (3). Isolated metabolites are given a designation, **SAMM** (**SAM**pangine **M**etabolite) followed by a number according to the chronological order of their isolation but not necessarily described in this order.

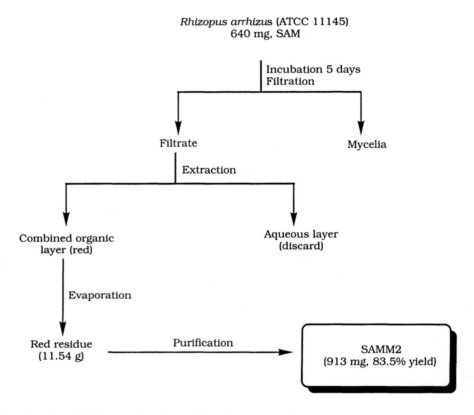

Fig. (3). Microbial biotransformation of SAM into SAMM2.

Structure Elucidation of the Metabolite SAMM2

The thermospray mass spectrum of SAMM2 indicated that its molecular weight (m/z 369; M$^+$) was the sum of sampangine (232) and a moiety of 164 mass units. High Resolution Fast Atom Bombardment (HRFAB) mass spectrum of SAMM2 showed the exact molecular weight which was consistent with the molecular formula $C_{21}H_{20}N_2O_6$. The IR spectrum showed a broad absorption at 3300 cm^{-1}, which suggested the presence of at least one hydroxyl group, while the absence of any strong absorption around 1670 cm^{-1} region indicated the conversion of the original sampangine carbonyl group into another functionality. The ^{13}C-NMR spectrum (75 MHz and 125 MHz) of SAMM2 showed 21 carbon resonances, one triplet, thirteen doublets, and seven singlets none of which was a carbonyl carbon. Five of those resonances were in the aliphatic oxygenated region (δ 60-80 ppm) (Table 4) and were consistent with the sugar glycoside. The carbon resonated at δ 107.7 ppm as a doublet was determined from Heteronuclear Correlation (HETCOR) and Heteronuclear Multiple Bond Correlation (HMBC) spectra to be consistent with an anomeric carbon. The other resonances were characteristic of the sampangine moiety with no substitution at any of its protons (see Table 3, 4).

Especially noteworthy are upfield shifts for C-4 (22.4 ppm), C-5 (11.6 ppm), C-6a (17.7 ppm), C-7 (51.6 ppm), H-4 (2.15 ppm), and H-5 (2.14 ppm) which indicated whatever moiety was added, it should have had a profound magnetic and electronic effect on that part (ring B and C-7 in ring C) of sampangine. The nuclear Overhauser effect (NOESY and ROESY) between the anomeric proton and H-8 (resonated at 8.46 ppm as a doublet with J=8.3 Hz), and the existence of 3-bond correlations (HMBC spectrum, Fig. (4)) between the anomeric proton and a singlet carbon (δ 129.9 ppm) which has another 3-bond correlation to H-8 led us to confidently conclude that SAMM2 is an O-glycoside in which the glycone part is coupled with sampangine through its presumably reduced carbonyl group, see Fig. (6).

From the COSY spectrum, H-9, H-10 and H-11 were determined. H-11 has a 3-bond correlation to two singlet carbons (δ 131.4 and δ 145.7). One of these singlet resonances (δ 131.4) was determined to be C-7a based on a 3-bond correlation to H-9. Therefore, the other singlet carbon (δ 145.7) should be C-11b. C-11b has another 3-bond correlation to a proton resonated at δ 8.27 ppm as a d (J=5.0 Hz). This proton should be H-2. Consequently, H-3 was determined from the COSY spectrum. The nuclear Overhasuer effect (NOESY and ROESY) connected H-3 to H-4 which resonated at δ 5.75 ppm as a d (J=7.2 Hz). Finally, H-5 was assigned from the COSY spectrum. The unambiguous assignments of H-2 and H-5 was crucial in distinguishing between the two tautomeric forms that can be drawn for SAMM2, see discussion on page 37 and Fig. (8). Comparing the coupling constants between protons 1' & 2' in the sugar part with those of the corresponding protons in some β-, and α-glycosides found in the literature [45-47] revealed that SAMM2 is a β-glycoside.

H-1' resonated at δ 4.69 as a d (J=7.9 Hz) and this large coupling constant indicated 1, 2-diaxial relationship with H-2' and proved the β-glycosidic linkage. Furthermore, application of the Insight and Discover program (Biosym®) with bond-energy minimizations on the proposed structure supported this conclusion. Since a careful comparison of the

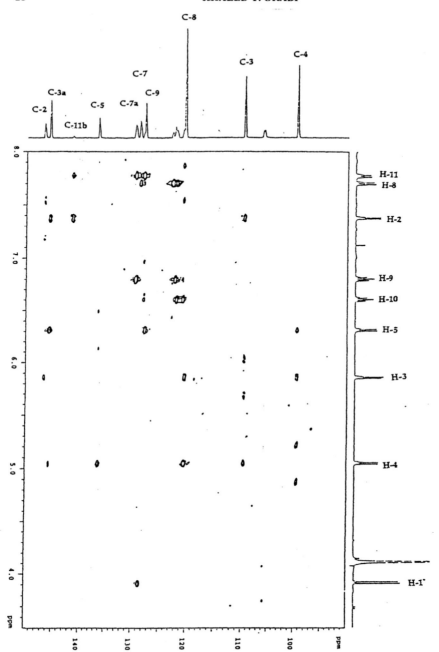

Fig. (4). HMBC spectrum of SAMM2.

^{13}C-NMR data for the sugar part in SAMM2 with those in an authentic β-D-glucopyranose [45-47] revealed identical resonances for each carbon, the conclusion that the sugar part is β-glucopyranose was made.

An additional piece of supportive evidence was drawn from the coupling constants. Coupling constants between protons 2' & 3', 3' & 4', and 4' & 5' were larger than 9.0 Hz (see Table 3) and that indicated the 1, 2-trans-diaxial relationship between each consecutive pair of those protons. This ultimately confirmed the identity of this hexose to be β-glucopyranose. It was further supported by the existence of nuclear Overhauser effects between protons 1' & 3', 1' & 5', 3' & 5', and 2' & 4' which proved the 1, 3-cis-diaxial interactions between those protons.

The complete unambiguous assignments of other carbon resonances, particularly the singlets, were made possible from HMBC and HMQC spectra. Independent assignments of proton and carbon resonances, and cross peaks in HMBC, HMQC, COSY and ROESY were also performed

Table 3. Comparative ^{1}H-NMR Assignments of SAM[a], SAMM1, SAMM2, Ac-SAMM2 and SAM MM1

Proton No.	SAM[b] δH	SAMM1[c] δH	SAMM2[c] δH	Ac-SAMM2[b] δH	SAM MM1[c] δH
2	8.84 (1H, d, J=5.8)	8.29 (1H, d, J=5.0)	8.27 (1H, d, J=5.0)	8.52 (1H, d, J=4.9)	8.20 (1H, d, J=6.6)
3	7.69 (1H, d, J=5.8)	6.68 (1H, d, J=5.0)	6.66 (1H, d, J=5.0)	6.69 (1H, d, J=4.9)	6.80 (1H, d, J=6.6)
4	7.90 (1H, d, J=5.5)	5.76 (1H, d, J=7.2)	5.75 (1H, d, J=7.2)	5.74 (1H, d, J=7.2)	6.05 (1H, d, J=7.3)
5	9.11 (1H, d, J=5.5)	6.97 (1H, d, J=7.2)	6.97 (1H, d, J=7.2)	6.89 (1H, dd, J=7.2, 6.5)	7.38 (1H, d, J=7.3)
8	8.43 (1H, dd, J=7.8,1.0)	8.46 (1H, dd, J=8.0, 0.8)	8.46 (1H, d, J=8.0)	7.89 (1H, d, J=8.1)	8.68 (1H, d, J=7.3)
9	7.67 (1H, ddd, J=7.8, 7.8,1.3)	7.56 (1H, dd, J=8.0, 8.0, 0.8)	7.56 (1H, ddd, J=8.0, 8.0, 1.2)	7.57 (1H, ddd, J=8.1, 8.1, 1.2)	7.75 (1H, dd, J=7.3, 7.3)
10	7.81 (1H, ddd, J=7.8, 7.8, 1.0)	7.36 (1H, ddd, J=8.0, 8.0, 0.8)	7.36 (1H, ddd, J=8.0, 8.0, 1.2)	7.43 (1H, ddd, J=8.1, 8.1, 1.2)	7.55 (1H, dd, J=7.3, 7.3)
11	8.78 (1H, dd, J=7.8, 1.3)	8.70 (1H , dd, J=8.0, 0.8)	8.70 (1H, d, J=8.0)	8.97 (1H, d, J=8.1)	8.62 (1H, d, J=7.3)
1'		4.67 (1H , d, J=7.8)	4.69 (1H, d, J=7.9)	5.02 (1H, d, J=8.1)	4.85 (1H, d, J=8.0)
2'		3.67 (1H, dd, J=9.3, 7.8)	3.68 (1H, dd, J=9.1, 7.9)	5.46 (1H, dd, J=9.5, 8.1)	3.80 (1H, dd, J=8.0, 8.0)

(Table 3). contd.....

Proton No.	SAM [b] δ H	SAMM1[c] δ H	SAMM2[c] δ H	Ac-SAMM2 [b] δ H	SAM MM1[c] δ H
3'		3.55 (1H, dd, J=9.3, 9.7)	3.47 (1H, dd, J=9.1, 9.1)	5.30 (1H, dd, J=9.5, 9.5)	3.55 (1H, dd, J=8.0, 8.0)
4'		3.34 (1H, dd, J=9.7, 9.7)	3.57 (1H, dd, J=9.1, 9.1)	5.19 (1H, dd, J=9.5, 9.5)	3.65 (1H, dd, J=8.0, 8.0)
5'		3.22 (1H, dt, J=9.7, 3.2)	3.25 (1H, m)	3.68 (1H, m)	3.65 (1H, d, J=8.0)
6'		3.73 (2H, m)	3.78 (2H, m)	4.25 (2H, m)	
OCH$_3$		3.59 (3H, s)			
N-H				8.07 (1H, d, J=6.5)[d]	
C̲H$_3$COO				2.29 (3H, s)	
C̲H$_3$COO				2.08 (3H, s)	
2C̲H$_3$COO				2.05 (6H, s)	

[a]Chemical shifts are reported in parts per million, J values are in Hertz.
[b]Spectra recorded in CDCl$_3$.
[c]Spectra recorded in CD$_3$OD.
[d]Exchangeable proton.

using a program called "AURELIA" which is a 2D and 3D analysis package to conduct such assignments. These assignments were indistinguishable from those obtained using normal assignment methods.

SAMM2 can be represented as sampangine-β-glucopyranside.

Other Strategies to Confirm the Structure SAMM2

Before the final conclusion about SAMM2 structure was reached and during the acquisition of some of the above data (particularly HMBC and HMQC spectra), other strategies that may be helpful in deducing the structure were pursued. Two main strategies were considered. Chemical degradation and single crystal X-ray analysis.

Chemical Degradation of SAMM2

The chemical degradation of SAMM2 was achieved by simple acid hydrolysis of the glucosidic bond followed by chromatographic identification of the hyrolysates. The sugar moiety was further confirmed to be glucose using this strategy.

Table 4. Comparative ^{13}C-NMR Assignments of SAM, SAMM1, SAMM2, Ac-SAMM2, SAM MM1, BZSAM and BZSAMM1

Carbon No.	SAM[a] δC	SAMM1[b] δC	SAMM2[b] δC	AC-SAMM2[a] δC	SAM MM1[b] δC	BZSAM[a] δC	BZSAMM1[c] δC
2	147.1 (1)[d]	150.2 (1)	150.7 (1)	151.9 (1)	145.3 (1)	148.7 (1)	149.9 (1)
3	118.9 (1)	111.8 (1)	111.9 (1)	112.9 (1)	110.1 (1)	115.4 (1)	110.5 (1)
3a	138.3 (0)	145.8 (0)	145.9 (0)	147.2 (0)	143.0 (0)	137.7 (0)	138.6 (0)
3b						123.3 (0)	115.9 (0)
4	123.2 (1)	100.9 (1)	100.8 (1)	101.9 (1)	100.8 (1)	122.8 (1)	124.4 (1)
5	148.1 (1)	136.7 (1)	136.5 (1)	135.3 (1)	140.0 (1)	130.2 (1)	120.4 (1)
6						131.5 (1)	131.9 (1)
6a	147.4 (0)	129.7 (0)	129.7 (0)	130.3 (0)	131.5 (0)		
7	181.5 (0)	129.8 (0)	129.9 (0)	130.8 (0)	135.0 (0)	132.9 (1)	115.6 (1)
7a	131.9 (0)	131.3 (0)	131.4 (0)	127.8 (0)	131.5 (0)	145.7 (0)	139.6 (0)
8	128.1 (1)	122.2 (1)	122.2 (1)	119.6 (1)	122.2 (1)		
8a						146.7 (0)	126.8 (0)
9	131.1 (1)	129.4 (1)	129.3 (1)	129.8 (1)	129.5 (1)	181.9 (0)	128.9 (0)
9a						132.4 (0)	130.4 (0)
10	134.3 (1)	123.9 (1)	123.9 (1)	124.4 (1)	124.3 (1)	128.6 (1)	121.3 (1)
11	125.1 (1)	124.5 (1)	124.6 (1)	126.2 (1)	122.5 (1)	131.1 (1)	128.3 (1)
11a	135.0 (0)	125.3 (0)	125.8 (0)	127.2 (0)	128.0 (0)		
11b	150.6 (0)	145.3 (0)	145.7 (0)	143.8 (0)	143.0 (0)		
11c	119.2 (0)	122.1 (0)	122.3 (0)	121.7 (0)	123.0 (0)		
12						134.8 (1)	122.6 (1)
13						125.7 (1)	123.5 (1)
13a						135.9 (0)	124.3 (0)
13b						150.4 (0)	145.1 (0)

(Table 4). contd.....

Carbon No.	SAM[a] δC	SAMM1[b] δC	SAMM2[b] δC	AC-SAMM2[a] δC	SAM MM1[b] δC	BZSAM[a] δC	BZSAMM1[c] δC
13c						116.9 (0)	117.6 (0)
1'		107.6 (1)	107.7 (1)	104.5 (1)	106.2 (1)		106.6 (1)
2'		75.7 (1)	75.6 (1)	72.8 (1)	74.0 (1)		74.3 (1)
3'		78.1 (1)	78.0 (1)	74.0 (1)	76.8 (1)		76.4 (1)
4'		80.1 (1)	70.8 (1)	69.5 (1)	75.8 (1)[e]		69.4 (1)
5'		77.1 (1)	78.1 (1)	74.1 (1)	72.0 (1)[e]		77.0 (1)
6'		61.8 (2)	62.1 (2)	62.6 (2)	176.8 (0)		60.8 (2)
OCH₃		60.8 (3)					
C̲H₃COO				21.8 (3)			
C̲H₃COO				21.9 (3)			
C̲H₃COO				22.0 (3)			
C̲H₃COO				22.2 (3)			
CH₃C̲OO				170.6 (0)			
CH₃C̲OO				170.7 (0)			
CH₃C̲OO				171.5 (0)			
CH3C̲OO				172.0 (0)			

[a]Spectra recorded in CDCl₃ with TMS as internal standard.
[b]Spectra recorded in CD₃OD with TMS as internal standard.
[c]Spectra recorded in DMSO-d₆ with TMS as internal standard.
[d]Number of attached protons determined by DEPTGL.
[e]Interchangeable.

Hydrolysis of SAMM2 using HCl: A known procedure [48] with minor modifications was applied. In this procedure SAMM2 was incubated with 6M hydrochloric acid for 2 days (till the red color of SAMM2 changed into yellow color). The acidic solution was made basic using 6M NaOH and extracted with chloroform. The combined extracts were washed with water, dried over anhydrous Na₂SO₄ and evaporated *in vacuo* at 40°C to afford a yellow residue. The aqueous alkaline solution was also evaporated to afford a colorless residue which appeared to have a lot of NaCl salt. TLC analysis of the chloroform extract residue revealed the presence of two major compounds, the more polar one was identified as sampangine and the less polar as 4-chlorosampangine (TLC, co-TLC comparison). Chromatographic purification of those compounds over a silica column, using chloroform as eluent afforded the two compounds in a pure form. NMR analysis of the more polar compound showed indistinguishable spectral data from those of sampangine. Likewise, the less polar compound showed identical spectral data to those of 4-chlorosampangine.

The nucleophilic chlorination of sampangine at position 4 is very expected since sampangine was left in contact with hydrochloric acid for about 50 hrs. and this reaction has been observed previously [6].

TLC analysis of the aqueous extract residue and comparison with an authentic sample of glucose, suggested that glucose was the sugar part. Both gave blue color with anisaldehyde and heat.

Hydrolysis of SAMM2 using H_2SO_4: Since the aqueous layer from the HCl-hydrolysis contained excessive NaCl, another method using H_2SO_4 was employed [49] to avoid the excessive amount of salt and chlorination of the produced sampangine as well. In this method SAMM2 was dissolved in methanol and treated with 0.1M H_2SO_4 at ambient temperature for about 12 hrs. The reaction mixture was extracted several times using chloroform. The combined extracts were washed with 5% $NaHCO_3$ and saturated solution of NaCl, then evaporated to dryness. Sampangine was detected using TLC as before. The aqueous layer was neutralized with saturated solution of $Ba(OH)_2$. The precipitate was filtered off and the solution evaporated to dryness. The presence of glucose was proven using TLC analysis as before.

Single Crystal X-ray Analysis

Single crystal X-ray analysis is considered the ultimate structure proof, particularly for the absolute stereochemistries of the sugar moiety. For this technique to be successful SAMM2, or a suitable derivative must be obtained in a suitable crystalline form.

Attempted Crystallization of SAMM2: A variety of solvents were used and afforded only solid powder. The uncrystallizability of SAMM2 was attributed to its instability, which was evident (by TLC) during these attempts.

SAMM2 Salt Formation: The hydrochloride salt was formed under anhydrous conditions, yet under these conditions partial hydrolysis of SAMM2 occurred as revealed by TLC. The product was not crystalline. Attempts to form other salts (oxalate, citrate, acetate) did not add any advantage neither to the stability nor crystallizability of SAMM2.

Single crystal X-ray analysis was unsuccessful, but the combined structural data and chemical degradation conclusively proved SAMM2 structure.

It is very important to appreciate the crucial role of the structure elucidation of SAMM2 in aiding the elucidation of that of SAMM1 and of the mammalian metabolites as will be seen below.

Preparative-scale Biotransformation of Sampangine by Beauvaria bassiana

A preparative-scale biotransformation of sampangine was performed with *Beauvaria bassiana* ATCC 7159 using the standard two-stage fermentation procedure. This afforded a pure red metabolite (33 mg, 9.7% yield) (Fig. (5)).

Structure Elucidation of the Metabolite SAMM1

The thermospray mass spectrum of SAMM1 indicated that its molecular weight (m/z 410; M$^+$) was the sum of sampangine (232) and a moiety of 178 mass units. A HRFAB mass spectrum of SAMM1 showed that exact molecular weight to be consistent with the molecular formula $C_{22}H_{22}N_2O_6$. The IR spectrum showed a broad absorption at 3280 cm^{-1}, which suggested the presence of at least one hydroxyl group, while the absence of any strong absorption around 1670 cm^{-1} region indicated the conversion of the original sampangine carbonyl group into another functionality.

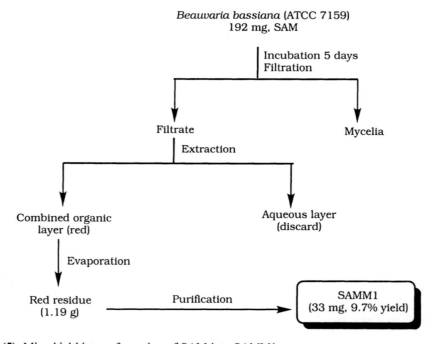

Fig. (5). Microbial biotransformation of SAM into SAMM1.

The ^{13}C-NMR spectrum of SAMM1 showed 22 carbon resonances none of which was a carbonyl, six of those resonances were in the aliphatic oxygenated region (δ 60-80 ppm) (Table 4). One of those six was a quartet (δ 60.8 ppm) which indicated the presence of a methoxyl group. The other resonances were characteristic of the sampangine moiety with no substitution at any of its protons.

The data, thus, led us to conclude that metabolite SAMM1 can be represented as O-methylhexose conjugate of sampangine. The resonance at δ 107.6 ppm was determined from the HETCOR spectrum to be an anomeric carbon. Especially noteworthy are upfield shifts for C-4 (22.3 ppm), C-5 (11.4 ppm), C-6a (17.7 ppm) and C-7 (51.7 ppm), and H-4 (2.14 ppm) and H-5 (2.14 ppm) which indicated whatever moiety was added, it should have had a profound magnetic and electronic effect on that part (ring B and C-7 in ring C) of sampangine.

This conjugate was determined to be a β-glycoside by comparing the coupling constants of protons 1', 2', 3', 4' and 5' in the sugar part with those of the corresponding protons in some β-, and α-D-glucosides found in the literature [45-47]. The H-1' resonated at δ 4.67 as d (J=7.8 Hz) (Table 3). This large coupling constant indicated 1, 2-diaxial relationship between H-1' and H-2' and proved the β-glycosidic linkage.

Coupling constants between protons 2' & 3', 3' & 4' and 4' & 5' were larger than 9.0 Hz which indicated the 1, 2-diaxial relationships between each consecutive pairs of those protons. This ultimately proved the identity of this hexose to be a glucopyranose derivative. A careful comparison of the ^{13}C-NMR data for the glucose part in SAMM1 with those of SAMM2 revealed nearly identical resonances except for C-4' which in SAMM1 showed a downfield shift (10 ppm) in relation to SAMM2. This downfield shift was attributed to the presence of an O-methyl group on C-4'. This methyl group resonated in the ^{13}C-NMR spectrum at δ 60.8, and in the ^{1}H-NMR spectrum at δ 3.6 as a singlet and integrated for three protons. This conclusion was further confirmed by comparing the ^{13}C resonances of the sugar part with those of the corresponding carbons in some 4'-O-methylglucopyranoside found in the literature [50].

The Proposed Mechanism Involved in the Formation of SAMM1 and SAMM2

The mass spectra for both metabolites (SAMM1 and SAMM2) revealed the presence of a common molecular ion at m/z 235; M[+]+1. The presence of such a common molecular ion in both samples led us to the conclusion that dihydrosampangine (MW=234), yet too unstable to isolate, is the common intermediate in the pathway leading to the formation of SAMM1 and SAMM2 as well. Thus, the proposed mechanism involved in the formation of SAMM1 and SAMM2 could be explained as follows (Fig. (6)):

1)   Reduction of sampangine into dihydrosampangine. This step could be accomplished through reductase or hydrogenase enzymes.
2)   Coupling of this "fragile" moiety through the produced hydroxyl group with glucose to form the more stable O-glucoside.
3)   Tautomerizm of H-7 to N-6.
4)   In the case of SAMM1, O-methylation of C-4' of glucose could be suggested to take place either before or after the coupling process.

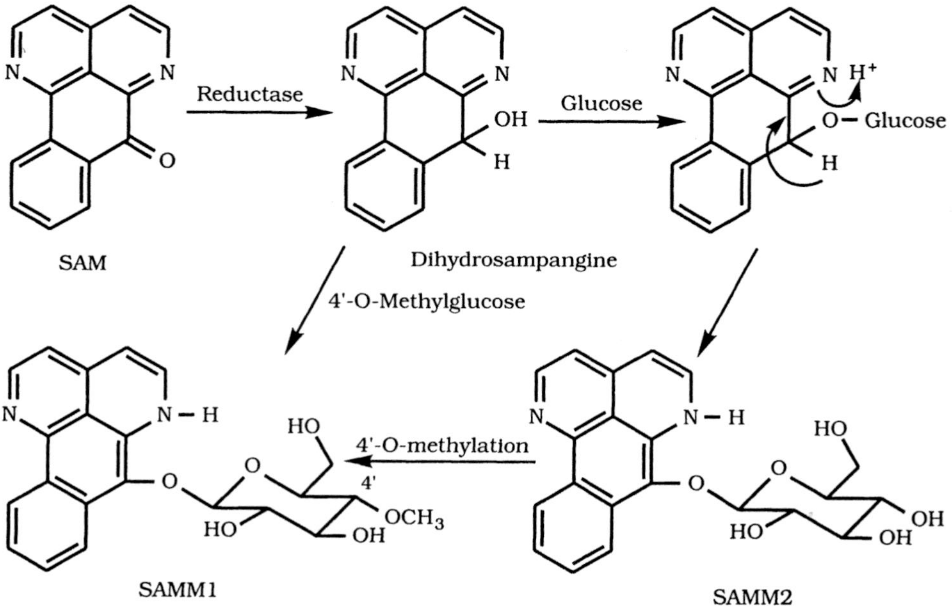

Fig. (6). The proposed mechanism involved in the formation of SAMM1 and SAMM2.

Tautomerism of SAMM1 and SAMM2

The structure of SAMM1/SAMM2 suggested a possible tautomerism of the proton on N-6 to N-1 and vice versa (Fig. (7)). To test the validity of this concept acetylation of the "N-H" moiety was pursued in SAMM2 since this metabolite was available in larger quantities.

N-acetylation of SAMM2

SAMM2 contains five acetylation-liable sites, four hydroxyl groups and one amino group. N-acetylation was pursued using neat acetic anhydride.

Theoretically, under these conditions N-acetylation would seem to be favored over O-acetylation. This procedure, after work up and purification, afforded a pure acetylated product (Ac-SAMM2).

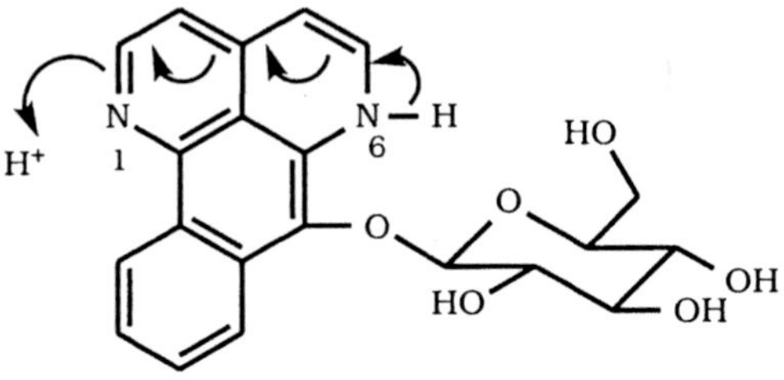

Fig. (7). The possible tautomeric structures of SAMM1/SAMM2.

Structure Elucidation of Ac-SAMM2

It was assumed that if N-acetylation took place it would be easy to prove it through NMR analyses. Strikingly, the ^{13}C-NMR spectrum showed four carbons resonating at δ 21.8, 21.9, 22.0 and 22.2 ppm each as a q, and another four carbons resonating at δ 170.6, 170.7, 171.5 and 172.0 ppm each as a singlet. Other carbon resonances were comparable to those of SAMM2 (Table 4). Likewise, the ^{1}H-NMR spectrum revealed the presence of three peaks, two of which integrated for three protons each, and the other one integrated for six protons (Table 3). HRFAB mass spectrum showed the exact molecular weight (m/z 565.1838; M+H$^+$) which was consistent with the molecular formula $C_{29}H_{28}O_{10}N_2$. This molecular weight was the sum of SAMM2 (396) and a moiety of 168 mass units (consistent with four units of CH_2CO). These findings proved the presence of four acetate groups, presumably on the four hydroxyl groups of glucose. Since H-2 and H-5 were unambiguously assigned (mentioned before), the splitting of H-5 into a dd (Table 3) suggested its coupling to an adjacent proton with a large coupling constant (J=6.5 Hz) and the only existing possibility would be the proton on N-6, which resonated at δ 8.07 as a d (J=6.5 Hz). This coupling pattern was confirmed by the collapse of H-5 into a d (J=7.2 Hz) and the disappearance of H-6 upon D_2O exchange. Especially noteworthy were the downfield shifts of H-2` (1.78 ppm), H-3' (1.83 ppm), H-4' (1.62 ppm), and H-6' (0.47 ppm) due to the deshielding effects of those acetates. Other protons were, more or less, the same except H-8 which showed an upfield shift (0.57 ppm). This upfield shift is due to the shielding effect of the carbonyl group of C-2' acetate. These findings proved two things; first, the inaccessibility of N-H group for acetylation, and second, SAMM2 exists in one tautomer in which the proton existed only on N-6. This could be explained by a possible hydrogen bonding between N-H and an oxygen atom (Fig. (8)).

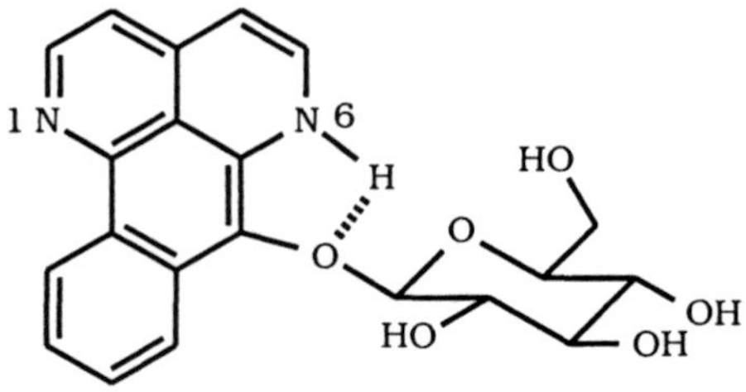

Fig. (8). The existing tautomeric structure of SAMM2, with H-bonding between NH and O.

Mammalian Metabolism Studies of Sampangine

The chronological order of both studies (microbial and mammalian) will determine the used approach (retrospective, parallel or prospective). However, in this case it is difficult to classify these studies into a distinct approach. The microbial metabolism studies were initiated after tentative information obtained from a limited mammalian metabolism study of sampangine. On the other hand, it is very important to note that complete structure elucidation of mammalian metabolites would not have been possible without the microbial metabolites having been identified first.

In the mammalian metabolism studies, two conjugate type metabolites have been produced and excreted in the urine after intraperitoneal administration of sampangine (80 mg/kg body weight) to rats. Remarkably, the urine was red in color. This observation led us to believe that whatever conjugate(s) were found in the urine would be similar if not identical to the microbial metabolites. The red rat urine was collected and extracted using water-saturated butanol. The combined extracts were evaporated under reduced pressure to afford a red residue. This residue was chromatographed over a reversed phase silica (C-18) column, using methanol-water (10: 0-8.5: 1.5) mixture as eluent to afford two red metabolites. These metabolites were given a designation, **SAM MM** (**SAM**pangine Mammalian Metabolite) followed by a number according to the chronological order of their isolation. Therefore, the isolated metabolites were designated SAM MM1 and SAM MM2. SAM MM2 was further purified by HPLC using a silica gel column, and methanol-chloroform-ammonia (19.5: 80: 0.5) mixture as an eluent to yield two closely related metabolites SAM MM2a and SAM MM2b. No spectral data were obtained and due to their low quantities and unstability, SAM MM2a and SAM MM2b structure elucidation is unlikely.

The solubility and stability of these metabolites have been assessed. The mammalian metabolites are freely soluble in dimethylsulfoxide and pyridine, sparingly soluble in water or methanol, and insoluble in non-polar solvents. They are unstable, particularly in methanol or pyridine.

Structure Elucidation of SAM MM1

The physiochemical properties indicated that SAM MM1 is similar, but not identical, to either one of the microbial metabolites. SAMM MM1 is red in color (in acidic medium) and more polar than both SAMM1 and SAMM2.

The thermospray mass spectrum of SAM MM1 showed no molecular ion peak, indicating that the conjugate decomposed thermally during analysis. High resolution electrospray impact (HRESI) mass spectrum of SAMM MM1 showed the exact molecular weight (m/z 411.1188; M+H$^+$) which was consistent with the molecular formula $C_{21}H_{18}N_2O_7$. This molecular weight was the sum of sampangine (232) and a moiety of 178 mass units (consistent with glucuronic acid). HRESI analysis and the fact that glucuronidation is one of the three common mammalian phase II metabolism led us to strongly believe that SAMM MM1 is an O-glucuronide of sampangine. Confirmation of this belief was achieved through NMR analyses.

The anomeric proton which resonated at δ 4.85 ppm (Table 3) had a large coupling constant (J=8.0 Hz) indicating that the glucuronic acid binds to the aglycone in a β-glycosidic linkage. These results are similar to the one seen with SAMM2 and to other published data [51]. Like SAMM2, the nuclear Overhauser effect (ROESY) between the anomeric proton and another proton at δ 8.68 ppm [(with J=7.3 Hz) (H-8)] led us to confidently conclude that SAM MM1 is an O-glycoside in which the aglycone part is coupled with sampangine through its reduced carbonyl group. Since the anomeric proton was distinguishable, other sugar protons were unambiguously assigned from the COSY spectrum. Likewise, other aromatic protons were assigned from the COSY spectrum starting from the preassigned H-8. Other ROESY cross peaks were seen for protons 2 & 3, 3 & 4, 4 & 5, 8 & 9, 9 & 10, and 10 & 11. The identity of glucuronic acid was confirmed from the coupling constants between protons 2' & 3', 3' & 4', and 4' & 5', which were larger than 7.0 Hz and that indicated the 1, 2-trans-diaxial relationship between each consecutive pair of those protons. It was further supported by the existence of nuclear Overhauser effects between H-1' and H-3', H-1' and H-5', H-3' and H-5', and H-2' and H-4' which proved the 1, 3-cis-diaxial interactions between those protons. The HMQC spectrum showed 21 carbon resonances. These resonances were comparable to those of SAMM2, except the one resonating at δ 176.8 ppm, which is consistent with the presence of a carbonyl of a carboxylic acid function. Unlike SAMM2, the HMQC spectrum showed the absence of any triplet carbon.

Antifungal Activity Evaluation of Sampangine and its Metabolites

Metabolic studies for sampangine have identified several metabolites and evaluation of their antifungal activities is very important. All of these metabolites were tested through one or more of the following stages; (1) *In vitro* qualitative evaluation, 2) *In vitro* quantitative evaluation, and 3) *In vivo* efficacy evaluation. Metabolite quantities dictated which one should go through which stage. An antifungal profile of these metabolites was determined against *Candida albicans* and *Cryptococcus neoformans*.

In vitro Qualitative Evaluation

In vitro qualitative evaluation was accomplished through bioautography [52] and/or the agar well-diffusion assay [53]. Both methods give qualitative, or more correctly, semiquantitative results expressed as zones of fungal inhibition measured in mm. Table 5 illustrates zones of inhibition of all sampangine metabolites against *C. albicans* NIH B311 and *C. neoformans* ATCC 52657 as compared to those of sampangine.

These results indicate that sampangine metabolites have comparable activities to sampangine, which was used as a positive control.

In vitro Quantitative Evaluation

In vitro quantitative evaluation was accomplished using two-fold serial broth macrodilution technique [54]. The activity expressed as minimum inhibitory concentration (μg/ml). Table 5 illustrates minimum inhibitory concentration (MIC) of sampangine metabolites against *C. albicans* B311 and *C. neoformans* ATCC 52657 as compared to those of sampangine.

SAMM1 was considered inactive against *C. albicans*, while it was as active as sampangine against *C. neoformans*.

In vivo Efficacy Evaluation

In vivo efficacy evaluation was accomplished by applying a "survival study" technique [55] in which a group of ICR Swiss female mice were heavily infected by *Cryptococcus neoformans* ATCC 52657. A daily injection of SAMM2 was given intraperitoneally (50 mg/kg) to those infected mice for 12 days. The survivors were counted daily. This study was conducted in comparison with a negative control in which mice were injected daily for 12 days with the vehicle (10% DMSO/saline). Since about 100 mg quantity was needed for this study, SAMM2 (available in these quantities) was the only metabolite to be tested. The results of this study showed that SAMM2 is inactive *in vivo* in the mouse model of cryptococcosis.

Table 5. *In vitro* Antifungal/Antibacterial Activities Evaluation of SAM and its Metabolites

Compound	Bioautography[a]		Agar Well Diffusion Assay[a]		Serial Broth Dilution Assay[b]		Alamar Blue Assay[b]			
	Ca[c]	Cn[d]	Ca	Cn	Ca	Cn	Sa[e]	Pa[f]	Bs[g]	Ec[h]
SAM	10	35	10	35	3.125	0.2	6.25	>25	3.12	>25
SAMM1	NT	20	5	26	100	0.2	NT	NT	NT	NT
SAMM2	NT	23	10	30	>25	0.78	>25	>25	>25	>25
SAM MM1	NT	35	6	26	>25	>25	>25	>25	>25	>25
Streptomycin	NT	NT	NT	NT	NT	NT	3.12	6.25	3.12	12.5

[a]Activity expressed as zone of inhibition measured in mm.
[b]Activity expressed as minimum inhibitory concentration (μg/ml).
[c]Ca; *Candida albicans* NIH B311.
[d]Cn; *Cryptococcus neoformans* ATCC 52657.
[e]Sa; *Staphylococcus aureus* ATCC 6538.
[f]Pa; *Pseudomonas aeruginosa* ATCC 15442.
[g]Bs; *Bacillus subtilus* ATCC 6633.
[h]Ec; *Escherichia coli* ATCC 10536.
NT; Not Tested.

Antibacterial Activity Evaluation of Sampangine and its Metabolites

In vitro quantitative antibacterial activity was evaluated using Alamar Blue Assay [56]. It is a colorimetric oxidation-reduction assay which involves the addition of Alamar blue dye as an indicator. It evaluates the metabolic activity of the microorganisms. The activity of the evaluated drug is expressed as minimum inhibitory concentration (μg/ml). Sampangine, SAMM2 and SAM MM1 were tested for their activity against *Staphylococcus aureus* ATCC 6538, *Pseudomonas aeruginosa* ATCC 15442, *Bacillus subtilus* ATCC 6633, and *E. coli* ATCC 10536. Streptomycin was used as a positive control.

Sampangine was only marginally to moderately active against *S. aureus* and inactive against *P. aeruginosa* and *E. coli*. Both SAMM2 and SAM MM1 had no antibacterial activities. Table 5 illustrates these results.

In conclusion, antifungal/antibacterial activities evaluation revealed that, sampangine metabolites have remarkable *in vitro* antifungal activities, whereas SAMM2 is inactive *in vivo* in the mouse model of cryptococcosis. This led us to conclude that metabolism *per se* is not the cause for the *in vivo* inactivity of sampangine, hence, further studies on the other pharmacokinetic parameters (bioavailability, distribution, clearance, etc) will be necessary. Furthermore, the absence of any antibacterial activity indicated that sampangine and its metabolites are selective in their action, which could be conceptualized as a "Host-Microorganism Selectivity", an important aspect in developing a "safe drug".

Benzosampangine

Chemically, benzosampangine is $9H$-benzo[f]naphtho[1,2,3,-ij][2,7] naphthyridin-9-one.

Acquisition of the Starting Material, Benzosampangine

All the benzosampangine used in this project was obtained through total synthesis via applying Peteron *et al.* method [2]. The yield was 38.3%.

Screening of Microorganisms for their Ability to Metabolize Benzosampangine

A total of fifty four microorganisms (Table 2) were evaluated for their ability to metabolize benzosampangine. TLC analysis of the culture extracts was used to identify the ability of these organisms to metabolize benzosampangine. Substrate control and culture control studies were utilized to ensure that these metabolites were a result of enzymatic activity and not a consequence of degradation or other non-metabolic changes and not related to the medium composition. Based on TLC analyses and control studies, some of these microorganisms were shown to be capable of complete conversion of benzosampangine to three polar compounds. Of those, *Cunninghamella blakesleeana* ATCC 8688a, *Mucor ramannianus* 1839 (CSih) and *Streptomyces rimosus* ATCC 23955 were chosen for preparative scale fermentation to produce sufficient quantities of the three compounds for structural elucidation.

Preparative-scale Biotransformation of Benzosampangine by Cunninghamella blakesleeana

A preparative-scale biotransformation of benzosampangine (192 mg) was performed with *Cunninghamella blakesleeana* ATCC 8688a. This afforded a partially pure metabolite (70.8 mg). This metabolite was further purified by HPLC over a silica gel column, using chloroform-methanol (8: 2) mixture as eluent and fractions were collected and combined based on retention time and TLC analysis. This afforded a pure metabolite (BZSAMM1, 6 mg, 2% yield). This yield is not a good representative of the actual conversion of benzosampangine into this metabolite, since rapid deterioration of this metabolite was evident during the purification process.

Structure Elucidation of the Metabolite BZSAMM1

Since both sampangine and benzosampangine are structurally related, it was logical to expect similar metabolites to be produced. Therefore, the glucoside of benzosampangine was highly expected.

The thermospray mass spectrum of BZSAMM1 showed no molecular ion peak, indicating that the conjugate decomposed thermally during analysis. A HRFAB mass spectrum showed the exact molecular weight (m/z 447.1553; M+H$^+$) which was consistent with the molecular formula $C_{25}H_{22}O_6N_2$. This molecular weight was the sum of benzosampangine (282) and a moiety of 164 mass units (consistent with glucose). NMR analysis (in DMSO-d$_6$) revealed similarities between SAMM2 and BZSAMM1 except for the additional benzene ring in BZSAMM1 (Tables 4 and 6).

The ^{13}C-NMR spectrum of BZSAMM1 showed 25 carbon resonances, one triplet, fifteen doublets, and nine singlets none of which were a carbonyl carbon. Five of those resonances were in the aliphatic oxygenated region (δ 60-80 ppm). The carbon resonating at δ 106.6 ppm as a doublet was determined from the HMQC spectrum to be the anomeric carbon, since it has a one-bond correlation with the anomeric proton (resonating at 4.62 ppm as a doublet with $J=7.9$ Hz). The presence of a 3-bond correlation (HMBC) between the anomeric proton and a singlet carbon (δ 128.9 ppm) which has another 3-bond correlation to H-10 (resonating at 8.48 ppm as a doublet with $J=7.8$ Hz) led us to confidently conclude that BZSAMM1 is an O-glycoside in which the glycogen part is coupled with benzosampangine through its presumably reduced carbonyl group. The mechanism involved in the formation of BZSAMM1 is similar to that involved in the formation of either SAMM1 or SAMM2, and is supported by the presence of a molecular ion at m/z 285; M+H$^+$, suggesting the presence of dihydrobenzosampangine (MW 284) as an intermediate. H-11, 12 and 13 were determined from the COSY spectrum. H-13 has a 3-bond correlation to two singlet carbons (δ 130.4 and 145.1). One of these singlet resonances (δ 130.4) was determined to be C-9a based on a 3-bond correlation to H-11. Therefore, the other singlet carbon (δ 145.1) should be C-13b. C-13b has another 3-bond correlation to a proton resonated at 8.77 ppm as a d ($J=5.3$ Hz) which is assigned to H-2. Consequently, H-3 was determined from the COSY spectrum. H-3 has a 3-bond correlation to two singlet carbons (δ 115.9 and 117.6 ppm). The singlet carbon at δ 115.9 ppm was determined to be C-3b based on 3-bond correlations to H-5 (resonating at δ 6.98 ppm as a double double doublet with $J=7.8, 7.8, 0.5$ Hz) and to H-7 (resonating at δ 7.06 ppm as a doublet with $J=7.8$ Hz). The other singlet carbon (δ 117.6 ppm) was determined to be C-13c since it has a 3-bond correlation to the proton on N-8 (exchangeable proton resonating at δ 10.3 ppm as a singlet) which in turn has a 3-bond correlation to C-3b.

Table 6.   Comparative ^{1}H-NMR Assignments [a] of BZSAM, BZSAMM1

Proton No.	BZSAM [b] δ H	BZSAMM1[c] δ H
2	8.97 (1H, d, J=5.7)	8.77 (1H, d, J=5.3)
3	8.30 (1H, d, J=5.7)	7.83 (1H, d, J=5.3)
4	8.55 (1H, dd, J=7.1, 1.4)	8.07 (1H, d, J=7.8)
5	7.84 (1H, ddd, J=7.1, 7.0, 1.4)	6.98 (1H, ddd, J=7.8, 7.8, 0.5)
6	7.93 (1H, ddd, J=7.1, 7.0, 1.4)	7.38 (1H, ddd, J=7.8, 7.8, 1.0)
7	8.55 (1H, dd, J=7.1, 1.4)	7.06 (1H, d, J=7.8)
10	8.44 (1H, dd, J=7.8, 1.0)	8.48 (1H, d, J=7.8)
11	7.66 (1H, ddd, J=7.8, 7.4, 1.0)	7.59 (1H, ddd, J=7.8, 7.8, 1.0)
12	7.80 (1H, ddd, J=7.8, 7.4, 1.0)	7.39 (1H, ddd, J=7.8, 7.8, 1.0)
13	8.79 (1H, dd, J=7.8, 1.0)	8.91 (1H, d, J=7.8)
1'		4.62 (1H, d, J=7.9)
2'		3.57 (1H, m)
3'		3.34 (1H, m[d])
4'		3.35 (1H, m[d])
5'		3.18 (1H, m)
6'		3.64 (1H, d, J=11.6)
6'		3.58 (1H, d, J=11.6)
N-H		10.3 (1H, s)
O-H		6.8 (1H, bs)
O-H		5.3 (1H, s)
O-H		5.1 (1H, s)
O-H		4.5 (1H, s)

[a]Chemical shifts are reported in parts per million, J values are in Hertz.
[b]Spectra recorded in CDCl$_3$.
[c]Spectra recorded in DMSO-d$_6$.
[d]Hidden under DMSO signal and were determined from HMQC spectrum.

The anomeric proton which resonates at δ 4.62 ppm had a large coupling constant, J=7.9 Hz, indicating that the glycone part binds to the aglycone through a β-glycosidic linkage. These results are similar to those seen with SAMM2. Other protons were assigned from the COSY spectrum. H-3' and H-4' were hidden under the DMSO impurity signal (δ 3.34 ppm) and were determined from HMQC spectrum.

Comparison of the [13]-C-resonances, [1]H-resonances, coupling patterns and coupling constants of the sugar part in both SAMM2 and BZSAMM1 revealed that sugar parts in both metabolites should be identical. Ultimately, the sugar part was determined to be β-glucose. Thus, BZSAMM1 is benzosampangine-β-glucopyranoside.

Preparative-scale Biotransformation of Benzosampangine by Mucor ramannianus and Streptomyces rimosus

Using the standard two-stage fermentation procedure, a preparative-scale biotransformation of benzosampangine (192 mg were fed to each microorganism) was performed with *Mucor rammannianus* 1839 (CSih) and *Streptomyces rimosus* ATCC 23955 to afford, after purification, BZSAMM2 (11.2 mg) and BZSAMM3 (20 mg), respectively. Those metabolites underwent rapid decomposition into benzosampangine as was shown by TLC analysis.

The physiochemical data showed that BZSAMM2 and BZSAMM3 are nearly identical to BZSAMM1. This similarity (not identicalness) was indicated since BZSAMM1, M2 and M3 were more polar than benzosampangine, orange in color and gave positive nitrogen test with Dragendorff's spray reagent. The electrospray mass spectra showed no characteristic molecular ion peak for those metabolites. Similar results were obtained with thermospray mass spectra indicating that BZSAMM2 and M3 were probably conjugate of benzosampangine. No other data was obtained and it has been speculated that these may be sulphate or phosphate conjugates.

3-Methoxysampangine

Although it was the original natural product of the sampangine class to be evaluated for its antifungal activity, 3-methoxysampangine metabolic fates were studied after sampangine and benzosampangine. Its scarcity was the primary reason behind this chronological order of study, and the reason why this study was never as comprehensive. Chemically, 3-methoxysampangine is 3-methoxy-7*H*-naphtho[1,2,3-*ij*][2,7]naphthyridin-7-one.

Due to its scarcity (0.000156% yield), 3-methoxysampangine used in this study was obtained through total synthesis.

Screening of Microorganisms for their Ability to Metabolize 3-Methoxysampangine

A total of thirty two microorganisms (Table 2) were evaluated for their ability to metabolize 3-methoxysampangine. *Streptomyces spectabilis* (ATCC 27465) and *Streptomyces griseus* (ATCC 13968) were chosen for screening studies to evaluate their demethylation ability based on a published study done by Hufford *et al*. [57]. In that study *S. griseus* and *S. spectabilis* were shown to be capable of demethylating colchicine into several demethylated metabolites. They were shown to do so, only when grown in medium θ which consists of (per liter of distilled water): glucose, 10 g; corn steep liquor, 10 g; soya flour, 10 g; dry malt extract, 5 g; CaCO$_3$, 1 g; NaCl, 5 g. Similar results were obtained in this study.

Based on TLC analyses and control studies, only *Streptomyces spectabilis* ATCC 27465 (grown in medium θ) was shown to be capable of complete conversion of 3-methoxysampangine into two more polar compounds, designated 3-MeOSAMM1 and 3-MeOSAMM2. Limitations in time and starting material for 3-methoxysampangine did not allow for preparative-scale fermentation studies.

Metabolite 3-MeOSAMM1 is thought to be the demethylated metabolite, 3-hydroxysampangine, since it gave blue color with ferric chloride spray reagent (specific to detect phenolic compounds), and positive nitrogen test with Dragendorff's spray reagent.

Metabolite, 3-MeOSAMM2, was much more polar than the first metabolite, it gave positive nitrogen test with Dragendorff's spray reagent, and negative ferric chloride test. No other data was obtained. Thus it is another highly polar conjugate.

It is needless to mention that this is only a tentative identification of those metabolites and further work has to be done in the synthesis of the starting material and the preparative-scale fermentation to produce enough quantities of those metabolites for structural elucidation and other biological studies.

ACKNOWLEDGEMENT

This chapter is a part of the author's Ph. D. dissertation. The author would like to acknowledge Dr. Charles D. Hufford, School of Pharmacy and National Center for the Development of Natural Products, The University Mississippi, University, MS 38677, USA, for providing the expertise, counseling and encouragement through this research project.

LIST OF ABBREVIATIONS

BZSAM:	Benzosampangine
BZSAMM:	Benzosampangine microbial metabolite

COSY:	Homonuclear correlated spectroscopy
g	Gram
HETCOR:	Heteronuclear correlated spectroscopy
HMBC:	Heteronuclear multiple bond correlation
HPLC:	High pressure liquid chromatography
hr:	Hour
HRESI:	High resolution electrospray impact
HRFAB:	High resolution fast atom bombardment
L:	Liter
MBC:	Minimum bacteriocidal concentration
3-MeOSAM:	3-Methoxysampangine
3-MeOSAMM:	3-Methoxysampangine microbial metabolite
MFC:	Minimum fungicidal concentration
MIC:	Minimum inhibitory concentration
NOESY:	Nuclear overhauser enhancement spectroscopy
OI:	Opportunistic infections
ROESY:	Rotating-frame overhauser enhancement spectroscopy
SAM:	Sampangine
SAMM:	Sampangine microbial metabolite
SAM MM:	Sampangine mammalian metabolite
TLC:	Thin layer chromatography

REFERENCES

[1]   Liu, S.C.; Oguntimein, B.; Hufford, C.D.; Clark, A.M. *Antimicrob. Agents Chemother.,* **1990**, *34*, 529.

[2]   Peterson, J.R.; Zjawiony, J.K.; Liu, S.; Hufford, C.D.; Clark, A.M.; Rogers, R.D. *J. Med. Chem.,* **1992**, *35*, 4069.

[3]   Bracher, F. *Liebigs Ann. Chem.,* **1989**, 87.

[4]   Zjawiony, J.K.; Khalil, A.A.; Hufford, C.D.; Clark, A.M.; Buolamwini, J.K. *J. Heterocyclic Chem.,* **1997**, *34*, 1233.

[5]   Rao, J.U.M.; Giri, G.S.; Hanumaiah, T.; Rao, K.V.J. *J. Nat. Prod.,* **1986**, *49*, 346.

[6]   Zjawiony, J.K.; Srivastava, A.R.; Hufford, C.D.; Clark, A.M. *Heterocycles,* **1994**, *39*, 779.

[7]   Conti, A.; Bickel, M. *Drug Metab. Rev.,* **1977**, *6*, 1.

[8]   Williams, R.T. *Detoxication Mechanisms*, Chapman & Hall: London, **1959**.

[9]   Testa, B.; Jenner, P. Drug Metabolism: Chemical and Biochemical Aspects, Dekker: New York, **1990**.

[10]   Testa, B.; Jenner, P. *Current Contents Life Sci.,* **1990**, *33*, 17.

[11]   Balant, L.P.; Roseboom, H.; Gundert-Remy, U.M. In *Advances in Drug Research;* Testa, B., Ed.; Academic Press Inc.: London, **1990**; Vol. *19*, pp. 1-138.

[12]   Testa, B.; Caldwell, J. In *The Metabolism of Drugs and Other Xenobiotics*; Testa, B.; Caldwell, J., Eds.; Academic Press Inc.: San Diego, CA, **1995**, pp. 1-40.

[13]   Rosazza, J.P.; Smith, R.V. *Adv. Appl. Microbial.,* **1979**, *25,* 169.
[14]   Clark, A.M.; Hufford, C.D. *Med. Res. Rev.,* **1991**, *11*, 473.
[15]   Lee, I.S.; Hufford, C.D. *Pharmac. Ther.,* **1990**, *48,* 345.
[16]   DelRaso, N.J. In *In Vitro Methods of Toxicology*; Watson, R.P., Ed.; CRC Press: Boca Raton, FL, **1992**, pp. 175-201.
[17]   Dagani, R. *Chem. Eng. News,* **1983**, Oct. 31, 7.
[18]   Rollin, B.E. *Animal Rights and Human Mortality*, Prometheus: Buffalo, **1981**.
[19]   Dagani, R. *Chem. Eng. News*, **1984**, Nov. 12, 25.
[20]   Dawson, M. *Pharm. J.,* **1984**, 607.
[21]   Holden, C. *Science,* **1982**, *215*, 35.
[22]   Orrenius, S.; Ernester, L. In *Molecular Mechanisms of Oxygen Activation*; Hayaishi O., Ed.; Academic Press Inc.: New York, **1974**, pp. 216-218.
[23]   Miller, L.L. In *Isolated Liver Perfusion and Its Applications*; Bartosik, I.; Guaitani, A.; Miller, L.L., Eds.; Raven: New York, **1973**, pp. 11-52.
[24]   Sies, H. In *Methods in Enzymology*; Fleischer, S.; Packer, L., Eds.; Academic Press Inc.: San Diego, CA, **1978**; Vol. *52*, pp. 48-59.
[25]   Begue, J.M.; LeBigot, J.F.; Guguen-Guillouzo, C.; Kiechel, J.R.; Guillouzo, A. *Biochem. Pharmacol.,* **1983**, *32*, 1643.
[26]   Davis, P.J. In *Antibiotics and Microbial Transformations*; Lamba, S.S.; Walker, C.A., Eds.; CRC Press: Boca Raton, FL, **1987**, pp. 47-70.
[27]   Smith, R.V.; Rosazza, J.P. *J. Pharm. Sci.,* **1975**, *64*, 1737.
[28]   Smith, R.V.; Rosazza, J.P. *Biotech. Bioeng.,* **1975**, *17*, 785.
[29]   Rosazza, J.P.; Smith, R.V. In *Applied Microbiology*; Perlman, D., Ed.; Academic Press Inc.: New York, **1979**; Vol. *25*, pp. 169-208.
[30]   Smith, R.V.; Rosazza, J.P. In *Microbial Transformations of Bioactive Compounds*; Rosazza, J.P., Ed.; CRC Press: Boca Raton, FL, **1982**; Vol. *2*, pp. 1-42.
[31]   Smith, R.V.; Rosazza, J.P. *J. Nat. Prod.,* **1983**, *46*, 79.
[32]   Smith, R.V.; Davis, J.P. In *Adv. Biochem. Eng.*; Ghasi, T.K.; Fiechter, A.; Blakebrough, N., Eds.; Springer-Verlag: Berlin, **1980**; Vol. *14*, pp. 61.
[33]   Smith, R.V.; Acosta, D.; Rosazza, J.P. *Adv. Biochem. Eng.,* **1977**, *5*, 70.
[34]   Rosazza, J.P. *Recent Results Cancer Res.,* **1978**, *63*, 58.
[35]   Ferris, J.P.; Fasco, M.J.; Stylianopoulou, F.L.; Jerina, D.M.; Dali, J.W.; Jeffrey, A.M. *Archs. Biochem. Biophys.,* **1973**, *156*, 97.
[36]   Rosazza, J.P. Microbial Transformations of Physiologically Active Substances, CRC Press: Boca Raton, FL, **1982**.
[37]   Goodhue, C.T. In *Microbial Transformations of Bioactive Compounds;* Rosazza, J.P., Ed.; CRC Press: Boca Raton, FL, **1982**; Vol. *2*, pp. 9-44.
[38]   Hufford, C.D.; Baker, J.K.; Clark, A.M. *J. Pharm. Sci.,* **1981**, *70*, 155.
[39]   Hufford, C.D.; Capiton, G.A.; Clark, A.M.; Baker, J.K. *J. Pharm. Sci.,* **1981**, *70*,   151.
[40]   Clark, A.M.; Hufford, C.D.; McChesney, J.D. *Antimicrob. Agents Chemother.,* **1981**, *19*, 337.
[41]   Kieshich, K. Microbial Transformations of Non-Steroid Cyclic Compounds, Wiley-Georg Thieme: Stuttgart, **1976**.
[42]   Beukers, R.; Marx, A.F.; Zuidweg, M.H.J. In *Drug Design;* Ariens, E.J., Ed.; Academic Press Inc.: San Diego, CA, **1972**; Vol. *3*, pp. 1-131.
[43]   Fonken, G.; Johnson, R.S. *Chemical Oxidation with Microorganisms*, Dekker: New York, **1972**.

[44]   Skryabin, G.K.; Golovleva, L.A.M. *Microorganisms in Organic Chemistry,* Nauka: Moscow, **1976**.

[45]   Nahrstedt, A.; Economou, D. *J. Nat. Prod.,* **1990**, *53*, 1387.

[46]   Sharma, P.; Alam, M. *J. Nat. Prod.,* **1989**, *52*, 395.

[47]   Agrawal, P.K. *Phytochemistry,* **1992**, *31*, 3307.

[48]   Hezari, M.; Davis, P.J. *Drug Metabolism and Disposition,* **1993**, *21*, 259.

[49]   Tanaka, T.; Tsukamoto, S.; Hayashi, K. *Phytochemistry,* **1990**, *29*, 229.

[50]   Seto, H.; Otake, N.; Koyama, M.; Ogino, H.; Kodama, Y.; Nishizawa, N.; Tsuruoka, T.; Inouye, S. *Tetrahedron Letters,* **1983**, *24*, 495.

[51]   Fujimaki, M.; Hakusui, H. *Xenobiotica,* **1990**, *20*, 1025.

[52]   Rahalison, L.; Hamburger, M.; Hostettmann, K.; Monod, M.; Frenk, E. *Phytochemical Analysis,* **1991**, *2*, 199.

[53]   Clark, A.M.; El-Feraly, F.S.; Li, W.-S. *J. Pharm. Sci.,* **1981**, *70*, 951.

[54]   Galgiani, J.M.; Bartlett, M.S.; Espinel-Ingroff, A.; Fromtling, R.A.; Pfaller, M.A.; Rinaldi, M.G. Reference method for broth dilution antifungal susceptibility testing of yeasts; proposed standard, NCCLS Document M27-P, **1992**, *12*, 1.

[55]   Rinaldi, M.G.; In *Recent Trends in the Discovery, Development and Evaluation of Antifungal Agents;* Fromtling, R.A., Ed.; J. R. Prous Science Publishers: Barcelona, Spain, **1987**, S1:11-S1:24.

[56]   Pfaller, M.A.; Grant, C.; Morthland, V.; Rhine-Chalberg, J. *J. Clin. Microbiol.,* **1994**, *32*, 506.

[57]   Hufford, C.D.; Collins, C.C.; Clark, A.M. *J. Pharm. Sci.,* **1979**, *68*, 1239.

Atta-ur-Rahman (Ed.) *Studies in Natural Products Chemistry, Vol. 23*
© 2000 Elsevier Science B.V. All rights reserved

ANSAMYCIN ANTIBIOTICS[A] DISCOVERY, CLASSIFICATION, BIOSYNTHESIS AND BIOLOGICAL ACTIVITIES

SHINJI FUNAYAMA*[1] *and* GEOFFREY A. CORDELL[2]

*Department of Bioscience and Biotechnology, Faculty of Engineering, Aomori University, 2-3-1- Kohbata, Aomori 030-0943, Japan
Department of Meeicinal Chemistry and Pharmacognosy, College of Pharmacy, University of Illinois at Chicago, 833 S. Wood Str., Chicago, IL 60612-7231, USA*

ABSTRACT: The discovery, classification, biosynthesis and biological activities of ansamycin antibiotics are reviewed.

INTRODUCTION - MACROCYCLIC LACTAMS AND ANSAMYCINS

Macrocyclic lactam antibiotics consist of the ansamycin antibiotics and a number of other compounds which are classified as non-ansamycin macrocyclic lactams (Scheme 1).

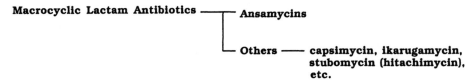

Scheme 1. Classification of macrocyclic lactam antibiotics.

The ansamycins constitute a class of antibiotics characterized by a cyclic structure consisting of an aromatic moiety (naphthalene or benzene derivative) and an aliphatic chain that forms a bridge linking two non-adjacent positions of an aromatic moiety. The word "ansa" is Latin and means grip. The name "ansamycin" was originally suggested by Prelog and Oppolzer [1], and the term "ansa" is derived from the "ansa compounds" coined by Lüttringhaus [2] who called the aliphatic chain of the compounds, such as indicated in Fig. 1, as an ansa moiety. The ansamycin antibiotics are structurally comprised of a naphthalene or a benzene derivative as a chromophore and an ansa chain moiety. To date, about 120

---

[a]Dedicated to the memory of Dr. Iwao Umezawa (1934-1988), the former director of the research division of the Kitasato Institute, Tokyo, Japan.

naturally occurring ansamycins have been discovered and characterized. In this review article, the discovery, classification, biosynthesis and biological properties of the ansamycin antibiotics are discussed.

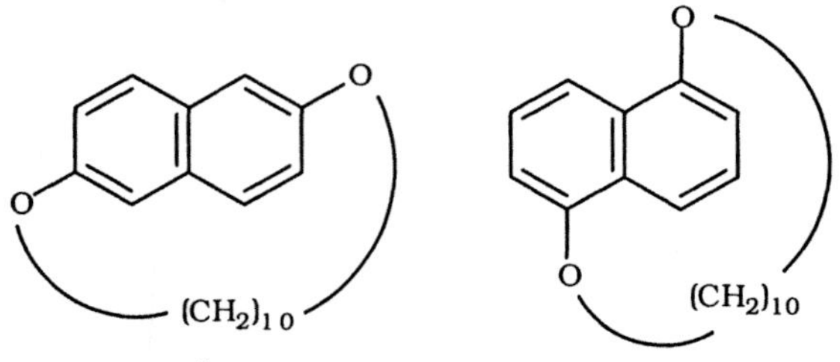

Fig. (1). The first ansa compounds.

The first report of an ansamycin antibiotic was the isolation of the rifomycin (the name rifomycin was changed into rifamycin subsequently) complex, and it was found later that rifamycin B (**1**) was the major product of this complex. Rifamycins are a family of antibiotics first isolated in the Lepetit Research Laboratories in 1957 [3].

The structures of the ansamycins which have provided the name for these antibiotics are remarkably complex, as we shall review in the following sections. However, it is the bioactivities of these compounds which first attracted the greatest attention.

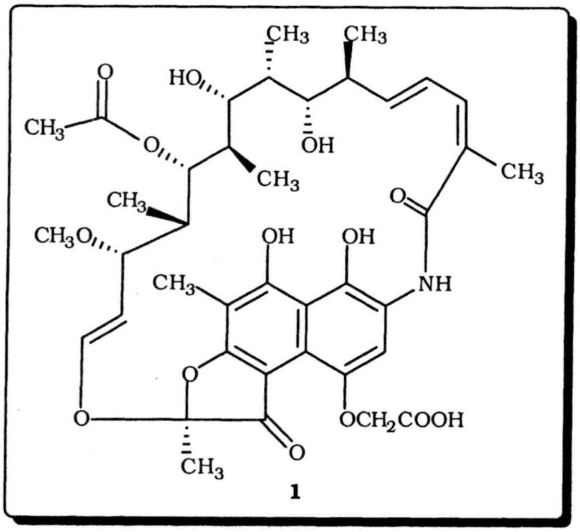

Among others, the ansamycin antibiotics have displayed antibacterial, antitumor and herbicidal activities. For many years, rifampicin, a semisynthesized antibiotic derived from rifamycin B, has been utilized for the treatment of tuberculosis.

As examples of the non-ansamycin macrocyclic lactams, stubomycin (hitachimycin) was first isolated as an antitumor antibiotic [4-8]. Ikarugamycin [9], capsimycin [10], rapamycin and demethoxy-rapamycin [11-13] are other examples. Through the investigation of the biosynthesis of stubomycin (hitachimycin), it was found that the benzene ring of the β-phenylalanine moiety was involved as a chromophore outside of the macrocyclic moiety [5,14].

For readers who are interested in the ansamycins, several review articles have appeared concerning various aspects of the ansamycin antibiotics [1, 15-24].

DISCOVERY AND CLASSIFICATION OF ANSAMYCINS

As described in the introduction, macrocyclic lactams can be classified as ansamycins and other compounds, and the ansamycins consist of an ansa-chain and a chromophore-containing moiety. The ansamycins are typically divided into two groups, i.e., the naphthalenoid ansamycins and the benzenoid ansamycins, according to their chromophores.

Among the naphthalenoid and benzenoid ansamycin antibiotics, some ansamycins possess a 1,4-naphthoquinone or a 1,4-quinone unit as a chromophore, and others possess a 1,4-hydroxynaphthalene or a 1,4-hydroquinone units (or their derivatives) as a chromophore. Since these two types of chromphore are in most cases reversible, it is not appropriate to classify by the difference of the oxidation stage of the chromophore. In this review, benzenoid and naphthalenoid ansamycins are further divided according to the difference of the length of their ansa chains. Thus, naphthalenoid ansamycins are divided into 3 groups, i.e., naphthalenoid ansamycins with C_{17} ansa chains, naphthalenoid ansamycins with C_{23} ansa chains, and naphthalenoid ansamycins with C_9 ansa chains. The benzenoid ansamycins are divided into 2 groups, i.e., benzenoid ansamycins with C_{15} ansa chains and benzenoid ansamycins with C_{17} ansa chains. The relationships between, and the antibiotics within these groups are indicated in Scheme 2.

The rifomycin complex was first isolated at the end of the 1950's [3], and the major product produced under certain conditions was named as rifomycin (rifamycin) B. The rifomycin complex is of interest its high degree of activity against Gram-positive bacteria and *Mycobacterium tuberculosis*, and for its low toxicity. The microbiological and pharmacological properties, as well as details of preliminary clinical investigations, of the rifomycin complex have been presented [25-27]. The name rifomycin was changed to rifamycin subsequently. Rifamycin B related compounds, such as rifamycin O [28] and rifamycin S [29], obtained by the chemical transformation of rifamycin B, and rifamycin L [30], rifamycin SV [31] and rifamycin Y [32,33], isolated from the cultured broth of *Streptomyces mediterranei* (which was later renamed as *Nocardia mediterranei*), were also described.

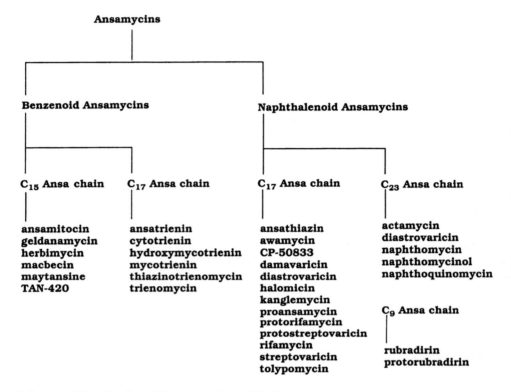

Scheme 2. Classification of the ansamycin antibiotics.

Naphthalenoid Ansamycin Antibiotics

As described above, the naphthalenoid ansamycins are divided into three groups according to the size of the macrocyclic ring, i.e., some of them possess C_{17} ansa chains, and other naphthalenoid ansamycins possess C_{23} and C_9 ansa chains, respectively.

Naphthalenoid Ansamycin Antibiotics with C_{17} Ansa Chains

Naphthalenoid ansamycins with C_{17} ansa chains can be further classified into three groups, namely, the rifamycin group, the proto-streptovaricin group and the streptovaricin group, based on the difference of their chromophores. Awamycin [34,35] is an example of the naphthalenoid ansamycin antibiotics and belongs to the protostrepto-varicin group with a sulphur atom in the molecule.

Rifamycin G [36] and 16,17-dehydrorifamycin G [37] are examples in the rifamycin series where C-1 is replaced by an oxygen atom. A γ-pyrone ring is therefore present, instead of the quinone or dihydro-quinone moiety of other rifamycin type ansamycin anibiotics.

Rifamycin B and Related Compounds

The structures of rifamycins B and S have been vigorously studied and finally the structure of rifamycin B (1) (as a *p*-iodoanilide) was determined by X-ray crystallographic analysis [38,39] and the configurations of 1 and related compounds were described [28,40]. As described above, the absolute configuration of 1 was investigated by X-ray crystallography, and it was defined with no mirror image problem [38,39,41,42].

Rifamycins S (2) and SV (3) were first obtained by the transformation of 1, and these antibiotics were also isolated from the fermentation broth of *Micromonospora lacustris* sp. *nova*, or its mutant, together with 3-thiomethylrifamycin SV (4) [43,44]. Rifamycin B (1) seemed to be formed from rifamycin S (SV) (2 (3)) by the addition of a C_3-precursor as the glycolic acid moiety [45]. As the synthetic precursor of the semisynthetic ansamycin antibiotics, 3-formylrifamycin SV (5) was prepared from rifamycin SV (3), as discussed in the following section.

Carbon-13 NMR spectral studies of rifamycins were reported and the unambiguous ^{13}C NMR assignments of rifamycin S (2) were given [47].

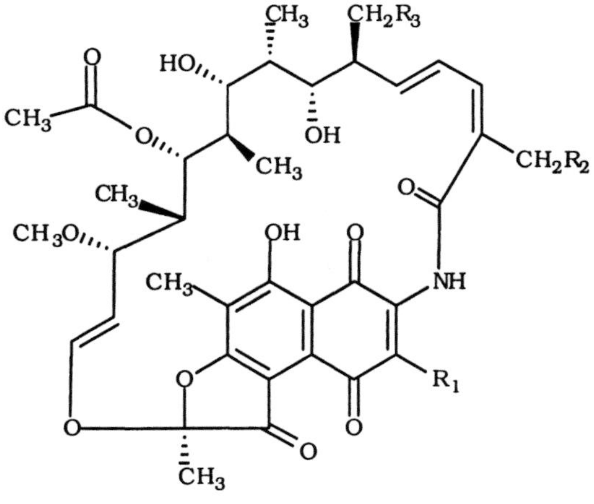

```
2   R_1 = R_2 = R_3 = H
13  R_1 = R_3 = H, R_2 = OH
15  R_1 = OH, R_2 = R_3 = H
16  R_1 = R_3 = OH, R_2 = H

3   R = H
4   R = SCH_3
5   R = CHO
```

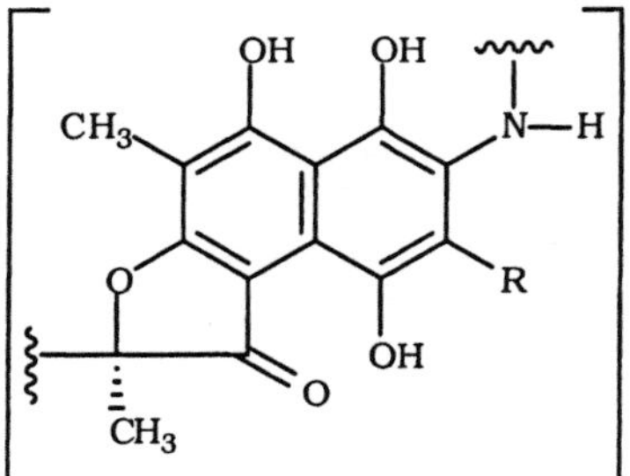

As a member of the rifamycin group of ansamycin antibiotics, tolypomycin Y (**6**) was isolated from the culture broth of *Streptomyces tolypophorus* and its structure was determined. Compound **6** consists of tolypomycinone and tolyposamine [48, 49].

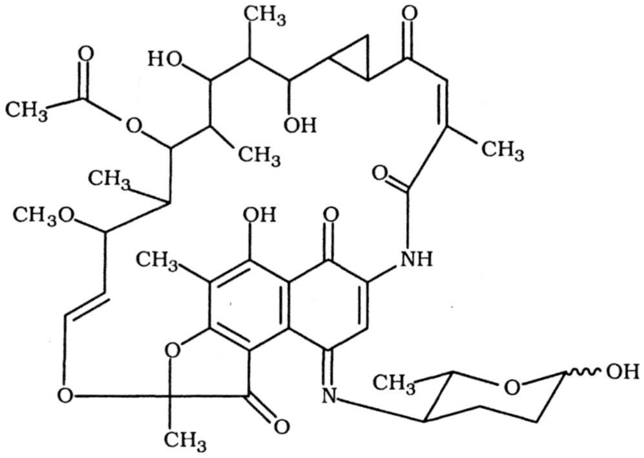

6

Varicinal A, with the same relative stereochemistry, was obtained by the degradation of both streptovaricin and rifamycin [15]. Rifamycin B (**1**) could be converted into rifamycin O (**7**) (and *vice versa*, Fig. 2) and, as shown above, rifamycin S (**2**) could be obtained from **1** [50]. An X-ray structural analysis and the NMR spectrum of **2** were reported [51]. The absolute configuration of **2** was deduced by X-ray crystallography and degradation studies [16,52]. It was reported that the OH groups attached to C-9 (C-23) and C-11 (C-25) of the rifamycin group antibiotics are oriented in the same direction as those of the two oxygens belonging to the chromophore [18].

Synthesis of the aliphatic building block of rifamycin S (**2**) was achieved in 1980 [53], and the total synthesis of racemic **2** was achieved subsequently [53-56].

Subsequently, 25-deacetyl-25-*epi*-hydroxy-rifamycin S (**8**), an example of a rifamycin S epimer, was synthesized with the aim of increasing the intrinsic activity of rifamycin S (**2**). However, the biological evaluation did not confirm this hypothesis [57]. 23-*Epi*-25-deacetylrifamycin S and 21-*epi*-rifamycin S were also prepared [58, 59].

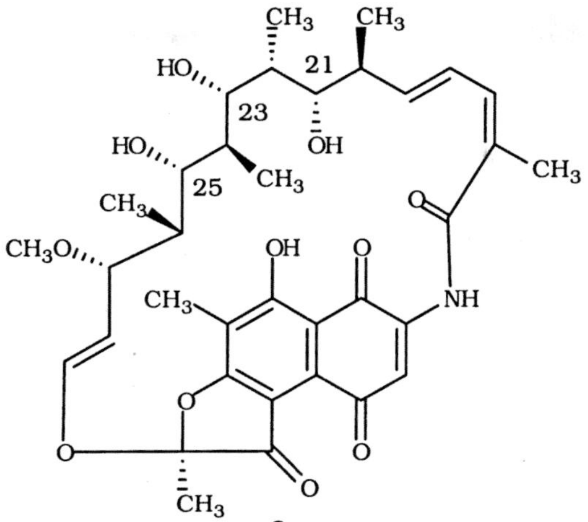

Several new rifamycin derivatives were synthesized and correlations of their NMR spectra indicate that C-3 substitution affects the overall conformation of the ansa bridge [60].

An antibiotic complex, consisting of at least four components, was isolated from the cultured broth of two new organisms belonging to the genus *Micromonospora*, specifically, *M. halophytica* sp. n. (NRRL 2998) and *M. halophytica* var. n. *nigra* (NRRL 3097), and named the halomicins. Halomicins showed inhibitory activity against Gram-positive organisms, including nine strains of clinically-isolated staphylococci which are resistant to penicillin [61]. The LD_{50} in mice is 5,650 mg/kg (sc), 1,250 mg/kg (ip) and >6,000 mg/kg (po).

Subsequently, the halomicins A (9), B (10) and C (11) were purified from the antibiotic complex and their structures determined. The structure of halomicin B (10) was quite similar to that of the rifamycin group, except that it possessed a pyrrolidine moiety at the C-4 position of the chromophore, and 10 possessed the same ansa moiety as some of the rifamycin group of antibiotics. In fact, the absolute configuration of the ansa moiety of 10 was determined by transforming rifamycin O (7) into 10 [62, 63].

As other rifamycin derivatives obtained from the culture broth of *Nocardia mediterranei*, rifamycin G (12) was isolated and characterized [36]. Rifamycin G (12) possesses a γ-pyrone ring as a partial structure of its chromophore instead of the quinone ring of rifamycin S (2), and the double bond at C_{16}-C_{17} of the ansa chain was hydrogenated.

On the other hand, rifamycin R (13) was isolated from the culture broth of a mutant of *Nocardia mediterranei* [64]. Rifamycin P (14) was also isolated from the fermentation of a mutant of *Nocardia mediterranei* or by

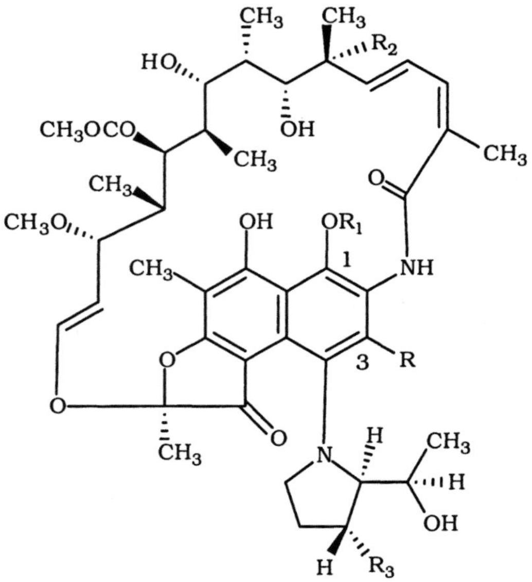

9    $R_1 = R_2 = H, R_3 = OH$
10   $R_1 = R_2 = R_3 = H$
11   $R_1 = R_3 = H, R_2 = OH$

the chemical modification of rifamycin S (**2**) [65]. The synthesis and biological activity of some derivatives of **14** were reported [66].

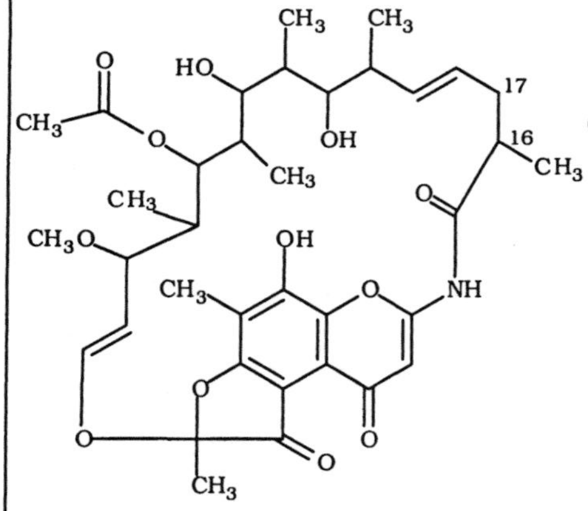

12

17 $\Delta^{16,17}$

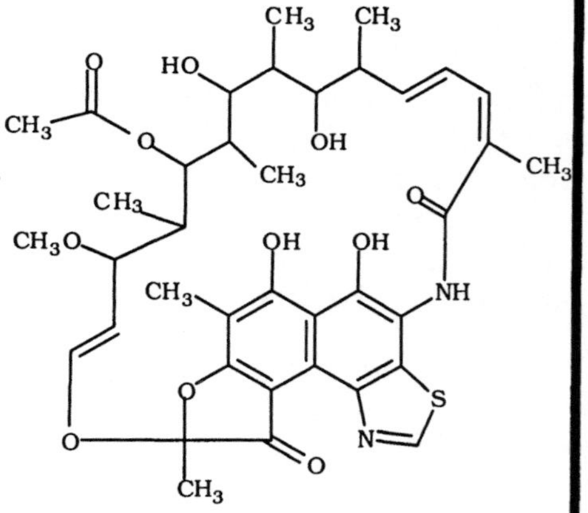

14

A recombinant strain of *Nocardia mediterranei* was found to produce a number of new rifamycins, in addition to the known rifamycins. The new ansamycins were structurally related to rifamycins S (2) and G (12), and were designated as 3-hydroxyrifamycin S (15), 3,31-dihydroxyrifamycin S (16), and 16,17-dehydrorifamycin G (17), respectively [67].

Kanglemycin A (18) was isolated from the culture broth of *Nocardia mediterranei* var. *kanglensis* 1747-64. Kanglemycin A (18) is one of the rifamycin-type ansamycin antibiotics which possesses inhibitory activity against Gram-positive bacteria [68].

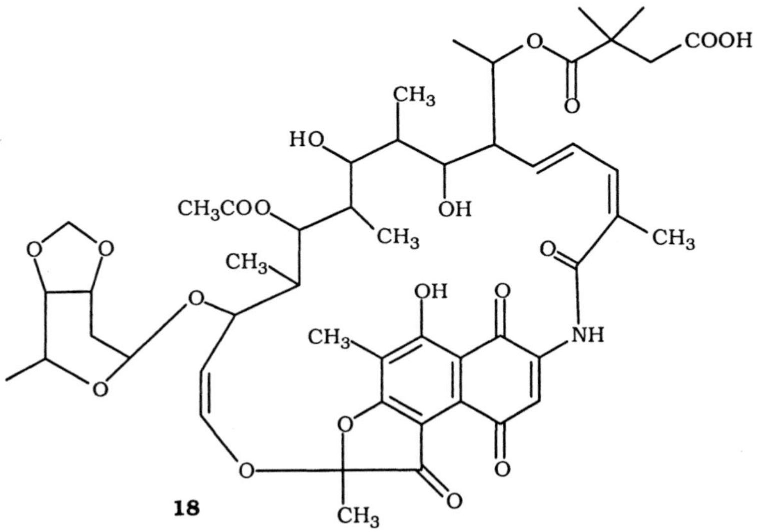

18

Rifampicin (Rifampin)

A semi-synthetic antibiotic, rifampicin (rifampin) (19), was prepared from compound 5, obtained by the formylation of the C-3 position of rifamycin SV (3), followed by the reaction of 5 with 1-amino-4-methylpiperazine in tetrahydrofuran (Fig. 2) [69].

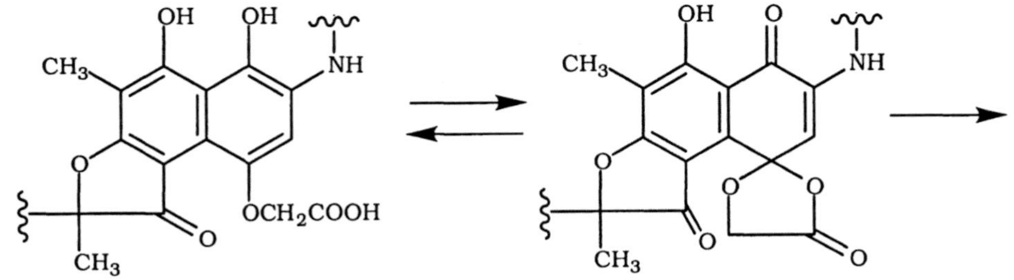

1   7

(Fig. 2). contd.....

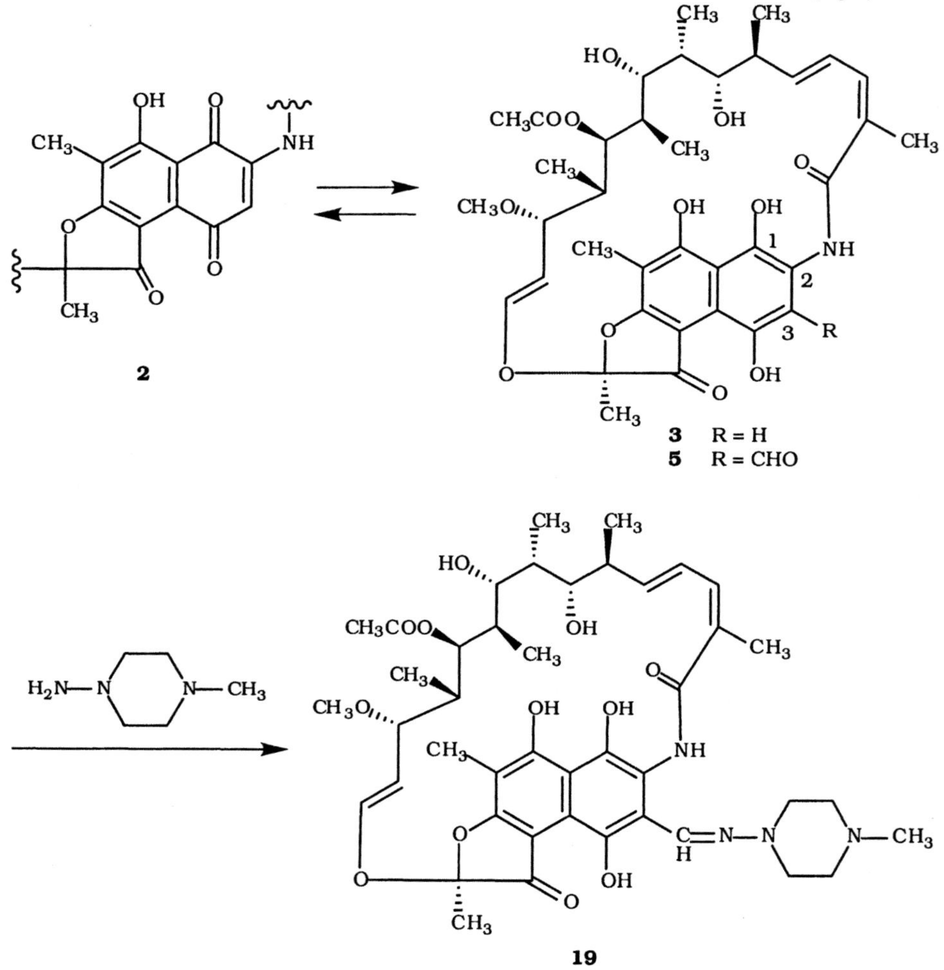

Fig. (2). Preparation of rifamycin S (2) and rifampicin (19) from rifamycin B (1).

Some other chemical reactivities of rifamycin S (SV), especially the reactivity of the C-3 position [60,70,71] and the ^{13}C NMR of these derivatives [72,73] were reported, and it was also reported that electronegative groups at C-3 of rifamycin S (2) enhanced its activity toward DNA-dependent RNA polymerase [74].

Damavaricins, Streptovaricins and Other Naphthalenoid Ansamycins

Rifamycin W (20), a naphthalenoid ansamycin antibiotic with a C_{17} ansa chain was isolated from a mutant strain of *Nocardia mediterranei*. Unlike rifamycins B (**1**) and S (**2**) and related compounds, rifamycin W (**20**) maintains a CH_3 moiety at the C_{16} position [75,76]. From *Amycolatopsis mediterranei*, 31-homorifamycin W (**21**) was isolated as a novel metabolite [77]. It possessed an ethyl moiety at the C_6 position instead of the methyl moiety of **20**.

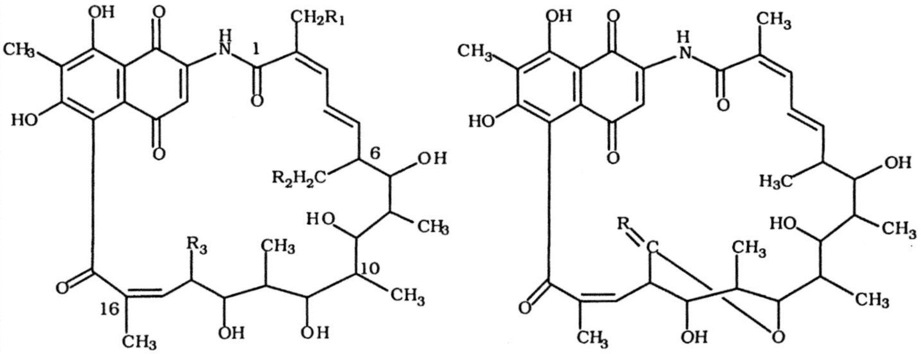

20	$R_1 = R_2 = H, R_3 = CH_2OH$
21	$R_1 = H, R_2 = CH_3, R_3 = CH_2OH$
24	$R_1 = OH, R_2 = H, R_3 = CH_2OH$
25	$R_1 = R_3 = OH, R_2 = H$

22	R = O
23	R = H, OH

A recombinant strain of *Nocardia mediterranei* produced rifamycin W-lactone (**22**), rifamycin W-hemiacetal (**23**), 30-hydroxyrifamycin W (**24**) and 28-dehydroxymethyl-28,30-dihydroxyrifamycin W (**25**), together with 3-hydroxyrifamycin S (**15**), 3,31-dihydroxyrifamycin S (**16**) and 16,17-dehydrorifamycin G (**17**) [67]. Rifamycin W-lactone (**22**) was also reported under the name of rifamycin Z [78].

Awamycin (**26**) was isolated from the cultured broth of *Streptomyces* sp. 80-217 [34,35] as an antitumor antibiotic. Awamycin (**26**) is one of the naphthalenoid ansamycin antibiotics containing a sulphur atom in the molecule at the C-3 position of the chromophore, and its structure is closely related to rifamycin W (**20**).

On the other hand, ansathiazin (**27**) was isolated from the culture filtrate of *Strepromyces albolongus* C-46366 [79] as a minor product, together with **26**. Compound **27** also contains a sulphur atom in the molecule, as in the case of **26**. Ansathiazin (**27**) was shown to possess inhibitory activity against Gram-positive bacteria and acid-fast bacteria.

Other examples of naphthalenoid ansamycin antibiotics with a sulphur atom are CP-50833 (**28**), obtained from the culture broth of *Streptomyces nigellus* subsp. *africanus* ATCC31496 [80], and the semisynthetic antibiotics derived from rifamycin S (**2**), i.e., 3-thiomethylrifamycin S and 3-thiomethylrifamycin SV (**4**) which were obtained by repeatedly treating rifamycin S with MeSH [81]. The double bond at the C_4-C_5 position of **28**

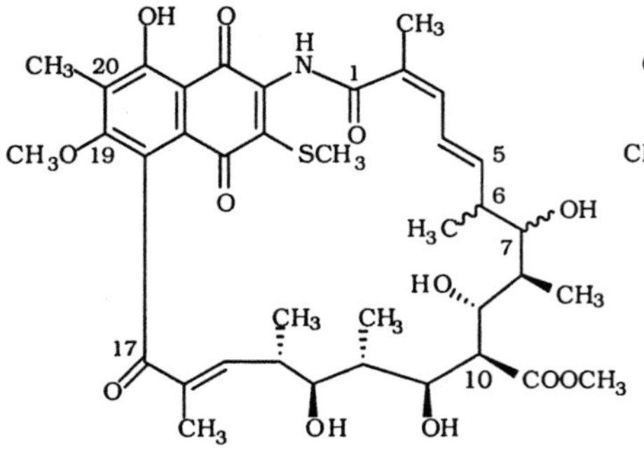

26

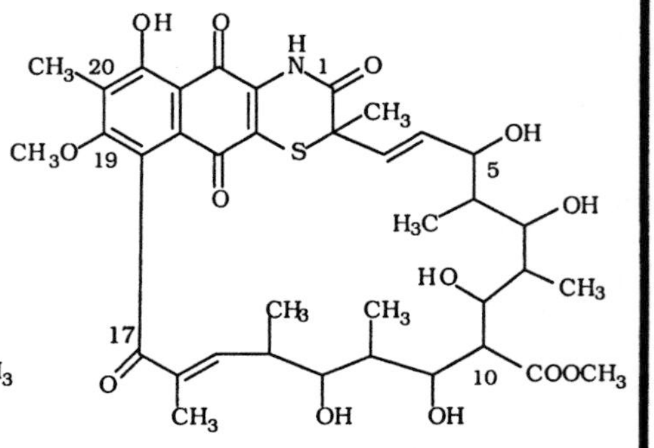

27

was reported to have a *Z* configuration instead of the *E* configuration in awamycin (**26**).

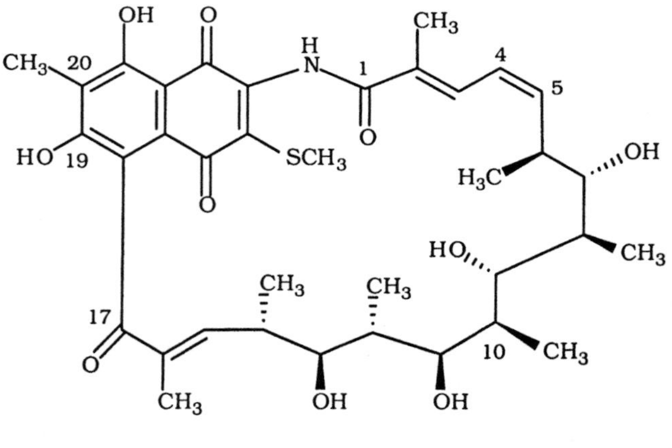

28

Protostreptovaricins I (**29**), II (**30**), III (**31**), IV (**32**) and V (**33**) were reported [82], and protorifamycin I (**34**) was isolated from a mutant strain derived from *Nocardia mediterranei* N813 [83]. Protorifamycin I (**34**) corresponds to 8-deoxyrifamycin W. Several 8-deoxy derivatives of rifamycin W (**20**), namely derivatives of protorifamycin I (**34**) were isolated from the cultured broth of *Nocardia mediterranei* [37,84,85].

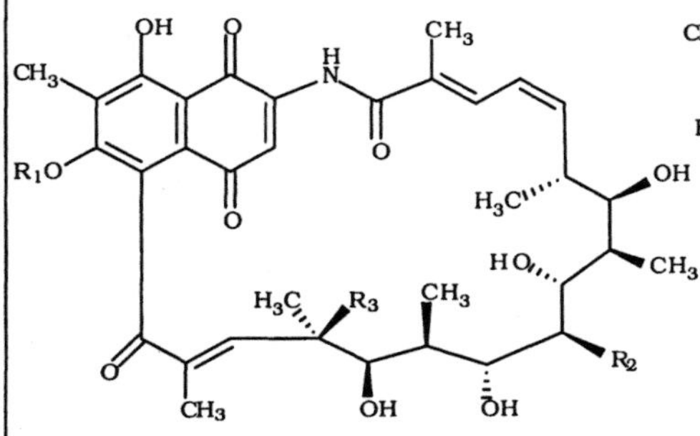

29	$R_1 = R_3 = H, R_2 = CH_3$
30	$R_1 = R_2 = CH_3, R_3 = H$
31	$R_1 = H, R_2 = CH_3, R_3 = OH$
32	$R_1 = R_2 = CH_3, R_3 = OH$
33	$R_1 = R_2 = R_3 = H$

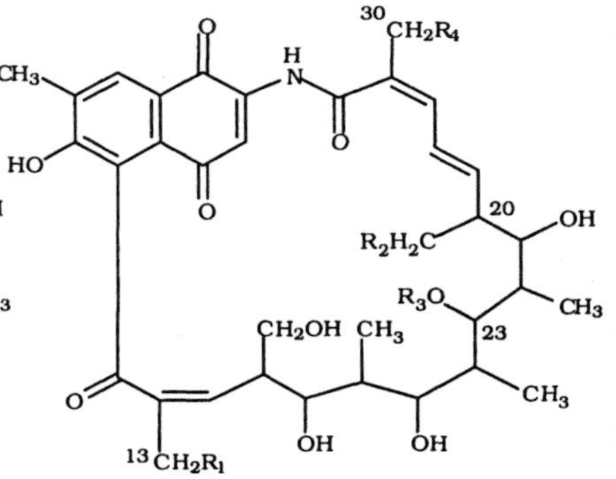

34	$R_1 = R_2 = R_3 = R_4 = H$
35	$R_1 = R_2 = R_4 = H, R_3 = COCH_3$
36	$R_1 = OH, R_2 = R_3 = R_4 = H$
37	$R_1 = R_3 = R_4 = H, R_2 = OH$
38	$R_1 = R_2 = R_3 = H, R_4 = OH$
39	$R_1 = R_2 = R_4 = H, R_3 = C_{23}$ keto

These compounds which are related to **34**, include 23-*O*-acetyl-protorifamycin I (**35**), 13-hydroxyprotorifamycin I (**36**), 20-hydroxyprotorifamycin I (**37**), 30-hydroxyprotorifamycin I (**38**) and 23-ketoprotorifamycin I (**39**). Carbon numbers indicated as a part of the names of these compounds are adopted from the original reports.

Damavaricins C (**40**) and D (**41**) are biologically active precursors of the streptovaricins and were isolated from the culture broth of a mutant of *Nocardia mediterranei* [86]. Damavaricin C (**40**) could be also obtained by

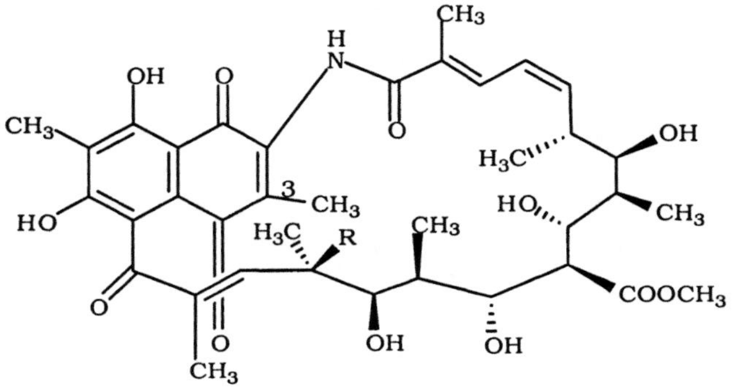

40	R = OH
41	R = H

the transformation of streptovaricin C, as will be mentioned subsequently. Damavaricins are considered to be the precursors of the streptovaricins, and both the damavaricins and streptovaricins possess a CH_3 moiety at the C-3 position [16].

Structures and field desorption mass spectra of of streptovaricins A (42), B (43), C (44), D (45), E (46), F (47), G (48) and J (49) were reported [87-89]. ^{13}C-NMR spectra of the streptovaricins and related compounds [90] and the ^{13}C-NMR spectroscopy and the biosynthesis of streptovaricin D (45) were discussed [91]. An X-ray crystallographic study of a streptovaricin C (44) derivative [92] was conducted.

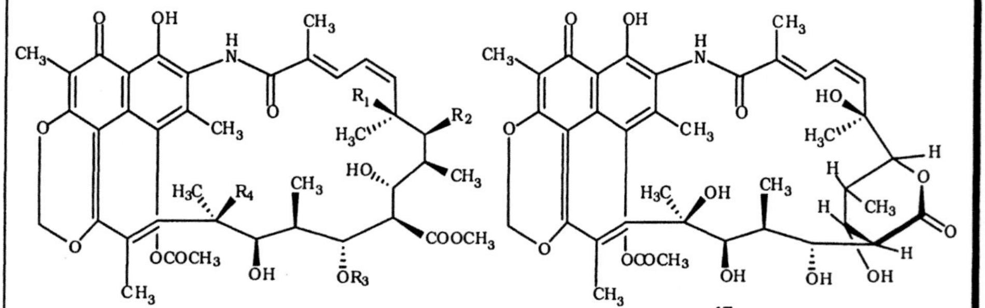

42	$R_1 = R_2 = R_4 = OH$, $R_3 = COCH_3$
43	$R_1 = H$, $R_2 = R_4 = OH$, $R_3 = COCH_3$
44	$R_1 = R_3 = H$, $R_2 = R_4 = OH$
45	$R_1 = R_3 = R_4 = H$, $R_2 = OH$
46	$R_1 = R_3 = H$, $R_2 = =O$, $R_4 = OH$
48	$R_1 = R_2 = R_4 = OH$, $R_3 = H$
49	$R_1 = R_3 = H$, $R_2 = OCOCH_3$, $R_4 = OH$

Unambiguous assignment of the ^{13}C NMR spectrum of streptovaricin C (44) was achieved and the biosynthesis of this compound was studied [93]. The configurations of the eight chiral centers, C-20 (C-6) to C-27 (C-23), and ansa system of the streptovaricins were reported [94], and the absolute configuration of 44 reported in 1971 [92] was revised in this paper [94].

When streptovaricin C (44) was heated in refluxing pyridine for 2.75 h, an atropisostreptovaricin C (50) was obtained, together with streptovaricin F_C (51) and atropisostreptovaricin F_C (52) (Fig. 3). The CD curves of the atropisomers were transformed into opposite to those of the original compounds [94].

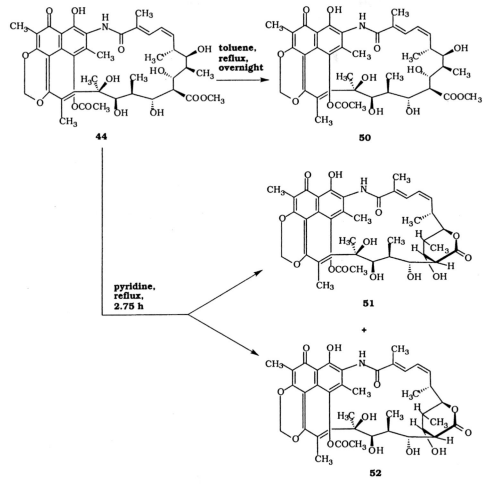

Fig. (3). Conversion of streptovaricin C (**44**) into atropisostreptovaricin C (**50**), streptovaricin F$_C$ (**51**) and atropisostreptovaricin F$_C$ (**52**).

Streptovaricin U (**53**) is an acyclic ansamycin and the configuration of this compound was studied by the comparison of its ^{13}C NMR spectrum with that of protostreptovaricin I (**29**) [95]. As other examples of such acyclic ansamycins, proansamycin B-M1 (**54**) and protorifamycin I-M1 (**55**) were isolated as minor compounds from the fermentation of the protorifamycin I (**29**) producing strain *Nocardia mediterranei* F 1/24 [96, 97]. These latter two compounds, **54** and **55**, lack a methyl moiety at the C-3 position and possess different geometries of the double bonds compared with those of **53**.

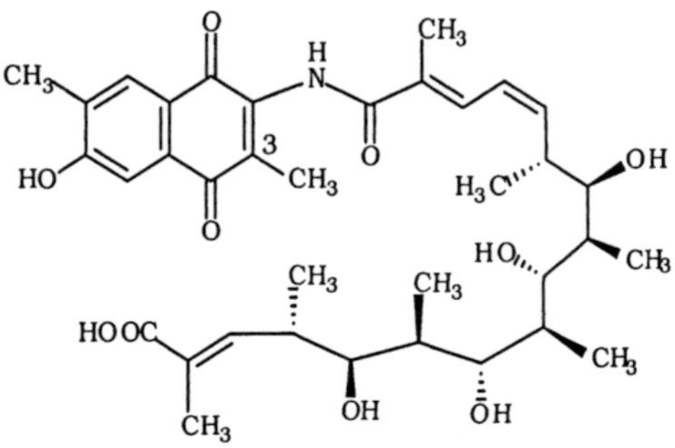

53

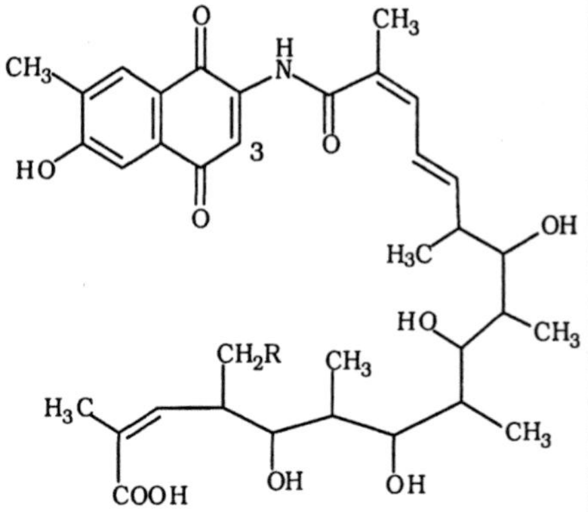

54   R = H

55   R = OH

Naphthalenoid Ansamycin Antibiotics with C_{23} Ansa Chains

Naphthalenoid ansamycins with C_{23} ansa chains, the naphthomycins, diastrovaricins, naphthoquinomycins, naphthomycinol and actamycin have been isolated and characterized.

Naphthomycin A and Related Compounds

Naphthomycin A (**56**) was isolated as a yellow pigment from the cultured broth of Strain Tü 105 of *Streptomyces collinus.*

Naphthomycin A was found to antagonize vitamin K [98] and the absolute configuration of this antibiotic was reported [99]. Subsequently, naphthomycins B (**57**) and C (**58**) were isolated from two different strains of *Streptomyces collinus* Tü 353 and Tü 1892 [100]. Four naphthomycin A (**56**) congeners, naphthomycins D (**59**), E (**60**), F (**61**) and G (**62**) were isolated from a strain of *Streptomyces* Tü 2357 and characterized [101]. Naphthomycins D (**59**) and E (**60**) are simple derivatives of **56**, possessing H at C-2 position instead of a CH_3 moiety, and OH and H at the C-30 position instead of a Cl atom, respectively. On the other hand, naphthomycins F (**61**) and G (**62**) contain an *N*-acetylcysteine residue linked to the aromatic moiety by a thioether group. Compound **61** shows some biological activity against Gram-positive bacteria and fungi, although much weaker than that of **56**. Compounds **59, 60** and **62** are inactive against microorganisms [101].

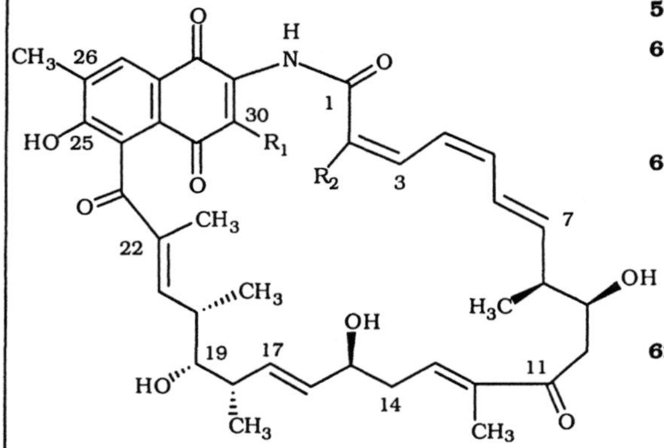

56 R_1 = Cl, R_2 = CH$_3$
59 R_1 = OH, R_2 = H
60 R_1 = H, R_2 = H

61 R_1 = -S-CH$_2$-C-COOCH$_3$, R_2 = H (NHCOCH$_3$, H)

62 R_1 = -S-CH$_2$-C-COOH, R_2 = H (NHCOCH$_3$, H)

65 R_1 = OCH$_3$, R_2 = H
66 R_1 = SCH$_3$, R_2 = H
67 R_1 = Cl, R_2 = H
68 R_1 = H, R_2 = CH$_3$, C$_{11}$-H, OH

Naphthomycin-related compounds, named diastrovaricins I (63) and II (64), were isolated from the culture broth of *Streptomyces diastochromogenes* subsp. *variabilicolor* n. subsp. as new inducers of Friend mouse erythroleukemia cells [102]. In addition, naphthoquinomycins A (65), B (66) and C (67) were isolated as inhibitors of fatty acid synthesis in *Escherichia coli* [103]. Naphthoquinomycin C (67) was found to be identical with the antibiotic given the name naphthomycin H [104,105].

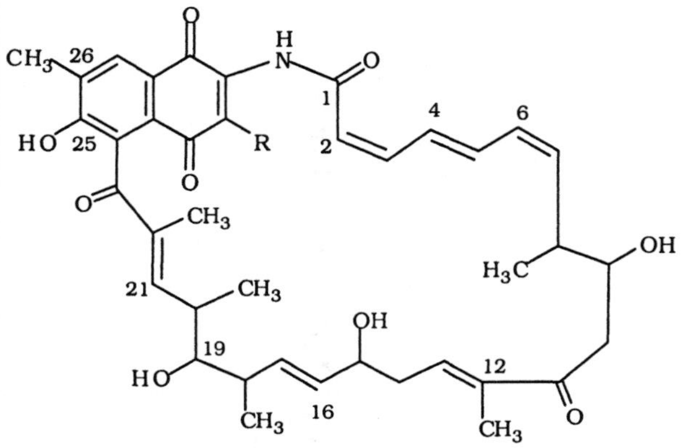

57 R = Cl
58 R = H
63 R = OH
64 R = -S-CH$_2$-CH-COOH (NHCOOCH$_3$)

Naphthomycinol (**68**), produced by *Streptomyces* sp. PF7, was isolated as a novel neuronal cell protecting substance. Naphthomycinol (**68**) is the first compound reported which has a hydroxyl function at C-11 [106].

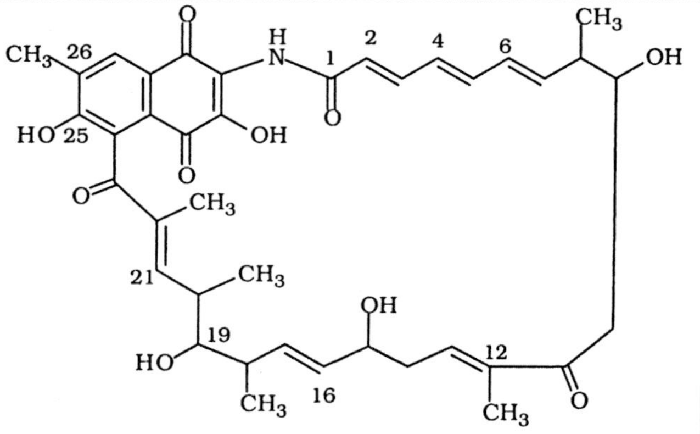

69

Actamycin (**69**) was isolated from the culture broth of *Streptomyces* sp. E784 and was active against Gram-positive bacteria. Actamycin (**69**) possesses the same plane structure as that of deastrovaricin I (**63**), whereas the geometries of the double bonds between these two antibiotics were different. Namely, the configurations of the double bonds of **69** were 2*E*, 4*E*, 6*E*, 12*E*, 16*E*, and 21*E* [107,108]. On the other hand, the configurations 2*Z*, 4*E*, 6*Z*, 12*E*, 16*E*, and 21*E* were established for diastrovaricin I (**63**). Consequently, it was concluded that **63** was a geometrical isomer of **69** [102].

Naphthalenoid Ansamycin Antibiotics with C₉ Ansa Chains

Naphthalenoid ansamycins with C_9 ansa chains consist of two compounds, rubradirin and protorubradirin. These antibiotics possess a C_9 ansa chain moiety and a *C*-nitroso-sugar or a *C*-nitro-sugar in the structure, respectively. As will be described below, rubradirin appears to be a photo-oxidation product of protorubradirin.

Rubradirin and Protorubradirin

Rubradirin (**70**) was isolated from the fermentation broth of *Streptomyces achromogenes* var. *rubradiris* [109-114]. Rubradirin (**70**) possesses pH indicator properties. Namely, it is red in acidic or neutral solutions and green in basic solutions, or in salt form. It is an extremely potent inhibitor of Gram-positive bacteria *in vitro*, and is effective in the treatment of experimental bacterial infections in mice [114]. The mechanism of the activity of this antibiotic has been shown to be the selective inhibition of ribosomal polypeptide biosynthesis [113].

The structure of **70** was reported [111,112], and the structure of its sugar moiety was revised subsequently. Thus, the sugar moiety of rubradirin was reported originally to be L-rubranitrose (2,3,6-trideoxy-3-C-methyl-3-nitro-L-hexopyranose), and was corrected to be D-rubranitrose by comparison with synthetic L- and D-rubranitorose [115,116].

Protorubradirin (**71**) was isolated in an attempt to improve the isolation of rubradirin from the same fermentation broth described above [117]. The product isolated was a mixture of rubradirin and a new antibiotic which was named as protorubradirin (**71**). Methanolysis of **71** in the dark yields an anomeric mixture of methyl glycosides of a C-nitroso-sugar instead of a C-nitro-sugar, as in the case of rubradirin. The methyl glycosides were converted into the methyl rubranitrosides by photo-oxidation. Thus, protorubradirin (**71**) is the C-nitroso-analogue of rubradirin (**70**) and it seems that protorubradirin (**71**) is the true secondary metabolite produced by S. achromogenes var. rubradiris and rubradirin (**70**) described earlier is the photo-oxidation product of **71**.

An early aromatic ansamycin-precursor containing the seven-carbon amino starter unit and designated as P8/1-OG (**72**) was isolated from the non-rifamycin producing UV-mutants derived from the *Nocardia mediterranei* strain N813 (a rifamycin B producer) [118]. This compound does not possess an ansa moiety, and can be considered to be an early ansamycin-precursor.

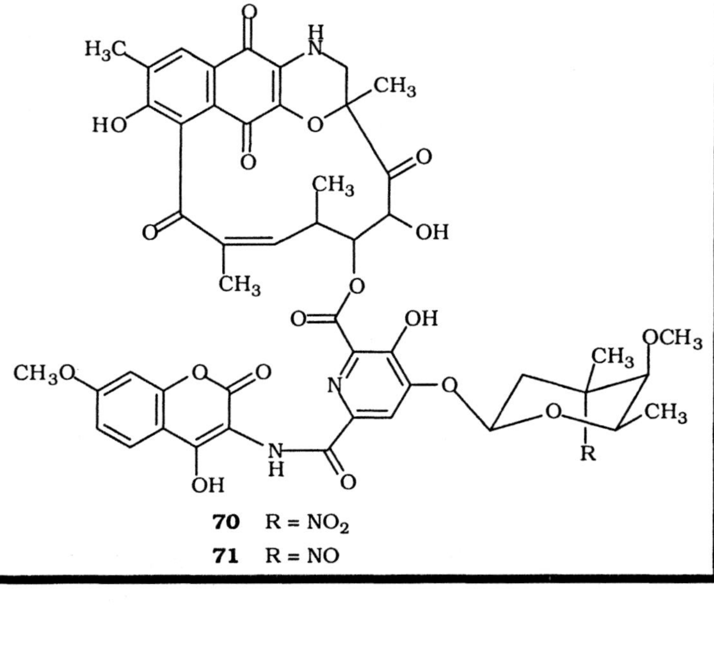

70 R = NO$_2$

71 R = NO

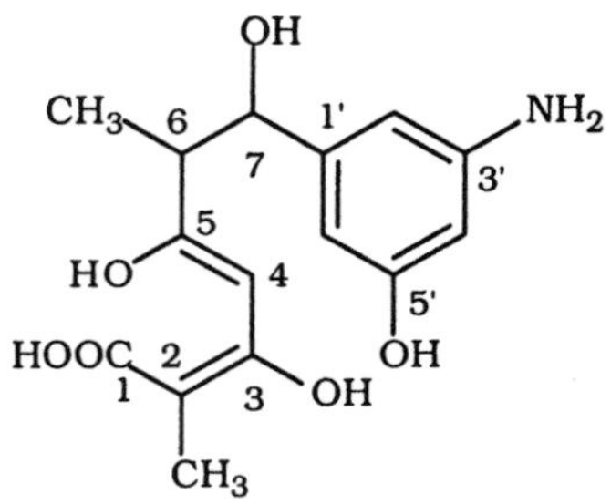

72

Benzenoid Ansamycin Antibiotics

Benzenoid Ansamycin Antibiotics with C_{15} Ansa Chains

Maytansine (**73**) was the first example of a benzenoid ansamycin with a C_{15} ansa chain. Maytansine was isolated originally from the higher plant *Maytenus ovatus* (Celastraceae). Subsequently, antibiotics related to maytansine were isolated from the fermentation broth of *Nocardia* sp. as well as other higher plants. Compounds belonging to the maytansinoid group possess a 1,3,5-trisubstituted benzene moiety as a chromophore.

Among other benzenoid ansamycins with C_{15} ansa chains, geldanamycin, the macbecins and the herbimycins are classified in this category. These compounds possess a 1,4-quinone or a 1,4-hydroquinone moiety as a chromophore.

Maytansine and Related Compounds

The first benzenoid ansa macrolide maytansine (**73**) was isolated from the bark of the Celastraceae plant *Maytenus ovatus* as an antileukemic principle [119, 120].

The crystal and molecular structure and the absolute configuration of maytansine (**73**) were studied using its 3-bromopropyl ether [121] and the cytotoxicity of **73** [122] was reported. Several other maytansinoids, such as colubrinol (**74**), normaytancyprine (**75**) [123], 10-epitrewiasine (**76**), trewiasine (**77**) and nortrewiasine (**78**) [124], were isolated from *Putterlickia verrucosa* and *Trewia nudiflora*, respectively, and the structure elucidation, and the structural requirements for antileukemic activity of natural and semisynthetic maytansinoids were reported [120, 125]. As described above, maytansine (**73**) was isolated initially from a higher plant, although it was evident that the content of this compound was extremely low.

On the other hand, a series of compounds possessing the same skeleton as maytansine (**73**) was isolated from the fermentation broth of *Nocardia* sp. No. C-15003 (N-1) and named as ansamitocins P-1 (**79**), P-2 (**80**), P-3 (**81**), P-3' (**82**) and P-4 (**83**).

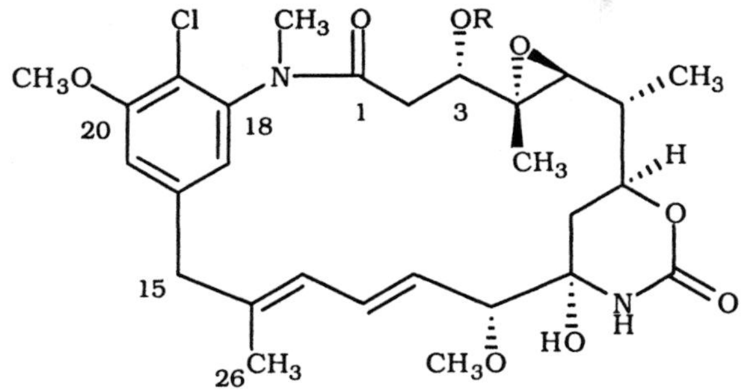

79	R = COCH$_3$
80	R = COCH$_2$CH$_3$
81	R = COCH(CH$_3$)$_2$
82	R = COCH$_2$CH$_2$CH$_3$
83	R = COCH$_2$CH(CH$_3$)$_2$
84	R = H

Among them, ansamitocins P-3 (**81**) and P-4 (**83**) were the most abundant [126]. Reductive hydrolysis of each antibiotic with LiAlH$_4$ at low temperature gave maytansinol (= ansamitocin P-0) (**84**). Acetylation of **84** with acetic anhydride in pyridine yielded an antitumor plant product maytanacine (ansamitocin P-1) (**79**). Maytansinol propionate, corresponding to ansamitocin P-2 (**80**), could be prepared by the acylation of **84**. Isolation procedures, chemical characterization and the structures of **79** - **83** were described [127].

Microbial conversion of ansamitocin P-3 (**81**) to other compounds, such as 20-*O*-demethylansamitocin P-3, maytansinol (**84**), 15-hydroxyansamitocin P-3, *N*-demethylansamitocin P-3 and 15-*epi*-15-hydroxyansamitocin P-3, were reported [128-131]. A mutant possessing a high ability to produce ansamitocins was derived from *Nocardia* sp. No. C-14482 (N-1001) by treating it with ethidium bromide. Thus, mutant N-1231 produced **81** and **83** as major components, and fifteen novel ansamitocin analogs as minor components. These analogues showed no activity against prokaryotic microorganisms.

The results of determining the activity inhibiting for cilia regeneration of deciliated *Tetrahymena pyriformis* suggest that hydroxylation of C$_{15}$ and C$_{26}$ and the presence of an acyl moiety at C$_3$ of the ansamitocins may cause a marked reduction of their antitubulin polymerization activities, whereas demethylation of the *N*-CH$_3$ only slightly affected their activities [132]. The antileukemic activity of ansamitocin P-3 (**81**) was reported [133].

Accordingly, the synthesis of *dl*-maytansinol [134] and (-)-maytansinol (**84**) [135] were conducted, and the total synthesis of maytansine (**73**)

[135,136] was achieved. The structure elucidation, chemistry and biological activities of maytansinoids were reviewed [21,23].

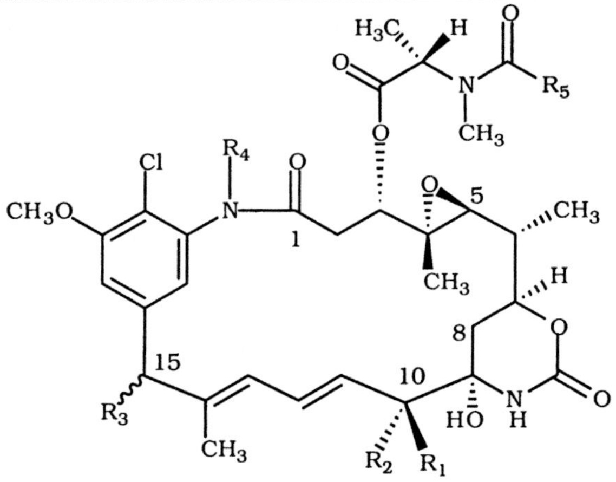

73 R_1 =H, R_2 = OCH$_3$, R_3 = H, R_4 = CH$_3$, R_5 = CH$_3$
74 R_1 =H, R_2 = OCH$_3$, R_3 = OH, R_4 = CH$_3$, R_5 = CH(CH$_3$)$_2$

75 R_1 = R_3 = H, R_2 = OCH$_3$, R_4 = CH$_3$, R_5 =

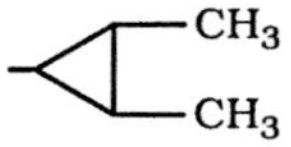

76 R_1 =R_3 = OCH$_3$, R_2 = H, R_4 = CH$_3$, R_5 = CH(CH$_3$)$_2$
77 R_1 = H, R_2 = R_3 = OCH$_3$, R_4 = CH$_3$, R_5 = CH(CH$_3$)$_2$
78 R_1 =R_4 = H, R_2 = R_3 = OCH$_3$, R_5 = CH(CH$_3$)$_2$

Geldanamycin and Related Compounds

As described above, the first benzenoid ansa macrolides were isolated from a higher plant. On the other hand, the first ansa macrolide from a microorganism was isolated from the culture filtrate of *Streptomyces hygroscopicus* var. *geldanus* and named as geldanamycin (**85**) [137]. Geldanamycin (**85**) showed an inhibitory effect on the growth and multiplication of protozoa and possessed potent cytocidal activities *in vitro*.

Geldanamycin (**85**) is extremely active against KB cells (<0.001 mcg/ml) and L1210 cells (<0.002 mcg/ml) and inhibits the growth and multiplication of protozoa (*Tetrahymena pyriformis* and *Crithida fasciculata*) at concentrations of 2 and 4 mcg/ml, respectively. Whereas the monoacetate derivative was inactive against *T. pyriformis* at a concentration of 2 mcg/ml. The structure of geldanamycin (**85**) was reported in 1970 [138] and the absolute configuration was determined subsequently [16]. The biosynthesis of **85** was studied [139].

Herbimycin A (**86**) was isolated as a herbicidal antibiotic produced by a strain of *Streptomyces hygroscopicus* AM-3672 [140,141]. Herbimycin A

(86) was found to possess herbicidal activity against most mono- and di-cotyledonous plants, as well as anti-tobacco mosaic virus activity, and its chemical structure was established, including relative configuration [140-142]. Through the screening for agents which convert the transformed morphology of Rous sarcoma virus-infected rat kidney cells to normal morphology, herbimycin A (86) was found to be active [143]. Herbimycins B (87) and C (88) were isolated and characterized subsequently [144].

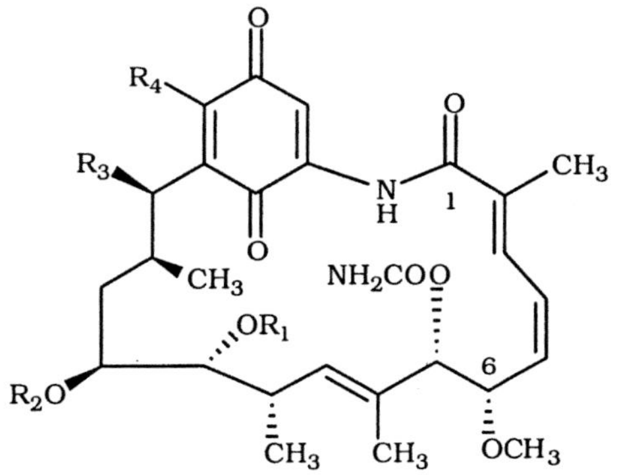

85	R_1 = H, R_2 = CH$_3$, R_3 = H, R_4 = OCH$_3$
86	R_1 = R_2 = CH$_3$, R_3 = OCH$_3$, R_4 = H
87	R_1 = R_2 = CH$_3$, R_3 = H, R_4 = H
88	R_1 = R_4 = H, R_2 = CH$_3$, R_3 = OCH$_3$
90	R_1 = R_2 = R_4 = H, R_3 = OCH$_3$

Antibiotics TAN-420A (89), B (90), C (91), D (88) and E (92) were isolated from the fermentation broth of *Streptomyces hygroscopicus* [145].

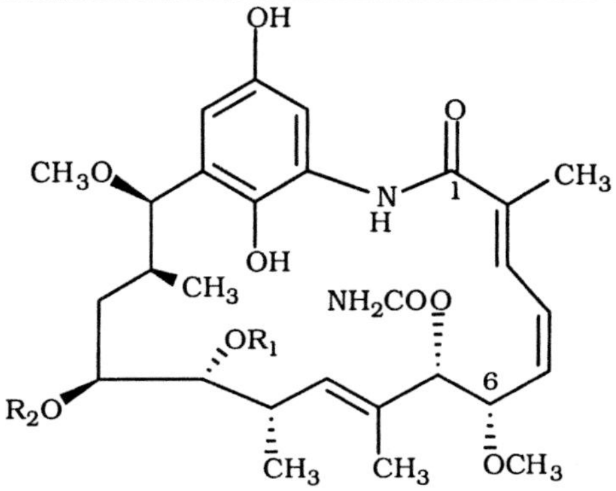

89	R_1 = R_2 = H
91	R_1 = H, R_2 = CH$_3$
92	R_1 = R_2 = CH$_3$

Antibiotics TAN-420B (90) and D (88) were oxidation products of the antibiotics TAN-420A (89) and C (91), respectively, and TAN-420D was identified to herbimycin C (88). On the other hand, the main product, TAN-420E (92), was a hydrogenated derivative of herbimycin A (86). ^{1}H-NMR assignments and cytotoxic activities against the P388 and KB lymphocytic leukemia cells of herbimycins A (86) and C (88) and TAN-420C (91) and E (92) were reported [146].

Other geldanamycin (85) related compounds, the macbecins I (93) and II (94), were isolated from the fermentation broth of *Nocardia* sp. No. C-14919 (N-2001) and their structures reported [147-150]. Macbecin I (93) possessed a quite similar chemical structure to that of herbimycin A (86), except that 93 possessed a methyl group at the C-6 position instead of a methoxyl group, as in 86. Macbecin II (94) possessed a hydroquinone moiety as a chromophoric unit.

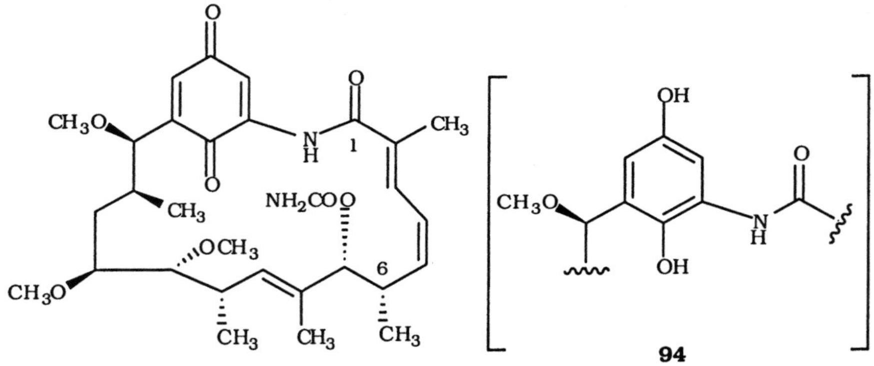

93   **94**

The macbecin-producing Actinomycete strain No. C-14919 was isolated from the leaf surface of a grass collected in Shiga Prefecture, Japan. It was reported that the macbecins were moderately active against several Gram-positive bacteria and fungi [148,149,151]. On the other hand, macbecin I (93) exhibited marked antitumor activity against intraperitoneally (ip) inoculated leukemia P388, B16 melanoma and for Ehrlich's carcinoma in mice on ip administration [152]. The biosynthesis of the macbecins was investigated by the combination of the incorporation of [14]C- and ^{13}C-labeled compounds into the antibiotic and an analysis of its ^{13}C-NMR spectra [153].

Benzenoid Ansamycin Antibiotics with C[17] Ansa Chains

Mycotrienins and ansatrienins also possess a 1,4-quinone or a 1,4-hydroquinone moiety as a chromophoric units, like geldanamycin and related compounds, but the former compounds possess a longer ansa moiety than the latter compounds. On the other hand, the trienomycins possess a quite similar ansa moiety to the mycotrienins and ansatrienins, although these antibiotics possess a 1,3,5-trisubstituted benzene moiety, like maytansinoids, as a chromophore. As well as these compounds, the cytotrienins and thiazinotrienomycins are classified within this group. This section deals with the ansamycin antibiotics with C_{17} ansa chains, which are sometimes referred to as triene ansamycins [24], because they all possess a triene moiety in the molecule.

Mycotrienin I and Related Compounds

Mycotrienins I (95) and II (96) were isolated from the fermentation broth of *Streptomyces rishiriensis* [154-158] and their structures determined. Mycotrienin II (96) corresponded to the mycotrienin isolated from the fermentation broth of a *Streptomyces* sp. [159].

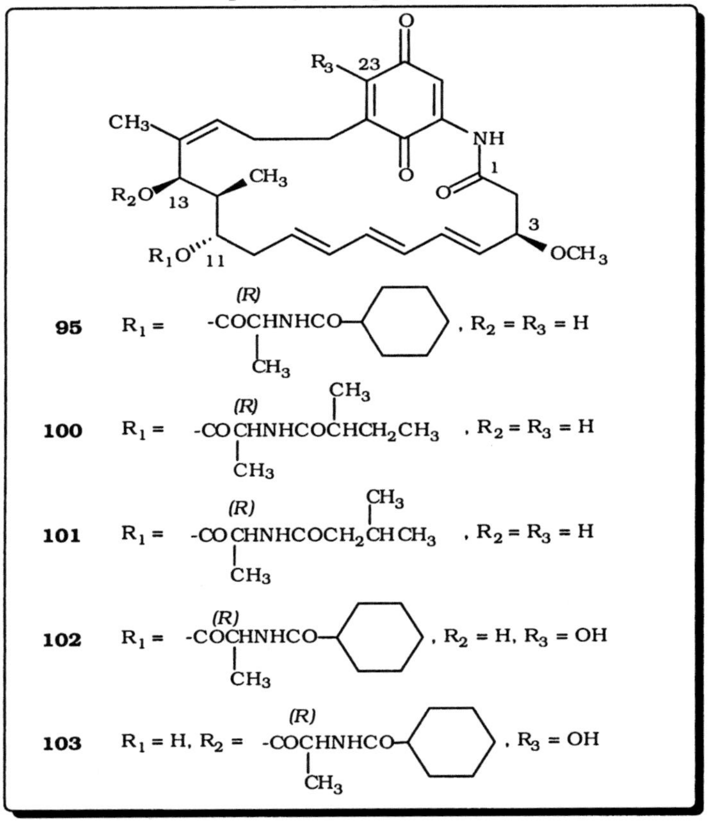

Mycotrienins I (95) and II (96) were reported to possess a cyclohexanecarboxyl-D-alanine moiety. Microbial conversion of mycotrienin II (**96**) to mycotrienol II (**97**), 34-hydroxymycotrienin II (**98**) and 22-*O*-β-D-glucopyranosylmycotrienin II (**99**) by *Bacillus megaterium* was reported [160].

Accordingly, it was shown that mycotrienins I (**95**) and II (**96**) possessed the same structures as those of the ansamycin antibiotics designated as ansatrienins A and B isolated from the fermentation broth of *Streptomyces collinus* [161, 162], except for the configuration of the alanine moiety. Thus, the alanine moiety of the ansatrienins was reported to be L, and was later revised to be D (the same configuration as those of mycotrienins). Ansatrienins A$_2$ (**100**) and A$_3$ (**101**) were isolated as minor congeners of ansatrienins A (**95**) and B (**96**) from the same fermentation broth [163].

Hydroxymycotrienin A (**102**), isolated from the culture broth of *Bacillus* sp. BMJ958-62F4 [164], is the 23-hydroxy derivative of mycotrienin I (**95**), and a congener hydroxymycotrienin B (**103**) was also isolated. The location of the cyclohexanecarboxyl-D-alanine moiety of the latter compound was at C$_{13}$ instead of C$_{11}$ in **102** [164].

Cytotrienins A (**104**) and B (**105**) were isolated from the culture broth of a *Streptomyces* sp. [165, 166]. Both **104** and **105** contain a 1-amino-

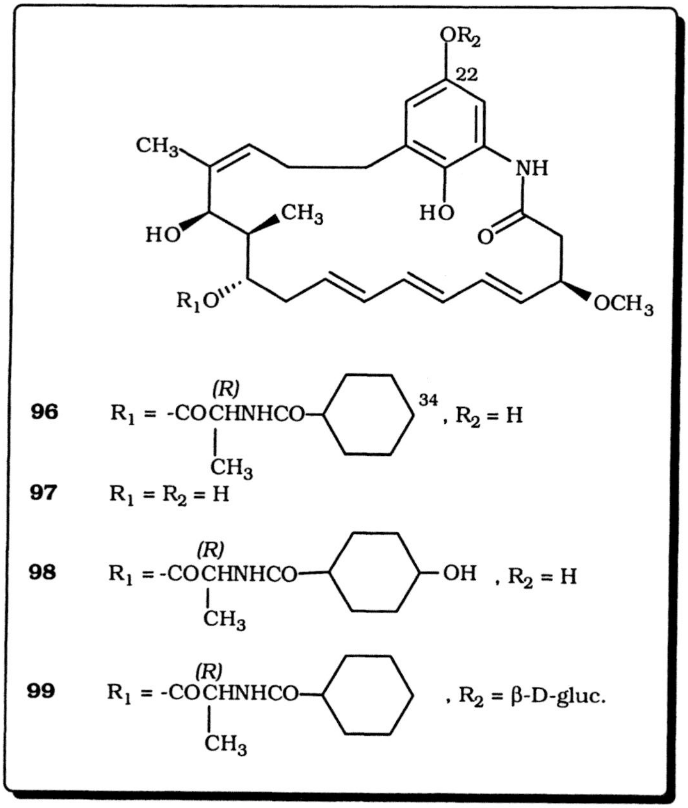

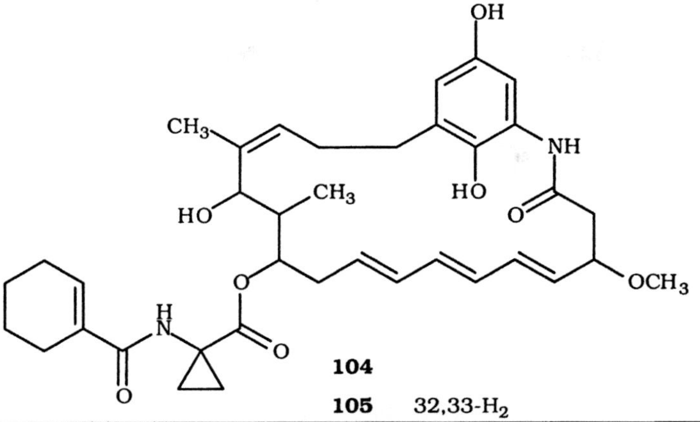

104

105 32,33-H$_2$

cyclopropane carboxylic acid moiety, and these antibiotics exhibit potent apoptosis-inducing activity on human leukemia HL-60 cells [165-167].

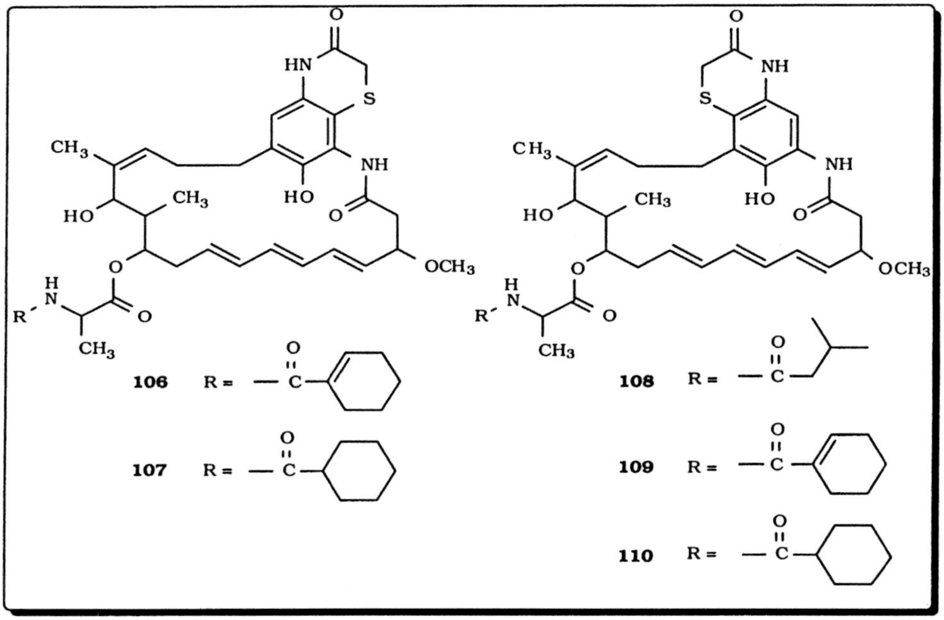

106 R = —C(=O)— (cyclohexenyl)

107 R = —C(=O)— (cyclohexyl)

108 R = —C(=O)—CH(CH$_3$)CH$_2$CH$_3$ (isobutyl)

109 R = —C(=O)— (cyclohexenyl)

110 R = —C(=O)— (cyclohexyl)

Thiazinotrienomycins A (**106**), B (**107**), C (**108**), D (**109**) and E (**110**) were isolated from the culture broth of *Streptomyces* sp. MJ672-m3. These antibiotics exhibited activity against cervical cancer cell lines [168].

Trienomycins A (**111**), B (**112**), C (**113**), D (**114**), E (**115**) and F (**116**), which possess cytotoxic and anti-tumor activity, were isolated from the fermentation broth of *Streptomyces* sp. 83-16 and their structures determined [169-172]. Unlike most of the other benzenoid ansamycin antibiotics of this group, the trienomycins possess a 1,3,5-trisubstituted benzene ring, and the partial structure (chromophoric unit) is rather similar to that of the maytansinoids.

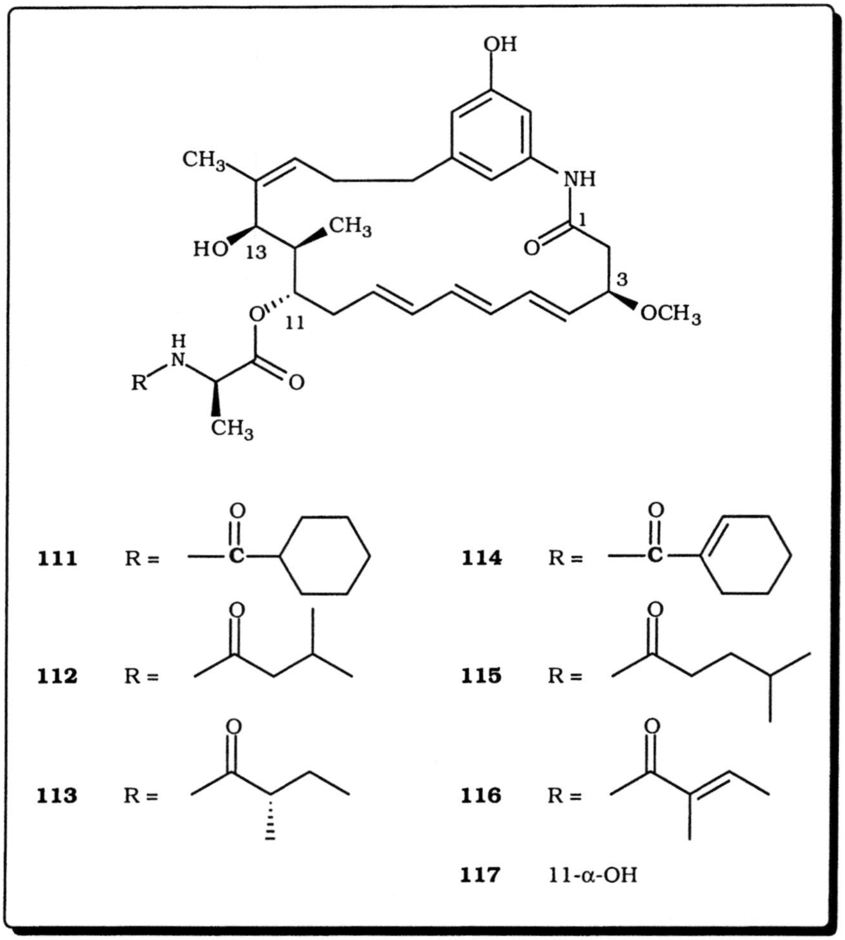

The relative and absolute stereochemistry of (+)-trienomycins A (111), B (112) and C (113) were established by chemical degradation studies and the total syntheses of 111 - 113 have been reported [173]. Also, the relative and absolute stereochemistry of (+)-mycotrienins I (95) and II (96) were established by chemical correlation with (+)-trienomycin A (111) [173,174]. Both trienomycinol (117), obtained by the reductive hydrolysis of 111 with LiAlH$_4$ at low temperature, and the hydrogenated derivative of 111 derived from H$_2$/Pd-C reduction, lacked cytotoxic activity [175].

BIOSYNTHESIS OF THE ANSAMYCINS

A description of the biosynthesis of the ansamycins can be divided into two parts, namely, the biosynthesis of the chromophoric units and the biosynthesis of the ansa moieties. It was established that the chromophoric moiety of both the benzenoid and the naphthalenoid ansamycins consists of a m-C_7N unit as the biosynthetic precursor. On the other hand, the ansa chains of these ansamycins are derived from polyketide units composed of such as acetic (C_2 unit) and propionic (C_3 unit) moieties through the polyketide biosynthesis pathway. The biosynthesis of the ansamycins has been reviewed [20,22].

As a prelude to the description of the biosynthesis of the ansamycins, the biosynthesis of the m-C_7N unit will be discussed, followed by a description of the biosynthesis of the ansa moiety.

Biosynthesis of the m-C_7N unit

With respect to the biosynthesis of the ansamycins, the incorporation of [1-^{13}C]glucose and [1-^{13}C]glycerate into the chromophoric moiety of rifamycin S was investigated [176].

The biosynthetic precursor of the chromophoric moiety was studied vigorously by several research groups. As a result it was established that during the biosynthesis of the ansamycins, a m-C_7N unit (Fig. 4) was incorporated into the chromophores of both the benzenoid ansamycins, such as geldanamycin (85) [177, 178] and ansamitocin P-3 (81) [179], and also the naphthalenoid ansamycins, such as the rifamycins [22,180-182] and actamycin (69) [183].

Fig. (4). m-C_7N unit.

As for the precursor(s) of the m-C_7N unit, it was reported that 3-amino-5-hydroxybenzoic acid (AHBA) (118) was incorporated into the ansa(myco)trienins (95 and 96). On the other hand, shikimic acid (119) was not incorporated into the m-C_7N unit of this compound, even though 118 was definitely derived via the shikimate pathway [184,185].

It was also found that when [1'-^{13}C]-AHBA (118) was administered to a streptovaricin C (44) producing culture, C-21 (the quinone methide carbonyl at 188.3 ppm) of 44 was specifically labelled (Fig. 5) [93], and the existence of 118 itself in the fermentation broth was reported [186,187]. It was also reported that the biosynthesis of the naphthalenoid ansamycin antibiotic actamycin (69) was markedly increased by the addition of AHBA (**118**) to the producing *Streptomyces* fermentation [187].

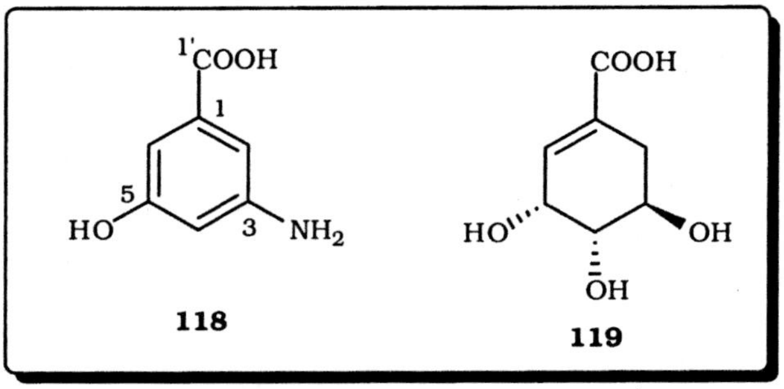

The biosynthesis of the rifamycins, including the fact that AHBA (**118**) was identified as the starter unit for the ansamycins, has been reviewed [22].

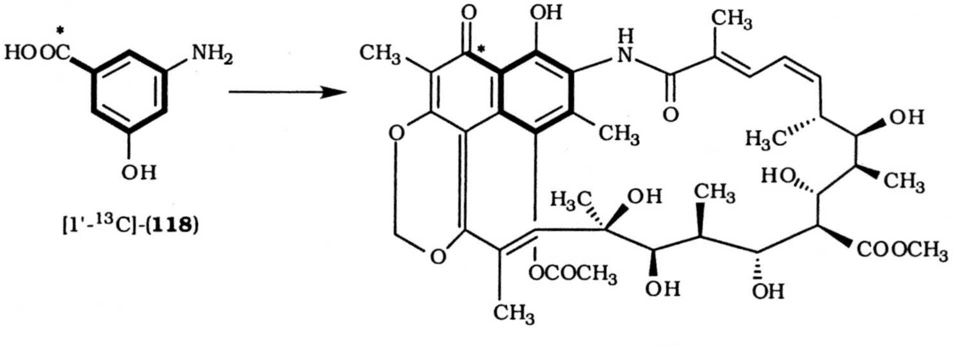

Fig. (5). Incorporation of [1'-^{13}C]-(**118**) into streptovaricin C (**44**).

Regarding the biosynthesis of the benzenoid ansamycin antibiotic ansamitocin P-3 (**81**), it was reported that the ^{13}C-NMR signal of the benzyl carbon of ansamitocin was enhanced by feeding [1'-^{13}C]-AHBA (**118**) [179].

The m-C$_7$N unit described was shown to be incorporated in the biosynthesis of both the ansamycins and also other related antibiotics, such as pactamycin and other compounds [188-190]. With respect to other naturally occurring compounds which contain a partial structure derived from the m-C$_7$N unit in the molecule, asukamycin [191] is a possible shunt metabolite from 3-dehydroquinic acid in the shikimate pathway. Incorporation of [1'-^{13}C]-AHBA (**118**) into the C-6 methyl group of porfiromycin (**120**) (a mitomycin group antibiotic) was reported (**Fig. 6**) [192], and the participation of a m-C$_7$N unit in this biosynthetic pathway and the antibiotics containing a m-C$_7$N unit were reviewed [178,188,190].

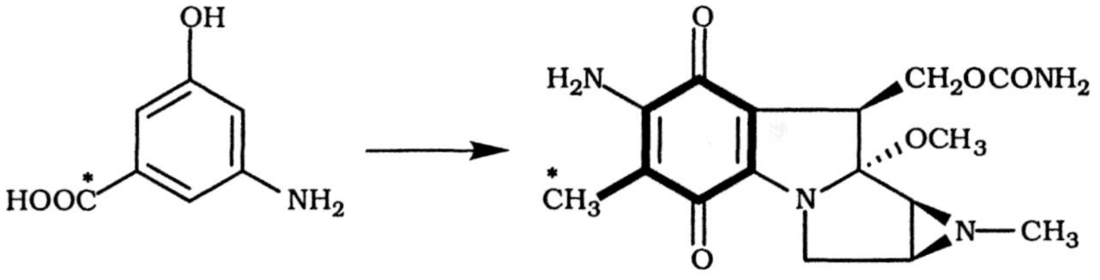

Fig. (6) Incorporation of $[1'-^{13}C]$-(118) into porfiromycin (120).

As for the other portions of the chromophoric unit, the origins of the substituent(s) at the C-3 and C-4 positions were discussed. A number of ansamycins bear different substituents at C-3 of the aromatic nucleus, e.g. hydroxyl on 3-hydroxyrifamycin S (15) and actamycin (69); thiomethyl in 3-thiomethylrifamycin SV (4) and awamycin (26); chlorine in naphthomycin A (56), the maytansinoids and the ansamitocins; and methyl in the streptovaricins, protostreptovaricins and damavaricins. Traxler and Ghisalba tested several 4-substituted 3-amino-5-hydroxybenzoic acids in mutasynthesis experiments as potential starter units for the biosynthesis of the 3-substituted rifamycins.

As a result, they concluded that the 3-substituents in the rifamycin and other ansamycin chromophoric units must be introduced as a late biosynthetic step because not even traces of the expected 3-hydroxyrifamycin S (15) or SV (when supplemented with 3-amino-4,5-dihydroxybenzoic acid (121)), 3-hydroxy-4-methoxyrifamycin S or SV (when supplemented with 3-amino-4-hydroxy-5-methoxybenzoic acid (122)), or 3-methylrifamycin S or SV (when supplemented with 3-amino-4-methyl-5-hydroxybenzoic acid (123)) were detected [84].

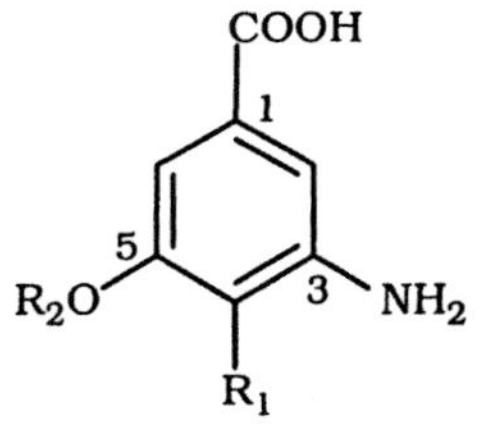

121	R_1 = OH, R_2 = H
122	R_1 = OH, R_2 = CH_3
123	R_1 = CH_3, R_2 = H

Because the C-8 phenolic hydroxy group and the C-8 carbon atom of rifamycins B (1), O (7) and S (2) were found to be derived from the carboxy group of AHBA (**118**), 8-deoxyansamycins were excluded as possible biosynthetic intermediates [193].

The biogenetic relationships of rifamycin S (**2**) and related compounds were discussed by Lancini *et al.* (1969). According to this research, incubation of **2** with the washed mycelium of *Streptomyces mediterranei* gives rifamycin B (**1**). The hypothesis that rifamycin O (**7**) was a progenitor in the biosynthesis of rifamycin B (**1**) was excluded because **2** (and/or **3**) was transformed into **1**. In addition, it was found that **1** and **7** are formed from **2** (and/or **3**) by different pathways using different C_3-precursors for the biosynthesis of their glycolic acid moieties [45,194].

P8/1-OG (**72**) produced by *Nocardia mediterranei* P14 is considered an early metabolite of rifamycin B (**1**). To investigate the origin of the C_7N-unit of **72**, racemic [1,2-$^{13}C_2$]glycerol was administered to growing cultures of *Nocardia mediterranei* P14 and the ^{13}C-NMR spectrum of labeled P8/1-OG (**72**) was studied [195]. Two incorporation patterns of racemic [1,2-$^{13}C_2$]glycerol were considered possible (Fig. 7).

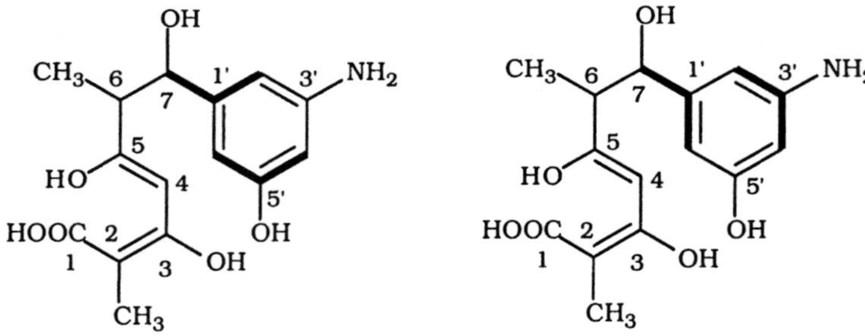

Fig. (7). Two possible coupling patterns (**A** and **B**) for P8/1-OG (**72**) after the $^{13}C_2$ incorporation of racemic [1,2-$^{13}C_2$] glycerol.

In the first case (**A**), phosphoenolpyruvate (**124**) derived from racemic [1,2-$^{13}C_2$]glycerol would label C-7, C-1', C-2' and erythrose 4-phosphate (**125**) would label C-4', C-5' and C-6', with C-3' remaining unlabelled. In the second case, **124** would label C-7, C-1' and C-6', and **125** would label C-2', C-3', C-4', with C-5' remaining unlabelled. This would suggest that in the preceding case, C-1 of **125** was aminated, and in the latter case, C-3 of **125** was aminated. The coupling pattern in the enriched sample of P8/1-

OG (72) revealed satellites resulting from the incorporation of intact C_2-units from racemic $[1,2^{-13}C_2]$glycerol showing two labelled C_3-units, and an additional C_2-unit in the side chain of 72. The fact that no satellites were observed for C-3' indicated that this C-atom was not labelled from a C_2-unit of racemic $[1,2^{-13}C_2]$glycerol. The missing satellites for C-3' were a clear confirmation of the first case (A) with C-7, C-1' and C-2' originating from phosphoenolpyruvate (124) and C-3', C-4', C-5' and C-6' originating from erythrose 4-phosphate (125). C-1 of erythrose 4-phosphate carries the N-atom or in the case of a cyclic intermediate, C-5 would be aminated.

With regard to the biosynthesis of the rifamycins, B-Factor [3'-(1-butylphosphoryl)adenosine] was isolated from a yeast extract as an inducer of rifamycin production in a non-rifamycin producing *Nocardia* mutant. Feeding of B-factor to the mutant culture demonstrated that the induction process was triggered during the early stationary phase. Rifamycin production in the mutant was also induced by an exogenous supply of AHBA (118), an intermediate in the antibiotic pathway, suggesting that a step upstream from the intermediate is regulated by B-factor. B-Factor analogues, *i.e.*, alkylesters of 3'-AMP with alkyl side chains of, C_2 - C_{12}, and *n*-butyl esters of 3'-GMP and 2'-AMP, all showed the B-factor activity [196].

Biosynthesis of the Ansa Chain of the Naphthalenoid Ansamycins

As mentioned in the introduction to this chapter, ansamycin antibiotics consist of a chromophoric unit and an ansa moiety. The main portion of the chromophore is derived from a *m*-C_7N unit, and the biosynthesis of this unit was described above. The biosynthesis of the ansa moiety (ansa chain) will be discussed below.

Thus, naphthalenoid ansamycin antibiotics with a C_{17} ansa moiety, naphthalenoid ansamycin antibiotics with a C_{23} ansa moiety, followed by the naphthalenoid ansamycin antibiotics with a C_9 ansa moiety will be discussed.

Naphthalenoid Ansamycin Antibiotics with a C_{17} Ansa Moiety

The biosynthetic pathway of the ansamycins was first studied in the formation of rifamycin S (2).

Through preliminary data derived from ^{14}C-precursor incorporation experiments, a scheme for rifamycin biosynthesis, in which the ansa chain was composed of five propionate and two acetate units, was proposed [197]. Rifamycin biosynthesis was also studied using ^{13}C-enriched precursors and ^{13}C-NMR spectroscopy, and the biosynthetic origin of the carbon atoms of the ansa chain of rifamycin S (2) was established

[197,198]. If the carbons derived from a propionate unit are indicated as P and carbons derived from an acetate unit are indicated as A, the common biosynthetic building units and their sequence in the naphthalenoid ansamycin antibiotics with a C_{17} ansa moiety (and a part of the chromophore) are presented as P-A-P-P-P-P-P-P-A-P (Fig. **8**).

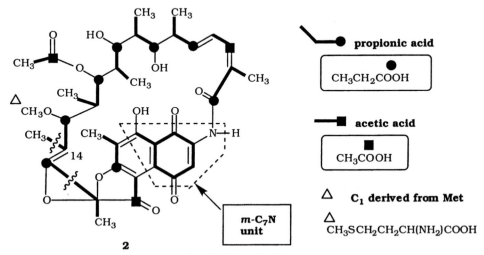

Fig. (8). The biosynthetic building units of rifamycin S (2).

An early aromatic ansamycin-precursor containing the seven-carbon amino starter unit designated as P8/1-OG (**72**) was isolated from the non-rifamycin producing UV-mutants derived from *Nocardia mediterranei* strain N813 (a rifamycin B producer). The *m*-C_7N unit (starter unit) and the three early units (P-A-P) incorporated are indicated in Fig. **9** [118].

Fig. (9). The biosynthetic building units of P8/1-OG (72).

It was proposed that the polyketide chain of rifamycin S (**2**) was formed by a linear condensation, in the appropriate sequence, of eight methylmalonates and two malonates [199] and it was also pointed out that the incorporation pattern of propionates and acetate for streptovaricin D (**45**) was in an amide-head direction. Namely, the amide carbonyl was derived from the carboxyl group of propionate by the incorporation study of sodium [1-^{13}C]propionate [91]. In this experiment, ^{13}C was incorporated at C-1, and was not incorporated at C-21 (Fig. 10). It appeared that all of the ansa chains of the rifamycins are derived from the same biosynthetic origin, and that a C_1 unit (CH_3 at C-14 position) is cleaved during the biosynthetic process in the case of the rifamycins.

On the other hand, the damavaricins are regarded as precursors of the streptovaricins [86], protorifamycin I (34), which is structurally related to the protostreptovaricins, and to rifamycin W (20) [83]. Washed mycelium of a rifamycin B (1) producing *Nocardia mediterranei* transformed ^{14}C-rifamycin W (20) into ^{14}C-rifamycin B (1). Consequently, it was suggested that rifamycin W (20) was the normal precursor of other rifamycins [75].

Celmer [200,201] reported a stereochemical and biogenetic model for the lactone ring of macrolide antibiotics such as oleandomycin, erythromycin and leucomycin. On the other hand, the structural similarity between the carbon skeleton (from C-5 to C-14) of the ansa chain of ansamycins, such as the rifamycins and the streptovaricins, and the lactone ring (from C-3 to C-12) of the macrolide antibiotics was also pointed out [1].

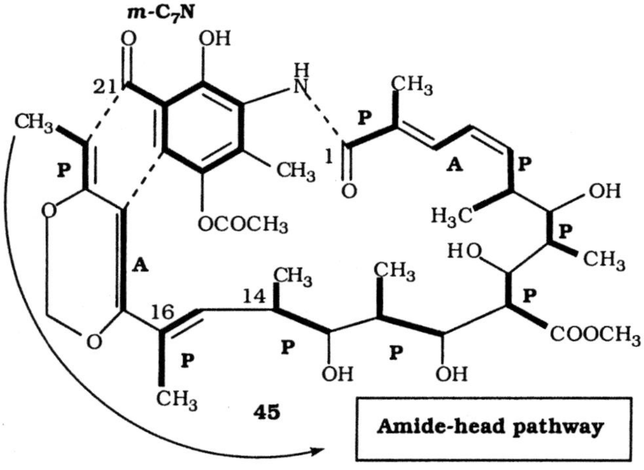

Fig. (10). The amide-head biosynthetic pathway of streptovaricin D (45).

Funayama *et al.* suggested the possibility of the application of this model for the determination of the stereochemistry and biosynthetic pathway of the known ansamycin antibiotics [202]. According to this proposal, it is important to consider that, of the C_{17} ansa chain naphthalenoid ansamycins, the absolute configurations of rifamycins S (2) [16, 194] and B (1) [33, 42], and streptovaricin C (44) [94], were established by X-ray analysis. The rigorous biosynthetic studies of rifamycin S (2) indicated P-A-P-P-P-P-P-P-A-P as the biosynthetic sequence of the ansa chain and a part of the naphthalenoid moiety [197].

This can be estimated to be common pathway to rifamycin B (1) and streptovaricin C (44). In addition to their biogenesis, the absolute configuration of these compounds can be speculated to be the same as each other, except for the configurations at the C-6 and C-7 positions (Fig. 11). Taking into consideration the configurational models tentatively established herein, the configuration of awamycin (26) was estimated to be as shown in the structure.

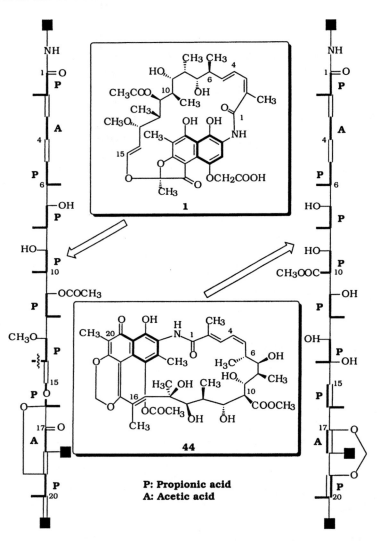

Fig. (11). The ansa moieties of rifamycin B (1) and streptovaricin C (44) according to Celmer's model.

About sixty naphthalenoid antibiotics possessing a C_{17} ansa chain have been isolated. However, the biosynthesis and/or absolute configuration of only a limited number of these antibiotics has been studied. Surprisingly, all of the naphthalenoid antibiotics possessing a C_{17} ansa chain appear to have essentially the same building unit sequence (P-A-P-P-P-P-P-P-A-P). The very few exceptions, such as kanglemycin (18) and 31-homorimanycin W (21), possess an additional C_1 unit at the C-31 position [68,77]. In addition, the absolute configurations of these compounds appear to be identical, except for the configurations at C-6 and C-7.

For example, the biosynthetic sequence and stereochemistry of the formation of rifamycin B (1) were the same as those of streptovaricin C (44), in spite of the occurrence of demethylation at the C-14 position (derived from a propionate unit) and the insertion of an oxygen atom between C-15 and C-16 in the former compound (Fig. 11).

Naphthalenoid Ansamycin Antibiotics with a C_{23} Ansa Moiety

Only small differences are found in the biosynthetic pattern of the ansa chains among the naphthalenoid ansamycin antibiotics possessing a C_{23} ansa chain. Among the C_{23} ansa chain-containing naphthalenoid ansamycin antibiotics, the absolute configuration of naphthomycin A (56) has been established by X-ray analysis [99]. On the other hand, the partial biosynthetic origin of actamycin (69) was reported [107, 108]. Using the previous conventions, the building unit of the ansa chain of actamycin (69) can be indicated as A-A-A-P-A-P-A-A-P-P-P-A-P. All of these C_{23} ansa chain naphthalenoid ansamycins, except for naphthomycin A (56) and naphthomycinol (68) [106], appear to possess essentially the same carbon skeleton as that of actamycin. Consequently, the building units of these compounds can be indicated as A-A-A-P-A-P-A-A-P-P-P-A-P, and that of naphthomycin A (56), which possesses a propionate unit at C-1, C-2 and a methyl at C-2 instead of an acetate unit, can be indicated as P-A-A-P-A-P-A-A-P-P-P-A-P. It is interesting that each starting building unit and the absolute configuration from C-11 to 20 (P-P-P-A-P) of rifamycins S (2) and B (1) and streptovaricin C (44) are the same as those of C-17 to C-26 (P-P-P-A-P) of the naphthalenoid ansamycins possessing a C_{23} ansa chain.

In the naphthalenoid ansamycins possessing a C_{23} ansa chain, an oxygen atom at C-25, derived from the carboxy group of AHBA (118), and a methyl group at C-26, derived from the propionate unit (corresponding to C-19 and C-20, respectively, in the naphthalenoid ansamycins possessing a C_{17} ansa chain), exists in all of these naphthalenoid ansamycin antibiotics. The structure of naphthomycin A (56), originally reported as possessing a 27-OH [203], was subsequenly revised to have a 25-OH [99,204]. Such a problem could be avoided if the structural and biogenetic observations described herein are considered.

Naphthalenoid Ansamycin Antibiotics with a C_9 Ansa Moiety

As examples of other naphthalenoid ansamycin antibiotics, rubradirin (**70**) and protorubradirin (**71**) were isolated from the fermentation broth of *Streptomyces achromogenes* var. *rubradiris* [109-114,117]. As indicated previously, protorubradirin (**71**) is the *C*-nitroso-analogue of **70**, and it seems that **71** is the true secondary metabolite produced by *S. achromogenes* var. *rubradiris*. Thus, rubradirin (**70**), reported earlier, is the photo-oxidation product of protorubradirin (**71**). The ansa moiety of these compounds possesses a C_9 chain, instead of a C_{17} or a C_{23} chain.

The ansa moiety and part of the chromophoric unit of **70** and **71** consist of nine carbons, and the biosynthetic building units of this part, and their sequence are considered to be P-A-P-P-A-P (Fig. 12). Through a comparison of the ansa moiety of this compound with that of the ansa chains of the naphthalenoid ansamycin antibiotics with a C_{17} ansa chain (P-A-P-P-P-P-P-P-A-P), it seems that the first two to four biosynthetic units (P-A, P-A-P or P-A-P-P), and the last two to four units (A-P, P-A-P or P-P-A-P), are involved in the formation of the antibiotic, whereas the four biosynthetic units (P-P-P-P) in the center of the sequence are missing.

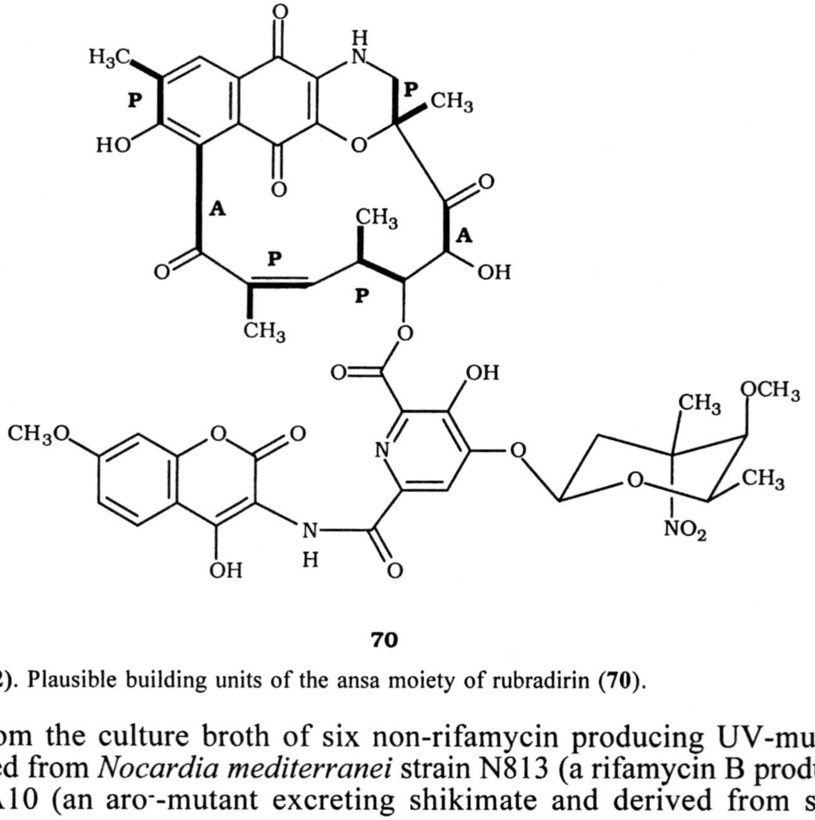

70

Fig. (12). Plausible building units of the ansa moiety of rubradirin (**70**).

From the culture broth of six non-rifamycin producing UV-mutants derived from *Nocardia mediterranei* strain N813 (a rifamycin B producer) and A10 (an aro⁻-mutant excreting shikimate and derived from strain

N813) P8/1-OG (72) was isolated. The structure of 72 was defined, and it could be easily recognized as a very early ansamycin precursor containing the seven-carbon amino starter-unit (m-C_7N unit) and the three initial acetate/propionate units of the ansa chain.

It appears that the starting three units in the biosynthetic process (P-A-P) are universal to the naphthalenoid ansamycin antibiotics which possess an ansa moiety consisting of the C_{17} and C_{23} ansa chains, and also the C_9 chains and even an early intermediate such as 72.

In the biosynthesis of compound 72 (Fig. 9) and other ansamycin antibiotics, 3-amino-5-hydroxyl-coenzyme A might act as a starter-molecule. To this seven-carbon amino unit the first propionate unit (*via* methylmalonyl-CoA), then an acetate unit (*via* malonyl-CoA) and finally another propionate unit are added by condensation and decarboxylation. In the case of 72, the resulting aromatic triketide is then converted into the product P8/1-OG (72) by hydrogenation of the keto group at C-7 and enolization of the keto groups at C-3 and C-5. The CoA is then split off, possibly during the excretion of the product [118].

Biosynthesis of the Ansa Chain of the Benzenoid Ansamycins

As described in the introduction to this chapter, the chromophoric moiety of the benzenoid ansamycin antibiotics are biosynthetically derived from a "m-C_7N unit" *via* the shikimate pathway, and the ansa chain moieties are derived from a "polyketide" chain *via* the condensation of acetate or glycolate (C_2 unit) and propionate (C_3 unit) moieties.

Because the biosynthesis of the m-C_7N unit was discussed previously, only the biosynthesis of the ansa moiety of the benzenoid ansamycin antibiotics with a C_{15} ansa moiety and the benzenoid ansamycin antibiotics with a C_{17} ansa moiety will be discussed here.

Benzenoid Ansamycin Antibiotics with C_{15} Ansa Moiety

The origin of the ansa chain and the direction of the synthesis of geldanamycin (85) were reported [139]. According to the report, the ansa chain consists of acetic and propionic acid moieties, and the direction of the biosynthesis was through the amide-head pathway, like that of streptovaricin D (45) (Fig. 10) [91]. It was also reported that two C_2 units of the ansa moiety of geldanamycin (85) at C-5-C-6 and C-11-C-12 arose from two glycolate (G) units [177]. Consequently, the biosynthetic building units and their sequence on 85 are indicated as *P-A-G-P-P-G-P (the *P unit corresponds to a C-1 carbonyl, C-2 and a methyl at C-2).

On the other hand, the absolute configuration [150] and bio-synthesis [153] of macbecins I (93) and II (94) was investigated through incorporation studies with ^{14}C- and ^{13}C-labeled substrates, followed by an analysis of the ^{13}C-NMR spectra of the ^{13}C-incorporated macbecins [153]. By feeding [2-^{13}C]-acetate, [1-^{13}C]-propionate, [3-^{13}C]-propionate, L-[methyl-^{13}C]-methionine, and L-[carbamoyl-^{14}C]-citrulline, the biosynthetic origin of the carbon skeleton of macbecin I (93), except for the C_2 unit at C-11-C-12 [estimated to be glycolate (G)], and the chromophoric moiety which was estimated to be derived from a m-C_7N unit, was defined [205].

In Fig. 13, the biosynthetic building units of macbecin I (93), including the C_2 unit at C-11-C-12 [estimated to be glycolate (G)] and the chromophoric moiety which was estimated to be derived from a m-C_7N unit, are shown.

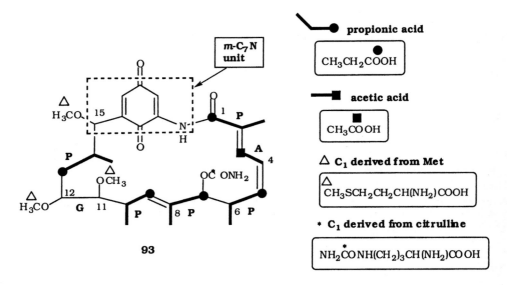

Fig. (13). The biosynthetic building units of macbecin I (93).

Thus, the biosynthetic sequence can be represented as P-A-P-P-P-G-P, which corresponds to that of geldanamycin, except for the C_2 unit at C-5 and C-6. The carbons at C-5 and C-6 of macbecin I (93) appear to arise from a glycolate (G) unit.

From these data, the absolute configurations and biosynthetic units of geldanamycin (85) and those of macbecin I (93) can be represented as shown in Fig. 14. Portions of the building units of both compounds are unclear, although it can be easily seen that both 85 and 93 possess quite similar biogenetic units and absolute configurations.

Funayama et al. reported the possibility for the application of their model to the stereochemistry and biosynthesis of the known benzenoid ansamycin antibiotics [202]. According to the analysis, herbimycin A (86) seemed to possess the same biosynthetic building units as those of

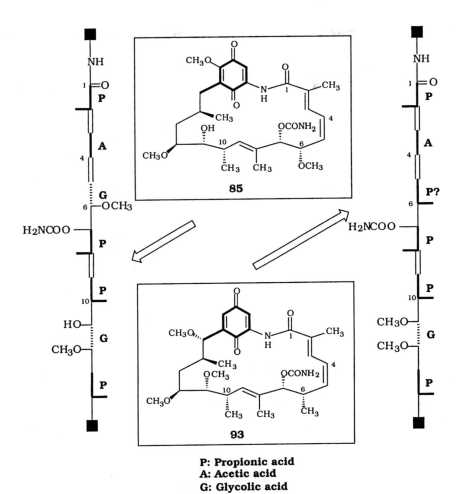

P: Propionic acid
A: Acetic acid
G: Glycolic acid

Fig. (14). The ansa moieties of geldanamycin (85) and macbecin I (93) by Celmer's model.

geldanamycin (85) because of the similarity of the ansa skeleton [202]. On the other hand, although only the relative stereochemistries were reported for 85 and 86, the absolute configurations of these compounds were estimated to be as shown, respectively from a comparison of the ansa moieties of these ansamycins with those of 93 for which the absolute configuration was defined [150].

Benzenoid Ansamycin Antibiotics with a C_{17} Ansa Moiety

It was shown that the chromophores of myco(ansa)trienins I (95) and II (96) originated from a m-C_7N unit and the ansa-chain from polyketide units (A-A-A-A-A-P-P-A). The cyclohexanecarboxylic acid moiety was derived from the sequential reduction of 2,5-dihydrobenzoic acid to cyclohexene-1-carboxylic acid [156,206]. Trienomycin A (111) appeared to be formed through the same biosynthetic pathway as the mycotrienins, except that 111 maintained an unoxidized *m*-C_7N unit as a chromophoric acid. The absolute configurations of the ansa moiety of trienomycin A (111) was established [173]. It is considered that the biosynthetic building units and their sequence are all A-A-A-A-A-P-P-A (Fig. 15) and are completely different from those of the naphthalenoid and the benzenoid ansamycin antibiotics. Thus, it appears that no universal rules exist for the biosynthesis of the benzenoid ansamycins with a C_{17} ansa chain, and naphthalenoid ansamycins with a C_{17}, a C_{23} or a C_9 ansa chain and the benzenoid ansamycins with a C_{15} ansa chain.

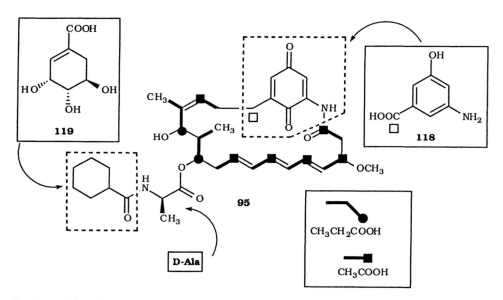

Fig. (15). The biosynthetic building units of mycotrienin I (95).

Cyclohexanecarboxylic acid moieties were previously observed in 11-cyclohexylundecanoic acid (126) and 13-cyclohexyltridecanoic acid (127) isolated from a thermophilic bacterium *Bacillus acidcaldarius* and *Curtobacterium pusillum* [207-209], asukamycin [191,210] and ketomycin (128) [211].

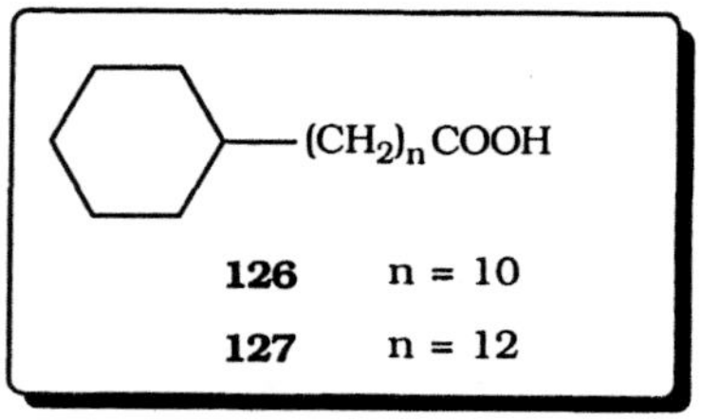

126 n = 10

127 n = 12

The biosynthesis of ketomycin (128) (R-3-cyclohexenylglyoxylic acid) was studied using D-[U-^{14}C]shikimic acid and D,L-[1,6-^{14}C$_2$]shikimic acid, and it was found that the cyclohexenyl moiety of this antibiotic was derived from shikimic acid (119), and that 1'- and 2'-C of ketomycin (128) were derived from carbon-1 and carbon-6 of 119 (Fig. 16) [211].

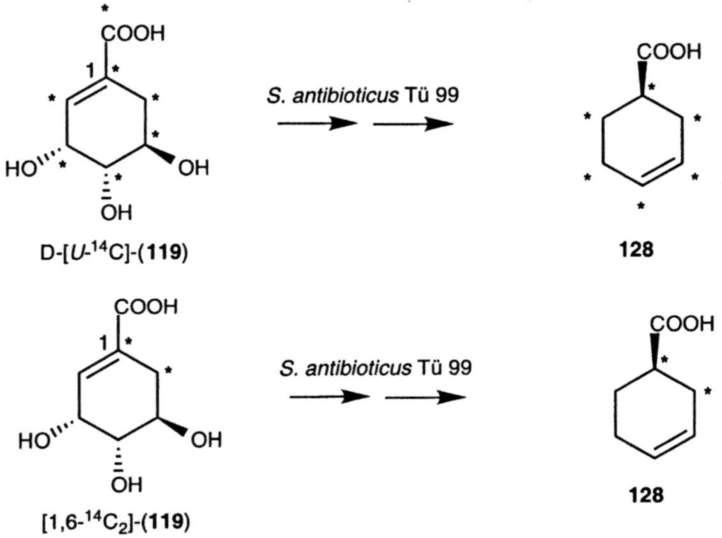

Fig. (16). The biosynthesis of ketomycin (128).

The biosynthetic origin of the cyclohexanecarboxylic acid moiety of the mycotrienins I (95) and II (96) was also considered to be shikimic acid (119) [184,204]. Cell-free extracts of *Streptomyces collinus* were tested with various cyclohexene- and cyclohexadiene-carboxylic acids in order to determine the latter stages of the conversion of 119 to cyclohexanecarboxylic acid. It was demonstrated that the final three steps of this process involve reduction of the α,β-double bond of 1,2(6)-cyclohexadienylcarbonyl CoA, an isomerization of the double bond of the resulting 2-cyclohexenylcarbonyl CoA to afford 1-cyclohexenylcarbonyl CoA, and a subsequent reduction of the newly formed α,β-double bond. Both of the reduction steps were shown to require NADPH as a cofactor [212].

BIOLOGICAL PROPERTIES OF THE ANSAMYCIN ANTIBIOTICS

Antimicrobial Activities

Rifamycin B (1), the main fermentation product from cultures of *Nocardia mediterranei*, displayed very poor biological activity. However on standing in aqueous solution, it was transformed into the highly biologically active rifamycin S (2). It was this compound which became the most important starting material for the production of semi-synthetic derivatives, including rifampicin (rifampin) (19).

Rifamycins are active against a large variety of organisms, including bacteria, eukaryotes and viruses, and for that reason they are sometimes called "wonder drugs". Clinically, they are particularly useful for the treatment of tuberculosis. For this purpose, a semi-synthetic ansamycin, rifampicin (19), is often used in the combination with INAH (isonicotinic acid hydrazide). A semi-synthetic rifamycin derivative, CGP-4832, is reported to be highly active against some Gram-negative bacteria [213].

The antimicrobial activity of these antibiotics is due to their prevention of the growth of bacterial cultures by specifically inhibiting the activity of DNA-directed RNA polymerase (DDRP). Any modification in the ansa bridge generally reduces this activity. The interaction of rifamycin with bacterial RNA polymerase was discussed [214]. It was reported that electronegative groups at C-3 of rifamycin S enhance its activity toward DDRP [215].

From all of the chemical and physical studies conducted on the rifamycins, it was concluded that some of the essential structural features necessary for activity are: 1) presence of a naphthalene ring carrying oxygen atoms at C(1) and C(8), either in the quinone or the hydroquinone form, 2) hydroxyl groups at positions C(21) and C(23) of the ansa chain, and 3) a well-defined spacial arrangement of the oxygen atoms at C(21) and C(23) [216].

Inactivation of rifampicin (19) by *Nocardia brasiliensis* and *N. otitidiscaviarum* [217,218] and by *Bacillus* sp. [219] was reported. Rifampicin (19) was glycosylated by a pathogenic species of *Nocardia*, i.e., *Nocardia brasiliensis*. The structures of the two glycosylated compounds isolated from the culture broth of the bacterium were determined to be 3-formyl-23-(O-[β-D-glucopyranosyl])rifamycin SV (129) and 23-(O-[β-D-glucopyranosyl])rifampicin (130), respectively. On the other hand, rifampicin was converted into 21-(O-phosphoryl)rifampicin (131) and 3-formyl-21-(O-phosphoryl)-rifamycin SV (132) by *Bacillus* sp. All these four compounds lacked antimicrobial activity against other Gram-positive bacteria, as well as the *Nocardia* species [215,218].

The molecular structure, conformation and interactions of rifamycin S (2) were investigated [220,221]. It had been previously reported that the activity of the antibiotics is dependent on the presence of a naphthalene ring carrying oxygen atoms at C-1 and C-8 in the chromophoric unit, either in the quinone or hydroquinone form, two unsubstituted hydroxyls at the C(21) and C(23) positions of the ansa chain, and a well defined spatial arrangement of these oxygen atoms. The chemical structure of rifamycin S (2), with the bond lengths along the ansa chain and torsion angles within the skeleton of the ansa chain in the active rifamycins, such as rifamycin SV (3), rifamycin S (2) and rifampicin (19), were presented [221]. The structure and conformation of halomicin B (10) in the solid state and in solution were also investigated in order to study the effect of substitution at the 4-position of the naphthohydroquinone on the conformation of the ansa chain [222].

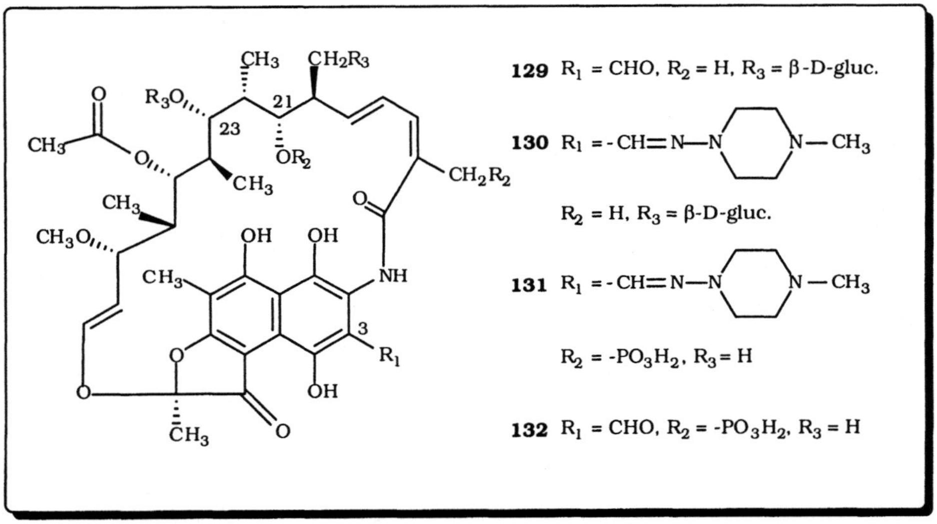

129 R_1 = CHO, R_2 = H, R_3 = β-D-gluc.

130 R_1 = - CH=N—N⟨⟩N—CH$_3$

R_2 = H, R_3 = β-D-gluc.

131 R_1 = - CH=N—N⟨⟩N—CH$_3$

R_2 = -PO$_3$H$_2$, R_3= H

132 R_1 = CHO, R_2 = -PO$_3$H$_2$, R_3 = H

The effect of rifamycin dimers on the activities of nucleic acid polymerases from various sources were tested and it was found that the chemical dimers of rifamycin SV (3) resembled the corresponding monomeric analogs with respect to their inhibitory properties *vs.* the nucleic acid polymerases [223].

Rifamycin P (14) was synthesized, and the biological activities of 14 and some derivatives of this compound were reported [66]. According to the report, the derivatives were more active than 14 against the *Mycobacterium avium* complex and other slowly and rapidly growing non-tuberculous mycobacteria. These organisms frequently cause systemic infections in patients with AIDS, and it was suggested that 2'-(diethylamino)rifamycin P appeared suitable for further investigation.

It was reported that the order of activity of streptovaricins A (42) - G (48) and J (49) against *E. coli* DDRP was A (42), G (48) > B (43), C (44) > D (45), J (49) > E (46) >> F (47) [88]. The acetate derivatives of streptovaricin G (streptovaricin G triacetate = streptovaricin A diacetate) and of streptovaricin C (streptovaricin C tri- and tetraacetates = streptovaricins B and J di- and triacetate) were inactive in this assay system.

Cytotoxic and Antitumor Activities

It was reported that some compounds which contained the chromophoric units of the ansamycins were cytotoxic. For example, hydroquinone isolated from Polemoniaceae plant *Ipomopsis aggregata* [224] and the naphthoquinone derivative, tricrozarin B (2,3,6-trimethoxynaphthazarin) isolated from the Iridaceae plant *Tritonia crocosmaeflora* [225] showed cytotoxic activity.

Among the naphthalenoid ansamycin antibiotics, the streptovaricin complex was reported to inhibit focus formation by MSV (MLV: murine leukaemia virions) and to inhibit the splenomegaly-induced Rauscher leukaemia by virus selectively [226,227]. On the other hand, awamycin (26) [34,35] was shown to possess activity against HeLa S3 cells and antitumor activity against experimental murine tumors. Antitumor activity tests on streptovaricin C (44) were conducted [226,227].

The antineoplastic activity of naphthomycin A (56) and its mode of action were reported. The antibiotic caused neither metaphase arrest nor prevented tubulin polymerization, and it was suggested that the mechanism of cytotoxicity of 56 was the inhibition of various SH enzymes, particularly those involved in nucleic acid biosynthesis [228,229].

Maytansine (73), the first benzenoid ansa macrolide isolated from the higher plant *Maytenus ovatus* was obtained as an antileukemic ansa macrolide [119,120]. The cytotoxic action of 73 was reported in a preliminary communication [122]. Several other maytansinoids, such as normaytancyprine (75) [123], 10-epitrewiasine (76) and nortrewiasine (78) [124], were isolated from *Putterlickia verrucosa* and *Trewia nudiflora*. The structural requirements for the antileukemic activity of naturally and semisynthetic maytansinoids were reported [120,125].

The growth inhibition of virus-transformed cells *in vitro* and antitumor activity *in vivo* of geldanamycin (85) and its derivatives [230,231], and the inhibition of DNA synthesis in murine tumor cells by 85 were reported [232,233]. Geldanamycin (85) is an antibiotic that preferentially inhibits G1/S transition and causes G2/M arrest in human leukemia HL-60 cells. Also, it was found that 85 selectively inhibited recombinant Src tyrosine kinase without significantly inhibiting protein kinase A. The perturbation of cell cycling by 85 was accompanied by marked suppression of c-MYC expression [234].

It was observed that macbecins I (93) and II (94) were moderately active against several Gram-positive bacteria and fungi, [149,205] whereas macbecin I (93) exhibited marked antitumor activity against intraperitoneally (ip) inoculated leukemia P388, B16 melanoma and Ehrlich's carcinoma in mice on ip administration [152]. According to this report, the maximum effect measured in terms of ILS (increase of life span) was 97% at a daily dose level of 10 mg/kg for leukemia P388, 103% at 5 mg/kg for B16 melanoma and 206% at 10 mg/kg for Ehrlich's carcinoma. The effect of macbecin I (93) on L1210 leukemia was slight (39% ILS) and no activity was observed against L5178Y or P388/P-3 leukemias (a line of P388 resistant to ansamitocin P-3), or the MOPC-104E myeloma. Cytotoxicity for 93 to cultured KB cells was observed at doses of 0.1 mcg/ml and more. Herbimycin A (86) was rediscovered through the screening of agents which convert the transformed morphology of Rous sarcoma virus-infected rat kidney cells to a normal morphology [143].

Halogenated and other related derivatives of herbimyicn A (86) were prepared and their antitumor activities against the Ehrlich ascites carcinoma were evaluated *in vivo*. Among these herbimycin derivatives, compounds modified at the C-4, C-5, C-6 and C-7 positions of the ansa chain showed comparatively high activities [235].

Hydroxymycotrienins A (102) and B (103), isolated from the culture broth of *Bacillus* sp. BMJ958-62F4, inhibited the growth of human cervical cancer cell lines of human papilloma virus (HPV) positive more strongly than that of HPV negative cell lines [164].

On the other hand, the trienomycins A (111) - C (113) were reported to have potent cytocidal activities [169,171,175] and antitumor activity was reported for trienomycin A (111) [236]. Cytotrienins A (104) and B (105) were isolated from the culture broth of *Streptomyces* sp., and these compounds exhibit potent apoptosis-inducing activity on human leukemia HL-60 cells with ED_{50} values of 7.7 nM, which were the same dose range as for their growth inhibitory activities [165-167].

Thiazinotrienomycins A (106) - E (110), isolated from the culture broth of *Streptomyces* sp. MJ672-m3, exhibited activities against cervical cancer cell lines, and the LD_{50} of thiazinotrienomycin B (107), the most active congener, was 125 mg/kg in mice (*ip*) [168]. Thiazinotrienomycins A (106) and B (107) showed about 10 times stronger activities against the cell lines of cervical cancers than against the cell lines of stomach, colon and breast cancers. Thiazinotrienomycin B (107) did not inhibit the growth of any bacteria, fungi and yeast at 100 mcg/ml *in vitro*. A preliminary study on the mode of action of 107 in HeLa cells showed that the antibiotic inhibited the membrane transport of thymidine and uridine, but not leucine, at concentrations where the cell growth was partially inhibited [168].

Inhibition of HIV Replication by 19-*O*-*n*-Pentyldamavaricin Fc *In Vitro*

Damavaricin Fc (133) is an atropisomeric mixture of two isomers produced by the alkaline degradation of streptovaricin C (44) [237]. Although the HIV-specificity is low as compared to that of HIV-selective inhibitors, including AZT (azidothymizine; zidovudine), *n*-pentyldamavaricin F_C (134) showed inhibitory activity on HIV replication with selective efficacy (ratio of CD_{50} to ED_{50}) in the range of about 3 to 5 [238, 239].

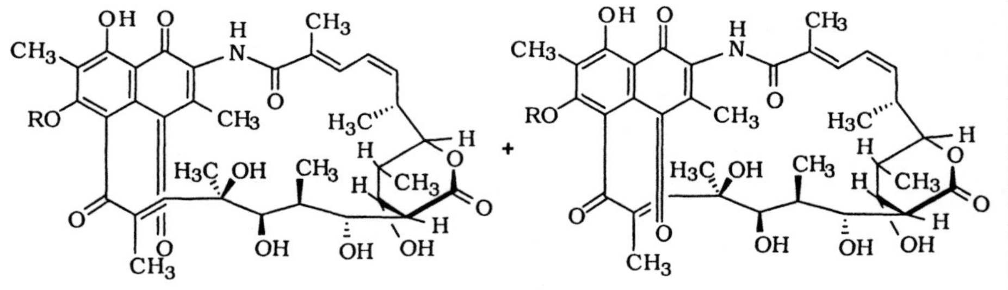

133 R = H
134 R = n-pentyl

Other Biological Activities

In the course of the screening for substances that protect neuronal hybridoma N18-RE-105 cells from L-glutamate toxicity, naphthomycinol (**68**) was isolated from the culture broth of *Streptomyces* sp. PF7 [106]. Naphthomycinol (**68**) is a member of the naphthomycin antibiotics with C_{23} ansa chain.

ACKNOWLEDGEMENTS

One of the authors (Shinji Funayama) is greatly indebted to Professor Nakao Ishida (Former President of the Tohoku University, Sendai, Japan) for his encouragement in preparing this manuscript.

REFERENCES

[1]   Prelog, V.; Oppolzer, W. *Helv. Chim. Acta* **1973**, *56*, 2279-2287.
[2]   Lüttringhaus, A.; Gralheer, H. *Ann. Chem.* **1942**, *550*, 67-98.
[3]   Sensi, P.; Greco, A. M.; Ballotta, R. *Antibiot. Ann.* **1959-1960**, 262-270.
[4]   Umezawa, I.; Takeshima, H.; Komiyama, K.; Koh, Y.; Yamamoto, H.; Kawaguchi, M. *J. Antibiot.* **1981**, *34*, 259-265.
[5]   Ōmura, S.; Nakagawa, A.; Tanaka, Y. *Trends in Antibiotic Research, Genetics, Biosyntheses, Actions and New Substances*, eds. Umezawa H, Demain AL, Hata T, Hutchinson CR, Japan Antibiotic Research Association, Tokyo **1982**, 135-145.
[6]   Oiwa, R.; Iwai, Y.; Takahashi, Y.; Kitao, K.; Ōmura, S. *Kitasato Arch. Exp. Med.* **1982**, *55*, 119-124.

[7]   Komiyama, K.; Edanami, K.; Yamamoto, H.; Umezawa, I. *J. Antibiot.* **1982**, *35*, 703-706.
[8]   Komiyama, K.; Edanami, K.; Tanoh, A.; Yamamoto, H.; Umezawa, I. *J. Antibiot.* **1983**, *36*, 301-311.
[9]   Jomon, K.; Kuroda, Y.; Ajisaka, M.; Sakai, H. *J. Antibiot.* **1972**, *25*, 271-280.
[10]   Aizawa, S; Akutsu, H.; Satomi, T.; Nagatsu, T.; Taguchi, R.; Seino, A. *J. Antibiot.* **1979**, *32*, 193-196.
[11]   Vezina, C.; Kudelski, A.; Sehgal, S. N. *J. Antibiot.* **1975**, *28*, 721-726.
[12]   Sehgal, S. N.; Baker, H.; Vezina, C. *J. Antibiot.* **1975**, *28*, 727-732.
[13]   Sehgal, S. N.; Baker, H.; Eng, C. P.; Singh, K.; Vezina, C. *J. Antibiot.* **1983**, *36*, 351-354.
[14]   Õmura, S.; Nakagawa, A.; Shibata, K.; Sano, H. *Tetrahedron Lett.* **1982**, *23*, 4713-4716.
[15]   Rinehart, K. L., Jr. *Acc. Chem. Res.* **1972**, *5*, 57-64.
[16]   Rinehart, K. L., Jr.; Shield, L. S. *Fortschr. Chem. Org. Naturst.* **1976**, *33*, 231-307.
[17]   Brufani, M. *Topics in Antibiotic Chemistry, vol. I,* ed. Sammes PG, John Wiley & Sons Inc., London **1977**, pp. 93-217.
[18]   Wehrli, W. *Topics in Current Chemistry* **1977**, *72*, 21-49.
[19]   Ganguly, A. K. *J. Chromatography Library* **1978**, *15*, 39-68.
[20]   Lancini, G.; Grandi, M. *Antibiotics IV Biosynthesis,* ed. J.W. Corcoran **1981**, pp. 12-40.
[21]   Reider, P. J.; Roland, D. M. *The Alkaloids vol. XXIII,* ed. A. Brossi, Academic Press, Inc., New York **1984**, pp. 71-156.
[22]   Ghisalba, O. *Chimiya* **1985**, *39*, 79-88.
[23]   Engvild, K. C. *Phytochemistry* **1986**, *25*, 781-791.
[24]   Tanaka, N.; Nakamura, S. In *"Koseibusshitu-Taiyo, 4th Ed.",* University of Tokyo Press **1992**, pp. 207-218.
[25]   Timbal, M. T. *Antibiot. Ann.* **1959-1960**, 271-276.
[26]   Maffii, G.; Timbal, M. T. *Antibiot. Ann. 1959-1960* **1959-1960**, 277-284.
[27]   Furesz, S.; Scotti, R. *Antibiot. Ann.* **1959-1960**, 285-292.
[28]   Oppolzer, W. O.; Prelog, V.; Sensi, P. *Experientia* **1964**, *20*, 336-339.
[29]   Oppolzer, W. O.; Prelog, V. *Helv. Chim. Acta* **1973**, *56*, 2287-2314.
[30]   Lancini, G.; Gallo, G. G.; Sartori, G.; Sensi, P. *J. Antibiot.* **1969**, *22*, 369-377.
[31]   Lancini, G.; Hengeller, C. *J. Antibiot.* **1969**, *22*, 637-638.
[32]   Leitich, J.; Prelog, V.; Sensi, P. *Experientia* **1967**, *23*, 505-507.
[33]   Brufani, M.; Fedeli, W.; Giacomello, G.; Vaciago, A. *Experientia* **1967**, *23*, 508-512.
[34]   Umezawa, I.; Oka, H.; Komiyama, K.; Hagiwara, K.; Tomisaka, S.; Miyano, T. *J. Antibiot.* **1983**, *36*, 1144-1149.
[35]   Funayama, S.; Okada, K.; Oka, H.; Tomisaka, S.; Miyano, T.; Komiyama, K.; Umezawa, I. *J. Antibiot.* **1985**, *38*, 1284-1286.
[36]   Lancini, G.; Sartori, G. *J. Antibiot.* **1976**, *29*, 466-468.
[37]   Traxler, P.; Schupp, T.; Fuhrer, H.; Richter, W. J. *J. Antibiot.* **1981**, *34*, 971-979.
[38]   Brufani, M.; Fedeli, W.; Giacomello, G.; Vaciago, A. *Rend. Acc. Naz. Lincei* **1964**, *36*, 113-117.
[39]   Brufani, M.; Fedeli, W.; Giacomello, G.; Vaciago, A. *Experientia* **1964**, *20*, 339-342.
[40]   Leitich, J.; Oppolzer, W.; Prelog, V. *Experientia* **1964**, *20*, 343-344.

[41] Brufani, M.; Fedeli, W.; Giacomello, G.; Vaciago, A. *Rend. Acc. Naz. Lincei* **1966**, *40*, 548.

[42] Brufani, M.; Cerrini, S.; Fedeli, W.; Vaciago, A. *J. Mol. Biol.* **1974**, *87*, 409-435.

[43] Celmer, W. D.; Sciavolino, F.; Routien, J. B.; Cullen, W. P. *Ger. Offen* **1975**a, 2,500,898 [*Chem. Abstr.* **1975**, *83*, 191340g].

[44] Celmer, W. D. *US Pat.* **1975**, 3,923,791 [*Chem. Abstr.* **1976**, *84*, 59610m].

[45] Ghisalba, O.; Roos, R.; Schupp, T.; Nüesch, J. *J. Antibiot.* **1982**, *35*, 74-80.

[46] Martinelli, E.; White, R. J.; Gallo, G. G.; Beynon, P. J. *Tetrahedron* **1973**, *29*, 3441-3448.

[47] Martinelli, E.; White, R. J.; Gallo, G. G.; Beynon, P. J. *Tetrahedron Lett.* **1974**, *15*, 1367-1368.

[48] Kishi, T.; Asai, M.; Muroi, M.; Harada, S.; Mizuta, E.; Terao, S.; Miki, T.; Mizuno, K. *Tetrahedron Lett.* **1969**, 91-95.

[49] Kishi, T.; Harada, S.; Asai, M.; Muroi, M.; Mizuno, K. *Tetrahedron Lett.* **1969**, 97-100.

[50] Prelog, V. *Pure Appl. Chem.* **1963**, *7*, 551-564.

[51] Gallo, G. G.; Martinelli, E.; Pagani, V.; Sensi, P. *Tetrahedron* **1974**, *30*, 3093-3097.

[52] Cellai, L.; Cerrini, S.; Segre, A.; Brufani, M.; Fedeli, W.; Vaciago, A. *J. Org. Chem.* **1982**, *47*, 2652-2661.

[53] Nagaoka, H.; Rutsch, W.; Schmid, G.; Iio, H.; Johnson, M. R.; Kishi, Y. *J. Am. Chem. Soc.* **1980**, *102*, 7962-7965.

[54] Iio, H; Nagaoka, H.; Kishi, Y. *J. Am. Chem. Soc.* **1980**, *102*, 7965-7967.

[55] Fox, J. L. *Chem. Eng. News* **1981**, 61-64.

[56] Paterson, I.; Mansuri, M. M. *Tetrahedron* **1985**, *41*, 3569-3624.

[57] Brizzi, V.; Brufani, M.; Cellai, L.; Segre, A. L. *J. Antibiot.* **1983**, *36*, 516-521.

[58] Brufani, M.; Cecchini, G.; Cellai, L.; Federici, M.; Guiso, M.; Segre, A. *J. Antibiot.* **1985**, *38*, 259-262.

[59] Brufani, M.; Cellai, L.; Cozzella, L.; Federici, M.; Guiso, M.; Segre, A. *J. Antibiot.* **1985**, *38*, 1359-1362.

[60] Dampier, M. F.; Chen, C. W.; Whitlock, H. W., Jr. *J. Am. Chem. Soc.* **1976**, *98*, 7064-7069.

[61] Weinstein, M. J.; Luedemann, G. M.; Oden, E. M.; Wagman, G. H. *Antimicrob. Agts. Chemother.* **1967**, 435-441.

[62] Ganguly, A. K.; Szmulewicz, S.; Sarre, O. Z.; Greeves, D.; Morton, J.; McGlotten, J. *Chem. Commun.* **1974**, 395-396.

[63] Ganguly, A. K.; Liu, Y. T.; Sarre, O. Z.; Szmulewicz, S. *J. Antibiot.* **1977**, *30*, 625-627.

[64] Martinelli, E.; Antonini, P.; Cricchio, R.; Lancini, G.; White, R. J. *J. Antibiot.* **1978**, *31*, 949-951.

[65] Cricchio, R.; Antonini, P.; Lancini, G. C.; Tamborini, G.; White, R. J.; Maritnelli, E. *Tetrahedron* **1980**, *36*, 1415-1421.

[66] Cavalleri, B; Turconi, M.; Tamborini, G.; Occelli, E.; Cietto, G.; Pallanza, R.; Scotti, R.; Berti, M.; Romanò, G.; Parenti, F. *J. Med. Chem.* **1990**, *33*, 1470-1476.

[67] Schupp, T.; Traxler, P.; Auden, J. A. L. *J. Antibiot.* **1981**, *34*, 965-970.

[68] Wang, N.-J.; Fu, Y.; Yan, G.-H.; Bao, G.-H.; Xu, C.-F.; He, C.-H. *J. Antibiot.* **1988**, *41*, 264-267.

[69] Maggi, N.; Pasqualussi, C. R.; Ballotta, R.; Sensi, P. *Chemotherapia* **1966**, *11*, 285-292.

[70] Kump, W.; Bickel, H. *Helv. Chim. Acta* **1973**, *56*, 2323-2347.

[71] Kump, W.; Bickel, H. *Helv. Chim. Acta* **1973**, *56*, 2348-2377.

[72] Fuhrer, H. *Helv. Chim. Acta* **1973**, *56*, 2377-2386.

[73] Casey, M. L.; Whitlock, H. W. *J. Am. Chem. Soc.* **1975**, *97*, 6231-6236.

[74] Whitlock, H. W.; Dampier, M. F. *J. Am. Chem. Soc.* **1975**, *97*, 6254-6256.

[75] White, R. J.; Martinelli, E.; Lancini, G. *Proc. Natl. Acad. Sci. USA* **1974**, *71*, 3260-3264.

[76] Martinelli, E.; Gallo, G. G.; Antonini, P.; White, R. J. *Tetrahedron* **1974**, *30*, 3087-3091.

[77] Wang, N.-J.; Han, B.-L.; Yamashita, N.; Sato, M. *J. Antibiot.* **1994**, *47*, 613-615.

[78] Cricchio, R.; Antonini, P.; Ferrari, P.; Ripamonti, A.; Tuan, G.; Marinelli, E. *J. Antibiot.* **1981**, *34*, 1257-1260.

[79] Tanida, S.; Shinagawa, S.; Takizawa, M.; Takahashi, T.; Harada, S.; Hasegawa, T. *Experientia* **1986**, *42*, 1167-1170.

[80] Celmer, W. D.; Cullen, W. P.; Oscarson, J. R.; Huang, L. H.; Shibakawa, R.; Tone, J. *US Pat.* **1980**, 4,225,674 [*Chem. Abstr.* **1981**, *94*, 28883e].

[81] Celmer, W. D.; Sciavolino, F.; Routien, J. B.; Cullen, W. P. *Japan. Kokai* **1975b**, 75-100,293, Aug. 8, 1975 [*Chem. Abstr.* **1975**, *83*, 191340g].

[82] Deshmukh, P. V.; Kakinuma, K.; Ameel, J. J.; Rinehart, K. L., Jr. *J. Am. Chem. Soc.* **1976**, *98*, 870-872.

[83] Ghisalba, O.; Traxler, P.; Nüesch, J. *J. Antibiot.* **1978**, *31*, 1124-1131.

[84] Traxler, P.; Schupp, T.; Wehrli, W. *J. Antibiot.* **1982**, *35*, 594-601.

[85] Traxler, P.; Ghisalba, O. *J. Antibiot.* **1982**, *35*, 1361-1366.

[86] Rinehart, K. L., Jr.; Antosz, F. J.; Deshmukh, P. V.; Kakinuma, K.; Martin, P. K.; Milavetz, B. I.; Sasaki, K.; Witty, T. R.; Li, L. H.; Reusser, F. *J. Antibiot.* **1976**, *29*, 201-203.

[87] Rinehart, K. L., Jr.; Maheshwari, M. L.; Antosz, F. J.; Mathur, H. H.; Sasaki, K.; Schacht, R. J. *J. Am. Chem. Soc.* **1971**, *93*, 6273-6274.

[88] Rinehart, K. L., Jr.; Antosz, F. J.; Sasaki, K.; Martin, P. K.; Maheshwari, M. L.; Reusser, F.; Li, L. H.; Moran, D.; Wiley, P. F. *Biochemistry* **1974**, *13*, 861-867.

[89] Rinehart, K. L., Jr.; Cook, J. C., Jr.; Maurer, K. H.; Rapp, U. *J. Antibiot.* **1974**, *27*, 1-13.

[90] Kakinuma, K.; Milavetz, B. I.; Rinehart, K. L., Jr. *J. Org. Chem.* **1976**, *41*, 1358-1364.

[91] Milavetz, B.; Kakinuma, K.; Rinehart, K. L., Jr.; Rolls, J. P.; Haak, W. J. *J. Am. Chem. Soc.* **1973**, *95*, 5793-5795.

[92] Wang, A. H. J.; Paul, I. C.; Rinehart, K. L., Jr.; Antosz, F. J. *J. Am. Chem. Soc.* **1971**, *93*, 6275-6276.

[93] Staley, A. L.; Rinehart, K. L., Jr. *J. Antibiot.* **1991**, *44*, 218-224.

[94] Rinehart, K. L., Jr.; Knöll, W. M. J.; Kakinuma, K.; Antosz, F. J.; Paul, I. C.; Wang, A. H. J.; Reusser, F.; Li, L. H.; Krueger, W. C. *J. Am. Chem. Soc.* **1975**, *97*, 196-198.

[95] Knöll, W. M. J.; Rinehart, K. L., Jr.; Wiley, P. F.; Li, L. H. *J. Antibiot.* **1980**, *33*, 249-251.

[96] Ghisalba, O.; Traxler, P.; Fuhrer, H.; Richter, W. J. *J. Antibiot.* **1979**, *32*, 1267-1272.

[97]   Ghisalba, O.; Traxler, P.; Fuhrer, H.; Richter, W. J. *J. Antibiot.* **1980**, *33*, 847-856.
[98]   Balerna, M.; Keller-Schierlein, W. K.; Martius, C.; Wolf, H.; Zähner, H. *Arch. Mikrobiol.* **1969**, *65*, 303-317.
[99]   Keller-Schierlein, W.; Meyer, M.; Cellai, L.; Cerrini, S.; Lamba, D.; Segre, A.; Fedeli, W.; Brufani, M. *J. Antibiot.* **1984**, *37*, 1357-1361.
[100] Keller-Schierlein, W.; Meyer, M.; Zeeck, A.; Damberg, M.; Machinek, R.; Zähner, H.; Lazar, G. *J. Antibiot.* **1983**, *36*, 484-492.
[101] Meyer, M.; Keller-Schierlein, W.; Megahed, S.; Zähner, H.; Segre, A. *Helv. Chim. Acta* **1986**, *69*, 1356-1364.
[102] Hotta, M.; Hayakawa, Y.; Furihata, K.; Shimazu, A.; Seto, H.; Õtake, N. *J. Antibiot.* **1986**, *39*, 311-313.
[103] Mochizuki, J.; Kobayashi, E.; Furihata, K.; Kawaguchi, A.; Seto, H.; Õtake, N. *J. Antibiot.* **1986**, *39*, 157-161.
[104] Mukhopadhyay, T.; Franco, C. M. M.; Reddy, G. C. S.; Ganguli, B. N.; Fehlhaber, H. W. *J. Antibiot.* **1985**, *38*, 948-951.
[105] Mochizuki, J.; Kobayashi, E.; Furihata, K.; Kawaguchi, A.; Seto, H.; Õtake, N. *Nihon Nogeikagakukai Koenyoshi-shu* **1986**, 384.
[106] Kim, J.-S.; Shin-ya, K.; Eishima, J.; Furihata, K.; Seto, H. *J. Antibiot.* **1996**, *49*, 1172-1174.
[107] Allen, M. S.; McDonald, I. A.; Rickards, R. W. *Tetrahedron Lett* **1981**, *22*, 1145-1148.
[108] McDonald, I. A.; Rickards, R. W. *Tetrahedron Lett.* **1981**, *22*, 1149-1152.
[109] Bhuyan, B. K.; Owen, S. P.; Dietz, A. *Antimicrob. Agts. Chemother. 1964* **1965**, 91-96.
[110] Meyer, C. E. Antimicrob. Agts. & Chemother. 1964 **1965**, 97-99.
[111] Hoeksema, H.; Mizsak, S. A.; Baczynskyj, L. *J. Antibiot.* **1979**, *32*, 773-776.
[112] Hoeksema, H.; Mizsak, S. A.; Baczynskyj, L.; Pschigoda, L. M. *J. Am. Chem. Soc.* **1982**, *104*, 5173-5181.
[113] Reusser, F. *Biochemistry* **1973**, *12*, 1136-1142.
[114] Reusser, F. *Antibiotics VI,* ed. Hahn FE **1983**, pp.187-198.
[115] Brimacombe, J. S.; Rahman, K. M. M. *J. Chem. Soc., Perkin Trans 1* **1985**, 1067-1072.
[116] Brimacombe, J. S.; Rahman, K. M. M. *J. Chem. Soc., Perkin Trans 1* **1985**, 1073-1079.
[117] Bannister, B.; Zapotocky, B. A. *J. Antibiot.* **1992**, *45*, 1313-1324.
[118] Ghisalba, O.; Fuhrer, H.; Richter, W. J.; Moss, S. *J. Antibiot.* **1981**, *34*, 58-63.
[119] Kupchan, S. M.; Komoda, Y.; Court, W. A.; Thomas, G. J.; Smith, R. M.; Karim, A.; Gilmore, C. J.; Haltiwanger, R. C.; Bryan, R. F. *J. Am. Chem. Soc.* **1972**, *94*, 1354-1356.
[120] Kupchan, S. M., Komoda, Y.; Branfman, A. R.; Sneden, A. T.; Court, W. A.; Thomas, G. J.; Hintz, H. P. J.; Smith, R. M.; Karim, A.; Howie, G. A.; Verma, A. K.; Nagao, Y.; Dailey, R. G., Jr.; Zimmerly, V. A.; Sumner, W. C., Jr. *J. Org. Chem.* **1977**, *42*, 2349-2357.
[121] Bryan, R. F.; Gilmore, C. J.; Haltiwanger, R. C. *J. Chem. Soc., Perkin Trans. 2* **1973**, 897-902.
[122] Wolpert-Defilippes, M. K.; Adamson, R. H.; Cysyk, R. L.; Johns, D. G. *Biochem. Pharmacol.* **1975**, *24*, 751-754.
[123] Sneden, A. T.; Sumner, W. C., Jr.; Kupchan, S. M. *J. Nat. Prod.* **1982**, *45*, 624-628.

[124] Powell, R. G.; Smith, C. R., Jr.; Plattner, R. D.; Jones, B. E. *J. Nat. Prod.* **1983**, *46*, 660-666.

[125] Kupchan, S. M.; Sneden, A. T.; Branfman, A. R.; Howie, G. A.; Rebhun, L. I.; McIvor, W. E.; Wang, R. W.; Schnaitman, T. C. *J. Med. Chem.* **1978**, *21*, 31-37.

[126] Higashide, E.; Asai, M.; Ootsu, K.; Tanida, S.; Kozai, Y,; Hasegawa, T,; Kishi, T.; Sugino, Y.; Yoneda, M. *Nature* **1977**, *270*, 721-722.

[127] Asai, M.; Mizuta, E.; Izawa, M.; Haibara, K.; Kishi, T. *Tetrahedron* **1979**, *35*, 1079-1085.

[128] Nakahama, K.; Izawa, M.; Asai, M.; Kida, M.; Kishi, T. *J. Antibiot.* **1981**, *34*, 1581-1586.

[129] Izawa, M.; Tanida, S.; Asai, M. *J. Antibiot.* **1981**, *34*, 496-506.

[130] Izawa, M.; Nakahama, K.; Kasahara, F.; Asai, M.; Kishi, T. *J. Antibiot.* **1981**, *34*, 1587-1590.

[131] Izawa, M.; Wada, Y.; Kasahara, F.; Asai, M.; Kishi, T. *J. Antibiot.* **1981**, *34*, 1591-1595.

[132] Tanida, S.; Izawa, M.; Hasegawa, T. *J. Antibiot.* **1981**, *34* 489-495.

[133] Hatano, K.; Higashide, E.; Yoneda, M. *Agr. Biol. Chem.* **1984**, *47*, 1889-1890.

[134] Meyers, A. I.; Reider, P. J.; Campbell, A. L. *J. Am. Chem. Soc.* **1980**, *102*, 6597-6598.

[135] Kitamura, M.; Ichikawa, Z.; Isobe, M.; Goto, T. *Nihon Yakugakukai Koen Yoshi-shu* **1984**, p.138.

[136] Corey, E. J.; Weigel, L. O.; Chamberlin, A. R.; Cho, H.; Hua, D. H. *J. Am. Chem. Soc.* **1980**, *102*, 6613-6615.

[137] DeBoer, C.; Meulman, P. A.; Wnuk, R. J.; Peterson, D. H. *J. Antibiot.* **1970**, *23*, 442-447.

[138] Sasaki, K.; Rinehart, K. L., Jr.; Slomp, G.; Grostic, M. F.; Olson, E. C. *J. Am. Chem. Soc.* **1970**, *92*, 7591-7593.

[139] Johnson, R. D.; Haber, A.; Rinehart, K. L., Jr. *J. Am. Chem. Soc.* **1974**, *96*, 3316-3317.

[140] Ōmura, S.; Iwai, Y.; Takahashi, Y.; Sadakane, N.; Nakagawa, A.; Oiwa, H.; Hasegawa, Y.; Ikai, T. *J. Antibiot.* **1979**, *32*, 255-261.

[141] Furusaki, A.; Matsumoto, T,; Nakagawa, A.; Ōmura, S. *J. Antibiot.* **1980**, *33*, 781-782.

[142] Ōmura, S.; Nakagawa, A.; Sadakane, N. *Tetrahedron Lett.* **1979**, *20*, 4323-4326.

[143] Uehara, Y.; Hori, M.; Takeuchi, T.; Umezawa, H. *Gann* **1985**, *76*, 672-675.

[144] Nakagawa, A.; Sadakane, N.; Ōmura, S. *J. Chem. Soc. Jpn., Chem. & Ind. Chem.* **1981**, 892-894.

[145] Tanida, S.; Muroi, M.; Hasegawa, T. *Eur. Pat. Appl. EP* **1984**, 110,710 [*Chem Abstr.* **1984**, *101*, 108940z].

[146] Lin, L.-Z.; Blaskó, G; Cordell, G. A. *J. Nat. Prod.* **1988**, *51*, 1161-1165.

[147] Muroi, M.; Haibara, K.; Asai, M.; Kamiya, K.; Kishi, T. *Symposium on the Chemistry of Natural Products, Fukuoka* **1979**, *21*, 386-393.

[148] Muroi, M.; Haibara, K.; Asai, M.; Kishi, T. *Tetrahedron Lett.* **1980**, *21*, 309-312.

[149] Muroi, M.; Izawa, M.; Kosai, Y.; Asai, M. *J. Antibiot.* **1980**, *33*, 205-212.

[150] Muroi, M.; Haibara, K.; Asai, M.; Kamiya, K.; Kishi, T. *Tetrahedron* **1981**, *37*, 1123-1130.

[151] Tanida, S.; Hasegawa, T.; Hatano, K.; Higashide, E.; Yoneda, M. *J. Antibiot.* **1980**, *33*, 192-198.

[152] Ono, Y.; Kozai, Y.; Ootsu, K. *Gann* **1982**, *73*, 938-944.

[153] Hatano, K.; Muroi, M.; Higashide, E.; Yoneda, M. *Agr. Biol. Chem.* **1982**, *46*, 1699-1702.

[154] Sugita, M.; Natori, Y.; Sasaki, T.; Furihata, K.; Shimazu, A.; Seto, H.; Õtake, N. *J. Antibiot.* **1982**, *35*, 1460-1466.

[155] Sugita, M.; Sasaki, T.; Furihata, K.; Seto, H.; Õtake, N. *Agr. Biol. Chem.* **1982**, *46*, 1111-1113.

[156] Sugita, M.; Sasaki, T.; Furihata, K.; Seto, H.; Õtake, N. *J. Antibiot.* **1982**, *35*, 1467-1473.

[157] Sugita, M.; Natori, Y.; Sueda, N.; Furihata, K.; Seto, H.; Õtake, N. *J. Antibiot.* **1982**, *35*, 1474-1479.

[158] Hiramoto, S.; Sugita, M.; Ando, C.; Sasaki, T.; Furihata, K.; Seto, H.; Õtake, N. *J. Antibiot.* **1985**, *38*, 1103-1106.

[159] Coronelli, C.; Pasqualucci, R. C.; Thiemann, J. E.; Tamoni, G. *J. Antibiot.* **1967**, *20A*, 329-333.

[160] Sugita, M.; Hiramoto, S.; Ando, C.; Sasaki, T.; Furihata, K.; Seto, H.; Õtake, N. *J. Antibiot.* **1985**, *38*, 799-802.

[161] Weber, W.; Zähner, H.; Damberg, M.; Russ, P.; Zeeck, A. *Zbl. Bakt. Hyg. I. Abt. Orig.* **1981**, *C2*, 122-139.

[162] Damberg, M.; Russ, P.; Zeeck, A. *Tetrahedron Lett.* **1982**, *23*, 59-62.

[163] Lazar, G.; Zähner, H.; Damberg, M.; Zeeck, A. *J. Antibiot.* **1983**, *36*, 187-189.

[164] Hosokawa, N.; Naganawa, H.; Hamada, M.; Takeuchi, T.; Ikeno, S.; Hori, M. *J. Antibiot.* **1996**, *49*, 425-431.

[165] Zhang, H.-P.; Kakeya, H.; Osada, H. *Tetrahedron Lett.* **1997**, *38*, 1789-1792.

[166] Kakeya, H.; Zhang, H.-P.; Kobinata, K.; Onose, R.; Onozawa, C.; Kudo, T.; Osada, H. *J. Antibiot.* **1997**, *50*, 370-372.

[167] Kakeya, H.; Osada, H. *Nippon Nogeikagaku Kaishi* **1998**, *72*, 525.

[168] Hosokawa, N.; Naganawa, H.; Iinuma, H.; Hamada, M.; Takeuchi, T.; Kanbe, T.; Hori, M. *J. Antibiot.* **1995**, *48*, 471-478.

[169] Umezawa. I.; Funayama, S.; Okada, K.; Iwasaki, K.; Satoh, J.; Masuda, K.; Komiyama, K. *J. Antibiot.* **1985**, *38*, 699-705.

[170] Funayama, S.; Okada, K.; Komiyama, K.; Umezawa, I. *J. Antibiot.* **1985**, *38*, 1107-1109.

[171] Funayama, S.; Okada, K.; Iwasaki, K.; Komiyama, K.; Umezawa, I. *J. Antibiot.* **1985**, *38*, 1677-1683.

[172] Nomoto, H.; Katsumata, S.; Takahashi, K.; Funayama, S.; Komiyama, K.; Umezawa, I.; Õmura, S. *J. Antibiot.* **1989**, *42*, 479-481.

[173] Smith, A. B., III; Wood, J. L.; Wong, W.; Gould, A. E.; Rizzo, C. J.; Funayama, S.; Õmura, S. *J. Am. Chem. Soc.* **1990**, *112*, 7425-7426.

[174] Smith, A. B., III; Wood, J. L.; Õmura, S. *Tetrahedron Lett.* **1991**, *32*, 841-842.

[175] Funayama, S.; Anraku, Y.; Mita, A.; Yang, Z.-B.; Shibata, K.; Komiyama, K.; Umezawa, I.; Õmura, S. *J. Antibiot.* **1988**, *41*, 1233-1230.

[176] White, R. J.; Martinelli, E. *FEBS Lett.* **1974**, *49*, 233-236.

[177] Haber, A.; Johnson, R. D.; Rinehart, K. L., Jr. *J. Am. Chem. Soc.* **1977**, *99*, 3541-3544.

[178] Rinehart, K. L., Jr.; Potgieter, M.; Wright, D. A. *J. Am. Chem. Soc.* **1982**, *104*, 2649-2652.

[179] Hatano, K.; Akiyama, S.; Asai, M.; Rickards, R. W. *J. Antibiot.* **1982**, *35*, 1415-1417.

[180] Ghisalba, O.; Nüesch, J. *J. Antibiot.* **1978**, *31*, 202-214.

[181] Ghisalba, O.; Nüesch, J. *J. Antibiot.* **1978**, *31*, 215-225.
[182] Ghisalba, O.; Nüesch, J. *J. Antibiot.* **1981**, *34*, 64-71.
[183] Kibby, J. J.; McDonald, I. A.; Rickards, R. W. *J. Chem. Soc., Chem. Commun.* **1980**, 768-769.
[184] Casati, R.; Beale, J. M.; Floss, H. G. *J. Am. Chem. Soc.* **1987**, *109*, 8102-8104.
[185] Meier, R.-M.; Tamm, C. *Helv. Chim. Acta* **1991**, *74*, 807-818.
[186] Kibby, J. J.; Rickards, R. W. *J. Antibiot.* **1981**, *34*, 605-607.
[187] Becker, A. M.; Herlt, A. J.; Hilton, G. L.; Kibby, J. J.; Rickards, R. W. *J. Antibiot.* **1983**, *36*, 1323-1328.
[188] Rinehart, K. L., Jr.; Weller, D. D.; Pearce, C.J. *J. Nat. Prod.* **1980**, *43*, 1-20.
[189] Rinehart, K. L., Jr.; Potgieter, M.; Delaware, D. L.; Seto, H. *J. Am. Chem. Soc.* **1981**, *103*, 2099-2101.
[190] Rinehart, K. L., Jr.; Potgieter, M.; Jin, W.; Pearce, C. J.; Wright, D. A.; Wright, J. L. C.; Walter, J. A.; McInnes, A. G. *Trends in Antibiotic Research, Genetics, Biosyntheses, Actions and New Substances,* eds. Umezawa H, Demain AL, Hata T, Hutchinson CR, Japan Antibiotic Research Association, Tokyo **1982**, 171-184.
[191] Kakinuma, K.; Ikekawa, N.; Nakagawa, A.; Ōmura, S. *J. Am. Chem. Soc.* **1979**, *101*, 3402-3404.
[192] Anderson, M. G.; Kibby, J. J.; Rickards, R. W.; Rothschild, J. M. *J. Chem. Soc., Chem. Commun.* **1980**, 1277-1278.
[193] Anderson, M. G.; Monypenny, D.; Rickards, R. W.; Rothschild, J. M. *J. Chem. Soc., Chem. Commun.* **1989**, 311-313.
[194] Roos, R.; Ghisalba, O. *Experientia* **1980**, *36*, 486-486.
[195] Meier, R.-M.; Tamm, C. *J. Antibiot.* **1992**, *45*, 400-410.
[196] Kawaguchi, T.; Azuma, M.; Horinouchi, S.; Beppu, T. *J. Antibiot.* **1988**, *41*, 360-365.
[197] White, R. J.; Martinelli, E.; Gallo, G. G.; Lancini, G.; Beynon, P. *Nature* **1973**, *243*, 273-277.
[198] Brufani, M.; Kluepfel, D.; Lancini, G. C.; Leitich, J.; Mesentseu, A. S.; Prelog, V.; Schmook, F. P.; Sensi, P. *Helv. Chim. Acta* **1973**, *56*, 2315-2323.
[199] Karlsson, A.; Sartori, G.; White, R. J. *Eur. J. Biochem.* **1974**, *47*, 251-256.
[200] Celmer, W. D. *Pure Appl. Chem.* **1971**, *28*, 413-453.
[201] Celmer, W. D. *Ann. N. Y. Acad. Sci.* **1986**, *471*, 299-303.
[202] Funayama, S.; Nakagawa, A.; Ōmura, S. *Symposium on the Chemistry of Natural Products, Sendai* **1986**, *28*, 73-80.
[203] Williams, T. H. *J. Antibiot.* **1975**, *28*, 85-86.
[204] Brufani, M.; Cellai, L.; Keller-Schierlein, W. *J. Antibiot.* **1979**, *32*, 167-168.
[205] Tanida, S.; Hasegawa, T.; Higashide, E. *J. Antibiot.* **1980**, *33*, 199-204.
[206] Wu, T. S.; Duncan, J.; Tsao, S. W.; Chang, C. J.; Keller, P. J.; Floss, H. G. *J. Nat. Prod.* **1987**, *50*, 108-118.
[207] Rosa, M. D.; Gambacorta, A.; Minale, L. *Chem. Commun.* **1971**, 1334-1334.
[208] Oshima, M.; Ariga, T. *J. Biol. Chem.* **1975**, *250*, 6963-6968.
[209] Suzuki, K.; Saito, K.; Kawaguchi, A.; Okuda, S.; Komagata, K. *J. Gen. Appl. Microbiol.* **1981**, *27*, 261-266.
[210] Ōmura, S.; Kitao, C.; Tanaka, H.; Oiwa, R.; Takahashi, Y.; Nakagawa, A.; Shimada, M.; Iwai, Y. *J. Antibiot.* **1976**, *29*, 876-881.
[211] Takeda, Y.; Mak, V.; Chang, C.; Chang, C.; Floss, H. G. *J. Antibiot.* **1984**, *37*, 868-875.

[212] Reynolds, K. A.; Wang, P.; Fox, K. M.; Floss, H. G. *J. Antibiot.* **1992**, *45*, 411-419.

[213] Wehrli, W.; Zimmermann, W.; Kump, W.; Tosch, W.; Vischer, W.; Zak, O. *J. Antibiot.* **1987**, *40*, 1733-1739.

[214] Wehrli, W.; Knusel, F.; Nüesch, J.; Staehelin, M. *Proc. Natl. Acad. Sci. USA* **1968**, *61*, 667-673.

[215] Dampier, M. F.; Whitlock, H. W., Jr. *J. Am. Chem. Soc.* **1975**, *97*, 6254-6256.

[216] Arora, S. K.; Main, P. *J. Antibiot.* **1984**, *37*, 178-181.

[217] Yazawa, K.; Mikami, Y.; Maeda, A.; Akao, M.; Morisaki, N.; Iwasaki, S. *Antimicrob. Agts. Chemother.* **1993**, *37*, 1313-1317.

[218] Morisaki, N.; Iwasaki, S.; Yazawa, K.; Mikami, Y.; Maeda, A. *J. Antibiot.* **1993**, *46*, 1605-1610.

[219] Dabbs, E. R.; Yazawa, K.; Tanaka, Y.; Mikami, Y.; Miyaji, M.; Andersen, S. J.; Morisaki, N.; Iwasaki, S.; Shida, O.; Takagi, H.; Kadowaki, K. *J. Antibiot.* **1995**, *48*, 815-819.

[220] Arora, S. K. *J. Med. Chem.* **1985**, *28*, 1099-1102.

[221] Arora, S. K.; Arjunan, P. *J. Antibiot.* **1992**, *45*, 428-431.

[222] Arora, S. K.; Kook, A. M. *J. Org. Chem.* **1987**, *52*, 1530-1535.

[223] Talpaert-Borlé, M.; Campagnari, F.; Discenza, G. *J. Antibiot.* **1975**, *28*, 580-589.

[224] Arisawa, M.; Funayama, S; Pezzuto, J. M.; Kinghorn, A. D.; Cordell, G. A.; Farnsworth, N. R. *J. Nat. Prod.* **1984**, *47*, 393-393.

[225] Masuda, K.; Funayama, S.; Komiyama, K.; Umezawa, I. *J. Nat. Prod.* **1987**, *50*, 958-960.

[226] Carter, W. A.; Brockman, W. W.; Borden, E. C. *Nature* **1971**, *232*, 212-214.

[227] Borden, E. C.; Brockman, W. W.; Carter, W. A. *Nature* **1971**, *232*, 214-216.

[228] Okabe, T.; Yuan, B. D.; Isono, F.; Sato, I.; Fukazawa, H.; Nishimura, T.; Tanaka, N. *J. Antibiot.* **1985**, *38*, 230-235.

[229] Okabe, T.; Suzuki, K.; Suzuki, H.; Inouye, Y.; Nakamura, S.; Tanaka, N. *J. Antibiot.* **1986**, *39*, 316-317.

[230] Sasaki, K.; Yasuda, H.; Onodera, K. *J. Antibiot.* **1979**, *32*, 849-851.

[231] Sasaki, K.; Inoue, Y. *Japan. Kokai* **1988**, 88-34866, July 12, **1998**.

[232] Yamaki, H.; Suzuki, H.; Choi, E. C.; Tanaka, N. *J. Antibiot.* **1982**, *35*, 886-892.

[233] Yamaki, H. *Studia Biophysica* **1984**, *104*, 313-316.

[234] Yamaki, H.; Nakajima, M.; Seimiya, H.; Saya, H.; Sugita, M.; Tsuruo, T. *J. Antibiot.* **1995**, *48*, 1021-1026.

[235] Shibata, K.; Satsumabayashi, S.; Sano, H.; Komiyama, K.; Nakagawa, A.; Ōmura, S. *J. Antibiot.* **1986**, *39*, 415-423.

[236] Komiyama, K.; Hirokawa, Y.; Yamaguchi, H.; Funayama, S.; Masuda, K.; Anraku, Y.; Umezawa, I.; Ōmura, S. *J. Antibiot.* **1987**, *40*, 1768-1772.

[237] Onodera, K.; Hiragun, A.; Sato, M.; Mitsui, H.; Sasaki, K. *J. Antibiot.* **1979**, *32*, 545-547.

[238] Ito, S.; Sasaki, K.; Yamamoto, N.; Onodera, K.; Nomoto, K. *Annual Meeting of the Aguricultural Chemical Society of Japan, Book of Abstracts* **1987**, p. 66.

[239] Ito, S.; Gilljams, G.; Wahren, B.; Wigzell, H.; Yamamoto, N.; Sasaki, K.; Onodera, K. *J. Antibiot.* **1990**, *43*, 1045-1046.

Atta-ur-Rahman (Ed.) *Studies in Natural Products Chemistry, Vol. 23*

(1→3,1→6)-β-D-GLUCANS OF YEASTS AND FUNGI AND THEIR BIOLOGICAL ACTIVITY

G. KOGAN

Institute of Chemistry, Slovak Academy of Sciences, 842 38 Bratislava, Slovakia
Phone: +421-7-5941 0274, Fax: +421-7-5941 0222
E-mail: kogan@savba.sk

ABSTRACT: Glucans, or polymers of D-glucose linked by (1→3)-β and (1→6)-β glycosidic linkages are the common polysaccharides of the fungal cell wall. They are usually located in the inner part of the wall and play the role of skeletal polysaccharide contributing to the shape and rigidity of the cell wall. The β-glucan molecules are interlinked by the hydrogen bonds and sometimes occur in a complex with other polysaccharides, such as chitin.

β-Glucans isolated from the various yeast and fungal species may have different molecular weight or other structural parameters such as degree of branching or the length of the side-chains which may affect their solubility in water. Many fungal β-glucans possess remarkable ability to enhance the immune system, *i.e.* act as immunomodulators. Numerous studies have demonstrated the activity of β-glucans as biological response modifiers that are able to exert beneficial effect in host by demonstrating antitumor, antibacterial, antiviral and antiparasitic activities. Some studies have attempted to correlate biological activity of individual β-glucans to their spatial structure or helical conformation. However neither such correlation nor the precise mechanism of the activation of the immune system by β-glucans have been unambiguously proven.

The present article reviews the existing knowledge on the immunological activity of the fungal β-glucans and provides some recent results on the radioprotective and antimutagenic activity of the β-glucan isolated from the baker's yeast.

INTRODUCTION

Glucans that contain the main chain of D-glucose units linked by (1→3)-β-glycosidic linkages and the side-chains constituted of single D-glucose units attached to the main chain by (1→6)-β-linkages are the important components of the fungal cell wall. As the skeletal components, (1→3,1→6)-β-glucans are located within the inner part of the wall and play the role of the structural carcass of the fungal wall that defines its shape and rigidity. They are often associated with other wall polymers, particularly polysaccharides, such as chitin. Some (1→3,1→6)-β-glucans can serve also as storage carbohydrates of some fungi, while those (1→3,1→6)-β-glucans that constitute the surface components or extracellular mucilages of certain fungi are important due to their involvement in the morphogenetic changes, as well as in the symbiotic or pathogenic interactions with higher plants and animals. This chapter will deal with the structural features of the fungal $(1\rightarrow3,1\rightarrow6)$-$\beta$-glucans (further called simply β-**glucans**) as well as with their remarkable immunological properties and with possible applications of their derivatives in pharmacy and medicine. For those seeking for an exhaustive review on all aspects of β-glucans, author can recommend an excellent book by Bruce A. Stone and Adrienne E. Clarke "Chemistry and Biology of $(1\rightarrow3)$-β-glucans", La Trobe University Press, Australia, 1992. A brief review on the structure-functional activity relationships of the $(1\rightarrow3)$-β-glucans was recently published: J.A. Bohn and J.N. BeMiller "$(1\rightarrow3)$-β-D-glucans as biological response modifiers: a review of structure-functional activity relationships" Carbohydrate Polymers, 28 (1995) 3-14. Some older reviews dealing partially with the structure and biological activities of fungal β-glucans are referred to in our review under reference numbers [28], [29] and [130].

OCCURRENCE OF β-GLUCANS IN YEASTS AND FUNGI

In fungi, polysaccharides usually account up to 75 % of the cell walls dry weight [1]. According to their function and localization in the cell wall, the polysaccharides may be subdivided in the two groups: the skeletal polysaccharides that are localized in the innermost layer of the cell wall, the polysaccharides of the amorphous wall matrix that overlies and surrounds the skeletal carcass, and the outermost layer of the mucilage. The mucilage polysaccharides are readily soluble in water, the amorphous matrix polysaccharides are generally soluble in dilute alkali, while the skeletal polysaccharides are more or less fibrillar or crystalline and are insoluble in alkali. The matrix polysaccharides are usually complexed with proteins and play a role of cementing substance that fills the space in the network made by the skeletal polysaccharides. It is quite difficult to separate the amorphous component from the fibrillar one and there is a possibility of the covalent link between them.

Chemically, the skeleton is generally composed of cellulose, chitin, and $(1\rightarrow3,1\rightarrow6)$-$\beta$-glucan, while the matrix contains various heteropolysaccharides and protein-polysaccharides [2]. The extracellular mucilage polysaccharides contain a large variety of neutral and acidic homo- and heteropolysaccharides, including $(1\rightarrow3,1\rightarrow6)$-$\beta$-glucans with the structure similar to that of the fibrillar β-glucan, but with the reduced molecular weight [3-6]. The crystalline character of skeletal β-glucan and its insolubility might be explained by its association with chitin through covalent bonds [7-9]. Recently some data were obtained showing that in yeast cell wall $(1\rightarrow3,1\rightarrow6)$-$\beta$-glucan is attached also to the glycosylphosphatidylinositol (GPI) anchor at the C-terminus of the cell wall mannoproteins [10-12], thus providing evidence that all matrix and skeletal polysaccharides are interconnected that indicates probable pathway of their biosynthesis during the cell wall assembly [13].

Table 1. Cell Wall Composition and Taxonomy of Fungi (Modified with permission from [6])

Group	Cell wall category		Taxon	Representative genera
	Alkali-insoluble	Alkali-soluble		
I	cellulose	glycogen?	MYXOMYCOTA (Acrasiomycetes)	*Polysphondylium, Dictyostelium*
II	cellulose, (1→3,1→6)-β- glucans	(1→3,1→6)-β-glucan	EUMYCOTA Mastigomycotina (Oomycetes)	*Phytophthora, Pythium, Saprolegnia*
III	cellulose, chitin	not determined	Mastigomycotina (Hyphochytridiomycetes)	*Rhyzidiomyces*
IV	chitin, chitosan, polyglucuronic acid	glucuronomanno-protein, polyphosphate	Zygomycotina (Zygomycetes)	*Mucor, Phycomyces, Zygorhynchus*
V	chitin, probably (1→3,1→6)-β-glucan	probably (1→3,1→6)-β-glucan	Mastigomycotina (Chytridiomycetes)	*Allomyces, Blastocladiella*
	chitin, (1→3,1→6)-β-glucans	(1→3,1→4)-α-glucan (1→3)-α-glucan, (galacto)-β-mannanprotein, (1→3,1→6)-β-glucan	Ascomycotina (Plectomycetes, Pyrenomycetes, Loculoascomycetes, Discomycetes)	*Neurospora, Histoplasma*
	chitin, (1→3,1→6)-β-glucans	(1→3)-α-glucan, xylomannanprotein, (1→3,1→6)-β-glucan	Basidiomycotina (Hymenomycetes, Teliomycetes)	*Schizophyllum, Fomes, Polyporus, Uromyces, Ustilago*
	chitin, (1→3,1→6)-β-glucans	(galacto)-α- mannanprotein, (1→3)-α-glucan	Deuteromycotina (Coelomycetes, Hyphomycetes)	*Aspergillus, Penicillium*
VI	(1→3,1→6)-β- glucans (chitin as minor component)	(galacto)-α- mannanprotein, α-glucans, (1→3,1→6)-β-glucan	Ascomycotina (Hemiascomycetes)	*Saccharomyces, Candida*
VII	chitin, β-mannan	α-glucans	Basidiomycotina (Basidiomycetous yeasts)	*Sporobolomyces, Rhodotorula*
VIII	polygalactosamine-galactan		Zygomycotina	*Amoebidium*

The relevance of the polysaccharide composition of the fungal cell wall for the taxonomic assignment of the fungi was initially stressed by Bartnicki-Garcia [14,15]. Stone and Clarke presented a table (Table I) demonstrating a correlation between the composition of the fungal cell walls and their taxonomy basing on the data of Bartnicki-Garcia [14,15] and Wessels and Sietsma [16].

ISOLATION OF β-GLUCANS FROM THE FUNGAL CELL WALLS

Since β-glucans occur in diverse locations in the fungal cells, the methods of their isolation greatly depend on their type and occurrence. If the polysaccharide is a part of extramural mucilage or capsule, or is extracellularly produced in the liquid culture during the fungal growth, it can be rather easily isolated by washing the cells and subsequent precipitation of the desired β-glucan from the water solution by conventional sedimentation techniques using organic solvents, *e.g.* ethanol, or ammonium sulfate. The subsequent purification procedures usually involve gel filtration or reprecipitation. Isolation of the water-insoluble but alkali-soluble β-glucans of the amorphous cellular matrix involves extraction with diluted alkali usually at elevated temperature and subsequent separation of other solubilized polysaccharides and other cellular components by the appropriate methods (*e.g.* α-mannans are separated by precipitation with Fehling's reagent [17]. However it should be kept in mind that exposing β-glucans to alkali for prolonged time at elevated temperatures may lead to the base-catalyzed degradation, so-called peeling-off reaction. Such degradation can be avoided by conversion of the reducing glucose residue to glucitol with sodium borohydride [18].

The most difficult problem represents isolation and purification of the alkali-insoluble β-glucans especially due to their association with other polysaccharides, such as chitin, α-glucans or mannans. The methods used usually involve disruption of the cell wall by physical methods or autolysis [19], separation of the alkali-soluble material (soluble glucans, mannans and proteins) by prolonged heating or boiling in dilute NaOH [20-22], subsequent digestion of the insoluble material with hot dilute HCl [21,23] or acetic acid [20]. Recently, Müller *et al.* [24] investigated the applicability of various protic acids for the extraction of water-insoluble β-glucan from *Saccharomyces cerevisiae*. They have found that the physicochemical parameters of the microparticulate glucans extracted with the use of acetic, formic or phosphoric acid were better than those of the glucan prepared with hydrochloric acid extraction. Besides, three examined acids were also less corrosive than HCl. Chitin can be separated from β-glucan by the action of chitinase [8] or by de-N-acetylation and subsequent degradation with nitrous acid [25], while concomitant α-glucans can be removed using α-amylase, glucoamylase or other appropriate enzymes [23]. For separation of the insoluble (1→3)-α-glucan, a method employing extraction with 4-methylmorpholine N-oxide and dimethyl sulfoxide (DMSO) was suggested [26]. The usual final yield of the insoluble microparticulate β-glucan isolated from the baker's yeast *Saccharomyces cerevisiae* does not exceed 2 % of the initial dry weight of the yeast mass [27].

CHEMICAL STRUCTURE OF THE FUNGAL β-GLUCANS

The extensive research that has been performed on the isolation and structural characterization of the fungal β-glucans starting from the late 1960s is accounted for remarkable biological properties (mostly antitumor) that are exerted by these polysaccharides. Hot water extracts from the edible mushrooms have been used for a long time as traditional medicine for cancer therapy as well as antiinfective and antiinflammatory drugs in Japan and China. Later research established that such biological activity is due to the presence of the polysaccharides belonging to the class of β-glucans having common structure of (1→3)-linked backbone with the single glucosyl units attached to the backbone through (1→6)-glycosidic linkage [28,29]. Structural investigations of the β-glucans has been usually performed using a combination of methylation analysis and nuclear magnetic resonance (NMR) spectroscopy. Although some early data indicated presence of low amount of (1→4)-β-linkages in the molecule of the β-glucan from *Lentinus edodes* [30], it might be attributed to the incorrect assignment of the ^{13}C NMR signals. Later results proved that lentinan has a common comb-like structure containing (1→3)-linked backbone with single glucosyl (1→6)-linked side-chains [31]. ^{13}C NMR spectra of fungal (1→3,1→6)-β-glucans because of their poor solubility in water are usually recorded in DMSO solution and contain broad signals and many overlapping resonances due to the backbone carbon atoms as well as minor signals due to the side-chain carbon atoms [32]. Fig. (1) shows ^{13}C NMR spectra of two important antitumor glucans from *Schizophyllum commune* (A) and *Lentinus edodes* (B).

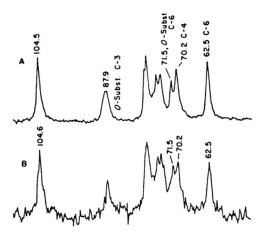

Fig. (1). ^{13}C NMR spectra of β-glucans from *Schizophyllum commune* (A) and *Lentinus edodes* (B). (Solvent DMSO-d$_6$; ambient temperature; chemical shifts expressed relative to external tetramethylsilane. Reproduced from [32] with permission of Academic Press Inc.)

The frequency of the occurrence of the (1→6)-linked glucosyl unit side-chains on the (1→3)-linked backbone (ratio side-chain units/backbone units - degree of branching - DB) can be determined by means of methylation analysis and varies greatly for different fungal β-glucans. Some β-glucans apparently may contain longer (1→6)-linked side-chains with more than one glucosyl units and some alternative structures containing both (1→3) and (1→6) linkages in the backbone have been suggested. Thus, Gomaa *et al.* proposed a structure for the major extracellular polysaccharide of *Glomerella cingulata* that involves side-chains composed of 1-4 (1→3)-linked glucosyl units attached through (1→6)-β-glycosidic linkage to the (1→3)-linked backbone [33]. Major difficulties with the determination of the chemical structures were encountered with the investigation of the alkali-insoluble glucans that are often associated with other polysaccharide components and their chemical analysis is further complicated due to their insolubility.

On the basis of methylation analysis, periodate oxidation and acetolysis Nakajima *et al.* suggested the possible structures for the alkali-insoluble glucan isolated from the cell walls of *Piricularia oryzae* [34]. The suggested structures contained (1→3)- and (1→6)-linkages in the backbone in the ratio ca. 7:1 and glucosyl or gentiobiosyl side-chains linked to the backbone through (1→6)-linkages on approximately every tenth (1→3)-linked unit of the backbone. Major effort of many researchers was associated with the determination of the structure of β-glucan of the cell wall of the baker's yeast *Saccharomyces cerevisiae*. The complexity of that task is due to the fact that contrary to the cell walls of Basidiomycetes, cell wall of *Saccharomyces cerevisiae* contains not one, but several β-glucans differing in molecular weight, solubility and relative ratio of (1→3) and (1→6) linkages. The structural studies of the *Saccharomyces cerevisiae* glucan started in the early 1950s, and the early work of Bell and Northcote [35] based solely on the results of methylation analysis suggested that it had a highly branched structure composed of (1→3)-linked glucooligosaccharides interconnected by ca. 11 % (1→2)-β-linkages. Later results assisted by the data on partial acid hydrolysis, according to Peat *et al.* however showed that alkali-insoluble component of baker's yeast glucan was linear and contained both (1→3) and (1→6) linkages [36]. The presence of ca. 10-20 % of (1→6) linkages was corroborated by the results of tosylation followed by iodination of the primary hydroxyl groups [37]. Using the newly developed procedure of Smith degradation for the structural elucidation of polysaccharides [38], Misaki *et al.* attempted to prove the location of (1→6) linkages and suggested the following structure of yeast glucan [39] (Fig.(2)).

components, one predominantly (1→3)-linked with minor portion of (1→6)-glycosidic bonds and the other one containing high proportion of (1→6)-linkages and smaller number of (1→3)-bonds. The major alkali-

$$....\rightarrow {}^6[\underset{\underset{G^1\rightarrow {}^3[G]_7}{\uparrow}}{G}]^1{}_x\rightarrow {}^6G^1\rightarrow {}^6G^1\rightarrow {}^6G^1\rightarrow {}^6[\underset{\underset{G^1\rightarrow {}^3[G]_7}{\uparrow}}{G}]^1{}_y\rightarrow ....$$

Fig. (2). Structure proposed by Misaki *et al.* G denotes β-glucopyranosyl unit, and x+y = 40-50. Modified from [39] with permission from Elsevier Science.

Manners *et al.* suggested that baker's yeast glucan contained two insoluble component that constituted ca. 85 % of whole glucan had high molecular weight of minimum 240,000 and contained 85 % (1→3)-bonds and 3 % interchain (1→6)-linkages [40]. The minor component that could be separated by a selective enzymolysis or acidic extraction contained 65 % (1→6)-linkages together with smaller amount of interresidual (5 %) and interchain (14 %) (1→3)-bonds. This glucan was referred to as (1→6)-β-glucan [41]. Regarding the structure of the major component, the authors emphasized that their results of methylation analysis could not discriminate between the suggested laminated structure with (1→6) interchain linkages and a comb-like structure. The authors also mentioned that different glucan preparations may greatly vary in structure depending on the microbial strain used and the conditions of cultivation, as well as on the procedure of glucan isolation. The structure suggested by Manners *et al.* for the alkali-insoluble glucan is schematically depicted in Fig. (3).

$$G^1\rightarrow [{}^3G^1]_a \rightarrow {}^3G^1$$
$$\downarrow$$
$$G^1\rightarrow [{}^3G^1]_b\rightarrow {}^6G^1\rightarrow [{}^3G^1]_c\rightarrow {}^3G^1$$
$$\downarrow$$
$$......... \rightarrow {}^3G^1\rightarrow {}^3G^1\rightarrow {}^3G^1 \rightarrow {}^3G^1 \rightarrow ....$$

Fig. (3). Structure of yeast glucan proposed by Manners *et al.* [40]. **a+b+c** comprise about 60 glucosyl units, although the exact values of **a,b** and **c** are unknown. Modified with permission from the Biochemical Society and Portland Press.

Such complex structure of yeast glucan, however, makes it highly improbable to adopt ordered helical structure as it is known to form in physiological conditions similarly to many other (1→3)-β-D-glucans of fungal and bacterial origin [31,42]. Kogan *et al.* [22] investigated the chemical structure of two yeast glucans and their ability to adopt helical conformation in solutions. According to their results, β-glucan isolated from *Saccharomyces cerevisiae* had a structure similar to that of other fungal glucans with (1→3)-β-linked backbone and (1→6)-β-linked side-chains that contained in average two glucosyl units. ^{13}C-NMR spectrum of yeast glucan (Fig.(4)) was very alike the spectra of other fungal glucans with classical comb-like structure (Fig. (1)).

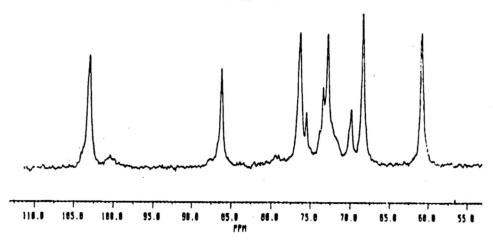

Fig. (4). ^{13}C-NMR spectrum of *Saccharomyces cerevisiae* glucan. (Solvent DMSO-d$_6$, temperature 28 °C, chemical shifts expressed relative to tetramethylsilane). Reproduced from [22].

Table 2 summarizes some of the investigated fungal β-glucans, together with their degree of branching and molecular weight values.

In contrast with rather extensively investigated glucans of the basidiomycetous and other macroscopic fungi, there is only limited information on the structure of the yeast glucans of the strains other than *Saccharomyces cerevisiae*. Gopal *et al.* isolated β-glucans from the regenerating spheroplasts of important human pathogen *Candida albicans* [56] as well as from its yeast, hyphal and germ-tube forming cells [57]. The results indicated that cell walls of *Candida albicans* contained at least two glucans. The acid-soluble glucan contained predominantly (67.0-76.6 %) (1→6)-linkages with smaller amount (11.7-14.3 %) of (1→3)-linkages and was highly branched (DB 5.3 - 8.4 %), while the alkali-insoluble skeletal glucan from intact yeast and hyphal cells contained 31.8 % and 29.6 % (1→3)- linkages and 53.2 % and 47.1 % (1→6)-linkages, respectively. Almost threefold excess of the determined amount of branching points over the non-reducing groups in the hyphal and germ-tube glucan preparations may imply that the (1→3)-linked chains might be interbound through (1→6)-linked fragments [57]. Kogan *et al.* investigated chemical structure and conformational behaviour of alkali-insoluble glucan isolated from the *Candida albicans* serotype A cell walls [22] and described that the glucan preparation that contained 72.4 % (1→3)- and only 7.7 (1→6)-linked glucosyl units, formed gel upon heating in neutral solutions and associated with Congo Red dye, indicating higher-order

Table 2.   Some Fungal β-Glucans and Their Structural Characteristics

Glucan	Source	DB	Molecular weight (Da)	References
Lentinan	*Lentinus edodes*	0.25-0.35	1,000,000	[43]
Schizophyllan	*Schizophyllum commune*	0.33	1,640,000	[44]
Grifolan	*Grifola frondosa*	0.35	1,200,000	[42]
Yeast glucan	*Saccharomyces cerevisiae*	0.03-0.2	240,000	[22,40]
VVG	*Volvariella volvacea*	0.2	450,000	[45]
Scleroglucan	*Sclerotium glucanicum*	0.33	1,400,000	[46]
SSG	*Sclerotinia sclerotiorum*	0.4-0.5	389,000	[47,48]
Glomerellan	*Glomerella cingulata*	0.25	681,000	[33]
AM-ASN	*Amanita muscaria*	0.3	260,000	[50]
Pleuran	*Pleurotus ostreatus*	0.25	not determined	[51]
Pichilan	*Pichia fermentans*	0.1	200,000	[52]
glucan	*Phytophthora parasitica*	0.33	9,000-200,000	[53]
G-A	*Ganoderma japonicum*	0.033	82,000	[54]
CI-6P	*Cordyceps cicadea*	0.04	21,000	[55]

structure formation. Katohda *et al.* studied alkali-soluble and alkali-insoluble glucan preparations from the yeast- and rod-like forms of the dimorphous yeast *Candida krusei* [58]. The analysis of the insoluble glucan showed that it contained very sparsely branched (1→3)-linked backbone (DB 0.03) with single glucosyl side-chain units linked through (1→6)-linkages. This glucan of both yeast- and rod-like forms was resistant to the action of β-1,6-glucanase due to the lack of consecutively (1→6)-linked portions. In contrast, alkali-soluble glucan preparations contained about 14 and 18 % (1→6)-linked units in yeast-like and rod-like forms, respectively, and it was possible to degrade certain amount of (1→6)-linked portions using β-1,6-glucanase [58].

Spatial supramolecular structure of β-glucans

In general, polysaccharides can exist in two conformational forms - random coils and ordered structures. In solutions, polysaccharides are expected to adopt random coil conformation since many flexible glycosidic linkages between the units in the polysaccharide chain allow rotation to occur around the glycosidic bonds involving little change in internal energy of the molecule. The number of the shapes the polymer can adopt depends on the internal freedom of the particular molecular fragments and is higher for the (1→6)-linked segments than for other types of glycosidic linkages since only (1→6)-linkage involves three torsional angles. In solutions, constant fluctuation between different possible conformations of the polysaccharide chain leads to the minimization of the total energy. However, if specific intramolecular hydrogen bonds exist in the molecule whose cooperative action can compete with the chain flexibility, they can provide some favoured ordered conformation with low internal energy. In concentrated solutions and gels additional intermolecular hydrogen bonds can contribute to energy minimization by creating supramolecular aggregates involving several polysaccharide molecules that adopt particular conformation in order to enable intermolecular association. However, even in gels these interactions are seldom strong enough to fix polysaccharide chain in one particular conformation, while in the solid state such fixation is enabled by greater molecular density and much stronger cooperative forces between different polysaccharide chains. However, even in solid state polysaccharides do not produce regular crystals that are required for precise determination of the conformation and three-dimensional structure by X-ray crystallography. For that reason, polysaccharides usually have to be obtained as highly oriented fibres from which fibre X-ray patterns can be obtained [59]. Thus, fibres of the linear (1→3)-β-D-glucans curdlan (from bacteria *Alcaligenes faecalis*) and paramylon (from algae *Euglena* spp.) were prepared and their triple-helical structure was established [59]. Using the computer-assisted methods for the calculation of the energy maps for the different polysaccharide conformations it has been shown, that for (1→3)- β-D-glucans the most favourable conformation is an open single helix with 6 residues per one turn of the helix [60] (Fig. (5).

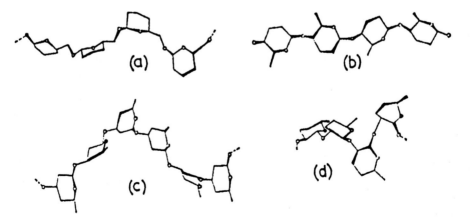

Fig. (5). Regular conformations of the β-D-glucans [60]. **a** (1→6)-β-glucan; **b** (1→4)-β-glucan; **c** (1→3)-β-glucan; **d** (1→2)-β-glucan. Reproduced with permission of the Royal Society of Chemistry.

When polysaccharides associate in solutions, gels or solid state, single helical chains may further interact through interchain hydrogen bonding and thus create multiple-helical aggregates. Calculations of the energy maps and predictions of the optimum conformation have been performed for $(1{\rightarrow}3)$-β-glucans as single helices, parallel and anti-parallel stranded double helices, as well as for right- and left-handed triple helices (60-63). Minimum energy calculations led to the conclusion that $(1{\rightarrow}3)$-β-glucans are highly probable to form triple helices [63].

Branched fungal $(1{\rightarrow}3)$-β-glucans generally having side-chains composed of the single $(1{\rightarrow}6)$- β-linked glucosyl units can also form triple-helical aggregates similarly as the linear $(1{\rightarrow}3)$-β-D-glucans which implies that short side-chains do not prevent adopting ordered conformation. Occurrence of schizophyllan, an extracellular glucan produced by *Schizophyllum commune*, in triple-helical conformation in neutral aqueous solutions was proved by means of sedimentation equilibrium, light scattering, and viscosity measurements [44]. Measurements of the weight-average and *z*-average molecular weight values of schizophyllan dissolved in water and in DMSO showed that the values determined for water were about three times as large as those for dimethyl sulfoxide. This observation and the investigation of schizophyllan behaviour in mixed water-DMSO solutions indicated that schizophyllan dissolved in water not as a single molecule, but as a triple helix, which underwent an abrupt "melting" when the content of DMSO added to aqueous solution exceeded about 85 %. Once broken with DMSO, the triple helix was not recoverable [44].

Using size-exclusion chromatography, Šoltés *et al.* studied the apparent differences in the molecular weight values of the native (M_{wN}) and DMSO-denatured (M_{wD}) samples of schizophyllan, lentinan and pleuran and discovered that in every case M_{wN} was significantly larger than M_{wD}, while the magnitude of this difference was indicative of the relative content of triple-helical associates for each glucan [64]. Presence of the triple-helical aggregates was demonstrated using different physico-chemical methods. Besides observing the apparent threefold difference in the molecular weight value of β-glucans in water and in DMSO, or upon the addition of chaotropic agents, such as urea and DMSO [65], changes in other parameters can be observed that indicate formation or dissolution of the triple-helical formations. Maeda *et al.* observed that addition of urea or DMSO decreased specific rotation $[\alpha]_D$ at 589 nm of lentinan in a concentration-dependent manner and that the removal of these denaturants resulted in recovery of $[\alpha]_D$ values. Measurements of optical rotatory dispersion in the spectral region between 600 and 200 nm showed the change in the higher ordered structure even more clearly. Lentinan was also denaturated by addition of NaOH in concentration-dependent manner. Denaturation of the triple-helical structure was associated with the loss of biological activity of lentinan [66].

Another prove of formation of the triple-helical structure is associated with the ability of β-glucans to form complexes with some dyes (Aniline blue, Congo red, Calcofluor). Although the precise mechanism of the dye-glucan interaction is not yet established, the evidence exists that an ordered form of β-glucans, possibly a triple helix, is responsible for complexing. Formation of the complex is evidenced by a shift in absorption maximum of the dye to the longer wave-lengths, a so-called metachromasy. Ogawa *et al.* [67] described complex formation of curdlan, a bacterial non- branched (1→3)-β-glucan, with Congo red that resulted in the shift of absorption maximum of the dye from 489 nm to 520 nm. As was previously established, the conformational transition of the curdlan structure from triple helix to random coil occurred at the NaOH concentration in the range 0.19-0.24 N [68].

Observation of the dependence of the absorption maximum of Congo red in the presence of curdlan on the alkali concentration showed that at low NaOH concentrations significant shift to the longer wave-lengths occurred. The concentration interval in which the conformational transition took place was also the region where the abrupt decrease of the Congo red absorption maximum occurred [66].

The study also showed that only the high molecular weight gel-forming curdlan (MW 80,000) was able to form triple helix and cause metachromasy, while water-soluble curdlan (MW 3,000) did not have such effect. This method of detection of the triple-helical structure was applied to the (1→6)-branched fungal β-glucans and the dependence of triple helix formation on the molecular weight of the glucan and alkali concentration was observed for grifolan [42,68]. β-Glucan PVG from ascomycotinous fungus *Peziza vesiculosa* formed a complex with Congo red in neutral and dilute alkali solution, but this complex dissociated at more than 0.2 N NaOH [49]. Glucan AM-ASN from the fruiting bodies of *Amanita muscaria* exhibited metachromasy with Congo red at NaOH concentrations below 0.25 N. Its molecular weight estimated by gel filtration in 0.1 N sodium chloride was 260,000, while in 0.5 N sodium hydroxide it was only 95,000 - almost three time less, which indicated triple helix formation [50]. Native β-glucan isolated from *Sclerotinia sclerotiorum* IFO 9395 (SSG) formed a complex with Congo red and induced metachromasy similarly as curdlan. When SSG was dissolved in 0.25 N NaOH, the solution contained only random-coil shaped glucan molecules and no complexing with Congo red was observed. However upon neutralization of the solution with hydrochloric acid, the helical structure of SSG was regenerated, and metachromasy was observed again [69]. In their study of the structure of the glucan isolated from the culture of the fungus *Glomerella cingulata*, Gomaa *et al.* [33] discovered that viscous gel-like structures the polysaccharide formed in neutral solutions were irreversibly destroyed by heating over 150 ºC, addition of > 0.05 N

Table 3.   Absorption Maximum of Congo red in Alkaline Solutions in the Presence and Absence of Yeast Glucans [22]

Sample	λ_{max} (nm)		
	0.1 N NaOH	0.2 N NaOH	0.3 N NaOH
Congo red only	484	486	484
Congo red + C. albicans glucan	490	500	494
Congo red + S. cerevisiae glucan	490	490	484

NaOH or over 80 % DMSO. The drop of viscosity was also accompanied by cessation of the ability of glucan solution to cause metachromasy of Congo red. These observations were in agreement with earlier findings that excessive heating [70,71], addition of DMSO [44,72] or treatment with alkali [67] destroy hydrogen bonding that is responsible for creation of the helical structures. Kogan et al. investigated the ability of two yeast glucans to produce complexes with Congo red [22]. Table 3 shows the data on the absorption maximum of Congo red obtained in the presence of the glucans at different alkali concentrations.

The results showed that both glucans induced metachromasy and in both cases the effect had its maximum at 0.2 N concentration of sodium hydroxide. In the case of S. cerevisiae glucan, the complex with Congo red decomposed at 0.3 N concentration of alkali, while C. albicans glucan was still able to induce metachromasy in these conditions. It was therefore concluded that S. cerevisiae glucan was less able to adopt helical conformation than C. albicans glucan. This conclusion was in agreement with the results of methylation analysis that revealed that the former glucan had longer side-chains that apparently prevented from adopting stable helical conformation.This was also corroborated by the fact that contrary to C. albicans glucan, the glucan from S. cerevisiae was unable to produce gel at neutral pH at the elevated temperature, since it is known that formation of ordered structures is the first step in the process of creating the spatial networks such as gels [73]. The schematic representation of the triple helix is shown in Fig. (6) as it was suggested for schizophyllan by Itou et al. [74].

Application of nuclear magnetic resonance spectroscopy, especially ^{13}C NMR methodology, became of a great advantage for detection and characterization of the ordered helical structures of β-glucans. In one of the earliest papers, Saitô et al. [75] described that in the ^{13}C NMR spectra of the gel-forming branched glucan A_3 isolated from Pleurotus ostreatus taken in DMSO-d_6 or in alkaline solution at pD 13.7 mainly signals of the $(1\rightarrow3)$-β-linked backbone were seen, while the signals assigned to the $(1\rightarrow4)$-α-linked side-chain glucose residues were buried in the more

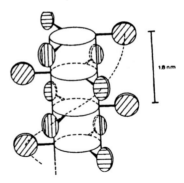

Fig. (6). Schematic representation of the schizophyllan triple helix. (Hatcheted disks represent the side-chain glucosyl residues, the central cylinder represents the helix core. The disks form a helical chain indicated by the dashed line.). Reproduced from [74] with permission from Elsevier Science.

intense signals of the backbone carbon atoms. The spectrum of the glucan recorded in the gel state contained poorly resolved broad signals chemical shifts of which corresponded solely to the carbon atoms of the $(1\rightarrow4)$-α-linked side-chains. Authors concluded that the complete disappearance of the backbone signals in the gel-state implied that the backbone adopted an ordered conformation which immobilized the segmental motion, while the side-chains remained in disordered conformation. Authors also calculated the values of spin-lattice relaxation times, nuclear Overhauser effects, and line-widths for the observable signals in the gel-state and in DMSO-d_6 solution. Taking into account the obtained data, the authors suggested that a multiple-stranded helix proposed by a computer modelling [76] as well as by the X-ray study of the solid-state conformation of xylan [77], was the most likely conformation to account for such immobility of the backbone. The similar almost complete suppression of the peak intensities was also reported to occur in the double-helical polynucleotides [78] and ι-carrageenan [79], and in triple-helical collagen [80].

^{13}C NMR investigation of the high molecular weight glucan lentinan and its acid-degraded fragments showed that the water soluble fraction with low molecular weight (3,640) gave the well-resolved spectrum from which the assignment of the backbone carbon signals including the C-6 branch points has been made. On the contrary, the spectra of the gel-form of intact lentinan (MW 1,000,000) and its degraded fragment with MW 16,200 contained signals of significantly suppressed amplitude [31]. The signals observed in the spectrum of the gel-state were unequivocally assigned to the $(1\rightarrow6)$-linked side-chains that were in disordered conformation. The only signal that belonged to the backbone and was still recognizable in the spectrum of the gel-state was the one of the unsubstituted C-6 because of the relatively higher mobility of the CH$_2$OH group of the $(1\rightarrow3)$-β-linked backbone units [81,82]. The suppressed peaks of the C-1 - C-5 atoms were partially recovered upon the addition of 0.03 N sodium hydroxide. From the comparison of the shift of the absorption maximum of Congo red complexed with lentinan with the data of the titration study of ^{13}C NMR spectrum, it was concluded that the ordered conformation of glucan was retained up to a concentration of sodium hydroxide of 0.20 N. Previously it was demonstrated that 20-30 % of the total gel-state signals (and 60 % of C-6 signal) were clearly observed in the resilient gel of a linear bacterial (1→3)-β-glucan curdlan [82]. These signals were ascribed to the single-helical portion of the glucan chain. Observation of the positions of the chemical shifts of C-1 and C-3 of the lentinan's backbone in different concentrations of NaOH and their comparison with the respective chemical shifts for the random-coil (1→3)-β-glucan laminaran showed that single-helical portions also exist in the branched glucans such as lentinan. These single-helical chains exist in addition to the triple-helical junction zones in the gel-state, ^{13}C NMR signals of which are not observed at the low alkali concentration [31]. Estimation of the size of peak areas of C-1, C-3 and C-6 observable for the single helix at pD 12.5 with the total area seen at pD 13.7 gave values of 30-40 % and 60 %, respectively which was in good agreement with those of resilient gel of linear curdlan [82].

Comparative study of the ^{13}C NMR spectra of the two branched glucans, lentinan and schizophyllan, on the concentration of sodium hydroxide allowed to define some general features of the conformational behaviour of fungal glucans in gel-state and in transition to solution [83]. Due to the higher degree of cross-linking in the branched glucans in comparison with the linear glucan curdlan, all ^{13}C NMR signals due to the (1→3)-β-linked glucosyl residues are completely suppressed in the gel-state. The multiple-helical forms that constitute the cross-link segments are converted into the random-coil form (gel-to-sol transition) at sodium hydroxide concentration around 0.2 N, which results in full peak areas in ^{13}C NMR spectra. Also, in contrast to the linear glucan, this helix-coil transition occurs in a non-cooperative way: the peak intensity and line width gradually chang with increasing concentration of sodium hydroxide. Such behaviour is best interpreted in terms of various degrees of cross-linking. Some loose cross-links are broken already in the lower concentrations of alkali (0.09 N), while others are resistant until complete transition to the random coil occurs (0.2 N). In the low concentration of alkali the signals of the single-helix form can be distinguished from the random-coil conformation by the characteristic downfield displacement of C-1 and C-3 signals by 3.0 and 2.8 ppm, respectively [31,75]. Thus, between 0.13 and 0.19 N NaOH, the distibution of the ^{13}C-peak positions of C-1 and C-3 reflects the sodium hydroxide-induced helix-random coil transition (or gel-to-sol transition, macroscopically). This transition is complete at almost the same concentration (0.2 N) of NaOH for both the linear [82] and the branched glucans. At the intermediate stage, at 0.13 N, a complex peak profile of C-3 reveals presence of two components, the broad peak **c** and the narrow peak **d**, which are readily assigned to the helical and random-coil forms, respectively. Similarly, the C-1 signals **a** and **b** correspond to the single-helix and random-coil forms (Fig. (7)).

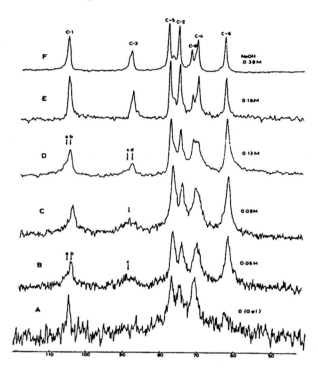

Fig. (7). ^{13}C NMR spectra of lentinan in the gel-state and its changes with the increased concentration of sodium hydroxide. Reprinted from [83] with peermission from Elsevier Science.

Interesting correlation was observed between the decreasing line-width of the C-3 signal and the change of the absorption maximum of Congo red complexed with lentinan. These two dependencies were almost identical that agrees well with the assumed role of the helical structure of glucan in the dye complexing. Authors had previously shown that the shift of absorption maximum of Congo red attached to $(1 \rightarrow 3)$-β-glucans is characteristic of the presence of the single-helix conformation [82,84]. This view was also supported by the results of Ogawa and Hatano who showed on the basis of circular dichroism measurements that single-helical part of the glucan chain produced the complex with Congo red [85]. Thus, the ^{13}C NMR measurements that revealed the relative proportions of the single-helix and random-coil conformations in different alkali concentrations satisfactorily agreed with the observation of metachromasy with Congo red.

The similar observations of the sodium hydroxide concentration dependencies of the peak positions, relative peak intensities, and line-widths of schizophyllan revealed that, like lentinan, its molecule adopted single-helical conformation at lower NaOH concentrations (<0.13 N), while the transition to the random-coil state took place at 0.19 N. Grifolan produced soft gel at neutral pH whose ^{13}C NMR spectrum was similar to those of lentinan and schizophyllan [42]. However, different preparations of grifolan possessed two kinds of solid-state structure and formed different gel types that were described as a helix type (similar to that of lentinan) and a native type that had different structure. Both structures were investigated by ^{13}C NMR spectroscopy, the accent being made on the differences in gel-to-sol transitions induced by addition of alkali, DMSO or urea [86].

Kogan et al. investigated ^{13}C NMR spectra of two yeast glucans in the DMSO solutions and in neutral aqueous gel-form [22]. Similarly to the branched glucans lentinan and schizophyllan, the ^{13}C spectrum taken in the DMSO-d_6 solution showed only the signals belonging to the (1→3)-β-linked backbone, while the less intense signals of the side-chains were buried in the backbone signals and were therefore indistinguishable. On the other hand, ^{13}C NMR spectrum of the aqueous gel produced upon heating at 55 °C contained different set of signals, that were almost identical with those observed in the spectrum of the linear (1→6)-β-D-glucan GE-3 from the fungus Gyrophora esculenta Miyoshi [31] (Table 4).

Table 4. ^{13}C NMR Signals Observed in the Spectra of the Glucan from *Candida albicans* (ppm) [22].

Sample	C-1	C-2	C-3	C-4	C-5	C-6
C. albicans glucan in DMSO	103.03	72.83	86.21	68.42	76.34	60.89
C. albicans glucan in aqueous gel-form	104.24	74.48	76.93	70.84	76.20	70.05
GE-3 [31]	104.2	74.2	76.8	70.7	76.1	70.0

The observed chemical shifts imply that while the signals of the (1→3)-β-linked backbone were completely suppressed in the gel-state due to the rigid helical conformation and the resulting restriction of the motion, (1→6)-linked side-chains preserved disordered conformation and in this manner their ^{13}C NMR signals became visible in the gel-state spectrum due to the almost complete loss of the major component - the backbone carbons signals. Since the signals of the side-chain residues were fully observed in the gel-state, there appears to be no specific role for (1→6)-linked units in gel formation. Therefore the gelation fenomenon may be discussed solely in terms of mutual association of (1→3)-linked chains.

Furthermore, the spectra of the yeast glucans taken at the different concentrations of NaOH showed interesting correlation with the measurements of Congo red absorption maximum displacement that implied that the optimum conditions for the metachromasy occured at 0.2 N NaOH, when apparently the maximum amount of single-helical conformers were present. Maximum suppression of the C-3 signal at this concentration of alkali corroborated this observation.

A new opportunity to study the conformation of fungal glucans in solid state using ^{13}C NMR spectroscopy was introduced by application of cross-polarization - magic angle spinning (CP/MAS) technique. This methodology allows to observe ^{13}C chemical shifts of the solid samples which avoids a requirement to dissolve usually badly soluble glucans. Saitô et al. have applied CP/MAS NMR spectroscopy to β-glucans, and suggested that the C-3 chemical shift was strongly influenced by the torsion angles [87,88]. Since C-3 is involved in the glycosidic linkage, its ^{13}C NMR signal is particularly sensitive to any conformantional change. Basing on the value of the C-3 chemical shift, authors distinguished four different types of solid-state conformations of β-glucans, namely, form I (laminaripentaose type), II (curdlan type), III (laminaran type), and IV (DMSO adduct). The first type was characterized by the lowermost position of the C-3 signal : 91.2 ± 0.2 ppm.It is clearly distinguishable from the signals of all other forms, since it was displaced downfield by 1.6 ppm from the second lowermost C-3 signal of curdlan and pachyman (type II - 89.6 ± 0.3 ppm). Interestingly, linear (1→3)-β-D-glucans curdlan and pachyman gave rise to almost identical spectra with the branched lentinan in spite of the differences in the primary structure. In a similar manner, linear laminaran and branched HA β-glucan from Pleurotus ostreatus (type III, C-3 signal at 85.7 ± 0.2 ppm) gave identical spectra patterns within the experimental error, that were distinct from those of the other abovementioned glucans. These data imply that presence of branching at C-6 does not affect secondary structures of the branched glucans in the solid state as compared with the linear glucans, although branching plays dominant role in gelation conditions [31]. Further, the chemical shifts were almost identical for all glucans lyophilized from DMSO solution (type IV containing a signal from CH_3 group at 39.5 ± 0.1 ppm and C-3 signal at 90.0 ± 0.3 ppm) independently on their molecular weight or branching [88]. The data obtained allowed to conclude that algal glucan paramylon from Euglena gracilis in the solid state did not consist of any specific conformation, but contained combination of all three types and additionally some unknown conformation that gave rise to C-3 signal at 82.4 ppm. Comparison of the results obtained by Ohno et al. who found two different conformations of the Grifola frondosa glucan [42,86], with the characteristical ^{13}C NMR spectra for all types of the glucans, implies that the helix form corresponds to the curdlan type, while the native form to laminaran type of solid-state conformation [70]. CP/MAS

NMR spectroscopy allowed to determine that the native SSG glucan from *Sclerotinia sclerotiorum* IFO 9395 and regenerated SSG obtained by neutralization of the alkaline solution had different conformations, since their C-3 signals appeared at 86 and 89 ppm, respectively. These values imply that while regenerated SSG possesses helical structure similar to that of curdlan, the native SSG in solid state has random-coil conformation [69].

Exhaustive examination of the gelation mechanism of different (1→3)-β-glucans by means of high-resolution CP/MAS ^{13}C NMR spectroscopy demonstrated that gelation of branched fungal glucans proceeds through association of the single helices into the triple-helical junction zones, while the gels of linear glucans such as curdlan consist almost entirely of single helices [89]. CP/MAS spectroscopy provided an excellent tool to get insight of the conformation of the fungal glucans in their native state in the fruiting bodies of the fungi or in the sclerotia. Ohno *et al.* demonstrated that in the sclerotia of *Sclerotinia sclerotiorum* IFO 9395 the nascent glucan was in random-coil conformation [70] and when the required chain length was achieved, the native and helix forms developed [69]. Majority of the sixteen basidiomycetous and ascomycetous fungi investigated by Ohno *et al.* by means of CP/MAS spectroscopy contained glucan in the native conformation (C-3 signal at 86 ppm), while the yeast cell walls of *Saccharomyces cerevisiae* produced a spectrum with C-3 signals at both 86 and 90 ppm. This observation allowed the authors to conclude that yeast glucan was present in native and helical conformations, and its primary structure was not similar to those of other glucans obtained from the culture broth, fruiting bodies or sclerotia of mushrooms [90]. This conclusion however might result from the fact that contrary to the rather homogenous β-glucans from the fungal sources, yeast glucan contains several components differing in primary structure and molecular weight.

An interesting phenomenon has been recently observed by Stokke *et al.* using electron microscopy. It was reported that denaturated samples of the branched β-glucans upon returning to the conditions favouring triple-helical association tend to form circular structures recognizable by electron microscopy. Such circular morphology detected for lentinan and schizophyllan [91,92] differs from that of other circular polysaccharides, *e.g.* cyclodextrins [93] and cyclic (1→2)-β-glucans [94] in that branched fungal glucans are much higher in molecular weight and that their cyclic form has multistranded structure that does not require covalent linkage to achieve the cyclization.

PREPARATION OF THE FUNCTIONAL DERIVATIVES OF β-GLUCANS

Since fungal β-glucans are often poorly soluble in aqueous media due to their high molecular weight as well as to interchain association, their clinical application is significantly limited. While topical administration of insoluble particulate glucan from *Saccharomyces cerevisiae* induced no toxicity, intravenous application of this insoluble form was associated with such adverse effects as granuloma formation, micro-embolization, hepatosplenomegaly, and enhanced endotoxin sensitivity [27]. For that reason, overcoming of this obstacle by means of preparation of the water-soluble derivatives of fungal glucans became the crucial task that would enable successful application of β-glucans as biological response modifiers (BRMs) in veterinary and human medicine. Miscellaneous biological activities of fungal β-glucans will be described in the next chapter. In this chapter the overview of different derivatives of β-glucans and their preparation will be presented.

In one of the first papers describing tumor-inhibitory activity of yeast glucan, DiLuzio *et al.* described the method of preparation of the soluble glucan using the modified procedure of Sasaki *et al.* [95]. The procedure included boiling of the particulate glucan with 90 % formic acid and provided water-soluble preparation that possessed antitumor and antibacterial activity comparable with that of the particulate glucan, but more appropriate for parenteral administration [21,96]. Although no characterization of the obtained soluble fraction was carried out, it can be suggested that simple acid hydrolysis took place that resulted in random degradation of the initial insoluble glucan and that the solubility was achieved due to reduced molecular weight.

Since due to (1→3)-linkages in the backbone of the fungal glucans it is resistant to the action of periodate oxidation, only (1→6)-linked side-chains can be oxidized by this procedure. Upon periodate oxidation and subsequent borohydride reduction of the oxidized product, Sone *et al.* prepared glucan-polyols from the β-glucan isolated from the culture mycelium of *Ganoderma lucidum* [97]. The authors reported that attachment of the polyol groups to the (1→3)-linked backbone significantly enhanced host-mediated antitumor effect of the glucan. The similar procedure was applied to grifolan [68], while the authors observed that introduction of the polyol groups into β-glucan structure did not affect its gel-forming ability or viscosity, while soluble polyol derivatives possessed higher antitumor activity than the original glucan. Periodate oxidation with subsequent borohydride reduction was also applied to schizophyllan and the resulting polyalcohol was investigated by ^{13}C NMR spectroscopy [98]. The obtained glucan-polyol was better soluble in water and produced more viscous solutions than schizophyllan. The solubility of the lyophilized polyalcohol increased with the prolongation of the time of oxidation. In the recent paper, Mizuno *et al.* applied the described procedure to the glucan isolated from the fruiting bodies of the mushroom *Tricholma giganteum* and observed that introduction of the increased number of hydrophilic alcohol groups and the decrease of the molecular weight of the glucan due to the degradation of the side-chains resulted in the enhanced antitumor effect [99].

Another approach to achieving better solubility and increased biological activity of fungal glucans involves introduction of the charged groups in the glucan molecules. Despite the fact that in this way chemically modified glucans lacked gel-forming and ordered conformation adopting ability, these derivatives possessed significant biological activity. Sulfated schizophyllan prepared by Kojima *et al.* by the treatment of schizophyllan solution in DMSO or pyridine with SO_3-pyridine complex caused proliferation of the murine spleen cells but lost the antitumor activity against sarcoma 180 that was characteristical for the non-derivatized schizophyllan. Interestingly, the derivative prepared in pyridine shifted absorption maximum of Congo red similarly as the underivatized schizophyllan, while the sulfoderivative prepared in DMSO did not exert metachromasy. It was therefore suggested, that the former sulfate formed a triple helix, while the latter one was in a random coil [100]. Using a sulfation procedure involving chlorosulfonic acid-pyridine, Cirelli *et al.* prepared a sulfoderivative of the glucan from the tree fungus *Cyttaria harioti* Fischer [101]. Interestingly, the sulfoderivative was devoid of antitumor activity but preserved mitogenic property similarly to the sulfated schizophyllan.

A very frequently used procedure for the preparation of the soluble derivatives of the insoluble fungal glucans is carboxymethylation. Ohno *et al.* prepared carboxymethylated derivatives of the highly branched glucan SSG from *Sclerotinia sclerotiorum* IFO 9395 with the degree of substitution (DS - number of carboxymethyl groups per anhydro glucose unit) 0.04-0.49 [102]. The derivatization reaction was carried out in isopropanol or water in the presence of sodium hydroxide and monochloroacetic acid. The DS of the resulting derivative was controlled by the amount of NaOH and monochloroacetic acid added. It was found that the DS values increased gradually with the addition of NaOH. The DS value was evaluated by a colorimetric method according to Eyler *et al.* [103] and the site of carboxymethylation was established using enzyme digestion, diethylaminoethyl-Sephadex chromatography, and ^{13}C NMR spectroscopy. The authors found that carboxymethylation took place at positions C-2, C-4 and C-6 of the glucose units. Carboxymethylated derivatives with higher DS > 0.14 were unable to form gels, and observation of the chemical shift of C-3 allowed to conclude that with the increasing DS a transition from the triple-helical through single-helical to random-coil conformation took place. Investigation of the carboxymethylated SSG (CM-SSG) in the solid state by means of CP/MAS ^{13}C NMR spectroscopy showed that CM-SSG had laminaran-type conformation. All CM-SSG derivatives possessed high antitumor activity that appeared to be not related to the ability to produce helical structures or to form gels [102]. Carboxymethylated grifolan also was not able to form gel, but possessed only weak antitumor activity that was detected only for a low DS compound (DS 0.25) at a high dose 250 µg/mouse administerd five times [68].

Using a similar procedure of carboxymethylation, Horváthová *et al.* prepared derivatized yeast glucan from *Saccharomyces cerevisiae* with DS 0.91 that was further fractionated by means of precipitation with acetone into three fractions with different molecular weight [104]. The fractions were characterized by gel permeation chromatography, light scattering, and viscometry. Native and DMSO-denaturated CM-glucan samples were investigated using high-performance gel permeation chromatography (HPGPC) [105]. The determined molecular characteristics of the native and denaturated polysaccharides were practically identical which implied that carboxymethylated samples did not associate in the multiple-stranded structures. Two high-molecular weight CM-derivatives of the glucan from *S. cerevisiae* with DS 0.56 (MW 346,000) and 0.91 (MW 366,000) were prepared by Machová *et al.* [106]. The obtained derivatives were subjected to the enzymic degradation with endo-1,3-β-glucanase from *Trichoderma reesei* as well as to ultrasonic treatment. The obtained fragments of CM-glucan were investigated by HPGPC and ^{13}C NMR spectroscopy. Ultrasonic treatment was proved to be an efficient tool for preparation of the water-soluble derivatives with reduced molecular weight for both samples, while enzymic depolymerization was not appropriate for the higher substituted CM-glucan. ^{13}C NMR spectra implied that carboxymethylation occured mainly at C-6 of the glucose units. Biological activity exerted by CM-glucan from *Saccharomyces cerevisiae* will be described in the next chapter.

Williams *et al.* described preparation of two soluble derivatives of the yeast glucan - glucan phosphate prepared by phosphorylation of the insoluble glucan with orthophosphoric acid in DMSO containing 8 M urea [27], and glucan sulfate prepared by a similar procedure using concentrated sulfuric acid [107]. Glucan phosphate contained two fractions, the major one having MW 110,000 constituted about 98 % of the polymer, while the minor high-molecular weight fraction had MW 3,570,000. It was found that in solutions glucan phosphate molecules associated into triple-helical arrangement. The prepared glucan sulfate also contained two fractions with the similar MW distribution as glucan phosphate. Contrary to glucan phosphate, glucan sulfate did not show any ordered conformation at low concentrations of NaOH. Physico-chemical parameters of both derivatives were determined using aqueous high-performance size-exclusion chromatography with on-line multi-angle laser light scattering photometry and differential viscometry. ^{13}C NMR did not provide information on the site of substitution, however it can be assumed that it occured preferentially at C-6. Both soluble derivatives possessed pronounced antibacterial, antitumor and other host-mediated activities that will be described in the next chapter.

Preparation of another biologically active derivative of yeast glucan - sulfoethyl glucan (SEG) has been described by Chorvatovičová *et al.* [108]. Particulate glucan was suspended in sodium hydroxide solution and treated with β-chloroethylsulfonate upon addition of isopropyl alcohol. The soluble derivative had molecular weight of 240,000 and revealed protective activity against the mutagenic effect of Cr^{VI} compounds in mice. Some other procedures used to obtain water-soluble biologically active derivatives of fungal glucans include oxidation with sodium hypochlorite [68,109], hydroxyethylation [68,110], formylmethylation and aminoethylation [111], as well as fragmentation of the high-molecular weight native glucans by means of ultrasonication resulting in production of the soluble lower-molecular weight polymers [112,113].

BIOLOGICAL ACTIVITY OF FUNGAL β-GLUCANS

Hot-water extracts from various mushrooms and tree fungi have been used for centuries as folk remedies in the Oriental countries and Eastern part of Russia as cancer prophylaxis or general health-stimulating preparations. It has been later discovered that the active component of the fungal extracts responsible for the medicinal action is β-glucan, which structure is very similar for many different types of fungi. Contrary to many synthetic drugs, β-glucans do not directly interact with the tumor cells, invading infective agents such as pathogenic fungi, bacteria or viruses, parasites or other causes of health impairment or disease. β-Glucans belong to the group of substances known as *biological response modifiers* and rather than attack the harmful agent they boost the defense mechanisms of the host, stimulating its intrinsic ability to resist disease-causing invaders. Such host-mediated defense is advantageous in comparison with the traditional approach of using synthetic or semi-synthetic therapeutics it that β-glucans are non-toxic to the host organism cells since their action is mediated through the organism's own tools of defense. Since protective action of β-glucans may usually involve many different pathways including T-cell stimulation, augmented macrophage participation and increased release of different cytokines, stimulation of reticulo-endothelial system (RES), activation of natural killer (NK) cells, activation of the classical and alternative complement pathways, increased antibody production, *etc*. β-glucans are also described as *non-specific immunomodulators*. Below we will provide the examples of different manifestations of the immunomodulating activity of the fungal β-glucans.

Antitumor Effect

About 30 years ago Chihara *et al.* described isolation of the polysaccharide with marked antitumor activity from the Japanese most popular edible mushroom shiitake (*Lentinus edodes*) and named it lentinan [114]. Since that time, numerous animal trials showed remarkable antitumor activity of lentinan against allogeneic, syngeneic and even autochtonous tumors [115-119]. Metastasis inhibiting action was described as well [120]. Hamuro *et al.* described the synergistic antimetastatic effect of lentinan and interleukin-2 (IL-2) with pre- and post-operative treatments of mice with metastatic methylcholanthrene (MC)-induced fibrosarcoma. 85 % reduction of the metastasis colony occurrence was observed with the combination pretreatment in comparison with lentinan alone or IL-2 alone pretreatment (7.1 and 28.4 % reduction, respectively) [121]. Lentinan proved also to be effective in combined endocrine therapy of chemically induced breast cancer in rats [122]. Table 5 demonstrates some antitumor activities of lentinan.

Table 5.   Some antitumor activities of lentinan (after Chihara *et al.* [120])

Tumors	Dose (mg/kg x days)	Tumor inhibition[1]	Complete regression[2]
Allogeneic			
Sarcoma 180	1 x 10	100	10/10
Ehrlich carcinoma	1 x 10	54.7	0/5
CCM adenocarcinoma	1 x 10	65.3	0/10
Syngeneic			
A/Ph.MC.S1 fibrosarcoma	1 x 10	100	18/18
MM-46 carcinoma	5 x 2	100	9/9
P-815 mastocytoma	5 x 4	89.0	2/8
Autochtonous			
MC-induced primary tumor	1 x 10	80	2/5
Metastases inhibition			
DBA/2.MC.CS-1 fibrosarcoma	1 x 10	94.2	
MH-134 hepatoma	1 x 10	100	
Prevention of oncogenesis			
Methylcholanthrene- induced	1 x 10	Tumor occurence[3]	
Adenovirus type 12-induced	10 x 3	83 - 33 % 79 - 40 %	

[1] In %. Tumor inhibition ratio = $(C - T)/C \times 100$ (C - tumor weight of control mice, T - tumor weight in lentinan treated mice.

[2] Number of tumor-free animals/number of treated tumor-bearing animals

[3] Tumor occurrence in the control group - tumor occurrence in lentinan-treated mice

Since the determined LD_{50} was very high (over 2,500 mg/kg intraperitoneally (i.p.) and 250-500 mg/kg intravenously (i.v.) in mice and rats [120], lentinan was found to be worthy for consideration in clinical application for cancer patients. Taguchi *et al.* [123] carried out and described phase I, II, and III clinical trials of lentinan administration to the gastrointestinal cancer patients. A statistically significant prolongation of the life-span was observed. The similar promising results were obtained with the combinational therapy of the stomach and colorectal cancer patients that included lentinan together with cytostatics mitomycin C plus 5-fluorouracyl or with tegafur [124].

Nanba and Kuroda studied the antitumor mechanisms of the orally administered *Lentinus edodes* to mice and observed significant inhibition of the tumor growth in carcinoma bearing animals. The increased superoxide anion production by macrophages, as well as twofold augmentation of the cytotoxic activities of NK-cells and killer T-cells was described [125]. Mechanisms of the immunological activities of lentinan and possible future trends of its application are extensively described by Chihara *et al.* [120,126,127].

Along with lentinan, schizophyllan, an extracellular polysaccharide from the culture filtrate of *Schizophyllum commune* represents the best investigated fungal glucan as far as antitumor and immunomodulating activities are concerned. Since the native schizophyllan has molecular weight of about 6,000,000 and its aqueous solutions are highly viscous, clinical grade schizophyllan termed SPG with molecular weight about 450,000 is produced by ultrasonic depolymerization and high-speed extrusion to the appropriate molecular weight [128]. Host-mediated antitumor action of schizophyllan against subcutaneously implanted tumors such as sarcoma-37, sarcoma-180, Ehrlich carcinoma, and Yoshida sarcoma which often resulted in complete regression of tumors was described already in 1969 [129]. SPG alone was effective against different allogeneic and syngeneic tumors in mice and rats, such as MM-46 and MH-134 carcinomas, BC-47 bladder tumor, AMC-60 fibrosarcoma, and A-755 mammary carcinoma. In association with chemotherapeutic agents, schizophyllan also inhibited L1210 leukemia, B-16 melanoma, and Meth-A fibrosarcoma [130,131]. Schizophyllan also proved to be effective on the metastases of Lewis lung cancer, in suppression of carcinogenesis by methylcholanthrene, as well as against AKR leukemia [131]. Since antitumor activity of schizophyllan disappeared in thymectomized mice, it is supposed to be T-cell-mediated [130]. Sakagami *et al.* described increased production of interleukin-2 and interferon-γ by the peripheral blood mononuclear cells. They concluded that antitumor effect of schizophyllan could be ascribed to the increased production of these cytokines [132]. Stimulative action of SPG on the production of interleukines 1, 2 and 3 by activated NK cells, spleen cells, and lymphoid cells, as well as on bone marrow cell stimulation was confirmed by Tsuchiya *et al.* [133]. SPG potentiated photodynamic therapy of squamous cell carcinoma [134] and fractionated X-ray treatment of B-16 melanoma bearing mice [135]. Both pronounced prolongation of life-span and survival rate were observed with the combined therapies involving pretreatment with SPG. Formylmethylated and aminoethylated derivatives of SPG showed increased antitumor activities, increased production of tumor regressing factor, as well as enhanced production of soluble cytotoxic factors, when compared with non-derivatized SPG [111]. Clinical use of SPG for cancer treatment has been approved in Japan since 1986. Although SPG is currently used only for the immunotherapy of stage II and III cervical cancers in combination with radiotherapy, recent clinical trials have demonstrated life-prolongation effects for lung and gastric cancer patients (combination therapy with mitomycin C and 5-fluorouracyl) [136].

Upon exhaustive pre-clinical investigation of the antitumor activity of protein-bound β-glucan from the basidiomycetous fungus *Coriolus versicolor* (PSK, Krestin) in the murine tumor models [137-141], Krestin was introduced for the adjuvant treatment of patients with lung cancer. The results showed that survival prognosis for the PSK-treated patients was significantly better than in the control group of patients that received radiotherapy alone. The results were even more significant in the group of elderly patients aged 70 years or more [142].

DiLuzio *et al.* described marked reduction in growth of the syngeneic anaplastic mammary carcinoma and melanoma B-16 in mice i.v. treated with soluble and particulate glucan isolated from baker's yeast *Saccharomyces cerevisiae*. Impressive results on the increased survival rate were obtained as well with both preparations. However, i.v. application of the particulate glucan led to granuloma formation in liver and therefore use of the soluble preparation was more favourable [21,96]. Olstad and Seljelid observed the increase in the number of T-cells, B-cells and macrophages after a single injection of particulate glucan to mice bearing syngeneic methylcholanthrene-induced sarcoma. The observed lack of the antitumor effect made authors to assume that the inappropriate dose or particulate form of glucan was applied [143]. Suspensions of particulate glucan proved to be inhibitory *in vivo* for allogeneic and syngeneic tumors: B-16 melanoma and adenocarcinoma [144], acute myelogenic leukemia [145], MBL2 leukemia [146], AKR leukemia [147], and methylcholanthrene-induced fibrosarcoma [148]. Interestingly, inhibition of B-16 melanoma could also be observed in nude athymic mice [149]. Sherwood *et al.* observed stimulation by glucan of the production of the cytolytic/cytostatic factors by splenic macrophages. The enhancement observed with glucan was higher than that obtained upon co-incubation of macrophages with bacterial endotoxin [150]. The authors concluded that enhanced secretion of the macrophage cytotoxic factor(s) could be an additional mechanism of glucan-induced antitumor activity. The results obtained in the same laboratory indicated that a combined therapy with cyclophosphamide and soluble glucan resulted in the reduced number of experimental hepatic metastases and in prolonged life-span in the group of the treated mice. A reduction of the weight of the primary tumor was observed as well [151]. Study of the tumoricidal activity of resting and glucan-activated Kupffer cells allowed to suggest that soluble glucan exerted antitumor activity by activation of Kupffer cell cytolytic activity and enhancement of cell-mediated immunity [151,152]. Soluble glucan was evaluated in combined administration with lymphokine-activated killer (LAK) cells for the therapy of experimental hepatic metastases in mice. Significant suppression of primary tumor growth as well as inhibited progress of hepatic metastases and increased survival in sarcoma-bearing mice were observed. Moreover, this therapy lacked the considerable toxicity associated with the systemic application of recombinant IL-2 that is used in the treatment of advanced neoplastic diseases [153]. Positive results obtained with the application of soluble glucan for antitumor therapy in animal models and the pre-clinical tests that showed that systemic administration of soluble glucan did not cause any adverse effects [154] indicate that soluble glucan can be considered for parenteral administration to humans.

Other observed and documented facts on antitumor activities of fungal β-glucans include, for example: glucan from the fruit body of Japanese edible mushroom Reishi, *Ganoderma lucidum,* that showed almost 100 % suppression of implanted sarcoma-180 [97,155] at i.v. application and a reduced antitumor effect at oral administration to mice [156]; grifolan from *Grifola frondosa* with potent antitumor effect against sarcoma-180 in mice [42,157-159]; PVG glucan from *Peziza vesiculosa* that also strongly inhibited growth of sarcoma-180 [49]; HA glucan from the hot-water extraction of *Pleurotus ostreatus* (Fr.) Quél. that showed marked antitumor activity at a dose of 0.1 mg/kg [160]; cold alkali-extracted glucan from *Volvariella volvacea* that exerted 97 % inhibition of sarcoma-180 at the dosage of 5 mg/kg x 10 days [45]; protein-bound glucan SN-C from *Cordyceps ophioglossoides* that exerted antitumor effect at intraperitoneal administration against sarcoma-180, Ehrlich sarcoma and syngeneic tumor X-5563 [161] as well as the glucan fraction isolated by ultrasonication from SN-C that exerted antitumor activity by oral administration against murine sarcoma-180 and syngeneic MM46 mammary carcinoma [162]; SSG, isolated from the culture filtrate of *Sclerotinia sclerotiorum* and its carboxymethylated derivative that inhibited growth of the various syngeneic tumors in mice by intraperitoneal, intravenous, and intralesional administration [102,163,164]; β-glucan As-I from the mycelial extract of *Cochliobolus miyabeanus* that inhibited growth of syngeneic tumors and enhanced their phagocytosis [165-167]; β-glucan isolated from the fruiting bodies of the Chinese mushroom Yū ěr (*Auricularia* species) that showed potent antitumor activity against the solid form of sarcoma-180 in mice and exhibited significant carbon clearance-enhancing activity [168,169]; β-glucans from the phytopathogenic fungus *Phytophthora parasitica* that caused inhibition and regression of the implanted allogeneic sarcoma-180 [53,170]. Some other sources from which antitumor-active β-glucans were isolated include such mushrooms and microscopic fungi as *Amanita muscaria* [50], *Cordyceps cicadae* [55], *Tricholma giganteum* [99], *Pythium aphanidermatum* [171], *Omphalia lapidescens* [172], *Tylopilus felleus* [173], etc.

Radioprotective Effect

It has been shown that application of baker's yeast glucan enhances survival of mice exposed to radiation less then 10 Gy. As with other immunostimulants that are radioprotective in this dose range (*e.g.* endotoxin, Bacillus Calmette Guerin, *Corynebacterium parvum, etc.*) such effect was ascribed to the enhanced hemopoietic recovery [174]. The similar hemopoietic-enhancing effect was observed among other soluble β-glucans only with schizophyllan, while scleroglucan, lentinan and krestin did not increase hemopoietic activity above radiation controls [174]. However, further experiments have shown that hemopoietic activity alone was unlikely to account for yeast glucan's ability to radioprotect animals, and the possibility that glucan may also enhance survival of irradiated animals by enhancing and prolonging the function and survival of already existing cell populations important in host defense against microbial invasion (such as granulocytes and macrophages) was considered. It was demonstrated that macrophages, but not granulocytes were responsible for prolonged radioprotection. In addition, evidence was obtained that glucan could also function as an effective free-radical scavenger [175]. Intraperitoneal administration of SSG caused a significant increase in the numbers of spleen and peripheral leukocytes, as well as of the macrophage progenitor cells. It was concluded that hemopoietic activity of SSG might make it valid as a drug for prevention of myelosuppression caused by radiation or chemotherapeutic agents [176].

Chirigos and Patchen evaluated a variety of synthetic and microbial biological response modifiers for their ability to enhance hemopoiesis and survival in irradiated mice or mice pretreated with cyclophosphamide, a known radiomimetic myelosuppressing agent [177]. Particulate glucan from *Saccharomyces cerevisiae* (Glucan-P), its lower molecular weight water-soluble derivative (Glucan-F) and lentinan demonstrated significant protection in mice against 9 Gy total-body irradiation. These BRMs modulated the macrophage and bone-marrow cell population and were capable of stimulating the production and secretion of colony-stimulating factor (CSF). Administration of BRMs induced the cytokine cascade that is crucial in maintaining hemopoiesis and authors have concluded that these BRMs can be therapeutically used as radioprotective drugs. Later results demonstrated that soluble Glucan-F used in combination with the exogenous radioprotector WR-2721 led to enhanced protection and survival of irradiated mice due to the glucan's ability to accelerate hemopoietic regeneration from the stem cells. Bone marrow and splenic

GM-CFC (granulocyte-macrophage colony-forming cells) numbers were greater in mice treated with both WR-2721 and glucan than in mice treated with either agent alone [178]. Another study in which the third radioprotective agent, selenium, was used corroborated the advantages of multiple-agent treatment [179]. It has been also demonstrated that glucan therapy was able to correct the adverse effect of the quinolone antibiotic pefloxacin that inhibited hemopoietic regeneration in critically myelosuppressed mice. Evaluation of GM-CFC recovery in mice receiving different treatments revealed that, compared with recovery in saline-treated mice, glucan stimulated GM-CFC recovery, pefloxacin suppressed GM-CFC recovery, while glucan administered in combination with pefloxacin could override pefloxacin's hemopoietic suppressive effect. Thus, the combined therapy led to the maximum protection and survival enhancement [180].

An extensive research on the radioprotective and hemopoiesis-stimulating activities of a soluble carboxymethyl derivative of baker's yeast glucan (CMG), prepared at the Institute of Chemistry, Slovak Academy of Sciences in Bratislava, was performed by the group of M. Popsíšil [181-185]. Administered prior to irradiation, CMG enhanced hemopoietic recovery in sublethally irradiated mice and increased survival of lethally irradiated ones [181]. Combined therapy by preirradiation cystamine and postirradiation CMG showed to have synergistic effect on radioprotection of lethally and sublethally irradiated mice [182]. Also combined use of diclofenac, an inhibitor of prostaglandin production, with CMG proved to be benefitiary at radioprotection due to an increased cell proliferation in the hemopoietic tissue [183]. Repeated CMG administration protected mice against fractionated irradiation and the effectivity of protection increased with the increasing number of CMG injections, however some inflammatory side effects were evoked by such procedure [184]. When the effect of single or multiple CMG applications after the sublethal 7-Gy irradiation was investigated, the best protective and enhanced hemopoietic effect was observed with multiple CMG injections, however, the most pronounced hepatic side effects were found in that group of mice as well [185]. Using CMG with two different degrees of substitution, Wagnerová et al. demonstrated its protective effect against the immunosuppressive action of cyclophosphamide, when CMG administration helped to overcome a significant decrease in peripheral blood cell count, as well as in spleen and bone marrow cellularity and accelerated the hemopoietic recovery in cyclophosphamide-treated mice [186].

Antibacterial, Antiinfective and Antiviral Effects

Activation of the complement system and RES upon administration of the fungal β-glucans leads to the increased resistance and protection against various infections of bacterial, viral, fungal, and parasitic origin. In some cases, administration of glucans prior to infection led to increased survival of the experimental animals, while post-infection application had a therapeutic effect. It should be emphasized that most of the work dealing with the antiinfective properties of fungal β-glucans were performed using the baker's yeast glucan. Song and DiLuzio presented a review of the early data obtained on the application of yeast glucan in the immunotherapy of infectious diseases [187]. They described prophylactic effect of *S. cerevisiae* glucan against systemic infections caused by extracellular bacteria *Staphylococcus aureus*, facultative intracellular parasite *Mycobacterium tuberculosis*, and opportunistic fungal pathogen *Candida albicans*, as well as against obligate intracellular parasite *Plasmodium berghei*. Therapeutic action of the glucan was described against the cutaneous and systemic infections caused by the facultative intracellular parasites *Cryptococcus neoformans* and *Sporotrichum schenkii*, as well as against the footpad infection by an obligate intracellular parasite *Mycobacterium leprae*. Mode of glucan's action involving stimulation of phagocytic function of RES by increasing the production of the macrophages and granulocytes, as well as lysosomes implication was discussed. Lahnborg *et al.* described the prophylactic effect of yeast glucan and antibiotic treatment on rats with intraabdominal sepsis. The best survival rate and the lowest mortality was observed in the group were combination of glucan with ampicillin was applied [188].

Initial studies on the antiviral effect of yeast glucan were undertaken by Williams and DiLuzio [189]. These studies revealed pronounced survival of mice lethally challenged with murine hepatitis virus (MHV) when glucan was administered prior, as well as after, viral challenge. The studies further suggest that maintenance of the phagocytic function and other macrophage secretory and metabolic activities resulted from glucan stimulation was responsible for increased survival and inhibition of hepatic necrosis. Employing either Venezuelan equine encephalomyelitis virus or Rift Valley fever virus, the i.v. administration of glucan prior to infection significantly enhanced survival of animals challenged with the viruses [190]. Significant protective effect of prophylactic administration of glucan against *Herpes simplex* Type II encephalitis in mice was demonstrated in the same laboratory as well as its effectiveness as a modifier of gram-negative infections [191]. In the latter case, glucan pretreatment resulted in 91 % survival of mice with *Escherichia coli*-induced sepsis, in comparison with 19 % survival in the control group. Glucan-pretreated rats showed increased intrapulmonary killing and phagocytosis of *Staphylococcus aureus*. Bactericidal effect against *Klebsiella pneumoniae* was even more pronounced and longlasting. However, only transient protection against pulmonary infection with group C *Streptococcus* was observed [192]. Pronounced increase of survival was observed in glucan-pretreated mice with post-operative

Candida albicans sepsis [193]. The observed effect allowed authors to suggest that BRMs such as glucan might be effectively employed in patients with increased risk of post-operative infections. Synergistic effect of glucan immunostimulation and antibiotic gentamicin therapy was evaluated in experimental *E. coli* peritonitis. In comparison with the control groups containing mice that received gentamicin alone or glucan alone administration, a combined therapy showed remarkable increase in long-term survival (56 % compared to 0 %) and significantly reduced bacteremia [194]. Investigation of the possible mechanisms mediating antiinfective activity of yeast glucan included observation of the enhanced IL-1 and IL-2 production [195] as well as stimulation of respiratory burst and superoxide anion production by macrophages [196]. It was also discovered that particulate glucan-P and soluble glucan-F activate macrophages by different pathways [197].

Later studies on the glucan protection against *E. coli* peritonitis showed that beneficial effect of glucan was mediated not only by activated macrophages, but also by neutrophilic leukocytes [198]. Glucan also provided protection of mice against protozoan parasite *Leishmania major* infection. A significant reduction of parasitic proliferation in liver and spleen of animals pretreated with glucan was demonstrated [199]. In another example of antiparasitic activity of glucan, a combination of glucan, porcine immunoglobulin and zinc exerted strong stimulatory effect on T- and B-lymphocytes, as well as on peritoneal macrophage phagocytic ability and inhibited experimental ascariosis (*Ascaris suum*) in cyclophosphamide-suppressed guinea pigs [200]. Buddle *et al.* described the protective effect of glucan against staphylococcal mammary infection in ewes. Glucan caused significant decrease in milk bacterial count and stimulated ovine mammary macrophages *in vitro* [201]. Another study demonstrated adjuvant effect of yeast glucan used together with *Trypanosoma cruzi* vaccine. Glucan's potential for application as an adjuvant for the immunoprophylactic trials has been stressed [202]. Particulate glucan as well as its soluble derivatives when applied intravenously or subcutaneously prior to intraperitoneal *Klebsiella pneumoniae* infection caused prominent increase of the mean survival time of experimentally infected mice [203]. Soluble glucan PGG protected mice against lethal peritonitis caused by *Escherichia coli* and *Staphylococcus aureus* [204] and has been evaluated in the phase I/II clinical trials in prevention of the post-operative sepsis in the high-risk patients who had undergone major abdominal or thoracic surgery [205]. PGG proved to be clinically safe, and the patients who received PGG had significantly fewer infectious complications, required less intravenous antibiotic administration and shorter intensive care unit length of stay. Soluble yeast glucan was evaluated also in the clinical trial involving prevention of the septic morbidity in trauma patients. Due to the enhanced macrophage function, benefitiary effect of glucan resulted in significant decrease of septic morbidity and total mortality rate in the group of glucan-treated patients [206]. Recent results demonstrated that many cells of the immune system [207-213] have specific receptors for $(1\rightarrow3)$-β-D-glucans and it is therefore possible that binding to these receptors is the first step in mediating the stimulating effects of glucans [214-217]. In the recent minireview [218], Williams *et al.* summarized the data obtained in the preclinical evaluation of glucan in septicaemia, clinical evaluation of glucan immunoprophylaxis in the prevention of infections in trauma patients, as well as in the clinical evaluation of PGG glucan in the prevention of infections in surgical patients. Cellular and molecular mechanisms of glucan antiinfective action is discussed and conclusions for the future directions of the glucan research regarding its antiinfective applications are being emphasized.

As has been mentioned above, apart from the yeast glucan, only few other fungal glucans have been evaluated for the antiinfective activity. The data obtained demonstrated protective activity of lentinan against bacteria *Mycobacterium tuberculosis* and *Listeria monocytogenes*, viruses Adenovirus type 12, Abelson virus, VSV-encephalitis virus, and parasites *Schistosoma mansoni* and *Schistosoma japonica* [116]. Lentinan was also shown to modify an experimentally induced septic process in rabbits [219]. Schizophyllan protected experimental animals against *E. coli*-induced peritonitis [116] and inhibited Sendai virus infection in mice when administered intraperitoneally or orally [220]. The first report of a β-glucan (SSG from *Sclerotinia sclerotiorum* IFO 9395) enhancement of the alveolar macrophage functions when administered by an oral route was published by Sakurai *et al.* [221]. Soluble scleroglucan from *Sclerotium glucanicum* increased resistance of mice to *Staphylococcus aureus*, *Candida albicans*, and mouse hepatitis virus A-59 [222].

Other miscellaneous activities of β-glucans

Already the early studies of modification of host defense mechanisms by glucans have shown that these properties are exerted also in invertebrates, where glucans proved to be protective against the fungal infections [223] as well as in plants, where they elicited protection against fungal and viral diseases [224]. In the latter paper, Griesbach and Ebel reported that fundamental elicitors of the substances in plants that possess antimicrobial properties, designated phytoalexins, are $(1\rightarrow3)$-β-D-glucans. Since all fungal cells have such glucan structures in their walls, plant probably recognizes the fungal attack by the presence of $(1\rightarrow3)$-β-D-glucans that activate production of plant $(1\rightarrow3)$-β-glucan hydrolases and phytoalexins that are involved in the plant defense and resistance. The structure and elicitor activity of the branched $(1\rightarrow3)$-β-D-glucan from *Phytophthora megasperma* var. *sojae* has been thoroughly investigated by Albersheim and coworkers [225-228]. Recently highly branched glucans isolated from the cell walls of *Phytophthora parasitica, Phytophthora megasperma* and *Fusarium oxysporum* were shown to effectively inhibit the early stages of infection of *Nicotiana* by numerous viruses of different taxonomic groups. It was found that very high antiviral activity was correlated with a high degree of branching at position 6 and with the length of the side-chains, while the molecular weight and ordered structure of the glucans were not essential [229]. A soluble derivative of yeast glucan obtained by its treatment with 33 % formic acid at 100 °C exerted 40 % inhibition of tobacco mosaic virus infection in bean [230]. Water-soluble derivatives of yeast glucan exerted protective effect against chemically-induced mutagenicity. Sulfoethyl glucan (SEG) was protective against the action of hexavalent chromium compounds in mice, when the group of mice pretreated with SEG showed lower frequency of micronuclei in bone marrow cells as well as decreased liver alkaline phosphatase activity in comparison with the control group to which only bichromate solution was administered [108]. Carboxymethyl glucan (CMG) with ultrasonically lowered molecular weight was applied both parenterally and orally to mice injected with cyclophosphamide and exerted protective effect against its mutagenic activity [113]. Reduction of serum cholesterol level was observed upon intravenous administration of yeast glucan to rats [231], as well as upon intraperitoneal application of high molecular branched glucan from the mushroom *Agrocybe cylindracea* to normal and diabetic mice [232].

STRUCTURE-FUNCTIONAL ACTIVITY RELATIONSHIPS

Already the early papers dealing with the pronounced biological activity of fungal β-glucans have also brought attention to the question, which structural feature of the glucans was responsible for the exerted properties. It has been observed that ordered conformation of lentinan was associated with its antitumor activity, since addition of urea, DMSO, or NaOH that destroys the higher structure led to disappearance of the antitumor activity, while renaturation of lentinan by removal of the chaotropic agents by dialysis caused recovery of antitumor, vascular dilatation and hemorrhage-inducing activity. The denaturation and renaturation of lentinan was associated with decrease and recovery of the specific rotation at 589 nm in a concentration-dependent manner, and the changes of $[\alpha]_D$ values and in the optical rotatory dispersion curves were correlated with the changes in the biological activity of lentinan [233]. The authors could not differentiate whether single or triple helical conformation was necessary for the expression of the biological activity.

One of the properties of β-glucans associated with their ordered structure is an ability to cause activation of coagulation factor G from *Limulus* amoebocyte lysate (LAL). The coagulation of amoebocyte lysate occurs in the presence of minor amounts of endotoxins (lipopolysaccharides, LPS) or $(1\rightarrow3)$-β-D-glucans. This property has been applied to determination of LPS in medicine and biological fluids. The *Limulus* test is one of the most sensitive methods for the determination of LPS [234]. The coagulation pathways involve several proteases and the cascades are similar to those in blood coagulation and the complement systems in mammals. While activation of the endotoxin-mediated pathway (factor C) is induced only in the presence of Gram-negative bacteria or synthetic derivatives of lipid A, activation of factor G can be caused by a variety of β-glucans that occur widely in the fruiting bodies and mycelia of Basidiomycetes and Ascomycetes, in weeds, lichens and plants. Ohno *et al.* have investigated reactivity of *Limulus* amoebocyte lysate towards different β-glucans and made conclusions on their structure-activity relationship [235]. β-Glucans that contained no $(1\rightarrow3)$-linkages induced little or no coagulation, while the $(1\rightarrow3)$-β-glucans curdlan (linear), grifolan and schizophyllan (*ca.* 33% branched), lentinan (*ca.* 40% branched), SSG (*ca.* 50% branched), and OL-2 from *Omphalia lapidescens* (*ca.* 66% branched) induced significant coagulation. The optimum concentration for coagulation was correlated with the degree of branching. The reactivity of LAL towards glucans decreased with the increase in the degree of branching. Also glucans with lower molecular weights (schizophyllan and laminarin) had lower coagulating activities. However, no direct correlation with the antitumor activity could be established, since laminarin and OL-2 that did not show any antitumor activity, also reacted with LAL. For carboxymethylated $(1\rightarrow3)$-β-glucans, the degree of substitution was important. The optimum dose for carboxymethylated grifolan (CM-GRN) with DS 0.78 was *ca.* 10 times higher than that of CM-GRN with DS 0.22, but the maximum reactivities of LAL to these derivatives were similar. These results reflected the suppressive effect of side-chains and polar substituent groups on the reactivity towards LAL. Using alkali-treated glucan samples that contained mainly single-helical form and the samples heated at 150 °C with the increased portion of triple-helical form, the authors have found that the single-chain form was more reactive towards LAL, than triple-helical one. Later, using high-resolution solid-state ^{13}C NMR spectroscopy it has been confirmed that single-helical form of β-glucan was responsible for the stimulation of factor G from LAL as well as for host-mediated antitumor activity [236]. The potency of activation of factor G was increased over 100-fold by treatment of glucan solution with NaOH which led to a complete or partial conversion from the triple to the single helix. Investigation of the ability of OL-2 glucan and its Smith-degraded derivatives to stimulate factor G and alternative complement pathway (ACP) led to the following observations: 1) the native OL-2 contained too many branches to exert stimulating activity; 2) several Smith-degraded derivatives that had lower degree of branching but preserved high molecular weight necessary for gel formation showed high stimulatory activity, and 3) low molecular weight less branched derivative that was unable to produce gel was inactive in LAL coagulation and ACP stimulation [237].

Cell wall of the oomycetous fungus *Pythium aphanidermatum* contained high molecular weight slightly branched storage (1→3,1→6)-β-D-glucan and acid-soluble highly branched (1→3)-β-D-glucan that contained 14 % of (1→6) and 8 % of (1→4) linkages. While the storage glucan showed high LAL coagulating activity and antitumor activity against Sarcoma-180 and DBA/2-MC.SC-1 fibrosarcoma in mice despite the fact that it did not adopt any ordered structure, the acid-soluble glucans showed no biological activity [171]. The authors have concluded that interruption of the (1→3)-linked backbone by appreciable proportions of (1→6) and (1→4) linkages were disadvantageous for antitumor activity. Antitumor activity of several linear and branched (1→3)-β-D-glucans was tested in relation to their conformation that was established by means of ¹³C NMR spectroscopy [238]. Linear glucans of triple-helical conformation were ineffective in tumor inhibition, while upon lyophilization from DMSO solutions they exerted antitumor activity, which was associated with a conformational change from triple-helical to single-helical form. Linear glucans, however, were not effective when assayed in DMSO solutions. In contrast, the branched glucans were active not only in either saline solutions of the triple-helical form or the lyophilized sample from DMSO, but also in DMSO solution.

Coagulation of LAL was used to assay conformation of β-D-glucans. Quantitative data on LAL coagulation with shizophyllan (SPG) and grifolan (GRN) and their derivatives allowed to determine that under neutral conditions alkaline-treated SPG was in a single-helical form, while polycarboxylated SPG had random coil conformation, and untreated SPG was in a triple-helical form. It was found that after heat treatment at 150 ºC the conformation of GRN was changed to predominantly triple helix, while upon NaOH treatment and dialysis, GRN conformation changed to single helix. About half of the single-helical conformer was gradually changed to triple-helical one upon staying over one week at 4 ºC. Some ambiguities about the fact that extensive heating of glucan could cause depolymerization and that decreased molecular weight rather than a change to the triple-helical conformation was responsible for inability to coagulate LAL, were ruled out by Aketagawa *et al.* [239]. The authors showed that molecular weight was not a dominant factor determining the biological effect, provided that it was large enough to allow formation of the single-helical conformation. Samples of β-D-glucans having same molecular weight differed significantly in their biological potency depending on whether they were in a single- or triple-helical form. Thus, the single-helical conformation of (1→3)-β-D-glucans was clearly demonstrated to be the dominant contributor to the activation of *Limulus* coagulation factor G. Using the LAL coagulation test Miura *et al.* investigated the dependence of the blood clearance of clinically used SPG (triple-helical form) and its alkali denatured product SPG-OH (single-helical conformer). It has been found that single-helical SPG-OH was removed from the blood faster than SPG following both i.p. and i.v. administration [240].

In a recent paper, Kulicke *et al.* examined several fungal glucans and their ability to stimulate production of the tumor necrosis factor-α (TNF-α) and superoxide-anion release from human blood monocytes depending on the molecular and structural characteristics [46]. SR-glucan from *Sclerotium rolfsii,* SG-glucan from *Sclerotium glucanicum,* MFN-glucan from *Monilinia fructigena* and MFL-glucan from *Monilinia fructicola* with degree of branching 0.31-0.39 and molecular weight from 280,000 to 550,000 were rendered water-soluble by ultrasonication. All investigated glucans showed pronounced stimulatory biological activity that was higher than that of the biomedical drug based on schizophyllan that is used in clinical practice in Japan. Interestingly, some glucans being in a random coil conformation gave much better results than those having partially or highly ordered structure. The increasing presence of helical conformation diminished the TNF-α release. These observations contradict the statement often cited in the literature that ordered structures are necessary for the release of the cytokines [241-243]. Moreover, lower molecular weight samples stimulated superoxide production better than those with high molecular weight. It seems therefore, that for the production of superoxide and for release of TNF-α neither helical conformation, nor the high molecular weight of β-glucan were important. Ohno *et al.* have summarized [244], that while triple-helical conformation of moderately branched glucans was required for the strong antitumor effect against solid tumors, nitric oxide synthesis in macrophages could not be induced by triple-helical conformers. On the other hand, interferon release was stimulated by both single- and triple-helical forms. Thus, apparently no general rules for the structure-activity relationship can be postulated, since for each type of biological activity probably different structural features of β-glucans are of importance.

ABBREVIATIONS

ACP	=	Alternative complement pathway
BRM	=	Biological response modifier
CM	=	Carboxymethyl
CP/MAS	=	Cross-polarization - magic angle spinning
CSF	=	Colony-stimulating factor
DB	=	Degree of branching
DS	=	Degree of substitution
DMSO	=	Dimethyl sulfoxide
GM-CFC	=	Granulocyte-macrophage colony-forming cells
GPI	=	Glycosylphosphatidylinositol
GRN	=	Grifolan

HPGPC	=	High-performance gel permeation chromatography
IL	=	Interleukin
i.p.	=	Intraperitoneal(ly)
i.v.	=	Intravenous(ly)
LAK-cells	=	Lymphokine-activated killer cells
LAL	=	Limulus amoebocyte lysate
LPS	=	Lipopolysaccharide(s)
MC	=	Methylcholanthrene
MHV	=	Murine hepatitis virus
MW	=	Molecular weight
NK-cells	=	Natural killer cells
NMR	=	Nuclear magnetic resonance
RES	=	Reticulo-endothelial system
SEG	=	Sulfoethyl glucan
SPG	=	Schizophyllan
TNF	=	Tumor necrosis factor

REFERENCES

[1]   Farkaš, V. In *Fungal Protoplasts. Applications in Biochemistry and Genetics*; Peberdy, J.F. and Ferenczy, L., Eds.; Marcel Dekker, Inc.: New York, Basel, **1985**, pp. 3-29.

[2]   Farkaš, V. *Microbiol. Rev.,* **1979**, *43,* 117.

[3]   Gander, J.E. *Annu. Rev. Microbiol.*, **1974**, *28,* 103.

[4]   Gorin, P.A.J.; Barreto-Bergter, E. In: *Polysaccharides*; Aspinall, G.O., Ed.; Academic Press, Inc.: Orlando, **1983**, Vol. *2,* pp. 365-409.

[5]   Barreto-Bergter, E. In: *Studies in Natural Products Chemistry, Structural Elucidation (part B)*; Atta-ur-Rahman, Ed.; Elsevier Science B.V.: Amsterdam, **1989**; Vol. *5,* pp. 275-340.

[6]   Stone, B.A.; Clarke, A.E. *Chemistry and Biology of (1→3)-β-Glucans*, La Trobe University Press, Australia, **1992**.

[7]   Sietsma, J.H.; Wessels, J.G.H. *J. Gen. Microbiol.*, **1979**, *114*, 99.

[8]   Hartland, R.P.; Vermeulen, C.A.; Klis, F.M.; Sietsma, J.H.; Wessels, J.G.H. *Yeast*, **1994**, *10*, 1591.

[9]   Kollar, R.; Petráková, E.; Ashwell, G.; Robbins, P.W.; Cabib, E. *J. Biol. Chem.*, **1995**, *270*, 1170.

[10]   Kapteyn, J.C.; Montijn, R.C.; Dijkgraaf, G.J.P.; Van den Ende, H.; Klis, F.M. *J. Bacteriol.*, **1995**, *177*, 3788.

[11]   Kapteyn, J.C.; Montijn, R.C.; Vink, E.; de la Cruz, J.; Llobel, A.; Douwes, J.E.; Shimoi, H.; Lipke, P.N.; Klis, F.M. *Glycobiology*, **1996**, *6*, 337.

[12]   Kollar, R.; Reinhold, B.B.; Petráková, E.; Yeh, H.J.C.; Ashwell, G.; Drgoňová, J.; Kapteyn, J.C.; Klis, F.M.; Cabib, E. *J. Biol. Chem.*, **1997**, *272*, 1772.

[13]   Klis, F.M. *Yeast*, **1994**, *10*, 851.

[14]   Bartnicki-Garcia, S. *Annu. Rev. Microbiol.*, **1968**, *22*, 87.

[15]   Bartnicki-Garcia, S. *Phytopathology*, **1969**, *59*, 1065.

[16]   Wessels J.G.H.; Sietsma, J.H. In *Plant Carbohydrates II. Extracellular Carbohydrates*; Tanner, W.; Loewus, F.A., Eds.; Springer Verlag: Berlin, **1981**; Vol. *13B*, pp. 352-394.

[17]   Jones, J.K.N.; Stoodley, R.J. *Methods Carbohydr. Chem.*, **1965**, *5*, 36.

[18]   Whistler, R.L.; BeMiller, J.N. *Adv. Carbohydr. Chem.*, **1958**, *13*, 289.

[19]   Shiota, M.; Nakajima, T.; Satoh, A.; Shida, M.; Matsuda, K. *J. Biochem.*, **1985**, *98*, 1301.

[20]   Manners, D.J.; Mason, A.J.; Patterson, J.C. *J. Gen. Microbiol.*, **1974**, *80*, 411.

[21]   DiLuzio, N.R.; Williams, D.L.; McNamee, R.B.; Edwards, B.F.; Kitahama, A. *Int. J. Cancer*, **1979**, *24*, 773.

[22]   Kogan, G.; Alföldi, J.; Masler, L. *Biopolymers*, **1988**, *27*, 1055.

[23]   Masler, L.; Babor, K.; Šandula, J.; Kogan, G. *Czechoslovak Patent* 274 030 (1992).

[24]   Müller, A.; Ensley, H.; Pretus, H.; McNamee, R.; Jones, E.; McLaughlin, E.; Chandley, W.; Browder, W.; Lowman, D.; Williams, D. *Carbohydr. Res.*, **1997**, *299*, 203.

[25]   Stagg, C.M.; Feather, M.S. *Biochim. Biophys. Acta*, **1973**, *320*, 64.

[26]   James, P.G.; Cherniak, R. *Carbohydr. Res.*, **1990**, *206*, 167.

[27]   Williams, D.L.; McNamee, R.B.; Jones, E.L.; Pretus, H.A.; Ensley, H.E.; Browder, I.W.; DiLuzio, N.R. *Carbohydr. Res.*, **1991**, *219*, 203.

[28]   Whistler, R.L.; Bushway, A.; Singh, P.R.; Nakahara, W.; Tokuzen, R. *Adv. Carbohydr. Chem. Biochem.*, **1976**, *32*, 235.

[29]   Witczak, Z.J.; Whistler, R.L. In *Indusrial Polysaccharides. The Impact of Biotechnology and Advanced Methodologies*; Stivala, S.S.; Crescenzi, V.; Dea, I.C.M. Eds.; Gordon and Breach Science Publishers: New York, **1987**; pp. 1-17.

[30]   Usui, T.; Yamaoka, N.; Matsuda, K.; Tuzimura, K.; Sugiyama, H.; Seto, S. *Agric. Biol. Chem.*, **1975**, *39*, 1071.

[31]   Saitô, H.; Ohki, T.; Takasuka, N.; Sasaki, T. *Carbohydr. Res.*, **1977**, *58*, 293.

[32]   Gorin, P.A.J. *Adv. Carbohydr. Chem. Biochem.*, **1981**, *38*, 13.

[33]   Gomaa, K.; Kraus, J.; Franz, G.; Röper, H. *Carbohydr. Res.*, **1991**, *217*, 153.

[34]   Nakajima, T.; Tamari, K.; Matsuda, K.; Tanaka, H.; Ogasawara, N. *Agric. Biol. Chem.*, **1972**, *36*, 11.

[35]   Bell, D.J.; Northcote, D.H. *J. Chem. Soc.*, **1950**, 1944.

[36]   Peat, S.; Whelan, W.J.; Edwards, T.E. *J. Chem. Soc.*, **1958**, 3862.

[37]   Peat, S.; Turvey, J.R.; Evans, J.M. *J. Chem. Soc.*, **1958**, 3868.

[38]   Goldstein, I.J.; Hay, G.W.; Lewis, B.A.; Smith, F. *Methods Carbohydr. Chem.*, **1965**, *5*, 361.

[39]   Misaki, A.; Johnson, Jr., J.; Kirkwood, S.; Scaletti, J.V.; Smith, F. *Carbohydr. Res.*, **1968**, *6*, 150.

[40]   Manners, D.J.; Masson, A.J.; Patterson, J.C. *Biochem. J.*, **1973**, *135*, 19.

[41]   Manners, D.J.; Masson, A.J.; Patterson, J.C. *Biochem. J.*, **1973**, *135*, 31.

[42]   Iino, K.; Ohno, N.; Suzuki, I.; Miyazaki, T.; Yadomae, T.; Oikawa, S.; Sato, K. *Carbohydr. Res.*, **1985**, *141*, 111.

[43]   Sasaki, T.; Takasuka, N. *Carbohydr. Res.*, **1976**, *47*, 99.

[44]   Norisuye, T.; Yanaki, T.; Fujita, H. *J. Polym. Sci., Polym. Phys. Ed.*, **1980**, *18*, 547.

[45]   Misaki, A.; Nasu, M.; Sone, Y.; Kishida, E.; Kinoshita, C. *Agric. Biol. Chem.*, **1986**, *50*, 2171.

[46]   Kulicke, W.-M.; Lettau, A.I.; Thielking, H. *Carbohydr. Res.*, **1997**, *297*, 135.

[47]   Ohno, N.; Suzuki, I.; Yadomae, T. *Chem. Pharm. Bull.*, **1986**, *34*, 1362.
[48]   Stokke, B.T.; Elgsaeter, A.; Hara, C.; Kitamura, S.; Takeo, K. *Biopolymers*, **1993**, *33*, 561.
[49]   Mimura, H.; Ohno, N.; Suzuki, I.; Yadomae, T. *Chem. Pharm. Bull.*, **1985**, *33*, 5096.
[50]   Kiho, T.; Katsuragawa, M.; Nagai, K.; Ukai, S.; Haga, M. *Carbohydr. Res.*, **1992**, *224*,237.
[51]   Karácsonyi Š. ; Kuniak, Š. *Carbohydr. Polym.*, **1994**, *24*, 107.
[52]   Desboeufs, J.; Ralambosoa, C.; Astoin, J.; Joram, C.; Salhi, S.L.; Bastide, J.-M. *Chem. Pharm. Bull.*, **1988**, *36*, 2766.
[53]   Bruneteau, M.; Fabre, I.; Perret, J.; Michel, G.; Ricci, P.; Joseleau, J.-P.; Kraus, J.; Schneider, M.; Blaschek, W.; Franz, G. *Carbohydr. Res.*, **1988**, *175*, 137.
[54]   Ukai, S.; Yokoyama, S.; Hara, C.; Kiho, T. *Carbohydr. Res.*, **1982**, *105*, 237.
[55]   Kiho, T.; Ito, M.; Yoshida, I.; Nagai, K.; Hara, C.; Ukai, S. *Chem. Pharm. Bull.*, **1989**, *37*, 2770.
[56]   Gopal, P.; Sullivan, P.A.; Shepherd, M.G. *J.Gen. Microbiol.*, **1984**, *130*, 1217.
[57]   Gopal, P.K.; Shepherd, M.G.; Sullivan, P.A. *J.Gen. Microbiol.*, **1984**, *130*, 3295.
[58]   Katohda, S.; Tsukinaga, M.; Tobinai, Y.; Sato, T. *Agric. Biol. Chem.*, **1982**, *46*, 1131.
[59]   Sarko, A.; Wu, H.C.; Chuah, C.T. *Biochem. Soc. Trans.*, **1983**, *11*, 139.
[60]   Rees D.A.; Scott, W.E. *Chem. Commun.*, **1969**, 1037.
[61]   Sathyanarayana, B.K.; Rao, V.S.R. *Biopolymers*, **1971**, *10*, 1605.
[62]   Whittington, S.G.; Glover, R.M. *Macromolecules*, **1972**, *5*, 55.
[63]   Bluhm T.L.; Sarko, A. *Carbohydr. Res.* **1977**, *54*, 125.
[64]   Šoltés, L.; Mislovičová, D.; Sébille, B. *Biomed. Chromatogr.*, **1996**, *10*, 53.
[65]   Yamada, Y.Y.; Watanabe, S.T.; Chihara, C.; Rokutanda, M. *Cancer Res.*, **1988**, *48*, 671.
[66]   Ogawa, K.; Tsurugi, J.; Watanabe, T. *Chemistry Lett.*, **1972**, 689.
[67]   Ogawa, K.; Watanabe, T.; Tsurugi, J.; Ono, S. *Carbohydr. Res.*, **1972**, *23*, 399.
[68]   Adachi, Y.; Ohno, N.; Ohsawa, M.; Sato, K.; Oikawa, S.; Yadomae, T. *Chem. Pharm. Bull.*, **1989**, *37*, 1838.
[69]   Ohno N.; Yadomae, T. *Carbohydr. Res.*, **1987**, *159*, 293.
[70]   Ohno, N.; Shinohara, H.; Yadomae, T. *Carbohydr. Res.*, **1987**, *168*, 110.
[71]   Kuge, T.; Suetsugu, N.; Nishiyama, K. *Agric. Biol. Chem.*, **1977**, *41*, 1315.
[72]   Norisuye, T. *Makromol. Chem. Suppl.*, **1985**, *14*, 105.
[73]   Rees, D.A. Adv. Carbohydr. Chem. Biochem.,**1969**, 24, 267.
[74]   Itou, T.; Teramoto, A.; Matsuo, T.; Suga, H. *Carbohydr. Res.*, **1987**, *160*, 243.
[75]   Saitô, H.; Ohki, T.; Yoshioka, Y.; Fukuoka, F. *FEBS Lett.*, **1976**, *68*, 15.
[76]   Rees, D.A.; Scott, W.E. J. *Chem. Soc. B.*, **1971**, 469.
[77]   Atkins, E.D.T.; Parker, K.D.; Preston, R.D. *Proc. Roy. Soc., B* **1969**, *173*, 209.
[78]   Smith, I.C.P.; Jennings, H.J.; Deslauriers, R. *Acc. Chem. Res.*, **1975**, *8*, 306.
[79]   Bryce, T.A.; McKinnon, A.A.; Morris, E.R.; Rees, D.A.; Thom, D. *Discuss. Faraday Soc.*, **1974**, 221.
[80]   Chien J.C.W.; Wise, E.B. *Biochemistry*, **1975**, *14*, 2785.
[81]   Colson, P.; Jennings, H.J.; Smith, I.C.P. *J. Am. Chem. Soc.*, **1974**, *96*, 8081.
[82]   Saitô, H.; Ohki, T.; Sasaki, T. *Biochemistry*, **1977**, *16*, 908.
[83]   Saitô, H.; Ohki, T.; Sasaki, T. *Carbohydr. Res.*, **1979**, *74*, 227.
[84]   Saitô, H.; Miyata, E.; Sasaki, T. *Macromolecules*, **1978**, *11*, 1244.

[85] Ogawa, K.; Hatano, M. *Carbohydr.Res.*, **1978**, *67*, 527.

[86] Ohno, N.; Adachi, Y.; Ohsawa, M.; Sato, K.; Oikawa, S.; Yadomae, T. *Chem. Pharm. Bull.*, **1987**, *35*, 2108.

[87] Saitô, H.; Tabeta, R.; Harada, T. *Chem. Lett.*, **1981**, 571.

[88] Saitô, H.; Tabeta, R.; Sasaki, T.; Yoshioka, Y. *Bull. Chem. Soc. Jpn.*, **1986**, *59*, 2093.

[89] Saitô, H.; Yoshioka, Y.; Yokoi, M.; Yamada, J. *Biopolymers*, **1990**, *29*, 1689.

[90] Ohno, N.; Adachi, Y.; Yadomae, T. *Chem. Pharm. Bull.*, **1988**, *36*, 1198.

[91] Stokke, B.T.; Elgsaeter, A.; Kitamura, S. *Int. J. Biol. Macromol.*, **1993**, *15*, 63.

[92] Stokke, B.T.; Elgsaeter, A.; Brant, D.A.; Kuge, T.; Kitamura, S. *Biopolymers*, **1993**, *33*, 193.

[93] Saenger, W. In *Inclusion Compounds*; Atwood, J.L.; Davies, J.E.D.; MacNicol, D.D., Eds. Academic Press: New York, **1984**; Vol. *II*, pp. 231-259.

[94] Rizzo, R.; Crescenzi, V.; Gasparrini, F.; Gargaro, G.; Misiti, D.; Sagre, A.L.; Zevenhuizen, L.P.T.M.; Fokkens, R.H. In *Biomedical and Biotechnological Advances in Industrial Polysaccharides*; Crescenzi, V.; Dea, I.C.M.; Paoletti, S.; Stivala, S.S.; Sutherland, I.W., Eds.; Gordon and Breach Science Publishers: New York, **1989**; pp. 485-493.

[95] Sasaki, T.; Takasuka, N.; Chihara, G.; Maeda, Y. *Gann*, **1976**, *67*, 191.

[96] DiLuzio, N.R.; Williams, D.L.; McNamee, R.B.; Malshet, V.G. In *Recent Results in Cancer Research*, Mathé, G.; Muggia, F.M., Eds.; Springer Verlag: Heidelberg, **1980**; Vol. *75*, pp. 165- 172.

[97] Sone, Y.; Okuda, R.; Wada, N.; Kishida, E.; Misaki, A. *Agric. Biol. Chem.*, **1985**, *49*, 2641.

[98] Schulz, D.; Rapp, P. *Carbohydr. Res.*, **1991**, *222*, 223.

[99] Mizuno, T.; Yeohlui, P.; Kinoshita, T.; Zhuang, C.; Ito, H.; Mayuzumi, Y. *Biosci. Biotech. Biochem.* **1996**, *60*, 30.

[100] Kojima, T.; Tabata, K.; Hirata, A.; Sugawara, I. *Agric. Biol. Chem.*, **1986**, *50*, 1635.

[101] Cirelli, A.F.; Covian, J.A.; Ohno, N.; Adachi, Y.; Yadomae, T. *Carbohydr. Res.*, **1989**, *190*, 329.

[102] Ohno, N.; Kurachi, K.; Yadomae, T. *Chem. Pharm. Bull.*, **1988**, *36*, 1016.

[103] Eyler, R.W.; Klung, E.D.; Diephuis, F. *Analyt. Chem.*, **1947**, *19* 24.

[104] Horváthová, M.; Mislovičová, D.; Šoltés, L.; Tuzar, Z.; Gemeiner, P.; Žúbor, V. *Carbohydr. Polym.*, **1991**, *15*, 79. L.;

[105] Horváthová, M.; Šoltés, L.; Lutonská, H.; Šandula, J.; Mislovičová, D. *J. Appl. Polym. Sci. Appl. Polym. Symp.*, **1991**, *48*, 33.

[106] Machová, E.; Kogan, G.; Alföldi, J.; Šoltés, L.; Šandula, J. *J. Appl. Polym. Sci.*, **1995**, *55*, 699.

[107] Williams, D.L.; Pretus, H.A.; McNamee, R.B.; Jones, E.L.; Ensley, H.E.; Browder, I.W.; DiLuzio, N.R. *Immunopharmacol.*, **1991**, *22*, 139.

[108] Chorvatovičová, D.; Kováčiková, Z.; Šandula, J.; Navarová, J. *Mutation Res.*, **1993**, *302*, 207.

[109] Miura, N.N.; Ohno, N.; Adachi, Y.; Yadomae, T. *Chem. Pharm. Bull.*, **1996**, *44*, 2137.

[110] Šandula, J.; Machová, E.; Hříbalová, V. *Int. J. Biol. Macromol.*, **1995**, *17*, 323.

[111] Usui, S.; Tomono, Y.; Sakai, M.; Kiho, T.; Ukai, S. *Biol. Pharm. Bull.*, **1995**, *18*, 1630.
[112] Stahmann, K.-P.; Monschau, N.; Sahm, H.; Koschel, A.; Gawronski, M.; Conrad, H.; Springer, T.; Kopp, F. *Carbohydr. Res.*, **1995**, *266*, 115.
[113] Chorvatovičová, D.; Machová, E.; Šandula, J. *Mutation Res.*, **1996**, *371*, 115.
[114] Chihara, G.; Maeda, Y.Y.; Hamuro, J.; Sasaki, T.; Fukuoka, F. *Nature*, **1969**, *222*, 687.
[115] Chihara, G.; Hamuro, J.; Maeda, Y.Y.; Arai, Y.; Fukuoka, F. *Cancer Res.*, **1970**, *30*, 2776.
[116] Chihara, G.; Maeda, Y.Y.; Hamuro, J. *Int. J. Tiss. Reac.*, **1982**, *IV*, 207.
[117] Suga, T.; Shiio, T.; Maeda, Y.Y.; Chihara, G. *Cancer Res.*, **1984**, *44*, 5132.
[118] Jeannin, J.F.; Lagadec, P.; Pelletier, H.; Reisser, D.; Olsson, N.O.; Chihara, G.; Martin, F. *Int. J. Immunopharmac.* **1988**, *10*, 855.
[119] Suzuki, M.; Takatsuki, F.; Maeda, Y.Y.; Hamuro, J.; Chihara, G. *Int. J. Immunopharmac.*, **1994**, *16*, 463.
[120] Chihara, G.; Hamuro, J.; Maeda, Y.Y.; Shiio, T.; Suga, T.; Takasuka, N.; Sasaki, T. *Cancer Detection and Prevention Supplement*, **1987**, *1*, 423.
[121] Hamuro, J.; Takatsuki, F.; Suga, T.; Kikuchi, T.; Suzuki, M. *Jpn. J. Cancer Res.*, **1994**, *85*, 1288.
[122] Kosaka, A.; Yamashita, A. *Int. J. Immunotherapy*, **1993**, *IX*, 111.
[123] Taguchi, T.; Furue, H.; Kimura, T.; Kondo, T.; Hattori, T.; Ogawa, N. In *Biological Response Modifiers in Human Oncology and Immunology*; Klein, K.; Spector, S.; Friedman, H.; Szentivanyi, A., Eds.; Plenum Press: New York, **1983**; pp. 181-187.
[124] Taguchi, T.; Furue, H.; Kimura, T.; Kondo, T.; Hattori, T.; Ito, I.; Ogawa, N. In *Rationale of Biological Response Modifiers in Cancer Therapy*; Tsubura, E.; Aoki, T.; Urushizaki, I., Eds.; Excerpta Medica: Amsterdam, **1985**; pp. 151-166.
[125] Nanba, H.; Kuroda, H. *Chem. Pharm. Bull.*, **1987**, *35*, 2459.
[126] Chihara, G.; Maeda, Y.Y.; Suga, T.; Hamuro, J. *Int. J. Immunotherapy*, **1989**, *V*, 145.
[127] Chihara, G. *Yakugaku Zasshi*, **1988**, *108*, 171.
[128] Tabata, T.; Ito, W.; Kojima, T.; Kawabata, S.; Misaki, A. *Carbohydr. Res.*, **1981**, *89*, 121.
[129] Komatsu, N.; Okubo, S.; Kikumoto, S.; Kimura, K.; Saito, G.; Sakai, S. *Gann*, **1969**, *60*, 137.
[130] Jacques, P. J. In *Current Concepts in Human Immunology and Cancer Immunomodulation*; Serrou, B., Ed.; Elsevier Biomedical Press B.V.: Amsterdam, **1982**; pp. 429-438.
[131] Chihara, G. In *Tissue Culture and RES*; Röhlich, P.; Bácsy, E., Eds.;, Akadémia Kiadó: Budapest, **1984**; pp. 179-191.
[132] Sakagami, Y.; Mizoguchi, Y.; Shin, T.; Seki, S.; Kobayashi, K.; Morisawa, S.; Yamamoto, S. *Biochem. Biophys. Res. Commun.*, **1988**, *155*, 650.
[133] Tsuchiya, Y.; Igarashi, M.; Inoue, M.; Kumagai, K. *J. Pharmacobio-Dyn.*, **1989**, *12*, 616.
[134] Krosl, G.; Korbelik, M. *Cancer Lett.*, **1994**, *84*, 43.
[135] Takai, Y.; Goodman, G.B.; Chaplin, D.J.; Grulkey, W.; Lam, G.K.Y. *Int. J. Oncology*, **1994**, *4*, 385.
[136] Misaki, A.; Kakuta, M. In *Fungal Cells in Biodefense Mechanism;* Suzuki, S.; Suzuki, M., Eds.; Saikon Publishing Co., Ltd.: Tokyo, **1997**; pp. 279- 300.

[137] Abe, S.; Yamazaki, M.; Mizuno, D. *Jpn. J. Cancer Res.*, **1978**, *69*, 223.
[138] Tsuru, S.; Nomoto, K. *J. Clin. Lab. Immunol.*, **1983**, *4*, 215.
[139] Tsukagoshi, S.; Hashimoto, Y.; Fujii, G.; Kobayashi, H.; Nomoto, K.; Orita, K. *Cancer Treat. Rev.*, **1984**, *11*, 131.
[140] Matsunaga, K.; Morita, I.; Iijima, H.; Endoh, H.; Oguchi, Y.; Yoshimura, M.; Fujii, T. Yoshikumi, C.; Nomoto, K. *J. Clin. Lab. Immunol.*, **1990**, *31*, 127.
[141] Ishii, K.; Kikuchi, Y.; Hirata, J.; Tode, T.; Kita, T.; Nagata, I.; Tsuru, S.; Rokutanda, M. *Cancer J.*, **1993**, *6*, 87.
[142] Hayakawa, K.; Mitsuhashi, N.; Saito, Y.; Takahashi, M.; Katano, S.; Shiojima, K.; Furuta, M.; Niibe, H. *Anticancer Res.*, **1993**, *13*, 1815.
[143] Olstad R.; Seljelid, R. *Acta Path. Microbiol. Scand. Sect. C*, **1980**, *88*, 97.
[144] DiLuzio, N.R.; Hoffmann, E.O.; Cook, J.A.; Browder, W.; Mansell, P.W.A. In *Control of Neoplasia by Modulation of the Immune System*; Chirigos, M.A., Ed.; Raven Press: New York, **1977**; pp. 475-499.
[145] DiLuzio, N.R. In *Kupffer Cells and Other Sinusoidal Cells*; Wisse, E.; Knook, D.L., Eds.; North Holland: Amsterdam, **1977**; pp. 397-406.
[146] Chirigos, M.A.; Jacques, P.J. In *Advances in Immunopharmacology*; Hadden, J.; Chedid, L.; Mullen, P.; Spreafico, F., Eds.; Pergamon Press: Oxford, **1981**; pp. 485-490.
[147] DiLuzio, N.R.; Cook, J.A.; Cohen, C.; Rodrigue, J.; Jones, E. In *The Macrophage and Cancer*; James, K.; McBride, B.; Stuart, A., Eds.; Edinburgh University Press: Edinburgh **1977**; pp. 188-201.
[148] Suit, H.D.; Elman, A.; Sedlacek, R.; Silobrcic, V. In *Immune Modulation and Control of Neoplasia by Adjuvant Therapy*; Chirigos, M.A., Ed.; Raven Press: New York, **1978**; pp. 235-241.
[149] Cook, J.A.; Taylor, D.; Cohen, C.; Rodrigue, J.; Malshet, V.; DiLuzio, N.R. In *Immune Modulation and Control of Neoplasia by Adjuvant Therapy*; Chirigos, M.A., Ed.; Raven Press: New York, **1978**; pp. 183-194.
[150] Sherwood, E.R.; Williams, D.L.; DiLuzio, N.R. *J. Biol. Response Modif.*, **1986**, *5*, 504.
[151] Williams, D.L.; Sherwood, E.R.; McNamee, R.B.; Jones, E.L.; Browder, I.W.; DiLuzio, N.R. *Hepatology*, **1987**, *7*, 1296.
[152] Sherwood, E.R.; Williams, D.L.; McNamee, R.B.; Jones, E.L.; Browder, I.W.; DiLuzio, N.R. *J. Leukocyte Biol.*, **1987**, *42*, 69.
[153] Sherwood, E.R.; Williams, D.L.; McNamee, R.B.; Jones, E.L.; Browder, I.W.; DiLuzio, N.R. *J. Biol. Response. Modif.*, **1988**, *7*, 185.
[154] Williams, D.L.; Sherwood, E.R.; Browder, I.W.; McNamee, R.B.; Jones, E.L.; DiLuzio, N.R. *Int. J. Immunopharmac.*, **1988**, *10*, 405.
[155] Mizuno, T.; Kato, N.; Totsuka, A.; Takenaka, K.; Shinkai, K.; Shimizu, M. *Nippon Nogeikagaku Kaishi*, **1984**, *58*, 871.
[156] Mizuno, T.; Suzuki, E.; Maki, K.; Tamaki, H. *Nippon Nogeikagaku Kaishi*, **1985**, *59*, 1143.
[157] Ohno, N.; Suzuki, I.; Sato, K.; Oikawa, S.; Miyazaki, T.; Yadomae, T. *Chem. Pharm. Bull.*, **1985**, *33*, 4522.
[158] Ohno, N.; Adachi, Y.; Suzuki, I.; Sato, K.; Ohsawa, M.; Yadomae, T. *J. Pharmacobio-Dyn.*, **1986**, *9*, 861.
[159] Mizuno, T.; Ohsawa, K.; Hagiwara, N.; Kuboyama, R. *Agric. Biol. Chem.*, **1986**, *50*, 1679.
[160] Yoshioka, Y.; Tabeta, R.; Saitô, H.; Uehara, N.; Fukuoka, F. *Carbohydr. Res.*, **1985**, *140*, 93.

[161] Ohmori, T.; Tamura, K.; Tsuru, S.; Nomoto, K. *Jpn. J. Cancer Res.*, **1986** *77*, 1256.

[162] Ohmori, T.; Tamura, K.; Wakaiki, A.; Kawanishi, G.; Tsuru, S.; Yadomae, T.; Nomoto, K. *Chem. Pharm. Bull.*, **1988**, *36*, 4512.

[163] Suzuki, I.; Hashimoto, K.; Yadomae, T. *J. Pharmacobio-Dyn.*, **1988**, *11*, 527.

[164] Suda, M.; Ohno, N.; Adachi, Y.; Yadomae, T. *Biol. Pharm. Bull.*, **1994**, *17*, 131.

[165] Nanba, H.; Kuroda, H. *Chem. Pharm. Bull.*, **1987**, *35*, 1285.

[166] Nanba, H.; Kuroda, H. *Chem. Pharm. Bull.*, **1987**, *35*, 1289.

[167] Nanba, H.; Kuroda, H. *Chem. Pharm. Bull.*, **1987**, *35*, 1523.

[168] Kiho, T.; Ito, M.; Nagai, K.; Hara, C.; Ukai, S. *Chem. Pharm. Bull.*, **1987**, *35*, 4286.

[169] Kiho, T.; Sakushima, M.; Wang, S.; Nagai, K.; Ukai, S. *Chem. Pharm. Bull.*, **1991**, *39*, 798.

[170] Franz, G.; Hensel, A.; Kraus, J. In *Biomedical and Biotechnological Advances in Industrial Polysaccharides;* Crescenzi, V.; Dea, I.C.M.; Paoletti, S.; Stivala, S.S.; Sutherland, I.W., Eds.; Gordon and Breach Science Publishers: New York, **1989**; pp. 241-249.

[171] Blaschek, W.; Käsbauer, J.; Kraus, J.; Franz, G. *Carbohydr. Res.*, **1992**, *231*, 293.

[172] Saito, K.; Nishijima, M.; Ohno, N.; Yadomae, T.; Miyazaki, T. *Chem. Pharm. Bull.*, **1992**, *40*, 261.

[173] Defaye, J.; Kohlmünzer, S.; Sodawiczny, K.; Wong, E. *Carbohydr. Res.*, **1988**, *173*, 316.

[174] Patchen, M.L.; DiLuzio, N.R.; Jacques, P.; MacVittie, T.J. *J. Biol. Response Mod.*, **1984**, *3*, 627.

[175] Patchen, M.L.; D'Alesandro, M.M.; Brook, I.; Blakely, W.F.; MacVittie, T.J. *J. Leukocyte Biol.*, **1987**, *42*, 95.

[176] Hashimoto, K.; Suzuki, I.; Ohsawa, M.; Oikawa, S,; Yadomae, T. *J. Pharmacobio-Dyn.*, **1990**, *13*, 512.

[177] Chirigos, M.A.; Patchen, M.L. *Pharmac. Ther.*, **1988**, *39*, 243.

[178] Patchen, M.L.; MacVittie, T.J.; Jackson, W.E. *Radiation Res.*, **1989**, *117*, 59.

[179] Patchen, M.L.; MacVittie, T.J.; Weiss, J.F. *Int. J. Radiation Oncology Biol. Phys.*, **1990**, *18*, 1069.

[180] Patchen, M.L.; Brook, I.; Elliott, T.B.; Jackson, W.E. *Antimicrob. Agents Chemother.*, **1993**, *37*, 1882.

[181] Pospíšil, M.; Šandula, J.; Pipalová, I.; Hofer, M.; Viklická, Š. *Physiol. Res.*, **1991**, *40*, 377.

[182] Pospíšil, M.; Netíková, J.; Pipalová, I.;. Jar'y, J. *Folia Biologica (Prague)*, **1991**, *37*, 117.

[183] Pospíšil, M.; Hofer, M.; Pipalová, I.; Viklická, Netíková, J.; Šandula, J. *Exp. Hematol.*, **1992**, *20*, 891.

[184] Hofer, M.; Pospíšil, M.; Pipalová, I.; Holá, J.; Šandula, J. *Folia Biologica (Prague)*, **1995**, *41*, 249.

[185] Hofer, M.; Pospíšil, M.; Viklická, Š.; Pipalová, I.; Holá, J.; Netíková, J.; Šandula, J. *Int. J. Immunopharmac.*, **1995**, *17*, 167.

[186] Wagnerová, J.; Líšková A.; Navarová, J.; Krištofová A.; Trnovec, T.; Ferenčík, M. *Immunopharmacol. Immunotoxicol.*, **1993**, *15*, 227.

[187] Song, M.; DiLuzio, N.R. In *Lysosomes in Biology and Pathology;* Dingle, J.T.; Jacques, P.J.; Shaw, I.B., Eds.; North Holland Press: Amsterdam, **1979**; Vol. *6*, pp. 533-547.

[188] Lahnborg, G.; Hedström, K.-G.; Nord, C.-E. *RES: J. Reticuloendothel. Soc.*, **1982**, 32, 347.

[189] Williams, D.L.; DiLuzio, N.R. *Science*, **1980**, *208*, 67.

[190] Reynolds, J.A.; Castello, M.D.; Harrington, D.G.; Crabbs, C.L.; Peters, C.J.; Jemski, J.V.; Scott, G.H.; DiLuzio, N.R. *Infect. Immun.*, **1980**, *30*, 51.

[191] DiLuzio, N.R. *Trends Pharmacol. Sci.*, **1983**, *4*, 344.

[192] Kimura, A.; Sherwood, R.L.; Goldstein, E. *RES: J. Reticuloendothel. Soc.*, **1983**, *34*, 1.

[193] Browder, I.W.; Williams, D.L.; Kitahama, A.; DiLuzio, N.R. *Int. J. Immunopharmac.*, **1984**, *6*, 19.

[194] Browder, W.; Williams, D.; Sherwood, E.; McNamee, Jones, R. E.; DiLuzio, N.R. *Surgery*, **1987**, *102*, 206.

[195] Sherwood, E. R.; Williams, D.L.; McNamee, R. B.; Jones, E. L.; Browder, I.W.; DiLuzio, N.R. *Int. J. Immunopharmac.*, **1987**, *9*, 261.

[196] Ferenčík, M.; Kotulová, D.; Masler, L.; Bergendi, Ľ.; Šandula, J.; Štefanovič, J. *Meth. Find. Exptl. Clin. Pharmacol.*, **1986**, *8*, 163.

[197] Gallin, E.K.; Green, S.W.; Patchen, M.L. *Int. J. Immunopharmac.*, **1992**, *14*, 173.

[198] Williams, D.L.; Sherwood, E. R.; Browder, I.W.; McNamee, R. B.; Jones, E. L.; Rakinic, J.; DiLuzio, N.R. *J. Surg. Res.*, **1988**, *44*, 54.

[199] Al Tuwajiri, A.S.; Mahmoud, A.A.; Al Mofleh, I.A.; Al Khuwaitir, S.A. *J. Med. Microbiol.*, **1987**, *23*, 363.

[200] Šolt'ys, J.; Benková, M.; Borošková, Z. *Vet. Immunol. Immunopathol.*, **1994**, *42*, 379.

[201] Buddle, B.M.; Pulford, H.D.; Ralston, M. *Vet. Microbiol.*, **1988**, *16*, 67.

[202] Williams, D.L.; Yaeger, R.G.; Pretus, H.A.; Browder, I.W.; McNamee, R. B.; Jones, E. L. *Int. J. Immunopharmac.*, **1989**, *11*, 403.

[203] Kogan, G.; Masler, L.; Šandula, J.; Navarová, J.; Trnovec, T. In *Biomedical and Biotechnological Advances in Industrial Polysaccharides*; Crescenzi, V.; Dea, I.C.M.; Paoletti, S.; Stivala, S.S.; Sutherland, I.W., Eds.; Gordon and Breach Science Publishers: New York, **1989**, pp. 485-493.

[204] Onderdonk, A.B.; Cisneros, R.B.; Hinkson, P.; Ostroff, G. *Infect. Immun.*, **1992**, *60*, 1642.

[205] Babineau, T.J.; Marcello, P.; Swails, W.; Kenler, A.; Bistrian, B.; Forse, R.A. *Ann. Surg.*, **1994** *220*, 601.

[206 Browder, W.; Williams, D.; Pretus, H.; Olivero, G.; Enrichens, F.; Mao, P.; Franchello, A. *Ann. Surg.*, **1990**, *211*, 605.

[207] Ross, G.D.; Cain, J.A.; Lachmann, P.J. *J. Immunol.*, **1985**, *134*, 3307.

[208] Konopski, Z.; Smedsrod, B.; Seljelid, R.; Eskeland, T. *Biochim. Biophys. Acta*, **1994**, *1221*, 61.

[209] Czop, J.K.; Austen, K.F. *J. Immunol.*, **1985**, *134*, 2588.

[210 Williams, J.D.; Topley, N.; Alobaidi, H.M.; Harber, M.J. *Immunology*, **1986**, *58*, 117.

[211] Goldman, R. *Exp. Cell. Res.*, **1988**, *174*, 481.

[212] Duan, X.; Ackerly, M.; Vivier, E.; Anderson, P. *Cell. Immunol.*, **1994**, *157*, 393.

[213] Müller, A.; Rice, P.J.; Ensley, H.E.; Coogan, P.S.; Kalbfleisch, J.H.; Kelley, J.L.; Love, E.J.; Portera, C.A.; Ha, T.; Browder, I.W.; Williams, D.L. *J. Immunol.*, **1996**, *156*, 3418.

[214] Konopski, Z.; Rasmussen, L.-T.; Seljelid, R.; Eskeland, T. *Scand. J. Immunol.*, **1991**, *33*, 297.

[215] Hoffman, O.A.; Olson, E.J.; Limper, A.H. *Immunol. Lett.*, **1993**, *37*, 19.

[216] Poutsiaka, D.D.; Mengozzi, M.; Vannier, E.; Sinha, B.; Dinarello, C.A. *Blood*, **1993**, *82*, 3695.

[217] Williams, D.; Müller, A.; Browder, W. *J. Endotoxin Res.*, **1995**, *2*, 203.

[218] Williams, D. L.; Müller, A.; Browder, W. *Clin. Immunother.*, **1996**, *5*, 392.

[219] Tsujinaka, T.; Yokota, M.; Kambayashi, J.; Ou, M.C.; Kido, Y.; Mori, T. *Eur. Surg. Res.*, **1990**, *22*, 340.

[220] Hotta, H.; Hagiwara, K.; Tabata, K.; Ito, W.; Homma, M. *Int. J. Immunopharmac.*, **1993**, *15*, 55.

[221] Sakurai, T.; Hashimoto, K.; Suzuki, I.; Ohno, N.; Oikawa, S.; Masuda, A.; Yadomae, T. *Int. J. Immunopharmac.*, **1992**, *14*, 821.

[222] Pretus, H.A.; Ensley, H.E.; McNamee, R.B.; Jones, E.L.; Browder, I.W.; Williams, D. L. *J. Pharmacol. Exp. Ther.*, **1991**, 257, 500.

[223] Unestam, T.; Söderhäll, K. *Nature*, **1977**, *267*, 45.

[224] H. Griesbach and J. Ebel, Angew. Chem., 17 (1978) 635-647.

[225] Ayers, A.R.; Ebel, J.; Finelli, F.; Berger, N.; Albersheim, P. *Plant. Physiol.*, **1976**, *57*, 751.

[226] Ayers, A.R.; Valent, B.; Ebel, J.; Albersheim, P. *Plant. Physiol.*, **1976**, *57*, 760.

[227] Ayers, A.R.; Valent, B.; Ebel, J.; Albersheim, P. *Plant. Physiol.*, **1976**, *57*, 766.

[228] Albersheim, P.; Valent, B. *J. Cell Biol.*, **1978**, *78*, 627.

[229] Rouhier, P.; Kopp, M.; Begot, V.; Brunetau, M.; Fritig, B. *Phytochemistry*, **1995**, *39*, 57.

[230] Slováková, L̈.; Šubíková, V.; Šandula, J. In *Works of the Institute of Experimental Phytopathology and Entomology 4;* Blahutiak, A., Ed. Vesna: Bratislava, **1993**, pp. 69-77.

[231] DiLuzio, N.R. US Patent Specification No. 3,081,226, **1963**.

[232] Kiho, T.; Sobue, S.; Ukai, S. *Carbohydr. Res.*, **1994**, *251*, 81.

[233] Maeda, Y.Y.; Watanabe, S.T.; Chihara, C.; Rokutanda, M. *Cancer Res.*, **1988**, *48*, 671.

[234] Iwanaga, S.; Morita, T.; Harada, T.; Nakanura, S.; Niwa, M.; Takada, K.; Kimura, T.; Sakakibara, S. *Haemostasis*, **1978**, *7*, 183.

[235] Ohno, N.; Emori, Y.; Yadomae, T.; Saito, K.; Masuda, A.; Oikawa, S. *Carbohydr. Res.*, **1990**, *207*, 311.

[236] Saitô, H.; Yoshioka, Y.; Uehara, N.; Aketagawa, J.; Tanaka, S.; Shibata, Y. *Carbohydr. Res.*, **1991**, *217*, 181.

[237] Saito, K.; Nishijima, M.; Ohno, N.; Nagi, N.; Yadomae, T.; Miyazaki, T. *Chem. Pharm. Bull.*, **1992**, *40*, 1227.

[238] Yoshioka, Y.; Uehara, N.; Saitô, H. *Chem. Pharm. Bull.*, **1992**, *40*, 1221.

[239] Aketagawa, J.; Tanaka, S.; Tamura, H.; Shibata, Y.; Saitô, H. *J. Biochem.*, **1993**, *113*, 683.
[240] Miura, N.M.; Ohno, N.; Adachi, Y.; Aketagawa, J.; Tamura, H.; Tanaka, S.; Yadomae, T. *Biol. Pharm. Bull.*, **1995**, *18*, 185.
[241] Kishida, E.; Sone, Y.; Misaki, A. *Carbohydr. Polym.*, **1992**, *17*, 89.
[242] Ohno, N.; Miura, N. N.; Chiba, N.; Adachi, Y.; Yadomae, T. *Biol. Pharm. Bull.*, **1995**, *18*, 1242.
[243] Ohno, N.; Asada, N.; Adachi, Y.; Yadomae, T. *Biol. Pharm. Bull.*, **1995**, *18*, 126.
[244] Ohno, N.; Miura, N. N.; Adachi, Y.; Yadomae, T. In *Fungal Cells in Biodefense Mechanism*; Suzuki, S.; Suzuki, M., Eds.; Saikon Publishing Co., Ltd.: Tokyo, **1997**; pp. 273- 278.

Atta-ur-Rahman (Ed.) *Studies in Natural Products Chemistry, Vol. 23*
© 2000 Elsevier Science B.V. All rights reserved

ORGANIC CHEMISTRY AND BIOLOGICAL ACTIVITY OF METABOLITES DERIVED FROM MARINE SEA PLUMES, PSEUDOPTEROGORGIA SPECIES OF GORGONIAN OCTACORALS

E. KATZ[†] *and* M. ADAMCZESKI[*1]

*Department of Chemistry, College of Arts and Sciences,
American University, 4400 Massachusetts Avenue, NW Washington,
DC 20016-8014, USA*

ABSTRACT: This review focuses on novel classes of secondary metabolites isolated from several identified and unidentified species of the *Pseudopterogorgia* genus of marine sea plumes. We conducted a thorough survey of the pertinent natural products literature, with the exception of the patent literature and conference abstracts, relating to the structures, stereochemistry, biological activity, biogenesis, total and/or partial synthesis of metabolites isolated from this genus. Although the pharmaceutical and other biological applications and evaluations of metabolites from these marine soft corals will be included, details of the biochemical and pharmacological studies as well as the screening strategies incorporated into the discovery process will be specifically omitted. Biosynthesis and mechanism(s) of chemical defense are reported when available from the literature.

INTRODUCTION

During the past several years, results described in the literature including those of our research [1] revealed that soft-bodied aquatic invertebrates have a high propensity for yielding novel biologically active natural products [2] that can benefit society [3] since many rely on chemical defense mechanisms [4] for their survival. *Pseudopterogorgia* (phylum Cnidaria, class Anthozoa, subclass Octocorallia, order Gorgonaceae, suborder Holaxonia, Family Gorgoniidae), a genus of colonial marine invertebrates known collectively as sea plumes or sea whips, are of great interest to natural products chemists as many of their secondary metabolites have shown to have both important biological activities and fascinating structures [5]. In fact a literature survey of gorgonians revealed

[†]Currently on Sabbatical from the Department of Organic Chemistry, Israel Institute for Biological Research, P.O. Box 19 Ness Ziona, Israel and *Department of Chemistry, American University, Washington, D.C. 20016-8014,USA
*New address: San José City College, 2100 Moorpark Ave. Dept. of Science and Mathematics, San José, CA 95128-2799, USA

that the *Pseudopterogorgia* genus is equipped with one of the best developed chemical defense mechanisms among marine invertebrates [6].

Sea whips are characterized by clusters of bushy, highly branched plumes and are widely distributed, although not restricted, to the Caribbean, Bahamas, and South Florida reef environments [7]. Since several individual species share similar morphological features, taxonomic identification often requires microscopic examination and/or taxonomic markers like secondary metabolites to distinguish among species. Still, many reports in the literature fail to taxonomically identify the species of *Pseudopterogorgia*. In those cases where secondary metabolites are isolated from unidentified *Pseudopterogorgia* species, their chemistry is discussed under the specie heading whose chemical distribution pattern and phenotypic descriptions are most closely matched. Thus, this review is organized in the following order according to the classes of metabolites from three major species of *Pseudopterogorgia*, namely, *P. elizabethae*, *P. americana* and *P. acerosa*. Furthermore the chemistry of the metabolites of four other species. *P. rigida*, *P. hummelinkii*, *P. kallos* and *P. bipinnata* is also included . Within these species, a comprehensive listing of all the secondary metabolites is reported in accord with their skeletal framework. While spectral evidence for the structures is not provided, their structures, stereochemistry, biological activity, proposed biosynthesis, biogenesis, total and/or partial synthesis are reported. We also allude to possible structure-activity-relationships (SAR) and discuss chemical defense mechanisms.

In addition, as part of a continuing study of bioactive metabolites from aquatic invertebrates we recently investigated the chemistry and biological activity of the solvent extracts of an unidentified *Pseudoptergorgia* specie collected from Indonesia [8]. We describe our findings under the genus *americana*.

PSEUDOPTEROGORGIA ELISABETHAE

A comprehensive study of *Pseudopterogorgia elisabethae* from different regions of the Caribbean Sea and tropical Atlantic areas unveiled a novel class of compounds, known collectively as the pseudopterosins. Pseudopterosins are metabolites which contain diterpene glycoside moieties. Polyhydroxy steroids are another class of compounds commonly isolated from solvent extracts of this specie.

Pseudopterosins

These polar lipid secondary metabolites belong to a family of tricyclic diterpene pentosides and have been thoroughly investigated by Fenical and his research group [9]. Most pseudopterosin-type compounds and derivatives thereof, possess exceptional anti-inflammatory and analgesic properties, with the most potent secondary metabolite being **1A**. Even more important is the fact that some of these compounds and other related pseudopterosin secondary metabolites extracted from *P. elisabethae* contain different sugar moieties (compounds **2-4**) [10] and are nontoxic in acute toxicity assays in mice ($LD_{50} > 300$ mg/kg) [10]. In fact, the first marine natural product from a sea plume to be commercialized is compound, **1A**, which was isolated from *Pseudopterogorgia elisabethae* solvent extracts [11]. The key ingredient in the skin care product Resilience™, recently developed as an anti-aging skin cream by Estee Lauder™, contains a pseudopterosin-type compound **1A** [12,13]. Hence, the marketplace success of natural products contributes to the impetus to screen aquatic invertebrates for novel structures with new (bio)chemical activities.

As pseudopterosins have proven to have a wide range of medical applications and while natural reserves are limited, extensive research was carried out towards their synthesis. Harrowven and coworkers [14] established an efficient synthesis of the tricyclic carbon skeleton. Two other research groups led by Corey [15] and Broka [16] developed a total enantiospecific synthesis of pseudopterosins. Broka and coworkers,

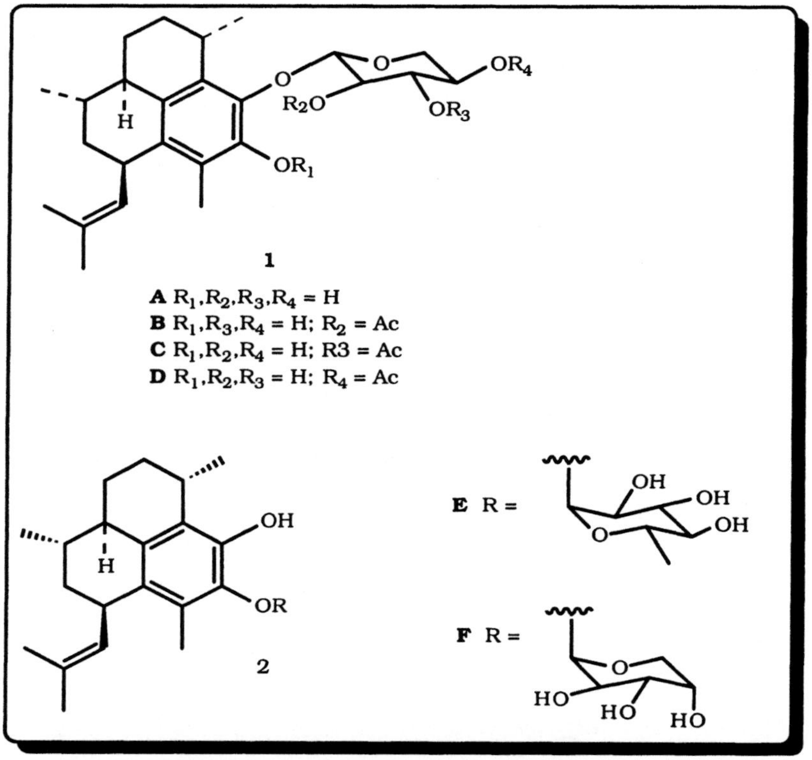

1

A $R_1, R_2, R_3, R_4 = H$
B $R_1, R_3, R_4 = H$; $R_2 = Ac$
C $R_1, R_2, R_4 = H$; $R3 = Ac$
D $R_1, R_2, R_3 = H$; $R_4 = Ac$

2

E R =

F R =

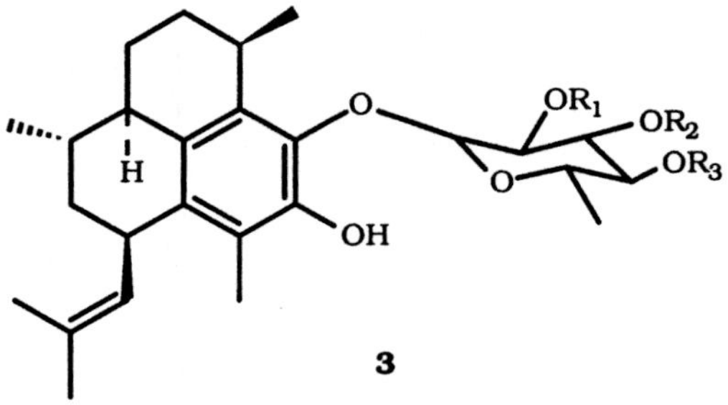

3

G $R_1=R_2=R_3=H$
H $R_1=Ac$, $R_2=R_3=H$
I $R_2=Ac$, $R_1=R_3=H$
J $R_3=Ac$, $R_1=R_2=H$

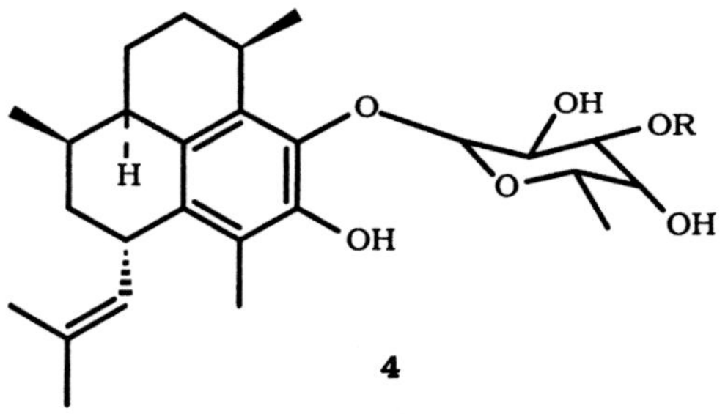

4

K $R=H$
L $R=Ac$

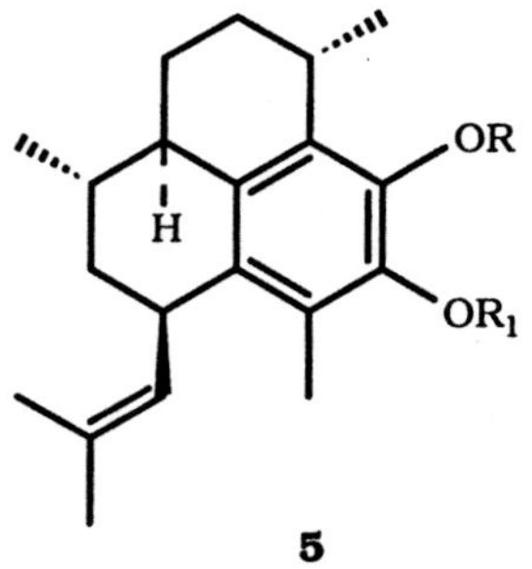

5

M $R=CH_3$, $R_1=H$
N $R=H$, $R_1=CH_3$

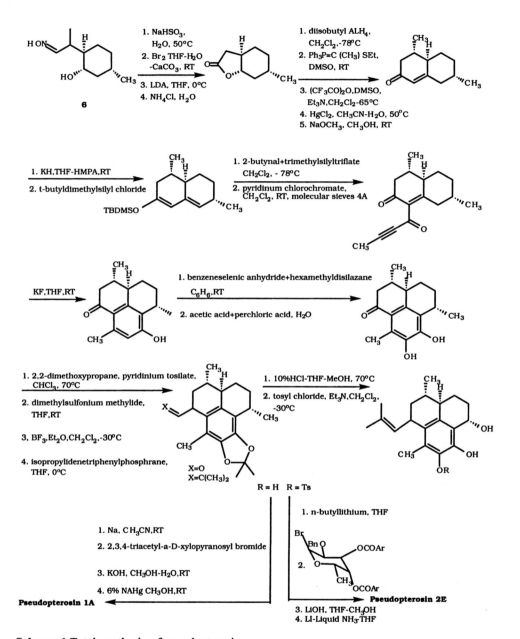

Scheme 1. Total synthesis of pseudopterosins.

synthesized pseudopterosin **1A** from (S)-(-)-limonene, in the optically active form, while Corey and Carpino [17] used (1S,2R,5S)-(+)-mentholoxime, **6,** (ie. easily prepared from (1S,2R,5S)-(+)-mentholnitrite ester) to prepare both pseudopterosins **1A** and **2E**. A detailed synthesis by Corey and Carpino [17] is presented in scheme 1.

Pseudopterosins

Pseudopterosin, **1A** exhibits potent anti-inflammatory and analgesic activity which appears to be pharmacologically different from typical cyclooxygenase inhibiting nonsteroidal anti-inflammatory drugs and from opioid analgetics [13]. Some of the prominent biological activities of pseudopterosin **1A** are featured in table 1. For comparative biological activity, indomethacin, a potent nonsteroidal anti-inflammatory drug is also included.

Table 1.   Biological Activity of Pseudopterosin 1A

compound	anti inflammatory[a] K	analgesia[b] ED_{50} (mg/kg)	toxicity (mice) LD_{50} (mg/kg)	IC_{50}^{c} (μM)
Pseudopterosin **1A**	8.93×10^{-4} M	3.12	~ 50	3.0, >80
Indomethacin	40 mM	~ 10	13	

[a]Blocking phorborol myristateacetate induced topical inflammation.

[b]Blocking stretch reflex response in mice induced by intraperitoneal injection of phenyl quinone.

[c]Inactivation of pancreatic phospholipase A_2.

Preliminary *in vitro* bioassay results demonstrate that pseudopterosin **1A** possesses additional significant biological activity [13]. For example **1A** has also been found to inhibit the synthesis of leukotrienes (human neutrophil cell studies), infering that it antagonizes lipogenase or enzymes higher in the arachidonic acid cascade. There are also indications that pseudopterosin **1A** inhibits degranulation in the same cells.

Seco-pseudopterosins

From a taxonomically distinct and still unidentified specie of the *Pseudopterogorgia* genus, a related series of pseudopterosins containing bicyclic diterpenoid glycosides, 7 seco **A-D,** were also isolated and their structures elucidated by Look and Fenical [18]. Compounds 7 seco **A-D** possess potent anti-inflammatory and analgesic activities that are similar to the pseudopterosins, and also show antimicrobial activity towards the bacteria, *Staphylococcus aureus*. For example: **7** seco-A, pseudopterosin exhibited a 69% reduction of inflammation in the mouse ear edema assay at a minimum inhibitory concentration (MIC) of 50 μg/ear.

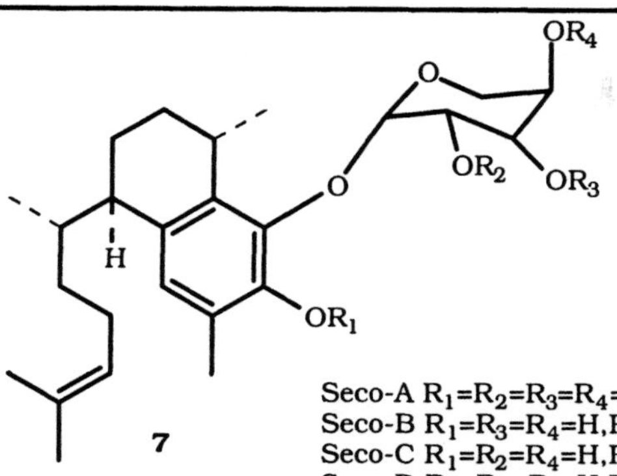

Seco-A R₁=R₂=R₃=R₄=H
Seco-B R₁=R₃=R₄=H,R₂=Ac
Seco-C R₁=R₂=R₄=H,R₃=Ac
Seco-D R₁=R₂=R₃=H,R₄=Ac

7

Fenical and coworkers [19] collected another unidentified, distinctly different taxonomic specie of this genus containing the novel diterpenoids, 8-11, which were also related to the pseudopterosins. The diterpenoids included a unique hydroperoxide, 10, and the quinone, 11, which contain the aglycon moiety found in the seco-pseudopterosins, 7 seco A-D.

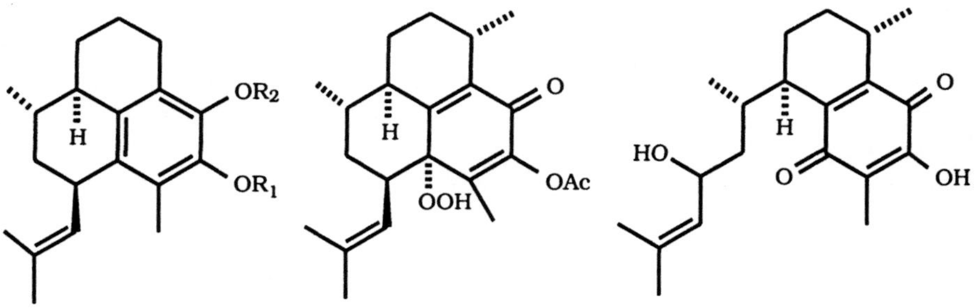

8 R₁=H, R₂=Ac
9 R₁=Ac, R₂=H

10

11

Polyhydroxy steroids

As described by D'Auria et al. 20] it is well known that many marine invertebrates are a rich source of polyhydroxylated steroids. One such metabolite, 5α-cholestane-3β,5,6β,9-tetrol, 12, initially isolated by Sjostrand et al. [21] and then later by Schmitz et al. [22], was the first polyhydroxy steroid to be isolated from the marine specie, *P. elisabethae*.

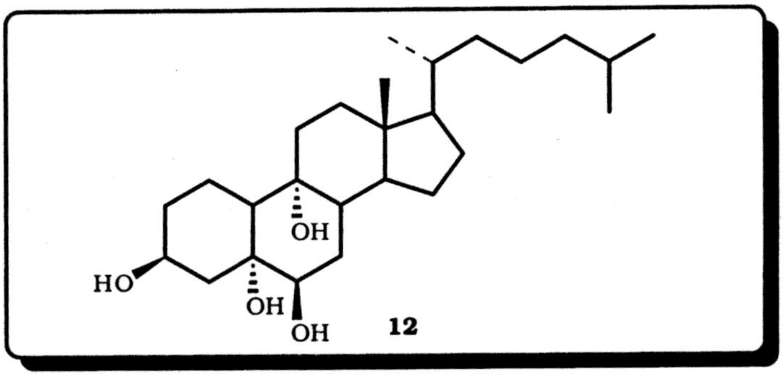

PSEUDOPTEROGORGIA AMERICANA

Pseudopterogorgia americana Gmelin is abundant in the Caribbean Sea and the chemistry of it's extract has been widely investigated. The secondary metabolites isolated as constituents in solvent extracts of *Pseudopterogorgia americana* were mainly sesquiterpene hydrocarbons and included: germacrenes, guaianines, gorgonenes, polysaccarides, sesquiterpenes, secogorgosterols, cholesterol and cetyl palmitate.

Germacrenes

Chan and coworkers [23] and Izac et al. [24] isolated germacrenes derivatives, **13-19**, as the major secondary metabolites from *Pseudopterogorgia americana*. Izac et al. [24] and Chan [23] isolated the new elemanolide **18** and elemanolide **19**, respectively along with C2 and C5 hydroxylated germacrenes.

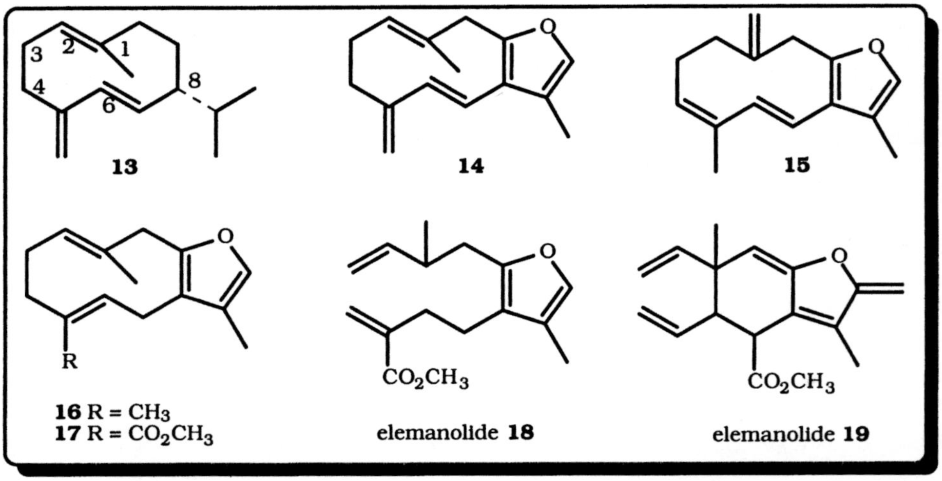

Guaianolides

Rodriguez and Boulanger [26] proposed that germacrenes isolated from *Pseudopterogorgia americana* are involved in the biogenesis of guaiane skeleton **20**. They suggested that rearrangement and further oxygenation of the germacrenes, **13-15**, can lead to the new guaiane series **20-31**. Interestingly, guaiane sesquiterpene-type compounds which constitute one of the largest families within the sesquiterpene lactones are mainly isolated from Compositae plants, as pointed out by Liu and Mabry [25]. The new guaiane metabolites **20-31**, which represent the first examples of guainine sesquiterpenes isolated from the gorgonian, *Pseudopterogorgia*

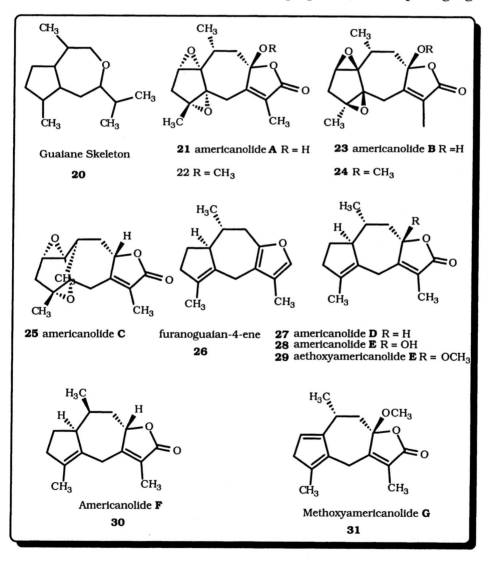

Guaiane Skeleton
20

21 americanolide **A** R = H
22 R = CH$_3$

23 americanolide **B** R =H
24 R = CH$_3$

25 americanolide **C**

furanoguaian-4-ene
26

27 americanolide **D** R = H
28 americanolide **E** R = OH
29 aethoxyamericanolide **E** R = OCH$_3$

Americanolide **F**
30

Methoxyamericanolide **G**
31

americana, were recently investigated by Rodriguez and Boulanger
[26,26]. From a biosynthetic standpoint, it is likely that americanolide
methyl ethers **22** and **24** originate from americanolides **21** and **23**
respectively, while americanolide **21** could arise from the oxidation of C8
of americanolide **25**.

Bioassay studies have been initiated on selected guaianolides.
Specifically, several guaianolides have been tested against human tumor
cell lines [27] and the results are summarized in table 2.

Table 2.   **Human Cytotoxicity of Americanolides**

compound	assay	IC_{50} (μg/ml)
21	MOLT-4 leukemia cells	0.1*
22	NCI panel	none
27	colon (KM-12) cancer cells	0.1
27	HeLa Cells	30
27	CHO-K1 cells	100

* selective inhibitor

Gorgonenes

Weinheimer and his research group [28] investigated a mixture of
sesquiterpene hydrocarbons which contained the well known 9-aristolene
32 and 1(10)-aristolene **33**, (+)-γ-maaliene **34**, both of which are
unprecendented in nature, and (+)-β-gorgonene **35**. The latter compound
35, possesses a new isoprenoid skeleton with a misplaced isopropenyl
residue.

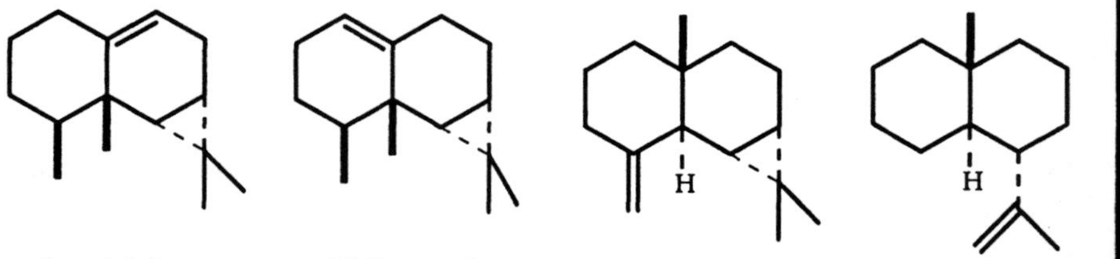

9- aristolene	1(10)-aristolene	(+)-γ-maaliene	(+)-β-gorgonene
32	**33**	**34**	**35**

The same research group [28] proposed a biogenetic process, whereby "migration" of the isopropenyl residue would result from hydride migration within the farnesyl-derived cyclic ion **36** and thus would lead to compound **37**. (+)-β-gorgonene, **35**, would result from a proton loss from **37**, leading to the monocyclic intermediate **38**, and followed by cyclization. The process is summarized in scheme 2.

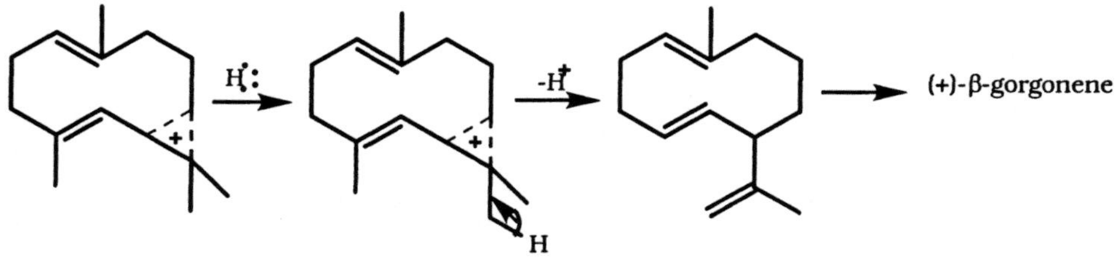

Scheme 2. Proposed biogenetic process towards (+)-β-gorgonene, **35**.

Polysaccharides

Another metabolite isolated from the specie, *Pseudopterogorgia americana,* is corallan, **39**, the chemistry of which was investigated by Molchanova and coworkers [29]. Corallan, **39**, a sulphated glycoprotein contains D-glucose, D- and L-galactoses, D-arabinose, L-fucose and D-glucoronic acid. Some structural features of the polysaccharide moiety of **39** are shown in scheme 3.

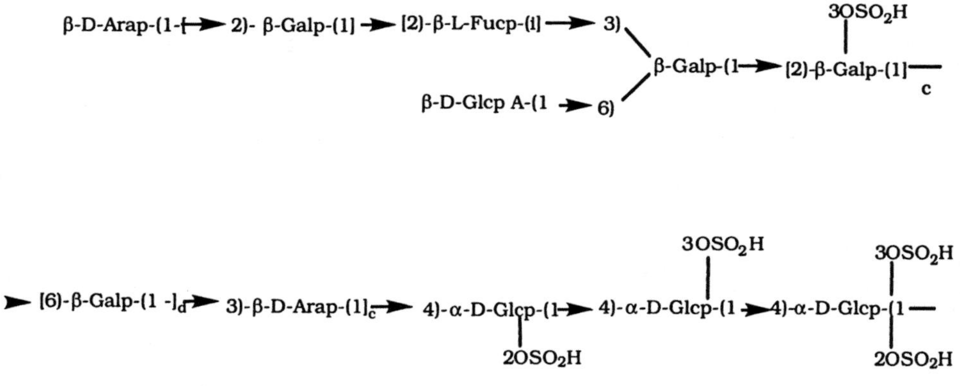

corallan **39**

Scheme 3. Structural features of the polysaccharide moiety in corallan **39**.

Sesquiterpenes

Besides the known bisabulene sesquiterpenes, (-)-curcuhydroquinone **41**, (-)-curcuphenolacetate **42**, and (-)-curcuquinone **44**, a new compound, (-)-curcuhydroquinone-1-monoacetate, **40**, was isolated by Miller and coworkers [30]. It is important to note that curcuhydroquinone was also isolated from *Pseudopterogorgia acerosa* by the same group [30].

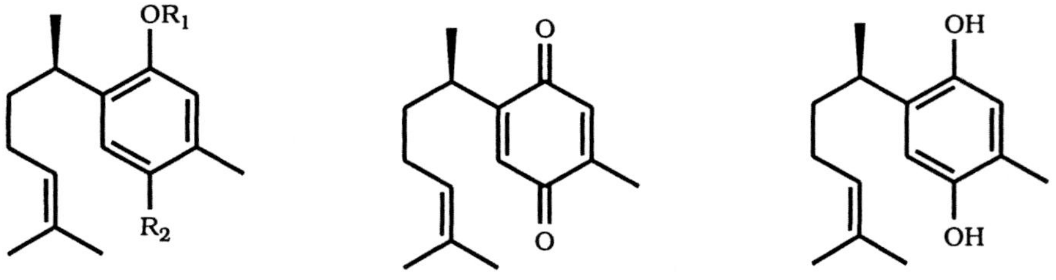

40	R$_1$=Ac, R$_2$=OH
41	R$_1$=H, R$_2$=OH
42	R$_1$=Ac, R$_2$=H
43	R$_1$=H, R$_2$=OH

curcuquinone

44

curcuphenol

45

As displayed in table 3, selected curcumene derivatives exhibited weak antibacterial activity against *Staphilococus aureus* and *Vibrio anguillarum*, with curcuphenol **45** being the most active [31].

Table 3.   **Antibacterial Activity* of Curcumene Derivatives. Zones of Inhibition (mm) against:**

Compound	µg (conc/disk)	Staphilococus aureus	*Vibrio* anguillarum
45	7	4	slight
45	70	11	3
44	18	3	0
44	180	3	0
41	9	slight	slight
41	90	3	2

* agar plates assay

A series of the same curcumene derivatives were isolated from *Pseudopterogorgia rigida* by McEnroe and Fenical [31]. Curcuphenol **45**, which was the most potent antibacterial metabolite was synthesized from 2-hydroxy-4-methyl acetophenone, **46**, as shown in scheme 4.

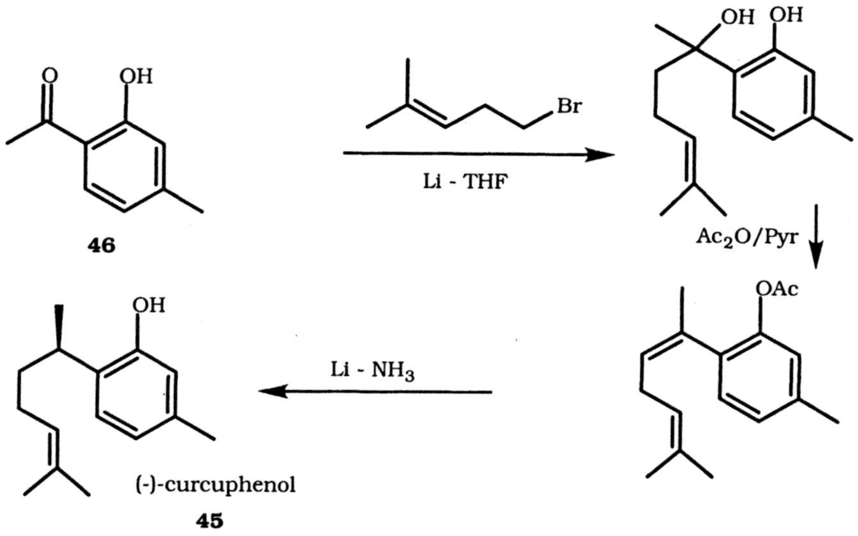

Scheme 4. Synthesis of Curcuphenol **45**.

Coincidentally, (+)-curcuphenol, **45**, which was also isolated from the marine sponge, *Didiscus flavus,* was investigated by Wright et al. [32] and found to be cytotoxic against several *in vitro* human tumor cell lines. A result of his study revealed that this metabolite possessed an IC_{50} of 7µg/mL against P-388 murine leukemia cells. The MIC's of (+)-curcuphenol against the growth of a human lung cancer cell line and the fungus, *Candida albicans* are presented in table 4.

Table 4.   **Minimum Inhibitory Concentration of (+)-Curcuphenol, 45 Against Growth of Cell Line and Fungus [32]**

Bioassay	MIC (µg/ml)
Human tumor cell lines A-549 (lung)	0.1, 10
Candida albicans	8

Rigidone **47**, another sesquiterpene *ortho*-quinone which is derived from the curcumenes **97**, was isolated from *Pseudopterogorgia rigida* by Freyer and coworkers [33]. It is important to mention that the structure of rigidone was verified by preparation from an authentic sample of the phenazine analog, **48**. Rigidone, **47**, exhibited a moderate Microphage

Scavenger Receptor down regulating inhibitory activity which is involved in the treatment of aertherosclerosis.

rigidone

47

phenazine analog **48**

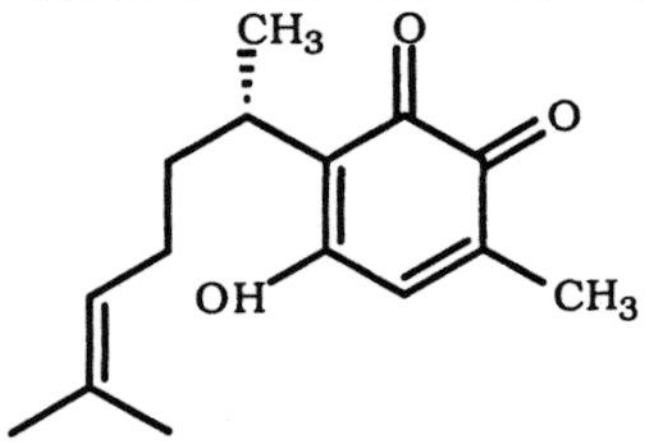

A new sesquiterpene alcohol, bisabulene derivative, **49,** was isolated by Look et al. [34], from a Caribbean sea plume of an undescribed species of the genus *Pseudopterogorgia.*

12-Hydroxy-E-γ-bisabolene **49**

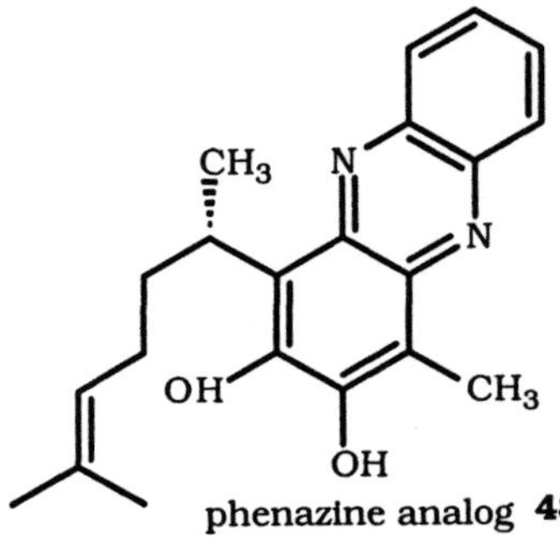

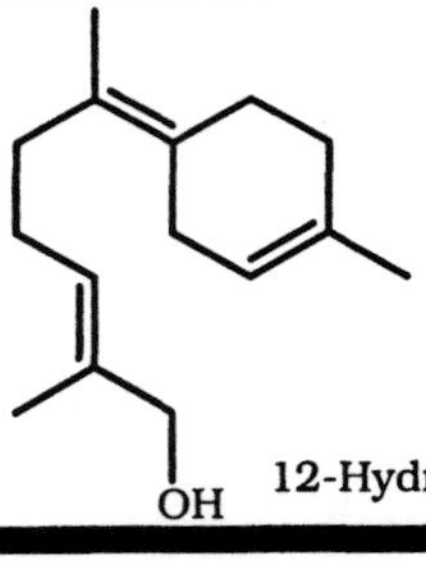

Secogorgosterols

Another class of compounds isolated from *Pseudopterogorgia americana* include the unusual secogorgosterols, **50, 52, 56-60,** which comprise a biogenetically unprecedented side chain containing a cyclopropyl group. A comprehesive study, including structural characterization of the side chain was carried out by several groups [35-37]. Enwall and van der Helm [38] isolated the secogorgosterol **50**, along with the related 6α-epoxide **52** [40]. The final structure proof of **50** was achieved by X-ray analysis of its 3-*p*-iodobenzoate derivative, **51**. Esterification of **50** to **51**, was accomplished by acetylation followed by selective saponification, and purification. The absolute configuration of the crystal structure of **51** was achieved by X-ray diffraction studies and was consistent with the presence of a cyclopropane ring in the side chain of the gorgosterol, **54,** class of compounds.

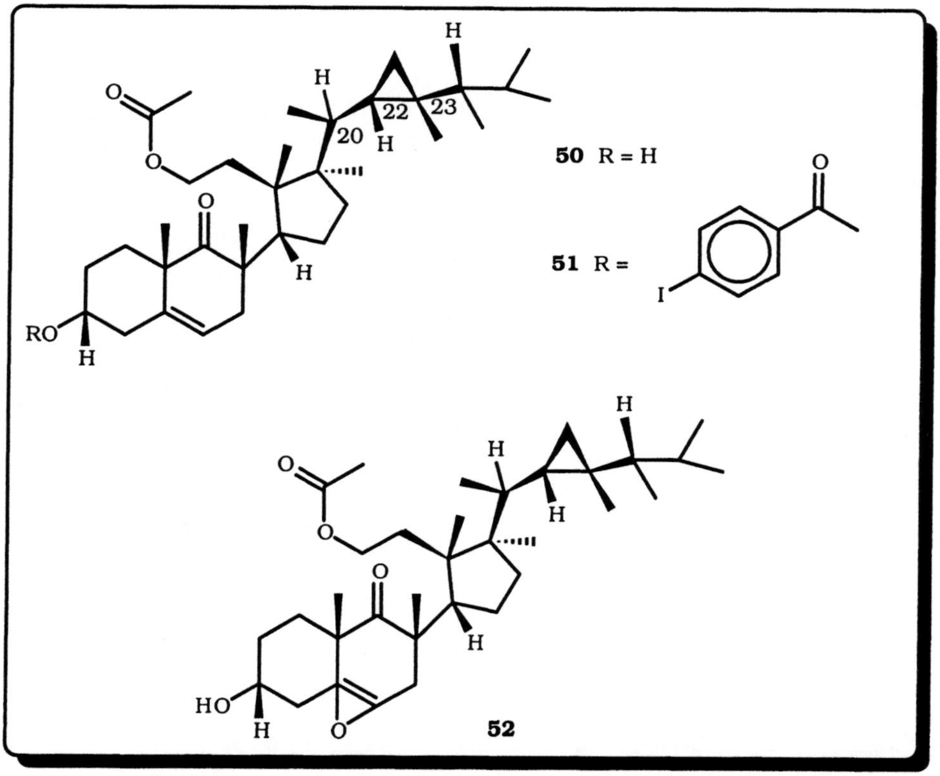

50 R = H

51 R =

The related secogorgosterol **53,** first isolated by Spraggins [39], was also isolated by Schultz and Clardy [40] as the major component from *Pseudopterogorgia hummelinkii*. The structure of this secogorgosterol, which has a different stereochemistry than **50** at C20, C22 and C23, was confirmed by the X-ray crystal structure of the unsubstituted compound.

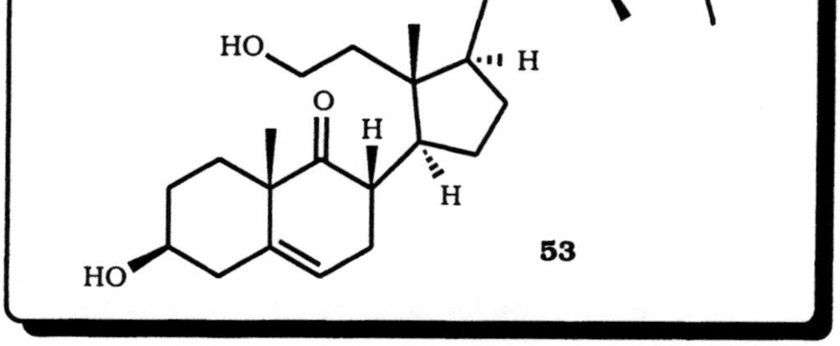

Secogorgosterols, **56-60**, were recently isolated from *Pseudopterogorgia americana* by Kerr and coworkers [41], while the closely related secogorgosterols, **57, 59**, and **60**, were isolated from an unidentified species of *Pseudopterogorgia sp.* by He and coworkers [42]. Besides the potent anti-proliferative and anti-inflammatory activities, the latter compounds exhibited inhibitory activity against protein kinase C (PKC). As discussed by Ker et al. [42] gorgosterol **54** and dinosterol **55** are the predominant sterols in crude extracts of *Pseudopterogorgia sp.* It has been suggested by Kerr et al. [42] that secosterols are produced by an enzymatic process. Kerr et al. [42] developed a fortified enzyme preparation of an extract of *Pseudopterogorgia americana* which transformed a variety of sterols to their 9(11)-secosterol derivatives. Results from these enzymatic studies confirmed the identity of gorgosterol **54**, as the biosynthetic precurser of 9(11)-secogorgosterol **57**.

Table 5.   **Cellular Activity, IC_{50} (μM), of Secosterols**

Compound	MCF-7a		NHEKb		Neutrophil	
	^{3}H-T[c]	MTT[d]	^{3}H-T	MTT		
57	7.8±1.1	79.3±2.5	1.1±0.2	12.8±0.9	>10	15.3±2.7
59	13.1±1.6	47.3±5.3	5.4±1.4	>100	3.9±0.5	>100
60	6.1±1.2	76.3±5.1	2.2±1.1	33.9±1.8	>10	not tested

[a]human breast carcinoma proliferation.

[b]normal human epidermal keratinocites proliferation.

[c]tritiated thymidine.

[d]direct cyto-lethality (MTT formazan dye reduction).

The occurrence of secosterols **59**, and **60**, in *Pseudopterogorgia sp.* appears to represent the first example of the 9,11-secosterols with the dinosterane skeleton. Furthermore, the side chain hydroxylation at C24 is very rare among marine sterols [41].

Secosterols **57, 59**, and **60** were found to inhibit human PKC enzymes with IC_{50} values from 12 to 50 μM. Since CPK has been implicated with both inflammatory and proliferative processes the compounds were tested in cells related to these processes [43] as shown in table 5. The results suggested that cellular activity may be due to a block in a cell cycle.

Sterols

The known sterol derivative **61**, containing a cyclopropane side chain was isolated in our lab (Adamczeski [8]) from *Pseudopterogorgia sp.* collection from Indonesia. This compound was first isolated from the soft coral

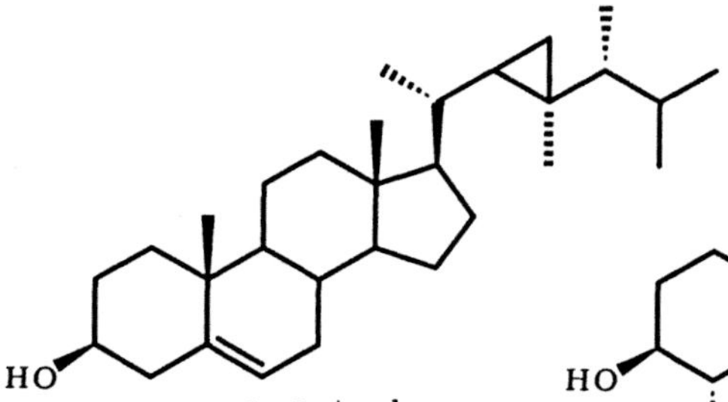

gorgosterol
54

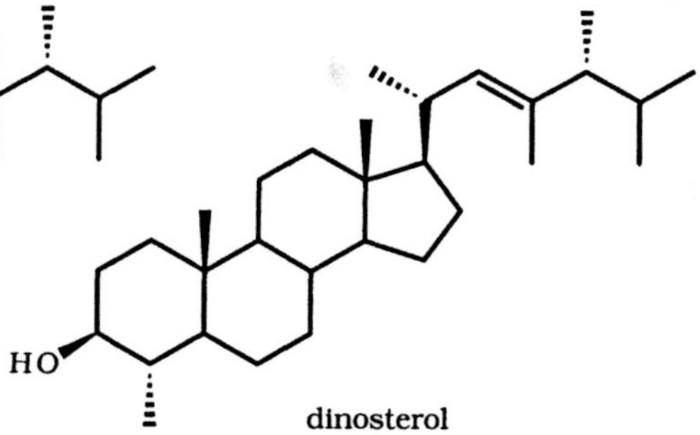

dinosterol
55

chemoenzymatic synthesis

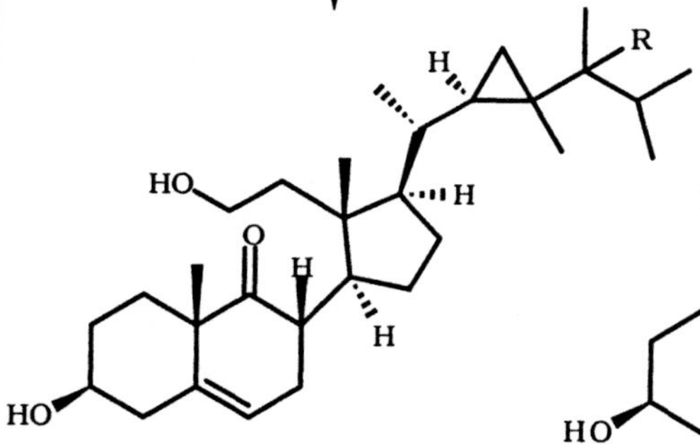

56 R = H
57 R = OH

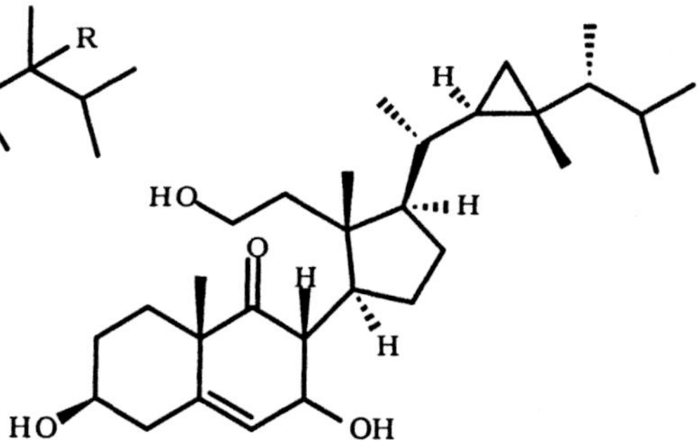

58

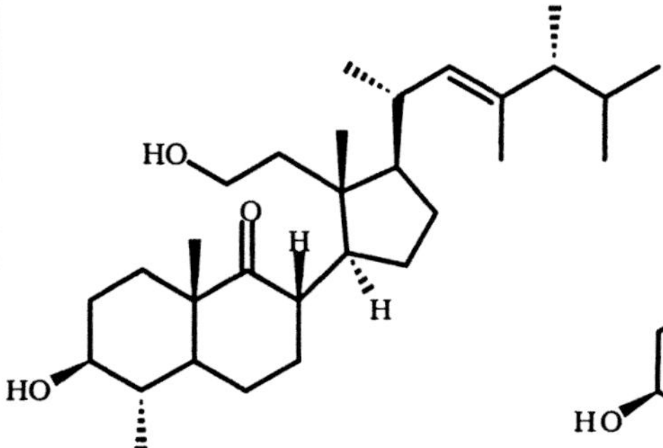

59

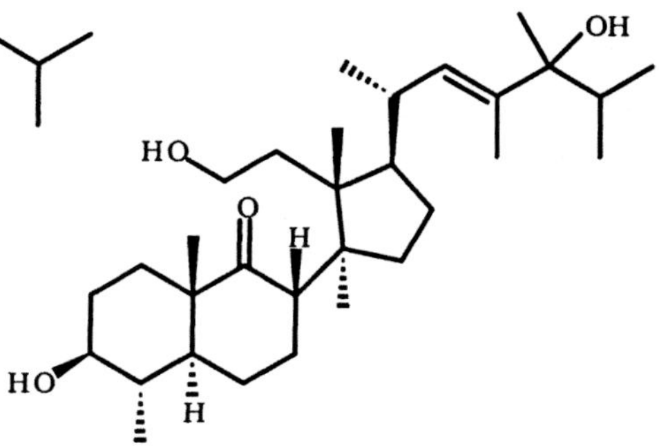

60

Sinularia dissecta, Jagodinska [43]. The structure and stereochemistry of **61** which possesses the polyoxygenated gorgosterol skeleton were characterized by comparison of the spectroscopic properties to the relevant data in the literature published by Tarchini et al. [44]. It is interesting to note that the stereochemistry of the side chain is similar to secosterol, **53**, which is thought to evolve from the basic gorgosterol structure, **54**. Using the brine shrimp assay (*Artemia salina*) [45], we tested the crude methanol extract whose NMR was consistent with sterol-type compounds related to **61**. Analysis of the data reveal that the crude extract was toxic to *Artemia salina* at an LD_{50} of 93 ppm.

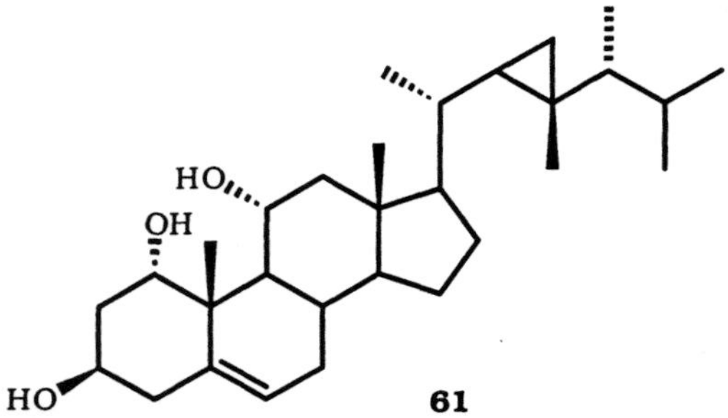

Miscellaneous

The eggs of *Pseudopterogorgia americana* contained lipids, with cetyl palmitate as the major product, glycerides, and sterols (i.e. mainly cholesterol). For more details see Wirahadikusumah [46].

PSEUDOPTREOGORGIA ACEROSA

Extracts of *Pseudopterogorgia acerosa* produced a family of diterpenoids derived from the pseudopterane skeleton **62** [5]. Other complex secondary metabolites produced by *P. acerosa* include cembranes, polyhydroxy steroids and carotenoids.

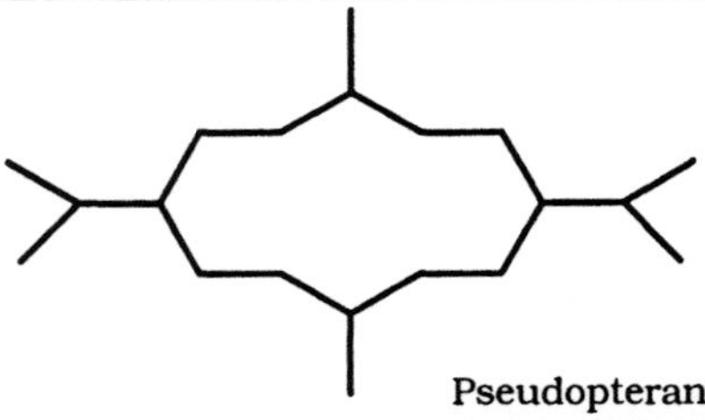

Pseudopterane skeleton **62**

Pseudopteranes

A variety of pseudopterane diterpenoids, **63-70**, based on the 12 memberd carbocyclic pseudopterane skeleton, **62**, and related to the pseudopterolide prototype, **63**, originally isolated by Fenical and and his research group [47], were recently isolated by Rodriguez et al. [48].

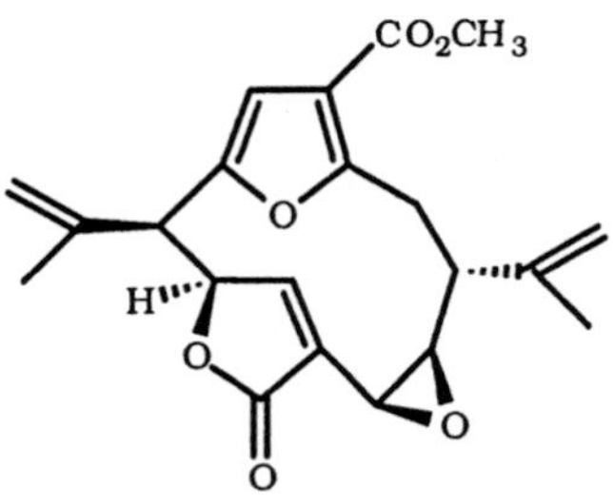

63

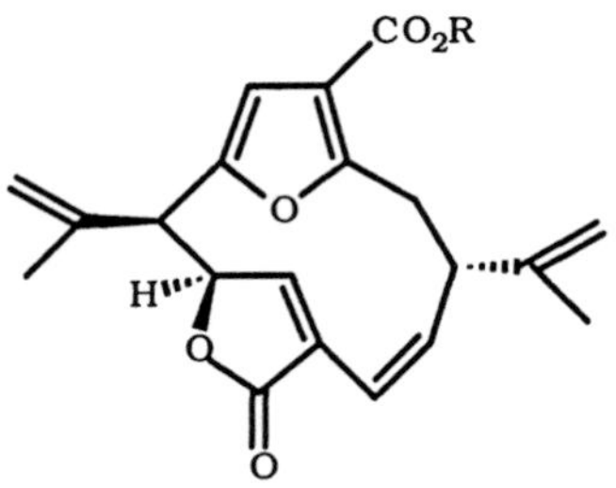

64   R = CH₃
65   R = H

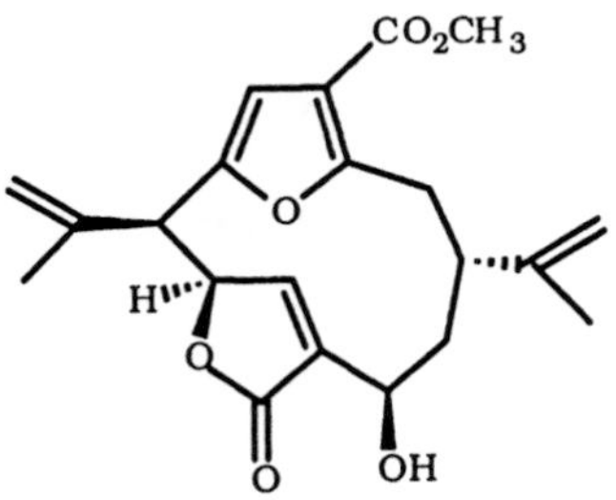

66

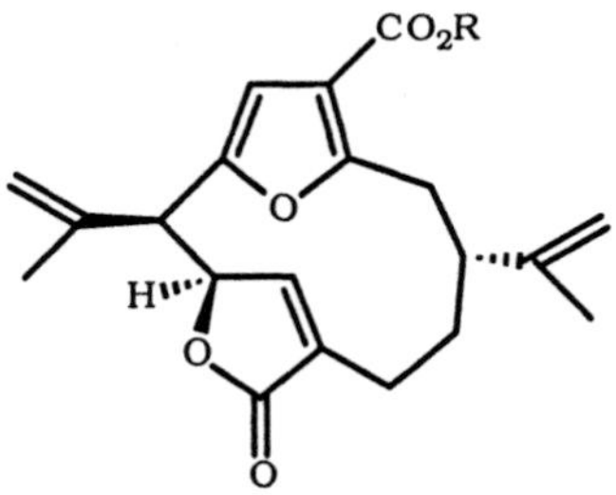

67   R = CH₃
68   R = H

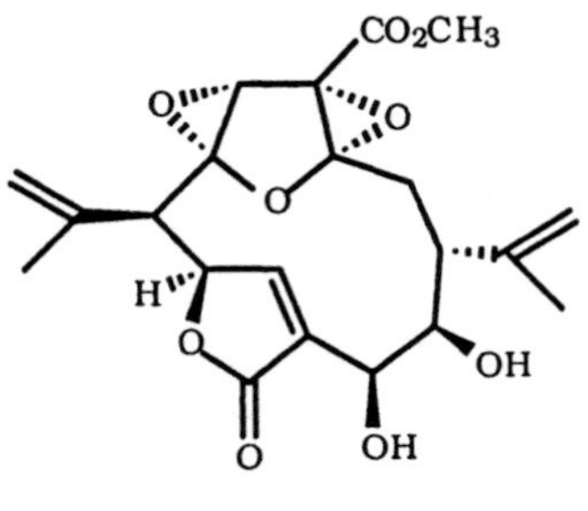

69

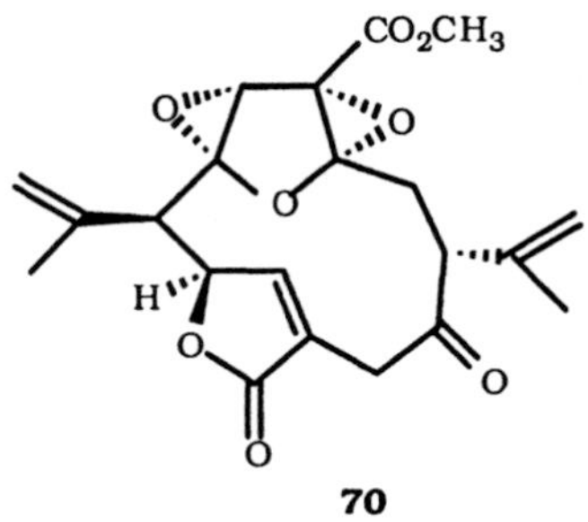

70

Cembranes

Pseudopterolide **71**, an irregular diterpenoid was isolated by Bandurraga et al. [47]. This metabolite exhibited unusual cytotoxic activity in that it inhibited overall cell cleavage while it did not inhibit nuclear division in the fertilized urchin egg assay.

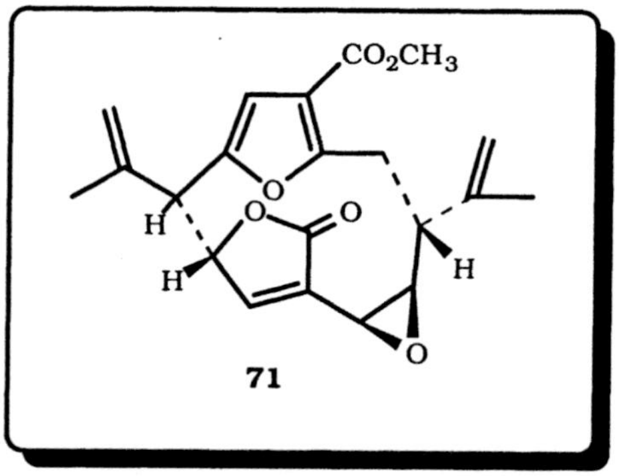
CO2CH3
O
O
O
H
H
H
O

71

Tobagolide **72**, another novel cembrane diterpenoid containing a dimethylamino moiety, was isolated by Tinto and Chan and coworkers [49,50]. As shown in scheme 5, an unusual reaction of **72** under acetylation conditions led to the replacement of the dimethylamino moiety with the acetate group to yield **73a**, the acetate product, and **73b** as the dehydro product.

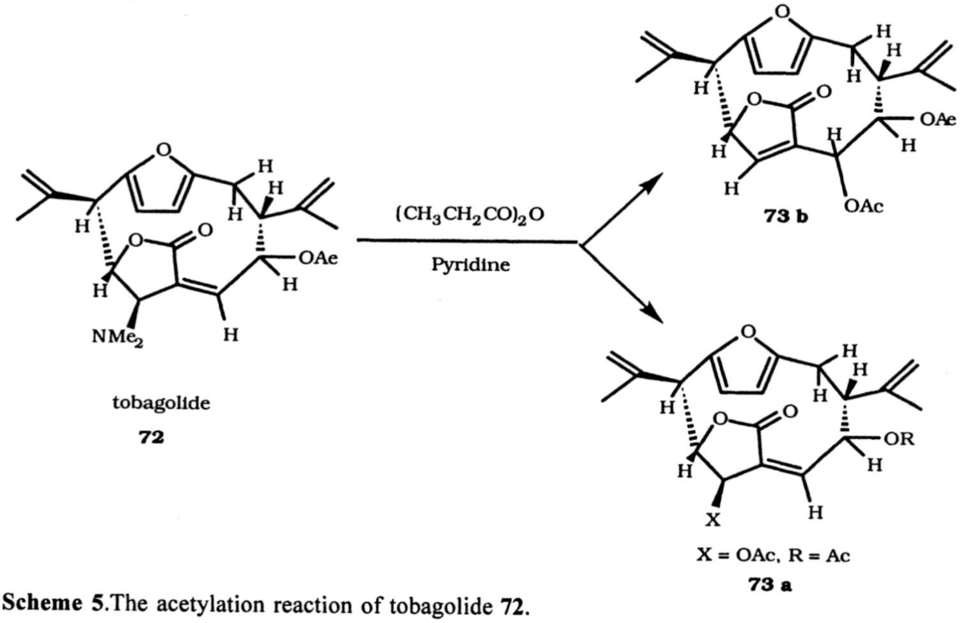
(CH3CH2CO)2O
Pyridine
NMe2
OAe
OAc
OR
X
tobagolide
72
73 b
X = OAc, R = Ac
73 a

Scheme 5. The acetylation reaction of tobagolide **72**.

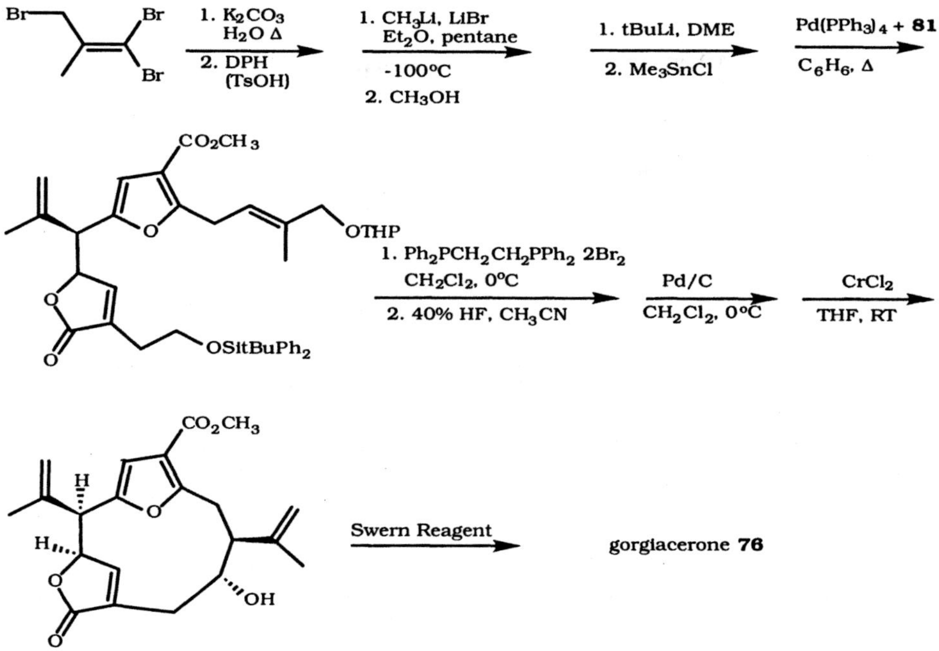

Scheme 6: Synthesis of gorgiacerone **76**.

Additional pseudopteranoids **74-80** were isolated by Tinto and coworkers [51,52]. Included among them are the unusually stable diepoxy furan derivative **79**, whose structural elucidation included X-ray analysis, and gorgiacerone, **76,** which was the most abundant metabolite isolated from the crude extract.

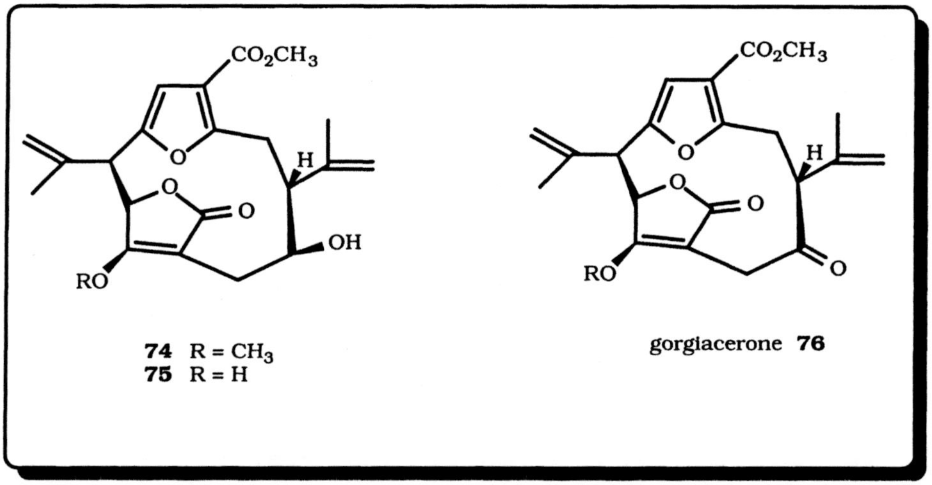

74 R = CH$_3$
75 R = H

gorgiacerone **76**

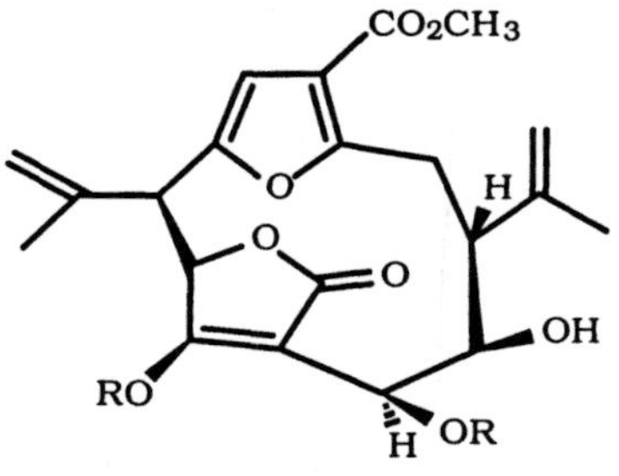

77 R = H
78 R = CH₃

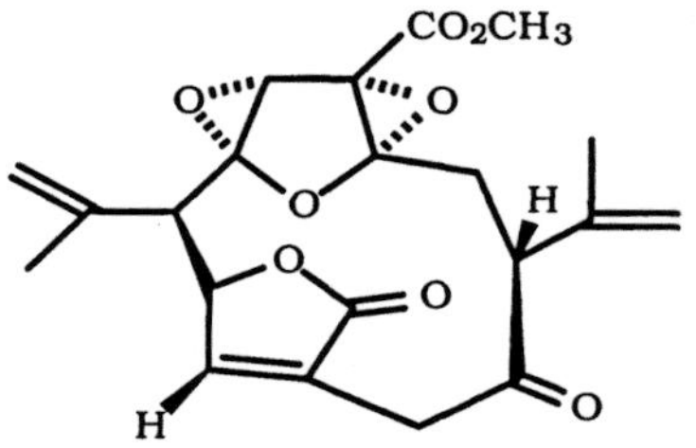

79

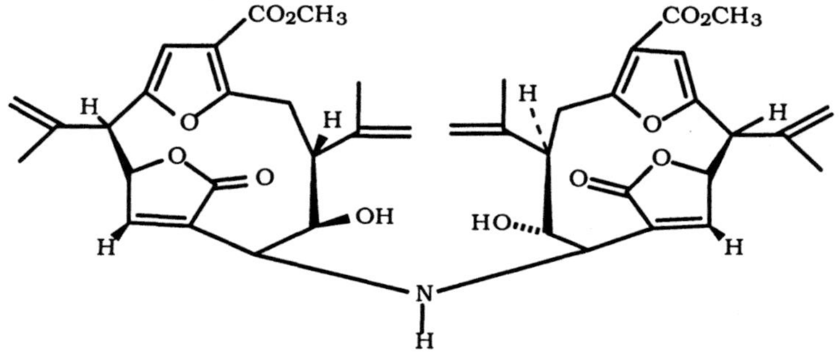

80

As shown in scheme 6, total synthesis of gorgiacerone, **76**, was reported by Rayner and coworkers [53]. The synthesis of gorgiacerone utilizes the geminal heterocyclic compound, **81** as a building block.

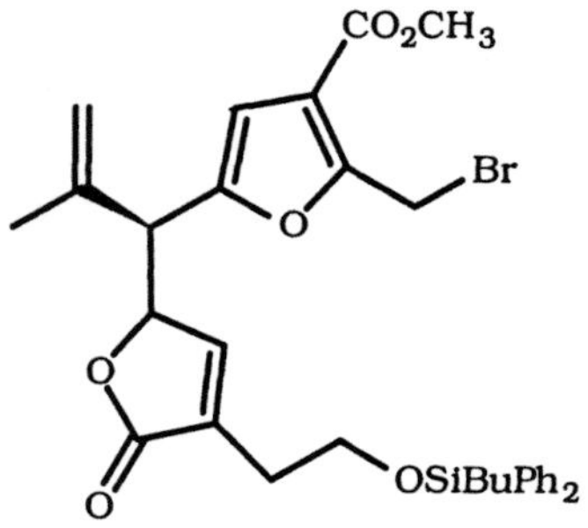

81

The four lactones, kallolides **82-85**, isolated from *Pseudopterogorgia kallos* (Bielschowsky) by Look et al. [54], are closely related to the cembrolide class of compounds and possess the pseudopterane skeleton (i.e. **62**). Among these diterpenoids, the major metabolite kallolide **83**, displays the same efficacy as the anti-inflammatory inhibitor, indomethacin (see table 1).

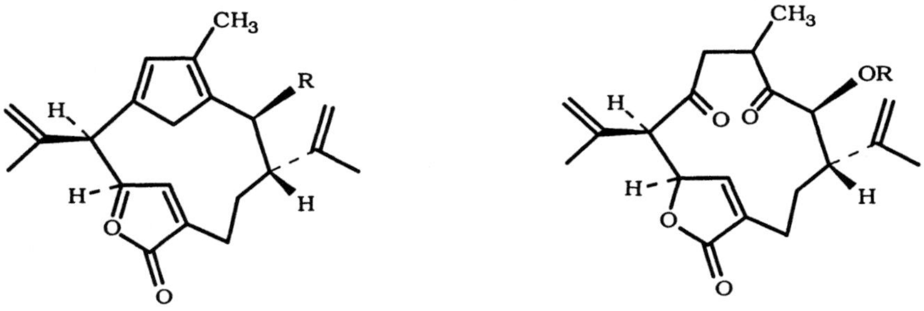

82	R = OH
83	R = OAc
84	R = H

R = H
85

Stemming from the fact that both cembrane and pseudopteranes skeletons are found in *Pseudopterogorgia acerosa*, Fenical [5] has suggested that the pseudopterane skeleton **62** might evolve from the ring contraction of a cembrane precursor **86**.

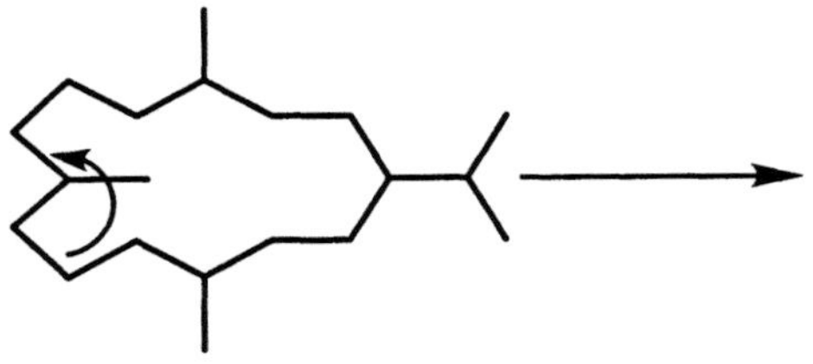

pseudopterane skeleton **62**

cembrane skeleton **86**

Also included among the cembrolides is the 14-membered furanocembrolide, namely acerolide, **87**, which was isolated and characterized by Chan et al [50]. Related 14-membered cyclic cembrolides include compounds **88 a-c** and **89** isolated from *Pseudopterogorgia bipinnata*. Shown in scheme 7 is a total synthesis of **87** proposed by Astles and Paquette [55].

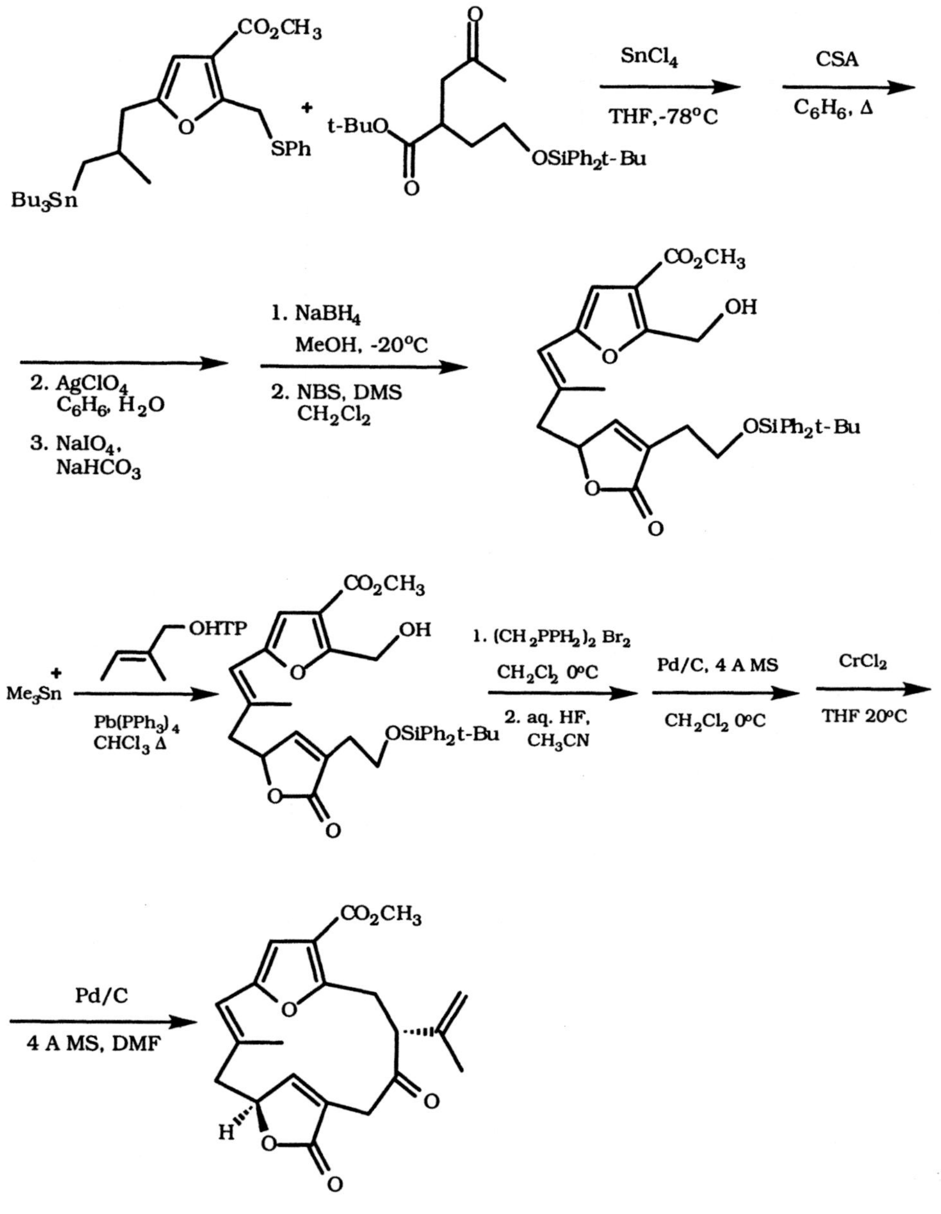

(+) acerolide **87**

Scheme 7. Total synthesis of racemic acerolide, **87**.

Investigation of the products of *Pseudopterogorgia bipinnata* by Wright [56] and coworkers yielded a new class of cembrenoids, namely, bipinnatins **88 a-c** and **89**.

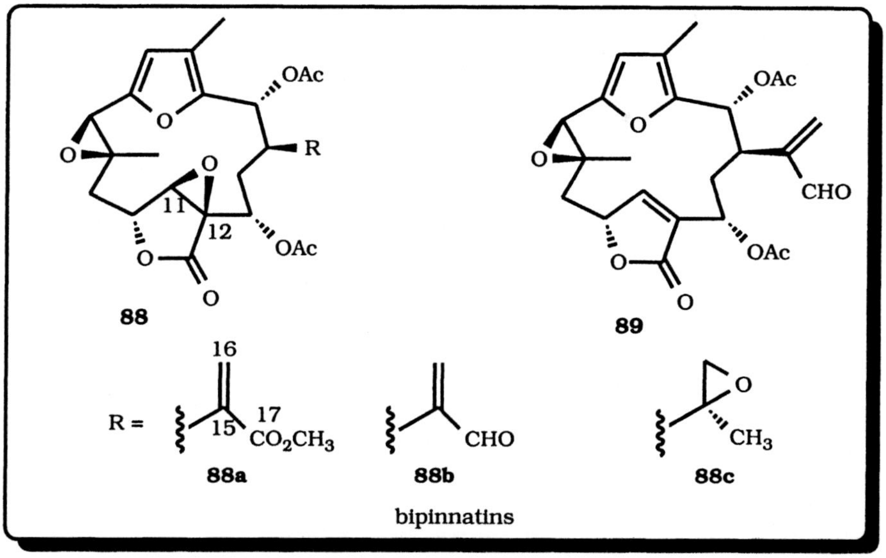

bipinnatins

Bipinnatins **88a, b**, and **89** exhibited *in-vitro* activity against murine tumor cell line as shown in table 6.

Table 6.   Inhibition of P388 Murine Tumor Cell Replication by Bipinnatins [56]

bipinnatin	88a	88b	88c	89
IC$_{50}$ (µg/ml)	0.9	3.2	46.6	1.5

Bipinnatins **88a, 88b** and **89** which exhibit very potent cytotoxic activity, share an α,β-unsaturated carbonyl functionality at C15-C17. Bipinnatin **88c** which lacks this moiety is a weak inhibitor, thus suggesting that the α,β-unsaturated carbonyl functionality at C15-C17 is essential for the observed cytotoxic activity of the bipinnatins.

On the other hand, the epoxide at C11-C12 which is not present in **89**, may not be a significant factor contributing to this activity.

Polyhydroxy Sterols

Another metabolite extracted and isolated from extracts of *Pseudopterogorgia acerosa* is acerosterol, **90**. This tetrahydroxy sterol was first isolated by John et al. [57] (see also references within). It should be noted that dinosterol **55**, and other similar derivatives having α-methyl substituents have been isolated from marine invertebrates containing a symbiotic zooxanthellae. Thus, it is speculated that acerosterol, **90**, may also have originated from a symbiotic zooxanthellae associated with this marine sea plume.

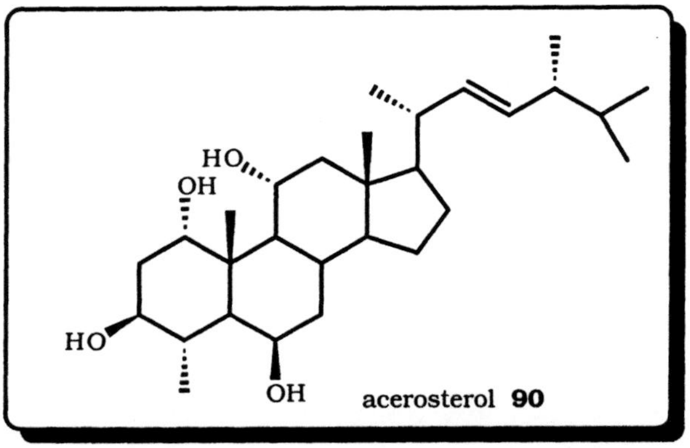

acerosterol **90**

Carotenoids

Skjenstad and coworkers [58] examined the carotenoids of unialgal cultures originating from zooxanthellae symbiotic with the following hosts; the giant clam, *Tridacna crocea*, the nudibranch, *Ptareolidiaianthina*, and *Pseudopterogorgia bipinnata*. The results showed that *P. bipinnata* produced a high total carotenoid content (0.62%). It should be noted that the dinoflagellates examined possessed a unique capacity to synthesize C37-norcarotenoids with peridinin **91**, as the main carotenoid (78%). Diadinoxanthin **96**, was the main C40-carotenoid (8%).
C37 carotenoids:

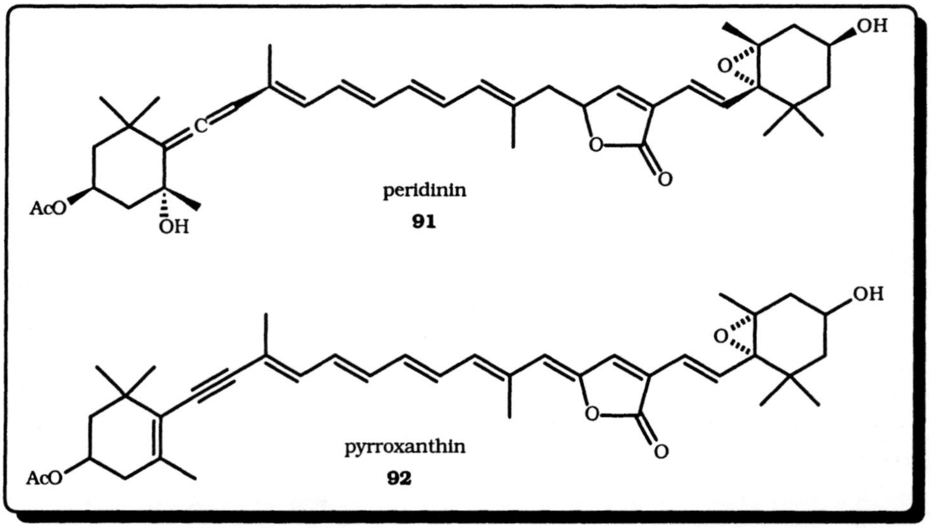

peridinin
91

pyrroxanthin
92

C40 Carotenoids:

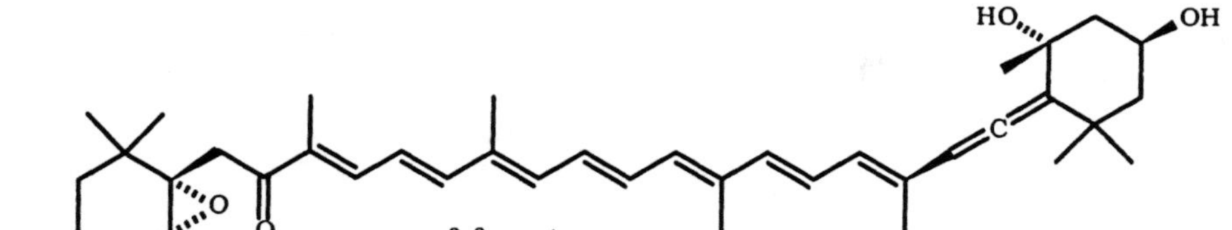

β, β-carotene
93

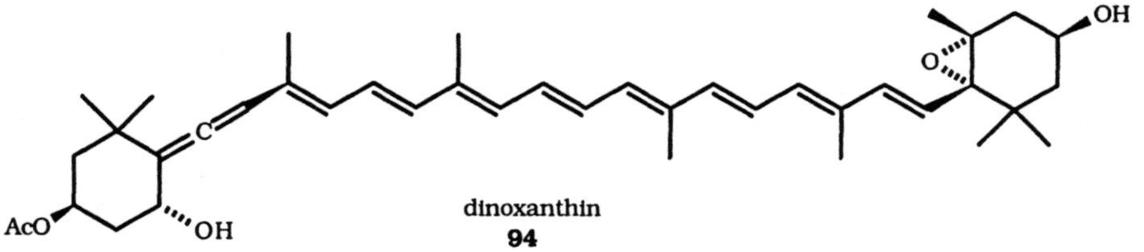

dinoxanthin
94

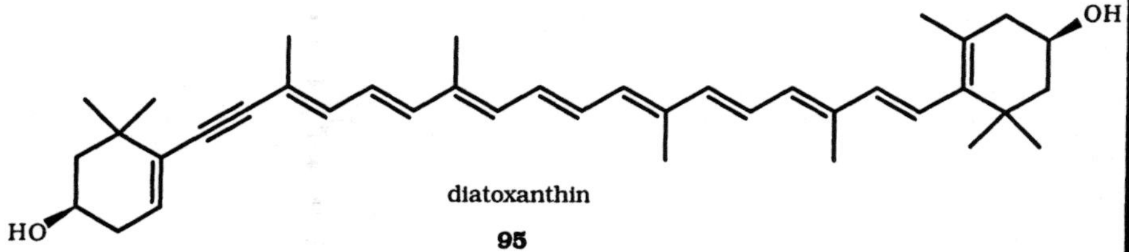

diatoxanthin
95

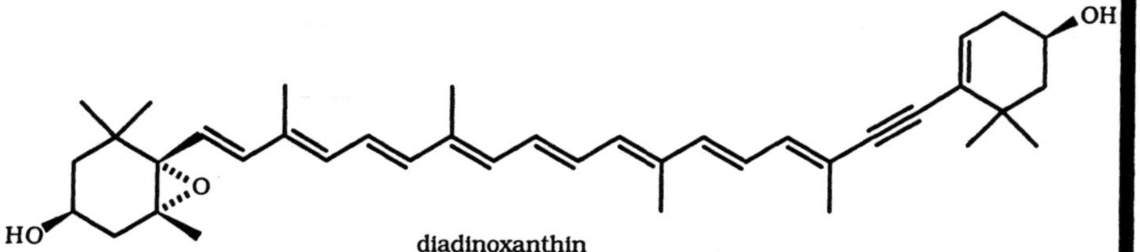

diadinoxanthin
96

BIOLOGICAL ACTIVITY OF METABOLITES FROM *PSEUDOPTEROGORGIA SP.*

As the metabolites of the different species of *Pseudopterogorgia* exihibit a wide scope of biological activities, we found it important to present an overview of the biological activity of its major groups of metabolite (table 7).

Table 7.   Biological Activities of Major Metabolites from *Pseudopterogorgia* Species

metabolite group	species	biological activity	references
pseudopterosins	*elisabethae*	anti-inflammatory, analgesic, nontoxic	[9],[10],[11], [12],[13]
secopseudopterosins	unidentified	anti-inflammtory, analegesic antimicrobial	[18]
guaianolides	*americana*	cytotoxic	[27]
sesquiterpenes (curcumenes)	*americana, acerosa, rigida*	cytotoxic, antibacterial, MSR inhibition	[30],[31] [32],[33]
secosterols	*americana*	protein kinase inhibitors, antiproliferative, anti- inflammatory	[43]
pseudopterolides, kallilolides	*acerosa* *kallos*	cytotoxic, cell and Sea Urchin eggs, anti-inflammatory, analgesic	[5], [47], [54],[55]
cembranes	*bippinata*	cytotoxic, anti-inflammatory	[5], [56]
metabolites in crude aqueous + mucus extracts	*americana,* unidentified	hemolytic, proteolitic, anti-cholineesterase	[59]
unknown metabolites in secretion extracts	*americana*	toxicity, exitatory + evolution depression in mice, edema , neuronal degeneration, kidney hemorrage, severe hepatic degeneration	[59]

DEFENSE MECHANISMS AGAINST PREDATION

Sea whips encompass widely diverse morphologies among soft-bodied marine invertebrates and are readily abundant in reef environments. As documented by Pawlik [4], and Fenical and coworkers [59], this genus of soft corals has fleshy tissues that appear to be physically unprotected from potential predators, yet seem to be unmolested and have only a few predators. Analysis of the wide spectrum of the potent biologically active metabolites of *Pseudopterogorgia* presented in table 7 may offer a possible explanation of the seemingly few predators of this specie.

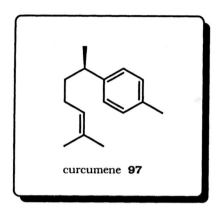

curcumene **97**

Ship-board bioassays were undertaken by Fenical and coworkers [60] to determine the palatabilty of food pellets coated with freshly extracted lipid soluble metabolites of Caribbean gorgoronians. Assays of extracts of one unidentified and three identified species of *Pseudopterogorgia,* namely, *americana, rigida* and *acerosa,* gave mixed results. While *Pseudopterogorgia rigida* and *P. acerosa* were very effective as food deterents (*rigida* yeilded the second most active extract assayed), extracts of *P. americana* and an unidentified specie of sea plume were found to be among the most palatable. The pure compounds isolated from the crude extracts from *Pseudopterogorgia rigida,* were assayed further by Harvell et al. [60]. Both curcuhydroquinone, **41,** and curcuquinone, **44,** were examined and found to deter feeding in aquarium and field assays at concentrations below those present in the gorgonian tissue. A hydrocarbon mixture containing mainly curcumene, **97,** and the monoacetate of curcuhydroquinone, **40,** did not deter feeding in either assay. Harvell et al. [61] suggested that the pseudopterolide, **63,** functions as the food deterent metabolite in *P. acerosa,* although no experimental details were presented.

CONCLUDING REMARKS

We presented a review of the natural product chemistry found in species of *Pseudopterogorgia*. These metabolites function as predator deterents and some of them proved to be (potential) therapeutic agents. The discovery of pseudopterosins and their potent anti-inflammatory activity should be especially highlighted as one of the most prominent (marine chemistry) scientific inventions of the last decade. We feel that the combination of the discoveries reviewed herein combined with drug design based on structure activity studies of these compounds will lead to therapeutic agents which although derived from marine life chemistry, will not rely on it as a source.

ACKNOWLEDGMENTS

Partial financial support for the research was provided by American University (AU) New Faculty, Senate and Mellon research awards. We should also like to acknowledge Mr. Wesley White at National Institutes of Health MD for support of our program and Mr. Martin Shapiro at AU for assistance with literature searches.

REFERENCES

[1]   (a) Faulkner, D.J. *Nat. Prod. Rep.* **1990**, 7, 269-309.; (b) Faulkner, D.J. *Nat. Prod. Rep.* **1991**, 8, 97-211.; (c) Faulkner, D.J. *Chem. Rev.* **1993**, 93(5), 1; and references cited therein.

[2]   Pettit, G.R.; Kamano, Y.; Aoyagi, R.; Herald, C.L.; Doubek, D.L.; Schmidt, J.M.; Rudloe, J.J. *Tetrahedron*, **1995**, 41, 986.

[3]   (a) Suffness, M.; Newman, D.J.; Sander, L. In: *Bioorganic Marine Chemistry*; Scheuer, P.J. Ed., Springer-Verlag; 1, pp. 132, **1 9 8 9** ; (b) Munro, M.H.G.; Luibrand, R.T.; Blunt, J.W. In: *Bioorganic Marine Chemistry*; Scheuer, P.H., Ed. Springer-Verlag; pp. 93 **1987**; and references cited therein.

[4]   Pawlik, J.R. *Chem. Rev.*, **1993**, 93(5), 1911.

[5]   Fenical, W. *J. Nat. Prod.* **1987**, 50, 1001.

[6]   Pawlik, J.R., Burch, M.T. and Fenical, W. J. *Exp. Mar. Biol. Ecol.*, **1987**, 108, 55.

[7]   Humann, P. In: *Reef Coral Identification.* New World Pub., Inc., Fl. **1993**.

[8]   Zhang. Z., Adamczeski, M. The 37th Annual Meeting of the American Society of Pharmacognosy at University of California, Santa Cruz, CA . July 28, **1996**. Poster number: 159.

[9]   Fenical, W. *J. Nat.Prod.*, **1987**, 50(6), 1001.

[10] Rousis, V., Wu, Z., Fenical, W., Strobel, S.A., Van Duan, G.D., Clardy, J. *J. Org. Chem.*, **1990**, 55, 4916.

[11] Rouhi, M. *C h e m . E n g . N e w s* , **1995**, November 20, 42.

[12] Look, S.A., Fenical, W., Matusumoto, G.K., Clardy, J. *J. Org. Chem.*, **1986**, 51, 5140.

[13] Look, S.A., Fenical, W., Jacobs, R.S., Clardy, J. *Proc. Natl. Acad. Sci. USA*, **1986**,83, 6238.

[14] Harrowven, D.C., Dennison, S.T., Howes, P. *Tetrahedron Letters,* **1994**, 35(24), 4243.

[15] Corey, E.J., Carpino, P. *Tetrahedron Letters*, **1990**, 31(27), 3857.

[16] Broka, C.A., Chan, S., Peterson, B. *J. Org. Chem.*, **1988**, 53, 1584.

[17] Corey, E.J., Carpino, P., *J. Am. Chem. Soc.*, **1989**, 111, 5472.

[18] Look, S.A., Fenical, W. *Tetrahedron*, **1987**, 43(15), 3363.

[19] Harvis, C.A., Burch, M.T., Fenical, W. *Tetrahedron Letters.*, **1988**, 29(35), 4361.

[20] D'Auria, M.V., Minale, L., Ricco, R. *Chem. Rev.*, **1993**, 93(5), 1839.

[21] Sjostrand, U., Bohlin, L., Fisher, L., Colin, M., Djerassi, C. *Steroids*, **1981**, 38(5), 347.

[22] Schmitz, F.J., Campbell, D.C., Kubo, I. *Steroids*, **1976**, 28(2), 211

[23] Chan, W.R., Tinto, W.F., Moore, R. *Tetrahedron*, **1990**, *46*(5), 1499.
[24] Izac, R.R., bandurraga, M.M., Wasylik, J.M., Dunn, F.W., Fenical, W. *Tetrahedron*, **1982**, *38*(2), 301.
[25] Liu, Y., Mabry, T.J. *J. Nat. Prod.*, **1981**, *44*, 722.
[26] Rodriguez, A.D., Boulanger, A. *J. Nat. Prod.*, **1996**, *59*(7), 653.
[27] Rodriguez, A.D., Boulanger, A. *J. Nat. Prod*, **1997**, *60*, 207.
[28] Weinheimer, A.J., Washechek, P.H., van Der Helm, D., Hossain, M.B. *Chem.Comm.*, **1968**, 1070.
[29] Molchanova, V.I., Ovodova, R.G., Ovodov, Y.S., Elkin, Y.N. *Carbohydrate Research*, **1985**, *141*, 289.
[30] Miller, S.L., Tinto, W.F., Mclean, S., Reyndols, W.F., Yo, M. *J. Nat. Prod.*, **1995**, *58*(7), 1116.
[31] McEnroe, F.J., Fenical, W. *Tetrahedron*, **1978**, *34*, 1661.
[32] Wright, A.E., Pomponi, S.A., McConnol, O.J., Kohmoto, S., McCarthy, P.J. *J. Nat. Prod.*, **1987**, *50*(5), 976
[33] Freyer, A.J., Patil, A.A., Kilmer, A.D., Zuber, G., Myers, C., Johnson, R.K. *J. Nat. Prod.*, **1997**, *60*(3), 309.
[34] Look, S.A., Buchoolz, K., Fenical, W. *Experiencia*, **1984**, *40*, 931.
[35] Hale, R.L., Leckleq, J., Turch, B., Djerssa, C., Gross, R.A. Junior, Weinhelm, A.J., Gupta, K., Scheur, P.J. *J. Am. Chem. Soc.*, **1970**, *92*, 2179.
[36] Ling, N.C., Hale, R.L., Djerassi, C. *J. Am. Chem Soc.*, **1970**, *92*, 5281.
[37] Schmitz, F.J., Pattabhiraman, T. *J. Am. Chem Soc.*, **1970**, *92*, 6074.
[38] Enwall, E.L., Van Der Helm, D. *Recueil*, **1974**, *93*, 53.
[39] Spraggins, R.L .(1970), *PhD thesis*, Univ. Of Oklahoma, Norman, Oklahoma, USA
[40] Schultz, L.W., Clardy, J. *Acta Crys.*, **1995**, *C51*, 415.
[41] Kerr, R.G., Rodriguez, L.C., Kellmann, J. *Tetrahedron Letters*, **1996**, *37*(46), 8301.
[42] He, H., Kulanthaivel, P., Baker, B.J., Kalter, K., Darges, J., Cofield, D., Wolf, L., Adams, L.*Tetrahedron*, **1995**, *51*(1), 51.
[43] Jagodinska, B.M., Trimmer, J.S., Fenical, W., Djerassi, C. *J. Org. Chem.*, **1985**, *50*, 1435.
[44] Tarchini, C., Rohmer, M., Djerassi, C. *Helvetica Chimica Acta*, **1979**, *62*(40), 1210.
[45] Colegate, S.M., Molyneux, R.J. *In:Bioactive Natural Products*, CRC Press, Inc., 441- 456, 1993.
[46] Wirahadikusumah, M. (1980), *PhD thesis*, Univ. Of Oklahoma, Norman, Oklahoma, USA
[47] Bandurraga, M.M., Fenical, W., Donovan, S.F. *J. Am. Chem. Soc.*, **1982**, *104*(23), 6463.
[48] Rodriguez, A.D., Soto, J. *J. Chem. Pharm. Bull.*, **1996**, *44*(1), 91.
[49] Tinto, W.F., Chan, W.R., Reynolds, W.F., McLean, S. *Tetrahedron Letters*, **1990**, *31*(4), 465.
[50] Chan, W.R., Tinto, W.F., Laydoo, R.S., Marchand, P.S., Reynolds, W.F., Maclean, S. *J. Org.Chem.*, **1991**, *56*, 1773.
[51] Tinto, W.F., John, L., Lough, A., J, MacLean, S. *Tetrahedron Letters*, **1991**, *32*(36), 4661.
[52] Tinto, W.F., John, L., Reynolds, W.F., McLean, S. *Tetrahedron*, **1991**, *47*(41), 8679.
[53] Rayner, C.M., Astles P.C., Paquette, L.A. *J. Am. Chem. Soc.*, 1992, *114*, 3926.

[54]   Look, S.A., Burch, M.T., Fenical, W., Qi-Tai, Z., Clardy, J. *J. Org. Chem.*, **1985**, *50*, 5741.
[55]   Astles, P. C., Paquette, L.A. *Syn. Lett.*, **1992**, 444.
[56]   Wright, E.W., Burres, N.S., Schulte, G.K. *Tetrahedron Letters*, **1989**, *30*(27), 349.
[57]   John, L.M.D., Tinto, W.F., MacLean, S., Reynolds, W.F. *J. Nat. Prod.*, **1993**, *56*(1), 144.
[58]   Skjenstad, T., Haxo, F.T., Liaan-Jensen, S.L. *Biochemical Systematics and Ecology*, **1984**, *12*(2), 149.
[59]   Pawlik, J.R., Burch, M.T., Fenical, W. J. *Exp. Mar. Biol. Ecol.*, **1987**, *108*, 55.
[60]   Harvell, C.D., Fenical, W., Greene, C.H. J. *Exp. Mar. Biol. Ecol.*, **1987**, *49*, 287.
[61]   Harvell, C.D., Fenical, W. *Limnol. Oceanogr.*, **1989**, *34*, 382.

Atta-ur-Rahman (Ed.) *Studies in Natural Products Chemistry, Vol. 23*

BIOACTIVE SECONDARY METABOLITES FROM OKINAWAN SPONGES AND TUNICATES

JUN'ICHI KOBAYASHI* *and* MASAMI ISHIBASHI†

Faculty of Pharmaceutical Sciences, Hokkaido University, Sapporo 060-0812 Japan and †Faculty of Pharmaceutical Sciences, Chiba University, Chiba 263-8522, Japan

ABSTRACT: During our studies on new bioactive substances from marine organisms, we have isolated a number of unique secondary metabolites having unprecedented chemical structures as well as unique bioactivities. In this review we describe our recent results of isolation and structural elucidation of novel bioactive secondary metabolites isolated from Okinawan marine sponges and tunicates. The review is subdivided into the following five sections, 1) pseudodistomins and penaresidins: aliphatic amino alcohols with piperidine and azetidine rings, respectively, isolated from a tunicate *Pseudodistoma kanoko* and a sponge *Penares* sp., respectively; 2) shimofuridins: seven new nucleoside derivatives from an Okinawan compound tunicate *Aplidium multiplicatum*; 3) manzamenones and *Plakortis* metabolites: new dimeric oxygenated fatty acid-derived substances and a series of other oxygenated lipid metabolites, isolated from sponges of the genus *Plakortis*; 4) taurospongin A and acetylene acids: a novel acetylene-containing natural product consisting of a taurine and two fatty acid residues isolated from the Okinawan marine sponge *Hippospongia* sp. and other acetylenic fatty acids isolated from sponges of the genera *Petrosia* and *Xestspongia*; and 5) theonezolides: novel 37-membered macrolides consisting of two principal fatty acid chains, isolated from a sponge of the genus *Theonella*.

During our continuing research projects on the search for new bioactive substances from marine organisms mostly collected at Okinawa islands, we have isolated a number of unique secondary metabolites having unprecedented chemical structures as well as unique bioactivities, some of which are contributable to the basic studies of various fields of life sciences and the development of clinically useful drugs. We have previously published several review articles describing our studies related to marine natural products chemistry including the subjects of "marine alkaloids" [1], "bioactive metabolites of symbiotic marine microorganisms" [2], "sphingosine-related marine alkaloids" [3], "unique macrolides from marine dinoflagellates" [4], and "marine natural products and marine chemical ecology" [5]. In this review we describe our recent results of isolation and structural elucidation of novel bioactive secondary metabolites isolated from Okinawan marine sponges and tunicates.

PSEUDODISTOMINS AND PENARESIDINS

In 1986 we investigated the bioactive substances from an Okinawan tunicate *Pseudodistoma kanoko*, which is an orange-colored compound tunicate and looks like a strawberry (Japanese name, 'ichigo-boya'). The material was collected off Ie Island, Okinawa, by SCUBA (-5 to -10 m). The methanol-toluene (3:1) extract of *P. kanoko* was partitioned between toluene and water. The aqueous layer was successively extracted with chloroform, ethyl acetate, and 1-butanol. By preliminary screening using mammalian muscle preparations, the chloroform-soluble fraction was found to exhibit marked antispasmodic activity on the isolated guinea-pig ileum; the contractile responses to carbachol and histamine were abolished by this fraction. The chloroform-soluble fraction was therefore subjected to bioassay-guided fractionations using silica gel flash column chromatography eluted with $CHCl_3/n\text{-}BuOH/H_2O/AcOH$ (1.5:6:1:1) followed by the reversed-phase HPLC separation (Develosil ODS-5, 50% MeCN with 0.1% TFA) to afford active fraction, which was positive on ninhydrin test on TLC plate. This active fraction was revealed to be a mixture of two components [pseudodistomins A (**1**) and B (**2**)], the separation of which was first carried out after converting them into acetates (**3** and **4**, respectively) by ODS-HPLC (YMC-Pack, AM) with 88% MeOH. The acetates (**3** and **4**) were used for characterizations and structural studies. A small amount of **1** and **2** (before acetylation) were obtained by careful HPLC (Develosil ODS-5) eluting with 37% MeCN with 0.2% TFA to supply for bioassays. In addition to antispasmodic activity, pseudodistomins A (**1**) and B (**2**) exhibited cytotoxic activity against murine leukemia cells, L1210 and L5178Y, in vitro (IC_{50} values: 2.5 and 0.4 μg/ml against L1210, respectively; 2.4 and 0.7 μg/ml against L5178Y, respectively). Both compounds **1** and **2** also exhibited calmodulin

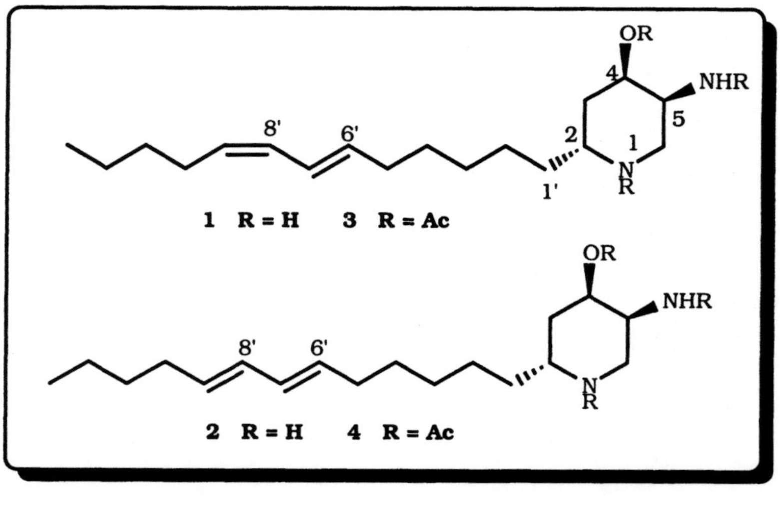

1 R = H   3 R = Ac

2 R = H   4 R = Ac

antagonistic activity; they both inhibited calmodulin-activated brain phosphodiesterase with IC_{50} values of 3 x 10^{-5} M, being approximately 3 times more potent than W-7, a well-known synthetic calmodulin antagonist [6].

Since the structures of pseudodistomins A and B (1 and 2) are reminiscent of sphingosine and marine aliphatic amino alcohols related to sphinganoids are of current interest of many scientists working in a broad range of biological sciences, the ninhydrin-positive fraction of the extract of *P. kanoko* was, after acetylation, further carefully examined by HPLC (Develosil ODS-5) eluting with 85% MeOH to afford a new piperidine alkaloid, named pseudodistomin C (5), as its acetate (6). Before acetylation, 5 was also obtained by preparative silica gel TLC (CHCl₃/MeOH/H₂O, 6:4:0.7), and was revealed to exhibit cytotoxicity against murine lymphoma L1210 and human epidermoid carcinoma KB cells *in vitro* (IC_{50} values, 2.3 and 2.6 µg/ml, respectively) [7].

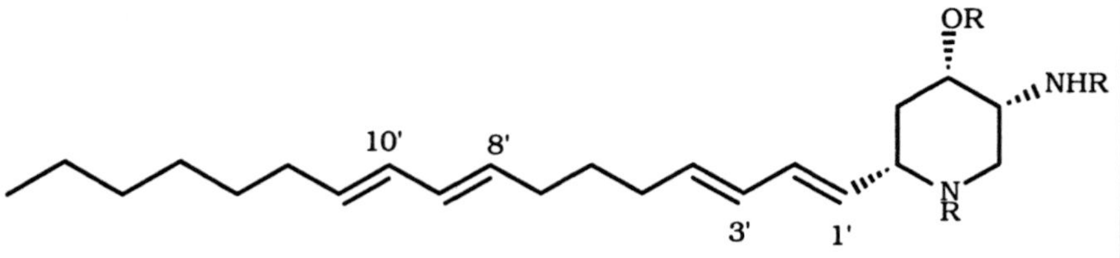

5 R = H   6 R = Ac

The ¹H NMR spectrum of the acetate (6) in CDCl₃ showed so broad signals that no signals were able to be assigned, and was quite different from the ¹H NMR spectrum of pseudodistomin B acetate (4). The 2D NMR experiments (¹H-¹H COSY, HSQC, and HMBC) of 6 were carried out in a CD₃OD solution, which showed relatively resolved signals. Since the ¹H NMR spectrum of natural compound (5) in a C₅D₅N solution appeared better in resolution, the ¹H-¹H COSY and HSQC spectra of 5 were recorded in this solution. From these spectral data, pseudodistomin C (5) was suggested to consist of a piperidine moiety and an unsaturated side-chain; the piperidine ring has the same substituents (4-hydroxyl and 5-amino groups) as those of pseudodistomins A and B (1 and 2), and the side-chain attached to C-2 contains two dienes. The positions of two dienes were clarified unambiguously by the following degradations. Pseudodistomin C (5) was treated with ozone followed by NaBH₄ reduction and acetylation to give a crude product, from which the tetraacetate [7, EIMS *m/z* 315 (M+H)⁺ and 255 (M − CH₃CONH₂)⁺] was obtained by HPLC purification, thus revealing one of the two dienes to be located at 1',3'-position. The second diene was deduced to be on 8',10'-

position since 1,5-pentanediol diacetate (**8**) was detected by reversed-phase TLC and HPLC analyses from the crude mixture of the ozonolysis products. These olefins were inferred to be all *E* from the coupling constants and the ^{13}C chemical shifts of the allylic methylenes.

Since the ^1H NMR spectrum of pseudodistomin C acetate (**6**) appeared quite different from that of pseudodistomin B acetate (**4**), a stereochemical evidence of the piperidine ring portion of pseudodistomin C (**5**) was required. Thus, the tetraacetate (**7**), which was obtained by ozonolysis of **5**, was prepared as an optically active form as shown in Scheme 1. Oxazolidine aldehyde (**9**), prepared from L-serine [8], was treated with allylmagnesium bromide to give a 1:1 diastereomeric mixture of allyl alcohols. After deprotection of acetonide group and conversion into pivaloyl ester, the unnecessary *threo*-isomer was removed by silica gel column chromatography. The *erythro*-monopivaloate (**10**) was transformed *via* 5 steps into a benzyl carbamate (**11**), which was subjected to amide mercuration to give 2*R*- and 2*S*-piperidine derivatives (**12** and **13**) in a ratio of 54:46. Oxidative demercuration of **12** and **13** gave primary alcohols (**14** and **15**, respectively), which were deprotected and acetylated to afford tetraacetates (L-7) and (**16**), respectively. The ^1H NMR spectrum of the tetraacetate (**7**) obtained from natural specimen of pseudodistomin C (**5**) was identical with that of the former [2*R*,4*R*,5*S*-derivative, L-7]. Since the latter tetraacetate (**16**) possessed the same relative configurations at C-2, 4, and 5 positions on the piperidine ring as those of pseudodistomins A (**1**) and B (**2**), relative configurations of pseudodistomin C (**5**) proved to be different from those of **1** and **2**. The sign of optical rotation of synthetic L-7 ($[\alpha]_D$ -19°) was opposite to that of the tetraacetate (**7**) ($[\alpha]_D$ +16°) derived from natural specimen of **5**. The absolute configuration of pseudodistomin C (**5**) was therefore revealed as 2*S*, 4*S*, and 5*R*. This result was, however, unexpected since the piperidine alkaloids isolated from the same tunicate possess different stereochemistries at C-4 and C-5 positions. To obtain further unambiguous confirmation of this conclusion, we prepared the enantiomer (D-7) from D-serine by the same procedures as above, and subjected it to chiral HPLC analysis (CHIRALPAK AD, Daicel Chemical Ind., Ltd.; 4.6 x 250 mm; flow rate: 0.5 mL/min; UV detection at 215 nm; eluent: hexane/2-propanol, 8:2), which established that the tetraacetate (7, t_R 16.5 min) derived from natural specimen (5) showed the same retention time as the enantiomer (D-7) prepared from D-serine (D-7: t_R 16.5 min; L-7: t_R 15.1 min), thus firmly establishing the 2S, 4S, and 5R-configurations for pseudodistomin C (5).

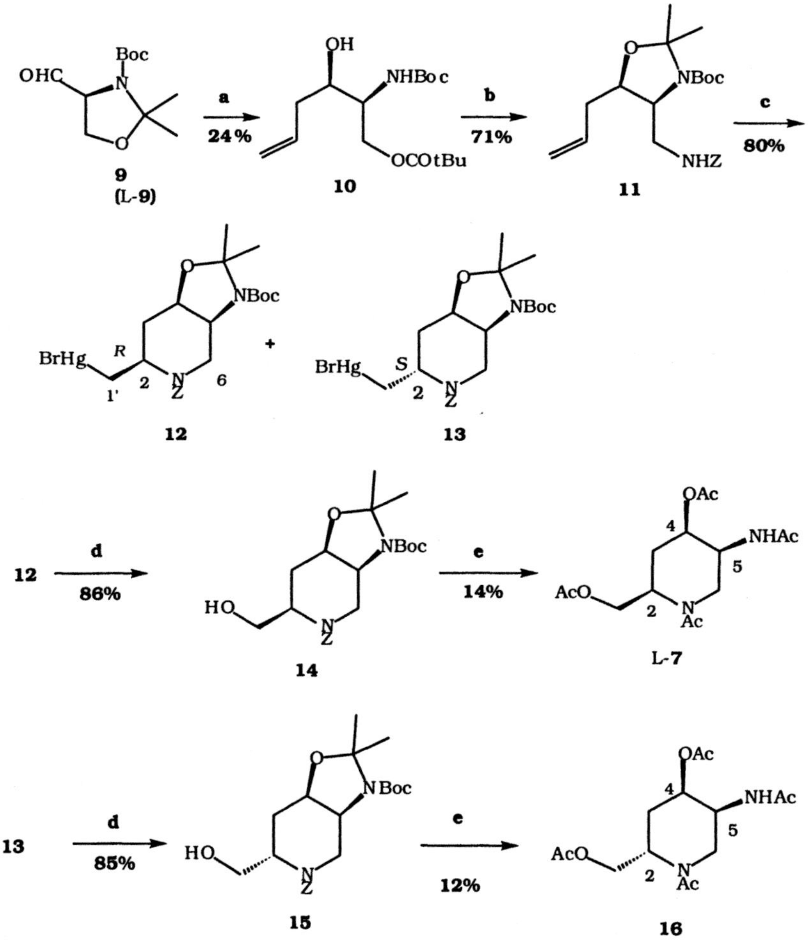

Scheme 1. (a) (1) CH_2=$CHCH_2MgBr$; (2) p-TsOH, MeOH; (3) PivCl, pyridine; (4) SiO_2 column, hexane/EtOAc (3:1); (b) (1) DMP, $BF_3.OEt$; (2) 2.5 N KOH, MeOH; (3) Phthalimide, DIAD, PPh3; (4) $H_2NNH_2.H_2O$, EtOH; (5) ZCl, 2N NaOH; (c) (1) $Hg(OCOCF_3)_2$, CHCl3; (2) NaHCO3; (3) NaBr; (d) NaBH4,O2, DMF; (f) (1) TFA, CH_2Cl_2; (2) Ac2O, pyridine; (3) H2, Pd/C, EtOH; (4) Ac2O, pyridine.

To provide a further unambiguous evidence for the whole structure of pseudodistomin C (**5**), total synthesis of **5** was investigated by us as follows [9]. Our synthesis began with Garner's aldehyde (D-**9**) derived from D-serine as summarized in Scheme 2. The Grignard reaction of allylmagnesium bromide with D-**9** afforded a 1:1 mixture of *erythro*- and *threo*-homoallyl alcohols. To obtain the *erythro*-alcohol (**18**) practically, the diastereomeric mixture was oxidized with Dess-Martin periodinane in DMF to give the ketone (**17**), which was reduced with $Zn(BH_4)_2$ to give the *erythro*-alcohol (**18**) in 96% de. The *erythro*-alcohol (**18**) was transformed into the *t*-butyl carbamate (**20**) via isomeric alcohol (**19**) in 7 steps by our previous method [7]; the terminal amine was protected by the Boc group to simplify the deprotection. Amide mercuration of **20** with $Hg(OAc)_2$ in $CHCl_3$ afforded (2*S*)-piperidine (**21**) and its (2*S*)-isomer in the ratio of 1.5:1. The (2*S*)-piperidine derivative (**21**) was oxidatively demercurated to give alcohol, which was treated with diphenyl disulfide

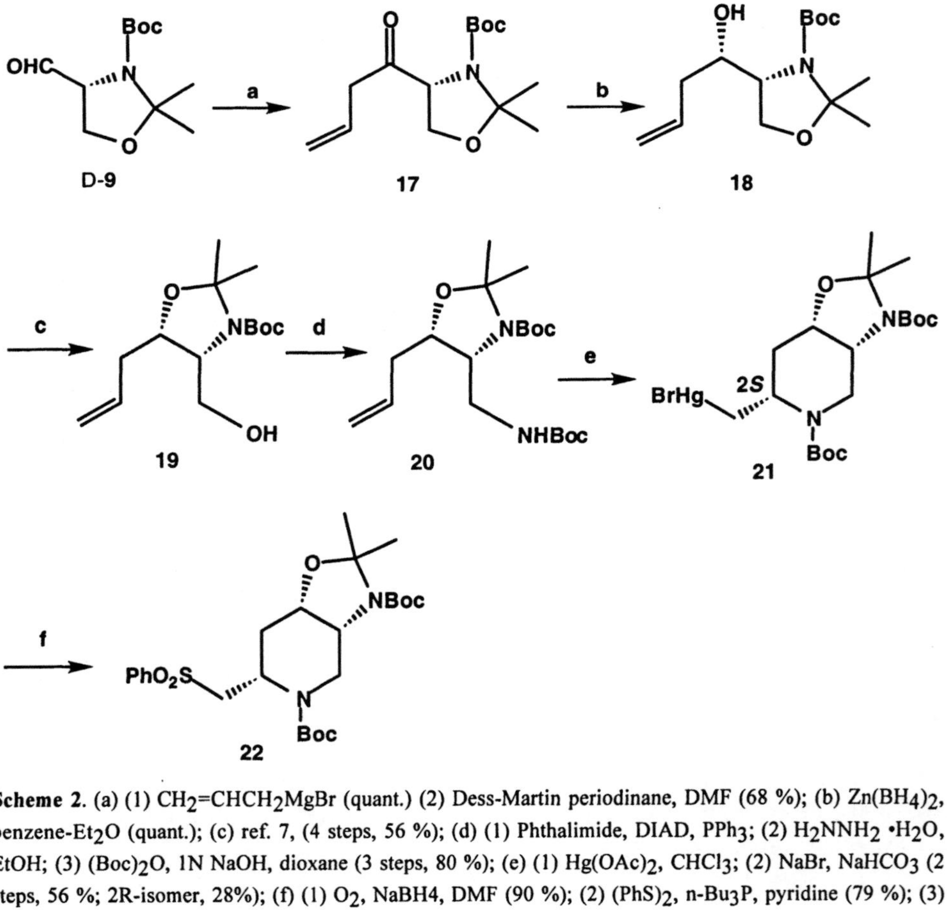

Scheme 2. (a) (1) CH_2=$CHCH_2MgBr$ (quant.) (2) Dess-Martin periodinane, DMF (68 %); (b) $Zn(BH_4)_2$, benzene-Et_2O (quant.); (c) ref. 7, (4 steps, 56 %); (d) (1) Phthalimide, DIAD, PPh_3; (2) $H_2NNH_2 \cdot H_2O$, EtOH; (3) $(Boc)_2O$, 1N NaOH, dioxane (3 steps, 80 %); (e) (1) $Hg(OAc)_2$, $CHCl_3$; (2) NaBr, $NaHCO_3$ (2 steps, 56 %; 2R-isomer, 28%); (f) (1) O_2, $NaBH_4$, DMF (90 %); (2) $(PhS)_2$, n-Bu_3P, pyridine (79 %); (3) Ph_2Se_2, 30% H_2O_2, CH_2Cl_2-Et_2O (73 %).

and tri-*n*-butylphosphine in pyridine followed by oxidation of the sulfide group with diphenyl diselenide and hydrogen peroxide to furnish phenylsulfone (**22**).

The side chain moiety of **5** was prepared as shown in Scheme 3, starting from known 1-bromo-3*E*,5*E*-decadienoyl bromide (**23**) [10]. Condensation of **23** and sodium dimethyl malonate afforded dimethyl ester (**24**), which was heated with sodium chloride in wet DMSO at 190 °C to afford the corresponding ester (**25**). Reduction of the ester group of **25** with DIBAL in toluene and Wittig reaction provided ethyl tetradecatrienoate (**26**) with all *E*-configurations revealed from coupling constants. The ester (**26**) was reduced with DIBAL and the resulting alcohol was oxidized with pyridinium chlorochromate (PCC) to afford the corresponding aldehyde, which was subjected to Julia olefination with the phenylsulfone (**22**). The sulfone (**22**) was treated with *n*-butyllithium in

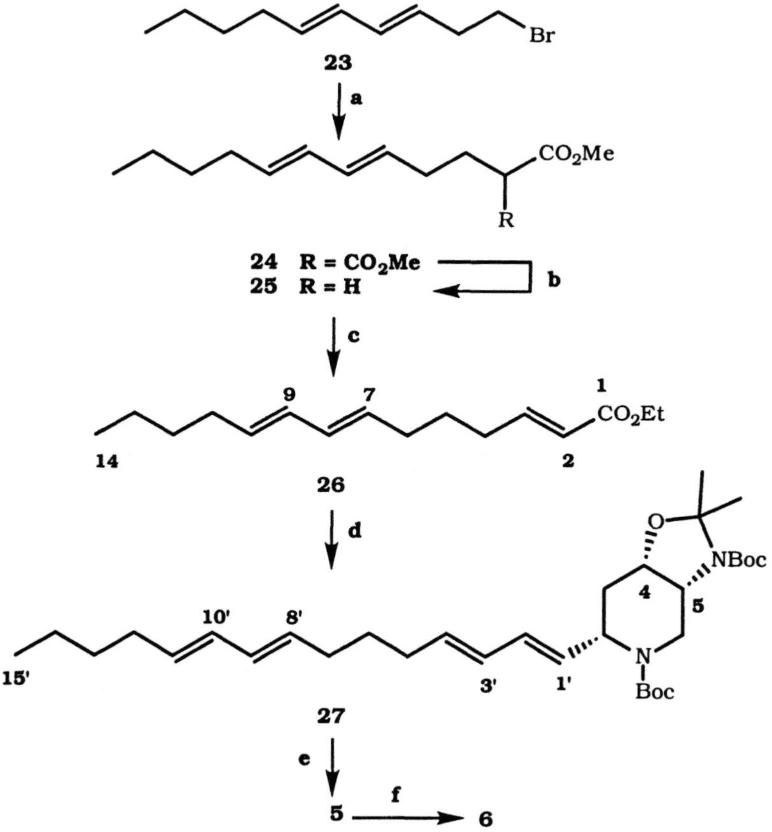

Scheme 3. (a) Na, MeOH, CH$_2$(CO$_2$CH$_3$)$_2$ (74 %); (b) (1) NaCl, DMSO, H$_2$O, 190 °C (74 %); (c) (1) DIBAL, toluene; (2) Ph$_3$P=CHCO$_2$Et, CH2Cl$_2$ (2 steps, 68 %); (d) (1) DIBAL, CH$_2$Cl$_2$ (80 %); (2) PCC, CH$_2$Cl$_2$ (74 %); (3) n-BuLi, 22, THF-HMPA; (4) Na-Hg, MeOH (13 % from 22); (e) 3N HCl, EtOAc (34 %); (f) Ac$_2$O, pyridine (56 %).

THF in the presence of HMPA at -78 °C to produce the orange sulfone anion, which was allowed to react with the aldehyde obtained from 26, and then quenched with benzoyl chloride to afford a diastereomeric mixture of β-benzoyloxy sulfones. Treatment of the crude mixture with sodium amalgam resulted in formation of the tetraene (27) possessing the backbone skeleton of pseudodistomin C (5). The ^{1}H NMR of 27 revealed that the last generated $\Delta^{1',2'}$-double bond was E ($J_{1',2'}$=14.0 Hz), and HPLC analysis of the tetraene (27) using reversed-phase column showed a single peak predominantly, suggesting that the tetraene (27) possesses all E-configurations. Removal of the protective groups of 27 with 3N HCl afforded pseudodistomin C (5), whose ^{1}H NMR and EIMS spectra as well as R_f values on TLC were completely identical with those of natural specimen [7]. The synthetic pseudodistomin C (5) was acetylated with acetic anhydride in pyridine to furnish pseudodistomin C triacetate (6), which was also identified with the triacetate (6) [7] derived from natural specimen of 5 on the basis of ^{1}H NMR and EIMS spectra as well as TLC and HPLC examinations. The sign of the optical rotation of synthetic triacetate (6) ($[\alpha]_D$ +43° (c 1, CHCl$_3$)) was also the same as that of natural one ($[\alpha]_D$ +85° (c 0.98, CHCl$_3$)).

Thus we found that our purpose of further structural confirmation of 5 by total synthesis was completed, although the absolute value of the optical rotation of synthetic triacetate (6) was smaller than that of natural specimen of 6. The optical purity of the synthetic compound was examined by means of chiral HPLC analysis after conversion of the corresponding alcohol derived from 21 into tetraacetate (7), which had been obtained by ozonolysis of 5 [7], to reveal that the synthetic tetraacetate (7) obtained in this study was 60% ee. The optical purity of synthetic 6 was estimated to be parallel to this result. This result may be attributable to partial racemization during oxidation-reduction process to obtain the *erythro*-alcohol (18). In our previous study [7], chiral HPLC had showed that no crucial racemization occurred since the *erythro*-alcohol (18) and its *threo*-isomer were separated after conversion into monopivaloyl esters (10 and its isomer) by 4-times repeated silica gel chromatographies.

Structures of pseudodistomins are suggestive that they are biogenetically classified as sphingosine derivatives, and a precursor of pseudodistomins A (1) and B (2) may be reasonably assumed to be D-(+)-*erythro*-sphingosine; the absolute configurations of 4R-hydroxyl and 5S-amino groups of 1 and 2 are coincident with those of corresponding positions of D-(+)-*erythro*-sphingosine containing 3R-hydroxyl and 2S-amino groups. It was, however, quite surprising that pseudodistomin C (5), isolated from the same tunicate as 1 and 2, possesses 4S,5R-configurations. Pseudodistomin C (5), therefore, has to be derived from unusual L-(-)-*erythro*-sphingosine (C$_{20}$-homolog), and the biosynthetic cyclization process of 1, 2, and 5 is assumed to give commonly the same stereochemistry at C-2 of the piperidine ring since all these compounds (**1**, **2**, and **5**) possess the same 2*S*-configuration.

During our studies on bioactive substances from marine sponges, we investigated extracts of an Okinawan sponge of the genus *Penares*, which was collected by netting at Unten Bay (-70 m), Okinawa island, in June 1987. The methanol extract of this sponge was partitioned between toluene and water, and the aqueous phase was subsequently extracted with chloroform, ethyl acetate, and *n*-butanol. From the toluene-soluble fraction a cytotoxic triterpene acid, penasterol [11], was isolated, while 6-bromoindole-3-acrylic acid (penaresin) was obtained from the chloroform-soluble fraction and was revealed to exhibit Ca-releasing activity in sarcoplasmic reticulum [12]. Further examination of the EtOAc-soluble fraction of this sponge by column chromatography on Sephadex LH-20 (MeOH/CHCl₃, 1:1) and silica gel (CHCl₃/*n*-BuOH/H₂O/AcOH, 1.5:6:1:1) resulted in the isolation of a mixture of two azetidine alkaloids, penaresidins A (**28**) and B (**29**) [13], in the ratio of *ca.* 1.5:1. Separation of this mixture using silica gel and ODS reversed-phase HPLC was unsuccessful even after conversion of them into tetraacetates (**30** and **31**, respectively). The mixture of penaresidins A and B (**28** and **29**) exhibited potent actomyosin ATPase-activating activity; the mixture elevated the ATPase activity of myofibrils from rabbit skeletal muscle to 181% of the control value at 3 x 10⁻⁵ mol/L. Very few substances are hitherto known as those which modulate the ATPase activities of myosin and actomyosin. Characterization and structural studies of penaresidins were carried out mostly using the mixture of tetraacetate (**30** and **31**). The presence of the azetidine skeleton was suggested by analyzing the ¹H-¹H COSY and HMBC spectra of the mixture of **30** and **31**. The relative stereochemistry of the azetidine ring portion was deduced from the coupling constants as 2,3-*trans* and 3,4-*cis*. The positions of the acetoxyl and secondary methyl groups were also indicated from the COSY and HMBC spectra.

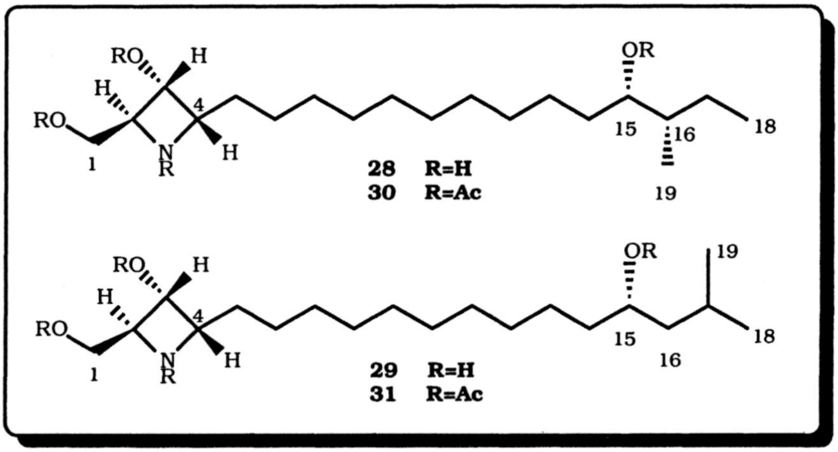

Penaresidins A and B (**28** and **29**) seems to be biogenetically derived from sphingosine or phytosphingosine through cyclization of N-2 to C-4. Natural products containing this type of sphingosine-related azetidine ring are unprecedented. The third compound belonging to this azetidine alkaloid group, named penazetidine A (**32**) [14], was isolated from a potato-shaped Indo-Pacific marine sponge, *Penares sollasi,* and was described to exhibit protein kinase C inhibition activity (IC_{50} 1 μM) and no inhibitory activity against protein tyrosine kinase. It seems interesting because sphingosine is known as a protein kinase C inhibitor. Penazetidine A (**32**) was reported to show in vitro cytotoxicity against human and murine cell lines as well.

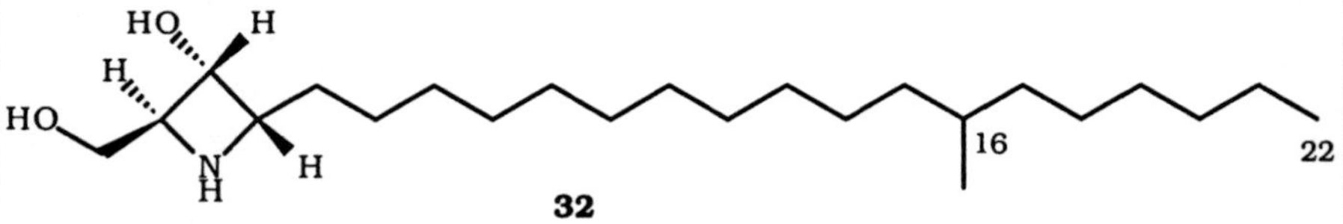

Because of their unprecedented structure and biological activity, synthesis of penaresidins is investigated by plural groups. Kamikawa and coworkers recently achieved the synthesis of a C16 analog (**33**) of penaresidins [15].

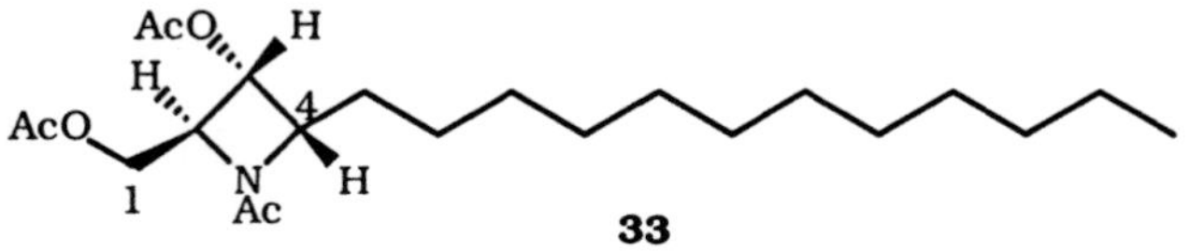

Mori and coworkers studied the synthesis of penaresidins [16], and showed that our initially proposed structure of penaresidin B (**29a**) with 14-hydroxyl and 16-methyl groups had to be revised to **29** possessing 15-hydroxyl and 17-methyl groups [17].

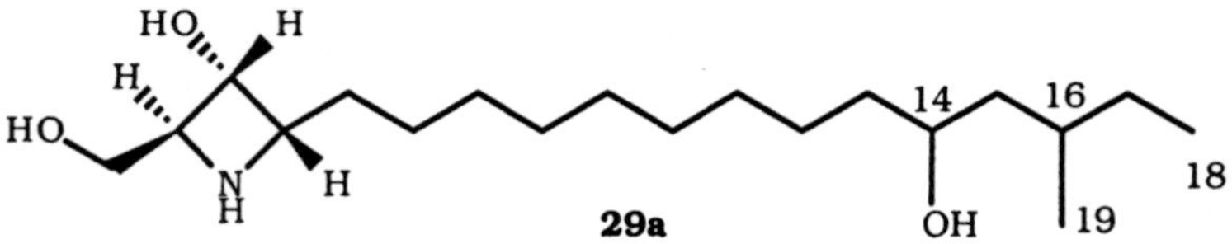

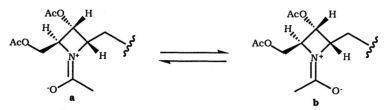

Fig. (1). Comparison of the ^{13}C NMR Chemical Shifts of Penaresidins A and B Acetates (**30** and **31**) with Those of Isoleucine and Leucine.

We reexamined the ^{13}C NMR data of mixture of penaresidin A and B acetates (**30** and **31**), and found that the ^{13}C NMR chemical shifts of the C-16 ~ C-19 positions of **30** and **31** corresponded well to those of corresponding positions of isoleucine and leucine [18], respectively (Fig. (1)). We therefore agreed to the conclusion given by Mori and coworkers. Mori and coworkers also encountered the phenomenon that signals for two isomers were observed in the ^{1}H NMR of the azetidine acetates; they proposed that this phenomenon were attributed to the geometrical isomers due to the amide bond (**a** and **b**, Fig. (2)).

Fig. (2). Isomerization of Azetidine Acetates.

Absolute stereochemistries of penaresidins A (**28**) and B (**29**) were determined to be 2S, 3R, 4S, 15S, and 16S for **28** and 2S, 3R, 4S, and 15S for **29**, respectively, on the basis of ^{1}H NMR data of their tri-O-MTPA esters [19]. Although the mixture of **28** and **29** as well as their acetates (**30** and **31**) was inseparable [13], the tri-O-MTPA derivatives of **28** and **29** were able to be completely separated by C_{18} HPLC (MeOH/H_2O). In addition, the spectral data of the tri-O-(R)-MTPA esters of natural penaresidins A (**28**) and B (**29**) proved to be identical with those of the tri-O-(R)-MTPA esters derived from synthetic (2S,3R,4S,15S,16S)-penarasidin A and (2S,3R,4S,15S)-penaresidin B, respectively.

In 1997 Yoda and coworkers reported another synthetic process for the preparation of an azetidine ring with contiguous stereogenic centers and the total synthesis of penaresidin B by featuring the elaboration of the functionalized homochiral lactam derived from D-glutamic acid [20]. Knapp and Dong also described the synthesis of several sphingosine-like azetidine alkaloids from the Garner aldehyde (**9**, see Scheme 1), including penaresidin A (**28**), penazetidine A (**32**), and their analogs such as allo-penaresidin A (**34**), 16-nor-penazetidine A (**35**), and mini-penazetidine (**36**) [21].

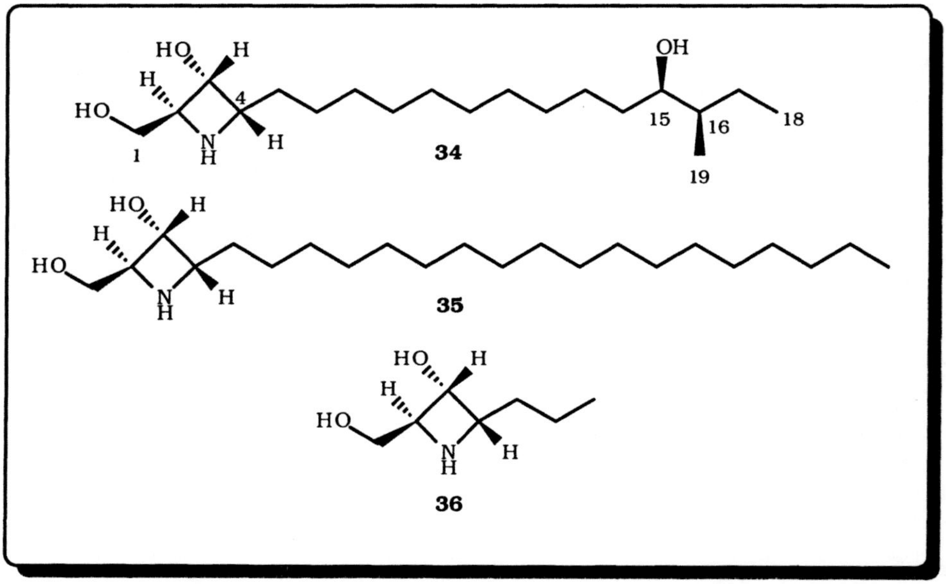

SHIMOFURIDINS

From the Okinawan marine tunicate *Aplidium multiplicatum*, shimofuridin A (**37**) [22], a nucleoside derivative embracing an acylfucopyranoside unit, was isolated and its structure including all absolute configurations was determined by spectral and chemical means, including degradation experiments and chiral HPLC and GC analyses. The tunicate *A. multiplicatum*, collected off Okinawa Island, and the Japanese name of this tunicate is "minami-shimofuri-boya", after which the compound is named. The EtOAc-soluble material of the MeOH extract of this tunicate was subjected to silica gel flash column chromatography (MeOH/CHCl$_3$, 30:70) followed by gel filtration on Sephadex LH-20 (MeOH/CHCl$_3$, 1:1) and reversed-phase HPLC (ODS, 70% MeOH) to give shimofuridin A (**37**, 0.002% wet weight). Extensive analysis of 2D NMR spectra (^{1}H-^{1}H COSY, HSQC, HMBC, and NOESY) suggested that shimofuridin A (**37**) consisted of a purine base, two sugar units, and two unsaturated fatty acids. A nucleoside unit (**38**) was obtained in the water-soluble fraction of

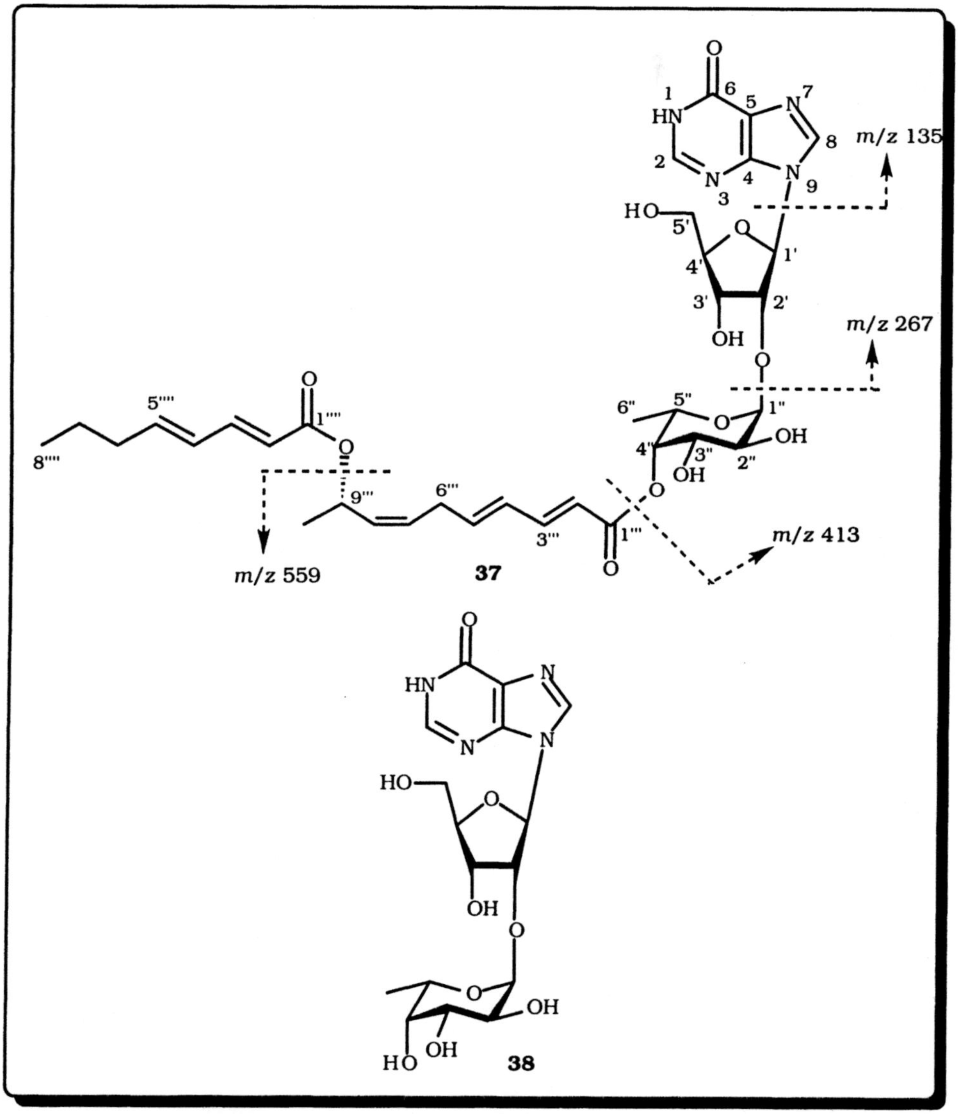

the alkaline hydrolysis product of **37**. The purine base of **37** was suggested as hypoxanthine from the UV absorption data of **38** [λ_{max} 244 nm (MeOH, pH 7); 252 nm (pH 2); 253 nm (pH 11)] as well as the fact that the ^{13}C NMR data of the purine moiety of **37** corresponded well to those of the purine base of inosine. The ^1H-^1H COSY spectrum of **37** suggested a pentafuranose and a 6-deoxyhexapyranose for the two sugar units, which were firmly identified as ribose and fucose, respectively, on the basis of GC analysis of the trimethylsilyl derivatives of the methanolysis product of the nucleoside moiety of **38**. The HMBC spectrum of **37** showed cross-peaks due to ^1H-^{13}C long-range correlations for H-1'/C-4 and H-1'/C-8, thus indicating that N-9 of hypoxanthine is connected to C-1' of ribose. The ^{13}C NMR chemical shifts for the ribose unit of **37** were analogous to those for the sugar part of inosine except for that of C-2' position (**37**: δ_C 80.7; inosine: δ_C 73.7), implying that the second sugar unit, fucose, is attached to this position, which was further confirmed by the HMBC correlations observed for H-2'/C-1" (the anomeric position of fucose) and H-1"/C-2'. The low-field resonance (δ_H 4.94) of H-4" of fucose implied that C-4" was acylated and this fact was supported by the HMBC connectivity between H-4" and C-1''' (ester carbonyl at δ_C 165.5).

The first unsaturated fatty acid chain contained in compound **37** was inferred to be 9-oxygenated deca-2,4,7-trienoyl group by analysis of the proton-connectivities observed in the ^1H-^1H COSY spectrum of **37**. H-9''' of this fatty acid unit resonated at δ_H 5.59 and this chemical shift suggested that C-9''' is further acylated by the second fatty acid chain, which was identified as octa-2,4-dienoyl group by the ^1H-^1H COSY data. The double bond geometries in the two unsaturated chains were elucidated as follows. The 2'''E- and 2''''E-configurations were deduced from the ^1H-^1H coupling constants ($J_{2''',3'''}$=15.2 Hz and $J_{2'''',3''''}$=15.6 Hz). The ^{13}C NMR chemical shifts for C-2'''' ~ C-8'''' positions of the second acyl chain agreed well with those for corresponding positions of 2E,4E-octadienoic acid. Particularly for the allylic position, C-6'''' of **37** was observed at δ_C 34.4, while C-6 of 2E,4E-octadienoic acid at δ_C 35.0. Thus 4''''E-configuration was revealed since the allylic carbons of cis-double bonds are known to resonate approximately at 27 ppm. The homo-spin decoupling experiment irradiating at H$_2$-6''' (the bis-allylic position of the first acyl chain) in CD$_3$OD revealed the $J_{7''',8'''}$-value to be 10.7 Hz, implying 7'''Z-configuration. This finding was supported by the NOESY cross-peak in CD$_3$OD observed between H$_2$-6''' and H-9''''. The bis-allylic carbon (C-6''') resonated at δ_C 30.8, which suggested that C-6''' is located between one cis and one trans olefins. The ^{13}C NMR chemical shifts for bis-allylic carbons between cis-cis, cis-trans, and trans-trans olefins are approximately 25, 30, and 35 ppm, respectively [23]. The $\Delta^{4''',5'''}$-double bond was therefore assigned as E.

The stereochemistries of the anomeric positions of the sugar moieties were assigned to be β for the ribofuranose and α for the fucopyranose, respectively, on the basis of the one-bond ^1H-^{13}C coupling constants recorded by INEPT experiment of **37** in CD$_3$OD [C-1' (ribose), $^1J_{C,H}$=150.7 Hz; C-1" (fucose), $^1J_{C,H}$=174.6 Hz], compared with the following data: (i) C-1' of adenosine (β-anomer), $^1J_{C,H}$=150.7 Hz; C-1' of α-adenosine (α-anomer), $^1J_{C,H}$=145.2 Hz; (ii) for hexapyranoses: ^1J (C, equatorial H), ca. 170 Hz; ^1J (C, axial H), ca. 160 Hz [24]. This finding for β-ribofuranose unit was consistent with the ROESY correlation observed between H-1' and H-4' in DMSO-d_6 solution of **37**. The absolute configurations of the ribose and fucose units were determined as D and L, respectively, by chiral HPLC (SUMIPAK OA-4000, 4.6 x 250 mm; flow rate, 1.0 mL/min; UV detection at 254 nm; hexane/EtOH, 70:1 ~ 199:1) and EIMS analysis using O-benzoyl derivatives of the methanolysis product of the nucleoside portion (**38**).

The EtOAc-soluble fraction of the alkaline hydrolysate of **37** containing a mixture of fatty acid units was treated with ozone followed by NaBH$_4$ reduction. Trimethylsilyl (TMS) derivatives of the ozonolysis product were subjected to chiral GC analysis (Chirasil-Val) to give a peak identical with the TMS-derivative of (S)-1,2-propanediol. Consequently, the absolute configuration of C-9''' position of **37** was assigned as S.

From all of these results the structure of shimofuridin A was concluded as inosine 2'-α-L-4''-[9'''S-(2''''E,4''''E-octadienoyloxy)-2'''E,4'''E,7'''Z-decatrie-noyl]-fucopyranoside (**37**). This structure was fully consistent with the fragment ions observed in the negative FABMS at m/z 559 (M - octadienoyloxy group)⁻, 413 (inosine α-L-fucopyranoside - H)⁻, 267 (inosine - H)⁻, and 135 (hypoxanthine - H)⁻. The conformation of the inosine unit of **37** was deduced to be *syn* from the cross-peak for H-8/H-1' observed in both of the ROESY in DMSO-d_6 and the NOESY in CD$_3$OD.

Shimofuridin A (**37**) exhibited cytotoxicity against murine lymphoma L1210 cells with an IC$_{50}$ value of 9.5 μg/mL in vitro, and antimicrobial activity against fungus *Trichophyton mentagrophytes* (MIC value, 133 μg/mL) and Gram-positive bacterium *Sarcina lutea* (MIC, 66 μg/mL). Compound **37** also showed endothelin converting enzyme (ECE) inhibition activity (31.2% at 100 μg/mL).

Further examination of the extract of this tunicate *A. multiplicatum* was carried out by HPLC using several types of reversed-phased columns (ODS, Ph, TMS, and NH$_2$) with various solvent systems, and separation of minor analogs was successfully achieved only when a column of Develosil ODS-HG-5 was used and eluted with 43 ~ 48% CH$_3$CN to give shimofuridins B (**39**, 0.001% wet weight), C (**40**, 0.003%), D (**41**, 0.002%), E (**42**, 0.002%), F (**43**, 0.0008%), and G (**44**, 0.0002%) [25] together with the major component, shimofuridin A (**37**). Shimofuridins B ~ E (**39** ~ **42**) were stereoisomers of **37** in the olefins of the acyl side-chains, while shimofuridins F (**43**) and G (**44**) were homologs of **37** with two more carbons in the second acyl chain.

The structures of shimofuridins B ~ G (**39** ~ **44**) are related closely to one another and the isolation procedure required careful operations to avoid air-oxidation or photoisomerization; all HPLC operations were carried out using well-degassed solvents under light-blocked conditions. Shimofuridin A (**37**) proved to be stable for several days in MeOH solution under room light. Formation of the isomers **39** ~ **42** was not detected by HPLC examination. We therefore concluded that **39** ~ **42** are not necessarily artificially produced compounds during isolation

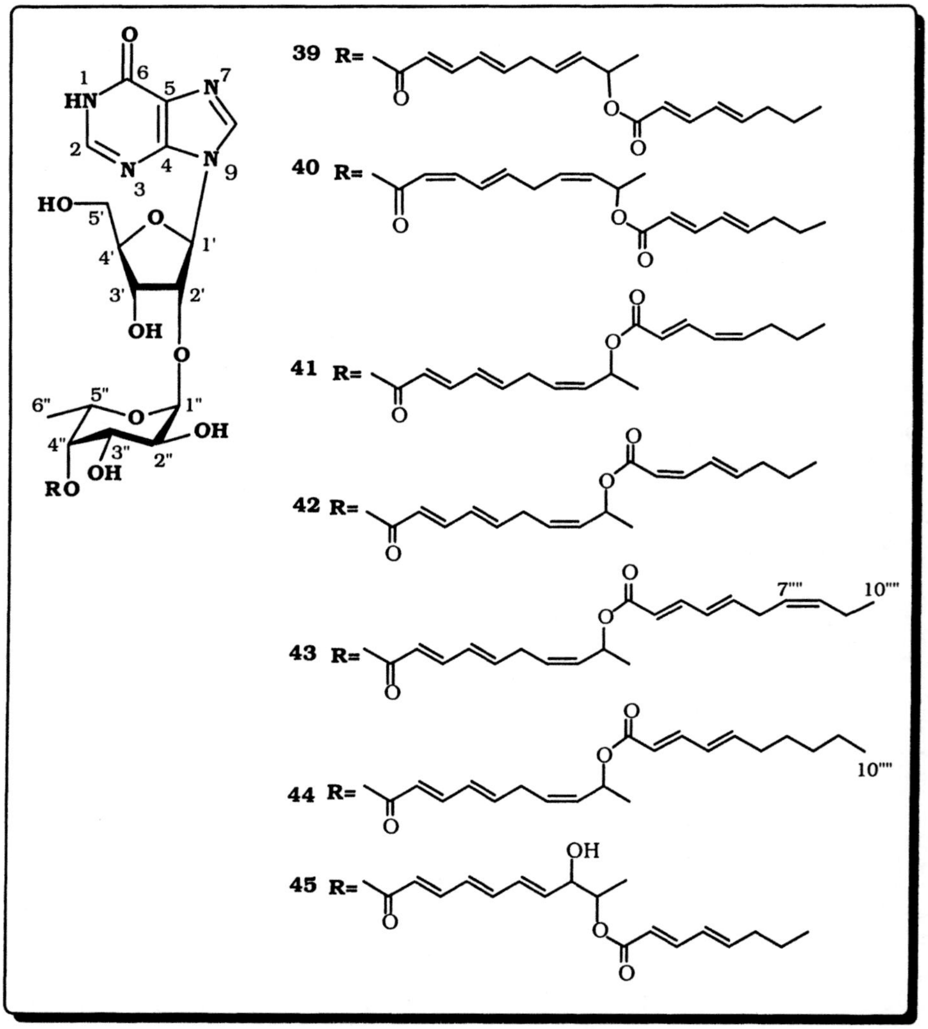

processes, while shimofuridins F (43) and G (44) are definitely natural products. Shimofuridin A (37), however, suffered from considerable decomposition under irradiation with 400 W high-pressure mercury lamp in MeOH solution. HPLC analysis showed formation of compounds 39 ~ 42, with the ratio of shimofuridins A ~ E (37 and 39 ~ 42) after 5 min to be 1:0.03:0.16:0.23:0.07. After 20-h irradiation, all of compounds (37 and 39 ~ 42) disappeared to yield a complex mixture, from which a peak due to an allyl alcohol (45) was detected by HPLC analysis. Compound 45 was isolated previously during HPLC separation of shimofuridins, in spite of taking care of air, and characterized as an air-oxidized product. Degradation experiments of 39 ~ 44 to elucidate the absolute stereochemistries were not carried out due to paucity of the samples. We, however, assume that the absolute stereochemistries of **39** ~ **44** were parallel to that of **37** because of the similarity of the spectral and optical data as well as formation of **39** ~ **42** from **37** via photoreaction.

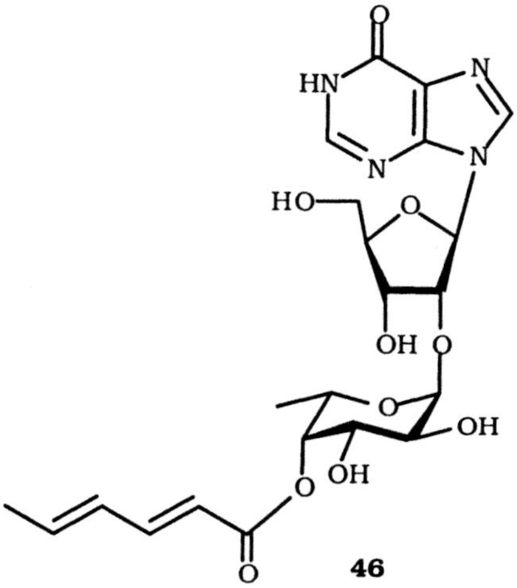

Studies on the synthesis of shimofuridin derivatives have been reported by two groups. van Boom and coworkers described a synthesis of a shimofuridin analog, 2'-*O*-[(4'-*O*-sorboyl)-α-L-fucopyranosyl]inosine (**46**), through a trimethylsilyl triflate mediated glycosylation as a key reaction [26]. Knapp and Gore reported the synthesis of shimofuridin nucleoside disaccharide (**38**) through fucosylation of a suitably protected inosine derivative by activation with *N*-iodosuccinimide and triflic acid [27].

MANZAMENONES AND *PLAKORTIS* METABOLITES

Marine sponges of the genus *Plakortis* have proven to be a rich source of a variety of unique bioactive compounds such as polycyclic aromatic alkaloids or peroxy aliphatic acids and esters [28]. The major peroxy ester contained in the sponge *Plakortis lita* was reported as chondrillin (**47**) [29, 30], which was initially isolated from a sponge of the genus *Chondrilla* [31], and exhibited cytotoxicity against P388 cells with IC_{50} value of 5 μg/mL [29]. We also isolated another peroxy ester, plakorin (**48**) [32], from the Okinawan marine sponge *Plakortis* sp., collected at Kerama Islands. Plakorin (**48**) was a sarcolasmic reticulum (SR) Ca^{2+}-ATPase activator; it activated SR Ca^{2+}-ATPase activity by 30% at 10^{-5} M. Plakorin (**48**) may provide a valuable chemical tool for studying the molecular mechanism of Ca^{2+} transport by Ca^{2+}-ATPase in the SR. Plakorin (**48**) also showed cytotoxicity against murine lymphoma L1210 cells and human epidermoid carcinoma KB cells in vitro with the IC_{50} values of 0.85 and 1.8 μg/mL, respectively. Syntheses of these peroxy esters were investigated [33], and in 1997 the absolute stereochemistry of chondrillin (**47**) was revised as 3*R* and 6*S* on the basis of the asymmetric total synthesis of its enantiomer [34].

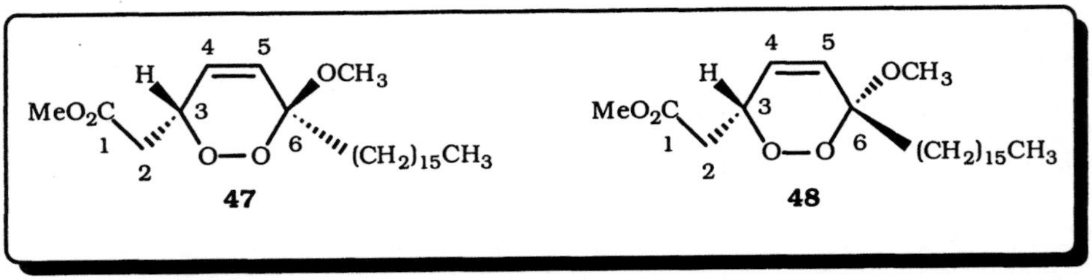

From the Okinawan *Plakortis* sponge, collected at Unten-harbor, we isolated a cytotoxic metabolite, named plakotenin (**49**) [35], which was active against L1210 and KB cells with the IC_{50} values of 5.4 and 7.4 μg/mL, respectively. Plakotenin (**49**) could be assumed as being biosynthetically derived from a linear precursor, which may be classified as a polypropionate and closely related to plakinic acid B (**50**) [36], through intramolecular [4+2] cycloaddition. Similar biosynthetic path has been proposed for ircinianin (**51**) [37], a furanosesterterpene from a marine sponge *Ircinia* sp.

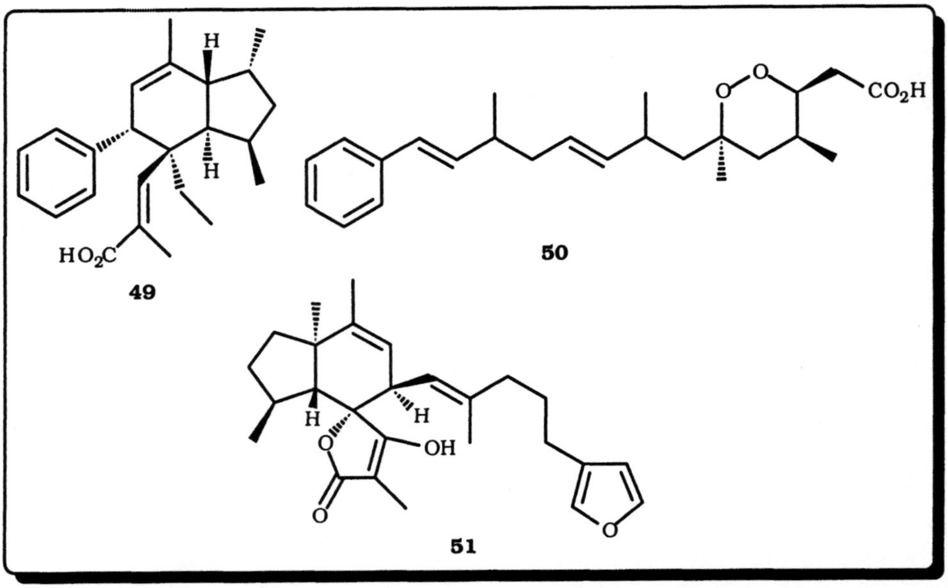

Our further investigations on the Okinawan sponges of the genus *Plakortis*, collected at Unten-harbor and off Manzamo at Okinawa Island, have resulted in isolation of a new class of lipid metabolites, manzamenones [38, 39, 40], which possess an unsymmetrical dimeric structure most likely composed of two molecules of fatty-acid derived precursors. Manzamenone A (**52**) contains unique bicyclo[4.3.0]nona-3,8-dien-7-one moiety with two long alkyl side chains (two hexadecyl groups), which was shown by extensive spectral studies using its methyl ester (**53**) prepared by treatment of **52** with diazomethane. The relative stereochemistry of manzamenone A (**52**) was revealed by difference NOE experiments of **53**, which showed NOE correlations for H-1/H-6, H-6/H-5, and H-5/H-4, implying that the H-1, H-6, and H-5 were on the same side of the cyclohexene ring and bridgehead protons (H-1 and H-6) were therefore oriented cis. Manzamenone B (**54**) was suggested as a stereoisomer of manzamenone A (**52**), and the proton-proton coupling constants of **53** and **55** (methyl ester from **54**) for H-1/H-6, H-6/H-5, and H-5/H-4 were almost the same, while those between H-1 and H-2 were quite different (6.0 and 0.5 Hz, respectively). In the difference NOE experiment of **55** irradiation of H-2 caused a significant NOE for H-1. Thus the configuration at C-2 position was different; H-1 and H-2 were trans for **53** and cis for **55**. These findings were coincident with the fact that treatment of **53** and **55** with 1 equiv of osmium tetraoxide afforded a 8,9-diol (**56**) and a 3,4-diol (**57**), respectively. The convex side of the $\Delta^{3,4}$-double bond of **53** was hindered by the methoxycarbonyl group on C-2, whereas this steric effect was small for **55**. To study the absolute stereochemistry of manzamenone A (**52**), the CD spectrum of the tetrahydroderivative (**58**) of the methyl ester (**53**) was recorded to show a negative Cotton effect [λ_{ext} 325 nm ($\Delta\varepsilon$ -0.27) and 286 nm ($\Delta\varepsilon$ +1.0)]. Based on the chiroptic data of *cis*-hexahydroinda-1-ones [41], 6*S*-configuration was inferred for **58**. Further investigations, however, were required to confirm the assignment since the structure of **58** is much different from those in the literature [41]. Recently Kusumi and coworkers demonstrated a new methodology to determine the absolute configurations of secondary carboxyl groups based on the ^{1}H NMR data of their amides of (*R*)- and (*S*)-phenylglycine methyl esters (PGME), which was evolved from modified Mosher's method [42]. Accordingly, manzamenone A (**52**) was treated with (*R*)- and (*S*)-PGME in the presence of BOP reagent and Et₃N in CH₃CN. Although partial racemization at the PGME moiety occurred during the reaction, (*R*)- and (*S*)-PGME amides of **52**, (*R*)- and (*S*)-**59**, respectively, were obtained after HPLC separation. The $\Delta\delta$ (δ_S - δ_R) values obtained from the ^{1}H NMR data of (*R*)- and (*S*)-**59** in CDCl₃ are affixed on the drawing of **59**, which implied 5*R*-configuration for **52**, being consistent with the inference from the CD data of **58**.

52 R=H
53 R=Me

54 R=H
55 R=Me

56

57

58

59

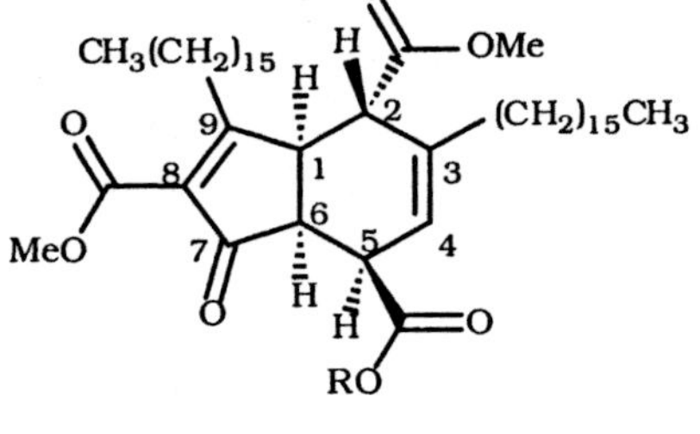

By further examination of the extracts of the same *Plakortis* sponges, manzamenones C (**60**), D (**61**), E (**62**), F (**63**) [38], G (**64**), H (**65**) [38], J (**66**), and K (**67**) [40] were also isolated. Manzamenones C ~ F (**60** ~ **63**) bore different esters or amides at C-5 carboxylic acid moiety, embracing ethyl ester, amide, L-valine, and *n*-butyl ester groups, respectively. Manzamenone G (**64**) possesses a previously unprecedented skeleton containing a bicyclo[4.4.0]decane ring system, while manzamenone H (**65**) contains a tyramine group at C-5 carboxylic acid moiety of manzamenone A (**52**) through an amide linkage.

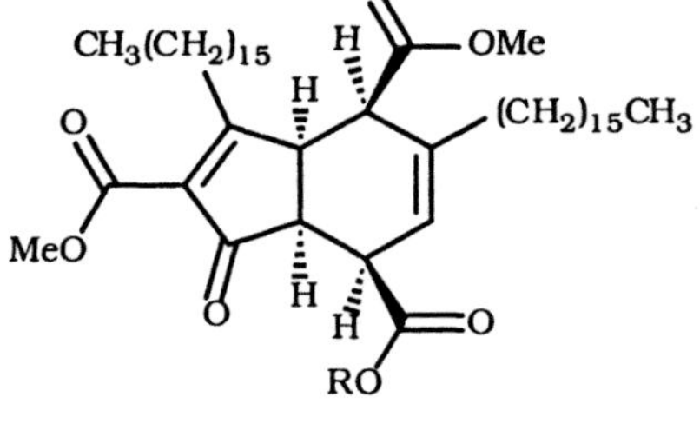

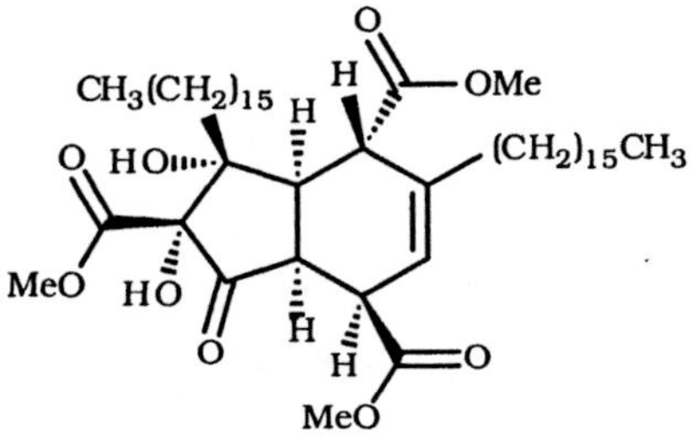

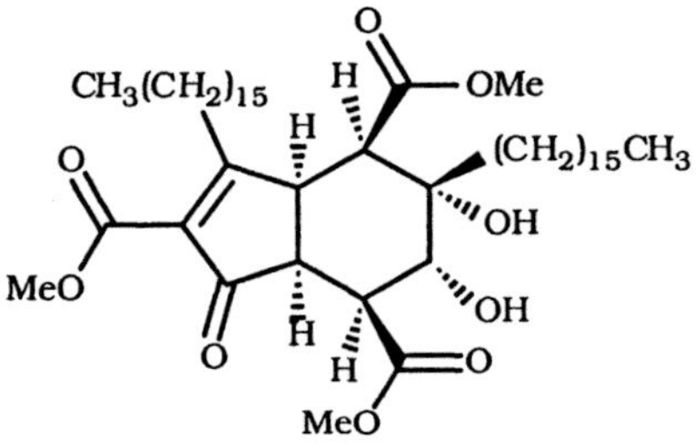

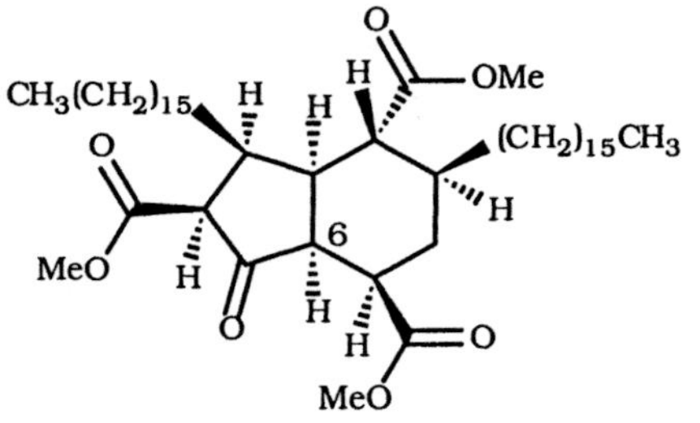

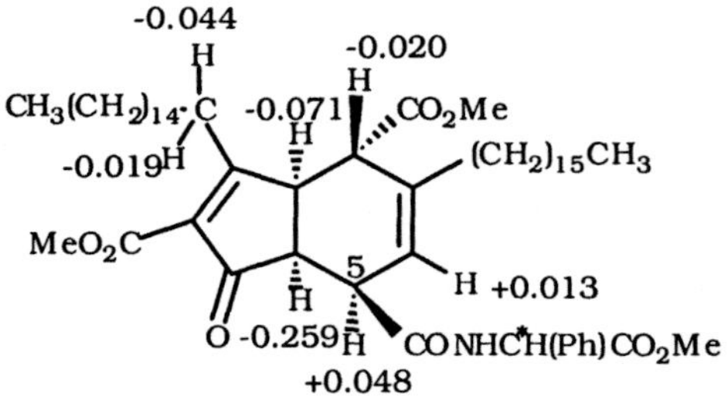

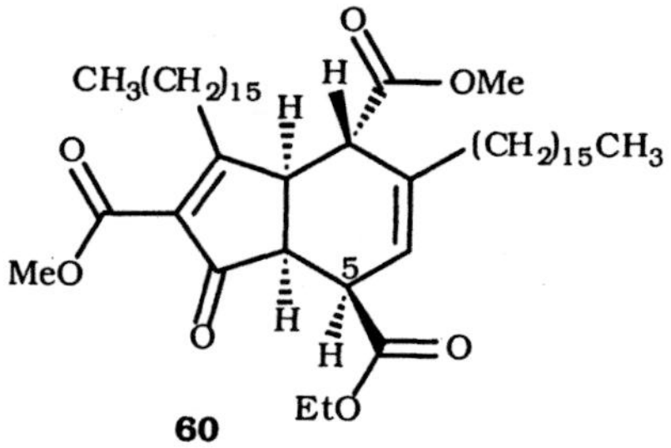

60

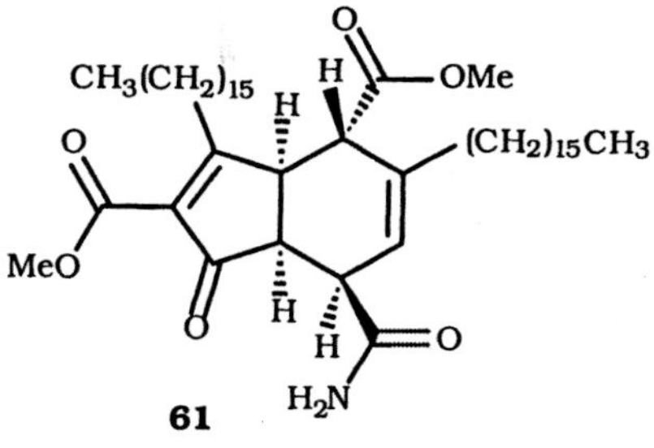

61

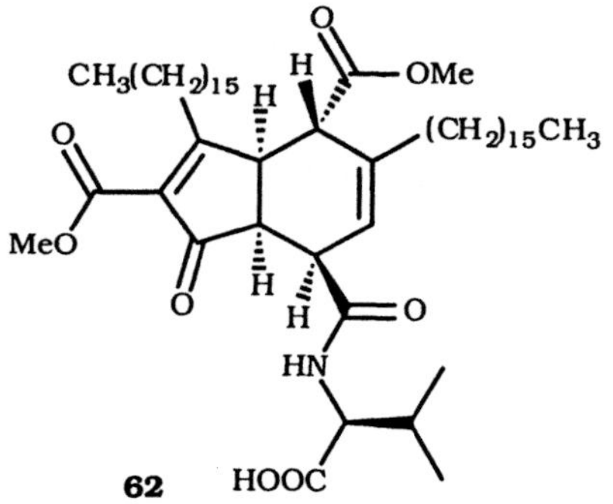

62

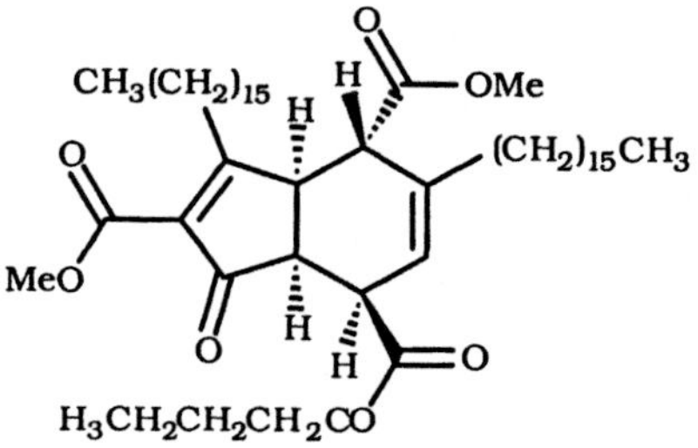

63

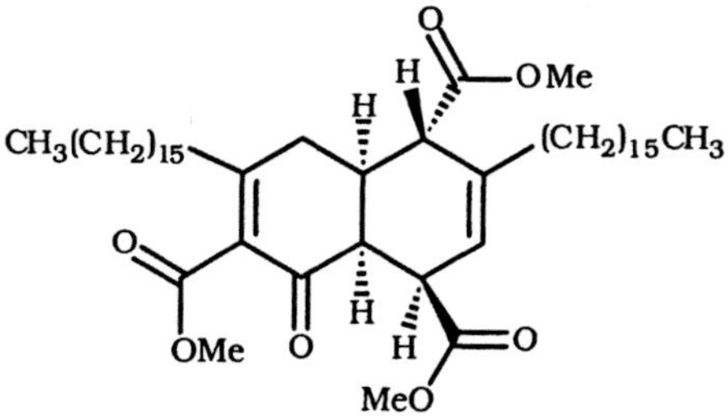

64

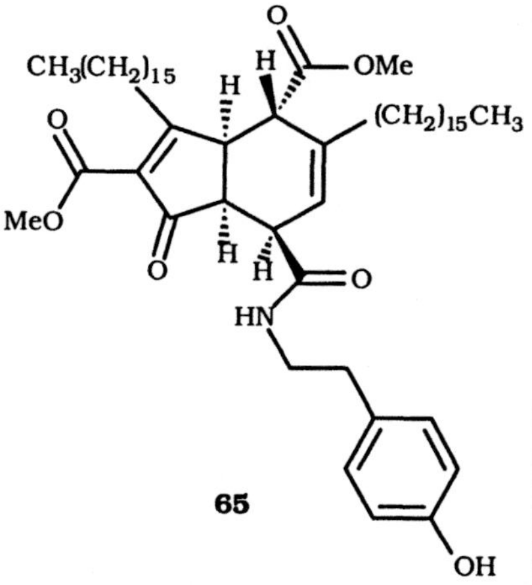

65

Manzamenone J (**66**) had the same composition as manzamenone D (**61**) with a conjugated dienone (λ_{max} 317 nm) functionality. Analysis of several types of 2D NMR spectra of **66** led to a bicyclo[4.3.0]nonane ring system involving a 1(6),2-dien-7-one chromophore. The stereochemistry of the vicinal methine protons on C-8 and C-9 were suggested to be anti from the NOESY spectrum of **66** showing substantial correlations from H-8 to H_2-26 and H_2-27. The stereochemistry of C-5 position relative to C-8 and C-9 portion remained unassigned.

Manzamenone K (**67**) was shown to be composed of two moieties: a cyclopentenone and a dioxabicylo[3.3.0]octane ring, which were linked between C-4 and C-3' by the HMBC correlations [H-4/C-3', H-5/C-3', and H-2'/C-4]. The EIMS of **2** showed intense peaks at m/z 638 (M – H_2O – MeOH – CH_2CO)$^+$ and 414, the difference of which (224 amu) corresponded to a hexadecyl group ($C_{16}H_{33}$ – H) assignable to the two aliphatic chains attached at C-6 and C-6' of **67**. The NOESY spectrum of **67** revealed a cross-peak between H-4 and H-4', indicating the C-3'/C-4' juncture to be *cis*. The NOESY correlations were also observed from H-4 to H-7'a as well as from H-4' to H-7'b, thus implying that H-4, H-4', and H_2-7' are located on the same side of the tetrahydrofuran ring. The relative stereochemistry of the C-4 position remains unassigned.

66   **67**

From the same *Plakortis* sponge collected at Unten-harbor, we have isolated two novel tyramine-containing pyrrolidine alkaloid, plakoridines A (**68**) [43] and B (**69**) [40], whose nitrogenated carbon framework is hitherto unknown. Plakoridine A (**68**) had a molecular formula of $C_{35}H_{57}O_5N$ as revealed by HRFABMS. The ^1H and ^{13}C NMR spectra of **68** suggested the presence of a *p*-substituted phenol group, and extensive 2D NMR experiments (^1H-^1H COSY, HSQC, HMBC, and NOESY) carried out in three solvents ($CDCl_3$, C_6D_6, and CD_3OD) gave rise to a tyramine unit. The ^{13}C NMR signals resonating at δ_C 166.6 (C-2), 90.2 (C-6), and 199.9 (C-7) were assignable to a β-amino substituted enone functionality, and the UV absorption maximum of **68** (λ_{max} 317 nm) is characteristic of the presence of a β-amino substituted enone. The HMBC correlations indicated linkage of C-2 and C-5 to the tyramine nitrogen atom to form a pyrrolidine ring, and an unbranched aliphatic chain ($C_{16}H_{33}$) was connected to ketone carbon (C-7). The FABMS and EIMS data for **68** and its diacetate afforded several fragment ion peaks indicating the presence of a heptadecanoyl ($COC_{16}H_{33}$) group. In order to further verify the length of the aliphatic chain, compound **68** was treated with ozone and subsequently reduced with dimethyl sulfide to give a lactam (**70**) and heptadecanoic acid (**71**), which were detected by EIMS analysis [**70**: m/z 321 (M$^+$); **71**: m/z 270 (M$^+$)]. From the NOESY spectral data, the relative stereochemistry of the substitutents on the pyrrolidine nucleus was deduced as 3,4-trans and 4,5-trans, and $\Delta^{2(6)}$-double bond was assigned as E-configuration. Plakoridine B (69) had a molecular formula of $C_{47}H_{81}O_5N$. The molecular weight of 69 was different from 68 by 168 amu, corresponding to twelve CH_2 units. Compound 69 was therefore reasonably assigned to a homologue of plakoridine A (68) with different length of the alkyl side chain. To elucidate the length of alkyl chains, plakoridine B (69) was treated with ozone and subsequently reduced with dimethyl sulfide to give a lactam (72) and heptadecanoic acid (71), which were detected by EIMS analysis [m/z 489 (M^+ for 72) and m/z 270 (M^+ for 71)]. The optical rotation of plakoridine B (69) as well as that of plakoridine A (68) was revealed as small as zero, and the CD spectra of 3 and 6 showed no characteristic curves. Plakoridine A (68) possesses a structurally unique fully-substituted pyrrolidine ring system. The hexadecanyl aliphatic chain was commonly embraced in manzamenones, while a tyramine unit was also contained in manzamenone H. Compound 68 was found to be weakly cytotoxic against murine lymphoma L1210 cells in vitro (IC_{50} 10 µg/mL).

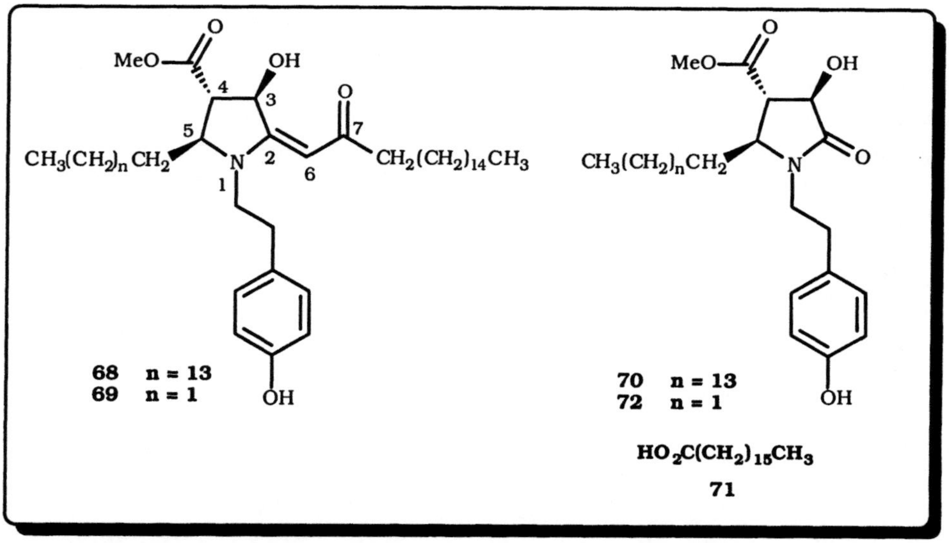

In connection with the biogenetic path of manzamenones and related *Plakortis* metabolites, we isolated 3-carboxy-5-(carboxymethyl)-4-tetradecyl-1-oxacyclopent-3-en-2-one (73) [38] from the same *Plakortis* sponge collected off Manzamo. The structure of compound 73 was defined by extensive spectroscopic analyses. On treatment with diazomethane compound 73 was converted into a dimethyl ester (74), indicating the presence of two carboxylic acids. From the 2D NMR studies, compound 73 was suggested to have a γ-butenolide structure with three substituents of a carboxyl group, a tetradecyl group, and a carboxymethyl group attached on C-3, C-4, and C-5, respectively. The absolute configuration of the chiral center at C-5 was deduced as R, since the di-*p*-bromophenacyl ester (75) exhibited a positive Cotton effect in the CD spectrum [λ_{ext} 247 nm ($\Delta\varepsilon$ +25) and 227 nm ($\Delta\varepsilon$ -17)], resulting from the positive chirality due to the two *p*-bromophenacyl groups.

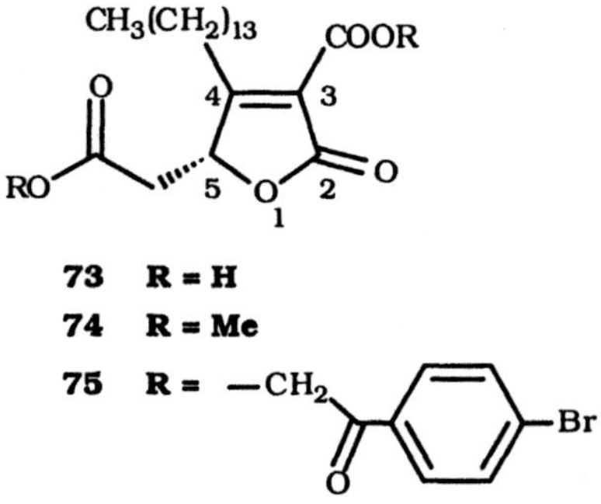

Manzamenone A (52) could be assumed as being biosynthetically generated from two fatty acid-derived precursors and we proposed a hypothetical biogenetical pathway (Scheme 4), in which manzamenone A (52) may be generated through an enantioselective intermolecular *endo*-type [4+2] cycloaddition between two precursors (c and d). Both c and d might be derived through condensation of malonate with 4-oxo-2,3-dehydrocarboxylic acid (e). A butenolide (76), which is an equivalent compound to e, was previously isolated from a Micronesian *Plakortis* sponge [30]. The butenolide compound (73) could be also assumed to be biosynthesized through a condensation of malonate and the hypothetical precursor e (a homologue with two less CH_2 units).

To substantiate the hypothetical route, we further investigated the minor constituents of the *Plakortis* sponge collected at Unten-harbor to result in isolation of a new cyclopentenone derivative, untenone A (77) [44], which is chemically equivalent to the hypothetical dienophile in biosynthesis of manzamenone A (52). Untenone A (77) may be assumed to be derived from the hypothetical intermediate (f) through a Dieckmann condensation. The structure of untenone A (77) was revealed by extensive spectroscopic analysis. The aliphatic chain on C-4 and the hydrogen on C-5 were *cis*-oriented since a distinct NOE was observed for the H-5 signal on irradiation of the methylene protons on C-6. Further verification of the structure of untenone A (77) has been provided by total syntheses of this compound (77) achieved by three groups [45, 46, and 47]. Asami and

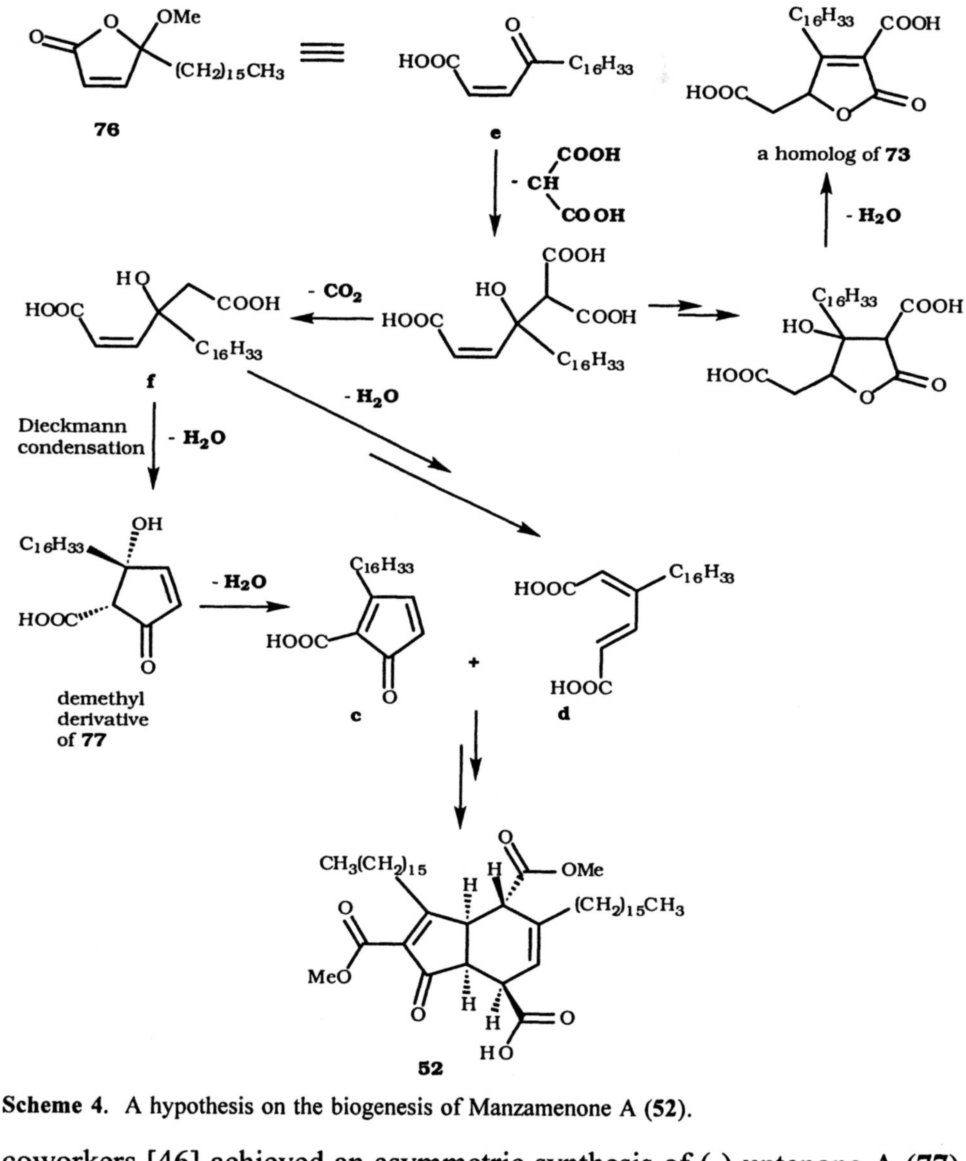

Scheme 4. A hypothesis on the biogenesis of Manzamenone A (**52**).

coworkers [46] achieved an asymmetric synthesis of (-)-untenone A (**77**) in the optically active form. As the synthesized (-)-untenone A (**77**) showed rather large specific rotation ($[\alpha]^{26}_D$ -73.3° (c 1.20, $CHCl_3$)) compared with natural untenone A ($[\alpha]^{19}_D$ +0.2° (c 2.1, $CHCl_3$)) [44], the isolated untenone A (**77**) was suggested to be almost racemic. Yamada and coworkers [47] prepared both of optically active (+)- and (-)-untenone A (**77**), and recorded the CD spectra of (-)-**77**, (+)-**77**, and natural **77**, which was provided by us. In (+)-**77**, the negative Cotton effect at 229 nm ($\Delta\varepsilon$ -6.2) and positive Cotton effect at 202 nm ($\Delta\varepsilon$ +4.8) were observed while the positive Cotton effect at 229 nm (Δε +5.4) and the negative Cotton effect at 202 nm (Δε -6.0) could be seen in (-)-77. However, natural untenone A (77) was observed not to exert the Cotton effect. Untenone A (77), isolated from the marine sponge, was therefore clearly shown to be racemic.

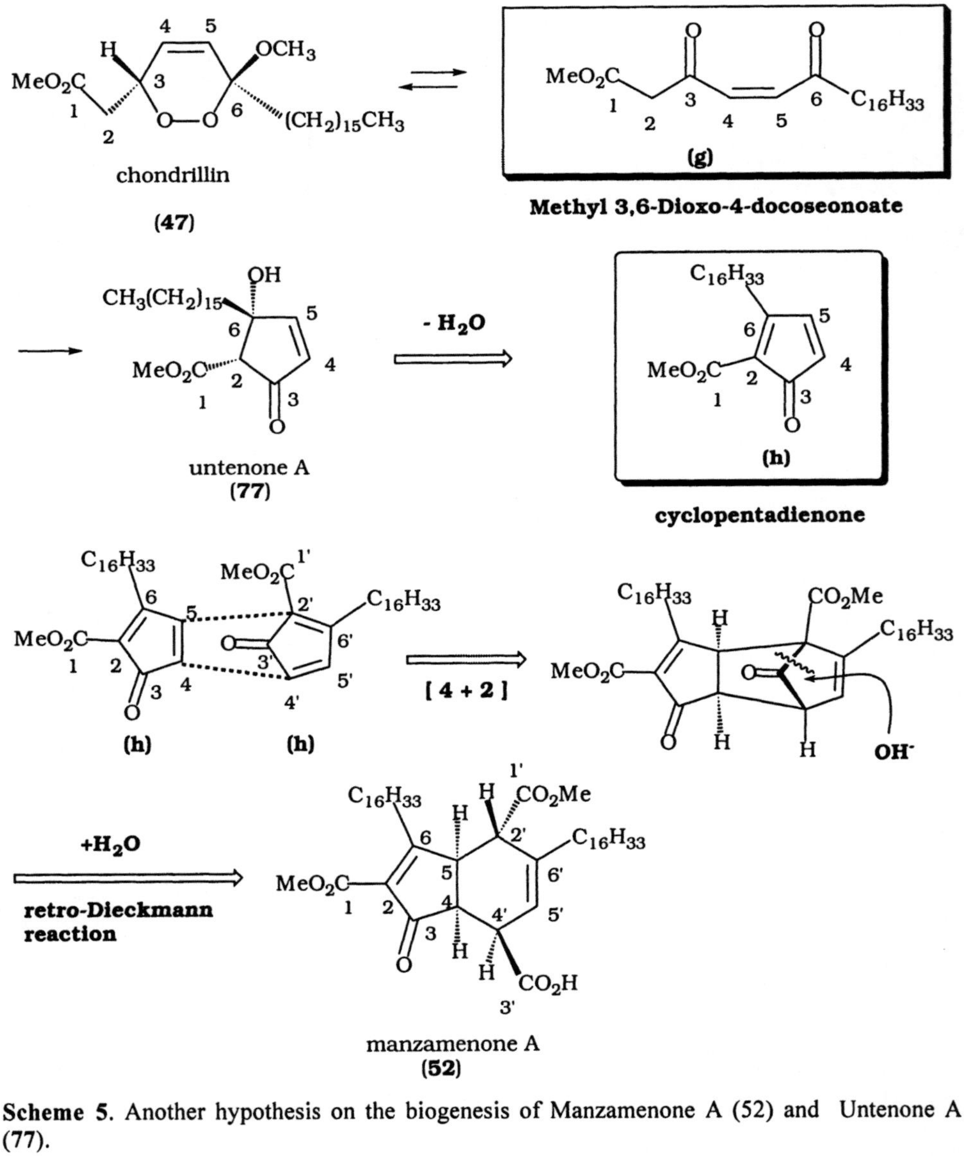

MeO2C
OCH3
(CH2)15CH3
chondrillin
(47)
MeO2C
C16H33
(g)
Methyl 3,6-Dioxo-4-docoseonoate
OH
CH3(CH2)15
- H2O
untenone A
(77)
(h)
cyclopentadienone
[4 + 2]
OH-
+H2O
retro-Dieckmann reaction
CO2Me
CO2H
manzamenone A
(52)

Scheme 5. Another hypothesis on the biogenesis of Manzamenone A (52) and Untenone A (77).

On the biogenesis of untenone A (77), it is another possibility that the cyclopentenone may be derived from a 1,4-dicarbonyl compound (3,6-

dioxo-4-docosenoic acid methyl ester (**g**), Scheme 5), which may be closely related to chondrillin (**47**). 1,4-Dicarbonyl compounds are known to be susceptible to intramolecular condensation to give cyclopentenones [48]. On the biogenesis of manzamenone A (**52**), another hypothetical pathway could be conceivable as shown in Scheme 5, where manzamenone A (**52**) is biogenetically derived from two molecules of hypothetical cyclopentadienone (**h**, Scheme 5) that may be an unstable intermediate generated by dehydration from untenone A (**77**). The *exo*-type [4+2] cycloaddition of two molecules of **h** followed by retro-Dieckmann reaction may furnish manzamenone A (**52**). Metabolites isolated from *Plakortis* sponges such as chondrillin (**47**), plakorin (**48**), manzamenones including manzamenone K (**67**), untenone A (**77**), and plakoridines A (**68**) and B (**69**) are all likely to be biogenetically related to one another, and methyl 3,6-dioxo-4-docoseonoate (**g**) may be a common key intermediate (Scheme 6).

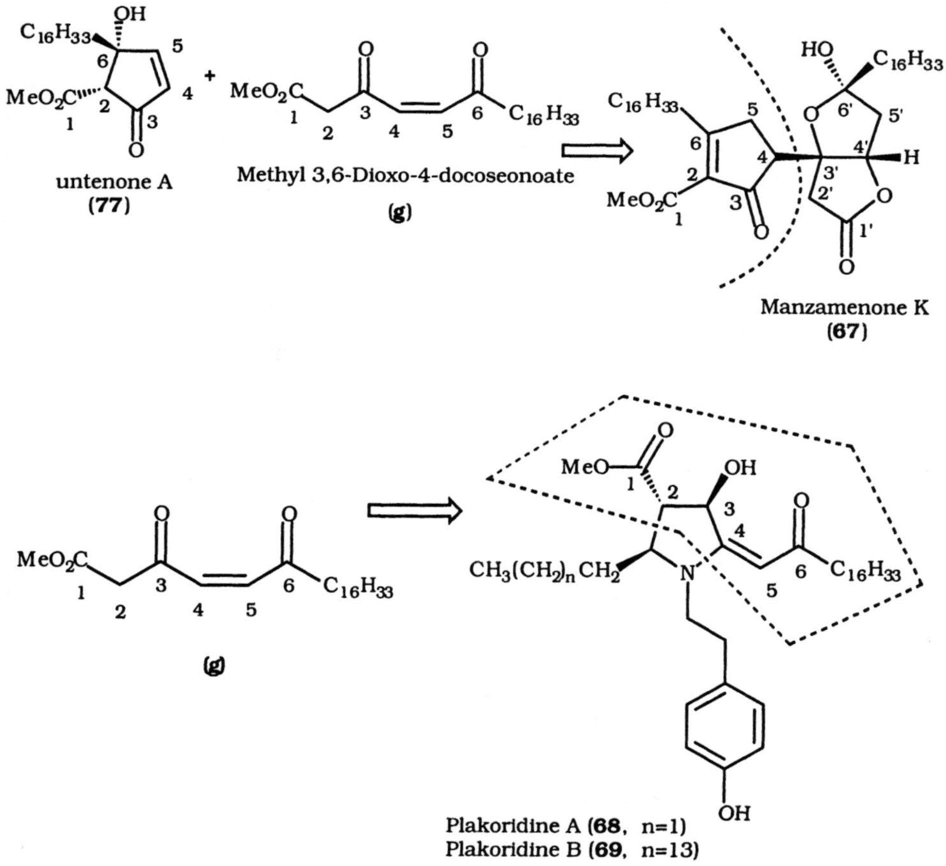

untenone A
(**77**)

Methyl 3,6-Dioxo-4-docoseonoate
(**g**)

Manzamenone K
(**67**)

(**g**)

Plakoridine A (**68**, n=1)
Plakoridine B (**69**, n=13)

Scheme 6. Hypothesis on the biogenesis of Manzamenone K (**67**) and Plakoridines A and B (**68** and **69**).

TAUROSPONGIN A AND ACETYLENE ACIDS

Acetylenic fatty acids mostly bearing bromine atoms were frequently isolated from marine sponges. We recently isolated a novel taurine-containing acetylenic lipid, taurospongin A (**78**) [49], from the Okinawan marine sponge *Hippospongia* sp. and determined its structure including the absolute stereochemistry on the basis of spectral data, chemical degradation, and the synthesis of a degradation product of **78**. Taurospongin A (**78**) was found to be a potent inhibitor of DNA polymerase β and HIV reverse transcriptase.

The purple-colored sponge *Hippospongia* sp., from which taurospongin A (**78**) was obtained, was collected off Okinawa Island. The sponge was extracted with MeOH, and the extract was partitioned between EtOAc and H$_2$O, and the aqueous phase was further extracted with *n*-BuOH. While the EtOAc-soluble fraction contained a sesquiterpenoid quinone, metachromin A [50], as a major constituent, the *n*-BuOH-soluble material was subjected to silica gel and gel filtration

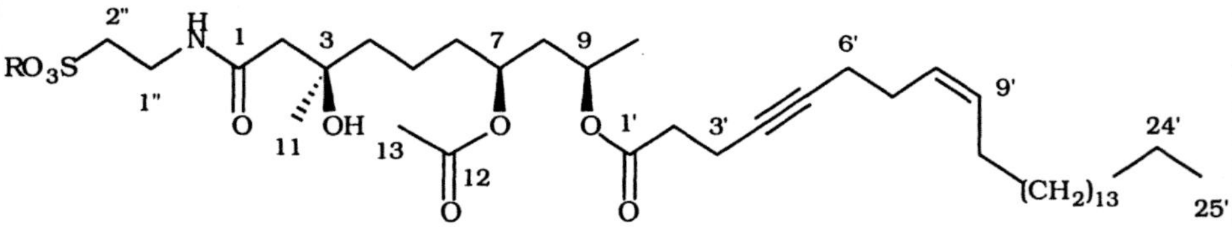

78 R = H
79 R = Me

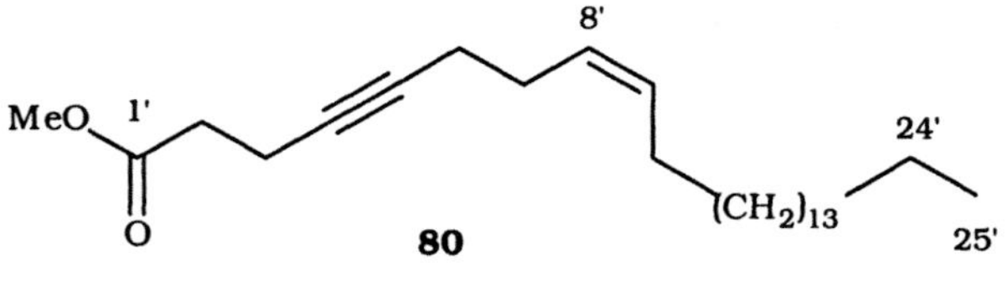

80

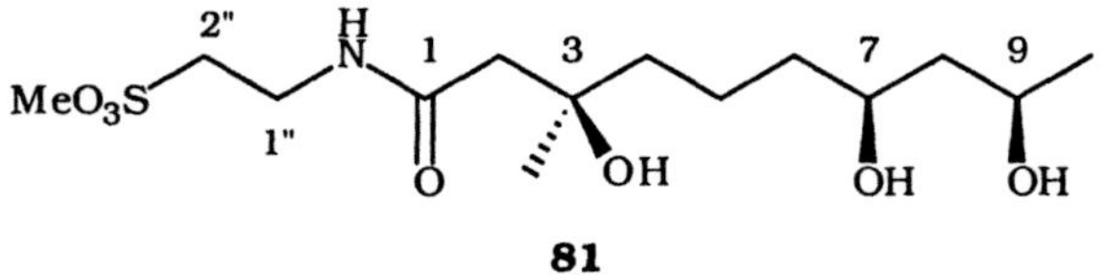

81

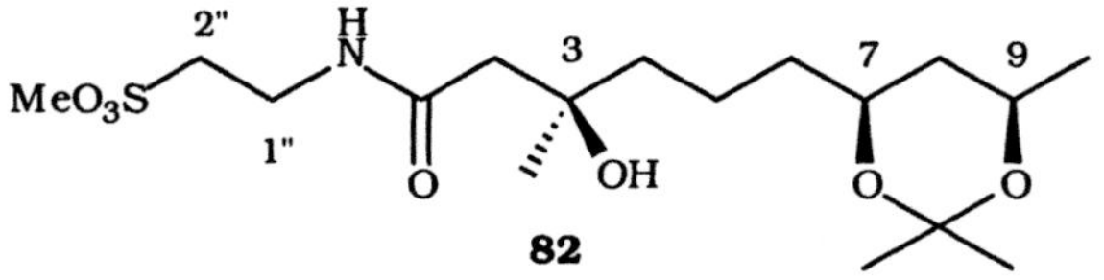

82

chromatographies to give taurospongin A (**78**), $C_{40}H_{70}NO_9S$, as colorless amorphous solids. The methyl ester (**79**), prepared from **78** by treatment with diazomethane, was used for the ^1H and ^{13}C NMR spectral studies, which revealed that taurospongin A (**78**) consisted of three elements, i.e., a taurine, a trihydroxyl fatty acid, and an unsaturated fatty acid. Amino acid analysis of the acid hydrolysate of **78** revealed the presence of taurine in **78**. The methyl ester (**79**) was therefore inferred to be a methyl sulfonate ester in the taurine moiety. The HMBC spectrum of **79** showed that the taurine unit was connected to C-1 of trihydroxy fatty acid moiety through an amide bond.

For the trihydroxy fatty acid moiety of **79**, 2D NMR data showed that a tertiary methyl and tertiary hydroxyl groups are located on C-3 quaternary carbon, and acetyl and unsaturated acyl groups were shown to be connected on C-7 and C-9 oxymethines, respectively, through ester linkages. However, the number of methylene carbons between C-3 and C-7 could not be firmly determined because of overlapping of the methylene signals. As to the unsaturated fatty acid portion, the ^1H-^1H COSY spectrum showed a cross peak for H$_2$-3'/H$_2$-6' due to long-range coupling through a triple bond. The relatively high-field resonances of C-3' (δ_C 15.2) and C-6' (δ_C 19.6) implied that these methylenes were adjacent to an acetylene group. The ^1H-^1H COSY revealed the connectivities from H$_2$-6' to H$_2$-10' through an olefin at C-8'/C-9' as well as a cross peak for the terminal part, H$_2$-24'/H$_3$-25'. Though the C-10' and C-24' had to be connected through sp^3 methylene carbons, the number of the methylenes between them was also unknown from spectral data. Methanolysis of **79** (MeOH/1N HCl, 10:1) was therefore carried out to afford two fragments, unsaturated fatty acid methyl ester [**80**, EIMS m/z 390 (M$^+$)] and trihydroxylamide [**81**, FABMS m/z 356 (M+H)$^+$]. Molecular weights of these methanolysis products (**80** and **81**) clearly showed the lengths of the methylene chains. The $\Delta^{8'}$-double bond was confirmed to be Z by the ^1H coupling constant for **80** ($J_{8',9'}$= 11.3 Hz).

Taurospongin A (**78**) contains three chiral centers at C-3, C-7, and C-9. The relative stereochemistry of 1,3-diol at the C-7/C-9 position was revealed to be *syn* by applying Rychnovsky's ^{13}C-acetonide analysis [51] for the acetonide (**82**: δ_C 19.9, 30.3, and 98.4) prepared from **81**. To elucidate the relative stereochemistry between C-3 and the C-7/C-9 position as well as the absolute stereochemistry of these three chiral centers, two possible diastereomers (**83** and **84**), which correspond to the acetonide derivative (**82**), were prepared as optically active forms as shown in Scheme 7. The methoxymethyl (MOM) ether (**86**) derived from (*R*)-3-hydroxy-3-methyl-5-pentanolide (**85**) was reduced with DIBAL to give a lactol, which was subjected to Wittig reaction with chloromethylenetriphenylphosphorane followed by protection with benzyloxymethyl group to yield a 2:1 mixture of the *E/Z* chlorovinyl

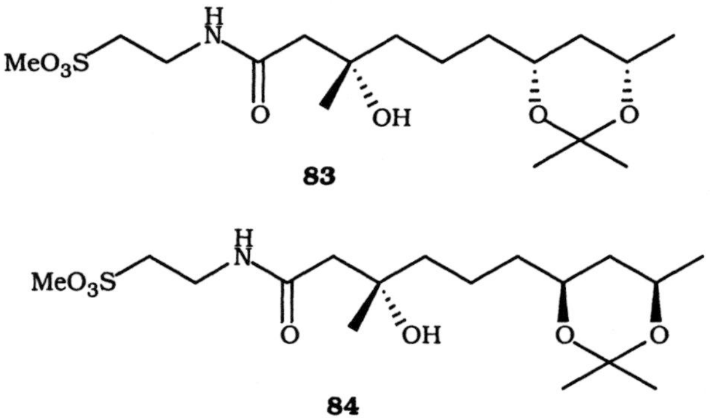

83

84

derivatives (**87**). On the other hand, (*S*)-3-*t*-butyldimethylsilyloxybutanal
((*S*)-**89**) was prepared from methyl (*S*)-(+)-3-hydroxybutylate ((*S*)-**88**) in
2 steps. The lithium acetylide generated from *E/Z* mixture of **87** (*n*-BuLi,
THF, 0 °C, 45 min) was coupled with the aldehyde ((*S*)-**89**) to afford a 1:1
mixture of the *anti/syn* alcohols (**90** and **91**). The diastereomeric mixture

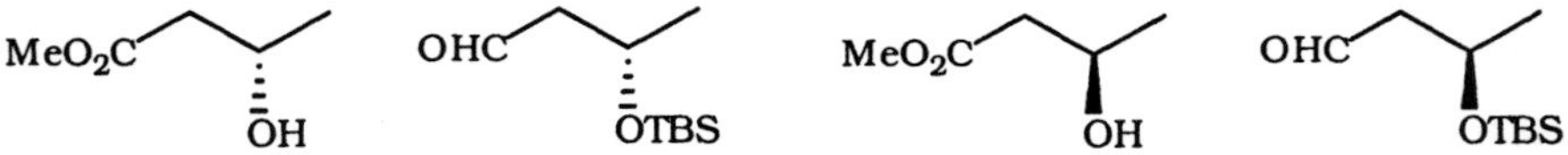

(*S*)-88   (*S*)-89   (*R*)-88   (*R*)-89

was separable by silica gel column chromatography, and each diastereomer
was then converted into its acetonide (**92** and **93**), respectively. Relative
stereochemistry of 1,3-diol moiety of **92** and **93** was elucidated from the
^{13}C-acetonide analysis [51] [**92** (*anti*): δ_C 23.8, 29.0, 99.8; **93** (*syn*): δ_C
19.4, 30.1, 99.0]. Since the 1,3-diol moiety at C-7 and C-9 of taurospongin
A (**78**) was already revealed as *syn*, the *syn* acetonide (**93**) was subjected
to hydrogenation in the presence of Raney Ni to furnish an alcohol (**94**).
RuO$_4$ oxidation of the alcohol (**94**) afforded an carboxylic acid, which was
coupled with taurine through an *N*-hydroxysuccinimide ester. Passing
through an ion exchange chromatography (Amberlite IR-122) resulted in
the deprotection of MOM ether and acetonide groups, and the resulting
sulfonic acid was treated with diazomethane to give trihydroxylamide
methyl ester (**95**), which was then converted into acetonide (**83**). The
diastereomeric *syn* acetonide (**84**) was also synthesized by coupling of the
acetylide from **87** with an enantiomeric aldehyde ((*R*)-**89**), prepared from

(R)-**88**, followed by the same reactions as described above. The ^{1}H NMR spectra of the diastereomers (**83** and **84**) could be discriminated from each other particularly by the AB quartet signals due to the C-2 methylene protons (**83**: δ_H 2.28 and 2.41; **84**: δ_H 2.29 and 2.40). The ^{1}H NMR

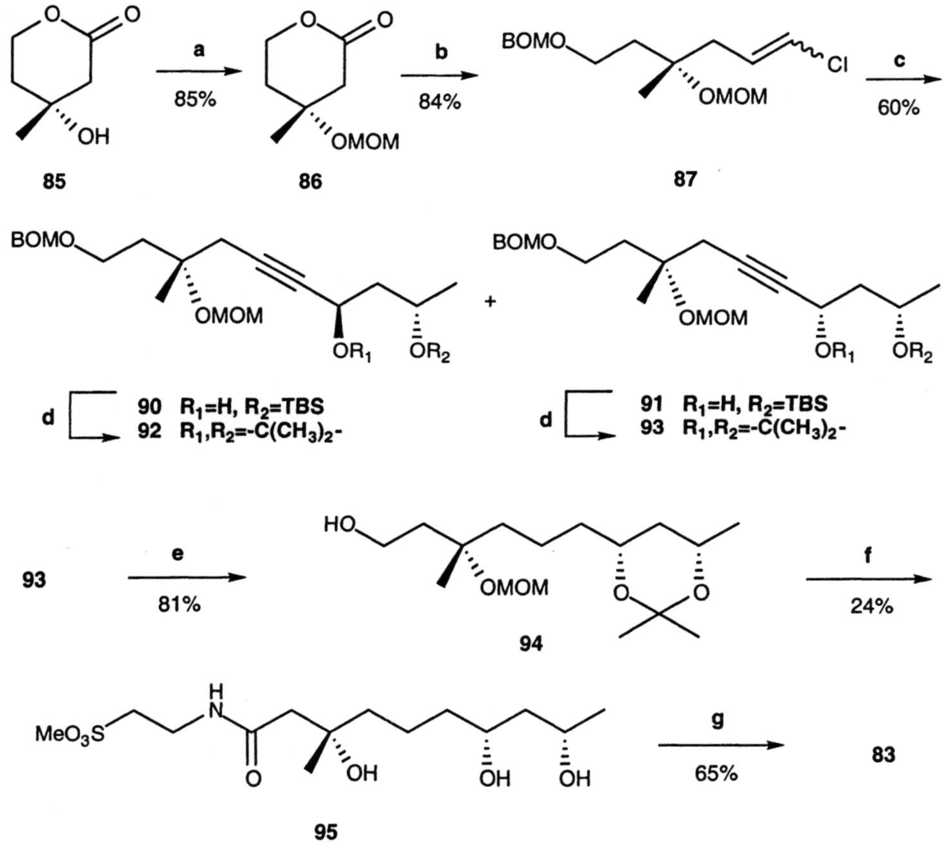

Scheme 7. (a) MOMCl, NEt(i-Pr)$_2$; (b) (1) DIBAL, Et$_2$O; (2) Ph$_3$P=CHCl; (3) BOMCl, NEt(i-Pr)$_2$; (c) n-BuLi, (S)-**89**; (d) (1) n-Bu$_4$NF; (2) DMP, PPTS; (e) H$_2$, Raney Ni (W-2); (f) (1) RuCl$_3$, NaIO$_4$; (2) HOSu, DCC; (3) Taurine, TEA; (4) Amberlite IR-122; (5) CH$_2$N$_2$; (g) DMP, PPTS.

spectrum of the acetonide (**82**) obtained from natural specimen of taurospongin A (**78**) was identical with that of the synthetic (3S,7R,9S)-derivative (**83**). Thus, the relative configuration of **82** was revealed as 3S*, 7R*, and 9S*. Synthetic sample (**83**), however, unfortunately had very small optical rotation ([α]$_D$ +0.65°), thus implying that determination of the absolute configurations of **82** based on the sign of the optical rotation was not convincing. This problem could be overcome by preparing Mosher's acid esters from synthetic and natural samples and distinguishing them by NMR spectroscopy. The synthetic triol (**95**) was treated with

(*S*)- and (*R*)-MTPACl to give (*R*)- and (*S*)-bis MTPA esters (**96** and **97**), respectively, while the triol **4** from natural specimen was treated with (*R*)-MTPACl to give bis-(*S*)-MTPA ester (**98**). The ^{1}H NMR spectra of synthetic samples, **96** and **97**, were easily distinguishable particularly by the methyl proton signals [**96**: δ_H 1.37 (H$_3$-10) and 1.11 (H$_3$-11); **97**: δ_H 1.22 (H$_3$-10) and 1.17 (H$_3$-11)]. The ^{1}H NMR of bis-(*S*)-MTPA ester (**98**) from natural specimen proved to be completely identical with that of synthetic bis-(*R*)-MTPA ester (**96**), implying that the synthetic and natural samples (**96** and **98**, respectively) are enantiomers. Thus the absolute configurations of taurospongin A (**78**) were unambiguously established as 3*R*, 7*S*, and 9*R*.

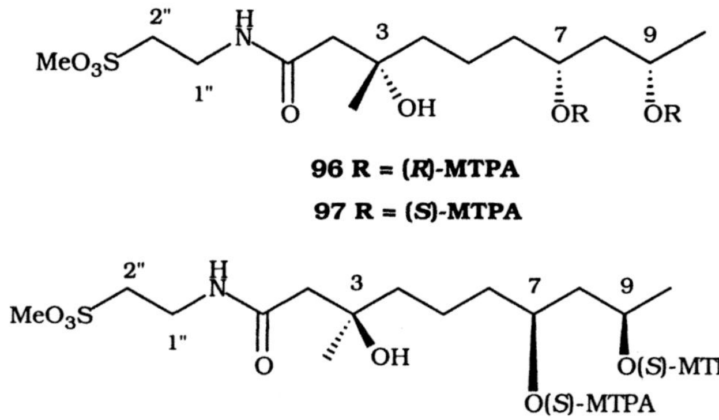

96 R = (*R*)-MTPA

97 R = (*S*)-MTPA

98

Taurospongin A (**78**) consisting of taurine, trihydroxy fatty acid, and unsaturated fatty acid residues appears to belong to an unprecedented class of marine natural products, whereas taurine-bearing mono-fatty acid derivatives were recently isolated from bivalve *Pinna muricata* [52] and protozoan *Tetrahymena thermophila* [53]. The trihydroxy and unsaturated C$_{25}$ fatty acid contained in taurospongin A (**78**) are also new fatty acids, whilst a C$_{25}$ acetylenic fatty acid with more unsaturations was recently isolated from Australian sponge *Phakellia cardus* [54]. Acetylenic fatty acids mostly bearing bromine atoms were frequently isolated from sponges of the genera *Petrosia* and *Xestspongia* (family Nepheliospongiidae) [55], and they were never obtained previously from *Hippospongia* sp.

Taurospongin A (**78**) exhibited potent inhibitory activity against DNA polymerase β and HIV reverse transcriptase with IC$_{50}$ values of 7.0 and 6.5 μM (Ki values of 1.7 and 1.3 μM), respectively. This inhibitory activity has been frequently associated with the sulfonic acid function. Furthermore, **78** showed weak inhibitory activity against c-erbB-2 kinase (IC$_{50}$ 28 μg/mL), while there was no cytotoxicity (IC$_{50}$ > 10 μg/mL) against murine lymphoma L1210 and human epidermoid carcinoma KB cells in vitro.

For other acetylenic compounds, we have isolated nepheliosyne A (99) [56], a new C_{47} acetylenic acid, from the sponge *Xestospongia* sp., collected off Kerama Islands, which also contained three cytotoxic and

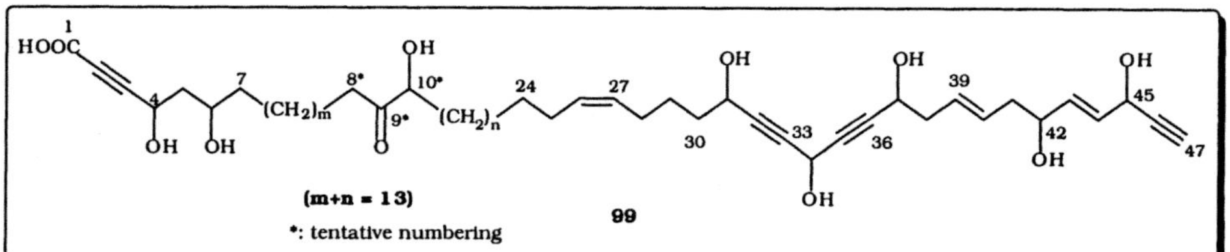

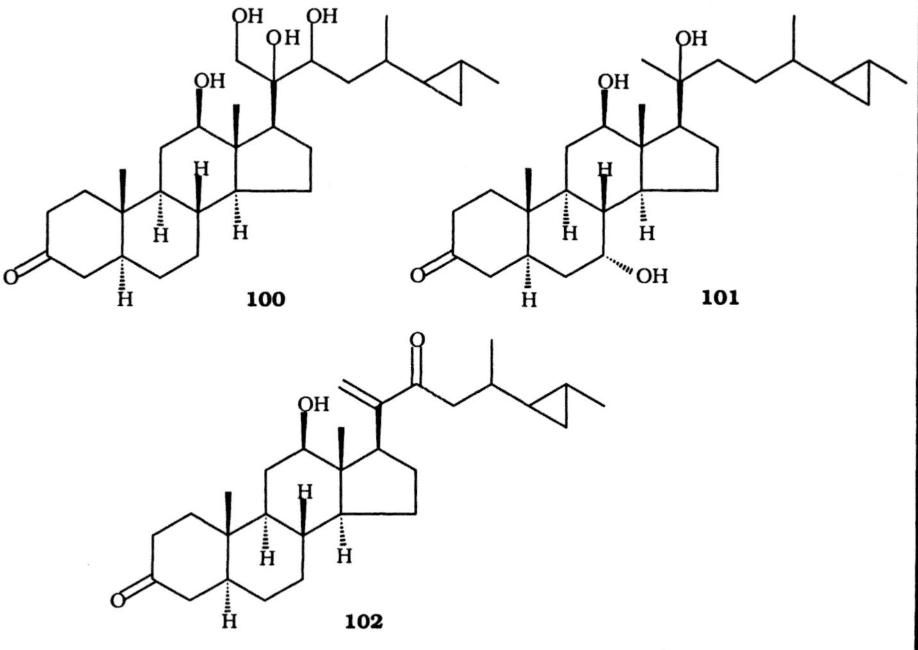

antimicrobial C_{29} sterols with cyclopropane moiety, named xestokerols A ~ C (100 ~ 102) [57]. Although the planar structure of nepheliosyne A (99) was elucidated by extensive spectral studies, the values of 'm' and 'n' (m+n = 13) were not able to be defined. The numbering for C-8* to C-10* was therefore tentative, and the positions of the carbonyl (C-9*) and oxymethine (C-10*) groups may be reversed. Nepheliosyne A (99) is a polyacetylene carboxylic acid with the highest molecular weight among those hitherto obtained from extracts of marine sponges, and is the first example of a C_{47} polyacetylene natural product. Nepheliosyne A (**99**), however, proved to possess a closely related structure to petrosolic acid (**103**) [58], a C_{44} oxo-octahydroxy-triene-tetraynoic carboxylic acid isolated from the Red Sea sponge *Petrosia* sp. The cytotoxicity of compound **99** proved to be weak; it showed only 40% and 14% inhibition at 20 μg/mL against lymphoma L1210 and human epidermoid carcinoma KB cells in vitro, respectively. On the other hand, petrosolic acid (**103**) was reported to be effective in inhibitory of the DNA polymerase activities of the reverse transcriptase of human immunodeficiency virus.

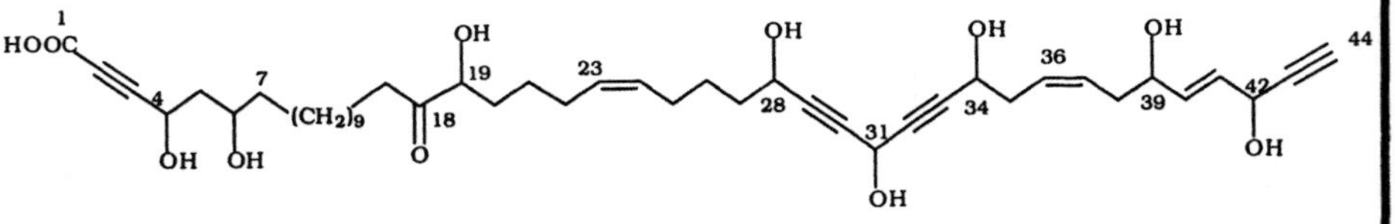

103

From another sponge of the genus *Xestospongia* collected off Motobu peninsula, Okinawa Island, seventeen brominated unsaturated fatty acid derivatives (**104 ~ 120**) were isolated [59]. These seventeen compounds (**104 ~ 120**) contained olefinic and/or acetylenic unsaturation as well as 1-3 vinylic bromine atoms, and they could be classified to two groups, short acids (C_{18} and C_{16}: **104 ~ 113**) and long acids (C_{27}, C_{28}, and C_{29}: **114 ~ 120**). The structures of these new compounds were elucidated extensive spectroscopic studies including several types of modern 2D NMR techniques and chemical correlations with diazomethane. These compounds are all first isolated as natural products, while three of them (**116, 117, and 119**) were previously identified as fatty acid components of the phospholipids of marine sponges [60]. Some of these new compounds (**104, 105, 106, 107, 114, and 117**) exhibited cytotoxicities against L1210 murine leukemia cell and KB human epidermoid carcinoma cell in vitro (IC_{50} values, 1 - 20 μg/mL).

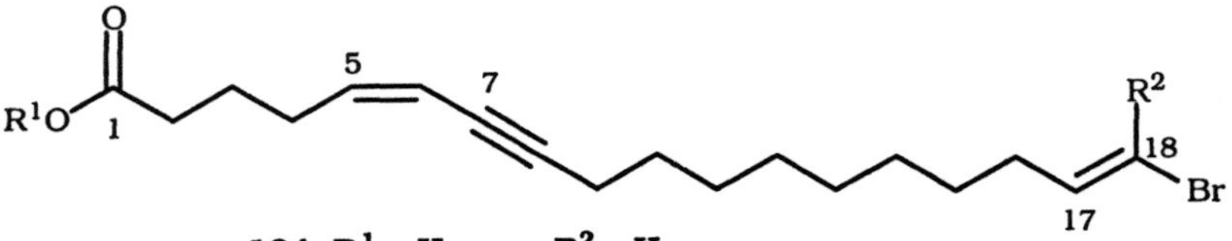

104	R^1 = H	R^2 = H
105	R^1 = H	R^2 = Br
106	R^1 = CH$_3$	R^2 = H
107	R^1 = CH$_3$	R^2 = Br

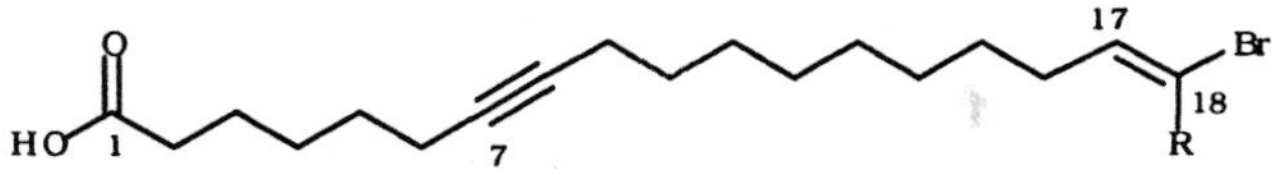

108 R = H
109 R = Br

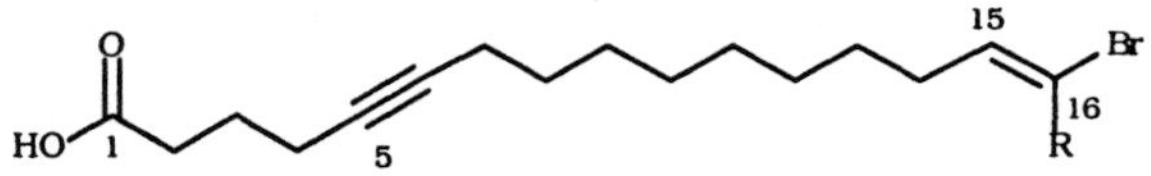

110 R = H
111 R = Br

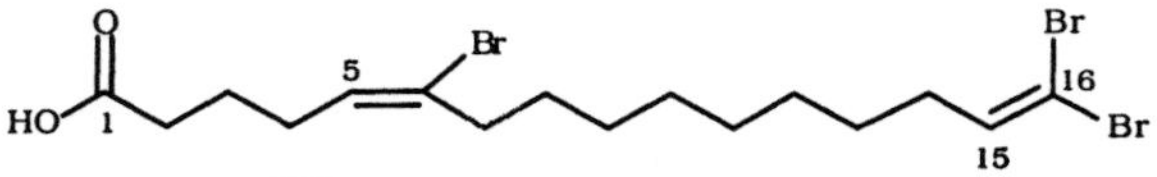

112

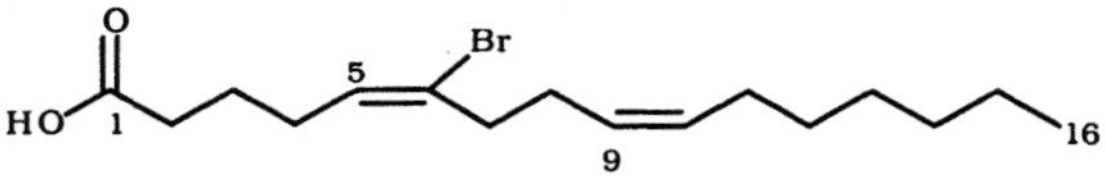

113

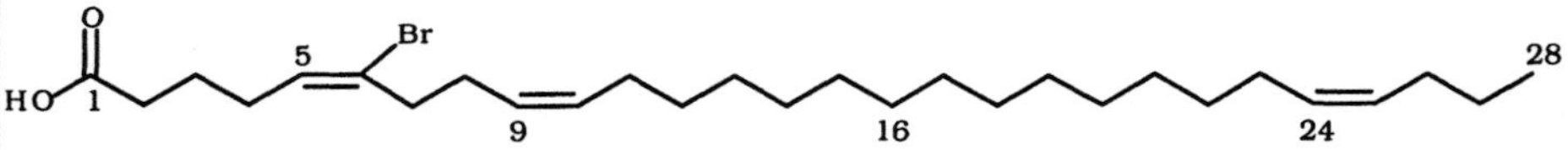

114

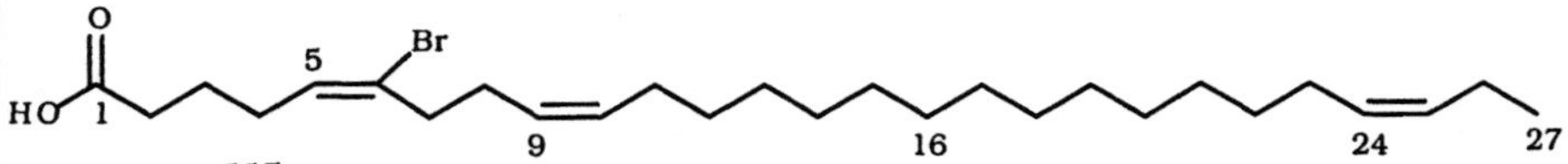

115

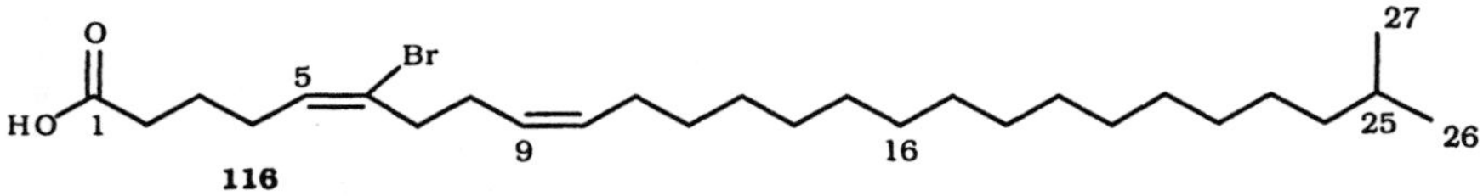

116

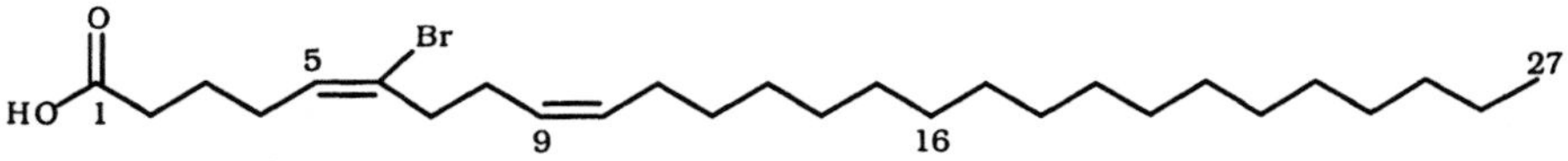

117

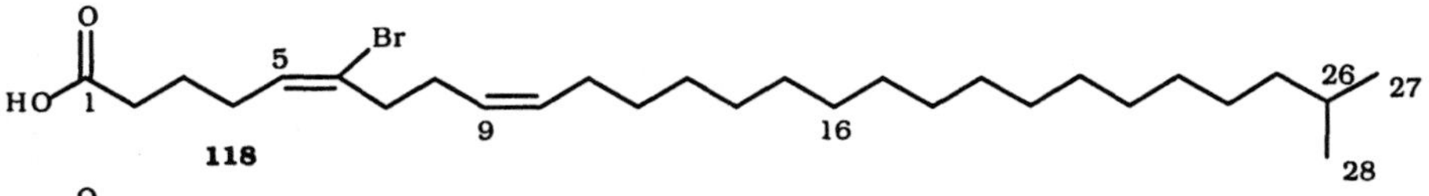

118

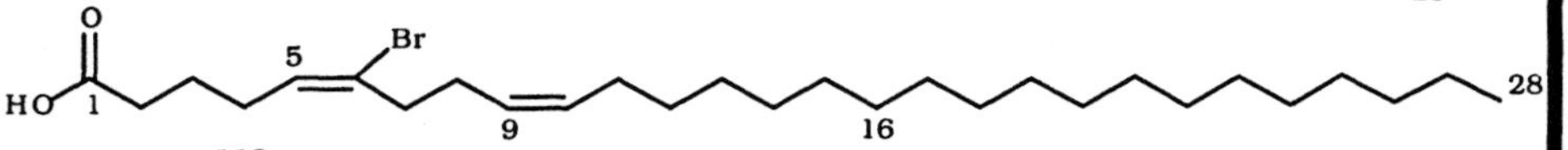

119

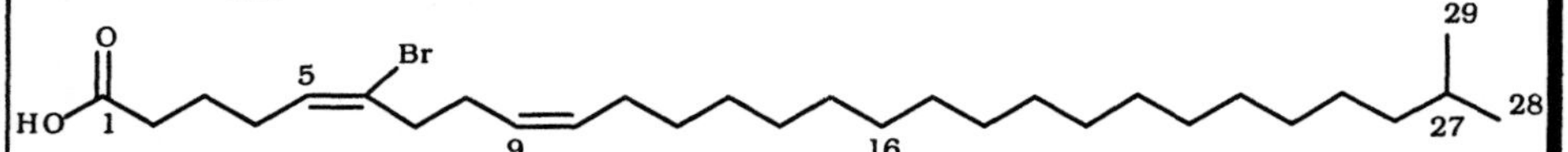

120

THEONEZOLIDES

Marine sponges of the genus *Theonella* frequently afford a variety of interesting secondary metabolites including polyoxygenated aliphatic compounds [61] as well as unusual cyclic peptides [62], most of which exhibit significant biological activities. We examined extracts of *Theonella* sponges of several collections and isolated a number of new bioactive substances such as pyridine alkaloids (theonelladins [63]), cyclic peptides with ureido linkage (konbamide [64] and keramamide A [65]), oxazole or thiazole containing cyclic peptides (keramamides B ~ D [66] and F [67]), congeners of macrolides or polyethers (swinholides [68], bistheonellides [69], and onnamides [70]), and a brominated benzyltetrahydro protoberberine alkaloid (theoneberine [71]).

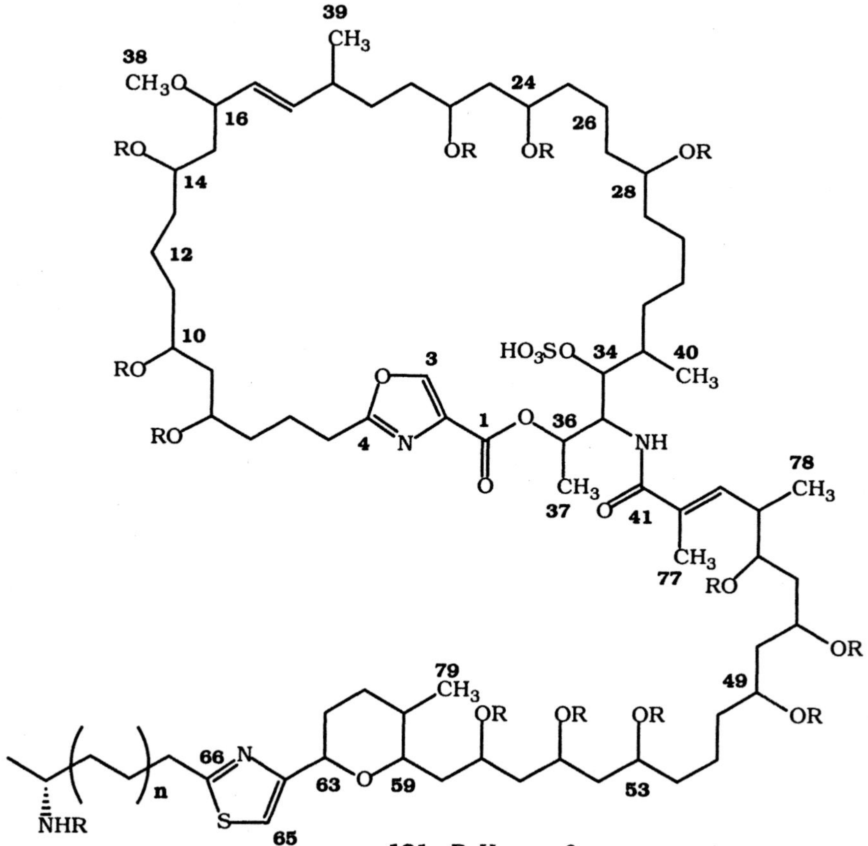

121 R=H,   n=3
122 R=Ac, n=3
127 R=H,   n=2
128 R=H,   n=4

During our further studies on extracts of *Theonella* sponges, theonezolide A (121) [72], a novel macrolide, was isolated from a sponge *Theonella* sp., collected off Ie Island, Okinawa. Theonezolide A (121) was the first member of a new class of polyketide natural products consisting of two principal fatty acid chains with various functionalities such as a sulfate ester, an oxazole, and a thiazole group, constituting a 37-membered macrocyclic lactone ring bearing a long side chain attached through an amide linkage. The planar structure of 121 was elucidated on the basis of extensive spectroscopic analyses and chemical degradation studies. The molecular formula of 121 was suggested as $C_{79}H_{140}N_4O_{22}S_2$ by negative HRFABMS and combustion analytical data. ^1H and ^{13}C NMR spectra of 121 revealed signals due to one di- and one trisubstituted olefins, seventeen oxymethines, two nitrogen-bearing methines, one vinyl and six secondary methyls, and many sp^3 methylenes. Acetylation of 121 afforded a tridecaacetate (122), and spectral comparison of 121 and 122 indicated the presence of twelve secondary hydroxyl and one amino groups for 121. The remaining five oxymethines were ascribed to those bearing a methoxy, an ester, a sulfate, and an ether-oxygen forming a tetrahydropyrane ring, whose ^1H resonances did not show the down-field shift by acetylation. The 2D NMR data of 121 including DQF-COSY, HOHAHA, ROESY, HSQC, and the recent techniques of gradient-enhanced HMBC and HSQC-HOHAHA spectra in DMSO-d_6 were extensively studied to give rise to five partial structures. The partial structure of the C-41 ~ C-69 moiety in the second fatty acid chain was further verified by the negative FABMS/MS experiment of 121 [parent ion *m/z* 1559 (M-H)⁻], which showed characteristic daughter ions generated by fissions at α and β positions to OH groups. The presence of the sulfate group proved to be desirable for the negative ion FABMS/MS analysis, and the presence of a sulfate group was also confirmed by the negative FABMS/MS experiment (parent ion *m/z* 1559), which exhibited intense daughter ions at *m/z* 97 and 80, assignable to HSO_4^- and SO_3^- ions, respectively.

Analyzing the DQF-COSY spectrum of 121 suggested the absence of 1,2-diols, while the presence of at least four 1,3-diols were inferred from the formation of the acetonides in which four acetone molecules were incorporated. The acetonides were obtained by treatment of 121 with 2,2-dimethoxypropane in DMF in the presence of *p*-toluenesufonic acid and found to be a mixture of four components with equal molecular weight based on HPLC and FABMS analysis. This result was coincident with the presence of two 1,3-diols and two 1,3,5-triols in 121. The ^{13}C NMR data of the major compound of the four acetonide products indicated that three 1,3-diols of the four acetonide groups in the major compound were *syn* and one 1,3-diol was *anti* on the basis of Rychnovsky's ^{13}C-acetonide analysis [51]. It was revealed that the ^{13}C NMR signals for sp^3 methylene carbons located between two hydroxy-bearing methines (viz., 2 position of 1,3-

diol) were observed at δ_C 43 ~ 45 ppm, whereas sp^3 methylene carbons between a hydroxy-bearing methine and another sp^3 methylene resonated at δ_C 35 ~ 38 ppm. The locations of secondary hydroxyls, a methoxy, and secondary methyl groups were elucidated mainly by the HMBC and HSQC-HOHAHA correlations. Applying the new technique of gradient-enhanced HMBC experiment afforded data of high sensitivity with almost no noise, while the HSQC-HOHAHA spectrum provided ^{1}H-^{13}C connectivity data through 5- or 6-bonds. It was, however, extremely difficult to obtain unambiguous evidence for connection of the partial structures through sp^3 methylenes by spectral means because of the heavy overlapping of the methylene signals in the NMR spectra. The chemical degradation experiments were therefore carried out to give solution to this problem. Treatment of the acetate (122) with ozone followed by NaBH$_4$ reduction and acetylation afforded a complex mixture, which was purified by reversed-phase HPLC (ODS) to give four useful products (123 ~ 126), corresponding to C-4 ~ C-17, C-18 ~ C-37, C-43 ~ C-64, and C-66 ~ C-76 moieties of 121, respectively. Thus the numbers of the sp^3 methylene carbons between the partial structures to connect them each other were clearly revealed by the molecular weights of these ozonolysis products (123 ~ 126) [FABMS, 123: m/z 490 (M+H)$^+$; 124: m/z 824 (M-H)$^-$; 125: m/z 788 (M+H)$^+$; 126: m/z 243 (M+H)$^+$]. The ^{1}H and ^{13}C NMR data facilitated by the ^{1}H-^{1}H COSY spectra of degradation products (123 ~ 126) provided additional proofs corroborating the total structure of 121. From all of these data the whole structure of theonezolide A was concluded as 121.

Theonezolide A (121) is the first member of an unprecedented class of polyketide natural products consisting of two principal fatty-acid chains, bearing several structural features of interest from the biogenetical viewpoint. The oxazole unit could be assumed to be derived from an amino acid, serine, whereas the origin of the thiazole ring is problematical. There are a number of oxazole- and thiazole-containing metabolites reported from marine origins, and in most cases cysteine is suggested as a precursor of the thiazole functionality [73]. It might be, however, possible that the thiazole ring of theonezolide A (121) was generated via backbone rearrangement from a nitrogen-involved polyketide intermediate as proposed for ulapualides [74] or kabiramides [75]. Theonezolide A (121) exhibited cytotoxicity against murine lymphoma L1210 and human epidermoid carcinoma KB cells in vitro with the IC$_{50}$ values of both 0.75 µg/mL.

Theonezolide A (121) caused a marked platelet shape change at low concentrations (0.2-0.6 µM) [76]. Increasing concentrations of 121 to 6 µM or more caused shape change followed by a small but sustained aggregation. In a Ca^{2+}-free solution, theonezolide A-induced aggregation was markedly inhibited, although the marked shape change was still observed. Aggregation stimulated by 121 increased in a linear fashion with increasing Ca^{2+} concentrations from 0.1 to 3.0 mM. Furthermore theonezolide A (121) markedly enhanced ^{45}Ca^{2+} uptake into platelets. Aggregation induced by theonezolide A (121) was inhibited by Arg-Gly-Asp-Ser, an inhibitor of fibrinogen binding to glycoprotein IIb-IIa, H-7 and staurosporine, protein kinase C inhibitors, or genistein and tyrphostin A23, protein tyrosine kinase inhibitors, whereas shape change was blocked by genistein and tyrphostin A23. H-7 or staurosporine did not affect the theonezolide A-induced shape change. These results suggested that theonezolide A-induced platelet shape change was not dependent on external Ca^{2+}, whereas theonezolide A-induced aggregation was caused by an increase in Ca^{2+} permeability of the plasma membrane. It was also suggested that both aggregation and shape change induced by theonezolide A (121) were associated with protein phosphorylation by protein kinase C and tyrosine kinase.

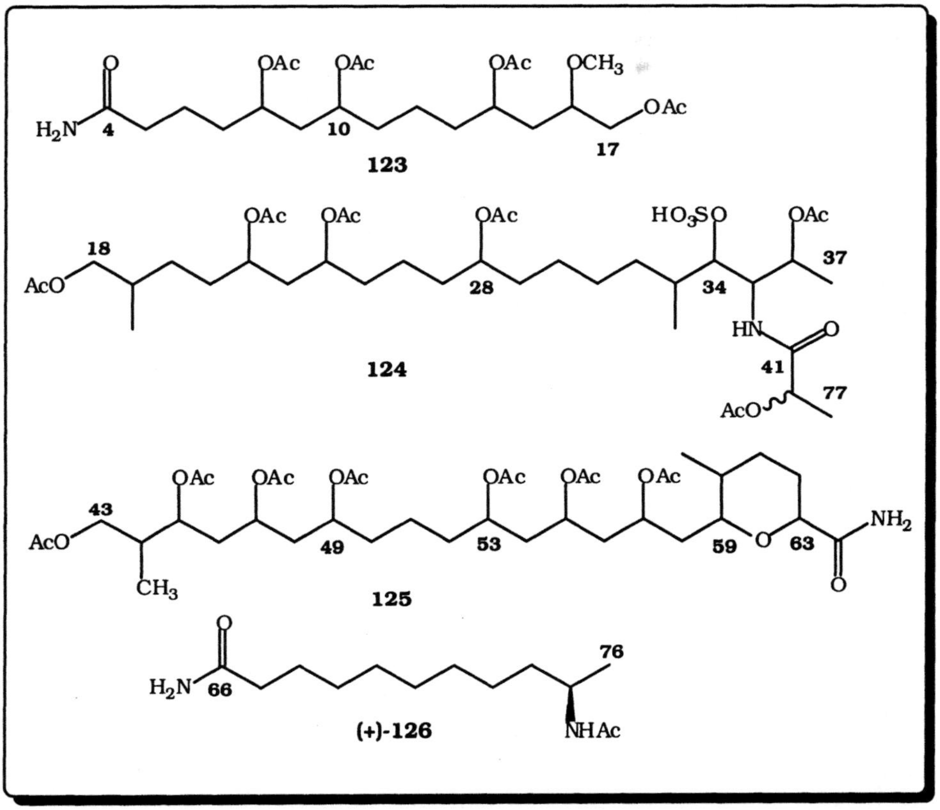

Two additional macrolides, theonezolides B (127) and C (128), were also isolated from the same Okinawan marine sponge *Theonella* sp., and their structures were elucidated on the basis of spectroscopic data as well as chemical degradation experiments [77]. Although the ^{1}H NMR spectra of theonezolides B (127) and C (128) appeared almost indistinguishable to that of theonezolide A (121), the negative FABMS spectra of 127 and 128 clearly showed the pseudomolecular ion peaks at m/z 1531 and 1587, respectively, being different from that of 121 (m/z 1559). The HRFABMS data of 127 and 128 suggested the molecular formulas as $C_{77}H_{136}N_4O_{22}S_2$ and $C_{81}H_{144}N_4O_{22}S_2$, respectively, which implied that 127 has two less CH_2 units than 121 while 128 includes two more CH_2 units than 121. Since information on the differences of the structures of 127 and 128 from that of 121 was hardly obtained from comparison of their ^1H and ^{13}C NMR data including several types of 2D NMR experiments (^1H-^1H COSY, HSQC, and HMBC), theonezolides B (127) and C (128) were subjected to degradation experiments by ozonolysis, which was also carried out for theonezolide A (121). Treatment of theonezolide B (127) with ozone followed by NaBH$_4$ reduction and acetylation afforded a complex mixture, from which four degradation products were isolated by careful HPLC separation. Three of the four products were revealed to be identical with compounds 123, 124, and 125, which had been obtained by ozonolysis of 121 and corresponded to C-4 ~ C-17, C-18 ~ C-37, and C-43 ~ C-64 moieties of 121, respectively, on the basis of comparison of their ^1H NMR, FABMS, and optical rotation data. The fourth product 129 obtained from theonezolides B (127) showed analogous ^1H NMR spectrum to that of 126 (C-66 ~ C-76 moiety) obtained from 121, and EIMS data indicated that 129 was a homologue of 126 with two less sp^3 methylenes [129: m/z 214 (M$^+$); 126: m/z 242 (M$^+$)]. Ozonolysis of theonezolide C (128) was also carried out by the same procedure described as above to afford three identical products (123, 124, and 125) and one homologous product 130 [EIMS, m/z 270 (M$^+$)] with two more CH_2 groups than 126. From these results, the structures of theonezolides B and C were concluded to be 127 and 128, respectively; viz., the numbers of sp^3 methylenes between the thiazole moiety (C-64 ~ C-66 position) and the terminal amino group of theonezolide A (121), B (127), and C (128) are 8, 6, and 10, respectively. Theonezolides B (127) and C (128) exhibited cytotoxicity against murine lymphoma L1210 and human epidermoid carcinoma KB cells in vitro (IC$_{50}$ values, 127: 5.6 and 11 µg/mL, respectively; 128: 0.3 and 0.37 µg/mL, respectively).

(+)-129   (+)-130

Theonezolides A (121), B (127), and C (128) contain 23 chiral centers, among which we first chose the chiral center at the terminal position bearing the primary amino and secondary methyl groups (C-75 of 121, C-73 of 127, and C-77 of 128) for the study of determination of the absolute stereochemistry. The fragments 126, 129, and 130 retaining the terminal chiral centers of 121, 127 and 128, respectively, were prepared in optically active forms (Scheme 8). The iodide 131 prepared from L-alaninol by the described procedure [78], was treated with Grignard reagent followed by hydroboration to give the alcohol (133), which was converted via 2 steps into the amide (135). Deprotection and acetylation of 135 afforded S-(-)-enantiomer of 126, the ^{1}H NMR and EIMS spectra of which were identical with those of 126 obtained by ozonolysis of 121. Since the sign of optical rotation of synthetic (-)-126 ($[\alpha]_D$ -10°) was opposite to that of (+)-126 ($[\alpha]_D$ +12°) from natural specimen, the absolute configuration of C-75 position of theonezolide A (121) was established as R. By the similar procedures, S-(-)-enantiomers of 129 and 130 were prepared and their ^{1}H NMR and EIMS spectral data were completely superimposable to those of 129 and 130 derived from natural theonezolides B (127) and C (128). The signs of their optical rotations, however, were also different (synthetic: (-)-129, $[\alpha]_D$ -8°; (-)-130, $[\alpha]_D$ -12°; natural: (+)-129, $[\alpha]_D$ +14°; (+)-130, $[\alpha]_D$ +10°). Theonezolides B (127) and C (128) were therefore revealed to have 73R- and 77R-configurations, respectively.

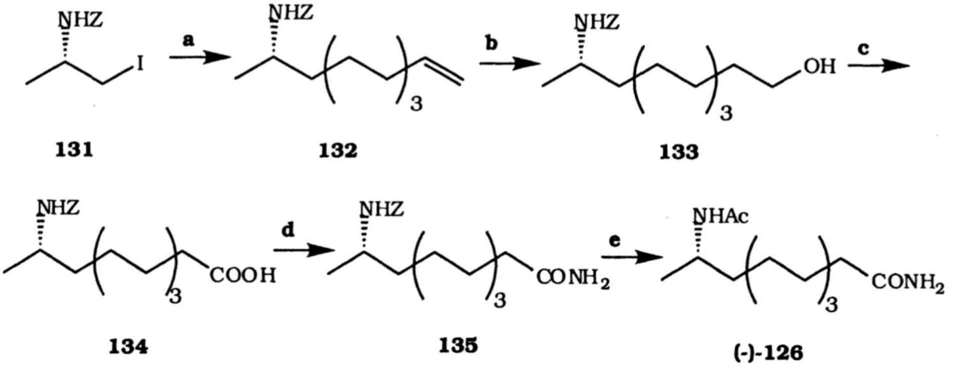

Scheme 8. (a) CH$_2$=CH(CH$_2$)$_6$MgBr, CuI, THF (74%) (b) BH$_3$-THF; H$_2$O$_2$-NaOH (69%) (c) PDC, DMF (60%) (d) 1) N,N'-Carbonyldiimidazole, THF, 2) NH$_3$ (98%) (e) 1) 10% Pd-C, H$_2$, MeOH 2) Ac$_2$O, Pyridine (74%) .

Subsequently we studied the stereochemistry of the C-4 ~ C-17 fragment (123), which were commonly obtained by ozonolysis of the three macrolides, theonezolides A (121), B (127), and C (128). This fragment (123) contains 4 chiral centers which comprise two 1,3-diol type moieties (14-OAc/16-OMe and 8-OAc/10-OAc); the numberings of the carbons for compound 123 and the following model compounds described here corresponded to those of the parent natural product, theonezolide A (**121**). The relative configurations of the two 1,3-diol type moieties of compound **123** were investigated by preparation of four model compounds (**136/137** and **138/139**) corresponding to *syn* and *anti* diastereomers for the 14-OAc/16-OMe and 8-OAc/10-OAc moieties, respectively.

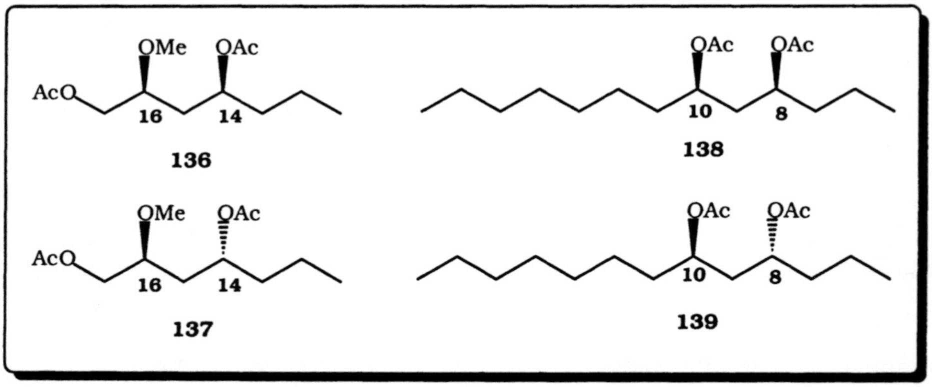

Synthesis of the four model compounds (*syn*: **136** and **138**; *anti*: **137** and **139**) began with the diastereoisomeric homoallylalcohols [**140** (*syn*) and **141** (*anti*)], respectively, which were both prepared from (-)-(*S*)-malic acid by literature procedures [79]. Preparation of the diastereomer (**136**) was outlined in Scheme 9. The secondary hydroxyl group of the *syn*

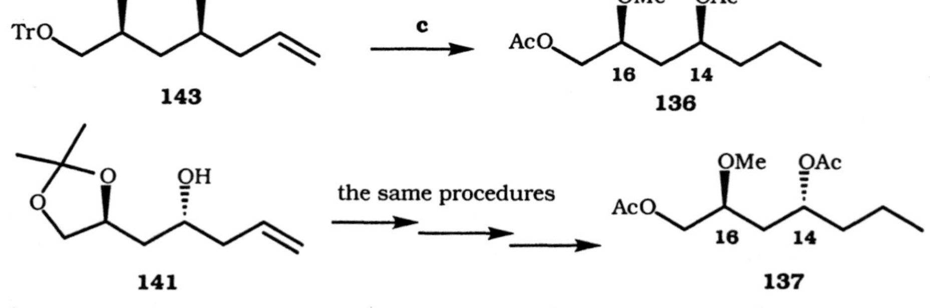

Scheme 9. (a) 1) BnBr, NaH, *n*-Bu4NI, DMF (79%); 2) 3N HCl, THF (99%). (b) 1) Ph3CCl, DMAP, pyridine (64%); 2) MeI, KH, THF (92%). (c) 1) H2, Pd(OH)2/C, EtOH (66%); 2) Ac2O, pyridine (85%).

homoallylalcohol (**140**) was protected with a benzyl ether, and the acetonide group was deprotected to give the 1,2-diol (**142**). The primary hydroxyl group of **142** was protected with triphenylmethyl (Tr) group, and the remaining hydroxyl group was then methylated to furnish the methyl ether (**143**). Hydrogenolysis of both benzyl and trityl groups of **143** followed by acetylation afforded the *syn* diastereomer (**136**). The *anti* diastereomer (**135**) was obtained by the same procedures for **136** (Scheme 9), starting from the *anti* homoallylalcohol (**141**). The *syn* diacetate (**138**) was prepared as shown in Scheme 10. After removal of the acetonide

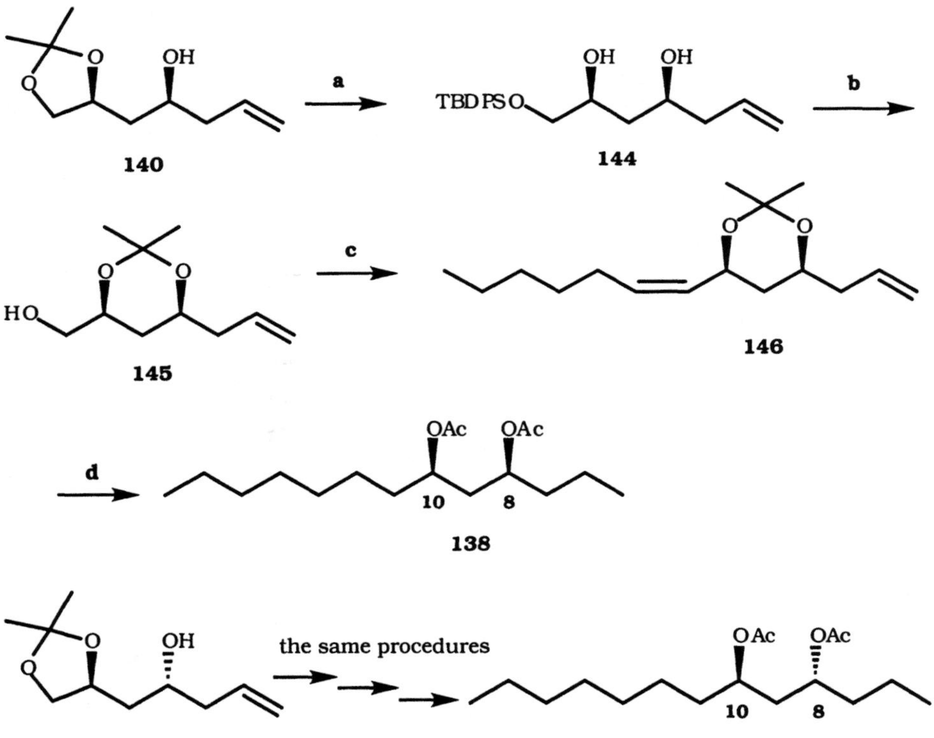

Scheme 10. (a) 1) 3N HCl, THF (80%); 2) *t*-BuPh$_2$SiCl, imidazole, CH$_2$Cl$_2$ (53%). (b) 1) (CH$_3$)$_2$C(OCH$_3$)$_2$, PPTS, CH$_2$Cl$_2$ (72%); 2) 2N NaOH, EtOH (50%). (c) 1) PCC, MS3Å, CH$_2$Cl$_2$ (72%); 2) *n*-BuLi, CH$_3$(CH$_2$)$_5$P+Ph$_3$Br-, THF (32%, 2 steps). (d) 1) H$_2$, Pd/C (10%), EtOH (84%); 2) 3N HCl, THF (99%); 3) Ac$_2$O, pyridine (65%).

group of the *syn* homoallylalcohol (**140**), the primary hydroxyl group was selectively protected with *t*-butyldiphenylsilyl (TBDPS) group to give the 1,3-diol (**144**). Protection of the 1,3-diol of **144** as isopropylidene ketal followed by desilylation afforded the acetonide (**145**), which was oxidized with pyridinium chlorochromate and the resulting aldehyde was subjected to Wittig reaction with *n*-hexyltriphenylphosphonium bromide to give the olefin (**146**). Hydrogenation of **146** followed by hydrolysis of the acetonide and acetylation furnished the *syn* diastereomer (**138**). Starting from the *anti* homoallylalcohol (**141**), the *anti* diacetate (**139**) was also prepared by the same methods as shown in Scheme 10. Comparisons of the ^1H and ^{13}C NMR data of synthetic *syn* and *anti* diastereomers (**136/137** and **138/139**) with those of the C-4 ~ C-17 fragment (**123**) derived from the natural specimens apparently showed that the ^1H and ^{13}C chemical shifts of *syn* diastereomers (**136** and **138**) corresponded to those of **123** quite better than those of *anti* diastereomers (**137** and **139**), as to both of the C-14 ~ C-17 positions and the C-8 ~ C-10 positions, respectively. Particularly, the ^1H NMR signals for methylene protons located between the two oxymethines were characteristic: the two methylene protons of *syn* diastereomers **136** [H_2-15: δ_H 1.66 and 1.89 (each 1H)] and **138** [H_2-9: δ_H 1.65 and 1.90 (each 1H)] were magnetically non-equivalent, while those of *anti* diastereomers **137** [H_2-15: δ_H 1.60 (2H)] and **139** (H_2-9: δ_H 1.80 (2H)] resonated equivalently [80]. The methylene proton signals of **123** (H_2-15 and H_2-9) were both magnetically non-equivalent [H_2-15: δ_H 1.64 and 1.90 (each 1H); H_2-9: δ_H 1.64 and 1.90 (each 1H)]. Thus, the two 1,3-diol type systems of the C-4 ~ C-17 fragment (**123**), the 14-OAc/16-OMe and 8-OAc/10-OAc moieties, were both suggested as *syn*. Consequently, out of sixteen possibilities, four feasible structures (**123a** and **123b** and their enantiomers) now remain for the C-4 ~ C-17 fragment (**123**) of theonezolides A (**121**), B (**127**), and C (**128**). Further investigation to establish the absolute stereochemistry of **123** is currently in progress by us on the basis of synthesis of **123a** and **123b**.

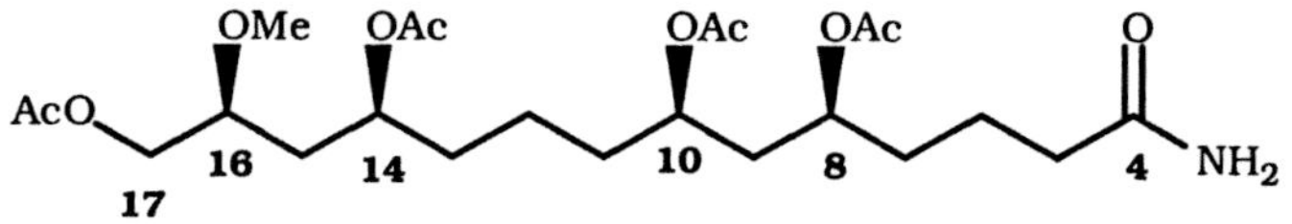

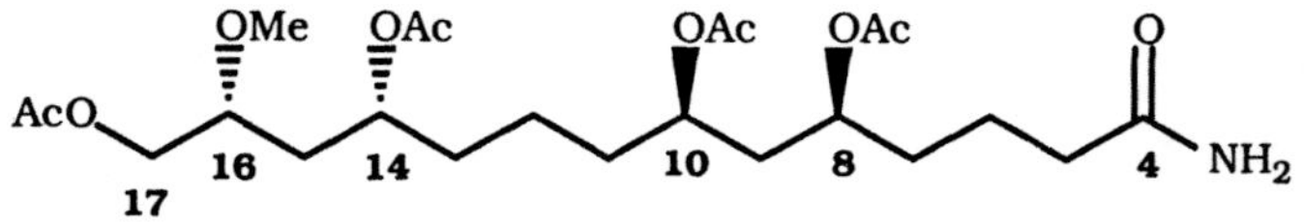

ACKNOWLEDGMENT

We thank Mr. Z. Nagahama for his help with collecting the sponges and tunicates. Financial support by a Grant-in-Aid for Scientific Research from the Ministry of Education, Science, Sports and Culture of Japan is gratefully acknowledged.

REFERENCES

[1]   Kobayashi, J.; Ishibashi, M. In *The Alkaloids*; Brossi, A.; Cordell, G. A. Eds.; Academic Press: San Diego, **1992**; Vol. 41, pp. 41-124.
[2]   Kobayashi, J.; Ishibashi, M. *Chem. Rev.*, **1993**, *93*, 1753-1769.
[3]   Kobayashi, J.; Ishibashi, M. *Heterocycles*, **1996**, *42*, 943-970.
[4]   Ishibashi, M.; Kobayashi, J. *Heterocycles*, **1997**, *44*, 543-572.
[5]   Kobayashi, J.; Ishibashi, M. In *Comprehensive Natural Product Chemistry*; Vol. 8, K. Mori, Ed.; Elsevier Science; Oxford, in press.
[6]   Ishibashi, M.; Ohizumi, Y.; Sasaki, T.; Nakamura, H.; Hirata, Y.; Kobayashi, J. *J. Org. Chem.* **1987**, *52*, 450-453.
[7]   Kobayashi, J.; Naitoh, K.; Doi, Y.; Deki, K.; Ishibashi, M. *J. Org. Chem.* **1995**, *60*, 6941-6945.
[8]   Garner, P.; Park, J. M. *Org. Synth.*, **1991**, *70*, 18-28.
[9]   Doi, Y.; Ishibashi, M.; Kobayashi, J. *Tetrahedron* **1996**, *52*, 4573-4580.
[10] Svirskaya, P. I.; Maiti, S. N.; Jones, A. J.; Khouw, B.; Leznoff, C. C. *J. Chem. Ecol.* **1984**, *10*, 795-807.
[11] Cheng, J.-F.; Kobayashi, J.; Ohizumi, Y.; Nakamura, H.; Hirata Y.; Sasaki, T. *J. Chem. Soc., Perkin Trans. 1* **1988**, 2403-2406.
[12] Kobayashi, J.; Cheng, J.-F.; Yamamura, S.; Sasaki, T.; Ohizumi, Y. *Heterocycles* **1990**, *31*, 2205-2208.
[13] Kobayashi, J.; Cheng, J.-F.; Ishibashi, M.; Wälchli, M. R.; Yamamura, S.; Ohizumi, Y. *J. Chem. Soc., Perkin Trans. 1* **1991**, 1135-1137.
[14] Alvi, K. A.; Jaspars, M.; Crews, P. *Bioorg. Med. Chem. Lett.* **1994**, *4*, 2447-2450.
[15] Hiraki, T.; Yamagiwa, Y.; Kamikawa, T. *Tetrahedron Lett.* **1995**, *36*, 4841-4844.
[16] Takikawa, H.; Maeda, T.; Mori, K. *Tetrahedron Lett.* **1995**, *36*, 7689-7692.
[17] Takikawa, H.; Maeda, T.; Seki, M.; Koshino, H.; Mori, K. *J. Chem. Soc., Perkin Trans. 1* **1997**, 97-111.
[18] Breitmaier, E.; Voelter, W. In ^{13}C *NMR Spectroscopy*; Verlag Chemie: Weinheim, **1978**; p. 276.
[19] Kobayashi, J.; Tsuda, M.; Cheng, J.-F.; Ishibashi, M.; Takikawa, H.; Mori, K. *Tetrahedron Lett.* **1996**, *37*, 6775-6776.
[20] Yoda, H.; Oguchi, T.; Takabe, K. *Tetrahedron Lett.* **1997**, *38*, 3283-3284.
[21] Knapp, S.; Dong, Y. *Tetrahedron Lett.* **1997**, *38*, 3813-3816.
[22] Kobayashi, J.; Doi, Y.; Ishibashi, M. *J. Org. Chem.* **1994**, *59*, 255-257.
[23] Gunstone, F. D.; Pollard, M. R.; Scrimgeour, C. M.; Vedanayagam, H. S. *Chem. Phys. Lipids* **1977**, *18*, 115-129.
[24] Bock, K.; Lundl, I.; Pedersen, C. *Tetrahedron Lett.* **1973**, 1037-1040.
[25] Doi, Y.; Ishibashi, M.; Kobayashi, J. *Tetrahedron* **1994**, *50*, 8651-8656.
[26] Duynstee, H. I.; Wijsman, E. R.; van der Marel, G. A.; van Boom, J. H. *Synlett* **1996**, 313-314.
[27] Knapp, S.; Gore, V. K. *J. Org. Chem.* **1996**, *61*, 6744-6747.

[28]   Gerwick, W. H.; Nagle, D. G.; Proteau, P. J. In *Topics in Current Chemistry*, Vol. 167, *Marine Natural Products - Diversity and Biosynthesis*; Scheuer, P. J., Ed.; Springer-Verlag: Berlin, **1993**; pp 167-180.

[29]   Sakemi, S.; Higa, T.; Anthoni, U.; Christophersen, C. *Tetrahedron* **1987**, *43*, 263-268.

[30]   De Guzman, F. S.; Schmitz, F. J. *J. Nat. Prod.* **1990**, *53*, 926-931.

[31]   Wells, R. J. *Tetrahedron Lett.* **1976**, 2637-2638.

[32]   Murayama, T.; Ohizumi, Y.; Nakamura, H.; Sasaki, T.; Kobayashi, J. *Experientia* **1989**, *45*, 898-899.

[33]   Snider, B. B.; Shi, Z. *J. Am. Chem. Soc.* **1992**, *114*, 1790-1800.

[34]   Dussault, P. H.; Woller, K. R. *J. Am. Chem. Soc.* **1997**, *119*, 3824-3825.

[35]   Kobayashi, J.; Takeuchi, S.; Ishibashi, M.; Shigemori, H.; Sasaki, T. *Tetrahedron Lett.* **1992**, *33*, 2579-2580.

[36]   Phillipson, D. W.; Rinehart, K. L., Jr. *J. Am. Chem. Soc.* **1983**, *105*, 7735-7736.

[37]   Hofheinz, W.; Schönholzer, P. *Helv. Chim. Acta* **1977**, *60*, 1367-1370.

[38]   Tsukamoto, S.; Takeuchi, S.; Ishibashi, M.; Kobayashi, J. *J. Org. Chem.* **1992**, *57*, 5255-5260.

[39]   Kobayashi, J.; Tsukamoto, S.; Takeuchi, S.; Ishibashi, M. *Tetrahedron* **1993**, *49*, 5955-5960.

[40]   Takeuchi, S.; Kikuchi, T.; Tsukamoto, S.; Ishibashi, M.; Kobayashi, J. *Tetrahedron* **1995**, *51*, 5979-5986.

[41]   Klyne, W. *Tetrahedron* **1961**, *13*, 29-47.

[42]   Nagai, Y.; Kusumi, T. *Tetrahedron Lett.* **1995**, *36*, 1853-1856.

[43]   Takeuchi, S.; Ishibashi, M.; Kobayashi, J. *J. Org. Chem.* **1994**, *59*, 3712-3713.

[44]   Ishibashi, M.; Takeuchi, S.; Kobayashi, J. *Tetrahedron Lett.* **1993**, *34*, 3749-3750.

[45]   Takeda, K.; Nakayama, I.; Yoshii, E. *Synlett* **1994**, 178.

[46]   Asami, M.; Ishizaki, T.; Inoue, S. *Tetrahedron Lett.* **1995**, *36*, 1893-1894.

[47]   Miyaoka, H.; Watanuki, T.; Saka, Y.; Yamada, Y. *Tetrahedron* **1995**, *51*, 8749-8756.

[48]   Piancatelli, G.; D'Auria, M.; D'Onofrio, F. *Synthesis* **1994**, 867-889.

[49]   Ishiyama, H.; Ishibashi, M.; Ogawa, A.; Yoshida, S.; Kobayashi, J. *J. Org. Chem.* **1997**, *62*, 3831-3836.

[50]   Ishibashi, M.; Ohizumi, Y.; Cheng, J.-F.; Nakamura, H.; Hirata, Y.; Sasaki, T.; Kobayashi, J. *J. Org. Chem.* **1988**, *53*, 2855-2858.

[51]   Rychnovsky, S. D.; Rogers, B.; Yang, G. *J. Org. Chem.* **1993**, *58*, 3511-3515.

[52]   Chou, T.; Kuramoto, M.; Otani, Y.; Shinkai, M.; Yazawa, K.; Uemura, D. *Tetrahedron Lett.* **1996**, *37*, 3871-3874.

[53]   Kouda, K.; Ooi, T.; Kaya, K.; Kusumi, T. *Tetrahedron Lett.* **1996**, *37*, 6347-6350.

[54]   Barrow, R. A.; Capon, R. J. *Aust. J. Chem.* **1994**, *47*, 1901-1918.

[55]   Fusetani, N.; Li, H.-Y.; Tamura, K.; Matsunaga, S. *Tetrahedeon*, **1993**, *49*, 1203-1210 and references cited therein.

[56]   Kobayashi, J.; Naitoh, K.; Ishida, K.; Shigemori, H.; Ishibashi, M. *J. Nat. Prod.* **1994**, *57*, 1300-1303.

[57]   Kobayashi, J.; Ishida, K.; Naitoh, K.; Shigemori, H.; Mikami, Y.; Sasaki, T. *J. Nat. Prod.* **1993**, *56*, 1350-1355.

[58]   Isaacs, S.; Kashman, Y.; Loya, S.; Hizi, A.; Loya, Y. *Tetrahedron* **1993**, *49*, 10435-10438.

[59]   Li, Y.; Ishibashi, M.; Sasaki, T.; Kobayashi, J. *J. Chem. Research (S)* **1995**, 126-127.
[60]   Carballeira, N. M.; Shalabi, F. *J. Nat. Prod.*, **1993**, *56*, 739-746.
[61]   Kitagawa, I.; Kobayashi, M. *Gazz. Chim. Ital.* **1993**, *123*, 321-327.
[62]   Fusetani, N.; Matsunaga, S. *Chem. Rev.* **1993**, *93*, 1793-1806.
[63]   Kobayashi, J.; Murayama, T.; Ohizumi, Y.; Sasaki, T.; Ohta, T.; Nozoe, S. *Tetrahedron Lett.* **1989**, *30*, 4833-4836.
[64]   Kobayashi, J.; Sato, M.; Murayama, T.; Ishibashi, M.; Wälchli, M. R.; Kanai, M.; Shoji, J.; Ohizumi, Y. *J. Chem. Soc., Chem. Commun.* **1991**, 1050-1052.
[65]   Kobayashi, J.; Sato, M.; Ishibashi, M.; Shigemori, H.; Nakamura, T.; Ohizumi, Y. *J. Chem. Soc., Perkin Trans. 1* **1991**, 2609-2611.
[66]   Kobayashi, J.; Itagaki, F.; Shigemori, H.; Ishibashi, M.; Takahashi, K.; Ogura, M.; Nagasawa, S.; Nakamura, T.; Hirota, H.; Ohta, T.; Nozoe, S. *J. Am. Chem. Soc.* **1991**, *113*, 7812-7813.
[67]   Itagaki, F.; Shigemori, H.; Ishibashi, M.; Nakamura, T.; Sasaki, T.; Kobayashi, J. *J. Org. Chem.* **1992**, *57*, 5540-5542.
[68]   Tsukamoto, S.; Ishibashi, M.; Sasaki, T.; Kobayashi, J. *J. Chem. Soc., Perkin Trans. 1* **1991**, 3185-3188.
[69]   Kobayashi, J.; Tsukamoto, S.; Tanabe, A.; Sasaki, T.; Ishibashi, M. *J. Chem. Soc., Perkin Trans. 1* **1991**, 2379-2383.
[70]   Kobayashi, J.; Itagaki, F.; Shigemori, H.; Sasaki, T. *J. Nat. Prod.* **1993**, *56*, 976-981.
[71]   Kobayashi, J.; Kondo, K.; Shigemori, H.; Ishibashi, M.; Sasaki, T.; Mikami, Y. *J. Org. Chem.* **1992**, *57*, 6682-6684.
[72]   Kobayashi, J.; Kondo, K.; Ishibashi, M.; Wälchli, M. R.; Nakamura, T. *J. Am. Chem. Soc.* **1993**, *115*, 6661-6665.
[73]   Groweiss, A.; Shmueli, U.; Kashman, Y. *J. Org. Chem.* **1983**, *48*, 3512-3516.
[74]   Roesener, J. A.; Scheuer, P. J. *J. Am. Chem. Soc.* **1986**, *108*, 846-847.
[75]   Matsunaga, S.; Fusetani, N.; Hashimoto, K.; Koseki, K.; Noma, M. *J. Am. Chem. Soc.* **1986**, *108*, 847-849.
[76]   Rho, M.-C.; Park, Y.-H.; Sasaki, S.; Ishibashi, M.; Kondo, K.; Kobayashi, J.; Ohizumi, Y. *Can. J. Physiol. Pharmacol.* **1996**, *74*, 193-199.
[77]   Kondo, K.; Ishibashi, M.; Kobayashi, J. *Tetrahedron* **1994**, *50*, 8355-8362.
[78]   Schlessinger, R. H.; Iwanowicz, E. J. *Tetrahedron Lett.* **1987**, *28*, 2083-2086.
[79]   Hanessian, S.; Ugolini, A.; Therien, M. *J. Org. Chem.* **1983**, *48*, 4427-4430.
[80]   Mynderse, J. S.; Moore, R. E. *Phytochemistry* **1979**, *18*, 1181-1183.

Atta-ur-Rahman (Ed.) *Studies in Natural Products Chemistry, Vol. 23*

AROMATIC ALKALOIDS FROM ASCIDIANS

BRUCE F. BOWDEN

School of Biomedical and Molecular Sciences, James Cook University, Townsville 4811 Qld., Australia

ABSTRACT: More than 200 aromatic alkaloids isolated from marine tunicates are grouped into structural types and discussed in terms of their reported pharmacological activity. The major groups of alkaloids which are discussed in detail include the benzopentathiepins and trithianes, lamellarins and related structures, ecteinascidins, β-carboline alkaloids, and the pyridoacridines. References to reported syntheses are also included. Effects of substituent changes within groups on observed activity (structure/activity relationships) are discussed where sufficient data is available to draw such conclusions. The current state of development and potential of the ecteinascidins and lamellarins in cancer treatment and control is summarised, together with details of current knowledge regarding their modes of action. The antiviral activity of the β-carboline alkaloids (eudistomins and eudistomidins) is discussed in structural terms. Cytotoxicity data for the benzopentathiepins (lissoclinotoxins and varacins) and for the pyridoacridine alkaloids are summarised, together with evidence for their modes of action. The high potential for some of these alkaloids to be developed as therapeutic agents for the treatment of cancer and viral infection is reinforced by the summarised data.

INTRODUCTION

Much interest and research over the past 15 or so years has been directed towards ascidian metabolites because of the high incidence of pharmacological activity that they display. Few comprehensive reviews exist on the topic; the only general ones being those produced by Davidson in 1993 [1,2] and the Marine Natural Product literature reviews by Faulkner which always contain a section on tunicate metabolites [3-13]. Although a number of non-nitrogenous metabolites have been isolated from the tunicates, the majority of compounds isolated have been derived from amino acids. In particular, the two most commonly encountered classes are cyclic peptides and polycyclic aromatic alkaloids.

SCOPE AND ORGANISATION OF THIS REVIEW

Although the division between the peptides and other nitrogen-containing metabolites derived from ascidians is not a clear cut one, the literature published on the non-peptidic, tunicate-derived alkaloids which contain aromatic (often in combination with heteroaromatic) rings has been summarised. The few linear peptides (which contain aromatic amino acid residues) have also been included because of their apparent structural relationship to some of the reported alkaloids. Cyclic peptides, nucleosides and simple heteroaromatic molecules such as pyrroles and polypyrroles are outside the scope of this review.

The emphasis is on the reported pharmacological activities of compounds and any leads to structure-activity relationships within groups of metabolites. For this reason, alkaloids are grouped in terms of structural similarity (or apparent biogenetic origin) rather than in groups based on their source species, genera, or families. This facilitates dealing with metabolites which have been isolated from two or more genera or families and avoids the question of whether such metabolites are produced by the host or by microbes (symbiotic algae, cyanobacteria, bacteria, fungi) which may be present. Some of the compounds which are discussed are strikingly similar to reported marine sponge metabolites. Indeed, in one or two cases the *same* metabolite has been isolated from members of these two very distinct groups of organisms. Such examples are likely to be the product of a symbiont which can co-exist with either organism, but proof of such a hypothesis is difficult. Although in sponges, some cell separation strategies have been used to locate metabolites within cells and to pin-point the likely producer of metabolites, to my knowledge, no comparable studies have been carried out with ascidians. I am also not aware of any authenticated examples of marine microorganism cultures which have yielded any of the chemicals isolated from extracts of their normal host.

Simple Derivatives of the Amino acids Phenylalanine and Tyrosine

The simplest aromatic alkaloids which have been isolated from ascidians are those derived directly or indirectly from phenylalanine, tyrosine, phenylethylamine or tyramine. Phenylethylamine (1) itself has been reported [14] from a *Lissoclinum* sp., as has its urea derivative (2) from *Didemnum ternatanum* [15]. The methyl ether of 3,5-diiodotyramine (3) and the corresponding urea derivative (4) were isolated from an unidentified Didemnid tunicate from Cocos lagoon, Guam, and (3) was reported to have marginal cytotoxic activity (IC_{50} 20 µg/mL against L1210

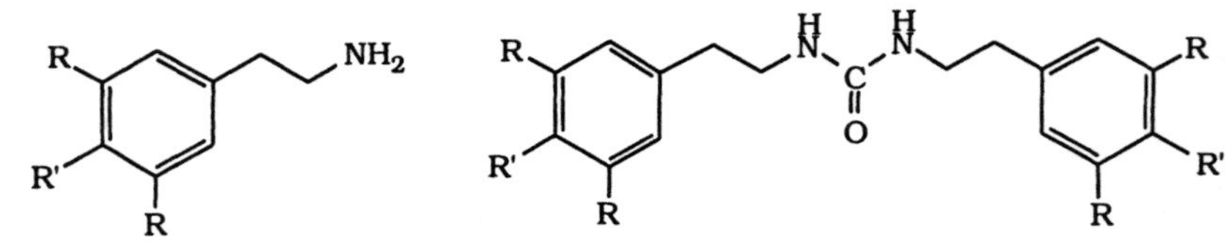

1. R = R' = H
3. R = I, R' = OCH$_3$

2. R = R' = H
4. R = I, R' = OCH$_3$

murine lymphoma cells in culture)[16] and antifungal activity against *Candida albicans*. No activity data was presented for the urea (4).

The genus *Aplidium* has afforded some metabolites where the amino group of what initially appears to have been a dopamine residue has ended up as part of a heteroaromatic ring. Three such metabolites: (5), (6), and (7), were reported from *Aplidium pliciferum* [17]. No activity data was presented for these metabolites, however the apparently related 1,2,3-trithiane (8) reported from the New Zealand *Aplidium* sp. D [18], exhibited antibacterial activity against gram positive bacteria (*Bacillis subtilis*) and fungi (*Candida albicans*), but no activity against gram negative bacteria (*Escherichi coli* and *Pseudomonas aeruginosa*). The trithiane ring underwent base-induced epimerisation to (9), and both isomers were marginally cytotoxic (IC_{50} 12-13 μg/mL) to P388 murine leukemia cells *in vitro*. No antiviral activity was observed, but both displayed cytotoxicity to the host basal cell carcinoma (BSC) cells in the antiviral assay [18].

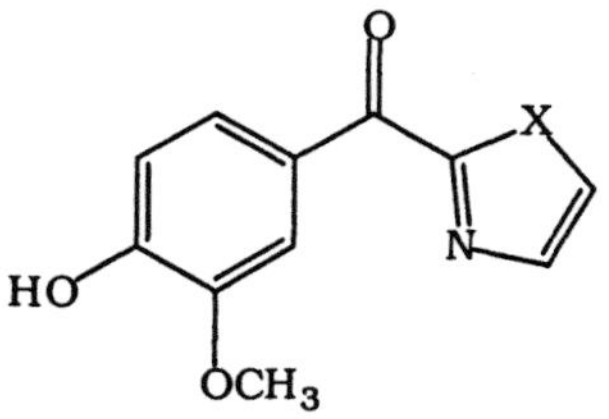

5. X = S
7. X = NH

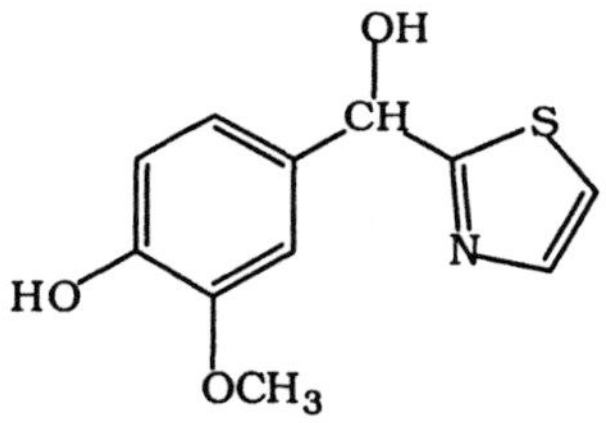

6.

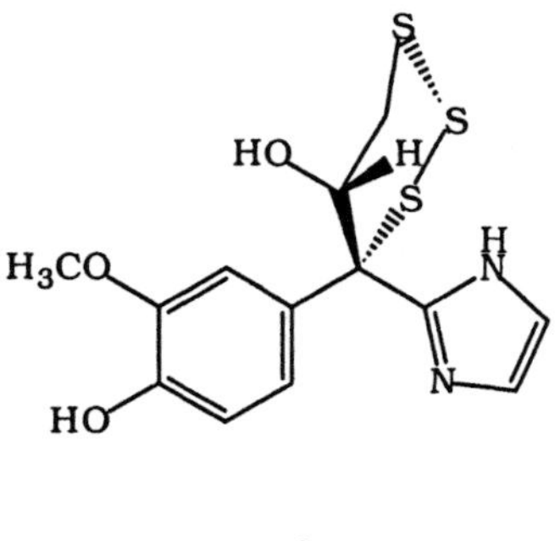

8.

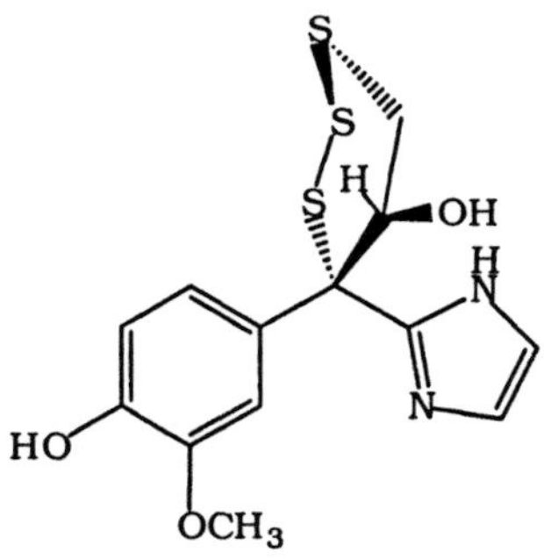

9.

Lissoclinotoxins, Varacins and Related Metabolites

A significant number of benzopentathiepin and benzotrithiane derivatives have been reported since the first report of a trithiane. The first reported example of these metabolites was lissoclinotoxin A [19], but the literature on this group is complicated by the instability of the free bases [20], the equilibrium between the pentathiepins and the trithianes [21], and the initial incorrect positional location of substituents in lissoclinotoxin A [19]. With the exception of the N-acetylated derivatives which have been reported [22], it appears that all benzopentathiepins which have been reported were in fact isolated as their protonated amine salts. The free bases are reported to be unstable; the instability is believed to result from attack by the free amino group on the pentathiepin ring [20].

Lissoclinotoxin A was initially reported [19] as the benzotrithiane (10) but the structure was subsequently corrected [21] to the benzopentathiepin (11). The chirality of such unsymmetrically substituted benzopentathiepins was noted by two authors [14,23] and optical activity has been reported [14]. Lissoclinotoxin A has been reported from *Lissoclinum perforatum* [19,21] from the Eastern Atlantic where it co-occurred with lissoclinotoxin B (12), and from a *Lissoclinum* sp. from the Great Barrier Reef (GBR), Australia [14].

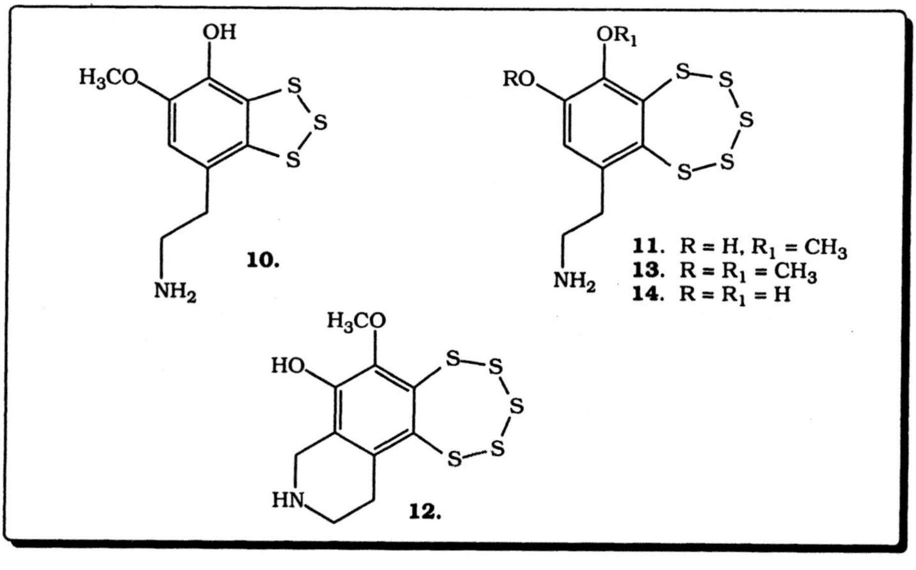

Varacin (13), the methyl ether of Lissoclinotoxin A, has been reported from *Lissoclinum vareau* [24], and from a "Far Eastern" *Polycitor* sp. (where it was also isolated as its N-acetyl derivative after acetylation of the crude extract to facilitate separation) [22]. A number of syntheses of varacin have been reported [20,25,26,27]. "Desmethylvaracin" (14) was reported as its trifluoracetate (TFA salt) from a *Eudistoma* sp. from Palau [28]. The salt of the thiomethyl derivative of varacin (15) was isolated as a 2:3 inseparable mixture with its trithiane analogue (16) from a *Lissoclinum* sp. from Pohnpei, while the N,N-dimethyl analogues (17), (18) of these two derivatives were isolated from *Lissoclinum japonicum* from Palau [28]. Lissoclinotoxins C (19) and D (20) were reported from an Australian collection of a *Lissoclinum* sp [14].

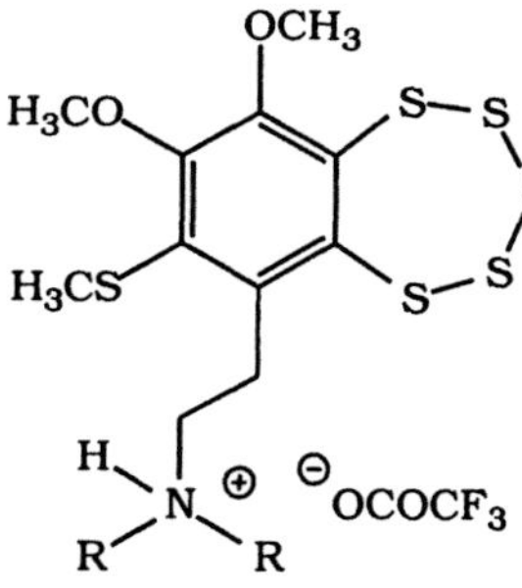

15. R = H
17. R = CH₃

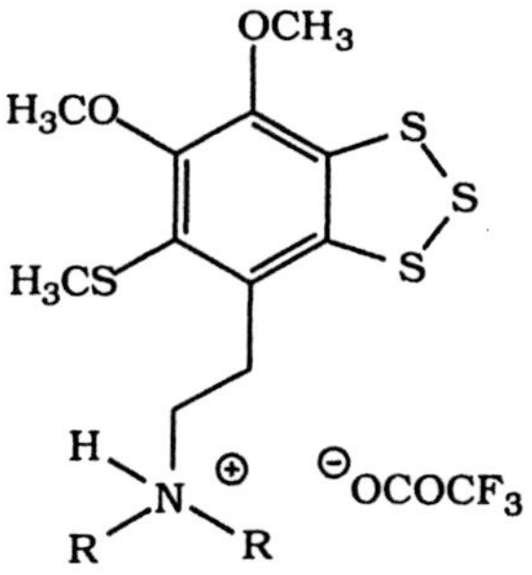

16. R = H
18. R = CH₃

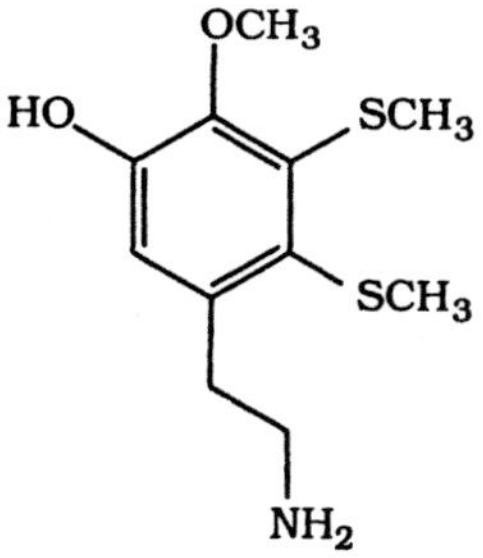

19.

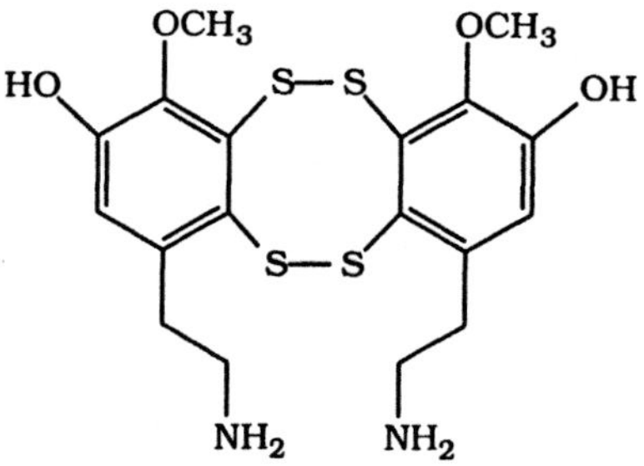

20.

The N-acetyl derivatives of varacins A (21), B (22) and C (23) were reported from fractionation of the acetylated extract of a "Far Eastern" *Polycitor* sp. [22].

Antimicrobial Activity

As a general rule, these sulfur-containing derivatives of tyramine exhibit good antimicrobial activity : lissoclinotoxin A is reported to be active against a wide range of both gram positive and gram negative bacteria, as well as yeasts and fungi [21]. Varacin (13) is reported to afford a 20mm zone of inhibition against *Bacillis subtilis* at 0.1 μg / disk [22].

Lissoclinotoxins A (**11**) and B (**12**) yield zones of 35mm and 30mm respectively against *Staphylococcus aureus* at 100 µg/disk [21], while the N,N-dimethyl metabolites (**17, 18**) afforded a 15 mm zone at the same concentration [28]. The reported antifungal activity of varacin against *Candida albicans* varies from zones of inhibition of 20 mm at 0.1 µg per disk [22] to 14mm at 2 µg/disk [24] while lissoclinotoxin A is reported to yield zones of inhibition of 27 mm at 40 µg/disk, 19 mm at 10 µg/disk and 8 mm at 1 µg/disk [14] with a minimum inhibitory concentration (MIC) of 40 µg/mL [21]. Lissoclinotoxin D was reported to be a little less active than A with inhibition zones of 19 mm at 40 µg/disk and 15 mm at 10 µg/disk [14], and the N,N-dimethyl derivatives (**17, 18**) are also reported to be active [28]. In order to produce comparable zones of inhibition to those observed with varamine using N-acetylvaramine against *B. subtilis* and *C. albicans*, 10 and 100 times as much per disk respectively were required. Similar results were observed for the other N-acetylated derivatives, varacins A-C (**14-16**), yet the authors stated: "This series does not depend upon the presence of the free amino group for its activity "[22].

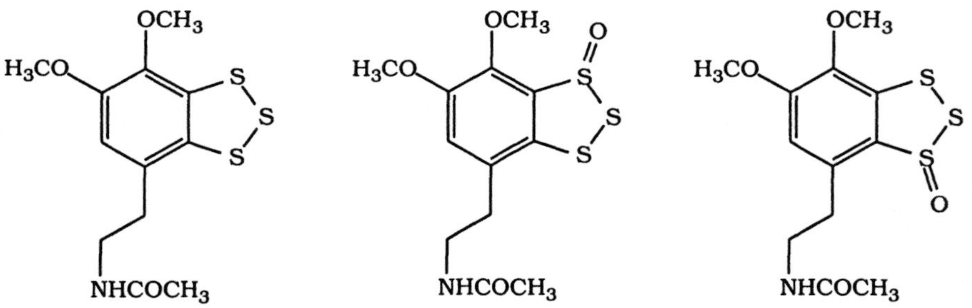

21.   22.   23.

Lissoclinotoxins A and B were active against ichthyopathogenic strains of *Aeromonas salmonicida* and *Vibrio anguillarum* , while lissoclinotoxin A also exhibited activity against resistant malaria, *Plasmodium faciparum* with an IC_{50} of 296 nM/L [21].

Cytotoxicity

As well as having antimicrobial activity, the benzopentathiepins and benzotrithianes are cytotoxic. The cytotoxicity of these metabolites was first reported for lissoclinotoxin A in the form of a sea urchin egg assay, where a concentration of 16 µg/mL inhibited cell division [21], then the

IC_{50} against L1210 murine leukemia cultured cells was reported in the range 1-4 µg/mL [19,21]. In *in vivo* screening against L1210 in mice, the survival rate of treated mice divided by controls and expressed as a percentage (T/C) was 125% at a dose of 3.125 mg/Kg, but dropped to 25% when a dose of 12.5 mg/Kg was utilised because of the toxicity of lissoclinotoxin A in mice (LD_{50} <50 mg/Kg) [21]. Varacin is more cytotoxic than lissoclinotoxin, with a reported IC_{50} of 0.05 µg/mL against HCT116 (human colon tumor cells) [24]. Preliminary assays suggested that the cytotoxicity was due to DNA damaging activity because of differential cytotoxicity towards the CHO cell line EM9 (chlordeoxyuridine sensitive) compared to BR1 (BCNU resistant) cells [24].

The N,N-dimethyl derivatives (17, 18) were tested against the American National Cancer Institute (NCI) 60 cell line panel but no useful selectivity was found [28].

Protein Kinase C Inhibition

Because Protein Kinase C-mediated phosphorylation of target proteins is thought to be critical to many aspects of cellular physiology that depend on control of mitosis or selective gene expression, inhibitors of PKC are sought as potential antineoplastic agents, as well as for their potential therapeutic value as cardiovascular, inflammatory, and CNS agents. The 2:3 mixture of 15 and 16 inhibited PKC with an IC_{50} of 0.3 µg/mL, and desmethyl varacin (14) showed comparable activity. The mixture of (15 and 16) was also separately tested against PKCα, PKCε and PKCζ (isozymes with different biochemical requirements); all IC_{50} values were in the range 0.8-2.1 µg/mL with the highest activity shown against PKCα. Although the N,N-dimethyl analogues (17, 18) had less activity than the other compounds against PKC, it was noted that the trithiane was twice as active as the pentathiepin, and this was interpreted as an indication that the activity is not merely due to release of elemental sulfur. The amino or dimethylamino group was concluded to be necessary for activity against PKC because the quaternary ammonium salt (24) which was prepared during structural elucidation was essentially inactive [28].

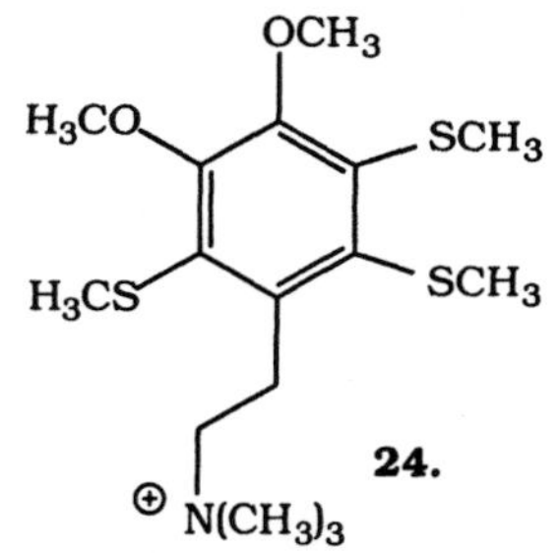

24.

Other Metabolites Containing Residues of One or Two Aromatic Amino Acids

Polycarpine (25), which was isolated from *Polycarpa aurata* collected at Chuk Atoll [29], inhibited inosine monophosphate dehydrogenase (an enzyme used to detect potential antiproliferative drugs) with an IC_{50} of 0.015μg/mL (0.03 μM). Inhibition was however reversible with excess dithiothreitol, which suggested polycarpine reacts with sulfhydryl groups on the enzyme.

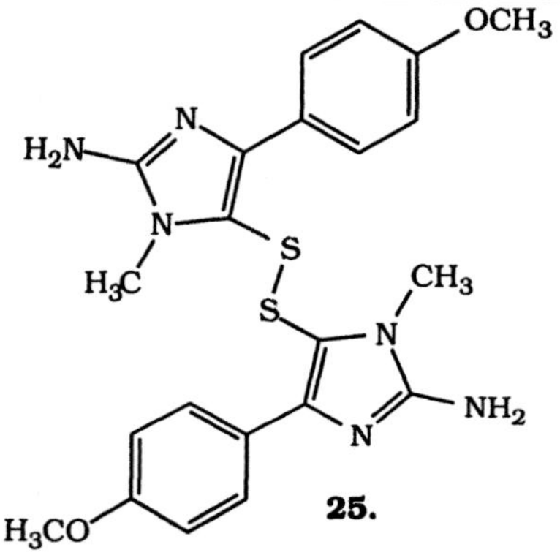

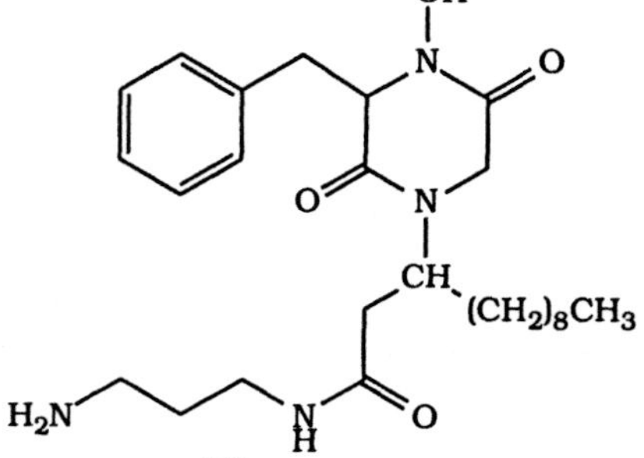

Etzionine (26) was isolated from an unidentified Red Sea tunicate [30] and reported to exhibit antifungal and antibacterial activity. The MIC for etzionine against *C. albicans* was 3 μg/mL in RPMI-1640 broth and 12.5 μg/mL in Sabouraud dextrose broth. Antibacterial activity was observed against gram positive microorganisms (*Aspergillis nidulans* and *B. subtilis*), but it was inactive against gram negative bacteria. Etzionine exhibited marginal cytotoxicity (IC_{50} 10 μg/mL against P388 murine leukemia cells), although the crude extract of the red ascidian was more active (IC_{50} of 5 μg/mL against P388), inferring that an unidentified cytotoxic agent has yet to be isolated [30].

Polyandrocarpamides A-D (27-30), from the marine ascidian *Polyandrocarpa* sp. [31], are metabolites which appear to be derived from tyrosine (bromotyrosine, or iodotyrosine) and tryptophane. No activity was reported for the polyandrocarpamides.

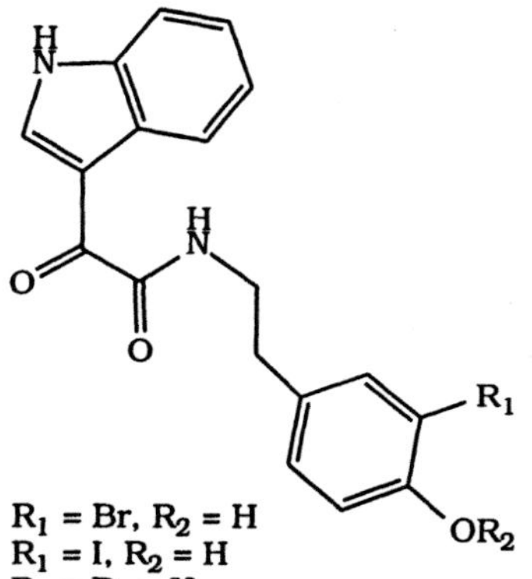

27. R_1 = Br, R_2 = H
28. R_1 = I, R_2 = H
29. R_1 = R_2 = H
30. R_1 = Br, R_2 = $COCH_3$

Similar metabolites which appear to be derived from tyrosine and brominated tyrosine include botryllamides A-D (31-34), which were isolated from brightly coloured styelid ascidians from the genus *Botryllus* collected in Fiji and in Australia [32]. Botryllamide D was reported to exhibit marginal cytotoxicity after 72 hour exposure to the human colon cancer cell line HCT 116 (IC_{50} 17 μg/mL), but were inactive *in vivo* . The rubrolides, which were isolated from *Riterella rubra* [33] may well be derived from precursors similar to the botryllamides, however, the rubrolides are not alkaloids so have been mentioned here only for comparative purposes.

Tunichromes

The tunichromes are a group of reducing blood pigments from tunicates which were initially implicated in metal ion accumulation (particularly vanadium and iron), and have been recently reviewed [34]. They are dipeptides constructed from 2β(3,4,5-trihydoxyphenyl)alanine (Topa) or 2β(3,4-dihydoxyphenyl)alanine (Dopa) and a C-terminally bound 3,4-dihydroxystyrylamine or 3,4,5-trihydroxy-styrylamine unit, and while it is now postulated that small proteins which are rich in Dopa and/or Topa units accumulate the metal ions rather than the tunichromes themselves,

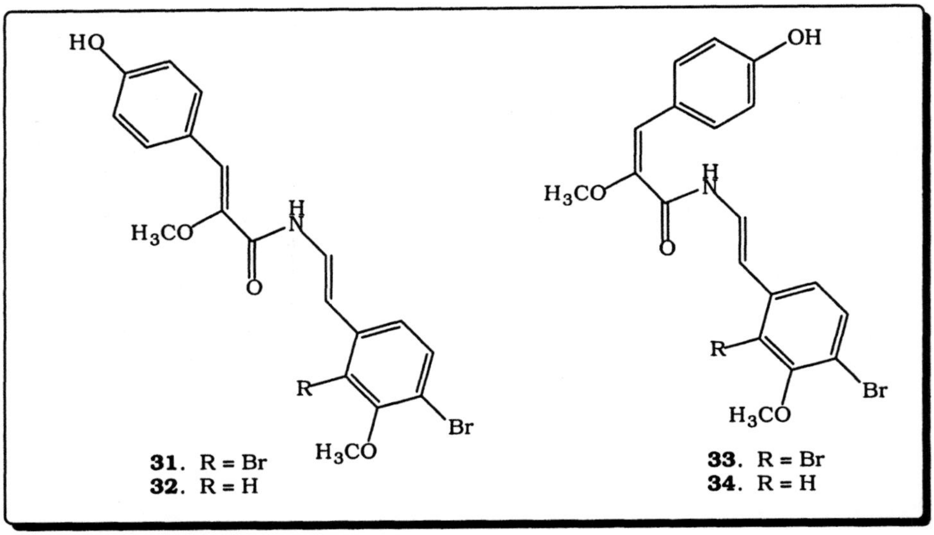

31. R = Br
32. R = H

33. R = Br
34. R = H

they are an interesting group of metabolites. Although in the strictest sense they are (decarboxylated) tripeptides, a brief overview is included in this review of alkaloids because of apparent structural similarities with rigidin, the polycitrins, polycitone A, the lukianols, ningalins, and lamellarins. The structure of Tunichrome B-1 (which is now referred to as Tunichrome An-1 because it was isolated from *Ascidia nigra*) was

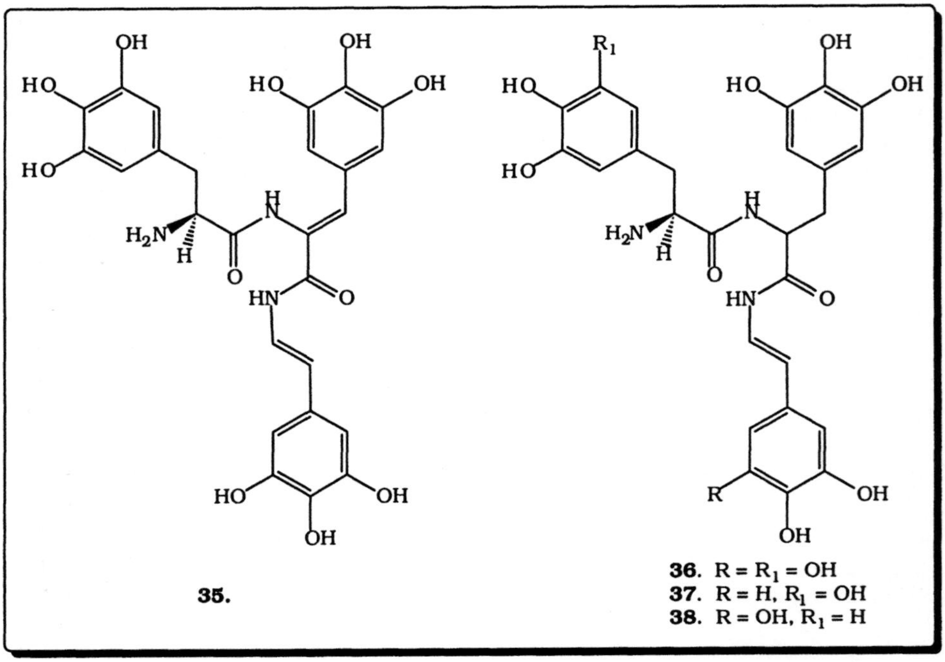

35.

36. R = R$_1$ = OH
37. R = H, R$_1$ = OH
38. R = OH, R$_1$ = H

reported [35] to be (35), and subsequently verified by synthesis [36]. The structure of tunichrome Pm-1 (36) (from *Phallusia mammosa*) and two related tunichromes, Pm-2 (37) and Pm-3 (38), which contain one less double bond than Tunichrome An-1, were subsequently reported, with evidence that tunichrome in tunicate cells is present as the free peptide, not bound to vanadium [37]. A third group of tunichromes, Mm-1 (39) and Mm-2 (40) were reported [38] from the iron accumulating stolidobranch ascidian *Morgula manhattensis*. These are composed of two Dopa residues and either a glycine (Mm-1) or a leucine (Mm-2).

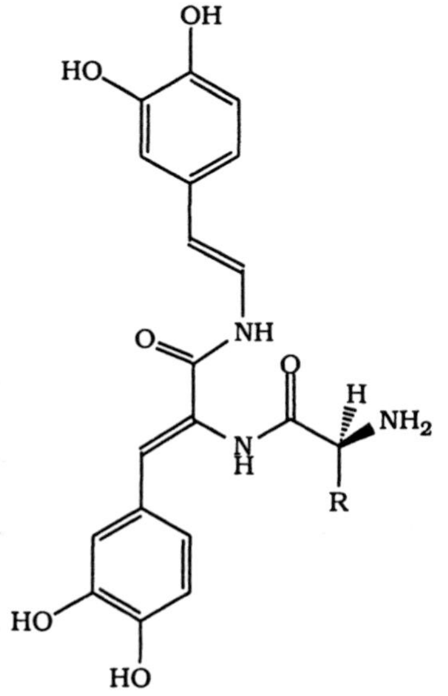

39. R = H
40. R = i-Bu

Related to the tunichromes are halocyamines A (41) and B (42) from *Halocynthia roretzi*. [39]. These are tetrapeptides which contain a Dopa residue (note: the authors have depicted a D-Dopa residue in their structure). Both halocyamines displayed antimicrobial activity against *B. subtilis* (MIC 50µg /mL), *B. megateruine* (MIC 50 µg/mL) , *B. cereus* (MIC 100 µg /mL) and the yeast *Cryptococcus neoformans* (MIC 100 µg /mL). Halocyamine A was also found to be cytotoxic to neuroblastoma N-18 cells at 160 µM for 24 hours. It caused degeneration of neurite and soma in cultured rat fetal brain cells at 100 µM and death of Hep-G2 cells at 200 µM for 24 hours.

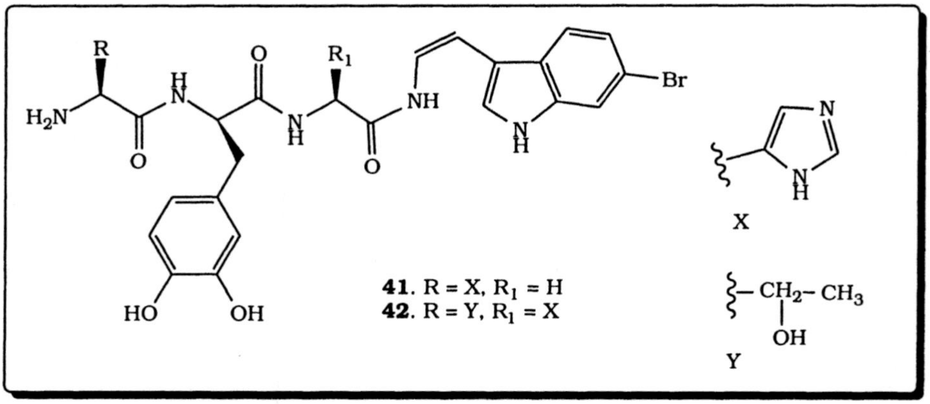

41. R = X, R₁ = H
42. R = Y, R₁ = X

The other modified linear peptides reported are virenamides A-E (43-47) from the didemnid ascidian *Diplosoma virens* [40,41]. Strictly speaking, these metabolites are apparently derived from 4 amino acids so should be regarded as tetrapeptides, but are more conveniently regarded as carboxy-protected tripeptides (where the carboxyl group has been

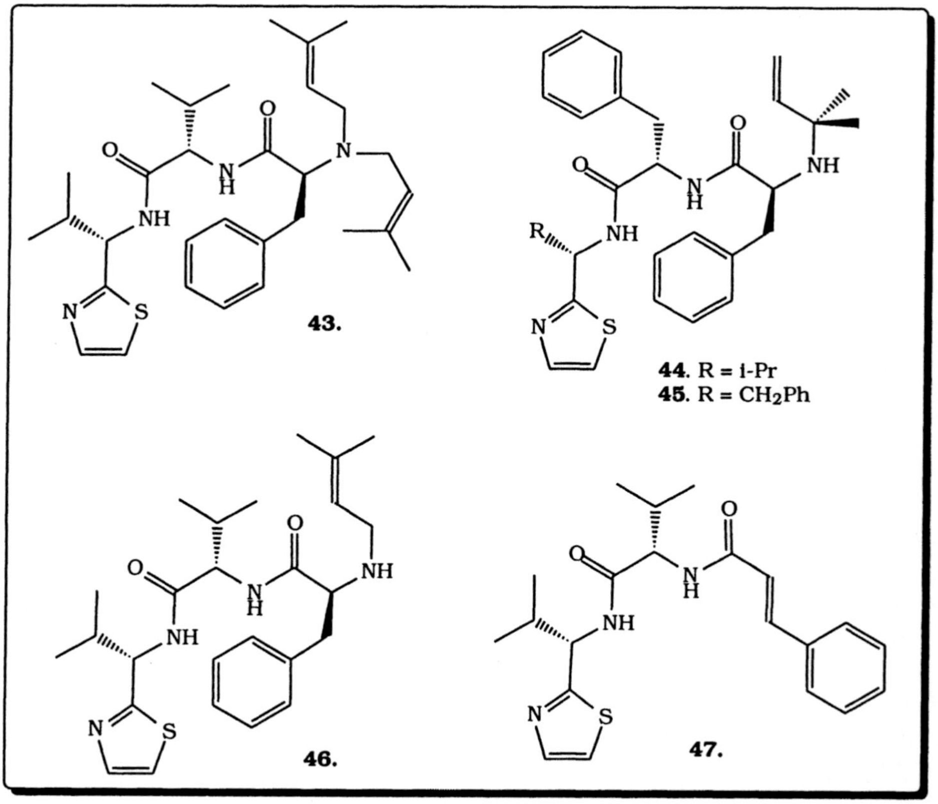

43.

44. R = i-Pr
45. R = CH₂Ph

46.

47.

converted, presumably via condensation with cysteine and subsequent decarboxylation, to a thiazole ring). While these metabolites do not contain Dopa or Topa residues they all contain at least one phenylalanine residue.

The virenamides showed modest cytotoxicity towards a panel of cultured cells: virenamide A exhibited IC_{50}'s of 2.5 μg/mL against P388 (murine leukemia), and 10 μg/mL against A549 (human lung carcinoma), HT29 (multidrug resistant human colon carcinoma) and CV1 (monkey kidney) cells. It exhibited Topoisomerase II activity (IC_{50} of 2.5 μg/mL). Virenamides B and C both exhibited IC_{50} values of 5 μg/mL against P388, A549, HT29 and CV1 cells.

A number of metabolites which appear to be derived from Tyrosine, Dopa, or Topa residues and to have similar biogenetic origins to the tunichromes have been reported. Rigidin (48), isolated from the Okinawan marine tunicate *Eudistoma c.f. rigida* exhibited calmodulin antagonistic activity with an IC_{50} of 5 x 10^{-5} M against calmodulin-activated brain phosphodiesterase [42]. Some other pyrrolo[2,3d]pyrimidine-2,4-diones which have been synthesised have shown weak affinity for the benzodiazepine receptor.

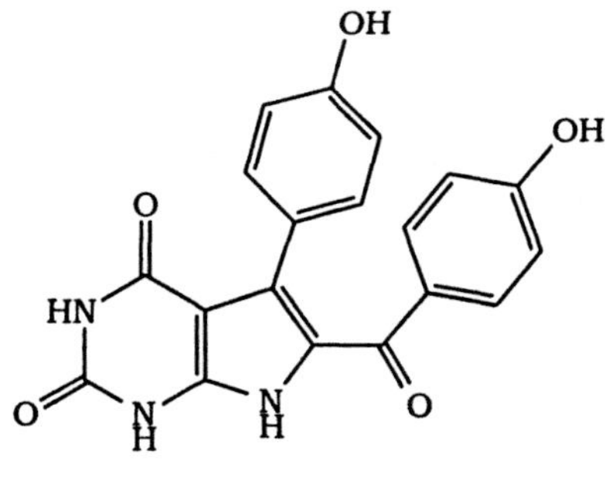

48.

Lukianols, Polycitrins, Lamellarins and Related Compounds

Lukianols A and B (49 and 50) were isolated from an unidentified ascidian collected at Palmyra atoll (N.E. of the Solomon Islands) [43]. Lukianol A exhibited moderate cytotoxicity (IC_{50} of 1 μg/mL against KB cell cultures), but lukianol B was 100 times less active [43]. The lukianols are closely related to polycitrins A and B (51 and 52), which were isolated from a *Polycitor* sp. [44]. Polycitrin A has been synthesised [45] by a proposed biomimetic route (based on known slime mould metabolites), but no activity data for the polycitrins has been published. These metabolites are some of the simplest of a group of highly substituted pyrrole derivatives which have been isolated from prosobranch molluscs, ascidians and sponges. The apparent biogenetic origin of these metabolites is from two, three or more tyrosine, Dopa or Topa moieties.

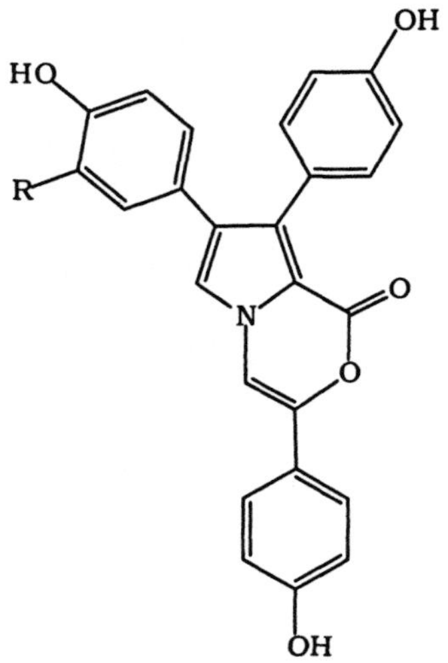

49. R = H
50. R = I

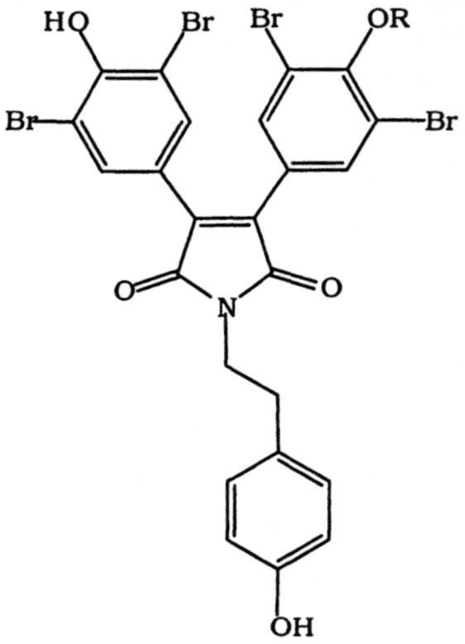

51. R = H
52. R = CH₃

The lukianols appear to arise by the same biogenetic pathways as for two metabolites (**53** and **54**) which were isolated from the southern Australian marine sponge *Dendrilla cactos* and called lamellarins O and P [46]. The choice of names is unfortunate, as these apparently tyrosine-derived tetracyclic metabolites are structurally closer to the lukianols and polycitrins than to the Dopa and Topa derived hexacyclic lamellarins. Some evidence for restricted rotation of the phenolic aromatic rings in these metabolites was reported. Subsequently, lamellarins Q and R (**55, 56**) were reported from a sample of *Dendrilla cactos* from southern New South Wales (Australia) [47]. Although both sponge extracts displayed mild antibiotic activity, the activity was not due to the lamellarins. Syntheses of lamellarin-O, lamellarin-Q, and lukianol-A have been reported [48]. Although lamellarins O - R were isolated from sponges the close structural association between these metabolites and those reported from ascidians is unquestionable. The ningalins were recently reported from an undescribed dark purple ascidian collected at Ningaloo Reef, Western Australia [49]. Ningalin A (**57**) appears to be a Dopa analogue closely allied to the tyrosine dimer lamellarin Q , while the pentacyclic ningalin B (**58**) fits into the biogenetic gap between the tetracyclic polycitrins, lamellarins O and P, and the more highly condensed hexacyclic lamellarins. No pharmacological activity has been reported for ningalins A and B.

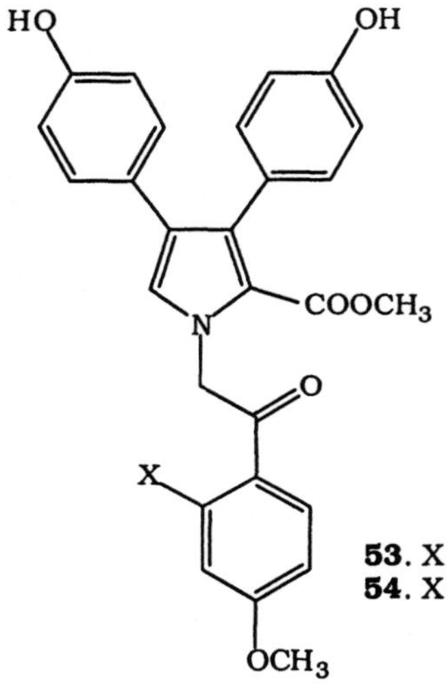

53. X = H
54. X = OH

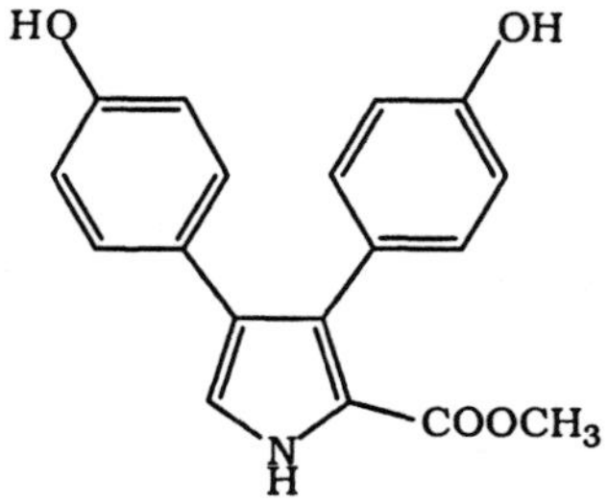

55.

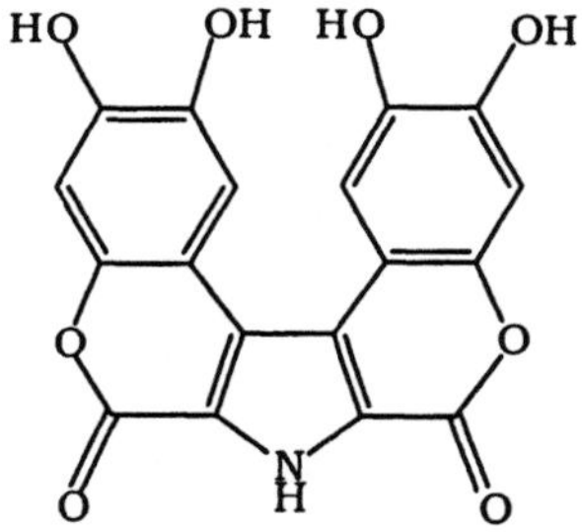

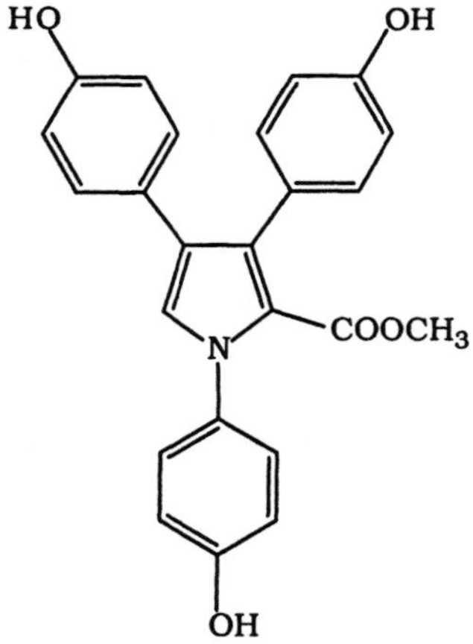

56.

57.

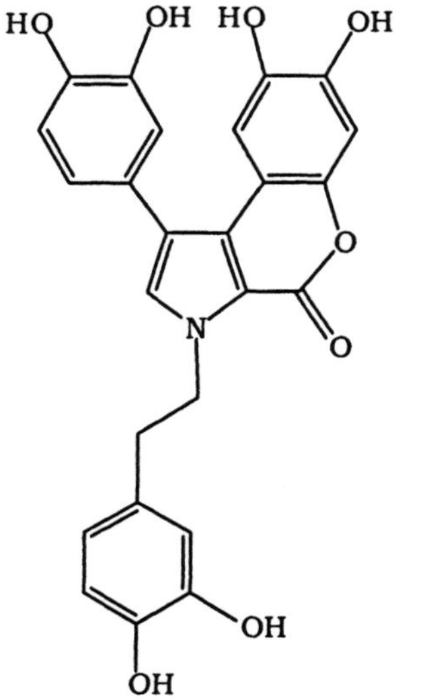

58.

Lamellarins A - D (**59-62**) were the first reported members [50] of what is now a fairly extensive group of alkaloids. Lamellarins A - D were initially isolated from prosobranch molluscs (which presumably fed on either ascidians or sponges). The structure of lamellarin A (**59**) was confirmed by x-ray analysis, and hindered rotation of the (oxygenated) phenyl ring was predicted [50]. Lamellarins E - H (**63 - 66**) were reported from the didemnid ascidian *Didemnum chartaceum* collected from the Republic of the Sechelles, atoll of Aldabra in the Indian Ocean [51], and two reports on the n.m.r. assignments for 5,6-dihydrolamellarin H (**67**) by 1,1-, 1,n-, n,1- and n,n-ADEQUATE experiments have been published [52,53]. No information on the origin of 5,6-dihydrolamellarin has been given other than that it was "from a marine ascidian of the genus *Didemnum* ." Presumably it was from the same source as lamellarins E-H. The synthesis of lamellarin-G trimethyl ether has been reported [54].

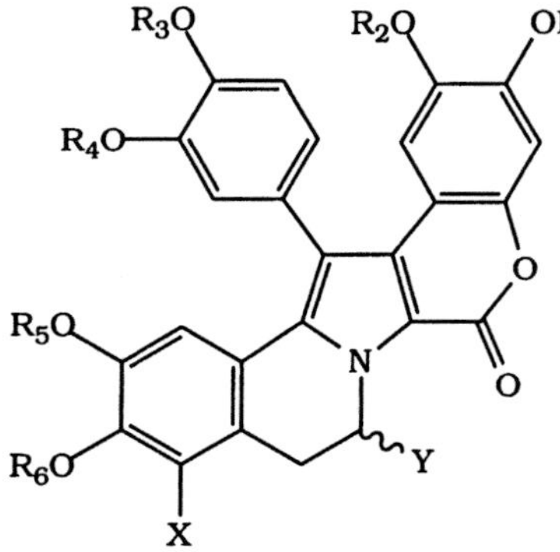

59. $R_1 = R_3 = H$, $R_2 = R_4 = R_5 = R_6 = CH_3$, $X = OCH_3$, $Y = OH$
61. $R_1 = R_3 = Y = H$, $R_2 = R_4 = R_5 = R_6 = CH_3$, $X = OCH_3$
63. $R_1 = R_4 = Y = H$, $R_2 = R_3 = R_5 = R_6 = CH_3$, $X = OH$
64. $R_1 = Y = H$, $R_2 = R_3 = R_4 = R_5 = R_6 = CH_3$, $X = OH$
65. $R_1 = R_3 = R_5 = CH_3$, $R_2 = R_4 = R_6 = X = Y = H$
67. $R_1 = R_2 = R_3 = R_4 = R_5 = R_6 = X = Y = H$
68. $R_1 = Y = H$, $R_2 = R_3 = R_4 = R_5 = R_6 = CH_3$, $X = OCH_3$
69. $R_1 = R_6 = X = Y = H$, $R_2 = R_3 = R_4 = R_5 = CH_3$
70. $R_1 = R_3 = Y = H$, $R_2 = R_4 = R_5 = R_6 = CH_3$, $X = OH$
71. $R_1 = R_4 = R_6 = X = Y = H$, $R_2 = R_3 = R_5 = CH_3$
74. $R_1 = R_2 = R_3 = R_4 = R_6 = X = Y = H$, $R_5 = CH_3$
75. $R_1 = R_4 = Y = H$, $R_2 = R_3 = R_5 = R_6 = CH_3$, $X = OCH_3$
76. $R_1 = R_4 = X = Y = H$, $R_2 = R_3 = R_5 = R_6 = CH_3$
77. $R_1 = R_4 = H$, $R_2 = R_3 = R_5 = R_6 = CH_3$, $X = OCH_3$, $Y = OH$
80. $R_1 = SO_3Na$, $R_2 = R_3 = R_5 = R_6 = X = OCH_3$, $R_4 = Y = H$
81. $R_1 = SO_3Na$, $R_2 = R_3 = R_5 = R_6 = CH_3$, $R_4 = X = Y = H$
82. $R_1 = SO_3Na$, $R_2 = R_3 = R_5 = R_6 = CH_3$, $R_4 = H$, $X = OCH_3$, $Y = OH$
83. $R_1 = SO_3Na$, $R_2 = R_3 = R_6 = OCH_3$, $R_4 = R_5 = X = Y = H$

Lamellarins I - N (**68 - 73**) were isolated along with the known lamellarins A - D from *Didemnum chartaceum* collected at Lihou Reef in the Coral Sea [55], while a sample of the same organism from the GBR yielded lamellarins A, B, C, I, K and M [55]. Lamellarin S (**74**) was isolated along with lamellarin K from a *Didemnum* sp. collected from the southern coast of N. S. W. (Australia) [56]. Although x-ray data and molecular modelling indicate a barrier of around 85 kJ/mol to rotation of the phenyl ring [50,56], no optical activity had been observed for the lamellarins until the transient atropism reported for lamellarin S, where the half-life for optical activity was estimated at around 90 days [56]. Lamellarin K reisolated from fresh material collected at Lihou Reef had no observable rotation despite the rotation being measured 17 days after collection of the living organism (author's unpublished results). A further five optically inactive lamellarins, lamellarins T-Y (**75-79**) were recently reported from an unidentified ascidian collected from the Arabian Sea. In addition, the 20-sulfate esters of lamellarin T (**80**), U (**81**), V(**82**) and the 20-sulfate of lamellarin Y(**83**) were reported [57]. Lamellarin N (**73**) was also isolated and characterised; it had previously only been characterised as its triacetate [55].

60. $R_1 = R_3 = R_4 = R_5 = CH_3$, $R_2 = H$, $X = OCH_3$
62. $R_1 = R_3 = R_4 = CH_3$, $R_2 = R_5 = X = H$
66. $R_1 = R_2 = R_3 = R_4 = R_5 = X = H$
72. $R_1 = R_3 = R_4 = R_5 = CH_3$, $R_2 = H$, $X = OH$
73. $R_1 = R_2 = R_4 = CH_3$, $R_3 = R_5 = H$, $X = H$
78. $R_1 = R_2 = R_4 = R_5 = CH_3$, $R_3 = H$, $X = OCH_3$
79. $R_1 = R_2 = R_4 = R_5 = CH_3$, $R_3 = H$, $X = OH$

Antimicrobial Activity and Cytotoxicity

The lamellarins generally do not exhibit much antimicrobial activity, but do show significant differential cytotoxicity. Lamellarin D was reported to cause 78% inhibition of development of fertilised sea urchin eggs at 19 μg/mL, while lamellarin C reportedly caused only 15% inhibition, and lamellarins A and B were inactive at the same concentration [50]. No activity data has been reported for lamellarins E - H, while from results of preliminary testing, lamellarins I, K and L were reported to show comparable and significant cytotoxicity against three different cell lines in culture with IC_{50} values of around 0.25 μg/mL [55]. Lamellarins K and L were also reported to exhibit moderate immunomodulatory activity [55]. A recent more extensive cytotoxicity assessment of a number of the lamellarins [58] is summarised in Table (1). Lamellarin D- triacetate and lamellarin K-triacetate displayed high potency against lung carcinoma cells (A549) in culture. Lamellarin N (73) was reported to show some selectivity in the NCI 60 cell line panel towards the melanoma cell lines SK-MEL-5 (LC_{50} 1.87 x 10^{-7} M) and UACC-62 (LC_{50} 9.88 x 10^{-6} M), but the lamellarin sulfates decomposed before they could be tested [57]. The selective cytotoxicity exhibited by several of the lamellarins makes them good candidates for further development as potential anticancer agents.

Table 1.   Cytotoxicity of Different Lamellarins Against a Panel of Tumor Cells [58]

Cell line:	Mean IC_{50} (μM)						
	P388	Schabel	AUXB1	CCH^RC5	A549	HT29	MEL28
Lamellarin A	0.89	0.91	0.36	0.71	0.9	2.1	0.93
Lamellarin B	10.1	10.4	5.5	18	5.2	>10	10.1
Lamellarin D-triacetate	0.11	0.14	0.05	0.06	0.008	0.8	0.16
Lamellarin I	4.9	4.8	0.38	2	5	4.7	5
Lamellarin I-acetate	9	9.2	4.1	9	9.3	>10	9.1
Lamellarin J	2.9	3.9	0.58	1.2	0.6	5.8	2.9
Lamellarin K	0.19	0.017	0.19	0.75	0.18	0.38	0.4
Lamellarin K-triacetate	0.09	0.16	0.15	0.16	0.005	0.47	0.93
Lamellarin L	1.2	1.4	0.8	1.3	0.6	6	1.2
Lamellarin L-triacetate	2.4	2.4	2.2	2.5	1.1	>3	2.3
Lamellarin M	0.15	0.17	0.07	0.17	0.06	0.56	0.54
Lamellarin M-triacetate	0.91	1.1	0.76	3.1	0.22	>1	0.9
Lamellarin N-triacetate	0.32	0.3	0.1	0.16	0.02	3.2	1.6

Perhaps even more interesting than the direct cytotoxicity is the reported ability of certain of the lamellarins to act on multidrug-resistant (MDR) cell lines, and to display resistance modifier activity [58]. Lamellarin I (68) in particular showed the highest chemosensitising activity of the lamellarins tested. Lamellarin I was 9-16 times more effective than verapamil (a well known chemosensitising agent), effectively increasing the cytotoxicity of doxorubin, vinblastine and daunorubicin in a concentration dependent manner in MDR cells. Results of *in vitro* measurements of rhodamine123 accumulation in multidrug-resistant Lo Vo/Dx cells indicated that lamellarin I reversed MDR by directly inhibiting the P-glycoprotein-mediated drug efflux [58]. Further studies to assess the potential of the lamellarins as therapeutic agents for the treatment of cancer and for their activity as chemosensitising agents in MDR cells are in progress.

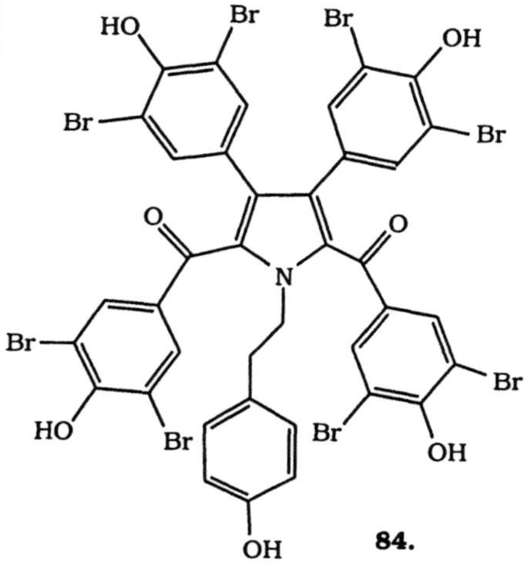

84.

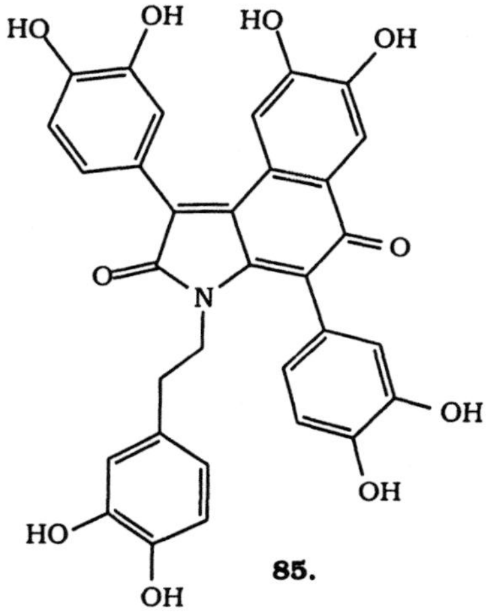

85.

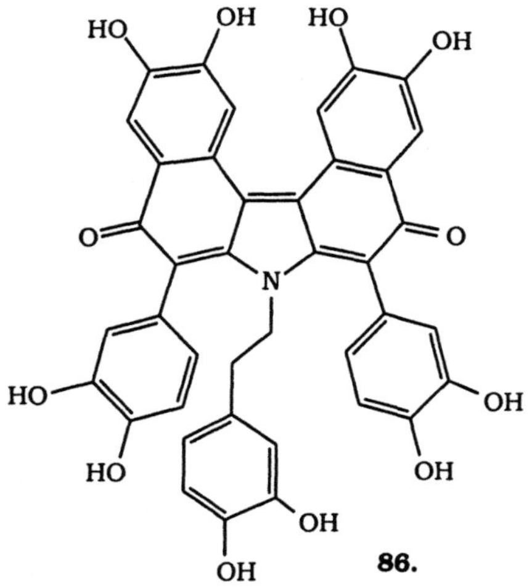

86.

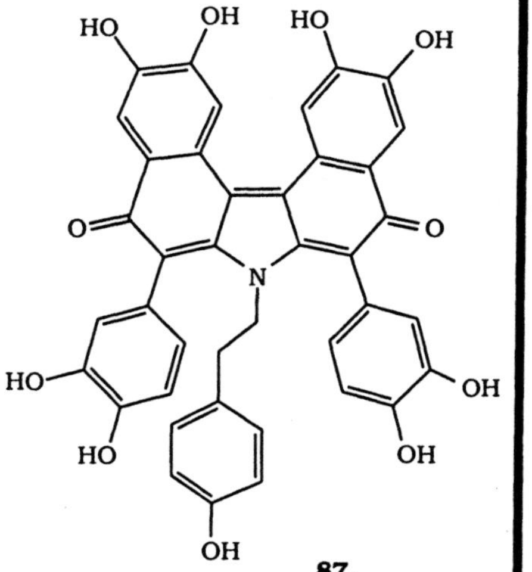

87.

Among the most highly substituted of the pyrrole derivatives reported from ascidians to date, is polycitone-A (**84**), which was isolated along with polycitrins A and B from a tunicate of the genus *Polycitor* [44]. The biosynthetic processes which lead to (substituted) phenacyl substituents at C2 and C5 of the pyrrole ring (in place of a carboxylic acid functionality, or hydrogen) are unclear. The penta-O-methyl derivative of polycitone A exhibited marginal cytotoxicity by inhibiting the growth of SV40 transformed fibroblast cells at 10 µg/mL. Polycitone A appears to be structurally similar to ningalins C (**85**) and D (**86**) as well as several reported sponge metabolites. Ningalins C and D co-occurred with ningalins A and B in the purple *Didemnum* sp. collected at Ningaloo Reef in Western Australia [49]. The origin of the additional carbon atoms needed to progress from a structure like ningalin B to the hexacyclic ningalin C or the octacyclic ningalin D has not been determined. The carbon skeleton for ningalin D is the same as for the sponge metabolite purpurone (**87**) which was reported [59] from an *Iotrochota* sp. Also apparently structurally related are storniamides A - D (**88-91**) from the Patagonian burrowing sponge, *Cliona* sp [60]. No pharmacological activity has been reported for any ningalins although speculation of the potential role these metabolites may play in iron accumulation is included in the report [49]. Purpurone is an inhibitor of ATP-citrate lyase with an IC_{50} value of 7 µg/mL [59], while the storniamides exhibited antibiotic activity against gram positive bacteria (*S. aureus, B. subtilis, Micrococcus luteus*) at 50 µg/disk, but no activity against gram negative bacteria or the yeast *C. albicans* [60].

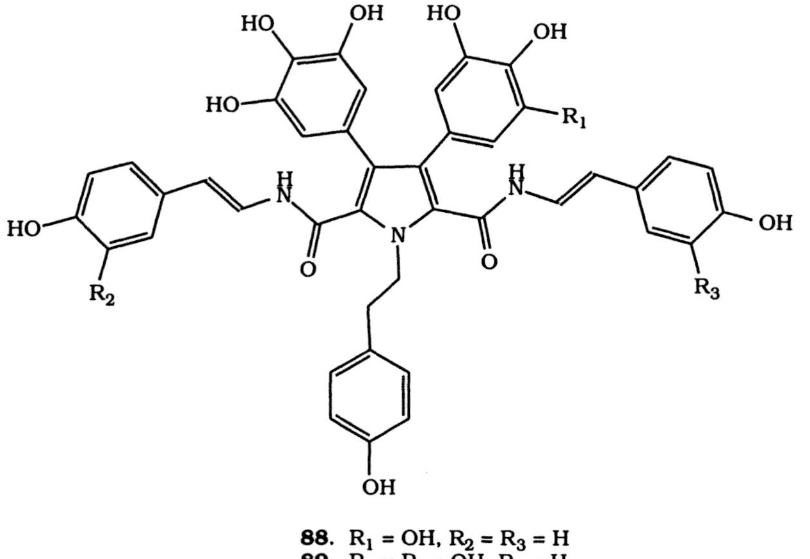

88. R_1 = OH, R_2 = R_3 = H
89. R_1 = R_3 = OH, R_2 = H
90. R_1 = H, R_2 = R_3 = OH
91. R_1 = R_2 = R_2 = OH

Tetrahydroisoquinoline Alkaloids - Ecteinascidins

This group of alkaloids is reasonably uncommon in ascidians, but the sole group of representatives, the ecteinascidins, are potentially one of the most useful group of anticancer agents found to date from marine sources.

The ecteinascidins are tetrahydroisoquinoline alkaloids which have been isolated from *Ecteinascidia turbinata*. Reports of the extraordinary cytotoxicity of extracts of *Ecteinascidia turbinata* date back to 1969 when *in vivo* screening of the crude extract against P388 leukemia in mice yielded T/C values of 272 with 4/6 cures. It was not however until 1990 that the first structures of the unstable cytotoxic constituents were eludidated and simulataneously published by two groups [61,62]. The ecteinascidins are similar in structure to the dimeric isoquinolinequinone alkaloids, the safracins and saframycins, which are microbial derived [63], and the sponge-derived renieramycins [64, 65]. The ecteinascidins have been named according to dominant ions in their mass spectra. Ecteinascidin 729 (**92**) = Et 729, Ecteinascidin 743 (**93**) = Et 743, Ecteinascidin 745 (**94**) = Et 745, Ecteinascidin 759A (**95**) = Et 759A, Ecteinascidin 759B (**96**) = Et 759B, and Ecteinascidin 770 (**97**) = Et 770. Several of the ecteinascidins have molecular weights 18 mass units higher than their names would suggest as a result of water loss in the mass spectrometer: the molecular weight of Et 729 is 747, while Et 743 and the two isomeric Et 759 structures have molecular weights of 761 and 777 respectively. X-ray

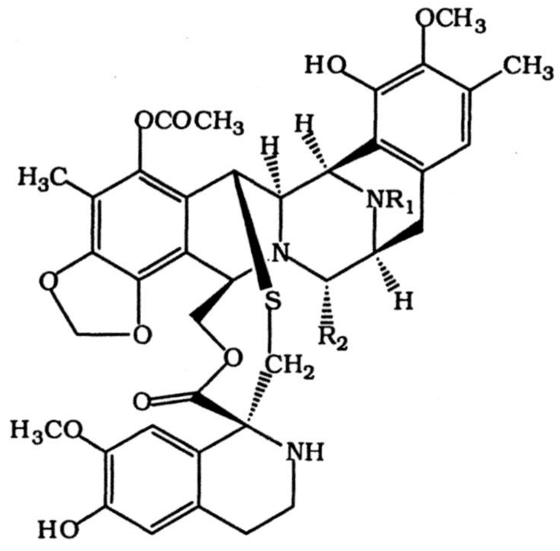

92. R_1 = H, R_2 = OH
93. R_1 = CH$_3$, R_2 = OH
94. R_1 = CH$_3$, R_2 = H
95. R_1 = CH$_3$, R_2 = OH
96. R_1 = CH$_3$, R_2 = OH ⎤
97. R_1 = CH$_3$, R_2 = CN ⎦ N-oxides

structures for 21-O-methyl-N12-formylecteinascidin 729, and for Et 759, the N-oxide of ecteinascidin 743 [66,67], confirmed the proposed relative stereochemistry. All members of this original group of ecteinascidins are composed of three tetrahydroisoquinoline units (A-C) .

A further six additional ecteinascidins have subsequently been reported. Two, ecteinascidin 722 (97) and ecteinascidin 736 (98), contain two tetrahydroisoquinoline units (A and B) but the C unit is a tetrahydro-β-carboline residue [66]. The remaining four ecteinascidins, Et 597 (99), Et 583 (100), Et 594 (101) and Et 596 (102) are also biologically active and have been described as "putative biosynthetic precursors" of the other ecteinascidins [68]. These ecteinascidins also have tetrahydroisoquinoline units A and B, but L-cysteine or its α-oxo analogue is the C unit. Identification of the absolute stereochemistry of that unit has led to confirmation that the absolute stereochemistries of the ecteinascidins are the same as for the safracins and saframycins [68].

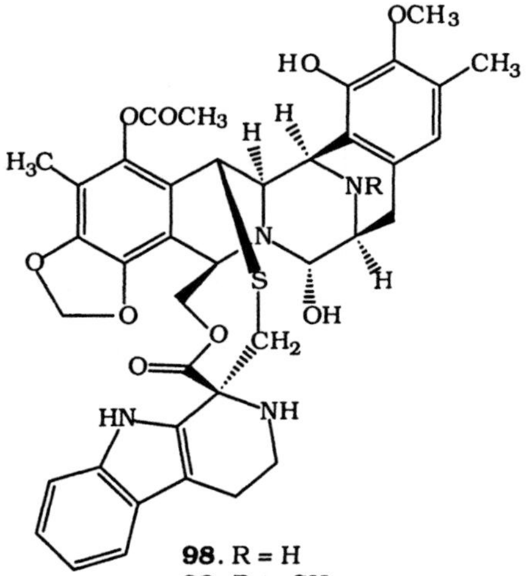

98. R = H
99. R₁ = CH₃

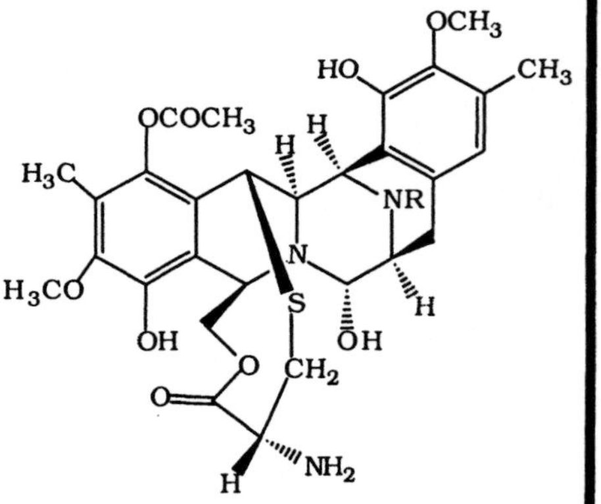

100. R = CH₃
101. R = H

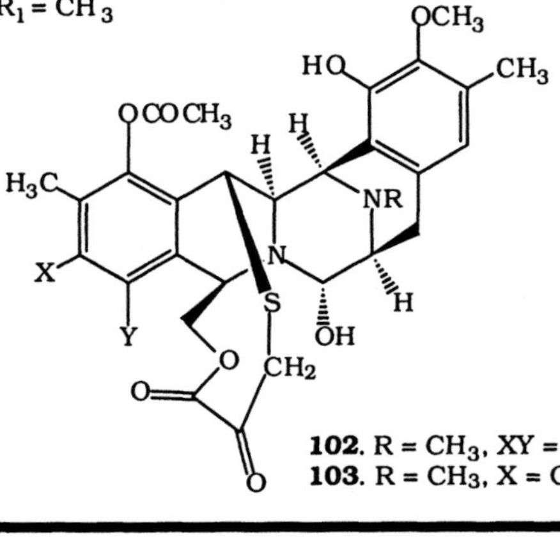

102. R = CH₃, XY = OCH₂O
103. R = CH₃, X = OCH₃, Y = OH

Biogenesis of the Ecteinascidins

Some biosynthetic studies involving incorporation of labelled tyrosine, cysteine, methionine, glycine and tryptophane [68,69], have already been performed on the ecteinascidins, to partially confirm the proposed biosynthetic origin of the A and B units from Dopa units. Further studies are in progress to test the hypothesis that the electrophilic ketone in an intermediate such as Et 594 or Et 596 can be condensed in a Pictet-Spengler reaction with a Dopa derivative to form the third tetrahydroisoquinoline group (C unit) in Et 729 or Et 743 [68].

Cytotoxicity

The ecteinascidins show very good *in vitro* and *in vivo* activity, and have been reported to have activity against P388 lymphoma, B16 melanoma, M5076 ovarian sarcoma, Lewis lung carcinoma, and the LX-1 and MX-1 human lung and human mammary xenografts [70]. Et 743 exhibited IC_{50} values of 1.3 and 0.5 ng/mL against P388 [62] and L1210 [61] respectively *in vitro* , and a T/C of 167 at 15 µg/Kg against P388 *in vivo* [61]. Et 729 exhibited an IC_{50} value of 0.93 ng/mL against P388 [62] and afforded T/C values of 214 at 3.8 µg/Kg against P388 and 246 at 10 µg/Kg against B16 melanoma *in vivo* [61] . An approximately 1:1 mixture of Et729 and 743 afforded a T/C value of 188 against B16 melanoma at 0.1mg/Kg administered daily for 9 days, and a T/C value of 147 against colon carcinoma 26 with the same dose regime [62]. Et 745 afforded an IC_{50} of 88 ng/mL against L1210 *in vitro* , and a T/C of only 111 at 250 mg/Kg against P388 *in vivo* [61]. Ecteinascidins 722 and 736 are both reported to be "highly active" against L1210 cells *in vitro* , and Et 722 is reported to be highly active *in vivo* against P388, B16 melanoma and Lewis lung carcinoma [71]. Comparative studies of Et 597, Et 583 and Et 594 against Et 743 and Et 729 reveal lower activities (4 to 100 times less) against P388 (murine lymphoma), A549 (human lung carcinoma), and HT29 (human colon carcinoma) [68]. Comparable activities were observed for Et 597 and Et 583 against MEL 28 (human melanoma) and for Et 597 against CV-1 (monkey kidney) cells. All strongly inhibit protein, DNA, and RNA synthesis, as well as RNA polymerase activity. The ecteinascidins also exhibit antimicrobial activity *(B. subtilis)* with MIC values (µg/disk) of 0.02 (Et 743), 0.08 (Et 729), 0.14 (Et 597), 0.74 (Et 583) and 0.37 (Et 594) [68]. Structure/activity relationships for the ecteinascidins are not clear at this point in time, but a comparison of reported activities for Et 743 and Et 745 clearly demonstrates the importance of the hydroxyl group to the cytotoxicity: its removal results in more than two orders of magnitude loss in activity. "The potent anticancer activity of ecteinascidins may, at least in part, be attributed to unit C, since the related saframycin A (104), for example, which lacks the C unit, has lower efficacy than Et 729 in comparable tumor models" [68].

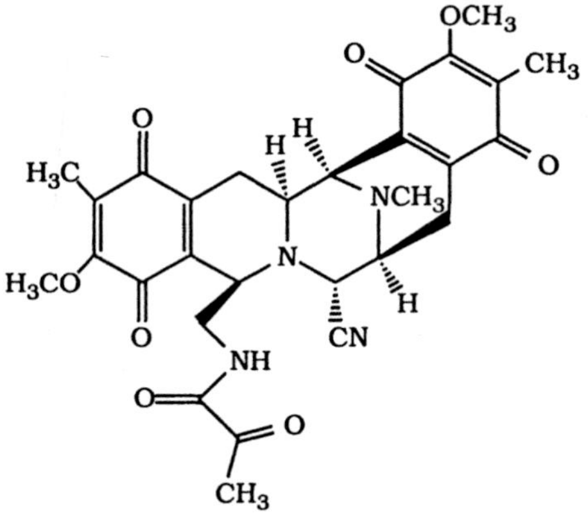

104.

Mode of Action

Ecteinascidin 743 has been demonstrated to bind in the minor groove of DNA by covalent attachment to the exocyclic amino group at position 2 of guanine. Comparison with anthramycin, another minor groove alkylating agent which alkylates at guanine N2 showed Et 743 covered 3-5 base pairs and exhibited a different sequence selectivity to anthramycin [72]. Computer modelling of the covalent adduct of Et 743 to DNA using the reactive carbinolamine group and N2 of guanine in the minor groove of the DNA double helix suggests that the A and B units stack against the DNA backbone [67].

Clinical Trials and Drug Supplies

The preclinical pharmacology of Et 729 has been reported [73]. The current clinical plan for the more available ecteinascidin, Et 743, calls for three 0.5 mg doses per patient [74]. Ecteinacidin 743 is currently in phase 1 clinical trials in three European countries and in the United States [68]. The ecteinascidins have currently been extracted from harvested tunicate. Some partial syntheses of the ecteinascidin structure have been reported [70, 75], and one enantioselective total synthesis of ecteinascidin 743 has been published [74].

Indole Derived Alkaloids

These are perhaps the most diverse group of aromatic alkaloids isolated from ascidians. Their structures span a range of complexity, starting with 6-bromoindole-3-aldehyde (105) (previously reported from a marine *Pseudomonas* sp.), and the brominated quinazolinedione (106) from *Pyura sacciformis* [76].

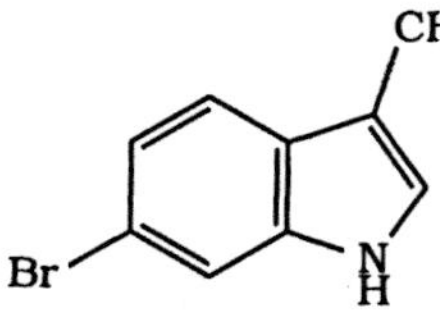

105.

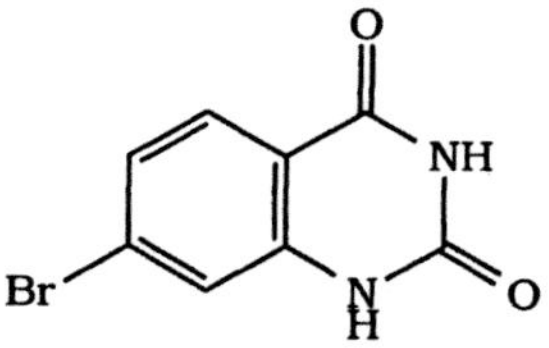

106.

6-Bromotryptamine (107) has been reported [14] from a collection of a *Lissoclinum* sp. from the GBR, and from *Didemnum candidum* from the Gulf of Mexico [77], while 5-bromo-N,N-dimethyl tryptamine (108) was reported from a New Caledonian collection of *Eudistoma fragum* where it co-occurred with woodinine [78]. 5-Bromo-N,N-dimethyltryptamine exhibited antimicrobial activity (with 12 mm and 17 mm zones of inhibition reported at 100 µg per 6 mm disk against *S.aureus* and *E coli* resp.; 17mm and 22 mm zones at 200 µg per disk against the same microorganisms). A synthesis of 5-bromo-N,N-dimethyltryptamine has been reported [79].

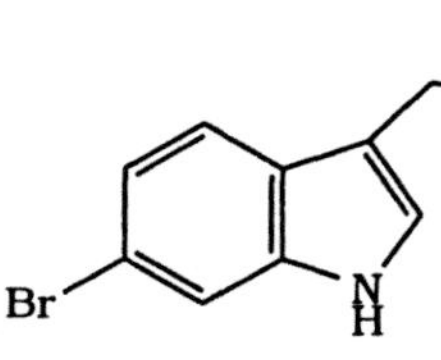

107.

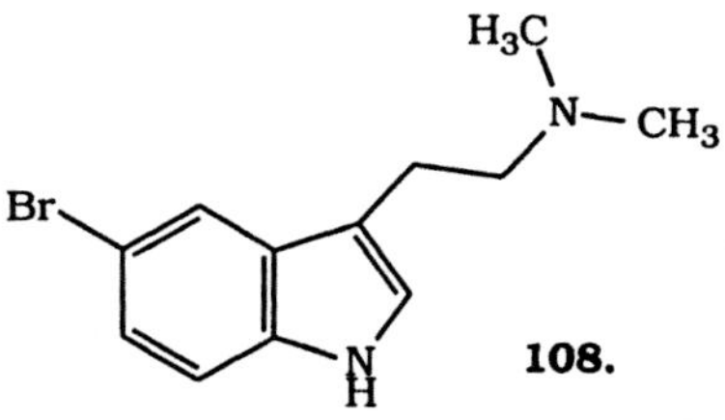

108.

Dendrodoine (109) [80,81], the related aminoimidazole (110) analogue [82] and the recently reported oxadiazinone, alboinon (111) [83] were reported from *Dendrodoa grossularia*. There are structural similarities between these tryptophane-derived metabolites, the phenylalanine-derived metabolites (5), (6) and (7), and the tyrosine-derived metabolite

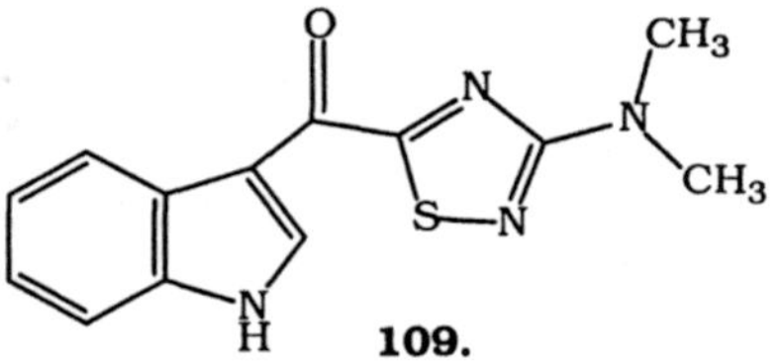

109.

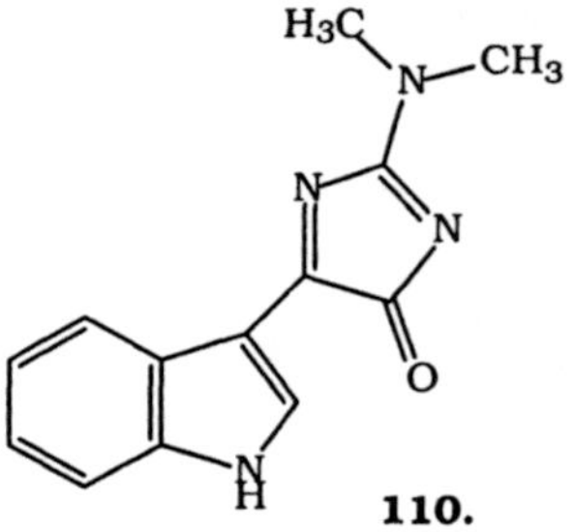

110.

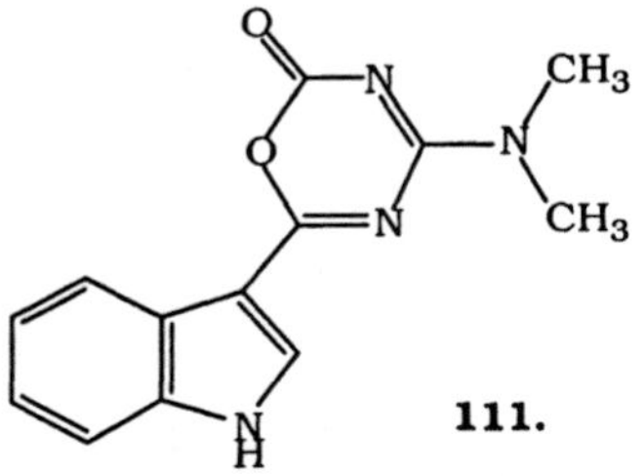

111.

polycarpine (**25**). Dendrodoine, whose synthesis has also been reported [84], is cytotoxic (ID$_{50}$ 10μg/mL against L1210 lymphoma cells) [85] and inhibits DNA synthesis *in vitro*, measured by incorporation of ^{3}H-thymidine by cultured L1210 cells [81]; no activity data has been reported for (**110**) but its structural relationship to methyl aplysinopsin, a powerful antidepressant, has been noted [82]. Didemnimides A-D (**112-115**) were isolated from the Caribbean mangrove ascidian, *Didemnum conchyliatum* [86]. The structure of didemnimide A was confirmed by x-ray analysis and the didemnimides were reported to be predator deterrents against carnivorous fish. Urochordamines A (**116**) and B (**117**) are larval settlement/metamorphosis-promoting alkaloids from the tunicate *Ciona savignyi* [87]. They also exhibited antibacterial activity against the gram positive marine microorganism *Bacillus marinus* but were inactive against gram negative marine bacteria and fungi.

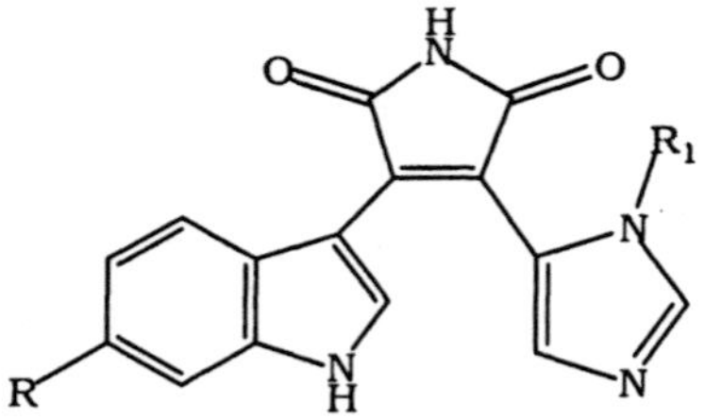

112. R = R$_1$ = H
113. R = Br, R$_1$ = H
114. R = H, R$_1$ = CH$_3$
115. R = Br, R$_1$ = CH$_3$

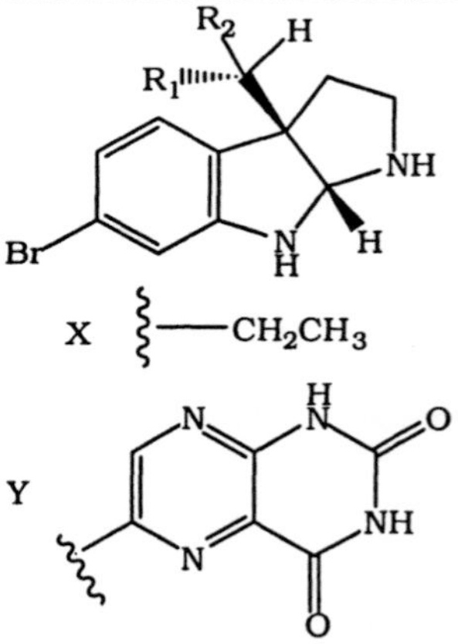

116. R$_1$ = X, R$_2$ = Y
117. R$_1$ = Y, R$_2$ = X

The symmetrical dimeric 6-bromoindole derivatives (118) and (119) have been reported from the Gulf of California tunicate *Didemnum candidum*, but no stereochemistry or activity data were reported [77]. Wakayin (120), a novel pyrroloiminoquinone alkaloid from an ascidian *Clavelina* sp., was reported to be cytotoxic (IC$_{50}$ of 0.5μg/mL against HCT116) [88]. It inhibits topoisomerase II at 250 μM and shows a 3-fold differential in cytotoxicity towards CHO cell line EM (which is DNA repair deficient) compared to BR1, indicating it exhibits cytotoxicity by interfering with or damaging DNA. It also exhibits antimicrobial activity (MIC 0.3μg/mL against *B. subtilis)*. The structural similarity between wakayin and some sponge metabolites such as isobatzelline C and the discorhabdins/prianosines has been noted.

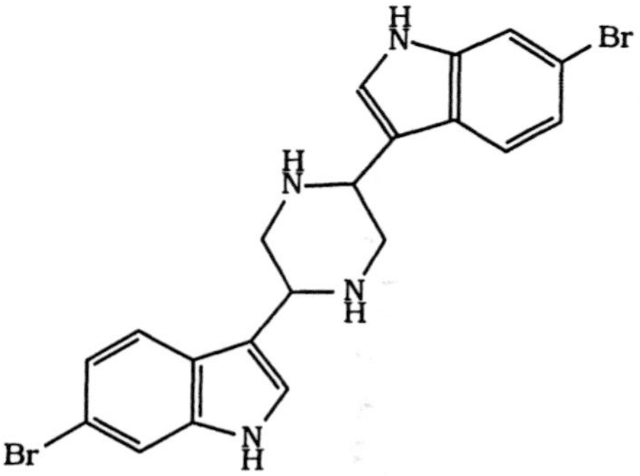

118.

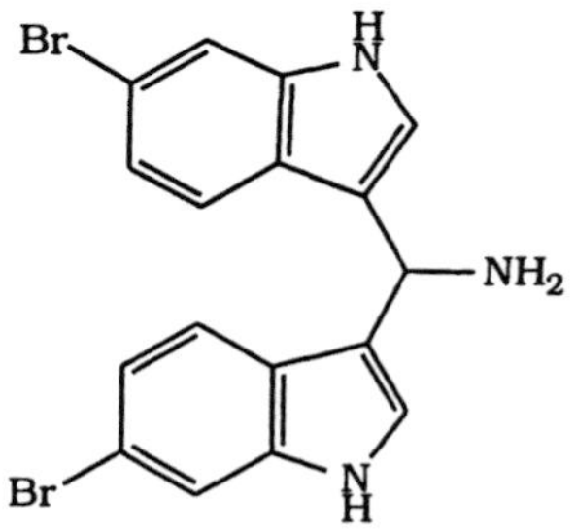

119.

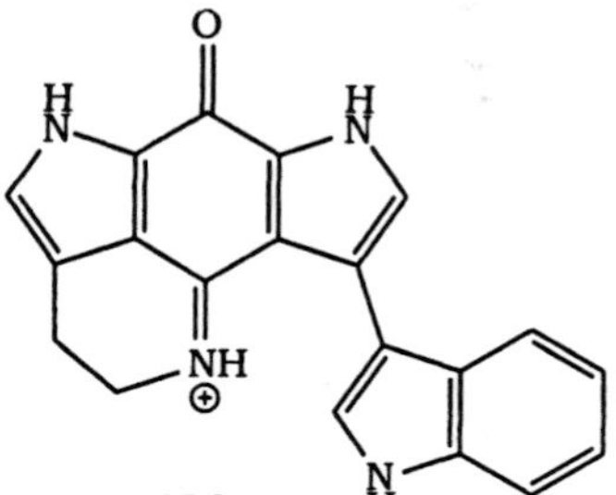

120.

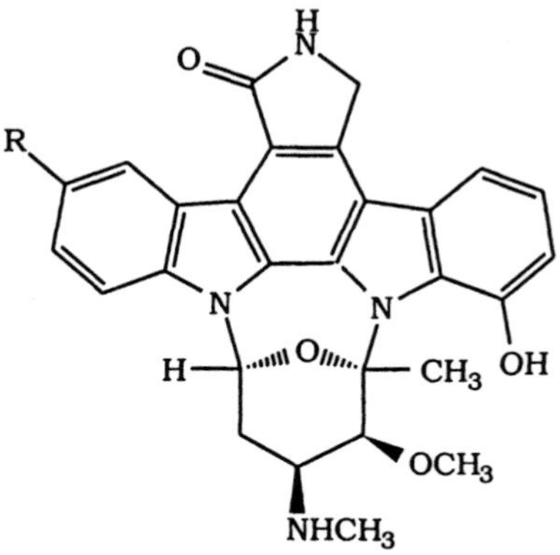

121. R = H
122. R = OH

11-Hydroxystaurosporine (121), another dimeric indole structure which was isolated from an unidentified tunicate [89], was reported to be a highly cytotoxic, powerful protein kinase C inhibitor. Cytotoxicity was characterised by an IC$_{50}$ of 0.7 nM against KB cells, while the IC$_{50}$ for

PKC inhibition was 2.2 nM. In a LoVo (MDR cytotoxicity) screen it exhibited 75% activity at 0.03μM. The 3,11-diol (**122**) was also reported to co-occur, but was incompletely characterised.

α-Carbolines

α-Carbolines are not common tunicate metabolites. The grossularines [80,85] from *Dendrodoa grossularia* are the sole representatives of this structural class. The initial structure reported for grossularine (= grossularine-1) was incorrect [80], but it was subsequently revised when data for both grossularine-1 (**123**) and -2 (**124**) were reported [85]. Both grossularines are cytotoxic (ID_{50} = 6μg/mL for grossularine-1 and 4μg/mL for grossularine-2 against L1210 murine lymphoma), and both grossularines cause accumulation of cells in the G1 phase (at 10μg/mL and 1.5μg/mL resp. by cell-flow cytofluorimetric analysis). Both are active down to 10 ng/mL in a cloning bioassay against WiDr (colon) and MCF7 (breast) solid human cancer cells. Grossularine-2 appears to act on DNA as a mono-intercalating agent [85]. Desmethyl grossularine-1 (**125**) has been reported from *Polycarpa aurata* from Chuk Atoll, but no activity data was presented [29].

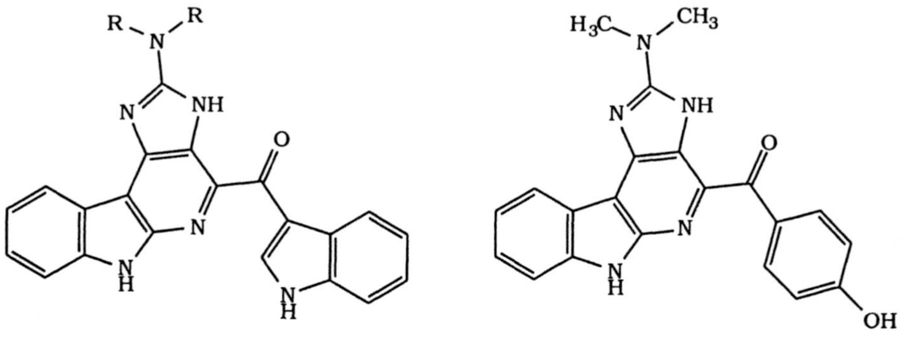

123. R = CH₃
125. R = H

124.

β-Carbolines and Tetrahydro- β-Carbolines

These form one of the largest group of reported metabolites from ascidians, and are notable for the antiviral activity which is displayed by some members. The majority of these alkaloids are either named as eudistomins or eudistomidins. For the purposes of this review, these have been grouped into C1-unsubstituted β-carbolines, β-carbolines with a C1-pyrrolyl group, β-carbolines with a C1-pyrrolinyl group, β-carbolines

Table 2.   Structures and Sources of Some β-Carboline Metabolites and Derivatives

Name	Structure	C5	C6	C7	C8	N9	Source [Reference]
		Substituents					
Simple β-carbolines which are unsubstituted at C1							
β-Carboline	126	H	H	H	H	H	*Riterella siginilloides* [119]
Eudistomin N	127	H	Br	H	H	H	*Eudistoma olivaceum* [91,120]
Eudistomin O	128	H	H	Br	H	H	*Eudistoma olivaceum* [91,120] *Riterella siginilloides* [119]
Eudistomin D	129	Br	OH	H	H	H	*Eudistoma olivaceum* [91,120] *Eudistoma glaucus* [96]
Eudistomidin D	130	Br	OH	H	H	CH$_3$	*Eudistoma glaucus* [96]
Eudistomin J	131	H	OH	Br	H	H	*Eudistoma olivaceum* [91]
Bromoeudistomin D	132	Br	OH	Br	H	H	Synthetic [120]
Bromoeudistomidin D	133	Br	OH	Br	H	CH$_3$	Synthetic [123]
β-Carbolines with a 2-pyrrolyl substituent at C1							
Eudistomin M	134	H	OH	H	H	H	*Eudistoma olivaceum* [91,120]
Eudistomin A	135	H	OH	Br	H	H	*Eudistoma olivaceum* [91,120]
β-Carbolines with a 2-(Δ^1-pyrrolinyl) substituent at C1							
Eudistomin I	136	H	H	H	H	H	*Eudistoma olivaceum* [91,120,99] *Eudistoma glaucus* [96]
Eudistomin G	137	H	H	Br	H	H	*Eudistoma olivaceum* [91,120,99]
Eudistomin H	138	H	Br	H	H	H	*Eudistoma olivaceum* [91,120,99] *Eudistoma glaucus* [96]
Eudistomin Q	139	H	OH	H	H	H	*Eudistoma olivaceum* [91,120]
Eudistomin P	140	H	OH	Br	H	H	*Eudistoma olivaceum* [91,120,99]
Eudistomidin A	141	H	Br	H	OH	H	*Eudistoma glaucus* [116]

with other C1 substituents, tetrahydro-β-carbolines and dihydro-β-carbolines. The structures and sources of isolated β-carboline metabolites and their pharmacologically active derivatives for C1-unsubstituted β-carbolines, β-carbolines with a C1-pyrrolyl group, and β-carbolines with a C1-pyrrolinyl group are tabulated in Table (2). Syntheses have been reported for eudistomins D [90, 91], H [91,92], I [91,93,92,94], N [91], M [94], O [91] and Q [91].

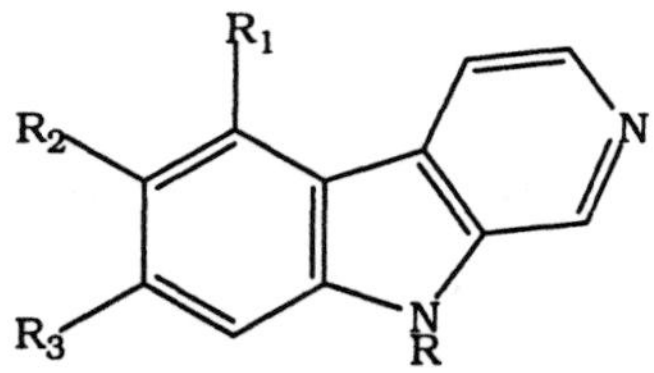

126. R = R_1 = R_2 = R_3 = H
127. R = R_1 = R_3 = H, R_2 = Br
128. R = R_1 = R_2 = H, R_3 = Br
129. R = R_3 = H, R_1 = Br, R_2 = OH
130. R = CH_3, R_1 = Br, R_2 = OH, R_3 = H
131. R = R_1 = H, R_2 = OH, R_3 = Br
132. R = H, R_1 = R_3 = Br, R_2 = OH
133. R = CH_3, R_1 = R_3 = Br, R_2 = OH

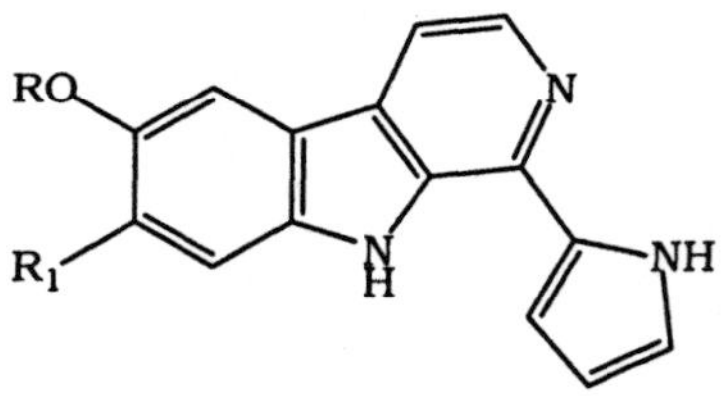

134. R = R_1 = H
135. R = H, R_1 = Br

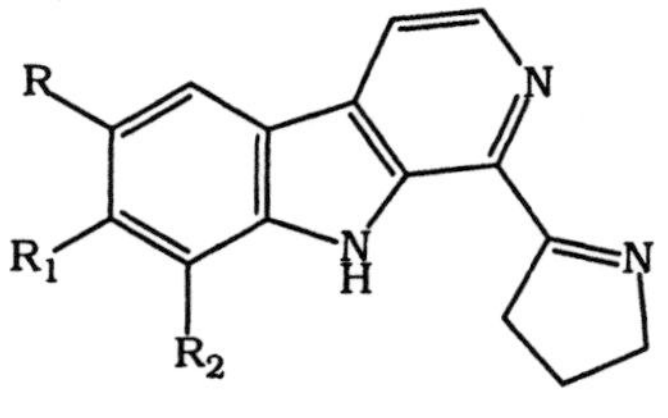

136. R = R_1 = R_2 = H
137. R = R_2 = H, R_1 = Br
138. R = Br, R_1 = R_2 = H
139. R = OH, R_1 = R_2 = H
140. R = OH, R_1 = Br, R_2 = H
141. R = Br, R_1 = H, R_2 = OH

β-Carbolines which have other substituents at C1 include eudistomin U (142) from the Caribbean didemnid *Lissoclinum fragile* [95], eudistomidins C (143) [96,97], E (144) [98] and F (145) [98] from the Okinawan ascidian *Eudistoma glaucus* , eudistomins R (146), S (147) and T (148) from the

Caribbean tunicate *Eudistoma olivaceum* [99], and eudistalbins A (149) and B (150) from the New Caledonian ascidian *Eudistoma album* [100]. The syntheses of eudistomins S [101], T [93,94,101,102,103,104] and U [90,105] have been reported.

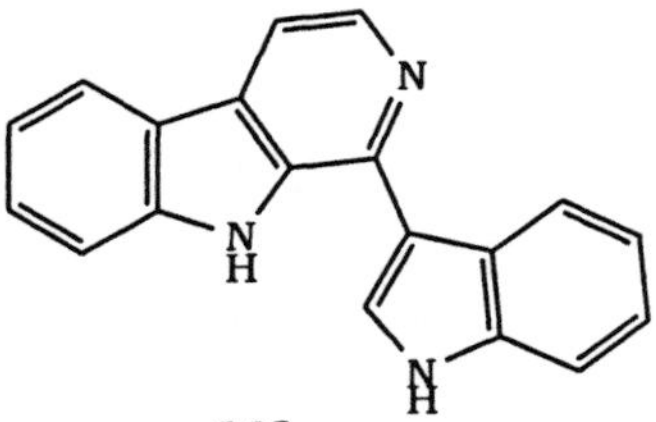

142.

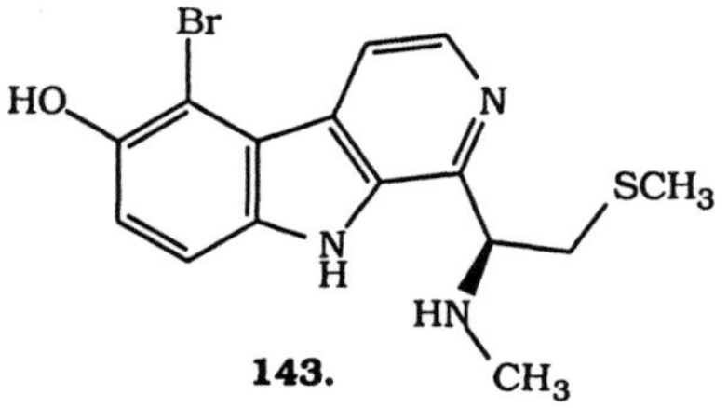

143.

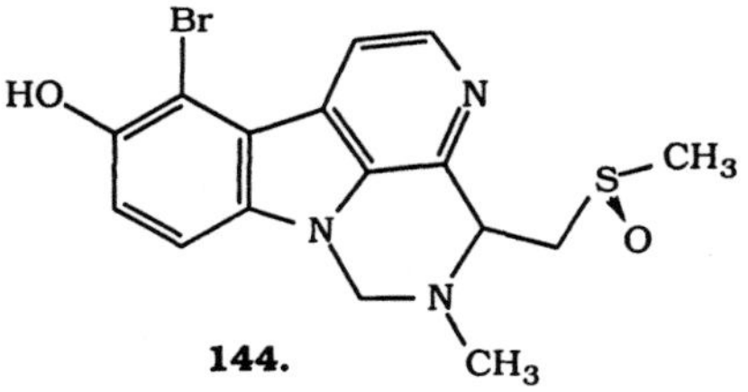

144.

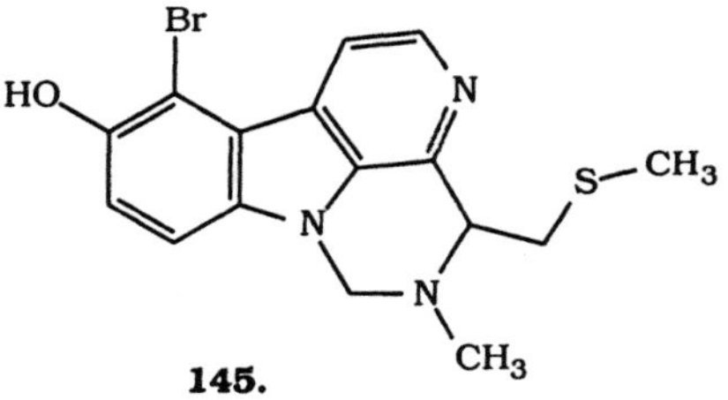

145.

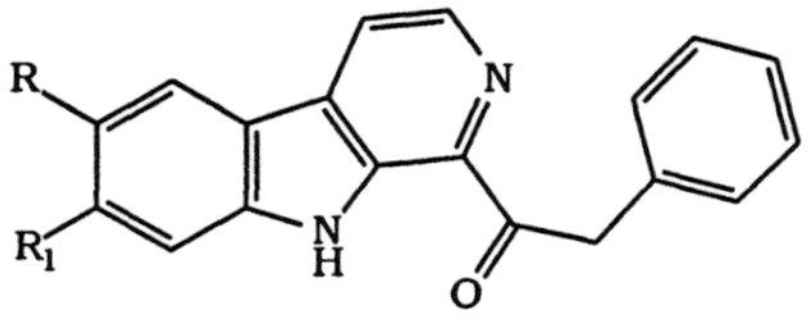

146. R = H, R₁ = Br
147. R = Br, R₁ = H
148. R = R₁ = H

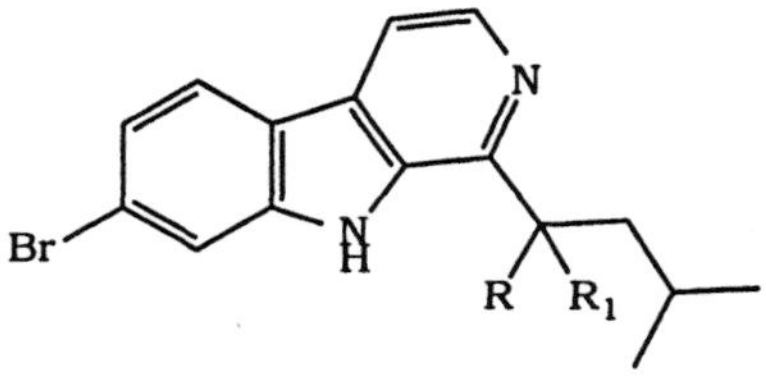

149. R = H, R₁ = NH₂
150. RR₁ = O

A number of tetrahydro-β-carbolines with various substituents at C1 have been reported. They include woodenine (151) [78], from the New Caledonian ascidian *Eudistoma fragum*, eudistomidin B (152) from *Eudistoma glaucus* [96,97] and lissoclin C (153) from a *Lissoclinum* sp. collected on the GBR [14]. The latter, together with the report of eudistomin U and isoeudistomin U from *Lissoclinum fragile* collected in the Caribbean [95], are the sole reports of isolation of β-carboline alkaloids from didemnid ascidians; they are more generally found in the families Polycitoridae and Polyclinidae. The structure of woodenine has been confirmed by total synthesis [92,102]. Arborescidimes A-D (154-157), examples of tetrahydro-β-carbolines which are brominated at C7, were reported from *Pseudodistoma arborescens* [106].

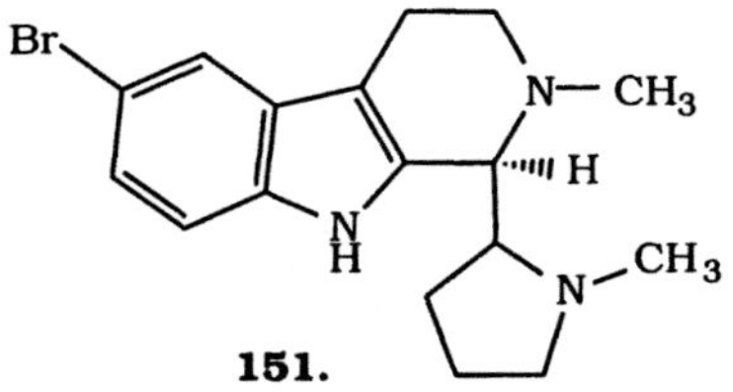

151.

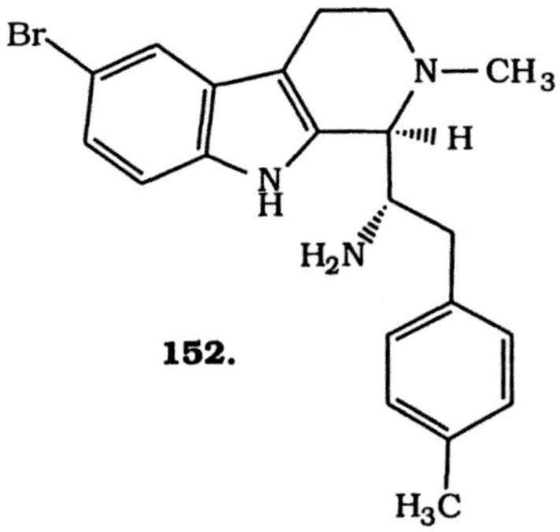

152.

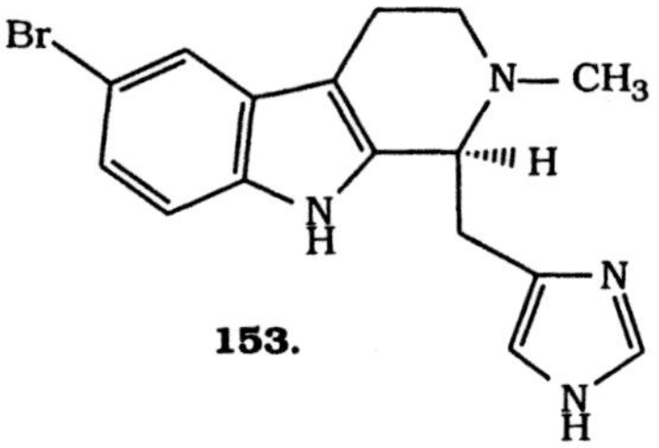

153.

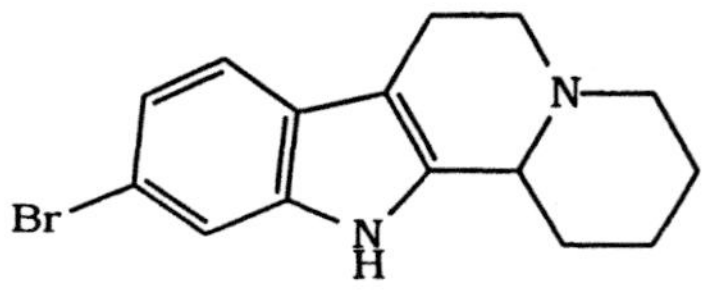

154.

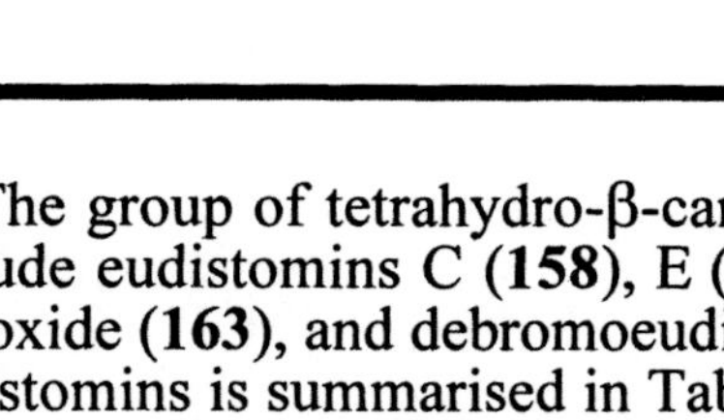

155.

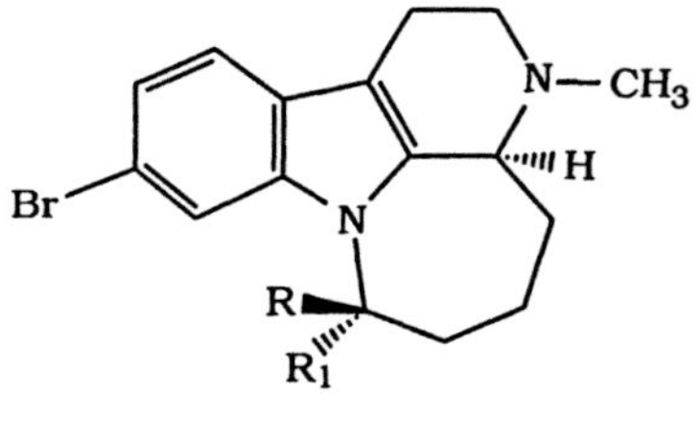

156. R = H, R₁ = OH
157. R = OH, R₁ = H

The group of tetrahydro-β-carbolines that have an oxathiazepine ring include eudistomins C (**158**), E (**159**), F (**160**), K (**161**), and L (**162**), K-sulfoxide (**163**), and debromoeudistomin K (**164**). The occurrence of these eudistomins is summarised in Table (**3**), and the highest levels of antiviral activity have been exhibited by members of this group. The stereochemistry of the N-O bond was initially suggested to be β [91], but later shown to be α [107]. The structure and absolute stereochemistry of eudistomin K has been confirmed by solution of its x-ray structure [108], while syntheses have been reported for eudistomin F [109], eudistomin L [110], N(10)-acetyleudistomin L [111,112] and debromoeudistomin L (= debromoeu-distomin K) [110, 113]. (-) O-Methyldebromoeudistomin E has also been synthesised [113].

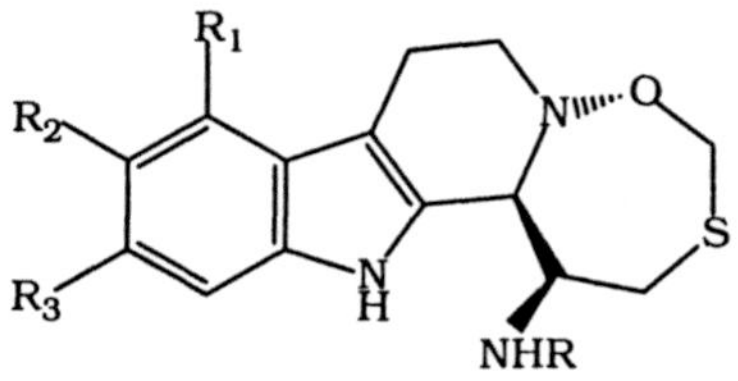

158. R = R$_1$ = H, R$_2$ = OH, R$_3$ = Br
159. R = R$_3$ = H, R$_1$ = Br, R$_2$ = OH
160. R = COCH$_3$, R$_1$ = H, R$_2$ = OH, R$_3$ = Br
161. R = R$_1$ = R$_2$ = H, R$_3$ = Br
162. R = R$_1$ = R$_3$ = H, R$_2$ = Br
164. R = R$_1$ = R$_2$ = R$_3$ = H

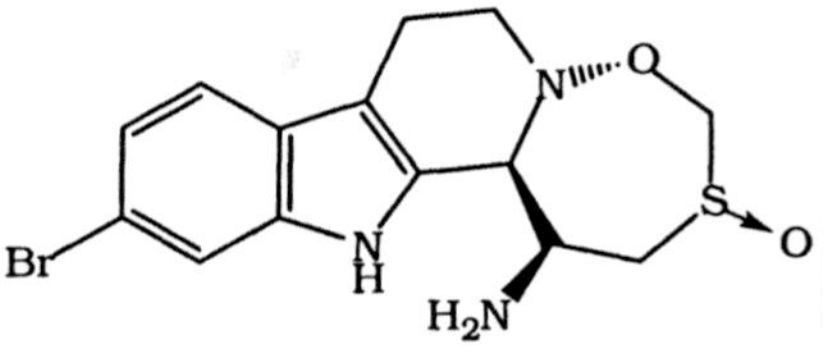

163.

Table 3.   Reported Occurrence of Eudistomins that Contain an Oxathiazepine Ring

Compound	Structure	Occurrence	Location [References]
Eudistomin C	158	*Eudistoma olivaceum* *Riterella sigillinoides*	Caribbean [91,125] New Zealand [119]
Eudistomin E	159	*Eudistoma olivaceum* *Eudistoma album*	Caribbean [91,125] New Caledonia [100]
Eudistomin F	160	*Eudistoma olivaceum*	Caribbean [91]
Eudistomin K	161	*Eudistoma olivaceum* *Riterella sigillinoides*	Caribbean [91,125] New Zealand [119]
Eudistomin K sulfoxide	163	*Riterella sigillinoides*	New Zealand [126,119]
Eudistomin L	162	*Eudistoma olivaceum*	Caribbean [91,125]
Debromoeudistomin K	164	*Riterella sigillinoides*	New Zealand [119]

There are only two reported examples of dihydro-β-carbolines. The first was isoeudistomin U which was initially reported to be a 4-substituted dihydro-α-carboline derivative [95], but whose structure was revised after total synthesis to 3,4-dihydroeudistomin U (**165**) [114]. Recently, the dihydro--β-carboline (**166**) was reported from an undescribed ascidian of the genus *Eudistoma* [115].

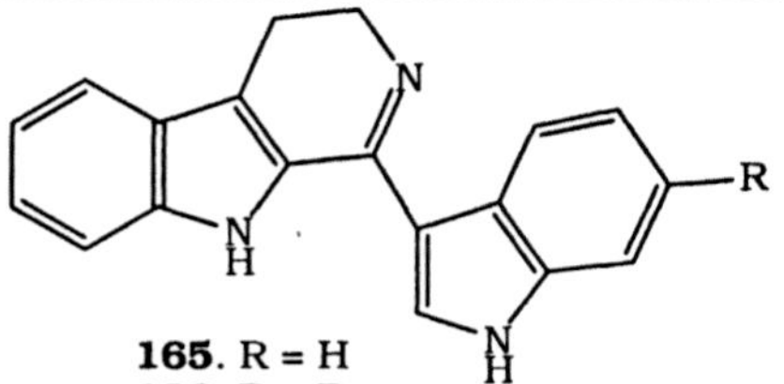

165. R = H
166. R = Br

Biosynthetic Studies

Biogenetic work to date has concerned only those β-carbolines which have a 5-membered N-containing ring substituted at C1. Although the biogenesis of eudistomidin A (141) was speculated to be from condensation of tryptophane with glutamate [116], biosynthetic studies with eudistomins H (138) and I (136) indicate a more direct route from tryptophane and proline [117,118]. Data which indicate a proposed order for bromination and decarboxylation of tryptophane have also been presented.

Activity of β-Carbolines, Tetrahydro-β-Carbolines and Dihydro-β-Carbolines

β-Carboline (126) is reported to inhibit HSV-1 and polio vaccine type 1 at a concentration of 2 μg per disk [119]. In preliminary antiviral assays eudistomins D (129), N (127), and O (128) displayed moderate inhibition of HSV-1 [120], and eudistomin O was subsequently reported to inhibit HSV-1 and polio vaccine type 1 *in vitro* at a concentration of 500 ng/disk[119]. Eudistomins D, N, and O also exhibited antibacterial activity against *B. subtilis* , but although N and O were initially reported to also inhibit the yeast *Saccharomyces cerevisiae* [120], the activity was not supported in a subsequent report [91]. Remarkable synergism was however observed with mixtures of N and O which produced antibiotic activity while none was observed with either of the pure compounds [91]. An investigation of the photoactive properties of N and O showed that N was the most active (similar to harmine) in the presence of UVA light (long wavelength UV), and that O was moderately phototoxic [121]. Eudistomidin D (130) is cytotoxic with IC_{50} values of 2.4 μg/mL and 1.8 μg/mL against the murine leukemias L1210 and L5178Y respectively. It also induces Ca^{2+} release from the sarcoplasmic reticulum, and is 10 times more potent than caffeine. Eudistomin J (131) was not reported to display any activity. The synthetic bromoeudistomin D (132) was reported to be a novel inducer of calcium release from fragmented sarcoplasmic reticulum that causes contraction of skinned muscle fibres at ≥10mM[122]. Contractions were completely blocked by 10mM procaine. It was reported that bromoeudistomin D may induce Ca^{2+} release from the sarcoplasmic reticulum through physiologically-relevant Ca^{2+} channels. Eudistomins D, J, N, and O were also tested; only D and J stimulated Ca^{2+} release but required 7 times higher concentrations than bromo eudistomin D, whose calcium-releasing effect was 400 times more potent than caffeine [91]. Results suggest that the 5-Br, 6-OH and 7-Br are all important for activity. Another semi-synthetic eudistomin derivative, N-methyl bromoeudistomin D (i.e. 7-bromo-eudistomidin D), was also subsequently reported to be a potent inducer of calcium release from the sarcoplasmic reticulum of skeletal muscle [123].

No activities have been observed for those eudistomins substituted at C1 with a pyrrole residue: eudistomins A (135) and M (134) [120,91], although eudistomin M was reported to be moderately phototoxic [121].

Eudistomins G (137), H (138), and I (136) in preliminary antiviral assays exhibited moderate inhibition of HSV-1, and an investigation of the photoactive properties of H and I showed that the presence of UVA (long wavelength UV) light had little effect on their activity [121]. Eudistomin I also displayed antibacterial activity against *B. subtilis* [120,91]. Eudistomins H and P (140) were reported to inhibit the yeast *S. cerevisiae* , while eudistomins P and Q (139) displayed antibacterial activity against *B. subtilis* [120,91]. Eudistomidin A (141) is a calmodulin agonist with an IC_{50} of 2×10^{-5} M against calmodulin-activated brain phosphodiesterase [116].

Eudistomin U (142) was not cytotoxic to CEM human leukemia lymphoblasts, nor did it exhibit antimicrobial activity against marine bacterial strains, although it exhibited strong activity against *Agrobacterium tumefaciens* [95]. In contrast to the activity of eudistomin A, eudistomidin C (143) is cytotoxic with IC_{50} values of 0.36 µg/mL and 0.42 µg/mL against L1210 and L5178Y cells *in vitro* and is a calmodulin antagonist, with an IC_{50} of 3×10^{-5} M [96]. No activities have been reported for eudistomidins E (144) and F (145) [98] . The original report of the structures of eudistomins R (146), S (147) and T (148) [99] contained no bioactivity data, but subsequent papers reporting the syntheses of eudistomin T have commented on its antimicrobial activity [124]. Eudistalbin A (149) is cytotoxic to KB human buccal carcinoma cells (ED_{50} 3.2 µg/mL) while eudistalbin B (150) is inactive in the same screen [100], suggesting that the amino group, a feature also present in eudistomidin C, is necessary for cytotoxicity.

Activity of Tetrahydro-β-Carbolines

The tetrahydro-β-carbolines generally exhibit higher levels of biological activity than their fully aromatic relatives. Antiviral activity is especially evident in the tetrahydro-β-carbolines that have an oxathiazepine ring, speculated to have been derived from cysteine [125].

Woodinine (151) displayed antimicrobial activity against *S. aureus* and *E.coli* , with zones of inhibition of 16mm and 8mm resp. at 100 µg per 6 mm disk, and 18mm and 11mm resp. at 200 µg / disk [78]. Eudistomidin B (152) was cytotoxic to both L1210 and L5178Y cells with reported IC_{50} values of 3.4 µg/mL and 3.1 µg/mL respectively. Eudistomidin B also activated rabbit heart muscle actomyosis ATPase by 93% at 3×10^{-5} M [96]. No activity data was reported for lissoclin C (153) [14]. The only arborescidime which was reported to be active was arborescidime D (157), which was cytotoxic to KB human buccal carcinoma cells (IC_{50} 3μg/mL) [106].

The group of eudistomins which contain an oxathiazapine ring exhibit the highest levels of antiviral activity. Those with a phenolic hydroxyl are active against HSV-1 down to 5-10ng/ disk [91]. The order of antiviral activity has been reported to be eudistomin E (150) (which strongly inhibited HSV-1 at 25-50 ng/12.5 mm disk) [125,91] followed by eudistomin C (158), (which has been reported to strongly inhibit HSV-1 [125,119] and Polio vaccine type 1 [119] at 40-50 ng /disk), followed by L (162) (some inhibition of HSV-1 at 100 ng/ disk) [125,91] and K (161) (some inhibition of HSV-1 at 250 ng/ disk [125,91] or both HSV-1 and polio vaccine type 1 at 40-50 ng/disk) [119]. These eudistomins generally also exhibit antimicrobial activity: C against *B. subtilis* , *E. coli* and *Penicillium atrovenetum* ; E against *B. subtilis* ; K and L against *B. subtilis* , *E. coli* , *S. cerevisiae* and *P. atrovenetum* [91]. No activity of eudistomin F (151) has been reported [91]. Eudistomin E was also reported to be cytotoxic to KB human buccal carcinoma cells with an ED_{50} of <5.0 ng/mL (100% cytotoxicity at 5 ng/mL) [100] , while eudistomin K was reported to be cytotoxic both *in vitro* against P388 cells (IC_{50} 0.01 μg/mL) and *in vivo* against L1210 and P388 murine leukemias as well as HCT-8, a human colon tumor. The T/C value of 137 at 100 μg/Kg was reported for P388 [119]. Debromoeudistomin K (164) is active against HSV-1 and Polio vaccine type 1 at a concentration of 400 ng/disk, while eudistomin K sulfoxide (163) was reported to be active at the same concentration [119] and also at 200 ng/disk [126].

The bioactivity of the dihydro-β-carboline isoeudistomin U (165) was reported to be similar to that of eudistomin U: it was not cytotoxic to CEM human leukemia lymphoblasts, nor did it exhibit antimicrobial activity against marine bacterial strains, although it exhibited strong activity against *Agrobacterium tumefaciens* [95]. No activity data was reported for (166) [115].

Quinolines

Two metabolites which were called trididemnic acids A (167) and B (168), were isolated from a Northeast Pacific ascidian *Trididemnum* sp. [127]. Xanthurenic acid (169) co-occurred with the trididemnic acids suggesting biosynthesis from tryptophane; no activity data was reported.

Pyridoacridine Alkaloids

Members of this group of alkaloids have been reported from ascidians (Urochordata), sponges (Porifora), anemones (Cnidaria) and one prosobranch (Mollusca), although the latter example is likely to be due to

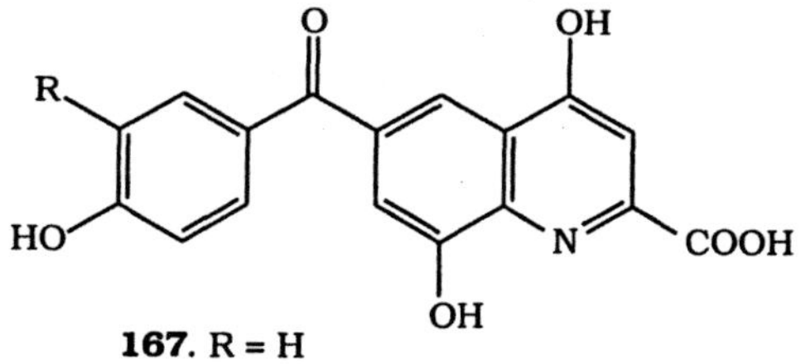

167. R = H
168. R = OH

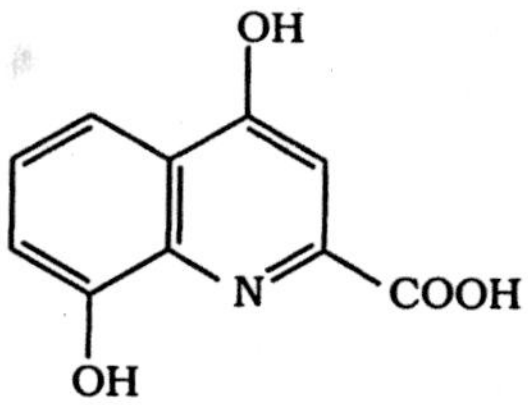

169.

accumulation of the alkaloids from dietary intake. In one or two cases the same metabolite has been reported from both tunicates and sponges. The discussion here will deal only with those alkaloids which are tunicate-derived, but is cross-referenced in cases where the same metabolite has also been isolated from another organism. The structure, synthesis and biological chemistry of marine-derived pyridoacridine alkaloids were reviewed by Molinski in 1993 [128].

Tetracyclic Pyridoacridine Alkaloids

The known tetracyclic pyridoacridine alkaloids from marine sources are dominated by those isolated from tunicates. All reported tetracyclic pyridoacridine alkaloids to date have oxygen functionality at C8, and the alkaloids can readily be divided into a group that has a carbonyl group at

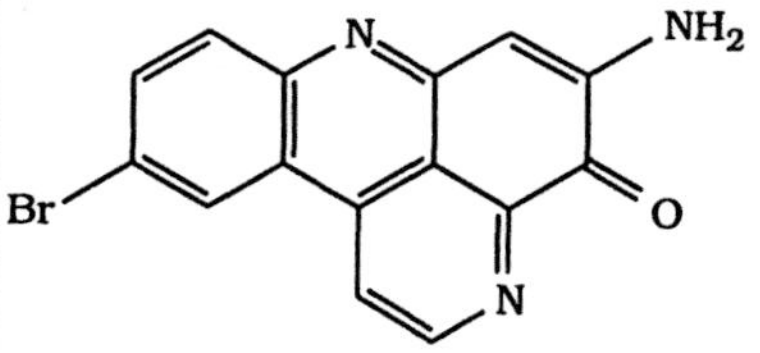

170.

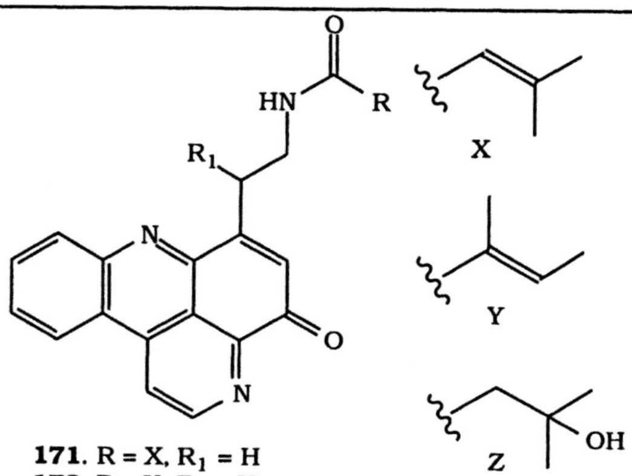

171. R = X, R_1 = H
172. R = Y, R_1 = H
173. R = Z, R_1 = H
174. R = X, R_1 = OH
175. R = Y, R_1 = OH
176. R = X, R_1 = OCH_3
177. R = Y, R_1 = OCH_3
178. R = X, R_1 = $OOC(CH_2)_7CH=CH(CH_2)_7CH_3$
179. R = Y, R_1 = $OOC(CH_2)_7CH=CH(CH_2)_7CH_3$
180. R = CH_3, R_1 = H

C8 (8*H*-pyrido[4,3,2-*mn*]acridones) and a group that has an ether at C8 (the tautomeric, 11*H*-pyrido[4,3,2-*mn*]acridines). The first group contains pantherinine **(170)** [129], cystodytins A-J **(171-180)** [130,131,132], diplamine **(181)** [133] and lissoclins A **(182)** and B **(183)** [14]. The second group contains norsegoline **(184)** [134,135], and varamines A **(185)** and B **(186)** [136].

Pantherinine **(170)**, the simplest example to date of the first group, was isolated from the South Australian ascidian *Aplidium pantherinum* [129]. Cystodytins A - I **(171-179)** were isolated from *Cystodytes dellechiajei* collected in Okinawa [130,131,137]. In each case the isomeric β,β-dimethyl acrylate and tiglate amides (R = X and R = Y) could not be separated and were characterised as mixtures. Cystodytin A was also reported from a Fijian *Cystodytes* sp., which in addition contained cystodytin J **(180)** along with a number of pentacyclic pyridoacridines [132]. Diplamine **(181)** was reported from a *Diplosoma* sp. [133], while lissoclins A **(182)** and B **(183)** (which only differ from diplamine by the nature of the acyl group in the amide side chain) were reported from a collection of a *Lissoclinum* sp. in Australia [14].

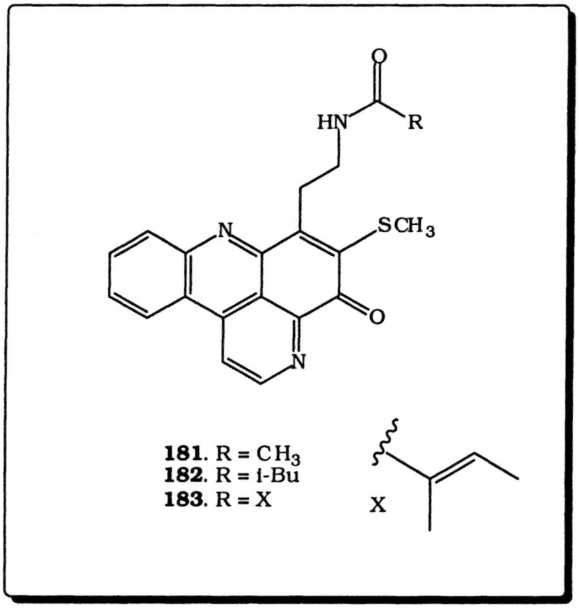

181. R = CH$_3$
182. R = *t*-Bu
183. R = X

The structure of norsegoline **(184)**, which was isolated from a *Eudistoma* sp. from the Red Sea, was disclosed in two reports [134,135], while varamines A **(185)** and B **(186)** were isolated from a Fijian collection of *Lissoclinum vareau* [136]. Syntheses of cystodytin J [138], other cystodytins [139], diplamine [138,140,141], and norsegoline [142,143] have been reported.

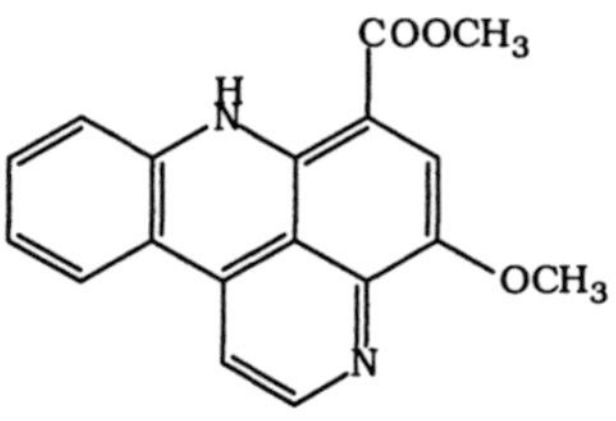

184

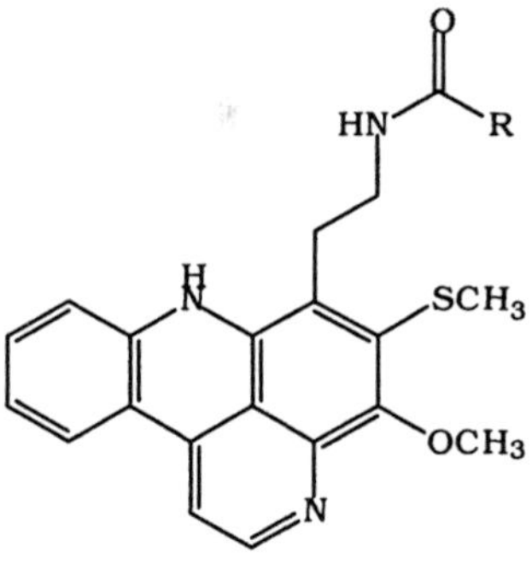

185. R = CH₂CH₃
186. R = CH₃

Biological Activity of the Tetracyclic Pyridoacridines

The cytotoxicity of some of the tetracyclic pyridoacridines is summarised in Table (4). Despite the published statement [136] that "the varamines are about 1 order of magnitude more cytotoxic than the cystodytins which contain the same aromatic nucleus but lack the S-methyl group", a comparison of the data for diplamine with that for cystodytin J does not support any significant cytotoxicity enhancement due to the thiomethyl group. The observed cytotoxicity difference between the varamines and comparable cystodytins is presumably due to their tautomeric aromatic nuclei. Based on limited data, among the cystodytins, the nature of the amide functionality on the side chain seems to have little effect on the cytotoxicity, but the presence of a hydroxyl group at the side chain benzylic position appears to marginally decrease the cytotoxicity while an ether functionality at the same position significantly enhances the activity.

A detailed assessment of the antileukemic properties of norsegoline has been published [144], but no activity data has been reported for lissoclins A and B [14].

As well as their cytotoxicity, varamines A and B were also reported to exhibit potent antifungal activity [136], while the cystodytin A/B mixture and cystodytin C were 36 times and 13 times repectively more powerful than caffeine in calcium-releasing activity in sarcoplasmic reticulum [130].

Mode of Action

Incorporation studies on a number of these alkaloids showed that pyridoacridines disrupt DNA and RNA synthesis with little effect on protein synthesis [132]. Results were consistent with DNA being the

Table 4.   **Reported Cytotoxicity of Some Tetracyclic Pyridoacridine Alkaloids**

Compound	Structure	IC_{50}	Cell line	Reference
Pantherinine	170	4.5 µg/mL	P388	129
Cystodytin J	180	0.5 µg/mL	HCT	132
Cystodytin A/B	171/172	0.22 µg/mL	L1210	130
Cystodytin C	173	0.24 µg/mL	L1210	130
Cystodytin D/E	174/175	1.1 µg/mL 1.4 µg/mL	L1210 KB	131 131
Cystodytins F/G	176/177	0.068 µg/mL 0.078 µg/mL	L1210 KB	131 131
Cystodytin H/I	178/179	0.080 µg/mL 0.092 µg/mL	L1210 KB	131 131
Diplamine	181	<0.5 µg/mL 0.02 µg/mL	HCT L1210	132 133
Varamine A	185	0.03 µg/mL	L1210	136
Varamine B	186	0.05 µg/mL	L1210	136
Shermilamine B	188	5 µg/mL 0.3-0.4 µg/mL	KB P388	148 145
Shermilamine C	189	7 µg/mL 0.3-0.4 µg/mL	HCT P388	132 145
Kuanoniamine A	190	1.0 µg/mL	KB	148
Kuanoniamine B	191	10 µg/mL	KB	148
Kuanoniamine D	193	5 µg/mL	KB	148
Dehydrokuanoniamine B	194	3.3 µg/mL	HCT	132

primary target of the pyridoacridine alkaloids, which appeared to be DNA intercalators. The effect of diplamine on Topoisomerase-I (Topo-I) was assessed using the supercoiled DNA relaxation assay and diplamine inhibited Topo-I at a much higher concentration than the Topoisomerase-II (Topo-II) decatenation reaction. The IC_{90} value for the Topo II decatenation assay was 9.2µM whereas Topo-I activity was still present at 14 mM. All tested pyridoacridines inhibited the Topo II -mediated decatenation of kinoplast DNA in a dose-dependent manner. In another study, shermilamine B was reported to inhibit Topo-II at 30 µM (compared to the reported [132] IC_{90} value of 118 µM) while shermilamine A was found to be inactive [145]. Disruption of the function of Topo II subsequent to intercalation is a probable mechanism by which pyridoacridines inhibit the proliferation of HCT (human colon tumor) cells.

Pentacyclic Pyridoacridine Alkaloids

Shermilamines A (187) [146,147] , B (188) [147,148], C (189) [132], kuanoniamines A-D (190-193) [148] and dehydrokuanoniamine B (194) [132] can be envisaged to arise from similar precursors to the cystodytins or varamines. Kuanoniamine C is identical to dercitamide which had previously been isolated from a sponge, *Dercitus* sp [149].

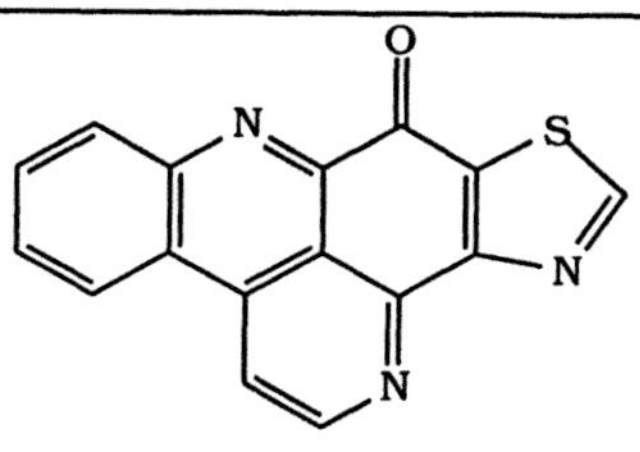

190.

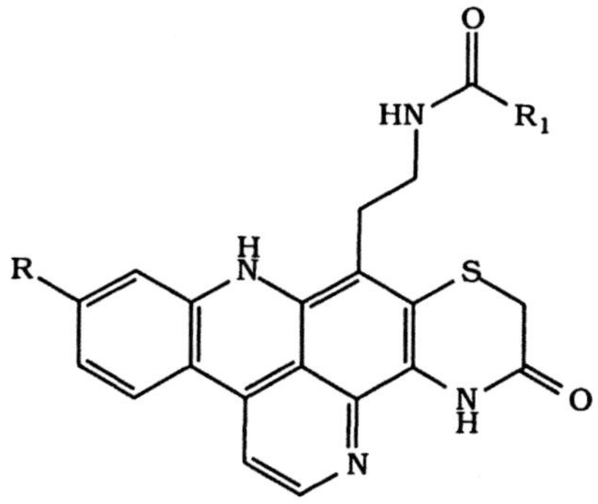

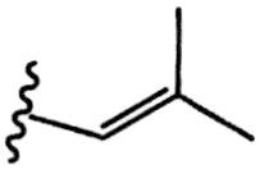

187. R = Br, R$_1$ = CH$_3$
188. R = H, R$_1$ = CH$_3$
189. R = H, R$_1$ = X

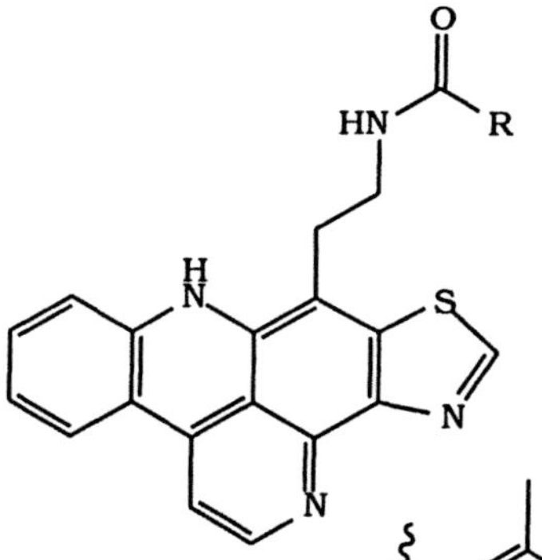

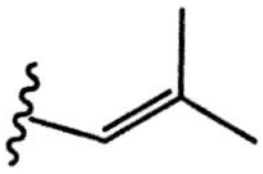

191. R = i-Bu   X
192. R = CH$_2$CH$_3$
193. R = CH$_3$
194. R = X

Shermilamines A and B were isolated from a purple *Trididemnum* sp. [146,147], while shermilamine C was reported from a purple fleshy *Cystodytes* sp. from Fiji [132]. Both shermilamines A and B were reported in a paper which described metabolites from *Amphicarpa meridiana* and *Leptoclinides* sp. [145], while shermilamine B was also reported under the name debromoshermilamine from a Red Sea *Eudistoma* sp. [135]. Biosynthetic studies on shermilamine B have been reported [150]. The kuanoniamines were isolated from the lamellarid mollusc, *Chelynotus semperi* and their unidentified tunicate food source from Pohnpei [148]. Dercitamide (**191**) and kuanoniamine D were subsequently reported from dark purple tunicates of the genus *Cystodytes* collected in Pohnpei [152], while dehydrokuanoniamine B [132] was isolated from a Fijian *Cystodytes* sp. Available cytotoxicity data on these metabolites, which is included in Table (**4**), indicate they are significantly less active than their tetracyclic analogues.

The remaining pentacyclic pyridoacridine alkaloids can be divided into two structural groups: those with an additional angular ring fused at C9,10 of the acridine ring C include ascididemnin (**195**), 11-hydroxyascididemnin (**196**), and 2-bromoleptoclinidinone (**197**), while those with a linear ring fusion at C8,9 of ring C include amphimedine (**198**), meridine (**199**) and cystodamine (**200**).

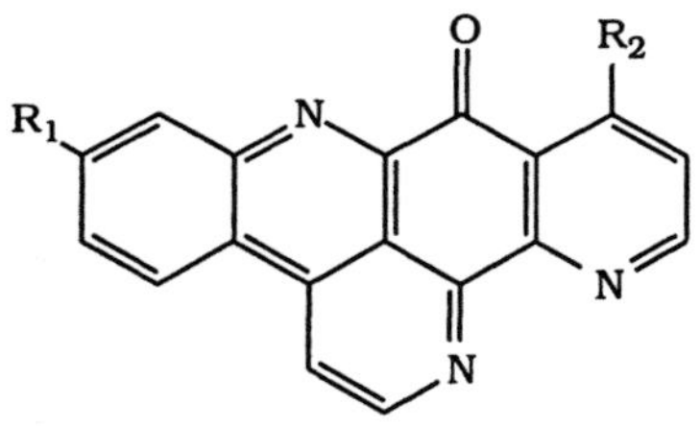

195. $R_1 = R_2 = H$
196. $R_1 = H$, $R_2 = OH$
197. $R_1 = Br$, $R_2 = H$

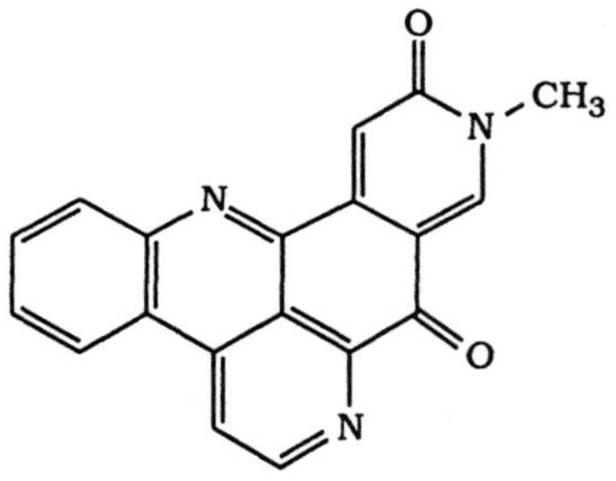

198.

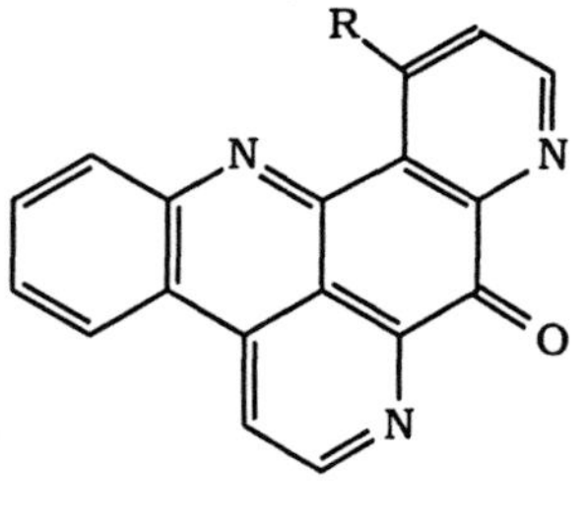

199. R = OH
200. R = NH$_2$

Ascididemnin (**195**), initially isolated from an Okinawan *Didemnum* sp., was reported to be cytotoxic (IC$_{50}$ of 0.39 µg/mL against L1210 murine leukemia), and to have calcium releasing activity in the sarcoplasmic reticulum (7 times more powerful than caffeine) [151]. It was subsequently reported from a *Eudistoma* sp. [153], and from a *Leptoclinides* sp. [145]. The latter confirmed the cytotoxicity (IC_{50} of 0.3 - 0.4 µg/mL against P388 murine leukemia cells), and reported Topoisomerase-II inhibition at 75µM for ascididemnin. In addition, the isolation of 11-hydroxyascididemnin (**196**) was reported, and said to be cytotoxic although no data was supplied to support the assertion. The structure of 2-bromoleptoclinidinone, isolated from an ascidian tentatively identified as a *Leptoclinides* sp., was initially incorrectly reported [154], but subsequently corrected to (**197**) [155]. 2-Bromoleptoclinidinone was reported to be cytotoxic with an IC_{50} of 0.4 µg/mL against P388 (PS) cells [154]. "Leptoclinidinone" is synonymous with ascididemnin. The synthesis of 2-bromoleptoclinidinone has been reported [156].

Amphimedine (**198**) was a known metabolite from an *Amphimedon* sp. of sponge [157] but it has also subsequently been reported from a *Didemnum* collected in Okinawa [151], and from a Mediterranean ascidian, *Cystodytes delle chiajei* [158]. Evidence was presented for the binding of amphimedine to DNA and it was proposed that inhibition of DNA synthesis (by intercalation) rather than inhibition of Topoisomerases caused the cytotoxicity [159]. A number of total syntheses of amphimedine have been reported [160,161,162,163,164,165]. The structure of meridine (**199**), which was isolated from the South Australian tunicate *Amphicarpa meridiana* [145, 166] and subsequently also reported from a sponge *Corticum* sp. [167], is supported by an x-ray structure. Meridine is reported to have antifungal properties [167] and two total syntheses have been reported [168,169]. A stable tautomer of meridine (**201**) was also reported to be cytotoxic to P388 (PS) cells (IC_{50} of 0.3-0.4 µg/mL) but to be inactive against Topoisomerase II [145]. Cystodamine (**200**) was reported from a Mediterranean ascidian, *Cystodytes delle chiajei* [158]. It has an amino group in place of the hydroxyl group in meridine and was reported to be cytotoxic (IC_{50} of 1.0 µg/mL against CEM human leukemic lymphoblasts).

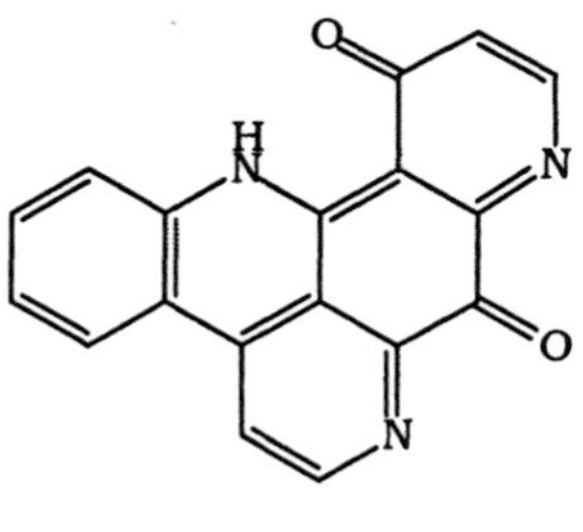

201.

More Complex Polycyclic Pyridoacridines

Among the more complex pyridoacridines from ascidians, the symmetrical dibenzotetraazaperylene eilatin (202) and eudistones A (203) and B (204) all are ring fused at both C8,9 and C9,10 of the acridine C-ring, while segoline A (205) and isosegoline A (206) represent yet another site of fusion. Eilatin (202) was isolated from a *Eudistoma* sp. from the Red Sea, along with segoline A (205), isosegoline A (206) and norsegoline (184) [170,135]. Eilatin was also subsequently reported from a Fijian *Cystodytes* sp. [132], where it was reported to be responsible for the positive biochemical prophage induction assay which was being used as a rapid test for antitumor agents that interact with DNA, and to be moderately cytotoxic. Segoline A, isosegoline and eilatin have been reported to be potent regulators of cellular growth and differentiation, and to affect cyclic

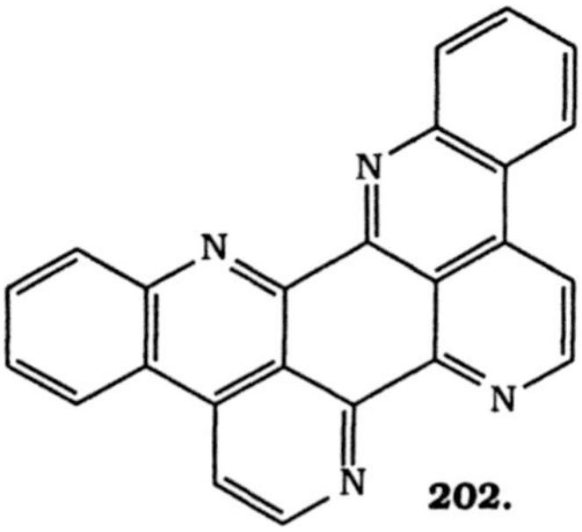

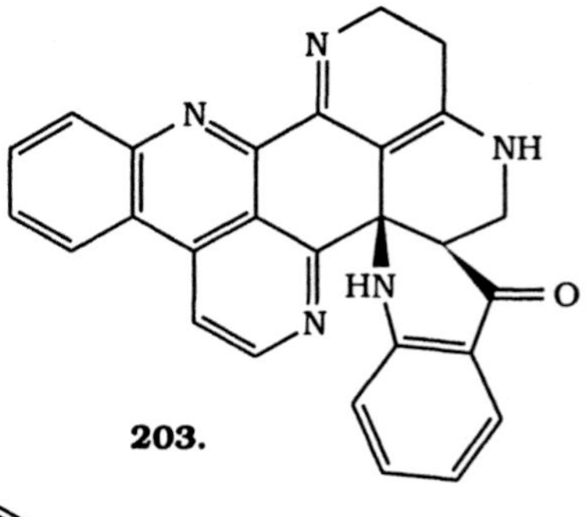

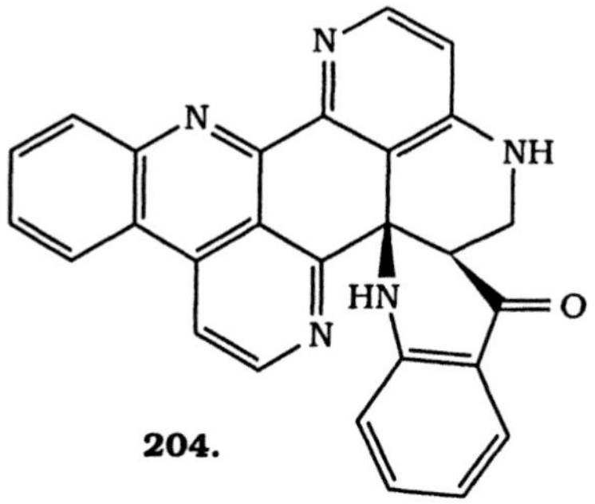

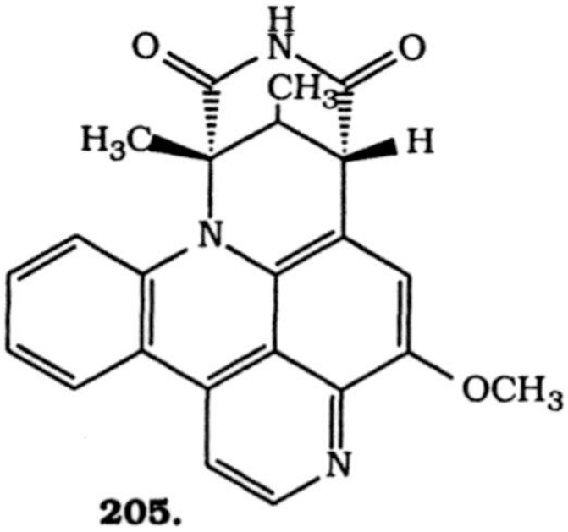

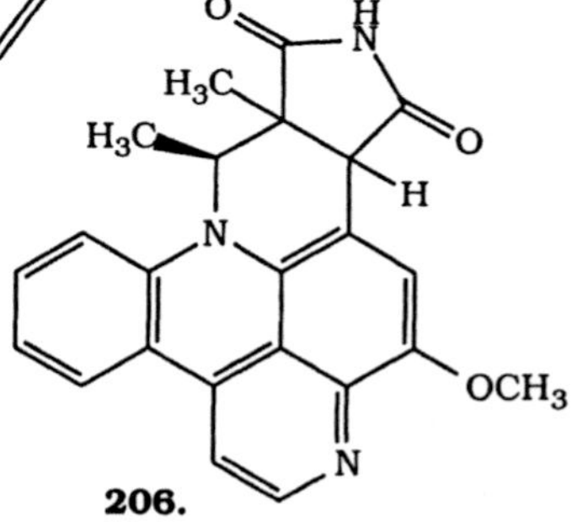

AMP-mediated processes [171]. No activity data has been published for eudistones A and B.

Ungrouped Alkaloids

Among those alkaloids whose biogenetic origins are not obvious are the polycarpamines. Polycarpamines A-F (207-212) were isolated from only one of a number of collection of *Polycarpa aurata* [172]. Only polycarpamine B exhibited significant antifungal activity (*S. cerevisiae* and *C. albicans*); the lack of optical activity for polycarpamine A and other polycarpamines with aliphatic methoxy groups suggests they may be artefacts of the isolation process.

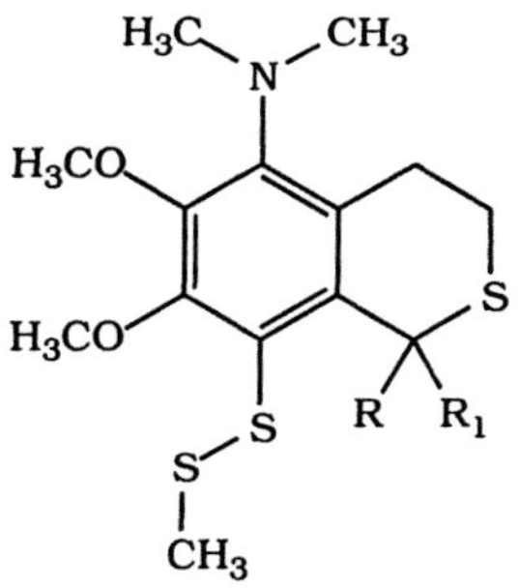

207. R = H, R₁ = OCH₃
208. R,R₁ = O
209. R,R₁ = S
210. R = COCH₃, R₁ = OCH₃

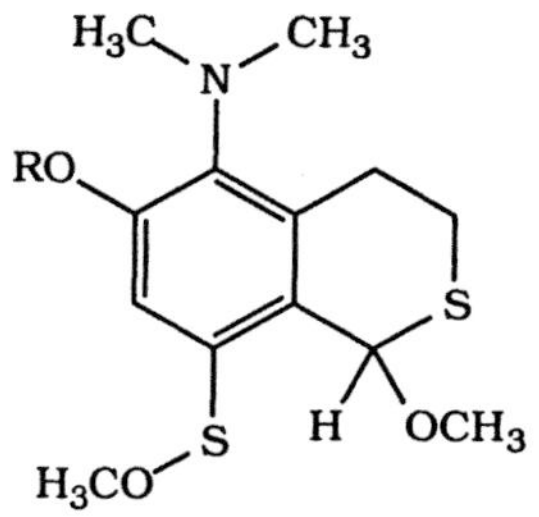

211. R = H
212. R = COCH₃

CONCLUDING REMARKS

Aromatic alkaloids from ascidians have provided a number of very promising leads especially in the area of cancer antiproliferative agents. The cytotoxic activity of the ecteinascidins and their application to clinical trials as anticancer agents stand out, but the potential of the lamellarins has yet to be fully investigated and exploited. A number of other metabolite groups also clearly hold potential: the antiviral activity of the eudistomins will undoubtedly be used as biochemical tools, whether or not they become clinical antiviral agents, and the mechanism of cytotoxicity of pyridoacridine alkaloids by suppression of DNA/RNA/protein synthesis will continue to be investigated.

With current emphasis on marine microorganism research it is conceivable that we may soon know the source of some of the metabolites which are common to certain ascidians and sponges.

ABBREVIATIONS

BSC	Basal cell carcinoma
Dopa	3,4-Dihydroxyphenylalanine
Et	Ecteinascidin
GBR	Great Barrier Reef, Queensland, Australia
HCT	Human colon tumor cell culture
ID_{50}/IC_{50}	Dose/Concentration required to reduce activity (or proliferation) by 50%
LD_{50}	Lethal dose for 50% of treated individuals
MDR	Multi-Drug Resistant
MIC	Minimum inhibitory concentration
NCI	American National Cancer Institute
T/C	Treated (usually days survival) divided by controls, expressed as %
TFA	Trifluoracetic acid
Topa	3,4,5-Trihydroxyphenylalanine

REFERENCES

[1]   Davidson, B.S. *Chem. Rev.*, **1993**, *93*, 1771.
[2]   Davidson, B.S. *Chem. Rev.*, **1994**, *94*, 1719.
[3]   Faulkner, D.J. *Nat. Prod. Rep.*, **1984**, *1*, 551.
[4]   Faulkner, D.J. *Nat. Prod. Rep.*, **1986**, *3*, 1.
[5]   Faulkner, D.J. *Nat. Prod. Rep.*, **1987**, *4*, 539.
[6]   Faulkner, D.J. *Nat. Prod. Rep.*, **1988**, *5*, 613.
[7]   Faulkner, D.J. *Nat. Prod. Rep.*, **1990**, *7*, 269.
[8]   Faulkner, D.J. *Nat. Prod. Rep.*, **1991**, *8*, 97.
[9]   Faulkner, D.J. *Nat. Prod. Rep.*, **1992**, *9*, 323.
[10]   Faulkner, D.J. *Nat. Prod. Rep.*, **1993**, *10*, 497.
[11]   Faulkner, D.J. *Nat. Prod. Rep.*, **1994**, *11*, 355.
[12]   Faulkner, D.J. *Nat. Prod. Rep.*, **1995**, *12*, 223.
[13]   Faulkner, D.J. *Nat. Prod. Rep.*, **1996**, *13*, 75.
[14]   Searle, P.A.; Molinski, T.F. *J. Org. Chem.*, **1994**, *59*, 6600.

[15]   Ireland, C.M.; Durso, A.; Scheuer, P.J. *J. Nat. Prod.*, **1981**, *44*, 360.
[16]   Sesin, D.F.; Ireland, C.M. *Tetrahedron Lett.*, **1984**, *25*, 403.
[17]   Arabshahi, L.; Schmitz, F.J. *Tetrahedron Lett.*, **1988**, *29*, 1099.
[18]   Copp, B.R.; Blunt, J.W.; Munro, M.H.G.; Pannell, L.K.*Tetrahedron Lett.*, **1989**, *30*, 3703.
[19]   Litaudon, M.; Guyot, M. *Tetrahedron Lett.*, **1991**, *32*, 911.
[20]   Behar, V.; Danishefsky, S.J. *J. Amer. Chem. Soc.*, **1993**, *115*, 7017.
[21]   Litaudon, M.; Trigalo, F.; Martin, M-T; Frapier, F.; Guyot, M.*Tetrahedron*, **1994**, *50*, 5323.
[22]   Makarieva, T.N.; Stonik, V.A.; Dmitrenok, A.S.; Grebnev, B.B.; Isakov, V.V.; Rebachyk, N.M.; Rashkes, Y.W. *J. Nat. Prod.*, **1995**, *58*, 254.
[23]   Davidson, B.S.; Ford, P.W.; Wahlman, M. *Tetrahedron lett.*, **1994**, *35*, 7185.
[24]   Davidson, B.S.; Molinsky, T.F.; Barrows, L.R.; Ireland, C.M. *J. Amer. Chem. Soc.*, **1991**, *113*, 4709.
[25]   Ford, P.W.; Davidson, B.S. *J. Org. Chem.*, **1993**, *58*, 4522.
[26]   Ford, P.W.; Narbut, M.R.; Belli, J.; Davidson, B.S. *J. Org. Chem.*, **1994**, *59*, 5955.
[27]   Toste, F.D.; Still, W.J. *J. Amer. Chem. Soc.*, **1995**, *117*, 7261.
[28]   Compagnone, R.S.; Faulkner, D.J. *Tetrahedron*, **1994**, *50*, 12785.
[29]   Abas, S.A.; Hossain, M.B.; van der Helm, D.; Schmitz, F.J.; Laney. M.; Cabuslay, R.; Schatzman, R.C. *J. Org. Chem.*, **1996**, *61*, 2709.
[30]   Hirsch, S.; Miroz, A.; McCarthy, P.; Kashman, Y. *Tetrahedron Lett.*, **1989**, *30*, 4291.
[31]   Lindquist, N.; Fenical, W. *Tetrahedron Lett.*, **1990**, *31*, 2521.
[32]   McDonald, L.A.; Swersey, J.C.; Ireland, C.M.; Carroll, A.R.; Coll, J.C.; Bowden, B.F.; Fairchild, C.R.; Cornell, L. *Tetrahedron*, **1995**, *51*, 5237.
[33]   Miao, S.; Andersen, R.J. *J. Org. Chem.*, **1991**, *56*, 6275.
[34]   Taylor, S.W.; Kammerer, B.; Bayer, E. *Chem. Rev.*, **1997**, *97*, 333.
[35]   1214 Bruening, R.C.; Oltz, E.M.; Furukawa, J.; Nakanishi, K.; Kustin, K. *J. Am. Chem. Soc.*, **1985**, *107*, 5298.
[36]   Horenstein, B.A.; Nakanishi, K. *J. Am. Chem. Soc.*, **1989**, *111*, 6242.
[37]   Bayer, E.; Schiefer, G.; Waidelich, D.; Scippa, S.; De Vincentiis, M. *Angew. Chem. Int. Ed. Eng.*, **1992**, *31*, 52.
[38]   Oltz, E.M.; Bruening, R.C.; Smith, M J.; Kustin, K.; Nakanishi, K. *J. Am. Chem. Soc.*, **1988**, *110*, 6162.
[39]   Azumi, K.; Yokosawa, H.; Ishii, S-i. *Biochemistry*, **1990**, *29*, 149.
[40]   Carroll, A.R.; Feng, Y.; Bowden, B.F.; Coll, J.C. *J. Org. Chem.*, **1996**, *61*, 4059.
[41]   Feng, Y.; Bowden, B.F. *Aust.J. Chem.*, **1997**, *50*, 337.
[42]   Kobayashi, J.; Cheng, J-f.; Kikuchi, Y.; Ishibashi, M.; Yamamura, S.; Ohizumi, Y.; Ohta, T.; Nozoe, S. *Tetrahedron Lett.*, *31*, 4617.
[43]   Yoshida, W.Y.; Lee, K.K.; Carroll, A.R.; Scheuer, P.J. *Helv. Chim. Acta*, **1992**, *75*, 1721.
[44]   Rudi, A.; Goldberg, I.; Stein, Z.; Frolow, F.; Benayahu, Y.; Schleyer, M.; Kashman, Y. *J. Org. Chem.*, **1994**, *59*, 999.
[45]   Terpin, A; Polborn, K.; Steglich, W. *Tetrahedron*, **1995**, *51*, 9941.
[46]   Urban, S.; Butler, M.S.; Capon, R.J. *Aust. J. Chem.*, **1994**, *47*, 1919.
[47]   Urban, S.; Hobbs, L.; Hooper, J.N.A.; Capon, R.J. *Aust. J. Chem.*, **1995**, *48*, 1491.

[48] Banwell, M.J.; Flynn, B.L.; Hamel, E.; Hockless, D.C.R. *J. Chem. Soc. Chem. Commun.*, **1997**, 207.
[49] Kang, H.; Fenical, W. *J. Org. Chem.*, **1997**, *62*, 3254.
[50] Andersen, R.J.; Faulkner, D.J.; Cun-heng, H.; Van Duyne, G.G.; Clardy, J. *J. Amer. Chem. Soc.*, **1985**, *107*, 5492.
[51] Lindquist, N.; Fenical, W.; Van Duyne, G.G.; Clardy, J. *J. Org. Chem.*, **1988**, *53*, 4570.
[52] Kock, M.; Reif, B.; Fenical, W.; Griesinger, C. *Tetrahedron Lett.*, **1996**, *37*, 363.
[53] Reif, B.; Kock, M.; Kerssebaum, R.; Kang, H.; Fenical, W.; Griesinger, C. *J. Mag. Res.* Ser. A, **1996**, *118*, 282.
[54] Heim, A.; Terpin, A.; Steglich, W. *Angew. Chem. Int. Ed. Eng.*, **1997**, *36*, 155.
[55] Carroll, A.R.; Bowden, B.F.; Coll, J.C. *Aust. J. Chem.*, **1993**, *46*, 489.
[56] Urban, S.; Capon, R.J. *Aust. J. Chem.*, **1996**, *49*, 711.
[57] Reddy, M.V.R.; Faulkner, D.J.; Venkateswarlu, M.; Rao, M.R. *Tetrahedron*, **1997**, *53*, 3457.
[58] Quesada, A.R.; Gravalos, M.D.G.; Puentes, J.L.F. *Br. J. Cancer*, **1996**, *74*, 677.
[59] Chan, G.W.; Francis, T.; Thureen, D.R.; Offen, P.H.; Pierce, N.J.; Westley, J.W.; Johnson, R.K.; Faulkner, D.J. *J. Org. Chem.*, **1993**, *58*, 2544.
[60] Palermo, J.A.; Brasco, M.F.R.; Seldes, A.M. *Tetrahedron*, **1996**, *52*, 2727.
[61] Rinehart, K.L.; Holt, T.G.; Fregeau, N.L.; Stroh, J.G.; Keifer, P.A.; Sun, F.; Li, L.H.; Martin, D.G. *J. Org. Chem.*, **1990**, *55*, 4512.
[62] Wright, A.E.; Forleo, D.A.; Gunawardana, G.P.; Gunasekera, S.P.; Koehn, F.E.; McConnell, O. *J. Org. Chem.*, **1990**, *55*, 4508.
[63] Kubo, A.; Saito, N. In *Studies in Natural Products Chemistry;* Atta-ur-Rahman, Ed.; Elsevier Science B.V: Amsterdam, 1992; Vol. *10*, pp. 77-145.
[64] He, H.; Faulkner, D.J. *J. Org. Chem.*, **1989**, *54*, 5822.
[65] Davidson, B.S. *Tetrahedron Lett.*, **1992**, *33*, 3721.
[66] Sakai, R.; Rinehart, K.L.; Guan, Y.; Wang, A.H.J. *Proc. Nat. Acad. Sci. USA*, **1992**, *89*, 11456.
[67] Guan, Y.; Sakai, R.; Rinehart, K.L.; Wang, A.H.J. *J. Biomol. Struct. Dyn.*, **1993**, *10*, 793.
[68] Sakai, R.; Jareserijman, E.A.; Manzanares, I.; Elipe, M.V.S.; Rinehart, K.L. *J. Am. Chem. Soc.*, **1996**, *118*, 9017.
[69] Kerr, R.G.; Miranda, N.F. *J. Nat. Prod.*, **1995**, *58*, 1618.
[70] Saito, N.; Tashiro, K.; Maru, Y.; Yamaguchi, K.; Kubo, A. *J. Chem. Soc. Perkin Trans. 1*, **1997**, 53.
[71] Rinehart, K., Sakai, R. PCT Int. Appl., **1992** (C.A. *117*, 205189z).
[72] Pommier, Y.; Kohlhagen, G.; Bailly, C.; Waring, M.; Mazumder A.; Kohn, K.W.*Biochemistry*, **1996**, *35*, 13303.
[73] Reid, J.M.; Walker, D.L.; Ames, M.M. *Cancer Chemother. Pharmacol.*, **1996**, *38*, 329.
[74] Corey, E.J.; Gin, D.Y.; Kania, R.S. *J. Am. Chem. Soc.*, **1996**, *118*, 9202.
[75] Corey, E.J.; Gin, D.Y. *Tetrahedron Lett.*, **1996**, *37*, 7163.
[76] Niwa, H.; Yoshida, Y.; Yamada, K. *J. Nat. Prod.*, **1988**, *51*, 343.
[77] Fahey, E.; Potts, B.C.M.; Faulkner, D.J.; Smith, K. *J.Nat. Prod.*, **1991**, *54*, 564.
[78] Debitus, C.; Laurent, D.; Pais, M. *J. Nat. Prod.*, **1988**, *51*, 799.
[79] Somei, M.; Kobayashi, K.; Tanii, K.; Mochizuki, T.; Kawada, Y.; Fukui, Y. *Heterocycles*, **1995**, *40*, 119.

[80] Moquin, C.; Guyot, M. *Tetrahedron Lett.*, **1984**, *25*, 5047.
[81] Heitz, S.; Durgeoat, M.; Guyot, M.; Brassy, C.; Bachet, B. *Tetrahedron Lett.*, **1980**, *21*, 1457.
[82] Guyot, M.; Meyer, M.*Tetrahedron Lett.*, **1986**, *27*, 2621.
[83] Bergmann, T.; Schories, D.; Steffan, B.*Tetrahedron*, **1987**, *53*, 2055.
[84] Hogan, I.T.; Sainsbury, M.*Tetrahedron*, **1984**, *40*, 681.
[85] Moquin-Pattey, C.; Guyot, M. *Tetrahedron*, **1989**, *45*, 3445.
[86] Vervoort, H.C.; Richards-Gross, S.E.; Fenical, W.; Lee, A.Y.; Clardy, J. *J. Org. Chem.*, **1997**, *62*, 1486.
[87] Tsukamoto, S, ; Hirota, H.; Kato, H.; Fusetani, N. *Tetrahedron Lett.*, **1993**, *34*, 4819.
[88] Copp, B.R.; Ireland, C.M.; Barrows, L.R. *J. Org. Chem.*, **1991**, *56*, 4596.
[89] Kinnel, R.B.; Scheuer, P.J. *J. Org. Chem.*, **1992**, *57*, 6327.
[90] Rocca. P.; Marsais, F.; Godard, A.; Queguiner, G.; Adams, L.; Alo, B. *Tetrahedron Lett.*, **1995**, *36*, 7085.
[91] Rinehart, K.L.; Kobayashi, J.; Harbour, G.C.; Gilmour, J.; Mascal, M.; Holt, T.G.; Shield, L.S.; LaFargue, F. *J. Amer. Chem. Soc.*, **1987**, *109*, 3378.
[92] McNulty, J.; Still, I.W.J. *Tetrahedron Lett.*, **1991**, *32*, 4875.
[93] Van Wagenen, B.C.; Cardellina, J.H. *Tetrahedron Lett.*, **1989**, *30*, 3605
[94] Wasserman H.H.; Kellt, T.A. *Tetrahedron Lett.*, **1989**, *30*, 7117.
[95] Badre, A.; Boulanger, A.; Abou-Mansour, E.; Banaigs, B.; Combaut, G.; Francisco, C. J. Nat. Prod., **1994**, *57*, 528.
[96] Kobayashi, J.; Cheng, J-f.; Ohta, T.; Nozoe, S.; Ohizumi, Y.; Sasaki, T. *J. Org. Chem.*, **1990**, *55*, 3666.
[97] Kobayashi, J.; Cheng, J-f.; Ishibashi, M.; Nakamura, H.; Ohizumi, Y.; Hirata, Y.; Ohta, T.; Nozoe, S.; Sasaki, T. *Tennen Yuki Kagobutsu Toronkai*, **1989**, *31*, 348.
[98] Marata, O.; Shigemon, H.; Ishibashi, M.; Sugama, K.; Hayashi, K.; Kobayashi, J. *Tetrahedron Lett.*, **1991**, *32*, 3539.
[99] Kinzer, K.F.; Cardellina, J.H. *Tetrahedron Lett.*, **1987**, *28*, 925.
[100] Adesanya, S.A.; Chbani, M.; Pais, M.; Debitus, C. *J. Nat. Prod.*, **1992**, *55*, 525.
[101] Still, I.W.J.; McNulty, J. *Heterocycles*, **1989**, *29*, 2057.
[102] McNulty, J.; Still, I.W.J. *J. Chem. Soc. Perkin Trans. 1*, **1994**, 1329.
[103] Bracher, F.; Hildebrand, D.; Ernst, L. *Arch. Pharm.* (Weinheim, Ger.), **1994**, *327*, 121.
[104] Rocca. P.; Marsais, F.; Godard, A.; Queguiner, G.; Adams, L.; Alo, B. *Synth.Commun.*, **1995**, *25*, 3373.
[105] Molina, P.; Fresneda, P.M.; Garcia-Zafra, S. *Tetrahedron Lett.*, **1995**, *36*, 3581.
[106] Chbani, M.; Pais, M.; Delaunaux, J-M.; Debitus, C. *J. Nat. Prod.*, **1993**, *56*, 105.
[107] Blunt, J.W.; Lake, R.J.; Munro, M.H.G.; Toyokuni, T. *Tetrahedron Lett.*, **1987**, *28*, 1825.
[108] Lake, R.J.; McCombs, J.D.; Blunt, J.W.; Munro, M.H.G.; Robinson, W.T. *Tetrahedron Lett.*, **1988**, *29*, 4971.
[109] Lui, J-J.; Nakagawa, M.; Harada, N.; Tsuruoka, A.; Hasegawa, A.; Ma, J.; Hino, T. *Heterocycles*, **1990**, *31*, 229.
[110] Nakagawa, M.; Lui, J.; Hino, T. *J. Am. Chem. Soc.*, **1989**, *111*, 2721.
[111] Still, I.W.J.; Strautmanis, J.R. *Tetrahedron Lett.*, **1989**, *30*, 1041.
[112] Still, I.W.J.; Strautmanis, J.R. *Can. J. Chem.*, **1990**, *68*, 1408.

[113] Hermkens, H.H.; v. Maarseveen, J.H.; Ottenheijm, H.C.J.; Kruse, C.G.; Scheeren, H.W. *J. Org. Chem.*, **1990**, *55*, 3998.
[114] Massiot, G.; Nazabadioko, S.; Bliard, C. *J. Nat. Prod.*, **1995**, *58*, 1636.
[115] Kang, H.; Fenical, W. *Nat. Prod. Lett.*, **1996**, *9*, 7.
[116] Kobayashi, J.; Nakamura, H.; Ohizumi, Y.; Hirata, Y. *Tetrahedron Lett.*, **1986**, *27*, 1191.
[117] Shen, G.C.; Baker, B.J. *Tetrahedron Lett.*, **1994**, *35*, 1141.
[118] Shen, G.C.; Baker, B.J. *Tetrahedron Lett.*, **1994**, *35*, 4923.
[119] Lake, R.J.; Blunt, J.W.; Munro, M.H.G. *Aust. J. Chem.*, **1989**, *42*, 1201.
[120] Kobayashi, J.; Harbour, G.C.; Gilmour, J.; Rinehart, K.L. *J. Am. Chem. Soc.*, **1984**, *106*, 1526.
[121] Hudson, J.B.; Saboune, H.; Abramowski. Z.; Towers, G.H.N.; Rinehart, K.L. *Photochem. Photobiology*, **1981**, *47*, 377.
[122] Nakamura, Y.; Kobayasi, J.; Gilmore, J.; Mascal, M.; Rinehart, K.L.; Nakamura, H.; Ohizumi, Y. *J. Biol. Chem.*, **1986**, *261*, 4139.
[123] Kobayasi, J.; Ishibashi, M.; Nagai, U.; Ohizumi, Y. *Experientia*, **1989**, *45*, 782.
[124] Molina, P.; Fresneda, P.M.; Garcia-Zafra, S. *Tetrahedron Lett.*, **1996**, *37*, 9353.
[125] Rinehart, K.L.; Kobayashi, J.; Harbour, G.C.; Hughes, R.G.; Mizsak, S.A.; Scahill, T.A. *J. Am. Chem. Soc.*, **1984**, *106*, 1524.
[126] Lake, R.J.; Brennan, M.M.; Blunt, J.W.; Munro, M.H.G.; Pannell, L.K. *Tetrahedron Lett.*, **1988**, *29*, 2255.
[127] de Silva; D.E., Miao, S.; Andersen, R.J.; Schultz, L.W.; Clardy, J. *Tetrahedron Lett.*, **1992**, *33*, 2917.
[128] Molinski, T.F. *Chem. Rev.*, **1993**, *93*, 1825.
[129] Kim, J.; Pordesimo, E.O.; Toth, S.I.; Schmitz, F.J.; van Altena, I. *J. Nat. Prod.*, **1993**, *56*, 1813.
[130] Kobayasi, J.; Cheng, J-f.; Walchli, M.R.; Nakamura, H.; Hirata, Y.; Sasaki, T.; Ohizumi, Y. *J. Org. Chem.*, **1988**, *53*, 1800.
[131] Kobayasi, J.; Tsuda, M.; Tanabe, A.; Ishibashi, M.; Cheng, J-f.; Yamamura, S.; Sasaki, T. *J. Nat. Prod.*, **1991**, *54*, 1634.
[132] McDonald, L.A.; Eldredge, G.S.; Barrows, L.R. ; Ireland, C.M. *J. Med. Chem.*, **1994**, *37*, 3819
[133] Charyulu, G.A.; McKee, T.C.; Ireland, C.M. *Tetrahedron Lett.*, **1989**, *30*, 4201.
[134] Rudi, A.; Benayahu, Y.; Goldberg, I.; Kashman, Y. *Tetrahedron Lett.*, **1988**, *29*, 3861.
[135] Rudi, A.; Kashman, Y. *J. Org. Chem.*, **1989**, *54*, 5331.
[136] Molinski, T.F.; Ireland, C.M. *J. Org. Chem.*, **1989**, *54*, 4256.
[137] Kobayasi, J.; Cheng, J-f.; Ishibashi, M.; Nakamura, H.; Ohizumi, Y.; Hirata, Y.; Walchli, M.R.; Sasaki, T. *Tennen Yuki Kagobutsu Toronkai*, **1988**, *30*, 268.
[138] Ciufolini, M.A.; Shen, Y.C. *Tetrahedron Lett.*, **1995**, *36*, 4709.
[139] Ciufolini, M.A.; Bryne, N.E. *J. Am. Chem. Soc.*, **1991**, *113*, 8016.
[140] Szczepankiewicz, B.G.; Heathcock, C.H. *J. Org. Chem.*, **1994**, *59*, 3512.
[141] Ciufolini, M.A.; Shen, Y.C.; Bishop, M.J. *J. Am. Chem. Soc.*, **1995**, *117*, 12460.
[142] Dunn, S.H.; McKillop A. *J. Chem. Soc. Perkin Trans. I*, **1993**, 879.
[143] Gellerman, G.; Rudi, A.; Kashman, Y. *Tetrahedron Lett.*, **1993**, *34*, 1823.
[144] Einat, M.; Nagler, A.; Lishner, M.; Amiel, A.; Yarkoni, S.; Rudi, A.; Gellerman, G.; Kashman, Y.; Fabian, I. *Clin. Cancer Res.*, **1995**, *1*, 823.
[145] Schmitz, F.J.; De Guzman, F.S.; Hossain, M.B.; van der Helm, D. *J. Org. Chem.*, **1991**, *56*, 804.

[146] Cooray, N.M.; Scheuer, P.J.; Parkanyi, L.; Clardy, J. *J. Org. Chem.*, **1988**, *53*, 4619.

[147] Carroll, A.R.; Cooray, N.M.; Poiner, A.; Scheuer, P.J. *J. Org. Chem.*, **1989**, *54*, 4231.

[148] Carroll, A.R.; Scheuer, P.J. *J. Org. Chem.*, **1990**, *55*, 4426.

[149] Gunawardana, G.P.; Kohmoto, S.; Burres, N.S. *Tetrahedron Lett.*, **1989**, *30*, 4359.

[150] Steffan, B.; Brix, K.; Putz, W. *Tetrahedron*, **1993**, *49*, 6223.

[151] Kobayasi, J.; Cheng, J-f.; Nakamura, H.; Ohizumi, Y.; Hirata, Y.; Sasaki, T.; Ohta, T.; Nozoe, S. *Tetrahedron Lett.*, **1988**, *29*, 1177.

[152] Gunawardana, G.P.; Koehn, F.E.; Lee, A.Y.; Clardy, J.; He, H.Y.; Faulkner, D.J. *J. Org. Chem.*, **1992**, *57*, 1523.

[153] He, H.Y.; Faulkner, D.J. *J. Org. Chem.*, **1991**, *56*, 5369.

[154] Bloor, S.J.; Schmitz, F.J. *J. Am. Chem. Soc.*, **1987**, *109*, 6134.

[155] De Guzman, F.S.; Schmitz, F.J. *Tetrahedron Lett.*, **1989**, *30*, 1069.

[156] Bracher, F. *Liebigs Ann. Chem.*, **1990**, 205.

[157] Schmitz, F.J.; Agarwal, S.K.; Gunasekera, S.P.; Schmidt, P.G.; Schoolery, J.M. *J. Am. Chem. Soc.*, **1983**, *105*, 4835.

[158] Bontemps, N.; Bonnard, I.; Banaigs, B.; Combaut, G.; Francisco, C. *Tetrahedron Lett.*, **1994**, *38*, 7023.

[159] Bonnard, I.; Bontemps, N.; Lahmy, S.; Banaigs, B.; Combaut, G.; Francisco, C.; Colson, P.; Houssier, C.; Waring, M.J.; Bailly, C. *Anti-Cancer Drug Des.*, **1995**, *10*, 333.

[160] Echavarren, A.M.; Stille, J.K. *J. Am. Chem. Soc.*, **1988**, *110*, 4051.

[161] Kubo, A.; Nakahara, S. *Heterocycles*, **1988**, *27*, 2095.

[162] Prager, R.H.; Tsopelas, C.; Heisler, T. *Aust. J. Chem.*, **1991**, *44*, 277.

[163] Prager, R.H.; Tsopelas, C.; Heisler, T. *Heterocycles*, **1989**, *29*, 847.

[164] Bracher, F.; Papke, T. *Liebigs Ann. Chem.*, **1996**, 115.

[165] Nakahara, S.; Tanaka, Y.; Kubo, A. *Heterocycles*, **1996**, *43*, 2113.

[166] Schmitz, F.J.; De Guzman, F.S.; Choi, Y-H.; Hossain, M.B.; Rizvi, S.K.; van der Helm, D. *Pure Appl. Chem.*, **1990**, *62*, 1393.

[167] McCarthy, P.J.; Pitts, T.P.; Gunawardana, G.P.; Kelly-Borges, M.; Pomponi, S. *J. Nat. Prod.*, **1992**, *55*, 1664.

[168] Bontemps, N.; Delforme, E.; Bastide, J.; Francisco, C.; Bracher, F. *Tetrahedron*, **1997**, *53*, 1743.

[169] Kitahara, Y.; Tamura, F.; Kubo, A. *Chem. Pharm. Bull.*, **1994**, *42*, 1363.

[170] Rudi, A.; Benayahu, Y.; Goldberg, I.; Kashman, Y. *Tetrahedron Lett.*, **1988**, *29*, 6655.

[171] Shochet, N.R.; Rudi, A.; Kashman, Y.; Hod, Y.; El-Maghrabi, M.R.; Spector, I. *J. Cell Physiol.*, **1993**, *157*, 481.

[172] Lindquist, N.; Fenical, W. *Tetrahedron Lett.*, **1990**, *31*, 2389.

Atta-ur-Rahman (Ed.) *Studies in Natural Products Chemistry, Vol. 23*

BIOACTIVE QUASSINOIDS

M. OKANO*, N. FUKAMIYA, *and* K. H. LEE[+]

Faculty of Integrated Arts and Sciences, Hiroshima University,
1-7-1 Kagamiyama, Higashi-Hiroshima 739-8521, Japan
[+]*School of Pharmacy, The University of North Carolina, Chapel Hill,*
NC 27599-7360, USA

ABSTRACT: This review covers various bioactive quassinoids in Simaroubaceous plants, which show cytotoxic antitumor, anti-tumor-promoter (= inhibitor against TPA-induced EBV-EA activation), anti-HIV, anti-tuberculosis, insect antifeedant, and other biological activities. Stability of quassinoids, the preparation of their derivatives, and their biological activities are also discussed.

INTRODUCTION

In a previous paper [1], we reviewed the isolation of quassinoids, alkaloids, and fatty acids from Simaroubaceous plants during the period 1972-1988 and discussed their biological activities, such as antitumor, antimalarial, and anti-inflammatory agents as well as insect antifeedants, amoebicides, and herbicides. This review will add new quassinoids isolated from Simaroubaceous plants from 1988 onwards and their biological activities, for example, antitumor, anti-tumor-promotor (= inhibitor against TPA-induced EBV-EA activation), anti-HIV, anti-tuberculosis, and insect antifeedant activities. Furthermore, the stability of quassinoids, lower activities of degradation products, and the synthesis of fluorinated quassinoids are also described.

ISOLATION OF QUASSINOIDS AND TESTING FOR VARIOUS BIOLOGICAL ACTIVITIES

From *Brucea antidysenterica*

An earlier investigation of the antileukemic constituents of the Ethiopian *Brucea antidysenterica* by Kupchan and associates [2] led to the isolation of quassinoids such as *bruceantin* (**1**), *bruceantarin* (**2**), *dehydrobruceantin* (**3**), *dehydrobruceantarin* (**4**), and *isobrucein-B* (**5**). Among these quassinoids, *bruceantin* (**1**) demonstrated potent antileukemic activity, including the P-388 and L-1210 lymphocytic leukemias, the Lewis lung carcinoma, and the B-16 melanoma. This compound was in Phase II clinical trial as an anticancer drug by the US National Cancer Institute (NCI).

We isolated the glycosides *bruceantinoside-A* (**6**) and *-B* (**7**) [3] from the same plant extract. We expected *B. antidysenterica* to contain glycosides of *bruceantin* (**1**), because *Brucea javanica* contains *bruceoside-A* (**9**) and *-B* (**10**) which are both glucosides of *brusatol* (**8**). *Brusatol* (**8**) is structurally identical with *bruceantin* (**1**) except for a slight difference in the C-15 ester side chain, namely, where *brusatol* (**8**) has a senecioate group, whereas *bruceantin* (**1**) bears a *trans*-3,4-dimethyl-2-pentenoate moiety. Careful examination of the thin layer chromatography

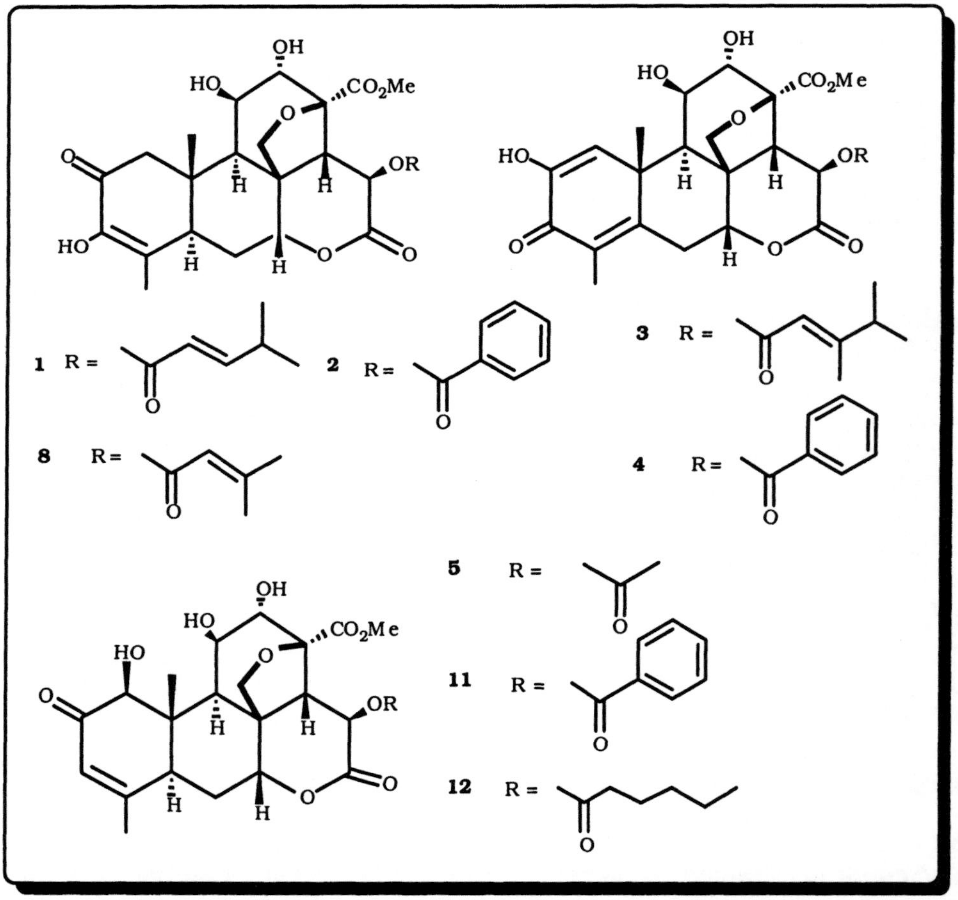

(TLC) of a chloroform soluble fraction of *B. antidysenterica* revealed the presence of spots differing from those previously isolated by Kupchan *et al* [2] Two of these spots showed Rf values slightly higher than those of *bruceoside-A* (**9**) and *-B* (**10**). Isolation of these two compounds by column chromatography (CC) on silica gel, followed by further purification of the appropriate fractions with preparative TLC and high performance liquid chromatography (HPLC), yielded the desired new glycosides, which were provisionally named *bruceantinoside-A* (**6**) and *-B* (**7**). Both were amorphous solids. Structural elucidation of *bruceantinoside-A* (**6**) and *-B* (**7**) was achieved by the direct comparison of NMR spectra of their hydrolyzed products with those of hydrolyzed product of *bruceoside-A* (**9**), *-B* (**10**), and related compounds.

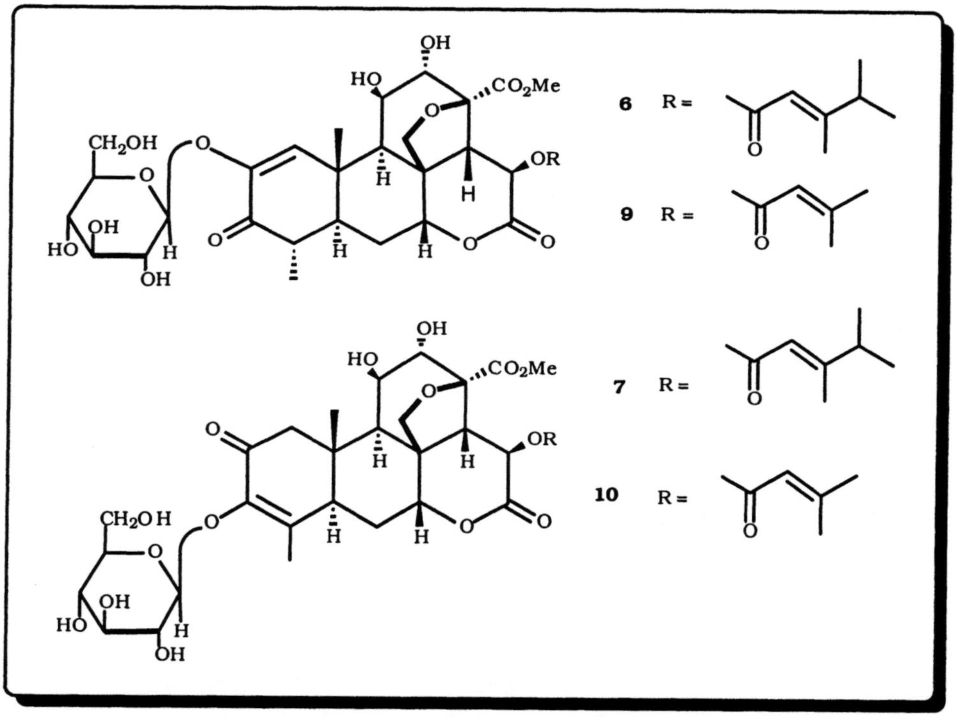

In vitro activity was assayed by the standard NCI procedure [4]. *Bruceantinoside-A* (**6**) and *-B* (**7**) showed significant (T/C%>125) antileukemic activity in P-388 lymphocytic leukemia (e.g. *bruceantinoside-A* (**6**): at 0.5, 1, 6, 10, and 20 mg/kg afforded T/C% values of 110, 106, 118, 152, and 172, respectively, and *bruceantinoside-B* (**7**): at 10 and 20 mg/kg afforded T/C% values of 132 and 160, respectively).

Sakaki *et al.*[5] also isolated *bruceantinoside-B* (**7**) from *B. javanica* but named it *yadanzioside-P*, because its NMR spectrum was different from our data [3]. On re-examination, our sample was found to be contaminated [6].

Further investigation of the *B. antidysenterica* extract [7] led to two new isobrucein type quassinoids, which were provisionally named *bruceanol-A* (**11**) and *-B* (**12**). Both were amorphous solids. *Bruceanol-A* (**11**) eluted after *isobrucein-B* (**5**) on preparative HPLC and *bruceanol-B*

(12) eluted after *bruceanol-A* (11). According to the NCI procedure [4], bruceanol-A (11) and -B (12) showed antileukemic activity in vitro against P-388 lymphocytic leukemia. *Bruceanol-A* (11): T/C%= 130 at 0.5 mg/kg, 129 at 1 mg/kg, 134 at 2 mg/kg; *bruceanol-B* (12): T/C%= 123 at 0.5 mg/kg. The control, *5-fluorouracil*, had T/C% = 135 at 200 mg/kg.

Further investigation of the *B. antidysenterica* extract [8] led to isolation of *yadanzioside-G* (13), *yadanzioside-N* (14) and a new compound, which was provisionally named as *bruceantinoside-C* (15). According to the standard NCI procedures [4], bruceantinoside-C (15), *yadanzioside-G* (13) and *yadanzioside-N* (14) showed significant ($ED_{50}<4.0$ μg/ml) cytotoxicity *in vitro* against P-388 and L-1210 lymphocytic leukemia tissue culture cells with ED_{50}=2.12 and 3.50, 1.25 and 2.58, and 2.24 and 4.56 μg/ml, respectively. The control drug, 5-fluorouracil, used in this assay showed ED_{50}=3.72 and 1.94 μg/ml, respectively.

By preparative HPLC of the $CHCl_3$ extract of *B. antidysenterica* and recrystallization from $EtOAc/CH_2Cl_2/Et_2O$, a colorless amorphous solid was obtained and was provisionally named *bruceanol-C* (16) [9]. According to the standard NCI procedure [4], *bruceanol-C* (16) demonstrated potent cytotoxicity against human KB, A-549 lung carcinoma, and HCT-8 colon tumor as well as murine P-388 lymphocytic leukemia with ED_{50} values of <0.04, <0.48, <0.40, and 0.56 μg/ml, respectively.

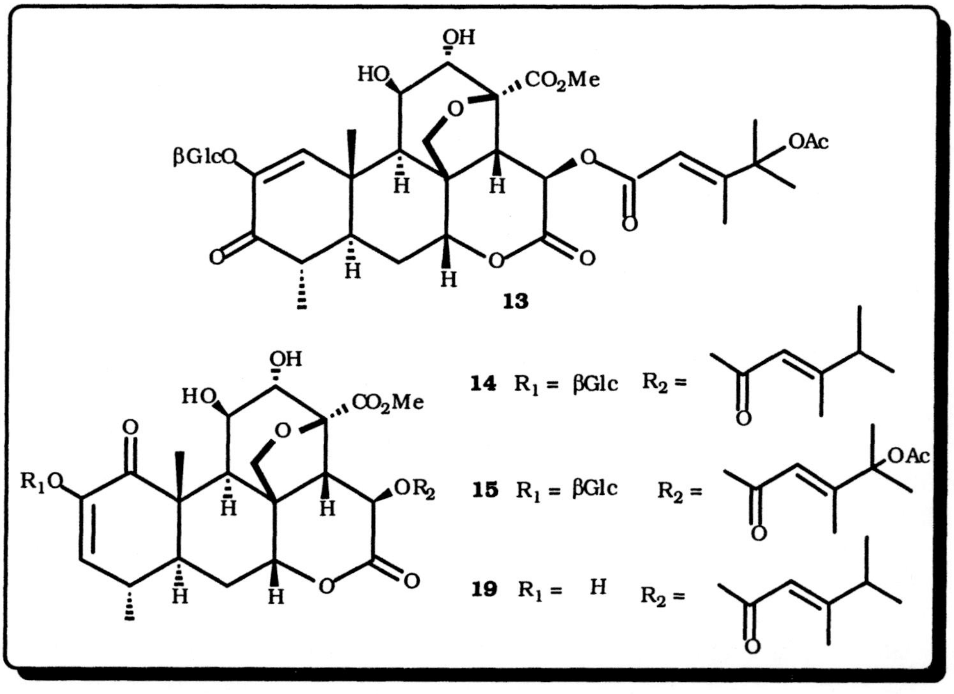

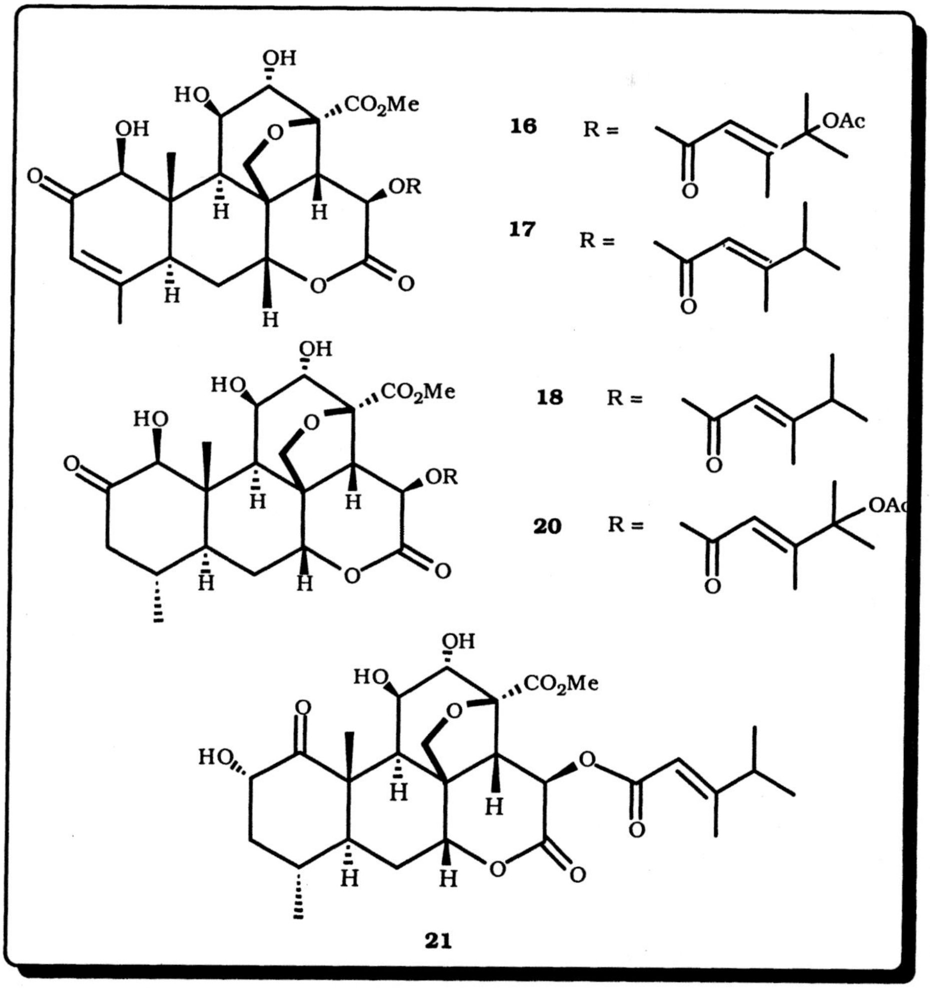

Further investigation of the *B. antidysenterica* extract [10] led to the isolation of three new quassinoids: *bruceanol-D* (17), *-E* (18), *-F* (19), and known quassinoids: *yadanzioside-N* (14), *bruceantin* (1), and *isobrucein-B* (5). *Bruceanol-D* (17) and *-F* (19) were both obtained as colorless amorphous solids and *bruceanol-E* (18) was obtained as colorless needles. They showed cytotoxicity for tumor cells such as KB (nasopharynx carcinoma), A-549 (human lung carcinoma), HCT-8 (human colon tumor), P-388 (murine lymphocytic leukemia), TE-671 (human medulloblastoma), and RPMI-7951 (human melanoma), as shown in Table 1.

Repeated preparative HPLC of the *B. antidysenterica* extract [11] gave *bruceanol-G* (20) and *-H* (21). Both were colorless needles. *Bruceanol-G* (20) and *-H* (21) were evaluated against three cancer cell lines, SK-MEL-5 (melanoma), COLO-205 (colon cancer), and KB (nasopharynx carcinoma).

Bruceanol-G (**20**) showed antitumor activity against the COLO-205 and KB cell lines, with ED_{50} values of 0.44 and 0.55 μM, respectively. *Bruceanol-G* (**20**) and -*H* (**21**) were only marginally cytotoxic in the SK-MEL-5 line, with ED_{50} values of 4.08 and 6.37 μM, respectively.

Table 1.   **Cytotoxicity of 17, 18, and 19**

Compound	Cell Line(ED_{50},μg/ml)					
	KB	A-549	HCT-8	P-388	TE-671	RPMI-7951
17	0.08	0.55	0.09	0.09	0.08	0.09
18	0.55	3.75	0.37	0.57	0.12	0.11
19	0.43	0.55	0.13	0.36	0.09	0.09

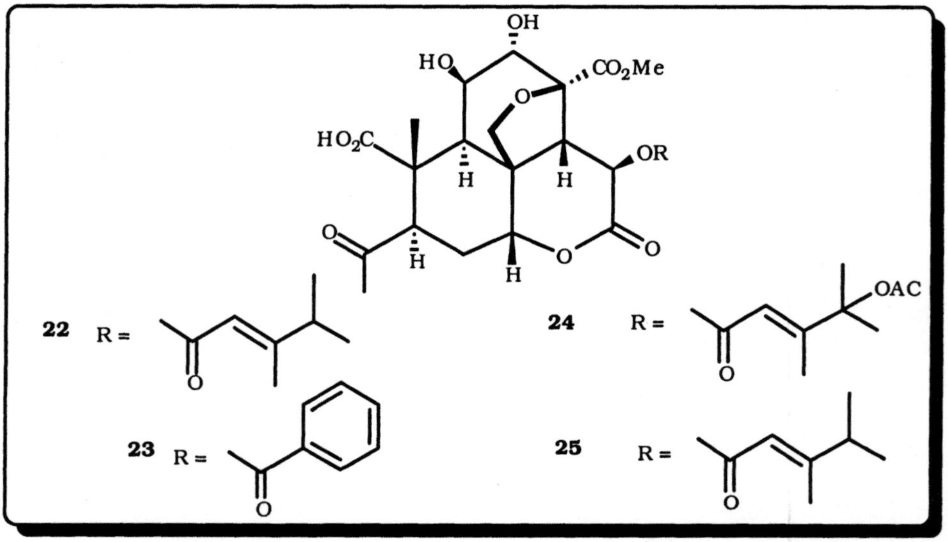

Detailed investigation of the *B. antidysenterica* extract [12] isolated *bruceanic acid-A* (**22**), -*B* (**23**), -*C* (**24**), and -*D* (**25**), which might be obtained by A-ring degradation of natural quassinoids with diosphenol structure, although this conversion should be investigated in more detail. They were isolated by CC, preparative TLC, and preparative HPLC from this plant extract, all as colorless amorphous solids. The methyl ester of *bruceanic acid-A* (**22**) was also isolated in the course of this study, as colorless amorphous solids. An *in vitro* cytotoxicity assay was carried out according to the procedure described by Geran *et al* [4] and Ferguson *et al* [13]. The assay against KB (nasal pharyngeal carcinoma), TE-671 (human medulloblastoma), A-549 (human lung carcinoma), HCT-8 (human colon carcinoma), and P-388 (murine leukemia) tumor cells was based on a method reported by Lee *et al* [14]. *Bruceanic acid-A* (**22**) demonstrated cytotoxicity against KB (ED$_{50}$ 4.16 µg/ml) and TE-671 (ED$_{50}$ 5.50 µg/ml) cell lines, and *bruceanic acid-D* (**25**) was cytotoxic against P-388 (ED$_{50}$ 0.77 µg/ml). Neither *bruceanic acid-A* (**22**) *nor -D* (**25**) were cytotoxic against A-549 and HCT-8 (ED$_{50}$=10 µg/ml) cells. Compounds **23**, **24**, and the methyl ester of **22** showed no cytotoxicity against KB, TE-671, A-549, HCT-8, and P-388 tumor cells. Thus, these acids were less active than the quassinoids with diosphenol structure.

From *Picrasma ailanthoides* (= *Picrasma quassioides*)

New quassinoid glucosides, *picrasinoside-A* (**26**), *-B* (**27**), *-C* (**28**), *-D* (**29**), *-E* (**30**), *-F* (**31**), *-G* (**32**) and two new quassinoid hemiacetals, *picrasinol-A* (**33**) and *-B* (**34**) were isolated from the stem wood of *P. ailanthoides* and their structures were elucidated by spectral evidence and chemical

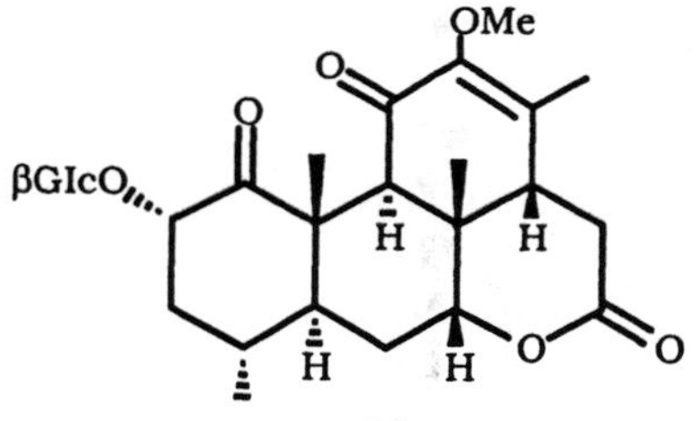

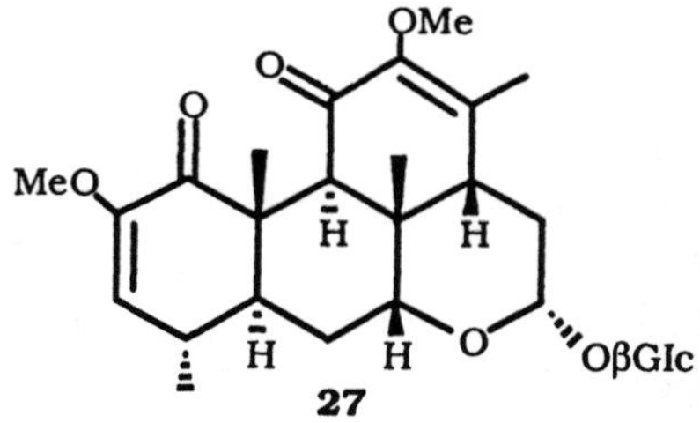

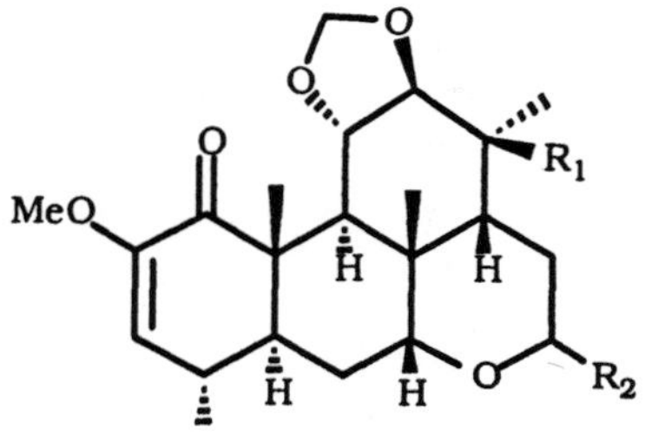

	R$_1$	R$_2$
28	H	····OβGlc
29	H	H, OH
35	OAc	····OβGlc

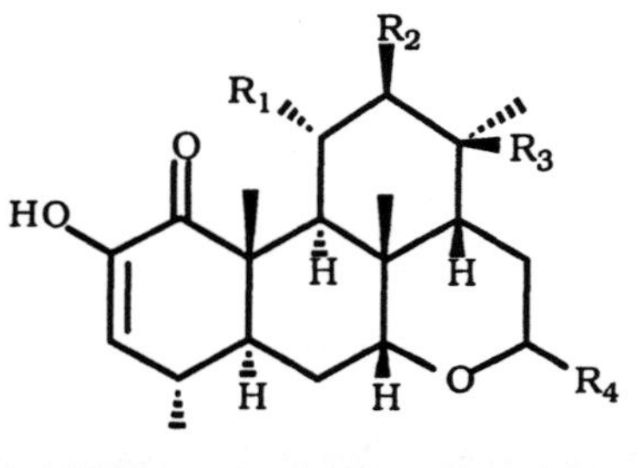

	R$_1$	R$_2$	R$_3$	R$_4$
30	OAc	OMe	H	····OβGlc
31	OAc	OMe	OH	····OβGlc
32	OH	OMe	H	····OβGlc
33	OH	OMe	OH	····OβGlc
34	OAc	OMe	H	H, OH

transformation into known compounds. *Picrasinoside-A* (**26**) and *-B* (**27**) were assayed for mean survival time (T/C%) in mice bearing P-388 lymphocytic leukemia, and both compounds were inactive (NCI) [15].

A new quassinoid glucoside, *picrasinoside-H* (**35**), was isolated from the same plant as a colorless amorphous solid. An *in vitro* cytotoxicity assay was carried out according to the NCI procedure [4] *Picrasinoside-A* (**26**), *-B* (**27**), *-C* (**28**), *-D* (**29**), *-E* (**30**), and *-G* (**32**) were tested against KB (nasal pharingeal carcinoma), TE-671 (human medulloblastoma), A-549 (human lung carcinoma), HCT-8 (human colon carcinoma), RPMI (human melanoma), and P-388 (murine leukemia) tumor cells. But, no significant (ED_{50} < 4 μg/ml) cytotoxicity was observed in these quassinoids [16].

From *Brucea javanica*

Bruceoside-C (**36**), a new quassinoid glucoside, and related compounds were isolated from *B. javanica*, and their structures were elucidated from spectral data. *Bruceoside-C* (**36**) demonstrated potent cytotoxicity against human epidermoid carcinoma of the nasopharynx (KB) (ED_{50}<0.1μg/ml), human lung carcinoma (A-549) (ED_{50}=9.44 μg/ml), human colon carcinoma (HCT-8) (ED_{50}=4.51 μg/ml), human melanoma (RPMI) (ED50<0.1 μg/ml), and CNS carcimoma (TE-671) (ED_{50}=0.29 μg/ml), as well as murine lymphocytic leukemia (P-388) (ED_{50}=5.11 μg/ml) [17].

Three new quassinoid glucosides, *bruceoside-D* (**37**), *-E* (**38**), and *-F* (**39**) were isolated from the same plant, and their structures were elucidated by spectral evidence and chemical transformation to known compounds. *Bruceoside-D* (**37**), *-E* (**38**), and *-F* (**39**) were evaluated for cytotoxicity in the NCI's *in vitro* human tumor cell line panel, which included 58 cell lines representing nine cancer types. These compounds showed selective cytotoxicity in cell lines of leukemia, non-small cell lung cancer, colon cancer, CNS cancer, melanoma, ovarian cancer, renal cancer, prostate cancer, and breast cancer. The log GI_{50} values (log concentration which reduced cell growth to 50%) ranged from >-4 to -5.72, as shown in Table 2 [18].

Table 2   Inhibition of *in vitro* Cancer Growth by Compounds 37-39 [Cytotoxicity Log $GI_{50}(M)$][a,b]

Cell Line	Compound		
	37	38	39
Leukemia			
CCRF-CEM	-4.44	-4.61	-4.36
HL-60(TB)	-4.45	-4.38	-4.32
K-562	-4.59	-4.62	-4.40

(Table 2) contd.....

Cell Line	Compound		
	37	38	39
MOLT-4	-4.65	-4.72	-4.47
SR	-4.51	-4.47	-4.31
Non-Small Cell Lung Cancer			
A549/ATCC	-4.26	-4.38	_c
EKVX	_c	_c	_c
HOP-62	-4.48	-4.35	-4.16
HOP-92	-4.50	-4.43	-4.18
NCI-H226	-4.55	-4.53	-4.14
NCI-H23	_c	_c	_c
NCI-H322M	_c	_c	_c
NCI-H460	-4.49	-4.56	-4.40
NCI-H522	-4.54	-5.66	-4.73
Colon Cancer			
COLO-205	_c	_c	_c
HCC-2998	-4.61	-4.69	-4.44
HCT-116	-4.53	-4.48	-4.42
HCT-15	_c	_c	_c
HT-29	-4.45	-4.47	_c
KM-12	-4.31	_c	_c
SW-620-	_c	_c	_c
CNS Cancer			
SF-268	_c	_c	_c
SF-539	-4.58	-4.69	-4.25
SNB-19	-4.50	-4.50	_c
SNB-75	_c	_c	_c
U-251	-4.40	-4.50	_c
Melanoma			
LOX1MVI	-4.39	-4.41	_c
MALME-3M	-4.62	-4.64	-4.25

(Table 2) contd.....

Cell Line	Compound		
	37	38	39
M-14	-4.49	_c	_c
SK-MEL-2	_c	_c	_c
SK-MEL-28	-4.43	-4.60	_c
SK-MEL-5	-437	-4.55	_c
UACC-257	_c	_c	_c
UACC-62	_c	_c	_c
Ovarian Cancer			
IGROV1	_c	-4.45	_c
OVCAR-3	-4.69	-4.44	-4.59
OVCAR-4	-4.60	-5.72	_c
OVCAR-5	_c	_c	_c
OVCAR-8	_c	_c	_c
SK-OV-3	_c	_c	_c
Renal Cancer			
786-0	_c	_c	_c
A-498	_c	_c	_c
ACHN	_c	_c	_c
CAKI-1	_c	_c	_c
RXF-393	-4.60	-4.97	-4.43
SN12C	_c	_c	_c
TK-10	_c	_c	_c
UO-31	-4.50	_c	_c
Prostate Cancer			
PC-3	_c	_c	_c
DU-145	_c	_c	_c
Breast Cancer			
MCF-7	-4.65	-4.55	-4.56
MCF-7/ADR-RES	_c	_c	_c
MDA-MB-231/ATCC	_c	_c	_c
HS-578T	_c	-4.40	_c

(Table 2) contd.....

	Compound		
Cell Line	37	38	39
MDA-MB-435	_c	_c	_c
MDA-N	_c	_c	_c
BT-549	_c	_c	_c
T-47D	_c	_c	_c

[a]Data obtained from NCI's *in vitro* disease oriented human tumor cells screen[seeGrever *et al.*[19]and Monksetal.[20]fordetails].

[b]Log concentrations which reduced cell growth to 50% of level at start of experiment.

[c]"_"means log GI_{50} is greater than-4.

Luyengi *et al.* isolated three known quassinoids, *brusatol* (**8**), *dehydrobrusatol* (**54**), and *yadanzioside-C* (**55**), from the same plant. In an HL-60 cell differentiation test, *brusatol* (**8**), *dehydrobrusatol* (**54**), and *yadanzioside-C* (**55**) exhibited ED_{50} values of 0.006, 0.8, and 0.6 µg/ml, respectively. Modification of the ring-A functionality of *brusatol* (**8**) as evident in *dehydrobrusatol* (**54**),led to almost a hundredfold diminution of the activity. *Bruceoside-B* (**10**), which is a glucoside of *brusatol* (**8**), had a value of 20 µg/ml; Thus, the sugar moiety appears to decrease the aglycon's activity [23].

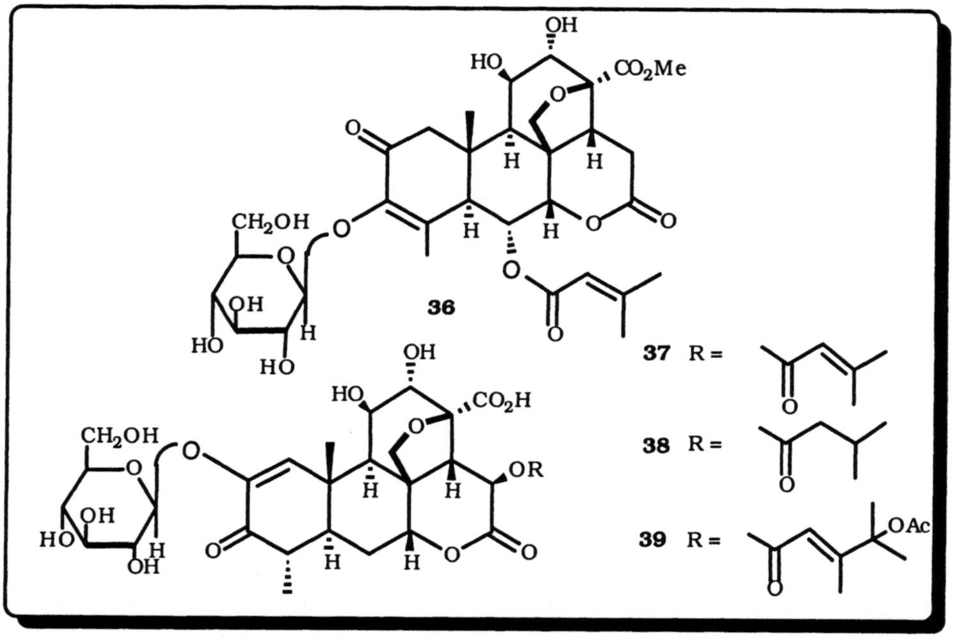

From *Ailanthus altissima and Castela tortuosa*

Four new quassinoids, *ailantinol-A* (**40**), *-B* (**41**), *-C* (**42**), *-D* (**43**), and related compounds were isolated from *A. altissima*, and their structures were elucidated from spectral evidence.[21] *Ailantinol-A* (**40**) showed anti-HIV activity as shown in Table 14.

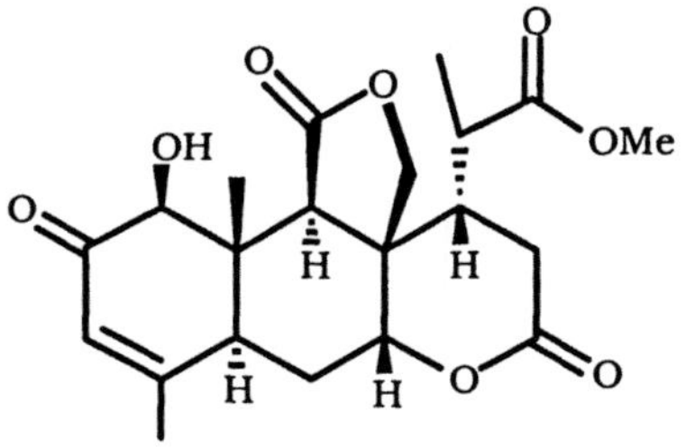

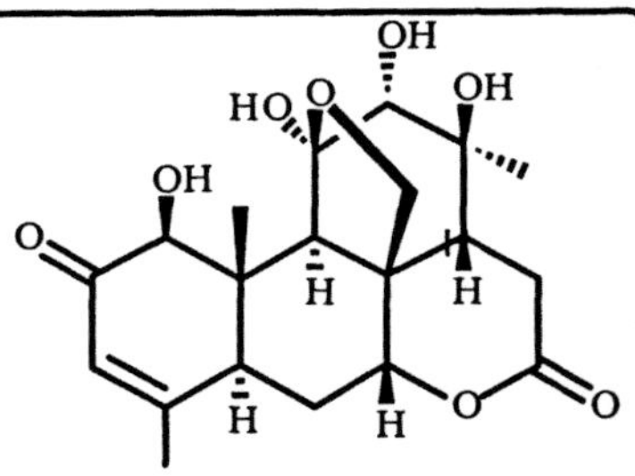

A methanol extract of *A. altissima* and a methylene chloride extract of *Castela tortuosa* exhibited plant growth inhibitory activity against *Brassica juncea, Eragrostis tef,* and *Lemna minor. Ailanthone* (**80**) and *chaparrinone* (**58**) were identified as the active constituents in *A. altissima* and *C. tortuosa*, respectively. In the bioassays with *B. juncea, E. tef,* and *L. minor, ailanthone* (**80**) inhibited growth by 50% (I_{50}) at 0.9, 2.6, and 21 μM, respectively, and the I_{50} values for *chaparrinone* (**58**) in these bioassays were 0.9, 2.1, and 4.2 μM, respectively. In a greenhouse test, *ailanthone* (**80**) exhibited inhibitory activity against several weed species, and it was more effective postemergent than preemergent [29].,

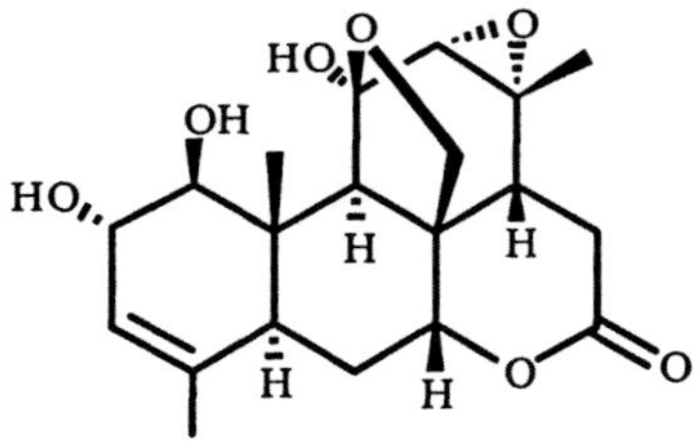

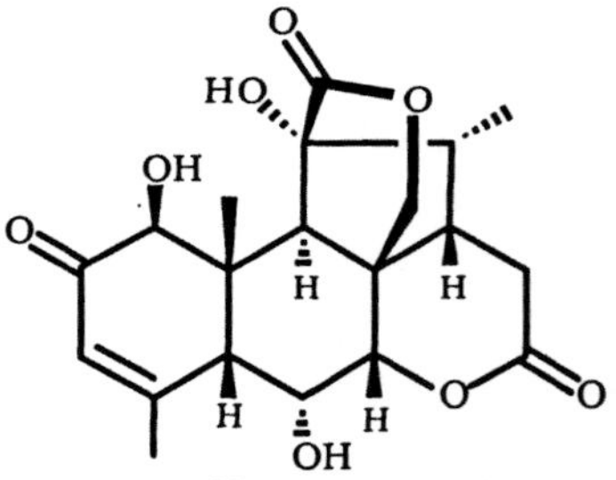

From *Quassia indica*

Kitagawa *et al.* isolated four new quassinoids named *samaderine-X* (**44**), *-Y* (**45**), *-Z* (**46**), *indaquassin-X* (**48**), and a new C_{19} quassinoid glycoside, 2-

O-glucosylsamaderine-C (**53**), together with five known quassinoids, *samaderine-B* (**50**), *-C* (**51**), *-E* (**47**), *indaquassin-C* (**49**), and *simarinolide* (**52**), from the stem of *Quassia indica* (Simaroubaceae), an Indonesian medicinal plant. The chemical structures of these quassinoids were elucidated on the basis of their chemical and physicochemical properties. *Samaderine-X* (**44**), *-Z* (**46**), *-E* (**47**), and *-B* (**50**) showed significant growth-inhibitory activity against the cultured malarial parasite *Plasmodium falciparum* (a chloroquin-resistant K1 strain), and **44**- **51** exhibited *in vitro* cytotoxicity (IC$_{50}$: 0.07 - 1.00 μg/ml) against KB cells (see Table 3). *Samaderine-X* (**44**), *-B* (**50**), and *-C* (**51**), as well as *indaquassin-X* (**48**), exhibited inhibitory activity in the *in vitro* endothelial cell-neutrophil leukocyte adhesion assay, whereas *samaderine-X* (**44**) and *-B* (**50**) exhibited significant anti-inflammatory activity [22].

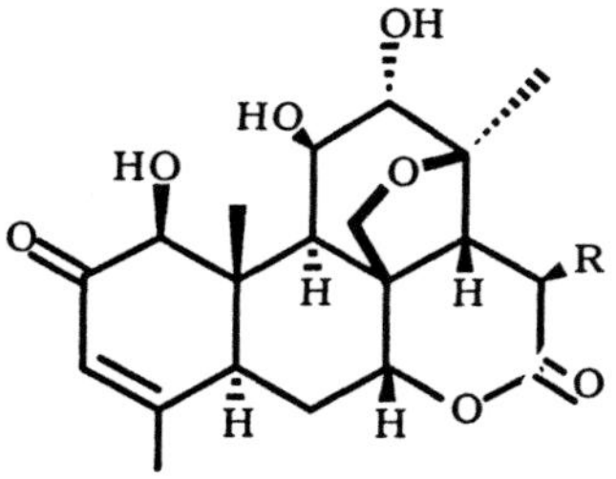

44 R = OAc
45 R = H
46 R = OH

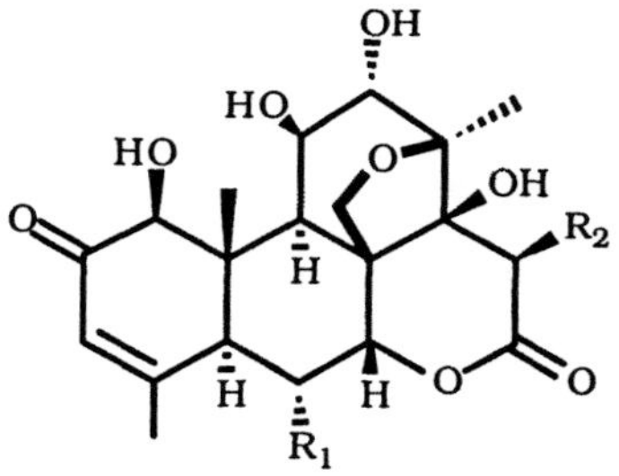

47 R$_1$ = H   R$_2$ = H
48 R$_1$ = OH   R$_2$ = OH
49 R$_1$ = OH   R$_2$ = H

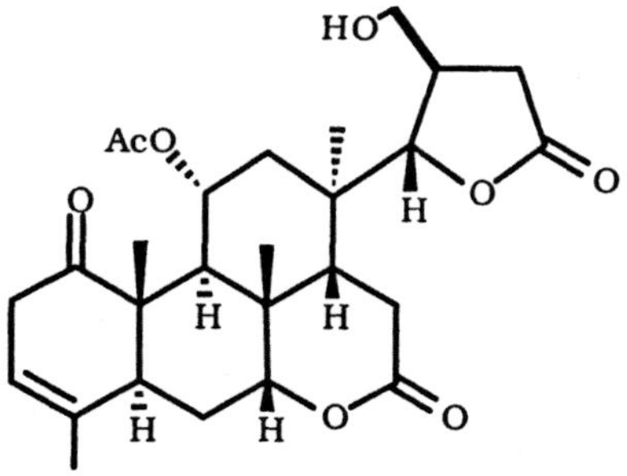

52

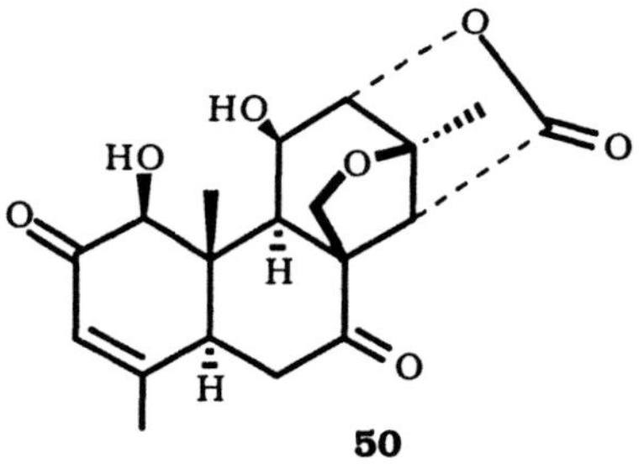

50

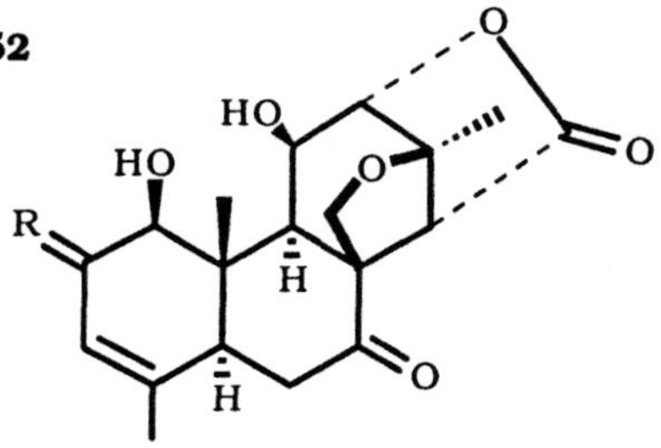

51 R = OH
53 R = OβGlc

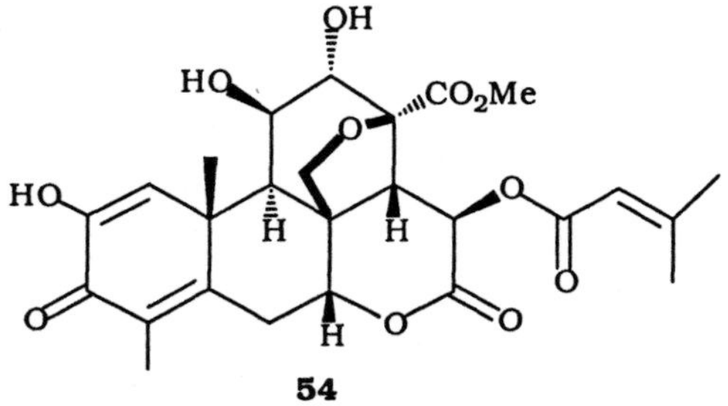

54

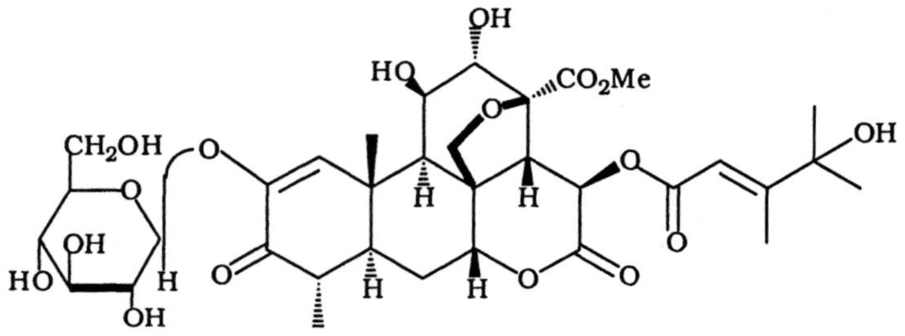

55

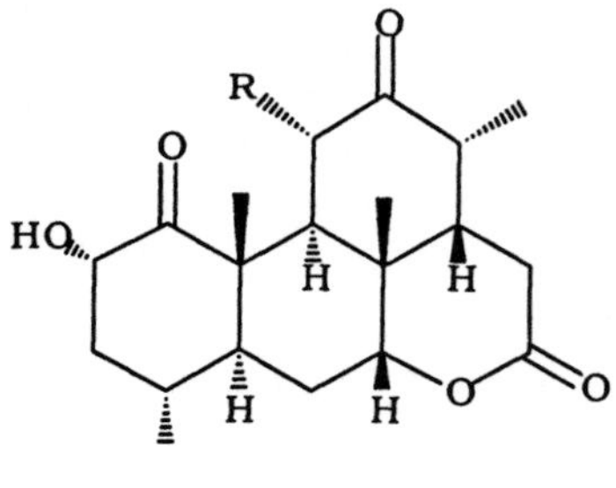

56   R =

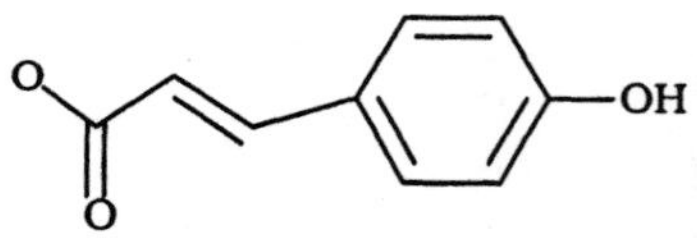

57   R = OH

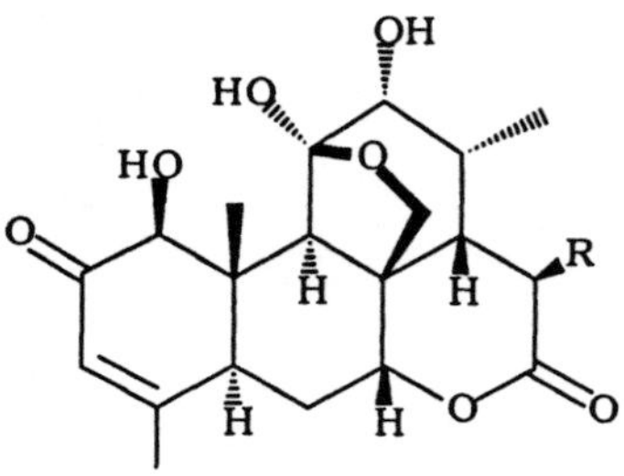

58   R = H

60   R = OH

61   R = OAc

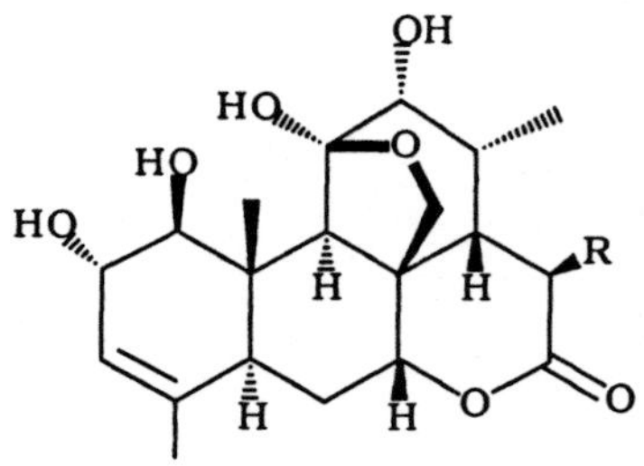

59   R = H

62   R = OβGlc

Table 3.   *In vitro* Cytotoxic Activity of Compounds 44-53 against KB Cells

Sample	IC$_{50}$(μg/ml)
Samaderine-X(44)	0.02
OSamaderine-Y(45)	0.10
Samaderine-Z(46)	0.20
Samaderine-E(47)	0.04
Indaquassin-X(48)	0.60
OIndaquassin-C(49)	1.00
Samaderine-B(50)	0.07
Samaderine-C(51)	0.40
Simarinolide(52)	>10
2-O-Glucosyl samaderine-C(53)	>10

From *Castela texana*

Dou *et al.* isolated a new quassinoid, *11-O-trans-p-coumaroylamarolide* (**56**), from *Castela texana*, and the structure was elucidated by spectroscopic analysis. The compound is the first coumaroyl quassinoid derivative to be isolated from nature. A new quassinoid, *11-O-trans-p-coumaroylamarolide* (**56**), and six known quassinoids, *amarolide* (**57**), *chaparrinone* (**58**), *chaparrin* (**59**), *glaucarubolone* (**60**), *holacanthone* (**61**), and *15-O-β-D-glucopyranosylglaucarubol* (**62**) were isolated, and tested for their cytotoxicity (see Table 4) and antimalarial activity (see Table 5).[24].

Table 4.   Cytotoxicity of Quassinoids 56-62 (IC$_{50}$ μg/ml)

Compound	OK562	KB	BT-549	SK-OV-3	Vero
11-*O-trans-p*-Coumaroylamarolide(56)	NA	NA	NA	NA	NA
Amarolide(57)	NA	NA	NA	NA	NA
Glaucarubolone(60)	0.60	1.00	0.40	0.80	1.50
Chaparrin(59)	NA	NA	NA	NA	NA
Holacanthone(61)	1.00	0.50	0.30	0.50	1.00
15-O-Glucopyranosylglaucarubol(62)	NA	NA	NA	NA	NA
Doxorubicin	0.035	0.12	0.008	0.01	1.5

NA=Noactivityatconcentration10 μg/ml or lower.

From *Eurycoma longifolia*

Itokawa *et al.* isolated three new quassinoids, *6α-hydroxy-eurycomalactone* (**63**), *longilactone* (**64**), and *14,15β-dihydroxy-klaineanone* (**65**), from the woods of *Eurycoma longifolia* (Simaroubaceae) along with three known quassinoids, *11-dehydro-klaineanone* (**66**), *eurycomalactone* (**67**), and *5,6-dehydro-eurycomalactone* (**68**). All showed potent cytotoxic activity against P-388 and KB cells (see Table 6) [25].

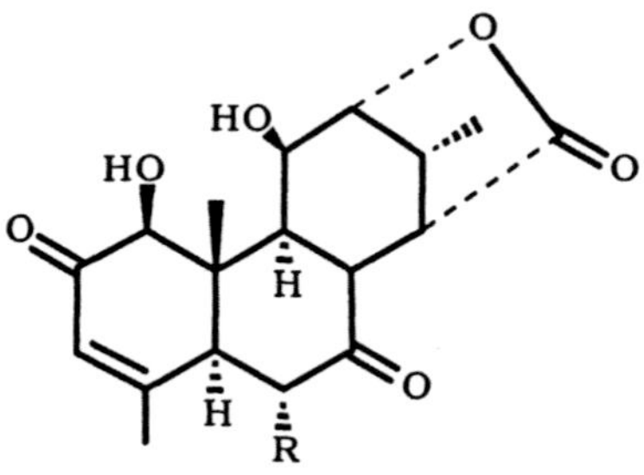

63   R = OH

67   R = H

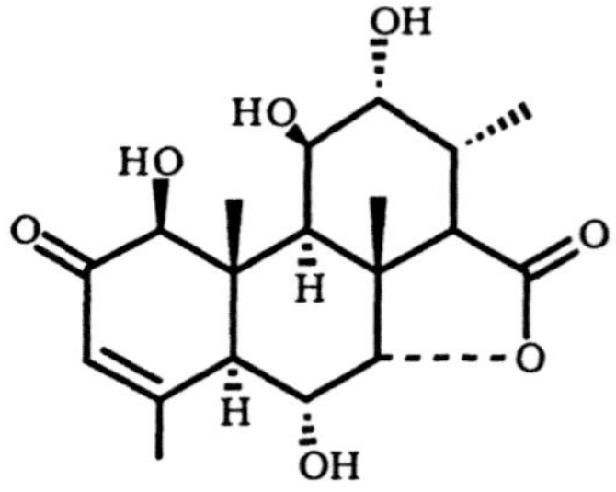

64

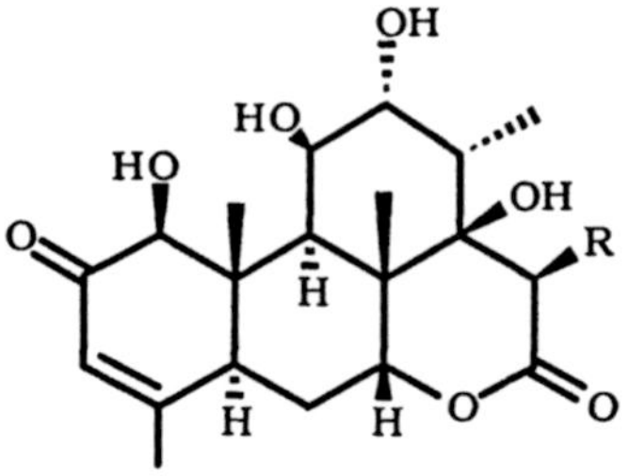

65   R = OH

66   R = H

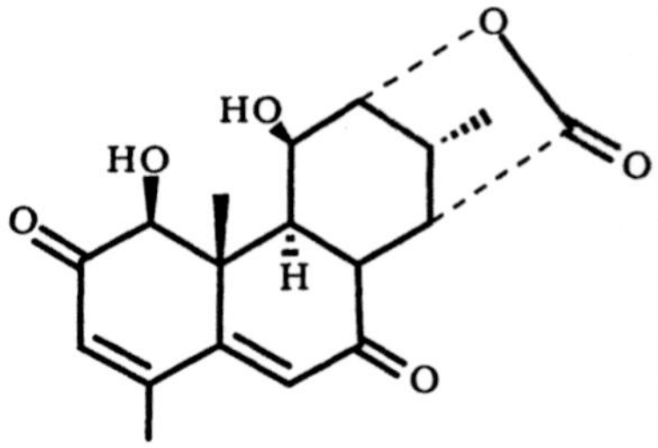

68

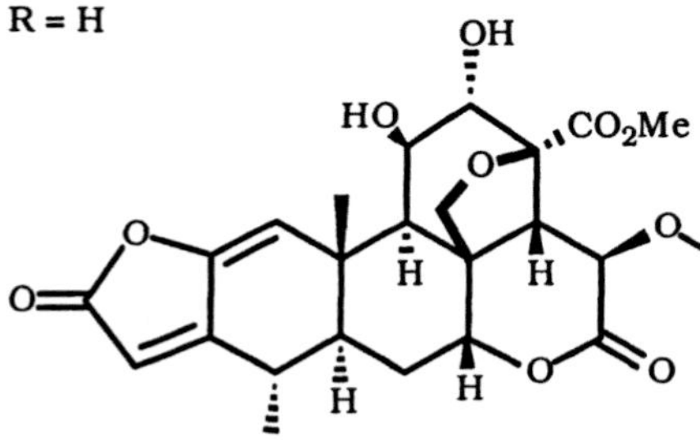

69

Table5.   **Anti malarial Activity of Quassinoids 56-62 (IC$_{50}$ µg/ml)**

Compound	P.falciparum(D6 clone)	P.falciparum(W2 clone)
11-O-trans-p-Coumaroylamarolide(56)	0.92	0.75
Amarolide(57)	NA	NA
Chaparrinone(58)	0.25	0.20
Glaucarubolone(60)	0.125	0.20
Chaparrin(59)	0.25	0.35
Holacanthone(61)	0.010	0.012
15-O-Glucopyranosylglaucarubol(62)	NA	NA
Chroroquine	0.006	0.1

NA= No activity at concentration 10 µg/ml or lower.

Table 6.   **Cytotoxic Activities of Compounds 63-68 against P-388 and KB Cells (IC$_{50}$ µg/ml)**

Compound	P-388	KB Cells
6α-Hydroxyeurycomalactone(63)	0.25	0.44
Longilactone(64)	1.3	3.4
14,15β-Dihydroxyklaineanone(65)	0.29	0.38
11-Dehydroklaineanone(66)	1.8	1.6
Eurycomalactone(67)	0.90	1.3
5,6-Dehydroeurycomalactone(68)	1.0	1.7

From *Cedronia granatensis*

The NCI *in vitro* primary disease-oriented antitumor screen was used to select and guide the fractionation of the organic and aqueous extracts of Cedronia granatensis. Two quassinoids, *sergeolide* (**69**) and *isobrucein-B* (**5**), were isolated, to which the screening panel cell lines exhibited up to a 1000-fold range of differential sensitivity, were isolated. The compounds typically produced IC$_{50}$-level responses at concentrations of 10^{-5} - 10^{-8} M against a majority of the melanoma lines and several of the colon, lung, and other solid tumor lines [26].

From *Hannoa klaineana*

Lumonadio *et al.* isolated five quassinoids from the root bark of *Hannoa klaineana*. Of these quassinoids, only *15-desacetyl-undulactone* (**70**) was active against P-388 mouse lymphocytic leukemia cells and colon 38 adenocarcinoma in mice. *Undulactone* (**71**) and especially *15-O-"-D-glucopyranosyl-21-hydroxy-glaucarubolone* (**72**) were more toxic, while *6α-tigloyloxy-glaucarubol* (**73**) and *21-hydroxy-glaucarubolone* (**74**) were inactive [27].

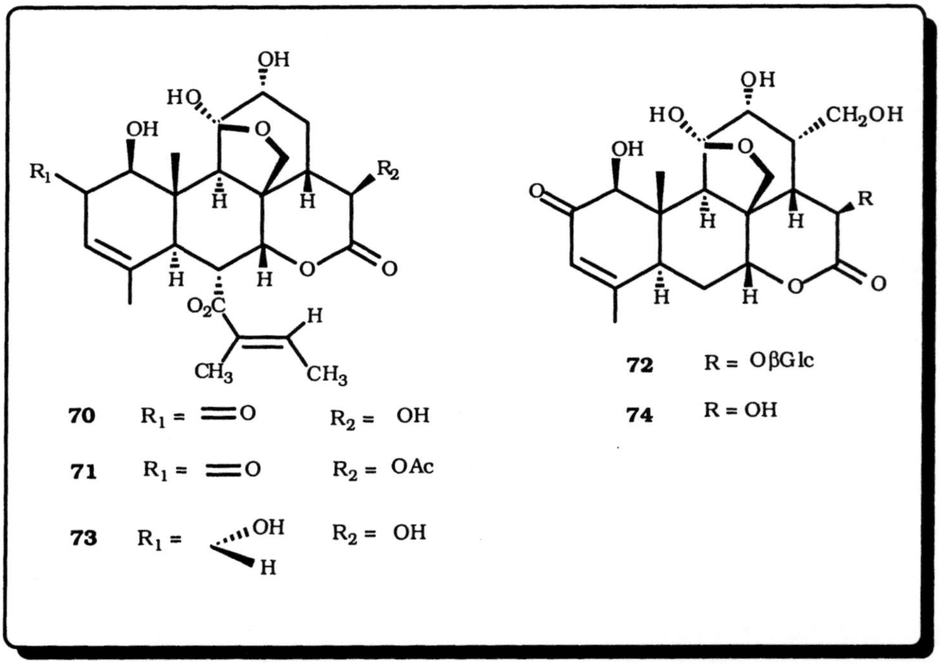

From *Myzus persicae*

A series of quassinoids was tested for antifeedant activity against the aphid *Myzus persicae* (Hemiptera, Aphididae) as shown in Table 7. *Isobrucein-B* (**5**), *brucein-B* (**75**), *brucein-C* (**76**), *glaucarubinone* (**77**), and *quassin* (**78**) decreased feeding at concentrations <0.05% and *isobrucein-A* (**79**) was effective at 0.01%. However, only *quassin* (**78**) showed no phytotoxic effects. Therefore, Polonsky *et al.* concluded that *quassin* (**78**) was the most promising compound for further development [28].

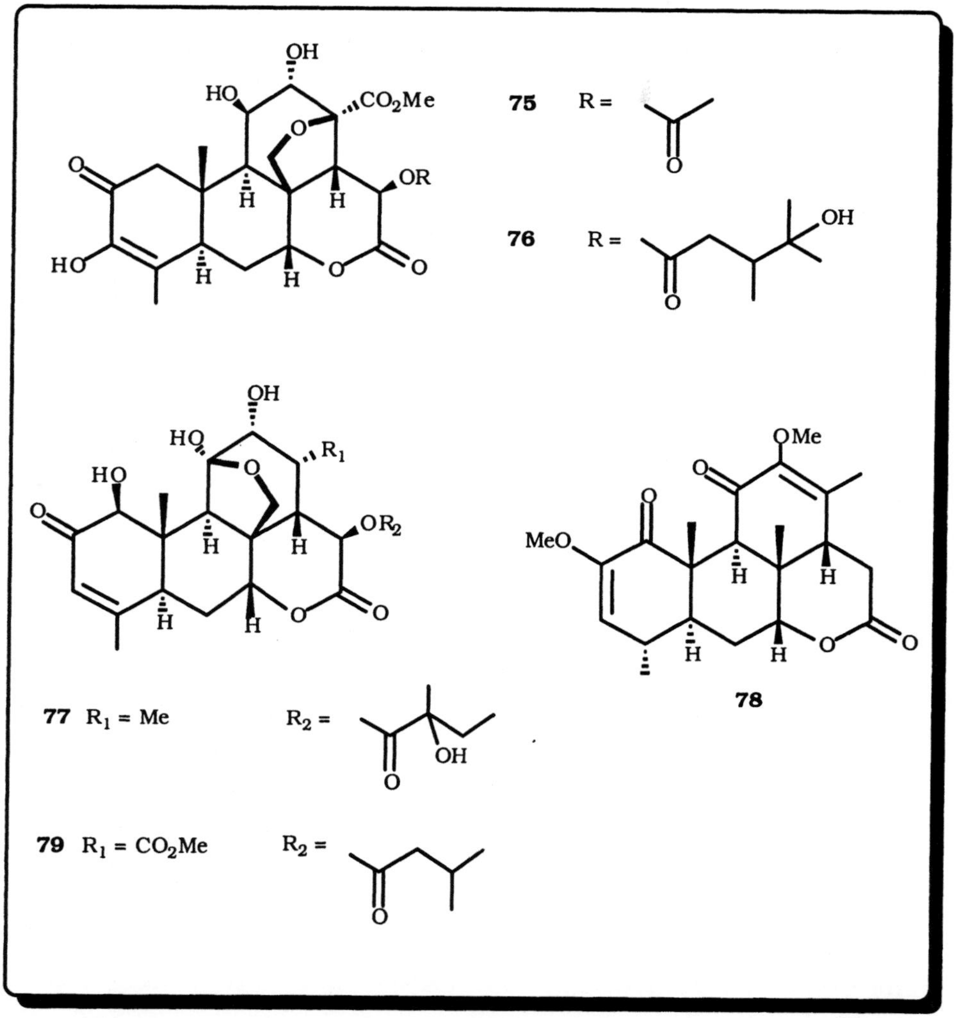

75　R =

76　R =

77　R_1 = Me　　　R_2 =

79　R_1 = CO_2Me　　　R_2 =

78

PREPARATION OF QUASSINOID DERIVATIVES FOR ANTITUMOR EVALUATION

A series of new bisbrusatolyl and brusatolyl esters and related compounds were synthesized and tested for *in vivo* antileukemic activity against a quassinoid sensitive strain of P-388 lymphocytic leukemia in BDF1 mice as shown in Table 8. *Bisbrusatolylmalonate* (**81**), *bisbrusatolylsuccinate* (**82**), *bisbrusatolylglutarate* (**83**), *bisbrusatolyladipate* (**84**), and *bisbrusatolyl sebacate* (**85**) were as active as, or more active than, *brusatol* (**8**). The C-3 esters of *brusatol* (**8**) or *bruceantin* (**1**) were also as active as, or more active than, the original *brusatol* (**8**) and *bruceantin* (**1**) in general.

Table 7.   Quassinoids Active as Aphid Antifeedants

Compound	Concentration(%)	MeanNo.ofAphidsSettled	
		Treated	Control
Isobrucein-A(79)	0.1	3.3	11.0
	0.05	4.8	10.7
	0.01	5.4	11.0
	0.005	8.9	9.0
Isobrucein-B(5)	0.1	3.7	12.0
	0.05	5.1	5.6
Brucein-B(75)	0.1	3.0	6.2
	0.05	7.4	8.5
Brucein-C(76)	0.1	2.8	7.4
	0.05	7.3	10.1
Glaucarubinone(77)	0.1	4.1	12.4
	0.05	4.2	12.5
	0.01	7.5	9.6
Quassin(78)	0.1	5.5	9.2
	0.05	4.8	9.2
	0.01	7.6	10.7

Free hydroxy groups at C-11 and C-12, as well as the enone double bond in ring-A of both bisbrusatolyl and brusatolyl esters, are required for antileukemic activity. The presence of a double bond in the ester side chain contributes slightly to the enhanced activity of these esters [30].

Table 8.   Anti leukemic Activity of Bisbrusatolyl and Brusatolyl Esters and Related Compounds against P-388 Lymphocytic Leukemia Cell Growth in BDF1 Mice

Compound	Dosemg/kg/dayintraperitoneally	AverageDaysSurvivedofTreated/Control	T/C,%[a]
Brusatol(8)	0.6	14.2/9.5	149
	0.3	14.3/9.5	150
Bisbrusatolylmalonate(81)	1.0	20.2/9.5"	213
	0.6	25.8/9.5	272[b]

(Table 8) contd.....

Compound	Dosemg/kg/dayintraperitoneally	AverageDaysSurvivedofTreated/Control	T/C,%[a]
Bisbrusatolylsuccinate(82)	0.3	20.4/9.5	215
Bisbrusatolylglutarate(83)	0.6	0.6/9.5	217
Bisbrusatoryladipate(84)	0.6	16.7/9.5	176
Bisbrusatolylsebacate(85)	0.6	16.8/9.5	176
12,12'-Diacetate of 82	0.6	13.6/9.5	143
12,12'-Diacetate of 83	0.6	11.2/9.5	118
12,12'-Diacetate of 84	0.6	11.6/9.5	122
22,23,22',23'-Tetrahydro-derivative of 81(131)	0.6	9.2/9.5	97
22,23,22',23'-Tetrahydro-derivative of 83	0.6	12.5/9.5	132
3,4,3',4'-Tetrahydro-derivarive of 131	0.6	13.9/9.5	147
3-Senecioate of 8(132)	0.6	10.7/9.5	113
3-Butanoate of 8(133)	0.6	17.6/9.5	185
3-Ethylsuccinate of 8(134)	0.6	15.7/9.5	166
3-3',4'-Dimethyl-2'-pentenoate of 8(135)	0.6	15.3/9.5	161
3-3',4',5'-Trimethoxybenzoate of 8(136)	0.6	14.2/9.5	149
3-Cinnamate of 8(137)	0.6	12.3/9.5	130
12-Acetate of 131	0.6	13.0/9.5	137
12-Acetate of 132	0.6	9.8/9.5	103
12-Acetate of 133	0.6	9.9/9.5	104
12-Acetate of 134	0.6	11.4/9.5	120
12-Acetate of 135	0.6	9.9/9.5	105
11,12-Diacetate of 131	0.6	10.6/9.5	112
11,12-Diacetate of 132	0.6	9.8/9.5	103
11,12-Diacetate of 133	0.6	10.3/9.5	109
11,12-Diacetate of 134	0.6	10.4/9.5	109
11,12-diacetate of 136	0.6	10.2/9.5	107

(Table 8) contd.....

Compound	Dosemg/kg/dayintraperitoneally	AverageDaysSurvivedofTreated/Control	T/C,%[a]
3,11,12-Triacetate of **8**	0.6	9.0/9.5	95
3,11-Dicinnamate of **8**	0.6	9.7/9.5	102
3,11,12-Tri-3',4'-dimethyl-2'-pentenoate of **8**	0.6	11.2/9.5	118
3-3',4'-Dimethyl-2'-pentenoate of **1**	0.6	12.6/9.5	133
22,23-Dihdrobrusatol	0.6	18.4/9.5	194
3,4,22,23-Tetrahydrobrusatol (**138**)	0.6	14.2/9.5	150
3,16-Diol of **138**	0.6	11.4/9.5	120
5-Fluorouracil	0.6	10.2/9.5	107
	2.5	15.7/9.5	166

[a]A compound is active if it exhibits a T/C>125%.These data were obtained at UNC.

[b]T/C% value of 197 was obtained at the other laboratory.[31]

Based on the fact that some known antineoplastic agents possess ester moieties within their structures, the esters of *helenalin* (sesquiterpene), *brusatol* (**8**), and *bruceantin* (**1**), were synthesized and tested for antileukemic activity in the P-388 screen. These agents gave different T/C% values depending on the P-388 lymphocytic strain and the host strain of mice used. Later studies demonstrated that the agents caused different degrees of inhibition of nucleic acid and protein synthesis in the various P-388 strains. The higher the degree of inhibition of precursor incorporation into the nucleic acid or protein, the higher the T/C% value obtained in a given P-388 strain. The study demonstrates the lack of consistency with P-388 lymphocytic leukemia cell lines used in various laboratories. The T/C% of *bisbrusatolylmalonate* (**81**) from one laboratory was 272% and from another laboratory was 197%. Thus, we concluded that the inbred strain of mice is a critical factor in the tolerance of drug toxicity.[31].

Thereafter, an in vitro test for human cancer was used as the bioassay system. For example, two new types of quassinoids (**86a, 86b, 87a, 87b, 88a, 88b**) obtained by the catalytic hydrogenation of dehydro-quassinoids [*dehydro-bruceantin* (**86**), *dehydro-bruceantinol* (**87**), and *dehydro-bruceantarin* (**88**)], which were isolated from *B. antidysenterica*, were

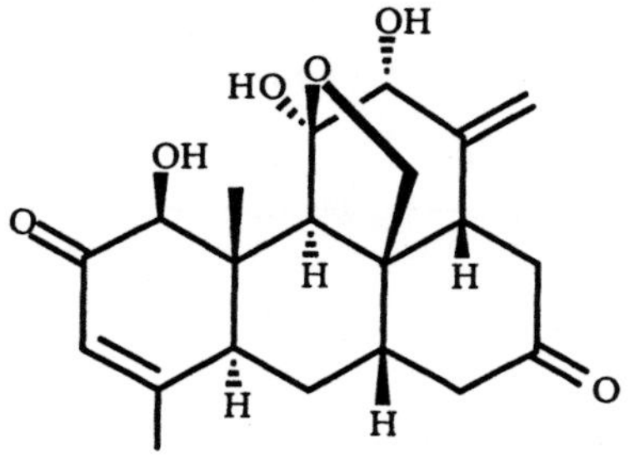

80

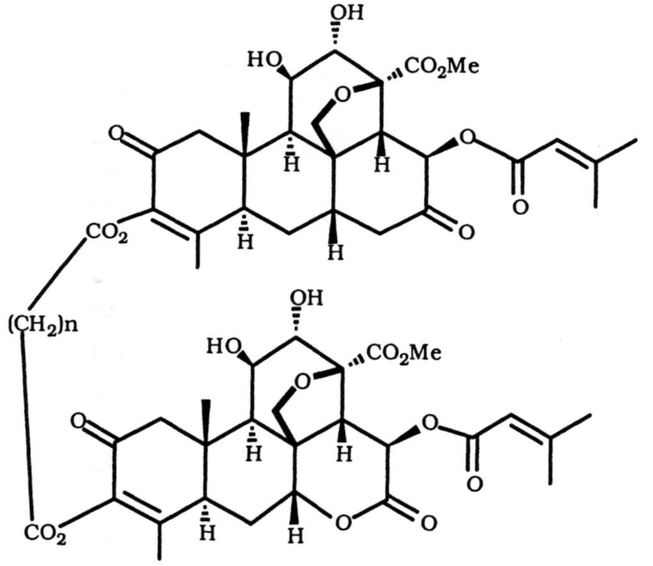

81 n = 1   82 n = 2   83 n = 3   84 n = 4   85 n = 8

tested against cancers mainly for human but including murine lymphocytic leukemia P-388, as shown in Table 9. Compounds **86a** and **86b** showed potent selective cytotoxicity against human medulloblastoma (TE-671). Significant cytotoxicity was also demonstrated by **88a** for TE-671 and for murine leukemia (P-388), and by **87b** for human melanoma (RPMI-7951) [32].

Table 9.   **Cytotoxicity [ED$_{50}$ (μg/ml)] of Quassinoids against Cancers**

Compound	Cancer					
	KB	A-549	HCT-8	RPMI-7951	TE-671	P-388
86[a]	5.50	6.22	>10	5.06	0.68	3.84
86[b]	>10	>10	>10	>10	5.56	>10
87[a]	5.50	>10	8.67	5.36	4.23	5.36
87[b]	5.50	>10	8.13	1.80	<0.1	4.83
88[a]	5.20	>10	>10	4.36	3.63	>10
88[b]	>10	>10	>10	>10	>10	>10

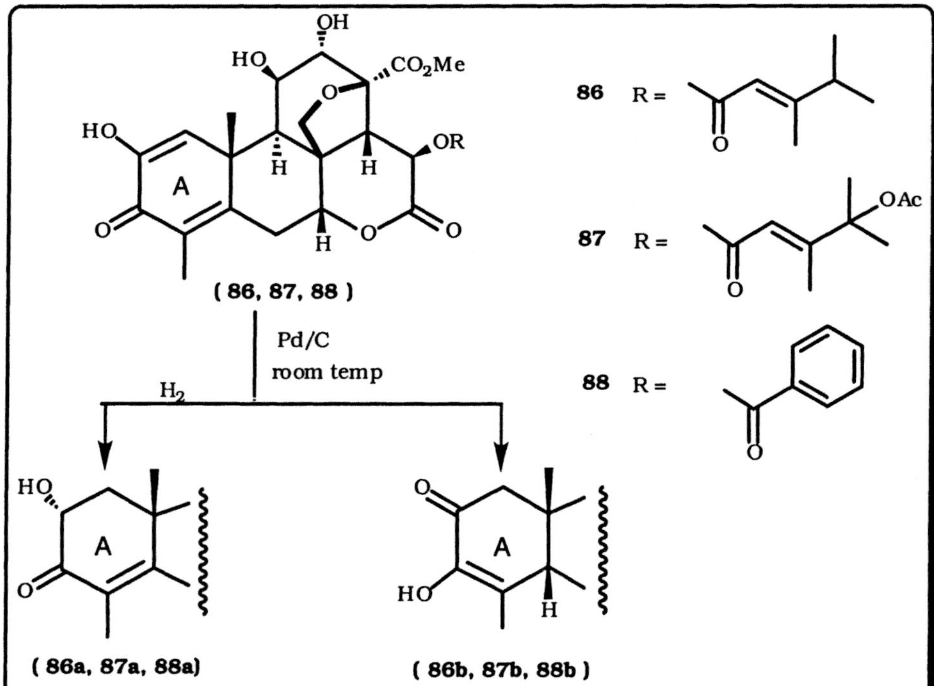

Catalytic Hydrogenation of Compounds **86, 87,** and **88**

Brucein-C (**76**), *bruceantin-4′,5′-epoxide,* and *bruceantin-5′-ol* were obtained as metabolites by incubation of *bruceantin* (**1**) in *Streptomyces griceus* [1]. This result revealed that the quassinoid side chain is biologically unstable. To overcome the instability, the side chain of quassinoid was halogenated in the following way. The C-15 senecioyl side chain of *brusatol* (**8**) was interchanged with fluorinated acyl groups, and C-3 hydroxy group of *bruceolide* (**89**) was esterified with fluorinated acyl chlorides. As shown in Table 10, these fluorinated quassinoids (**90 - 93**) showed significant cytotoxic antitumor activity against eight human cancer cell lines including small and non-small cell lung, colon, CNS, ovarian, and renal cancers, leukemia, and melanoma [33].

Table 10. Log GI_{50} Values from the Human Disease-oriented Cancer Cell Line Screening Panel for Compounds 90-93[GI_{50}=50% Growth Inhibition(mol/l)]

Panel/Cell Line	Compound			
	90	**91**	**92**	**93**
Leukemia				
CCRP-CEM	-5.41	-5.06	-5.39	-7.92
HL-60(TB)	-6.44	-5.94	-6.57	-8.43
K-562	-5.31	-5.03	-5.22	-7.80
MOLT-4	-6.00	-5.67	-6.11	-8.33
RPMI-8226	-5.81	-5.69	-6.23	-7.94
SR	-6.18	-5.48	-5.56	-8.37
Non-Small Cell Lung Cancer				
A549/ATCC	-5.36	-4.78	-4.76	-7.58
EKVX	-4.92	-4.41	-4.58	-6.89
HOP-18	-5.20	-4.52	-5.70	-6.51
HOP-62	-5.22	-4.91	-4.82	-7.01
HOP-92	-5.67	-5.23	-6.05	-8.02
NCI-H226	-5.69	-5.19	-5.69	-7.97
NCI-H23	-5.83	-5.34	-5.86	-7.90
NCI-H322M	-4.94	-4.52	-4.65	-6.40
NCI-H460	-5.99	-5.61	-5.75	-8.27
NCI-H522	-5.62	-5.20	-5.46	-8.05
LXFL-529	-5.58	-5.13	-5.42	-8.14

(Table 10) contd.....

Panel/Cell Line	Compound			
	90	91	92	93
Small Cell Lung Cancer				
DMS-114	-5.38	-5.37	-5.84	-8.02
DMS-273	-6.01	-5.46	-6.42	-8.16
Colon Cancer				
COLO-205	-5.18	-4.95	-5.17	-7.00
DLD-1	-5.27	-5.13	-5.56	-7.06
HCC-2998	-5.75	-5.46	-5.70	-7.98
HCT-116	-5.61	-5.06	-5.61	-8.00
HCT-15	-5.10	-4.62	>-4.20	-6.78
HT-29	-5.66	-5.17	-5.16	-7.88
KM-12	-5.24	-4.84	-4.95	-7.49
KM-20L2	-5.41	-5.07	-5.42	-7.71
SW-620	–	–	-5.92	-7.66
CNS Cancer				
SP-268	-5.06	-4.95	-5.16	-7.06
SF-295	-5.38	-5.10	-5.29	-7.35
SF-539	-6.00	-5.42	-5.82	-7.98
SNB-19	-5.44	-5.33	-6.09	-8.10
SNB-75	-5.11	-5.40	-5.68	-7.35
SNB-78	-5.04	-4.95	-4.53	-6.69
U-251	-5.36	-5.18	-5.32	-7.69
XP-498	-5.65	-5.01	-6.89	-8.28
Melanoma				
MALME-3H	-5.94	-5.10	-6.67	-8.00
M-14	-5.45	-5.06	-5.70	-7.46
M-19-MEL	-5.92	-5.29	-6.36	-7.79

(Table 10) contd.....

Panel/Cell Line	Compound			
	90	91	92	93
SK-MEL-2	-5.08	-5.10	-5.53	-7.11
SK-MEL-28	-5.33	-5.27	-5.82	-7.92
SK-MEL-5	-5.48	-5.32	-5.79	-7.90
UACC-257	-5.15	-5.06	-5.53	-7.12
UACC-62	-5.69	-5.63	-5.95	-8.13
Ovarian Cancer				
IGROV1	-5.67	-5.41	-5.65	-7.77
OVCAR-4	-5.20	-4.75	-4.83	-7.17
OVCAR-5	-5.06	-4.66	-4.71	-6.84
OVCAR-8	-4.98	-4.92	-5.33	-7.24
SK-OV-3	-5.14	-5.10	-5.28	-7.29
Renal Cancer				
786-0	-5.10	-5.04	-5.72	-7.19
A498	-5.74	-6.28	-6.81	-7.67
ACHN	-5.38	-4.86	-5.08	-7.64
CAKI-1	-5.12	-4.70	-5.03	-7.25
RXF-393	-5.22	-4.77	-5.33	-7.04
SN12C	-5.06	-5.10	-5.70	-6.84
TK-10	-5.11	-5.00	-5.26	-7.18
UO-31	-5.17	-4.45	-4.97	-6.83
MG-MID	-5.44	-5.13	-5.54	-7.56

INHIBITORY EFFECTS OF NATURAL AND SEMI-SYNTHETIC QUASSINOIDS ON TPA-INDUCED EBV-EA ACTIVATION

Short-term *in vitro* assays for tumor promoters and anti-tumor promoters (Epstein-Barr virus activation test) were carried out for 45 quassinoids, as shown in Table 11. As a result, some quassinoids showed potent activity, more than 50% inhibition at a molar ratio of 1:1 (TPA/quassinoids). These results led to the following structure-activity relationships: (1) a methyleneoxy bridge and side chain enhance the activity: and (2) a sugar moiety reduces the activity [34].

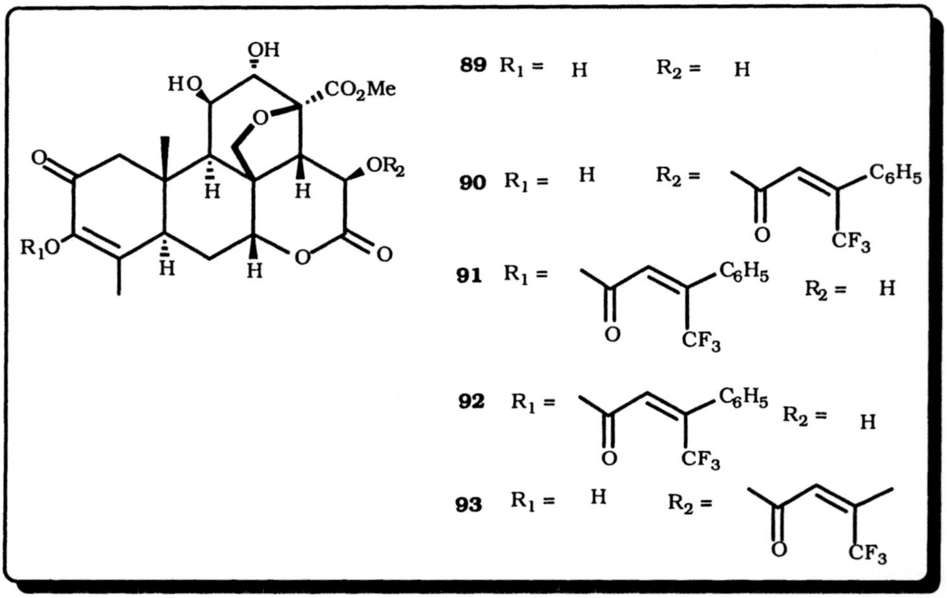

Table 11.   Inhibitory Effects of Quassinoids on TPA-induced EBV-EA Activation

Compound	%EBV-EA Positive Cells					Control	Plant
	Compound Concentration mol ratio/32 pmol TPA)						
	1000	500	100	10	1		
Brucein-D(94)	0	0	6	11	40	40	a
Yadanziolide-C(95)	0	0	30	40	40	40	a
Brucein-E(96)	3	7	16	40	40	40	a
Yadanzioside-F(97)	0	14	32	40	40	40	a
Yadanzioside-L(98)	0	0	11	26	40	40	a
Bruceoside-C(36)	0	0	16	30	40	40	a
Bruceoside-B(10)	0	0	7	32	40	40	a
Yandanzioside-C(55)	0	0	10	22	35	40	a
Yadanzioside-B(99)	0	0	9	27	40	40	a

(Table 11) contd.....

Compound	%EBV-EA Positive Cells					Control	Plant
	Compound Concentration mol ratio/32 pmol TPA)						
	1000	500	100	10	1		
Bruceoside-A(9)	0	0	8	28	40	40	a
Yadanzioside-E(100)	0	0	24	40	40	40	a
Yadanzioside-M(101)	0	0	22	40	40	40	a
Yadanzioside-A(102)	0	4	28	40	40	40	a
Yadanzioside-G(13)	0	6	25	36	40	40	a
Bruceantinoside-A(6)	0	0	9	29	40	40	a
Isobrucein-B(5)	0	0	0	6	17	40	b
Dehydrobruceantarin(88)	0	0	5	33	40	40	b
Bruceantinoside-C(15)	0	7	18	35	40	40	b
Bruceanol-A(11)	0	0	0	7	19	40	b
Bruceantinoside-B(7)	0	4	12	30	40	40	b
Bruceanol-C(16)	0	0	0	0	11	40	b
Yadanzioside-N(14)	0	0	7	16	32	40	b
Bruceanol-G(20)	0	0	0	8	12	40	b
Dehydrobruceantinol(87)	0	0	0	0	33	40	b
Bruceanol-D(17)	0	0	0	0	8	40	b
Dehydrobruceantin(86)	0	0	0	0	12	40	b
Bruceanol-B(12)	0	0	0	0	10	40	b
Bruceantin(1)	0	0	0	7	12	40	b
Bruceanol-F(19)	0	7	15	22	31	40	b
Bruceanol-E(18)	0	0	0	0	4	40	b
Bruceanol-H(21)	0	0	11	19	25	40	b
Picrasin-B(103)	0	0	26	40	40	40	c
Nigakilactone-F(104)	14	27	35	40	40	40	c
Nigakilactone-H(105)	0	7	34	40	40	c	

(Table 11) contd.....

Compound	%EBV-EA Positive Cells					Control	Plant
	Compound Concentration mol ratio/32 pmol TPA)						
	1000	500	100	10	1		
Nigakilactone-L(106)	0	7	25	34	40	40	c
Picrasin-A(107)	9	19	29	40	40	40	c
Nigakilactone-E(108)	0	12	27	34	40	40	c
Quassin(78)	0	19	34	40	40	40	c
Picrasinol-D(109)	7	14	34	40	40	40	c
Nigakihemiacetal-A(110)	6	13	29	40	40	40	c
Picrasin-E(111)	13	27	37	40	40	40	c
Picrasinol-C(112)	0	11	30	38	40	40	c
Nigakihemiacetal-D(113)	0	0	28	40	40	40	c
Picrasin-D(114)	0	13	27	37	40	40	c
Neoquassin(115)	0	10	26	40	40	40	c

a:*Brucea javanica*,b:*Brucea antidysenterica*,c:*Picrasma ailanthoides*

Short-term *in vitro* assays for tumor-promoters and anti-tumor-promoters (Epstein-Barr virus activation test) were carried out for 14 quassinoids isolated from *A. altissima*, as shown in Table 12. Some quassinoids, including *ailantinol-B* (41), *ailantinol-C* (42), *ailanthone* (80) and *shijulactone-A* (116), showed moderate activity at a molar ratio of 1:100 (TPA/quassinoids), and the results led to the following structure-activity relationships: (1) an epoxy group enhances the activity; (2) an OH group at C-6 and/or at C-11 enhances the activity; (3) an OH group at C-13 and/or C-15 weakens the activity; and esterification of the OH group at C-15 may enhance the activity [35].

The same assay was carried out for five semi-synthetic quassinoids, which were obtained by esterification of the C-15 OH group of *desacetyl-isobrucein-B* (124) as shown in Table 13. All ester derivatives showed higher antitumor promoting activity than that of the starting quassinoid, *desacetyl-isobrucein-B* (124). A compound containing a fluorinated aliphatic ester showed the highest potency [36].

Table 12.   Inhibitory Effects of Quassinoids[a] on TPA-induced EBV-EA Activation

Compound	%EBV-EA Positive Cells%				
	Compound Concentration (mol ratio/ TPA[b])				
	1000	500	100	10	Control
Ailantinol-A(**40**)	0 (70)[c]	7	27	37	40
Ailantinol-B(**41**)	0 (70)	0	14	33	40
Ailantinol-C(**42**)	0 (60)	0	10	27	40
Ailantinol-D(**43**)	0 (70)	0	21	37	40
Ailanthone(**80**)	0 (70)	7	13	27	40
Shinjulactone-A(**116**)	0 (70)	0	7	33	40
Shinjudilactone(**117**)	7 (70)	18	30	40	40
Amarolide(**57**)	0 (70)	9	34	40	40
Amaloride11-acetate(118)	9 (70)	19	34	40	40
Shinjulactone-K(**119**)	0 (70)	10	27	35	40
$\Delta^{13(18)}$-Dehydroglaucarubinone(**120**)	0 (70)	10	27	35	40
$\Delta^{13(18)}$-Dehydroglaucarubolone(**121**)	0 (70)	7	32	40	40
Shinjulactone-B(**122**)	0 (70)	10	35	40	40
Shinjulactone-C(**123**)	17 (70)	25	35	40	40

[a]Obtained from *Ailanthus altissima.*
[b]32 pmol was used.
[c]Viability percentage of Raji cells

Table 13.   **Inhibitory Effects of Quassinoid Derivatives on TPA-induced EBV-EA Activation**

Compound	EBV-EA Positive Cells(%)				
	Compound Concentration (mol ratio/TPA[a])				
	1000	500	100	10	Control
Deacetylisobrucein-B(124)	0 (80)[b]	0	19	36	40
3-3'Mmethylbutanoate of 124(125)	0 (80)	0	11	33	40
3-Senecioate of 124(126)	0 (80)	0	13	36	40
3-3'-Trifluoromethylbutanoate of 124(127)	0 (80)	0	4	12	40
3-3'-Fluorobenzoate of 124(128)	0 (80)	0	6	19	40
3-3'-Trifluoromethylbutanoate of 124(129)	0 (80)	0	8	24	40
Curcumine	0 (70)	10	29	40	40
β-Carotene	3 (60)	12	32	40	40
Glycyrrhizin	5 (60)	21	32	40	40

[a]TPA=32 pmol.
[b]Viability percentages of Raji cells.

ANTI-HIV ACTIVITY OF NATURAL QUASSINOID GLYCOSIDES AND THEIR AGLYCONES

Eighteen quassinoid glycosides and nine known quassinoids were tested for inhibitory activity against HIV replication in H9 lymphocytic cells, as shown in Table 14. Of the compounds tested, *shinjulactone-C* (**123**) demonstrated the highest anti-HIV activity (EC_{50}=10.6 μM) in the absence of cytotoxicity with a therapeutic index of >25 [37].

Table 14.   Cytotoxicity and Anti-HIV Activity of Quassinoids

Compound	$IC_{50}(\mu M)$	$EC_{50}(\mu M)$	Therapeutic Index
Yadanzioside-F(97)	156	109	1.4
Bruceoside-A(9)	29	29	1
Yadanzioside-A(102)	146	110	1.3
Bruceantinoside-A(6)	4	a	a
Yadanzioside-G(13)	37	23	1.6
Yadanzioside-C(55)	55	a	a
Yandanzioside-M(101)	7	a	a
Bruceoside-D(37)	52	a	a
Bruceoside-E(38)	59	74	0.8
Bruceoside-F(39)	>132	112	>1.2
Bruceoside-B(12)	3	3	1
Yadanzioside-B(99)	4	5	0.8
Yadanzioside-P(7)	18	18	1
Yadanzioside-L(98)	10	5	2
Bruceoside-C(36)	44	a	a
Yadanzioside-N(14)	49	a	a
Bruceantinoside-C(15)	26	46	0.6
Yadanzioside-E(100)	29	a	a
$\Delta^{13(18)}$-Dehydroglaucarubinone(120)	b	b	b
Shinjulactone-C(123)	>264	10.6	>25
Shinjulactone-B(122)	>287	28	>10
Shinjulactone-A(116)	5	5	1
Ailantinol-A(40)	>246	30	>8.2
Ailanthone(80)	b	b	b
Amarolide(57)	>275	179	>1.5
Amarolide11-acetate(118)	>246	246	>1
Shinjudilactone(117)	>266	43	>6.2
ddC[c]	500	0.03	16.667

[a]No suppression.

[b]Toxic at all concentrations tested.

[c]ddC is the control drug.

ANTI-TUBELCULOSIS   ACTIVITY   OF   NATURAL   QUASSINOID GLYCOSIDES AND THEIR AGLYCONES

In vitro evaluation of anti-tuberculosis activity was conducted for fifty-six quassinoids isolated in our laboratory from the Simaroubaceous plants *A. altissima* (10 compounds), *B. antidysenterica* (16 compounds), *P. ailanthoides* (14 compounds), and *B. javanica* (16 compounds), as shown in Table 15. Of the compounds tested, *shinjulactone-K* (**119**)

Table 15.   Inhibitory Effects of Quassinoids as Anti-tuberculosis Agents

Compond	%Inhibitionat12.5µg/ml	Plant
Shinjulactone-K(**119**)	19	a
Ailanthone(**80**)	17	a
Shinjudilactone(**117**)	15	a
Dehydrobruceantin(**86**)	15	b
Nigakihemiacetal-D(**113**)	12	c
Amarolide(**57**)	12	a
Bruceantin(**1**)	9	b
Nigakilactone-L(**106**)	9	c
Dehydrobruceantarin(**88**)	8	b
Neoquassin(**115**)	8	c
Nigakihemiacetal-A(**110**)	8	c
Bruceoside-D(**37**)	7	d
Quassin(**78**)	7	c
Nigakilactone-H(**105**)	5	c
Nigakilactone-E(**108**)	5	c
Picrasin-A(**107**)	4	c
Bruceanol-F(**19**)	3	b
Dehydrobruceantinol(**87**)	1	b

a:*Ailanthus altissima*,b:*Brucea antidysenterica*,c:*Picrasma ailanthoides*, d:*Brucea javanica*

ailanthone (**80**), *shinjudilactone* (**117**), and *dehydrobruceantin* (**3**) were the most potent. Although the activities were very low (0-19%), the resulting data provided a picture of structure-activity relationships: (1) the carbonyl group in ring-A may play an important role in the emergence of the activity; (2) some quassinoids with a side chain at C-15 are active and an aliphatic side chain is preferred rather than an aromatic one; (3) acetylation of the aliphatic side chain decreases the activity; (4) a hemiacetal group in ring-D is preferred rather than a lactone group; and (5) a sugar moiety decreases the activity [38].

INSECT ANTIFEEDANT ACTIVITY OF NATURAL QUASSINOIDS, THEIR GLYCOSIDES, AND SEMI-SYNTHETIC QUASSINOID DERIVATIVES

Using 3rd instar larvae of the diamondback moth (*Plutella xylostella*), the antifeedant and insecticidal activities of sixteen quassinoids were compared with those of known insect antifeedant *chlorodimeform* (= *galecron*), as shown in Table 16. The insecticidal activity of *quassin* (**78**) was higher than that of *chlorodimeform*, although its antifeedant activity was nearly identical to that of the reference compound [39].

Table 16. Antifeedant Activity of Quassinoids against the Diamondback Moth, *Plutellaxylostella*

Compound	$63.7\mu g/cm^2$ Application"			$31.9\mu g/cm^2$ Application			$16.0\mu g/cm^2$ Application		
	1 Day	2 Days	5 Days	1 Day	2 Days	5 Days	1 Day	2 Days	5 Days
Chlorodimeform	+++	+++	+++	+++	+++	+++	+++	+++	+++
Quassin(78)	+++	+++	+++	+++	+++	+++	+++	+++	+++
Picrasin-B(103)	+++	+++	+++	++	++	+++	++	++	++
Picrasin-D(114)	+++	+++	+++	+++	++	+++	+++	++	++
Nigakilactone-E(108)	+	++	++	+	++	++	+	++	++
Nigakilactone-F(104)	++	++	++	+	++	++	+	++	++
Picrasin-E(111)	+	++	++	+	++	++	+	++	++
Neoquassin(115)	++	++	++	++	++	++	+	++	++
Picrasinol-B(34)	+++	++	++	+	++	++	+	++	++

(Table 16) contd.....

Compound	63.7µg/cm^2 Application"			31.9µg/cm^2 Application			16.0µg/cm^2 Application		
	1 Day	2 Days	5 Days	1 Day	2 Days	5 Days	1 Day	2 Days	5 Days
Picrasin-A(107)	+++	+++	+++	+++	+++	+++	++	++	+++
Nigakilactone-H(105)	-	+	+	-	+	++	-	+	-
Nigakilactone-L(106)	-	++	++	-	++	++	-	++	++
Picrasin-G(130)	-	++	++	-	+	++	-	-	-
Picrasinoside-C(28)	+	++	++	+	+	+	-	+	+
Picrasinoside-E(30)	-	-	-	-	-	-	-	-	-
Picrasinoside-G(32)	-	-	-	-	-	-	-	-	-

Four semi-synthetic and fourteen additional natural quassinoids were also tested in these same assays (Table 17). Chemical conversion of the methoxy group in ring-A and methylenedioxy group in ring-C into hydroxy groups (derivatives from **78, 103, 114,** and **130**) resulted in decreased activity [40]. Therefore, the methoxy groups in these quassinoids may be very important for the activities. However, some quassinoids such as *isobrucein-B* (**5**) also showed potent activities, although these contain some hydroxy groups. Therefore, protection of these hydroxy groups by methylation or esterification may enhance the activity.

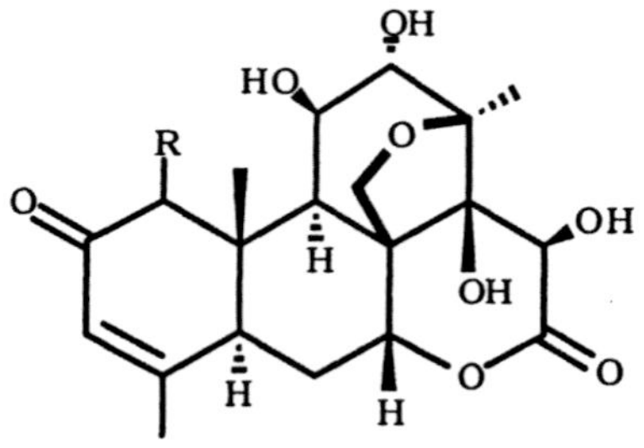

94 R = ─ OH
95 R = ···· OH

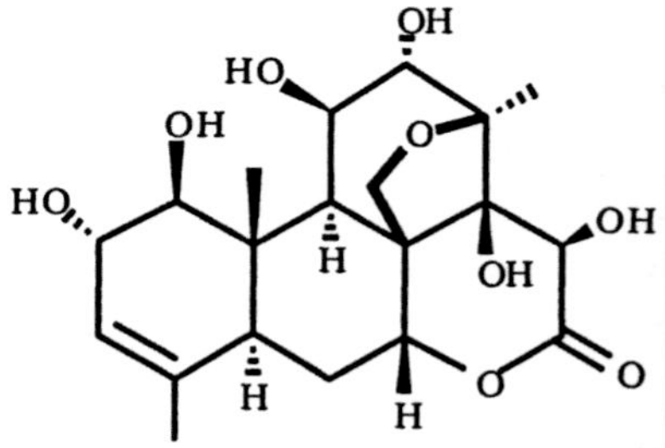

96

Table 17. **Antifeedant Activity of Quassinoids Against the Diamondback Moth,** *Plutella xylostella*

Compound	63.7 µg/cm^2 Application			31.9 µg/cm^2 Application			16.0 µg/cm^2 Application		
	1 Day	2 Days	5 Days	1 Day	2 Days	5 Days	1 Day	2 Days	5 Days
Quassin(78)	+++	+++	+++	+++	+++	+++	+++	+++	+++
Picrasin-B(103)	+++	+++	+++	++	+++	+++	++	++	++
Picrasin-D(114)	+++	+++	+++	+++	++	+++	+++	++	++
Picrasin-G(130)	-	++	++	-	+	++	-	-	-
Derivative from 78	-	++	++	-	++	+	-	++	++
Derivative from 103	+	+	++	-	-	-	-	-	-
Derivative from 114	-	++	++	-	++	+	-	+	-
Derivative from 130	+	-	-	-	+	-	-	-	-
Nigakiheniacetal-A(110)	-	++	+	+	+	-	-	++	-
Nigakihemiacetal-D(113)	+	++	++	-	++	+	+	+	+
Bruceantin(1)	+++	+++	+++	+	++	++	+++	+++	+++
Isobrucein-B(5)	++	+++	+++	+++	+++	+++	++	++	+++
Bruceanol-A(11)	+++	+++	+++	+++	+++	+++	++	++	+++
Bruceoside-A(9)	+++	+++	+++	-	++	++	-	++	++
Yadanziolide-A(102)	++	++	+++	+	++	++	-	++	++
Yadanziolide-C(95)	++	+++	+++	++	++	+++	++	++	+++
Yadanzioside-F(97)	++	+++	+++	-	++	++	-	++	++
Yadanzioside-G(13)	+	++	++	++	++	+++	-	++	++
Bruceoside-B(12)	++	+++	+++	+++	++	+++	++	++	+++
Yadanzioside-B(99)	+++	+++	+++	+++	+++	+++	+++	+++	+++
Yadanzioside-L(98)	+++	+++	+++	+++	+++	+++	++	++	+++
Brucein-E(96)	-	++	++	-	++	++	-	++	++

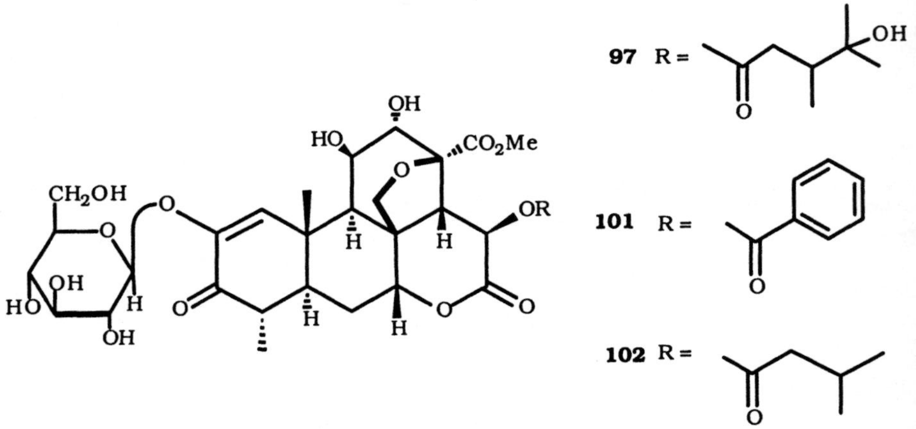

97 R =

101 R =

102 R =

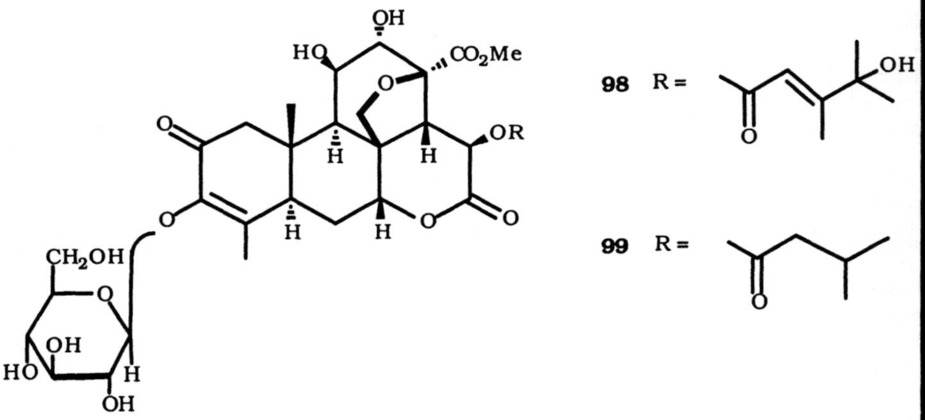

98 R =

99 R =

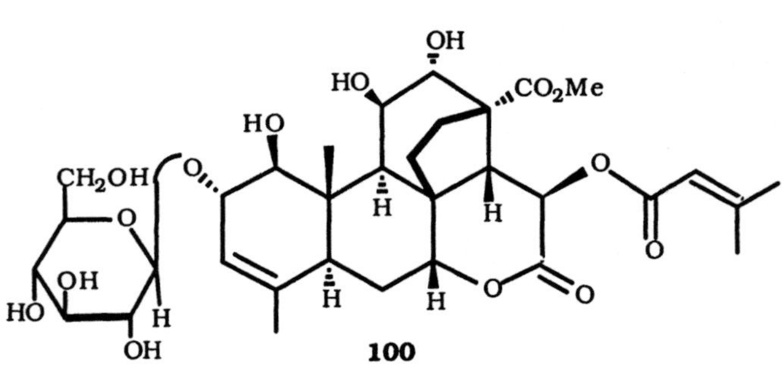

100

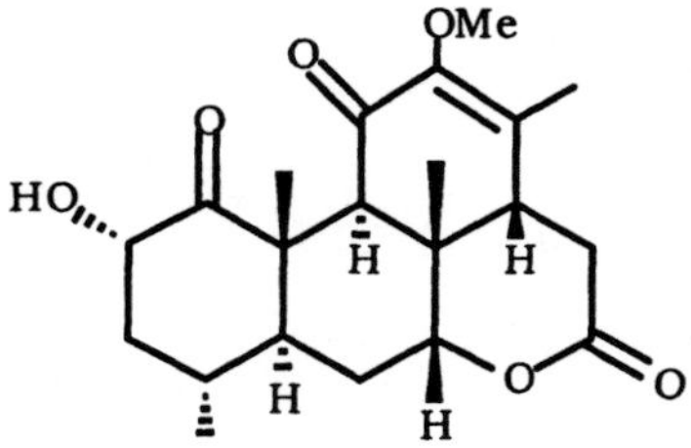

103

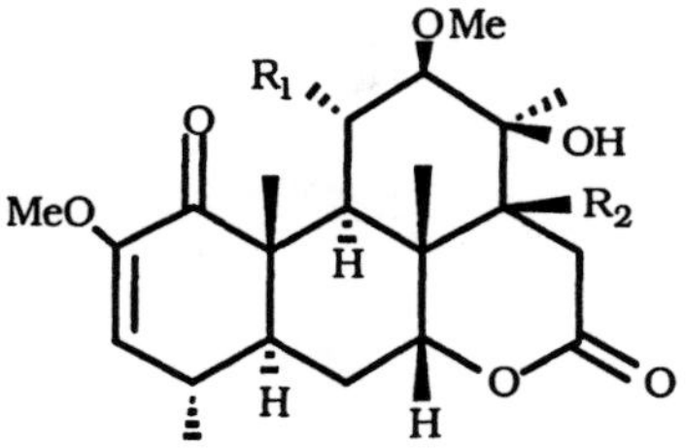

104 R_1 = OH R_2 = H

105 R_1 = OH R_2 = OH

108 R_1 = OAc R_2 = H

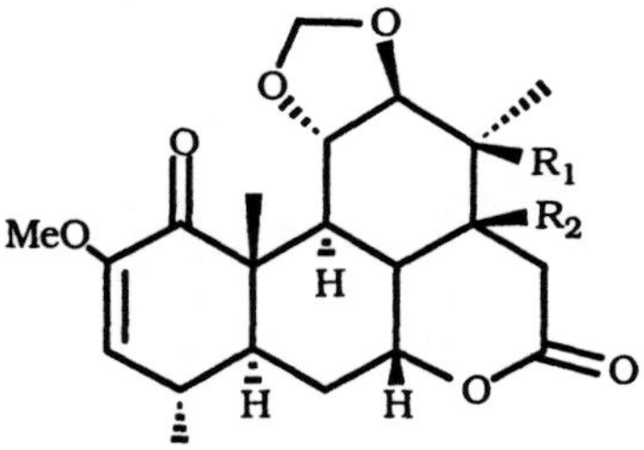

106 R_1 = OH R_2 = H

111 R_1 = OH R_2 = OH

114 R_1 = H R_2 = H

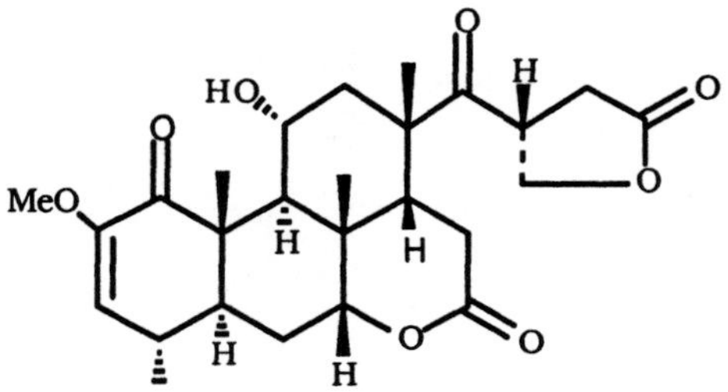

107

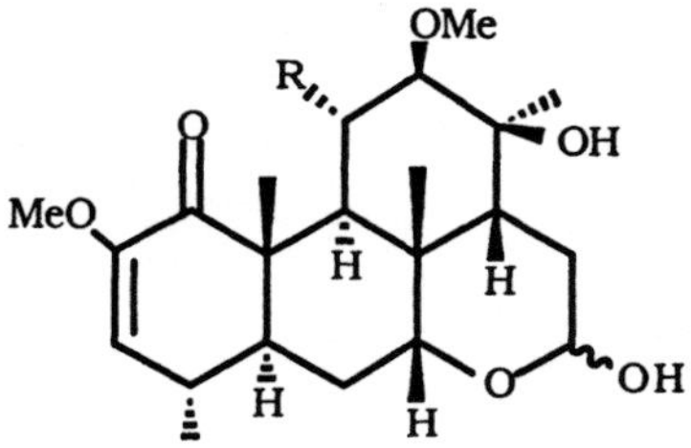

110 R = OH

113 R = OAc

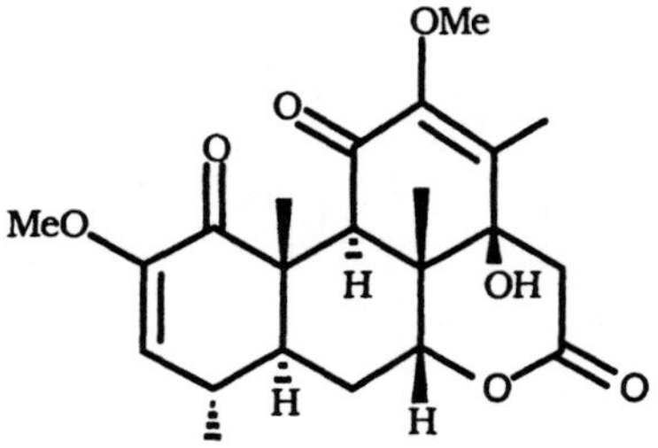

112

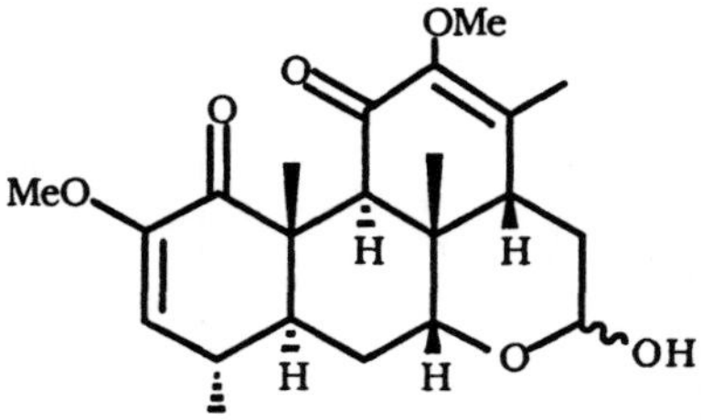

115

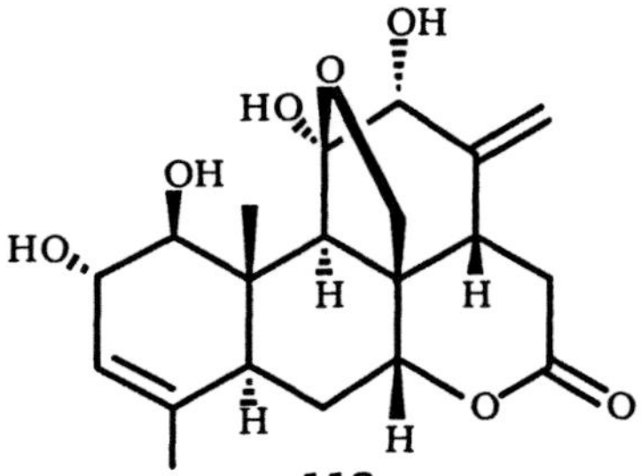

116

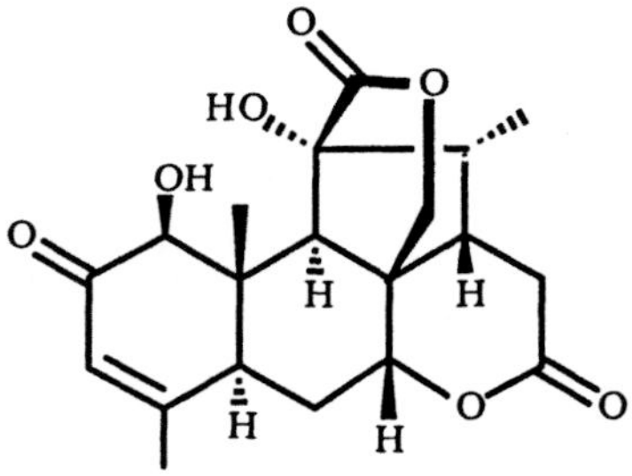

117

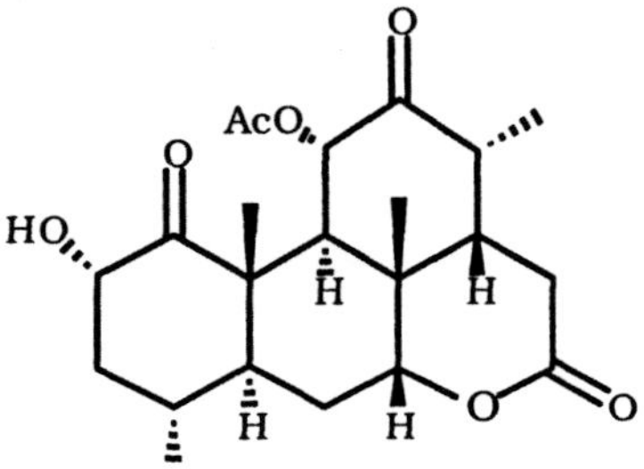

118

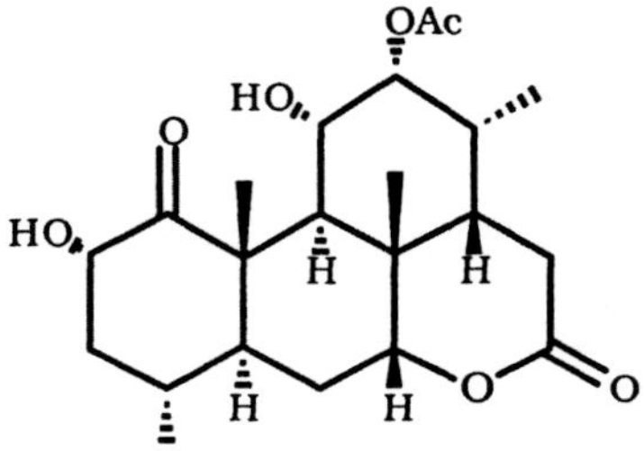

119

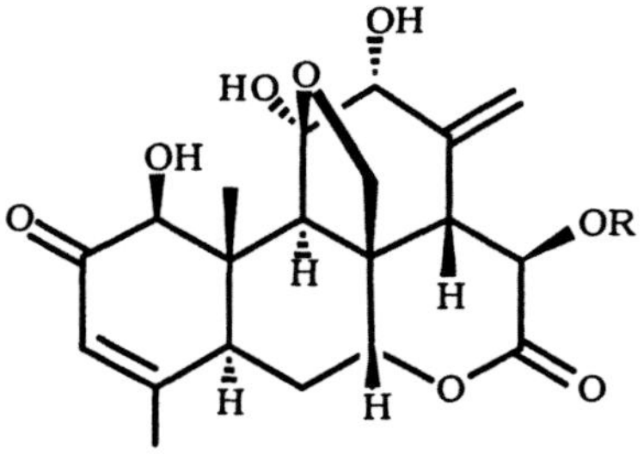

120   R =

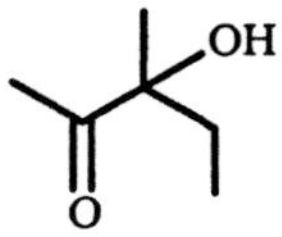

121   R = H

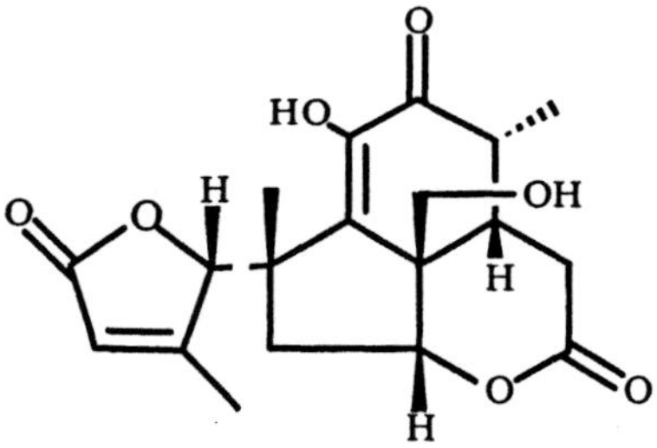

122

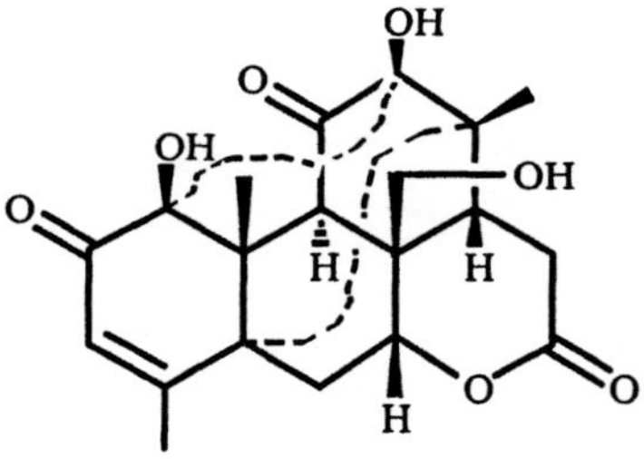

123

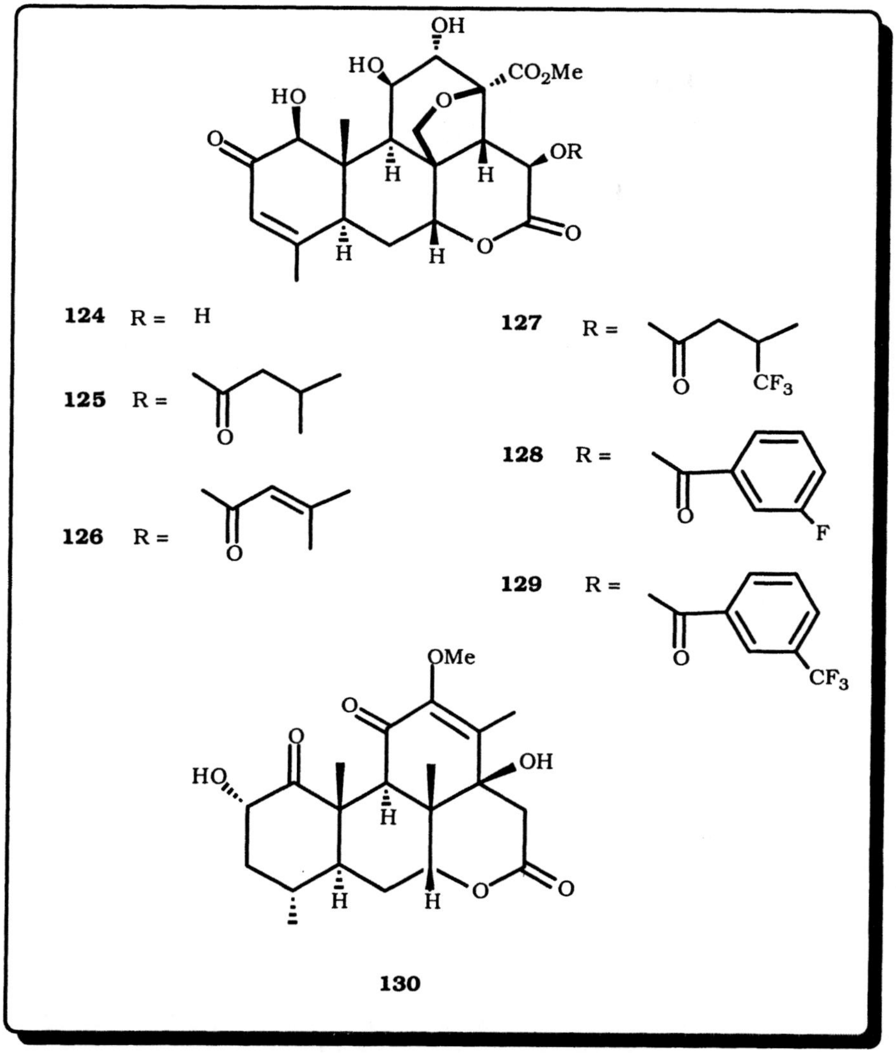

124 R = H

125 R = [structure]

126 R = [structure]

127 R = [structure] CF₃

128 R = [structure] F

129 R = [structure] CF₃

130

STABILITY OF QUASSINOIDS

From the correlated structures of **1**, **86**, and **22**, it seems reasonable to assume that *bruceantin* (**1**) ($ED_{50}=0.59$ µg/ml) [1] might be converted to *dehydrobruceantin* (**86**) ($ED_{50}=1.0$ µg/ml) [1] and **86** might further converted to *bruceanic acid-A* (**22**) ($ED_{50}=4.16$ µg/ml) (the values in the parentheses are antitumor activity for KB cells) [12]. Thus, on the basis of these data, the activity drops with degradation of the quassinoid.

Luengi *et al* [23]. also made this point clear. In an HL-60 cell differentiation assay, *brusatol* (**8**) showed ED_{50} value of 0.006 μg/ml, but *dehydrobrusatol* (**54**) showed a value of 0.8 μg/ml.

In the P-388 assay of *bisbrusatolylmalonate* (**81**) [30], different values of T/C (272% and 197%, see Table 8) were obtained at different laboratories. Therefore, the stability of ring-A structure in **81** should also be questioned considering these data.

Bruceantin (**1**) was converted into *brucein-C* (**76**), *bruceantin-4', 5'-epoxide*, and *bruceantin-5'-ol* by its incubation in *S. griceus* [1]. Thus, the quassinoid side chain is biologically unstable.

ACKNOWLEGMENTS

It is with great pleasure that we acknowlege the contributions of all colleagues, most of whom are cited in the references. In particular, we wish to thank Drs. K. Tagahara (Kobe Pharmaceutical University, Japan), H. Tokuda (Kyoto Prefectural University of Medicine, Japan), and H. Yamazaki (Agricultural Chemicals Research Laboratory, Sumitomo Chemical Co. Ltd., Japan).

ABBREVIATIONS

NCI	=	National Cancer Institute of USA
CC	=	Column Chromatography
TLC	=	Thin Layer Chromatography
HPLC	=	High Performance Liquid Chromatography
NMR	=	Nuclear Magnetic Resonance
T/C%	=	Ratio of test group survival to control group survival in tumored animals
ED_{50}	=	50% Effective Dose
GI_{50}	=	50% Growth Inhibition
P-388	=	Murine lymphocytic leukemia
L-1210	=	Murine leukemia
B-16	=	Murine melanoma
KB	=	Nasopharynx carcinoma or nasal pharingeal carcinoma
A-549	=	Human lung carcinoma
HCT-8	=	Human colon tumor

TE-671	=	Human medulloblastoma
RPMI-7951	=	Human melanoma
SK-MEL-5	=	Human melanoma
COLO-205	=	Human colon cancer
CNS	=	Human medulloblastoma
Quassinoids **1**	=	Bruceantin
2	=	Bruceantarin
3	=	Dehydrobruceantin
4	=	Dehydrobruceantarin
5	=	Isobrucein-B
6	=	Bruceantinoside-A
7	=	Bruceantinoside-B
8	=	Brusatol
9	=	Bruceoside-A
10	=	Bruceoside-B
11	=	Bruceanol-A
12	=	Bruceanol-B
13	=	Yadanzioside-G
14	=	Yadanzioside-N
15	=	Bruceantinoside-C
16	=	Bruceanol-C
17	=	Bruceanol-D
18	=	Bruceanol-E
19	=	Bruceanol-F
20	=	Bruceanol-G
21	=	Bruceanol-H
22	=	Bruceanic acid-A
23	=	Bruceanic acid-B
24	=	Bruceanic acid-C
25	=	Bruceanic acid-D
26	=	Picrasinoside-A
27	=	Picrasinoside-B

28	=	Picrasinoside-C
29	=	Picrasinoside-D
30	=	Picrasinoside-E
31	=	Picrasinoside-F
32	=	Picrasinoside-G
33	=	Picrasinol-A
34	=	Picrasinol-B
35	=	Picrasinoside-H
36	=	Bruceoside-C
37	=	Bruceoside-D
38	=	Bruceoside-E
39	=	Bruceoside-F
40	=	Ailantinol-A
41	=	Ailantinol-B
42	=	Ailantinol-C
43	=	Ailantinol-D
44	=	Samaderine-X
45	=	Samaderine-Y
46	=	Samaderine-Z
47	=	Samaderine-E
48	=	Indaquassin-X
49	=	Indaquassin-C
50	=	Samaderine-B
51	=	Samaderine-C
52	=	Simarinolide
53	=	2-*O*-Glucosylsamaderine-C
54	=	Dehydrobrusatol
55	=	Yadanzioside-C
56	=	11-*O*-trans-p-Coumaroylamarolide
57	=	Amarolide
58	=	Chaparrinone
59	=	Chaparrin

60	=	Glaucarubolone
61	=	Holacanthone
62	=	15-*O*-β-D-Glucopyranosylglaucarubol
63	=	6-α-Hydroxyeurycomalactone
64	=	Longilactone
65	=	14,15-β-Dihydroxyklaineanone
66	=	11-Dehydroklaineanone
67	=	Eurycomalactone
68	=	5,6-Dehydroeurycomalactone
69	=	Sergeolide
70	=	15-Desacetylundulactone
71	=	Undulactone
72	=	15-*O*-β-D-Glucopyranosyl-21-hydroxy-glaucarubolone
73	=	6-α-Tigloyloxyglaucarubol
74	=	21-Hydroxyglaucarubolone
75	=	Brucein-B
76	=	Brucein-C
77	=	Glaucarubinone
78	=	Quassin
79	=	Isobrucein-A
80	=	Ailanthone
81	=	Bisbrusatolylmalonate
82	=	Bisbrusatolylsuccinate
83	=	Bisbrusatolylglutarate
84	=	Bisbrusatolyladipate
85	=	Bisbrusatolylsebacate
86	=	Dehydrobruceantin
87	=	Dehydrobruceantinol
88	=	Dehydrobruceantarin
89	=	Bruceolide
90	=	15-3'-Trifluoromethylcinnamate of **89**
91	=	3-3'-Trifluoromethylcinnamate of **89**

92	=	3-3'-Trifluoromethylbutanoate of **89**
93	=	15-3'-Trifluoromethylbutanoate of **89**
94	=	Brucein-D
95	=	Yadanziolide-C
96	=	Brucein-E
97	=	Yadanzioside-F
98	=	Yadanzioside-L
99	=	Yadanzioside-B
100	=	Yadanzioside-E
101	=	Yadanzioside-M
102	=	Yadanzioside-A
103	=	Picrasin-B
104	=	Nigakilactone-F
105	=	Nigakilactone-H
106	=	Nigakilactone-L
107	=	Picrasin-A
108	=	Nigakilactone-E
109	=	Picrasinol-D
110	=	Nigakihemiacetal-A
111	=	Picrasin-E
112	=	Picrasinol-C
113	=	Nigakihemiacetal-D
114	=	Picrasin-D
115	=	Neoquassin
116	=	Shinjulactone-A
117	=	Shinjudilactone
118	=	Amarolide 11-acetate
119	=	Shinjulactone-K
120	=	$\Delta^{13(18)}$-Dehydroglaucarubinone
121	=	$\Delta^{13(18)}$-Dehydroglaucarubolone
122	=	Shinjulactone-B
123	=	Shinjulactone-C

124	=	Deacetylisobrucein-B
125	=	3-3'-Methylbutanoate of 124
126	=	3-Senecioate of 124
127	=	3-3'-Trifluoromethylbutanoate of 124
128	=	3-3'-Fluorobenzoate of 124
129	=	3-3'-Trifluromethylbutanoate of 124
130	=	Picrasin-G
131	=	22,23,22',23'-Tetrahydro-derivative of 81
132	=	3-Senecioate of 8
133	=	3-Butanoate of 8
134	=	3-Ethylsuccinate of 8
135	=	3-3',4'-Dimethyl-2'-pentenoate of 8
136	=	3-3',4',5'-Trimethoxybenzoate of 8
137	=	3-Cinnamate of 8
138	=	3,4,22,23-Tetrahydrobrusatol

REFERENCES

[1]   Okano, M.; Fukamiya, N.; Lee, K. H. In *Studies in Natural Products Chemistry*; Atta-ur-Rahman, Ed.; Elsevier Science Publishers B. V.: Amsterdam, **1990**; Vol.7, pp.369-404.

[2]   Kupchan, S. M.; Britton, R. W.; Lacadie, J. A.; Ziegler, M. F.; Sigel, C. W. 0 *J. Org. Chem.* **1975**, *40*, 648-654.

[3]   Okano, M.; Lee, K. H., Hall, I. H., Boettner, F. E. *J. Nat. Prod.* **1981**, *44*, 470-474.

[4]   R. I. Geran, R. I.; Greenberg, N. H.; MacDonald, M. M.; Schumacher, A. M.; Abbott, B. *J. Cancer Chemocether. Rep.*, **1972**, Part *3*, 1-88.

[5]   Sakaki, T.; Yoshimura, S.; Tsuyuki, T.; Takahashi, T.; Honda, T. *Chem. Pharm. Bull.* **1986**, *34*, 4447-4450.

[6]   Okano, M.; Fukamiya, N.; Toyota, T.; Tagahara, K.: Lee, K. H. *J. Nat. Prod.* **1989**, *52*, 398-401.

[7]   Okano, M.; Fukamiya, N.; Aratani, T., Ju-ichi, M.; Lee, K. H. *J. Nat. Prod.* **1985**, *48*, 972-975.

[8]   Fukamiya, N.; Okano, M.; Aratani, T., Lee, K. H. *J. Nat. Prod.* **1987**, *50*, 1075-1079.

[9]   Fukamiya, N.; Okano, M.; Aratani, T.; Lee, K. H. *J. Nat. Prod.* **1988**, *51*, 349-352.

[10]   Imamura, K.; Fukamiya, N.; Okano, M.; Tagahara, K.; Lee, K. H. *J. Nat. Prod.* **1993**, *56*, 2091-2097.

[11]   Imamura, K.; Fukamiya, N.; Nakamura, M.; Okano, M.; Tagahara, K.; Lee, K. H. *J. Nat. Prod.* **1995**, *58*, 1915-1919.

[12]   Toyota, T.; Fukamiya, N.; Okano, M.; Tagahara, K.; Chang, J. J.; Lee, K. H. *J. Nat. Prod.* **1990**, *53*, 1526-1532.

[13]   Ferguson, P. J.; Fisher M. H.; Stephenson, J.; Li, D. H.; Zhau, B. S.; Chen, Y. C. *Cancer Res.* **1988**, *48*, 5956-5964.

[14]   Lee, K. H.; Lin, Y. M.; Wu, T. S., Zhang, D. C.; Yamagishi, T.; Hayashi, T.; Hall, I. H.; Chang, J. J.; Wu, R. Y.; Yang, T. H. *Planta Med.* **1988**, *54*, 308-311.

[15]   Okano, M.; Fujita, T.; Fukamiya, N.; Aratani, T. *Bull. Chem. Soc. Jpn.* **1985**, *58*, 1793-1800.

[16]   Matsuzaki, T.; Fukamiya, N.; Okano, M.; Fujita, T.; Tagahara, K.; Lee K. H. *J. Nat. Prod.* **1991**, *54*, 844-848.

[17]   Fukamiya, N.; Okano, M.; Miyamoto, M.; Tagahara, K.; Lee, K. H. *J. Nat. Prod.* **1992**, *55*, 468-475.

[18]   Ohnishi, S.; Fukamiya, N.; Okano, M.; Tagahara, K.; Lee, K. H. *J. Nat. Prod.* **1995**, *58*, 1032-1038.

[19]   Grever, M. R.; Schepartz, S. A.; Chabner, B. A. *Seminars Oncol.* **1992**, *19*, 622-638.

[20]   Monks, A.; Scudiero, D.; Skehan, P.; Shoemaker, R.; Paul, K.; Vistica, D.; Hose, C.; Langley, J.; Cronise, P.; Vaigro-Wolff, A.; Gray-Goodrich, M.; Campbell, H,; Mayo, J.; Boyd, M. *J. Natl. Cancer Inst.* **1991**, *83*, 757-766.

[21]   a) Kubota, K.; Fukamiya, N.; Hamada, T.; Okano, M.; Tagahara, K.; Lee, K. H. *J. Nat. Prod.* **1996**, *59*, 683-868. b) Kubota, K.; Fukamiya, N.; Okano, M.; Tagahara, K.; Lee, K. H. *Bull. Chem. Soc. Jpn.* **1996**, *69*, 3613-3617.

[22]   Kitagawa, I.; Mahmud, T.; Yokota, K.; Nakagawa, S; Mayumi, T.; Kobayashi, M.; Shibuya, H. *Chem. Pharm. Bull.* **1996**, *44*, 2009-2014.

[23]   Luyengi, L.; Suh, N.; Fong, H. H. S; Pezzuto, J. M.; Kinghorn, A. D. *Phytochemistry* **1996**, *43*, 409-412.

[24]   Dou, J.; Mcchesney, J. D.; Sindelar, R. D.; Goins, D. K.; Walker, L. A. *J. Nat. Prod.* **1996**, *59*, 73-76.

[25]   Itokawa, H.; Kishi, E.; Morita, H.; Takeya, K. *Chem. Pharm. Bull.*, **1992**, *40*, 1053-1055.

[26]   Tischler, M.; Cardellina, J. H., II; Boyd, M. R.; Cragg, G. M. *J. Nat. Prod.*, **1992**, *55*, 667-671.

[27]   Lumonadio, L.; Attassi, G.; Vanhealen, M.; Vanhealen-Fastre, R. *J. Ethnopharmacol.* **1991**, *31*, 59-65.

[28]   Polonsky, J.; Bhatnager, S. C.; Griffiths, D. C.; Pickett, J. A.; Woodcock, C. M. *J. Chem. Ecol.*, **1989**, *15*, 993-998.

[29]   Lin, L. J.; Peiser, G.; Ying, B. P.; Mathias, K.; Karasina, F.; Wang, Z. Itatani, J.; Green, L.; Hwang, Y. S. *J. Agric. Food Chem.*, **1995**, *43*, 1708-1711.

[30]   Lee, K. H.; Okano, M.; Hall, I. H., Brent, D. A.; Soltmann, B. *J. Pharm. Sci.* **1982**, *71*, 338-345.

[31]   Hall, I. H.; Lee, K. H.; Okano, M.; Sims, D.; Ibuka, T.; Liou, Y. F.; Imakura, Y. *J. Pharm. Sci.* **1981**, *70*, 1147-1150.

[32]   Imamura, K.; Fukamiya, N.; Okano, M.; Tagahara, K.; Lee, K. H. *Bull. Chem. Soc. Jpn.* **1992**, *65*, 2684-2689.

[33]   Ohno, N.; Fukamiya, N.; Okano, M.; Tagahara, K.; Lee, K. H. *Bio-Org. & Med. Chem.* **1997**, *5*, 1489-1495.

[34]   Okano, M.; Fukamiya, N.; Tagahara, K.; Tokuda, H.; Iwashima, A.; Nishino, H.; Lee, K. H. *Cancer Lett.* **1995**, *94*, 139-146.

[35] Kubota, K.; Fukamiya, N.; Tokuda, H.; Nishino, H.; Tagahara, K.; Lee, K. H, Okano, M. *Cancer Lett.* **1997**, *113*, 165-168.

[36] Rahman, S.; Fukamiya, N.; Ohno, N.; Tokuda, H.; Nishino, H.; Tagahara, K.; Lee, K. H.; Okano, M. *Chem. Pharm. Bull.* **1997**, *45*, 675-677.

[37] Okano, M.; Fukamiya, N.; Tagahara, K.; Cosentino, M.; Lee, T. T. Y.; Natschke, S. M.; Lee, K. H. *Bio-Org. & Med. Chem. Lett.*, **1996**, *6*, 701-706.

[38] Rahman, S.; Fukamiya, N.; Okano, M.; Tagahara, K.; Lee, K. H. *Chem. Pharm. Bull.* **1997**, *45*, 1527-1529.

[39] Daido, M.; Fukamiya, N.; Okano, M.; Tagahara, K.; Hatakoshi, M.; Yamazaki, H. *Biosci. Biotech. Biochem.* **1993**, *57*, 244-246.

[40] Daido, M.; Ohno, M.; Imamura, K.; Fukamiya, N.; Hatakoshi, M.; Yamazaki, H.; Tagahara, K.; Lee, K. H.; Okano, M. *Biosci. Biotech. Biochem.* **1995**, *59*, 974-979.

Atta-ur-Rahman (Ed.) *Studies in Natural Products Chemistry, Vol. 23*

COUMARINS, THE BIOACTIVE STRUCTURES WITH ANTIFUNGAL PROPERTY

S. SARDARI,[1] S. NISHIBE,[2] *and* M. DANESHTALAB[1]

[1] Faculty of Pharmacy and Pharmaceutical Sciences, University of Alberta, Edmonton, AB, Canada T6G 2N8
[2] Faculty of Pharmaceutical Sciences, Health Sciences University of Hokkaido, Ishikari-Tobetsu, Hokkaido, Japan 061-02

ABSTRACT: The emergence of fungal infections due to the increase in the number of immunocompromised patients has become a matter of concern. Among the available sources for treating fungal infections, natural products can play a significant role. In this article, we have tried to categorize the different sources and chemical classes of fungitoxic agents as following: (1) animal sources such as defensins and other peptides, (2) plant sources such as terpenoids, flavonoids, and coumarins, (3) fungal metabolites like fatty acids, sugar derivatives and polyenes and (4) bacterial sources such as amino acids. The potentials of coumarins, which are mainly distributed in four plant families, Umbelliferae, Rutaceae, Leguminosae and Compositae, have been uncovered in this respect. In addition to antifungal property, the other bioactivities of coumarins include: anticoagulant, antibacterial, antiparasitic, anticancer, photodynamic, edema relieving, anti-inflammatory, radical scavenging, lipolytic, antiplatelet, and anti-allergic properties. Studies in structure-antifungal activity relationships of coumarins, which is presented here, reveal the essential strucutral features for their activity. In order to have a better understanding of the mode of action of coumarins, the cellular sites and the interactions of coumarins with biological systems have been explained. Effects on the enzymes gyrase and topoisomerase, and photodynamic reaction with lipids, proteins and DNA are among the main issues discussed. In this respect, morphological changes of *Candida albicans* cells due to coumarins, which are studied by scanning electron microscopy (SEM), support their biochemical effects.

INTRODUCTION

Background

Despite the presence of ~100,000 known species of fungi [1] that inhabit planet Earth, serious illness in humans due to fungal infections is relatively rare. Most fungi are virtually no threat to humans because of their special growth requirement and/or limited ecological niches. A broad array of host defense mechanisms, both immune [2] and non-immune, have evolved in humans to protect them against fungal invasion. Indeed, the majority of serious fungal infections occur in patients who have defects in one or more of these defenses. Moreover, concomitant with the increasing numbers of severely immunosuppressed patients, there has been an increment in the number of reports of infections caused by fungi that were not previously considered pathogens.

With different immunodeficiency states, the sequential appearance of different diseases serves as a herald of clinical disease progression. The role of the immunocompromised state on the medical history, physical findings, and natural history of fungal infections is important. The appearance and increased frequency of rare fungal infections is observed in immunosuppressed individuals such as patients with bone marrow transplant and those infected by HIV [3].

In addition, with the employment of modern chemotherapeutic modalities in 1960's, susceptible hosts were created for a broad range of opportunistic pathogenic fungi. The role of commensals, such as *Candida* species, is changing. Ranging from a 75% increase in small hospitals to over 400% increase in some large care centers, *Candida* is now ranked as the third most common causative agent of nosocomial systemic infections in most hospitals [4].

The development of azoles has revolutionized the treatment of many fungal infections, however, treatment of many of them necessitates application of the highly toxic drug, amphotericin B or a combination of drugs. Emergence of new resistant species of fungi in addition to the poor safety profile and pharmacokinetics challenges the clinicians in their way to handle the fungal infections.

Although combination therapy, improving the delivery systems, immunotherapy and synthesis of new analogues of the present antifungal drugs has improved the fungal disease management, the increasing demand in finding novel antifungal agents capable of overcoming deep-seated mycoses, and resistance induction has diverted scientists attention toward natural products. Over 25% of the present drugs are of natural origin. In addition, nature has indirectly provided more than 60% of whole drugs [5,6]. Echinocandins, pneumocandins, pradimycins, and nikkomycins are representatives of novel classes of antifungals with novel mechanism of activity. These agents, which have been isolated as fermentation products, have been semi-synthetically modified to possess favorite pharmacokinetic profiles. The structural complexity and narrow spectrum of activity are the major drawbacks for the future development of these compounds. Since plants are considered as an important source of simple bioactive molecules, we have tried to study the antifungal properties of the specific plant-derived chemical species. This article is a contribution to promote and acknowledge the potential of small molecules like coumarins in the therapy of new challenging diseases.

History

The isolation of coumarin was first reported by Vogel in Munich in 1820. He associated the pleasant odor of the Tonka bean from Guiana with that of clover, *Melilotus officinalis*, which give rise to the characteristic aroma of new-mown hay. The name coumarin originates from a Caribbean word 'coumarou' for the Tonka tree, which was known botanically at one time as *Coumarouna odorata* Aubl. By 1868, coumarin was known to have the molecular formula $C_9H_6O_2$. W.H.Perkin Snr, in his first synthesis of a vegetable perfume, treated the sodium salt of *o*-hydroxybenzaldehyde with acetic anhydride and obtained synthetic coumarin identical with that isolated from the Tonka bean [7,8].

Furanocoumarins have been used in folk medicine for a long time. Indian sacred book "Athara Veda" describes the treatment of leukoderma (vitiligo) using a poultice from a plant now known as *Psoralea corylifolia*. Ancient Egyptians had used *Ammi majus* for vitiligo. In the 13th century, Ibn El Bitar described the usefulness of *Ammi majus* in his book "Mofradat El Adwiya". The first furanocoumarin, 5-methoxypsoralen, was isolated in 1838 by Kalbrunner from bergamot oil [9].

Chemistry

Coumarins are lactones with the basic structure of 1,2-benzopyrone. Biosynthetically, they are mainly originated from shikimic acid pathway, with the intermediate of cinnamic acid. By the addition of a furan ring, the furanocoumarins are resulted, which are represented by psoralen (linear) and angelicin (angular) [10].

The phenylpropanoid biosynthetic pathway controls the synthesis of lignin, flower pigments, signaling molecules, and a variety of compounds involved in plant defense against pathogenic organisms like coumarins. The key steps in this pathway mostly include P_{450} enzymes so that more than 15 P_{450}-dependent reactions have been characterized in this pathway. One enzyme of the main pathway, cinnamate 4-hydroxylase (C4H), is the most extensively studied plant P_{450} and is suspected of playing an essential role in the regulation of the whole pathway [11].

A feature common to most coumarins is oxygenation at C-7 position. The 7-hydroxycoumarin, commonly known as umbelliferone (1), is often regarded as the parent, both structurally and biogenetically, of the more complex coumarins [8]. Another commonality among coumarins is the presence of isoprenoid chains, frequently of one, but often of two or three units, attached to a carbon, or oxygen, or both. The prenyl group may be found as the simple 3-methylbut-2-enyl unit, but it is often encountered as the corresponding epoxide or vicinal glycol, or in a variety of oxidized and skeletally rearranged forms [8]. Biogenetically, an additional heterocyclic ring can be formed when the prenyl group interacts with an *o*-phenolic group. The structural variations of this type encountered in the natural coumarins mostly include dihydrofuran, hydroxydihydropyran and their derivatives, furan and dihydropyran.

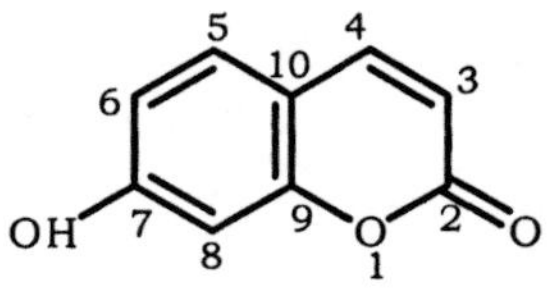

(1) 7-Hydroxycoumarin

Prenylation at C-6 and C-8 can lead to linear coumarins, such as psoralen (2), and xanthyletin (3), or angular forms, like angelicin (4) and seselin (5), respectively. In this regard, coumarins are classified as following [8]:

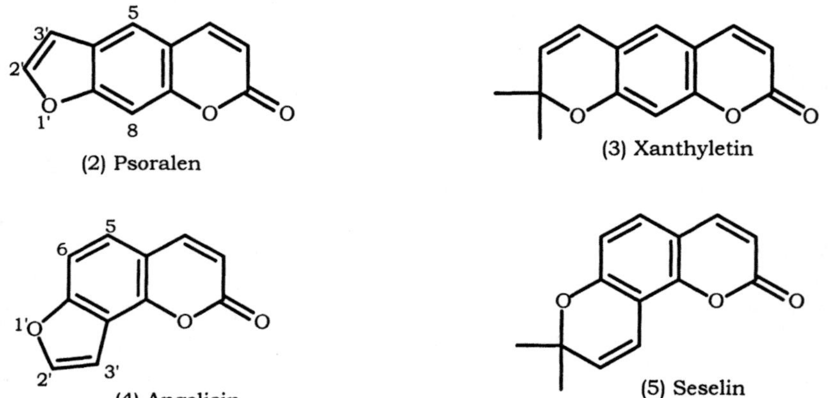

(2) Psoralen

(3) Xanthyletin

(4) Angelicin

(5) Seselin

a)   simple, which includes coumarin, its hydroxylated, alkoxylated, and alkylated derivatives and their glycosides
b)   furanocoumarins, including the typical linear form, psoralen, and angular type, angelicin
c)   pyranocoumarins, which have a six membered ring attached to the benzoid part like seselin and xanthyletin
d)   pyron-ring substituted coumarins, like 4-hydroxycoumarins, and 3-phenylcoumarins.

IDENTIFICATION AND SPECIFIC SPECTROSCOPIC FEATURES

To date, at least 1300 coumarins have been identified and many new coumarin structures are being reported each year [12,13]. A variety of methods have been developed for the identification of known and novel coumarins. The most obvious physical property of most coumarins is the fluorescence they display in UV light (365 nm), which can readily be examined by applying them on TLC plates, or paper chromatographs [10,14,15]. Spectroscopic methods are among the most useful techniques in identifying coumarins. UV absorption spectra are useful for distinguishing coumarins from chromones. Chromones normally have a strong absorption at 240-250 nm (log ε 3.8) whereas coumarins usually show a weak absorption at this wavelength [8]. Coumarin shows absorption bands at 274 and 311 nm, which have been attributed to the benzene and pyrone rings, respectively. Methyl substitution at C-5, C-7, or C-8 leads to a bathochromic shift of the 274 nm maximum but leaves the 311 nm maximum practically unchanged [16]. The introduction of a hydroxyl group into the coumarin nucleus causes a bathochromic shift of the major absorption bands. 7-Oxygenated coumarins show strong absorption bands at about 217 and 315-330 nm with weak peaks or shoulders at 240-250 nm. Linear furanocoumarins show four zones of absorption at 205-225, 240-255, 260-270, and 298-316 nm. Angular furanocoumarins can readily be distinguished from linear forms since the maxima at 242-245 and 260-270 are absent [17].

Another spectrometric method is IR, which apart from identifying functional groups is useful in revealing the conjugated lactone functionality. Coumarins are isomeric with chromones, but the two classes differ considerably in their IR spectra. The carbonyl stretching frequency in coumarins (α-pyrones) is observed in the region 1700-1750 cm^{-1} whereas in chromones (γ-pyrones) it is found at 1650 cm^{-1} [8].

A wide range of NMR techniques has been applied to the structural elucidation of naturally occurring coumarins. The structure-spectra correlation rules were established first in 1972 [18], and by applying modern calculation techniques reached more completion afterwards [19]. In ring-proton analysis, a pair of doublets, J 9.5 Hz, centered at δ 6.1-6.4 and 7.5-8.3, strongly indicates a coumarin unsubstituted in the pyrone ring. These characteristic signals arise from the H-3 and H-4 protons, respectively. The majority of natural coumarins have an oxygen function at C-7, which through electron donating property, leads to an increase in the electron density at C-3. Therefore, this group of coumarins has their H-3 and C-3 moved to higher field [20]. H-4 resonance is found in the region δ 7.5-7.9 in coumarins lacking a C-5 oxygen function [18]. An oxygen or alkyl at C-5 shifts the resonance downfield by about 0.3 ppm to δ 7.9-8.2 [21,22]. The presence of an unsubstituted furan ring is easily recognizable from the pair of doublets, J ~ 2.5 Hz, which arise from H-2' and H-3'. The doublets from H-3' are frequently coupled, by long range coupling of J ~ 1Hz, to H-8 in psoralens and H-6 in angelicins [18].

In ^{13}C-NMR, the chemical shift of the carbonyl-carbon atom has been found to be approximately the same (about 160 ppm) for most coumarins [19,23]. The effect of hydroxyl and methoxyl on the benzoid ring is quite

Human (HNPl): ACYCRIPACIAGERRYGTCIYQGRLWAFCC

Rabbit (NPl):   VVCACRRALCLPRERRAGFCRIRGRIHPLCCRR

Rat (RtNPl):   TCYCRRTRCGFRERLSGACGYRGIYRLCCR

Mouse (MuCrl): LRDLVCYCRTRGCKRRERMNGTCRKGHLMYTLCCR

(6) Amino acid sequence of several mammalian defensins.

characteristic in that the signal from the newly formed quaternary carbon is found approximately 30 ppm downfield from the value observed in coumarin, while the carbons *ortho* and *para* to the substituent move upfield by about 13 and 8 ppm, respectively [24,25].

STRUCTURE OF NATURAL ANTIFUNGALS

Animal Sources

To date, structures and properties of about a hundred peptide antibiotics from various mammals (including man), birds, frogs, and many types of insects have been published. Peptide antibiotics exert a bacteriostatic effect on the normal flora of the skin and of the oral cavity and other orifices. It is also possible that peptide antibiotics protect us against many pathogenic organisms [26,27].

Insects compose a huge number of species. Nearly one million species are recorded, and estimates show that the number of individual insects living at a given time could be as high as 10^{18}. These high numbers indicate that insects must have very effective means to defend themselves against infections. The adaptation ability of insects to different ecological habitats, frequently rich in microflora like decomposing organic material, is indicative of fundamentally different immune system. Moreover, since insects have never been found to posses lymphocytes or immunoglobulins, and since their immunity lacks the high degree of specificity that is found in vertebrates, one can expect their immune system to be designed in an entirely different way. In fact, insects use a cell-free immunity, by defense substances, to defend against a wide verity of organisms that may attack them. Attacins, diptericins, cecropin are among the inducible peptides that insects use for protecting themselves against microbial invasion [28,29]. Several other defense-designed chemicals in insects like defensins (6) have been found as well (Table 1).

Table 1. **Structural Category of Natural Antifungals and Their Sources**

Chemical Group	Source	Example	Ref.
1. Animals			
Cyclic peptide	Marine sponge, *Discodermia* sp.	Discobahamins A and B	[55]
Cyclic peroxides	Plakortis sp.	Peroxyplakoric acid A1 methyl ester (8), and B1 methyl ester	[56]
Depsipeptide	Marine sponge, *Halichondria cylindrata*	Halicylindramides A-C	[57]
Epoxyfatty acid	Human	Leukotoxin (9)	[58]
Halogenated cyclic peptides	*Jaspis* sp. and *Theonella* sp.	Jaspamide (10), and theonellamide	[59]
Peptide	Human, insects	Defensins (6)	[36]
Peptide	Insect, *Drosophila* sp.	Drosomycin	[60]
Peptide	Frog, *Phyllomedusa sauvagii*	Dermaseptin I	[61]
Peptide	Toad, *Bombina variegata*	Bombinins	[62]
Peptide	Scorpion, *Androctonus australis*	Androctonin,	[63]
2. Plants			
Acetylenes	*Heteromorphia trifoliata, Artemisia capillaris, Vicia faba*	Capillin, wyeronic acid (11)	[45, 64]
Acid terpenes	*Chamaecaparis nootkatensis*	Chamic acid (12), chaminic acid (13)	[64]
Aliphatic Alcohols	*Camellia sinensis*	1-Octanol	[65]
Aliphatic aldehydes	*Ducrosia anethifolia*	Decanal and dodecanal	[66]
Aliphatic sulfur derivatives	*Allium cepa, Allium sativum*	Allicin, alliin, and diallyl trisulfide	[64,67]
Alkaloid	*Cleistopholis patens*	Sampangines	[68,69]
Amines	*Musa* sp. (banana)	Dopamine derivatives	[70]
Amino acids	Legumes	L-canavanine	[64]
Anthranilic acid	*Avena sativa*	Avenalumin I (15)	[71]
Benzodioxin-2-one	*Lillium maximowczii*	Yurenolide	[72]
Benzofuarn	*Cotoneaster lactea*	Cotonefuran	[73]
Benzopyrans	*Crotalaria madurensis*	Crotmadine and crotmarine	[74]
Benzoquinones	*Croton lacciferus*	2,6-Dimethyoxybenzoquinone	[75]

(Table 1). contd.....

Chemical Group	Source	Example	Ref.
Benzoxazinoids	Rye, wheat, corn (*Zea mays*)	2,4-Dihydroxy-7-methoxy-2*H*-1,4-benzoxazin-3(4*H*)-one	[76]
Benzoxazolones	Rice, wheat, and corn	2(3*H*)-benzoxazolone	[64]
Biphenyls	*Fischerella ambigua*	Ambigols A and B	[77]
Chromenes	*Hypericum revolutum*		[45]
Coumarins	*Polygala fruticosa*	Scopoletin, xanthyletin, psoralen, and luvangetin	[45,78]
Diterpenes	*Pseudolarix kaempferi*	Pseudolaric acid B	[79]
Fatty acids	*Oryza sativa* (rice)	Epoxy- and hydroxylinoleic and linolenic acids	[80,81,82]
Flavonoids	*Psiadia trinervia, Pyrus* sp.	Apigenin 7- and 4'-glucoside, Luteoline 7- and 4'-glucoside, and quercetin 3-glucoside	[45,78]
Halogenated monoterpenes	Red algae, *Portieria hornemannii*	6-Bromo-3-chloromethyl-1,7-dichloro-7-methylocta-2-gene	[83, 84]
Isoflavonoids	*Phaseolus vulgaris*	Phaseollinisoflavan	[78]
Ketone terpenes	*Humulus lupulus*	Humulone (16), lupulone	[64]
Lactone terpenes	*Sorbus aucuparia*	Papsorbic acid	[64]
Long chain alcohol	*Persea americana* (avocado)	2-Hydroxy-4-oxo-heneicosadiene derivative	[85]
Monoterpenes	Many sources	Isopulegol and carvone	[86]
Naphthoquinones	*Sesamum angolense*	Naphthoxirenes	[87]
Peptide	*Mirabilis jalapa*	Mj-AMP1-2	[88]
Phenanthrenes	*Dioscorea batatas* and Orchids	Batasin I (17), orchinol, hircinol acid	[89,64]
Phenolic glycosides	*Serindea warneckei*	Tannic acid	[90]
Phenols	*Pinus silvestris*	Stilbene	[64]
Phenylpropanoids	*Kigelia pinnata*	Phenylpropanoic acid	[91]
Polyphenols	*Cordia goetzei, Brackenridgea zanguebarica*	Cordigone, cordigol	[45]
Polysaccharides	*Zea mays* (corn)	Methanol-soluble polysaccharides	[92]
Protein	*Raphanus sativus*	Rs-AFP1-2	[93]
Quinones	*Juglans regia*	Juglone	[64]

(Table 1). contd.....

Chemical Group	Source	Example	Ref.
Quinonoid	*Sesamum angolense, Clerodendrum uncinatum*	Naphthoxinrene derivatives	[45]
Saponin glycosides	*Trillium grandiflorum*		[94]
Sesquiterpenes	*Nicotiana glutinosa*	Glutionosone	[64]
Steroid glycoalkaloids	*Solanum acaule*	Tomatin	[95]
Sulphur-containing amides	*Glycosmis cyanocarpa*	Sinharine (18)	[96,97]
Triterpenes	*Medicago sativa* (Alfalfa)	2-β-Hydroxy-3-b-*O*-(β-D-glucopyranosyl)-delta12-oleanene-23,28-dionic acid	[98]
Tropolones	*Thuja plicata*	Thujaplicins	[64]
Xanthones	*Hypericum brasiliense*	Hyperbrasilone	[40]
3. Fungi			
Azaphilones	*Aspergillus deflectus*	Deflectins	[99]
Fatty acids	*Cantharellus cibarius* (chanterelle) and *C. tubaeformis*	Cibaric acid and 10-hydroxy-8-decenoic acid	[100]
Guanidine	*Streptomyces riseochromogenes*	Blasticidin	[64]
Guanidine-type alkaloid	*Actinomadura* sp.	Sch 40873 (19)	[68]
Halogenated terpenoid	*Penicillium griseofulvum*	Griseofulvin	[101]
Isocoumarin	*Cladosporium cladosporioides*	Cladosporin (20)	[102]
Lipopeptides	*Zalerion arboricola, Aspergillus aculeatus,* and *A. sydowi*	Pneumocandins Aculeanic acid, mulundocandin	[103]
Nucleosides	*Streptomyces cacaoi*	Polyoxins	[64]
Peptide (macrocyclic)	*Aurebasidium pullulans*	Aureobasidin A-R	[68]
Polyenes	*Streptomyces nodosus*	Amphotericin B	[50]
Quinoid	*Nematoctonus robustus, Streptomyces exfoliatus*	Pleurotin, dihydropleurotinic acid, yurinelide	[104, 105]
Quinoid alkaloid	*Actinomadura hibisca*	Pradimicin A	[68]
Sugar derivatives	*Streptomyces kasugaensis, S. kagawaensis*	Kasugamycin (22), prumycin	[64]

(Table 1). contd.....

Chemical Group	Source	Example	Ref.
Terpenoids	*Leptosphaeria oraemaris*	Culmorin	[106]
Tricarboxylic acid	*Sporormiella intermedia, Leptodontium elatius*, and *Mollisia* sp.	Zaragozic acids	[51,52,107]
4. Bacteria			
Amino acid	*Bacillus cereus*	Cispentacin (23)	[108]
Cyanogens	*Pseudomonas* sp.	Hydrogen cyanide	[109]
Phenazines	*Pseudomonas* sp.	Phenazine-1-carboxylic acid	[109]
Phenols	*Pseudomonas* sp.	C-acetyl phloroglucinols	[109]
Phosphono-oligopeptides	*Bacillus subtilis*	Rhizocticin A	[110]
Pyroles	*Pseudomonas* sp.	Pyoluteorin (24), and pyrrolnitrin (25)	[109]
5. Cyanobacteria			
Cyclic peptides and depsipeptides	*Lyngbya majuscula*	Majusculamide C (26)	[111]

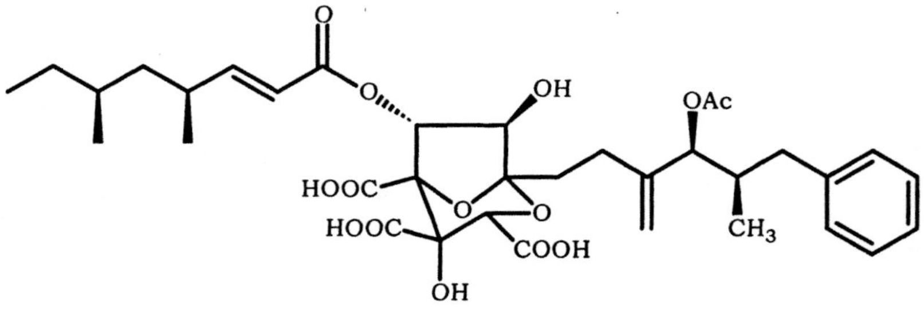

(7) Zaragozic acid A

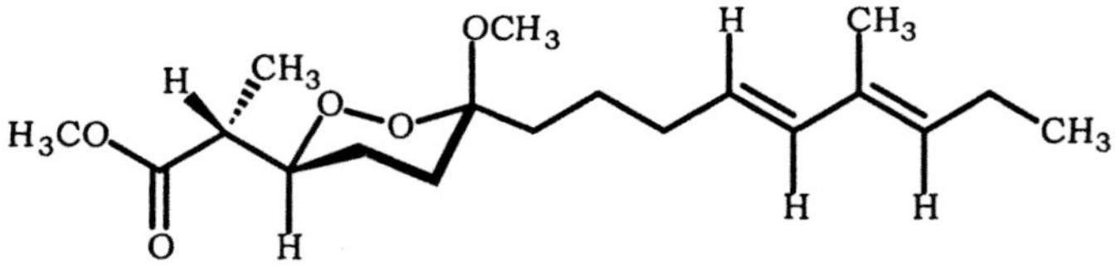

(8) Peroxyplakoric acid A 1 methyl ester

Human immune system also exhibits interesting aspects in this regard. The most abundant phagocytic cell circulating in human blood, the polymorphonucleated neutrophil (PMN) has potent oxidative and non-oxidative microbicidal mechanisms. The non-oxidative mechanisms are mediated by endogenous antimicrobial polypeptides located in PMN's cytoplasmic azurophil granules, whose contents are delivered to phagocytic vacuoles containing ingested microorganisms. These antimicrobial molecules include defensins [30].

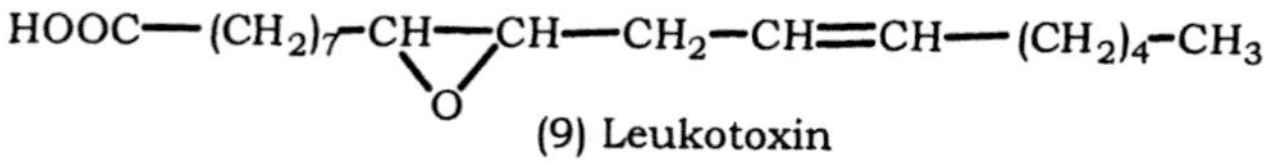

$$HOOC-(CH_2)_7-CH-CH-CH_2-CH=CH-(CH_2)_4-CH_3$$

(9) Leukotoxin

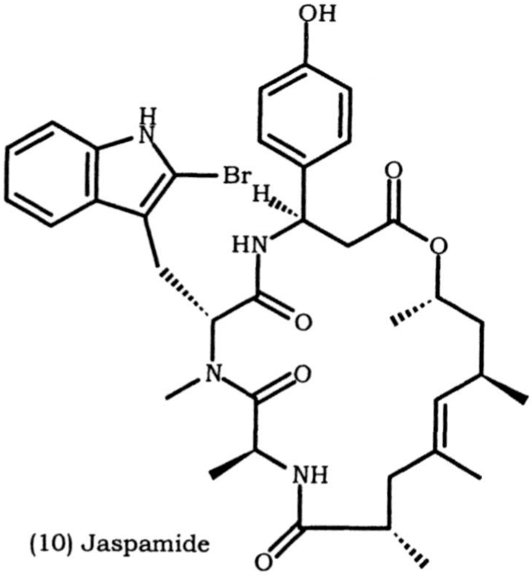

(10) Jaspamide

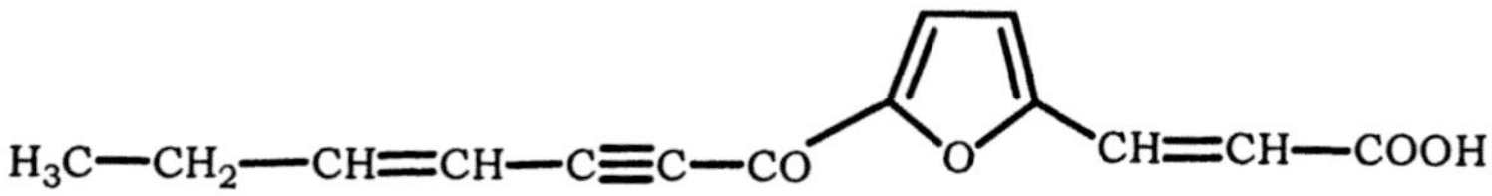

$$H_3C-CH_2-CH=CH-C\equiv C-CO \quad CH=CH-COOH$$

(11) Wyeronic acid

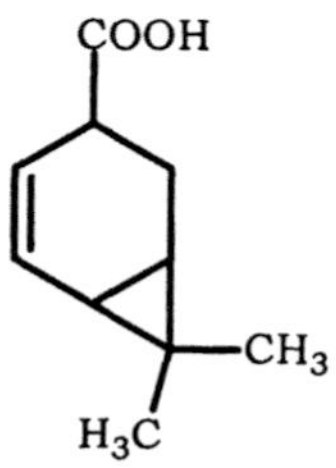

(12) Chamic acid

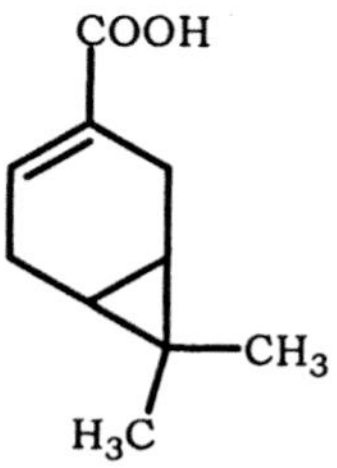

(13) Chaminic acid

Defensins are variably cationic, relatively arginine-rich, nonglycosylated peptides comprised of 29-34 amino acid residues. Human (HNP-1) and rabbit (NP-5) defensins have been studied by NMR techniques and showed that they are rich in beta sheet structure. Although both types of defensins permeabilize the outer membrane of *E. coli* [31], they differ from other peptides with similar action but rich in alpha helices such as magainin, mellitin and cecropin.

Some defensins are broadly antimicrobial, antiviral and cytotoxic, while others are chemotactic, or may modulate hormonal responses. The three-dimensional fold of defensins forms a highly amphiphilic molecule. Defensins are synthesized by phagocytes or their precursors as a 94-95 amino acid charge-neutralized preprodefensin, an arrangement that may avoid cytotoxic injury to the phagocyte. Although defensins were recognized only recently, the existence of homologues in certain invertebrates suggests that they are ancestral components of the host defense system [32,33].

Microbicidal and cytotoxic properties of defensins are most likely a consequence of their ability to insert into biological membranes and to generate voltage-sensitive channels [34,35]. Defensins exerted antimicrobial effects *in vitro* against many Gram-positive and Gram-negative bacteria, fungi, mycobacteria and some enveloped viruses, and were cytotoxic to a wide range of normal and malignant targets [36,37,38].

Plant Sources

There are many antifungal compounds with plant origin. These compounds may be constitutive or induced. From the former category, fatty acids and long chain alcohols, which also can be found in the human body, are reminded. In the latter group, coumarins can be named [39]. Some simple and interesting structures which exhibit antifungal activity and to some extent may resemble those of coumarins are indicated in Table 1 [39,40,41,42,43]. In addition to coumarins, still some natural products with very simple structure like sarisan (14) [44], or complex molecules like zeamatin or permeatin [126], which are peptides, exhibit fungicidal activity.

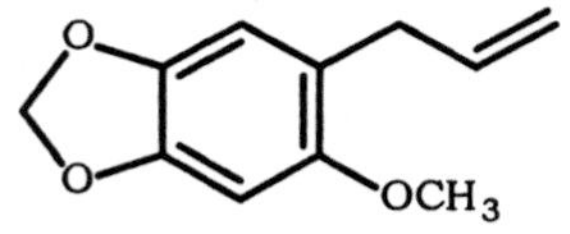

(14) Sarisan

Hostettmann *et al.*, [41,45,46] in a number of publications has referred to a variety of biologically active natural products. By the use of simple bioassays in conjunction with an approach involving bioactivity-guided fractionation, novel compounds with interesting properties have been isolated. Several hundred plant extracts have been submitted to different screenings, including bioautography on TLC plates with the fungi *Cladosporium cucumerinum* and *Candida albicans*. Using a variety of modern separation techniques, a number of bioactive compounds have been isolated and identified. These include prenylated xanthones, tetracyclic phenols and saponins [47]. A purified preparation of trypsin and chymotrypsin inhibitors from cabbage foliage *in vitro* inhibited germ tube elongation and germination of phytopathogens [48]. Essential oil ingredients such as anethole, cinnamaldehyde, cinnamic acid, eugenol, thymol, d-limonene inhibited mycotoxin production from a group of *Aspergillus* sp. Also oil of bitter orange, cinnamon, clove, ginger, lime, rosemary, nutmeg, orange, and pepper had the same effect [49]. Grayer and Harborne [39] have reviewed a few hundred other antifungal compounds from plants; antibiotics and antifungal agents from algae have also been listed by Glombitza and Guven [113,124].

Fungal Sources

Amphotericin B is the most important clinically and best defined chemically of the macrolide heptaene antibiotics. A fermentation product of the soil actinomycetes *Streptomyces nodosus*, amphotericin B binds selectively to ergosterol in the cell membrane of susceptible fungi, inducing changes in permeability that can produce lethal cell injury [50].

A number of chemically diverse agents including fatty acid derivatives have been found to possess antifungal activity [127]. The zaragozic acids (ZAs), a family of fungal metabolites containing 4,6,7-trihydroxy-2,8-dioxobicyclo[3.2.1]octane-3,4,5-tricarboxylic acid core, were discovered recently [51]. All these compounds contain the same core but differ in their 1-alkyl and 6-acyl side chains. Four novel zaragozic acids, F-10863A, B, C and D, were isolated from a culture broth of the fungus *Mollisia* sp. [52]. The ZAs are produced in an extensive taxonomic range of Ascomycotina or their anamorphic states. These compounds are very potent inhibitors of squalene synthase that inhibit cholesterol synthesis and lower plasma cholesterol levels in primates. They also inhibit fungal ergosterol synthesis and are potent fungicidal compounds. These natural products have the potential for development as cholesterol lowering agents and/or as antifungal agents [51,52].

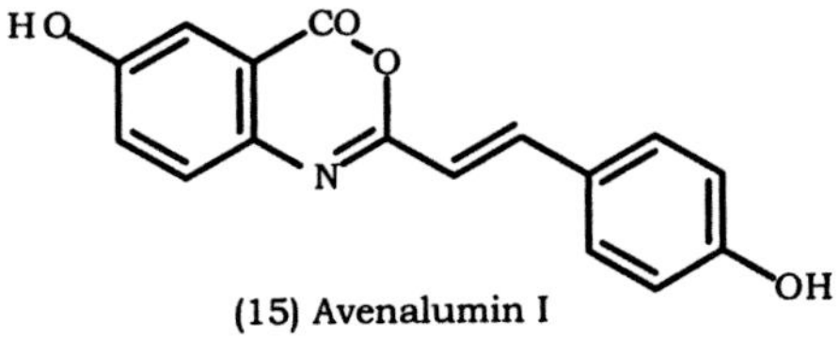

(15) Avenalumin I

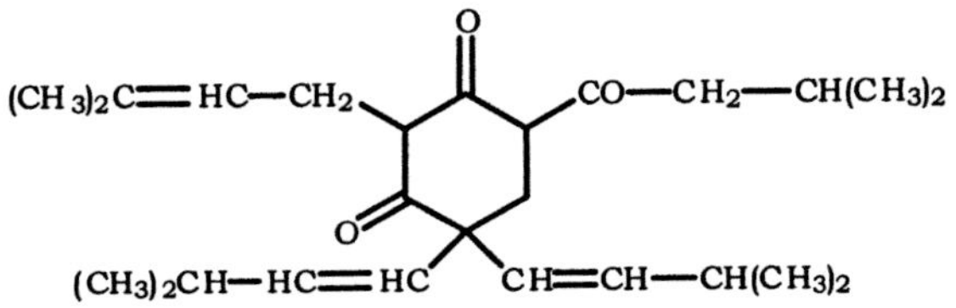

(16) Humulone

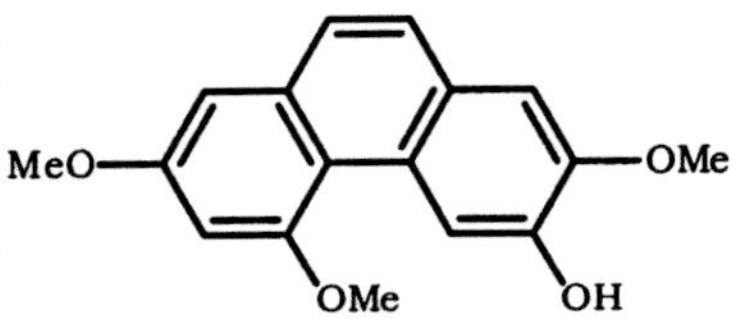

(17) Batasin I

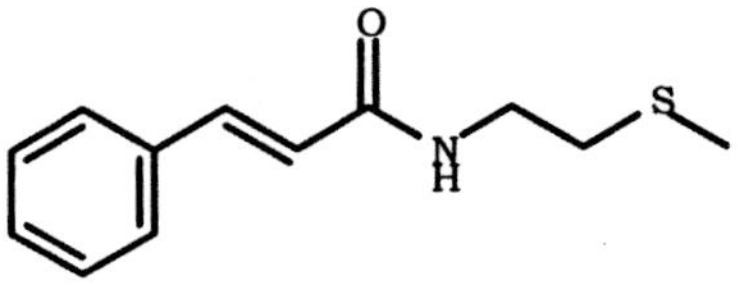

(18) Sinharine

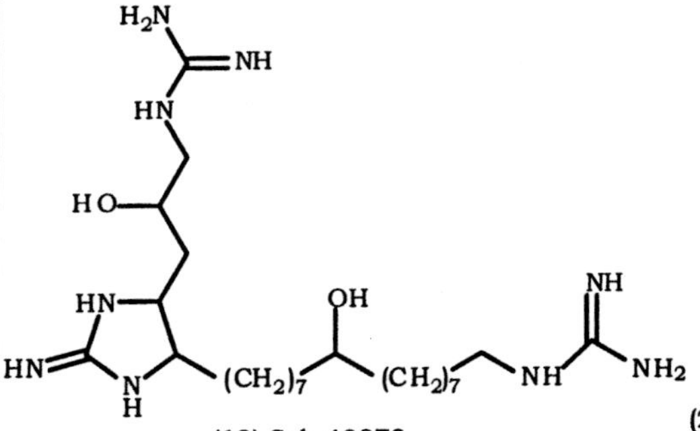

(19) Sch 40873

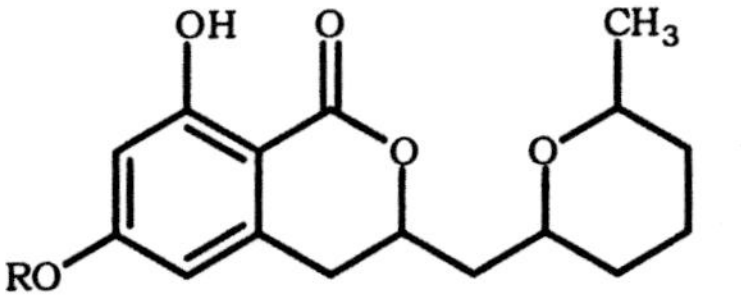

(20) Cladosporin, R=H
(21) Cladosporin methyl ester, R = COCH$_3$

Our group has evaluated the antifungal activity of myristic acid analogues. Synthesis and biological evaluation of about 30 different analogues led to identification of very potent compounds. Structure-activity relationship indicated a role for inhibition of the enzyme involved in N-myristoylation of proteins in fungal cells by these compounds. The selectivity of these compounds was tested by toxicity test on KB cell line, and the proposed mechanism of action was supported. This has brought up the possibility of their usage as antifungal compounds in agriculture [53].

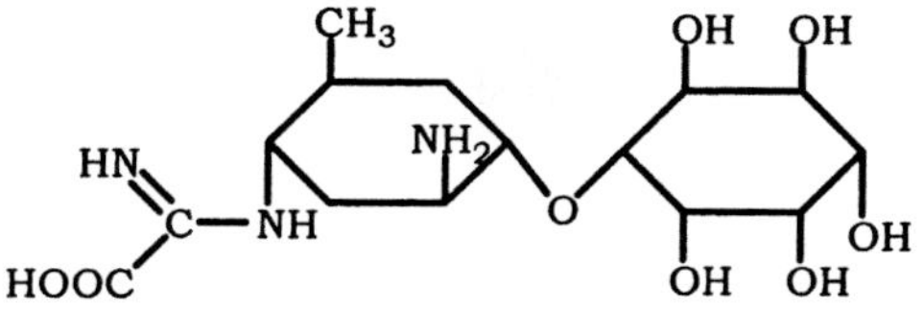

(22) Kasugamycin

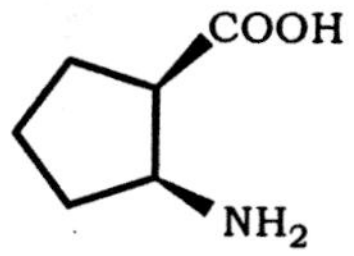

(23) Cispentacin

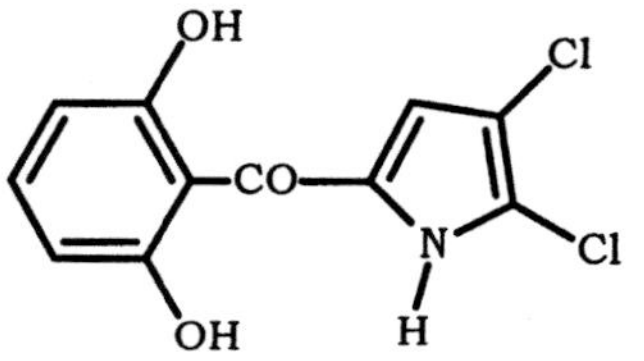

(24) Pyoluteorin

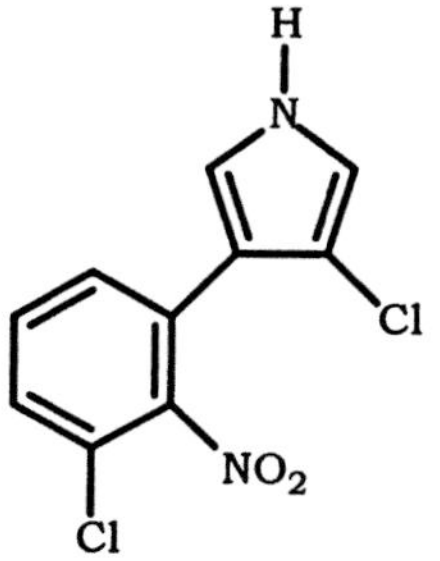

(25) Pyrrolnitrin

A number of synthetic iron (II) complex of vitamin C analogues (L-threo-hexo-2-enaro-1,4-lactone) showed antibacterial and antifungal actions against *Aspergillus niger, Geotrichum candidum, Aureobasidium pullulams, Penicillium steckii* and *Trichoderma* sp. [54].

Prokaryotic Sources

In addition to bacteria, which produce antifungal compounds [109,112], other prokaryotes were identified as rich sources of these agents, as well.

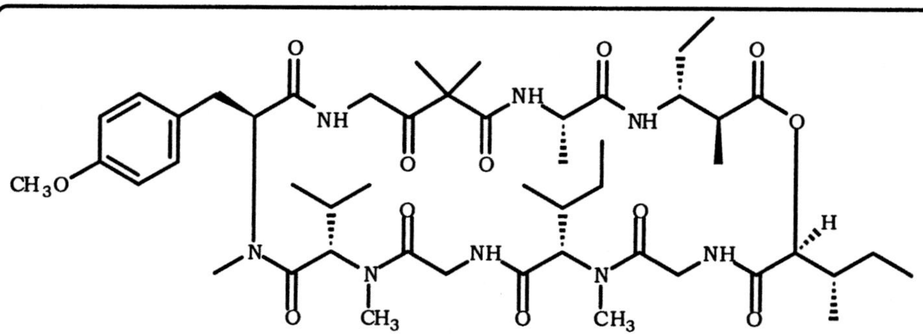

(26) Majusculamide C

An elaborate array of structurally novel and biologically active cyclic peptides and depsipeptides are found in blue-green algae (cyanobacteria). Most of these cyclic peptides and depsipeptides, such as the microcystins and the lyngbyatoxins, will probably be useful only as biochemical research tools. A few, however, have the potential for development into useful commercial products. For example, majusculamide C (26), a microfilament-depolymerizing agent from *Lyngbya majuscula*, shows potent fungicidal activity and may have use in the treatment of resistant fungal-induced diseases of domestic plants and agricultural crops [111].

COUMARINS IN NATURE: DISTRIBUTION AND SOURCES

Coumarins are almost unknown in the animal kingdom. On the other hand, they are widely distributed in the plant kingdom, especially in families Umbelliferae, Rutaceae, Leguminosae and Compositae [9,125]. Generally, the highest level of coumarins is occurred in the fruits, followed by roots, stems and leaves, respectively.

Umbelliferae plants, which are very rich in coumarins, compose a large family with 240-300 genera and over 3000 species. More than 410 species of Umbelliferae family belonging to 87 genera have been known to contain coumarins. These figures represent only about 14% of species and 36% of the genera in this family [8]. The occurrence of simple coumarins like aesculetin, scopoletin, scoparone (27) and umbelliferone (1) in this family have been reported to be less than 10% of the taxa except umbelliferone (1). Osthole (28) from *Angelica archangelica* [114] and imperatorin (29)

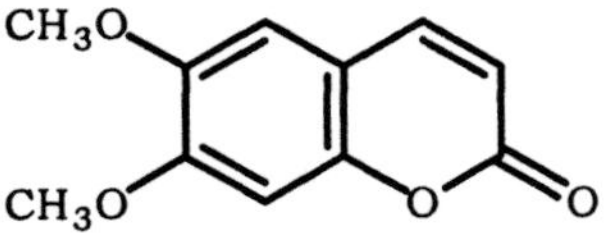

(27) Scoparone

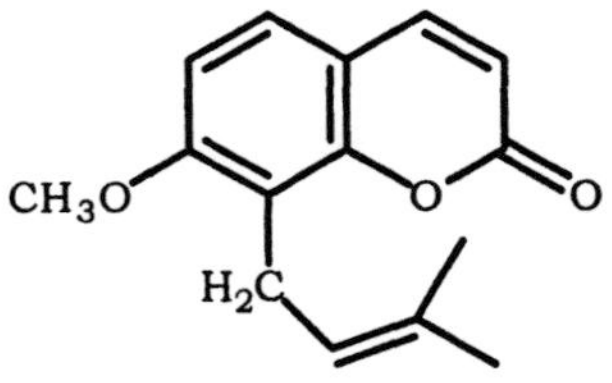

(28) Osthole

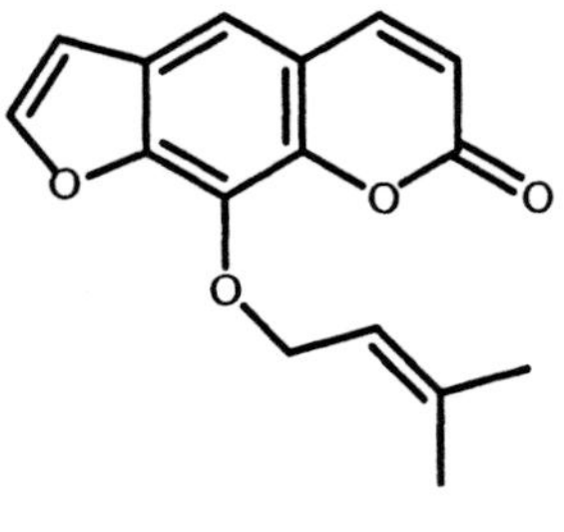

(29) Imperatorin

from *Angelica dahurica* [115,116] are known as simple coumarins possessing phenyl group.

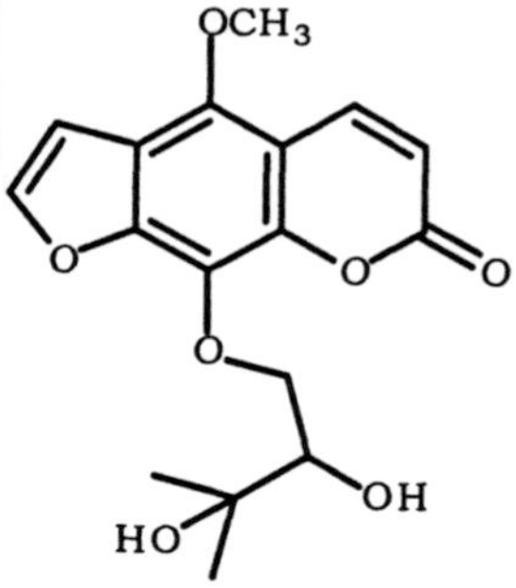

(30) Byakangelicin
(31) Nodakenetin 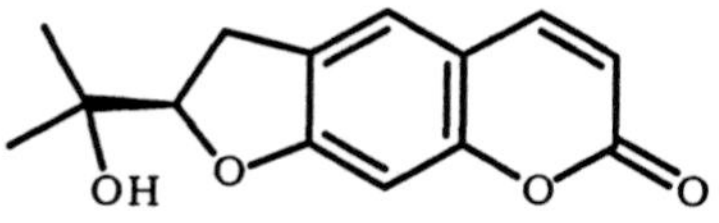
(32) Archangelicin 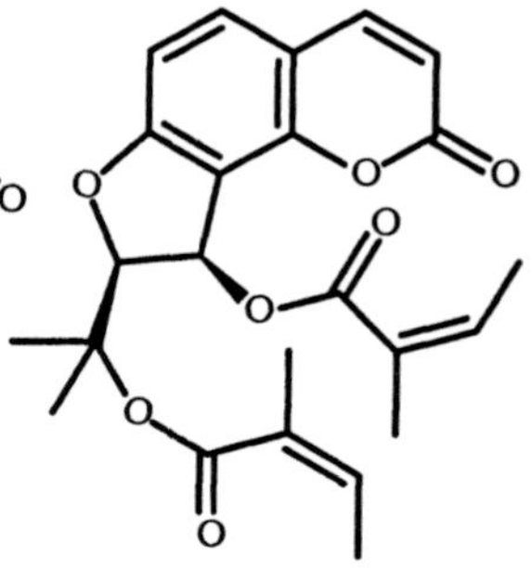

Furanocoumarins occur in Umbelliferae and are found only in the most evolved genera. The angular furanocoumarins are mainly present in genera *Heracleum*, *Pastinaca* and *Pimpinella* and are almost absent in genera *Angelica*, *Peucedanum*, *Prangos* and *Seseli*. On the other hand, the linear furanocoumarins are present in genera *Prangos*, *Angelica* and *Peucedanum*. One of the typical linear furanocoumarins is byakangelicin (30) from *Angelica* [115,116]. Generally, dihydrofuranocoumarins co-occur with furanocoumarins. Nodakenetin (31) (linear type) from *Angelica decursiva* [117] and archangelicin (32) (angular type) from *Angelica kieskei* [118] are representatives of dihydrofuranocoumarins.

The angular pyranocoumarin seselin (5), which was first isolated from *Seseli indicum* and later from *Foeniculum vulgare*, occurs also in *Apium graveolens*, *A. leptophyllum*, *Carum roxburghianum* and *Pimpinella heyeana*. The linear pyranocoumarins do not occur in Umbelliferae.

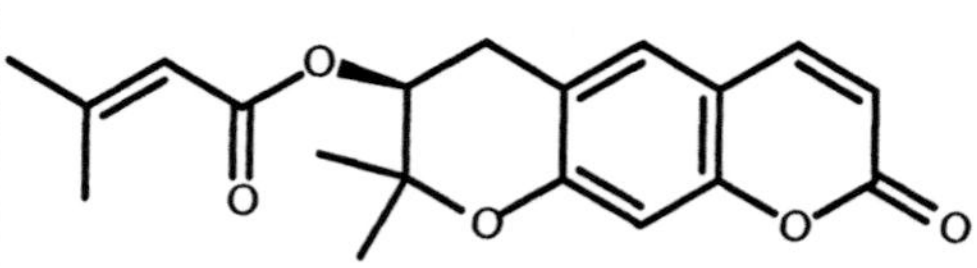

(33) Decursin
(34) Visnadin

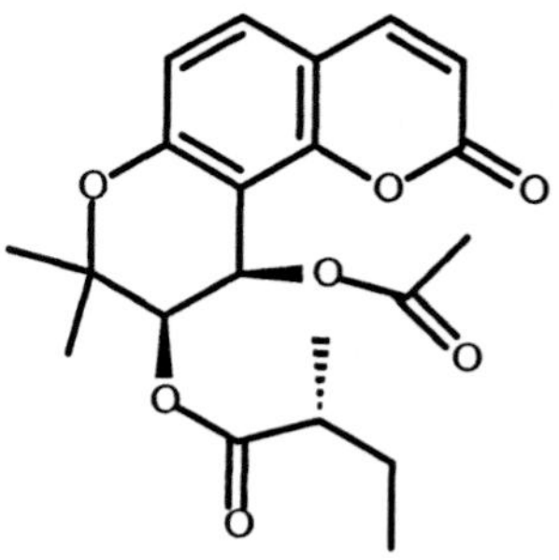

Dihydropyranocoumarins occur in *Angelica* as a complex mixture with dihydrofuranocoumarins. Some of the typical dihydropyranocoumarins are decursin (33) (linear type) from *Angelica decursiva* [117] and visnadin (34) (angular type) from *Ammi visnaga* [119,120].

Rutaceae is a family that consists of 150 genera and around 1600 species. Coumarins have been characterized in 178 rutaceous species from 50 genera, representing 11% of the family's total species and 33% of its genera [8]. These percentages are quite similar to those in Umbelliferae.

In Rutaceae, the commonest simple coumarins are umbelliferone (1), aurapten and scoparone (27). Generally, coumarins having free phenolic groups are relatively rare in Rutaceae. Bergapten occurs as the commonest furanocoumarin and xanthyletin (3) as the commonest pyranocoumarin. Isopimpinellin (35) and xanthotoxin (36) are known as the common methoxylated furanocoumarins. The high occurrence of the pyranocoumarin seselin (5) is observed in Rutaceae. Unusual 5,7-dioxygenated prenyl coumarins were isolated from *Toddalia aculeata* [121]. Bergaptol and limettin (37) are typical rutaceous coumarins in *Citrus* spp. Coumarins having 1,1-dimethylallyl group at C-3 position are very common in *Ruta* spp.

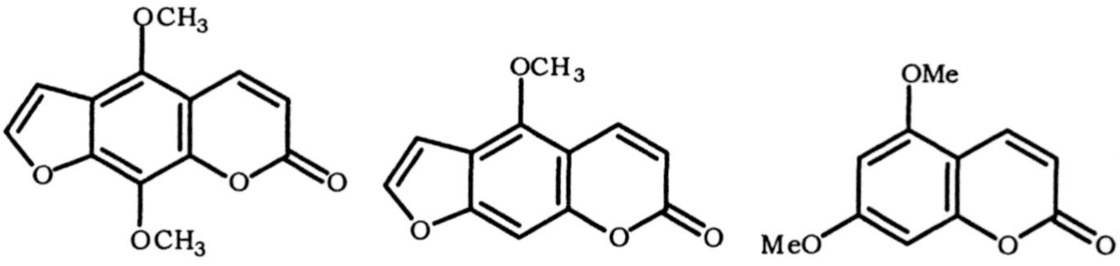

(35) Isopimpinellin   (36) Xanthotoxin   (37) Limettin

Leguminosae consists of over 600 genera and 12,000-17,000 species but the occurrence of coumarins has been known in only 155 species [8].

4-Aryl-6,7-dioxygenated coumarins are characteristic of Leguminosae. In addition, sisafolin (38), which is 5,7-dioxygenated 4-arylcoumarin, was isolated from Leguminosae.

Compositae is a very large family with over 1000 genera and around 15,000 species. Only 172 species have been known to contain coumarins.

In Guttiferae, 4-alkyl- and 4-aryl-5,7-dioxygenated coumarins are characteristic of this family.

The simple coumarins like aesculetin, fraxetin and isofraxinol also occur in families, Apocynaceae, Oleaceae and Araliaceae *etc.*

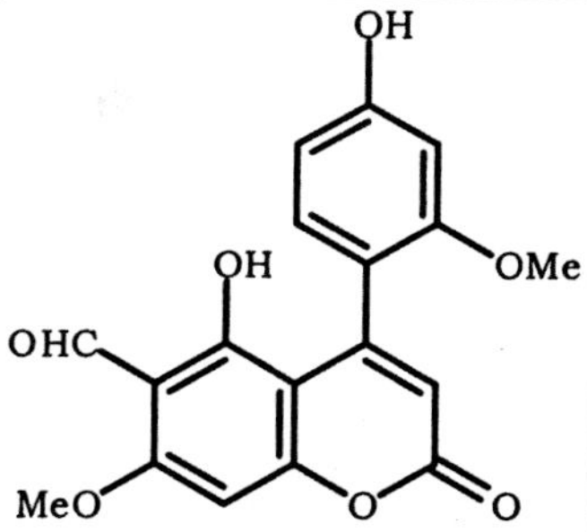

(38) Sisafolin

The food sources of furanocoumarins include figs (contains xanthotoxin), parsley [has bergapten (39)], and fruits of *Ammi majus* and *Ficus carica* (include imperatorin) [122]. They have also been reported in limes, mustard, carrots and celery [123].

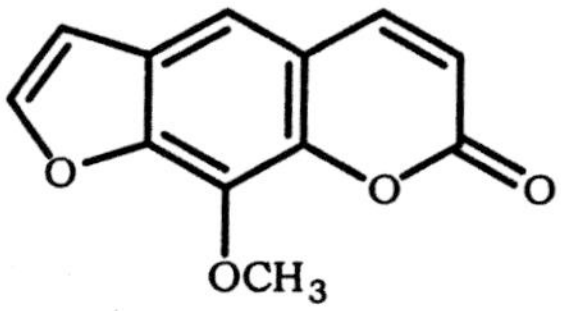

(39) Bergapten

BIOACTIVITIES OF COUMARINS

Coumarins exhibit many bioactivities including: anticoagulant, estrogenic, dermal photosensitizing activity, antimicrobial, vasodilator, molluscacidal, anthelmintic, sedative and hypnotic, analgesic and hypothermal activity [10, 128,129]. In fact, the diversity of bioactivities among coumarin compounds is so huge that the term "pharmacological promiscuity" has been applied on their case [12]. Notable among the physiological effects exerted by coumarins are the acute hepatotoxicity and carcinogenicity of certain aflatoxins, the anticoagulant action of dicoumarol (40), and the antibiotic activity of novobiocin and coumermycin A_1. The photosensitizing effect on the cells by certain linear furanocoumarins, which is of great biochemical interest, has been reported to be associated with cross-linking of DNA strands. These biological actions are described as the following:

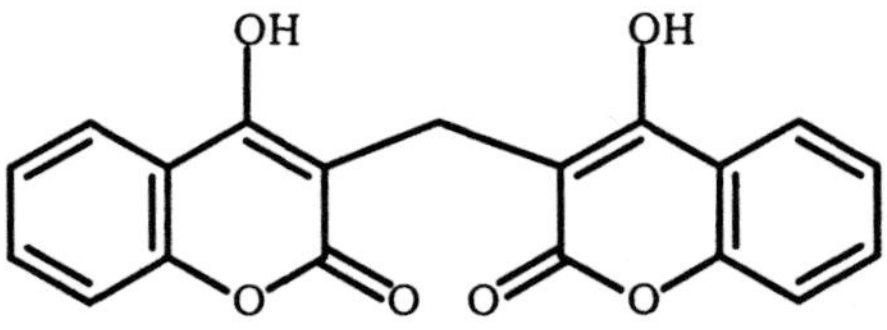

(40) Dicoumarol

Antifungal Activities

Herniarin (7-methoxycoumarin, or ayapanin) (41) has shown activity against *Aspergillus glaucus* and *A. flavus* in both diffuse light and in complete darkness [130]. Coumarin inhibits the germination of spores of *Aspergillus niger, Penicillium glaucum,* and *Rhizopus nigricans.* The effect of coumarin is fungistatic not fungicidal, and a physiological adaptation, which ends in tolerance to this compound, occurs [131]. Novobiocin and other 4-hydroxycoumarins are generally ineffective against fungi [130].

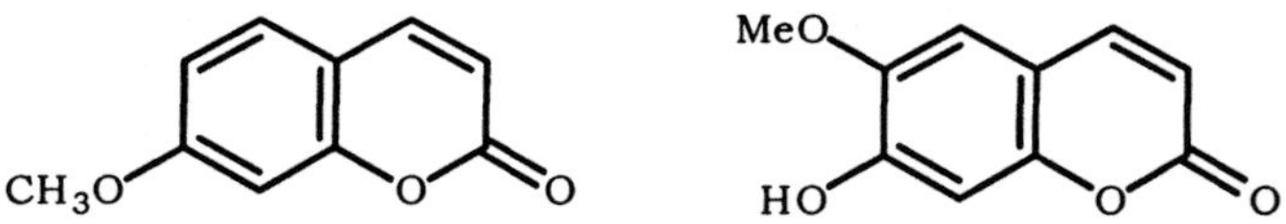

(41) Herniarin   (42) Scopoletin

Furanocoumarins also have photosensitizing effect on dermatophytes [132]. Xanthotoxin (36) showed activity in darkness against *Candida albicans* and *Cryptococcus laurentii* [133].

The fungus *Cladosporium cladosporioides* produces isocoumarin cladosporin (20). This compound and its monoacetyl derivative inhibit growth of a number of *Trichophyton* species in addition to *Microsporum canis* and *Epidermophyton floccosum* at a concentration of 75 µg/ml on the agar medium. *Aspergillus niger, A. versicolor* and *A. nidulans* spore germination was inhibited by cladosporin and its acetylated derivative at a concentration of 40 µg/ml [102].

The isocoumarin, oosponol (43), which is produced by Basidiomycete, *Gloeophyllum sepiarium,* possesses marked antifungal activity against *Candida albicans* and *Trichophyton asteroides* [134]. In addition, many examples of synthetic isocoumarins have been reported to possess antifungal activity [135,136,137].

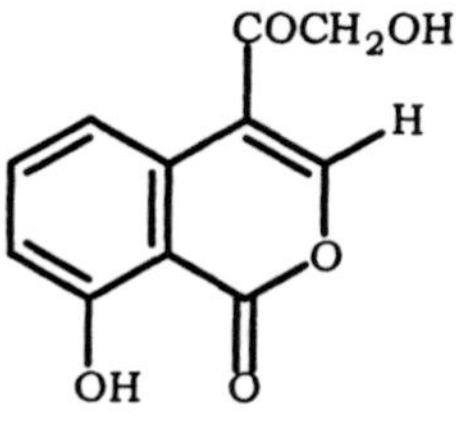

(43) Oosponol

The photosensitizing effect of 8-methoxypsoralen on the inactivation and mutation of *Aspergillus* conidia by near ultraviolet light has been studied [138]. Treatment of conidia with either 8-MOP, UV, visible light, 8-MOP plus visible light, or UV followed by 8-MOP show no inactivating or mutagenic effect, whereas treatment with UV in the presence of 8-MOP produces inactivation and mutation under similar conditions. The prior treatment of conidia with 8-MOP, before exposure to UV, produces a more rapid inactivating effect presumably because 8-MOP reaches a higher concentration in the conidia [138].

The effect of coumarins on fungal spore germination has been studied. Van sumere *et al.* [139] showed that coumarin is inhibitory to the germination of wheat stem rust. Knypl [140] concluded that coumarin at concentrations of 100-1000 ppm inhibited germination of spores in *Aspergillus niger*, and *Rhizopus nigricans*. Coumarin at lower concentrations of 5-10 ppm dissolved in yeast water considerably stimulated germination of *R. nigricans* spores [140]. In the case of soil molds, *Penicillium jenseni* and *P. nigricans*, presence of coumarin and sucrose in their growth medium causes vigorous growth [141].

Antifungal Activity and its Ecological Importance

There are numerous reports on the antifungal activity of compounds with plant origin. These compounds may be constitutive or induced. Coumarins are classified in the latter group [39]. This concept that coumarins are inducible antifungal compounds in plants has been supported by a number of observations. When sweet potato roots are attacked by fungi *Ceratocistis fimbriata* and *Fusarium oxysporum*, umbelliferone (1), scopoletin (42), aesculetin, and scopolin are produced in the plant tissue [142]. After invasion of *Phytophthora infestans* (organism causing potato blight) the affected tissues surrounding the lesion become fluorescent under UV mainly due to accumulation of coumarins [143,144]. When roots of celery (*Apium graveolens*) and parsnip (*Pastinaca sativa*) were inoculated with *Sclerotinia sclerotiorum*, a raise in the level of furanocoumarins was observed at the early stage of infection [145].

Coumarins have biologic effects in the plant cells. These compounds may well be considered as a defense tool for plants against pathogenic fungi [146]. They are phytoalexins and in a study that was done by Zobel, *et al.* [147], many coumarins were found in the extract resulted from washing the leaf surface of the plants that belong to five species of the Umbelliferae. In fact, the co-occurrence of the coumarins on the surface of leaves was related to their role in communication between the plant and its environment [146,147]. Presence of coumarins on the plant surface and the fact that their antifungal activity generally increases by the application of UV light, which is due to photodynamic activity of coumarins, can be correlated.

Other Activities

The search for useful pharmaceutical leads from plants, used in various cultures as medicine, led to a resurgence of interest in coumarins and related compounds. In general, since the purified substance display potent and desired pharmacological activities, while at the same time appearing to lack toxicity in mammalian systems, it is sensible to relate the biological activity observed in the crude extracts to specific pure compounds isolated from the same plant. In this context, coumarins can be classified as compounds with potential pharmacological profile. Although in this review, we focused on antifungal property of various coumarins, many of which are derived from natural sources especially plants, it is necessary to look at the potentials of this group of natural products from a wider angle. In Table 2 selected bioactivities of coumarin compounds have been listed.

Table 2.   Selected Bioativites of Coumarin Compounds

Bioactivity	Coumarin	Comment	Ref.
Adrenergic	Coumarin	Possibly due to inhibition of catechol-O-methyltransferase	[206]
Anthelmintic	Psoralen (2)	Isolated from fruits of *Psoralea corylifolia*	[207]
Anti-arrhythmic	Coumarin	It did not affect a- and b-receptors	[208]
Antibiotic	Streptonivicin	Produced by *Streptomyces nivens*	[209]
Antileukoderma	Methoxsalen (8-MOP)	Increased tyrosinase activity	[123]
Antioxidant	Fraxetin, esculetin, 4-methyldaphnetin	Inhibitor of microsomal lipid peroxidation	[202,210,211]
Choleretic	Isofraxidine, scopoletin (42), umbelliferone (1)	Isolated from *Artemisia abrotanum*	[212]

(Table 2). contd.....

Bioactivity	Coumarin	Comment	Ref.
DNA gyrase inhibition	Coumermycin A_1	Antibacterial	[157]
Eicosanoid inhibition	Umbelliferone (1)	Inhibitor of prostaglandin synthesis	[213]
Fish poison	5-methoxypsoralen (36)	Angelicin does not have this activity	[214]
Gonadotropin inhibition	Byakangelicin (30)	Interacted with sulfhydryl group in pregnant mare's serum gonadotropin, possibly through addition of protein –SH group to the 3,4-double bond of the coumarin	[215,216]
Histamine release inhibition	Coumarin		[206]
Hypolipidemic	Scoparone (27)		[217]
Immunostimulant	Coumarin	Stimulator of macrophages	[161,181]
Immunosuppressive	Scoparone (27)	Maybe caused by inhibition of tyrosine kinase	[218,219]
Inhibition of depolarization induced Ca^{2+} uptake	Columbianadin (46)	May act on voltage-dependent calcium channels	[220]
Lens aldose reductase (bovine-LAR) inhibition	Scoparone (27) and scopoletin (42)	Isolated from *Artemisia capillari* and tested on bovine enzyme	[221]
Mitodepressant	Coumarin, umbelliferone (1) and 4-hydroxycoumarin, psoralen and xanthotoxin (36)	All phases of mitosis were affected in the same proportions, except umbelliferone; plant cells were studied	[167]
Molluscacidal	Bergapten (39) and isopimpinellin (35)	Showed a potency comparable to pentachlorophenate	[78]
Narcosis induction	Coumarin	Due to effect on CNS and not heart or lungs; affected both cold- and warm-blooded animals	[222]
Parkinson's syndrome	Visnadin (34)	Clinical status and EEG both improved	[223]
Platelet aggregation inhibition	Osthole (28), cloricromene, scoparone (27) and scopoletin (42)	Perhaps prevented thromboxane generation and phosphoinositide turn over	[221,224,225]
Preventing dental caries	Coumarin	Used in dentifrice	[226]
Psoriasis (treatment)	Methoxsalen (8-MOP)	Inhibited DNA synthesis and cell division	[123]
Reduction of capillary fragility	Aesculin	Tested in pigs	[155]

(Table 2). contd.....

Bioactivity	Coumarin	Comment	Ref.
Reduction of swelling and edema in man	Coumarin	Clinically used in elephantiasis free of side effect	[227,228]
Spasmolytic and smooth muscle relaxant	Osthole (28), cloricromene, scoparone (27) and imperatorin (29)	Effective on trachea and aorta	[12,229, 230,231,23 2]
Suppression of seed germination in plants	Angelicin (4), psoralen (2) and coumarin	Inhibitor of root and sprout formation in potato	[155,146]
Treatment of HPLO (high protein lymphoedema)	Coumarin	Possibly by attaching to proteins and acting like a hapten made proteins antigenic and increased phagocytosis and proteolysis	[233,234]
Vasodilator	Scoparone (27), scopoletin (42)	Maybe mediated through radical scavenging effect thus potentiating c-GMP-mediated effects of internal nitric oxide	[217,235]

Anticoagulant

One of the most well known activities of coumarins is their anticoagulant property. This action was first observed through poisonous effects of contaminated hay in animals. Later in 1941 dicoumarol (40) was identified as the active compound in this regard. It was found that clover or the fungal contaminants by themselves are non-toxic to animals but their mixture is deadly. The reason was shown to be the conversion of *o*-coumaric acid to dicoumarol by fungal contaminants [7,148,149]. None of the 60 natural coumarins tested were pathologic as moldy hay. Coumarin can be oxidized to 4-hydroxycoumarin and coupled to formaldehyde and another coumarin, all by the contaminating *Aspergillus*, to form dicoumarol [150]. In 1948, Link by searching in the compounds synthesized by Ikawa, discovered a powerful coumarin analogue effective on coagulation [151,152]. This compound, 3-(α-acetonylbenzyl)-4-hydroxycoumarin, was named warfarin (44) as its patent right was given to the Wisconsin Alumni Research Foundation (from the first letters of the foundation plus suffix –arin from coumarin) [153]. The discovery of oral anticoagulants has not only provided physicians with potent antithrombotic drugs but also helped elucidate much basic science of factors involved in blood coagulation [150,154]. It is interesting that several coumarins show the opposite activity to warfarin. For example, while the 4-hydroxycoumarins have anticoagulant activity, herniarin (41) and ayapin (45) are shown to possess a hemostatic activity both *in vitro* and *in vivo* [155].

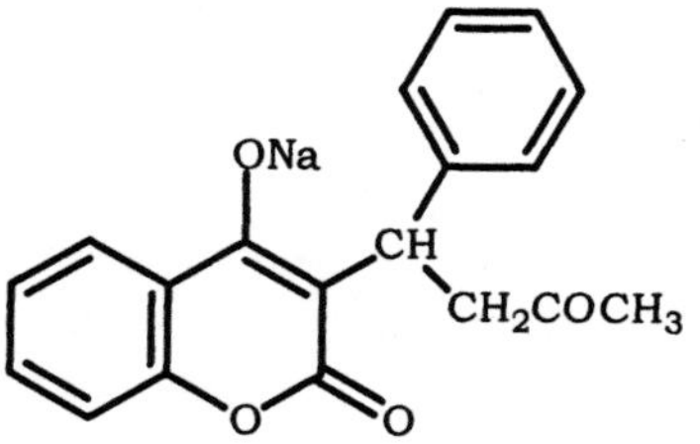

(44) Warfarin (sodium salt)

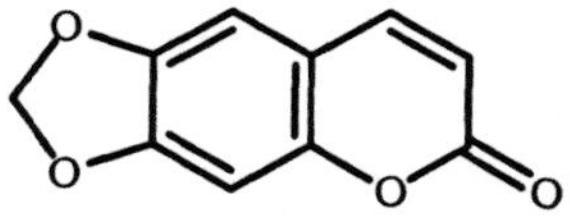

(45) Ayapin

Antibacterial and Antiparasitic

Coumermycin is one of the coumarins that have been used clinically as an antibiotic [156,157]. Dicoumarol (40) has also excellent activity against many bacteria including *Bacillus anthracis*, *Staphylococcus aureus*, *S. albus*, *Streptococcus pyogenes* and *Pasteurella avicida* [78,158]. Furanocoumarins show more photosensitized lethal effect in gram-positive bacteria than gram negatives [9]. Coumarin has been used in the treatment of brucellosis in humans [159,160]. This usage led to its application in other chronic infections like mononucleosis, mycoplasmosis, toxoplasmosis, Q fever, and psittacosis [161,162,163].

The coumarin group of antibiotics, including coumermycin, targets the bacterial enzyme DNA gyrase. The drugs bind to the B subunit of gyrase and inhibit DNA supercoiling by blocking the ATPase activity. The mode of inhibition of the gyrase ATPase reaction by coumarins is probably not simple competitive inhibition, and the drugs may act by stabilizing a conformation of the enzyme with low affinity for ATP [164,165].

Coumarin derivatives also act on type II topoisomerases, which are enzymes that break and re-legate DNA phosphodiester bonds while crossing over DNA strands and altering DNA topology. Therefore, they can interfere with the spatial organization of chromatin and are involved in several crucial biological functions, such as DNA replication and transcription, and chromosome segregation [166].

Antimitotic and Anticancer

Coumarin and its derivatives, including 4-hydroxy coumarins and furanocoumarins inhibit mitosis in *Allium cepa* L. bulb cells [167]. In human, a case of mycosis fungoides that was treated with topical PUVA therapy using 4,6,3'-TMA (trimethylangelicin) exhibited no side effects or changes in the normal values of the laboratory examinations [168].

In the treatment of cancer, coumarin has been used in a multicenter trial in the chemotherapy of melanoma [169,170]. In the untreated group the reoccurrence rate was 10 out of 14, while in coumarin-treated group was 2 out of 13 patients after 2 years.

Several other reports have demonstrated that coumarin in combination with cimetidine can produce objective antitumor responses in some patients with advanced melanoma [171,172]. There are reports that review the clinical development of coumarin, with or without cimetidine, with special reference to renal cell carcinoma (RCC) [173,174]. While objective tumor regressions were observed in renal carcinoma, no symptomatic or organ dysfunction toxicity was observed in any of the trials [175]. Some results suggest osthole as a selective antiproliferative agent in vascular smooth muscle cells. The antiproliferative effect occurs at the early G1 phase of the cell cycle and is due to the increase in cyclic AMP and cyclic GMP contents [176].

Coumarin has undergone a double-blinded, placebo-controlled trial for RCC and 7-hydroxycoumarin (1) is in phase I clinical trial for this purpose [177]. Recently, xanthotoxin has also been used in the treatment of cancer [178, 179]. Studies have been initiated in two multi-center trials of 1 g coumarin daily. Metastatic prostatic carcinoma patients of normal performance status were treated in a phase II trial [180].

While the exact mechanism(s) of action of coumarin is unknown, it has been shown that this drug possesses immunomodulatory activity *in vitro* and *in vivo* [181,182]. Both coumarin and its 7-hydroxyl derivative were found to be growth-inhibitory (cytostatic) for several human malignant cell lines. The growth inhibition was dependent on dose and time and was reversible upon removal of cells from medium containing the drug. Coumarin and 7-HC inhibited [3H]thymidine, [3H]uridine and [3H]leucine incorporation. In addition, coumarins stimulated apoptosis in HL-60 cells but not in other cell lines tested. Therefore, it can be concluded that coumarin and 7-hydroxycoumarin (1) have a direct antitumor (cytostatic) activity [183]. As a cancer chemotherapeutic agent, in general, it could be explained that coumarin acts on three levels. First, inhibition of tumor cells directly or via its metabolites. Second, it activates the immune system and prevents metastasis. Third, it acts as an antimutagenic agent [161].

The number of studies in the field of antimutagenesis is increasing. Some naturally occurring flavorings such as vanillin, cinnamaldehyde, and coumarin have been reported to inhibit mutagenesis induced by mutagens in bacterial and mammalian cells [184]. The chemical carcinogen, 7,12-dimethylbenz(α)anthracene (DMBA) has been administered before and after coumarin. No sign of tumor formation was observed in mice when given after coumarin treatment [185]. Coumarin is considered to act as antimutagen by modifying DNA replication and/or DNA repair systems after cellular DNA was damaged by mutagens [184,186].

Photodynamic Activity

Furanocoumarins are known to have photosensitizing activity in biological systems [187]. Skin photosensitizing ability among angular furanocoumarins is much less than linear ones like psoralen (2). In the psoralen group also this activity varies depending on the structure [122]. This relationship among several derivatives of psoralen has been shown in figure (1). As it is observed, psoralen (2) is the strongest skin photosensitizer in this group, while the hydroxyl derivatives and those with saturated 3,4- or 2',3'-double bond are almost inactive.

Photochemotherapy of psoriasis by the application of 8-methoxypsoralen (methoxsalen) and long wave UV in patients has ended in promising results [188,189,190]. Clinical application of these compounds has been successful in clearing psoriasis and lack of phototoxicity particularly for 4,6,3'-TMA is promising [191]. Methyl derivatives of angelicin have been tested clinically in psoriasis using their antiproliferative property. 4,6,3'-Trimethylangelicin showed a high antiproliferative effect and reduced genotoxicity in comparison with 8-methoxypsoralen (8-MOP) [191]. 8-MOP inhibits the DNA synthesis in the epidermal cells [190].

The mode of photosensitizing action of coumarins has been studied before [192]. An important mechanism in this action is coupling of the compounds to thymine of DNA [188]. Some plant cells especially the

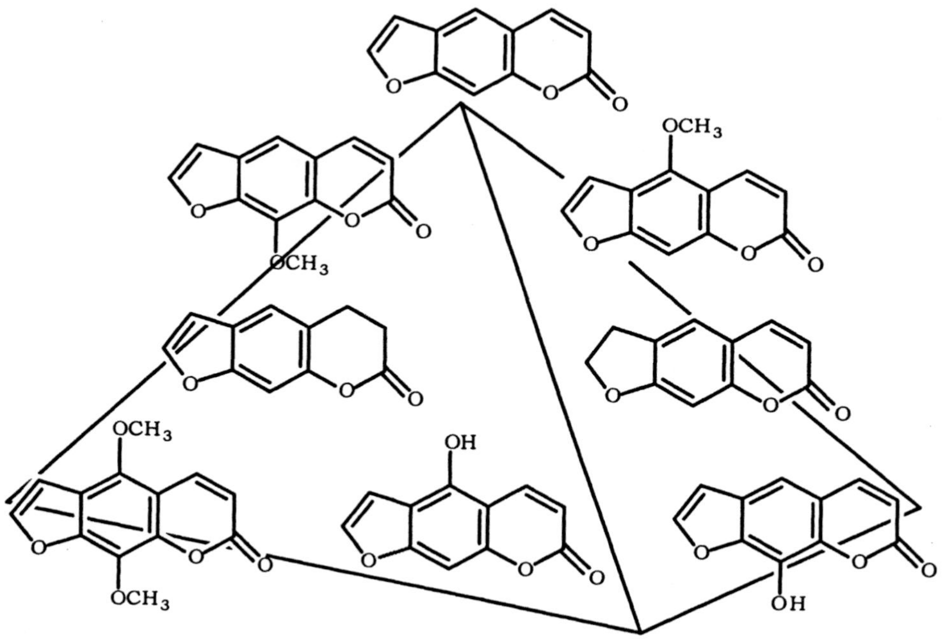

Fig. (1). The pyramid of relative skin photosensitizing activity among several psoralens. The furanocoumarins at the bottom are almost inactive.

dividing cells seem to be sensitive to furanocoumarin phototoxicity, while some others like *Heracleum* are unaffected [133]. Utilizing high intensity monochromator, the biologic action spectrum for 8-methoxypsoralen (8-MOP) was derived from threshold energy values required for erythema production in sensitized guinea pigs, and inhibition of growth of *C. albicans* sensitized with 8-MOP. Greatest efficiency in long ultraviolet was observed between 320 and 340 nm, and no activity was found at wavelengths longer than 380 nm. These findings correlate more closely with the known absorption spectrum of 8-MOP [193]. The capacity of some linear and angular furanocoumarins to induce DNA-protein cross-links by UVA (320-400 nm) irradiation has been evaluated in Chinese hamster ovary cells. Two linear furanocoumarins, psoralen (2) and 8-methoxypsoralen appeared to be capable of inducing DNA-protein cross-links to a noticeable extent [194].

3'-Methylangelicin and 4,3'-dimethylangelicin formed only reduced amounts of DNA-protein cross-links, while angelicin (4) and 4,6,3'-trimethylangelicin seemed to be unable to induce significant levels of this lesion. This might have an important role in furanocoumarin sensitization. The capacity for inducing DNA-protein cross-links appears to be a property of the skin phototoxic furanocoumarins. These results suggest the hypothesis of a connection between this damage and the formation of skin erythemas [194]. Other reports have noticed that skin and soft tissue necrosis is a rare complication of anticoagulation therapy with coumarins [195,196,197].

Edema Relief

Coumarins reduce all high-protein edemas, including lymphedema and elephantiasis. Coumarin itself has an established efficacy in slow-onset long-term reduction of lymphedema in man, as was confirmed in recent double-blind trials against elephantiasis and postmastectomy swelling of the arm [12]. Coumarin and oxerutins have also been used in many clinical trials on a variety of high-protein edemas. The mechanism of action is not quite clear, but may involve macrophage-induced proteolysis of edema protein. It has been suggested that they increase the numbers of macrophages and their normal proteolytic capability, which results in the removal of the excess protein, and the edema. They can also relieve the stimulus that the excess protein provides for chronic inflammation and fibrosis, and its action as a culture medium for bacteria [12,198].

Anti-inflammatory and Radical Scavenging Activities

The effect of coumarins on inflammation and reactive oxygen has been related to their action on eicosanoids. We have already reported the radical scavenging activity of several coumarins and phenolic compounds [199]. 5-Lipoxygenase and cyclo-oxygenase control stereospecific free radical peroxidation of arachidonic acid at their active sites. Therefore, coumarins and phenolic compounds that scavenge peroxyradicals might be expected to interfere with this process. Our finding in regard to the inhibition of the proinflammatory enzyme, 5-lipoxygenase, at micromolar levels by coumarins is a clear indication of this phenomenon. In addition to fraxetin and fraxin, other coumarins have been reported to inhibit lipoxygenase and cyclo-oxygenase pathways [200].

Although only a few coumarins are potent inhibitors of eicosanoid pathways, they show significant selectivity in this regard [201]. For example, dihydroxylated coumarins selectively inhibit 5-lipoxygenase, whereas only 5,7-dihydroxy-4-methylcoumarin shows high potency against cyclooxygenase. Although it is a potent inhibitor of cyclo-oxygenase, it is not pro-oxidant [12]. Both 5,7- and 6,7-dihydroxy-4-methylcoumarin have been reported to decrease the duration of ventricular fibrillation in which oxygen-derived free radicals are implicated. This shows that these antioxidant coumarins possess beneficial properties in this pathophysiological condition [12]. The coumarins with ortho-dihydroxy functions, like fraxetin and 4-methyldaphnetin, are potent inhibitors of lipid peroxidation and scavengers of superoxide anion radicals and of aqueous alkylperoxyl radicals. However, they may be pro-oxidant (enhancing generation of hydroxyl radicals) in the presence of free iron ions. For scavenging of reactive oxygen species, they act through their iron-chelating and iron ion-reducing abilities [202]. These properties of coumarins could be used for the design of potential non-toxic anti-inflammatory agents for treating disease in which eicosanoid generation and the production of reactive oxygen species are involved [203].

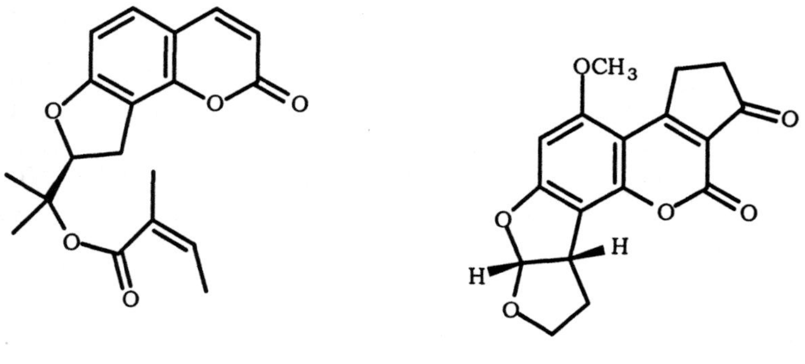

(46) Columbianadin

(47) Aflatoxin B_2

Lipolytic Activity

Scoparone (6,7-dimethoxycoumarin) has been purified from the hypolipidemic Chinese herb *Artemisia scoparia* and shown to reduce total cholesterol and triglycerides, and to retard the characteristic pathomorphological changes in hypercholesterolemic diabetic rabbits [12]. 5-Methoxypsoralen also can release and reduce cholesterol in atheromatous plaques. These activities have been attributed to various properties of coumarins including ability to scavenge reactive oxygen species, inhibition of tyrosine kinases, and potentiation of prostaglandin generation. The change in autoxidation of cholesterol can lead to an increase in its polarity and thus solubility, thereby facilitating the uptake by lipoproteins [12,204,205].

Antiplatelet Activity

GU-7, a 3-arylcoumarin derivative, has been isolated from *Glycyrrhizae radix*, which is a crude herbal medicine. GU-7 caused inhibition of platelet aggregation, phosphorylation of 40K and 20K dalton proteins, inositol 1,4,5-trisphosphate production, intraplatelet calcium increase and phosphodiesterase activity *in vitro*. The data indicate that GU-7 inhibits platelet aggregation by increasing intraplatelet cAMP concentration. Anti-platelet action may also explain the mechanism by which traditional medicines are effective in diabetic neuropathy [236]. Osthole causes hypotension *in vivo*, and inhibits platelet aggregation and smooth muscle contraction *in vitro*. It may interfere with calcium influx and cyclic nucleotide phosphodiesterases [12]. Cloricromene, a synthetic coumarin derivative, also possesses antithrombotic-antiplatelet activity [237]. Some of these properties of cloricromene have been attributed to the inhibition of arachidonate release from membrane phospholipids [12].

The *in vitro* effect of AD6 (8-monochloro-3-β-diethylamino-ethyl-4-methyl-7- ethoxycarbonylmethoxycoumarin) is reported on washed human platelet aggregation. Acetylsalicylic acid was ineffective on the same parameters, while the Ca^{2+} antagonist diltiazem and the cyclo-oxygenase-lipoxygenase inhibitor nordihydroguaiaretic acid inhibited aggregation. The results indicate that AD6 is an inhibitor of platelet aggregation *in vitro* with a mode of action different from cyclo-oxygenase blockade [238].

Anti-allergic Activity

Picumast dihydrochloride (PDH), (3,4-dimethyl-7-[4-chlorobenzyl)piperazine-1-yl]propoxycoumarin dihydrochloride) is a prophylactically active anti-allergic compound, which combines inhibition of mediator release and action. The activity profile of PDH differs clearly from that of known prophylactic anti-allergic drugs such as DSCG and ketotifen. PDH has H1-antagonistic property. This effect plus other inhibitory actions of PDH may be the cause of the suppression of immediate and late phase allergic reactions in animals and also allergic rhinitis and bronchial responsiveness in asthmatic patients [239,240].

Toxicity

Although many coumarins have been safely used in clinics, there are a number of reports about their unwanted side effects [241,242]. Coumarin mycotoxins in food products can pose health hazards in animals and human. At high levels in feed, these mycotoxins may cause animal toxicoses, such as aflatoxicosis. At lower levels in feed these mycotoxins may have no apparent effect on livestock production, but their residues and related substances may move up the food chain [243]. Investigations from several European countries indicate that, in former days, there was a connection between mycotoxins, mortality and reduced birth rate. In addition, many mycotoxins can interfere with the immune system, even in small concentrations, and increase susceptibility to tuberculosis and infant mortality [244].

Metabolism and hepatotoxicity of coumarins have been reviewed previously [245,246,247]. The differences in coumarin hepatotoxicity among mammals appear to be metabolism-related. In rat, coumarin is highly hepatotoxic and is primarily metabolized to 3-hydroxyl derivative and the product of the heterocyclic ring cleavage. Coumarin is less toxic in the baboon, gerbil and certain strains of mice, which resemble man in their formation of the 7-hydroxy metabolite. Liver toxicity, in patients receiving relatively high daily doses of coumarin is very rare [12]. In a study on human subjects, only 0.37% of 2173 patients with cancer or chronic infections that were treated with coumarin developed elevated liver enzyme levels during therapy with coumarin. It has been proposed that this hepatitis was probably a form of idiosyncratic hepatotoxicity and may have been of immune origin [248].

Metabolism

It has been shown that in insects' cytochrome P_{450} monooxygenases are involved in metabolism of host plant allelochemicals by *Lepidoptera* larvae. Expression of the CYP6B1 alleles in lepidopteran cell lines has demonstrated that both P_{450} isoforms metabolize substantial amounts of linear furanocoumarins, such as xanthotoxin and bergapten, but not angular furanocoumarins, such as angelicin (4) and sphondin [249].

Coumarin is 7-hydroxylated by the P_{450} isoform CYP2A-5 in mice and CYP2A6 in humans. The effect of 90 chemicals including coumarins, was evaluated as possible inhibitors of coumarin 7-hydroxylase (COH) activity in mouse microsomes [250]. The furanocoumarin derivatives methoxsalen and psoralen (2) proved to be the most potent inhibitors of mouse COH activity. Although methoxsalen was a potent inhibitor of COH activity also in human liver microsomes, bergapten, isopimpinellin and imperatorin had no effect [250]. These results not only suggest that the active sites of CYP2A-5 and CYP2A6 are structurally different, but also drew a more complex picture of the coumarins' effect on their own metabolism.

8-MOP induces mixed-function oxidase in mouse liver. This compound clearly induces *p*-nitroanisol-*O*-demethylase, but psoralen (2) and 4,2',8-trimethylpsoralen failed to induce the enzyme [251,252]. Since coumarins have phosphodiesterase inhibitory activity, it is expected that the intracellular level of cAMP increases after their administration. This situation might induce COH activity and CYP2A5 mRNA levels [253].

STRUCTURE-ANTIFUNGAL ACTIVITY RELATIONSHIP

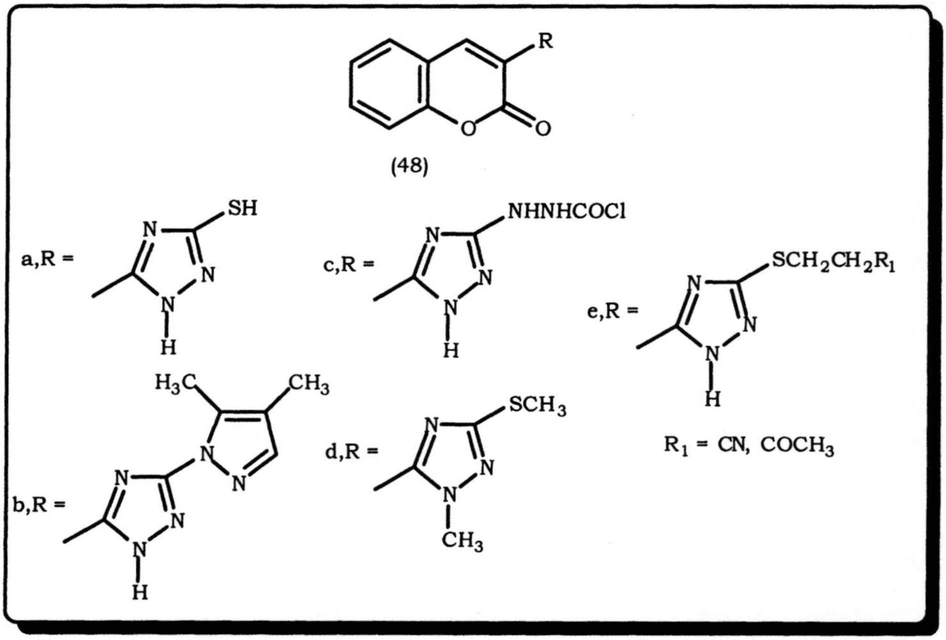

Coumarins are among the well-studied classes of natural products, which have been documented as compounds with diverse biological activity profiles. The nature of the expected biological activities may vary depending on the structural variances, which may contribute to the difference in binding characteristics, and receptor affinity. Cell biological and biochemical studies on the coumarin targets have resulted in the in-depth knowledge of cellular behavior toward this interesting nucleus. Synthetic modification on the parent compounds, on the basis of the above studies, have introduced novel molecules with improved activities, and biological potential. In this section we will try to introduce the available and current information on the coumarin derivatives with potential antifungal activity.

Among the non-fused coumarin derivatives those with the substitution on C-3 had shown superior *in vitro* antifungal activity to compare with the unsubstituted congeners. For example, 3-(1,2,4-triazol-5-yl)coumarins (48_{a-e}) have been reported to exhibit more than 75% activity in concentrations lower than 250 ppm on *A. niger*, *P. digitatum*, and *T. viride* [254].

In the same context, 3-(2-benzimidazolyl) coumarins ($49_{a,b}$) and (50) have shown *in vitro* activity against *Aspergillus* sp., *Fusarium*, and *Trichoderma*, respectively. Among the above compounds, (49_a) exhibited activity against *Fusarium* and *Trichoderma*, while (49_b) was only active against *Fusarium*. Compound (50) was showing superior activity against wide range of fungi [255].

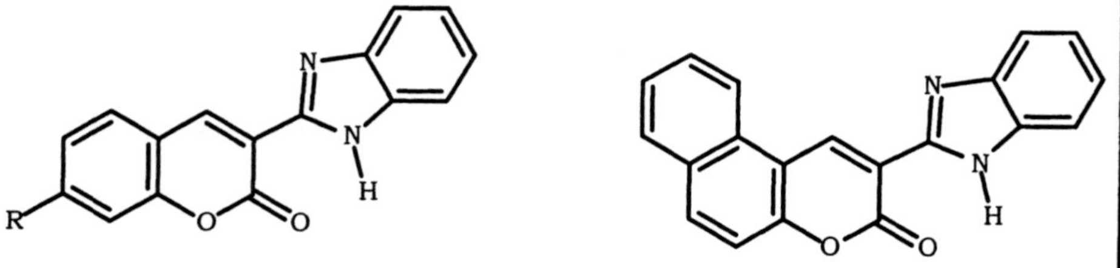

(49), a, R = OH
    b, R = H

(50)

Hydroxylation of the aromatic ring in coumarins has substantial effect on their antifungal activities. For example, *in vitro* activity of 3-phenyl

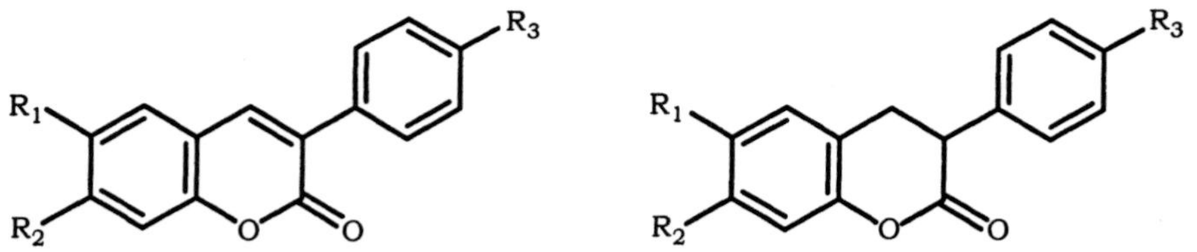

(51)

a, $R_1 = R_2 = OH$; $R_3 = H$
b, $R_1 = R_2 = R_3 = OH$
c, $R_1 = R_2 = OH$; $R_3 = OCH_3$

(52)

substituted coumarin (51_{a-c}) and its dihydro analogue (52) against seed contaminating molds, such as *F. oxysporum* and *T. harzionum*, appear to be dependent upon the number of hydroxyl groups substituted on the aromatic ring. Namely, compound (51_a) shows significant activity against both molds, while (51_b) is less effective. Methoxylation of the para position of the phenyl (51_c) has a minor effect on the activity. The dihydro analogues (52_{a-c}) do not show antifungal activity against the above fungi [256].

The same effect is reported by Verma *et al.* [257]. In this report the antifungal activity of 2,3-dihydrocyclopenta[*c*][*1*]benzopyran-4(H)-ones (53_{a-g}) against *D. oryzae* (rice foliage), *P. aphanidermatum* (tomato seedling), *F. solani* (cowpea seedling), and *M. phaseolina* (cowpea seedling) have been evaluated. The *in vitro* results indicate that all compounds exhibit strong activity against above fungi, with compound (53_e) being the most active. In this series, substitution of a hydroxyl group at C-7 position is essential for the activity, and an alkyl group at C-8 generally potentiates the activity remarkably.

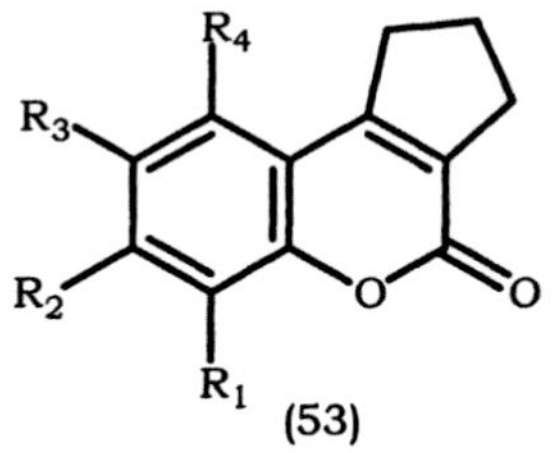

	R_1	R_2	R_3	R_4
a,	H	H	OH	H
b,	H	OH	H	H
c,	OH	OH	H	H
d,	H	OH	H	OH
e,	H	OH	Et	H
f,	H	OH	n-Pr	H
g,	H	OH	n-Pt	H

Substitution of a carbonyl containing functional group generally decreases the antifungal activity in coumarins. For example carbamodithioic acid esters of 3-acetylcoumarins (54) do not show activity against *Trichophyton* sp. and *Microsporum* sp. [259].

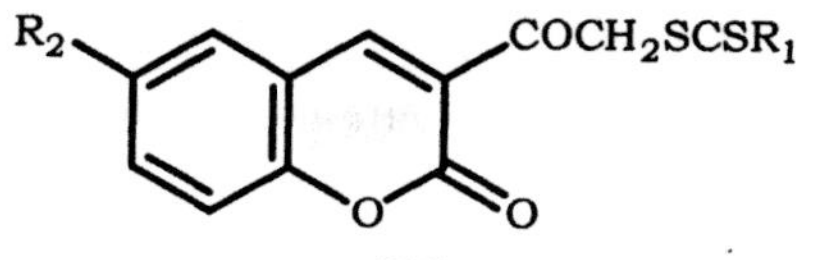

(54)

R_2 = H, Br
R_1 = Sec. -Amine, Pyrrolidinyl, Piperazinyl, N-Morpholinyl, etc

The same effect has been reported for the coumarins hydroxylated at C- 7 and/or C- 8 positions [260]. Namely, the 3-carboxylic acids and ester derivatives ($55_{b,c}$) were less active than their 3-unsubstituted congeners umbelliferone and daphnetin ($55_{a,d}$) against phytopathogenic fungi. Acetylation of the hydroxyl groups generally increased the activity, probably due to the increase in lipophilic characteristics of these compounds.

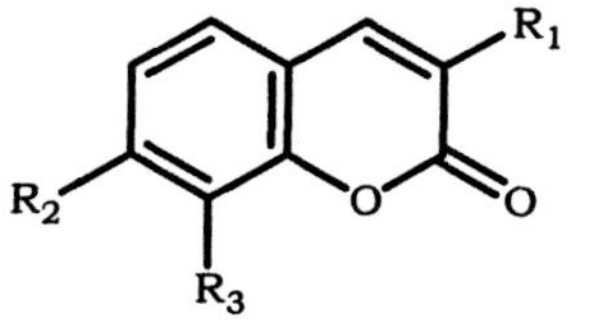

(55)

a, R_1, R_3 = H; R_2 = OH (Umbeliferone)
b, R_1 = CO_2Et; R_2 = OH; R_3 = H
c, R_1, R_2 = CO_2H; R_3 = H
d, R_1 = H; R_2, R_3 = OH (Daphnetin)

The same trend has been reported for polymethylene coumarins [258]. While the 3-alkylcoumarins (56_{a-d}) and their acylamino analogue (57_a) were showing significant activity against *A. niger*, compounds (57_b and 58) were inactive against the same fungus.

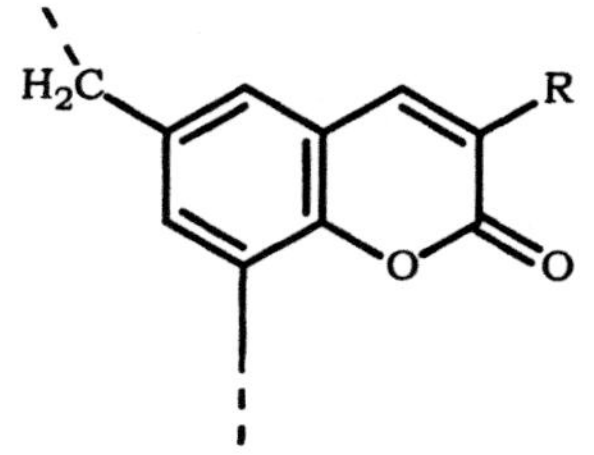

(56); a, R = Me   (57); a, R = Me   (58); R = Acetyl
      b, R = Et   b, R = Ph
      c, R = Allyl

Considering the bio-isosterism phenomena, and the lack of activity in 3-carbonyl containing coumarins, one can predict the lack of antifungal activity in the 3-sulfonyl substituted congeners. Nofal *et al.* [261] have reported an opposite observation in this regard. The 3-sulfonimine derivatives of 4-hydroxycoumarins (59_{a-c}) have been reported to exhibit strong to moderate *in vitro* activity against *C. albicans* and *A. niger*. The same range of activity has been observed for 3-amino, and 3-mercapto analogues (60_{a-f}), respectively.

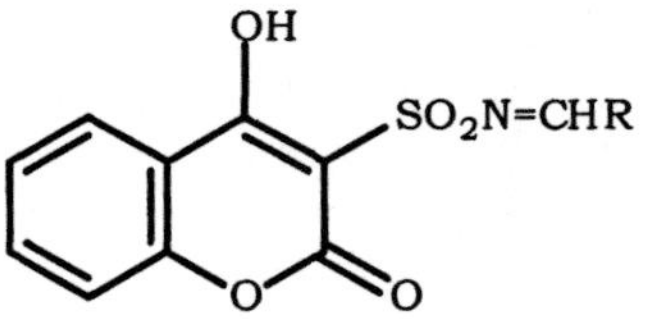

(59); a, R = *p*-Methoxyphenyl
      b, R = 2-Thienyl
      c, R = *p*-Nitrophenyl

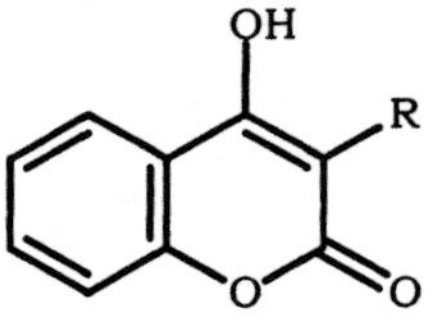

(60); a, R = ——NHCH(CH₂)₃N(Et)₂
                      |
                      CH₃
      b, R = Piperidinyl
      c, R = Morpholinyl
      d, R = 4,4,6-Trimethylpyrimidine-2-thiol
      e, R = 2-Thiazolylamino
      f, R = Mercaptophenyl

In the same context, the hydrazone derivatives of 3-acetylcoumarine (61_{a-c}) have been reported to show activity against *C. albicans* at the same level as Griseofulvin, while compound (61_d) exhibits higher activity than the standard [262].

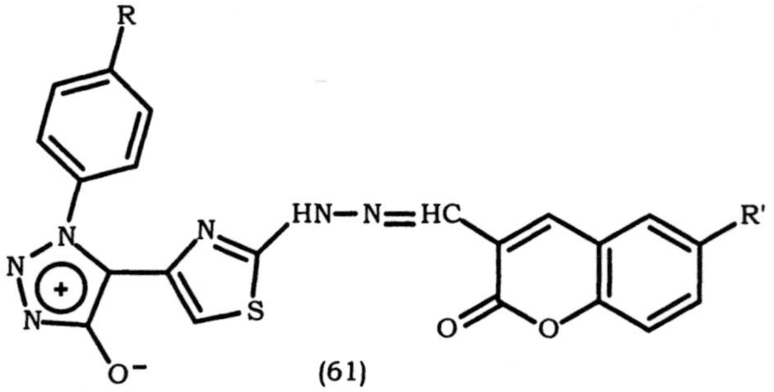

(61)

a, R, R' = Cl
b, R = Me R' = Cl
c, R = Cl, R' = Br
d, R = OMe, R' = Cl

Complex products of uranium (IV), dioxouranium (VI), and thorium (IV) with coumarin biheterocycles (62_{a-c} and 63_{a-c}) have been reported to exhibit antifungal activity against human pathogenic fungi [263]. In this regard, the CTBz ligands were highly active against *C. albicans*, but weakly active against *A. niger*, while CIP ligands were moderately active against both fungi. The activity of both ligands was less than standard Griseofulvin.

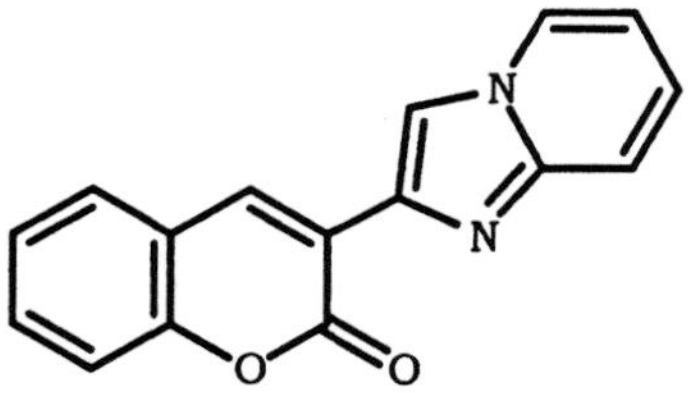

(62)
2-(3-Coumarinyl)imidazo-
[1,2-a]pyridine (CIP)
a, [U(CIP)Ac$_4$]
b, [UO$_2$(CIP) Ac$_2$]
c, [Th(CIP)$_2$(NO$_3$)$_4$]

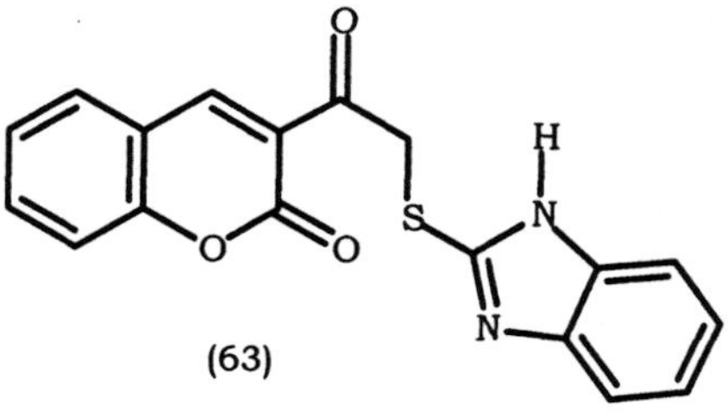

(63)
2-(3-Coumarinyl)thoacetyl]-
benzimidazole (CTBZ)
a, [U(CTBZ)Ac$_4$]
b, [UO$_2$(CTBZ Ac$_2$]
c, [Th(CTBZ)$_2$(NO$_3$)$_4$]

Substitution on the C-8 position of protected and unprotected 7-hydroxycoumarins have been reported to affect the antifungal activity of these compounds. For example 7-(β-*t*-alkylaminoalkoxy)-8-benzoyl-4-methylcoumarins (64_{a-g}) have been reported to show reasonable activity against *Fusarium oxysporum* and *Curvularia lunata*. Among the above compounds the diethylaminoethoxy (64_c) and *N*-piperidinoethoxy (64_e) have shown the highest activity at low, as well as high concentrations [264].

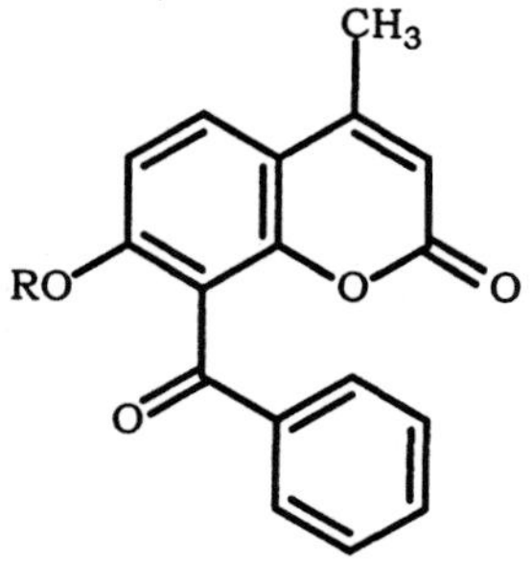

(64); a, R = Dimethylaminoethyl
b, R = 1-Dimethylamino-2-propyl
c, R = Diethylaminoethyl
d, R = 1-Pyrrolidinoethyl
e, R = 1-Piperidinoethyl
f, R = 1-Piperidinopropyl
g, R = 4-Morpholinoethyl

Following the same trend, compounds (65_{a-d}) and (66_{a-d}) have been reported to exhibit remarkable activity against *A. niger* in the same range as the standard drug Carbendazim [265].

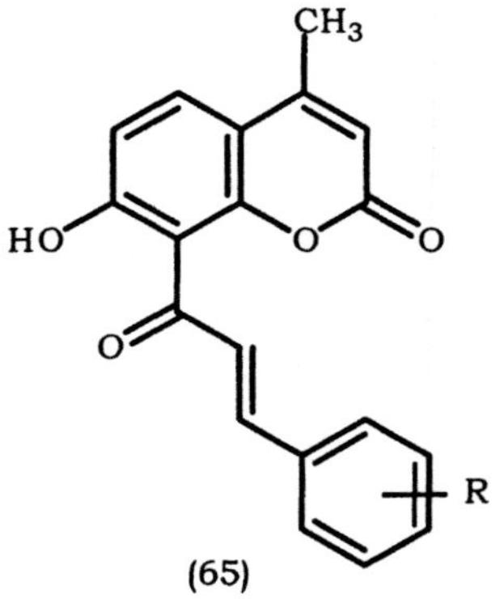

(65)

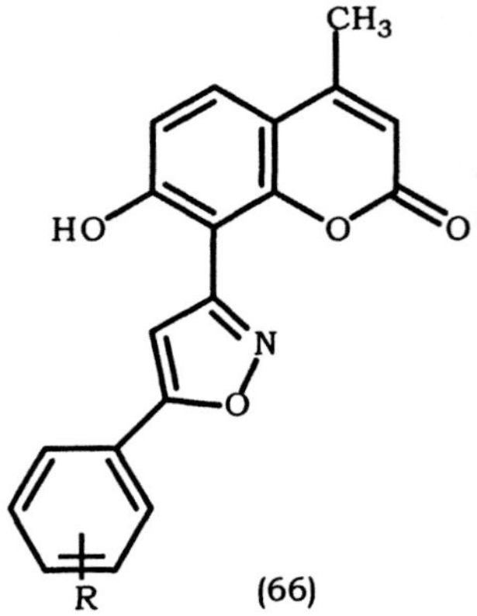

a, R = 4-Methyl
b, R = 3-Nitro
c, R = 4-Et, 3-OMe
d, R = 4-Cl

(66)

Among the naturally occurring coumarins the hydroxyl substituted congeners are usually exhibiting a better antifungal activity against human pathogenic fungi, compared to those lacking the hydroxyl substitution. For example, among 10 coumarin derivatives isolated from *Cyperus incompeltus* only four compounds (67_{a-d}) have been reported to show remarkable *in vitro* activity against *C. albicans* and *A. niger* [129].

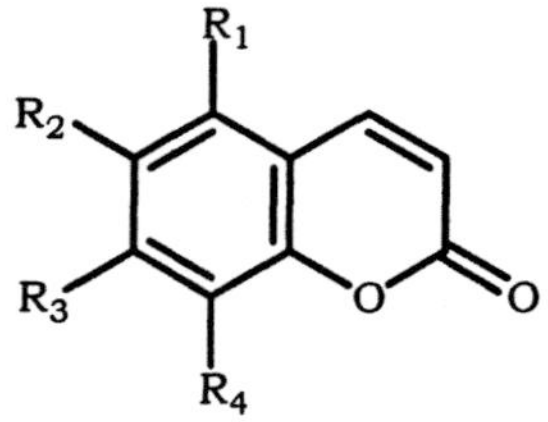

(67); a, R_1, R_2, R_4 = H; R_3 = OH
b, R_1, R_2 = H; R_3, R_4 = OMe
c, R_1, R_4 = H; R_2 = OH; R_3 = Dimethylallyloxy
d, R_1, R_4 = H; R_2 = OMe; R_3 = OH

Alkylation of the hydroxyl groups at C-4, and C-7 positions of coumarin nucleus has a dramatic effect on the overall antifungal activity of

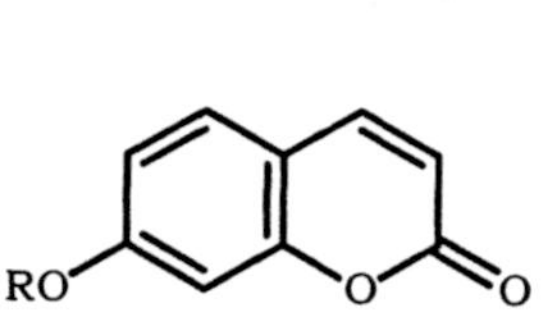

(68)

a, R = H
b, R = Me
c, R = Et
d, R = Geranyl
e, R = Allyl
f, R = Benzyl
g, R = Benzoyl

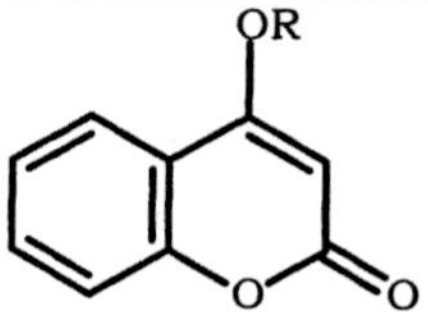

(69)

a, R = H
b, R = Me
c, R = Allyl
d, R = Geranyl

these compounds. Compounds (68$_a$) and (69$_a$) are naturally occurring coumarins with remarkable activity against molds and yeasts. Methylation of the hydroxyl groups, in both cases, has increased the activity of these compounds (68$_b$ and 69$_b$) against 5 *Aspergillus* species, and 8 of 9 species of yeasts tested. The ethoxy derivatives have been as effective as their methoxy counterparts. The geranyloxy derivatives were as active as the parent compounds, probably due to the hydrolysis and transformation to their parent molecule. Psoralen (2), and its methoxylated derivative showed activity after exposure to the light. The benzoylated derivatives of 4-hydroxycoumarins are less active than their 7-hydroxyl congeners. The activity of psoralen and xanthyletin (3) may be accounted for by the above generalization, since the cyclic alkoxy groupings (the furano and chromeno rings) presented in these compounds are stable to hydrolytic ring fusion [130].

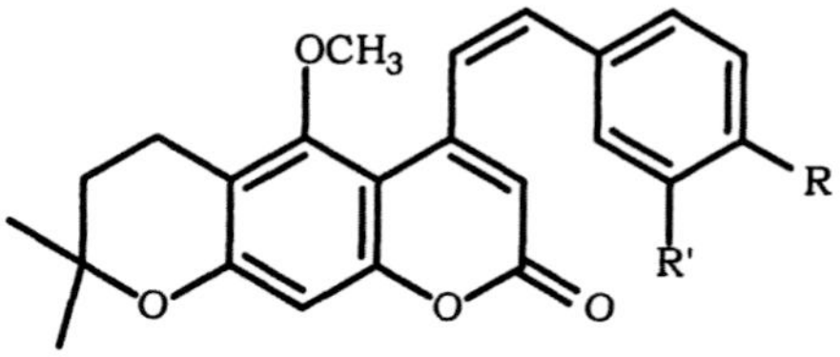

(70-74)

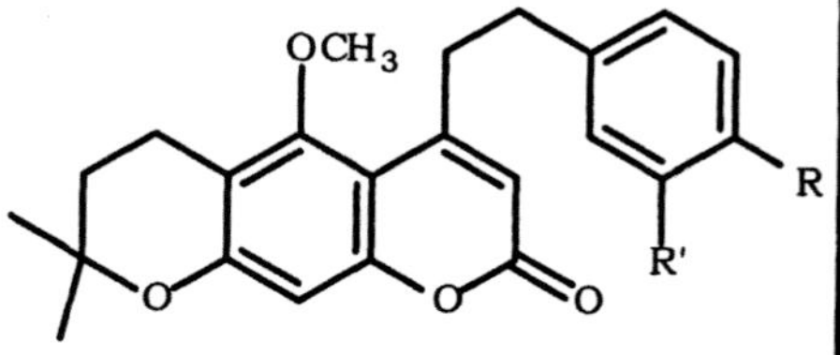

(75-79)

Compound No.	R	R'
70, 75	H	H
71, 76	Me	H
72, 77	OMe	H
73, 78	OMe	OMe
74, 79	Methylenedioxy	

Antifungal activity of pyrano- and furanocoumarins (natural and synthetic) has been studied and reported by different authors. For example, 4-arylethenyl and 4-arylethyl-6,7-dihydropyranocoumarin derivatives (70-79) have been reported to show *in vitro* antifungal activity against *A. niger* [72 (48.6%), 73 (55.3%), 74 (54.7%)], and *A. flavus* [74 (58.1%), 75 (49.9%)] in comparison with the reference compound Bavistin (80.0%) [266].

From the above data, it is evident that substitution on the phenyl ring has a critical effect on the activity. The unsubstituted, as well as mono methylsubstituted phenyl derivatives do not exhibit antifungal activity (compounds 70, 71, 75, 76). The disubstituted derivatives (73, 74, 78, 79) are showing almost equal activity. Among the monomethoxylted derivatives, only compound (72) shows reasonable potency, indicating the important role of the double bond on the activity.

Among four furanocoumarins tested against *C. albicans, C. neoformans, S. cerevisiae,* and *A. niger* in our laboratory, Angelicin (4_a) exhibited the strongest activity against all microorganisms tested [267]. Although compounds (80), (81), and (82) exhibited remarkable activity against all four fungi, they were less active than angelicin. Modification on the Angelicin molecule (4_{b-c}) resulted in various activity profiles, among which compound (4_e) appeared to be the most active one [328].

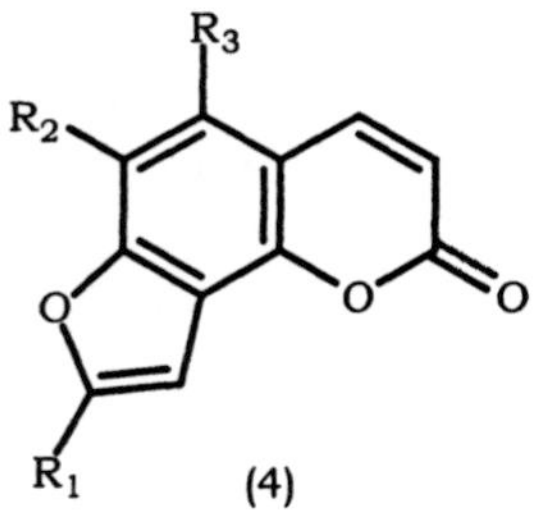

a, R_1, R_2, R_3 = H (Angelicin)
b, R_1 = COR_4 (R_4 = OH, OEt, NHR_5); R_2, R_3 = H
c, R_1, R_3 = H; R_2 = OR_6 (R_6 = H, Acyl)
d, R_1 = H; R_2 = Acetyl; R_3 = Benzyl
e, R_1 = NO_2; R_2, R_3 = H

Although most of the above mentioned coumarin derivatives have been reported to show strong *in vitro* activity against a wide range of fungal species, no *in vivo* efficacy data on the mammalian fungal disease models is reported in the literature. To evaluate the real potential of this class of compounds it is necessary to perform appropriate *in vivo* studies, and determine the efficacy of the compounds.

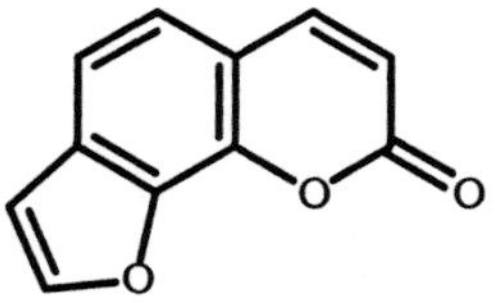

(80) Pseudo-isopsoralen

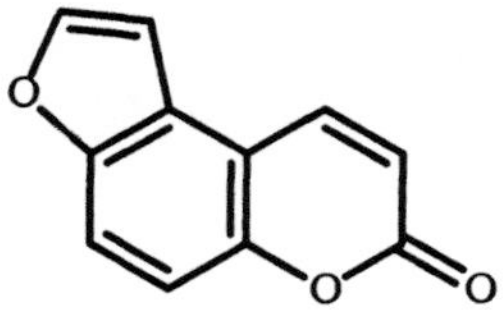

(81) Isopsoralen

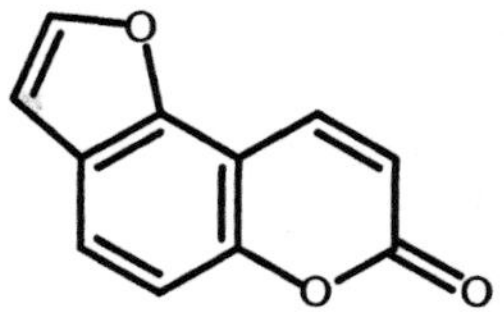

(82) Allopsoralen

MECHANISM OF ACTION

The study of mechanism of biologic action and especially antifungal property among coumarins is a very important issue. The mechanistic study would provide scientists with the understanding of the subcellular events that lead to the final bioactivity. In this way, not only manipulation of the agent and the activity becomes easier, but also discovery of new targets that are urgently needed to overcome resistance to older generations of antifungals becomes a real possibility. In addition, study of the mode of action of both natural and synthetic substances can conveniently be used for clarifying metabolism, genetic aspects, morphogenesis of cell structures, life cycle and differentiation of fungi [268].

The mechanism of action of coumarins are too complex and far from being completely understood. However, they generally can be divided into two light-dependent and independent processes. Since there are already many reports on the light-dependent action of coumarins and furanocoumarins [10,241,269,270,271,272,273] that can explain their antifungal activity, here we try to focus on the light-independent mechanisms.

Uncoupling Activity and Effect on Respiration

It was demonstrated that the rate of aerobic phosphorylation in rat liver mitochondria is depressed by dicoumarol [274]. Antibacterial activity of dicoumarol has been suggested to be related to its uncoupling characteristic [10]. This compound was reported to be a powerful uncoupler of oxidative phosphorylation, which inhibits phosphorylation at every step in the electron transport chain [275].

Kaiser, et al. [276] studied another example of inhibition of phosphorylation process by coumarins. In that study, it was shown that imperatorin, a furanocoumarin from *Imperatoria ostruthium* L. and *Ammi majus* L., inhibited respiration and phosphorylation of isolated mitochondria of the liver in the presence of succinate [276,326].

Coumarin has been reported to inhibit glycolysis and oxidative phosphorylation [277,278,279]. Coumarin and xanthotoxin can lower the oxygen uptake by meristematic cells of *Apium cepa* root tip [280]. They have also induced morphologic changes on the mitochondrial matrix to make it dense. Since the mentioned coumarins show antimitotic activity at the concentrations in which they occur on the plant surface [280], they may act as a defense barrier against phytopathogenic fungi. Inhibition of mitosis may be caused by a lack of intracellular energy or changes in mitochondrial structure [281,282,283].

Effect on DNA

Interaction between furanocoumarin and DNA occurs in two steps: (a) intercalation of coumarin between two base pairs of the DNA, (b) covalent photoconjugation through 3,4 and 2',3' double bonds of bifunctional furanocoumarins. In the case of monofunctional furanocoumarins, 2',3' double bond can participate in the reaction [269]. Psoralen photoaddition to thymine (83) is also performed in two steps under UV light of 365 nm wavelength, which provides the final product (84) [9].

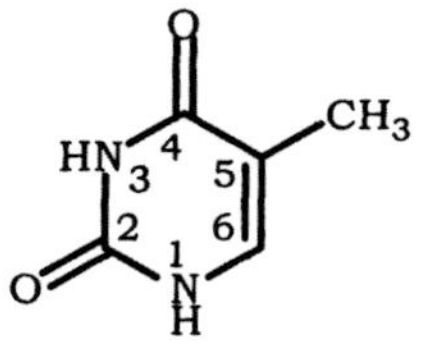

In a study on haploid and diploid cells of *Saccharomyces cerevisiae*, it was shown that the haploid cells are more susceptible to the action of furanocoumarins under 365 nm light, which is probably due to different repair capacity [284].

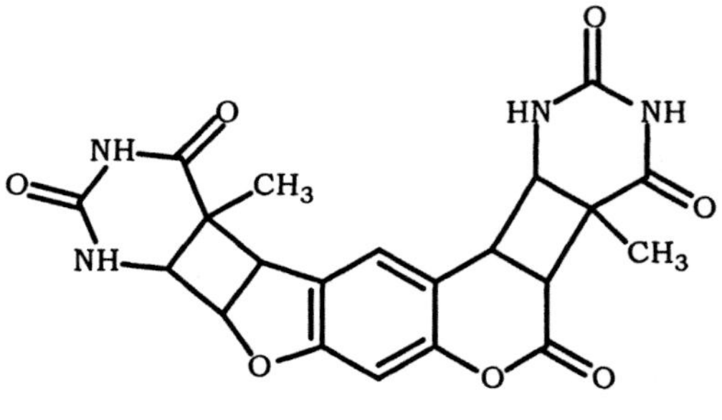

Furanocoumarins like angelicin (4) that only form monoadduct (involving the 3,4 double bond), induce mutation at a frequency comparable to ultraviolet light of 254 nm. No effect can be observed at the wavelengths above 254 nm [284].

It was shown previously that 8-MOP inhibits the DNA synthesis in the epidermal cells [190]. Angelicin (4) and 4,6,3'-TMA are unable to induce significant level of DNA-protein cross-link. Although the DNA-protein cross linkage may well be related to the phototoxic reaction of furanocoumarins, Bussey, *et al.* [285] synthesized and tested two furanocoumarin derivatives (85) and (86) that were able to cross link DNA and protein, but did not show phototoxicity in mouse [194,285].

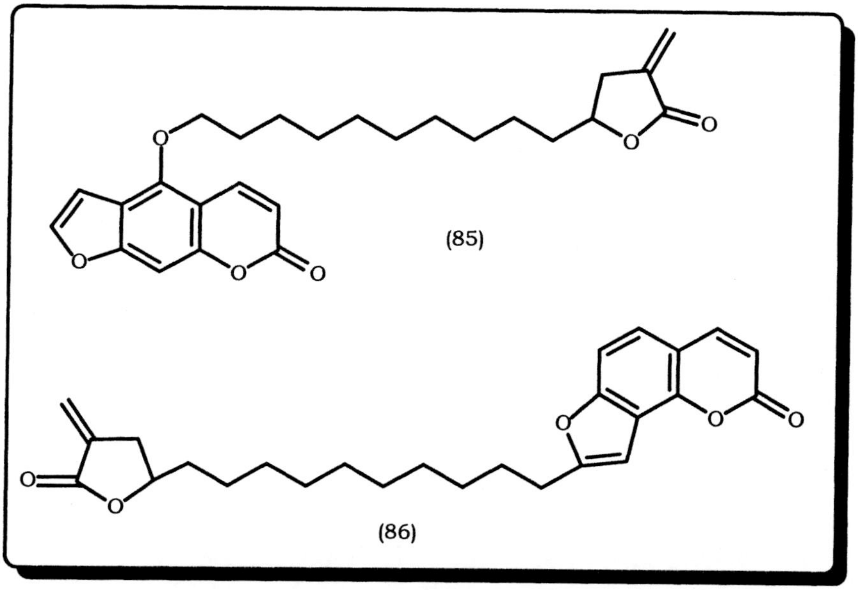

(85)

(86)

Effect on Enzymes: Gyrase and Topoisomerase

Two classes of antimicrobial gyrase inhibitors i.e., 4-quinolones like oxolinic acid (87), and coumarins (novobiocin and coumermycin A1) have been used to represent the different aspects of gyrase and topoisomerase reaction. Coumarins are inhibitors of the ATP hydrolysis and DNA supercoiling reactions catalyzed by DNA gyrase [286]. They block ATP hydrolysis at the Gyr B subunit [287]. Organisms like *Borrelia burgdorferi*, have gyrase enzyme resistant to nalidixic acid or oxolinic acid (87) type inhibitors, but show a distinct sensitivity to coumarin antibiotics like coumermycin. This phenomenon opens a prospect of

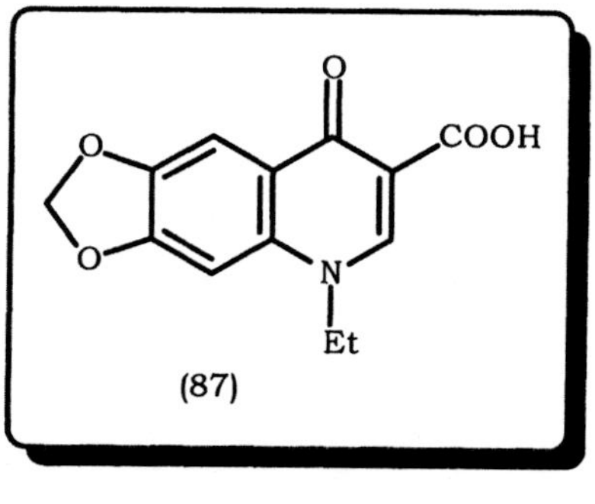

(87)

designing new drugs, which can specifically be utilized for the treatment of Lyme disease and probably fungal infections [288]. The mode of inhibition of the gyrase ATPase activity by coumarins is unlikely to be a competitive inhibition [164].

Topoisomerase is another enzyme that acts on DNA. Inhibition of topoisomerase can lead to cell toxicity and death. The mechanism of cell killing by topoisomerase poisons is unclear. These compounds include coumarins that act at the level of ATP binding or hydrolysis. They interfere with binding of ATP to the DNA-enzyme complex or ATP hydrolysis. In any case, this enzyme seems to be a good target for drug design [289,290,291]. But it should be noted that many effective compounds do not exhibit any activity in the *in vitro* tests. The reason for this is that factors like primary structure of DNA, spatial organization of enzyme-DNA complex, and the structure of chromatin fiber play critical roles in the overall reaction. These factors determine which locus of DNA will be cleaved, but in the *in vitro* tests only pure enzyme and naked DNA are being involved [166,292].

Interaction with Lipids and Proteins

Furanocoumarins can photoreact with unsaturated fatty acids. For example, angelicin (4) can form the adduct (88) with linolenic acid. It has been suggested that since the unsaturated fatty acids have important role in the phosphatidyl inositol system, the furanocoumarin adducts may change the regulatory function of this system. Also these adducts may inhibit phospholipase and then prevent the activation of protein kinase C [269].

The importance of the photoreaction with fatty acids becomes more evident if we know the comparative and quantitative results of furanocoumarin photoreaction with different cell fractions. Caffieri, *et al.* [293] showed that the highest proportion of photobinding of 8-MOP was to lipid fraction (60%) followed by protein and nucleic acid fractions (20% each). Similar results have been reported for 4,6,3'-TMA [269,294].

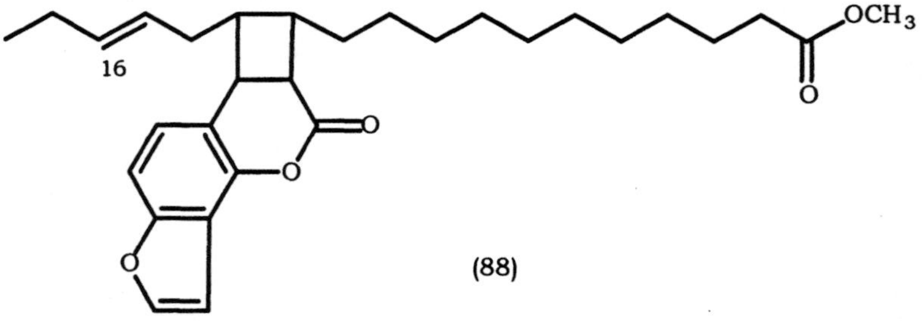

(88)

Furanocoumarin 8-methoxypsoralen (8-MOP) (1-100 μg/ml) in the dark showed a protective affect against hypotonic hemolysis of the erythrocyte membrane. However, the effect against heat-induced hemolysis was dependent on the concentration of 8-MOP; lower concentrations of 8-MOP showed an inhibiting effect, whereas higher concentrations caused acceleration of hemolysis. The erythrocytes reacted with 8-MOP in the dark were shrunk and had altered shapes. It can be deduced that modification of erythrocyte membrane by 8-MOP is via the reaction with membrane lipids and proteins [295]. From these results, it could be concluded that the effect on the cell membrane by coumarins could have an important role in their bioactivity.

Effect on Transport of Metabolites

Coumarin reduced labeled glucose incorporation and carbohydrate and protein content of the cytoplasm of *Pythium,* but enzymes related to the metabolism of cell wall polysaccharides were not affected [296]. It has been reported that moulds are more sensitive to coumarin when they are cultured on the simple synthetic nutrient medium than on the yeast water [140]. This suggests that coumarin blocks the synthesis of an unknown metabolite of fundamental importance. Coumarin was found to affect the level of a variety of free amino acids present in a cell. In addition, some nucleic acid precursors partially reverse the coumarin-induced inhibition of maize mesocotyle [131]. The stimulation of germination and growth caused by coumarin in low concentrations may be assumed as a super-compensation mechanism.

Receptors for Coumarin Action

In psoriasis, PUVA therapy (psoralen and UVA) leads to an inhibition of abnormally high rates of epidermal cell growth associated with the disease [297]. It has been shown that photoactivated psoralens are also potent inhibitors of epidermal cell growth *in vitro.* There are several lines of evidence that indicate the psoralen receptor is not associated with the DNA [298]. Studies on the distribution of compounds in cells based on their fluorescence properties demonstrated that psoralen (2) was present in the cytoplasm and cell surface membranes [299]. When cell membrane fraction of HeLa cells were prepared, psoralen receptor binding could be detected [297]. Using [3H]-8-MOP (methoxypsoralen), the cellular distribution of covalently bound psoralen in HeLa cells was also examined [300]. In these experiments, cells were treated with [3H]-8-MOP, then pulsed with UV light. It was indicated that labeled 8-MOP was distributed in cytoplasm and membrane fractions. These results have been repeated in several laboratories and all of them implicate targets outside of the nucleus for the action of psoralens [301,302,303,304,305]. In addition, saturable, specific receptor sites for psoralen have been found in HeLa cells that are non-responding to angular coumarin, trimethyl angelicin [298]. It has been hypothesized that these receptors mediate, at least in part, the biological actions of these coumarins. One receptor is reported to be a 22,000 Dalton protein and presents in membrane and cytoplasmic fractions of responsive cells [306]. In addition, another potential receptor is EGF (epidermal growth factor). Photoactivated psoralens inhibit binding of EGF [297]. TPA (12-*O*-tetradecanoyl-phorbol-13-acetate) and TCDD (2,3,7,8-tetrachlorodibenzo-*p*-dioxin) modulate cell growth. In fact, inhibition of EGF binding in cells by TPA and psoralens occurs almost in an identical manner, suggesting a similar mechanism for both compounds [307,308,309]. It has been shown that TPA binds to the calcium- and phospholipid-dependent protein kinase, protein kinase C [325]. Laskin, *et al.* [297] has proposed that photoactivated psoralen receptor modulates the TGF receptor by inducing its phosphorylation. This may occur indirectly through the activator of protein kinase C. A similar phenomenon can happen in fungi. We have observed the effect of coumarins in inhibiting fungal morphogenesis. In our previous report, the role of protein kinase C in fungal morphogenesis has also been described [310]. Since morphological studies are normally performed in sub-inhibitory concentrations of compounds, higher concentrations of coumarins can inhibit fungal growth in a similar way through action on protein kinase C. A furanocoumarin-fatty acid adduct can inhibit a phospholipase and then prevent the activation of protein kinase C or other regulatory proteins. Inhibition of protein kinase C could also account for the antiproliferative effects of PUVA therapy [311].

It is possible that there are other proteins and enzymes that coumarins react with. As mentioned before they can react with thiol group of enzymes and hormones. In fact this has been shown in the case of byakangelicin that through its 3,4-double bond forms a 1,4-addition product with -SH group of HCG (human chorionic gonadotropin) [215,216].

Morphological Changes

Electron microscope examination of the fungal cells is a powerful mean of studying the morphological changes due to different bioactive agents. Scanning electron microscopy (SEM) is a technique that is utilized to look at the surface of these organisms. SEM pictures of *C. albicans* cells treated with amphotericin B (AmB), and 2'-nitroangelicin, synthesized in our group, are shown in figures (2-3). As can be seen in figure (3), cells are in a phase of multibudding. In the cells treated with subinhibitory concentrations of nitroangelicin, cell wall is rugged. However, the AmB group has very few cells with wrinkled surface. AmB group cells do not exhibit marked changes, figure (**3**), as the minimum inhibitory concentration (MIC) of AmB is very sharp and subinhibitory concentration practically does not trail.

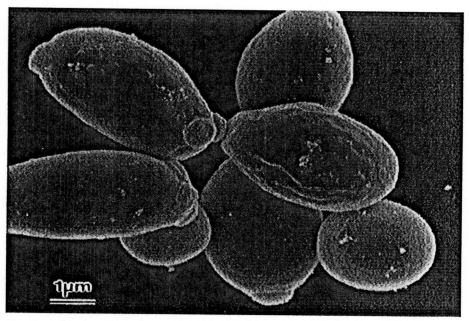

Fig. (2). SEM Picture of the *C. albicans* cells exposed to AmB, grown in RPMI 1640 at 37°C for 24 h.

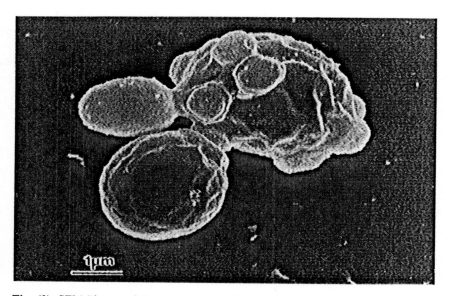

Fig. (3). SEM Picture of the *C. albicans* cells exposed to nitroangelicin, grown in RPMI 1640 at 37°C for 24 h.

cAMP has important roles in fungal physiology, such as yeast to hyphae transition [312,313,314,315]. Also, cAMP can induce translation of some genes and can affect cell cycle in fungi, and vertebrates [316]. IBMX (a phosphodiesterase inhibitor) at concentrations 1 and 10 μM caused an increase in intracellular cAMP and stimulated the production of endoglucanase in fungal cells [317]. A change in the sterol composition of the membrane is shown to lower overall specific activity of the enzyme, chitin synthase [318].

cAMP in *C. albicans* rises during yeast to mycelial form transition, even at lower temperatures. Cysteine, which suppressed germination, also reversed the increase in intracellular cAMP (possibly through activation of cyclic nucleotide phosphodiesterase as in *E. coli*). Cysteine promotes yeast like growth via, for example, reduction of disulfide linkages in glucomannan protein. In *M. rouxii* addition of db-cAMP or cAMP induced the yeast like morphology. *Mucor racemosus* is similar to *M. rouxii*, however, *Histoplasma capsulatum,* like *C. albicans* tends to be in mycelial form in this condition [314]. Although no mycelial form was observed in our study, the increased cAMP could be responsible for some of the morphologic changes seen in SEM study.

Since the coumarin compounds show phosphodiesterase inhibitory activity, the observed change in the morphology of *Candida albicans* could be partly due to this property. Each *Candida* yeast cell can normally have a budding scar or the daughter cell still attached to it [319]. The possible cause of multibudding phenomenon could be PD inhibition. A similar observation was done in treatment with caffeine, which at the subinhibitory concentrations caused an increase in unusual modes of proliferation with signs of multiple budding in *Candida albicans* [320].

CONCLUSION

The importance of natural products become more clear by looking at their current status in human society. It is estimated that 80% of the world's population rely solely on indigenous systems of medicines that are mainly plant-based [321]. The World Health Organization estimates that some 20,000 species of higher plant are used medicinally throughout the world. On the other hand, it also has been estimated that 25% of all plant species (i.e., 60,000) will be disappeared by mid-21 century [321], which urges a worldwide protection-development program on them.

Higher plants, many of which are threatened with extinction, are used as sources of pharmaceuticals. Plants have been an integral part of traditional medicines and are of value in new drug discovery systems. Artemisinin, taxol and camptothecin are examples of natural products, which are undergoing clinical and commercial development. Several natural products isolated from plants used in traditional medicine have potent antifungal action. Furanocoumarins as a group of photoactive drugs were found in a search on traditional medicines used by ancient Egyptians and Middle East people.

The focus of this review was to look at the position of natural products in antifungal research and role of coumarins in this respect. In addition, we tried to consider other aspects of coumarin compounds with major bioactivities as well. Natural products have served as an important source of drug for centuries, as about half of the pharmaceuticals in today's market are derived from natural products [322]; In addition, they have had a significant role in identifying new biomedical cellular/molecular targets and act as biochemical probes, therefore, more attention should be paid to this field of research by scientists and pharmaceutical companies. In this article, we tried to present potentials of coumarins for such attentions. Moreover, Mother Nature has provided us with many new ideas in drug discovery itself. Combinatorial chemistry, wherein a molecular scaffold is substituted with a variety of substituents, was initiated as a close mimic of synthesis of different antibodies by the immune system [323].

Coumarins show diverse biological activities. The individual coumarins could also have a wide range of activities, which may not necessarily limit their application. In fact, it could be argued that several of the indicated properties could be beneficial if combined. However, semisynthetic derivatives of some coumarins could offer improvements in activity, toxicology, or pharmacokinetic behavior [324]. In view of the established low toxicity, relative cheapness and availability, presence in the diet and occurrence in various herbal remedies of coumarins, it is tempting to evaluate properties and applications of coumarins further.

REFERENCES

[1]   Rippon, J. W. Medical Mycology: the Pathogenic Fungi and the Pathogenic Actinomycetes. W. B. Saunders: Philadelphia, **1988**.
[2]   Levitz, S. M. *Clin. Infect. Dis.*, **1992**, *14*, S37.
[3]   Conant Ma. *Dermatologic Clinics*, **1996**, *14*, 155.
[4]   Pfaller, M. A.; Wenzel. R. P. *Eur. J. Clin. Microbiol. Infec. Dis.*, **1992**, *11*, 287.
[5]   Farnsworth, N. R. In *Medicinal Resources of the Tropical Forests*, Balick, M. J.; Elisabetsky, E.; Laird, S. A. Eds.; Columbia University Press: N.Y., **1996**; pp. IX-X.
[6]   Cragg, G. M.; Newman, D. J.; Snader, K. M. *J. Nat. Prod.*, **1997**, *60*, 52.
[7]   Loew, F. M. Bulletin of the History of Medicine, **1981**, 55, 263.
[8]   Murray, R. H.; Mendez, J.; Brown, S. A. *The Natural Coumarins; Occurrence, Chemistry and Biochemistry*, Johns Wiley & Sons Ltd.: Chichester, **1982**.
[9]   Scott, B. R.; Pathak, M. A.; Mohn, G. R. *Mutat. Res.*, **1976**, *39*, 29.
[10]   Soine, T.O. *J. Pharm. Sci.*, **1964**, *53*, 231.
[11]   Werck-Reichhart, D. *Drug Metab. Drug Interactions*, **1995**, *12*, 221.
[12]   Hoult, J. R.; Paya, M. *Gen. Pharmacol.*, **1996**, *27*, 713.
[13]   Murray, R. D. *Nat. Prod. Rep.*, **1989**, *6*, 591.

[14]   Beyrich, T. *J. Chromatogr.* **1964,***13,* 181.
[15]   Seshadri, T. R.; Vishwapaul, *J. Sci. Ind. Res.* **1973,** *32,* 227.
[16]   Masrani, K. V.; Rama, H. S.; Bafna, S. L. *J. Applied Chem. Biotechnol.* **1974,** *24,* 331.
[17]   Lee, K. H.; Soine, T. O. *J. Pharm. Sci.,* **1969,** *58,* 681.
[18]   Steck, W.; Mazurek, M. *J. of Nat. Prod.,* **1972,** *35,* 418.
[19]   Mikhova, B.; Duddeck, H. In *Studies in Natural Products Chemistry*, Part K, Atta-ur-Rahman, Ed.; Elsevier: Amsterdam, **1996,** Vol.*18,* pp.971-1080.
[20]   Dean, F. M., Costa, A. M. B. S. R. C. S., Harborne, J. B., Smith, D. M. *Phytochemistry,* **1978,** *17,* 505.
[21]   Fisher, J. F.; Nordby, H.E. *Tetrahedron,* **1966,** *22,* 1489.
[22]   Kutney, J. P.; Young, R. N.; Verma, A. K. *Tetrahedron Lett.,* **1969,** 1845.
[23]   Elgamal, M. H. A.; Elewa, N. H.; Elkhrisy, E. A. M.; Duddeck, H. *Phytochem.,* **1979,** *18,* 139.
[24]   Cussans, N. J.; Huckerby, T. N. *Tetrahedron,* **1975,** *31,* 2719.
[25]   Bose, A. K.; Fujiwara, H. *Tetrahedron,* **1979,** *35,* 13.
[26]   Boman, H. G. *Nordisk Medicin,* **1996,** *111,* 176.
[27]   Warr G. W.; Cohen, N. *Phylogenesis of Immune Functions*, CRC Press: Boca Raton, **1991.**
[28]   Boman, H. g.; Hultmark, D. *Ann. Rev. Microbiol.,* **1987,** *41,* 103.
[29]   Sima, P.; Vetvicka, V. *Evolution of Immune Reactions.* CRC Press: Boca Raton, **1990.**
[30]   Lehrer, R. I.; Ganz, T.; Selsted, M. E. *Cell,* **1991,** *64,* 229.
[31]   Lehrer, R. I.; Barton, A.; Daher, K. A.; Harwig, S. S. L.; Ganz, T.; Selsted, M. E. *J. Clin. Invest.,* **1989,** *84,* 553.
[32]   Ganz, T.; Selsted, M. E.; Lehrer, R. I. *Eur. J. Haematol.,* **1990,** *44,* 1.
[33]   Lehrer, R. I.; Ganz, T. *Blood,* **1990,** *76,* 2169.
[34]   Lehrer, R. I.; Ganz, T. *Ciba Foundation Symp.,* **1992,** 171, 276.
[35]   Ganz, T.; Lehrer, R. I. *Curr. Opinion in Immunol.,* **1994,** *6,* 584.
[36]   Kagan, B. L.; Ganz, T.; Lehrer, R. I. *Toxicol.,* **1994,** *87,* 131.
[37]   Ganz, T.; Lehrer, R. I. *Pharmacol. Ther.,* **1995,** *66,* 191.
[38]   Porter, E. M.; van Dam, E.; Valore, E. V.; Ganz, T. *Inf. Immun.,* **1997,** *65,* 2396.
[39]   Grayer, R. J.; Harborne, J. B. *Phytochem.,* **1994,** *37,* 19.
[40]   Rocha, L.; Marston, A.; Kaplan, M.; Stoeckli-Evans, H.; Thull, U.; Testa, B.; Hostettmann, K. *Phytochem.,* **1994,** *36,* 1381.
[41]   Hostettmann, K.; Marston, A.; Maillard, M.; Wolfender, J. L. *Adv. Exp. Med. Biol.,* **1996,** *404,* 117.
[42]   Bashir, A.; Hamburger, M.; Rahalison, L.; Monod, M.; Gupta, M. P.; Solis, P.; Hostettmann K. *Planta Med.,* **1991,** *57,* 192.
[43]   Griffin, D. H. *Fungal Physiology*, Wiley-Liss: N.Y., **1994**; pp.399-423.
[44]   Harborne, J. B.; Baxter, H. *Phyochemical Dictionary.* Taylor and Francis: London, **1993**; pp. 351.
[45]   Hostettmann, K.; Marston, A. In *Studies in Natural Product Chemistry*; Atta-ur-Rahman, Ed.; Elsevier: Amsterdam, **1990**; Vol.7, part A., pp. 405-437.
[46]   Hostettmann, K.; Potterat, O.; Wolfender, J-L., *Pharm. Ind.,* **1997,** *59,* 339.
[47]   Marston, A.; Maillard, M.; Hostettmann, K. *J. Ethnopharmacol.,* **1993,** *38,* 215.
[48]   Lorito, M.; Broadway, R. M.; Hayes, C. K.; Woo, S. L.; Noviello, C.; Williams, D. L.; Harman, G. E. *Mo. Plant-Microbe Interact.,* **1994,** *7,* 525.
[49]   Janssen, A.; Scheffer, J.; Baerheim Svendsen, A. *Planta Med.,* **1987,** *53,* 395.

[50] Abu-Salah, K. M. *Br. J. Biomed. Sci.,* **1996,** *53,* 122.
[51] Bergstrom, J. d.; Dufresne, C.; Bills, G. F.; Nallin-Omstead, M.; Byrne, K. *Ann. Rev. Microbiol.,* **1995,** *49,* 607.
[52] Tanimoto, T.; Hamano, K.; Onodera, K.; Hosoya, T.; Kakusaka, M.; Hirayama, T.; Shimada, Y.; Koga, T.; Tsujita, Y. *J. Antibiot.,* **1997,** *50,* 390.
[53] Parang, K.; Knaus, E. E.; Wiebe, L. I.; Sardari, S.; Daneshtalab, M.; Csizmadia, F. *Arch. Pharmazie.* **1996,** *329,* 475.
[54] Kasai, T.; Inoue, K.; Tsujimura, M.; Hasegawa, T. *Vitamines,* **1994,** *68,* 275.
[55] Gunasekera, S. P.; Pomponi, S. A.; McCarthy, P. J. *J. Nat. Prod.,* **1994,** *57,* 79.
[56] Kobayashi, M.; Kondo, K.; Kitagawa, I. *Chem. Pharm. Bull.,* **1993,** *41,* 1324.
[57] Li, H. Y.; Matsunaga, S.; Fusetani, N. *J. Med. Chem.,* **1995,** *38,* 338.
[58] Akabane, H. Hokkaido Igaku Zasshi - Hokkaido (J. Med. Sci.), **1991,** *66,* 510.
[59] Woerdenbag, H. J.; Koulman, A.; Proksch, P.; Ebel, R.; Van Uden, W.; Pras, N.; Beijnen, J. H. *Pharm. Weekbl.,* **1995,** *130,* 1290.
[60] Fehlbaum, P.; Bulet, P.; Michaut, L.; Lagueux, M.; Broekaert, W. F.; Hetru, C.; Hoffmann, J. A. *J. Biol. Chem.,* **1994,** *269,* 33159.
[61] Mor, A.; Nicolas, P. *Eur. J. Biochem.,* **1994,** *219,* 145.
[62] Simmaco, M.; Barra, D.; Chiarini, F; Noviello, L.; Melchiori, P.; Kreil, G.; Richter, K. *Eur. J. Biochem.,* **1991,** *199,* 217.
[63] Ehret-Sabatier, L.; Loew, D.; Goyffon, M.; Fehlbaum, P.; Hoffmann, J.; van Dorsselaer, A.; Bulet, P. *J. Biol. Chem.,* **1996,** *271,* 29537.
[64] Kramer, W. In *Chemistry of Pesticides*; K. H. Buchel, Ed.;John WIley and Sons: New York, **1983;** , pp. 227-321.
[65] Kubo, I. In *Bioactive Volatile Compounds from Plants*; R. Teranishi, R. G. Buttery and H. Sugisawa, Eds.; American Chemical Society: Washington DC, **1993;** p. 57.
[66] Janssen, A.; Scheffer, J.; Baerheim Svendsen, A.; Aynehchi, Y. *Pharm. Weekb. - Scientific Edition.,* **1984,** *6,* 157.
[67] Lun, Z. R.; Burri, C.; Menzinger, M.; Kaminsky, R. *Ann. Soc. Belge Med. Trop.,* **1994,** *74,* 51.
[68] Barrett, J. F.; Klaubert, D. H. In *Annual Reports in Medicinal Chemistry*; J. A. Bristol, Ed.; Academic Press: San Diego, **1992;** Vol. *27,* pp. 149-158.
[69] Liu, S.; Oguntimein, B.; Hufford, C.; Clark, A. *Antimicrob. Agents Chemother.,* **1990,** *34,* 529.
[70] Muirhead, J. F.; Deverall, B. J. *Aust. J. Botany,* **1984,** *32,* 575.
[71] Crombie, L.; Mistry, J. *Tetrahedron Lett.,* **1990,** *31,* 2647.
[72] Monde, K.; Kishimoto, M.; Takasugi, M. *Tetrahedron Lett.,* **1992,** *33,* 5395.
[73] Burden, R. S.; Kemp, M. S.; Wiltshire, C. W.; Owen, J. D. *J. Chem. Soc. Perkin Trans.,* **1984,** *I,* 1445.
[74] Bhakuni, D.; Chaturvedi, R. *J. Nat. Prod.,* **1984,** *47,* 585.
[75] Ratnayake Bandara, B. M.; Wimalasiri, W. R. *Phytochem.,* **1988,** *27,* 225.
[76] Hashimoto, Y.; Ishizaki, T.; Shudo, K. *Yakugaku Zasshi (J. Pharm. Soc. Jpn),* **1995,** *115,* 189.
[77] Falch, B. S.; Konig, G. M.; Wright, A. D.; Sticher, O.; Angerhofer, C. K.; Pezzuto, J. M.; Bachmann, H. *Planta Med.,* **1995,** *61,* 321.
[78] Mabry, T. J.; Ulubelen, A. *J. Agric. Food Chem.,* **1980,** *28,* 188.
[79] Li, E.; Clark, A.; Hufford, C. *J. Nat. Prod.,* **1995,** *58,* 57.
[80] Kato, T.; Yamaguchi, Y.; Uyehara, T.; Yokoyama, T. *Tetrahedron Lett.,* **1983,** *24,* 4715.

[81]   Kato, T.; Yamaguchi, Y.; Namai, T.; Hirukawa, T. *Biosc. Biotechn. Biochem.*, **1993**, *57*, 283.

[82]   Namai, T.; Kato, T.; Yamaguchi, Y.; Hirukawa, T. *Biosc. Biotechn. Biochem.*, **1993**, *57*, 611.

[83]   Konig, G. M.; Wright, A. D.; Sticher, O.; Angerhofer, C. K.; Pezzuto, J. M. *Planta Med.*, **1994**, *60*, 532.

[84]   Wright, A. D.; Konig, G. M.; Sticher, O. *Tetrahedron*, **1991**, *47*, 5717.

[85]   Pursky, D.; Keen, N. T.; Eaks, I. *Physiol. Plant Path.*, **1983**, *22*, 189.

[86]   Naigre, R.; Kalck, P.; Roques, C.; Roux, I.; Michel, G. *Planta Med.*, **1996**, *62*, 275.

[87]   Potterat, O.; Stoeckli-Evans, H.; Msonthi, J. D.; Hostettmann, K. *Helv. Chim. Acta*, **1987**, *70*, 155.

[88]   Cammue, B. P.; De Bolle, M. F.; Terras, F. R.; Proost, P.; Van Damme, J.; Rees, S. B.; Vanderleyden, J.; Broekaert, W. F. *J. Biol. Chem.*, **1992**, *267*, 2228.

[89]   Takasugi, M.; Kawashima, S.; Monde, K.; Katsui, N.; Masamune, T.; Shirata, A. *Phytochem.*, **1987**, *26*, 371.

[90]   Salako, N. *Odonto-Stomatologie Trop.*, **1990**, *13*, 117.

[91]   Binutu, O. A.; Adesogan, K. E.; Okogun, J. I. *Planta Med.*, **1996**, *62*, 352.

[92]   Neucere, J. N.; Godshall, M. A. *Mycopathol.*, **1991**, *113*, 103.

[93]   Terras, F. R.; Schoofs, H. M.; De Bolle, M. F.; Van Leuven, F.; Rees, S. B.; Vanderleyden, J.; Cammue, B. P.; Broekaert, W. F. *J. Biol. Chem.*, **1992**, *267*, 15301.

[94]   Hufford, C. D.; Liu, S. C.; Clark, A. M. *J. Nat. Prod.*, **1988**, *51*, 94.

[95]   Kuc, J. In. *Phytochemical resources for medicine and agriculture.* H. N. Nigg and D. Seigler, Ed.; Plenum Press: New York,.**1992**; pp. 159-183.

[96]   Greger, H.; Hofer, O.; Kahlig, H.; Wurz, G. *Tetrahedron*, **1992**, *48*, 1209.

[97]   Greger, H.; Hofer, O.; Zechner, G.; Wurz, G.; Hadacek, F. *Planta Med.*, **1992**, *58*, A721.

[98]   Polacheck, I.; Zehavi, U.; Naim, M.; Levy, M.; Evron, R. Zentralblatt Fur Bakteriologie, Mikrobiologie, Und Hygiene – Ser. A, Med. Microbiol., Inf. Dis., Virol., Parasitol., **1986**, 261, 481.

[99]   Anke, H.; Kemmer, T.; Hofle, G. *J. Antibiot.*, **1981**, *34*, 923.

[100] Anke, H.; Morales, P.; Sterner, O. *Planta Medica*, **1996**, *62*, 181.

[101] Sato, Y. Yakugaku Zasshi (J. Pharm. Soc. Jpn), **1995**, 115, 892.

[102] Scott, P. M.; Walbeek, W. V.; McLeac, W. M. *J. Antibiot.*, **1971**, *24*, 747.

[103] Iwamoto, T.; Fujie, A.; Sakamoto, K.; Tsurumi, Y.; Shigematsu, N.; Yamashita, M.; Hashimoto, S.; Okuhara, M.; Kohsaka, M. *J. Antibiot.*, **1994**, *47*, 1084.

[104] Stadler, M.; Sheldrick, W. S.; Dasen-Brock, J.; Steglich, W.; Anke, H. *Nat. Prod. Lett.*, **1994**, *4*, 209.

[105] Potterat, O.; Zahner, H.; Volkmann, C.; Zeeck, A.; Monde, K.; Kishimoto, M.; Takasugi, M. *J. Antibiot.*, **1993**, *46*, 346.

[106] Konig, G. M.; Wright, A. D. *Planta Med.*, **1996**, *62*, 193.

[107] Bergstrom, J. D.; Kurtz, M. M.; Rew, D. J.; Amend, A. M.; Karkas, J. D.; Bostedor, R. G.; Bansal, V. S.; Dufresne, C.; VanMiddlesworth, F. L.; Hensens, O. D.; et al. *Proc. Nat. Acad. Sci. USA.*, **1993**, *90*, 80.

[108] Konishi, M.; Nishio, M.; Saitoh, K.; Miyaki, T.; Oki, T.; Kawaguchi, H. *J. Antibiot.* **1989**, *42*,1749.

[109] Dowling, D. N.; O'Gara, F. *TIBTECH*, **1994**, 12(April), 133.

[110] Kugler, M.; Loeffler, W.; Rapp, C.; Kern, A.; Jung, G. *Arch. Microbiol.*, **1990**, *153*, 276.
[111] Moore. R. E. *J. Ind. Microbiol.*, **1996**, *16*, 134.
[112] Rapp, C. *Angew. Chem. Int. Ed. Eng.* **1988**, *27*, 1733.
[113] Glombitza, K. -W. In *Marine Algae in Pharmaceutical Sciences*; H. A. Hoppe and T. Levring, Eds.; Walter de Gruyter: Berlin, **1979**; pp. 303-342.
[114] Corcillius, F. Arch. Pharm. Ber. Disch. Pharm. Ges., **1956**, 289, 81.
[115] Hata, K.; Kozawa, M.; Yen, K.-Y. *Yakugaku Zasshi*, **1963**, *83*, 606.
[116] Fujita, Y. *Shokubutsu Kenkyu Zasshi*, **1963**, *38*, 359.
[117] Hata, K.; Sano, K. *Tetrahedron Lett.*, **1966**, 1461.
[118] Kozawa, M.; Morita, N.; Baba, K.; Hata, K. *Yakugaku Zasshi*, **1978**, *98*, 636.
[119] Badran, N.; Starkowsky, N. A. *Proc. Pharm. Soc. Egypt.*, **1956**, *38*, 93.
[120] Smith, E.; Hosansky, N.; Bywater, W. G. *Chem. Ind. (London)*, **1956**, 718.
[121] Combes, G.; Pernet, R.; Pierre, R. *Bull. Soc. Chim. Fr.*, **1961**, 1609.
[122] Musajo, L.; Rodighiero, G. *Experientia*, **1962**, *18*, 153.
[123] United States pharmacopeia dispensing information, *Drug Information for the Health Care Professional*. United States Pharmacopeial Convention: Rockville, MD, vol. *I*; pp. 2025-2028, **1996**.
[124] Guven, K. C.; Guler, E. In *Marine Algae in Pharmaceutical Sciences*; H. A. Hoppe and T. Levring, Eds.; Walter de Gruyter: Berlin, **1979**; pp. 693-710.
[125] Murray, R. D. H. In *Progress in the Chemistry of Organic Natural Products (Fortschritte der Chemie organischer Naturstoffe)*; W. Herz, G. W. Kirby and W. Steglich, Eds.;Springer-Verlag: Wien, **1991**; Vol. *58*, pp. 84-316.
[126] Selitrennikoff, C. P.; Vigers, A. J.; Roberts, W. K. In *New Approaches for Antifungal Drugs*; P. B. Fernandes, Ed.; Birkhauser: Boston, **1992**; pp. 88-113.
[127] Worthington, P. A. *Nat. Prod. Rep.*, **1988**, *5*, 47-66.
[128] Edelson, R. L. *Sci. Am.*, **1988**, August, 68.
[129] Dini, A.; Ramundo, E.; Saturnino, P.; Scimone, A.; Stagno, d'Alcontres, I. *Boll. Soc. Ital. Biol. Sper.*, **1992**, *68*, 453.
[130] Jurd, D. L.; King, J. R. A. D.; Mihara, K. *Experientia.* **1970**, *26*, 1281.
[131] Knypl, J. S. Ph.D. Thesis. **1963**, Univ. Lodz.
[132] Mikkelson, V.E.; Fowlks, E.W.; Griffith, D. G. *Arch. Phys. Med. Rehabil.*, **1961**, *42*, 609.
[133] Camm, E. L.; Wat, C. K.; Towers, G. H. N. *Can. J. Bot.*, **1976**, *54*, 2562.
[134] Nakijima,K., Kawai, K. Yamada, S. *Phytochem.*, **1976**, *15*, 327.
[135] Nozawa, K.; Yamada, M.; Tsuda, Y.; Kawai, K.; Nakajima, S. *Chem. Pharm. Bull.*, **1981**, *29*, 2491.
[136] Nozawa, K.; Yamada, M.; Tsuda, Y.; Kawai, K.; Nakajima, S. *Chem.Pharm. Bull.*, **1981**, *29*, 3486.
[137] Nozawa, K.; Yamada, M.; Tsuda, Y.; Kawai, K.; Nakajima, S. *Chem. Pharm. Bull.* **1981**, *29*, 2689.
[138] Alderson, T.; Scott B. R. *Mutat. Res.*, **1970**, *9*, 569.
[139] Van Sumere, C. F.; Van Sumere, d. P. C.; Vining, I. C.; Ledingham, G. A. *Can. J. Microbiol.*, **1957**, *3*, 847.
[140] Knypl, J. S. *Nature.* **1963**, *200*, 800.
[141] Bellis, D. M. *Nature*, **1958**, *182*, 806.
[142] Uritani, I. *J. Assoc. Offic. Agr. Chemists*, **1967**, *50*, 105.
[143] Hughes, J. C.; Swain, T. *Phytopathol.*, **1960**, *50*, 398.
[144] Austin, D. J.; Clarke, D. D. *Nature*, **1966**, *210*, 1165.
[145] Uecker, S.; Jira, T.; Beyrich, T. *Pharmazie*, **1991**, *46*, 599.

[146] Scheel, L.D. *Microb. Toxins*, **1972**, *8*, 47.
[147] Zobel, A. M.; Wang, J.; March, R. E.; Brown, S. A. *J. Chem. Ecol.*, **1991**, *17*, 1859.
[148] Perone, V. B. *Micropial Toxins*, **1972**, *8*, 67.
[149] Stahman, M. A.; Huebner, C. F.; Link, K. P. *J. Biol. Chem.*, **1941**, *138*, 513.
[150] Mueller, R. l.; Scheidt, S. *Circulation*, **1994**, *89*, 432.
[151] Ikawa, M.; Stahmann, M. A.; Link, K. P. *J. Am. Chem. Soc.*, **1944**, *66*, 902.
[152] Link, K. P. *Circulation*, **1959**, *19*, 97.
[153] Gershen, B. *Maryland Med. J.*, **1993**, *42*, 393.
[154] Pineo, G. f.; Hull, R. D. *Drug Safety*, **1993**, *9*, 263.
[155] Bose, P. K. *J. Indian Chem. Soc.*, **1958**, *35*, 367.
[156] Godfrey, J. C.; Price, K. E. *Adv. Appl. Microbiol.*, **1972**, *15*, 231.
[157] Godfrey, J. C. *Adv. Appl. Microbiol.*, **1981**, *27*, 125.
[158] Duddeck, H.; Snatzke, G.; Yemul, S. S. *Phytochem.*, **1978**, *17*, 1369.
[159] Thornes, R. D. *Irish Med. J.*, **1977**, *70*, 480.
[160] Thornes, R. D. *Vet. Record*, **1977**, *101*, 27.
[161] Egan, D.; O'Kennedy, R.; Moran, E.; Cox, D.; Prosser, E.; Thornes, R. D. *Drug Metab. Rev.*, **1990**, *22*, 503.
[162] Thornes, R. D.; Wall, P. G. *Vet. Record*, **1991**, *129*, 496.
[163] Thornes, R. D.; Early, A. M.; Hogan, B.L.; Reen, D. *Irish Med. J.*, **1982**, *75*, 423.
[164] Maxwell, A. *Mol. Microbiol.*, **1993**, *9*, 681.
[165] Drlica, K.; Coughlin, S. *Pharmacol. Ther.*, **1989**, *44*, 107.
[166] Paoletti, C. *Pharmacol. Ther.*, **1993**, *60*, 381.
[167] Podbielkowska, A.; Kupidlowska, E.; Walesa, M.; Dobrzynska, K.; Louis, S. A.; Keightley, A.; Zobel, A. M. *Int. J. Pharmacogn.*, **1994**, *32*, 262.
[168] Morita, A.; Takashima A.; Nagai M.; Dall'Acqua F. *J. Dermatol.*, **1990**, *17*, 545.
[169] Thornes, D.; Daly, L.; Lynch, G.; Browne, H.; Tanner, A.; Keane, F.; O'Loughlin, S.; Corrigan, T.; Daly, P.; Edwards, G.; *et al. Eur. J. Surg. Oncol.*, **1989**, *15*, 431.
[170] Thornes, R. D.; Daly, L.; Lynch, G.; Breslin, B.; Browne, H.; Browne, H.Y.; Corrigan, T.; Daly, P.; Edwards, G.; Gaffney, E.; *et al. J. Cancer Res. Clin. Oncol.*, **1994**, *120*, S32.
[171] Thornes, R.D.; Lynch, G.; Sheehan, M. V. *Lancet*, **1982**, *8293*, 328.
[172] Marshall, M. E.; Butler, K.; Cantrell, J.; Wiseman, C.; Mendelsohn, L. *Cancer Chemother. Pharmacol.*, **1989**, *24*, 65.
[173] Marshall, M. E.; Mendelsohn, L.; Butler, K.; Riley, L.; Cantrell, J.; Wiseman, C.; Taylor, R.; Macdonald, J. S. *J. Clin. Oncol.*, **1987**, *5*, 862.
[174] Marshall, M. E.; Mendelsohn, L.; Butler, K.; Cantrell, J.; Harvey, J.; Macdonald, J. *Cancer Treatment Rep.*, **1987**, *71*, 91.
[175] Gallicchio, V. S.; Hulette, B. C.; Harmon, C.; Marshall, M. E. *J. Biol. Response Modifiers*, **1989**, *8*, 116.
[176] Guh, J. H.; Yu, S. M.; Ko, F. N.; Wu, T.S.; Teng, C. M. *Eur. J. Pharmacol.*, **1996**, *298*, 191.
[177] Marshall, M. E.; Mohler, J. l.; Edmonds, K.; Williams, B.; Butler, K.; Ryles, M.; Weiss, L.; Urban, D.; Bueschen, A.; Markiewicz, M.; *et al. J. Cancer Res. Clin. Oncol.*, **1994**, *120*, S39.
[178] Knobler, R. M.; Edelson, R. l. *Med. Clin. North Am.*, **1986**, *70*, 109.

[179] Edelson, R. L.; Berger, C.; Gasparro, E.; Jegasothy, B.; Heald, P.; Wintroub, B.; Vonderheid, E.; Knobler, R. *New Eng. J. Med.*, **1987**, *316*, 297.
[180] Mohler, J. l.; Williams, B. T.; Thompson, I. M.; Marshall, M. E. *J. Cancer Res. Clin. Oncol.*, **1994**, *120*, S35.
[181] Marshall, M. E.; Conley, D.; Hollingsworth, P.; Brown, S.; Thompson, J.S. *J. Biol. Response Modifiers*, **1989**, *8*, 70.
[182] Marshall, M. E.; Riley, L. K.; Rhoades, J.; Eichhorn, T.; Jennings, C.D.; Cibull, M.; Thompson, J. *J. Biol. Response Modifiers*, **1989**, *8*, 62.
[183] Marshall, M. E.; Kervin, K.; Benefield, C.; Umerani, A.; Albainy-Jenei, S.; Zhao, Q.; Khazaeli, M. B. *J. Cancer Res. Clin. Oncol.*, **1994**, *120*, S3.
[184] Ohta, T. *Crit. Rev. Toxicol.*, **1993**, *23*, 127.
[185] Feuer, G.; Kellen, J. A.; Kovacs, K. *Oncol.*, **1976**, *33*, 35.
[186] Talalay, P. *Adv. Enzyme Regul.*, **1989**, *28*, 237.
[187] Song, P. S.; Tapley, K. J. Jr. *Photochem. Photobiol.*, **1979**, *29*, 1177.
[188] Baden, H.; Parrington, J.; Delhanty, J.; Pathak, M. *Biochim. Biophys. Acta*, **1972**, *262*, 247.
[189] Warin, A. P.; Carruthers, J. A. *Clin. Exp. Dermatol.*, **1976**, *1*, 181.
[190] Parrish, J.; Fitzpatrick, T.; Tanenbaum, L.; Pathak, M. *New Engl. J. Med.*, **1974**, *291*, 1207.
[191] Cristofolini, M.; Recchia, G.; Boi, S.; Piscioli, F.; Bordin, F.; Baccichetti, F.; Carlassare, F.; Tamaro, M.; Pani, B.; Baburdi, N.; Guiotto, A.; Rodighiero, P.; Vedaldi, D.; Dall'Acqua, F. *Br. J. Dermatol.*, **1990**, *122*, 513.
[192] Musajo, L.; Rodighiero, G. *Photophysiol.*, **1972**, *7*, 115.
[193] Owens, D.; Glicksman, J.; Freeman, R.; Carnes, R. *J. Invest. Dermatol.*, **1968**, *51*, 435.
[194] Bordin, F.; Carlassare, F.; Busulini, L.; Baccichetti, F. *Photochem. Photobiol.*, **1993**, *58*, 133.
[195] Cole, M. S; Minifee, P. K; Wolma, F. J. *Surgery*, **1988**, *103*, 271.
[196] Chambers, M.; Thornes, R.D.; McKernan, M. *Br. Med. J.*, **1989**, *298*, 755.
[197] Marshall, M. E.; Butler, K.; Metcalfe, M.; Tate, M. *Surgery*, **1989**, *105*, 237.
[198] Casley-Smith J. R; Casley-Smith J. R. *Aust. J. Dermatol.*, **1992**, *33*, 69.
[199] Nishibe, S.; Sardari, S.; Kodama, A.; Horita, K.; Kudo, M.; Koike, K.; Nikaido, T. *Nat. Med.*, **1997**, *51*, 482.
[200] Kimura, Y.; Okuda, H.; Arichi, S.; Baba, K.; Kozawa, M. *Biochim. Biophys. Acta*, **1985**, *834*, 224.
[201] Hoult, J. R.; Moroney, M. A.; Paya, M. *Methods Enzymol.*, **1994**, *234*, 443.
[202] Paya, M.; Halliwell, B.; Hoult, J. R. *Biochem. Pharmacol.*, **1992**, *44*, 205.
[203] Hoult, J. R.; Forder, R. A.; de las Heras, B.; Lobo, I. B.; Paya, M. *Agents Actions*, **1994**, *42*, 44.
[204] Kimura, Y.; Ohminami, H.; Arichi, H.; Okuda, H.; Baba, K.; Kozawa, M.; Arichi, S. *Planta Med.*, **1982**, *45*, 183.
[205] Baker, J.; Cros, M.; Lafont, H.; Hauton, J. C.; Aycard, J. P. In. *Psoralens: past, present and future of photochemoprotection and other biological activities.* T. B. Fitzpatrick, P. Forlot, M. A. Pathak and F. Urbach, Ed.; John Library Eurotext: Paris,.**1988 (pub. 1989)**; pp. 201-206.
[206] Middleton, E. *Trends Pharmacol. Sci.*, **1984**, *615*, 335.
[207] Rangari, V. D.; Agrawal, S. R. *Indian Drugs*, **1992**, *29*, 662.
[208] Sandor, P.; Kovach, A. G. B. *Arzneim. Forsch.*, **1968**, *18*, 1287.
[209] Smith, C. G.; Dietz, A.; Sokolski, W. T.; Savage, G. M. *Antibiot. Chemother.*, **1956**, *6*, 135.

[210] Paya, M.; Goodwin, P. A.; De Las Heras, B.; Hoult, J. R. *Biochem. Pharmacol.*, **1994**, *48*, 445.

[211] Paya, M.; Halliwell, B.; Hoult, J. R. *Free Radical Res. Commun.*, **1992**, *17*, 293.

[212] Nieschulz, v., O.; Schmersahl, P. *Arzneim. Forsch.*, **1968**, *18*, 1330.

[213] Lee, R. E.; Bykadi, G.; Ritschel, W. A. *Arzneim. Forsch.*, **1981**, *31*, 640.

[214] Spath, E.; Kuffner, F. *Monatsh. Chem.*, **1936**, *69*, 75.

[215] Pakrashi, A. *Ind. J. Exp. Biol.*, **1968**, *6*, 204.

[216] Pakrashi, A. *Ind. J. Exp. Biol.*, **1968**, *6*, 212.

[217] Huang, H.-C.; Weng, Y.-I.; Lee, C. R.; Jan, T.-R.; Chen, Y.-L.; Lee, Y.-T. *Br. J. Pharmacol.*, **1993**, *110*, 1508.

[218] Huang, H. -C.; Chu, S. H.; LeeChao, P. D. *Eur. J. Pharmacol.*, **1991**, *198*, 211.

[219] Huang, H.-C.; Huang, Y.-L.; Chang, J.-L.; Chen, C.-C.; Lee, Y. T. *Eur. J. Pharmacol.*, **1992**, *217*, 143.

[220] Tornquist, K.; Vuorela, H. *Planta Med.*, **1990**, *56*, 127.

[221] Okada, Y.; Miyauchi, N.; Suzuki, K.; Kobayashi, T.; Tsutsui, C.; Mayuzumi, K.; Nishibe, S.; Okuyama, T. *Chem. Pharm. Bull.*, **1995**, *43*, 1385.

[222] Ellinger, A. *Arch. Exptl. Pathol. Pharmakol.*, **1908**, Suppl., 150.

[223] Tzavellas, v., O. *Arzneim. Forsch.*, **1968**, *18*, 1027.

[224] Prosdocimi, M.; Zatta, A.; Gorio, A.; Zanetti, A.; Dejana, E. *Naunyn Schmiedebergs Arch. Pharmacol.*, **1986**, *332*, 305.

[225] Ko, F. N.; Wu, T. S.; Liou, M. J.; Huang, T. F.; Teng, C. M. *Thromb. Haemostasis*, **1989**, *62*, 996.

[226] Stralfors, C. E. A. *Brit. Patent.*, **1968**, *1*,130,566, 2pp.

[227] Casley-Smith, J. R., Jamal, S., Casley-Smith, J. R. *Ann. Trop. Med. Parasitol.*, **1993**, *8*, 247.

[228] Casley-Smith, J. R., Morgan, R. G., Piller, N. B. *N. Engl. J. Med.*, **1993**, *329*, 1158.

[229] Aporti, F.; Finesso, M.; Granata, L. *Pharmacol. Res. Commun.*, **1978**, 10, 469.

[230] Khadzhai, Y. I.; Kuznetsova, V. F. *Farmatsevt. Zh.*, **1962**, *17*, 57.

[231] Ko, F. N.; Wu, T. S.; Liou, M. J.; Huang, T. F.; Teng, C. M. *Eur. J. Pharmacol.*, **1992**, *219*, 29.

[232] Teng, C. M.; Lin, C. H.; Ko, F. N.; Wu, T. S.; Huang, T. F. *Naunyn Schmiedebergs Arch. Pharmacol.*, **1994**, *349*, 202.

[233] Bolton, T.; Casley-Smith, J. R. *Experientia*, **1975**, *31*, 271.

[234] Foldi-Borlsok, E.; Bedall, F. K.; Fahlfs, V. W. *Arzneim-Forsch (Drug Research)*, **1971**, *21*, 2025.

[235] Huang, H.-C.; Lee, C. R.; Weng, Y.-I.; Lee, M.-C.; Lee, Y.-T. *Eur. J. Pharmacol.*, **1992**, *218*, 123.

[236] Tawata, M.; Yoda, Y.; Aida, K.; Shindo, H.; Sasaki, H.; Chin, M.; Onayam T. *Planta Med.*, **1990**, *56*, 259.

[237] Del Maschio, A.; Bazzoni, G.; Zatta, A.; Chen, Z. M.; Dejana, E.; Prosdocimi, M. *Eur. J. Pharmacol.*, **1990**, *187*, 541.

[238] Zanetti, A.; Zatta, A.; Prosdocimi, M.; Dejana, E. *Eur. J. Pharmacol.*, **1986**, *128*, 119.

[239] Wilhelms, O. H.; Roesch, A.; Schaumann, W. *Agents Actions Supp.*, **1991**, *34*, 335.

[240] Danno, K.; Toda, K.; Horio, T. *J. Invest. Dermatol.*, **1985**, *85*, 110.

[241] Grigg, G. W. *Mutat. Res.*, **1977**, *47*, 161.

[242] Rodighiero, G. *Prog. Biochem. Pharmacol.,* **1978,** *14,* 94.

[243] Kuiper-Goodman, T. *Vet. Hum. Toxicol.,* **1991,** *33,* 325.

[244] Stormer, F. C. *Tidsskr. Nor. Laegeforen.,* **1993,** *113,* 1061.

[245] Fentem, J. H.; Fry, J. R. Comp. Biochem. Physiol. C: Comp. Pharmacol. Toxicol., **1993,** 104, 1.

[246] Cohen, A. J. *Food & Cosmetics Toxicology,* **1979,** *17,* 277.

[247] Mandula, B. B.; Pathak, M. A. *Biochem. Pharmacol.,* **1979,** *28,* 127.

[248] Cox, D.; O'Kennedy, R.; Thornes, R. D. *Human Toxicol.,* **1989,** *8,* 501.

[249] Ma, R.; Cohen, M.; Berenbaum, M.; Schuler, M. *Arch. Biochem. Biophys.,* **1994,** *310,* 332.

[250] Maenpaa, J.; Sigusch, H.; Raunio, H.; Syngelma, T.; Vuorela, P.; Vuorela, H.; Pelkonen, O. *Biochem. Pharmacol.,* **1993,** *45,* 1035.

[251] Mandula, B. B.; Pathak, M. A.; Dudek, G. *Science,* **1976,** *193,* 1131.

[252] Mandula, B. B.; Pathak, M. A.; Nakayama, Y.; Davidson, S. J. *Brit. J. Dermatol.,* **1978,** *99,* 687.

[253] Salonpaa, P.; Pelkonen, O.; Kojo, A.; Pasanen, M.; Negishi, M.; Raunio, H. *Biochem. Biophys. Res. Commun.,* **1994,** *205,* 631.

[254] Farargy, A. E.; Yassin, F.; Abdel-Chani, E.; El-Said, N.; Saleh, R. *Heterocycles,* **1992,** *34,* 25.

[255] Hammad, A.; El-Sayed, E. S.; Islam, I. E.; Shafik, N. *J. Chem. Soc. Pak.,* **1990,** *12,* 292.

[256] Weidenbornen, M.; Hem Chandra Jha, *Chem. Mikrobiol. Technol. Lebensm,* **1995,** *17,* 22.

[257] Verma, B. S.; Abrol, V.; Sanguan, N. K.; Malik, O. P. *Chim. Acta Turcica,* **1989,** *17,* 433.

[258] Brahmbhatt, D. I.; Singh, S.; Patel, K. C. *Int. J. Polymeric Mat.,* **1997,** *35,* 145.

[259] Gursoy, A.; Ates, O.; Karali, N.; Cesur, N.; Kiraz, M. *Eur. J. Med. Chem.,* **1996,** *31,* 643.

[260] Niedmann, C.; Kumerlin, R.; Sylva, M. *Anal. Real. Acad. Farm.,* **1990,** *56,* 171.

[261] Nofal, Z. M.; Mandour, A. H.; Nassar, M. I. *Egypt. J. Chem.,* **1990,** *33,* 509.

[262] Hosmani, C. K.; Mallur, S. G.; Badami, B. V. *Indian J. Heterocyclic Chem.,* **1996,** *6,* 107.

[263] Gudasi, K. B.; Dhumwad, S. D.; Goudar, T. R. *Indian J. Chem.,* **1994,** *33A,* 782.

[264] Geethanjali, Y.; Nirmala, K.; Rao, M. K. *Indian J. Heterocyclic Chem.,* **1995,** *4,* 307.

[265] Sharan, P.; Giri, Nizamuddin, S. *Indian J. Chem., Soc.,* **1989,** *66,* 393.

[266] Ahluwalia, V. K.; Mann, R. R.; Baba, S. *Indian J. Chem.,* **1989,** *28B,* 608.

[267] Daneshtalab M.; Sardari, S., *et al.,* unpublished data

[268] Betina V. *Folia Microbiol.,* **1985,** *30,* 80.

[269] Bordin, F.; Dall'Acqua, F.; Guiotto, A. *Pharmac. Ther.,* **1991,** *52,* 331.

[270] Fowlks, W. L. *J. Invest. Dermatol.,* **1959,** *32,* 233.

[271] Rodighiero, G.; Dall'Acqua, F.; Averbeck, D. In. *Psoralen DNA Photobiology.* F. P. Gasparro, Ed.; CRC Press: Boca Raton, **1988;** Vol. *I:* 37-114.

[272] Juttermann, R.; Averbeck, D.; Averbeck, C. *Il Farmaco,* **1985,** *40,* 3.

[273] Gasparro, F. P. In. *Psoralen DNA Photobiology.* F. P. Gasparro, Ed.; CRC Press: Boca Raton, **1988;** Vol. *I:* 5-36.

[274] Maritus, C.; Nitz-Litzow, D. *Biochem. Biophys. Acta,* **1953,** *12,* 134.

[275] Judah, J.D. *J. Pharam. Pharmacol.*, **1959**, *11*,1.
[276] Kaiser, E.; Kramar, R.; Farkouh, E.; Podbielkowska, M.; Kupidlowska, E.; Waleza, M.; Dobrzynska, K.; Louis, S. A.; Keightley, A.; Zobel, A. M. *Enzymologia*, **1966**, *30*, 64.
[277] Stenlid, G.; Saddik, K. *Physiol. Plant.*, **1962**, *15*, 369.
[278] Knypl, J. S. *Acta Soc. Bot. Pol.*, **1968**, *37*, 51.
[279] Knypl, J. S. *Biol. Plant.*, **1969**, *12*, 199.
[280] Kupidlowska, E.; Dobrzynska, K.; Parys, E.; Zobel, A. M. *J. Chem. Ecol.*, **1994**, *20*, 2471.
[281] Podbielkowska, M.; Zarska-Maciejewska, B.; Kacperska-Palacz, A. *Protoplasma*, **1975**, *83*, 201.
[282] Podbielkowska, M.; Kupidlowska, E. *Acta Soc. Bot. Pol.*, **1976**, *45*, 239.
[283] Podbielkowska, M.; Waleza, M.; Zobel, A. M. *Acta Soc. Bot. Pol.*, **1981**, *50*, 563.
[284] Averbeck, D.; Chandra, P.; Biswas; R. K. *Radiat. Environ. Biophys.*, **1975**, *12*, 241.
[285] Bussey, C.; Lepoittevin, J. P.; Benezera, C. *Bioorg. Med. Chem. Lett.*, **1993**, *3*, 1283.
[286] Contreras, A.; Maxwell, A. *Mol. Microbiol.*, **1992**, *6*, 1617.
[287] Fisher, L. M.; Austin, C. A.; Hopewell, R.; Margerrison, E. E.; Oram, M.; Patel, S.; Plummer, K.; Sng, J. H.; Sreedharan, S. *Philos. Trans. R. Soc. London. Ser. B.*, **1992**, *336*, 83.
[288] Samuel, D. S.; Garon, C. F. *Antimicrob. Agents Chemother.*, **1993**, *37*, 46.
[289] Goldman, G. H.; Yu, W. H. Y.; Sanders, M. M.; La Voie, E. J.; Liu, L. F. *Biochem.*, **1997**, *36*, 6488.
[290] Fostel, J.; Montgomery, D.; Lartey, P. *FEMS Microbiol. Lett.*, **1996**, *138*, 105.
[291] Fosse, P.; Rene, B.; Charra, M.; Paoletti, C.; Saucier, J-M. *Mol. Pharmacol.*, **1992**, *42*, 590.
[292] Fosse, P.; Charra, M.; Paoletti, C.; Saucier, J. M.; Rene, B. *Bull. Cancer*, **1994**, *81*,194.
[293] Caffieri, S.; Schoonderwoerd, S. A.; Daga A. Dall'Adcqua, F.; Beijersbergen, Van H. G. M. J. *Med. Biol. Environ.*, **1989**, *17*, 797.
[294] Schoonderwoerd, S. A.; Beijersbergen, van Henegouwen G. M.; Persons, C. C.; Caffieri, S.; Dall'Acqua, F. *J. Photochem. Photobiol. B – Biol.*, **1991**, *10*, 257.
[295] Gawron, A.; Pawlikowska, B. *Journal of Pharm. Pharmacol.*, **1993**, *45*, 1087.
[296] Corderio Neto, F.; Dietrich, S. *Appl. Environ. Microbiol.* **1977**, *34*, 258.
[297] Laskin, J. D.; Laskin, D. L. In. *Psoralen DNA Photobiology*. F. P. Gasparo, Ed.; CRC Press: Boca Raton,.**1988**; Vol. *II*: 135-148.
[298] Laskin, J. D.; Lee, E.; Yurkow, E. J.; Laskin, D.L.; Gallo, M. A. *Proc. Nat. Acad. Sci. USA*, **1985**, *82*, 6158.
[299] Moreno, G.; Salet, C.; Kohen, C.; Kohen, E. *Biochim. Biophys. Acta*, **1982**, *721*, 109.
[300] Yurkow, E. J.; Laskin, J. D. *J. Biol. Chem.*, **1987**, *262*, 8439.
[301] Ben-Hur, E.; Song, P. S. *Adv. Radiat. Biol.*, **1984**, *11*, 131.
[302] Bertaux, B.; Dubertret, L.; Moreno, G. *Acta Derm. Venereol.*, **1981**, *61*, 481.
[303] Frederiksen, S.; Hearst, J. E. *Biochim. Biophys. Acta*, **1979**, *563*, 343.
[304] Pathak, M. A.; Kramer, D. M. *Biochim. Biophys. Acta*, **1969**, *195*, 197.
[305] Toda, K.; Danno, K.; Tachibana, T.; Horio, T. *J. Invest. Derm.*, **1986**, *87*, 113.
[306] Laskin, J. D. *J. Am. Coll. Toxicol.*, **1989**, *8*, 797.

[307] Lockyer, J. M.; Bowden, G. T.; Matrisian, L. M.; Magun, B. E. *Cancer Res.*, **1981**, *41*, 2308.

[308] Magun, B. E.; Matrisian, L. M.; Bowden, G. T. *J. Biol. Chem.*, **1980**, *255*, 6373.

[309] Laskin, J. D.; Lee, E.; Laskin, D.L.; Gallo, M. A. *Proc. Nat. Acad. Sci. USA*, **1986**, *83*, 8211.

[310] Sardari, S.; Mori, Y.; Daneshtalab, M. *Pharm. Res.*, **1997**, *14*, S344.

[311] Midden, R. In. *Psoralen DNA Photobiology*. F. P. Gasparo, Ed.; CRC Press: Boca Raton,.**1988**; Vol. *II*: 1-50.

[312] Egidy, G.; Paveto, C.; Passeron, S.; Galvagno, M. A. *Cell Biol. Int. Rep.*, **1990**, *14*, 59.

[313] Larsen, A. D.; Sypherd, P. S. *J. Bacteriol.* **1974**, *117*, 432.

[314] Nimi, M.; Nimi, K.; Tokunaga, J.; Nakayma, H. *J. Bacteriol.* **1980**, *142*, 1010.

[315] Cho, T.; Hamatake, H.; Kaminishi, H.; Hagihara, Y.; Watanabe, K. *J. Med. Vet. Mycol.*, **1992**, *30*, 35.

[316] Lane, H. A.; Nigg, E. A. *Trend in Cell Biol.*, **1997**, 7(Feb.), 63.

[317] Sestak, S.; Farkas, V. *Can. J. Microbiol.*, **1993**, *39*, 342.

[318] Walker-Caprioglio; H. M.; Casey, W. M.; Parks, L. W. *Appl. Environ. Microbiol.*, **1990**, *56*, 2853.

[319] Yoshii, Z.; Tokunaga, J.; Tawara, J. *Atlas of Scanning Electron Microscopy in Microbiology*. Igaku Shoin Ltd.: Tokyo, **1976**; pp.116-137.

[320] Mittag, H. *Mycoses.*, **1994**, *37*, 337.

[321] Phillipson, J. D. *Trans. Royal Soc. Trop. Med. Hyg.*, **1994**, *88*, S1/17.

[322] Clark, A. M. *Pharm. Res.*, **1996**, *13*, 1133.

[323] Plunkett, M. J.; Ellman, J. A. *Sci. Am.*, **1997**, April, 68.

[324] Hector, R. F. *Clin. Microbiol. Rev.*, **1993**, *6*, 1.

[325] Nishizuka, Y. *Nature*, **1984**, *308*, 693.

[326] Kaiser, E.; Kramer, R. E. *Enzymologia*, **1965**, *30*, 65.

[327] Bertaux, B.; Moreno, G.; Vinzens, F.; Dubertret, L. *Skin Pharmacol.*, **1988**, *1*, 250.

[328] Sardari, S.; Abel, M.; Micetich, R.G.; Daneshtalab, M. *Pharmazie*, **1999**, *54*, 156.

Atta-ur-Rahman (Ed.) *Studies in Natural Products Chemistry, Vol. 23*

CHEMICAL AND BIOLOGICAL PERSPECTIVES OF ELLAGITANNIN OLIGOMERS FROM MEDICINAL PLANTS

TAKASHI YOSHIDA*, TSUTOMU HATANO, HIDEYUKI ITO *and* TAKUO OKUDA

Faculty of Pharmacognosy, Faculty of Pharmaceutical Sciences, Okayama University, Tsushima, Okayama 700, Japan

ABSTRACT: The phytochemical study of hydrolyzable tannins, naturally occurring complex polyphenols, has been facilitated by recent developments in modern analytical and spectroscopic techniques. Among numerous tannins characterized during the last two decades, those classified as oligomeric ellagitannins have attracted widespread attention as to their diverse structures and biological activities including antioxidative, antitumor and antiviral effects. We describe herein the molecular structures of representative oligomeric hydrolyzable tannins and general methods for their extraction, separation, and structural elucidation based on MS, NMR and CD spectra. Biological activities specific to each oligomer structure are also summarized.

INTRODUCTION

Hydrolyzable tannins are widely distributed in dicotyledonous plants of angiospermae. Early studies in tannin chemistry focused mostly on the characterization of components of plant extracts used in the leather industry [1]. However, the number of reports dealing with the isolation and structural elucidation of this class of natural polyphenols from medicinal plants and foods has increased markedly during the last two decades [1,2]. The remarkable progress in the chemistry of hydrolyzable tannins has been mostly based on developments of modern analytical and isolation techniques and various spectroscopies including high-field NMR. In addition, interest in the tannin constituents of medicinal plants has grown as a results of recent findings that some tannins exhibit various biological, including antitumor, antiviral and antioxidant, activities [3]. Among the numerous hydrolyzable tannins characterized to date, oligomeric ellagitannins have received considerable attention because of their vast structural diversity and biological activities specific to each structure [2].

The purpose of this paper is to describe perspectives of chemistry of ellagitannin oligomers including structural elucidation by spectroscopic analyses, and of their biological activities.

DISCOVERY OF ELLAGITANNIN OLIGOMERS

In 1970, Hillis and Seikel isolated an ellagitannin from a *Eucalyptus* species, and tentatively suggested it to be a dimer composed of two glucoses, two galloyl and three hexahydroxydiphenoyl (HHDP) groups based on ^{1}H-NMR data and chromatographic behavior [4]. This was the first report implying the occurrence of dimeric ellagitannin in nature although the compound was left uncharacterized. The ^{1}H-NMR spectrum reported in that paper is now considered to be identical with that of a monomeric *C*-glucosidic tannin, casuarinin. Agrimoniin (**1**), the first structurally defined dimeric ellagitannin, was isolated in 1982 together with its related monomeric tannins, agrimonic acids A (**2**) and B (**3**), and potentillin (**4**) from the roots of *Agrimonia pilosa* (Rosaceae) which have been used as anticarcinogenic crude medicine in China [5].

The structure of agrimoniin was established as a dimer of **4** based on extensive ^{1}H-NMR and ^{13}C-NMR spectral analyses and methanolysis of its methylated derivative yielding dimethyl penta-*O*-methyldehydrodigallate (**5**) and dimethyl hexamethoxydiphenate (**6**). The absolute configurations of chiral HHDP groups in **1** were determined to be all (*S*) by application of the empirical rule in the CD spectrum [6].

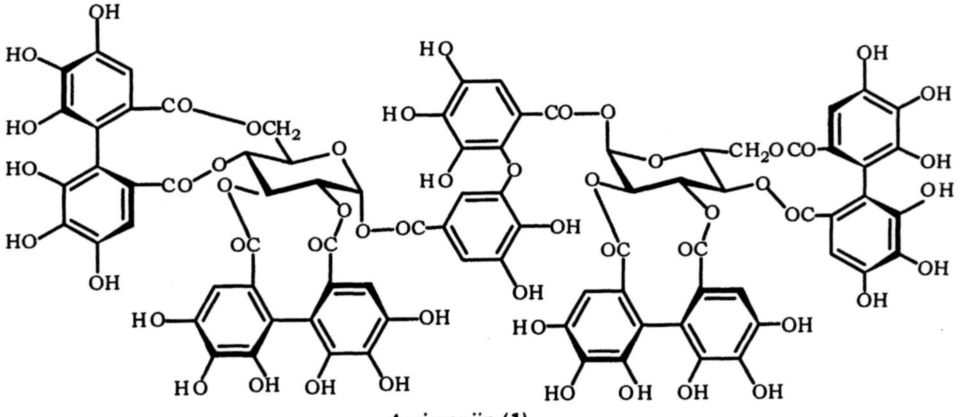

Agrimoniin (**1**)

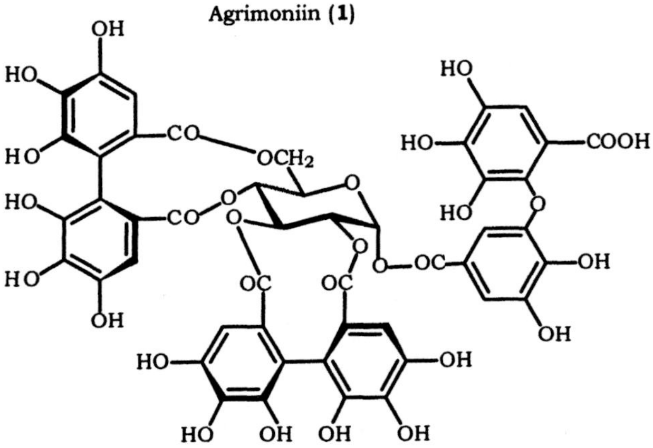

Agrimonic acid A (**2**)

(Chart 1). contd.....

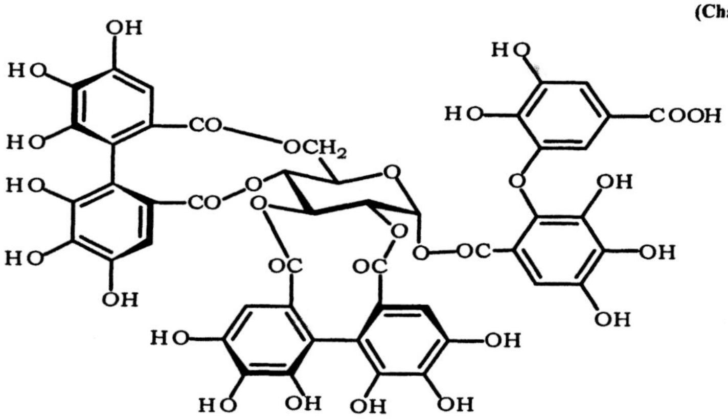

Agrimonic acid B (3)

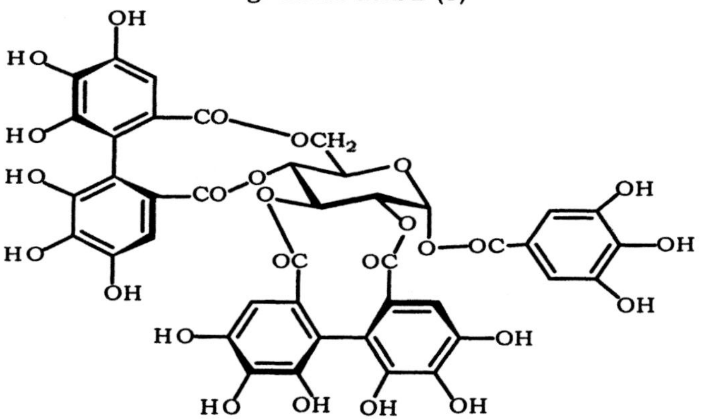

Potentillin (4)

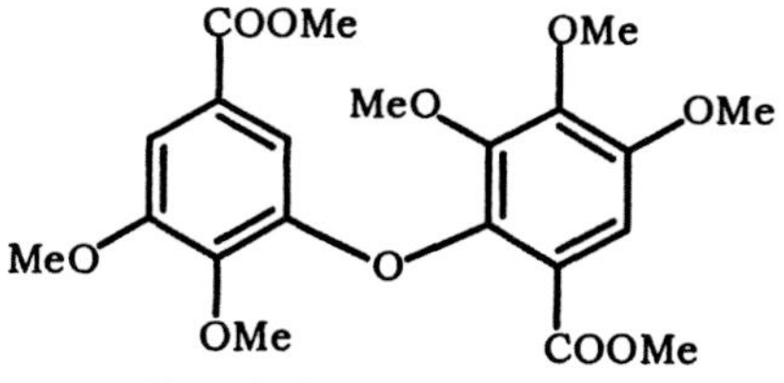

Dimethyl penta-O-
methyldehydrodigallate (5)

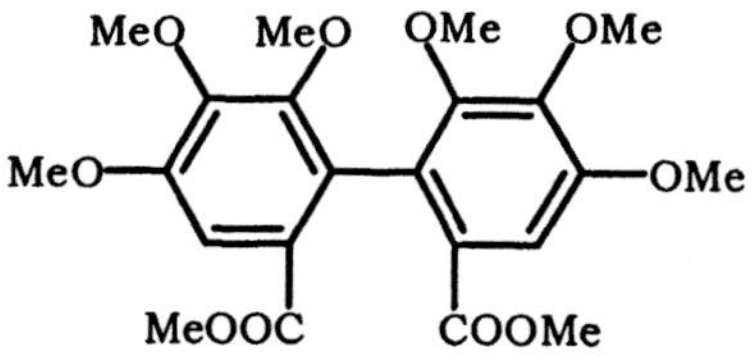

Dimethyl hexamethoxy
diphenate(6)

Chart 1.

In addition to agrimoniin (1), several other dimers and oligomers such as rugosins, sanguiins and gemins were found in the Rosaceous plants, *Rosa rugosa* [7,8], *Filipendula ulmaria* [1], *Sanguisorba officinalis* [9] and *Geum japonicum* [10], respectively. Among them, gemins B and C having the dehydrodigalloyl (DHDG) group as a linking unit were dimers structurally analogous to **1**, while rugosins and sanguiins were different types of oligomers with the valoneoyl and sanguisorboyl group, respectively, as a linking unit.

Miyamoto *et al.* revealed that an antitumor component isolated from the *A. pilosa* leaves upon fractionation guided by antitumor effect against Sarcoma 180 in mice was identical with agrimoniin [11]. Discovery of agrimoniin (**1**) with such biological activity and other new oligomers from the Rosaceous plants encouraged us to develop chemical and biological techniques with which to study of this type of hydrolyzable tannin, leading to the isolation of many new dimeric and oligomeric ellagitannins from various species of other plant families. To date, more than 160 new hydrolyzable tannin oligomers have been characterized [2,12]. Most of these are dimeric ellagitannins, and only six are tetramers, which are the largest hydrolyzable tannins characterized so far.

STRUCTURAL DIVERSITY OF OLIGOMERIC ELLAGI-TANNINS

Biogenesis of oligomers

Oligomeric hydrolyzable tannins are now regarded as metabolites biogenetically produced through intermolecular oxidative coupling(s) between the galloyl or HHDP group of one monomer and the HHDP or galloyl group of the other [1,12]. This biogenesis of oligomers is based on the hypothesis [13,14] proposed by Schmidt and Haslam for the formation of a chiral HHDP group in monomeric ellagitannins through intramolecular C-C coupling between galloyl groups in spatially suitable proximity on the glucopyranose core. It is implied from the structures of hydrolyzable tannins characterized so far that the C-C oxidative coupling between polyphenolic acyl groups occurs intramolecularly in the monomers, whereas the C-O coupling preferentially takes place intermolecularly to form a DHDG or valoneoyl group as the linking unit of monomers [2], although an exceptional example was recently reported [15].

Classification of Oligomers

Oligomeric hydrolyzable tannins can be classified in several ways based on the difference of 1) structures of the unit linking monomers, 2) structures of constituent monomers, 3) conformations of constituent glucose cores ($^4C_1$, $^1C_4$ conformation or open-chain form), and 4) degree of oligomerization. Of these, we believe the first way, associated with biogenesis, best facilitates comprehensive understanding of not only the structural diversity of oligomers but also their chemotaxonomical significance [2,12]. The types of oligomers are as follows:

Oligomers Having a Dehydrodigalloyl Group as a Linking Unit

Dimers of this type are biogenetically produced by intermolecular C-O oxidative coupling between galloyl groups of two monomers. They include agrimoniin (1) and gemins A (7) and B (8) from *Agrimonia* and *Geum* species in Rosaceae as described earlier, coriariin A (9) [16] from *Coriaria japonica* (Coriariaceae) and nupharin C (10) [17] from *Nuphar japonicum* (Nymphaeaceae). Tamaricaceous plants also produce this type of oligomer, for example, hirtellins A (11) and C (12), the latter of which possesses the dehydrodigalloyl (DHDG) group and its isomer (iso-DHDG) as linking units [18,19]. It is noteworthy that the position and/or configuration of the galloyl group on each glucose core, which participates in formation of the linking unit, seem to be specific to the plant family. Rosaceous and Coriariaceous plants produce dimers having a DHDG group at C-1 of each glucose residue. However they are distinguishable

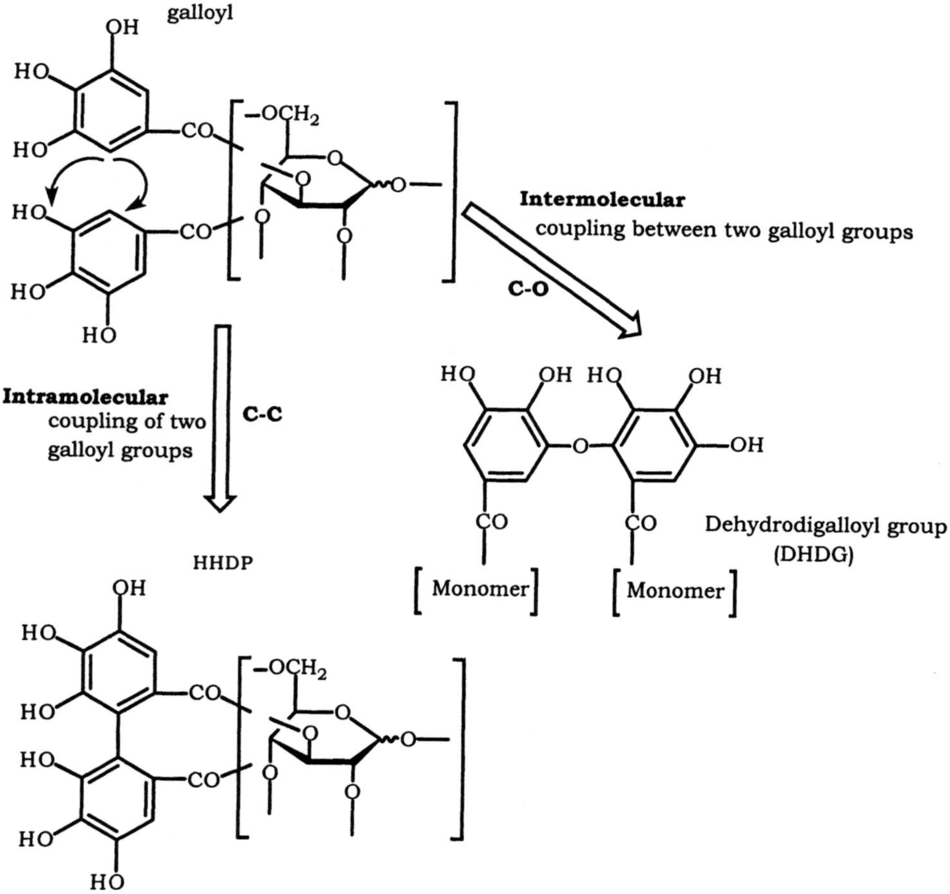

(Chart 2). contd.....

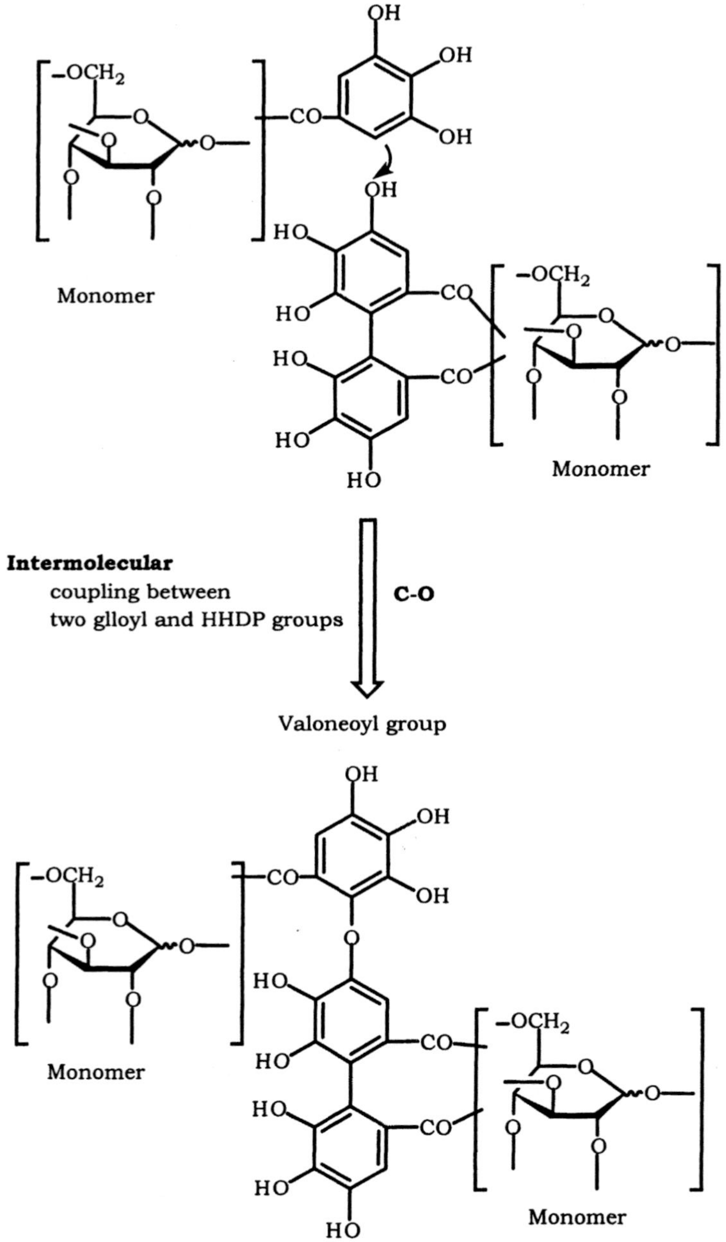

Chart 2.

from each other in the configuration of DHDG at C-1 (C-1'): which is
being α-oriented for the former (e.g. **1** and **7**) and β-oriented for the latter
(e.g. **9**). The oligomers of Tamaricaceae and Nymphaeaceae are produced
through similar coupling between galloyl groups at C-1 of one monomer

and C-2 of another. The constituent monomers of these oligomers do however differ, e.g. tellimagrandin II (or I) in Tamaricaceae and 3,6-HHDP-1, 2,4-trigalloyl-β-D-glucose in Nymphaeaceae. These structural features might thus be of chemotaxonomical significance.

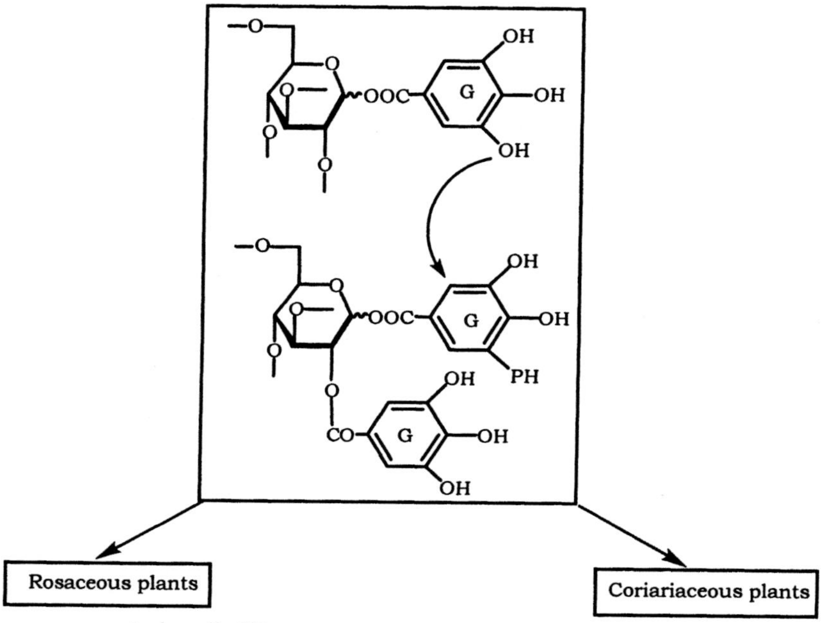

Agrimoniin (1)

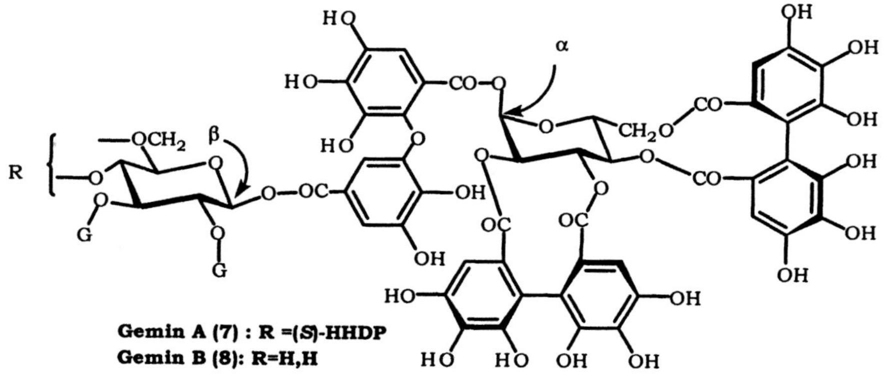

Gemin A (7) : R =(S)-HHDP
Gemin B (8): R=H,H

(S)-HHDP =

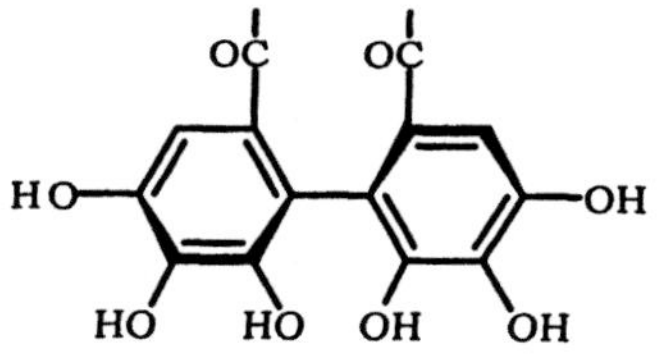

G =

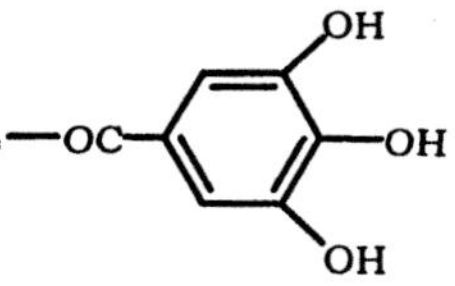

(Chart 3). contd.....

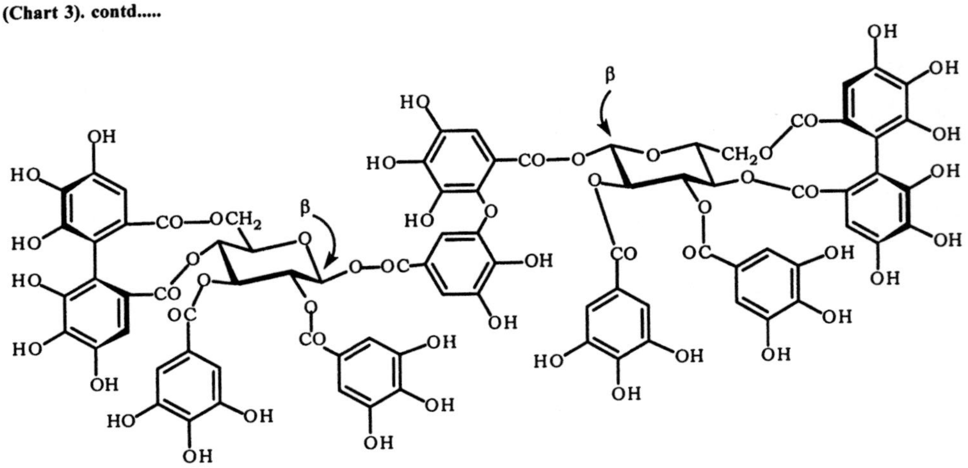

Coriariin A (**9**)

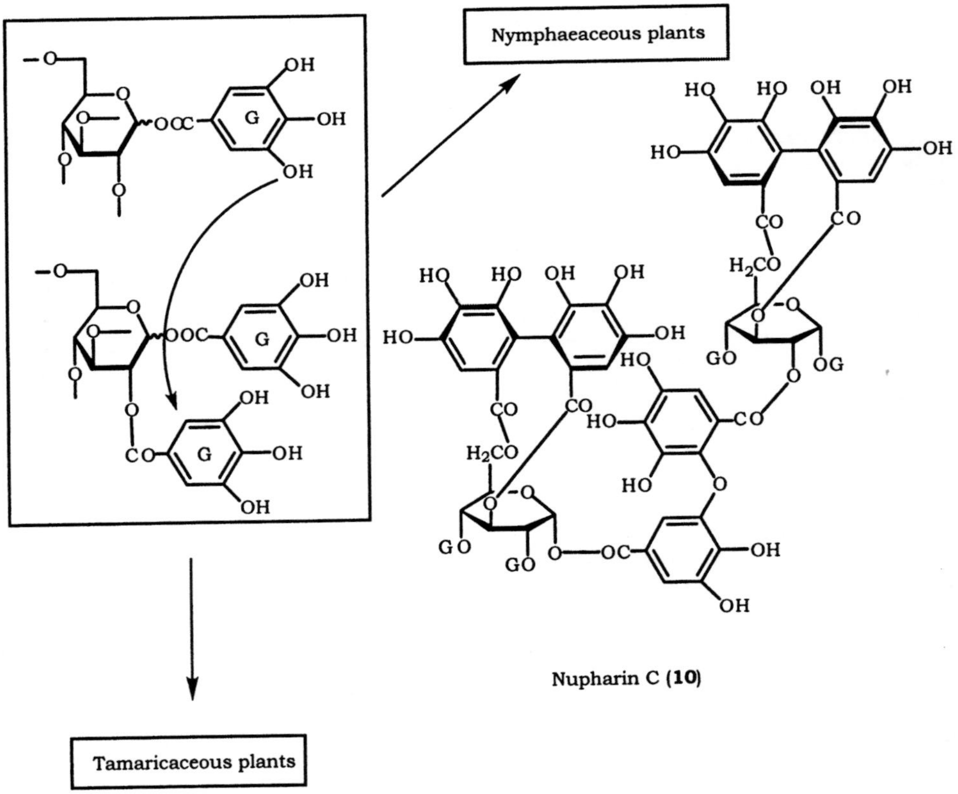

Nymphaeaceous plants

Tamaricaceous plants

Nupharin C (**10**)

(Chart 3). contd.....

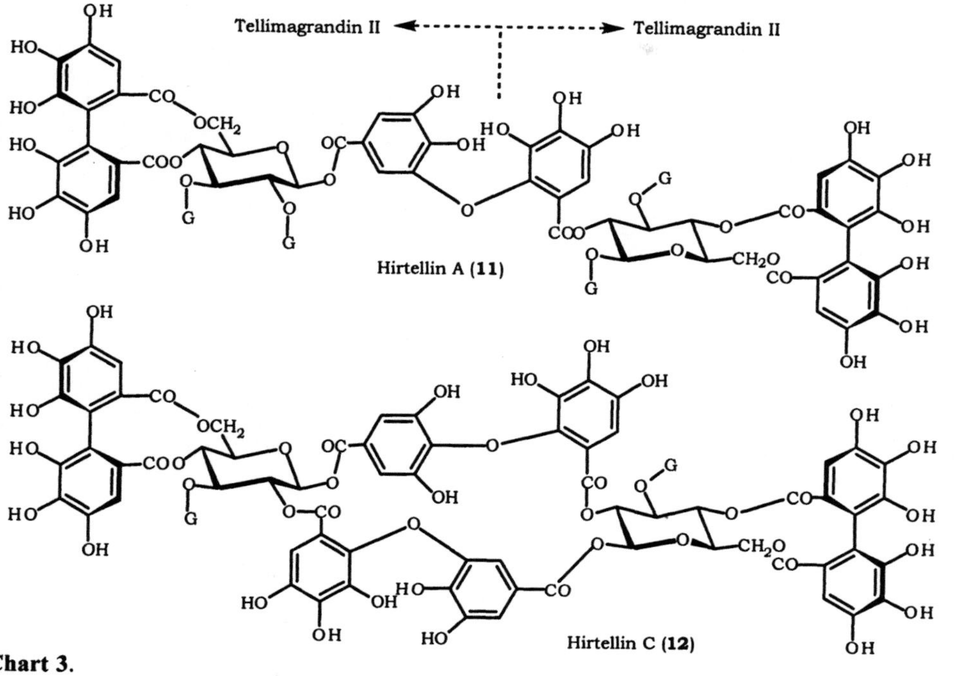

Chart 3.

Oligomers Having a Valoneoyl Group and its Isomeric Acyl Group as a Linking Unit

Oligomers of this type constitute the largest group of oligomeric hydrolyzable tannins and distribute in a relatively wide range of plant families. They include approximately 75 dimers, 15 trimers and 3 tetramers. Plant species producing oligomers of this type are shown in Table 1.

Table 1.   Plant Species Producing Oligomers Having a Valoneoyl Group and its Equivalent

Family	Species	Compound	Oligomerization	Ref.
Oligomers based on 4,6-valoneoyl-glucose core				
Fagaceae	*Quercus phillyraeoides*	Phillyraeoidin A-D	dimer	20
Betulaceae	*Corylus heterophylla*	Heterophylliin A,B (22)-F	dimer	21,22
		Heterophylliin G	trimer	22
		Rugosins D (13), F	dimer	22

(Table 1). contd.....

Family	Species	Compound	Oligomerization	Ref.
Rosaceae	*Rosa rugosa*	Rugosin D (**13**), E (**14**), F	dimer	8
		Rugosin G	trimer	8
	R. roxburghii	Roxbin A	dimer	23
	Rosa davurica	Davuriciin D_1	dimer	24
		Davuriciin T_1	trimer	24
Coriariaceae	*Coriaria japonica*	Coriariins D, E	dimer	25
Melastomataceae	*Tibouchina semidecandra*	Nobotanin A (**18**), F (**19**)	dimer	26
		Nobotanin C, E	trimer	27
	Heterocentron roseum	Nobotanin A (**18**), F (**19**)	dimer	28
		Nobotanin E, J	trimer	28
		Nobotanin K	tetramer	28
	Medinilla magnifica	Medinillin A	dimer	29
	Melastoma malabatricum	Malabatrin B, C, D	dimer	30
	Melastoma normale	Nobotanin A (**18**)	dimer	31
Onagraceae	*Oenothera erythrosepala*	Oenothein B (**23**)	dimer	32
	Oenothera biennis	Oenothein A (**42**), B (**23**)	trimer, dimer	33
	Oenothera laciniata	Oenothein A (**42**), B (**23**)	trimer, dimer	34
		Oenothein D, F	dimer	34
	Epilobium capense and other *Epiloboium* spp.	Oenothein A (**42**), B (**23**)	trimer, dimer	35
Lythraceae	*Lythrum anceps*	Oenothein B (**23**)	dimer	36
	Woodfordia fruticosa	Woodfordin A-C (**24**)	dimer	37
		Woodfordin D (**43**), E	trimer	33
		Woodfordin F	tetramer	38
Theaceae	*Camellia japonica*[a)]	Camelliin A (**17**), B (**26**)	dimer	39

(Table 1). contd.....

Family	Species	Compound	Oligomerization	Ref.
	Camellia oleifera	Camelliin B (26)	dimer	40
		Camellioferin A	dimer	40
	Schima walichii	Schimawalin B	dimer	41
Myrtaceae	*Eucalyptus alba*	Eucalbanin B, C (31)	dimer	42
	Eucalyptus robusta	Eurobustin A, B	dimer	43
Myrtaceae	*Eugenia uniflora*	Eugniflorin D$_1$ (25), D$_2$ (44)	dimer	44
Cornaceae	*Cornus officinalis*	Cornusiin A, D, E	dimer	45
		Cornusiin C (20)	trimer	45
Trapaceae	*Trapa japonica*	Trapanin A (21)	trimer	46
		Trapanin B	tetramer	46
Nyssaceae	*Camptotheca acuminata*	Camptothin A, B	dimer	47
Combretaceae	*Terminalia calamansanai*	Calamanin B, C	dimer	48
Hamamelidaceae	*Liquidamber formosana*	Isorugosin D (15), E (16)	dimer	49,50
		Isorugosin G	trimer	51
	Loropetalum chinense	Loropetalin B	trimer	52

Oligomers based on 2,3-valoneoyl-glucose core

Family	Species	Compound	Oligomerization	Ref.
Melastomataceae	*Tibouchina semidecandra*	Nobotanin B (27)	dimer	27
	Heterocentron roseum	Nobotanin B (27), G, H, I	dimer	53
Lythraceae	*Lagerstoroemia flos-regina*	Reginin D (28)	dimer	54
Theaceae	*Camellia japonica*[b)]	Camelliatannin H (29)	dimer	55
		Camelliatannin D (67)	dimer	56

Oligomers based on 3,6-valoneoyl-glucose core

Family	Species	Compound	Oligomerization	Ref.
Euphorbiaceae	*Euphorbia hirta*	Euphorbin A (30), B (32)	dimer	57
		Euphorbin C (40), D (41)	dimer	58,59

(Table 1). contd.....

Family	Species	Compound	Oligomerization	Ref.
		Euphorbin E	dimer	60
	Euphorbia tirucalli	Euphorbin F	dimer	61
		Tirucallin B	dimer	61
	Euphorbia prostrata	Euphorbin G, H	dimer	62
	Euphorbia helioscopia	Euphorherin	dimer	63
	Euphorbia maculata	Euphorbin A (30), B (32)	dimer	64
		Eumaculin A-C	dimer	64,65
	Euphorbia watanabei	Euphorbin A (30), B (32)	dimer	66
		Euphorbin I (33)	dimer	66
	Excoecaria kawakamii	Excoecarianin A, B	dimer	67
	Antidesma pentandrum var. *barbatum*	Antidesmin A	dimer	68
	Bishofia javanica	Bishofianin	dimer	69
	Mallotus japonicus	Mallotannin A, B	dimer	70

a) Flowers. b) Leaves and fruits

The vast structural diversity comes largely from the variation in the positions of the HHDP and galloyl groups that participate in formation of the valoneoyl group through intermolecular C-O oxidative coupling. Even in the C-O oxidative coupling between a galloyl group of one monomer and a 4,6-HHDP group of another, the coupling modes are grouped into 5 subtypes (a-e) as illustrated in Chart 4, and they seem to be characteristic of the plant families.

Oenothein B (23) and woodfordin C (24) are members of a unique class of oligomers with macrocylic structures and diverse pharmacological activities. Oenothein B (23) was first isolated from *Oenothera erythrosepala* [32] and has been shown to be characteristic of Onagraceae [2,12,35]. Oenothein B (23) and its galloyl congeners, woodfordin C (24) and eugeniflorin D$_1$ (25), have also been found in species of Lythraceae [37] and Myrtaceae [44]. The macro-ring in these tannins might be formed by double coupling between galloyl and HHDP groups of each monomer in the modes of b) and e) (Chart 4). Camelliin B (26) [39], an another member of this class, differs from 23 and 24 in the position of a galloyl part of one of the valoneoyl groups in the molecule [modes, a) and e)].

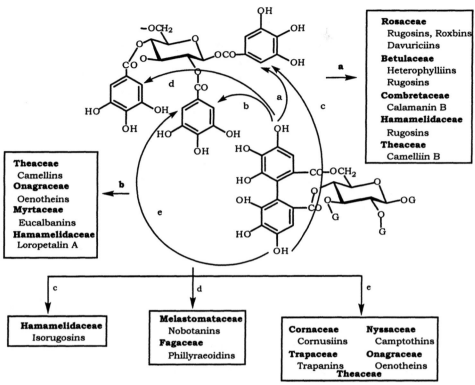

Chart 4. Coupling modes leading to oligomers with a valoneoyl group at O-4/-6 of the glucose core.

Rugosin D (**13**) : R=(β)-OG
Rugosin E (**14**) : R=OH

(Chart 5). contd.....

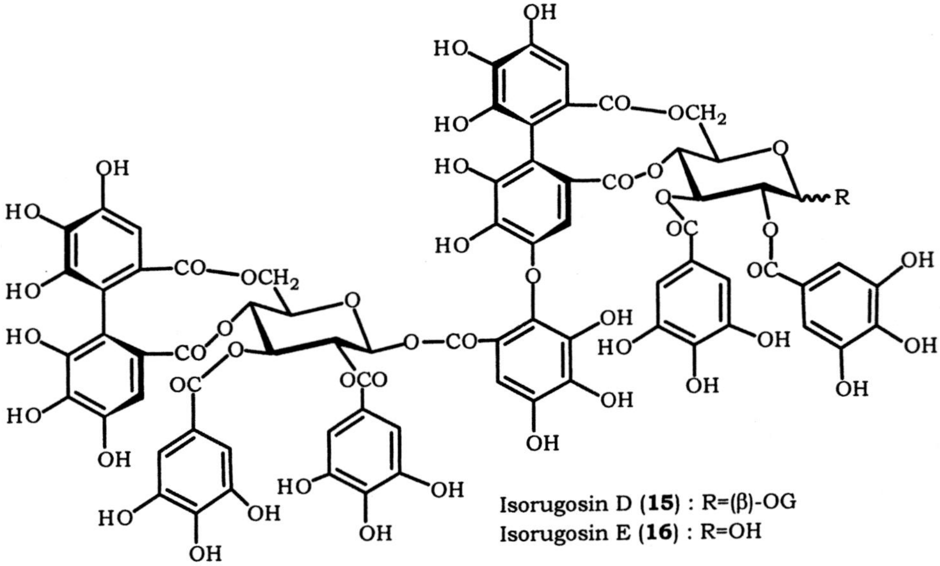

Isorugosin D (**15**) : R=(β)-OG
Isorugosin E (**16**) : R=OH

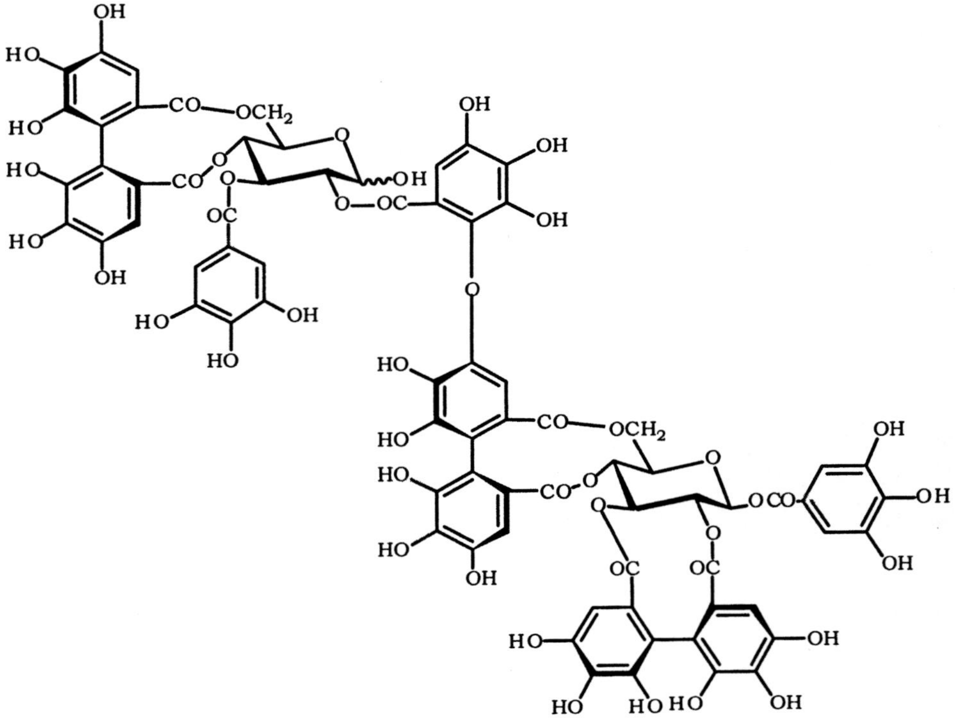

Camelliin A (**17**)

(Chart 5). contd.....

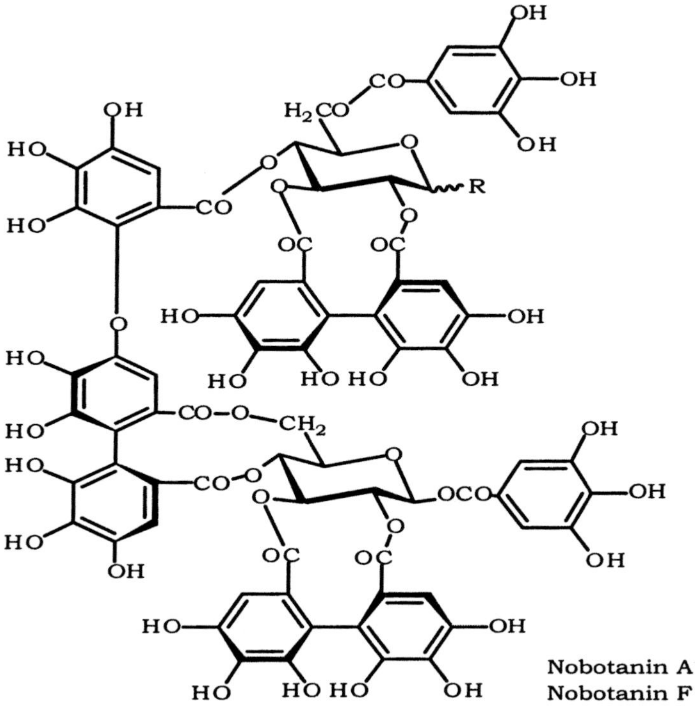

Nobotanin A (18) : R=OH
Nobotanin F (19) : R=(β)-OG

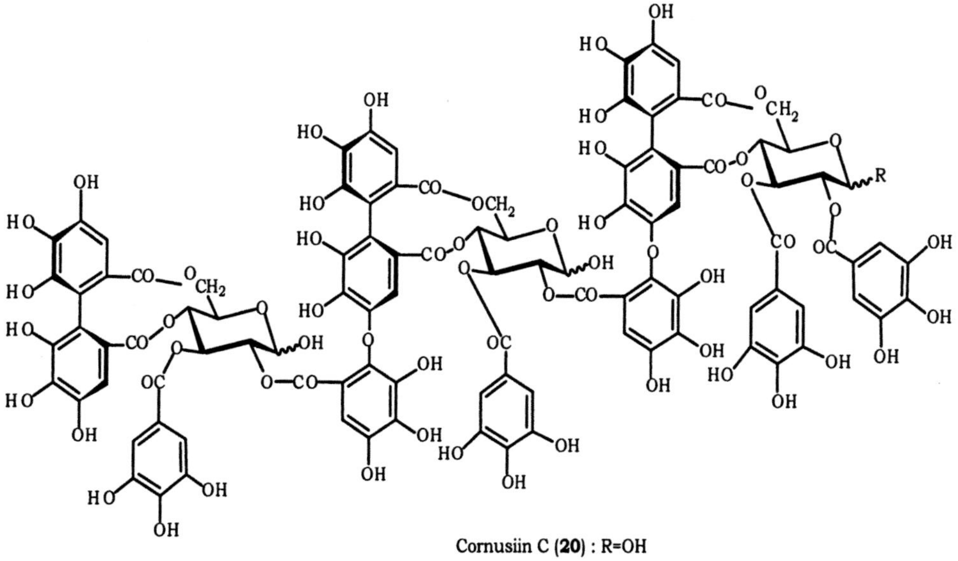

Cornusiin C (20) : R=OH
Trapanin A (21) : R=β)-OG

(Chart 5). contd.....

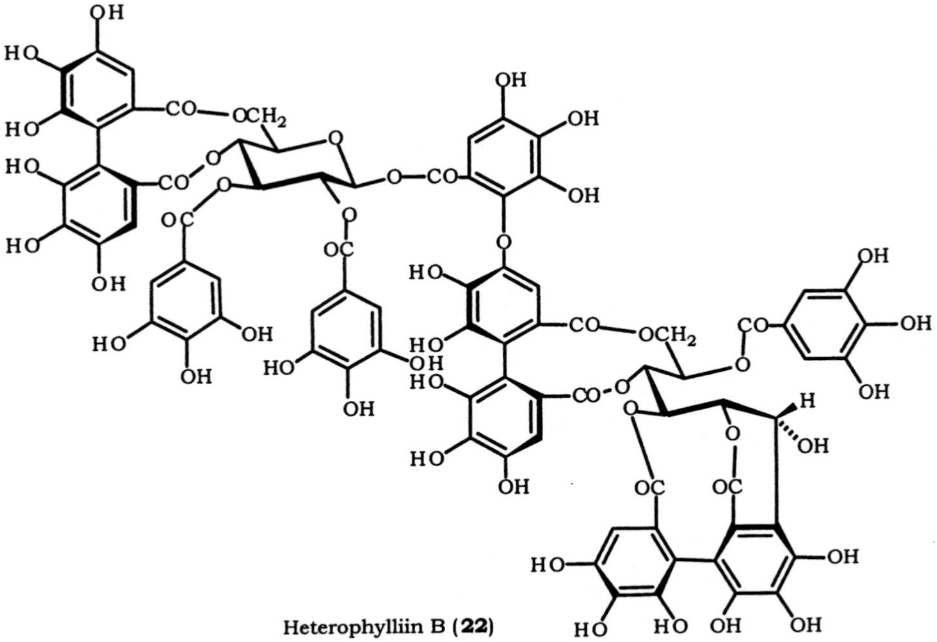

Heterophylliin B (**22**)

Chart 5.

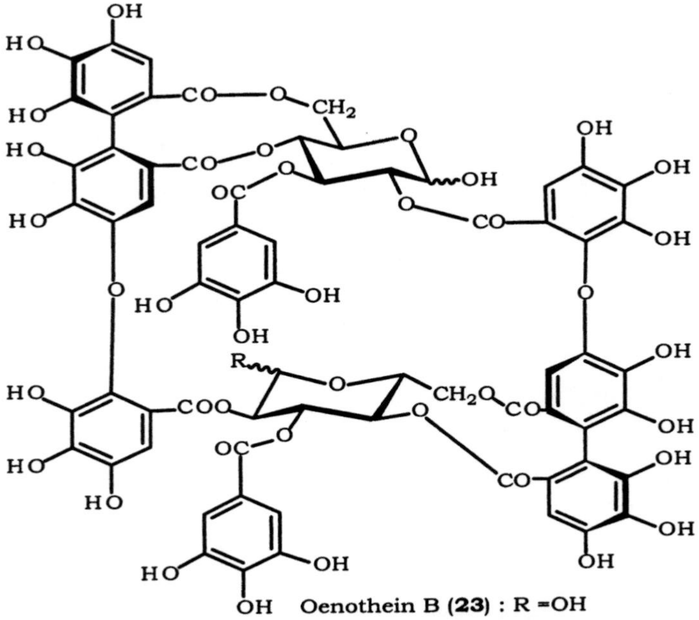

Oenothein B (**23**) : R =OH
Woodfordin C (**24**) : R = (α)-OG
Eugeniflorin D_1 (**25**) : R = (β)-OG

(Chart 6). contd.....

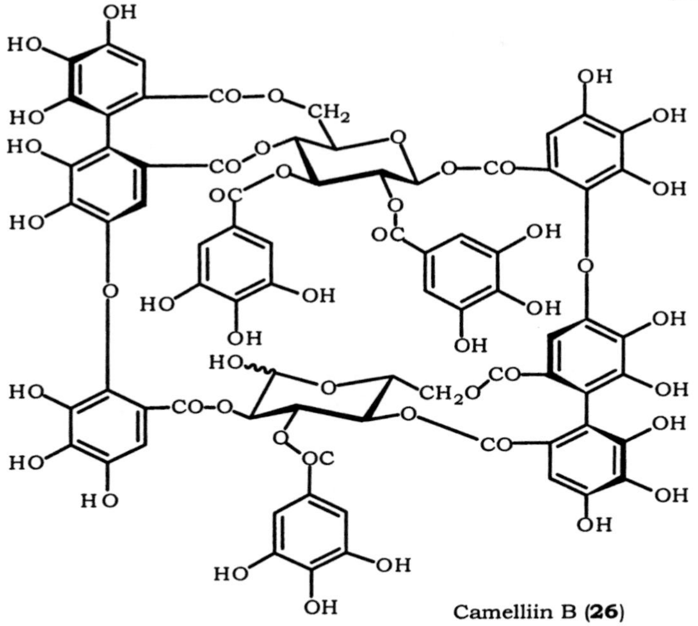

Camelliin B (**26**)

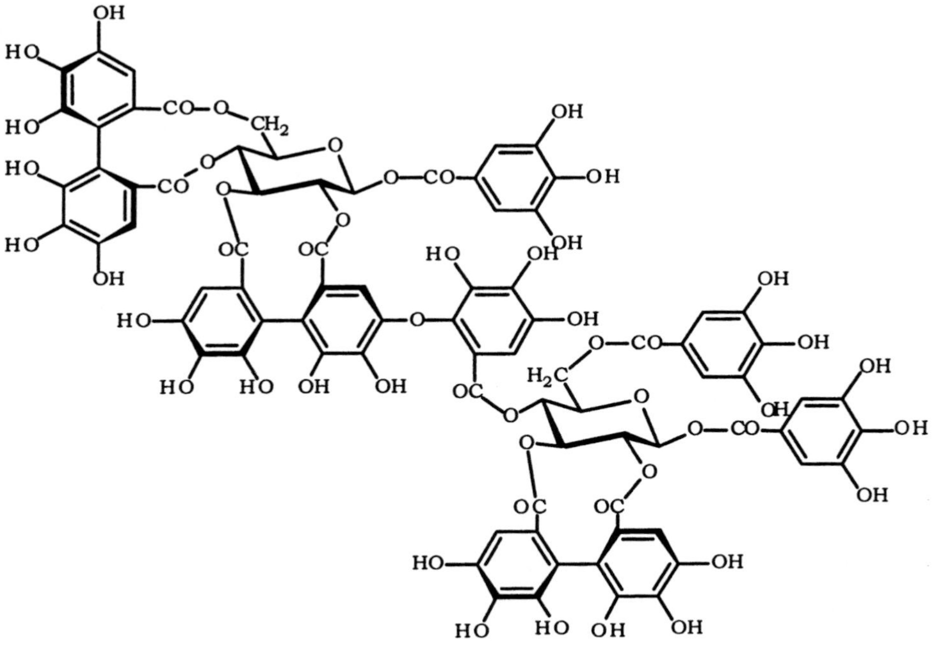

Nobotanin B (**27**)

(Chart 6). contd.....

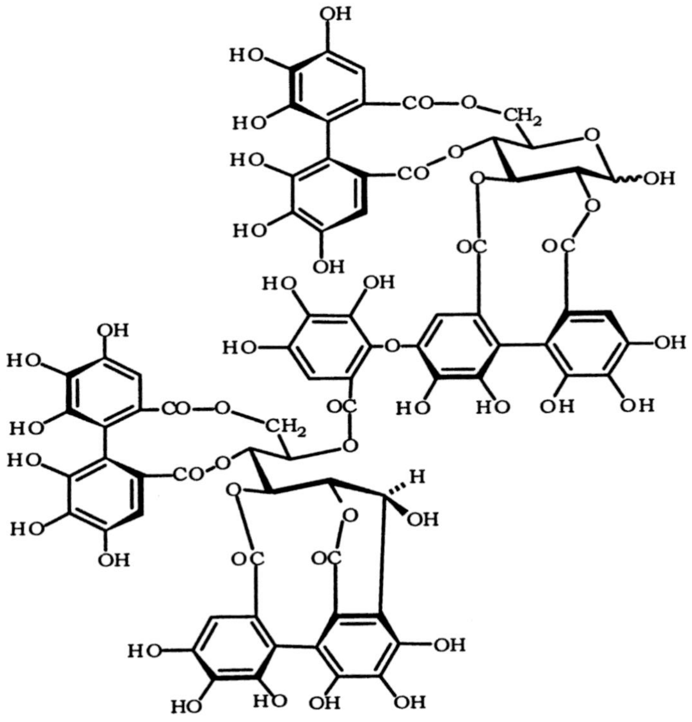

Reginin D (**28**)

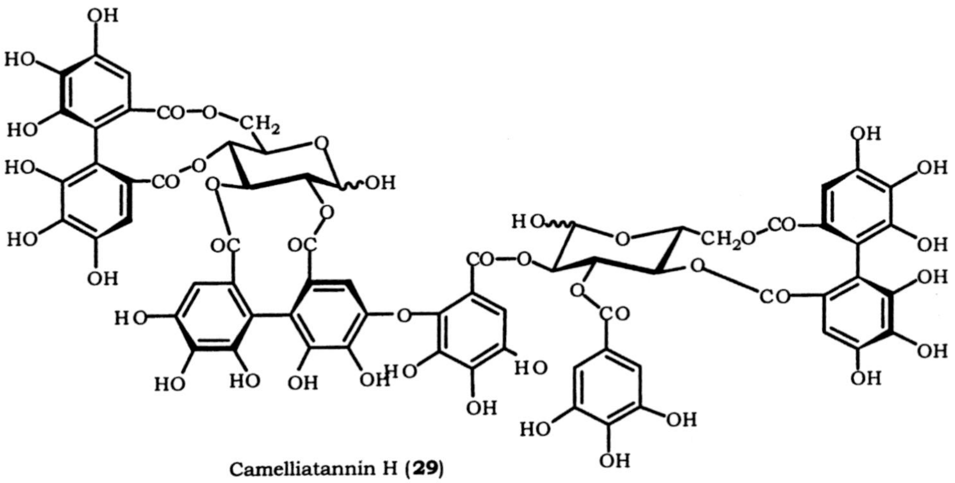

Camelliatannin H (**29**)

(Chart 6). contd.....

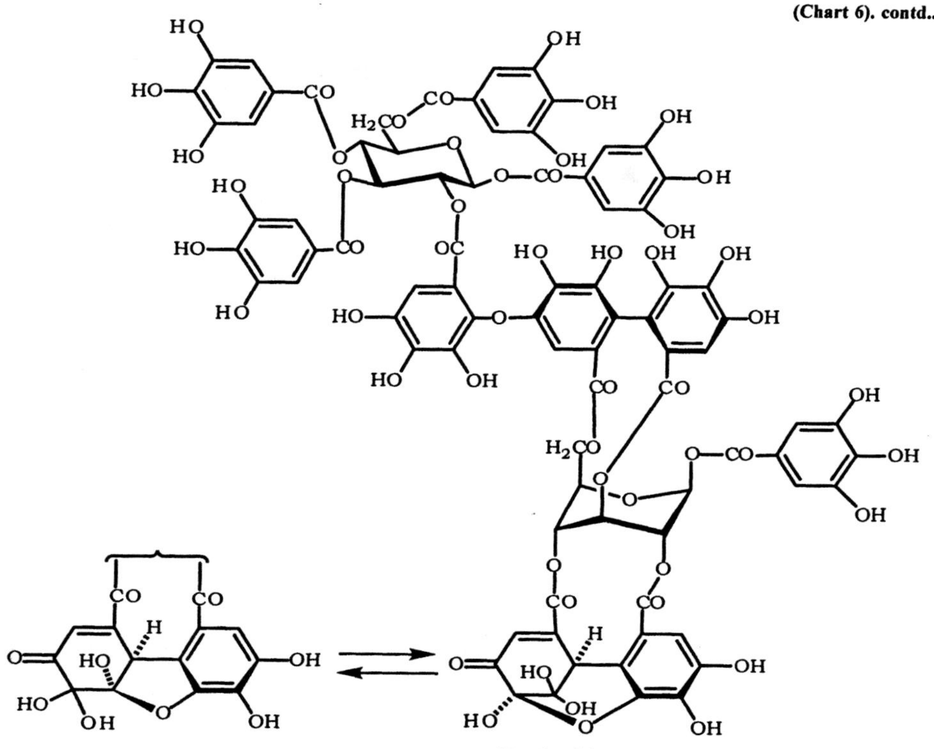

Euphorbin A (30)

Chart 6.

In contrast to the wide distribution of the oligomers having a valoneoyl group at positions 4,6 on the glucose core, those having 2,3-valoneoyl and 3,6-valoneoyl groups have been found in only a few families. Oligomers of the former type are represented by nobotanin B (27) [27], reginin D (28) [54] and camelliatannin H (29) [55] from the species of Melastomataceae, Lythraceae and Theaceae, respectively, while examples of the latter include euphorbins from Euphorbiaceae. It should be noted that euphorbin A (30) and its analogs based on the $^4C_1$ and $^1C_4$ glucose cores [57] are characteristic of Euphorbiaceae, and have not been found in any other plant family.

Tergalloyl group, which is an isomer of the valoneoyl group in terms of the position of the ether bond on the HHDP group, was first found as a component of monomeric tannin, tergallagin isolated from *Terminalia catappa* [71]. Dimers possessing a tergalloyl group as a linking unit of monomers are rather rare and have been found hitherto only in the *Eucalyptus* and *Euphorbia* species. Examples of such dimers are eucalbanin C (31) from *Eucaryptus alba* [42] and eurobustin A from *Eucalyptus robusta* [43]. Whereas these tannins have an (S)-tergalloyl group, those

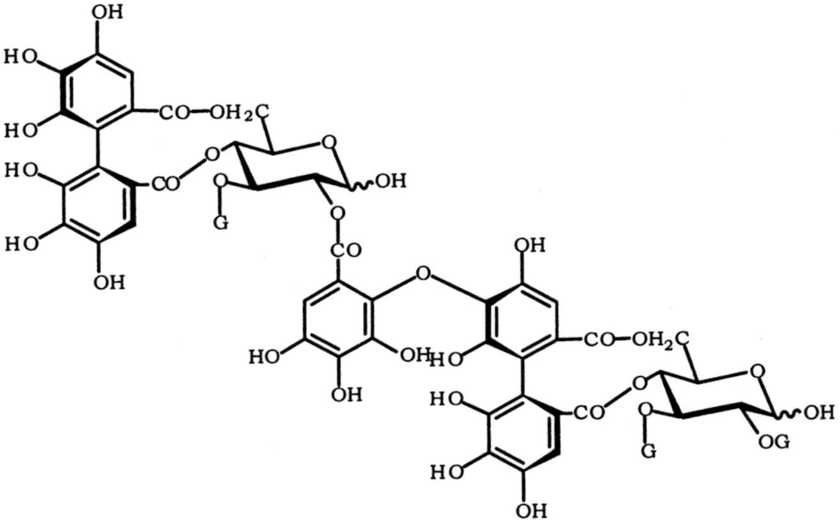

Eucalbanin C (**31**)

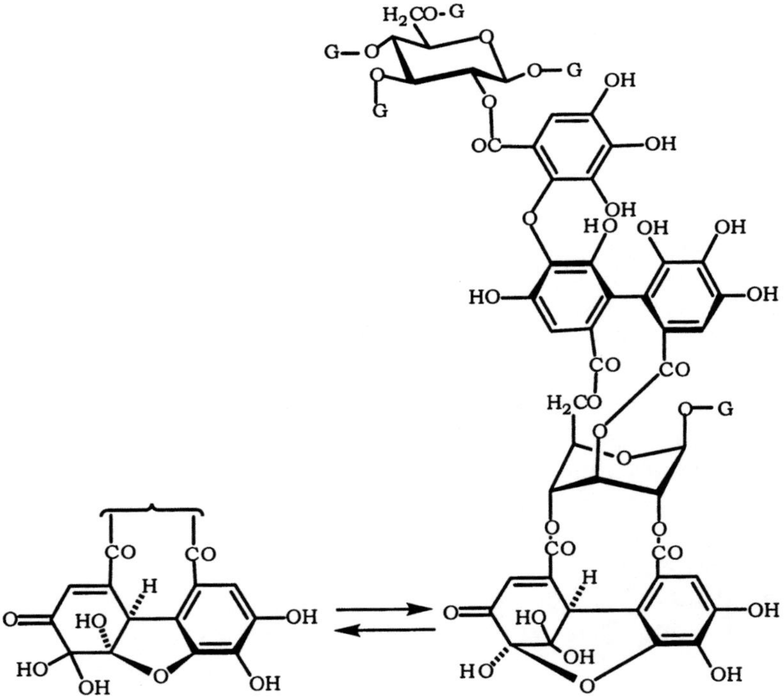

Euphorbin B (**32**)

(Chart 7). contd.....

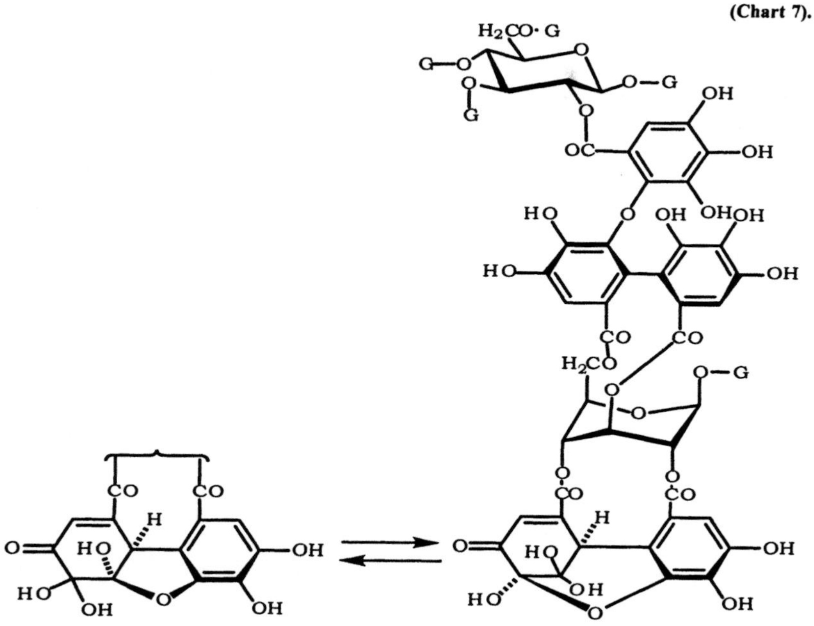

Euphorbin I (33)

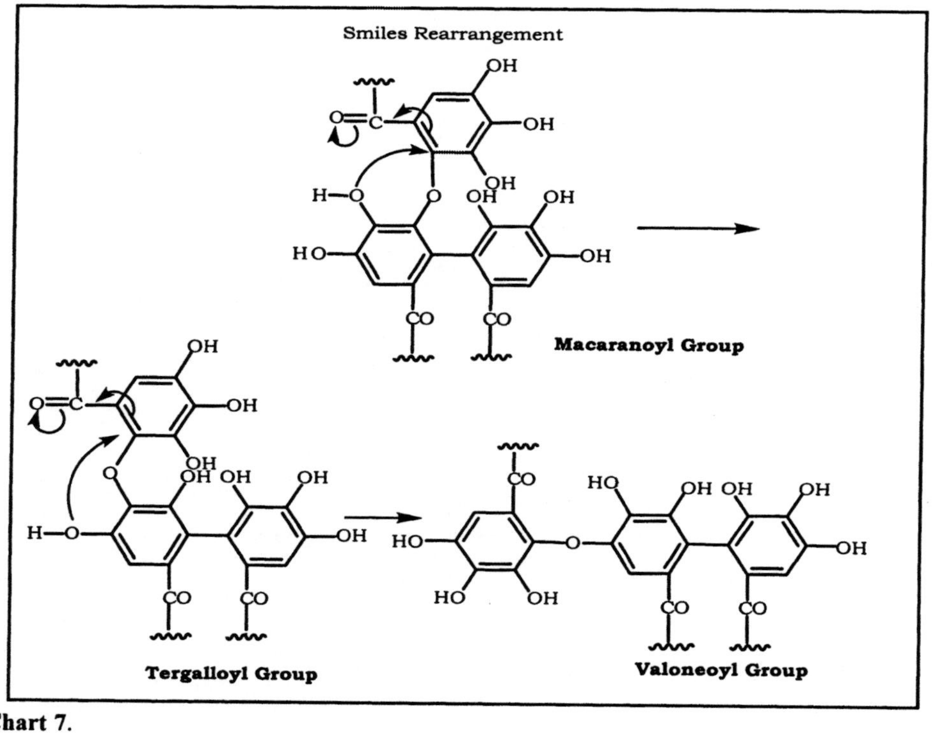

Smiles Rearrangement

Macaranoyl Group

Tergalloyl Group

Valoneoyl Group

Chart 7.

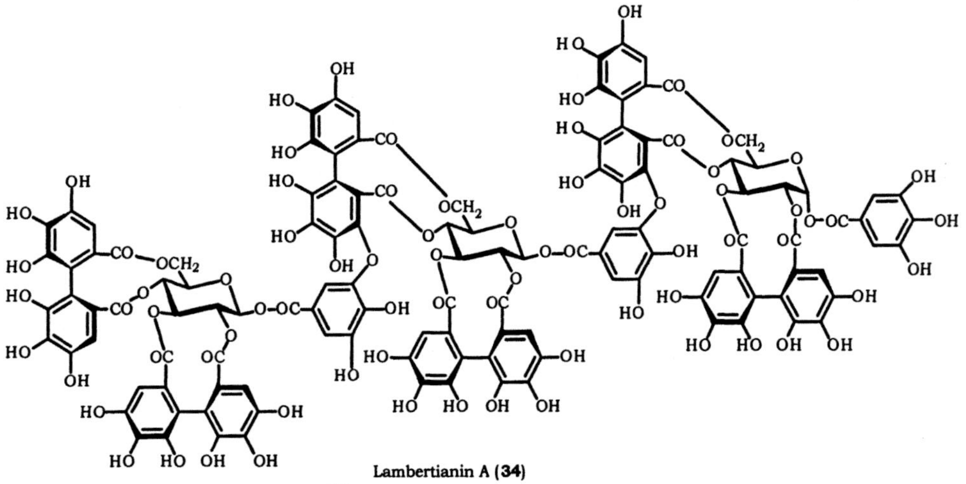

Lambertianin A (34)

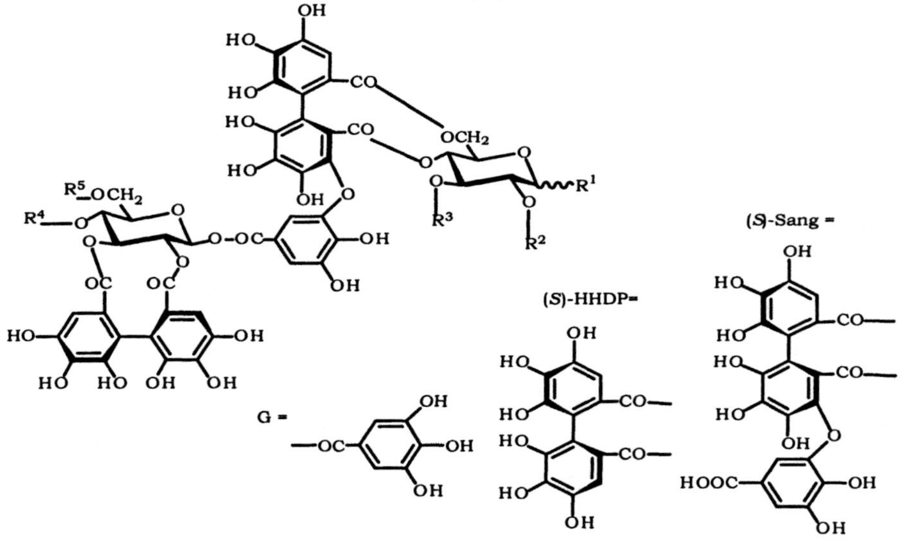

(S)-Sang =

(S)-HHDP=

G =

	R^1	R^2	R^3	R^4	R^5
Sanguiin H-6 (35)	(α)-OG	(S)-HHDP			(S)-HHDP
Roshenin A (36)	(α)-OG	(S)-HHDP			(S)-Sang
Roshenin B (37)	OH	(S)-HHDP			(S)-HHDP
Roshenin C (38)	(α)-OG	H	H		(S)-HHDP
Roshenin D (39)	(α)-OG	(S)-HHDP	H	G	

Chart 8.

produced by *Euphorbia* species have an (*R*)-tergalloyl group as exemplified by euphorbin B (**32**) [66]. Euphorbin I (**33**) which was obtained from *Euphorbia watanabei* [66] is a dimer composed of geraniin and pentagalloylglucose as constituent monomers like euphorbin B, but with a macaranoyl group as a linking unit that is an alternative isomer of the valoneoyl group. The tergalloyl and macaranoyl groups have a much bulkier structure than the valoneoyl group and are easily isomerized to the latter through Smiles rearrangement [72] upon treatment with aqueous solution containing a small amount of phosphate buffer (pH 6.0). Thus the structures of euphorbins B (32) and I (33) were confirmed by their chemical correlations with euphorbin A (30) through Smiles rearrangement.

Oligomers Having a Sanguisorboyl Group as a Linking Unit

Oligomers of this type, sanguiins H-6 (35) and H-8 were first isolated from *Sanguisorba officinalis* and were found to be tannins characteristic of the genera *Sanguisorba* [9]. Sanguiin H-6 (35) and its congeners, lambertianins A (34) and B also widely distribute in the *Rubus* species [73]. Besides the genera *Sanguisorba* and *Rubus*, *Rosa henryi* was found to contain lambertianin A and its analogues, roshenins A (36)-D (39) [74].

Oligomers Having a Highly Oxidative Valoneoyl Group as a Linking Unit

Euphorbins C (40) and D (41) which were obtained from *Euphorbia hirta* contain an euphorbinoyl group which might be produced through C-O oxidative coupling of an additional galloyl group with a valoneoyl group [58.59]. This tetraquis galloyl unit is an isomer of a woodfordinoyl unit, which is one of the polyphenolic components in unique macrocyclic trimers, e.g., oenothein A (42), woodfordin D (43) and a tetramer, woodfordin F. These oligomers were isolated from *Woodfordia fruticosa* (Lythraceae), a Jamu medicine called Sidowaya, which is popularly used for treatment of dysentary and bronchial complaints in Indonesia [33,38]. Highly oxidative ellagitannin dimers also include euphrobin E [60] from *Euphorbia hirta* and eugeniflorin D_2 (44) [44] from *Eucalyptus eugeniflora*, in which one galloyl part of the tris- and tetraquisgalloyl group linking monomers is oxidized to a dehydrogalloyl unit as shown in Chart 9.

C-Glucosidic Ellagitannin Dimers

Condensation of *C*-glucosidic ellagitannins involves intermolecular C-C coupling in addition to C-O oxidative coupling as mentioned above. Such condensation takes place between an anomeric carbon of open-chain glucose and the HHDP group on the other sugar unit as represented by alienanins A and B (45) from *Quercus aliena* [75], which are composed of casuarinin (60) and stachyurin (61). Roburins from an European oak, *Quercus robur* are analogs of these dimers composed of castalagin (62) and vescalagin (63) [76]. Dimeric tannins of this class have been found in Fagaceae, Theaceae, Combretaceae, Myrtaceae, Melastomataceae and

Lythraceae. Other types of *C*-glucosidic dimers are formerly produced through intermolecular coupling between a C-1 of the sugar in one *C*-glucosidic tannin monomer and a galloyl group in another as exemplified by casuglaunin A (46) from *Casuarina glauca* [77] and *Elaeagnus umbellata* [78].

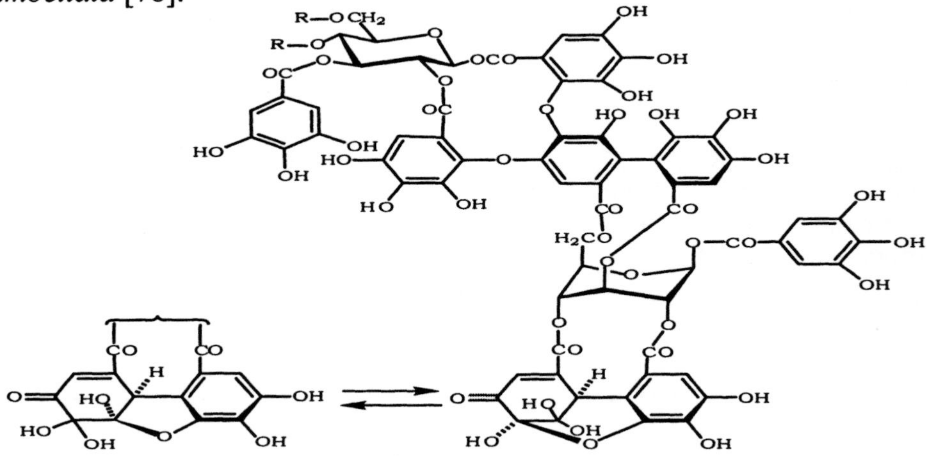

Euphorbin C (40) : R ~ R' = (*S*)-HHDP
Euphorbin D (41) : R = R' = G

(*S*)-HHDP =

G =

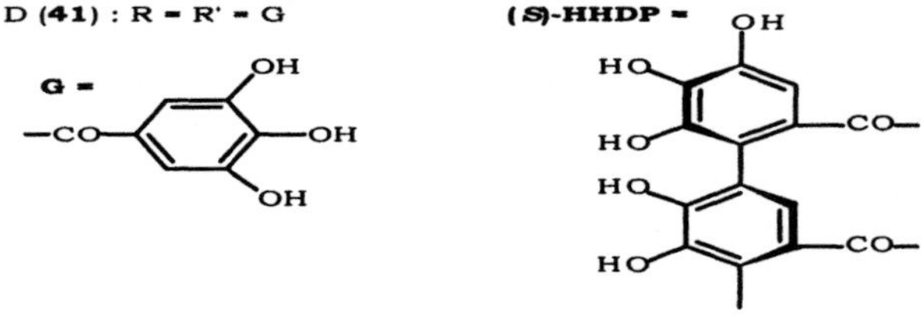

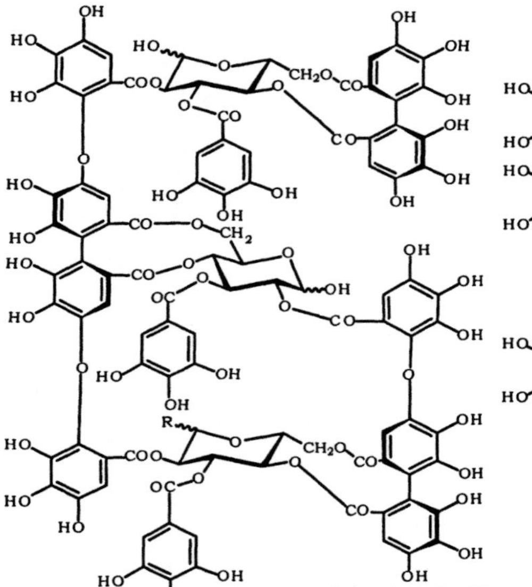

Oenothein A (42) : R =OH
Woodfordin D (43) : R = (α)-OG

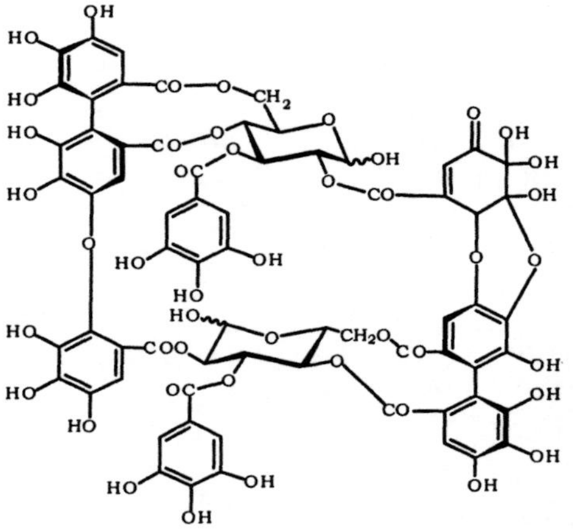

Eugeniflorin D$_2$ (44)

Chart 9.

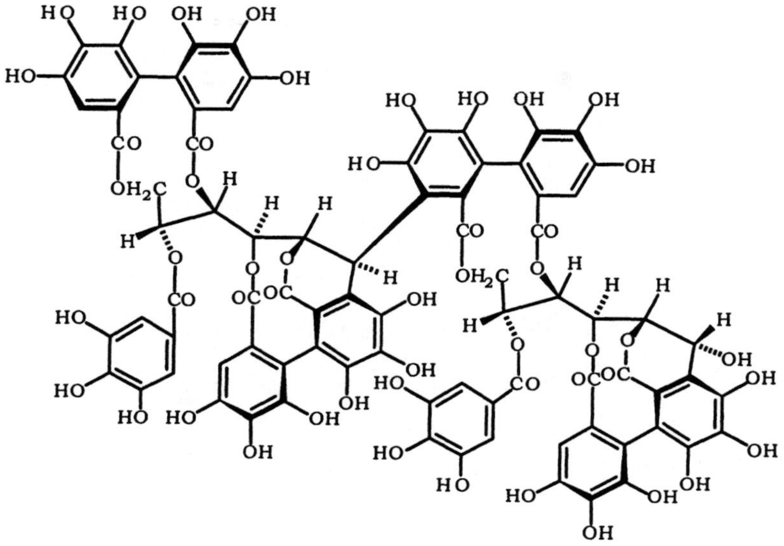

Alienanin B (45)

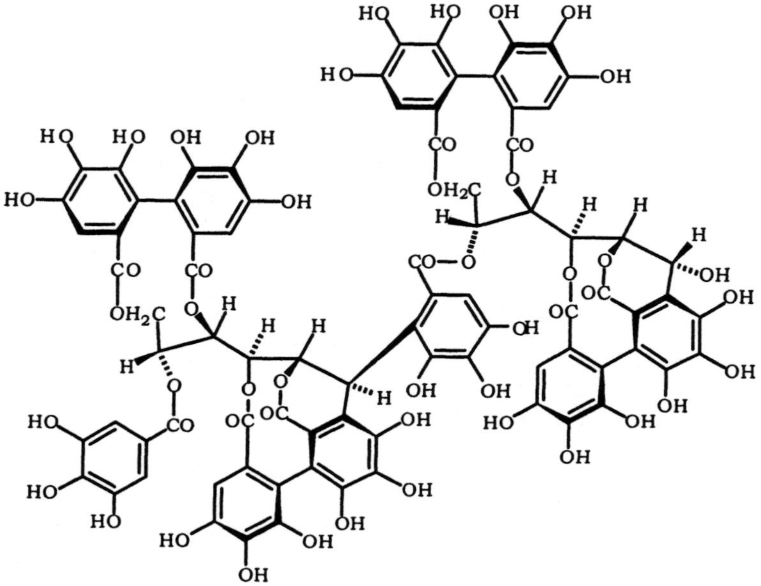

Casuglaunin A (46)

(Chart 10). contd.....

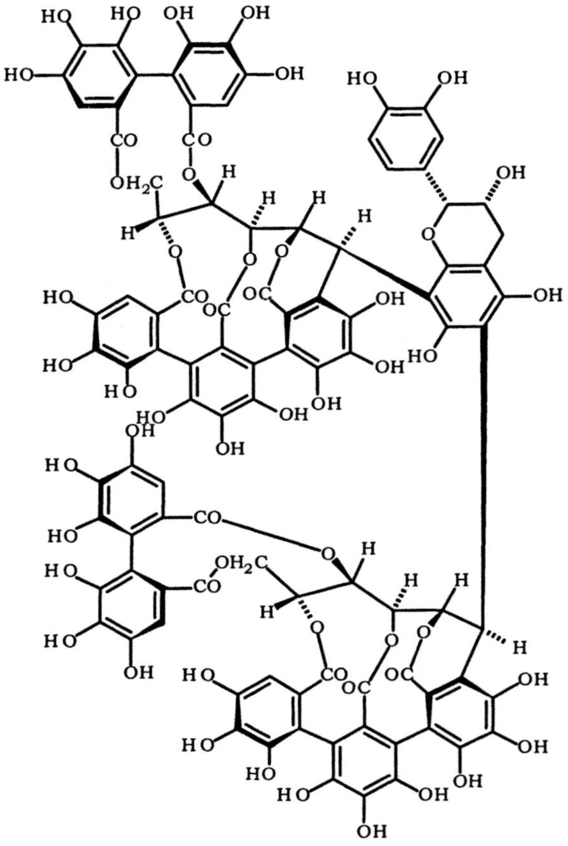

Anogeissinin (**47**)

Chart 10.

Examples of complex tannin dimers which have a flavan-3-ol unit attached through C-C linkage to C-1 of a *C*-glucosidic monomer include anogeissinin (**47**) and anogeissusins [79].

ISOLATION OF TANNINS

Hydrolyzable tannins, high polar polyesterified glucoses, usually occur in complex mixtures of analogues with related structures in a plant, and are sensitive to heat and alkaline conditions.

Although several qualification and quantitation methods for tannins have been developed, only a few are specific towards hydrolyzable or condensed tannins. Among colorimetric methods, the Folin-Denis or Folin Ciocalteu reagents (for total phenolics), and vanillin-HCl and *n*-BuOH-HCl methods are widely used for quantification of condensed tannins [80].

The *n*-BuOH-HCl color test is quite specific for condensed tannins, and is based on acid hydrolysis of interflavan linkages to yield anthocyanidins. The term "proanthocynidins" for condensed tannins is thus preferred these days. Of the hydrolyzable tannins, gallotannins (galloylglucoses) give a transient red color with KIO_3 in aqueous or aqueous organic solutions. On the other hand, ellagitannins are distinguished from gallotannins and/or condensed tannins (proanthocynidins) by color change from reddish orange to dark blue with the sodium nitrate-acetic acid reagent in the absence of oxygen [81].

Table 2. HPLC Conditions for Analyzing Hydrolyzable Tannins

Column	Solvent
Normal phase Superspher Si-60 (4.0 mm I.D. x 120 mm) (Merck) YMC-Pack SIL-A003 (4.6 mm I.D. x 250 mm) (YMC Co., Ltd.)	(A) *n*-hexane-MeOH-THF-HCOOH (55 : 33 : 11 : 1) + $(COOH)_2$ 450 mg / L (B) *n*-hexane-MeOH-THF-HCOOH (60 : 45 : 15 : 1) + $(COOH)_2$ 500 mg / 1.2L
Reversed phase LiChrospher RP-18 (4.0 mm I.D. x 250 mm) (Merck) YMC-Pack ODS-A312 (4.6 mm I.D. x 150 mm) (YMC Co., Ltd.) Inertsil ODS-3 (4.0 mm I.D. x 250 mm) (GL Science)	(C) 10 mM H_3PO_4-10mM KH_2PO_4-EtOH-EtOAc (42.5 : 42.5 : 10 : 5) (D) 10 mM H_3PO_4-10mM KH_2PO_4-CH_3CN (42.5 : 42.5 : 15)

Among solvents such as methanol, aqueous alcohol, acetone and water used for extraction of polar natural products, aqueous acetone or aqueous MeOH containing 30-50% water is the most suitable solvent for the extraction of both hydrolyzable and condensed tannins [82]. Plant materials are thus homogenized or soaked in aqueous acetone at room temperature and the filtered solution is evaporated or concentrated under reduced pressure below 40ºC. To date, separation and purification procedures have been based on a combination of chromatographic and partition techniques. Silica gel, the most common adsorbent for natural products, is not suitable for separation of tannins because of strong adsorption to the support preventing recovery from the column with organic solvent. Repeated column chromatography utilizing different adsorbents such as polystyrene (Dia-ion HP-20, MCI-gel CHP-20P), and polyvinyl (Toyopearl (TSK) HW-40W, coarse or fine grades) gels and Sephadex LH-20 has successfully resulted in the purification of each component of the complex mixture. Elution of the tannins from such columns is effected in a stepwise gradient mode using aqueous alcohol or aqueous alcohol-acetone as solvent. Modern liquid-liquid partition chromatography such as droplet countercurrent chromatography (DCCC) and centrifugal partition chromatography (CPC) has also been successfully applied to initial and/or final fractionation of hydrolyzable tannins [82,84,85]. These methods have the advantage that they prevent inevitable loss of sample due to adsorption to solid support. CPC separation, which is usually achieved within a few hours, is superior to other methods in the quick separation of labile oligomeric hydrolyzable tannins [86]. In the fractionation of tannins by DCCC or CPC, the following solvent systems are used: n-BuOH-AcOH-H_2O (4:1:5) and n-BuOH-PrOH-H_2O (4:1:5) with normal- or reversed-phase development.

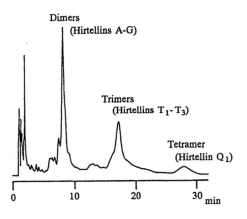

Fig. (1). HPLC profile [normal phase : solvent (B)] of the *Raumuria hirtella* extract.

The purity of each fraction obtained with these separation methods and the isolated tannins may be monitored by means of analytical HPLC in normal phase and reversed-phase modes under the conditions outlined in Table 2. The molecular sizes of hydrolyzable tannins can be estimated by normal phase HPLC and, as retention time increases, rank in the order of monomer→dimer→trimer→tetramer as seen in Fig. 1.

Hydrolyzable tannin oligomers are mostly obtained in a pure state as amorphous powders, while monomers are sometimes isolable in a crystalline state. Although traces of solvents may be removed by exposure to a high vacuum overnight at 40°C, complete removal of water is usually difficult because of the extent of hydration to polyphenolic hydroxyl groups. The stability of tannins with such a hydration structure is evidenced by the recent X-ray analysis of crystalline geraniin, which revealed that 7 moles of water are included as hydrated forms in a molecule [87]. Condensed tannins are known to be easily oxidized or polymerized under exposure to air upon storage, while hydrolyzable tannins are rather more stable to air. It is noteworthy that the stability of oligomeric hydrolyzable tannins largely depends on their structure and extent of condensation. Dimers having a galloyl part of the valoneoyl group at the anomeric center as represented by rugosins D-G are less stable than those having the same group at C-2 of a glucose core like cornusiins A and C, and oenothein B etc. Trimers and tetramers are also less stable than dimers. This unstability of oligomers on storage might result from hydrolysis of the ester linkages on the glucose core with water included in the hydrated molecule.

PHYSICAL METHODS FOR STRUCTURAL ELUCIDATION OF OLIGOMERIC ELLAGITANNINS

Mass Spectroscopy

Mass spectroscopy has played a major role in the determination of the structure of natural products including tannins and related polyphenols. Among several ionization methods which enable us to detect the molecular ion species of nonvolatile, high polar and thermally unstable compounds, fast atom bombardment (FAB) is applicable without derivatization to a wide range of natural products including biopolymers (peptide, polysaccharide and nucleic acids etc). Its usefulness over the whole range of polyphenolics classified as hyrolyzable tannins, proanthocyanidins and flavonoid glycosides was reviewed by R. Self et al. [88]. Upon application of FAB mass spectrometry to oligomeric hydrolyable tannins, molecular ion species could be measured to up to 3300 of tetramer (woodfordin F). The observed molecular ion species in FAB-MS are usually $[M+H]^+$ in the positive ion mode and $[M-H]^-$ in the negative mode, from both of which a molecular weight can be unequivocally determined. An alternative way of discriminating and confirming the molecular ion in the positive FAB-MS is by adding a small amount of NaCl or KCl to a matrix agent such as m-nitrobenzyl alcohol, which often gives a predominant ion peak due to $[M+Na]^+$ or $[M+K]^+$.

Electrospray ionization (ESI) recently developed in connection with LC/MS is superior to FAB in the soft ionizations of labile high molecular weight compounds, and has the advantage of producing predominant pseudo-molecular ions such as $[M+H]^+$ or $[M+Na]^+$ for the ionic minute compounds. The ionization effect in ESI-MS depends on whether the solvent contains acetic acid, formic acid or trifluoroacetic acid etc. This technique with direct injection has been successfully applied to hydrolyzable tannins up to tetramers when 50% aqueous MeOH or 50% aqueous acetonitrile containing a small amount of $AcONH_4$ was used as a solvent. In this condition, the observed molecular ion species for hydrolyzable tannins are an $[M+NH_4]^+$ with high sensitivity.

The usefulness of ESI-MS, used either directly or on line after HPLC separation, in the analyses of proanthocyanidin oligomers of apple, grape seed and litchi pericarp extracts, was recently demonstrated [89]. Pseudo-molecular ion [M-H]⁻ in the negative mode of ESI-MS and characteristic fragmentation ions generated by increasing the orifice voltage can be used to determine the nature of constituent units, degree of polymerization and type (A-or B-type) of interflavonoid linkages as well as molecular size. However, it is emphasized that some difficulty in discriminating a [M-H]⁻ ion from doubly and triply charged ions is encountered in the case of high molecular weight oligomers and their mixture. Nevertheless, it is likely that the ESI technique in LC/MS and MS/MS will be useful in the structural elucidation of this class of polyphenols and also the microanalysis of polyphenol mixtures in the extracts of crude drugs.

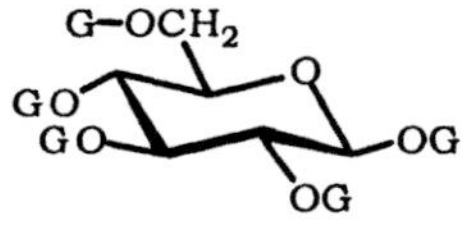

1,2,3,4,6-Pentagalloyl-β-D-glucose (48)

G =

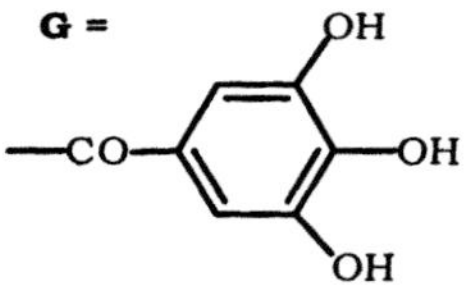

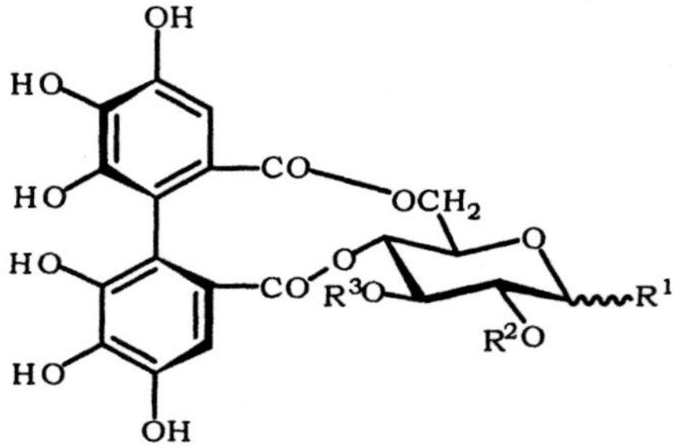

Tellimagrandin I (49) : R^1 = OH, R^2 = R^3 = G
Tellimagrandin II (50) : R^1 = (β)-OG, R^2 = R^3 = G
Strictinin (51) : R^1 = (β)-OG, R^2 = R^3 = H

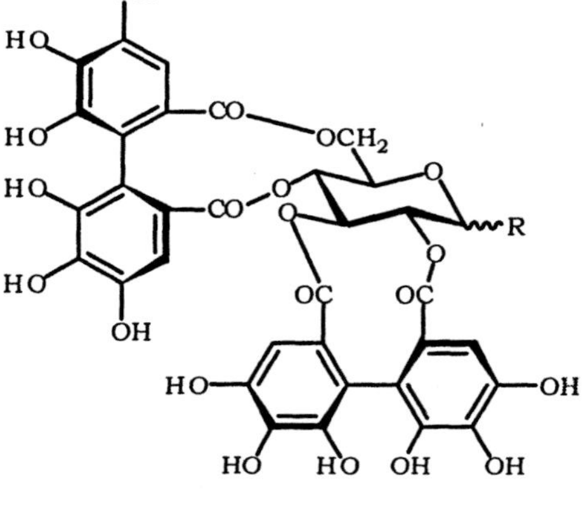

Pedunculagin (52) : R = OH
Casuarictin (53) : R = (β)-OG
Potentillin (4) : R = (α)-OG

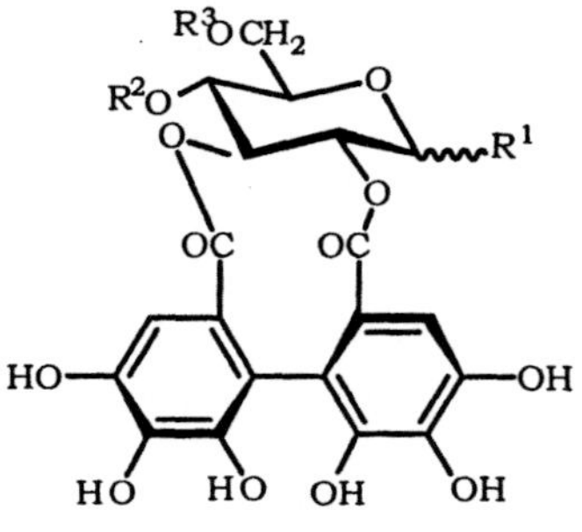

Praecoxin B (54) : R^1 = OH, R^2 = R^3 = G
Pterocaryanin C (55) : R^1 = (β)-OG, R^2 = G
Isostrictinin (56) : R^1 = (β)-OG, R^2 = R^3 = H

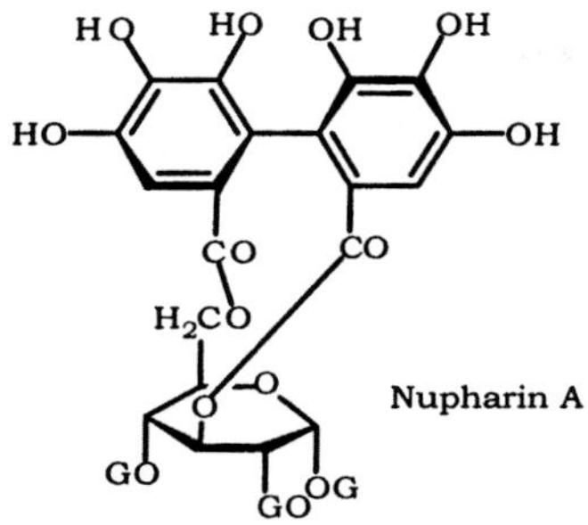

Nupharin A (57)

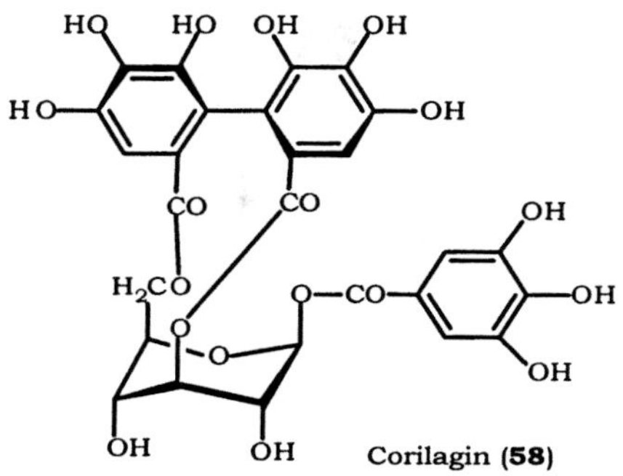

Corilagin (58)

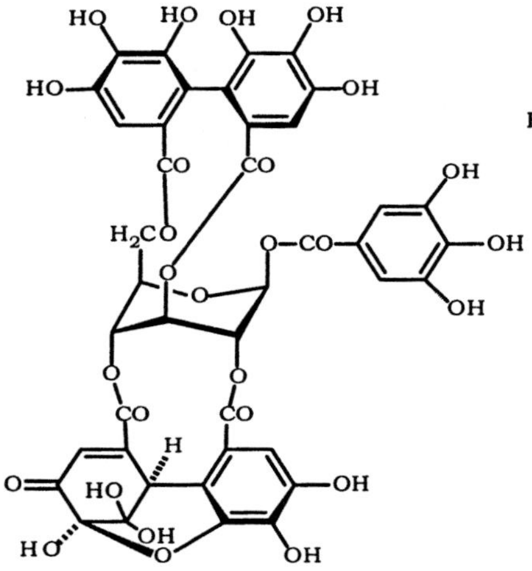

Geraniin (59)

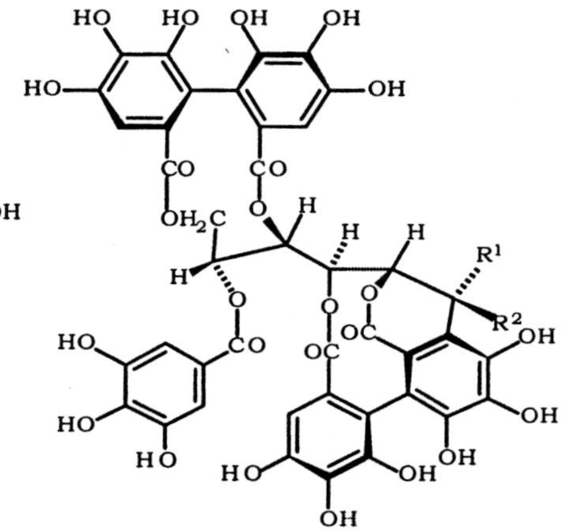

Casuarinin (60) : R^1 = OH, R^2 = H
Stachyurin (61) : R^1 = H, R^2 = OH

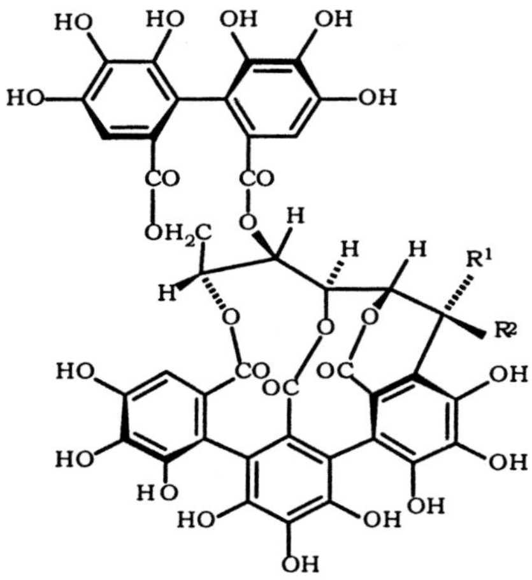

Castalagin (62) : R^1 = OH, R^2 = H
Vescalagin (63) : R^1 = H, R^2 = OH

Chart 11.

Table 3 ^{1}H-NMR Data of Monomers (500 MHz, Acetone-d_6+D$_2$O, J in Hz)

Compounds	Glucose							Acyl groups	
	H-1	H-2	H-3	H-4	H-5	H-6	H-6	Galloyl-H (2H, s)	HHDP-H (1H, s)
1, 2, 3, 4, 6-Pentagalloyl-β-D-glucose (48)	6.29, d (8.5)	5.67, dd (8.5, 10)	6.00, t (10)	5.56, t (10)	4.54, ddd (2, 5, 10)	4.58, dd (2, 12)	4.31, dd (5, 12)	7.15, 7.10, 7.05, 6.99, 6.96	
Tellimagarandin I (49) (α-anomer)	5.57, d (4)	5.11, dd (4, 10)	5.88, t (10)	5.11, t (10)	4.67, ddd (1.5, 6.5, 10)	5.28, dd (6.5, 13)	3.77, dd (1.5, 13)	7.05, 6.98	6.65, 6.58
(β-anomer)	5.13, d (8)	5.24, dd (8, 10)	5.61, t (10)	5.11, t (10)	4.27, ddd (1.5, 6.5, 10)	5.30, dd (6.5, 13)	3.84, dd (1, 13)	7.04, 6.93	6.64, 6.47
Tellimagarandin II (50)	6.18, d (8)	5.59, dd (8, 10)	5.82, t (10)	5.20, t (10)	4.53, dd (6, 10)	5.34, dd (6, 13)	3.87, d (13)	7.10, 6.99, 6.96	6.67, 6.47
Strictinin (51)	5.71, d (8)	3.67, dd (8, 9)	3.80, dd (9, 10)	4.88, t (10)	4.08, dd (6.5, 10)	5.30, dd (6.5, 13.5)	3.75, d (13.5)	7.17	6.70, 6.58
Pedunculagin (52) (α-anomer)	5.46, d (4)	5.06, dd (4, 10)	5.46, t (10)	5.08, t (10)	4.60, ddd (1.5, 7, 10)	5.27, dd (7, 13)	3.75, dd (1.5, 13)		6.65, 6.58, 6.54, 6.31
(β-anomer)	5.10, d (8)	4.87, dd (8, 10)	5.22, t (10)	5.07, t (10)	4.21, ddd (1.5, 6.5, 10)	5.30, dd (6.5, 13)	3.83, dd (1, 13)		6.66, 6.57, 6.49, 6.31
Casuarictin (53)	6.18, d (8.5)	5.16, t (8.5)	5.41, dd (8.5, 10.5)	5.13, t (10.5)	4.47, dd (6.5, 10.5)	5.31, dd (6.5, 13)	3.85, d (13)	7.16	6.65, 6.54, 6.45, 6.36
Potentillin (4)	6.61, d (4)	5.40, dd (4, 10)	5.56, t (10)	5.21, t (10)	4.64, dd (6, 10)	5.31, dd (6, 13)	3.81, d (13)	7.22	6.66, 6.58, 6.46, 6.35
Praecoxin B (54) (α-anomer)	5.48, d (3.5)	5.07, dd (3.5, 10)	5.62, t (10)	5.50, t (10)	4.54, ddd (2, 4, 10)	4.49, dd (2, 12)	4.27, dd (4, 12)	7.14, 7.13	6.60, 6.39
(β-anomer)	5.17, d (8)	4.88, dd (8, 9.5)	5.35, t (9.5)	5.46, t (9.5)	4.22, ddd (2, 5, 9.5)	4.49, dd (2, 12)	4.26, dd (5, 12)	7.14, 7.10	6.59, 6.39

(Table 3), cond.....

Compounds	Glucose							Acyl groups	
	H-1	H-2	H-3	H-4	H-5	H-6	H-6	Galloyl-H (2H, s)	HHDP-H (1H, s)
Pterocaryanin C (55)	6.30, d (8.5)	5.21, t (8.5)	5.55, dd (8.5, 9)	5.59, t (9)	4.50, dd (5, 9)	4.55, d (12)	4.32, dd (5, 12)	7.13, 7.12, 7.12	6.44, 6.43
Isostrictinin (56)	6.07, d (8.5)	4.98, dd (8.5, 9.5)	5.16, t (9.5)	3.87, t (9.5)	3.70, ddd (2, 4.5, 9.5)	3.87, dd (2, 12.5)	3.77, dd (4.5, 12.5)	7.12	6.69, 6.40
Nupharin A (57)	6.69, d (3)	5.50, dd (3, 8)	5.71, d (8)	5.10, s	4.39, brs	5.32, d (13)	3.90, d (13)	7.24, 7.06, 6.96	7.25, 7.15
Corilagin (58)	6.36, brs	4.05, brs	4.82, brs	4.45, brs	4.52, dd (9.5, 11.5)	4.96, t (11.5)	4.09, dd (9.5, 11.5)	7.10	6.81, 6.67
Geraniin[a] (59)	6.60, brs	5.60, brs	5.50, brs	5.56, brs	4.81, m	4.93, t (11)	4.33, dd (8, 11)	7.22	7.13, 6.71, 7.25,[b] 6.56,[b] 5.16[b]
Casuarinin (60)	5.60, d (5)	4.64, dd (1.5, 5)	5.41-5.44, m	5.41-5.44, m	5.31, dd (3.5, 8.5)	4.83, dd (3.5, 13)	4.04, d (13)	7.09	6.81, 6.54, 6.49
Stachyurin (61)	4.93, d (2)	4.86, t (2)	4.98, t (2)	5.62, dd (2, 9)	5.36, dd (3, 9)	4.84, dd (3, 13)	4.02, d (13)	7.14	6.82, 6.55, 6.50
Castalagin (62)	5.68, d (4.5)	5.01, d (4.5)	5.00, dd (4.5, 7)	5.22, t (7)	5.58, dd (3, 7)	5.08, dd (3, 13)	3.98, d (13)		6.80, 6.76, 6.62
Vescalagin (63)	4.86, d (2)	5.21, t (2)	4.55, dd (2, 8)	5.19, t (8)	5.62, dd (3, 8)	5.07, dd (3, 13)	3.98, d (13)		6.74, 6.72, 6.57

a) Measured in acetone-d_6. b) DHHDP unit

Table 4. ^{13}C-NMR Data of Monomers (126 MHz, Acetone-d_6+D$_2$O)

	C-1	C-2	C-3	C-4	C-5	C-6
1, 2, 3, 4, 6-Pentagalloyl-β-D-glucose (48)	93.4	71.9	73.5	69.5	74.1	62.9
Tellimagrandin I (49) (α-anomer)	91.2	72.9	71.1	71.1	67.2	63.5
(β-anomer)	96.7	74.1	73.5	71.1	72.0	63.5
Tellimagrandin II (50)	93.8	71.8	73.3	70.8	73.1	63.1
Strictinin (51)	95.9	74.7	75.6	72.8	73.2	63.7
Pedunculagin (52) (α-anomer)	91.8	75.6	75.8	69.9	67.4	63.6
(β -anomer)	95.4	78.3	77.6	69.6	72.5	63.6
Casuarictin (53)	92.4	76.0	77.3	69.3	73.5	63.1
Potentillin (4)	90.7	74.1	76.0	69.1	71.0	63.2
Praecoxin B (54) (α -anomer)	91.3	75.1	75.4	68.7	68.4	63.1
(β -anomer)	94.9	77.7	77.6	68.5	73.1	63.2
Pterocaryanin C (55)	91.9	75.3	77.4	69.2	73.9	62.7
Isostrictinin (56)	92.9	71.1	72.9	73.4	69.2	65.9
Nupharin A (57)	89.5	70.0	73.5	74.9	77.9	65.1
Corilagin (58)	95.0	69.2	70.7	62.5	76.0	64.8
Geraniin (59)	91.6	76.6	68.7	67.6	76.8	65.2
Casuarinin (60)	67.6	76.7	69.8	74.2	71.2	64.6
Stachyurin (61)	65.5	81.0	70.9	73.3	72.0	64.5
Castalagin (62)	67.2	74.0	66.2	69.2	71.1	65.3
Vescalagin (63)	66.1	78.3	71.7	69.1	70.0	66.0

NMR Spectroscopy

High-field NMR spectral analyses including a variety of two-dimensional
techniques are the most powerful tools for the structural determination of
oligomers. The ^{1}H- and ^{13}C-NMR spectra provide useful information on
the nature and number of polyphenolic acyl units, e.g. galloyl, HHDP,
DHHDP, valoneoyl units and others, and on the substitution modes of
glucose residues.

In spite of the structural diversity of oligomers as mentioned above, the
types of constituents monomeric units of oligomers are relatively limited.
Such monomers include pentagalloylglucose (**48**), tellimagrandin I (**49**), II
(**50**), strictinin (**51**), pedunculagin (**52**), casuarictin (**53**), corilagin (**58**) and

geraniin (**59**) etc. Most of the *C*-glucosidic ellagitannin dimers or trimers are composed of casuarinin (**60**), stachyurin (**61**), castalagin (**62**) and/or vescalagin (**63**). The ^{1}H- and ^{13}C-NMR data, in particular, for the glucose moiety of these monomers thus often facilitate the assignment of the constructing monomers of oilgomers by NMR spectral comparison [90] (Tables 3-5).

Table 5. ^{13}C-NMR Data of Dimers (126 MHz, Acetone-d_6+D_2O)

	Monomer Units	C-1 C-1	C-2 C-2	C-3 C-3	C-4 C-4	C-5 C-5	C-6 C-6
Agrimoniin (1)	4	91.5	75.1	76.5$^{a)}$	69.7$^{b)}$	71.5$^{c)}$	63.8$^{d)}$
	4	91.9	75.1	76.7$^{a)}$	69.3$^{b)}$	71.5$^{c)}$	63.2$^{d)}$
Laevigatin B	4	91.0	93.6	78.0	67.5	76.3	61.4
	(A)	90.8	74.2	75.9	68.9	71.3	63.2
Gemin A (7)	4	91.3	75.1	76.7	69.4	71.5	63.5
	50	94.3	72.2	74.1	71.3	73.6	63.8
Rugosin D (13)	50	93.1	71.7	73.1$^{e)}$	70.5$^{f)}$	72.8$^{g)}$	62.9$^{h)}$
	50	93.5	71.7	73.2$^{e)}$	70.6$^{f)}$	72.9$^{g)}$	63.0$^{h)}$
Rugosin F	50	93.4	71.9	73.5	70.7	73.1	63.2
	53	92.3	74.9	77.2	69.3	73.5	63.2
Woodfordin A	50	93.5	71.8	73.1	70.7	72.7	63.2
	(B)	93.1	72.0	75.5	69.4	76.2	64.5
Nobotanin F (19)	55	92.0	75.1	77.4	67.7	73.9	62.5
	53	92.1	75.6	77.0	69.1	73.4	63.0
Nobotanin G	53	92.3	76.6	77.0	69.5	73.3	63.2
	51	92.6	74.2	75.2	70.9	73.1	63.2
Euphorbin F	50	93.6	71.6	73.9	70.4	72.7	62.8
	59	91.0	69.9	63.3	66.9	73.4	64.0
Heterophylliin B (22)	50	93.1	71.7	73.2	70.5	72.8	62.9
	60	67.6	76.7	69.8	74.2	71.2	64.6
Sanguiin H-6 (35)	4	90.6	73.8	75.2	69.2	71.1	63.0
	53	92.4	75.8	77.2	68.9	73.4	62.7
Sanguiin H-10	4	90.9	75.1	73.7	71.4	69.2	63.4
	51	95.8	75.1	74.0	73.0	72.5	63.4
Alienanin B (45)	60	67.1	77.1	69.2	74.5	71.4	65.7
	61	40.2	80.9	69.5	74.6	73.2	64.3

(A) Sanguiin H-4, (B) 1,2,3,6-Tetragalloyl-β-D-glucose

a)-h) Interchangeable

A crucial step in the structural determination of new oligomeric hydrolyzable tannins as well as monomers is to define the position of each acyl group on the glucose cores in the molecules, even though convenient assignment of monomers could be made in the way described above. Partial hydrolysis of oligomers under mild conditions often provides conclusive chemical evidence for the structures by identification of the monomeric partial hydrolyzates of known structure. However, upon determining the attached positions of the linking units, i.e., the DHDG, valoneoyl and their analogous groups, in the oligomer molecules, great difficulty is inevitably encountered. For this purpose, 2D long-range correlation spectrum, which enables us not only to assign the ^{13}C resonances but also to establish the connectivity of a phenolic aromatic proton with a glucose proton through an ester carbonyl carbon, has been applied. Three-bond couplings among a variety of C-H long-range couplings are thus of most practical value. These coupling constants are

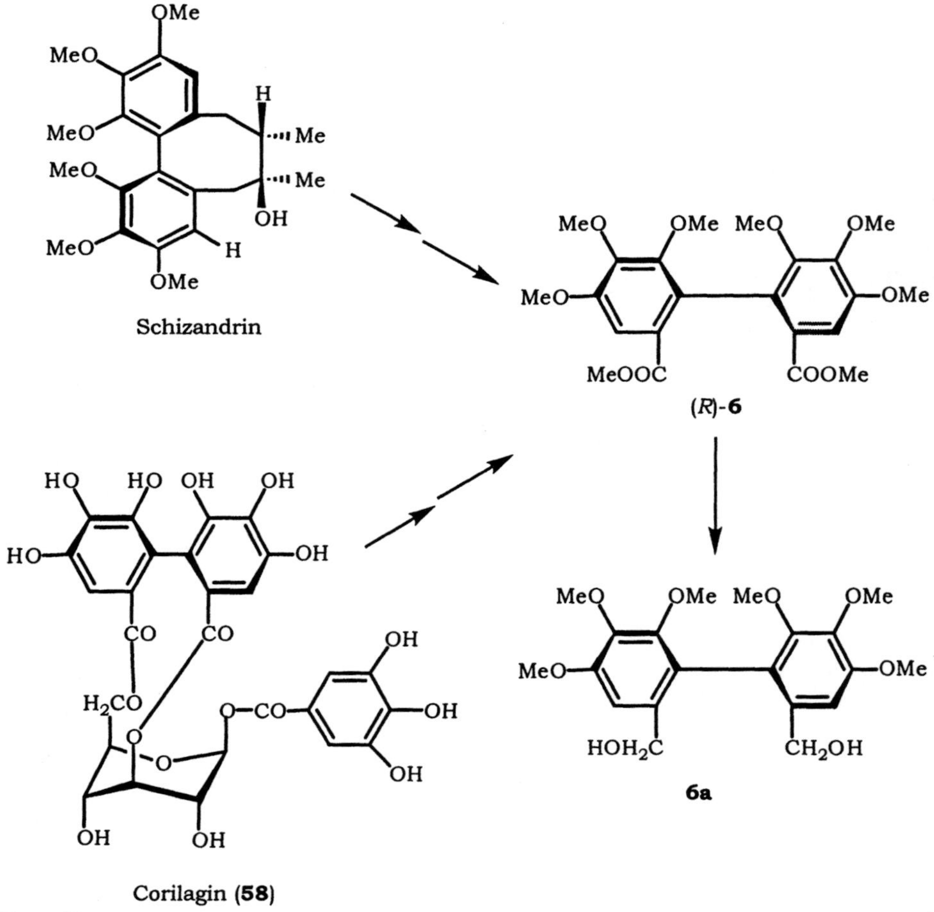

Schizandrin

(*R*)-**6**

6a

Corilagin (**58**)

Chart 12.

recognized to depend on several factors such as bond distance, bond order, angle and electronegativity of the substituents. Based on accumulated data, we found that a ^{1}H-^{13}C COLOC or HMBC measurement set at ^3J=5-7 Hz usually permits straightforward assignments of the position of each acyl group on the glucose core(s).

CD Spectroscopy

Ellagitannin monomers have one or two chiral biphenyl (HHDP) groups per molecule. Up to 1982, the absolute configuration of the HHDP group remained undetermined although the sign of specific rotation of dimethyl hexamethoxydiphenate (6) obtained by methylation and subsequent methanolysis of each ellagitannin was reported. Dextrorotatory 6 was demonstrated to have (R)-configuration [91], based on the ORD comparison of the dihydroxy derivative (6a) prepared by LiAlH$_4$ reduction of (+)-6 with that derived from schizandrin whose biphenyl portion had (R)-configuration [92]. The CD spectra of (R)-6 and (S)-6 exhibit strong Cotton effects at 225 and 250 nm, of the sign opposite to each other, as shown in Fig. 2 [6]. These Cotton effects are associated with the biphenyl conjugation bands (222 nm and 260 nm) in the UV spectra.

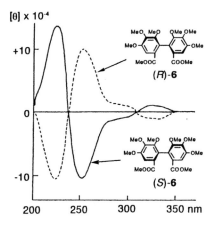

Fig (2). CD spectra of (R)-6 and (S)-6 in MeOH

A convenient empirical method [6] for determining the absolute configurations of the biphenyl moieties in ellagitannins without chemical degradation was exploited based on the accumulated CD spectra of the compounds whose atropisomerism of the chiral HHDP group in each molecule was determined by specific rotation of **6** after chemical degradation.

The CD spectra of mono-HHDP glucoses are similar to those of **6** except for a small bathochromic shift of each Cotton effect (235 and 265 nm). The positive and negative Cotton effects at a longer wavelength are thus related with (R) and (S)-configuration of the HHDP group, respectively.

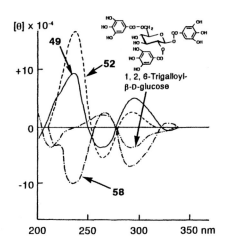

Fig. (3). CD spectra of monomeric hydrolyzable tannins in MeOH.

Ellagitannins such as tellimagrandin I (**49**), tellimagrandin II (**50**), pedunculagin (**52**) and corilagin (**58**) which have one or two HHDP and galloyl groups showed Cotton effects at around 235, 265 and 285 nm (Fig. **3**). The third extremum was regarded as a longer wavelength component of the couplet Cotton effect centered at ca. 277 nm which corresponds to the UV maxima (278 nm) of galloyl chromophore, and to have arisen from intramolecular charge-transfer of galloyl groups in a way similar to that of benzoylglucoses [93]. The negative first Cotton effect (284 nm) of 1,2,6-trigalloylglucose is consistent with anti-clockwise chirality of 1,2-digalloyl esters in the application of the exciton chirality rule [93]. The extremum around 265 nm in the CD spectra of tellimagrandin I (**49**) and II (**50**) is thus an overlapping peak composing of a longer wavelength Cotton effect of the HHDP group and a shorter wavelength extremum of couplet Cotton effect due to galloyl groups. Although this extremum may be variable depending on the galloyllation pattern on the glucose core, the Cotton effect at around 235 nm is unaffected by the presence or absence of galloyl groups and also by conformational change of the glucose core. Therefore the shorter wavelength Cotton effect is a diagnostic in determining the absolute configuration of the HHDP group in ellagitannins. That is, the positive and negative Cotton effects at around 235 nm indicate (*S*)- and (*R*)-configuration, respectively. Its amplitude is also indicative of the number of HHDP groups in the molecule, e.g., those of pedunculagin (**52**) and casuarictin (**53**) are two-fold of that of tellimagrandin I (**49**) or II (**50**) (Table 6).

Table 6.   CD Data of Monomeric and Oligomeric Hydrolyzable Tannins

Compounds	HHDP	Chiral biphenyl Valoneoyl or its equiv.	CD maxima [θ] x 10^{-4} (nm) (MeOH)		
(*R*)-6	1		-5.5(225)	+4.8 (250)	
(*S*)-6	1		+6.7 (225)	-5.6 (250)	
(*R*)-64		1	-3.8 (221)	+2.4 (251)	
(*S*)-64		1	+5.9 (220)	-2.8 (251)	
(*R*)-65		1	-1.2 (222)	+1.5 (252)	
(*S*)-65		1	+1.5 (225)	-1.7 (256)	
(*R*)-66		1	-3.1 (231)	+1.6 (262)	
(*S*)-66		1	+1.7 (231)	-1.1 (263)	
Monomers					
1, 2, 6-Trigalloyl-β-D-glucose			-3.3 (215)	+2.2 (261)	-3.5 (284)
1, 2, 3- Trigalloyl-β-D-glucose				-0.1 (257)	+1.1 (280)
1, 2, 3, 4, 6-Pentagalloyl-β-D-glucose (**48**)			+1.2 (225)	+0.2 (0.2)	+1.5 (270)
Tellimagrandin I (**49**)	1		+8.8 (232)	-3.7 (259)	+4.0 (284)
Tellimagrandin II (**50**)	1		+8.1 (235)	-1.0 (264)	+0.8 (285)
Pedunculagin (**52**)	2		+16.8(233)	-5.4 (259)	+2.1 (282)
Casuarictin (**53**)	2		+20.2 (234)	-5.3 (261)	+0.7 (281)
Corilagin (**58**)	1		-10.3 (237)	+1.7 (262)	-6.6 (284)
Geraniin (**59**)	1	*	-7.1 (236)	+4.6 (261)	-6.2 (291)
Dimers					
Agrimoniin (**1**)	4		+36.5 (235)	-10.6 (260)	+9.7 (282)
Rugosin D (**13**)		2	+25 (225)	-3.8 (258)	+8.1 (280)
Oenothein B (**23**)		2	+15 (236)	-4.3 (258)	+11 (280)
Nobotanin F (**19**)	2	1	+19.7 (235)	-7.1 (262)	+3.4 (282)

(Table 6). contd.....

Compounds	Chiral biphenyl		CD maxima [θ] x 10⁻⁴ (nm) (MeOH)		
	HHDP	Valoneoyl or its equiv.			
Eucalbanin B		1	+11.3 (235)	-7.5 (263)	+7.8 (287)
Eumaculin A		1	-11.0 (245)	+6.3 (261)	-8.2 (288)
Eumaculin B		1	-6.4 (236)	+3.6 (263)	-5.6 (293)
Eumaculin C	1	1	-3.0 (238)	+7.4 (265)	-2.2 (297)
Trimers					
Heterophylliin G	3	2	+62 (229)	-26 (261)	+18 (282)
Cornusiin C (20)	1	2	+14 (238)	-21 (259)	+15 (285)
Oenothein A (42)	1	2	+21.0 (238)	-11.5 (259)	+19.1 (283)
Nobotanin C	2	2	+35.6 (235)	-12.4 (262)	+4.5 (283)
Tetramers					
Nobotanin K	3	3	+37.9 (234)	-18.3 (260)	+7.6 (280)
Woodfordin F	1	3	+18.0 (237)	-14.9 (258)	+20.5 (282)
Trapanin B	1	3	+56 (220)	-33 (259)	+25 (285)

* DHHDP

This CD empirical rule is applicable to the assignment of atropisomerism of the valoneoyl group and its isomeric acyl groups as the linking units. The Cotton effects of dimethyl (*R*)- and (*S*)-octa-*O*-methylvaloneates (64) showed hypsochromic shift by ca. 10 nm with reduced amplitude relative to the corresponding Cotton effects of 6 [(Fig. 4)]. The dimers or oligomers possessing a valoneoyl as well as HHDP group thus exhibit a diagnostic Cotton effect at around 225 nm with a shoulder at 235 nm (Fig, 5).

(*S*)- and (*R*)-Trimethyl ocat-*O*-methyltergallate (65) were obtained by methylation followed by methanolysis of eucalbanin C (31) and euphorbin B (32), respectively. Similarly the enantiomers of the methylated derivative (66) of macaranoic acid were prepared from eurobustin A [42] and euphorbin I (33) [66]. The amplitudes of the Cotton effect at around 230 nm in the CD spectra of these methylated derivatives were smaller than that of the valoneoyl derivative [Fig. (4), Table 5]. This imbalance of amplitudes might be due to a difference in the angle between biphenyl rings [93, 94]. As mentioned earlier, the tergalloyl and macaranoyl group in the molecule is easily isomerized to the valoneoyl group through

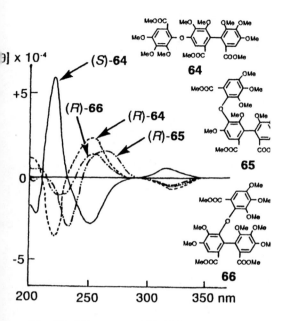

Fig. (4). CD spectra of **64-66** in MeOH.

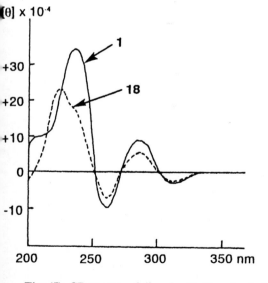

Fig. (5). CD spectra of dimeric ellagitannins in MeOH.

Smiles rearrangement even during the methylaion. Thus CD data which provide a straightforward assignment of chirality without derivatization would particularly be useful for dimers having labile linking units.

Structure Determination of Oligomeric Hydrolyzable Tannins

Practical examples of structural determination achieved with the aid of spectral analyses including two-dimensional NMR techniques are presented below.

Structure of Camelliin B (26)

Camelliin B (26) first isolated from the flowers of *Camellia japonica* (Theaceae) [39,41] is a unique dimer with a macrocyclic structure like oenothein B (23). This tannin forms an equilibrium mixture of α- and β-anomers (*ca.* 5:1) as revealed by duplication of each proton signal in the ¹H-NMR spectrum (Fig. 6). The spectrum of the major anomer indicated the signals are ascribable to three galloyl groups (δ 6.81, 7.11 and 7.17, each a 2H singlet) and two valoneoyl groups (δ 5.97, 6.48, 6.72, 6.86, 7.21 and 7.44, each a 1H singlet). These acyl units were confirmed by their characteristic signals in the ¹³C-NMR spectrum, and methanolysis of the methylated derivative yielding methyl tri-*O*-methylgallate and dimethyl octa-*O*-methylvaloneate (64). The atropisomerism of the valoneoyl groups in the molecule was evidenced to be all (*S*)-series by a strong positive Cotton effect at 220 nm ([θ] +28 x 10⁴) and 237 nm ([θ] +4.9 x 10⁴) in the CD spectrum.

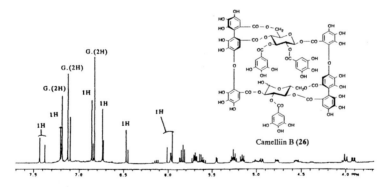

Camelliin B (26)

Fig. (6). ¹H-NMR spectrum of camellin B (26) (500 MHz, acetone-d_6 + D$_2$O).

The ¹H-¹H COSY spectrum of 26 revealed that all of the hydroxyl groups on the glucose residues are acylated except for an anomeric center. These spectral characteristics along with the molecular formula $C_{75}H_{52}O_{48}$ which was deduced from the FAB-MS (m/z 1743 [M+Na]⁺) implied a macrocircular structure. The binding modes of the acyl groups were determined as follows. In order to reduce the complexity of the signals due to mixture of α- and β-anomers, 26 was treated with NaBH₄ in aqueous

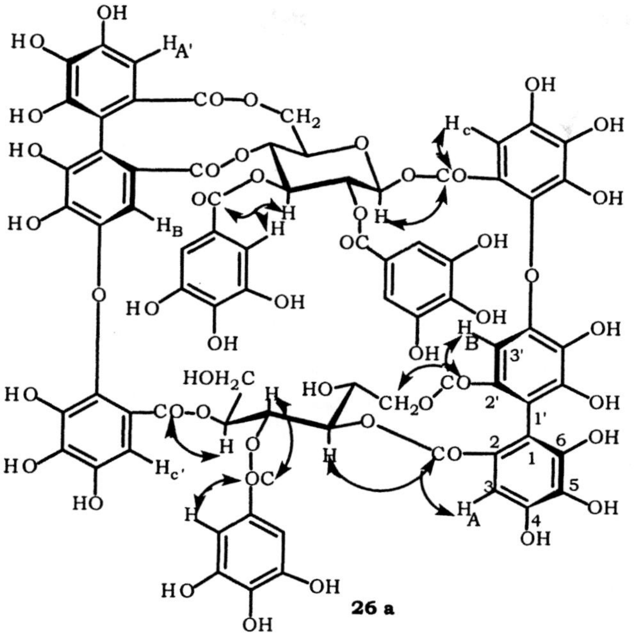

26 a

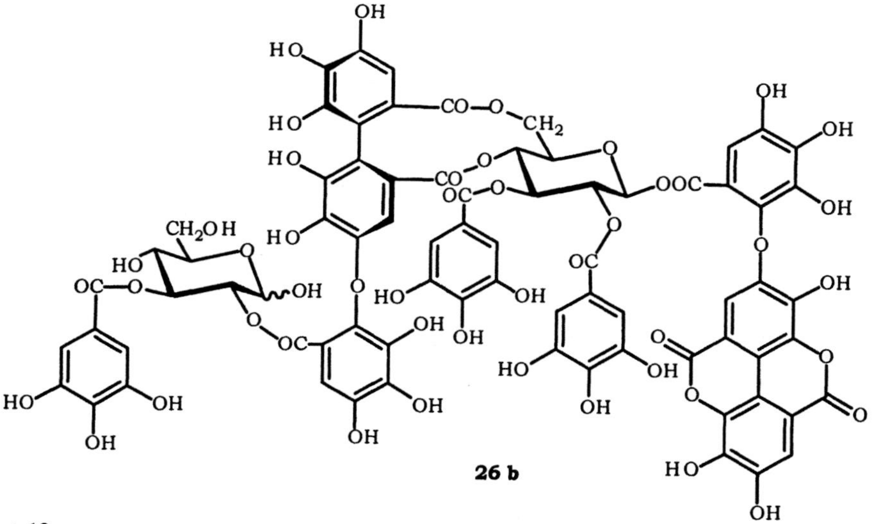

26 b

Chart 13.

MeOH to afford a dihydrocamelliin B (**26a**), FAB-MS m/z 1745 [M+Na]$^+$, which gave simplified ^{1}H- and ^{13}C-NMR spectra without duplication of signals. The signals due to glucose-II (glucitol) were unequivocally distinguished from those of glucose-I by the ^{1}H-^{1}H and ^{1}H-^{13}C COSY spectra. Among the valoneoyl proton signals (δ 6.18, 6.32, 6.64, 6.79, 6.83 and 7.19), the H$_A$(H$_A$') and H$_B$ (H$_B$') signals were assigned by the correlations with C-1 and C-1' (δ 115.7-117.3) through three-bond couplings in the ^{1}H-^{13}C long-range COSY spectrum measured at $^3J_{CH}$=7 Hz and 8 Hz. The signals at δ 6.18 and 6.32 were then assigned to valoneoy $H_{B(B')}$ signals based on the two-bond correlations with ethereal carbon signals at δ 146.2 and 146.3, respectively, which resonated at the lowest region for oxygen-bearing aromatic carbons. The H_A and $H_{A'}$ were thus assigned to the singlets at δ 6.79 and 6.64. The H_B signal at δ 6.18 was correlated with an ester carbonyl carbon resonance at δ 169.4 which also showed a cross peak with the glucitol H-6' signal (δ 4.55, dd, J=5.0, 12.5 Hz). Similarly, the connectivity of H_A with the glucitol H-4' was shown by respective three-bond couplings with a common ester carbonyl carbon (δ 168.8). The other long-range correlations are illustrated in Chart 13, establishing the orientation of the valoneoyl group at O-4'/O-6' in camelliin B (**26**). Although the connectivities between the $H_{A'}$ ($H_{B'}$) of the other valoneoyl group and the glucose-I protons were not observed, the orientation at O-4/O-6 was assigned by characterization of partial hydrolysate (**26b**) of **26**.

Structure of Camelliatannin D (67)

Camelliatannin D (**67**) isolated from the leaves and fruits of *Camellia japonica* [56] is a complex tannin dimer having a flavan-3-ol unit. Its FAB-MS spectra showed an (M+H)$^+$ ion peak at m/z 1859 (positive-ion mode) and (M-H)$^-$ ion peak at m/z 1857 (negative-ion mode), corresponding to the molecular formula $C_{83}H_{62}O_{50}$. Although the ^{1}H-NMR spectrum of this compound was complicated by duplication of signal like **26**, the aliphatic protons were attributed to a glucose residue adopting an open-chain form and a glucopyranose residue of two anomeric forms (α-anomer: β-anomer = 1:2), with the aid of the ^{1}H J-resolved NMR spectrum (Fig. **7**) and the ^{1}H-^{1}H COSY spectrum. C-Ring protons of a flavan-3-ol residue of 2,3-*cis* structure [δ 4.61 (br s, H-2), 4.25 (m, H-3), 2.56 (dd, J=2.5, 17 Hz, H-4) and 2.25 (dd, J=3.5, 17 Hz) (α-anomer); δ 4.61 (br s, H-2), 4.23 (m, H-3), 2.56 (dd, J=2.5, 17 Hz, H-4) and 2.16 (dd, J=3.5, 17 Hz) (β-anomer)] were also observed in the aliphatic region. Chemical shifts and coupling patterns of the protons of A- and B-rings of the flavan residue [δ 5.71 (s, A-ring H), 6.97 (d, J=1.5 Hz, H-2'), 6.75 (d, J=8.5 Hz, H-5'), 6.78 (dd, J=1.5, 8.5 Hz, H-6') (α-anomer); δ 5.70 (s, A-ring H), 6.97 (d, J=1.5 Hz, H-2'), 6.74 (d, J=8.5 Hz, H-5'), 6.78 (dd, J=1.5, 8.5 Hz, H-6') (β-anomer)], indicated that the flavan-3-ol is 8- or 6-substituted epicatechin. The chemical shifts of the A-ring proton are similar to those of camelliatannin C (**68**) (δ 5.84) having a 6-substitued epicatechin residue rather than those of camelliatannin E (**69**) (δ 6.03) having an 8-substitued epicatechin residue [95]. Remaining aromatic singlets were assigned to protons of galloyl, valoneoyl and/or HHDP groups, and the presence of these acyl groups was clearly evidenced by methanolysis of the

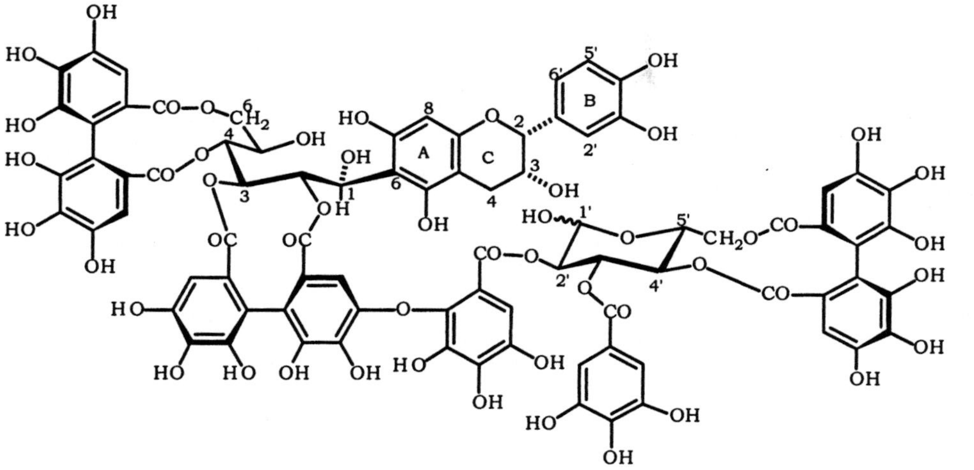

Camelliatannin D (**67**)

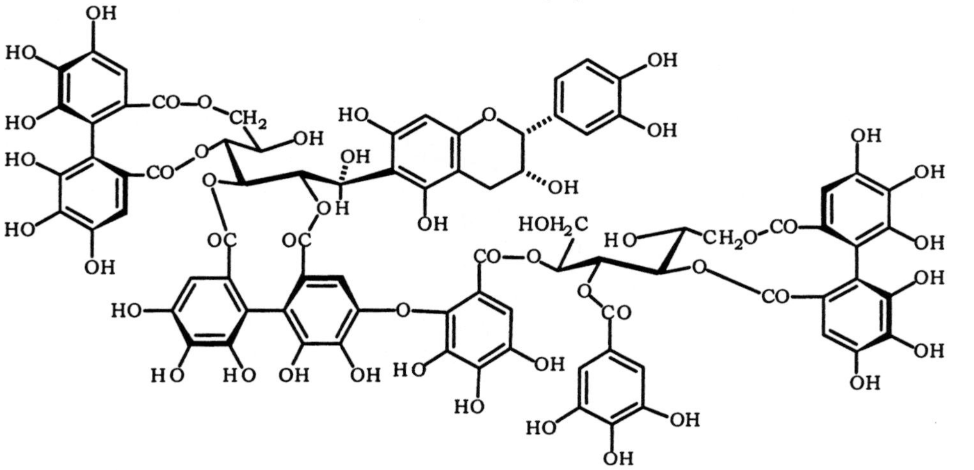

Dihydrocamelliatannin D (**67a**)

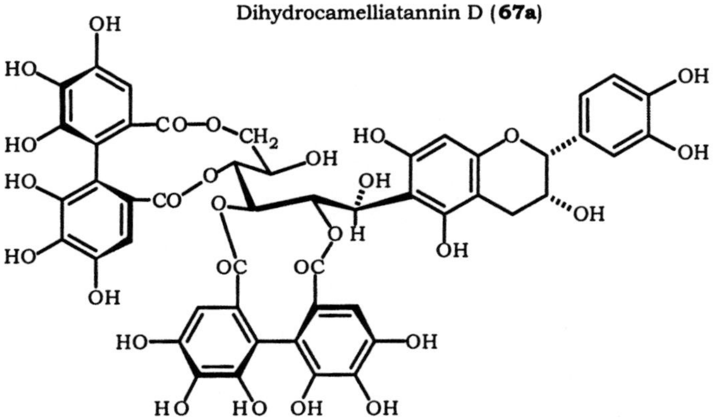

Camelliatannin C (**68**)

(Chart 14). contd.....

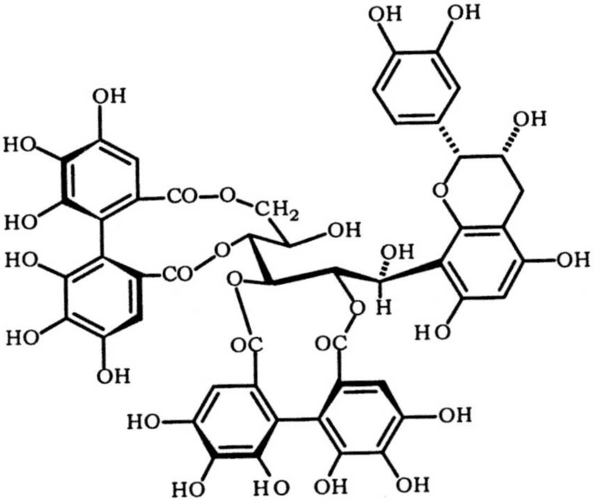

Camelliatannin E (**69**)

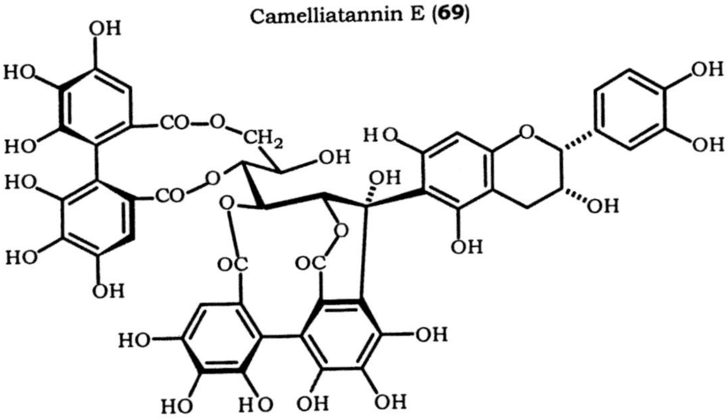

Camelliatannin B (**70**)

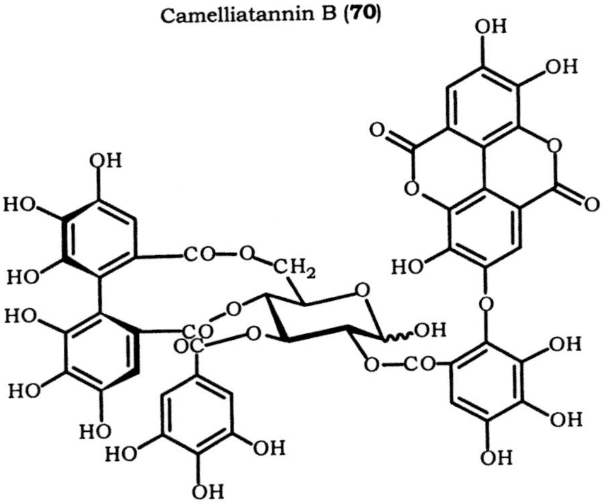

Cornusiin B (**71**)

Chart 14.

methylated tannin to give methyl tri-*O*-methylgallate, dimethyl hexamethoxydiphenate (**6**) and trimethyl octa-*O*-methylvaloneate (**64**) in a molar ratio of 1:2:1. Therefore, this tannin is composed of one galloyl, two HHDP, one valoneoyl, one 6-substitued epicatechin and two glucose residues.

Chemical shifts and coupling constants of the open-chain glucose and the glucopyranose residue were similar to those of camelliatannin C (**68**) and cornusiin B (**71**) [45], respectively, suggesting that the patterns of substitution of the acyl groups on the two glucose residues are the same as those of these two monomers. The configuration at C-1 of the open-chain glucose of camelliatannin D (**67**) was also assigned the same as that of **68**, based on the comparison of the coupling constants of the glucose protons. The configuration of **68** was determined on the basis of its ROESY spectrum.

The locations of the acyl groups on the two glucose residues were assigned by the long-range ^{1}H-^{13}C COSY spectra of camelliatannin D (**67**) and its dihydro derivative (**67a**) obtained by treatment with NaBH$_4$. The structure of **67** was chemically substantiated by its partial degradation to give camelliatannins C (**68**) and B (**70**) along with cornusiin B (**71**). The CD spectrum of camelliatannin D (**67**) showed a positive Cotton effect with a large amplitude in the short wavelength region ([θ]$_{234}$ +2.3 x 10^5), indicating that the configurations of the valoneoyl and HHDP groups are all *S*. The structure of camelliatannin D, composed of camelliatannin C (**68**) and tellimagrandin I (**49**), was thus established as **67** .

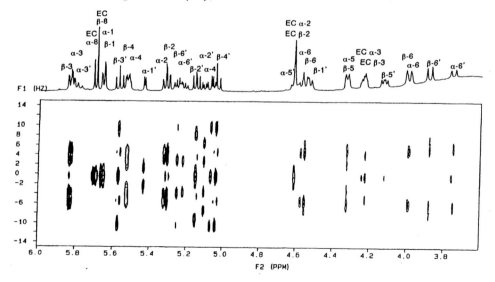

Fig. (7). 2D J-resolved NMR spectrum of cammeliatannin D (**67**) (500 MHz, acetone-*d*$_6$ + D$_2$O). α-3 means glucose H-3 of α-anomer. EC means (-)-epicatechin residue.

BIOLOGICAL ACTIVITY

Recent progress in the characterization of numerous tannins has enabled the investigation of the biological activities of these compounds on the basis of structural differences in a way similar to that used for the other types of natural organic compounds such as alkaloids, terpenoids and flavonoids. Besides astringency, a fundamental property of tannins, based on strong affinity to biopolymers such as proteins and polysaccharides, various biological and pharmaceutical activities beneficial to human health, have been found.

Antioxidative Activity

Oxidative cellular damage by reactive oxygen species such as superoxide anion, hydroperoxy and hydroxyl radicals etc. is known to be associated with various human chronic diseases, e.g. cancers, inflammation, arthritis, atherosclerosis and also with the process of ageing. Claims that diet and increased intake of nutrients exhibiting antioxidative activity have a preventative effect on chronic diseases have increased in recent years. In this context, polyphenolic compounds such as tannins, flavonoids, coumarins, lignans and caffeic acid derivatives, which are abundantly contained in a large number of medicinal plants, foods and beverages, are of particular interest for human health care because of the antioxidative properties widely found in plant phenolics. The antioxidative activity of tannins has been extensively studied in various *in vitro* and *in vivo* experimental systems and summarized in reviews [96, 97]. Such activity includes the inhibition of lipid peroxidation induced by NADPH-ADP and ascorbic acid-ADP in rat liver microsomes and mitochondria, respectively [98]. Most of the polyphenolics tested including ellagitannins, condensed tannins and related low molecular weight polyphenols showed stronger inhibitory effect in these experimental systems than did α-tocopherol. Inhibition of over 90% was observed with pentagalloylglucose (**48**), pedunculagin (**52**) and isoterchebin at a concentration of 5 μg/ml. These antioxidative properties of tannins are based on their potent radical scavenging ability as revealed by mechanistic studies for autoxidation of methyl linoleate based on kinetic, static and *in situ* ESR measurements [99]. In this radical-chain autoxidation, the radical scavenging activity of the HHDP group was stronger than that of the galloyl group. That is, tannins with many HHDP groups exhibited greater levels of scavenging activity than tannins having galloyl groups instead of HHDP group.

Radical scavenging effects of tannins and related polyphenols were also evidenced by ESR studies for the effects on the superoxide anion radical generated in the hypoxanthine-xanthine oxidase system and on the 1,1-diphenyl-2-picrylhydrazyl radical which is often used as a model of free radicals in lipid peroxidation [100]. Investigation of structure-activity relationships in these systems suggested that the scavenging effects of tannins are generally dependent on the molecular size and hence the number of phenolic hydroxyl groups with an *ortho*-trihydroxyl structure in the molecule. Dimeric hydrolyzable tannins possessing more than twenty hydroxyl groups thus have higher activity levels than those of monomers and related low molecular polyphenols as shown in Table 7.

Table 7. Scavenging Effects of Tannins and Related Polyphenols on Superoxide Anion Radicals Generated by Hypoxanthine-xanthine Oxidase

Compound	Galloyl	HHDP (+Other)	Number of phenolic hydroxyl groups	EC_{50}[a] (μM)
Hydrolyzable tannins				
Monomers				
1,2,6-Trigalloyl-β-D-glucose	3	0	9	4.1
1,2,3,6-Teragalloyl-β-D-glucose	4	0	12	2.8
1,2,3,4,6-Pentagalloyl-β-D-glucose(**48**)	5	0	15	3.4
Tellimagrandin II (**50**)	3	1	15	3.1
Pedunculagin (**52**)	0	2	12	2.8
Casuarictin (**53**)	1	2	15	2.8
Rugosin A	3	1+1(b)	17	2.7
Geraniin (**59**)	1	1+1(c)	11	3.0
Dimers				
Cornusiin A	3	1+1(b)	23	1.6
Rugosin D (**13**)	5	1+1(b)	29	1.7
Coriariin A (**9**)	4	1+1(d)	29	1.3
Proanthocyanidins				
Procyanidin B-2	0	0	8	14.0
Procyanidin C-1	0	0	12	8.8
Epicatechin gallate dimer	2	0	10	5.0
Low molecular phenols				
Gallic acid	1	0	3	6.8
Ellagic acid	0	0	4	19.0
(-)-Epicatechin	0	0	4	23.0
(-)-Epicatechin gallate	1	0	5	4.8

a) Concentration giving a 50% decrease of superoxide anion radical.
b) valoneoyl group ; c) DHHDP group ; d) DHDG group

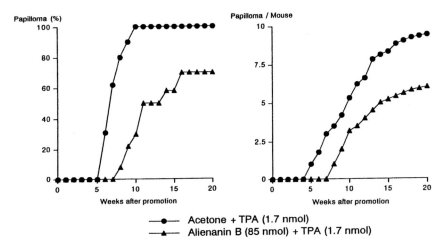

—●— Acetone + TPA (1.7 nmol)
—▲— Alienanin B (85 nmol) + TPA (1.7 nmol)

Fig. (8). Inhibitory effect of alienanin B **(45)** on tumor promotion induced with DMBA and TPA.

Although the antioxidative effects of tannins vary depending on the character of the radicals or on the experimental system, this property underlies the medicinal usages of polyphenol-rich plants and folk medicines.

Anti-tumor Promoting Activity

Cancer chemoprevention is one of the most important strategies for cancer prevention [101]. Extensive studies on multi-stage carcinogenesis in humans have revealed that the tumor-promoting step is a long and reversible process, and growth of benign tumors can be arrested at this stage by preventive agents, although it is difficult to prevent the tumor initiation which is a rapid and irreversible process. Therefore, inhibitors of tumor promotion, rather than initiation or progression of tumor from benign to malignant stage, are expected to be effective for cancer prevention in humans. In an *in vitro* survey of preventive agents against tumor promotion from medicinal plants and drugs, polyphenols such as (-)-epigallocatechin gallate (EGCG), pentagalloylglucose **(48)**, pedunculagin **(52)** and chebulinic acid etc. were found to exhibit a competitive binding activity to TPA receptor in a particulate fraction of mouse skin [102,103]. In a two-stage experiment for mouse skin carcinogenesis, EGCG, the main tannin in green tea (*Camellia sinensis*), showed a remarkable inhibition against tumor promotion induced with 7,12-dimethylbenz(*a*)anthracene (DMBA) and okadaic acid [102]. Inhibitory effects of EGCG on various systems of chemical carcinogenesis have then been studied extensively in Japan and America, and EGCG and green tea extract are now considered to be one of the most promising chemopreventive agents for human cancers [101,104].

Among screening tests developed to discover possible anti tumor-promoting agents from natural sources, an assay of activation of Epstein-Barr virus early antigen (EBV-EA) induced by 12-O-tetradecanoylphorbol-13-acetate (TPA) has been widely used as a short-term *in vitro* primary screening [105]. Upon the bioassay-guided fractionation of the aqueous acetone extract of *Cowania mexicana* (Rosaceae), *C*-glucosidic ellagitannin monomer and dimers, stenophyllanin B, and alienanin B (**45**) and casuglaunin A (**46**), were isolated as potent inhibitors of the EBV-EA activation. Their inhibitory effects were 70.1%, 81.5% and 68.4 %, respectively, at a concentration of 500 mol ratio/TPA (32 pmol), stronger than those of EGCG (65.1 %) and glycyrrhetinic acid (45.7%) as positive controls [106]. In 2-steps *in vivo* carcinogenesis on mouse skin using TPA as promoter and DMBA as initiator, dimeric tannins, alienanin B (**45**) and casuglaunin A (**46**) significantly decreased the number of tumor-bearing mice and reduced the number of papilloma in individual mice as shown in Fig. (**8**).

Host-mediated Antitumor Activity

This type of activity of tannins was first found for agrimoniin (**1**) [11]. A remarkable antitumor activity of agrimoniin was exhibited on intraperitoneal injection several days before inoculation of MM2 ascite tumors or sarcoma-180 tumor cells into mice.

The *in vivo* assay for evaluation of antitumor activity of other tannins and related polyphenols was thus routinely performed by administration of tannins with a single intraperitoneal injection (10 mg/kg) to female ddY mice 4 days before inoculation of sarcoma 180 cells (1×10^5). Sixty days after inoculation of the tumor cells, survivors were killed and autopsied. The antitumor activity of each tannin was estimated by the number of regressors and the % increases in the life span (% ILS) [107]. Among approximately 100 polyphenols of different types tested, most of the oligomeric hydrolyzable tannins exhibited a potent activity in contrast to the negligible activity of the monomeric hydrolyzable tannins, proanthocyanidins and their related low-molecular weight polyphenols such as gallic acid, catechins and caffeoyl derivatives.

The most potent activity was shown by oenothein B (**23**) which gave 196% ILS and 4 survivors out of 6 mice after 60 days: none of these survivors had a tumor. Hirtellin B (**72**) [50] and its degalloylated congener, tamarixinin A (**73**) [19], in which two monomeric units are doubly linked to form a macrocyclic structure like oenothein B, also showed strong activity comparable to that of oenothein B (**23**) (survivors; 4 and 3 mice out of six, respectively). Other active oligomers include macrocyclic tannins, e.g. oenothein A (**42**), woodfordin C (**24**) and camelliin B (**26**), and the oligomers with linear structure, e.g. coriariin A (**9**) and rugosin D (**13**) etc. These active oligomers are mostly composed of tellimagrandin I (**49**) and/or tellimagrandin II (**50**). The potency of the activity is, however, not proportional to the degree of oligomeriztion, as revealed by much weaker activity of oenothein A (**42**) and woodfordin F, which are trimer and tetramer of tellimagrandin I (**49**), respectively. These results suggest that the presence of tellimagrandin I or II units in a molecule as well as the molecular size including the conformation of the whole molecule is important for antitumor activity in this assay system.

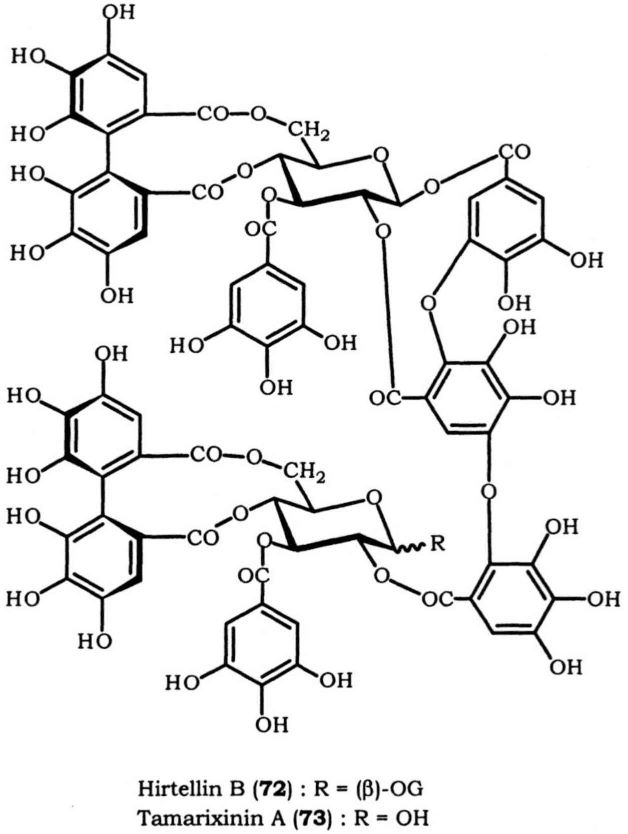

Hirtellin B (**72**) : R = (β)-OG
Tamarixinin A (**73**) : R = OH

Chart 15.

This activity is exhibited by peritoneal administration of tannin once 4days before inoculation of tumor cells in mouse, hence indicating that it is attributable to enhancement of the immune response of host animals. In fact, these tannins significantly stimulated macrophages *in vitro* and *in vivo,* resulting in induction of interleukin-1 and increase of cytotoxic immunocytes [108, 109].

Other Significant Activities of Oligomeric Ellagitannins

Besides the above-mentioned activities, antitumoric properties have been reported for oligomeric hydrolyzable tannins. For example, woodfordin C (**24**) is a potent inhibitor of human DNA-topoisomerase II showing an IC_{50} value of 2.5 μg/mL [110]. This activity of **24** was 10-fold stronger than that of adriamycin and etoposide when compared at molar concentration. On the other hand, in a study of inhibitors against 5α-reductase and aromatase which convert testosterone into dihydrotestosterone and 17β-estradiol, respectively, the macrocyclic oligomers, oenotheins B (**23**) and A (**42**) were isolated from the *Epilobium* extract as potent inhibitors (IC_{50} 0.44 μM for **23** and 1.24 μM for **42** against 5α-reductase: inhibition at 50 μM, 33% (**23**) and 70% (**42**) against aromatase) [35]. Testosterone and its metabolites have been shown to be associated with development of benign prostatic hyperplasia (BPH) [111], and thus the inhibitors against the above enzymes are expected to have a potential role in the treatment of BPH.

Oenothein B (**23**) was also found to be a potent and specific *in vitro* inhibitor of poly(ADP-ribose)glycohydrolase which is suggested to be involved in regulation of DNA repair, replication and transcription [112]. This activity was exhibited only by hydrolyzable tannins, among which oligomeric ellagitannins such as nobotanin B (**27**), coriariin A (**9**) and nobotanin K (tetramer) as well as **23**, were more potent, with IC_{50} values ranging from 0.38 to 8.5 μM, than monomers (IC_{50} >11 μM). Oenothein B (**23**) suppressed the expression of glucocorticoid-sensitive mouse mammary tumor virus (MMTV) genes integrated into 34I cells [113]. This suppression was suggested to result from *in vivo* inhibition of de-poly(ADP-ribosyl)ation on chromosomal proteins by **23**.

Antiviral activities of oligomeric ellagitannins have also been reported. For example, oenothein B (**23**) and coriariin A (**9**) are efficient inhibitors of *Herpes simplex* virus with EC_{50} values in the nano molar range [114]. Significant anti-HIV activity was observed for the oligomers, camelliin B (**26**), nobotanin B (**27**) and trapanin B. This activity was suggested to involve the inhibition of virus absorption to MT-4 cells, and of HIV reverse transcriptase [115].

CONCLUSIONS

Remarkable recent advances in isolation and spectroscopic analyses have
led to the characterization of numerous hydrolyzable tannins including
newly found ellagitannin oligomers. Details as to the chemical diversity
and potency in various biological and pharmaceutical activities, especially
with respect to antioxidant, antitumor and antiviral activities, of tannins
have consequently been elucidated. From a chemical and biological point of
view, hydrolyzable tannins are of interest to organic chemists, biochemists
and pharmacologists as well as to chemists searching for natural bioactive
products. Several monomeric ellagitannins such as tellimagrandins I (**49**)
and II (**50**), pedunculagin (**52**), praecoxin B (2,3-HHDP-4,6-di-*O*-
galloylglucose) and its gallate, pterocaryanin C (**55**) have already been
chemically synthesized as a results of developments in methodology for
building up a chiral biphenyl acyl group on the glucose template [116-
119]. However, the biosynthesis of ellagitannins still remains to be
clarified, although that of galloylglucoses has been extensively studied
[120]. More importantly, studies on the absorption and metabolism of
polyphenols of this class in human bodies are needed before they can be
applied to the practical prevention of diseases and to improving human
health care.

ACKNOWLEDGEMENTS

The authors wish to acknowledge the efforts of our coworkers cited in the
references.

REFERENCES

[1]   Haslam, E. *Plant Polyphenols. Vegetable Tannins revisited*, Cambridge
       University Press, Cambridge, **1989**.
[2]   Okuda, T.; Yoshida, T.; Hatano, T. *Fortschritt. Chem. Org. Naturstoffe*, **1995**,
       66, 1.
[3]   Okuda, T.; Yoshida, T.; Hatano, T. In *Economic and Medicinal Plant Research*.
       Wagner, H; Farnsworth, N.R., Eds.; Academic Press: London, **1991**, Vol. *5*, pp.
       129-165.
[4]   Hillis, W.E.; Seikel, M.K. *Phytochemistry*, **1970**, *9*, 1115.
[5]   Okuda, T.; Yoshida, T.; Kuwahara, M.; Memon, M.U.; Shingu, T. *J. Chem.
       Soc., Chem. Commun.*, **1982**, 163.
[6]   Okuda, T.; Yoshida, T.; Hatano, T.; Koga, T.; Toh, N.; Kuriyama, K.
       Tetrahedron Lett., **1982**, *23*, 3937.
[7]   Okuda, T.; Hatano, T.; Ogawa, N. *Chem. Pharm. Bull.*, **1982**, *30*, 4234.
[8]   Hatano, T.; Ogawa, N.; Shingu, T.; Okuda, T. *Chem. Pharm. Bull.*, **1990**, *38*,
       3341.
[9]   Tanaka, T.; Nonaka, G.; Nishioka, I. *J. Chem. Res. (M)*, **1985**, 2001.

[10] Yoshida, T.; Okuda, T.; Memon, M.U.; Shingu, T. *J. Chem. Soc., Perkin Trans. 1*, **1985**, 315.
[11] Miyamoto, K.; Koshiura, R.; Ikeya, Y.; Taguchi, H. *Chem. Pharm. Bull.*, **1985**, *33*, 3977.
[12] Okuda, T.; Yoshida, T.; Hatano, T. *Phytochemistry*, **1993**, *32*, 507.
[13] Haslam, E. Fortschritt. Chem. Org. Naturstoffe, **1982**, *41*, 1.
[14] Schmidt, O. Th. Fortschritt. Chem. Org. Naturstoffe, **1956**, *13*, 570.
[15] Nonaka, G.; Sakai, T.; Mihashi, K.; Nishioka, I. *Chem. Pharm. Bull.*, **1991**, *39*, 884.
[16] Hatano, T.; Hattori, S.; Okuda, T. *Chem. Pharm. Bull.*, **1986**, *34*, 4092.
[17] Ishimatsu, M.; Tanaka, T.; Nonaka, G.; Nishioka, I.; Nishizawa, M.; Yamagishi, T. *Chem. Pharm. Bull.*, **1989**, *37*, 1735.
[18] Yoshida, T.; Ahmed, A.F.; Memon, M.U.; Okuda, T. *Chem. Pharm. Bull.*, **1991**, *39*, 2849.
[19] Yoshida, T.; Ahmed, A.F.; Okuda, T. *Chem. Pharm. Bull.*, **1993**, *41*, 672.
[20] Nonaka, G.; Nakayama, S.; Nishioka, I. *Chem. Pharm. Bull.*, **1989**, *37*, 2030.
[21] Yoshida, T.; Jin, Z.-X.; Okuda, T. *Chem. Pharm. Bull.*, **1991**, *39*, 49.
[22] Jin, Z-X.; Ito, H.; Yoshida, T. *Phytochemistry*, **1998**, *48*, 333.
[23] Yoshida, T.; Chen, X.-M.; Hatano, T.; Fukushima, M.; Okuda, T. *Chem. Pharm. Bull.*, **1987**, *35*, 1817.
[24] Yoshida, T.; Jin, Z.-X.; Okuda, T. *Phytochemistry*, **1991**, *30*, 2747.
[25] Hatano, T.; Hattori, S.; Okuda, T. *Chem. Pharm. Bull.*, **1986**, *34*, 4533.
[26] Yoshida, T.; Ohbayashi, H.; Ishihara, K.; Ohwashi, W.; Haba, K.; Okano, Y.; Shingu, T.; Okuda, T. *Chem. Pharm. Bull.*, **1991**, *39*, 2233.
[27] Yoshida, T.; Ohwashi, W.; Haba, H.; Ohbayashi, H.; Ishihara, K.; Okano, Y.; Shingu, T.; Okuda, T. *Chem. Pharm. Bull.*, **1991**, *39*, 2264.
[28] Yoshida, T.; Haba, K.: Arata, R.; Nakata, F.; Shingu, T.; Okuda, T. *Chem. Pharm. Bull.*, **1995**, *43*, 1101.
[29] Yoshida, T.; Ikeda, Y.; Ohbayashi, H.; Ishihara, K.; Ohwashi, W.; Shingu, T.; Okuda, T. *Chem. Pharm. Bull.*, **1986**, *34*, 2676.
[30] Yoshida, T.; Nakata, F.; Hosotani, K.; Nitta, A.; Okuda, T. *Phytochemistry*, **1992**, *31*, 2829.
[31] Yoshida, T.; Arioka, H.; Fujita, T.; Chen, X.-M.; Okuda, T. *Phytochemistry*, **1994**, *37*, 863.
[32] Hatano, T.; Yasuhara, T.; Matsuda, M.; Yazaki, K.; Yoshida, T.; Okuda, T. *J. Chem. Soc., Perkin Trans. 1*, **1990**, 2735.
[33] Yoshida, T.; Chou, T.; Matsuda, M.; Yasuhara, T.; Yazaki, K.; Hatano, T.; Nitta, A.; Okuda, T. *Chem. Pharm. Bull.*, **1991**, *39*, 1157.
[34] Yoshida, T.; Chou, T.; Shingu, T.; Okuda T. *Phytochemistry*, **1995**, *40*, 555.
[35] Ducrey, B.; Marston, A.; Gohring, S.; Hartmann, R.W.; Hostettmann, K. *Planta Med.*, **1997**, *63*, 111.
[36] Okuda, T.; Yoshida, T.; Hatano, T. *Centrifugal Partition Chromatography* (Foucault, A.P., Ed.), Marcel Dekker, New York, **1995**, pp. 99-131.
[37] Yoshida, T.; Chou, T.; Nitta, A.; Miyamoto, K.; Koshiura, R.; Okuda, T. *Chem. Pharm. Bull.*, **1990**, *38*, 1211.
[38] Yoshida, T.; Chou, T.; Nitta, A.; Okuda, T. *Chem. Pharm. Bull.*, **1992**, *40*, 2023.
[39] Yoshida, T.; Chou, T.; Maruyama, Y.; Okuda, T. *Chem. Pharm. Bull.*, **1990**, *38*, 2681.

[40]   Yoshida, T.; Nakazawa, T.; Hatano, T.; Yang, R.-C.; Yang, L.-L.; Yen, K.-Y.; Okuda, T. *Phytochemistry*, **1994**, *37*, 241.

[41]   Yoshida, T.; Chou, T.; Nitta, A.; Okuda, T. *Chem. Pharm. Bull.*, **1991**, *39*, 2247.

[42]   Yoshida, T.; Maruyama, T.; Nitta, A.; Okuda, T. *Chem. Pharm. Bull.*, **1992**, *40*, 1750.

[43]   Kondo, S.; Hatano, T.; Yoshida, T.; Okuda, T.; Lu, C.-F.; Yang, L.-L.; Yen, K.- Y. Abstract Papers, 32nd Annual Meeting of the Chugoku-Shikoku Branch of the Pharmaceutical Society of Japan, **1993**, p. 44.

[44]   Lee, M.-H.; Nishimoto, S.; Yang, L.-L.; Yen, K.-Y.; Hatano, T.; Yoshida, T.; Okuda, T. *Phytochemistry*, **1997**, *44*, 1343.

[45]   Hatano, T.; Ogawa, N.; Kira, R.; Yasuhara, T.; Okuda, T. *Chem. Pharm. Bull.*, **1989**, *37*, 2083.

[46]   Hatano, T.; Okonogi, A.; Yazaki, K.; Okuda, T. *Chem. Pharm. Bull.*, **1990**, *38*, 2707.

[47]   Hatano, T.; Ikegami, Y.; Shingu, T.; Okuda, T. *Chem. Pharm. Bull.*, **1988**, *36*, 2017.

[48]   Tanaka, T.; Morita, A.; Nonaka, G.; Lin, T.-C.; Nishioka, I.; Ho, F.-C. *Chem. Pharm. Bull.*, **1991**, *39*, 60.

[49]   Hatano, T.; Kira, R.; Yasuhara, T.; Okuda, T. *Chem. Pharm. Bull.*, **1988**, *36*, 3920.

[50]   Yoshida, T.; Hatano, T.; Ahmed, F.A.; Okonogi, A.; Okuda, T. *Tetrahedron*, **1991**, *47*, 3575.

[51]   Hatano, T.; Okonogi, A.; Okuda, T. In *Plant Polyphenols: Synthesis, Properties, Significance* ; Hemingway, R.W. ; Laks, P.E., Eds.; Plenum Press: New York, **1992**. pp. 195-207.

[52]   Yoshida, T.; Tanei, S.; Liu, Y.-Z.; Yuan, K.; Ji, C.-R.; Okuda, T. *Phytochemistry*, **1993**, *32*, 1287.

[53]   Yoshida, T.; Haba, K.; Nakata, F.; Okano, Y.; Shingu, T.; Okuda, T. *Chem. Pharm. Bull.*, **1992**, *40*, 66.

[54]   Xu, Y.-M.; Tanaka, T.; Nonaka, G.; Nishioka, I. *Chem. Pharm. Bull.*, **1991**, *39*, 647.

[55]   Han, L.; Hatano, T.; Yoshida, T.; Okuda. T. *Chem. Pharm. Bull.*, **1994**, *42*, 1399.

[56]   Hatano, T.; Han, L.; Taniguchi, S.; Okuda, T.; Kiso, Y.; Tanaka, T.; Yoshida, T. *Chem. Pharm. Bull.*, **1995**, *43*, 2033.

[57]   Yoshida, T.; Chen, L.; Shingu, T.; Okuda, T. *Chem. Pharm. Bull.*, **1988**, *36*, 2940.

[58]   Yoshida, T.; Namba, O.; Chen, L.; Okuda, T. *Chem. Pharm. Bull.*, **1990**, *38*, 86.

[59]   Yoshida, T.; Namba, O.; Yokoyama, K.; Okuda, T. Symposium Papers of *The 31st Symposium on the Chemistry of Natural Products*. **1989**, P.601.

[60]   Yoshida, T.; Namba, O.; Chen, L.; Okuda, T. *Chem. Pharm. Bull.*, **1990**, *38*, 1113.

[61]   Yoshida, T.; Yokoyama, K.; Namba, O.; Okuda, T. *Chem. Pharm. Bull.*, **1991**, *39*, 1137.

[62]   Yoshida, T.; Namba, O.; Kurokawa, K.; Amakura, Y.; Liu, Y.-Z..; Okuda, T. *Chem. Pharm. Bull.*, **1994**, *42*, 2005.

[63] Lee, S.-H.; Tanaka, T.; Nonaka, G.; Nishioka, I. *Chem. Pharm. Bull.*, **1991**, *39*, 630.

[64] Amakura, Y.; Kawada, K.; Hatano, T.; Agata, I.; Sugaya, T.; Nishibe, S.; Okuda, T.; Yoshida, T. *Can. J. Chem.*, **1997**, *75*, 727.

[65] Agata, I.; Hatano, T.; Nakaya, Y.; Sugaya, T.; Nishibe, S.; Yoshida, T.; Okuda, T. *Chem. Pharm. Bull.*, **1991**, *39*, 881.

[66] Amakura, Y.; Yoshida, T. *Chem. Pharm. Bull.*, **1996**, *44*, 1293.

[67] Lin, J.-H.; Tanaka, T.; Nonaka, G.; Nishioka, I.; Chen, I.-S. *Chem. Pharm. Bull.*, **1990**, *38*, 2162.

[68] Yoshida, T.; Namba, O.; Lu, C.-F.; Yang, L.-L.; Yen, K.-Y.; Okuda, T. *Chem. Pharm. Bull.*, **1992**, *40*, 338.

[69] Tanaka, T.; Nonaka, G.; Nishioka, I.; Kouno, I.; Ho, F.-C. *Phytochemistry*, **1995**, *38*, 509.

[70] Saijo, R.; Nonaka, G.; Nishioka, I.; Chen, I.-S.; Hwang, T.-H. *Chem. Pharm. Bull.*, **1989**, *37*, 2940.

[71] Tanaka, T.; Nonaka, G.; Nishioka, I. *Chem. Pharm. Bull.*, **1986**, *34*, 1039.

[72] Maki, Y.; Suzuki, M.; Masugi, T. *Chem. Pharm. Bull.*, **1968**, *16*, 559.

[73] Tanaka, T.; Tachibana, H.; Nonaka, G.; Nishioka, I.; Hsu, F.-L., Kohda, H.; Tanaka, H. *Chem. Pharm. Bull.*, **1993**, *41*, 1214.

[74] Yoshida, T.; Feng, W.-S.; Okuda, T. *Chem. Pharm. Bull.*, **1992**, *40*, 1997.

[75] Nonaka, G.; Sakai, T.; Mihashi, K.; Nishioka, I. *Chem. Pharm. Bull.*, **1991**, *39*, 884.

[76] Penhoat, C.L.M.H.; Michon, V.M.F.; Peng, S.; Viriot, C.; Scalbert, A.; Gage, D. *J. Chem. Soc. Perkin Trans. 1*, **1991**, 1653.

[77] Shimokawa, H.; Nonaka, G.; Nishioka, I. Abstract Papers, 111th Annual Meeting of the Pharmaceutical Society of Japan, **1991**, p. 147.

[78] Ito, H.; Miki, K.; Yoshida, T. *Chem. Pharm. Bull.*, **1999**, in press.

[79] Lin, T.-C.; Tanaka, G.; Nonaka, G.; Nishioka, I.; Young, T.-J. *Chem. Pharm. Bull.*, **1991**, *39*, 1144.

[80] Lueng, J.; Fenton, T.W.; Mueller, M.M.; Clandinin, D.R. *J. Food Sci.*, **1979**, *44*, 1313.

[81] Bate-Smith, E.C. *Phytochmiestry*, **1981**, *20*, 211.

[82] Scalbert, A. In: *Plant Polyphenols. Synthesis, Properties, significance.* Hemigway, R. W.; Laks, P. E. Eds.: Plenum Press, New York; **1992**, pp. 259-280.

[83] Okuda, T.; Yoshida, T.; Hatano, T.; Yazaki, K.; Kira, R.; Ikeda, Y. *J. Chromatogr.*, **1986**, *362*, 375.

[84] Okuda, T.; Yoshida, T.; Hatano, T.; Ikeda, Y.; Shingu, T.; Inoue, T. *Chem. Pharm. Bull.*, **1986**, *34*, 4075.

[85] Okuda, T.; Yoshida, T.; Hatano, T.; Mori, K.; Fukuda, T. *J. Liq. Chromatogr.*, **1990**, *13*, 3637.

[86] Yoshida, T.; Hatano, T.; Okuda, T. *J. Chromatogr.*, **1989**, *467*, 139.

[87] Luger, P.; Weber, M.; Kashino, S.; Amakura, Y.; Yoshida, T.; Okuda, T.; Beurskens, G.; Dauter, Z. *Acta Crystallographica B*, **1998**, *54*, 687.

[88] Self, R.; Eagles, J.; Galletti, G.C.; Mueller-Harvey, I.; Hartley, R.D.; Lea, A.G.H.; Magnolato, D.; Richli, U.; Gujer, R.; Haslam, E. *Biomed. Envir. Mass Spectr.*, **1986**, *13*, 449.

[89] Cheynir, V.; Doco, T.; Fulcrand, H.; Guyot, S.; Rouz, E. Le; Souquet, J.M.; Rugaud, J.; Moutounet, M. *Analusis*, **1997**, *25*, M32.

[90]   Yoshida, T.; Hatano, T.; Okuda, T.; Memon, M.U.; Shingu, T.; Inoue, K. *Chem. Pharm. Bull.*, **1984**, *32*, 1790.

[91]   Okuda, T.; Yoshida, T.; Hatano, T. *J. Chem. Soc., Perkin Trans. 1*, **1982**, 9.

[92]   Ikeya, Y.; Taguchi, H.; Yoshioka, I.; Kobayashi, H. *Chem. Pharm. Bull.*, **1979**, *27*, 1383.

[93]   Harada, N.; Nakanishi, K. *Acc. Chem. Res.*, **1972**, *5*, 257.

[94]   Chen, S.; Harada, N.; Nakanishi, K. *J. Amer. Chem. Soc.* **1974**, *96*, 7352.

[95]   Hatano, T.; Han, L.; Taniguchi, S.; Shingu, T.; Okuda, T.; Yoshida, T. *Chem. Pharm. Bull.*, **1995**, *43*, 1629.

[96]   Okuda, T.; Yoshida, T.; Hatano, T. In *Phenolic Compounds in Food and Their Effects on Health II., Antioxidant & Cancer Prevention* ; Huang, M.-T.; Ho, C.-T.; Lee, C.Eds.; American Chemical Society: Washington DC, **1992**; pp.87-97.

[97]   Okuda, T.; Yoshida, T.; Hatano, T. In *Food Phytochemicals for Cancer Prevention II, Teas, Spices and Herbs.* ; Ho.; C.-T.; Osawa, T.; Huang, M.-T.; Rosen, R. T. Eds.; American Chemical Society: Washington DC, **1994**; pp.133-143.

[98]   Okuda, T.; Kimura, Y.; Yoshida, T.; Hatano, T.; Okuda, H. *Chem. Pharm. Bull.*, **1983**, *31*, 1625.

[99]   Fujita, Y.; Komagoe, K.; Niwa, Y.; Uehara, I.; Hara, R.; Mori, H.; Okuda, T. *Yakugaku Zasshi*, **1988**, *108*, 528.

[100]   Hatano, T.; Edamatsu, R.; Hiramatsu, M.; Mori, A.; Fujita, Y.; Yasuhara, T.; Yoshida, T.; Okuda, T. *Chem. Pharm. Bull.*, **1989**, *37*, 2016.

[101]   Fujiki, H.; Komori, A.; Suganuma, M. In *Comprehensive Toxicology, Vol. 12, Chemical Carcinogens and Anticarcinogens* ; Bowden, G.T.; Fischer, S.M., Eds; Pergamon: New York, **1997**; p. 453.

[102]   Yoshizawa, S.; Horiuchi, T.; Fujiki, H.; Yoshida, T.; Okuda, T.; Sugimura, T. *Phytotherapy Res.*, **1987**, *1*, 44.

[103]   Yoshizawa, S.; Horiuchi, T.; Suganuma, M.; Nishiwaki, S.; Yatsunami, J.; Okabe, S.; Okuda, T.; Muto, Y.; Frenkel, K.; Troll, W.; Fujiki, H. In *Phenolic Compounds in Food and Their Effects on Health II., Antioxidant & Cancer Prevention*; Huang, M.-T.; Ho, C.-T.; Lee, C., Eds.; American Chemical Society: Washington DC, **1992**; pp.316-325.

[104]   Jankum, J.; Selman, S. H.; Swiercz, R. *Nature*, **1997**, *387*, 561.

[105]   Ito, Y.; Kawanishi, M.; Harayama, T.; Takabayashi, S. *Cancer Letters*, **1981**, *12*, 175.

[106]   Itoh, H.; Hatano, T.; Miyake, M.; Yoshida, T.; Kozuka, M.; Tokuda, H.; Okuda, T. In *Polyphenols 94*; INRA: Paris, **1995**; pp.419-420.

[107]   Miyamoto, K.; Nomura, M.; Murayama, T.; Furukawa, T.; Hatano, T.; Yoshida, T.; Koshiura, R.; Okuda, T. *Biol. Pharm. Bull.*, **1993**, *16*, 379.

[108]   Miyamoto, K.; Murayama, T.; Nomura, M.; Hatano, T.; Yoshida, T.; Furukawa, T.; Koshiura, R.; Okuda *Anticancer Res.*, **1993**, *13*, 37.

[109]   Miyamoto, K.; Murayama, T.; Yoshida, T.; Hatano, T.; Okuda, T. In: *Antinutrients and Phytochemicals in Food* ; Shahidi, F., Ed.; p.245, American Chemical Society: Washington DC, **1997**; pp.245-259.

[110]   Kuramochi-Motegi, A.; Kuramochi, H.; Kobayashi, F.; Ekimoto, H.; Takahashi, K.; Kadota, S.; Takamori, Y.; Kikuchi, T. *Biochem. Pharmacology*, **1992**, *44*, 1961.

[111]   Bluestein, D.L.; Oesterling, J. In: *Prostate Diseases*, p.182, W.B.Sanders Company, Philadelphia, **1993**.

[112] Aoki, K.; Nishimura, K.; Abe, H.; Maruta, H.; Sakagami, H.; Hatano, T.; Okuda, T.; Yoshida, T.; Tsai, Y.-J.; Uchiumi, F.; Tanuma, S. *Biochim. Biophys. Acta*, **1993**, *1158*, 251.

[113] Aoki, K.; Maruta, H.; Uchiumi, F.; Hatano, Yoshida, T.; Tanuma, S. *Biochem. Biophys. Res. Commun.*, **1995**, *210*, 329.

[114] Fukuchi, K.; Sakagami, H.; Okuda, T.; Hatano, T.; Tanuma, S.; Kitajima, K.; Inoue, Y.; Inoue, S.; Ichikawa, S.; Nonomiya, M.; Konno, K. *Antiviral Res.*, **1989**, *11*, 285.

[115] Nakashima, H.; Murakami, T.; Yamamoto, N.; Sakagami, H.; Tanuma, S.; Hatano, T.; Yoshia, T.; Okuda, T. *Antiviral Res.*, **1992**, *18*, 91.

[116] Quideau, S.; Feldman, K.S. *Chem. Rev.* **1996**, *96*, 475, and references cited therein.

[117] Nelson, T. D.; Mayers, A. I. *J. Org. Chem.*, **1994**, *59*, 2577.

[118] Itoh, T.; Chika, J.; Shirakami, S.; Ito, H.; Yoshida, T.; Kubo, Y.; Uenishi, J. *J. Org. Chem.*, **1996**, *61*, 3700.

[119] Khanbabaee, K.; Lotzerich, K. *Liebigs Ann./Recueil*, **1997**, 1571.

[120] Gross, G.G. In *Plant Polyphenols. Synthesis, Properties, Significance* ; Hemingway, R.W.; Lukas, P.E., Eds.; Plenum Press: New York, **1992**; pp.43-60.

Atta-ur-Rahman (Ed.) *Studies in Natural Products Chemistry, Vol. 23*

RECENT STUDIES ON THE CHEMISTRY AND BIOLOGICAL ACTIVITIES OF THE ORGANOSULFUR COMPOUNDS OF GARLIC (ALLIUM SATIVUM)

A. KAMEL* and M. SALEH

Department of Chemistry, Texas Southern University, Houston, Texas 77004, USA.

Abstract: Garlic, *Allium sativum*, is known to contain a wide range of biologically active compounds. The organosulfur compounds are mainly the biologically active components of garlic. Many health benefits has been ascribed to them, possessing biological activities including antibacterial, antifungal, antiatherosclerotic, antihypertensive, lipid and cholesterol lowering effects, inhibition of carcinogenesis, enhancing the immune system and other biological activities. The chemistry of the organosulfur compounds in garlic varies with the method it has been processed. For example, the composition of the organosulfur compounds change when garlic is cooked, crushed, ingested, or powdered. This review summarizes the composition of these compounds in garlic's different forms and discusses the recent studies on the biological activities of the organosulfur compounds of garlic from 1996- January 1998, with special attention to the role of these compounds in preventing cancer. Garlic, in any of its forms, would be effective to the human body in spite of the differences in the chemical composition in each form. It is shown in this review that the water soluble S-alkyl cysteines, the thiosulfinates, ajoenes and vinyl dithiins, and the oil soluble allyl sulfides, which are found in the different forms of garlic, possess the biological activities known for garlic. It is therefore advisable to consider including garlic in our daily diets to benefit from its effects, especially in cancer prevention.

Garlic, *Allium sativum*, is a member of the Liliaceae family. It's name may have derived from the Celtic word "all", which means pungent. Garlic is of such antiquity as a cultivated plant, that it is difficult with any certainty to trace the country of its origin. Garlic has been found in King Tut-ankh-Amon's tomb Fig. **(1)** and pictured on the walls of other ancient Egyptian tombs from 3200-1500 B.C. The hieroglyphic name of garlic *khidjana*, found in a papyrus among a list of offering presented to the pharaoh Ramses III, was later changed to *shegen* by the Coptic Christians then to *thoum*, the current name for garlic, in the Arabic language [1]. Garlic names in other languages were previously discussed [2].

Garlic is one of the most ancient healing plants. It has been esteemed in the Chinese medical tradition for thousands of years. No doubt the early Egyptians were aware of garlic's antibiotic properties, too. Hieroglyphic records show that laborers consumed large amounts of garlic while constructing the pyramids to gain strength and fight malaria, common cold

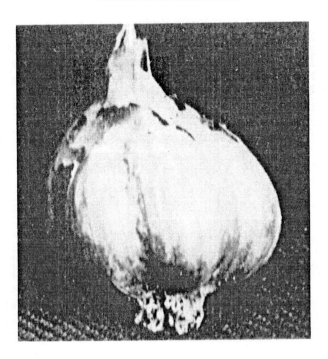

Fig. (1). Garlic found in the tomb of King Tut-ank-Amon exhibited at the Cairo Museum in Egypt. (adapted from Derby *et.al.,* 1977 [1]).

and other diseases. Other Egyptian medical papyrus dating to about 1550 B.C., of the more than 800 therapeutic formulas, 22 mention garlic for a variety of ailments including heart problems, headaches, bites, worms and tumors. The Phoenicians were also known to take garlic on long voyages as a valuable remedy for various ailments common to sailors. The Greek, who knew garlic from the Egyptian, also used garlic as a vermifuge and expectorant. During the first Olympic games the athletes were said to have employed garlic as a stimulant. Garlic has also been used in the Indian medicine in the form of an antiseptic lotion for washing wounds and ulcers. More recently, garlic was the remedy of choice for treating wounded soldiers during World War I. Although penicillin was discovered in 1928 and antibiotics replaced garlic after then, when the Russian army ran out of penicillin in World War II they turned again to garlic. After that, garlic was nicknamed the Russian penicillin [3].

Current research reveals that the therapeutic value of garlic is mainly due to its high content of organosulfur compounds which is the highest sulfur content (3.2 mg/g) among all vegetables and fruits [4]. The organosulfur compounds of garlic are responsible for the antibacterial, anti-hypertensive, cholesterol and lipid reducing, anti-thrombotic and anti-platelet aggregation, hypoglycemic, anti-cancer, immune system stimulation and many other activities of garlic. Recent studies show that individuals and populations consuming moderate quantities of garlic in their food have a lower risk for cancer. For example, Northern Italy inhabitants who consume less amounts of garlic than the southern provinces have 3 folds risk of stomach and colon cancer [5]. It therefore seems to be important to start prevention of cancer and other diseases by garlic nutrition as early as possible and adhere to it over a long period. More than one thousand research papers, review articles and books on the chemistry and therapeutic value of garlic has been published to date [3, 6-10]. One of the most recent reviews was that of Koch and Lawson, in 1996 [2] in which most previous research was collected in a book.

This review compiles recent studies on the chemistry and biological activities of the organosulfur compounds of garlic with special attention to the role of these compounds in preventing cancer.

CHEMISTRY OF THE ORGANOSULFUR COMPOUNDS OF GARLIC

The total sulfur content of garlic bulbs is 0.35% of its fresh weight and about 1% of its dry weight [11]. The nature of the organosulfur compounds differ from one form of garlic to another as will be discussed in details in the next sections. In garlic bulbs, before they are being cut, more than 75% of the organosulfur compounds are γ-glutamylcysteines and S-alk(en)ylcysteine sulfoxides which neither possess the pungent odor of garlic nor possess biological activities. Once the garlic cloves are cut and in the presence of water, the enzyme allinase (alliin lyase) acts on S-allylcysteine sulfoxide (alliin) converting it to diallyl thiosulfinate (allicin) which is the organosulfur compound responsible for its strong odor and is the parent of the compounds of most of the biological activities of garlic

γ-Glutamylcysteines

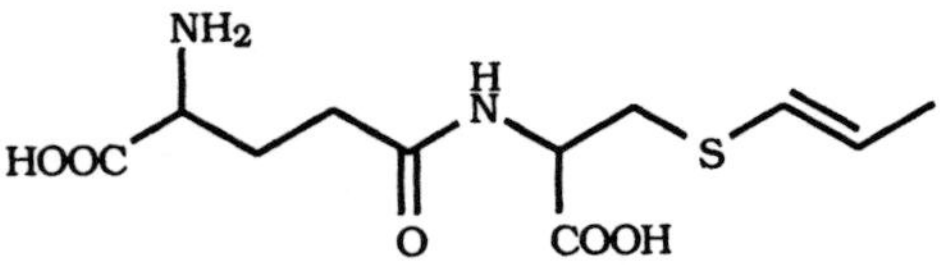

S-*trans*-1-propenyl-γ-glutamylcysteine

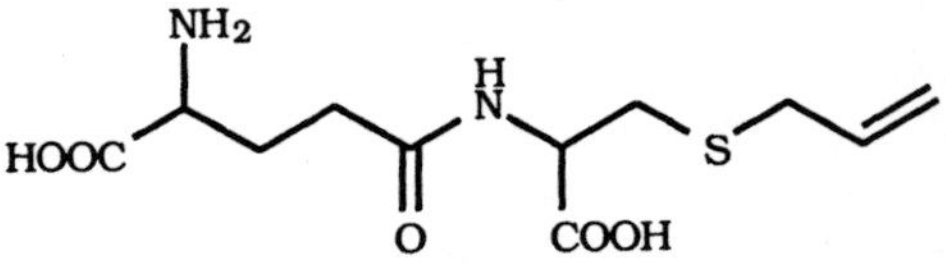

S-*trans*-2-propenyl-γ-glutamylcysteine

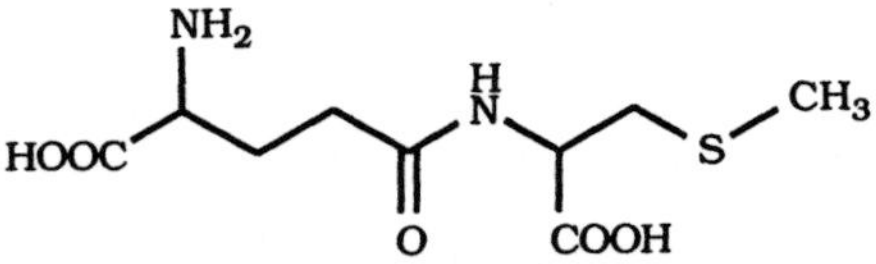

S-methyl-1-γ-glutamylcysteine

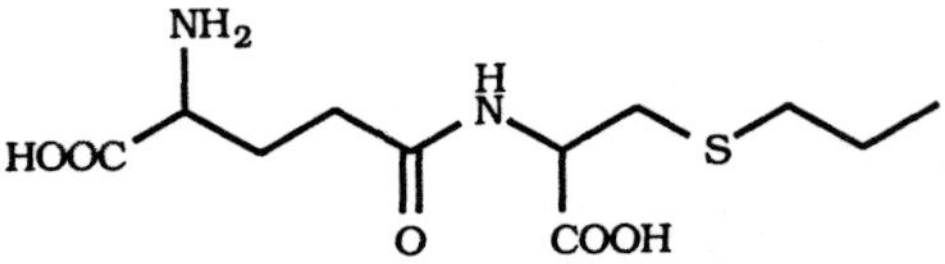

S-propyl-γ-glutamylcysteine

S-Alkyl cysteines

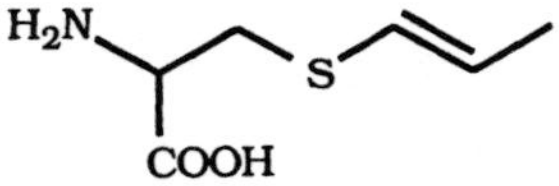

trans-1-propenyl cysteine

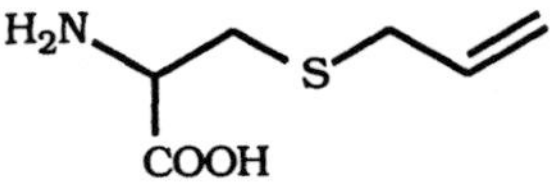

trans-2-propenyl cysteine
(S-allyl cysteine, SAC)

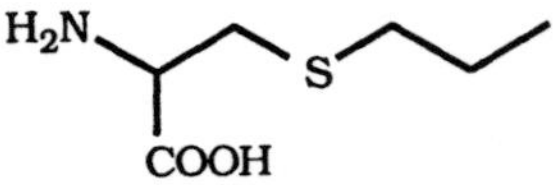

S-propyl cysteine
(SPC)

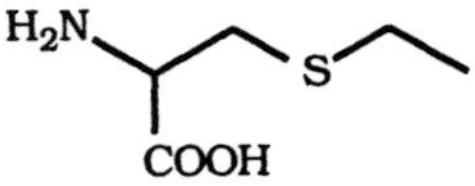

S-methyl cysteine
(SMC)

Alkylcysteine sulfoxides

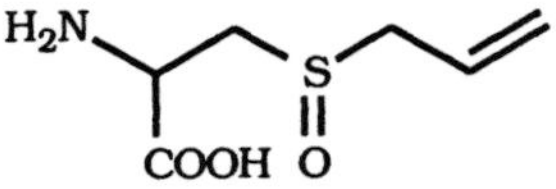

S-allylcysteine sulfoxide
(alliin)

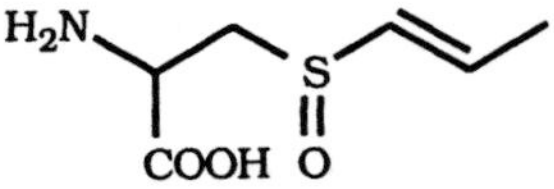

S-1-propenylcysteine
sulfoxide (isoalliin)

S-methylcysteine
sulfoxide (methiin)

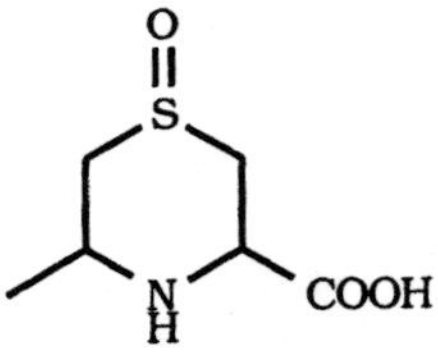

cycloalliin

(Fig. 2). contd.....

Thiosulfinates

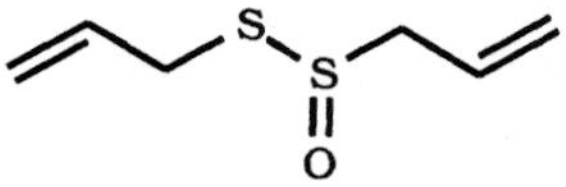

allyl thiosulfinate
(diallyl thiosulfinate)
(allicin)

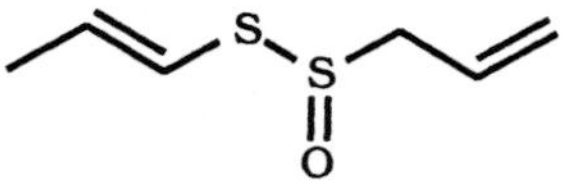

trans-1-propenyl
2-propene thiosulfinate

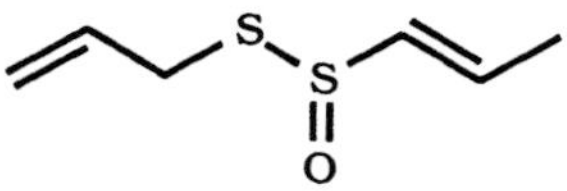

allyl *trans*-1-propenyl
thiosulfinate

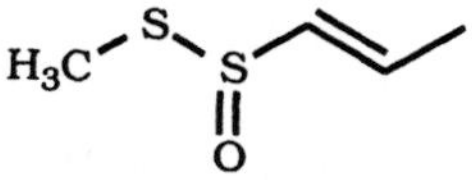

methyl *trans*-1-propenyl
thiosulfinate

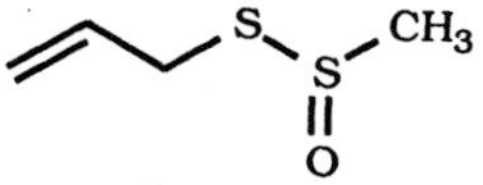

allyl methane
thiosulfinate

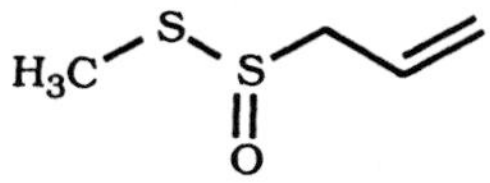

methyl *trans*-2-propenyl
thiosulfinate

Vinyl dithiins

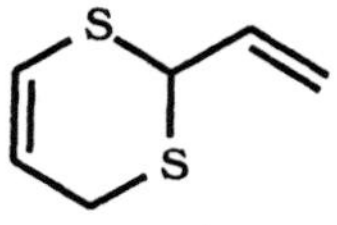

2-vinyl 4H-1,3 dithiin

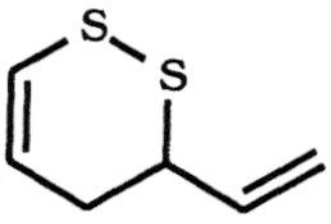

3-vinyl 4H-1,2 dithiin

Ajoenes

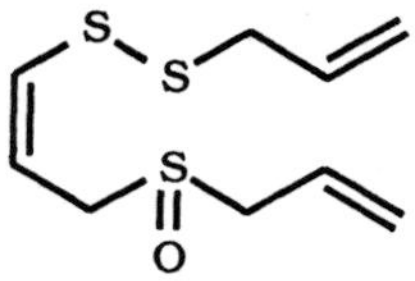

Z-ajoene

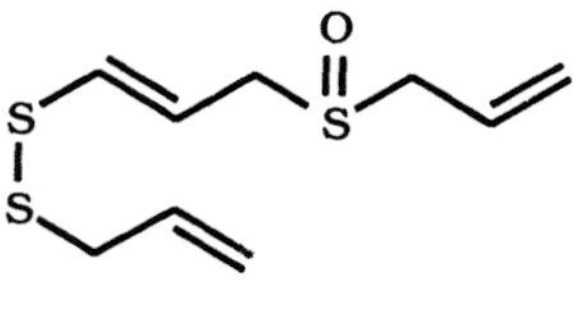

E-ajoene

Allyl sulfides

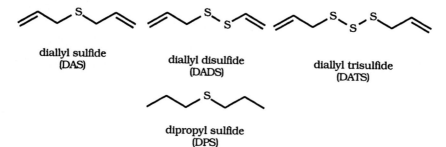

diallyl sulfide
(DAS)

diallyl disulfide
(DADS)

diallyl trisulfide
(DATS)

dipropyl sulfide
(DPS)

Fig. (2). Chemical structures of the organosulfur compounds found in garlic.

and is itself a strong antibacterial substance [12]. Allicin is an unstable compound and undergoes degradation and transformations into ajoenes, vinyldithiins, and allylsulfides [13]. Steam distillation or cooking of crushed garlic cloves reveals the degradation products of allicin. The exact composition of the organosulfur compounds in each stage are discussed below. Figure (**2**) demonstrates some of the chemical structures of the organosulfur compounds found in garlic.

GARLIC BULBS

The major organosulfur compounds (82%) found in garlic bulbs (whole garlic cloves) are the γ-glutamylcysteines; S-*trans*-1-propenyl γ-glutamylcysteine (3-9 mg/g), S-*trans*-2-propenyl γ-glutamylcysteine (S-allyl γ-glutamylcysteine) (2-6 mg/g), and S-methyl γ-glutamylcysteine (0.1-0.4 mg/g) [10, 14-15], and the cysteine sulfoxides; S-allylcysteine sulfoxide (alliin) (5-14 mg/g), S-methylcysteine sulfoxide (methiin) (0.5-2 mg/g), S-*trans*-1-propenylcysteine sulfoxide (isoalliin) (0.2-1.2 mg/g) and cycloalliin (0.5-1.5 mg/g) [10, 15-18]. Small percentages of other organosulfur compounds have also been detected; γ-glutamyl methionine (0.02-0.12 mg/g), S-2-carboxypropylglutathione (0.09 mg/g), γ-glutamyl-S-mercaptocysteine (0.01-0.03 mg/g) and scordinins [2].

S-alkyl cysteines, such as S-allyl cysteine (SAC), S-methyl cysteine (SMC), and S-propyl cysteines are slowly formed during sprouting from the action of γ–glutamyl transpeptidase on γ–glutamyl cysteines. S-allyl cysteine is also rapidly formed in the body by hydrolysis of γ–glutamyl cysteine particularly in the kidney which contains an abundant amount of γ–glutamyl transpeptidase activity [2].

The composition ratios of the γ-glutamylcysteines and the cysteine sulfoxides vary during different stages of growth of the garlic plant and during storage or refrigeration. In 7-week premature plants, γ-glutamylcysteines were not detected until shortly before harvest time. [2]. Cool storage of garlic bulbs for a few months resulted in steady decreases of the γ-glutamylcysteines with almost equimolar increases in the cysteine sulfoxides [15, 19].

The γ-glutamylcysteines were first discovered in the early 1960s [20,21], however the most abundant, S-*trans*-1-propenyl γ-glutamylcysteine was discovered about 30 years later [14-15, 19]. These compounds serve as a reserve pool that can increase its levels during sprouting and storage [19]. They play an important function as reserve compounds for producing additional alliin and isoalliin during wintering and sprouting, increasing the antibiotic capacity of the young plants [22] S-allylcysteine sulfoxide (alliin) was discovered earlier in 1947 and was synthesized by Stoll and Seebeck [23,24]. Alliin was the first compound in nature to possess both chiral sulfur and chiral carbon atoms [24]. Ten years later the other cysteine sulfoxides S-methylcysteine sulfoxide (methiin), S-trans-1-propenylcysteine sulfoxide (isoalliin) and cycloalliin were discovered.[25,26].

Alliin and its analogues are more likely formed from the actions of both γ-glutamyl transpeptidase and oxidase on the corresponding γ-glutamylcysteine (Scheme 1)[27]. The γ-glutamyl peptides are closely connected to glutathione metabolism and may have related biological roles in the plant [28].

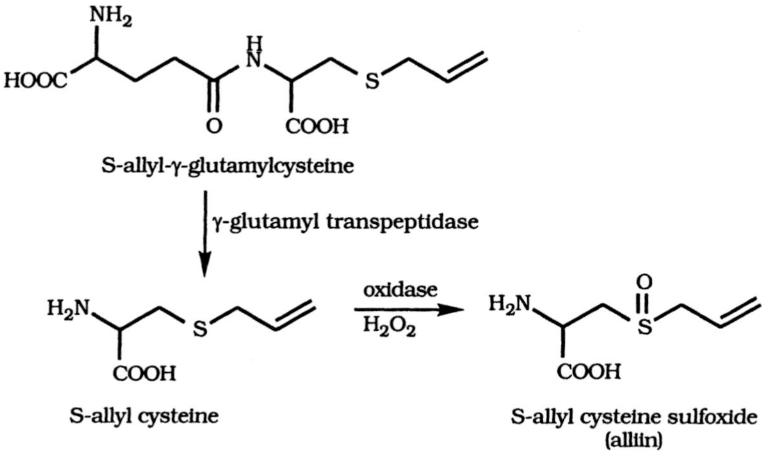

S-allyl-γ-glutamylcysteine

γ-glutamyl transpeptidase

S-allyl cysteine

oxidase
H₂O₂

S-allyl cysteine sulfoxide
(alliin)

Scheme (1). Formation of alliin from S-allyl γ-glutamylcysteine

CRUSHED GARLIC

Upon transversely cutting a garlic clove, small yellow spots appear. These are vascular bundle sheath cells located around the veins. Inside the cytoplasmic globular granules of these cells the enzyme allinase (alliin lyase or alliin alkyl-sulfenate-lyase) was detected [29,30]. Its structure was established by cloning. The mRNA encodes a 486-amino-acid polypeptide with a molecular mass of 55,623 Da. [31]. On the other hand, alliin has been found to be concentrated in the very abundant storage mesophyll cells, with none present near the bundle sheath cells where allinase is present (Fig. (3)). When a garlic clove is cut, crushed or chewed, alliin and allinase come into contact and react giving rise to diallyl thiosulfinate (allicin), via the formation of allylsulfenic acid, as well as the distinct odor of garlic. The allylsulfenic acid have extremely short lifetimes before condensation to form allicin. The formation of allicin from alliin is very rapid and is complete in less than 0.5 minutes at pH 5-8 and 30-35°C [32]. It has been found that over half of the alliin is converted to allicin in

6 seconds with the formation of pyruvic acid and ammonia (Scheme 2) [24, 36]. In a similar manner, allicin's analogues allyl methyl thiosulfinate, allyl *trans*-1-propenylthiosulfinate, methyl *trans*-1-propenylthiosulfinate, and methyl methanethiosulfinate are formed with their isomers from the corresponding alliin analogues by the action of the same enzyme allinase [33]. Allicin is however the major product of the lysis and comprises about 80% of the thiosulfinates.

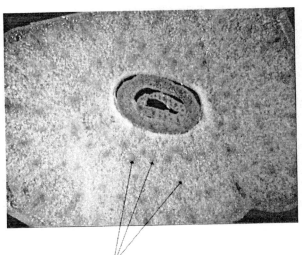

Yellow vascular bundle sheath cells containing allinase

Mesophyll storage cells containing alliin

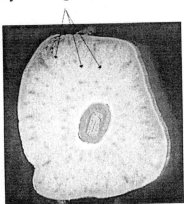

Fig. (3). Cross-section of a garlic clove showing the locations of alliin and allinase

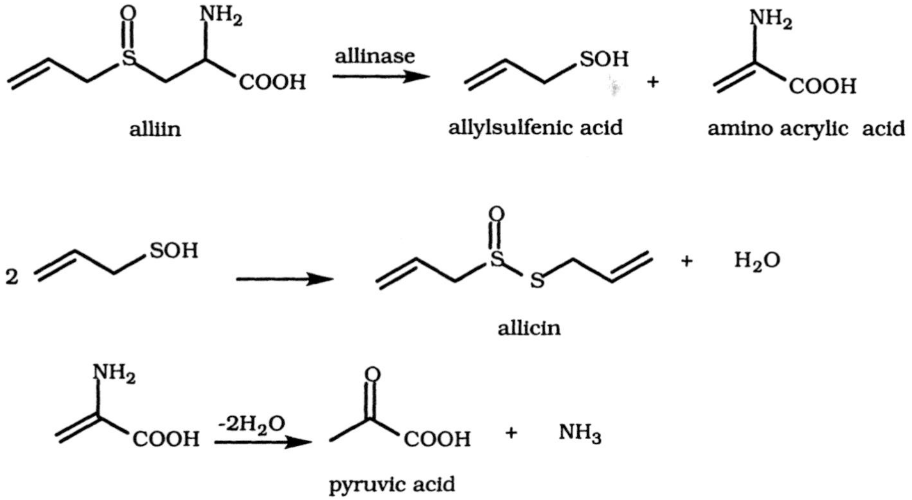

Scheme (2). Formation of allicin from the action of allinase on alliin. (modified from Block *et.al.* 1984 [13]).

Allicin is unstable and undergoes degradation and transformations into, ajoenes, vinyl dithiines, thioacroleines and allyl sulfides within 24 hours [13, 34-35]. Different schemes for the transformations have been suggested. Block [13, 35] suggested that β- elimination of allicin should afford allylsulfenic acid and thioacrolein which would dimerize to vinyl dithiins, while S-thioallylation of allicin followed by Cope-type elimination and re-addition of allylsulfenic acid should give ajoene or could undergo hydrolysis giving diallyl trisulfide and allyl alcohol. Allicin is also transformed into diallyl disulfide via the attack of allyl mercaptan, which is supported by the formation of sulfur dioxide and allyl alcohol (Scheme 3).

S-allyl mercaptocysteine (SAMC) results from the reaction of allicin with cysteine. It is undetected in whole garlic, but as chopped garlic ages, protein hydrolysis yields cysteine which then reacts with allicin to yield SAMC whose amount increases as chopped garlic ages [37].

STEAM DISTILLATION OR COOKING

The transformation products of allicin and other thiosulfinates present in crushed garlic is greatly accelerated and expanded upon heating such as during steam distillation or cooking. As many as 20 different sulfides have been identified in the steam distillate and include diallyl mono to hexasulfides, allyl methyl mono to hexasulfides dimethyl mono to hexasulfides, and allyl 1-propenyl di and trisulfides [38].

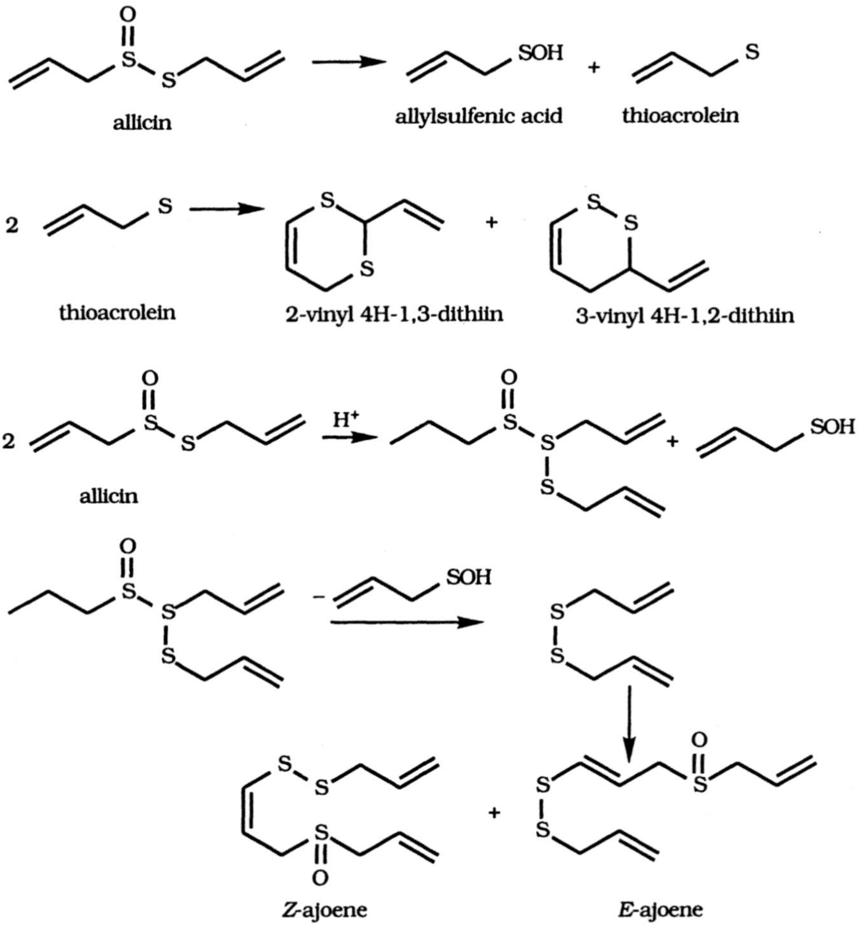

Scheme (3). Formation of ajoenes and vinyl dithiins from allicin. (modified from Block *et.al.*, 1986 [35]).

CHEMISTRY OF DIFFERENT GARLIC FORMS AND PREPARATIONS

Cooked Garlic

Boiling of whole cloves inactivates the enzyme allinase and only 0.5-1% of alliin is converted to allicin which is rapidly converted to diallyl trisulfide. Also 12% of the γ-glutamylcysteines are hydrolyzed to S-allyl cysteine (SAC) and S-1-propenyl cysteine. A similar loss of alliin is found [37]. Boiling of crushed or tiny chopped garlic enables the formation of allicin which undergoes degradation to yield mainly diallyl trisulfide most of which evaporates during the cooking process.

Frying garlic in oil or microwaving show almost similar results, however about 16% of the sulfides are retained in the oil [37].

Powdered Garlic

Since alliin and allinase are stable to temperatures up to 60°C, garlic powder produced by oven drying of whole garlic cloves at intermediate temperatures retain the content of its alliin and allinase which is to a great extend stable for years upon storage under 4-7% humidity conditions. Once water gets in contact with the garlic powder the enzyme is activated and the conversion to allicin is initiated. Lyophilization shows similar results, however, the amount of alliin in lyophilized garlic powder is a little higher.

If garlic powder is obtained by cutting the cloves in small pieces and then drying, the conversion of alliin will be initiated. An industrial and economic method to dry garlic is to cut the cloves, followed by immediate freezing, then drying in the oven at 60°C. However, the yield of alliin in whole dried cloves is always higher than drying of the cut cloves. Several reviews has discussed the processing of garlic [39-44]. A detailed study on the allicin yield of commercially available garlic powder and the effect of the size of cut cloves has been carried out [37].

Odorless Garlic

Various pharmaceutical preparations of garlic are available containing odorless garlic. There are several methods to prepare odorless garlic that mainly depend on the deactivation of allinase. This could be done by boiling garlic cloves, spray-drying, or the addition of enzyme inhibitors. The therapeutic value of this odorless garlic will depend on the activity of allinase in the human body. In some preparations deodorized garlic powder is provided with another pill of allinase to ensure the initiation of the chemical reaction producing allicin. It is important to consider the instability of allinase in the acidic medium of the stomach, therefore, garlic powder tablets or pills should be protected from the stomach gastric acid by an enteric coat, usually made of cellulose esters. In this case, allicin will be produced in the intestine. The quality of many brands of garlic tablets has been compared in several publications [10, 45-50].

Chewed or Swallowed Garlic

It has been prescribed in traditional medicine that chewing or swallowing 1 or 2 medium size garlic cloves with water will obtain the maximum health benefits of garlic. The chemistry here is dependent on the stomach contents and varies from one person to another. Allinase is immediately inactivated at low pH levels usually found in the stomach, however, a high protein meal can activate the enzyme [45, 46]. It is therefore advisable to swallow garlic when the acidity in the stomach is somehow neutralized.

Oil-macerated Garlic

When garlic cloves are minced or macerated in vegetable oil, allicin transformation compounds are mainly vinyldithiins and ajoenes with little allyl sulfides. These compounds remain stable in the oil for at least 3 years, the vinyldithiins being considerably more stable than the ajoenes. Refrigeration of these oils prolong the lifetime of the ajoene content [49].

Compounds in Human Breath

Recently, a variety of sulfur- and selenium-bearing compounds were identified through the use of GC/MS techniques [51-53], element specific technique of gas chromatography with atomic emission detection (GC-AED) [54], and by proton-transfer-reaction mass spectrometry (PTR-MS)[55] in the breath of persons after the ingestion of garlic.

Thermal desorption of purged and trapped garlic breath was studied by GC/MS [53] in which allyl methyl sulfide, diallyl sulfide, diallyl disulfide, *p*-cymene and D-limonene were found consistently in all subjects, however, allyl thiol was detected occasionally. Preliminary evidence suggest that stomach acid caused increased evolution of hydrogen sulfide, a potential breath odor compound, during digestion.

Among the main garlic-related compounds detected in the breath by GC-AED were allyl methyl sulfide, allyl methyl disulfide, diallyl sulfide, diallyl disulfide, 2-propenethiol, dimethyl disulfide, and diallyl trisulfide [54]. Measurements taken shortly after the ingestion of garlic and at four hourly intervals thereafter, indicating a more rapid decrease with time of diallyl disulfide than allyl methyl sulfide. Allyl methyl disulfide, diallyl sulfide, and diallyl trisulfide were reported to decrease slowly, while 2-propenethiol vanished rapidly.

The breath of a test person was analyzed over a time period of about 30 hours by means of proton-transfer-reaction mass spectrometry [55]. After ingestion of cut raw garlic, the components allyl methyl sulfide, allyl methyl disulfide, diallyl sulfide, diallyl disulfide, diallyl trisulfide, dimethyl sulfide, and acetone were detected. While the concentrations of allyl methyl disulfide, diallyl sulfide, diallyl disulfide, and diallyl trisulfide reached maximum concentration shortly after ingestion of garlic and declined to baseline values within the next 2-3 h, concentrations of allyl methyl sulfide, dimethyl sulfide, and acetone increased much more slowly and showed enhanced values even 30 h after garlic consumption. The strong increase of the concentration of acetone might be indicative of enhanced metabolism of serum cholesterol, triglycerides, and total lipids in the blood stream.

Supercritical Fluid Extraction of Garlic

Interest in supercritical fluid (SF) technology has increased because of its effectiveness in the extraction and analysis of natural products and concern related to the use and disposal of conventional organic solvents. Several laboratories have employed supercritical fluid extraction (SFE) methodologies in the extraction of garlic and onion [56-58].

Homogenized garlic was extracted by $SF-CO_2$ and the major thiosulfinates were characterized with liquid chromatography/ atmospheric pressure chemical ionization mass spectroscopy (LC/APCI/MS) [59]. In addition to the thiosulfinates, small quantities of ajoene were also found in the SF extracts of garlic homogenates. The identified thiosulfinates are listed in Table (1).

Table (1). Thiosulfinates found in $SF-CO_2$ extract of garlic*

Compound	mole %
Dimethyl thiosulfinate	0.9
Methyl 2-propenethiosulfinate + Allyl methanethiosulfinate	18.5
S1-propenyl methanethiosulfinate-(E,Z) + Methyl 1-propene thiosulfinate-(E)	9.4
Diallyl thiosulfinate (allicin)	59.0
n-Propyl 2-propenethiosulfinate	traces
S1-propenyl allyl thiosulfinate-(E,Z) + Allyl 1-propenyl thiosulfinate-(E)	19.6

* Modified from Calvey et al., 1994 [57].

Extraction of these organosulfur compounds in garlic by $SF-CO_2$ and applying the LC/APCI/MS technique for analysis provided an effective alternative method for extraction and analysis of these thermally labile compounds in garlic. The low UV absorbance, low concentration and thermal liability of these compounds made detection by other means difficult. While APCI-MS may give ambiguous data based solely on the protonated molecule and adduct ions, the MS-MS spectra show that the regioisomers of the various sulfur containing compounds present can be uniquely distinguished [59].

BIOLOGICAL ACTIVITIES OF THE ORGANOSULFUR COMPOUNDS OF GARLIC

Garlic is known to contain a wide range of biologically active compounds. The organosulfur compounds[1] are the biologically active components of garlic. Many health benefits has been ascribed to them, possessing biological activities including antibacterial [12], antifungal [63], antiatherosclerotic [64], antihypertensive [65], lipid and cholesterol lowering effects [66], inhibition of carcinogenesis [67], enhancing the immune system [68] and other biological activities. Most research has been conducted on garlic extracts, garlic powders and raw garlic rather than on the organosulfur compounds. Studies on the biological activities of garlic and its organosulfur compounds including anti-cancer effects, were previously discussed [3]. Recent studies on the biological activities of the organosulfur compounds of garlic are presented below and are listed in (Table 2).

Table 2.   Recently Reported Biological Activities of Organosulfur Compounds of Garlic [1996-1997]

Compound	Biological Activity	Reference
S-alkyl cysteines		
S-allyl cysteine (SAC)	neurotrophic, prevention of atherosclerosis, depression of liver carcinogenesis, blocking NMOR formation, blocking NF-kappa B activation, GST activiation, prevention of colon cancer, reduction of BaP-DNA adduct formation	71-73, 76-78.
S-methyl cysteine (SMC)	GST activiation	75
Cysteine	GST activation, reduction of NMOR formation,	75, 76
S-propyl cysteine (SPC)	reduction of NMOR formation	76
S-allylmercaptocysteine (SAMC)	depression of erythroleukemia cells, inhibition of breast and prostate cancer cell lines	79, 80
Thiosulfinates		
S-2-propenylcysteine thiosulfinate (allicin)	antibacterial, antifungal, antiprotozoal, antiparasitic, antiviral, inhibition of RNA synthesis, inhibition of cholesterol biosynthesis, iron oxidation in hemoglobin.	12, 85, 87-89, 91, 112-113
Ajoenes and vinyl dithiines		
(*E,Z*) Ajoene	antithrombotic, inhibition of platelet aggregation, lipid biosynthesis inhibitor, antibacterial, antimycotic, antifungal, antimutagenic, protection of immune cells.	13, 35, 86, 94-98,100.

[1] Organoselenium compounds have recently proven to possess some of the biological activities of garlic [60-62].

(Table 2) contd.....

Compound	Biological Activity	Reference
2-Vinyl-4H-1,3-dithiin	antithrombotic	35, 94
Allyl sulfides		
Diallyl sulfide (DAS) Diallyl disulfide (DADS) Diallyl trisulfide (DATS)	inhibition of BP-induced fore-stomach, lung and liver cancer in mice, increase hepatic and fore-stomach GST and EROD activities, inhibition of tumors in human colon cell line (DADS only), anti-mutagenic, antihepatotoxic,	102, 105, 107-111.
Dipropyl sulfide (DPS) Dipropyl disulfide (DPDS)	increase in hepatic EROD activity in BP-induced cancer, anti-mutagenic	102, 109

The S-alkylcysteine sulfoxides (alliin and its analogues) has been shown not to possess any of the biological activities of garlic in absence of the enzyme alliin lyase (allinase) [32]. The ability of sulfur compounds in garlic to depress nitrosamine formation and bioactivation is consistant with epidemiologic evidence that higher intake of *Allium* plants is associated with a reduction in the risks of some cancers. For instance, it has long been recognized that people in the countries that border the Mediterranean Sea have a reduced risk for colon and breast cancers. Although the observed protection is not solely attributable to *Alliums*, they do contribute to it [69].

It has been suggested that differences in the ability of organosulfur compounds to modulate glutathione S-transferase (GST) activity may, at least in part, account for their differential chemopreventive efficacy against benzo(a)pyrene induced cancer in mice [70].

S-ALKYL CYSTEINES

Alkyl cysteines are water soluble organosulfur compounds in garlic and are formed from the hydrolysis of γ-glutamylcysteines. These compounds were found to possess activities in cancer prevention, atherosclerosis and neurotrophic disorders.

Neurotrophic Activity

Several organosulfur compounds found in garlic extract promoted the survival of rat hippocampal neurons *in vitro*. From the analysis of structure-activity relationship, the thioallyl group in these compounds is essential for the manifestation of neurotrophic activity. S-Allyl-L-cysteine (SAC), one of the organosulfur compounds having thioallyl group in garlic extract, also promoted the axonal branching of cultured neurons. These results suggest that thioallyl compounds make a unique group of neurotrophic factors [71].

Atherosclerosis

S-allyl cysteine (SAC) may be useful for prevention of atherosclerosis and can protect vascular endothelial cells from injury caused by oxidized LDL. Pulmonary artery endothelial cells pre-incubated with S-allylcysteine (0.1, 1, 10 and 20 mM) at 37°C and 5% CO_2 for 24 h, washed, and then exposed to 0.1 mg/ml oxidized LDL for 24 h, significantly prevented membrane damage, loss of cell viability and lipid peroxidation. This indicates that SAC can protect vascular endothelial cells from injury caused by oxidized LDL, and suggest that it may be useful for prevention of atherosclerosis [72].

Carcinogenesis

Organosulfur compounds are the biologically active components of allium vegetables. Many health benefits have been ascribed to them, including inhibition of carcinogenesis. Inasmuch as several of these compounds are quite unstable and others are rapidly inactivated in the body. One of the hypotheses explaining the mechanisms of the chemopreventive activity of these substances is that they activate detoxification systems such as glutathione S-transferase (GST) [73]. A mechanism(s) other than GST induction is likely to be responsible for the differential effects of diallyl sulfide (DAS) and diallyl trisulfide (DATS) on benzo(a)pyrene (BP)-induced lung cancer in mice. It has been suggested that relatively lower efficacies of the organosulfur compounds (OSCs) against BP-induced lung cancer than against fore-stomach neoplasia may be attributed to (a) a lack of expression in the lung of an isoenzyme corresponding to fore-stomach GST 9.5 and (b) a comparatively lower level of induction of pi type GST in the lung than in the fore-stomach by these OSCs [74]. SAC may be useful in the prevention of BP-associated tumorigenesis and that further evaluation of their preventive potential in humans at risk appears feasible.

S-Methyl cysteine (SMC) and cysteine exerted significant inhibitory effects to decrease the number of glutathione S-transferase placental form (GST-P)-positive foci when given in the promotion stage of the Ito test.

When given during the initiation stage, these two OSCs also significantly inhibited focus formation. Regarding the mechanism underlying the inhibitory effects of SMC and cysteine, measurement of ornithine decarboxylase in SMC and cysteine treated liver tissues after partial hepatectomy (PH) revealed a significantly reduced activity, and the proportion of hepatocytes positive for proliferating cell nuclear antigen was significantly decreased by SMC or cysteine administration.

Moreover, examination of the expression of the early response proto-oncogenes, c-fos, c-jun, and c-myc, after PH demonstrated down-regulated induction of c-jun mRNA transcripts by SMC, sustained for an eight-hour period [75].

S-allyl cysteine (SAC), (20, 40, and 80 mM) depressed N-nitrosomorpholine (NMOR) formation, a known liver carcinogen, by 16%, 27%, and 43%, respectively ($p < 0.05$). The ability of SAC to block NMOR formation decreased as the $NaNO_7$ and morpholine concentrations increased. SAC and its non-allyl analog S-propyl cysteine (SPC) effectively blocked NMOR formation. SAC and SPC were less effective than isomolar cysteine in reducing NMOR formation ($p < 0.05$). The oil-soluble sulfur compounds diallyl disulfide (DADS), dipropyl disulfide (DPS), and diallyl sulfide (DAS) were ineffective inhibitors of NMOR generation ($p > 0.05$). SAC and DADS reduced the mutagenicity of NMOR in *Salmonella typhimurium* TA100 ($p < 0.05$). SAC at 70 mumol/plate reduced the number of histidine revertants per plate by 51% ($p < 0.05$), whereas DADS at 0.12 mumol/plate reduced mutant colony number by 76% ($p < 0.05$). SAC and DADS were more effective than isomolar cysteine in reducing NMOR mutagenicity ($p < 0.05$) [76].

S-allyl cysteine (SAC), was found to protect bovine pulmonary artery endothelial cells from oxidant injury induced by hydrogen peroxide (H_2O_2). Also, SAC may act via antioxidant mechanisms to block NF-kappa B activation in Jurkat cells in which it consistently exhibited a dose-dependent inhibition of NF-kappa B activation induced by both TNF-alpha and H_2O_2. Supershift with specific antibodies to NF-kappa B subunits confirmed that the inducible retarded bands observed in the EMSA and p65-p50 heterodimer of the NF-kappa B/Rel protein [77].

S-allyl cysteine (SAC) decreased the number of aberrant crypt foci, which are the most likely precursors of colon cancers, by 33 and 54% in groups given 40 or 80% maximum tolerated dose (MTD) of SAC respectively. The number of aberrant crypt foci, however, was not changed when SAC was given during the promotion period. GST activity in the liver was increased significantly by 41% 12 h after a single oral administration of 3.5 mmol/kg SAC and this elevated GST level was maintained over a 72 h period. GST levels were increased significantly by the administration of SAC (1.8 mmol/kg/ day for 3 days) not only in the liver but also in the proximal and middle small bowel. Hepatic GST-α and GST-μ isoenzymes were significantly increased by 35 and 42% respectively after oral administration of SAC using Western blotting. GST-π levels were lower than the detection limit (130 ng/mg/protein) in both the control and SAC-treated groups [73].

SAC significantly decreased BaP-DNA adduct formation at concentrations of 0.01 and 0.1 mg/ml while no significant reduction in BaP-DNA adduct formation was found for diallyl sulfide in stimulated human peripheral blood lymphocytes *in vitro*. BaP-DNA adduct formation was not associated with cell viability or proliferation of peripheral blood lymphocytes after the various treatments. No clear scavenging activity was detected for the garlic constituents. Aryl hydrocarbon hydroxylase activity was not decreased, nor was formation of sulfate and glucuronide conjugates of 3-hydroxy-BaP increased in the presence of SAC, indicating that increased glutathione S-transferase activity or a more efficient repair of BaP-DNA adducts may explain the observed effects. In addition, reactive oxygen species-induced 8-oxodeoxygunosine in DNA was reduced in the presence of SAC [78].

S-allylmercaptocysteine (SAMC) was found to be an effective antiproliferative agent against erythroleukemia cells that induces cell death by apoptosis. The antiproliferative potential of SAMC induces a dose-dependent inhibition of cell growth with an LD_{50} of 0.046 mM for OCIM-1 cells and 0.093 mM for HEL cells. [3H]thymidine incorporation was reduced in cells treated with this thioallyl compound, and analysis of high-molecular-weight DNA showed fragmentation compatible with apoptosis. Flow cytometric analyses of DNA revealed an abnormal cell cycle progression in both types of erythroleukemia cells, with the major portion of the unsynchronized cells in the G2/M phase. Measurement of acid-soluble free sulfhydryl groups showed an initial increase in response to SAMC followed by a progressive dose-dependent decrease with extended incubation of cells [79].

S-allylmercaptocysteine (SAMC), was susceptible to the growth-inhibitory influence of the two hormone-responsive cancer cell lines of breast and prostate MCF-7 and CRL-1740, respectively. The antiproliferative effect of SAMC was limited to actively growing cells. Human umbilical vein endothelial cells that had reached confluence escaped the reduction in viability so noticeable in the cancer cell lines tested [80].

THIOSULFINATES

The thiosulfinates, mainly allicin, were subjected to extensive research from which it was thought that allicin is the most important organosulfur compound of garlic and that most of the known biological activities of garlic are due to it. Allicin has shown strong antibacterial effects in which 1 mg of allicin is equivalent to 15 IU of penicillin (10 μg) [12,81]. The thiosulfinate structure of allicin (S(=O)S) is thought to be responsible for the antibacterial activity, since upon transformation to diallyl disulfide the activity is abolished [82]. Ajoene, which also contains the S=O group is almost as active as allicin, while the other allyl sulfides are much less active. It has been suggested that allicin inhibits SH-containing enzymes in microorganisms by the rapid reaction of the thiosulfinates with SH-groups [83, 84]. This is confirmed by the structural differences of the various bacterial strains in which they differ in the lipid content of their cell membranes [84]. It has also been found that allicin caused only partial inhibition of DNA and protein synthesis, but the RNA synthesis was completely inhibited [85]. Another explanation for the mechanism by which thiosulfinates and ajoene inhibit bacterial growth suggests that the disulfide bond (S-S) is necessary for the antibacterial activity since reduction by cysteine, which reacts with disulfide bonds, abolished its antibacterial activity [86].

Allicin not only shows antibacterial activity, but also antifungal [63], antiprotozoal [87], anti-parasitic [88] and antiviral activities [89]. It has recently been shown to inhibit the ability of *Entamoeba histolytica* trophozoites to destroy monolayers of baby hamster kidney cells [90]. Cysteine proteinases, an imporatant contributor to amebic virulence, as well as alcohol dehydrogenase, are strongly inhibited by allicin [90].

Most of the studies on allicin have been conducted *in-vitro*, however, *in-vivo* studies on the activity of allicin carried out prove that allicin has been transformed to other sulfides, S-allylmercaptocysteine (SAMC), or allyl mercaptan [37, 92]. Later, it has been demonstrated not to be an active compound in the body [91]. Neither allicin nor any other thiosulfinate has been detected in the human blood circulation [92] nor in the human breath [55] which shows that it reacts very rapidly with other species in the stomach, or intestine and its metabolites are thus considered the actual active compounds [93]. Surprisingly, allicin was shown that it might be harmful and that the iron in hemoglobin may rapidly be oxidized by the interaction with allicin in a dose of 0.2 mg/ml [91].

AJOENES AND VINYL DITHIINES

Ajoenes and vinyl dithiines are the compounds formed from the first step in allicin's transformations. They are well-known for their strong antithrombotic and their powerful ability to inhibit platelet aggregation [13, 35]. It has been reported that 2-vinyl-4H-1,3-dithiin and (*E,Z*)-4,5,9-trithiadodeca-1,6,11-triene 9-oxide (*E,Z*-ajoene) are potent antithrombotic agents derived from garlic extract [35, 94]. Unlike several other inhibitors of platelet aggregation, those compounds have been found to inhibit aggregation induced by all known inductors. Ajoene impairs platelet aggregation by inhibiting the functional exposure of platelet integrins GPIIb/IIIa [95].

Ajoene was shown to have potent antiproliferative activity against epimastigotes and amastigotes of *Trypanosoma cruzi in vitro*; this activity was associated with a significant alteration of the phospholipid composition of the cells with no significant effects on the sterol content. This shows the potential of lipid biosynthesis inhibitors as useful therapeutic agents in the treatment of leishmaniasis and Chagas' disease [96]

Ajoene also exhibited broad-spectrum antimicrobial activity. Growth of Gram-positive bacteria, such as *Bacillus cereus, Bacillus subtilis,*

Mycobacterium smegmatis, and Streptomyces griseus, was inhibited at 5 μg/ml of ajoene. *Staphylococcus aureus* and *Lactobacillus plantarum* also were inhibited below 20 μg/ml of ajoene. For Gram-negative bacteria, such as *Escherichia coli, Klebsiella pneumoniae*, and *Xanthomonas maltophilia*, MICs were between 100 and 160 μg/ml. Ajoene also inhibited yeast growth at concentrations below 20 μg/ml. The microbicidal effect of ajoene on growing cells was observed at slightly higher concentrations than the corresponding MICs. *B. cereus* and *Saccharomyces cerevisiae* were killed at 30 μg/ml after 24 h of cultivation when cultivation was started at 10(5) cells per ml. However, the minimal microbicidal concentrations for resting cells were at 10 to 100 times higher concentrations than the corresponding MICs [86].

Ajoene was also studied for short-term therapy of *Tinea pedis*. The use of ajoene as a 0.4% (w/w) cream resulted in complete clinical and mycological cure in 27 of 34 patients (79%) after 7 days of treatment. The remaining seven patients (21%) achieved complete cure after seven additional days of treatment. All patients were evaluated for recurrence of mycotic infections 90 days after the end of treatment, yielding negative cultures for fungus. These results show that ajoene is an alternative, efficient and low-cost antimycotic drug for short-term therapy of *Tinea pedis* [97].

(E,Z)-ajoene also showed inhibition of *Paracoccidiodes bransiliensis*, a dimorphic fungus pathogenic for humans. Ajoene induced alterations in phospholipid and fatty acid proportions such that phosphatydylcholine was reduced to about 18% in both morphological phases and phosphatidylethanolamine increased to 38% (Y phase) or 44% (M phase), suggesting inhibition of phosphatydylcholine synthesis. Ajoene also reduced saturated fatty acids (16:O and 18:O) from 67 to 35% in the Y phase, with a corresponding increase in the saturated components. This effect was not seen in the M phase [98].

Immunoblotting-based experiments using an anti-phosphotyrosine antibody showed that incubation of platelets with ajoene enhanced the phosphorylation of at least four proteins (estimated molecular weights 76, 80, 84 and 120 kDa), both in resting platelets and in platelets subsequently stimulated with thrombin (0.1 U/ml). This effect was both dose- and incubation-time-dependent. High concentrations of ajoene (50 μM) or long periods of incubation (10 min) led to nonselective 'hyperphosphorylation' of numerous proteins. Also, protein tyrosine phosphatase activity was inhibited when platelets were incubated with ajoene before lysis, but not when ajoene was added to lysates of platelets which had not been pre-exposed to ajoene [99].

Ajoene has been found a potent modulator of membrane-dependent functions of immune cells. It strongly inhibited the proliferation induced in human lymphocytes by the mitogens phytohemagglutinin (PHA), phorbol myristate acetate (PMA) and anti-CD3, and the capping formation induced in B lymphocytes by anti-IgM antibodies. On macrophages, ajoene was also found to partially inhibit the lypopolysaccharide-induced production of Tumor Necrosis Factor (TNF), and to decrease the phagocytic activity of thioglycolate-elicited mouse peritoneal macrophages for IgG-opsonized human erythrocytes. Ajoene also partially prevented the lytic effect of human and rabbit TNF on Actinomycin D-treated WEHI 164 cells [95].

The antimutagenic effects of ajoene were investigated by the Ames test. Ajoene inhibited mutagenesis induced by both benzo[a]pyrene (B[a]P) and 4-nitro-1,2-phenylenediamine (NPD) in a dose-dependent manner. In particular, NPD-induced mutagenesis was more effectively suppressed by ajoene than the B[a]P-induced type. Furthermore, the inhibition of mutagenesis by ajoene was more effective for transition-type mutations than for the frame shift type. HPLC analysis of B[a]P metabolism in the presence of the rat liver microsomal fraction (S-9) showed that ajoene dose-dependently inhibited the metabolic activation of B[a]P which suggests that ajoene affected the metabolic enzymes in the S-9 fraction [100].

ALLYL SULFIDES

Allyl sulfides are the last series of transformation organosulfur compounds of allicin and are stable. These organosulfur compounds are formed during cooking, steam distillation, storing of crushed garlic, or found in aged garlic extracts. These compounds have shown powerful chemopreventive activities as compared to other organosulfur compounds in garlic.

Anti-cancer Effects

The chemoprotective effects of diallyl sulfide (DAS), have been attributed to its inhibitory effects on CYP2E1-mediated bioactivation of certain carcinogenic chemicals. In addition to being a competitive inhibitor of CYP2E1 *in vitro*, DAS is known to cause irreversible inhibition of CYP2E1 in rats *in vivo*. The latter property is believed to be mediated by the DAS metabolite diallyl sulfone (DASO$_2$), which is thought to be a mechanism-based inhibitor of CYP2E1, although the underlying mechanism remains unknown [101]. While DAS and DASO undergo extensive oxidation *in vivo* at the sulfur atom, the allylic carbon, and the terminal double bonds, CYP2E1 preferentially catalyzes oxidation of the sulfur atom to form the sulfoxide and the sulfones (DASO and DASO$_2$). However, it appears that the end product of this sequence, namely, DASO$_2$, undergoes further CYP2E1-mediated activation of the olefinic pi-bond, a reaction which transforms many terminal olefins to potent mechanism-based P450 inhibitors. It is therefore hypothesized, that it is this final metabolic event with $DASO_2$ which leads to autocatalytic destruction of CYP2E1 and which is mainly responsible for the chemoprotective effects of DAS *in vivo* [101]. Diallyl sulfide (DAS), diallyl disulfide (DADS), diallyl trisulfide (DATS), dipropyl sulfide (DPS) and dipropyl disulfide (DPDS) caused a small but significant increase (37-44%) in hepatic ethoxyresorufin O-deethylase (EROD) activity in benzo(a)pyrene (BP)-induced cancer in mice with the exception of DATS. However, the fore-stomach EROD activity did not differ significantly between control and treated groups. Only DAS treatment caused a modest but statistically significant reduction (about 25%) in pulmonary EROD activity. These results suggest that while reduction of EROD activity may, at least in part, contribute to the DAS-mediated inhibition of BP-induced lung cancer, anticarcinogenic effects of OSCs against BP-induced fore-stomach carcinogenesis seems to be independent of this mechanism [102]. Treatment of mice with DAS, DADS and DATS which are potent inhibitors of BP-induced fore-stomach cancer in mice, resulted in a significant increase, as compared with control, in both hepatic (3.0-, 3.2-and 4.4-fold, respectively) and fore-stomach (1.5-, 2.7- and 2.7-fold, respectively) glutathione transferase (GST) activity toward anti-7β,8α-dihydroxy-9α,10α-oxy-7,8,9,10-tetrahydrobenzo(a)pyrene (anti-BPDE), which is the ultimate carcinogen of BP [102, 107]. On the contrary, this activity was not increased in either organ by dipropyl sulfide (DPS), which is ineffective against BP-induced fore-stomach cancer. The pulmonary GST activity was not increased by any of the tested OSCs. Even though epoxide hydrolase (EH) activity was differentially altered by these OSCs, a correlation between chemopreventive efficacy of OSCs and their effects on EH activity was not apparent [102]. The chemopreventive efficacy of these OSCs correlated with their ability to increase the expression of GST π. For example, DAS treatment resulted in approximate increases of 1.7- and 2.2-fold in hepatic and fore-stomach GST π expression, respectively, over the control. Treatment of mice with DATS, which is a relatively more potent inhibitor of BP-induced fore-stomach cancer than DAS, resulted in about 3.8- and 3.2-fold increases, respectively, in hepatic and fore-stomach GST π expression over the control. On the contrary, the expression of hepatic and fore-stomach GST π was increased only marginally (10-20%) upon DPS administration [107].

Diallyl sulfide (DAS), diallyl trisulfide (DATS) and dipropyl sulfide (DPS), significantly differ in their efficacy against BP-induced lung cancer in mice toward the ultimate carcinogenic metabolite of benzo(a)pyrene (BP), (+)-anti-7 β, 8 α-dihydroxy-9 α, 10 α-oxy-7,8,9,10-tetrahydro-benzo(a)pyrene ((+)-anti-BPDE). The GST activity in the lung was due to two α-class (pI 9.4 and 6.0), two μ-class (pI 8.7 and 8.6), and one π-class (pI 8.9) isoenzyme. The GST isoenzyme profile of the lung was different from that of the A/J mouse fore-stomach, which also is a target organ for

BP-induced cancer in mice. Noticeably, an α-class heterodimeric isoenzyme (pI 9.5) present in the fore-stomach of A/J mouse, which is exceptionally efficient in the glutathione (GSH) conjugation of (+)-anti-BPDE could not be detected in the lung [103]. The specific activities of the lung GSTs in the GSH conjugation of (+)-anti-BPDE were in the order of GST 8.9 > GST 8.7 > GST 9.4 > GST 6.0. While DPS treatment did not increase the levels of any pulmonary GST isoenzyme, the expression of π-class GST 8.9 was significantly increased in response to both DAS and DATS administrations. Interestingly, DATS, an OSC which lacks activity against BP-induced lung cancer in mice, was a relatively more potent inducer of π–class GST isoenzyme than DAS, which is a potent inhibitor of BP-induced lung tumorigenesis [104].

Diallyl disulfide (DADS) was found to be more cytotoxic than diallyl sulfide (DAS) (showing a Dq of 1.6 µg/ml and a D0 of 0.6 µg/ml as opposed to values of 295 and 90 µg/ml, respectively) in a Chinese hamster ovary cell line. The two compounds were found to induce both chromosome aberrations and sister chromatid exchanges (SCEs) with DADS again being more active on a weight-for-weight basis, exhibiting activity at concentrations below 10 µg/ml compared with the levels of 300 µg/ml and above required for DAS to show any effect. The addition of rat liver S-9 activation fraction to the assays modified the effects of the two compounds in a non-consistent manner. It reduced the induction of SCEs by both compounds, enhanced the generation of aberrations by DADS (but not by DAS) and radically altered the parameters of both survival curves, reducing the Dq values almost to zero but increasing the D0 values [105].

By means of ionspray LC-MS/MS, ten GSH conjugates were identified in bile collected from rats dosed with DAS, namely: S-[3-(S'-allyl-S'-dioxomercapto)-2-hydroxypropyl]glutathione (M1, M2; diastereomers), S-[3-(S'-allyl-S'-dioxomercapto)-2-hydroxypropyl]-glutathione (M5), S-[2-(S'-allyl-S'-dioxomercapto)-1-(hydroxymethyl)ethyl]glutathione (M3, M4; diastereomers), S-[3-(S'-allylmercapto)-2-hydroxypropyl]glutathione (M6), S-(3-hydroxypropyl)-glutathione (M7), S-(2-carboxyethyl)-glutathione (M8), allyl glutathionyl disulfide (M9), and S-allylglutathione (M10) in order to investigate the nature of the reactive intermediate(s) responsible for the inactivation of CYP2E1 by DAS and its immediate metabolites. With the exception of M6, all of the above GSH conjugates were detected in the bile of rats treated with DASO, while only M3, M4, M5, M7, M8, and M10 were found in the bile of rats treated with DASO$_2$. *In vitro* experiments showed that GSH reacted spontaneously with DASO to form conjugates M9 and M10, and with DASO$_2$ to form M10. In the presence of NADPH and GSH, incubation of DAS with cDNA-expressed rat CYP2E1 resulted in the formation of metabolites M6, M9, and M10, while incubation with DASO led to the formation of

M3, M4, M5, M9, and M10. When $DASO_2$ acted as substrate, CYP2E1 generated only conjugates M3, M4, M5, and M10 [101].

Female rats (n = 5) fed diets containing sodium selenite (2.0 mg Se/kg diet), garlic powder (20 g/kg diet), water-soluble S-allyl cysteine (SAC; 5.2 μmol/kg diet), and oil-soluble diallyl disulfide (DADS; 5.2 μmol/kg diet) inhibited (p < 0.05) total DNA adducts by 45%, 40%, 80%, and 75%, respectively. Combining selenite with garlic powder, SAC, or DADS further inhibited DNA adducts. Selenite, but not garlic powder, SAC, or DADS, enhanced liver glutathione S-transferase and uridine diphosphate-glucuronosyltransferase activities. Selenite, garlic powder, SAC, or DADS did not affect liver cytochrome P-450 1A1 activities. This provides evidence that synergistic protection against the initiation of DMBA carcinogenesis occurs when selenite is supplemented in conjunction with garlic or its allyl sulfur components [106].

DADS has been shown to be an effective antitumorigenic agent against xenografts resulting from an established human colon tumor cell line [108]. Intraperitoneal injection of 1 mg diallyl disulfide DADS thrice weekly in 6-wk-old female NCr v/v mice with an initial body weight of 20-22 g reduced tumor volume of human colon tumor cell line, HCT-15, xenografts by 69% (P < 0.05) without apparent ill consequences such as altered growth of the host. Providing this quantity of DADS intragastrically also inhibited growth of the HCT-15 tumor. At equivalent DADS dosages, intraperitoneal treatment was proportionately more effective (P < 0.05) in reducing tumor growth than gastric intubation. Tumor inhibition caused by DADS (0.5 mg thrice weekly) was similar to that occurring with 5-fluorouracil (5-FU) treatment (0.5 mg thrice weekly). Combining DADS and 5-FU was no more effective in inhibiting tumor growth than using either compound alone. However, concurrent DADS treatment significantly (P < 0.05) inhibited the depression in leukocyte counts and spleen weight and prevented the elevated plasma urea caused by 5-FU treatment which suggests that DADS reduces the toxicity of 5-FU [108].

Anti-mutagenic Effects

Three heterocyclic aromatic amines, 2-amino-3-methyl-imidazo[4, 5-f]quinoline (IQ),2-amino-3,4-dimethylimidazo[4,5-f]quinoxaline and 2-amino-3,4-dimethylimidazo[4,5-f]quinoline, have been found in boiled pork juice [109]. Six organosulfur compounds - diallyl disulfide (DAD), dipropyl disulfide (DPD), diallyl sulfide (DAS), allyl methyl sulfide (AMS), allyl mercaptan (AM) and cysteine - were added separately to the pork juice before reflux boiling and then the mutagenicity of each sample was examined with the *Salmonella typhimurium* strain TA98 in the presence of S9 mix. All six compounds were found to inhibit the mutagenicity of boiled pork juice. The greatest inhibitory effect was observed with DAD and DPD, and this was 111-fold higher than that of the lowest, cysteine. To elucidate the inhibitory effect of DAD on mutagen formation in boiled pork juice, the major mutagenic fractions were monitored after HPLC separation by their mutagenicity with S. typhimurium TA98. By comparing the retention times of authentic IQ compounds from boiled pork juice with those following the addition of DAD, the mutagenicity of three major fractions was significantly inhibited compared with those same fractions in boiled pork juice alone. In addition, the Maillard reaction products (MRPs) in the boiled pork juice with and without the addition of DAD were quantified and identified by capillary gas chromatography and gas chromatography-mass spectrometry. The results show that the reduction in the total amount of MRPs (pyridines, pyrazines, thiophenes and thiazoles) in boiled pork juice after boiling for 12 h is correlated with their mutagenicity. Among the MRPs, tetrahydrothiophene-3-one exhibited the strongest correlation suggesting that the inhibition of IQ mutagen formation by DAD is mediated through the reduction of MRPs production [109].

Anti-hepatotoxic Effects

Diallyl sulfide (DAS), is known to modulate drug metabolism and may protect animals from chemically induced toxicity and carcinogenesis. In the hepatotoxicity evaluation of Fischer 344 rats there was a dose-dependent increase in the odds of mortality rate by acetaminophen (APAP) (P = 0.009); DAS treatment significantly protected rats from APAP-related mortality (P = 0.026). Liver toxicity determined by lactate dehydrogenase activity was significantly increased by APAP treatment (0.75 g/kg). Pretreatment with DAS protected animals from APAP-induced liver toxicity in a time- and dose-dependent fashion. Treatment of DAS (50 mg/kg) 3 hr after APAP dosing significantly (P < 0.05) protected rats from APAP-induced liver toxicity. The metabolism of APAP (50 μM) *in vitro* was significantly inhibited by DAS (0.3-1 mM) in liver microsomes isolated from F344 rats. As the effect of DAS on APAP-induced hepatotoxicity *in vivo* was observed only when DAS was administered before or shortly after (< 3 hr) APAP dosing. This suggests that the protective effect of DAS is mainly at the metabolic activation step of APAP. However, the possibility that DAS may also have effects on other drug metabolism systems, such as glutathione (GSH) and glutathione S-transferases, cannot be ruled out [110].

According to the results of lactate dehydrogenase (LDH) leakage and microscopic examination, 0.5 or 1 mM DAS treatment did not have any adverse effects on the viability of hepatocytes. Intracellular GSH contents of cells treated with 0.5 and 1 mM DAS (58.6 and 66.4 nmol GSH/mg protein, respectively) were higher than in the controls (54.2 nmol GSH/mg protein), around 8-23%, at 24 hr of incubation; a significant difference (P < 0.05) was observed for 1 mM DAS treatment at 48 hr. This phenomenon is beneficial to the detoxification and antioxidation capabilities of hepatocytes. Further, when the hepatocytes were treated with 0.5 or 1 mM DAS, the activities of glutathione S-transferase (GST), glutathione peroxidase (GPx) and glutathione reductase (GRd) were almost the same as those of the controls. On the other hand, treatment with 5 mM DAS was associated with a significant decrease ($P < 0.05$) in cell viability, namely in increased LDH leakage (50% at 24-hr treatment), significant changes in the morphology of the hepatocytes, low intracellular GSH level (45% lower than in the controls at 24-hr treatment), and low activities of GST, GPx and Grd [111].

Inhibition of Cholesterol Biosynthesis

Using primary rat hepatocyte cultures, the potency of several garlic-derived organosulfur compounds to inhibit cholesterol biosynthesis *in toto* as well as at early and late steps of this metabolic pathway was compared [112]. Concerning early steps, allicin significantly inhibited incorporation of [^{14}C] acetate into nonsaponifiable neutral lipids already at concentrations as low as 10 µM, while diallyl disulfide and allyl mercaptan were effective above 100 µM only. Likewise, inhibition in response to two vinyl-dithiins (1,3-vinyl-dithiin and 1,2-vinyldithiin) started at 500 µM. If [^{14}C] acetate was replaced by [^{14}C] mevalonate, inhibition due to allicin, diallyl disulfide, and allyl mercaptan disappeared suggesting that HMGCoA-reductase was the target of inhibition. In contrast, for the vinyl-dithiins a stimulation of mevalonate incorporation was found. Concerning the late step, the potency to exert accumulation of lanosterol presumably by inhibiting lanosterol 14 α-demethylase decreased in the order allicin > diallyl disulfide > allyl mercaptan = 1,3-vinyl-dithiin >> 1,2-vinyldithiin, the effect of the latter compound being close to zero. With respect to the total inhibition of [^{14}C] acetate labeling of cholesterol, the half-maximal effective concentration-value of allicin was determined to be 17 +/- 2 µM compared to 64 +/- 7 µM for diallyl disulfide and to 450 +/- 20 µM for allyl mercaptan. Cytotoxicity as determined by the lactate dehydrogenase leakage assay was slightly higher for the two vinyl-dithiins than for diallyl disulfide and allyl mercaptan, but was apparent only at concentrations higher than 10 mM and, consequently, was irrelevant for the effects described. These results demonstrate that different garlic-derived organosulfur compounds interfere differently with cholesterol biosynthesis and, thus, may provoke multiple inhibition of this metabolic pathway in response to garlic consumption [112]. The fact that allicin was the most effective inhibitor argues against the possibility that its degradation products, namely diallyl disulfide or allyl mercapatan, might mediate its effects, a possibility that might be true, however, in the case of the vinyl-dithiins.

Antifungal Effects

Although amphotericin B remains the drug of choice for systemic fungal infections, its use is limited by considerable side effects. In The Peoples' Republic of China, commercial *Allium sativum* derived compounds are widely used as an antifungal drug to treat systemic fungal infections. To evaluate the scientific merit of using *A. sativum* derived compounds as antifungal agents, a Chinese commercial preparation, allitridium was studied [113]. This preparation contained mainly diallyl trisulfide as confirmed by high performance liquid chromatography. Allitridium, with and without amphotericin B, was tested to determine its efficacy in killing three isolates of *Cryptococcus neoformans*. The minimum inhibitory concentration of the commercial preparation was 50 μg/ml and the minimum fungicidal concentration was 100 μg/ml against 1×10^5 organisms of *C. neoformans*. In addition, the commercial preparation was shown to be synergistic with amphotericin B in the *in vitro* killing of *C. neoformans*. This study demonstrates that diallyl trisulfide and other polysulfides possess potent *in vitro* fungicidal effects and their activity is synergistic with amphotericin B. These observations lend laboratory support for the treatment of cryptococcal infections with both amphotericin B and the Chinese commercial preparation [113].

CONCLUSION

Previous and recent studies show the role of garlic and its organosulfur compounds to prevent many body disorders. Garlic, in any of its forms, would be effective. If fresh whole garlic cloves are cut and swallowed with water, the evolved allicin would start its transformation reactions and supply the body with compounds that would reduce, or prevent, many health risks.   If fresh garlic, or garlic powder is being cooked or oil macerated, the allyl sulfides will provide cancer prevention and other health benefits. Available pharmaceutical formulations of garlic that claim to duplicate these benefits would also be effective. Further evaluation of the preventive potential of these organosulfur compounds in humans at risk appears feasible. It therefore seems to be important to start prevention of cancer and other diseases by garlic nutrition as early as possible and adhere to it over a long period.

REFERENCES

[1]   Darby W.J.; Ghalioungui, P.; Grivetti, L. *Food: The Gift of Osiris*, vol.2, Academic Press, London, , **1977**.

[2]   Koch, H.P.; Lawson, L. Garlic, The Science and Therapeutic Application of Allium sativum L. and Related Species. 2nd edition, Williams and Wilkins, Maryland, **1996**.

[3]   Block, E. In *Folk Medicine: The Art and the Science*; ACS book series, Steiner, R. American Chemical Society; Washington, DC, **1986** p.134.

[4]   Nielson K.K; Mahooney, A.W.; Williams, L.S.; Rogers, V.C. *J. Food Comp. Analysis*, **1991**, *4*, 39.

[5]   Buiatti, E.; Palli, D.; Decarli, A.; Amadori, D.; Avellini, C.; Bianchi, S.; Biserni, R.; Cipriani, F.; Cocco, P.; Giacosa, A., *Int. J. Cancer* **1989**, *44*, 611.

[6]   Koch, H.; Hahn, G. *Knoblauch* Urban and Schwarzenberg, Baltimore, **1988**.

[7]   Fenwick, G.; Hanley, A. *CRC Crit. Rev. Food. Sci. Nutr.* **1985**, *23*, 1.

[8]   Abdullah, T. H.; Kandil, O.; Elkadi, A.; Carter, J. *J. Natl. Med. Assoc.* **1988**, *80*, 439.

[9]   Petkov, V. *J. Ethnopharmacol.* **1986**, *15*, 121.

[10]   Lawson, L. In *Human Medicinal Agents from Plants* ACS book series, Kinghorn, A.D.; Balandrin, M.F. American Chemical Society; Washington, DC, **1993**, pp. 306-330.

[11]   Pentz, R.; Guo, Z.; Kress, G.; Müller, D.; Müller, B.; Sigers, C.P. *Planta Med.***1990**, *56*, 691.

[12]   Cavallitto, C.J.; Baily, J.H. *J. Am. Chem. Soc.* **1944**, *66*, 1950.

[13]   Block, E.; Ahmad, S.; Catalfamo, J.; Mahendra, K.J.; Apitz-Castro, R. *J. Am. Chem. Soc.* **1986**, *108*, 7045.

[14]   Mütsch-Eckner, M.; Meier, B.; Wright, A.D.; Sticher, O. *Phytochemistry* **1992**, *31*, 2389.

[15]   Lawson, L.D.; Wang, Z.Y.J.; Hughes, B.G. *J. Nat. Prod.* **1991**, *54*, 436.

[16]   Ueda, Y.; Kawajiri, H.; Miyamura, N.; Miyajima, R. *J. Jap. Soc. Food Sci. Technol.* **1991**, *38*, 429.

[17]   Sendl, A.; Elbl, G.; Steinke, B.; Redl, K.; Breu, W.; Wagner, H. *Planta Med.* **1992**, *58*, 1.

[18]   Block, E.; Naganathan, S.; Putnam, D.; Zhao, S.H. *J. Agric. Food Chem.*, **1992**, *40*, 2418.

[19]   Lawson, L.D.; Wood, B.G.; Hughes, B.G. *Planta Med.* **1991**, *57*, 263.

[20]   Suzuki, T.; Sugii, M.; Kakimoto, T. *Chem. Pharm. Bull.* **1961**, *9*, 251.

[21]   Virtanen, A.I.; Mattila, I. *Suom. Kemistil.* **1961**, *B34*, 73.

[22]   Lancaster, J.E.; Shaw, M.L. *Phytochemistry* **1991**, *30*, 2857.

[23]   Stoll, A.; Seebeck, E. *Experientia* **1947**, *3*, 114.

[24]   Stoll, A.; Seebeck, E. *Helv. Chim. Acta* **1951**, *56*, 481.

[25]   Fujiwara, M.;Yoshimura, M.; Tsuno, S.; Murakami, F. *J. Biochem.* **1958**, *45*, 141.

[26]   Granroth, B. *Acta Chem. Scand.* **1968**, *22*, 3333.

[27]   Ceci, L.N.; Curzio, O.A.; Pomilio, A.B. *Phytochemistry*, **1992**, *31*, 441.

[28]   Rennenburg, H. *Phytochemistry*, **1982**, *21*, 2771.

[29]   Ellmore, G.S.; Feldberg, R.S. *Am. J. Bot.*, **1994**, *81*, 89.

[30]   Wen, J.Y.; Mato, A.; Malik, M.N.; Jenkins, E.C.; Sheikh, A.M.; Kim, K.S. *J. Cell. Biochem.*, **1995**, *58*, 481.

[31]   Van Damme, E.J.M.; Smeets, K.; Torrekens, S.; Van Leuven, F.; Peumans, W.J. *Eur. J. Biochem.*, **1992**, *209*, 751.

[32]   Stoll A.; Seebeck, E. *Adv. Enzymol.*, **1951**, *11*, 377.

[33]   Lawson, L.D.; Wood, S.G.; Hughes, B.G. *Planta Med.* **1991**, *57*, 263-270.

[34]   Brodnitz, M.H.; Pascale, J.V.; Van Derslice, L. *J. Agric. Food Chem.* **1971**, *19*, 273.

[35]   Block, E.; Ahmad, S.; Jain, M.K.; Crecely, R.W.; Apitz-Castro, R.; Cruz, M.R. *J.Am.Chem.Soc.,* **1984**, *106*, 8295.

[36]   Lawson, L.D.; Hughes, B.D. *Planta Med.*, **1992**, *58*, 345

[37]   Lawson, L.D.; Wang, C.J. *Planta Med.*, **1994**,

[38]   Lawson, L.D.; Wang, Z.G.; Hughes, B.G. *Planta Med.* **1991**, *57*, 363.

[39]   Raghavan, B.; Abraham, K.O.; Shankaranaryana, M.L. *PAFAI J.*, **1986**, *8*, 11.

[40]   Fenwick, G. R.; Hanely, A. B. *CRC Crit. Rev. Food Sci. Nutr.,* **1985**, *22*, 199.

[41]   Fenwick, G. R.; Hanely, A. B. *CRC Crit. Rev. Food Sci. Nutr.,* **1985**, *22*, 273.

[42]   Koch, H. P. *Dtsch. Apoth. Ztg.,* **1987**, *127*, 367.

[43]   Sticher, O. *Dtsch. Apoth. Ztg.,* **1991**, *131*, 403.

[44]   Gassmann, B. *Ernähr. Umsch.*, **1992**, *39*, 444.

[45]   Blania, G.; Spangenburg, B. *Planta Med.*, **1991**, *57*, 371.

[46]   Lawson, L. D.; Hughes B. G. *Planta Med.*, **1992**, *58*, 345.

[47]   Winkler G.; Lohmüller, E.M.; Landshuter, J.; Weber, W.; Knobloch, K. *Dtsch. Apoth. Ztg.,* **1992**, *132*, 2312.

[48]   Pentz, R.; Guo, Z.; Müller, B.; Aye, R. D.; Siegers, C.P. *Dtsch. Apoth. Ztg.,* **1992**, *132*, 1779.

[49]   Voigt, M.; Wolf, E. *Dtsch. Apoth. Ztg.,* **1986**, *126*, 591.

[50]   Schardt, D.; Liebman, B. *Nutrition Action Health Lett.*, **1995**, *22*, 4.

[51]   Laakso, I.; Seppänan-Laasko, T.; Hiltunen, R.; Müller, B.; Jansen, H.; Knobloch, K. *Planta Med.,* **1989**, *55*, 257.

[52]   Minami, T.; Boku, T.; Katsuhiro, I; Masanori, M.; Okazaki, Y. *J. Food Science,* **1989**, *54*, 763.

[53]   Ruiz, R.; Hartman, T.G.; Karmas, K.; Lech, K.; Rosen, R.T. In *Food Phytochemicals I: Fruits and Vegetables* ACS Symp. Ser., American Chemical Society; Washington, DC, **1994**, *546*, pp.102-119.

[54]   Cai, X.J.; Block E.; Uden, P.C.; Quimbey, B.D.; Sullivan, J.J. *J. Agric. Food Chem.,* **1995** 43, 1751.

[55]   Taucher, J.; Hansel, A.; Jordan, A.; Lindinger, W. *J. Agric. Food Chem.* **1996**, *44(12)*, 3778.

[56]   Miles, W.S.; Quimby, B.D. *Am. Lab.,* **1991**, July, 28F.

[57]   Calvey, E.M.; Matusik, J. E.; White, K. D.; Betz, J. M.; Block, E.; Littlejohn, M. H.; Naganathan, S.; Putman, D. *J. Agric. Food Chem.*, **1994**, 42, 1335.

[58]   Calvey E. M.; Block, E. In *Spices Flovor Chemistry and Antioxidant Properties,* Risch, S. J.; Ho, C.-T. ACS Symp. Ser. 660, American Chemical Society, Washington D. C. **1997**, pp. 113-124.

[59]   Calvey, E. M.; Matusik, J. E.; White, K. D.; DeOrazio, R.; Sha, D.; Block, E. *J. Agric. Food Chem.*, **1997**, *45*(11), 4406.

[60]   Ip, C.; Lisk, D. J.; Thompson, H. J. *Carcinogenesis,* **1996**, *17*(9),1979.

[61]   El-Bayoumi, K.; Chae, Y. H.; Upadhayaya, P.; Ip, C. *Anticancer Res.,***1996**, *16*(5A), 2911.

[62]   Lu, J.; Pei, H.; Ip, C.; Lisk, D. J.; Ganther, H.; Thompson, H. J. *Carcenogenesis,* **1996**, *17*(9), 1903.

[63]   Schmidt, P. W.; Marquardt, U. Zentrabl. Bakteriol. Parasitenkd. Infektionskrankh.Hyg., **1936**, 138, 104.

[64]   Silber, W. *Klin. Wochenschr.*, **1933**, *12*, 509.

[65]   De Torrescacasna, E. U. *Rev. Espan. Fisiol.*, **1946**, *2*, 6.

[66]   Kamanna, V. S.; Chandrasekhara, N. *Ind. J. Med. Res.*, **1984**, *79*, 580.

[67]   Sumiyoshi, H.; Wargovich, M. J. *Asia Pac. J. Pharmacol.*, **1989**, *4*, 133.
[68]   Kandil, O. M.; Abdullah, T. H.; Elkadi, A. *Fed. Proc.*, **1987**, *46*, 441.
[69]   Wargovich, M. J.; Uda, N.; Woods, C.; Velasco, M.; McKee, K. *Biochem. Soc. Trans.*, **1996**, *24*, 811.
[70]   Srivastava, S. K.; Hu, X.; Xia, H.; Zaren, H. A.; Chatterjee, M. L.; Agarwal, R.; Singh, S. V. *Cancer Lett.*, **1997**,*118*(1), 61.
[71]   Moriguchi, T.; Matsuura, H.; Kodera, Y.; Itakura, Y.; Katsuki, H.; Saito, H.; Nishiyama, N. *Neurochem. Res.* **1997**, *22*(12),1449.
[72]   Ide, N.; Lau, B. H. *J. Pharm. Pharmacol.*, **1997**, *49*(9), 908.
[73]   Hatono, S.; Jimenez, A.; Wargovich, M. J. *Carcingenesis,* **1996**, *17*(5), 1041.
[74]   Hu, X.; Singh, S. V. *Arch. Biochem. Biophys.* **1997**, *340*(2), 279.
[75]   Takada, N.; Yano, Y.; Wanibuchi, H.; Otani, S.; Fukushima, S. *Jpn. J. Cancer Res.,* **1997**, *88*(5), 435.
[76]   Dion, M. E.; Agler, M.; Milner, J. A. *Nutr. Cancer,* **1997**, *28*(1), 1.
[77]   Geng, Z.; Rong, Y.; Lau, B. H. *Free Radic. Biol. Med.,* **1997**, *23*(2), 345.
[78] Hageman, G. J.; Van Herwijnen, M. H.; Schilderman, P. A.; Rhijnsburger, E. H.; Moonen, E. J.; Kleinjans, J. C. *Nutr. Cancer,* **1997**, *27*(2), 177.
[79]   Sigounas, G.; Hooker, J. L.; Li, W.; Anagnostou, A.; Steiner, M. *Nutr. Cancer, 1997, 28*(2), 153.
[80]   Sigounas, G.; Hooker, J.; Anagnostou, A.; Steiner, M. *Nutr. Cancer,* **1997**, *27*(2), 186.
[81]   Zwergal, A. *Pharmazie*, **1952**, *7*, 245.
[82]   Bailey, J. H.; Cavallito, C. J. *J. Bacteriol.*, **1948**, *55*, 175.
[83]   Tynecka, Z.; Gos, Z. *Ann. Univ. Mariae Curie-Sklodowska*, **1975**, *30D*, 5.
[84]   Hörhammer, L.; Wagner, H.; Seitz, M.; Vejdelek, Z. J. *Pharmazie*, **1968**, *23*, 462.
[85]   Feldberg, R. S.; Chang, S. C.; Kotik, A. N.; Nadler, M.; Neuwirth, Z.; Sundstrom, D. C.; Thompson, N. H. *Antimicrob. Agents Chemother.*, **1988**, *32*, 1763.
[86]   Naganawa, R.; Iwata, N.; Ishikawa, K.; Fukuda, H.; Fujino, T.; Suzuki, A. *Appl Environ. Microbiol.*, **1996**, *62*(11), 4238.
[87]   Mirelman, D.; Varon, S. *Neue Aerztliche*, **1987**, *59*, 10.
[88]   Araki, M.; Yokota, Y.; Kuga, M.; Chin, S.; Fujikawa, F.; Nakajima, K.; Fujii, H.; Tokuoka, A.; Hirota, Y. *J. Pharm. Soc. Japan*, **1952**, *72*, 979.
[89]   Holzhey, M.; Roth, H. H.; Höpfner, V. *Germany Patent*, **1992**, 4 024 155.
[90]   Ankri, S.; Miron, T.; Rabinkov, A.; Wilchek, M.; Mirelman, D. *Antimicrob. Agents Chemother.,* **1997**, *41*(10), 2286.
[91]   Freeman, F.; Kodera, Y. *J. Agric. Food Chem.*, **1995**, *43*, 2332.
[92]   Lawson, L. D.; Ransom, D. K.; Hughes, B. G. *Thromb. Res.*, **1992**, *65*, 141.
[93]   Freeman, F.; Kodera, Y. *J. Agric. Food Chem.*, **1997**, *45*, 3709.
[94]   Apitz-Castro, R.; Cabrera, S.; Cruz, M. R.; Ledezma, E.; Jain, M. K. *Throm. Res.*, **1983**, *32*, 155.
[95]   Romano, E. L.; Montano, R. F.; Brito, B.; Apitz, R.; Alonso, J.; Romano, M.; Gebran, S.; Soyano, A. *Immunopharmacol. Immunotoxicol.*, **1997**, *19*(1), 15.
[96]   Urbina, J. A. *Parasitology*, **1997**, *Suppl:S91-S99*, 114.
[97]   Ledezma, E.; DeSousa, L.; Jorquera, A.; Sanchez, J.; Lander, A.; Rodriguez, E.; Jain, M. K.; Apitz-Castro, R. *Mycoses*, **1996**, *9-10*, 393.
[98]   San-Blas, G.; Urbina, J. A.; Marchan, E.; Contreras, L. M.; Sorais, F.; San-Blas, F.   *Microbiology*, **1997**, *143*(5), 1583.

[99]   Villar, R.; Alvarino, M. T.; Flores, R. *Biochim. Biophys. Acta* **1997**, *1337*(2), 233.

[100]   Ishikawa, K.; Naganawa, R.; Yoshida, H.; Iwata, N.; Fukuda, H.; Fujino, T.; Suzuki, A.; *Biosci. Biotechnol. Biochem.,* **1996**, *60*(12), 2086.

[101]   Jin, L.; Baillie, T. A. *Chem. Res. Toxicol.* **1997**, *10*(3), 318.

[102]   Srivastava, S. K.; Hu, X.; Xia, H.; Zaren, H. A.; Chatterjee, M.L.; Agarwal, R.; Singh, S.V. *Cancer Lett.* **1997**, *118*(1):61.

[103]   Hu, X.; Srivastava, S. K.; Xia, H.; Awasthi, Y. C.;Singh, S. V. *J. Biol. Chem.,* **1996**, *271*, 32684.

[104]   Hu, X.; Singh, S. V. *Arch. Biochem. Biophys.,* **1997**, *340*(2):279.

[105]   Musk, S. R.; Clapham, P.; Johnson, I.T. *Food Chem. Toxicol.,* **1997**, *35*(3-4), 379.106. Schaffer, E. M.; Liu, J. Z.; Milner, J. A. *Nutr. Cancer* **1997**, *27*(2), 162.

[106]   Hu, X.; Benson, P. J.; Srivastava, S. K.; Mack, L. M.; Xia, H.; Gupta, V.; Zaren, H. A.; Singh, S. V. *Arch. Biochem. Biophys.,* **1996**, *336*(2), 199.

[107]   Sundaram, S. G.; Milner, J. A. *J Nutr.,* **1996**, *126*(5), 1355.

[108]   Tsai, S. J.; Jenq, S. N.; Lee, H. *Mutagenesis,* **1996** , *11*(3), 235.

[109]   Hu, J. J.; Yoo, J. S.; Lin, M.; Wang, E. J.; Yang, C. S. *Food Chem. Toxicol.,* **1996**, *34*(10), 963.

[110]   Sheen, L. Y.; Lii, C. K.; Sheu, S. F.; Meng, R. H.; Tsai, S. J. *Food Chem. Toxicol.,* **1996**, *34*(10), 971.

[111]   Gebhardt, R.; Beck, H. *Lipids,* **1996**, *31*(12), 1269.

[112]   Shen, J.; Davis, L. E.; Wallace, J. M.; Cai, Y.; Lawson, L. D. *Planta Med.,* **1996**, *62*(5), 415.

Atta-ur-Rahman (Ed.) *Studies in Natural Products Chemistry, Vol. 23*
© 2000 Elsevier Science B.V. All rights reserved

CHEMISTRY OF SOME NATURAL PRODUCTS OF BIOLOGICAL INTEREST

S.B. MAHATO

Indian Institute of Chemical Biology, 4, Raja S.C. Mullick Road, Jadavpur, Calcutta - 700032, India

ABSTRACT: Chemistry of some natural products of biological interest is reviewed. Cleomeolide, a diterpene lactone with unprecedented structural features is isolable in quantity from renewable regions of *Cleome icosandra* widely distributed in India. Its chemical convertibility to a tricarboxylic system analogous to that found in taxane indicated its potential for use as a base material for the synthesis of taxol analogues. Novel triterpene acids isolable by acid hydrolysis of the saponins from *Mimusops elengi* are potential bioactive compounds. Acaciasides A, B and C, the acylated triterpenoid bisglycosides from the pericarps of *Acacia auriculiformis*, abundantly available throughout India hold promise for their use as spermicidal and antifilarial agents. Unambiguous structure elucidation of four new saponins from *Bacopa monniera* reputed for its use as a nervine tonic for memory improvement has been achieved and presence of at least eight other saponins has been confirmed. The results disprove the prevailing notion of occurrence in the plant of only bacoside A_1 and bacoside A_3. Saikosaponins are biologically interesting compounds reported so far from *Bupleurum* species. Saikosaponins-like glycosides corchorusins A, B, C, D, C_1, D_1, D_2 and D_3 have been isolated from *Corchorus acutangulus*, a plant of different genus and family occurring throughout the hotter parts of India. Interesting acid-catalysed rearrangemednts of some aglycones are described.
Spermidine and spermine alkaloids which belong to the class of polyamine conjugates possess diverse biochemical profiles. Three spermidine alkaloids, Caesalpinines A, B and C have been isolated from the prickly scandent shrub growing in eastern India. Caesalpinine A possessses a novel skeleton of a 13-membered lactum ring fused to a five-membered lactum ring. Flavonoid-, and phenylpropanoid glycosides from polar fraction of *Lantana camara* widely occurring in many tropical and subtropical parts of the world are of interst for their use for therapeutic intervention in a wide range of diseases including cancer.

INTRODUCTION

Natural Products Chemistry has contributed significantly towards the development of modern medicines. Many of the modern medicines are modifications or survivals of ancient herbalism. Natural products often serve as chemical models for the design and total synthesis of new drugs.The structure elucidation of natural products which was once an attractive area of research is no longer the spearhead of the subject. The application of the powerful spectroscopic techniques has tremendously eased the problem. However, although the routine adoption of these techniques in structure elucidation studies may appear to have created a limitation on the generation of new chemical knowledge, these methodologies have nevertheless opened up new vistas and research activities are now moving forward into areas which were otherwise inaccessible. While much of our basic chemical knowledge was previously derived from degradative studies, its extension today rests primarily on the synthesis and modification of natural products and their analogues. The real spearhead of natural products chemistry now lies in studies associated with biological investigations leading to the understanding of structure-activity relationship. It may be emphasized, however, that the structural study has not lost its importance from a practical standpoint and plants continue to be important sources of new drugs. We have seen in recent years the discovery and industrial stimulus provided by such natural products as taxol [1], a diterpene (anticancer), artemisinin [2], a sesquiterpene lactone (antimalarial), azadirachtin [3,4], a tetranortriterpenoid (antifeedant) and forskolin [5-8], a diterpene (hypotensive, cardioactive). Bioactive natural products often have highly complex structures with many chiral centres which are sometimes necessary for biological activity. Such complex compounds cannot be synthesized economically with the present state of chemical knowledge and as such development of processes for large scale isolation from potential natural sources or their partial synthesis from widely available natural products becomes imperative. This article covers the chemistry and potential of some biologically interesting natural products with novel structural features isolated from plants used in traditional medicine.

DITERPENES

Cleomeolide

The macrocyclic diterpene lactone, cleomeolide (1) was first isolated by us [9,10] in the late 1970s and its complete structure and stereochemistry were determined by means of NMR, X-ray and CD methods. Burke et al. [11] isolated and characterised it subsequently and described some interesting reactions encountered during the structural investigation. The host plant is Cleome icosandra (Syn. Cleome viscosa) which is a sticky herb widely distributed in India. It has yellow flowers and strong penetrating odour. The leaves are rubefacient, vesicant, sudorifice and is also used in external applications for wounds and ulcers. The plant is used by the poorer people as vegetable after discarding the flowers and pods [12].

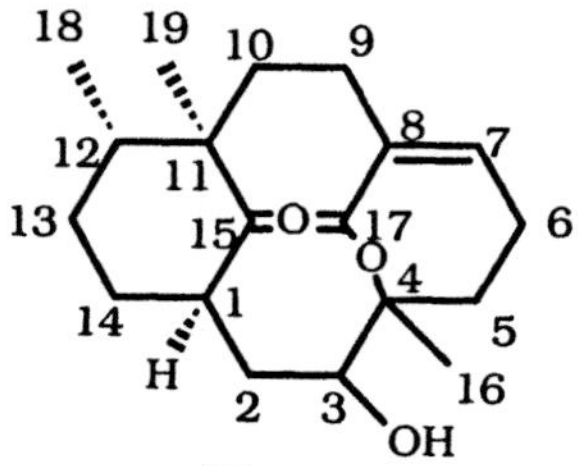

(1)
cleomeolide

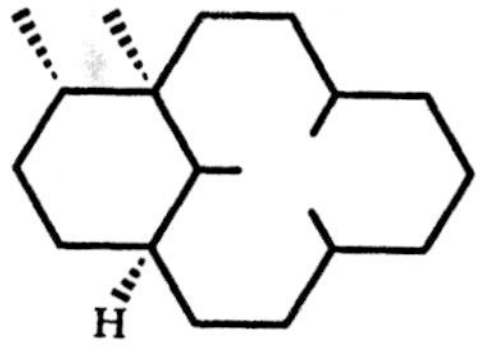

(2)
cleomane

The unprecedented structural features of cleomeolide encompass a 12-membered carbocyclic ring *cis*-fused to a methylene cyclohexane ring as well as a 7-membered α, β–unsaturated lactone whose double bond is at the bridgehead position. The unique carbocyclic framework arises by a nine-membered carbon chain *cis*-fused in 2,6-fashion to a methylenecyclohexane core.

The novelty of the carbon skeleton (2) named cleomane [10] present in cleomeolide was revealed by its ^{13}C NMR spectral data which demonstrated the occurrence of only one sp^3 quaternary carbon atom, a structural feature unprecedented in naturally occurring diterpenes. A somewhat similar carbocyclic framework containing diterpene, verticillol (3) was isolated by Karlsson et al. [13] from *Sciadopitys verticillata* more or less during the same period of time.

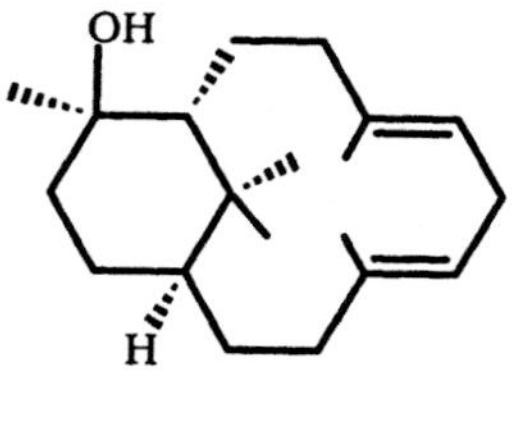

(3)
verticillol

The biogenetic pathway of formation of cleomeolide may be envisaged to arise either by the oxidative cyclization of geranyl linalool expoxide [10] as outlined in Scheme 1 or via head-to-tail cyclization of geranyl geranyl pyrophosphate [11] as shown in Scheme 2. However, one would be tempted to prefer the former considering the oxygenation pattern in the molecule.

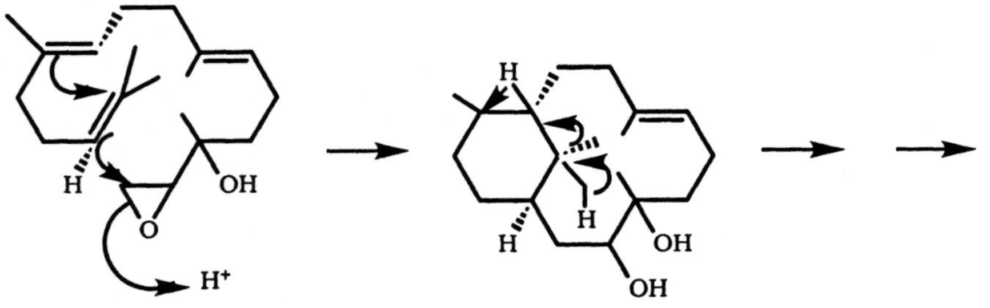

Scheme 1 Oxidative cyclisation of geranyl linalool epoxide.

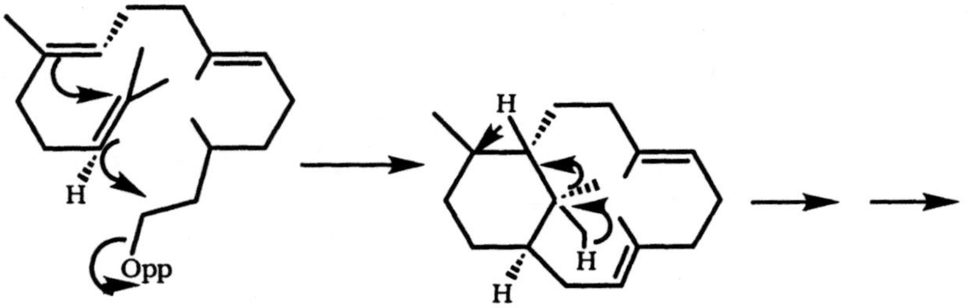

Scheme 2 Head-to-tail cyclisation of geranyl geranyl pyrophosphate.

The macrocyclic diterpenes, cembrene [14] and casbene [15] are regarded as the prototypes of a variety of natural products showing significant biological activity [16]. Phorbol esters are known cocarcinogens [17] and jatrophone [18], gnidimacrin [19] and asperdiol [20] are reported for their antileukemic activity whereas taxol (4) has been approved for use against metastatic ovarian and breast cancers and is undergoing evaluation against a variety of other ailments [21,22]. Indeed taxol (4) has attracted much attention in recent years for ameliorating the problem of its availability by the phytochemical route as well as by total synthesis. However, the compound (4) tends to be localized in non-renewable domains of most taxol-containing plants in very low amounts. On the other hand baccatin III (5) and 10-deacetylbaccatin III (6), themselves not possessing useful biological function are isolable in quantity from renewable regions of variety of plants. Successful semisynthesis of taxol (4) from rather more accessible base materials (5) and (6) has been achieved [23,24]. Total synthesis of (4) via total synthesis of (5) has been reported by the research groups of Holton [25], Nicolaou [26] and Danishefsky [27]. Much interesting chemistry has been developed following synthetic explorations pertinent to taxol which is covered in an excellent review [28]. However, availability problem for taxol remains and tremendous efforts are on to solve it. The synthetic explorations will bring with them new nuclei which may be used as starting materials for the synthesis of medicinally promising variants of taxol. Potential precursors to analogues of taxol are also obtainable from certain easily available natural products.

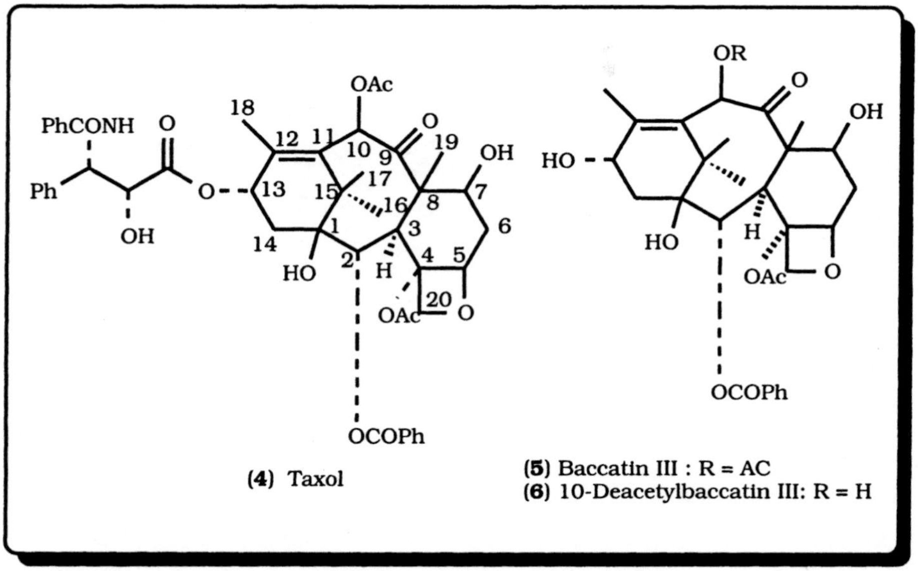

(4) Taxol

(5) Baccatin III : R = AC
(6) 10-Deacetylbaccatin III: R = H

Besides cleomeolide itself having a great potential for its biological activity its chemical convertibility to a tricarbocyclic system analogous to that found in taxane has created much interest. Cleomeolide (1) in benzene or acetone on oxidation with Jones reagent at room temperature furnished [10] the ketone (7) which on treatment with 10% KOH and MeOH at room temperature afforded [11] two products (8) and (9). The interesting

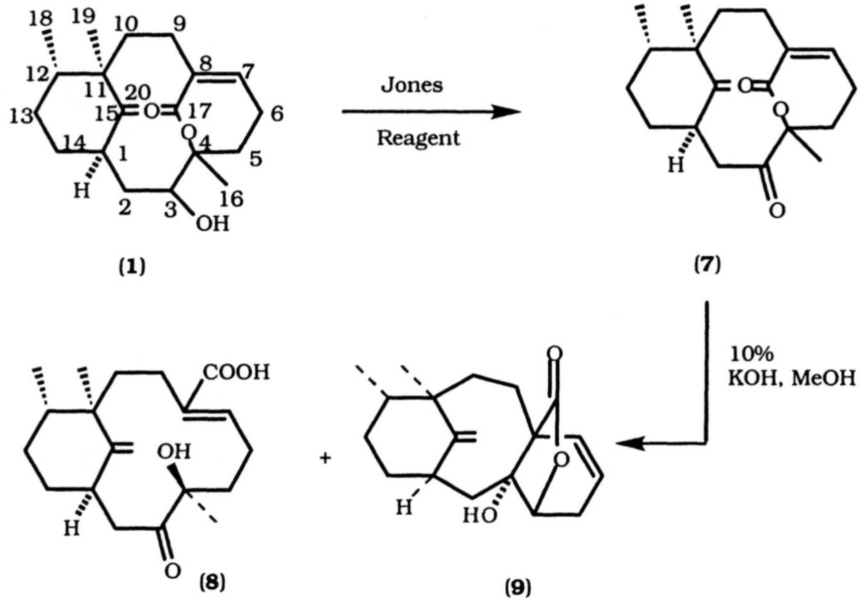

Chart 1. Transformation of cleomeolide to lactone (9).

isomeric lactone (9) which was characterized by X-ray crystallography [11] has structural relationship to the taxane diterpenes such as taxol (4) (Chart-1).

It is noteworthy that taxol is isolated from *Taxus* yews which are rare plant species requiring special agroclimatic conditions to grow. Moreover, these yews take longer period to get matured. The increasing demand for taxol has prompted the investigators to search for alternative source for this diterpene or its potential analogues. *Cleome icosandra* which is widely distributed in India and probably in other tropical countries may be a potential source for cleomeolide which may be required as a precsursor to analogues of taxol.

The first total synthesis of (+)-cleomeolide has been accomplished in enantioselective fashion from optically pure Wieland Miescher ketone (10) by Paquette and his co-workers [29,30]. Their well designed and ultimately successful stratagem was dependent on effective use of a diagnosable interpendence of functional group deployment and adoption of serviceable global energy conformational minima. The successful use of

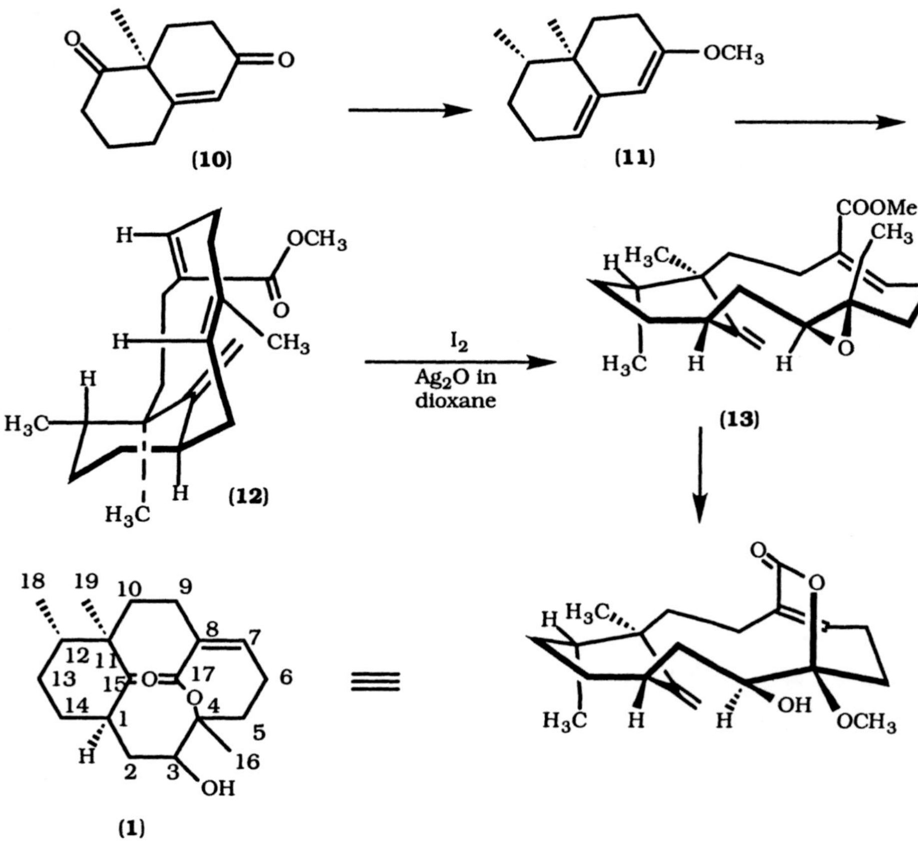

Chart 2. Key steps in the synthesis of cleomeolide (1).

these tacticts will encourage the adoption of related possibilities for stereocontrolled macrocyclic ring assembly in future synthetic ventures.

The key steps in the synthesis were (a) improved conversion of optically pure Wieland-Miescher ketone (10) into dienol ether (11) and oxidative cleavage of the latter to aldehydo ester (12) ; (b) avoidance of complications arising from steric blockade of C-15 for introduction of the methylene group at that site; (c) employment of an intramolecular Wadsworth-Emmons cyclization for macroclyclic ring construction; (d) modulation of the conformation adopted by the medium ring by diastereofacial control of epoxidation of the C-3/C-4 double bond; and (e) intramolecular cyclization of the epoxy acid derived from (12) by nucleophilic capture at the more substituted oxiranyl carbon. The deep-seated topographical change that occurs during the formation of (13) projects the macrocyclic ring quasiaxially from the methylenecyclohexane subunit. Thus the intramolecular cyclization to set the bridgehead double bond was facilitated. The compound (12) on treatment with the combination of iodine and Ag_2O in aqueous dioxane provided the desired product (13) in modest yield (Chart-2).

Verticillol

The macrocylic diterpene alcohol, verticillol (3) was isolated from the wood of *Sciadopitys verticillata* Sieb. Et zucc. (Taxodiaceae) [13]. The structure of verticillol (3) was elucidated from NMR-LIS studies on it as well as its correlation to the diepoxide (14) whose structure was established by direct single crystal X-ray analysis. The absolute configuration of verticillol (3) was determined by CD data of the verticillol norketodiepoxide (15). The carbocyclic framework of verticillol (3) is somewhat akin to that of cleomeolide (1). However, the presence of the substituents(3-OH and C-8→C-4 lactone) in appropriate positions in (1) facilitates its conversion to compounds analogous to taxane diterpenes.

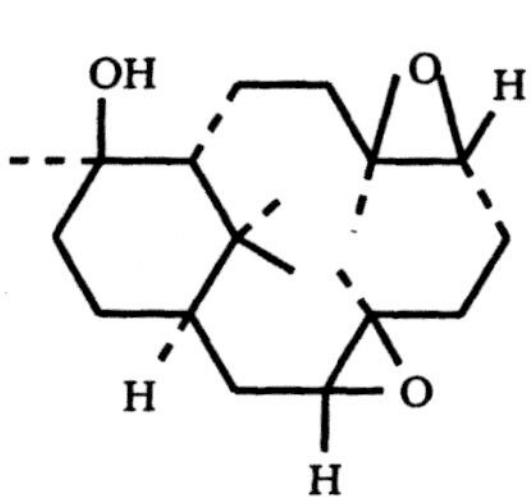

(14)
Verticillol diepoxide

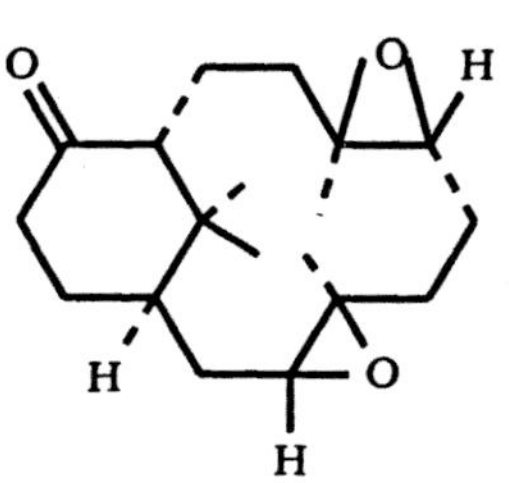

(15)
Verticillol norketodiepoxide

TRITERPENES

Mimusopic and Mimusopsic Acids

Triterpenes of diverse structural types are widely distributed in prokaryotes, as well as eukaryotes. Some plants contain large quantities of triterpenes in their latex and resins and the physiological function of these metabolites is generally believed to be a chemical defence against pathogens and herbivores. It is expected, therefore, that triterpenes should act against certain pathogens causing human and animal diseases. Application of triterpenes as successful therapeutic agents is limited thus far. However, the widespread reports in recent years on useful biological activities of triterpenes, particularlly of triterpene acids [31-33] and indeed some practical applications [34-36], have made these products more relevant and interesting. *Mumusops elengi* is a widely distributed tree in India and is held in high repute in Indian traditional medicine. Bassic acid (17) which is considered to be a chemical marker for the presence of saponins of protobassic acid (16) was isolated previously from the plant

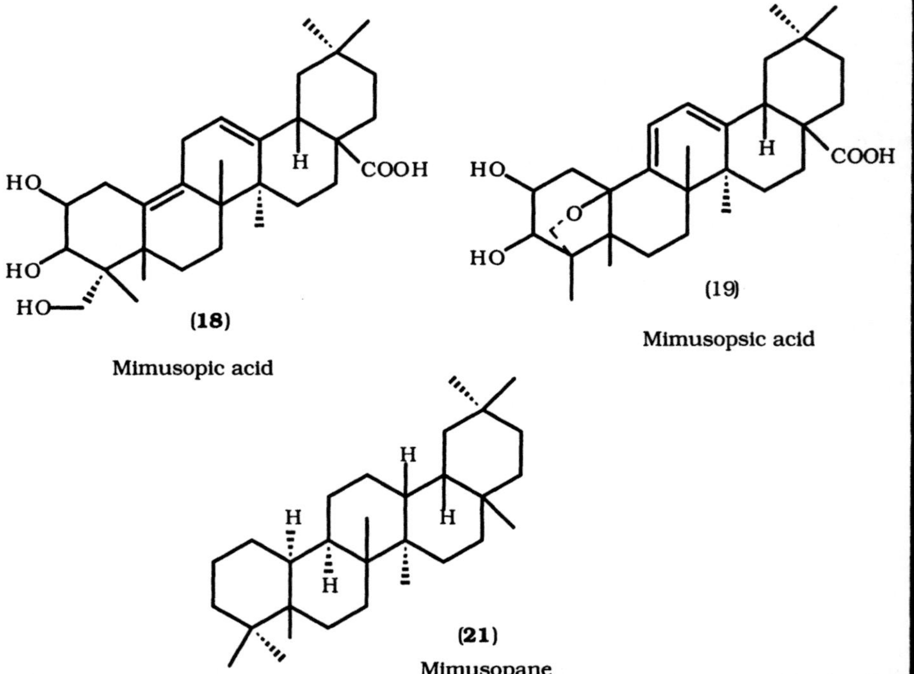

(18)

Mimusopic acid

(19)

Mimusopsic acid

(21)

Mimusopane

[37,38]. Two new pentacyclic triterpene acids, mimusopic acid (18) and mimusopsic acid (19) possessing the novel migrated oleanane skeleton, mimusopane (21) have recently been isolated from the acid hydrolysate of the MeOH extract of the seeds of the plant [39]. Their structures were elucidated by spectroscopic methods and chemical transformations.

The co-occurrence of acids(17), (18) and (19) strongly indicated that both acids (18) and (19) could be artefacts formed from bassic acid (17) during acid hydrolysis of the saponin fraction. In fact, when a solution of bassic acid (17) in MeOH-HCl (aq.) was boiled under reflux for 4h and worked up as usual for the isolation of the products, acid (18) (40%), (19)

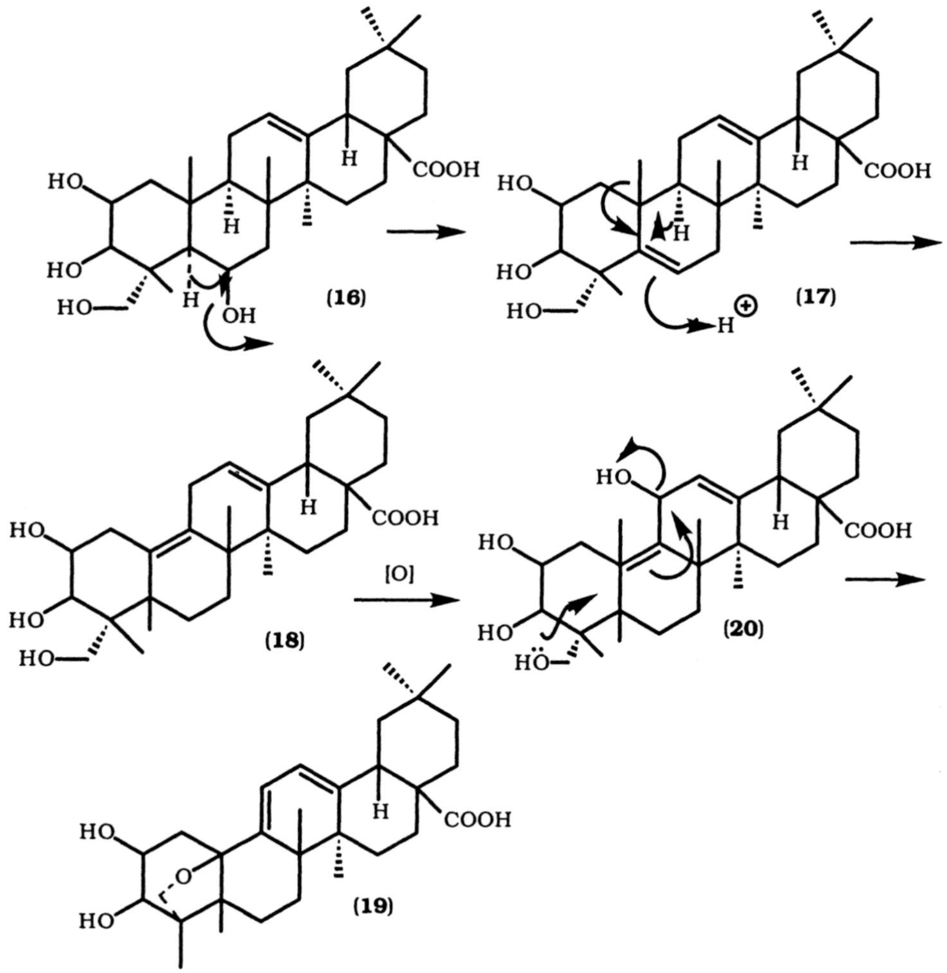

Scheme 3. Mechanism of formation of mimusopic acid (18) and mimusopsic acid (19) from basic acid (17).

(10%) and unconverted acid (17) (50%) were obtained. Although transformation of protobassic acid (16) to bassic acid (17) is known to occur by facile dehydration of 6 β (axial)-hydroxyl group under acidic condition which is under strong steric 1,3-diaxial interactions with 4β-, 8β- and 10β-axial methyls, further conversions of acid (17) to acids (18) and (19) are unprecedented. The mechanism of formation of acids (18) and (19) is rationalised as shown in Scheme 3. It is apparent from Dreiding model inspection that 1,3-diaxial interaction exists in acid (17) involving 10β-, 4β-methyls and 2β-hydroxyl group. Protonation of the 5,6-double bond followed by migration of the 10β-methyl to 5β - position, resulting in reversal of conformation of the A-ring substituents, release the strains and lead to the formation of mimusopic acid (18) with the novel 5,10-friedooleanane skeleton. It is conceivable that the diallylic nature of 11-CH_2 of acid (18) it is prone to aerial oxidation leading to the intermediate formation of 11-hydroxy compound(20). Elimination of the 11-hydroxy of acid (20), migration of the 9:10 double bond to 9:11 position and formation of 23 → 10 oxido ring furnish the novel 2β,3β-dihydroxy-(23→10) oxido-5,10-friedooleane-9(11),12-diene-28-oic acid(mimusopsic acid) (19). The acids (18) and (19) represent the first members of a hitherto unknown 5,10-friedooleanane (mimusopane) skeleton (21).

The other new triterpenes isolated [40] so far from the acid hydrolysate of the MeOH extract of the seeds of the plant are mimusopgenone (22) and mimugenone (23).

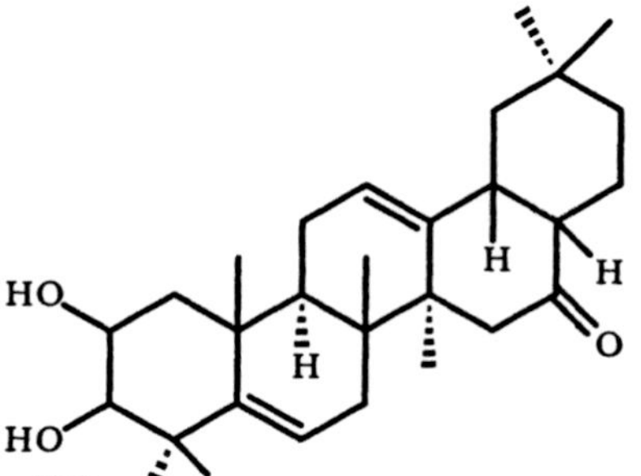

(22)
Mimusopgenone

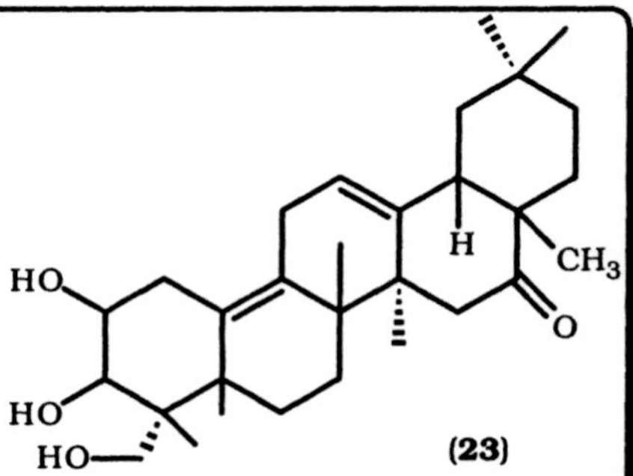

(23)
Mimugenone

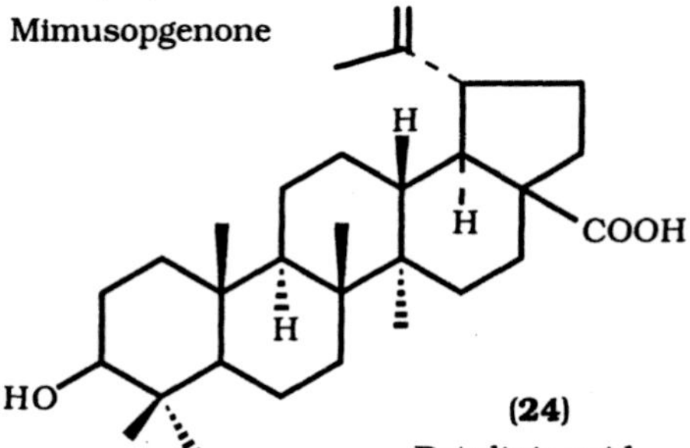

(24)
Betulinic acid

Saponins containing protobassic acid as the aglycone have been isolated from a number of plants [41-43]. Considerable attention is being given in recent years for evaluation of antitumor activity of triterpene acids and encouraging results in this regard have been published [44-46]. Betulinic acid (24), a pentacyclic triterpene has been identified as a melanoma-specific cytotoxic agent [47]. A variety of cellular responses indicated that the antitumor activity was mediated by the induction of apoptosis. It is noteworthy that betulinic acid is abundantly available from common natural sources and is inexpensive. The compound is currently undergoing preclinical development for the treatment or prevention of malignant melanoma. The triterpene acids (16) and (17) with new skeleton which are derivable from widely available protobassic acid containing saponins are of much interest to study their structure activity relationship.

SAPONINS

Acaciasides A, B and C

As an extension of chemical investigation on the pharmaceutically important naturally occurring saponins [48-57], Mahato and his co-workers isolated three acylated triterpenoid bisglycosides, acaciasides A (25), B (26) and C (27) from the water soluble saponin fraction obtained

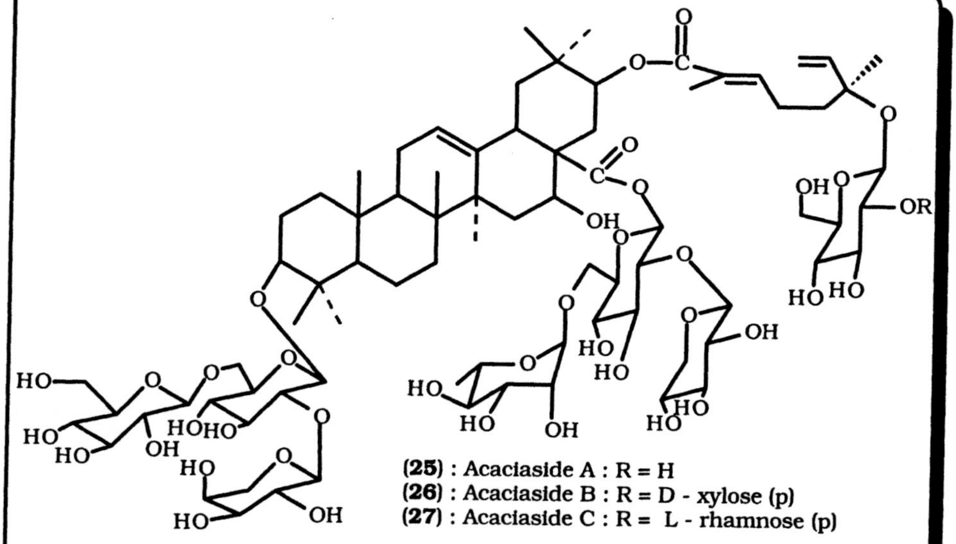

(25) : Acaciaside A : R = H
(26) : Acaciaside B : R = D - xylose (p)
(27) : Acaciaside C : R = L - rhamnose (p)

from the pericarps of *Acacia auriculiformis* A. Cunn. (Leguminosae)
[58,59]. This plant species which produces large amounts of fruits is
abundantly available throughout India. The saponins which are present as
a complex mixture are the main constituents of the pericarps of the plant.
The water soluble saponin fraction from the fruits exhibited strong
spermicidal property [60] as well as antifilarial activity [61]. The
structural features of the sponins were elucidated by a combination of fast
atom bombardment mass spectrometry, ^{1}H and ^{13}C NMR spectroscopy
and strategic chemical degradations. These useful biological properties of
the saponins as well as their easy availability from a widely occurring
plant source make them the ideal target for their large scale isolation and
beneficial exploitation. The four new sparingly water soluble triterpenoid
saponins isolated from the plant are acaciaside (**28**) [52], proacaciaside-I
(**29**), proacaciaside-II (**30**) and acaciamine (**31**) [62]. The last one is an
amino sugar containing saponin and such saponins rarely occur in the plant
kingdom.

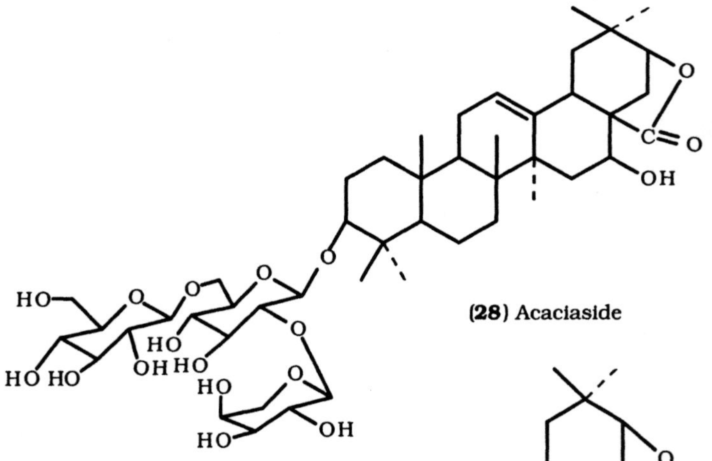

(28) Acaciaside

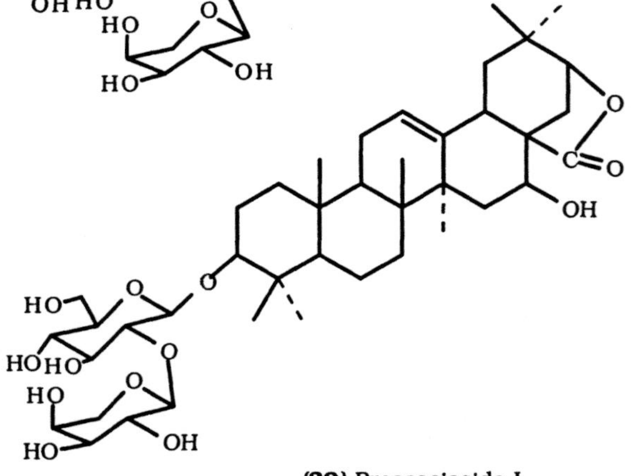

(29) Proacaciaside-I

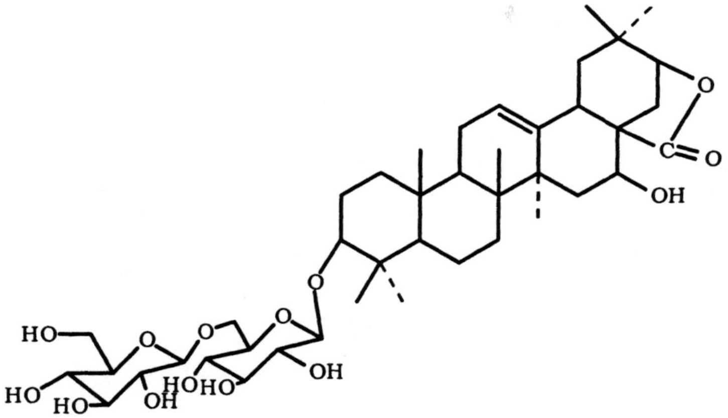

(30) Proacaciaside-II

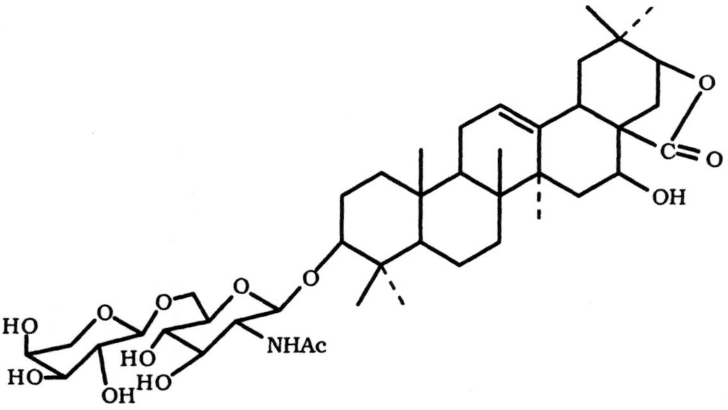

(31) Acaciamine

Saponins from *Bacopa monniera*

Bacopa monniera Wettst, popularly known as "Brahmi" is widely used as a nervine tonic for memory improvement, cardio tonic and diuretic in Indian traditional medicine [63]. The alcoholic leaf extract of the plant was reported to improve the performance of rats in various learning experiments as manifested by better acquisition, consolidation and retention of newly acquired behavioural responses [64-66]. The activities are believed to be associated with the polar fraction of the leaf extract, which contains mainly saponins as a complex mixture.

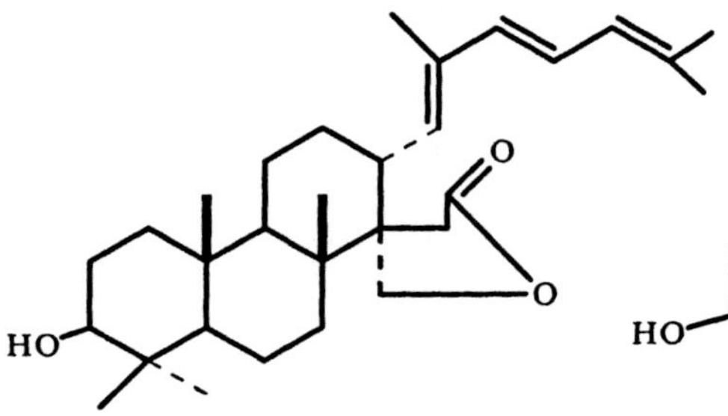

(32) Ebelin lactone

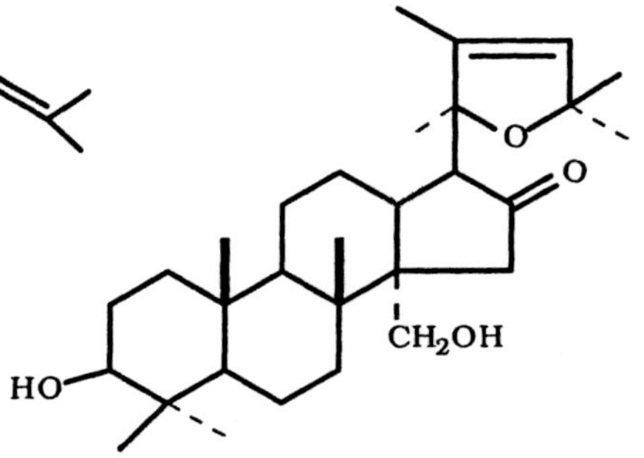

(33) Bacogenin A$_1$

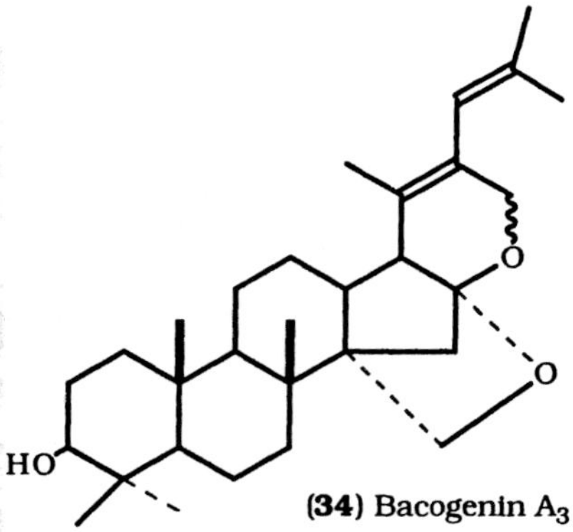

(34) Bacogenin A$_3$

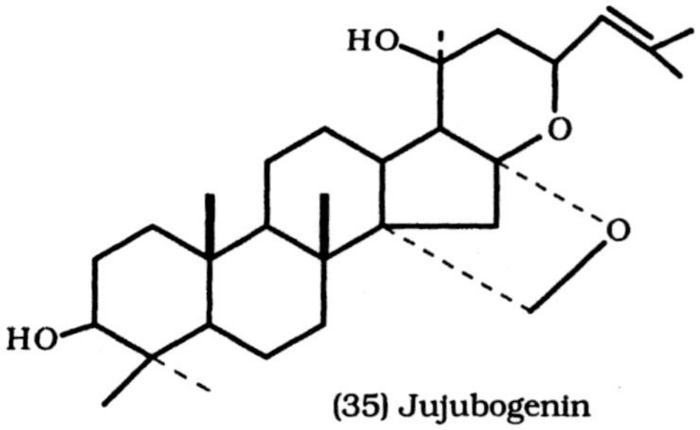

(35) Jujubogenin

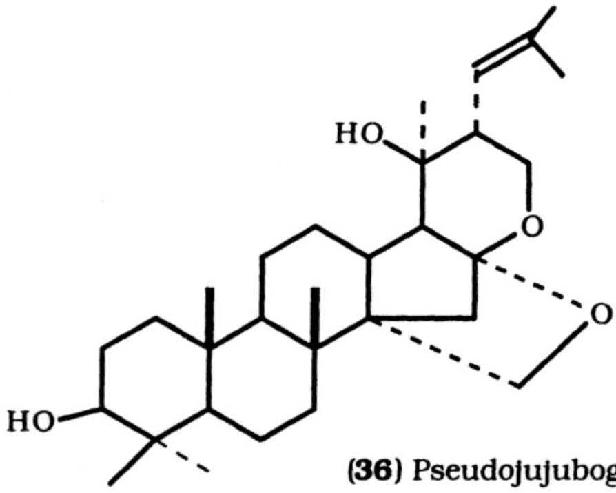

(36) Pseudojujubogenin

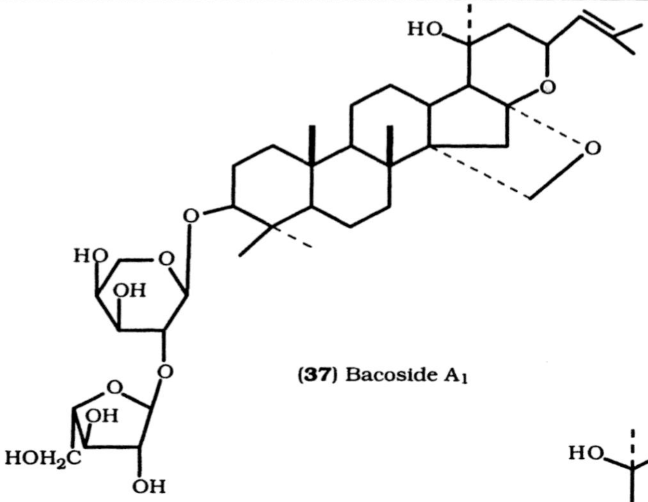

(37) Bacoside A$_1$

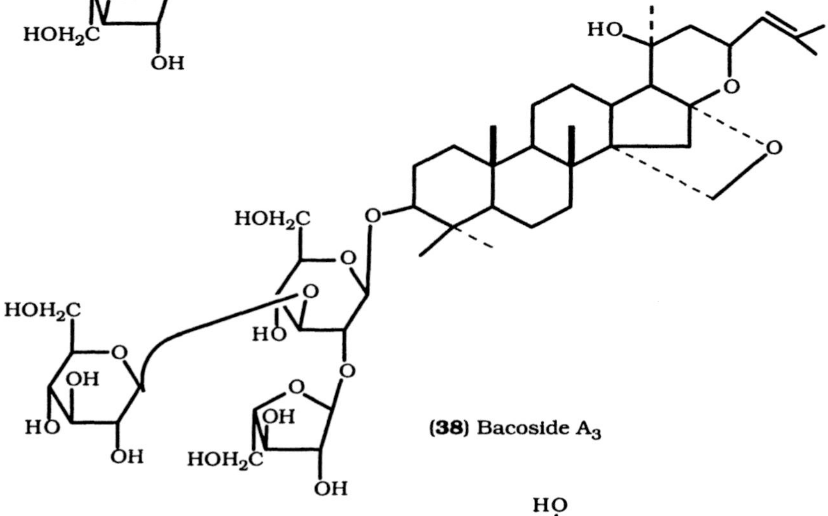

(38) Bacoside A$_3$

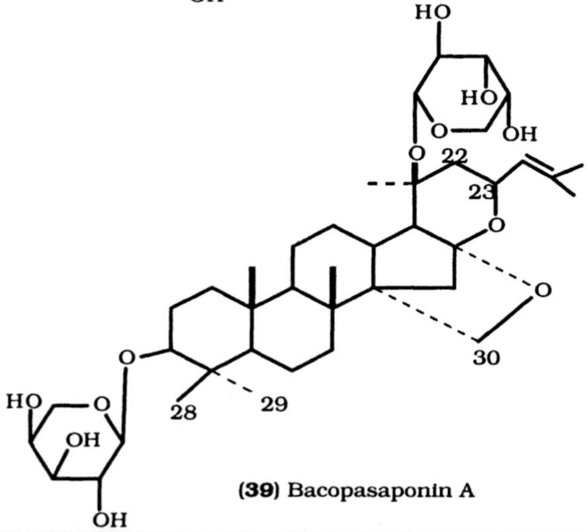

(39) Bacopasaponin A

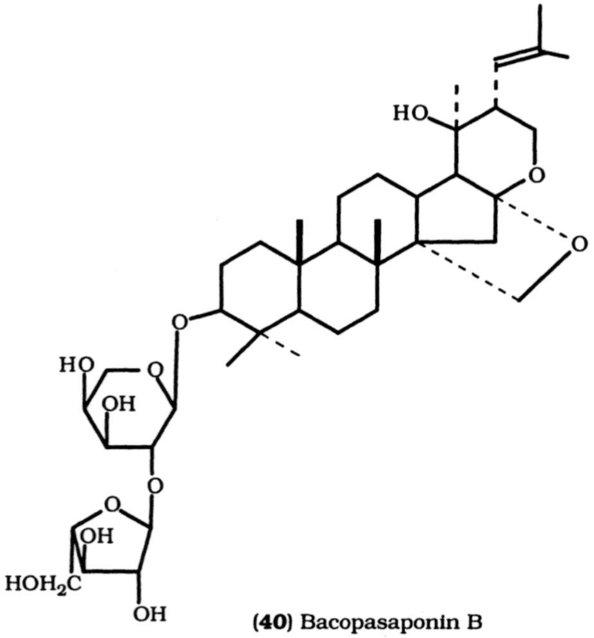

(40) Bacopasaponin B

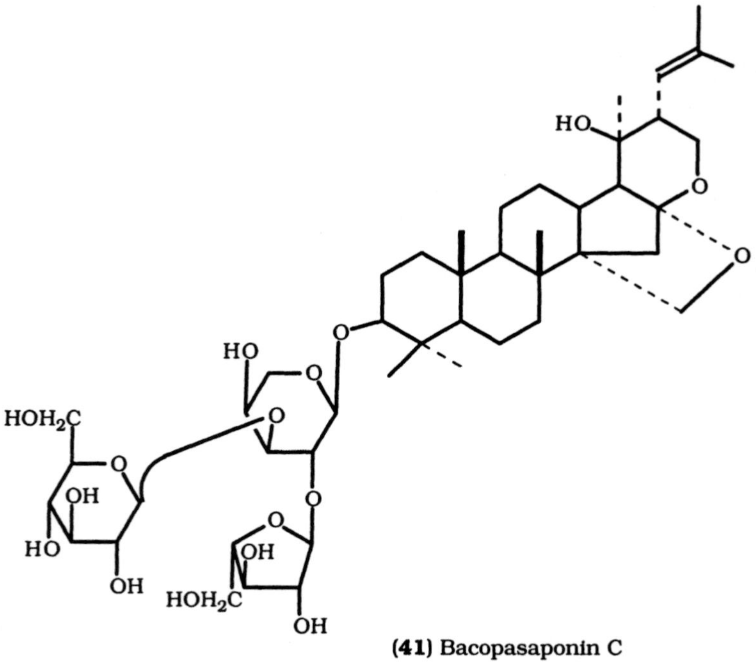

(41) Bacopasaponin C

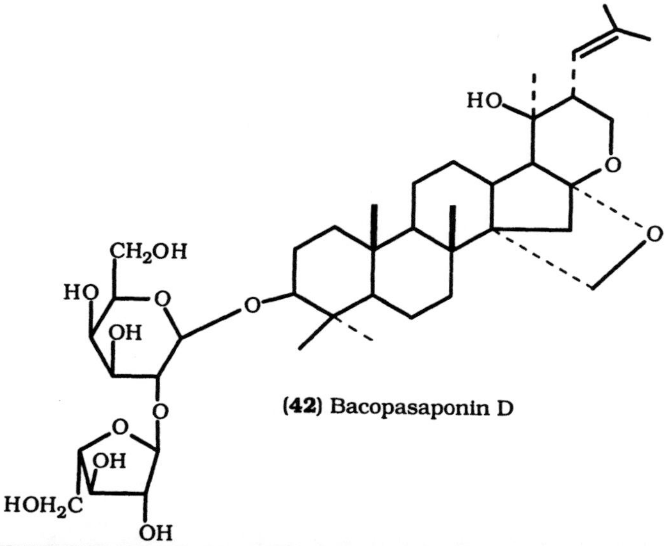

(42) Bacopasaponin D

In earlier chemical investigations the isolation was reported of three aglycones ebelin lactone (32) [67], bacogenin A₁ (33) [68] and bacogenin A₃ (34) [69] which are artefacts formed during acid hydrolysis, and two genuine sapogenins, jujubogenin (35) [70] and pseudojujubogenin (36) [71]. However, the structure of pseudojujubogenin was not completely elucidated, the configurations at its C-20 and C-22 remaining undetermined. The isolation of two jujubogenin saponins designated bacoside A₁(37) [72] and bacoside A₃(38) [73] was also reported. However, it is to be particularly mentioned that these two saponins are not the only saponin constituents present as is wrongly claimed in some commercial advertisements of the drug. It has been reported recently by the author's group [74,75] the isolation and structure elucidation of four new dammarane-type saponins, bacopasaponins A, B, C and D (39-42). Of these four saponins bacopasaponin A (39) is a jujubogenin bisarabinoside and the other three are pseudojujubogenin glycosides. The hitherto undetermined configurations at C-20 and C-22 of pseudojujubogenin were elucidated by phase-sensitive Rotating frame Overhauser Enhancement Spectroscopy (ROESY). The ROEs observed in phase-sensitive ROESY suggested 20(S) and 22(R) configurations. The ¹H and ¹³C signals of bacopasaponins A-D (39-42) were assigned with the help of DEPT, ¹H - ¹H COSY, HSQC and HMBC techniques which are shown in Tables 1-5. Thus the structures of the four saponins were elucidated unambiguously. Besides these four saponins, the presence of at least eight other saponins was disclosed by TLC and HPLC examination of the purified saponin mixture whose isolation and characterization are yet to be accomplished.

Acid hydrolysis of bacopasaponins yields artefacts instead of the genuine aglycones which are acid labile. While jujubogenin glycosides furnish ebelin lactone (32) as the major aglycone, pseudojujubogenin saponins liberate bacogenin $A_1(33)$ as the main sapogenin. The mechanism of acid catalyzed transformation of jujubogenin (35) to ebelin lactone (32) is rationalised as shown in Scheme 4. Dehydration and retro Diels' Alder collapse of the dihydropyran ring yields the lactone (32). The isolation of bacogenin A_3 (34) [69], the intermediate dehydrated product lends support to this proposition.

Table 1.   **^{1}H NMR Chemical Shifts of Bacopasaponin A (39)**

Proton		39	Proton		39
1	α	0.75(ddd, J=3.5,13.4,13.4Hz)	29		0.86(s)
	β	1.50(ddd, J=2.7,2.7,13.4Hz)	30	α	4.17(d,J=7.5Hz)
2	α	1.80(m)		β	4.04(dd,J=1.4,7.5Hz)
	β	2.10(dddd,J=2.7,3.5,4.4,13.9Hz)	3-0-Ara		
3		3.24(dd,J=4.4,11.7Hz)	1		4.70(d,J=7.0Hz)
4		-	2		4.33(dd,J=7.0, 8.8Hz)
5		0.62(dd,J=1.7,11.2Hz)	3		4.09(dd,J=2.2, 8.8Hz)
6	α	1.23(m)	4		4.25(m)
	β	1.36(m)	5	a	4.23(m)
7	α	1.40(m)		b	3.75(dd,J=2.5, 12.7Hz)
	β	1.26(m)			
8		-	20-0-Ara		
9		0.80(dd,J=2.4,12.4Hz)	1		4.78(d,J=6.6Hz)
10		-	2		4.29(dd,J=6.6, 8.6Hz)

Table 1). contd.....

Proton		39	Proton		39
11	α	1.82(*m*)	3		4.08(*dd,J*=2.4, 8.6Hz)
	β	1.17(*m*)	4		4.23(*m*)
12	α	1.67(*dddd,J*=3.8,13.2,13.2,13.2Hz)	5	a	4.14(*dd,J*=3.0,12.2Hz)
	β	1.77(*m*)		b	3.65(*dd,J*=2.0,12.2Hz)
13		2.95(*dddd,J*=1.4,4.4,5.4,13.2Hz)			
14		-			
15	α	2.32(*d, J*=8.5Hz)			
	β	1.33(*d, J*=8.5Hz)			
16		-			
17		1.31(*d, J*=5.4Hz)			
18		0.96(*s*)			
19		0.66(*s*)			
20		-			
21		1.35(*s*)			
22	α	1.97(*dd, J*=1.7,13.9Hz)			
	β	1.39(*dd, J*=11.0,13.9Hz)			
23		5.27(*ddd, J*=1.7,7.3,11.0Hz)			
24		5.43(*dsept, J*=7.3,1.3Hz)			
25		-			
26		1.62(*d, J*=1.3Hz)			
27		1.75(*d, J*=1.3Hz)			
28		1.17(*s*)			

Ara : arabinose

Table 2. ^{1}H NMR Chemical Shifts of Bacopasaponin B (40)

Proton		40	Proton		40
1	α	0.76(ddd,J=3.9,13.5,13.5Hz)	29		0.96(s)
	β	1.50(ddd,J=2.0,3.9,13.5Hz)	30	α	4.24(d,J=7.6Hz)
2	α	1.82(m)		β	4.16(d,J=7.6Hz)
	β	2.10(dddd,J=3.9,3.9,4.4,13.7Hz)	3-0-Ara		
3		3.24(dd,J=4.4,11.7Hz)	1		4.89(d,J=6.3Hz)
4		-	2		4.40(dd,J=6.3,8.0Hz)
5		0.65 (dd,J=1.7,11.7Hz)	3		4.18(dd.J=6.6,8.0Hz)
6	α	1.32(m)	4		4.26(m)
	β	1.45(m)	5	a	4.28(m)
7	α	1.47(m)		b	3.77(dd,J=3.4,12.7Hz)
	β	1.35(m)	Ara(f)		
8		-	1		6.10(d,J=1.2Hz)
9		0.85(dd,J=2.5,12.7Hz)	2		4.98(dd,J=1.2,2.7Hz)
10		-	3		4.81(dd,J=2.7,4.6Hz)
11	α	1.53(m)	4		4.89(m)
	β	1.33(m)	5	a	4.26(m)
12	α	1.75(m)		b	4.17(m)
	β	1.90(m)			
13		2.82(dddd,J=1.4,4.4,5.4,13.2Hz)			
14		-			
15	α	2.43(d,J=8.3Hz)			
	β	1.33(dd,J=1.2,8.3Hz)			
16		-			
17		1.63(dd,J=1.3,7.8Hz)			
18		1.05 (s)			

(Table 2) contd.....

Proton		40	Proton	40
19		0.72 (*s*)		
20		-		
21		1.36 (*s*)		
22		2.58(*ddd,J*=2.2,2.2,10.5Hz)		
23	α	3.85(*dd,J*=2.2,10.7Hz)		
	β	4.68(*dd,J*=2.2,10.7Hz)		
24		5.82(*dsept,J*=10.7,1.2Hz)		
25		-		
26		1.68(*d,J*=1.2Hz)		
27		1.60(*d,J*=1.2Hz)		
28		1.22 (*s*)		

Ara : arabinose, f : furanose

Table 3. ^{1}H NMR Chemical Shifts of Bacopasaponin C (41)

Proton		41	Proton		41
1	α	0.71(*ddd,J*=3.4,13.6, 13.6Hz)	29		1.00 (*s*)
	β	1.48(*m*)	30	α	4.21(*d,J*=7.6Hz)
2	α	1.82(*m*)		β	4.14(*d,J*=7.6Hz)
	β	2.10(*dddd,J*=3.4,3.4,4.1,13.7Hz)	3-0-Ara		
3		3.19(*dd,J*=4.1,11.7Hz)	1		4.76(*d,J*=6.8Hz)
4		-	2		4.43(*dd,J*=6.8,8.8Hz)
5		0.65(*brd,J*=11.5Hz)	3		4.18(*dd,J*=2.7,8.8Hz)
6	α	1.27(*m*)	4		4.47(*ddd,J*=2.7,2.7,2.7Hz)
	β	1.42(*m*)	5	a	4.16(*dd,J*=2.7,10.0Hz)

(Table 3) contd.....

Proton		41	Proton		41
7	α	1.47(*m*)		b	3.73(*dd,J*=2.7,10.0Hz)
	β	1.32(*m*)	Ara(f)		
8		-	1		6.00(*d,J*=2.7Hz)
9		0.80(*dd,J*=2.0,12.6Hz)	2		4.93(*dd,J*=2.7,5.1Hz)
10		-	3		4.77(*dd,J*=5.4,7.3Hz)
11	α	1.47(*m*)	4		4.71(*ddd,J*=3.4,3.7,7.3Hz)
	β	1.28(*m*)	5	a	4.25(*dd,J*=3.4,12.0Hz)
12	α	1.70(*m*)		b	4.16(*dd,J*=3.7,12.0Hz)
β		1.86(*m*)	Glc		
13		2.78(*ddd,J*=4.6,5.4,12.7Hz)	1		5.04(*d,J*=7.8Hz)
14		-	2		3.89(*dd,J*=7.8,9.0Hz)
15	α	2.38(*d,J*=8.3Hz)	3		4.14(*dd,J*=9.0,9.0Hz)
	β	1.46(*brd,J*=8.3Hz)	4		4.04(*dd,J*=9.0,9.5Hz)
16		-	5		3.87(*ddd,J*=2.4,5.6,9.5Hz)
17		1.58(*brd,J*=7.8Hz)	6	a	4.40(*dd,J*=2.4,11.7Hz)
18		1.02(*s*)		b	4.17(*dd,J*=5.6,11.7Hz)
19		0.65(*s*)			
20		-			
21		1.33(*s*)			
22		2.60(*ddd,J*=2.2,2.2,10.2Hz)			
23	α	3.79(*dd,J*=2.2,10.7Hz)			
	β	4.63(*dd,J*=2.2,10.7Hz)			
24		5.77(*dsept,J*=10.2,1.2Hz)			
25		-			
26		1.64(*d,J*=1.2Hz)			
27		1.58(*d,J*=1.2Hz)			
28		1.21(*s*)			

Ara : arabinose, Glc : glucose; f: furanose

Table 4. ^{13}C **NMR Chemical Shifts of Bacopasaponins A (39) and B (40) in Pyridine-d5**

Carbon	39	40	Carbon	39	Carbon	40
1	38.3	38.8	3-0-Ara		3-0-Ara	
2	26.7	26.7	1	107.3	1	105.8
3	88.6	88.9	2	72.9	2	76.3
4	39.7	39.6	3	74.6	3	83.1
5	56.2	56.2	4	69.4	4	68.5
6	18.3	18.3	5	66.6	5	65.4
7	36.0	36.1	20-0-Ara		Ara(f)	
8	37.5	37.5	1	98.8	1	109.9
9	53.0	53.1	2	73.1	2	81.2
10	37.3	37.3	3	75.1	3	78.8
11	21.8	21.8	4	69.3	4	88.1
12	28.4	28.6	5	66.6	5	62.8
13	36.1	37.2				
14	53.7	53.5				
15	37.4	37.0				
16	110.2	110.3				
17	55.2	51.4				
18	18.8	18.9				
19	16.3	16.3				
20	75.8	71.9				
21	25.1	27.2				
22	41.6	46.3				
23	68.8	66.1				
24	127.4	124.2				
25	133.8	132.9				
26	25.7	26.1				
27	18.3	18.5				
28	28.1	28.1				
29	16.7	16.7				
30	65.9	65.9				

Table 5. ^{13}C NMR Chemical Shifts of Bacopasaponins C (41) and D (42) in Pyridine-d$_5$

Carbon	41	42	Carbon	41	Carbon	42
1	38.7	38.9	3-0-Ara		Glc(p)	
2	26.6	26.8	1	105.3	1	105.9
3	88.6	89.1	2	76.7	2	78.3
4	39.7	39.7	3	83.1	3	89.1
5	56.1	56.2	4	68.2	4	71.9
6	18.2	18.3	5	65.4	5	78.2
7	35.9	36.1	Ara(f)		Ara(f)	
8	37.4	37.6				
9	52.9	53.0	1	110.0	1	109.7
10	37.1	37.2	2	83.5	2	80.9
11	21.6	21.8	3	77.7	3	78.8
12	28.4	28.7	4	84.7	4	88.5
13	37.0	37.2	5	61.9	5	62.9
14	53.3	53.6				
15	36.8	37.0	Glc			
16	110.2	110.4	1	104.6		
17	51.1	51.4	2	74.9		
18	18.7	18.9	3	77.6		
19	16.2	16.3	4	71.2		
20	71.7	71.9	5	78.1		
21	26.8	27.2	6	62.2		
22	45.9	46.3				
23	66.0	66.2				
24	124.0	124.2				
25	132.9	132.9				
26	25.9	26.1				
27	18.3	18.5				
28	27.7	28.1				
29	16.4	16.9				
30	65.8	63.0				

The mechanism of formation of bacogenin $A_1(33)$ from pseudojujubogenin (36) is proposed as shown in Scheme 5. In (36) the 20-hydroxy group and the isobutenyl side chain are in a geometrically favourable disposition to cyclise to tetrahydrofuran ring. Thus, initial cyclization followed by cleavage of the 16-ketal group lead to the formation of aglycone(33).

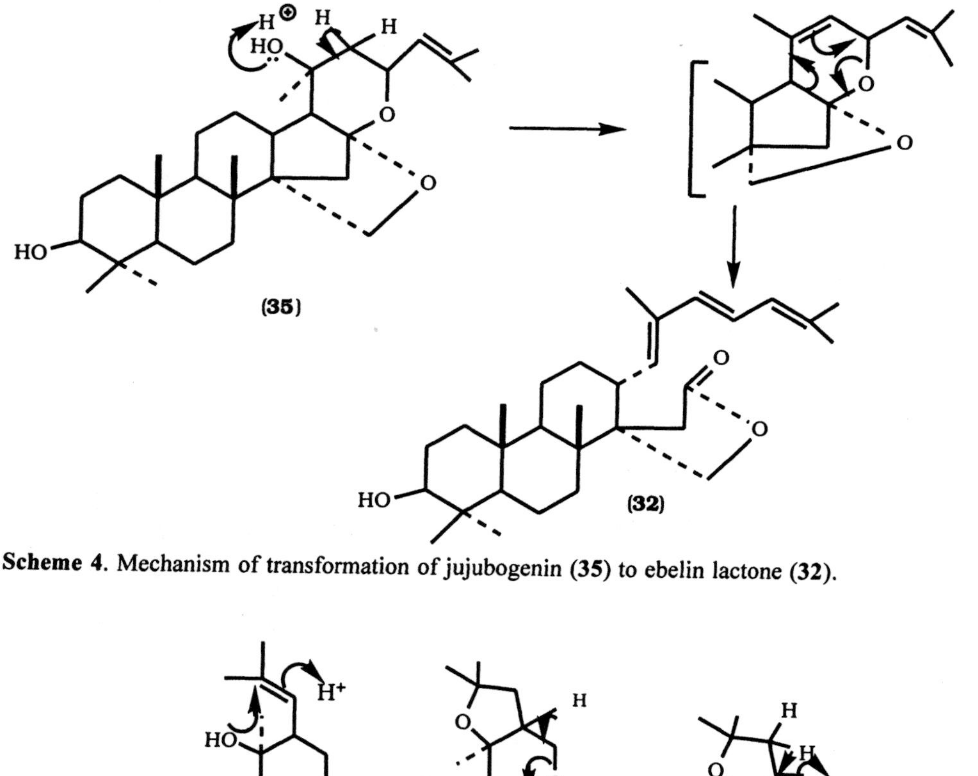

Scheme 4. Mechanism of transformation of jujubogenin (35) to ebelin lactone (32).

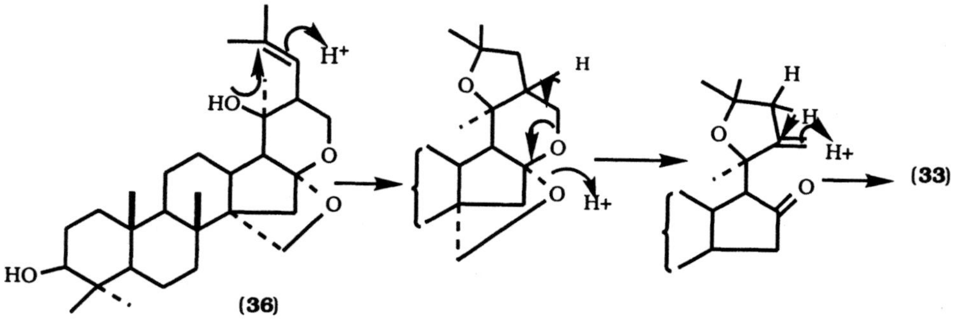

Scheme 5. Mechanism of formation of bacogenin A1 (33) from pseudojujubogenin (36).

Saponins of *Corchorus acutangulus* Lam

Corchorus acutangulus Lam. (Syn. *C. aestuans* Linn. (Tiliaceae) is a medicinal plant occurring throughout the hotter parts of India. Unlike its sister species *C. capsularis* and *C olitorious* which yield the jute of commerce in India, it is a wild species rarely used for extracting fibre [76]. Eight new saponins corchorusins A (**43**), B (**44**), C (**45**) and D (**46**) [49] and C_1 (**47**), D_1 (**48**), D_2 (**49**) and D_3 (**50**) [77] were isolated and

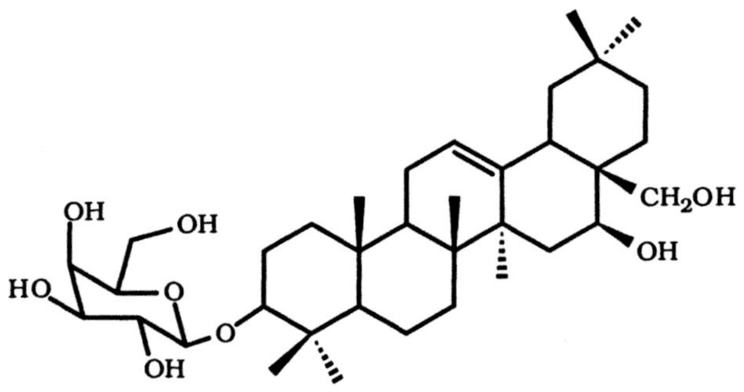

(**43**) Corchorusin A

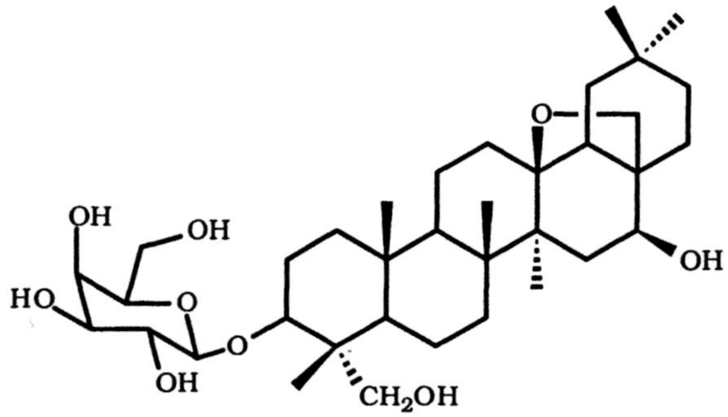

(**44**) Corchorusin B

characterised from <u>n</u>-BuOH-soluble fraction of MeOH extract of the leaves of the plant. Corchorusins B (**44**), D (**46**), C_1 (**47**), D_1 (**48**) and D_3 (**50**) possess saikogenins F, E, C, B and C respectively as their aglycones and are saikosaponins-like compounds. It is noteworthy that saikosaponins containing saikogenins as aglycones and different combinations of fucose, glucose and rhamnose have been reported so far only from medicinally used *Bupleurum* species (Umbelliferae) occurring in Japan, China and Korea [78,79]. Some of the saikosaponins have been reported to have antiviral [80], antiinflammatory [81], haemolytic [82] and plasma-cholesterol lowering [83] activities. *C. acutangulus* growing wild in India appears to be the only alternative plant of different genus and family so far which contains saikosaponins-like compounds. The activity profile of the corchorusins seems to be interesting because of their structural similarity with the saikosaponins.

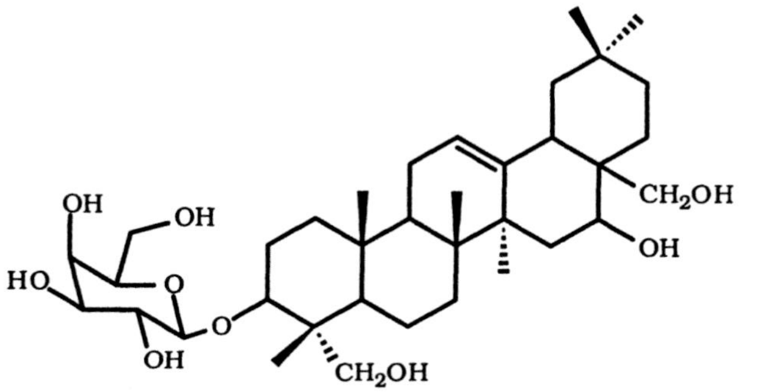

(45) Corchorusin C

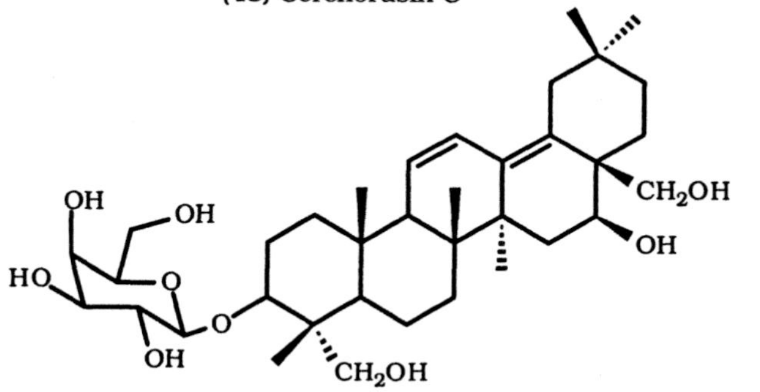

(46) Corchorusin D

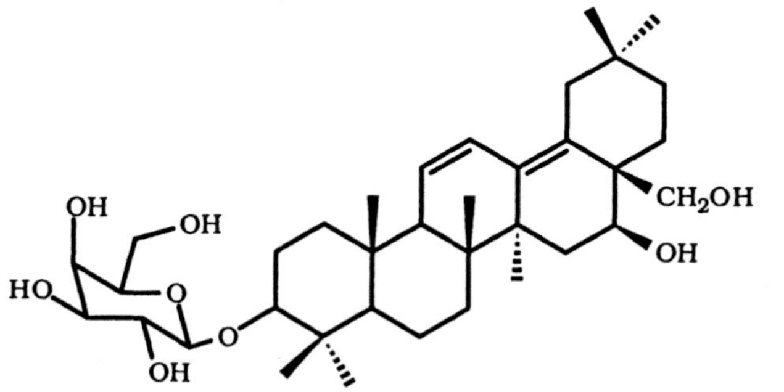

(47) Corchorusin C_1

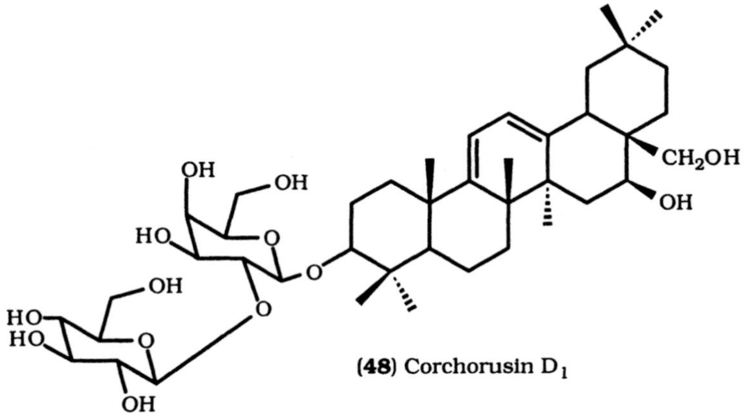

(48) Corchorusin D$_1$

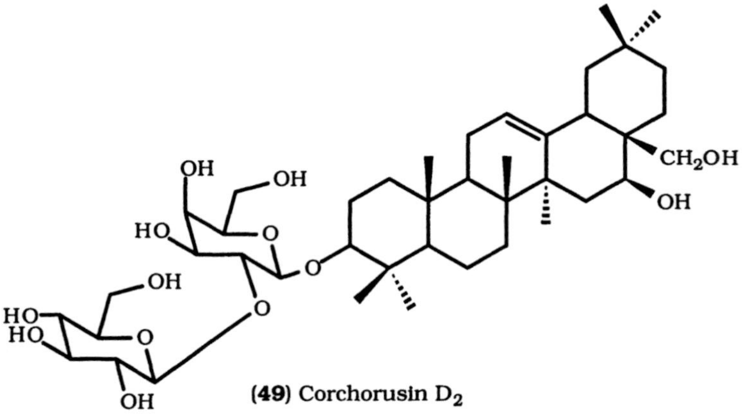

(49) Corchorusin D$_2$

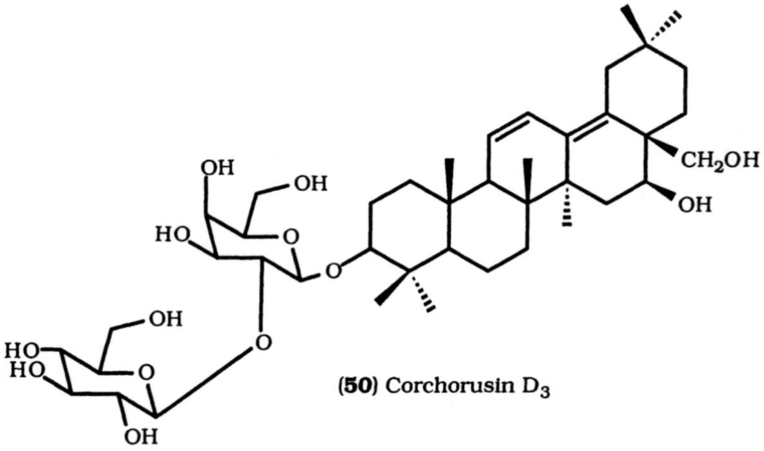

(50) Corchorusin D$_3$

Acid-catalysed Rearrangement of Some Aglycones

The saponins are composed of aglycones and sugar moieties which are recovered after hydrolysis and separately investigated for structure elucidation. Acid hydrolysis is often used. However, if a saponin contains a acid labile aglycone, artefacts are obtained. Arjunglucoside-1 (51) [84] on alkaline hydrolysis yielded arjungenin (52). Acid hydrolysis of the glucoside, however, generated a major aglycone which was characterised as tomentosic acid (53) by spectroscopic and X-ray analysis [85]. The acid (53) was isolated by Row and Rao [86] from *Terminalia tomentosa* not as a sapogenin but as a naturally occurring triterpene. In an attempt to

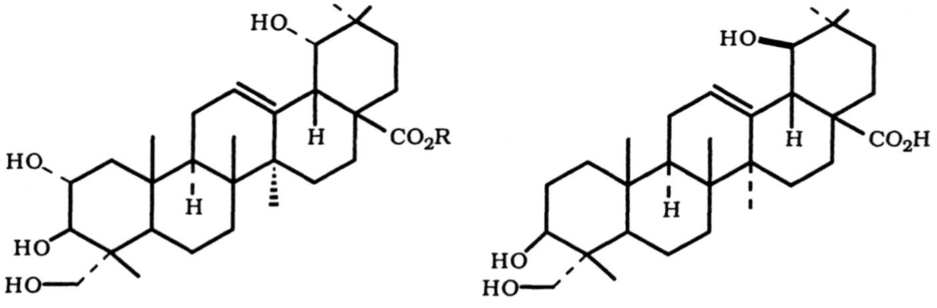

(51) Arjunglucoside - 1 : R = D - glucose (p)
(52) Aujugenin : R = H

(53) Tomentosic acid

elucidate the mechanism of transformation of arjungenin(52) to tomentosic acid (53) it was observed that although arjungenin (52) on being boiled with 5% methanolic hydrocholic acid produced tomentosic acid (53), the methyl ester of arjungenin was recovered unchanged after similar acid treatment. Consequently, the epimerisation of 19α-OH was reasonably thought to be *via* lactonisation. Isolation of the intermediate lactone (54) strongly supported this presumption and mechanism of the transformation was rationalised [85] as shown in Scheme 6. Aoki and Suga reported [87] acid catalysed epimerisation of the 16β-OH group of cochalic acid. They ascribed this phenomenon to thermodynamic stability. However, longispinogenin glycoside [49] which contains a 28-CH$_2$OH instead of a 28-COOH yielded, on acid hydrolysis, the genuine aglycone longispinogenin instead of its 16-epimer, primulagenin A. This experiment demonstrated that epimerisation of the 16β-OH group to 16α-OH does not take place on prolonged treatment with acid if there is no 28-COOH group available to form the 28 → 16 lactone. Inspection of Dreiding model also suggested that the 16α-OH (axial) group is not thermodynamically more stable than the 16β-OH(equatorial) group. As such this epimerisation was also suggested to occur *via* lactonisation between 28-COOH and 16β-OH as shown in Scheme 7. Thus in both cases the kinetically controlled products are formed and it was suggested that where the lactonisation and delactonisation are possible in a triterpene the kinetically controlled product is obtained exclusively [85].

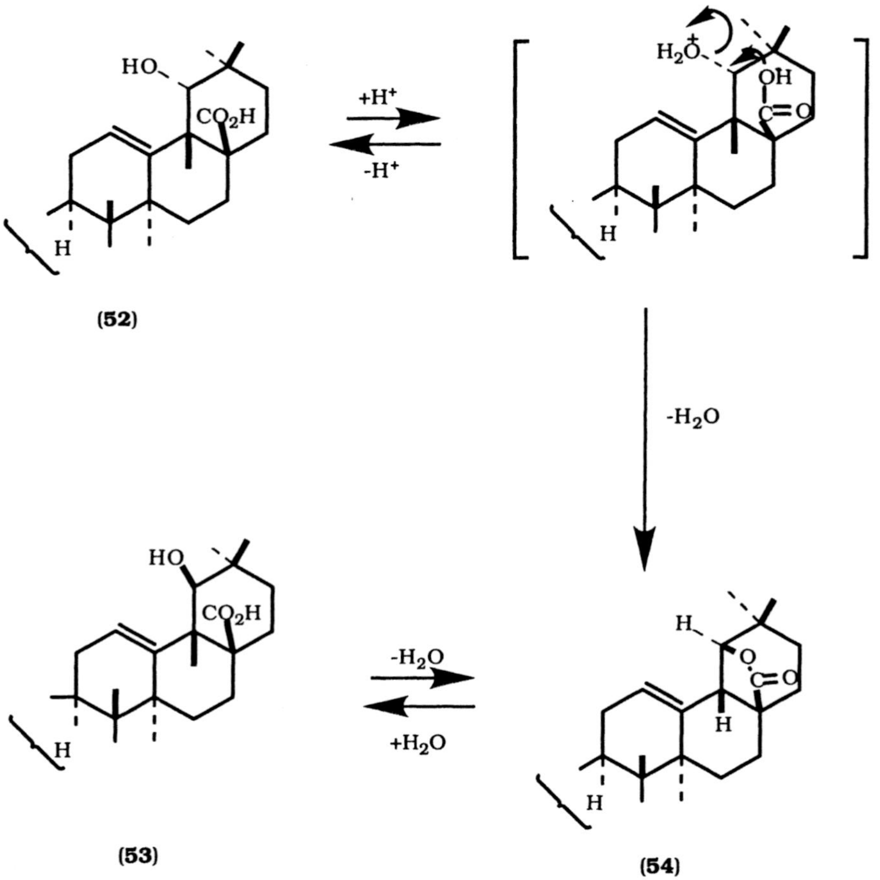

Scheme 6. Mechanism for transformation of arjungenin (52) to tomentosic acid (53).

During isolation of sapogenols by acid hydrolysis of the saponins from *Medicago hispida* [54] it was observed that soyasapogenol B (55) which is a genuine aglycone is converted to soyasapogenol D (56) and soyasapogenol F (57). Jurzysta [88] and Price *et al.* [89] reported that compounds (56) and (57) are artefacts of compound (55). However, the mechanism of formation of these products which apparently seemed to be

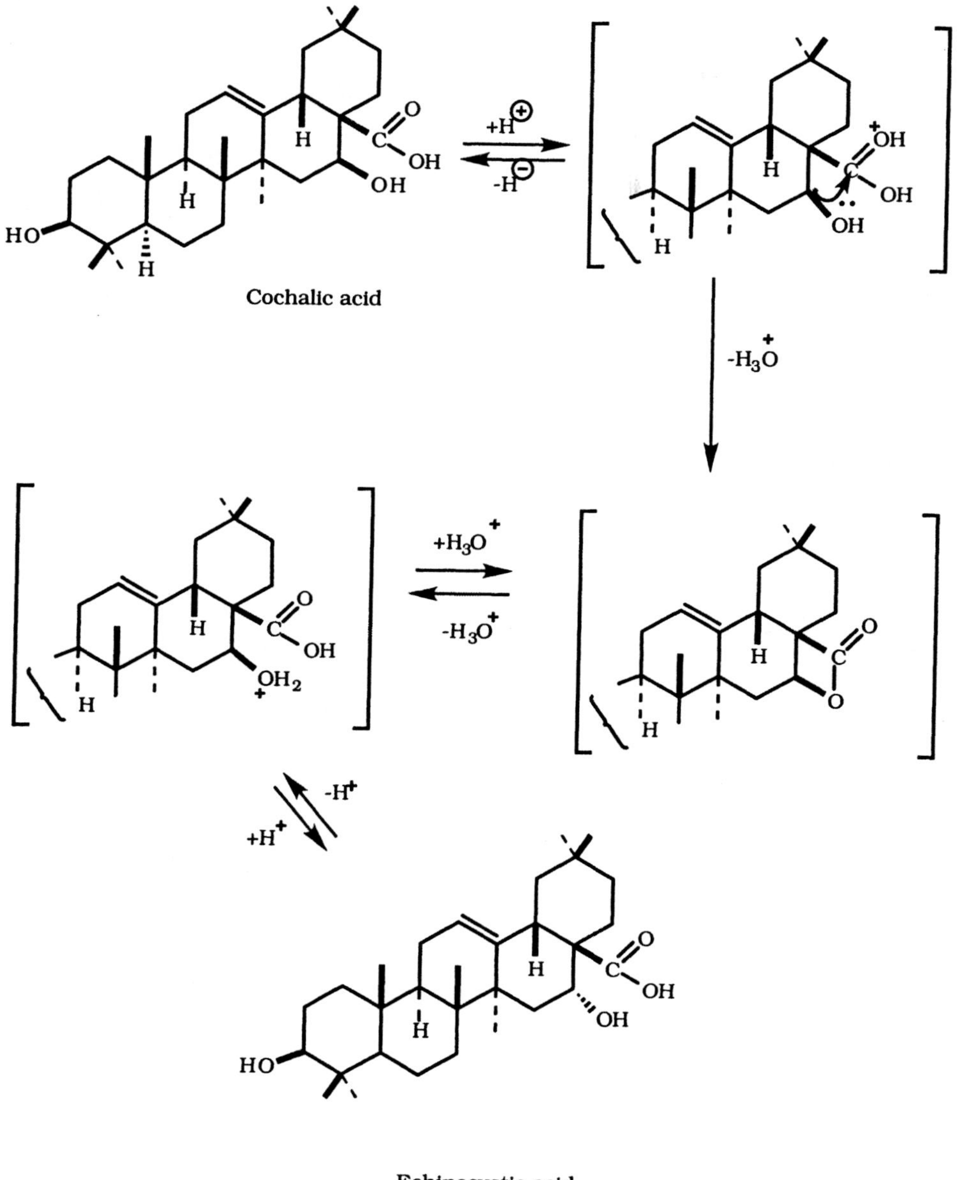

Cochalic acid

Echinocystic acid

Scheme 7. Proposed mechanism for transformation of cochalic acid to echinocystic acid.

rather unusual had not been rationalised. It was observed [49] that longispinogenin or its glycosides corchorusins A and D_2 containing 16-hydroxy group did not yield any $\Delta^{13(18)}$ - aglycone on similar treatment with aqueous methanolic hydrochloric acid. It was, therefore, indicated that the 22β(axial)-hydroxy group present in soyasapogenol B (**55**) or its glycosides might have some role in inducing this transformation. Assuming

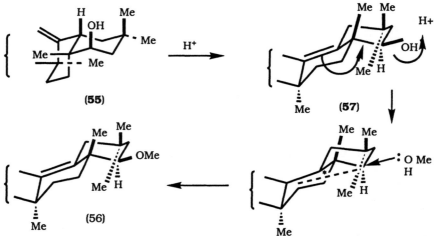

all-chair conformation of (**55**) having D/F *cis*-fusion, inspection of its Dreiding model disclosed that the 22β -axial hydroxyl experiences strong steric interaction with the 20β-axial methyl, in addition to that with the 17-methyl group. Migration of the double bond from the 12:13 to 13:18 position transforms the 22β-hydroxyl from axial to equatorial orientation, thereby releasing the 1,3-diaxial interaction. Substitution of the 22β-hydroxy group in (**57**) with a methoxy group may then occur as shown in Scheme 8 *via* (a) its protonation, (b) elimination of the protonated group through participation of the 13:18 double bond which is favourably disposed to form a protonated cyclopropane as intermediate and (c) attack by a molecule of methanol on the intermediate carbocation, to generate a methoxy derivative with retention of configuration [54].

Scheme 8. Mechanism of transformation of soyasapogenol B (**55**) to soyasapogenol D (**56**) and soyasapogenol F (**57**).

An Overview on Saponins

The saponins occurring widely in nature possess a broad spectrum of biological properties [90-96]. Although the saponins are highly toxic when given intravenously to higher animals, their toxic effects are very much lower when they are administered orally [97]. Many of the most common saponins of foods and feedingstuffs are apparently without significant oral toxicity. However, saponins from non-edible plants are known to be highly toxic to man and other species. Genseng, the widely known plant drug, has been used as an expensive traditional medicine and the genseng saponins have been the subject of much discussion [98-101]. The root extract of the plant has been used for centuries for increasing mental efficiency recovering physical balance and stimulating metabolic function in oriental countries. The possible antiaging effect of genseng stem-leaf saponins was studied [102,103] and the results indicated that the antiaging activity of the saponins is related to its free radical scavenging effect. There are evidences which show that saponins from a number of plant species can reduce plasma cholesterol levels in human. A recent review deals with hypocholesterolemic effects of dietary saponins and mechanism of these effects [104]. There is little doubt that saponins can be included in human diets at level, which give a beneficial effect but do not entail a risk of acute toxicity. Saponins of some medicinal plants exhibit diverse biological activities. For example the saponins from *Gymnema sylvestre*, a reputed Indian medicinal plant, are attracting much attention for their varied beneficial biological activities [105]. A recent review [106] incorporates the host of biological activities of triterpenoid saponins along with newer trends in isolation and structure elucidation reported during the period late 1989 - mid 1996. Although the saponins often occur as complex mixtures, it is possible sometimes to isolate individual major saponins by simplified nonchromatographic methods. For example, a simplified method for isolation of asiaticoside, a potential antileprotic agent [107] and whose complete structure has been determined by single cystal X-ray analysis [108] has been patented [109].

Macrocyclic Spermidine Alkaloids

Macrocyclic spermidine and spermine alkaloids which belong to the class of polyamine conjugates are comparatively a new group of plant bases. These are formed by a polyamine (spermidine or spermine) and a properly functionalised aliphatic or cinnamoylic acid or one of their derivatives. The free bases putrescine, spermidine and spermine are believed to play important roles in cellular differentiation and proliferation [110]. The polyamine levels in dividing cells, e.g., cancer cells are much higher than the resting cells [111] and this phenomenon has been exploited in chemotherapy and diagnostic test for malignant tumors [112,113]. Since early 1970s many unique polyamine conjugates have been isolated which possess diverse biochemical profiles [114].

We isolated three spermidine alkaloids designated caesalpinine A (**58**) [115], B (**59**) [116] and C (**60**) [117] from the leaves of *Caesalpinia digyna* Rottl. (Leguminosae). The plant is a prickly scandent shrub growing in eastern India, Burma, and Ceylon. It is reputed for its use in phthisis, scrofula, and diabetes [118]. caesalpinine B (**59**) and C (**60**) were found to be identical with celacinnine and celallocinnine isolated from *Maytenus arbutifolia* [119]. Caesalpinine A (**58**) turned out to be a novel macrocyclic spermidine alkaloid with a new skeletone. The structure and stereochemistry of the alkaloid were determined by mass, NMR and single-cystal X-ray analysis. It possesses a 13-membered lactam ring fused to a five-membered lactam ring. It also contains two monosubstituted benzene rings. The X-ray investigations of (**58**) demonstrated the presence of two independent molecules linked by a hydrogen bond in the assymmetric unit. In spermidine alkaloids the formation of a 13-membered lactam ring is a normal phenomenon. However, the unique structural feature of caesalpinine A (**58**) is its 5-membered lactam ring formed by an interesting biogenetic pathway.

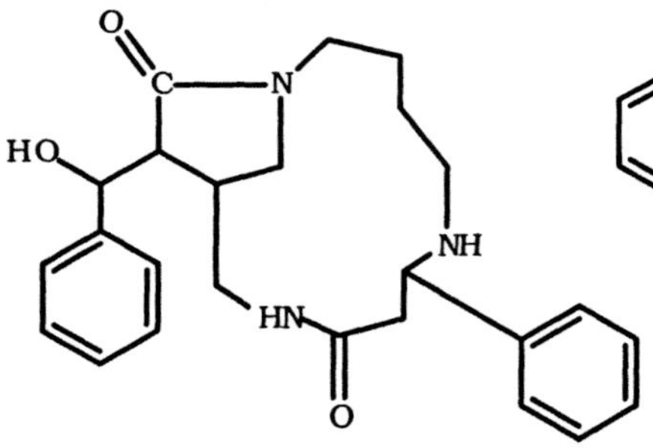

(**58**) Caesalpinine A 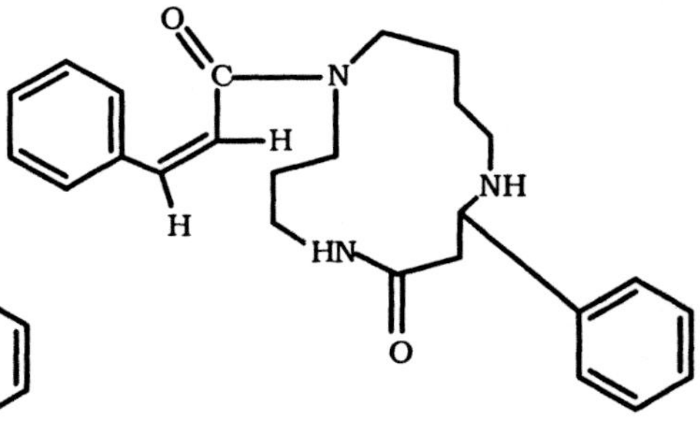   (**59**) Caesalpinine B

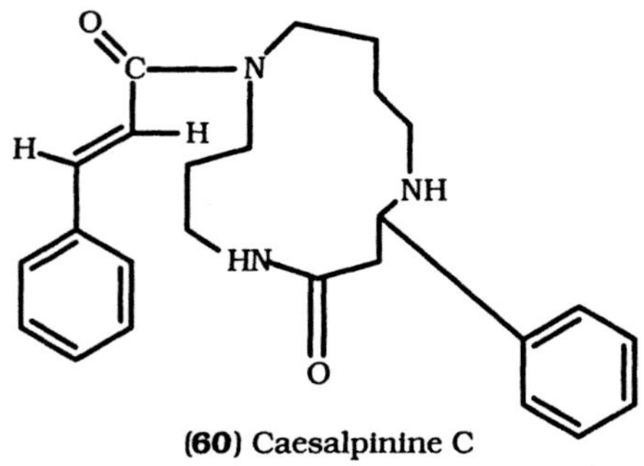

(**60**) Caesalpinine C

A proposition for the biosynthesis of caesalpinine A (58) envisages participation of two cinnamic acid moieties and a dehydrospermidine. Periphylline, a dehydrospermidine moiety containing alkaloid has, in fact, been isolated from a plant source [120]. The C-C bond is believed to arise *via* protonation of the dehydrospermidine as proposed in Scheme 9 and subsequent attack by the cennamoyl bond with concomitant hydroxylation.

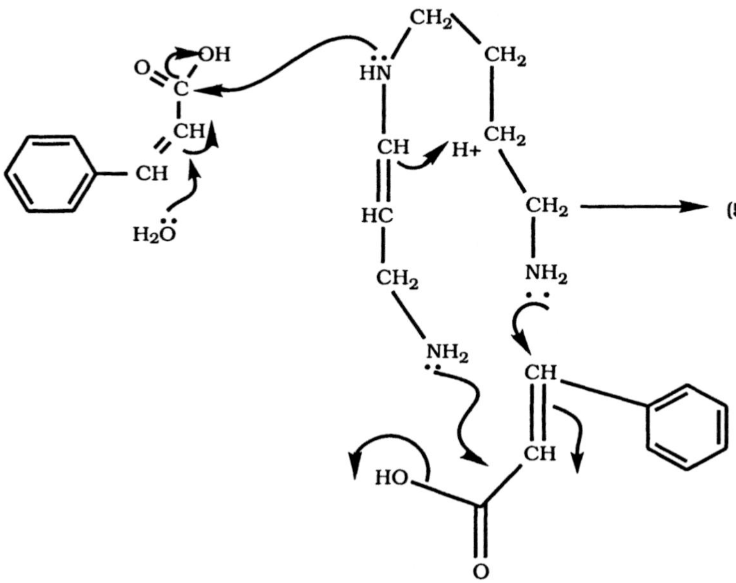

Scheme 9. A proposition for the biosynthesis of caesalpinine A (58).

Another proposition envisages involvement of one molecule of cinnamic acid, a molecule each of benzoylacetic acid and an appropriately substituted spermidine rather than spermidine itself. Elimination of the spermidine substituent (possibly involving the aziridinium ion) is followed by an attack from the carbanion derived from benzoylacetic acid with the formation of a new C-C bond. The other reactions involved in the biogenesis are of well-established nature *viz*, Michael addition, carbonyl reduction and amide formation. The carbonyl reduction step (and possibly even the cyclization to the γ −lactam) necessarily follows the C-C bond formation and the generation of the other C-N bonds may either precede or succeed the above step (Scheme 10).

Flavonoid-, and Phenylpropanoid Glycosides

Lantana camara Linn. Var *aculeata* (Verbenaceae) is a woody straggling plant with a number of flower colours viz. red, pink, white, yellow and

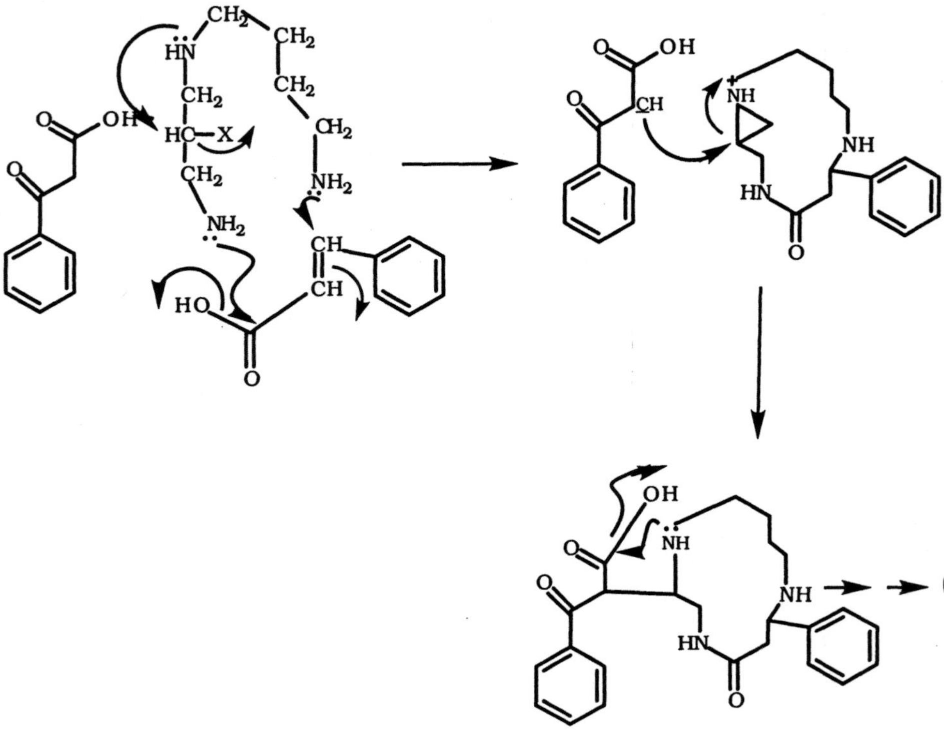

Scheme 10. An alternative proposition for the biosynthesis of caesalpinine (**58**).

violet. The plant has encroached upon a vast expanse of pastures, orchards and forest areas in many tropical and subtropical parts of the world [121,122]. The plant has the notoriety of causing hepatotoxicity and photosensitization in grazing animals [121-123]. It has also the reputation of being used in traditional medicine [124]. Considerable phytochemical work on this plant mainly on nonpolar constituents led to the isolation of a number of triterpenoids, a partially methylated flavonoid, umuhengerin and essential oils which have been covered in an excellent review [125]. The plant also contains allelochemicals possessing insecticidal, pesticidal and weedicidal activities which have been suggested to be polar [126]. Herbert and his co-workers [127] recently isolated from the plant, verbascoside (**61**), a polar phenylpropanoid glycoside possessing antimicrobial, immuno-suppressive and antitumor activities [128-130].

Mahato *et al.* [131] very recently isolated besides verbascoside (**61**) and a flavone glycoside (**62**), a novel flavonol glycoside, camaraside (**63**) and a new phenylpropanoid glycoside lantanaside (**64**) from the polar fraction of the MeOH extract of the leaves of the plant. The glycoside (**62**) was previously isolated from *Eria ravanica* [132]. Lantanaside (**64**) contains *cis*-caffeoyl moiety instead of trans-caffeoyl as in verbascoside (61) (Chart-3).

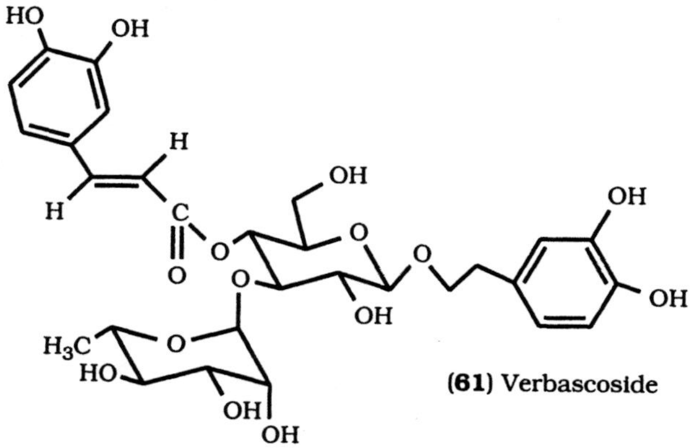

(61) Verbascoside

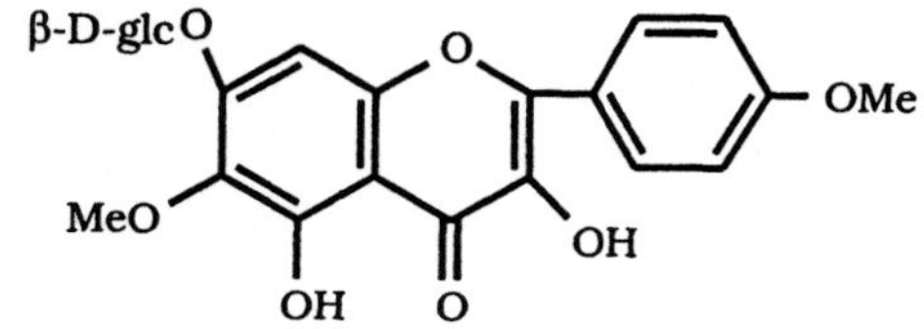

(63) Camaraside

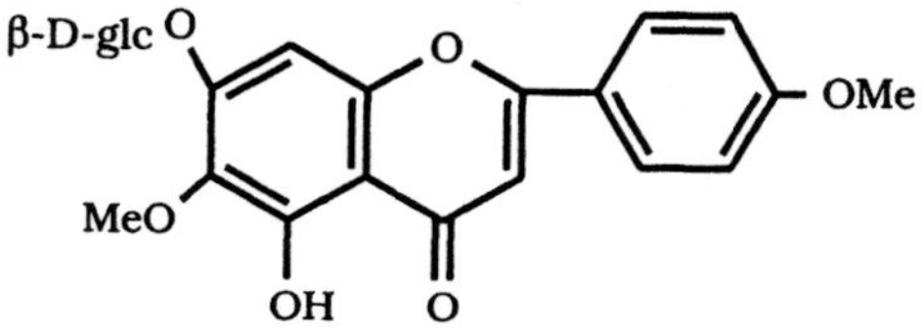

(62) Pectolinarigenin- 7- O - glucoside

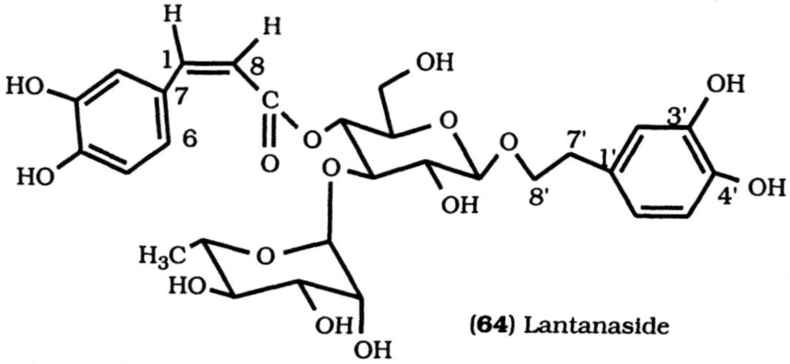

(64) Lantanaside

Chart 3. Polar constituents from *Lantana camara*.

Vervascoside isolated from *L. camara* has been disclosed to be an inhibitor of protein kinase C (PKC) which plays a key role in cellular growth and differentiation [133]. The compound also showed a potent antiproliferative effect *in vitro* against L-1210 cells and the authors [127] suggested for exploring the possibility of using verbascoside for therapeutic intervention in a wide range of disease including cancer, inflammation and immune disorders. Moreover, flavonol glycosides have also been reported to have inhibitory effects on 12-0-tetradecanoylphorbol-13-acetate (TPA)-induced tumor promotion and the effects have been attributed at least partly to activation of immune responses against tumors [134]. Yasukawa *et al.* also reported inhibitory effect of flavonol glycosides on TPA-induced inflammation in mice [135]. These reports strongly suggest that the flavone and flavonol glycosides reported [123] may also be responsible for the antitumor and possibly other medicinal properties of *L. camara* [124]. The reports of investigation on the plant thus far available suggest the strong possibility of beneficial utilization of the huge biomass of *L. camara* available.

PERSPECTIVES

The advances in extraction technology, separation science, and analytical and spectroscopic techniques have greatly increased the prospect of discovery of complex natural products of novel carbon frameworks possessing useful biochemical profiles. Natural products are also expected to play a significant role in agriculture as pesticides and herbicides. Insect, pest or weed control by using synthetic chemicals results in toxic residues in soil and plants. The crops contaminated with such residues are health hazard and as such natural plant chemicals will undoubtedly play a greater role in the future of pest control. A number of chemical companies are evaluating plant extracts and chemicals for the commercial use in agriculture. The complex structures of many plant chemicals preclude their economical chemical synthesis. However, genetically engineered synthetic technique has the potential to solve this problem. In fact, the feasibility of multi-enzyme, one-flask total synthesis of natural products has already been demonstrated by the preparation of tetramethyl corphinoid structure and of the alkaloid strictosidine by gene transfer and overexpression of the corresponding gene products in *E. coli* [136]. Continued attempts to synthesize complex natural products by the application of this technology is anticipated.

REFERENCES

[1]   Kingston, D.G.I.; Molinero, A.A.; Rimoldi, J.M. *Prog. Chem. Org. Nat. Prod.*, **1993**, *61*, 156.
[2]   Klayman, D.L. *Science*, **1985**, *228*, 1049.
[3]   Nakanishi, K. *J. Nat. Prod.*, **1982**, *45*, 15.
[4]   Balandrin, M.F.; Klocke, J.A.; Wurtele, E.S.; Bollinger, W.H. *Science*, **1985**, *228*, 1154.
[5]   Bhat, S.V.; Bajawa, B.S.; Dornauer, H.; De Souza, N.J. *Tetrahedron Lett.*, **1977**, 1669.
[6]   Inamdar, P.K.; Kanitkar, P.V.; Reden, J.; De Souza, N.J. *Planta Med.*, **1984**, *50*, 30.
[7]   Corey, E.J.; Jardine, P.D.; Rohloff, J.C. *J. Am. Chem. Soc.* **1988**, *110*, 3672.
[8]   Sen, J.; Sharma, A.K.; Sahu, N.P.; Mahato, S.B. *Planta Med.* **1992**, *58*, 325.

[9] Mahato, S.B.; Pal, B.C.; Kawasaki, T.; Miyahara, K.; Tanaka, O.;Yamasaki, K. *Proceedings Annual Convention of Chemists, Indian Chemical Society*, Waltair, **1978.**

[10] Mahato, S.B.; Pal, B.C.; Kawasaki, T.; Miyahara, K.; Tanaka, O.; Yamasaki, K. *J. Am. Chem. Soc.* **1979**, 101, 4720.

[11] Burke, B.A.; Chan, W.R.; Honkan,V.A.; Blount, J.F.; Manchand, P.S. *Tetrahedron*, **1980**, *36*, 3489.

[12] Sastri, B. N., Ed, In *The Wealth of India, Raw Materials*, CSIR, Delhi, **1950,** Vol.2, p.231.

[13] Karlsson, B.; Pilotti,A.M.; Soderholm, A.C.; Norin, T.; Sundin, S.; Sumimoto, M. *Tetrahedron*, **1978**, *34*, 2349.

[14] Dauben, W.G.; Hubbell, J.P.; Oberhansli, P.; Thiessen, W.E. *J.Org.Chem.* **1979**, *44*, 669.

[15] Sutton, D .; West, C.A. *Phytochemistry*, **1975**, *14*, 1921.

[16] Weinheimer, A.G.; Chang, C,W.J.; Matson, J.A. *Prog.Chem.Org.Nat.Prod.* **1979**, *36*, 285.

[17] Hecker, E. *Pure and Appl. Chem.*, **1977** *49*, 1423.

[18] Kupchan, S.M.; Sigel, C.W.; Matz, M.J.; Gilmore, C.J.; Bryan, R.F. *J. Am. Chem. Soc.*, **1976**, *98*, 2295.

[19] Kupchan, S.M.; Shizuri, Y.; Murae, T.; Sweeny, J.G.; Hayes, H.R.; Shen, M.S.; Barrick, J.C.; Bryan, R.F.; Holm, D.; Wu, K.K. *J. Am. Chem. Soc.*, **1976**, *98*, 5719.

[20] Weinheimer, A.J.; Matson, J.A.; Helm, D.; Poling M. *Tetrahedron Lett.*, **1977**, 3481.

[21] Chmurny, G.N.; Hilton, B.D.; Brobst, S.; Look, S.A.; Whitherup, K.M.; Beutler, J.A. *J. Nat. Prod.*, **1992**, *55*, 414.

[22] Georg, G.I.; Chen, T.T.; Ojima, I.; Vyas, DM. *Taxane Anticancer Agents*, American Cancer Society, San Diego, **1995**.

[23] Denis, J.N.; Greene, A.E.; Guenard, D.; Gueritte-Voegelein, F.; Mangatal, L.; Potier, P. *J. Am. Chem. Soc.*, **1988**, *110*, 5917.

[24] Ojima, I.; Sun, C.M.; Zucco, M.; Park, Y.M.; Duclos, O.; Kuduk, S. *Tetrahedron Lett.*, **1993**, *34*, 4149.

[25] Holton, R.A.; Somoza, C.; King, H.B.; Liang, F.; Biediger, R.J.;Boatman, D.; Shindo, M.; Smith, C.C.; Kim, S.; Nadizadeh, H.; Suzuku, Y.; Tao, C.; Vu, P.; Tang, S.; Zhang, P.; Murthi, K.K.; Gentile, L.S.; Liu, J.H. *J. Am. Chem. Soc.*, **1994**, *116*, 1597.

[26] Nicolaou, K.C.; Zang, Z.; Liu, J.J.; Ueno, H.; Nantermet, P.G.; Guy, R.K.; Claiborne, C.F.; Renaud, J.; Couladouros, E.A.; Paulvannan, K.; Sorensen, E.J. *Nature*, **1994**, *367*, 630.

[27] Danishefsky, S.J.; Masters, J.J.; Young, W.B.; Link, J.T.; Snyder, L.B.; Magee, T.V.; Jung, D.K.; Isacs, R.C.A.; Bornmann, W.G.; Alaimo, C.A.; Coburn, C.A.; Di Grandi, M.J. *J. Am. Chem. Soc.*, **1996**, *118*, 2843.

[28] Nicolaou, K.C.; Dai, W.M.; Guy, R.K. *Angew Chem., Int. Ed. Engl.*, **1994**, *33*, 15.

[29] Paquette,L.A.; Want, T.Z.; Wang, S.; Phillippo,C.M.G. *Tetrahedron Lett.*, **1993**, *34*, 3523.

[30] Paquette, L.A.; Wang, T.W.; Philippo, C.M.G.; Wang, S. *J. Am. Chem.Soc.*, **1994**, *116*, 3367.

[31] Mahato, S.B.; Nandy, A.K.; Roy, G. *Phytochemistry*, **1992**, *31*, 2199.

[32]   Umehara, K.; Takagi, R.; Kuroyanagi, M.; Ueno, A.; Taki, T.; Chen, Y.J.
        Chem. Pharm. Bull., 1992, 40, 401.
[33]   Mahato, S.B.; Sen, S. Phytochemistry, 1997, 44, 1185.
[34]   Shanbrom, E. US Patent, 1992, 5, 128, 149 (CA 117, 178292u).
[35]   Shanbrom, E US Patent, 1992, 5, 128, 149 (CA 117, 178293v).
[36]   Suzuki, H.; Watano, S.; Sasazuka, T.; Tutumi,T. PCT Int. Patent WO, 1992,
        9209, 533 (CA 117, 178142v).
[37]   Misra, G.; Mitra, C.R. Phytochemistry, 1968, 7, 501.
[38]   Heywood, B.J. Kon, G.A.R. J. Chem. Soc., 1940, 713.
[39]   Sen, S.; Sahu, N.P.; Mahato, S.B. Tetrahedron, 1993, 49, 9031.
[40]   Sen, S.; Sahu, N.P.; Mahato, S.B. Phytochemistry, 1995, 38, 205.
[41]   Toyota, M.; Nsonthi, J.D.; Hostettmann, K. Phytochemistry, 1990, 29, 2849.
[42]   Varshney, I.P.; Logani, M.K. Indian J. Appl. Chem, 1969, 32, 173.
[43]   Massiot, G.; Lavaud, C.; Delalude, C.; Binst, G.V.; Millers, S.P.F.; Fales,
        H.M. Phytochemistry, 1990, 29, 3291.
[44]   Liu, J.-S.; Tao, Y. Tetrahedron, 1992, 48, 6793.
[45]   Kaneda, N.; Pezzuto, J.M.; Kinghorn, A.D.; Farnsworth, N.R.; Santisuk, T.;
        Tuchinda, P.V.J.; Rentrakul, V. J. Nat. Prod., 1992, 55, 654.
[46]   Nozaki, H.; Matsuura, Y.; Hirono, S.; Kasai, R.; Chang, J.J.; Lee, K.H. J. Nat.
        Prod., 1990, 53, 1039.
[47]   Pisha, E.; Chai, H.; Lee, I.S.; Chagwedera, T.E.; Farnsworth, N.R.;Cordell,
        G.A.; Beecher, C.W.W.; Fong, H.H.S.; Kinghorn, A.D.; Brown, D. M.; Wani,
        M.C.; Wall, M.E.; Hieken, T.J.; Das Gupta, T.K.; Pezzuto, J. M. Nature
        Medicine, 1995, 1, 1046.
[48]   Mahato, S.B.; Sahu, N.P.; Ganguly, A.N.; Miyahara, K.; Kawasaki, T. J. Chem.
        Soc., Perkin Trans.I, 1981, 2405.
[49]   Mahato, S.B.; Pal, B.C. J. Chem. Soc., Perkin Trans.I, 1987, 629.
[50]   Pal, B.C.; Mahato, S.B. J. Chem. Soc., Perkin Trans I, 1987, 1963.
[51]   Waltho, J.P.; Williams, D.H.; Mahato, S.B.; Pal, B.C.; Barna, J.C.J. J. Chem.
        Soc., Perkin Trans I, 1986, 1527.
[52]   Mahato, S.B.; Pal, B.C.; Price, K.R. Phytochemistry, 1989, 28, 207.
[53]   Mahato, S.B.; Sahu, N.P.; Roy, S.K.; Pramanik, B.N. J. Chem. Soc., Perkin
        Trans I, 1989, 2065.
[54]   Mahato, S.B. Phytochemistry, 1991, 30, 3389.
[55]   Mahato, S.B.; Sahu, N.P.; Roy, S.K.; Sen, S. Tetrahedron, 1991, 28, 5215.
[56]   Mahato, S.B.; Nandy, A. K.; Kundu, A.P. Tetrahedron, 1992, 48, 2483.
[57]   Kundu, A.P.; Mahato, S.B. Phytochemistry, 1993, 32, 999.
[58]   Mahato, S.B.; Pal, B.C.; Nandy, A.K. Tetrahedron, 1992, 48, 6717.
[59]   Mahato, S.B. In Saponins Used in Traditional and Modern Medicine, Adv. Exp.
        Med. Biol., Waller, G.R.; Yamasaki, K. Eds., Plenum Publishing Corporation,
        New York, 1996, vol. 404, 173.
[60]   Pakrashi, A.; Ray, H.; Pal, B.C.; Mahato, S.B. Contraception, 1991, 43. 475.
[61]   Ghosh, M.; Sinha Babu, S.P.; Sukul, N.C.; Mahato, S.B. Indian J.
        Experimental Biology, 1993, 31, 604.
[62]   Garai, S.; Mahato, S.B. Phytochemistry, 1997, 44, 137.
[63]   Chopra, R.N.; Chopra, I.C.; Varma, B.S. In Glossary of Indian Medicinal
        Plants, CSIR, Rafi Marg, New Delhi, 1969, p. 10.
[64]   Singh, H.K.; Dhawan, B.N. Indian J. Pharmacol., 1978, 10, 72.
[65]   Singh, H.K.; Dhawan, B.N. J. Ethnopharmacol., 1982, 5, 205.

[66] Singh, H.K.; Rastogi, R.P.; Srimal, R.C.; Dhawan, B.N. *Phytother.Res.*, **1988**, *2*, 70.

[67] Kulshrestha, D.K.; Rastogi, R.P. *Phytochemistry*, **1973**, *12*, 887.

[68] Kawai, K.; Iitaka, Y.; Shibata, S.; Kulshrestha, D.K.; Rastogi, R.P. *Acta. Cryst.*, **1973**, *B29*, 2947.

[69] Chandel, R.S.; Kulshrestha, D.K.; Rastogi, R.P. *Phytochemistry*, **1977**, *16*, 141.

[70] Kawai, K.; Akiyama, T.; Ogihara, Y.; Shibata, S. *Phytochemistry*, **1974**, *13*, 2829.

[71] Kawai, K.;Shibata, S. *Phytochemistry*, **1978**, *17*, 287.

[72] Jain,P.; Kulshrestha, D.K. *Phytochemistry*, **1993**, *33*, 449.

[73] Rastogi, S.; Pal, R.; Kulshrestha, D.K. *Phytochemistry*, **1994**, *36*, 133.

[74] Garai, S.; Mahato, S.B.; Ohtani, K.; Yamasaki, K. *Phytochemistry*, **1996** *42*, 815.

[75] Garai, S.; Mahato, S.B.; Ohtani, K.; Yamasaki, K. *Phytochemistry*, **1996**, *43*, 447.

[76] Sastri, B.N., Ed. In *The Wealth of India, Raw Materials*, C.S.I.R., Delhi, **1950**, Vol. *2*, p. 326.

[77] Mahato, S.B.; Pal, B.C.; Sarkar, S.K. *Phytochemistry*, **1988**, *27*, 1433.

[78] Ding, J.K.; Fujino,H.; Kasai, R.; Fuzimoto, N.; Tanaka, O.; Zhou, J.; Matsuura, H.; Fuwa, T. *Chem. Pharm. Bull.*, **1986**, *34*, 1158.

[79] Kimata, H.; Himaya, C.; Yahara, S.; Tanaka, O.; Ishikawa, O.; Aiura, M. *Chem. Pharm. Bull.*, **1979**, 27, 1836.

[80] Rao, G.S.; Sinsheimer, J.E.; Cochran, K.W. *J. Pharm. Sci.*, **1974**, *63*, 471.

[81] Yamamoto, M.; Kumagai, A.; Yamamura, Y. *Arzneim-Forsch.*, **1975**, *25*, 1021.

[82] Abe, H.; Sakaguchi, M.; Konishi, H.; Tani, T.; Arichi, S. *Planta Med.*, **1978**, *34*, 160.

[83] Yamamoto, M.; Kumagai, A.; Yamamura, Y. *Arzneim-Forsch.*, **1975**, *25*, 1240.

[84] Nandy, A.K.; Podder, G.; Sahu, N.P.; Mahato, S.B. *Phytochemistry*, **1989**, *28*, 2769.

[85] Mahato, S.B.; Nandy, A.K.; Luger P.; Weber, M. *J. Chem. Soc., Perkin Trans II*, **1990**, 1445.

[86] Row, L.R.; Subba Rao, G.S.R. *Tetrahedron*, **1962**, *18*, 827.

[87] Aoki, T.; Suga, T. *Phytochemistry*, **1978**, *17*, 771.

[88] Jurzysta, M. *14th Int. Symp. Nat. Prod.*, **1984**, 127.

[89] Price, K.R.; Fenwick, G.R.; Jurzysta, M. *J. Sci. Food Agric.*, **1986**, *37*, 1077.

[90] Mahato, S.B.; Ganguly,A.N.; Sahu, N. P. *Phytochemistry*, **1982**, *21*, 959.

[91] Mahato, S.B.; Sarkar, S.K.; Podder, G. *Phytochemistry*, **1988**, *27*, 3037.

[92] Mahato, S.B.; Nandy, A.K. *Phytochemistry*, **1991**, *30*, 1357.

[93] Price, K.R.; Johnson,I.T.; Fenwick, G.R. *CRC Crit. Rev. Food Sci. Nutr.* **1987**, 26, 27.

[94] Bader, G.; Hiller, K. *Pharmazie*, **1987**, *42*, 577.

[95] Schopke, T.; Hiller, K. *Pharmazie*, **1990**, *45*, 313.

[96] Hiller, K. In Hostettman, K.; Lea, P.J. Eds. *Biologically Active Natural Products*, Clarendon Press, Oxford, **1987**.

[97] George, A.J. *Fd. Cosmet. Toxicol.* **1965**, *3*, 85.

[98] Sibata, S. In Barton, D.H.R.; Ollis, W.D. Eds. *Advances in Medicinal Phytochemistry*, John Libbey, **1986**.

[99] Tanaka, O. *Pure & Appl. Chem.* **1990**, *62*, 1281.

[100] Tanaka, O.; Kasai, R.; Morita, T. *Abstracts of Chinese Medicines I*, **1986**, 130.

[101] Mahato, S.B. In Sainsbury, M. Ed. *Second Supplement to the* 2nd Edition of Rodd's Chemistry of Carbon Compounds, Vol. 2 B,C,D and E, Elsevier, London, **1994**, Chapter 16, pp 509-554.

[102] Wu, C.; Yu, Q.; Liu, W.; Guo, Y.; Zhang, G. *Shenyang Yaoxueyuan Xue- bao,* **1992**, *9*, 37.

[103] Pan, H.; Huang, L.; Zhou, Z.; Liang, G.; Hong, X.; Huan, Z. *Zhongcaoyao,* **1993**, *24*, 581.

[104] Amarowicz, R.; Shimoyamad, M.; Okubo, K. *Rocz. Panstw. Zakl. Hig.* **1994**, *45*, 125.

[105] Mahato, S.B. In Rahman, A. Ed. *Studies in Natural Products Chemistry,* Elsevier, Amsterdam, **1996**, Vol. *18*, pp. 649-676.

[106]   Mahato, S.B.; Garai, S. *Prog. Chem. Org. Nat. Prod.* **1998**, *74, 1.*.

[107] Medda, S.; Mahato, S.B.; Mahadevan, P.R.; Basu, M.K. *Indian J. Biochem. Biophys.* **1995**, *32*, 147.

[108] Mahato, S.B.; Sahu, N.P.; Luger, P.; Muller, E. *J. Chem. Soc. Perkin Trans II,* **1987**, 1509.

[109] Mahato, S.B.; Sahu, N.P. An improved process for the isolation of asiaticoside, useful as an antileprotic agent from *Centella asiatica,* Indian patent filed, **1993**.

[110] Heby, O. *Differentiation* (Berlin), **1981**, *19*, 1.

[111] Flink, I.; Pittijohn, D.E. *Nature* (London), **1975**, *253*, 62.

[112] Morris, D.R.; Morton, L.J. *Polyamines in Biology and Medicine*, Dekkar, New York, **1981**.

[113] Bergeron , R.J. *Acc. Chem. Res.* **1986**, *19*, 105.

[114] Ganem, B. *Acc. Chem. Res.* **1982**, *15*, 280.

[115] Mahato, S.B.; Sahu, N.P.; Luger, P. *J. Am. Chem. Soc.* **1983**, *105*, 4441.

[116] Mahato, S.B.; Sahu, N.P.; Muller, E.; Luger, P. *J. Chem. Soc. Perkin Trans II,* **1985**, 193.

[117] Mahato, S.B. Annual Convention of Chemists, Bodh Gaya, 1990.

[118] Sastri, B.N. Ed. In *The Wealth of India Raw Materials*, CSIR, Delhi, **1950**, Vol. *2*, p.4.

[119] Guggisberg, A.; Hessee, M. In Brossi, A. Ed. The Alkaloids, Academic Press, New York, **1983**, Vol. *22*, pp. 85-188.

[120] Hocquemiller, R.; Leboeuf, M.; Das , B.C.; Husson, H.P.; Potier, P.: Cave, A.; Hebd, C.R. *Seances Acad. Sci.*, Ser. C **1974**, *278*, 525.

[121] Sharma, O.P.; Makkar, H.P.S.; Dawra, R.K. *Toxicon,* **1988**, *26*, 975.

[122] Pass, M.A.; *Vet. J.* **1986**, *63*, 169.

[123] Sharma, O.P.; Makkar, H.P.S.; Dawra, R.K.; Negi, S.S. *Clin. Toxicol.* **1981**, *18*, 1077.

[124] Duke, J.A. In *Handbook of medicinal Herbs*, CRC Press, Boca Raton, Florida, **1985**, p.226.

[125] Sharma, O.P.; Sharma, P.D. *J. Sci. Industr. Res.* **1989**, *48*, 471.

[126] Acchireddy, N.R.; Sing, M.; Acchireddy, L.L.; Nigg, H.N.; Nagy, S. *J. Chem. Ecol.* **1985**, *11*, 979.

[127] Herbert, J.M.; Maffrand, J.P.; Taoubi, K.; Augereau, J.M.; Fouraste, I.; Gleye, J. *J. Nat. Prod.* **1991**, *54*, 1595.

[128] Endo, K.; Hikino, H. *Hetrocycles,* **1982**, *19*, 2033.

[129] Sasaki, H.; Nishimura, H.; Morota, T.; Chin, M.; Wei, H.; Yu-Lang, X. *Planta Med.* **1989**, *55*, 458.

[130] Pettit, G.; Niumata, A.; Takemura, T.; Ode, R.; Narula, A.; Schmidt, J.; Cragg, G.; Pase, C. *J. Nat. Prod.* **1990**, *53, 456.*

[131] Mahato, S.B.; Sahu, N.P.; Roy, S.K.; Sharma, O.P. *Tetrahedron*, **1994**, *50*, 9439.

[132] Williams, C.A. *Phytochemistry*, **1979**, *18*, 803.

[133] Nisbizuka, Y. *Science*, **1986**, *233*, 305.

[134] Yasukawa, K.; Takido, M.; Takeuchi, M.; Sato, Y.; Nitta, K.; Nakagawa, S. *Chem. Pharm. Bull.* **1990**, *38*, 774.

[135] Yasukawa, K.; Takido, M.; Takeuchi, M.; Nakagawa, S. *Chem. Pharm. Bull.* **1989**, *37*, 1071.

[136] Scott, A.I. *Tetrahedron*, **1992**, *48*, 2559.

Atta-ur-Rahman (Ed.) *Studies in Natural Products Chemistry, Vol. 23*

BIOACTIVE CONSTITUENTS OF THE FAMILY DIPTEROCARPACEAE

EUN-KYOUNG SEO[1] *and* A. DOUGLAS KINGHORN*

Program for Collaborative Research in the Pharmaceutical Sciences and Department of Medicinal Chemistry and Pharmacognosy, College of Pharmacy, University of Illinois at Chicago, Chicago, Illinois 60612, U.S.A.

ABSTRACT: The family Dipterocarpaceae consists three subfamilies, Dipterocarpoideae, Monotoideae, and Pakaraimoideae. Plants of the largest subfamily, Dipterocarpoideae, are the sources of resveratrol oligomers (oligostilbenoids), sesquiterpenes, and triterpenes, in particular. Resveratrol oligomers, the most characteristic compound class of this subfamily, have been associated with several types of biological activity, such as antibacterial, antiviral, and cytotoxic effects. The two remaining subfamilies, Monotoideae and Pakaraimoideae, which are distinguished from the Dipterocarpoideae by the absence of resin, do not elaborate resveratrol oligomers. Compounds of the oligostilbenoid type seem worthy of more extensive biological investigation than has been the case to date.

INTRODUCTION

The Dipterocarpaceae is moderate sized family of higher plants represented typically by resinous buttressed tropical rain forest trees. The leaves are alternate and simple with entire or sinuate margins. The stipules are paired and the inflorescences are paniculate and racemose. The flowers are secund or distichous, bisexual, actinomorphic, scented, and nodding. The stamens are 5-10, and the ovary is superior or semi-inferior. The fruit is indehescent and one seeded [1]. It is estimated that 16 genera and more than 500 species occur in the Dipterocarpaceae [2]. This predominantly Asian family is divided into three subfamilies, Dipterocarpoideae (Old World), Monotoideae (Africa), and Pakaraimoideae (New World, South America) [1,3-6]. The subfamily Dipterocarpoideae (13 genera and about 470 species) is restricted to tropical Asia [1,4,7], while the subfamily

---

*Address correspondence to this author at Program for Collaborative Research in the Pharmaceutical Sciences and Department of Medicinal Chemistry and Pharmacognosy, College of Pharmacy, University of Illinois at Chicago, Chicago, Illinois 60612, U.S.A.

[1]Current address: Chemistry and Life Sciences Group, Research Triangle Institute, P.O. Box 12194, Research Triangle Park, North Carolina 27709.

Monotoideae (two genera and more than 30 species) occurs in mainland Africa and Madagascar [3,7,8], and the subfamily Pakaraimoideae (one genus and one species) is indigenous to northern South America [6]. All members of the Dipterocarpaceae grow in tropical climates, and survive approximately 1000 mm of annual rainfall and a dry season of less than six months per year [1]. Most species in this family are found below 1000 m, but a few are confined only to altitudes above 700 m [1].

The genera represented in the subfamily Dipterocarpoideae are *Anisoptera, Cotylelobium, Dipterocarpus, Dryobalanops, Hopea, Neobalanocarpus, Parashorea, Shorea* (*Doona*), *Stemonoporus, Upuna, Vateria, Vateriopsis*, and *Vatica* [1,2]. The stipules of the plants in this subfamily are often conspicuously large, and the anthers are basifixed. The petals are longer than the sepals. The wood rays are multiserate, and the wood, leaves and ovaries have resin or secretory ducts [5]. Several species in these genera are considered as important timber and/or oleoresin-producing trees in tropical Asia [3,9]. For example, the oleoresin from *Dipterocarpus alatus* is used for illumination, waterproofing baskets and boats, and for making paint, varnish and lacquer [10]. Therefore, *D. alatus* is considered as a highly important economic plant in Southeast Asia [9,10].

The subfamily Monotoideae is composed of two genera, *Monotes* (more than 30 species) and *Marquesia* (three to four species) [2,3,8,10]. The anthers of plants in this subfamily are basi-versatile and the stipules are small and caducous. The petals are longer than the sepals, and the wood rays are uniseriate. The wood, flowers, and leaves of these trees do not produce a resin or have secretory ducts. The first record of a plant of African origin in the Monotoideae was *Vatica africana* Welwitsch, which proved to be different from Asiatic *Vatica* species in its stamen characteristics [8,11]. Accordingly, *V. africana* was separated from the main genus *Vatica* and established as part of a new genus, *Monotes* [8]. Gilg suggested the institution of the subfamily Monotoideae inclusive of the genus *Monotes*, which diverged from the Dipterocarpaceae and had secretory cavities instead of resin canals as found in the plants in the Dipterocarpoideae [8].

The third subfamily, Pakaraimoideae, has only one genus, *Pakaraimaea* [1,5]. The first *Pakaraimaea* species (and so far the only one discovered) was found in 1951 as a tree on the Imbaimadai Savanna of the Pakaraima Uplands of Guyana [5]. Its morphological characteristics in terms of the anthers (basi-versatile), stipules (small and caducous), and the lack of resin ducts are similar to those of the Monotoideae. The petals are shorter than the sepals, and the wood rays are dominantly biserate [5]. The subfamily Pakaraimoideae seems to be closer to the Monotoideae than the Dipterocarpoideae taxonomically. However, the lack of resin canals in the wood of both the Monotoideae and the Pakaraimoideae constitutes a major distinction from the Dipterocarpoideae [5]. It has been suggested that the subfamily Monotoideae be treated as a separate family Monotaceae along with the Pakaraimoideae, to constitute a taxonomic group closely related to the Tiliaceae [12-15].

A preliminary chromatographic analysis of phytochemical constituents of selected species the three subfamilies in the Dipterocarpaceae was reported in 1977 [5], and indicated the presence of triterpenoids similar to dipterocarpol as constituents common to the three subfamilies. It was recommended that further studies on phenolic compounds such as flavonoids, which might be used as chemotaxonomic markers, should be carried out [5]. The present chapter focuses on the biologically active constituents of the plants in the family Dipterocarpaceae.

BIOACTIVE CONSTITUENTS OF THE SUBFAMILY DIPTEROCARPOIDEAE

The most characteristic secondary metabolites from plants in the Dipterocarpaceae are the oligostilbenoids, which are of only limited distribution in the plant kingdom [24-62]. This type of compound has been isolated only from a limited number of families, namely, the Cyperaceae, Dipterocarpaceae, Gnetaceae, Leguminosae, and Vitaceae [62]. The resveratrol oligomers in the Dipterocarpaceae which have been structurally characterized so far are balanocarpol (2) [44,45], canaliculatol (7) [46], copalliferol A (8) [47-49], copalliferol B (9) [50], dibalanocarpol (12) [51], distichol (10) [52], hopeaphenol (13) [25,53,54], malibatols A (3) and B (4) [51], shoreaphenol (5) [55], stemonoporol (11) [48,49], vatdiospyroidol (14) [56], vaticaffinol (15) [48,57,58], vaticaphenol A (16) [56], and ε-viniferin (6) [45,52]. Interestingly, resveratrol itself has never been reported as a constituent of any species in the Dipterocarpaceae, although recently (E)-resveratrol 3-O-β-D-glucoside (1) was isolated from *Vatica diospyroides* Sym. as the first resveratrol monomer from this family [61].

As may be seen from Table 1, biological activity has been shown in recent years for several oligostilbenoid constituents of the subfamily Dipterocarpoideae. The resveratrol trimers copalliferol A (8) [48], copalliferol B (9) [50], and distichol (10) [52] showed antibacterial activity against Oxford *Staphylococcus* and *Escherichia coli* when tested by a filter paper disk method using Mueller Hinton Agar medium, although the MIC values were not disclosed [52]. Canaliculatol (7) [46] and vaticaffinol (15) [58,59] exhibited antifungal activity on TLC plates that were sprayed with *Cladosporium cladosporiodes* in Czapex-Dox nutrient solution and incubated at room temperature for 48 hours [46]. Balanocarpol (2) and dibalanocarpol (12) were reported very recently as modest HIV-inhibitors (EC_{50} values of 46 and 20 µg/mL, respectively), and, in the same study, malibatols A (3) and B (4) [51] showed cytotoxicity to the host cells (CEM-SS lymphocytes; IC_{50} values of 13 and 21 µg/mL, respectively) without having anti-HIV activity themselves [51]. In recent work from our laboratory, vatdiospyroidol (**14**) from *Vatica diospyroides* displayed significant cytotoxic activity against the KB (human oral epidermoid carcinoma; IC_{50} 1.0 µg/mL), Col2 (colon cancer; IC_{50} 1.9 µg/mL), and BC1 (breast cancer; IC_{50} 3.8 µg/mL) cell lines in an *in vitro* tumor cell panel [56]. From this plant, a known compound, (*E*)-resveratrol 3-*O*-β-D-glucoside (**1**) was also isolated as the first resveratrol monomer found in the Dipterocarpaceae [56]; this compound is known to have antioxidant activity in test systems assessing radical scavenging [1,1-diphenyl-2-picryl-hydrazyl (DPPH) was used as a free radical: IC_{50} 198 ± 16.8 µM] and inhibition of lipid peroxidation [(Cu^{+2}-catalyzed lipid peroxidation in human low-density lipoprotein): IC_{50} 19.1 ± 3.0 µM] [61]. ε-Viniferin (**6**) has been found to show antihepatotoxic activity employing primary cultured rat hepatocytes in a carbon tetrachloride- and D-galactosamine-induced cytotoxicity model system [62,63]. Resveratrol oligomers from plants in other families than the Dipterocarpaceae have also exhibited interesting biological activities such as inhibiting protein kinase C [64] and inhibition of the L-DOPA oxidase activity of tyrosinase [65].

Table 1.   **Biologically Active Constituents of Plants in the Subfamily Dipterocarpoideae**

Compound class/name/code	Activity	Reference(s)
OLIGOSTILBENOIDS		
Resveratrol Monomer		
(*E*)-Resveratrol 3-*O*-β-D-glucoside (**1**)	Antioxidant activity (scavenging of free radicals, and inhibition of lipid peroxidation)[a]	61
Resveratrol Dimers		
Balanocarpol (**2**)	Anti-HIV activity *in vitro*	51
Malibatol A (**3**)	Cytotoxicity to the host cells (CEM SS) in a HIV antiviral assay	51
Malibatol B (**4**)	Cytotoxicity to the host cells (CEM SS) in a HIV antiviral assay	51
ε-Viniferin (**6**)	Protective effect in mice against carbon tetrachloride-induced hepatic injury[a]	63
Resveratrol Trimers		
Canaliculatol (**7**)	Antifungal activity against *Cladosporium cladosporioides*	46
Copalliferol A (**8**)	Antibacterial activity against Oxford *Staphylococcus* and *E. coli*	48
Copalliferol B (**9**)	Antibacterial activity against Oxford *Staphylococcus* and *E. coli*	50

(Table 1). contd.....

Compound class/name/code	Activity	Reference(s)
Distichol (10)	Antibacterial activity against Oxford *Staphylococcus* and *E. coli*	52
Resveratrol Tetramers[b]		
Dibalanocarpol (12)	Anti-HIV activity *in vitro*	51
Vatdiospyroidol (14)	Cytotoxicity against three human tumor cell lines	56
Vaticaffinol (15)	Antifungal activity against *Cladosporium cladosporioides*	58
TERPENOIDS		
Sesquiterpenes		
Caryophyllene (17)	Antibacterial activity against *Propionibacterium acnes*[a]	68
Copaene (18)	Male medfly chemoattractant[a]	69
α-Gurjunene (19)	Antibacterial and antifungal activities[a]	70
Triterpenoids		
β-Amyrin (20)	Cytotoxicity against the CAMA-1 human tumor cell line[a]	71
Betulinic acid (21)	Selective cytotoxic activity against melanoma cell lines; antitumor effect in melanoma murine xenograft system[a]	72
	Induction of apoptosis in neuroectodermal tumors[a]	73
	Selective cytotoxicity toward starving resting L1210 leukemia cells[a]	74
Oleanolic acid (22)	Antioxidant activity by inhibition of the copper-catalyzed oxidation of LDL (low density lipoprotein)[a]	75
	Anti-tumor promoter by inhibition of TPA-induced Epstein-Barr virus activation in Raji cells[a]	76
	Protective effect against hepatotoxicity produced by certain chemicals[a]	77
ψ-Taraxasterol (23)	Antibacterial activity against *Staphylococcus aureus*[a]	78
Ursolic acid (24)	Anti-tumor promoter by inhibition of TPA-induced Epstein-Barr virus activation in Raji cells[a]	77
	Inhibition of HIV-1 protease dimerization[a]	79
	Toxicity and feeding deterrence towards the aphid, *Schizaphis graminum*[a]	80
Ursonic acid (25)	Anti-HSV-1 activity *in vitro*[a]	81

(Table 1). contd.....

Compound class/name/code	Activity	Reference(s)
MISCELLANEOUS		
Coumarins		
Bergenin (26)	Anti-HIV activity *in vitro*[a]	82
Scopoletin (27)	Inhibition of the growth of Chinese cabbage seedlings[a]	83
	Photoreceptor and phototransducer[a]	84
	Protective effects against histamine aerosol-induced bronchospasm[a]	85
Ellagic acid derivative		
Ellagic acid (28)	Anticarcinogenic activity by detoxification of diol-epoxide of benzo[α]pyrene[a]	86
	Inhibition of *E. coli* DNA gyrase supercoiling[a]	87
	Inhibition of peroxidase-mediated oxidation of 4-hydroxystilbene[a]	88
	Inhibition of protein kinase C (inhibitory only towards the particulate enzyme obtained after carcinogen treatment)[a]	89
Flavonoids		
Cyanidin (29)	Antioxidant activity by inhibition of the copper-catalyzed oxidation of LDL (low density lipoprotein)[a]	90-92
Delphinidin (30)	Anti-inflammatory effects in carrageenan paw edema and croton oil ear edema models in rats[a]	93, 94
Kaempferol (31)	Antibacterial activity against *Streptococcus mutans*, *Actinomyces viscosus*, and *Porphyromonas gingivalis*[a]	95
	Inhibition of protein kinase C[a]	89
	Inhibition of topoisomerase 1-catalyzed DNA religation[a]	96
	Modulation of the carcinogenicity of aflatoxin B1[a]	97, 98
Myricetin (32)	Antibacterial activity against *Streptococcus mutans*, *Actinomyces viscosus*, and *Porphyromonas gingivalis*[a]	95
	Antioxidant activity by inhibition of lipid peroxidation[a]	99, 100
	Inhibition of 4-nitroquinoline-1-oxide (4NQO)-induced mutagenicity[a]	101
	Inhibition of eukaryotic topoisomerases I and II[a]	97

(Table 1). contd.....

Compound class/name/code	Activity	Reference(s)
Quercetin (33)	Antioxidant activity by inhibition of oxidation of the fluorochrome dichlorofluorescin in HepG2 cells[a]	102
	Antioxidant activity by reduction of oxidative DNA damage and lipid peroxidation, and ability to quench free radicals[a]	103
	Inhibition of protein kinase C[a]	89
	Inhibition of topoisomerase 1-catalyzed DNA religation[a]	96
Phenolics		
Caffeic acid (34)	Antioxidant activity by inhibition of lipid peroxidation[a]	104
	Chemosensitizing activity with selectivity for drug-resistant MCF-7/Dox human breast carcinoma cells[a]	105
p-Coumaric acid (35)	Antioxidant activity by inhibition of LDL oxidation through interaction with ascorbate[a]	106
	Inhibition of growth of contaminating yeast cells isolated from fresh white cheese[a]	107
Ferulic acid (36)	Antioxidant activity by extending the induction time of lipid oxidation in the Rancimat test[a]	108
	Inhibition of the growth of contaminating yeast cells isolated from fresh white cheese[a]	107
Gallic acid (37)	Antioxidant activity by inhibition of the copper-catalyzed oxidation of LDL (low density lipoprotein)[a]	109
	Hepatoprotective activity against carbon tetrachloride-induced alterations in the liver[a]	110
Quinones		
2,6-Dimethoxy-1,4-benzoquinone (38)	Cytotoxic activity to parental L5178Y cells[a]	111
Chrysophanic acid (= Chrysophanol) (39)	Brine shrimp toxicity[a]	112
	Molluscicidal activity against the snails Oncomelania hupensis and Biomphalaria glabrata[a]	113
	Tumor-promoting effect by enhancement of transformation of C3H/M2 mouse fibroblasts[a]	114

[a]Activity was found as a constituent from another plant family rather than the Dipterocarpaceae. In such cases, only the more recent literature has been surveyed.

[b]Although not formally included in this table, our group has obtained preliminary evidence of the biological activity of hopeaphenol (13) and vaticaphenol A (16) in a cell-based anticancer assay and in an antiviral assay, respectively.

Oligostilbenoids occur in the subfamily Dipterocarpoideae as resveratrol dimers, trimers, and tetramers, and are formed from resveratrol [(*E*)-3,5,4'-trihydroxystilbene] monomeric units by oxidative reactions [24]. As indicated above, (*E*)-resveratrol 3-*O*-β-D-glucopyranoside (1) was found for the first time from a plant in the Dipterocarpaceae, namely, *Vatica diospyroides* [56]. This species was also found to elaborate the novel resveratrol tetramers vatdiospyroidol (14) and vaticaphenol A (16) [56]. Resveratrol oligomers can be recognized as such from their elemental formulas, which are multiplets of the elemental formula of resveratrol, $C_{14}H_{12}O_3$. These compounds are relatively non-polar, and may be extracted with methanol, and then concentrate into ethyl acetate when partitioned with water [56]. In general, the UV spectrum displayed by an oligostilbenoid exhibits an absorption maximum in the range 281-286 nm (log $\in$ 4.0-4.4), consistent with the presence of one or more substituted phenyl rings [48,52,56,57,66]. The hydroxyl groups present are readily evident as a strong absorption band at 3200-3360 cm^{-1} in the IR spectrum of a resveratrol oligomer [52,56,57,66]. In the ^{1}H- and ^{13}C-NMR spectra of resveratrol oligomers, *ortho*- and *meta*- coupled aromatic protons appear in downfield regions (δ_H 6.0-8.8 and δ_C 95.0-130.0, respectively), and signals at δ_H 3.1-5.8 and δ_C 35.0-95.0 for the aliphatic methine groups are also found [52,56,67].

Hopeaphenol (13), the first resveratrol oligomer to be structurally characterized from a species in the subfamily Dipterocarpoideae, was isolated initially from the heartwood of *Hopea odorata* and *Balanocarpus heimii*, and its structure was determined in 1966 by X-ray crystallography on its dibromo-derivative [24,53,60]. In the case of other oligostilbenoids isolated from this subfamily reported until the early 1990's, mass spectrometry, ^{1}H NMR, ^{13}C NMR, and biogenetic rationalizations, supported occasionally by NOE and homo-decoupling NMR experiments, were employed to assign their structures [44,46-48,50,52,57,58]. Methylation and/or acetylation of the hydroxyl groups of the oligostilbenoids were also performed in support of these structural proposals [44,46-48,50,52,57,58]. In the following paragraphs, descriptions of the stages in the structure elucidation of two complex biologically active resveratrol tetramers, vaticaffinol (15) and vatdiospyroidol (14) (Table 1), will be given as examples.

The oligostilbenoid vaticaffinol (15) was isolated initially from *Vatica affinis* [57]. High-resolution mass spectrometry on the molecular ion was unsuccessful for this compound, but its decamethyl ether showed a molecular ion at *m/z* 1056.4253 ($C_{66}H_{62}O_{12}$). Its UV absorption maximum was observed at λ_{max} 282 nm (EtOH, log $\in$ 4.4) [48], and the IR spectrum showed prominent hydroxyl group absorption at 3200 cm^{-1}. The ^{13}C-NMR spectrum of vaticaffinol (15) at 75 MHz displayed 56 carbons, and the ^{1}H-NMR spectrum at 270 and 300 MHz showed six aliphatic protons

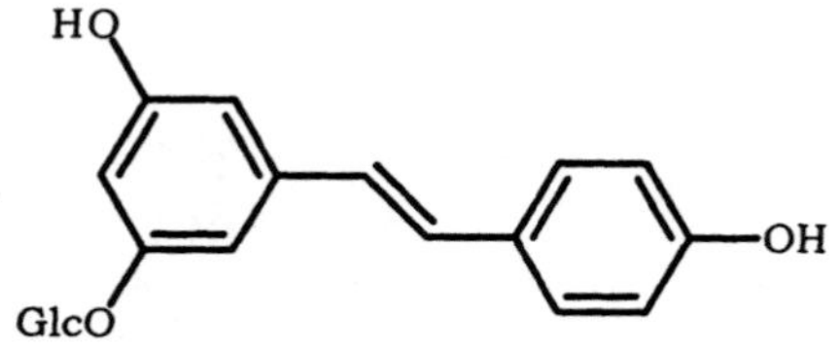

1

(Glc = β-D-glucopyranosyl)

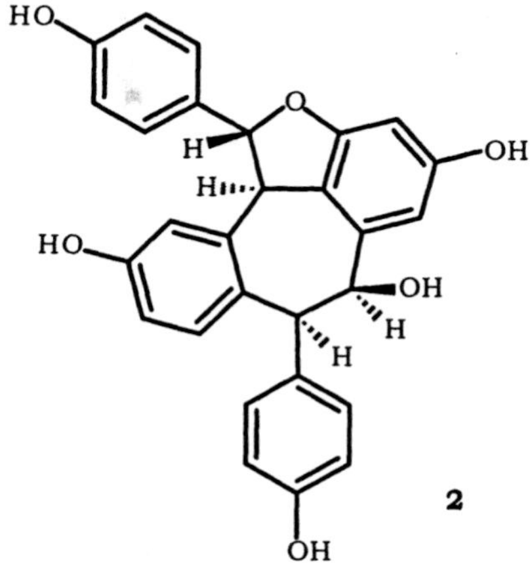

2

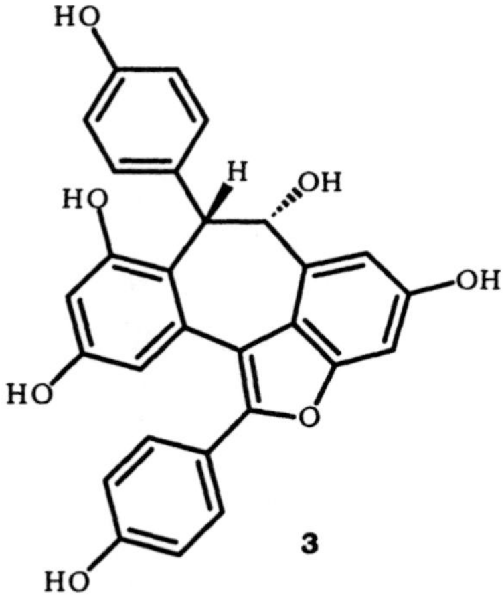

3

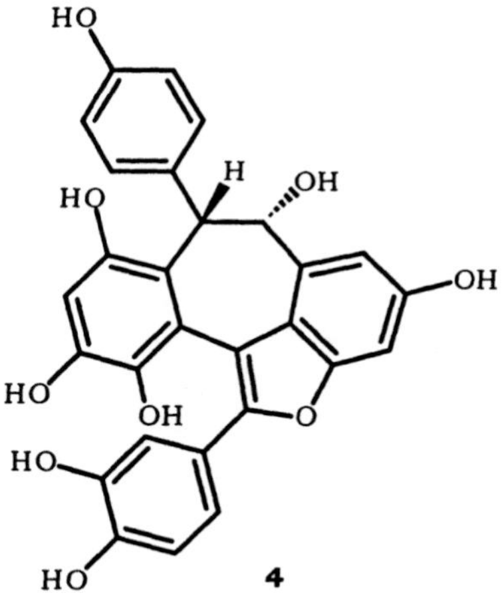

4

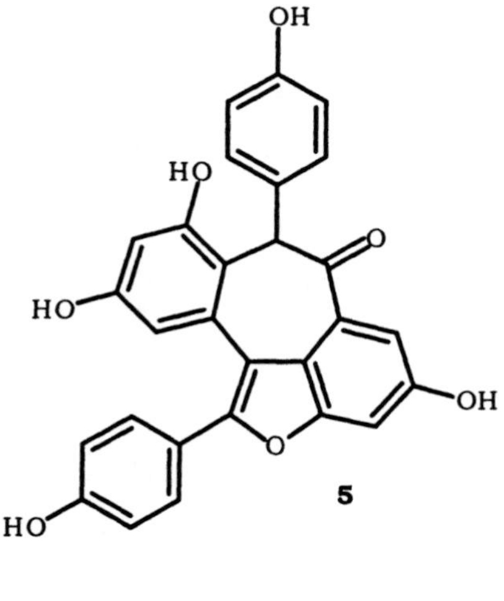

5

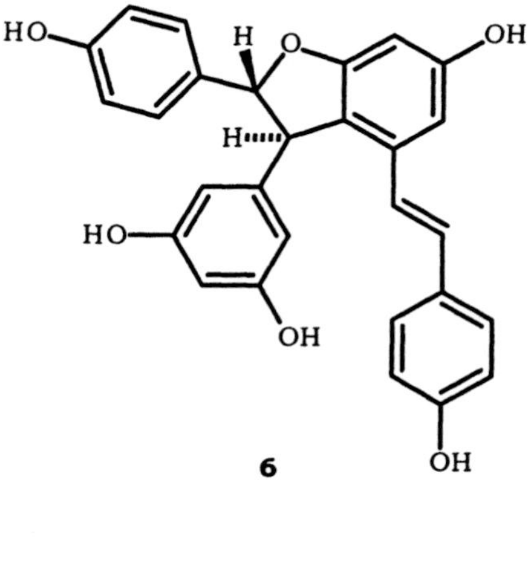

6

Fig. (1). contd.....

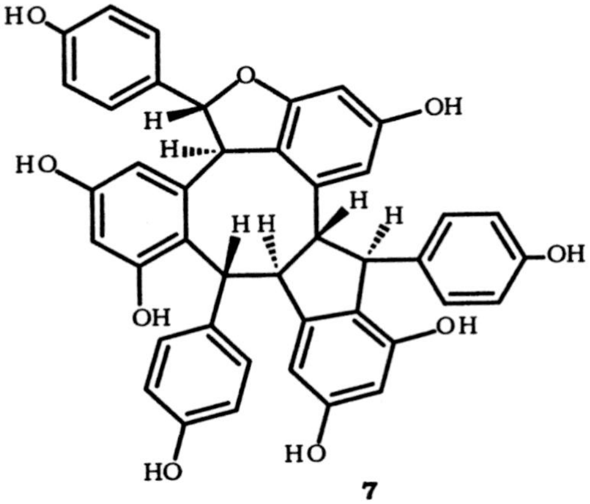

7

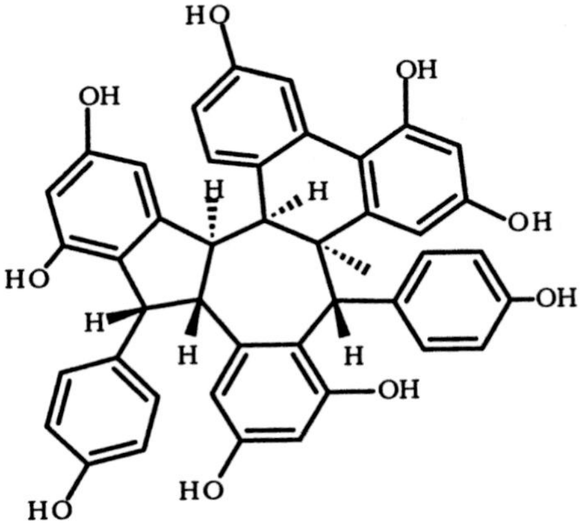

8

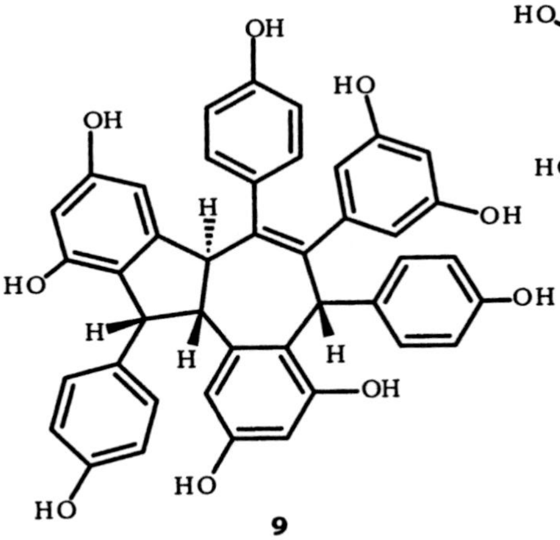

9

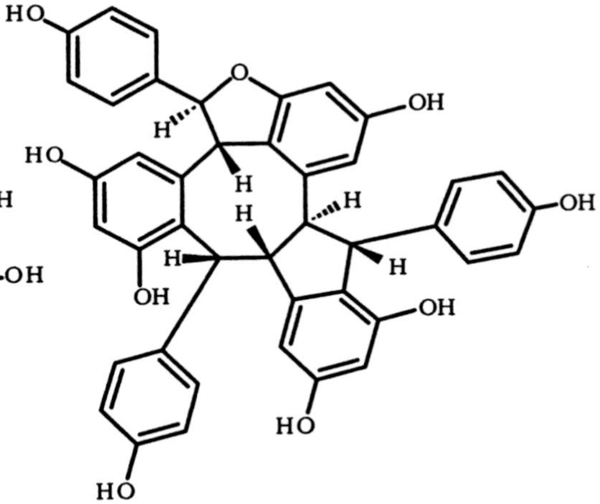

10

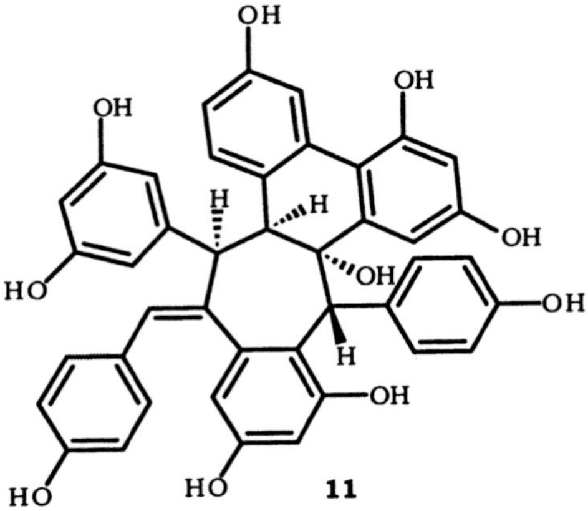

11

Fig. (1). contd.....

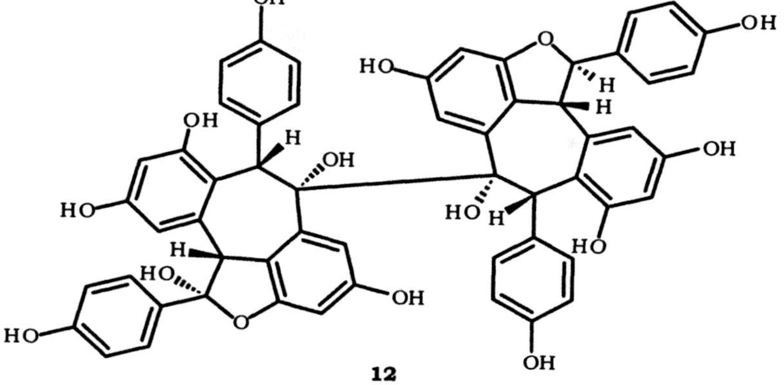

12

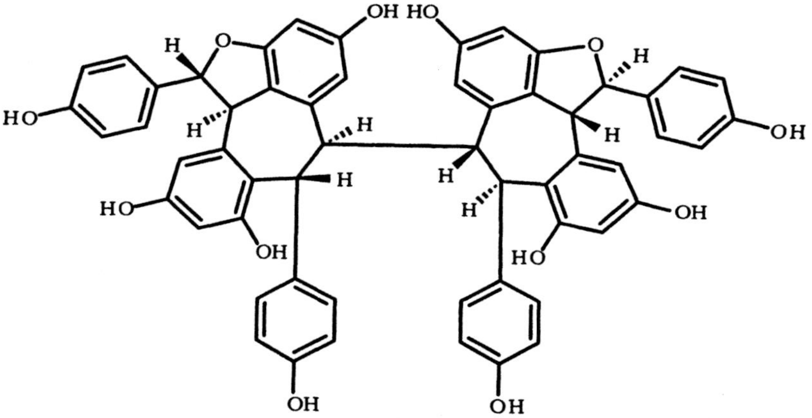

13

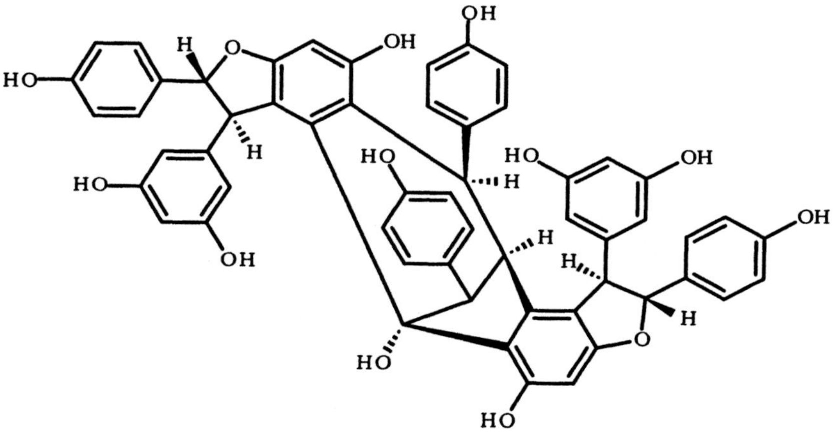

14

Fig. (1). contd.....

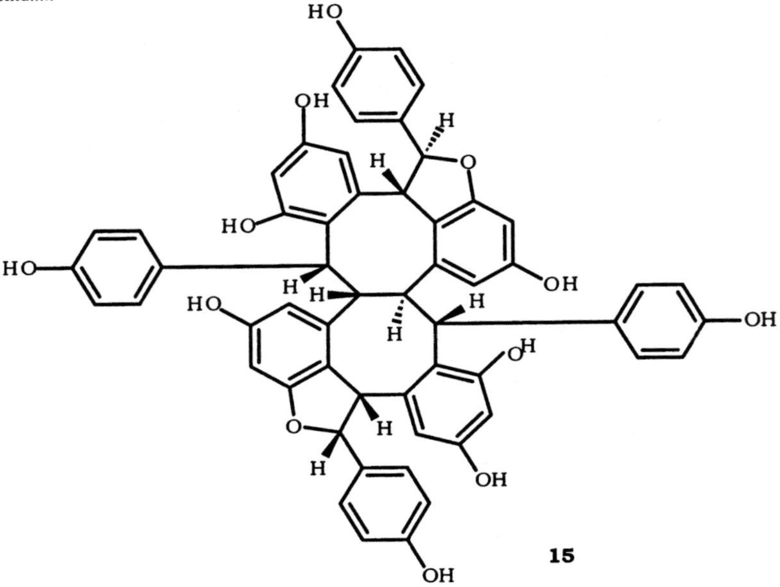

15

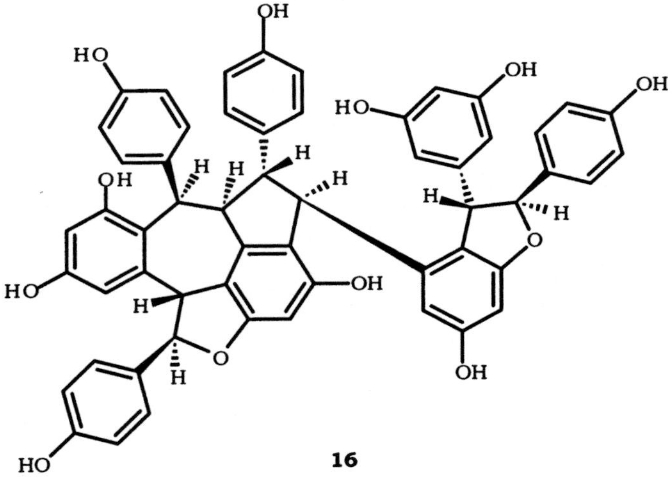

16

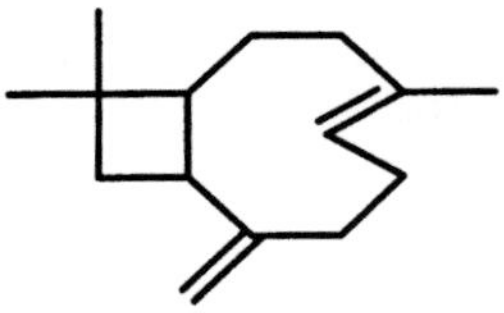

17

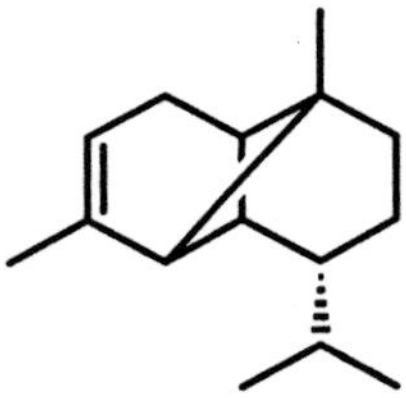

18

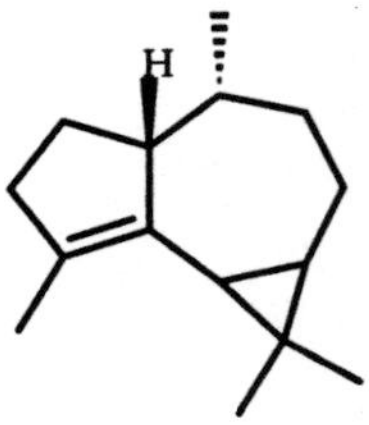

19

Fig. (1). contd.....

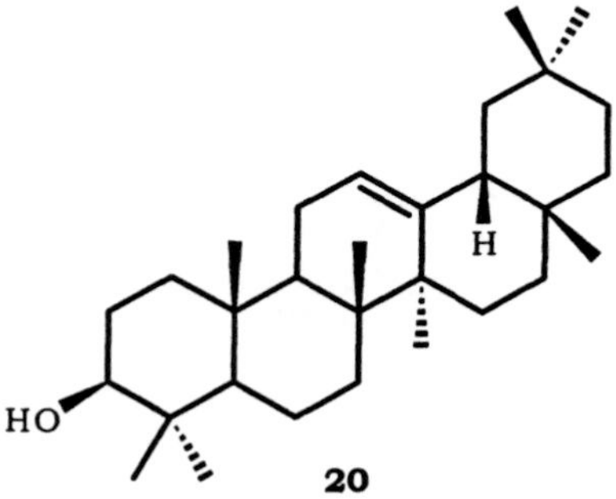

20

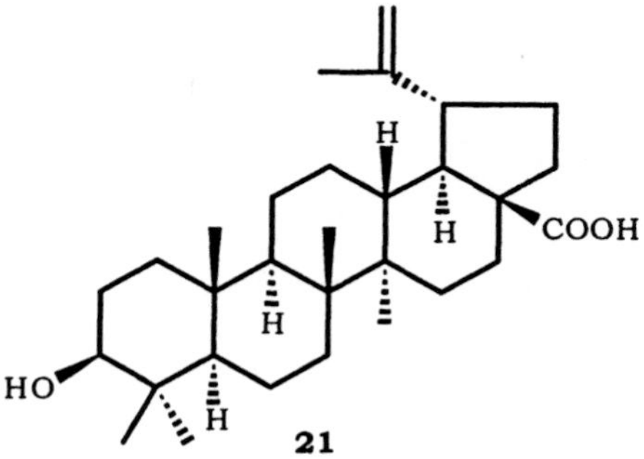

21

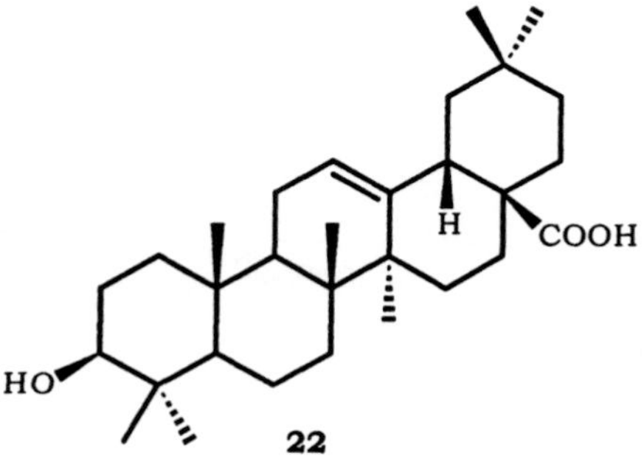

22

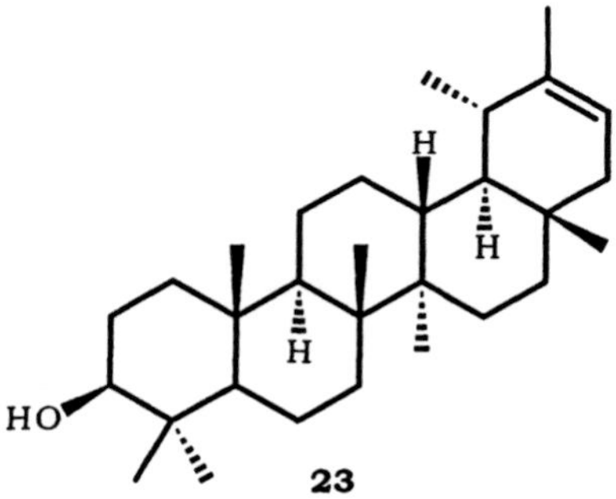

23

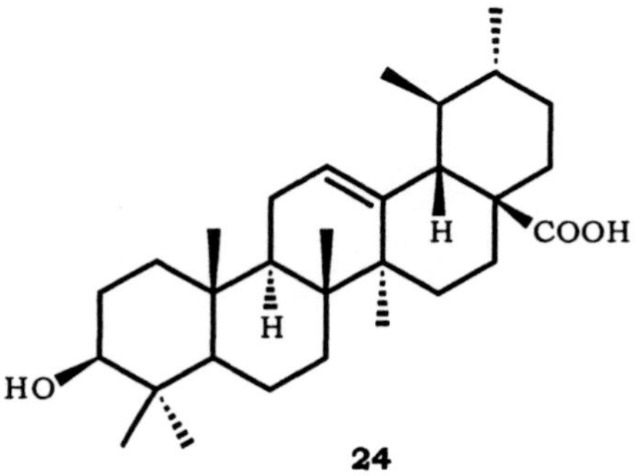

24

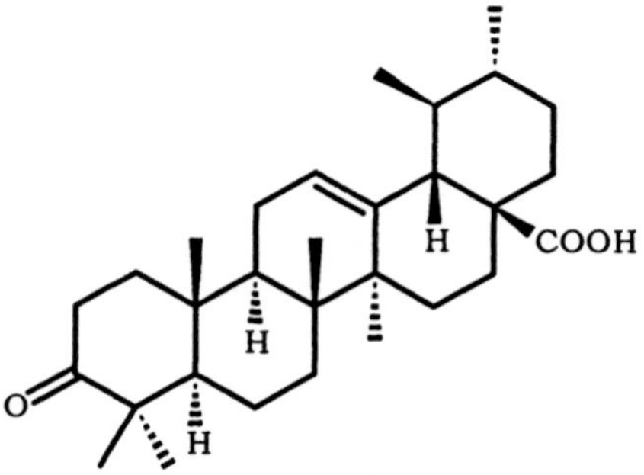

25

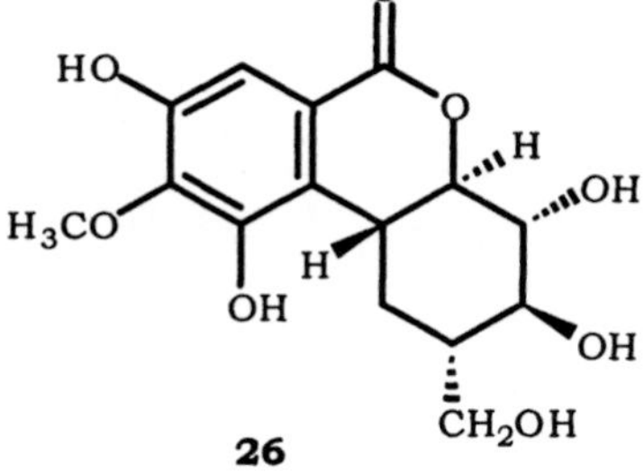

26

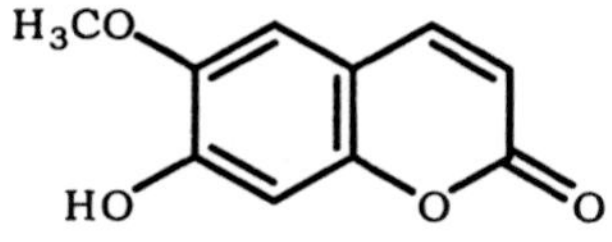

27

Fig. (1). contd.....

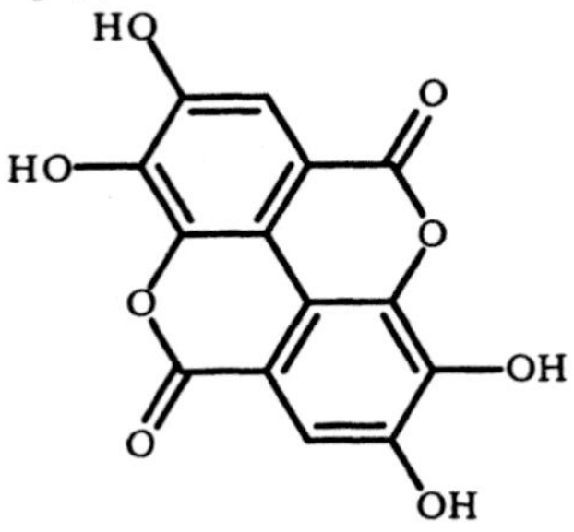

28

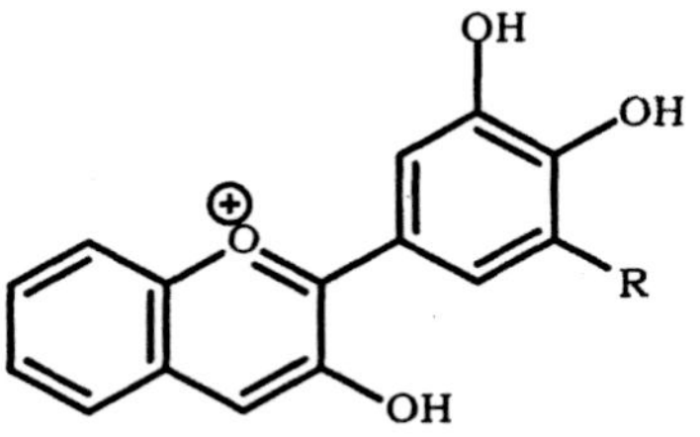

29 R = H
30 R = OH

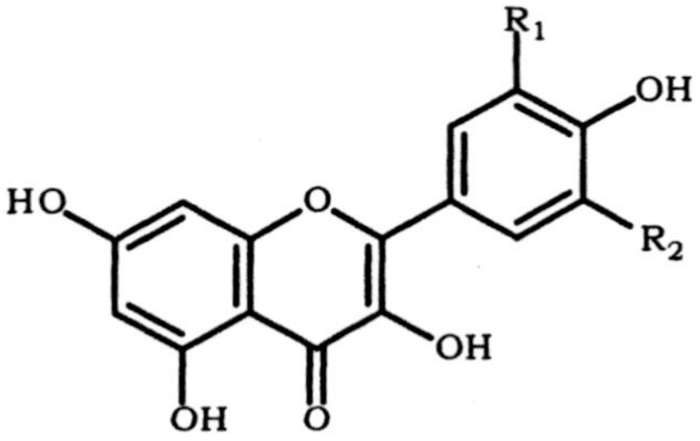

31 R$_1$ = R$_2$ = H
32 R$_1$ = R$_2$ = OH
33 R$_1$ = OH, R$_2$ = H

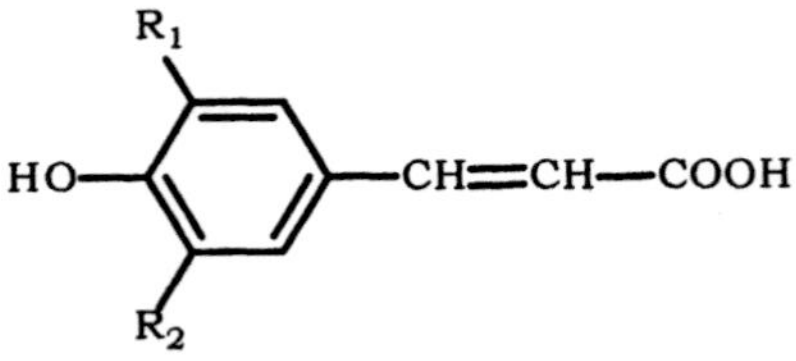

34 R$_1$ = R$_2$ = OH
35 R$_1$ = R$_2$ = H
36 R$_1$ = OCH$_3$, R$_2$ = OH

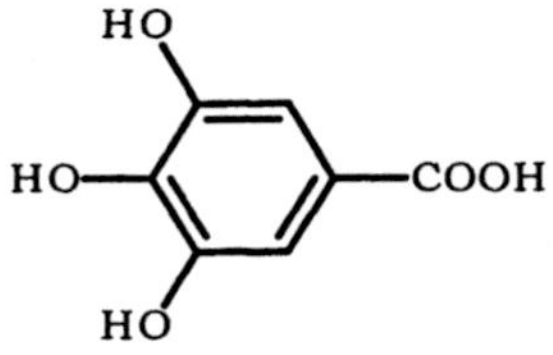

37

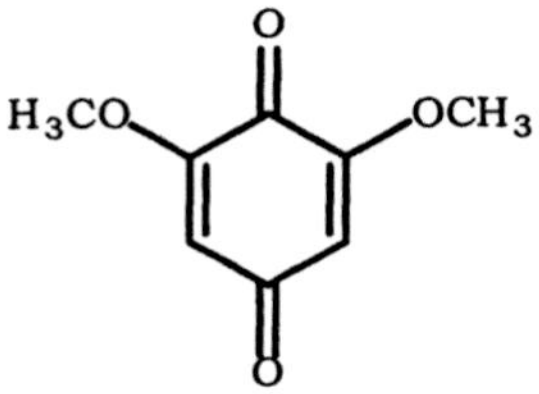

38

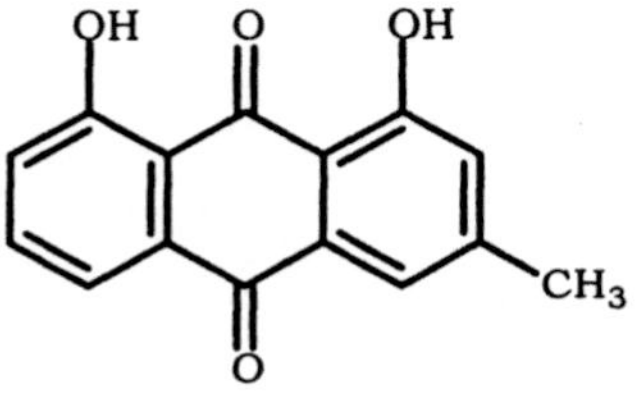

39

Fig. (1). Structures of constituents of plants in the subfamily Dipterocarpoideae with known biological activity.

and complex aromatic signals. The structure of vaticaffinol (15) was proposed from an analysis of its ^{1}H- and ^{13}C-NMR data and mass fragmentation together with biogenetic considerations. It was further supported by NOE and homo-decoupling NMR experiments, confirming the relationship of the aliphatic protons [58]. However, owing to the lack of supportive 2D-NMR techniques available when this study was performed, assignments of the ^{1}H- and ^{13}C-NMR signals for the complex aromatic region of vaticaffinol (15) remained ambiguous [48,57,58].

More recently, the structures of oligostilbenoids such as malibatols A (3) and B (4) [51], dibalanocarpol (12) [51], vatdiospyroidol (14), and vaticaphenol A (16) [56] were proposed using two inverse-detected NMR methods, namely, ^{1}H-^{13}C HMQC for one-bond correlations and ^{1}H-^{13}C HMBC for two- and three-bond correlations. For instance, in our own work on the structural assignment of vatdiospyroidol (14), the molecular formula of this resveratrol tetramer was deduced as $C_{56}H_{42}O_{12}$ by high-resolution negative-ion fast-atom bombardment mass spectrometry (found m/z [M-H]$^-$ 905.2638). A strong IR hydroxyl absorption band at 3361 cm$^-1$ and the UV absorption maximum at 285 nm (log $\in$ 4.0) were indicative of this compound being an oligostilbenoid [56]. Its ^{1}H-NMR spectrum showed ortho- or meta-coupled aromatic signals in the region of δ_H 6.0-7.3, and aliphatic methine functionalities appeared between δ_H 2.8 and 5.7. Each proton signal was correlated to adjacent carbons in the HMQC NMR spectrum. Ten hydroxyl protons of vatdiospyroidol (14) were also observed in the ^{1}H-NMR spectrum when in acetone-d_6, and a decamethylated product was obtained on the permethylation of this compound. In the HMBC NMR spectrum of vatdiospyroidol (14), all of these proton signals showed cross peaks with carbon atoms positioned two or three bonds away so that the complex structure of the skeleton could be postulated. The ^{1}H-^{1}H COSY and NOESY NMR spectral data of 14 provided confirmatory evidence for the overall structure hypothesized. To propose the relative stereochemistry of this compound, the NOESY NMR spectrum was analyzed in detail, with support by computer-aided molecular modeling. The energy-minimized stereostructure of the molecule was obtained by a molecular mechanics calculation method using HyperChem™ 4.0 software as shown in Fig. 2. From this energy-minimized stereostructure, the dihedral angles between the adjacent aliphatic methine protons were calculated and incorporated into a Karplus correlation graph to calculate their coupling constants. The expected values between protons were found to correspond to the observed coupling constants from the ^{1}H-NMR spectrum of vatdiospyroidol (14). For example, a dihedral angle of 79.80° calculated for H-7d and H-8d would be expected to result in only small couplings from the vicinal Karplus correlation graph, and experimentally these two protons (H-7d, δ_H 4.17; H-8d, δ_H 3.10) both appeared as broad singlets in the ^{1}H-NMR spectrum of vatdiospyroidol (14). Experimental values obtained for proton resonances throughout the whole molecule provided strong evidence for the proposed stereochemistry [56].

The resins of a number of species in the subfamily Dipterocarpoideae were investigated and found to contain several sesquiterpenes and triterpenes. Thus, the resins from five species of the genus Doona (Shorea) were examined and reported to contain sesquiterpenes such as caryophyllene (17), copaene (18), β-elemene, and humulene, and

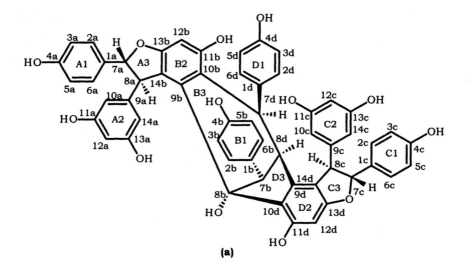

(a)

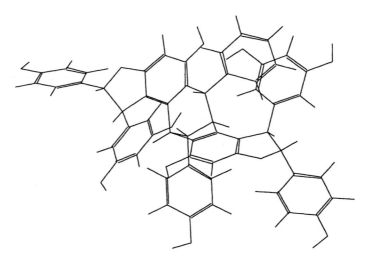

(b)

Fig. (2). (a) Structure of vatdiospyroidol (**14**) showing numbering of carbon atoms; (b) Energy-minimized stereostructure of vatdiospyroidol (**14**) (MM$^+$ calculation using the HyperChem™ 4.0 molecular modeling program).

triterpenes such as β-amyrin (**20**), dammarenediol-I, hydroxydammarenone-I, ψ-taraxasterol (**23**), and ursolic acid (**24**) [17]. Similarly, resins from species of the genus *Dipterocarpus* were investigated, and betulinic acid (**21**) [18], calarene, caryophyllene (**17**), copaene (**18**), cyperene, farnesane, α-gurjunene (**19**), γ-gurjunene, humulene, dipterocarpol (hydroxydammarenone-II), dammaradienone, dammarenediol-II, 20*R*-ocotillone, and 20*S*-ocotillone [19] were identified.

Certain species in other genera of this family, including *Anisoptera*, *Cotylelobium*, *Dryobalanops*, *Hopea*, *Stemonoporus*, *Upuna* and *Vateria*, and were analyzed in a similar manner, and found to contain similar types of sesquiterpenes and triterpenes to *Shorea* (*Doona*) and *Dipterocarpus*, as shown in Table 2 [20-23].

Many of these terpenoids have been reported to be biologically active as indicated in Table 1, although such work has occurred on these compounds when isolated as constituents of plant families other than the Dipterocarpaceae [68-81]. For example, the common triterpenoid, betulinic acid (21), was reported as having selective cytotoxic activity for several human melanoma cell lines, and also demonstrated highly effective tumor growth inhibition in athymic (nude) mice with human melanoma cells at non-toxic doses [72].

Table 2.   Plant Sources of Compounds with Known Biological Activity in the Subfamily Dipterocarpoideae

Compound class/name/code	Plant source (part) [reference citation]
OLIGOSTILBENOIDS	
Reveratrol Monomer	
(*E*)-Resveratrol 3-*O*-β-D-glucoside (1)	*Vatica diospyroides* Syn. (st) [56]
Resveratrol Dimers	
Balanocarpol (2)	*Balanocarpus zeylanicus* Trimen (b) [44], *Hopea brevipetiolaris* (b) [44], *H. jucunda*[b] (b) [44], *H. malibato* Foxw. (l) [51]
Malibatol A (3)	*Hopea malibato* (l) [51]
Malibatol B (4)	*Hopea malibato* (l) [51]
ε-Viniferin (6)	*Shorea disticha*[b] (b) [52], *Vatica affinis* Thr. (b) [45]
Resveratrol Trimers	
Canaliculatol (7)	*Stemonoporous canaliculatus* (b) [46]

(Table 2). contd.....

Compound class/name/code	Plant source (part) [reference citation]
Copalliferol A (8)	*Hopea cordifolia* (Thw.) Alston (b) [49], *H. brevipetiolaris* (b) [49] *Shorea stipularis*[b] (b) [49], *Stemonoporus affinis* (b) [48], *S. elegans* (Thw.) Alston (b) [48], *S. kanneliensis*[b] (b) [48], *S. oblongifolius* (b) [48], *Vateria copallifera* (Retz.) Alston (b) [47]
Copalliferol B (9)	*Vateria copallifera* (b) [50]
Distichol (10)	*Shorea disticha* (Thw.) Ashton (b) [52]
Resveratrol Tetramers	
Dibalanocarpol (12)	*Hopea malibato* (l) [51]
Vatdiospyroidol (14)	*Vatica diospyroides* (st) [56]
Vaticaffinol (15)	*Vatica affinis* (b) [57], *Stemonoporus canaliculatus* Thw. (b) [58], *S. cordifolius* (b) [48], *S. lancifolius* (b) [48]
TERPENOIDS **Sesquiterpenes**	
Caryophyllene (17)	*Anisoptera aurea* (r) [21], *A. curtisii* Dyer (r) [21], *A. glabra* Kurz (r) [21], *A. grossivenia* v. Sl. (r) [21], *Anisoptera laevis* (r) [21], *A. megistocarpa* v. Sl. (r) [21], *A. oblonga* (r) [21], *A. thurifera* (r) [21], *Colytelobium burckii* (Heim) Heim (r) [21], *C. malayanum* v. Sl. (r) [21], *C. melanoxylon* (Hook. f.) Pierre (r) [21], *Dipterocarpus acutangulus* (r) [19], *D. artocarpifolius* Pierre (r) [19], *D. borneensis* v. Sl. (r) [19], *D. chartaceus* (r) [19], *D. geniculatus* ssp. *grandis* Ashton (r) [19], *D. granduloflorus* (r) [19], *D. hispidus* (r) [18], *D. palembanicus* (r) [19], *D. appendiculatus* Scheff (r) [19], *D. caudiferus* Merr. (r) [19], *D. concavus* Foxw. (r) [19], *D. confertus* v. Sl. (r) [19], *D. costatus* Roxb. (r) [19], *D. cf. fagineus* (r) [19], *D. glandulosus* (r) [19], *D. gracilis* Bl. (r) [19], *D. hasseltii* Bl. (r) [19], *D. indicus* (r) [19], *D. oblongifolius* Bl. (r) [19], *D. pilosus* Roxb. (r) [19], *D. pseudofagineus* Foxw. (r) [19], *D. rigidus* Ridl. (r) [19], *D. sarawakensis* (r) [19], *D. trinervis* Gaertner f. (r) [19], *D. tuberculatus* (r) [19], *D. warburgii* (r) [19], *D. verrucosus* Foxw. (r) [19], *D. zeylanicus* (r) [19], *Doona congestifolia* Thw. (r) [17], *D. gardneri* (r) [17], *D. macrophylla* Thw. (r) [17], *D. oblonga* Thw. (r) [17], *D. zeylanica* Thw. (r) [17], *Dryobalanops aromatica* (r) [21], *D. becarii* Dyer (r) [21], *Upuna borneensis* (r) [21], *Vateria indica* (r) [72]
Copaene (18)	*Anisoptera aurea* (r) [21], *A. costata* (r) [21], *A. curtisii* (r) [21], *A. glabra* (r) [21], *A. grossivenia* (r) [21], *A. laevis* (r) [21], *A. megistocarpa* (r) [21], *A. oblonga* (r) [21], *A. thurifera* (r) [21], *Cotylelobium burckii* (r) [21], *C. malayanum* (r) [21], *Dipterocarpus caudiferus* (r) [19], *D. cf. fagineus* (r) [19], *D. obtusifolius* Teysm. (r) [19], *D. cf. palembanicus* (r) [19], *D. turbinatus* (r) [19], *D. warburgii* (r) [19], *Doona congestifolia* Thw. (r) [17], *D. gardneri* (r) [17], *D. macrophylla* Thw. (r) [17], *D. oblonga* Thw. (r) [17], *D. zeylanica* Thw. (r) [17], *Dryobalanops becarii* (r) [21], *Upuna borneensis* (r) [21]

(Table 2). contd.....

Compound class/name/code	Plant source (part) [reference citation]
α-Gurjunene (19)	*Dipterocarpus alatus* (r) [19], *D. artocarpifolius* (r) [19], *D. baudii* (r) [19], *D. borneensis* (r) [19], *D. cornutus* (r) [19], *D. dyeri* (r) [19], *D. fagineus* (r) [19], *D. granduloflorus* (r) [19], *D. humeratus* (r) [19], *D. intricatus* (r) [19], *D. jourdainii* (r) [19], *D. obtusifolius* (r) [19], *D. pseudofagineus* (r) [19], *D. sublamellatus* (r) [19], *D. tuberculatus* (r) [19], *D. turbinatus* (r) [19], *Upuna borneensis* (r) [21]
Triterpenoids	
β-Amyrin (20)	*Cotylelobium scariusculum* (Thw.) Brandis (tm) [23], *Doona congestifolia* (r) [17,18], *D. gardneri* (r) [17], *D. macrophylla* (r) [17], *D. oblonga* (r) [17], *D. zeylanica* (r) [17], *D. zeylanicus* (tm) [23], *Hopea brevipetiolaris* (tm) [23], *H. cordifolia* (b, tm) [23], *H. jucunda* var. *modesta* (tm) [23], *Shorea affinis* (Thw.) Ashton (tm) [23], *S. agami* Ashton (r) [22], *S. albida* (r) [22], *S. andulensis* Ashton (r) [22], *S. argentifolia* (r) [22], *S. assamica* Dyer ex Wood var. *globifera* (Ridl.) Sym. (r) [22], *S. balanocarpoides* (r) [22], *S. bracteolata* (r) [22], *S. congestiflora* (tw) [23], *S. faguetiana* (r) [22], *S. faguetioides* (r) [22], *S. flava* (r) [22], *S. gibbosa* (r) [22], *S. glaucescens* W. Meijer (r) [22], *S. gratissima* Dyer (r) [22], *S. guiso* (r) [22], *S. hypochra* Hance (r) [22], *S. isoptera* (r) [22], *S. laevis* (r) [22], *S. leprosula* (r) [22], *S. macrophylla* (r) [22], *S. magistophylla* (b) [23], *S. mecistopteryx* (r) [22], *S. cf. obscura* W. Meijer (r) [22], *S. ochracea* Sym. (r) [22], *S. ovalis* (r) [22], *S. aff. parvifolia* Dyer (r) [22], *S. cf. patoiensis* Ashton (r) [22], *S. quadrinervis* v. Sl. (r) [22], *Shorea resina-nigra*[b] (r) [134], *S. rubella* (r) [22], *S. scaberrima* (r) [22], *S. seminis* (r) [22], *S. sericeiflora* Fischer et Hutch. (r) [22], *S. smithiana* Sym. (r) [22], *S. stipularis* (b, tm) [23], *S. symingtonii* Wood (r) [22], *S. venulosa* (r) [22], *S. virescens* Parijs (r) [22], *S. worthingtonii*[b] (b, tm) [119], *S. xanthophylla* (r) [22], *Vateria copallifera* (b) [23], *Vatica obscura* (b, tm) [65], *V. affinis* (tm) [23], *Upuna borneensis* Sym. (r) [21]
Betulinic acid (21)	*Cotylelobium scariusculum* (b) [23], *Dipterocarpus glandulosus* (tm) [23], *D. hispidus* (tm) [18,23], *D. insignis* (tm) [23], *D. zeylanicus* (tm) [23], *Hopea brevipetiolaris* (tm) [23], *H. cordifolia* (tm) [23], *H. jucunda* var. *modesta* (tm) [23], *Vatica diospyroides* (st) [56], *Shorea affinis* (tm) [23], *S. congestiflora* (tm) [23], *S. dyerii* (tm) [23], *S. magistophylla* (b) [23], *S. oblongifolia* (tm) [23], *S. ovalifolia* (tm) [23], *S. wangtianshuea*[b] (rb) [80], *S. trapezifolia* (tm) [23], *Stemonoporus affinis* (tm) [23], *S. canaliculatus* (tm) [23], *S. cordifolius* (tm) [23], *S. oblongifolius* (tm) [23], *S. petiolaris* (tm) [23], *S. reticulatus* (tm) [23], *Vatica affinis* (tm) [23]
Oleanolic acid (22)	*Hopea pubescens* (hw) [138], *Dryobalanops aromatica* (r) [131]
ψ-Taraxasterol (23)	*Doona congestifolia* (r) [17], *D. gardneri* (r) [17], *D. macrophylla* (r) [17], *D. oblonga* (r) [17], *D. zeylanica* (r) [17], *Shorea resina-nigra*[b] (r) [133]

(Table 2). contd.....

Compound class/name/code	Plant source (part) [reference citation]
Ursolic acid (24)	*Cotylelobium scabriusculum* (tm) [23], *Dipterocarpus glandulosus* (tm) [23], *D. hispidus* (tm) [23], *D. insignis* (tm) [23], *D. zeylanicus* (tm) [23], *Doona congestiflora* (r) [17,18], *D. gardneri* (r) [17], *D. macrophylla* (r) [17,18], *D. oblonga* (r) [17], *D. zeylanica* (r) [17], *Hopea brevipetiolaris* (tm) [23], *H. cordifolia* (b, tm) [23], *H. jucunda* var. *modesta* (tm) [23], *Shorea affinis* (tm) [23], *S. congestiflora* (tm) [23], *S. dyeri* Thw. (tm) [23], *S. lysophylla* Ashton (tm) [23], *S. magistophylla* (b) [23], *S. oblongifolia* (tm) [23], *S. ovalifolia* (tm) [23], *S. robusta* (r) [67], *S. stipularis* (tm) [23], *S. trapezifolia* (tm) [23], *S. worthingtonii* (b, tm) [119], *S. zeylanica* (tm) [23], *Stemonoporus affinis* (b, tm) [20,23], *S. canaliculatus* (b, tm) [20,23], *S. cordifolius* (b, tm) [20], *S. elegans* (b) [20], *S. lancifolius* (b, tm) [23], *S. oblongifolius* (b, tm) [20,23], *S. petiolaris* (b, tm) [20,23], *S. reticulatus* (b, tm) [20,23], *Vateria copallifera* (tm) [23], *Vatica affinis* (b, tm) [23]
Ursonic acid (25)	*Stemonoporus affinis* (b, tm) [20], *S. canaliculatus* (tm) [20], *S. lancifolius* (b, tm) [20], *S. petiolaris* (b, tm) [20], *S. reticulatus* (b, tm) [20]
MISCELLANEOUS **Coumarins**	
Bergenin (26)	*Dipterocarpus indicus* Bedd. (b) [117], *Shorea leprosula* Miq. (hw) [132], *Stemonoporus affinis* Thw. (b) [20,48], *S. canaliculatus* (Thw.) (b) [20] *S. kanneliensis*[b] (b) [48], *S. lancifolius* (Thw.) Ashton (b) [20,48], *S. petiolaris* Thw. (b) [20], *S. reticulatus* (Thw.) (b) [20], *Vateria indica* L. (l, r) [54], *Vatica affinis* Thw. (b) [45], *V. diospyroides* Sym. (st) [56]
Scopoletin (27)	*Cotylelobium scabriusculum* (Thw.) Brandis (tm) [23], *Shorea lysophylla* Thw. (tm) [23], *S. magistophylla* Ashton (tm) [23], *S. trapezifolia* (Thw.) Ashton (tm) [23], *Stemonoporus affinis* (tm) [23], *S. oblongifolius* Thw. (tm) [23], *S. petiolaris* (tm) [23], *S. reticularis* Thw. (tm) [23], *Vatica affinis* (tm) [23], *V. obscura* Trimen (tm) [119]
Ellagic acid derivative	
Ellagic acid (28)	*Balanocarpus heimii* King (l) [16], *Dipterocarpus baudii* Korth. (l) [16], *D. crinitus* Dyer (l) [16], *Dryobalanops aromatica* Gaertner f. (l) [16], *D. oblongifolia* Dyer (l) [16], *Hopea beccariana* Burck (l) [16], *H. mengarawan* Miq. (l) [16], *H. nutans* Ridl. (l) [16], *H. odorata* Roxb. (l) [16], *H. sangal* Korth. (l) [16], *H. subalata* Sym. (l) [16], *Shorea bracteolata* Dyer (l) [16], *S. foxworthyi* Sym. (l) [16], *S. hemsleyana* (King) King ex Foxw. (l) [16], *S. maxima* (King) Sym. (l) [16], *S. maxwelliana* King (l) [16], *S. multiflora* (Burck.) Sym. (l) [16], *S. robusta* (s) [120], *Vatica stapfiana* (King) v. Sl. (l) [16], *V. wallichii* Dyer (l) [16]

(Table 2). contd.....

Compound class/name/code	Plant source (part) [reference citation]
Flavonoids	
Cyanidin (29)	*Dipterocarpus baudii* (l) [16], *D. kerrii* (l) [16], *Dryobalanops aromatica* (l) [16], *D. oblongifolia* (l) [16], *Shorea bracteolata* (l) [16], *S. curtisii* (l) [16], *S. kunstleri* (l) [16], *S. leprosula* (l) [16], *S. maxwelliana* (l) [16], *S. multiflora* (l) [16], *Vatica nitens* (l) [16]
Delphinidin (30)	*Dipterocarpus baudii* (l) [16], *D. costulatus* (l) [16], *D. crinicus* (l) [16], *D. kerrii* (l) [16], *Dryobalanops aromatica* (l) [16], *D. oblongifolia* (l) [16], *Shorea foxworthi* Sym. (l) [16], *S. curtisii* (l) [16], *S. leprosula* (l) [16], *S. singkawang* (Miq.) Burck. (l) [16], *Vatica nitens* (l) [16]
Kaempferol (31)	*Anisoptera laevis* Ridl. (l) [16], *Hopea beccariana* (l) [16], *Shorea bracteolata* (l) [16], *S. leprosula* (l) [16], *S. kunstleri* King (l) [16], *S. maxima* (l) [16], *S. multiflora* (l) [16]
Myricetin (32)	*Dipterocarpus baudii* (l) [16], *D. costulatus* v. Sl. (l) [16], *D. crinicus* (l) [16], *D. kerrii* King (l) [16], *Dryobalanops aromatica* (l) [16], *D. oblongifolia* (l) [16], *Hopea beccariana* (l) [16], *Shorea maxwelliana* (l) [16], *S. curtisii* Dyer ex King (l) [16], *Vatica nitens* King (l) [16]
Quercetin (33)	*Dipterocarpus costulatus* (l) [16], *D. kerrii* (l) [16], *Dryobalanops aromatica* (l) [16], *Hopea beccariana* (l) [16], *H. nutans* (l) [16], *H. subalata* (l) [16], *Shorea bracteolata* (l) [16], *S. curtisii* (l) [16], *S. kunstleri* (l) [16], *S. leprosula* (l) [16]
Phenolics	
Caffeic acid (34)	*Balanocarpus heimii* (l) [16], *Hopea odorata* (l) [16], *Hopea sangal* (l) [16]
p-Coumaric acid (35)	*Dipterocarpus macrocarpus*[b] (hw) [121]
Ferulic acid (36)	*Anisoptera laevis* (l) [16], *Hopea mengarawan* (l) [16]
Gallic acid (37)	*Shorea robusta* (s) [120]
Quinones	
2,6-Dimethoxy-1,4-benzoquinone (38)	*Shorea* species[b] (w) [125]
Chrysophanic acid (= Chrysophanol) (39)	*Shorea affinis* (tm) [23], *S. congestiflora* (tm) [23], *S. dyerii* (tm) [23], *S. lysophylla* (tm) [23], *S. magistophylla* (tm) [23], *S. stipularis* (tm) [23], *S. worthingtonii* Ashton (tm) [119], *S. zeylanica* (tm) [23]

[a]Plant part code: b = bark, eo = essential oil, ep = entire plant, hw = heart wood, l = leaf, ns = not stated, r = resin, rb = root bark, s = seed, so = seed oil, st = stem, tb = trunk bark, tm = timber, w = wood. [b]Authority was not stated.

Miscellaneous constituents of the subfamily Dipterocarpoideae, such as coumarins, ellagic acid, flavonoids, and other phenolics were identified in phytochemical screening carried out on the fresh leaves of 28 species from Malaysia by TLC comparison with authentic samples [16]. Many of these compounds have well defined biological activities, as indicated in Table 1 [84-114]. For instance, bergenin (26) showed weak anti-HIV activity in C8166 cells infected with HIV-1$_{MN}$ (EC$_{50}$ 40 μg/mL, selectivity index >25) [82]. Ellagic acid (28) has exhibited a variety of anticarcinogenic activities. It inhibited the CYP1A1-dependent activation of benzo[α]pyrene, detoxified the diol-epoxide of benzo[α]pyrene, and induced the phase II detoxification enzyme, glutathione-*S*-transferase [86]. Biological activities for these and other compounds are summarized in Table 1.

In Table 2, the occurrence of constituents of the plants in the subfamily Dipterocarpoideae with known biological activity is summarized.

BIOACTIVE CONSTITUENTS OF THE SUBFAMILY MONOTOIDEAE

Recently, five new flavanones [150] were reported in our work from the leaves of *Monotes engleri*, namely, 6-(1,1-dimethylallyl)naringenin (40), 6-(1,1-dimethylallyl)eriodictyol (41), 3'-*O*-methyl-6-(1,1-dimethylallyl)erio-dictyol (42), 6-[(2*RS*)-hydroxy-3-methyl-3-butenyl]-8-prenyleriodictyol, and 5,4'-dihydroxy-4",4",5"-trimethyl-5"*H*-dihydrofurano[2",3":6,7] flavanone along with the known flavanones 6,8-diprenyleriodictyol (43) [151] and hiravanone (44) [152]. 6- (1,1-Dimethylallyl)naringenin (40), 6-(1,1-dimethylallyl)eriodictyol (41), 3'-*O*-methyl-6-(1,1-dimethylallyl)eriodictyol (42), 6,8-diprenyleriodictyol (43) and hiravanone (44) were broadly cytotoxic when evaluated against a panel of human cancer cells at the University of Illinois at Chicago [56,150]. These flavanones exhibited their most potent cytotoxic activity against U373 cells (human glioma, IC$_{50}$ 3.3, 5.3, 1.9, 3.1 and 2.8 μg/mL, respectively), thereby showing some selectivity for this cell line [56,150].

Flavanones 43 and 44, which contain two prenyl side-chains at C-6 and C-8, exhibited cytotoxic activity, whereas 6-[(2*RS*)-hydroxy-3-methyl-3-butenyl]-8-prenyleriodictyol, was regarded as inactive even though it has the same structure as 43 except for the side chain at C-6. From these observations, it is suggested that the prenyl group affixed to C-6 in compounds 43 and 44 is required for their cytotoxic activity. Flavanones 40-42, containing a 1,1-dimethylallyl group at C-6, demonstrated cytotoxic activity. However, 5,4'-dihydroxy-4",4",5"-trimethyl-5"*H*-dihydro-furano[2",3":6,7]flavanone, resulting from the cyclization of the 1,1-dimethylallyl group at C-6 of compound 40, was found to be a non-cytotoxic compound. Therefore, it is apparent that the 1,1-dimethylallyl functionality at C-6 in flavanones 40-42 is important for the mediation of cytotoxic activity. In addition, 6-(1,1-dimethylallyl)naringenin (40) demonstrated weak cytotoxicity against several cell lines such as breast cancer [HS 578T -$\log_{10}$ GI$_{50}$ (4.82), TGI (4.31), LC$_{50}$ (>4.00)], leukemia [CCRF-CEM-$\log_{10}$ GI$_{50}$ (5.03), TGI (4.61), LC$_{50}$ (>4.21)], and renal cancer [A498 -$\log_{10}$ GI$_{50}$ (5.20), TGI (4.70), LC$_{50}$ (>4.34)] in the National Cancer Institute 60 human tumor cell line panel [150,153].

Table 3.   **Bioactive Compounds from *Monotes engleri* in the Subfamily Monotoideae**

Compound class/name/code	Activity	Reference
FLAVONOIDS[a]		
6-(1,1-Dimethylallyl)naringenin (40)	Cytotoxicity against human cancer cell panel	150
6-(1,1-Dimethylallyl)eriodictyol (41)	Cytotoxicity against human cancer cell panel	150
3'-O-Methyl-6-(1,1-dimethylallyl)-eriodictyol (42)	Cytotoxicity against human cancer cell panel	150
6,8-Diprenyleriodictyol (43)	Cytotoxicity against human cancer cell panel	150
Hiravanone (44)	Cytotoxicity against human cancer cell panel	150
COUMARIN[b]		
5-(4-Hydroxyphenethenyl)-4,7-dimethoxycoumarin (45)	Cytotoxicity against HT (fibrosarcoma) and LNCaP (human hormone-dependent prostate) cell lines	56

[a]Isolated from the leaves of *Monotes engleri.*
[b]Isolated from the combined extracts of the root wood, root bark, and stem bark of *M. engleri.*

From the root wood, root bark, and stem bark of *M. engleri*, a new compound, 5-(4-hydroxyphenethenyl)-4,7-dimethoxycoumarin (45) was isolated [56], and showed weak cytotoxic activity against fibrosarcoma (HT: IC$_{50}$ 5.3 μg/mL) and human hormone-dependent prostate (LNCaP, IC$_{50}$ 1.9 μg/mL) cancer cell lines [150]. The structures of the constituents of species in the subfamily Monotoideae are shown in Fig. 3. Active compounds are summarized in Table 3.

CONSTITUENTS OF THE SUBFAMILY PAKARAIMOIDEAE

Only one preliminary study has been published so far on the single species known in the subfamily Pakaraimoideae, namely, *Pakaraimaea dipterocarpacea* Maguire & Ashton. Ellagic acid which is also reported as a bioactive compound in the subfamily Dipterocarpoideae in Table 1, was identified in the wood extract of this species by comparison with an authentic sample [149]. UV analysis was used to detect the anthocyanidin leuco-7-O-methyl pelargonidin [149]. 3-O-Glycosides of flavonols such as

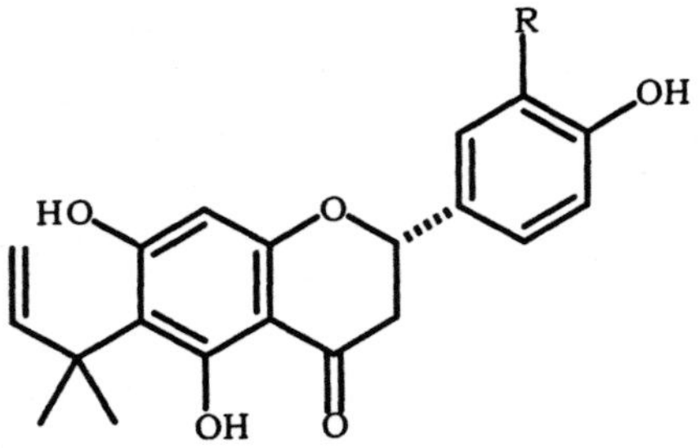

40 R = H
41 R = OH
42 R = OCH$_3$

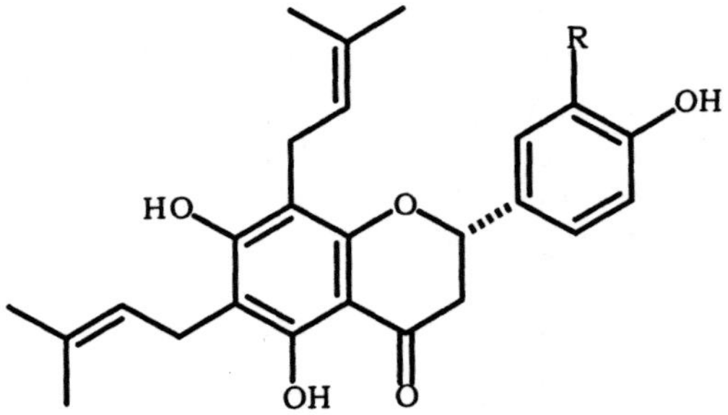

43 R = H
44 R = CH$_3$

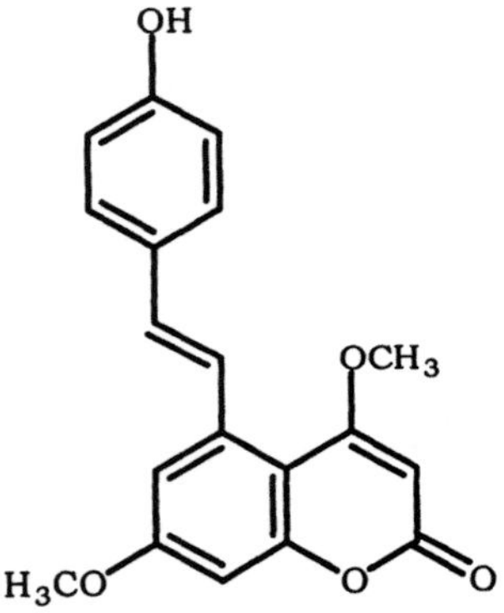

45

Fig. (3). Structures of reported bioactive constituents of plants in the subfamily Monotoideae.

myricetin, quercetin, and kaempferol were reported from *P. dipterocarpacea* without any identification of specific glycosides. Analysis by GC/MS showed sesquiterpenes, probably based on the elemane, farnesane, and humulane skeletons, as well as triterpenes similar to hydroxydammarenone-I, dipterocarpol, and dammarenediol, although these compounds were not identified conclusively [149].

CONCLUSIONS

The plant family Dipterocarpaceae is comprised of three subfamilies, Dipterocarpoideae, Monotoideae, and Pakaraimoideae. Species of the largest subfamily, Dipterocarpoideae, are of most interest because of their economic use in producing timber and oleoresins. Plants of the Dipterocarpoideae are typified as sources of oligostilbenoids (resveratrol oligomers) which have shown various biological activities such as antibacterial, antifungal, anti-HIV antiviral, and cytotoxic effects. This class of plant secondary metabolite seems worthy of more detailed biological and chemical investigation. The considerable structural complexity of members of this compound class has resulted in difficulties in confirming their structures and in assigning their ^{1}H- and ^{13}C-NMR signals unambiguously. In the past few years, 2D-NMR techniques such as ^{1}H-^{13}C HMQC and ^{1}H-^{13}C HMBC and molecular modeling have facilitated the structure elucidation of these complex oligostilbenoids. The recent identification of the first resveratrol monomer, (E)-resveratrol 3-O-β-D-glucoside (1), from a plant in the subfamily Dipterocarpoideae (*Vatica diospyroides*) is noteworthy from both a chemotaxonomic and biogenetic point of view. Oligostilbenoids have not yet been characterized from any plant in the subfamilies Monotoideae and Pakaraimoideae. Many phenolic compounds, sesquiterpenes and triterpenoids which were reported as active substances in other families, have been reported also in the Dipterocarpoideae. Several triterpenes, which are considered as chemotaxonomic markers of the family Dipetrocarpaceae, such as dipterocarpol and dammaradienone, need to be studied for their potential biological activities.

Prenylated flavanones obtained from a plant of the subfamily Monotoideae, showed significant cytotoxic activity against several human cancer cell lines. According to preliminary phytochemical screening data, plants of the subfamily Monotoideae and Pakaraimoideae also seem to contain sesquiterpenes and triterpenes similar to those of the Dipterocarpoideae, although such compounds have only been identified tentatively using chromatographic separatory techniques. Prenylated flavanones have been found as major constituents of the leaves of *Monotes engleri* (subfamily Monotoideae), but have not been reported from any other plant in the Dipterocarpaceae so far. Prenylated flavanones, in particular flavanones containing a 1,1-dimethylallyl group or prenyl group at C-6, may have a role as chemotaxonomic markers for the subfamily Monotoideae.

As discussed in some depth in this chapter, information on the biological activities of constituents of plants in the tropical, mid-sized family Dipterocarpaceae has been emerging steadily over the last few years. Additional aspects of biological activity may be expected in the future, particularly because some of the species in the subfamily Dipterocarpoideae have considerable economic importance for their oleoresin and timber. The resveratrol oligomers stand out as unusual secondary metabolites of this subfamily, and it will be of interest to determine if these compounds as polyphenols are similar in biological activity to the "vegetable tannins" (gallotannins and proanthocyanidins). The tannins are well-known to complex with proteins and polysaccharides through the formation of hydrogen bonds, and have a wide range of biological activities and practical uses as a consequence [154]. Perhaps the oligostilbenoids from the Dipterocarpaceae and other plant families will ultimately prove to exhibit an even wider array of biological activities than the vegetable tannins, owing to the antioxidant, antimutagenic, and cancer chemopreventive activity of their monomer, resveratrol, as recently described [155].

ACKNOWLEGMENT

The studies mentioned in this chapter which were carried out in our laboratory were supported by NIH grant U01/U19 CA 52956. We are grateful to the individual contributions of many scientific colleagues associated with this project in carrying out this collaborative work.

REFERENCES

[1]   Ashton, P.S. In *Flora Malesiana*, Ser. I, *Spermatophyta*; Steenis, C.G.G.J., Ed.; Martinus Nijhoff-Dr. W. Junk Publishers: Boston, **1983**; Vol. *9*, pp. 237-552.
[2]   Mabberley, D.J. *The Plant-Book*, Cambridge University Press: New York, **1993**; p.188.
[3]   Verdcourt, B. In *Flora of Tropical East Africa*, Polhill, R.M., Ed.; A. A. Balkema: Rotterdam, **1989**; pp. 1-11.
[4]   Ashton, P.S. *Gard. Bull. Soc.*, **1963**, *20*, 229-288.
[5]   Maguire, B.; Ashton, P.S. *Taxon*, **1977**, *26*, 343-368.
[6]   Maguire, B.; Ashton, P.S. *Taxon*, **1980**, *29*, 225-231.
[7]   Van Slooten, D.F. In *Bulletin du Jardin Botanique. Buitenzorg*, Van Leeuwen, W.M.D.; Von Faber, F.C.; Beumée, J.G.B., Eds.; Archipel Drukkerij-Buitenzorg: Buitenzorg, Netherlands, **1927**; Vol. *9*, pp. 67-136.
[8]   Bancroft, H. *Am. J. Bot.*, **1935**, *22*, 505-519.
[9]   Ismam, M.B.; Gunning, P.J.; Spollen, K.M. In *Phytochemicals for Pest Control*, Hedin, P.A.; Hollingworth, R.M.; Masler, E.P.; Miyamoto, J.; Thompson, D.G., Eds.; American Chemical Society: Washington, DC, **1997**; ACS Symp. Ser. 658, pp. 27-37.
[10]   Duvigneaud, P. In *Flora Zambesiaca, Vol. I.*; Exell, A.W.; Wild, H. Eds.; Crown Agents for Overseas Governments and Administrations: London, **1960**; pp. 407-420.
[11]   Oliver, D. *Flora of Tropical Africa*. Vol. 1, **1868**.
[12]   Kostermans, A.J.G.H. *Taxon*, **1978**, *27*, 357-359.
[13]   Kostermans, A.J.G.H. *Taxon*, **1985**, *34*, 426-435.
[14]   Kostermans, A.J.G.H. *Taxon*, **1989**, *38*, 123-124.
[15]   Kostermans, A.J.G.H. *A Handbook of the Dipterocarpaceae of Sri Lanka*. Wildlife Heritage Trust of Sri Lanka: Colombo, Sri Lanka, **1992**.
[16]   Bate-Smith, E.C.; Whitmore, T.C. *Nature*, **1959**, *184*, 795-796.
[17]   Diaz, M.A.; Ourisson, G.; Bisset, N.G. *Phytochemistry*, **1966**, *5*, 855-863.
[18]   Bandaranayake, W.M.; Gunasekera, S.P.; Karunanayake, S.; Sotheeswaran, S.; Sultanbawa, M.U.S. *Phytochemistry*, **1975**, *14*, 2043-2048.
[19]   Bisset, N.G.; Diaz, M.A.; Ehret, C.; Ourisson, G.; Palmade, M.; Patil, F.; Pesnelle, P.; Streith, J. *Phytochemistry*, **1966**, *5*, 865-880.

[20] Bandaranayake, W.M.; Karunanayake, S.; Sotheeswaran, S.; Sultanbawa, M.U.S.; Balasubramaniam, S. *Phytochemistry*, **1977**, *16*, 699-701.

[21] Bisset, N.G.; Diaz-Parra, M.A.; Ehret, C.; Ourisson, G. *Phytochemistry*, **1967**, *6*, 1395-1405.

[22] Bisset, N.G.; Chavanel, V.; Lantz, J.-P.; Wolff, R.E. *Phytochemistry*, **1971**, *10*, 2451-2463.

[23] Gunawardana, Y.A.G.P.; Sultanbawa, M.U.S.; Balasubramaniam, S. *Phytochemistry*, **1980**, *19*, 1099-1102.

[24] Sotheeswaran, S.; Pasupathy, V. *Phytochemistry*, **1993**, *32*, 1083-1092.

[25] Madhav, R.; Seshadri, T.R.; Subramanian, G.B.V. *Phytochemistry*, **1967**, *6*, 1155-1156.

[26] Pryce, R.J.; Langcake, P. *Phytochemistry*, **1977**, *16*, 1452-1454.

[27] Nakajima, K.; Taguchi, H.; Endo, T.; Yosioka, I. *Chem. Pharm. Bull.*, **1978**, *26*, 3050-3057.

[28] Lins, A.P.; Ribeiro, M.N.De.S.; Gottlieb, O.R.; Gottlieb, H.E. *J. Nat. Prod.*, **1982**, *45*, 754-761.

[29] Khan, M.A.; Nabi, S.G.; Prakash, S.; Zaman, A. *Phytochemistry*, **1986**, *25*, 1945-1948.

[30] Powell, R.G.; Bajaj, R.; McLaughlin, J.L. *J. Nat. Prod.*, **1987**, *50*, 293-296.

[31] Baba, K.; Maeda, K.; Tabata, Y.; Doi, M.; Kozawa, M. *Chem. Pharm. Bull.*, **1988**, *36*, 2977-2983.

[32] Hegnauer, R. *Chemotaxonomie der Pflanzen*, Birkhauser Verlag: Boston, **1989**; Band 8, pp. 392-398.

[33] Boralle, N.; Gottlieb, H.E.; Gottlieb, O.R.; Kubitzki, K.; Lopes, L.M.X.; Yoshida, M.; Young, M.C.M. *Phytochemistry*, **1993**, *34*, 1403-1407.

[34] Kurihara, H.; Kawabata, J.; Ichikawa, S.; Mishima, M.; Mizutani, J. *Phytochemistry*, **1991**, *30*, 649-653.

[35] Oshima, Y.; Ueno, Y. *Phytochemistry*, **1993**, *33*, 179-182.

[36] Ohyama, M.; Tanaka, T.; Iinuma, M.; Goto, K. *Chem. Pharm. Bull.*, **1994**, *42*, 2117-2120.

[37] Kawabata, J.; Mishima, M.; Kurihara, H.; Mizutani, J. *Phytochemistry*, **1991**, *30*, 645-647.

[38] Oshima, Y.; Kamijou, A.; Moritani, H.; Namao, K.-I.; Ohizumi, Y. *J. Org. Chem.*, **1993**, *58*, 850-853.

[39] Korhammer, S.; Reniero, F.; Mattivi, F. *Phytochemistry*, **1995**, *38*, 1501-1504.

[40] Ohyama, M.; Tanaka, T.; Iinuma, M. *Phytochemistry*, **1995**, *38*, 733-740.

[41] Li, W.W.; Ding, L.-S.; Li, B.G.; Chen, Y.Z. *Phytochemistry*, **1996**, *42*, 1163-1165.

[42] Ourtoule, J.-C.; Bourhis, M.; Vercauteren, J. *Tetrahedron Lett.*, **1996**, *37*, 4697-4700.

[43] Reniero, F.; Rudolph, M.; Angioni, A.; Bernreuther, A.; Cabras, P.; Mattivi, F. *Vitis*, **1996**, *35*, 125-127.

[44] Diyasena, M.N.C.; Sotheeswaran, S.; Surendrakumar, S.; Balasubramanian, S.; Bokel, M.; Kraus, W. *J. Chem. Soc., Perkin Trans. I*, **1985**, 1807-1809.

[45] Sotheeswaran, S.; Sultanbawa, M.U.S.; Surendrakkumar, S.; Balasubramaniam, S.; Bladon, P. *J. Chem. Soc., Perkin Trans. I*, **1985**, 159-162.

[46] Bokel, M.; Diyasena, C.; Gunatilaka, A.A.L.; Kraus, W.; Sotheeswaran, S. *Phytochemistry*, **1988**, *27*, 377-380.

[47] Sultanbawa, M.U.S.; Surendrakkumar, S. *J. Chem. Soc., Chem. Commun.*, **1980**, 619-620.

[48]   Samaraweera, U.; Sotheeswaran, S.; Sultanbawa, M.U.S. *Phytochemistry*, **1982**, *21*, 2585-2587.

[49]   Sotheeswaran, S.; Sultanbawa, M.U.S.; Surendrakumar, S.; Bladon, P. *J. Chem. Soc., Perkin Trans. I*, **1983**, 699-702.

[50]   Gunawardana, Y.A.G.P.; Sotheeswaran, S.; Sultanbawa, M.U.S.; Surendrakumar, S.; Bladon, P. *Phytochemistry*, **1986**, *25*, 1498-1500.

[51]   Dai, J.-R.; Hallock, Y.F.; Cardellina II, J.H.; Boyd, M.R. *J. Nat. Prod.*, **1998**, *61*, 351-353.

[52]   Sultanbawa, M.U.S.; Surendrakkumar, S.; Bladon, P. *Phytochemistry*, **1987**, *26*, 799-801.

[53]   Coggon, P.; Janes, N.F.; King, F.E.; King, T.J.; Molyneux, R.J.; Morgan, J.W.W.; Sellars, K. *J. Chem. Soc.*, **1965**, 406-409.

[54]   Dayal, R. *J. Indian Chem. Soc.*, **1987**, *64*, 259.

[55]   Saraswathy, A.; Purushothaman, K.K.; Patra, A.; Dey, A.K.; Kundu, A.B. *Phytochemistry*, **1992**, *31*, 2561-2562.

[56]   Seo, E.K. Structure Elucidation of Bioactive Constituents of Two Plants in the Dipterocarpaceae. Ph.D. Thesis, University of Illinois at Chicago, Chicago, IL, **1997**.

[57]   Sultanbawa, M.U.S.; Surendrakkumar, S.; Wazeer, I.M. *J. Chem. Soc., Chem. Commun.*, **1981**, 1204-1206.

[58]   Sotheeswaran, S.; Diyasena, M.N.C.; Gunatilaka, A.A.L.; Bokel, M.; Kraus, W. *Phytochemistry*, **1987**, *26*, 1505-1507.

[59]   Ito, J.; Niwa, M. *Tetrahedron*, **1996**, *52*, 9991-9998.

[60]   Coggon, P.; King, T.J.; Wallwork, S.C. *Chem. Commun.*, **1966**, 439-440.

[61]   Teguo, P.W.; Fauconneau, B.; Deffieux, G.; Huguet, F.; Vercauteren, J.; Merillon, J.-M. *J. Nat. Prod.*, **1998**, *61*, 655-657.

[62]   Ohyama, M.; Ichise, M.; Tanaka, T.; Iinuma, M.; Bab, K.; Doi, M.; Burandt, C.L., Jr. *Tennen Yuki Kagobutsu Toronkai Koen Yoshishu*, **1995**, *37*, 469-474.

[63]   Oshima, Y.; Namao, K; Kamijou, A.; Matsuoka, S.; Nakano, M.; Terao, K.; Ohizumi, Y. *Experientia*, **1995**, *51*, 63-66.

[64]   Kulanthaivel, P.; Janzen, W.P.; Ballas, L.M.; Jiang, J.B.; Hu, C.-Q.; Darges, J.W.; Seldin, J.C.; Cofield, D.J.; Adams, L.M. *Planta Med.*, **1995**, *61*, 41-44.

[65]   Kang, B.-S.; Shin, N.-H.; Lee, S.-H.; Min, K.P.; Kim, Y. *Med. Sci. Res.*, **1998**, *26*, 235-237.

[66]   Silverstein, R.M.; Bassler, G.C.; Morrill, T.C. *Spectroscopic Identification of Organic Compounds*, 5th edn., John Wiley & Sons: New York, **1991**.

[67]   Hota, R.K.; Bapuji, M. *Phytochemistry*, **1993**, *32*, 466-468.

[68]   Kubo, I.; Muroi, H.; Himejima, M. *J. Agric. Food Chem.*, **1992**, *40*, 245-248.

[69]   Flath, R.A.; Cunningham, R.T.; Mon, T.R.; John, J.O. *J. Chem. Ecol.*, **1994**, *20*, 2595-2609.

[70]   Fournier, G.; Hadjiakhoondi, A.; Leboeuf, M.; Cavé, A.; Charles, B.; Fourniat, J. *Phytother. Res.*, **1994**, *8*, 166-169.

[71]   Ling, H.C.; King, M.L.; Chen, C.F.; Hsu, K.P.; Su, M.H.; Lin, M.H. *Chunghua I Hsueh Tsa Chih* (Taipei), **1982**, *29*, 308-315.

[72]   Pisha, E.; Chai, H.-B.; Lee, I.-S.; Chagwedera, T.E.; Farnsworth, N.R.; Cordell, G.A.; Beecher, C.W.W.; Fong, H.H.S.; Kinghorn, A.D.; Brown, D.M.; Wani, M.C.; Wall, M.E.; Hieken, T.J.; Das Gupta, T.K.; Pezzuto, J.M. *Nature Med.*, **1995**, *1*, 1046-1051.

[73]   Fulda, S.; Friesen, C.; Los, M.; Scaffidi, C.; Mier, W.; Benedict, M.; Munez, G.; Krammer, P.H.; Peter, M.E.; Debatin, K.-M. *Cancer Res.*, **1997**, *57*, 4956-4964.

[74]   Noda, Y.; Kaiya, T.; Kohda, K.; Kawazoe, Y. *Chem. Pharm. Bull.*, **1997**, *45*, 1665-1670.

[75]   Liu, J.; Liu, Y.; Parkinson, A.; Klaassen, C.D. *J. Pharmacol. Exp. Ther.*, **1995**, *275*, 768-774.

[76]   Balanehru, S.; Nagarajan, B. *Biochem. Int.*, **1991**, *24*, 981-990.

[77]   Ohigashi, H.; Takamura, H.; Koshimizu, K.; Tokuda, H.; Ito, Y. *Cancer Lett.*, **1986**, *30*, 143-151.

[78]   Villarreal, M.L.; Alvarez, L.; Alonso, D.; Navarro, V.; Garcia, P.; Delgado, G. *J. Ethnopharmacol.*, **1994**, *42*, 25-29.

[79]   Quere, L.; Wenger, T.; Schramm, H.J. *Biochem. Biophys. Res. Commun.*, **1996**, *227*, 484-488.

[80]   Varanda, E.M.; Zuniga, G.E.; Salatino, A.; Roque, N.F.; Corcuera, L.J. *J. Nat. Prod.*, **1992**, *55*, 800-803.

[81]   Ryu, S.Y.; Lee, C.K.; Ahn, J.W.; Lee, S.H.; Zee, O.P. *Arch. Pharmacal Res.*, **1993**, *16*, 339-342.

[82]   Piacente, S.; Pizza, C.; De Tommasi, N. *J. Nat. Prod.*, **1996**, *59*, 565-569.

[83]   Shimomura, H.; Sashida, Y.; Nakata, H.; Kawasaki, J.; Ito, Y. *Phytochemistry*, **1982**, *21*, 2213-2215.

[84]   Mueller-Enoch, D. *Biosci.*, **1994**, *49*, 763-771.

[85]   Ojewole, J.A.O. *Fitoterapia*, **1983**, *54*, 153-161.

[86]   Barch, D.H.; Rundhaugen, L.M.; Stoner, G.D.; Pillay, N.S.; Rsoche, W.A. *Carcinogenesis*, **1996**, *17*, 265-269.

[87]   Weider-Wells, M.A.; Altom, J.; Fernandez, J.; Fraga-Spano, S.A.; Hilliard, J.; Ohemeng, K.; Barrett, J.F. *Bioorg. Med. Chem. Lett.*, **1998**, *8*, 97-100.

[88]   Zapata, J.M.; Calderon, A.A.; Barcelo, A.R. *Wein-Wiss.*, **1994**, *49*, 114-116.

[89]   Mistry, K.J.; Krishna, M.; Bhattacharya, R.K. *Cancer Lett.*, **1997**, *121*, 99-104.

[90]   Rice-Evans, C.A.; Miller, N.J.; Bolwell, P.G.; Bramley, P.M.; Pridham, J.B. *Free Radical Res.*, **1995**, *22*, 375-383.

[91]   Teissedre, P.-L.; Waterhouse, A.L.; Frankel, E.N. *J. Int. Sci. Vigne Vin.*, **1995**, *29*, 205-212.

[92]   Frankel, E.N.; Waterhouse, A.L.; Teissedre, P.L. *J. Agric. Food. Chem.*, **1995**, *43*, 890-894.

[93]   Gabor, M.; Razga, Z. *Acta Physiol. Hung.*, **1991**, *77*, 197-207.

[94]   Gabor, M.; Razga, Z. *Hung. Kiserl. Orvostud.*, **1990**, *42*, 57-66.

[95]   Cai, L.; Wu-Yuan, C.D. *J. Nat. Prod.*, **1996**, *59*, 987-990.

[96]   Boege, F.; Straub, T.; Kehr, A.; Boesenberg, C.; Christiansen, K.; Andersen, A.; Jakob, F.; Koehrle, J. *J. Biol. Chem.*, **1996**, *271*, 2262-2270.

[97]   Aboobaker, V.S.; Balgi, A.D.; Bhattacharya, R.K. *In Vivo*, **1994**, *8*, 1095-1098.

[98]   Firozi, P.F.; Bhattacharya, R.K. *J. Biochem. Toxicol.*, **1995**, *10*, 25-31.

[99]   Chung, T.Y.; Kim, M.A.; Jones, A.D. *Han'guk Nonghwa Hakhoechi*, **1996**, *39*, 320-326.

[100]   Chen, Z.Y.; Chan, P.T.; Ho, K.Y.; Fung, K.P.; Wang, J. *Chem. Phys. Lipids*, **1996**, *79*, 157-164.

[101]   de Flora, S.; Rosenkranz, H.S.; Klopman, G. *Mutagenesis*, **1994**, *9*, 39-45.

[102]   Musonda, C.A.; Helsby, N.; Chipman, J.K. *Hum. Exp. Toxicol.*, **1997**, *16*, 700-708.

[103]   Cai, Q.; Rahn, R.O.; Zhang, R. *Cancer Lett.*, **1997**, *119*, 99-107.

[104] Milic, B.L.; Djilas, S.M.; Canadanovic-Brunet, J.M. *Food Chem.*, **1998**, *61*, 443-447.
[105] Ahn, C.-H.; Choi, W.C.; Kong, J.Y. *Anticancer Res.*, **1997**, *17*, 1913-1917.
[106] Vieira, O.; Laranjinha, J.; Madeirat, V.; Almeida, L. *Biochem. Pharmacol.*, **1998**, *55*, 333-340.
[107] Rosenthal, I.; Rosen, B.; Bernstein, S. *Milchwissenschaft.*, **1997**, *52*, 134-138.
[108] Chen, J.H.; Ho, C.-T. *J. Agric. Food Chem.*, **1997**, *45*, 2374-2378.
[109] Heinonen, I.M.; Lehtonen, P.J.; Hopia, A.I. *J. Agric. Food Chem.*, **1998**, *46*, 25-31.
[110] Anand, K.K.; Singh, B.; Saxena, A.K.; Chandan, B.K.; Gupta, V.N.; Bhardwaj, V. *Pharmacol. Res.*, **1997**, *36*, 315-321.
[111] Halinska, A.; Belej, T.; O'Brien, P.J. *Br. J. Cancer, Suppl.*, **1996**, *74*, S23-S27.
[112] Demirezer, O.L.; Kuruuzum, A. *Biosci.*, **1997**, *52*, 665-669.
[113] Liu, S.Y.; Sporer, F.; Wink, M.; Jourdane, J.; Henning, R.; Li, Y.L.; Ruppel, A. *Trop. Med. Int. Health,* **1997**, *2*, 179-188.
[114] Woelfle, D.; Schmutte, C.; Westendorf, J.; Marquardt, H. *Cancer Res.*, **1990**, *50*, 6540-6544.
[115] Brown, S.O.; Hamilton, R.J.; Shaw, S. *Phytochemistry*, **1975**, *14*, 2726.
[116] Kojima, Y.; Djamal, S.; Kayama, T. *Mokuzai Gakkaishi*, **1985**, *31*, 312-315.
[117] Desai, P.D.; Dutia, M.D.; Ganguly, A.K.; Govindachari, T.R.; Joshi, B.S.; Kamat, V.N.; Prakash, D.; Rane, D.F.; Sathe, S.S.; Viswanathan, N. *Indian J. Chem.*, **1967**, *5*, 523-524.
[118] Carruthers, W.R.; Hay, J.E.; Haynes, L.J. *Chem. Ind.*, **1957**, Jan. 19, 76-77.
[119] Gunawardana, Y.A.G.P.; Kumar, N.S.; Sultanbawa, M.U.S. *Phytochemistry*, **1979**, *18*, 1017-1019.
[120] Kumar, D. *Heterocycles*, **1979**, *12*, 1281-1284.
[121] Krishnamurty, H.G.; Parkash, B.; Parthasarathy, M.R.; Seshadri, T.R. *Indian J. Chem.*, **1974**, *12*, 520-522.
[122] Saha, P.K.; Ganguly, S.N. *Fitoterapia*, **1979**, *50*, 7-9.
[123] Mustafa, A.I.; Bashar, A.; Huq, M.S. *Dacca Univ. Stud. Part B* (Bangladesh), **1981**, *29*, 55-61.
[124] Beri, R.M. *Indian Oil Soap J.*, **1970**, *35*, 274-275.
[125] Hausen, B.M. *Contact Dermatitis*, **1978**, *4*, 204-213.
[126] Grover, G.S.; Rao, J.T. *Perfum. Flavor*, **1982**, *7*, 15-17.
[127] Purushothaman, K.K.; Sarada, A.; Kalyani, D. *Indian Drugs*, **1984**, *21*, 515-516.
[128] Rücker, G.; Hefendehl, F.W. *Phytochemistry*, **1978**, *17*, 809-810.
[129] Ehret, C.; Ourisson, G. *Tetrahedron*, **1969**, 1785-1799.
[130] Klein, E.; Rojahn, W. *Tetrahedron Lett.*, **1970**, 279-282.
[131] Cheung, H.T.; Feng, M.C. *J. Chem. Soc.*, **1968**, 1047-1051.
[132] Kolhe, J.N.; Bhaskar, A.; Brongi, N.V. *Lipids*, **1982**, *17*, 166-168.
[133] Cheung, H.T.; Yan, T.C. *Aust. J. Chem.*, **1972**, *25*, 2003-2012.
[134] Purushothaman, K.K.; Saraswathy, A.; Sasikala, E. *Indian Drugs*, **1989**, *26*, 146-150.
[135] Pavanasasivam, G.; Sultanbawa, M.U.S. *Phytochemistry*, **1974**, *13*, 2002-2006.
[136] Zheng, Z.B.; Zhao, S.X.; Deng, J.Z.; Zhao, H.U.; Ye, W.C.; Wang, M.S. *Zhongguo Yaoke Daxue Xuebao*, **1994**, *25*, 262-264.
[137] Gupta, A.S.; Dev, S. *Tetrahedron*, **1971**, *27*, 823-834.
[138] Chan, K.C. *Phytochemistry*, **1969**, *8*, 1051-1052.
[139] Rahman, L.; Ahmed, M. *J. Indian Chem. Soc.*, **1973**, *50*, 751.
[140] Cheung, H.T.; Wong, C.-S.; Yan, T.C. *Tetrahedron Lett.*, **1969**, *58*, 5077-5080.

[141] Cheung, H.T.; Tökés, L. *Tetrahedron Lett.*, **1968**, 4363-4366.

[142] Cheung, H.T. *Tetrahedron Lett.*, **1967**, 2807-2809.

[143] Cheung, H.T. *J. Chem. Soc.*, **1968**, 2686-2689

[144] Hirose, Y.; Yanagawa, T.; Nakatsuka, T. *J. Jap. Wood Res. Soc.*, **1968**, *14*, 36.

[145] Ikeda, T.; Kitao, K. *Mokuzai Gakkaishi*, **1974**, *20*, 460-466.

[146] Hota, R.K.; Bapuji, M. *Phytochemistry*, **1994**, *35*, 1073-1074.

[147] Cheung, H.T.; Wong, C.S. *Phytochemistry*, **1972**, *11*, 1771-1780.

[148] Mills, J.S. *J. Chem. Soc.*, **1956**, *3*, 2196-2202.

[149] Giannasi, D.E.; Niklas, K.J. *Taxon*, **1977**, *26*, 380-385.

[150] Seo, E.K.; Silva, G.L.; Chai, H.B.; Chagwedera, T.E.; Farnsworth, N.R.; Cordell, G.A.; Pezzuto, J.M.; Kinghorn, A.D. *Phytochemistry*, **1997**, *45*, 509-515.

[151] Harborne, J.B.; Greenham, J.; Williams, C.A.; Eagles, J.; Markham, K.R. *Phytochemistry*, **1993**, *34*, 219-226.

[152] Ito, C.; Sato, K.; Oka, T.; Inouse, M.; Ju-Ichi, M.; Omura, M.; Furukawa, H. *Phytochemistry*, **1989**, *28*, 3562-3564.

[153] Boyd, M.R.; Paull, K.D. *Drug Dev. Res.*, **1995**, *34*, 91-109.

[154] Haslam, E. (Ed.). *Plant Polyphenols. Vegetable Tannins Revisited*. Cambridge University Press: Cambridge, U.K., **1989**.

[155] Jang, M.; Cai, L.; Udeani, G.O.; Slowing, K.V.; Thomas, C.F.; Beecher, C.W.W.; Fong, H.H.S.; Farnsworth, N.R.; Kinghorn, A.D.; Mehta, R.G.; Moon, R.C.; Pezzuto, J.M. *Science*, **1997**, *275*, 218-220.

Atta-ur-Rahman (Ed.) *Studies in Natural Products Chemistry, Vol. 23*
© 2000 Elsevier Science B.V. All rights reserved

THE VERATRUM ALKALOIDS: NATURAL TOOLS FOR STUDYING EMBRYONIC DEVELOPMENT

W. GAFFIELD

*Western Regional Research Center, U.S. Department of Agriculture,
Albany, CA 94710, USA*

ABSTRACT: Holoprosencephaly is a malformation sequence in which impaired midline cleavage of the embryonic forebrain is the defining feature and cyclopia is the most severe manifestation. Of the teratogenic agents known to induce holoprosencephaly in mammals, the *Veratrum* alkaloids (*e.g.*, jervine and cyclopamine) are the most prominent. Malformations observed in a variety of mammals include cyclopia, hypotelorism, cebocephaly, cleft palate, exencephaly, and encephalocele. Animal models currently favored for studying the defect include hamsters and chick embryos.

Structure-terata research has revealed that the presence of C-5, C-6 unsaturation in the steroidal framework of jervanes, solanidanes, and spirosolanes is a critical structural factor in the induction of teratogenicity. Enhanced teratogenicity induced when C-5, C-6 is an olefinic linkage may imply either that this functionality contributes significantly to complex stability at a receptor site or that an α,β-unsaturated ketone metabolically generated from a C-5, C-6 unsaturated steroidal alkaloid reacts more readily at the relevant receptor than a saturated ketone of a cyclohexanone moiety.

Recently, proteins secreted by the *Sonic hedgehog* gene have been implicated as being essential for the proper development of the mammalian neural tube and forebrain. Another malformation related to holoprosencephaly, but that is less severe, is the Smith-Lemli-Opitz syndrome that is associated with a lack of cholesterol in embryonic development. The application of the *Veratrum* alkaloids in revealing and defining the intricacies of Hedgehog signalling may permit manipulation of the signalling in the identification of therapeutic approaches where Hedgehog-responsive cell types are the targets of disease processes.

INTRODUCTION

The *Veratrum* alkaloids occupy a position of importance among the extensive group of natural products generally known as steroidal alkaloids and occur in certain perennial herbs that are distributed throughout the temperate regions of the Northern Hemisphere [1-4]. The structural complexity of these molecules, which contain six or even seven carbocyclic and heterocyclic rings, a variety of functional groups, and a large number of chiral carbon atoms, has stimulated numerous structural investigations for over a century [5]. Because of their relatively rich pharmacological history and the fact that many classes of drugs are based upon natural product prototypes, both the jerveratrum and ceveratrum groups of alkaloids derived from species of *Veratrum* have served as targets for diverse synthetic efforts that have continued to the present [6-8].

Several uses of *Veratrum* by Native Americans are listed in Table 1 including preparation of a tea from fresh root that was taken as a form of birth control [9]. Native American tribes of California and Nevada (listed at the bottom of Table 1) ate the bulb and stem of *Veratrum* plants after roasting in hot ashes and peeling away the outer covering. *Veratrum viride* is still used as a medicinal herb and was cited as a dangerous plant to be avoided in a recent review article [10]. Over twenty *Veratrum* alkaloids are known to exhibit important cardiovascular, neuromuscular, and respiratory effects [11]. Most significantly, the *Veratrum* alkaloids reduce considerably the blood pressure in hypertensive patients. For many years, *Veratrum* was used in treating eclampsia (convulsions occurring during pregnancy and parturition) and in the 19th Century the plant and its alkaloids were used in treating mania and epilepsy. However, toxic effects (*e.g.*, nausea, fainting, vomiting, substernal constriction, convulsions, and death) have made these compounds unattractive for human usage despite their vasodilator properties. Occasionally, *Veratrum* poisonings are still reported in France, usually resulting from erroneous identification of *Veratrum* as *Gentiana*, a plant used in the preparation of herbal tea and homemade liquors [12].

Table 1.   **Native American Usage of Veratrum Californicum**

Purgative

Anti-Epileptic

Birth control

Treatment of venereal disease

Treatment of snakebite

Poultice on wounds

Treatment for sore throat, inflamed tonsils, and colds

Snuff

Poison used on arrow heads

Maidu/Miwok/Washo

One aspect of *Veratrum* alkaloid toxicity, namely the activity of veratradine, has been exploited earlier as a pharmacological tool in studying the mechanism of action of ion channels [13]. Veratridine is one of a group of four types of lipid-soluble polycyclic compounds that serve as agonists of the sodium ion channel receptors that mediate the excitability of heart, nerve, and skeletal muscle. This review focuses primarily on the potential application of teratogenic *Veratrum* alkaloids such as jervine and cyclopamine (11-deoxojervine) as probes or tools to aid in the elucidation of pathways that control embryonic development of the mammalian forebrain [14]. Current advances in the chemistry and pharmacology of *Veratrum* alkaloids are reviewed periodically in *Natural Product Reports* [15,16].

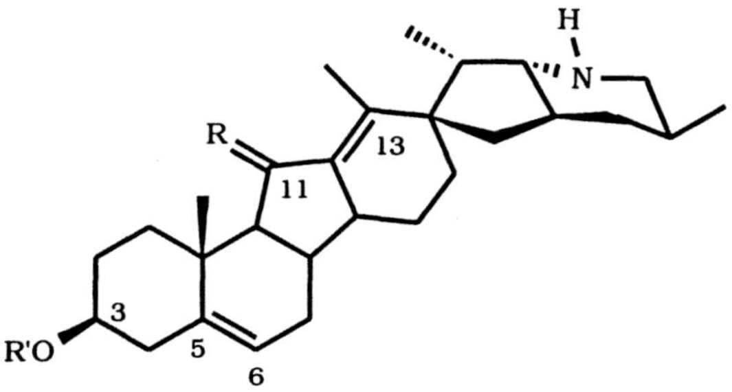

	R'	R
JERVINE (1)	H	O
CYCLOPAMINE (2)	H	H_2
CYCLOPOSINE (3)	D-GLC	H_2

HISTORICAL PERSPECTIVE

Cyclopia has fascinated man since antiquity. The Cyclopes, one-eyed giants who lived as shepherds on the coast of Sicily, were described in detail in Homer's *The Odyssey* [17]. Descriptions of cyclopia are common in the writings of the natural philosophers of the 16th and 17th Centuries and cyclopia has been the subject of artistic expression throughout the 20th Century [18]. Holoprosencephaly is a malformation sequence in which impaired cleavage of the embryonic forebrain is the defining feature and cyclopia is the most severe manifestation [18]. Various gradations of facial dysmorphism are commonly associated with holoprosencephaly in humans. For example, the spectrum of facial types associated with holoprosencephaly in order of decreasing severity includes; Fig (1), (A) cyclopia with a single median eye and lack of proboscis, (B) single median eye with varying degrees of doubling of the ocular structures, (C) formation of an abnormal proboscis, (D) closely-spaced eyes (hypotelorism) with an abnormal proboscis, (E) cebocephaly (a misshapen nasal passage) and nose with a single nostril, and (F) cleft lip with closely spaced eyes [19].

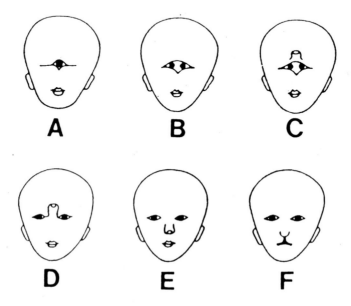

Fig. (1). Holoprosencephaly in humans in order of decreasing severity; *cf*, description in text. Nishimura, H.; Okamoto, N. *Sequential Atlas of Human Congenital Malformations: Observations of Embryos, Fetuses, and Newborns,* University Park Press: Baltimore, **1976.**

Of the teratogenic agents known to induce holoprosencephaly in animals, the *Veratrum* alkaloids are the most prominent [20,21]. During the early portion of the 20th Century, epidemics of congenital craniofacial malformations in newborn lambs were common on sheep ranches in the Boise, Sawtooth, and Challis National Forests of the western United States. During the 1950's, ranchers in certain areas of Idaho experienced incidences of congenitally deformed lambs of nearly 25% of live births. The most common malformation was called 'monkey-face' lamb disease by Basque sheepherders. This double-globe cyclopia was expressed as two closely-spaced corneas in a single distorted, often dumbbell shaped, sclera. Many afflicted animals were truly cyclopic, Fig (2), having a single median eye and in severe instances of the disease, the eyes and nose were absent and the cerebrum rudimentary [22-24]. Related malformations

Fig. (2). Cyclopic lamb from a ewe dosed with cyclopamine on the fourteenth day of gestation.

included; anophthalmia (the complete absence of ocular tissue), cebocephaly, and occasionally a pronounced curvature of the lower jaw. 'Monkey-face' lambs carried to full term were usually born alive and occasionally lived for a short time. Often, gestation was prolonged to over 225 days and the newborn lambs weighed over 25 lbs, which was remarkable because sheep normally give birth after approximately 150 days of pregnancy and lambs normally weigh 8-12 lbs [22,25]. Careful examination of the cyclopic fetuses showed that the base of their brains was deformed in the region of the hypothalamus and pituitary. Thus, the observation of prolonged pregnancy in sheep that had consumed *Veratrum* became a very dramatic clue regarding the fetal signals that initiated parturition. An underlying theme of these observations suggested involvement of the fetal hypothalamus, pituitary, and adrenal gland in the birth process. Subsequent medical research proved that the fetus secretes more ACTH, cortisol, and neuropeptide Y producing changes that lead to birth [26,27].

The cyclopic lambs generally died soon after birth due to their inability to breathe and feed properly and most deformed lambs were humanely killed at birth. A genetic basis for the disease was eliminated by careful breeding experiments in which 'carrier' ewes were mated to 'carrier' rams [22]. The incidence of the disease appeared to correlate both with the range area grazed by afflicted ewes and certain plant species present in those areas. About 1960, field study investigations and experimental feeding trials began which established that the malformations resulted when ewes consumed the range plant *Veratrum californicum* during the second and third weeks after conception [25].

Veratrum californicum, commonly known as false hellebore or corn lily, is a coarse, weedy plant that thrives between 6500-7000 feet of elevation and is avoided by most livestock, but not sheep [28]. The numerous erect stems of the plant are connected to each other by underground stems or rhizomes. Some of the stems bear open clusters (panicles) of dull greenish-white flowers, Fig. (3), whereas others end in a somewhat smaller leaf. The latter, Fig. (4), are a fascinating botanical curiosity. Each leaf blade is supported by a sheath that encircles the

Fig. (3). *Veratrum californicum* (flowering).

Fig. (4). *Veratrum californicum* (non-flowering).

apparent 'stem'. What appears to be a tall, leafy stem is actually only a cluster of leaves and sheaths wrapped around one another. True stems are found only on those shoots that bear flowers and seed capsules.

Retrospective analysis reveals that the 'monkey-face' lamb disease resulted from grazing protocol. Freshly bred ewes were often grazing the teratogenic *Veratrum* plants precisely during the period (fourteenth day of gestation) that they were most susceptible to induction of terata [29]. A practical solution to the problem was formulated when ranchers were advised to keep their pregnant ewes away from areas containing *Veratrum californicum* until after the ewes had passed the fourteenth day of gestation [29,30]. Subsequently, the incidence of 'monkey-face' lamb disease became virtually nonexistent [31]. Currently, several chemical and mechanical methods are used to control *Veratrum* in alpine areas [32].

Although *Veratrum* no longer poses a problem for sheep in the lower 48 states of the United States, a recent outbreak of cyclopia was reported in domestic alpacas in 1995 and 1996 in Alaska (K.E. Panter; J. Leach, personal communication). Suspect plants in the vicinity grazed by the alpacas included *Veratrum album*, from which several jerveratrum alkaloids have been isolated and which are likely responsible for the induction of cyclopia in the alpaca llamas [33-35].

The culmination of the extensive research efforts of R. F. Keeler and his colleagues at the Poisonous Plant Research Laboratory in Logan, Utah was the isolation and identification of three jerveratrum alkaloids from *Veratrum californicum*; jervine (**1**) [36], cyclopamine (11- deoxojervine) (**2**) [37], and its 3-glucosyl derivative cycloposine, (**3**) [38], all of which induced 'monkey-face' lamb disease when orally administered to ewes on the 14th day of gestation, which includes the primitive streak/neural plate stage of embryonic development in sheep [30].

Further experiments established that the teratogenic activity of *Veratrum* alkaloids was not restricted to sheep. Cattle and goats also produced malformed offspring upon ingestion of *Veratrum californicum*; however, similar to sheep, they possess rumen microorganisms that might or might not metabolize the jerveratrum alkaloids into teratogens [39]. Experiments in non-ruminants revealed that rabbits, mice, rats, and hamsters were also susceptible to terata-induction by cyclopamine, without possible conversion of the alkaloid to other compounds by ruminal metabolism, although other metabolic processes associated with oral administration cannot be ruled out with certainty [40]. For logistical reasons, hamsters became the assay model of choice for plant alkaloid teratogen research [41].

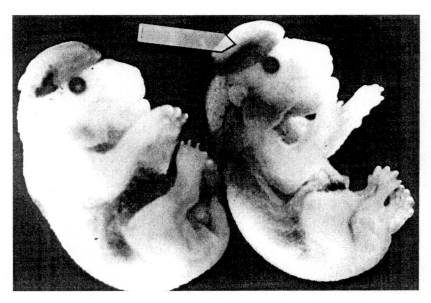

Fig. (5). Exencephalic hamster fetuses.

Three major types of craniofacial malformations were induced in hamsters upon oral adminstration of *Veratrum* alkaloids during the primitive streak/neural plate development phase which is on the eighth day of the hamsters' 16 day gestation period [42]. These primary malformations include: exencephaly, Fig. (**5**), which is a fully-exposed brain that appears developmentally similar to anencephaly in humans; encephalocele, Fig. (6), which is herniation of the brain that may appear as

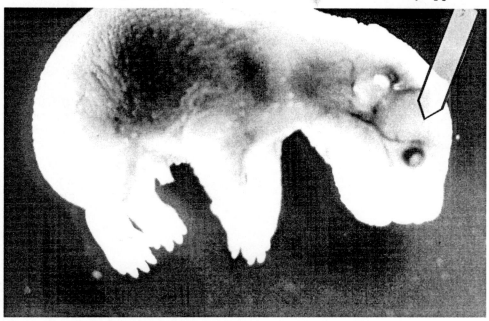

Fig (6). Encephalocele in a hamster fetus.

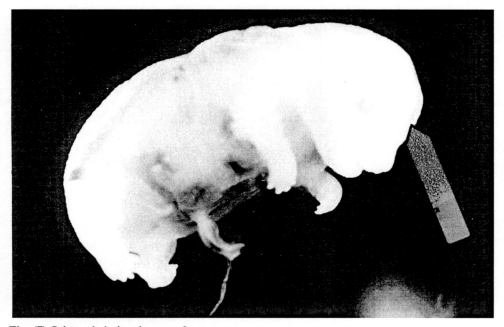

Fig. (7).Cebocephaly in a hamster fetus.

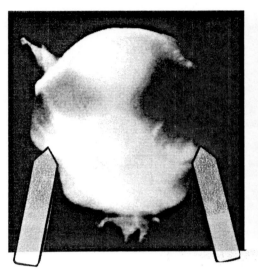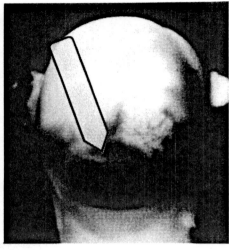

Fig (8). Ventral view of the head of a six-day old chick embryo; left, bilateral eyes of a control embryo; right, cyclopia induced in a cyclopamine-treated embryo.

a protrusion of meningeal or skin-covered brain tissue (i.e., partially exposed brain) ; and cebocephaly, Fig. (7), which is expressed as a misshapen nasal chamber having an absent or incompletely formed nasal septum and is part of the holoprosencephaly manifold of forebrain and midface malformations. Often closely-spaced eyes (hypotelorism) with a single nostril nose is associated with cebocephaly. Cyclopia is only rarely observed in hamsters [43]; thus, induction of forebrain or midline craniofacial malformations in hamsters is usually expressed as cebocephaly or cleft lip and/or palate (premaxillary dysgenesis). Currently, chick embryos are a favored experimental animal model for the investigation of cyclopia and related malformations [44]. In Fig. (8) is shown a ventral view of the 'face' of a pair of chick embryos; one normal and one cyclopamine-treated cyclopic embryo. The embryos are at 6.5 days of development with hatching occurring at 21.5 days (J. P. Incardona, personal communication).

STRUCTURE - TERATA CORRELATIONS

Each of the teratogenic expressions induced by *Veratrum* in sheep, which include not only craniofacial, but also limb, palate, and tracheal malformations that are induced at later gestational periods [45], has related counterparts in humans [46]. Although *Veratrum* species are not eaten by humans, many *Solanum* alkaloid-containing species are used as human foods [47]. For these reasons, numerous investigations of steroidal alkaloid-induced teratogenesis have been conducted over the past quarter century [48]. One of the primary focuses of this research has centered on whether or not stereoselective teratogenicity could be induced by plant alkaloids.

An interesting example of stereospecific teratogenicity concerns the enantiomeric 2 n- propyl-4-pentynoic acids that are structural analogs of the anticonvulsant drug valproic acid. No enantioselectivity of the anticonvulsant activity of these compounds has been observed; however, stereoselective dysmorphology for the enantiomeric valproic acid analogs has been demonstrated across rodent species, Fig. (**9**) [49]. Thus, the *S*-enantiomer is dysmorphogenic, embryolethal, and retards embryonic development in both rats and mice whereas exposure of the rodents to the *R*-enantiomer failed to produce any of these effects. These data demonstrate that certain biological mechanisms associated with closure of the neural tube are stereoselective and that the dysmorphogenicity induced by the valproic acid analogs is not a species-specific phenomenon derived from maternal factors such as preferential absorption, metabolism, or biliary excretion [50].

<table>
<tr><td>H,,,, CO₂H</td><td>H,,,, CO₂H</td></tr>
<tr><td>(S)
DYSMORPHOGENIC</td><td>(R)
NON-DYSMORPHOGENIC</td></tr>
</table>

Andrews, Ebron-McCoy, Bojic, and Nau (1997)

Fig. (9). Enantiomeric analogs of the anticonvulsant drug valproic acid.

The significance to hamster teratogenesis of both the presence of nitrogen in the F-ring of spirosolanes and its configurational placement appeared to be demonstrated by the absence of terata induction by tomatidine (**4**), which bears a nitrogen atom that projects above the steroidal plane and by the non-nitrogenous steroidal sapogenin, diosgenin (**5**), in contrast to the isomeric spirosolane, solasodine (**6**), which induced brain malformations in hamsters, and whose imino group projected below the steroidal plane [51].

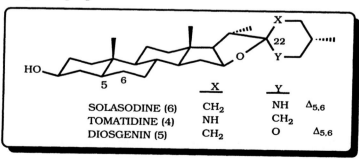

	X	Y	
SOLASODINE (6)	CH_2	NH	$\Delta_{5,6}$
TOMATIDINE (4)	NH	CH_2	
DIOSGENIN (5)	CH_2	O	$\Delta_{5,6}$

In other studies, solanidane-induced teratogenicity was proposed to depend upon C-22 configuration, *cf* Fig. **(10)**, and the teratogenicity of epimeric solanidanes was suggested to correlate with the direction in which the unshared pair of electrons on the nitrogen atom projects relative to the steroidal plane [52]. Earlier structure - terata relations of plant steroidal alkaloids have been summarized [53].

SOLANIDINE	R=H
α-SOLANINE	R= GLC
α-CHACONINE	R= RHA

Fig. (10). Structure of solanidane alkaloids.

However, extensive studies, of terata induction upon oral administration of jervanes, solanidanes, and spirosolanes to hamsters have shown that the primary structural feature controlling hamster teratogenesis was not the configurational placement of the nitrogen atom, but instead was the level of saturation at C-5, C-6 [54]. Thus, all three structural types (jervanes, solanidanes, and spirosolanes) were significantly less teratogenic upon saturation of the C-5, C-6 linkage. Fetuses obtained from animals administered jervine were malformed to the extent of 92%, whereas terata occurred in only 14% of fetuses derived from tetrahydrojervine-dosed hamsters. Similarly, the percentage of malformed fetuses derived from solanidine-treated hamsters declined from 24% to 3% and the percentage of deformed fetuses from solasodine-treated animals from 29% to 6% upon conversion of C-5, C-6 from an unsaturated to a saturated linkage. Clearly, hamster teratogenicity induced upon oral administration of jervanes, solanidanes, and spirosolanes appears to correlate closely with the presence or absence of C-5, C-6 unsaturation in these steroidal alkaloids although the effect of molecular configuration of stereocenters near the amino group plays a role in the teratogenicity both of solanidanes and spirosolanes [54,55]. Studies of steroidal alkaloids in appropriate *in vitro* assays, where dose- response functions may be established, should clarify the importance of these functionalities to structure-terata relationships.

With this structure-terata information available, a compilation of relative teratogenic potencies was proposed (Table 2) based upon extrapolation of recent data and older literature data to equivalent oral dosage in hamsters [56]. The data on teratogenic potencies were based upon litters rather than fetuses according to teratological protocol [57,58]. This tabulation is only an approximate correlation and is intended as a qualitative rather than a quantitative assessment. Teratogenicity data were obtained from differently controlled animal experiments at two research laboratories using hamsters from the same supplier but whose relative teratogenic susceptibilities may have varied. However, the relative teratogenic potencies shown in Table 2 should serve as a guide for comparison of structure-terata relations and for estimating the mammalian teratogenicity of newly-discovered steroidal alkaloid glycosides and their aglycones.

The difference in potency for different structural classes of steroidal alkaloids (for chemical structures of the alkaloids listed in Table 2, see

Table 2.   Relative Teratogenic Potencies of Steroidal Alkaloids

Alkaloid	Relative Teratogenic Potency
Jervine	100
Dihydrojervine	65
S,R-Solanidanes	50
Cyclopamine	47
Chaconine	43
Tetrahydrojervine	40
R,S-Solanidine	32
Solanidine-N-oxide	32
Solanine	32
Dihydrosolanidine	9
Muldamine	9
Solasodine	6
Dihydrosolasodine	4
Tomatine	1
Tomatidine	0

W. Gaffield and R.F. Keeler, *J. Nat. Toxins*, **5**: 25-38 (1996).

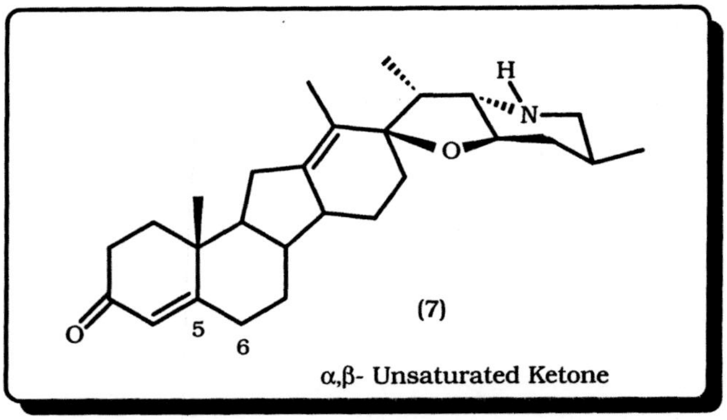

(7)

α,β- Unsaturated Ketone

Appendix) is graphically emphasized in Fig. (11) with values of the relative teratogenic potency of individual steroidal alkaloids from Table 2 placed adjacent to their structural type. The variance in values for the jervanes, solanidanes, and spirosolanes primarily reflects the level of saturation at C-5, C-6; e.g., 100 vs. 40 for jervanes, 32 vs. 9 for solanidanes, and 6 vs. 4 for spirosolanes, In each instance the higher value

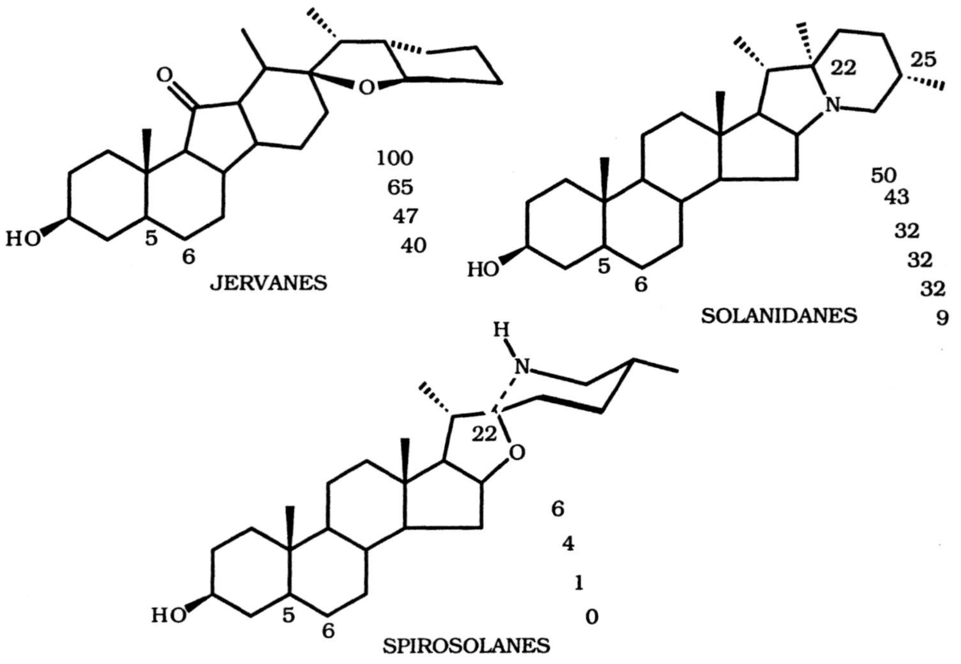

JERVANES

100
65
47
40

SOLANIDANES

50
43
32
32
32
9

SPIROSOLANES

6
4
1
0

RELATIVE TERATOGENIC POTENCIES OF STEROIDAL ALKALOIDS

Fig. (11). Correlation of relative teratogenic potencies (from Table 2) with steroidal alkaloid structure. For example, values for jervine (100), dihydrojervine (65), cyclopamine (47), and tetrahydrojervine (40) are shown adjacent to the tetrahydrojervine structure that represent jervanes, in general.

represents the relative teratogenic potency (RTP) of an alkaloid with C-5, C-6 as an unsaturated bond and the lower value represents an RTP of the same alkaloid with the C-5, C-6 linkage saturated. With rare exceptions, the jervane structure is the only one that induces forebrain and midface malformations in hamsters that are associated with holoprosencephaly; the spirosolanes and solanidanes generally induce neural tube malformations that are expressed as brain exposure [56]. The importance of C-5, C-6 unsaturation in structure- terata relations may imply that unsaturation at this position contributes significantly to complex stability at a receptor site, although it is uncertain whether this effect is a result of favorable interactions of the polarizable double bond with a receptor or due to differences in the steric fit inside a receptor cavity. Another speculative suggestion concerns potential oxidation products derived from the C-5, C-6 saturated and unsaturated steroidal alkaloids. If metabolic oxidation converts the former to a cyclohexanone and the latter to an electrophilic α,β-unsaturated ketone, the possibility exists that the latter is a much more reactive substrate at the relevant receptor site. This view is supported by the high teratogenicity induced in hamsters upon administration of the α,β- unsaturated ketone derived from cyclopamine, 11-deoxojervine-4-en-3-one (7) [59]. The teratogenicity of this compound was claimed to be greater than jervine and was attributed to the hormone-like A-B ring system that might facilitate greater interaction with hormone receptor sites. A summary of the embryolethality and teratogenicity of jervanes, spirosolanes, and solanidanes is provided in Table 3 .

BIOLOGICAL IMPLICATIONS

Holoprosencephaly is a disorder involving the development of forebrain and midface. The most severe form, alobar holoprosencephaly, is usually incompatible with postnatal life, involving complete failure of division of the forebrain into right and left hemispheres and is characteristically associated with facial anomalies involving cyclopia and primitive nasal structure and/or midfacial clefting. At the mild end of the spectrum, lobar holoprosencephaly, malformations may include microcephaly and mild hypotelorism (closely-spaced eyes). Clinical manifestations include severe mental retardation, seizures, temperature imbalance, and abnormal facies, [18,19]. In addition to the various defects of the craniofacial malformation complex (cf, Fig. (1)), several other syndromes are believed to be associated with holoprosencephaly, including Martin syndrome, the CHARGE association, and the Pallister-Hall syndrome [60].

Holoprosencephaly occurs in approximately 1 of 16000 live human births although the overall incidence is appreciably greater because 99% of affected embryos abort spontaneously [61]. Thus, holoprosencephaly is considerably more common in early embryogenesis, occurring in about 1 of

Table 3. The Embryolethality and Teratogenicity of Jervanes, Spirosolanes, and Solanidanes

	Dose (mmol/kg)	Dams fed	Dams with 100% resorption [c]	Maternal overdose death [c]	Malformed litters (malformed fetuses)	Percent abnormal litters (p) [d]	Percent abnormal fetuses (p) [d]	Total fetuses [c]	Fetuses/litter [c]	Total resorbed fetuses [c]	Resorbed Fetuses/litter [c]	Mean litter size [c]	Mean fetal weight (g) [c] (p) [e]	Percent dead fetuses [c]
Jervanes and Spirosolanes														
Jervine[a]	0.28	13	2	1	10(47)	100 (<0.0001)	92 (<0.0001)	52	5.2	46	4.6	9.8	1.44 (0.0002)	36.5
Tetrahydrojervine[a]	0.28	10	0	0	4(11)	40 (0.02)	14 (<0.0001)	80	8.0	20	2.0	10.0	1.64 (0.069)	7.5
Cyclopamine[b]	0.43	16	0	1	11(36)	73 (<0.0001)	49 (<0.0001)	73	4.9	46	3.1	8.0	-	-
Solasodine[a]	3.38	10	0	2	6(18)	75 (0.0004)	29 (<0.0001)	59	7.4	27	3.4	10.8	1.61 (0.046)	6.8
Dihydrosolasodine[a]	3.37	11	0	0	5(6)	45 (0.009)	6 (0.005)	96	8.7	19	1.7	10.4	1.70 (0.30)	1.0
Controls	-	14	0	0	0(0)	0	0	131	9.4	18	1.3	10.7	1.74	1.5
Solanidanes														
Solanidine[a]	0.44	12	0	0	6(13)	50 (0.003)	24 (<0.0001)	55	4.6	46	3.8	8.4		
Solanidine-N-oxide[b]	0.43	15	1	0	7(14)	50 (0.0003)	16 (<0.0001)	87	6.2	46	3.3	9.5		
Dihydrosolanidine[a]	0.44	15	1	0	2(2)	14 (0.22)NS	3 (0.17)NS	71	5.1	34	2.4	7.5		
Controls	-	42	0	0	0(0)	0	0	303	7.2	86	2.0	9.2		

[a] Data from ref. 54; [b] Data from ref. 55; [c] Data from ref.56; [d] Probability values (p) of the occurrence of abnormal litters or fetuses by comparison of treatment with alkaloid vs. carrier controls obtained by chi-square analysis using two-tailed Fisher exact results; [e] One-tailed p-values, students' t vs. controls.

250 spontaneous abortions [62]. Holoprosencephaly is derived from heterogeneous etiologies from environmental to genetic. An estimated 50% of all cases are associated with major chromosomal defects. For several years, attention has focused on certain human chromosomal regions believed to contain genes crucial for normal forebrain and midface development (Fig. (**12**)) [60,63]. Recent examination of families with autosomal dominant holoprosencephaly has demonstrated linkage to a gene in the 7q 36 region [64].

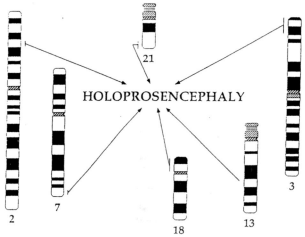

Fig. (12). Chromosomal regions involved in holoprosencephaly. *Cf,* Muenke, M. *Semin. Dev. Biol.,* **1994,** *5,* 293.

The 19th century poet William Ernest Henley wrote, 'I am the master of my fate, I am the captain of my soul' [65]. However, for the cells of the developing embryo, the situation is far more complex. Indeed, just when and how an unidentified precursor cell becomes 'fated' to fulfill its destiny as a specialized cell in the body is one of the major unresolved questions in developmental biology [66].

Secreted signalling proteins encoded by the *Hedgehog* gene family induce specific patterns of differentiation in a variety of tissues and structures during vertebrate and invertebrate development [67]. Vertebrate homologs of such a secreted protein have been implicated recently in the induction of ventral cell types in the neural tube comprising the developmental pathway from notochord to floorplate to adult organs [68-70]. One member of this gene family, *Sonic Hedgehog*, is expressed in embryonic structures such as the zone of polarizing activity, by cells of the notochord prechordal plate and floorplate over the period that these cell groups exhibit their inductive activities (Fig. (**13**)) [71].

The notochord has several distinct inductive properties; it is the source of a contact-dependent floor plate-inducing activity and a long-range motor neuron-inducing activity. The floor plate itself shares both of these inductive properties with the notochord. *Sonic hedgehog* appears to mediate all ventralizing activities of the notochord and the floor plate [72].

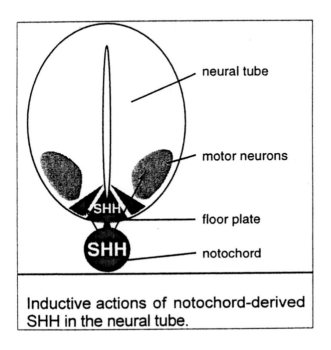

Fig. (13). Inductive actions of notochord-derived Sonic hedgehog in the neural tube.

Sonic hedgehog is not only sufficient, but also is required for the correct differentiation of ventral cell types in the neural tube. Targeted disruption or deletion of the *Sonic hedgehog* gene in mice [73] and humans [74] leads to defects in the maintenance of the notochord and establishment of the floor plate, with subsequent development of severe malformations of forebrain and cranium characteristic of human holoprosencephaly. Thus, failure of the induction of the optic stalk and other cells in the forebrain of *Sonic hedgehog* null mouse embryos produces holoprosencephaly and cyclopia [73].

The protein that the *Sonic hedgehog* gene encodes is required for the correct patterning of many organs or organ parts and it guides the development of the embryo's hands and feet, distinguishing clearly between the different sorts of fingers and toes. Biochemical experiments have revealed an interesting facet of the Sonic hedgehog protein; it is synthesized as an inactive precursor (45 kDa) which must be autoproteolytically cleaved by intramolecular processing into two similar sized fragments, a 20 kDa amino-terminal fragment and a 25 kDa carboxy-terminal fragment [75,76]. Of the two fragments, only the amino-terminal protein carries the developmental signal [77]. The C-terminal region of Sonic hedgehog serves both as the endopeptidase for cleavage or processing and as a lipid transferase [78]. After the cleavage has occurred, the signalling protein must be delivered to the proper part of the body in the correct concentration for normal development to occur [71].

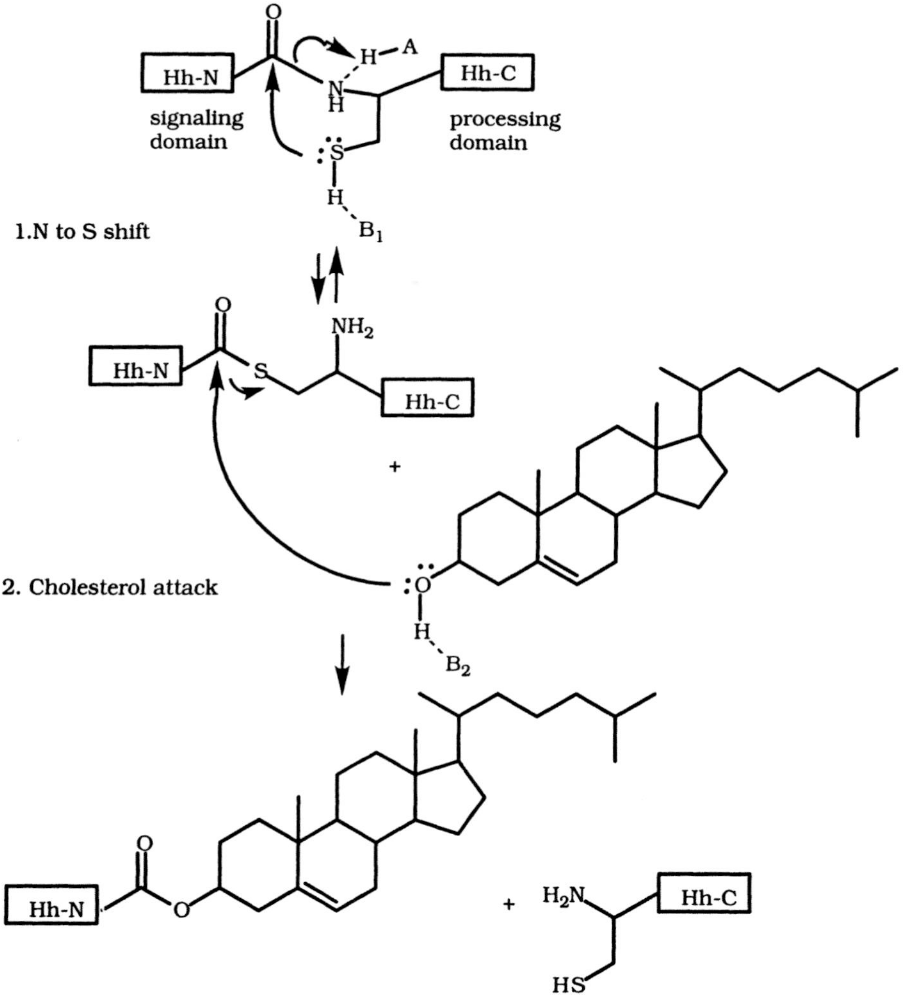

Fig. (14). Intramolecular autoprocessing reactions of Sonic hedgehog proteins.

Hedgehog autoprocessing proceeds through two steps as shown in Fig. **(14)** [78]. In the first step, the thiol group of a cysteine residue acts as a nucleophile to attack the carbonyl group of the preceding amino acid residue, forming a thioester linkage in place of the peptide bond and resulting in a nitrogen to sulfur shift. In the second step, the activated thioester intermediate is subjected to nucleophilic attack from the 3-β-hydroxyl group of a cholesterol molecule, resulting in cleavage of the thioester with release of the carboxy-terminal Sonic hedgehog fragment and formation of an ester linkage between cholesterol and the carboxyl terminus of the amino-terminal Sonic hedgehog fragment [79]. The significance of the lipid modification relates to the increase in the hydrophobic character of the modified amino-terminal Sonic hedgehog fragment which influences the subsequent spatial and subcellular distribution of the fragment. Different concentrations of amino-terminal Sonic hedgehog protein affect cells differently; high concentrations induce neural plate explants to form floor plate whereas low concentrations promote motor neuron formation [80]. The need for cell-cell contact likely reflects a requirement for a very high concentration of Sonic hedgehog protein, localized in or near the cells that produce it; modification by cholesterol causes association of Sonic hedgehog with the cell membrane. Conversely, long range-signalling may be achieved by lower concentrations of protein that has diffused away from producing cells [81]. Possibly, Sonic hedgehog signalling by an individual cell may be regulated by modulation of the amount of lipid-linked and unmodified Sonic hedgehog protein that is produced.

Until recently, cholesterol has been known to serve primarily as a precursor in steroid hormone and bile component biosynthesis and as an important structural component of biological membranes in animals [82]. A fourth major function for cholesterol in animals is suggested by its formation of a covalent linkage to amino-terminal Sonic hedgehog fragment and its modulation of spatial and subcellular distribution that affects mammalian patterning activities [79].

Another link between cholesterol and vertebrate development is the cholesterol synthesis defect associated with Smith-Lemli-Opitz syndrome, which is a milder form of holoprosencephaly [83,84]. Estimates of the prevalence of this autosomal recessive disorder are approximately 1 in 9000 births (alive and stillborn). Severe Smith-Lemli-Opitz syndrome is characterized by some of the malformations associated with holoprosencephaly, including microcephaly, lack of pituitary development, and midline facial anomalies as well as defects of other organs. The genetic defect in Smith-Lemli-Opitz syndrome inactivates the last enzyme of the cholesterol biosynthetic pathway, 7-dehydrocholesterol-Δ^7-reductase. Thus, impaired conversion of 7-dehydrocholesterol to cholesterol limits the formation of this critical lipid, instead allowing 'toxic' cholesterol precursors to accumulate in tissues and body fluids. Hypocholesterolemia-inducing chemicals (*e.g.*, AY9944 and BM15.766) might cause the holoprosencephaly pathogenesis by disrupting Hedgehog signalling proteins either by depriving them of cholesterol or by substituting aberrant sterols [85]. Another less severe form of holoprosencephaly occurred in mouse embryos null for gp

330/megalin, a member of the low density lipoprotein receptor gene family [86].

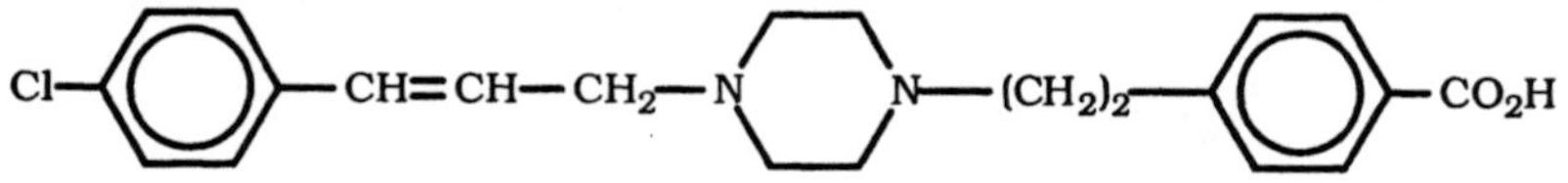

BM 15.766

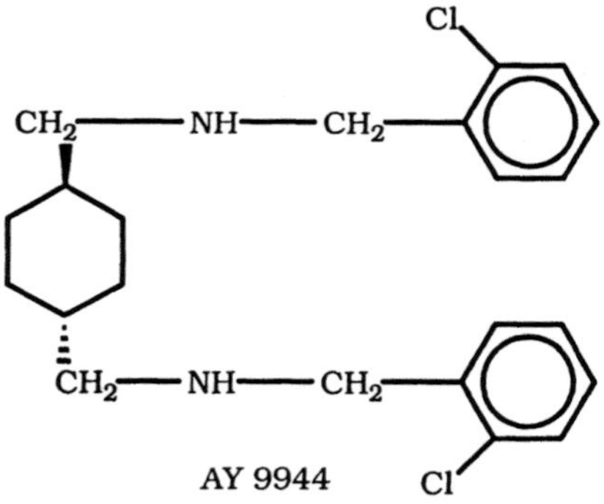

AY 9944

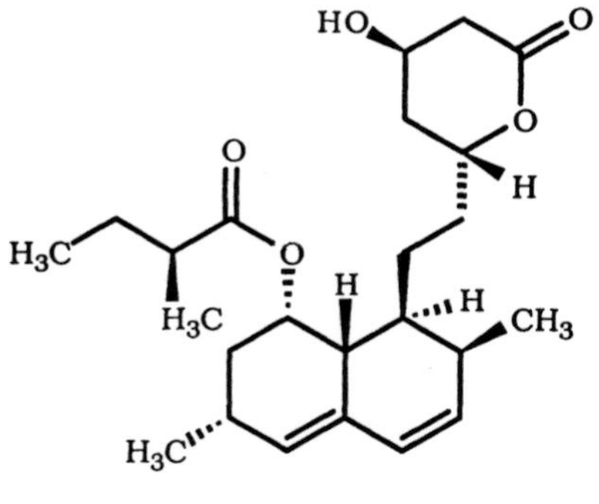

LOVASTATIN

Several levels exist in the Hedgehog pathway that could be adversely affected by the jerveratrum alkaloids: interference of initial establishment of Sonic hedgehog transcription in the notochord and prechordal plate; disruption of post translational processing of Sonic hedgehog including autoproteolytic cleavage; or inhibition of transduction of Sonic hedgehog signals in Sonic hedgehog-responsive cells. Furthermore, although cyclopamine incorporates a modified steroidal ring system, its similarity to the structure of cholesterol, Fig. (15), suggests that it might function as an enzyme inhibitor of cholesterol biosynthesis, cf [87].

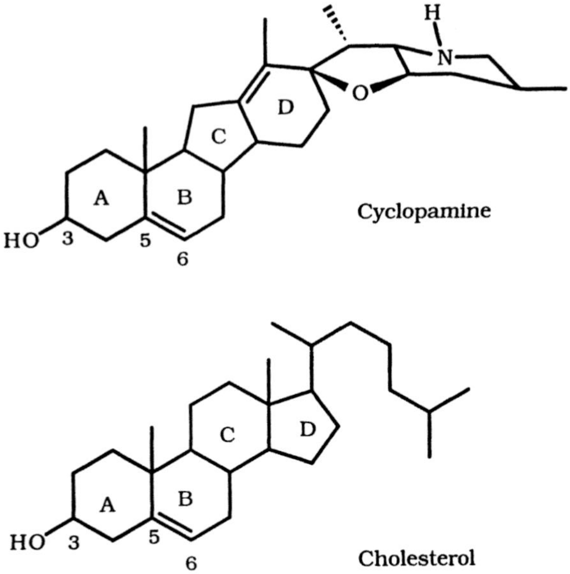

Fig. (15). Structure of cyclopamine and cholesterol.

The transcription factor *HNF-3β* regulates the expression of other genes that are essential for normal development of rostral embryonic structures, including the forebrain [88,89]. *HNF-3β* is expressed during gastrulation along the midline in the notochord and later in the floor plate of the neural tube. Treatment of hamsters with cyclopamine caused a transient loss of *HNF-3β* immunoreactivity in prechordal mesenchyme, floor plate, and notochord [90]. This transient down-regulation of *HNF-3β* might underlie the craniofacial malformations induced in cyclopamine-exposed hamster embryos. The normal pattern of *HNF-3β* expression in notochord, prechordal mesenchyme, and floor plate depends upon expression of the signalling protein Sonic hedgehog [80,81]. Because mutations that negate Sonic hedgehog expression cause cyclopia in mice [73] and are associated with holoprosencephaly in humans [74], reduced *HNF-3β* expression may reflect impaired Sonic hedgehog mediated signals rather than a primary effect of cyclopamine treatment [90].

Current molecular biological research underway in several research laboratories on the effect of jerveratrum alkaloids on Sonic hedgehog processing and signalling and as potential inhibitors of cholesterol biosynthesis should provide a clearer understanding of the molecular events that underlie the induction of mammalian teratogenicity by cyclopamine and structurally related alkaloids. Furthermore, these efforts should enhance our understanding of the regulation and evolution of other patterning mechanisms during mammalian development. In addition to Hedgehog, several other protein families appear to employ autoprocessing mechanisms; these proteins are known as self-splicing proteins [91].

Finally, recent research has indicated that other human diseases, such as the commonest form of skin cancer, basal cell carcinoma, are caused by mutations in genes involved in the Sonic hedgehog signalling pathway. The *Patched* (*Ptc*) gene encodes a Sonic hedgehog receptor [92,93] and a tumor suppressor protein that is defective in basal cell nevus syndrome [94].

The hedgehog receptor Patched (Ptc) possesses multiple membrane-spanning domains and forms a non-covalent complex with Smoothened, a member of the serpentine family of G-protein-coupled receptors [95]. Several of Patched's transmembrane domains are homologous to sterol-sensing domains present in several proteins that are involved in regulating cholesterol metabolism [96-98]. If inhibition of Sonic hedgehog signal transduction by cyclopamine was observed, then a sterol-sensing domain of Patched could serve as a potential target of the jerveratrum alkaloids. The application of the *Veratrum* alkaloids as molecular probes in helping to reveal the intricacies in Hedgehog signalling may eventually permit manipulation of the signalling in order to derive therapeutic approaches where Hedgehog-responsive cell types are the targets of disease processes.

APPENDIX

Structures of the steroidal alkaloids whose relative teratogenic potencies are listed in Table 2 and whose embryolethality and teratogenicity is recorded in Table 3 are provided as follows; jervine (**1**), cyclopamine (**2**), tomatidine (**4**), and solasodine (**6**) are listed in the text. Structures for 22*R*, 25*S*-solanidine, α-solanine, and α-chaconine are provided in Fig. (**10**). In solanidine-N-oxide, the electron pair on the nitrogen atom of solanidine in Fig. (**10**) is replaced by an oxygen atom. Dihydrosolanidine refers to solanidine with a reduced C-5, C-6 linkage. Dihydrojervine refers to the jervine structure (**1**) with the C-12, C-13 bond saturated and tetrahydrojervine has both C-5, C-6 and C-12, C-13 linkages saturated. Dihydrosolasodine is shown as structure **6** having C-5, C-6 saturated.

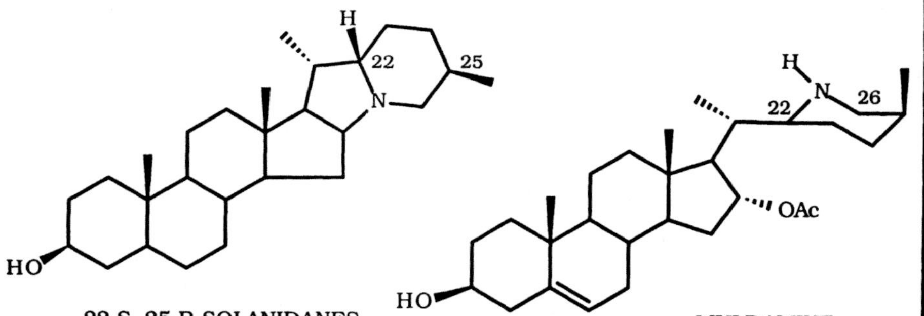

22 S, 25 R-SOLANIDANES   MULDAMINE

Tomatine is tomatidine-3β-lycotetrose. The lycotetrose moiety has a galactose attached to the 3-hydroxyl-group of tomatidine that is substituted on its 4-hydroxyl by a glucose. This glucose is further substituted at its 2-hydroxyl by another glucose and at its 4-hydroxyl by a xylose. Structures for muldamine and 22S, 25R- solanidanes (representing solanidanes both saturated and unsaturated at C-5, C-6) are shown below.

ADDED IN PROOF:

Examination of a suite of Sonic hedgehog-dependent cell types in the neural tube and somites of chick embryos with cyclopamine-induced malformations has shown that essentially all aspects of Sonic hedgehog signaling in these tissues are interrupted by cyclopamine treatment [99,100]. Cell types normally induced in the ventral neural tube by sonic hedgehog either are absent or appear aberrantly at the ventral midline, whereas dorsal cell types normally repressed by Sonic hedgehog appear ventrally. The alkaloid appears to exert its effects on developing embryos primarily by inhibiting signal transduction,

Although cyclopamine and the teratogenic inhibitor of cholesterol AY 9944 inhibit Sonic hedgehog response in the absence of added cholesterol, only cyclopamine blocks the Sonic hedgehog signal when exogenous cholesterol is present. The failure to rescue Sonic hedgehog signaling with exogenous cholesterol implies that cyclopamine-induced teratogenesis results from a more direct interaction with certain elements in the Sonic hedgehog signal transduction cascade [100].

REFERENCES

[1]   Prelog, V.; Jeger, O. In *The Alkaloids: Chemistry and Physiology*; Manske, R.H. F.; Holmes, H.L., Eds.; Academic Press: New York, **1953**; Vol. *3*, pp. 247-312.

[2]   Prelog, V.; Jeger, O. In *The Alkaloids: Chemistry and Physiology*; Manske, R.H.F., Ed.; Academic Press: New York , **1960**; Vol. *7*, pp. 363-417.

[3]   Kupchan, S.M.; By, A.W. In *The Alkaloids: Chemistry and Physiology*; Manske, R.H.F., Ed.; Academic Press: New York, **1968**; Vol *10*, pp. 193-285.

[4]   Tomko, J.; Voticky, Z. In *The Alkaloids: Chemistry and Physiology*; Manske, R.H.F., Ed.; Academic Press: New York, **1973**; Vol. *14*, pp. 1-82.

[5]   Fieser, L.F.; Fieser, M. *Steroids*, Reinhold: New York, **1959**; pp. 867-895.

[6]   Kutney, J.P. *Bioorg. Chem.*, **1977**, 6, 371.

[7]   Brown, E.; Ragault, M. *Tetrahedron*, **1979**, *35*, 911.

[8]   Jones, K.; Newton, R.F.; Yarnold, C.J. *Tetrahedron*, **1996**, *52*, 4133.

[9]   Sweet, M. *Common Edible and Useful Plants of the West*, Naturegraph Co.: Healdsburg, CA, **1962**.

[10]   Yarnell, E.; Meserole, L. *Altern. Complemen. Therap.*, **1997**, *3*, 13.

[11]   Hollman, A. *Br. Heart J.*, **1991**, *65*, 286.

[12]   Quatrehomme, G.; Bertrand, F.; Chauvet, C.; Ollier, A. *Hum. Experiment. Toxicol.*, **1993**, *12*, 111.
[13]   Codding, P.W. *J. Am. Chem. Soc.*, **1983**, *105*, 3172.
[14]   Gaffield, W.; Keeler, R.F. *J. Toxicol., Toxin Revs.*, **1996**, *15*, 303.
[15]   Atta-ur-Rahman; Choudhary, M.I. *Nat. Prod. Rep.*, **1995**, *12*, 361.
[16]   Atta-ur-Rahman; Choudhary, M.I., *Nat. Prod. Rep.*, **1997**, *14*, 191.
[17]   Mandelbaum, A. *The Odyssey of Homer, Book IX*, University of California Press: Berkeley, CA, **1990**, pp. 171-192.
[18]   Cohen, Jr., M.M.; Sulik, K.K. *J. Craniofac. Genet. Dev. Biol.*, **1992**, *12*, 196.
[19]   Nishimura, H.; Okamoto, N. Sequential Atlas of Human Congenital Malformations: Observations of Embryos, Fetuses, and Newborns, University Park Press: Baltimore, **1976**.
[20]   Keeler, R.F. In *Isopentenoids in Plants: Biochemistry and Function*; Nes, W.D.; Fuller, G.; Tsai, L.-S., Eds.; Dekker: New York, **1984**, pp. 531-562
[21]   Keeler, R.F. In *Alkaloids: Chemical and Biological Perspectives*; Pelletier, S.W., Ed.; Wiley - Interscience: New York, **1986**; Vol. *4*, pp. 389-425.
[22]   Binns, W.; Thacker, E.J.; James, L. F.; Huffman, W.T. *J. Am. Vet. Med. Assoc.*, **1959**, *134*, 180.
[23]   Binns, W.; Anderson, W.A.; Sullivan, D. J. *J. Am. Vet. Med. Assoc.*, **1960**, *137*, 515.
[24]   Binns, W.; James, L. F.; Shupe, J.L.; Thacker, E.J. *Arch. Environ. Health*, **1962**, *5*, 106.
[25]   Binns, W.; James, L.F.; Shupe, J.L.; Everett, G. *Am. J. Vet. Res.*, **1963**, *24*, 1164.
[26]   Nathanielsz, P.W. *Amer. Sci.*, **1996**, 84, 562.
[27]   Hamilton, G. *New Sci.*, **1998**, *157*, 24.
[28]   Hutchinson, J.L.; Stebbins, G.L. *A Flora of the Wright's Lake Area*, Cal Central Press: Sacramento, CA, **1986**, p. *136.*
[29]   Binns, W.; Shupe, J.L.; Keeler, R.F.; James, L.F. *J. Am. Vet. Med. Assoc.*, **1965**, *147*, 839.
[30]   Keeler, R.F.; Binns, W. *Teratology*, **1968**, *1*, 5.
[31]   Keeler, R.F. *J. Range Manage.*, **1978**, *31*, 355.
[32]   Anderson, V.J.; Thompson, R.M. *USDA Forest Service Intermountain Res. Station Res. Paper*, Ogden, UT, **1993**, No. 469, pp. 1-6.
[33]   Atta-ur-Rahman; Ali, R.A.; Parveen, T.; Choudhary, M.I.; Sener, B.; Turkoz, S. *Phytochemistry*, **1991**, *30*, 368.
[34]   Atta-ur-Rahman; Ali, R.A.; Choudhary, M.I.; Sener, B.; Turkoz, S. *J. Nat. Prod.*, **1992**, *55*, 565.
[35]   Atta-ur-Rahman; Ali, R.A.; Ashraf, M.; Choudhary, M.I.; Sener, B.; Turkoz, S. *Phytochemistry*, **1996**, *43*, 907.
[36]   Keeler, R.F.; Binns, W. *Can. J. Biochem.*, **1966**, *44*, 819.
[37]   Keeler, R.F. *Phytochemistry*, **1969**, *8*, 223.
[38]   Keeler, R.F. *Steroids*, **1969**, *13*, 579.
[39]   Binns, W; Keeler, R.F.; Balls, L.D. *Clin. Toxicol.*, **1972**, *5*, 245.
[40]   Keeler, R. F. *Lipids*, **1978**, *13*, 708.
[41]   Keeler, R.F. *Proc. Soc. Exp. Biol. Med.*, **1975**, *149*, 302.
[42]   Gaffield, W.; Keeler, R.F.; Baker, D.C. In *Natural Toxins; Toxicology, Chemistry, and Safety*; Keeler, R.F.; Mandava, N.B.; Tu, A.T., Eds.; Alaken, Inc.: Fort Collins CO, **1992**, pp. 18-34.
[43]   Kalter, H. *Teratology*, **1992**, *46*, 207.

[44] Bryden, M.M.; Perry, C.; Keeler, R.F. *Teratology*, **1973**, *8*, 19.

[45] Keeler, R.F. *J. Anim. Sci.*, **1988**, *66*, 2414.

[46] Warkany, J. *Congenital Malformations*, Year Book Medical Publ.: Chicago, IL, **1971**.

[47] Facciola, S. *Cornucopia: A Source Book of Edible Plants*, Kampong Publ.: Vista CA, **1990**, pp. 207-209.

[48] Keeler, R.F.; Baker, D.C.; Gaffield, W. In *Handbook of Natural Toxins, Toxicology of Plant and Fungal Compounds*; Keeler, R.F.; Tu, A.T., Eds.; Dekker: New York, **1991**; Vol. *6*, pp. 83-99.

[49] Nau, H.; Hauck, R.-S.; Ehlers, K. *Pharmacol. Toxicol.*, **1991**, *69*, 310.

[50] Andrews, J.E.; Ebron-McCoy, M.T.; Bojic,U.; Nau, H.; Kavlock, R.J. *Teratology*, **1997**, *55*, 314.

[51] Keeler, R.F.; Young, S.; Brown, D. *Res. Commun. Chem. Pathol. Pharmacol.*, **1976**, *13*, 723.

[52] Brown, D.; Keeler, R.F. *J. Agric. Food Chem.*, **1978**, *26*, 566.

[53] Gaffield, W.; Keeler, R.F. In *Plant Associated Toxins: Agricultural, Phytochemical and Ecological Aspects*; Colegate, S.M.; Dorling, P.R., Eds.; CAB International: Wallingford, UK, **1994**, p. 333-338.

[54] Gaffield, W.; Keeler, R.F. *Experientia*, **1993**, *49*, 922.

[55] Gaffield, W.; Keeler, R.F. *Chem. Res. Toxicol.*, **1996**, *9*, 426

[56] Gaffield, W.; Keeler, R.F. *J. Nat. Toxins*, **1996**, *5*, 25.

[57] Staples, R.E.; Haseman, J.K. *Teratology*, **1974**, *9*, 259.

[58] Healy, M.J.R. *Appl. Stat.*, **1972**, *21*, 155.

[59] Brown, D.; Keeler, R.F. *J. Agric. Food Chem.*, **1978**, *26*, 561.

[60] Muenke, M. *Semin. Dev. Biol.*, **1994**, *5*, 293.

[61] Roach, E.; DeMyer, W.; Conneally, P.M.; Palmer, C.; Merritt, A.D. *Birth Defects*, **1975**, *11* (2), 294.

[62] Matsunaga, E.; Shiota, K. *Teratology*, **1977**, *16*, 261.

[63] Belloni, E.; Muenke, M.; Roessler, E.; Traverso, G.; *et al. Nature Genet.*, **1996**, *14*, 353.

[64] Roessler, E.; Ward, D.E.; Gaudenz, K.; Belloni, E.; Scherer, S.W.; Donnai, D.; Siegel- Bartelt, J.; Tsui, L.-C.; Muenke, M. *Hum. Genet.*, **1997**, *100*, 172.

[65] Bartlett, J. *Familiar Quotations*, Little, Brown: Boston, **1980**, Ed. *15*, p. 663.

[66] Johnson, R.L.; Tabin, C.J. *Cell*, **1997**, *90*, 979.

[67] Perrimon, N. *Cell*, **1995**, *80*, 517.

[68] Bitgood, M.J.; McMahon, A.P. *Dev. Biol.*, **1995**, *172*, 126.

[69] Roberts, D.J.; Johnson, R.L.; Burke, A.C.; Nelson, C.E.; Morgan, B.A.; Tabin, C. *Development*, **1995**, *121*, 3163.

[70] Urase, K.; Mukasa, T.; Igarashi, H.; Ishii, Y.; Yagusi, S.; Momoi, M.Y.; Momoi, T. *Biochem. Biophys. Res. Commun.*, **1996**, *225*, 161.

[71] Hammerschmidt, M. *Trends Genet*, **1997**, *13*, 14.

[72] Weed, M.; Mundlos, S.; Olsen, B.R. *Matrix Biol.*, **1997**, *16*, 53.

[73] Chiang, C.; Litingtung, Y.; Lee, E.; Young, K.E.; Corden, J.L.; Westphal, H.; Beachy, P.A. *Nature*, **1996**, *383*, 407.

[74] Roessler, E.; Belloni, E.; Gaudenz, K.; Jay, P.; Berta, P.; Scherer, S.W.; Tsui, L.-C.; Muenke, M. *Nature Genet.*, **1996**, *14*, 357.

[75] Lee, J.J.; Ekker, S.C.; von Kessler, D.P.; Porter, J.A.; Sun, B.I.; Beachy, P.A. *Science*, **1994**, *266*, 1528.

[76] Bumcrot, D.A.; Takada, R.; McMahon, A.P. *Mol. Cell. Biol.*, **1995**, *15*, 2294.

[77]   Porter, J.A.; von Kessler, D.P.; Ekker, S.C.; Young, K.E.; Lee, J.J.; Moses, K.; Beachy, P.A. *Nature*, **1995**, *374*, 363.

[78]   Porter, J.A.; Ekker, S.C.; Park, W.-J.; von Kessler, D.P.; Young, K.E.; Chen, C.-H.; Ma, Y.; Woods, A.S.; Cotter, R.J.; Koonin, E.V.; Beachy, P.A. *Cell*, **1996**, *86*, 21.

[79]   Porter, J.A.; Young, K.E.; Beachy, P.A. *Science*, **1996**, *274*, 255.

[80]   Roelink, H.; Porter, J.A.; Chiang, C.; Tanabe, Y.; Chang, D.T.; Beachy, P.A.; Jessell, T.M. *Cell*, **1995**, *81*, 445.

[81]   Ericson, J.; Morton, S.; Kawakami, A.; Roelink, H.; Jessell, T.M. *Cell*, **1996**, *87*, 661.

[82]   Gibbons, G.F.; Mitropoulos, K.A.; Myant, N.B. *Biochemistry of Cholesterol*, Elsevier: Amsterdam, **1982**.

[83]   Tint, G.S.; Irons, M.; Elias, E.R.; Batta, A.K.; Frieden, R.; Chen, T.S.; Salen, G. *N. Engl. J. Med.*, **1994**, *330*, 107.

[84]   Salen, G.; Shefer, S.; Batta, A.K.; Tint, G.S.; Xu, G.; Honda, A.; Irons, M.; Elias, E.R. *J. Lipid Res.*, **1996**, *37*, 1169.

[85]   Kolf-Clauw, M.; Chevy, F.; Siliart, B.; Wolf, C.; Mulliez, N.; Roux, C. *Teratology*, **1997**, *56*, 188.

[86]   Willnow, T.E.; Hilpert, J.; Armstrong, S.A.; Rohlmann, A.; Hammer, R.E.; Burns, D.K.; Herz, J. *Proc. Natl. Acad. Sci., USA*, **1996**, *93*, 8460.

[87]   Kusano, G; Takahashi, A.; Sugiyama, K.; Nozoe, S. *Chem. Pharm. Bull.*, **1987**, *35*, 4862.

[88]   Ang, S.-L.; Rossant, J. *Cell*, **1994**, *78*, 561.

[89]   Weinstein, D.C.; Ruiz i Altaba, A.; Chen, W.S.; Hoodless, P.; Prezioso, V.R.; Jessell, T.M.; Darnell, Jr., J.E. *Cell*, **1994**, *78*, 575.

[90]   Coventry, S.; Kapur, R.J.; Siebert, J.R. *Pediatric and Developmental Pathology*, **1998**, *1*, 29.

[91]   Hall, T.M.T.; Porter, J.A.; Young, K.E.; Koonin, E.V.; Beachy, P.A.; Leahy, D.L. *Cell*, **1997**, *91*, 85.

[92]   Marigo, V.; Davey, R.A.; Zuo, Y.; Cunningham, J.M.; Tabin, C.J. *Nature*, **1996**, *384*, 176.

[93]   Stone, D.M.; Hynes, M.; Armanini, M.; Swanson, T.A.; *et al. Nature*, **1996**, *384*, 129.

[94]   Goodrich, L.V.; Milenkovic, L.; Higgins, K.M.; Scott, M.P. *Science*, **1997**, *277*, 1109.

[95]   van den Heuvel, M.; Ingham, P.W. *Nature*, **1996**, *382*, 547.

[96]   Carstea, E.D.; Morris, J.A.; Coleman, K.G.; Loftus, S.K.; *et al. Science*, **1997**, *277*, 228.

[97]   Hua, X.; Nohturfft, A.; Goldstein, J.L.; Brown, M.S. *Cell*, **1996**, *87*, 415.

[98]   Loftus, S.K.; Morris, J.A.; Carstea, E.D.; Gu, J.Z.; *et al. Science*, **1997**, *277*, 232.

[99]   Cooper, M. K.; Porter, J. A.; Young, K. E.; Beachy, P. A.. *Science*, **1998**, *280*, 1603.

[100]   Incardona, J. P.; Gaffield, W.; Kapur, R. P.; Roelink, H. *Development*, **1998**, *125*, 3553.

Atta-ur-Rahman (Ed.) *Studies in Natural Products Chemistry, Vol. 23*

CHEMICAL AND BIOLOGICAL EVALUATION OF GENUS TEUCRIUM

AYHAN ULUBELEN*,**, GÜLAÇTI TOPÇU** *and*
UFUK SÖNMEZ*

**Faculty of Pharmacy, University of Istanbul, 34452, Istanbul, Turkey;
**TUBİTAK, Marmara Research Center, P.O. Box 21, 41470, Gebze,
Kocaeli, Turkey*

ABSTRACT: The genus *Teucrium* is represented by about 300 species in the world and 27 species in Turkey, 8 of them being endemic.
Teucrium species are rich source of neo-clerodane diterpenoids. Although several furanoid diterpenes with the neo-clerodane skeleton have been isolated from many plants, the genus *Teucrium* is the most abundant natural source of these compounds, therefore *Teucrium* species are accepted as chemotaxonomic markers for neo-clerodanes. Chemical investigations on this genus showed that some of the species also contain rearranged neo-clerodane or abietane diterpenes, sesquiterpenes, triterpenes and steroids. Flavonoids and aromatic compounds, although not as abundant as in the genus *Salvia* from the same family, are also found in the genus *Teucrium*. In this review, general chemical properties of neo-clerodanes isolated from *Teucrium* species are given with their stereochemical studies.
Teucrium species have been used for more than 2000 years as medicinal plants. They showed diuretic, diaphoretic, antiseptic and antipyretic activities. Antifeedant activities of some species previously studied against *Spodeptera littoralis* were presented herein. *Teucrium* species are used as hypoglycemic agents in North Africa and Saudi Arabia. They also possess antispasmolitic activity and used in the treatment of stomach aches. Some of the *Teucrium* extracts exhibited central nervous system depressant action.

INTRODUCTION

Genus *Teucrium* is represented by approximately 300 species in the world. There are 27 species in Turkey, 8 of them being endemic [1]. The plant is distributed mainly in South-East Asia, Central and South America and around the Mediterranean district.

Teucrium species are a rich source of *neo*-clerodane diterpenes which can be used as chemotaxonomic markers for the genus. They possess therapeutic properties and are used as folk medicine in many parts of the world.

The chemical studies on the genus started about three decades ago and since then many compounds have been isolated and their structures have been established. The first review article about genus *Teucrium* written by Fujita et al. [2] which was published in 1976, followed by the articles of

Piozzi [3-5]. Furthermore, in the review articles of Hanson [6-13] some information was given about *Teucrium* species.

In the present review article, the chemistry and biological uses of the genus will be discussed.

Teucrium species are perennial, rarely annual or biennial herbs or small shrubs. Leaves are entire dentate or deeply dissected flowers are in axils of upper leaves or born in racemes, panicles or heads, calyx is often gibbous at base, regular or bilabiate, with five equal or unequal teeth.

Although *Teucrium* species contain triterpenes, sesquiterpenes and steroidal compounds, as well as flavonoids, the main secondary metabolites are clerodane and/or 19-norclerodane diterpenoids.

According to Rowe's nomenclature [14], the hydrogen at C-10α being and methyl at C-5 β(Me-19) indicate clerodane, the reverse *ent*-clerodane structure. Roger [15] assigned *ent-neo*-clerodane for the former and neo-clerodane for the latter type of structures. In the present day, clerodane and *neo*-clerodane names are used commonly. Most of the diterpenoids isolated from *Teucrium* species have *neo*-clerodane structures.

CHEMISTRY OF THE SECONDARY METABOLITES

Neo-clerodane Diterpenoids

Neo-clerodanes consist of three or four ring systems. Either a five-membered spiro-lactone or a lactol ring attached to ring B at C-9 is considered as the third ring (ring C) or in some other cases where this five-membered ring is missing and an ethylene group is present.

A furan ring, attached to either ring C (**1**) or to ethylene group (**2**), is almost always present, forming the third or the fourth ring of the molecule. Additional rings could be formed as lactone or ether between C$_4$-C$_5$ (**3**), C$_4$-C$_6$ (**4**), C$_5$-C$_{20}$ (**5**) or at some other positions.

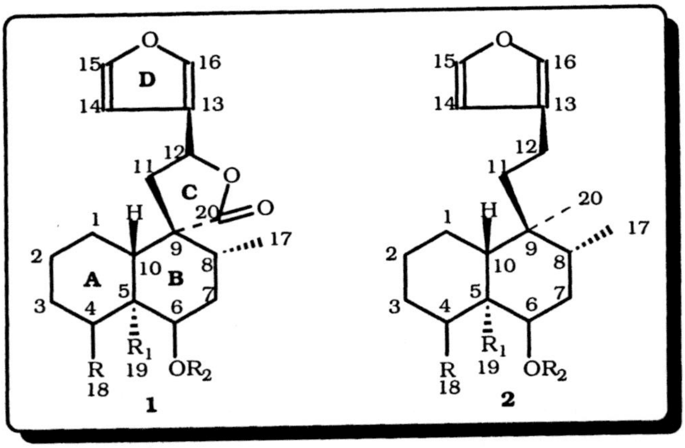

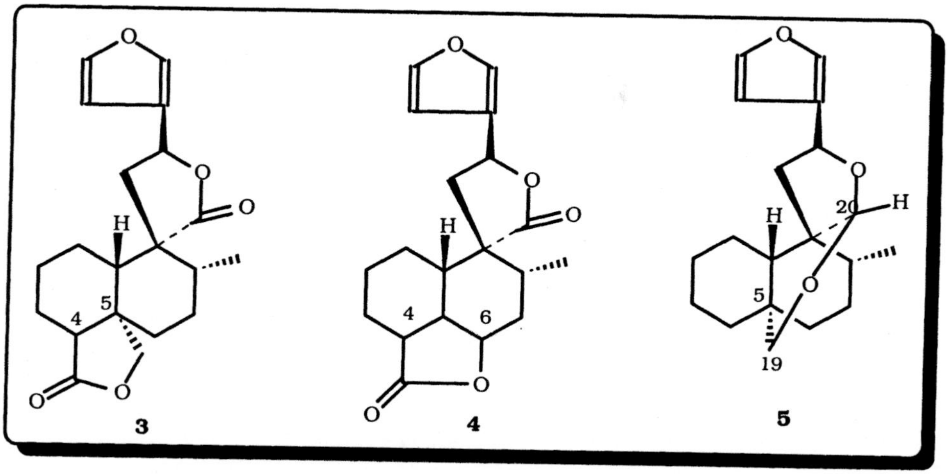

3   4   5

Instead of a five-membered ring for ring-C, rarely a six-membered pyrane ring can be present(**6**), however, this type of compounds are mostly found in *Salvia* species [16].

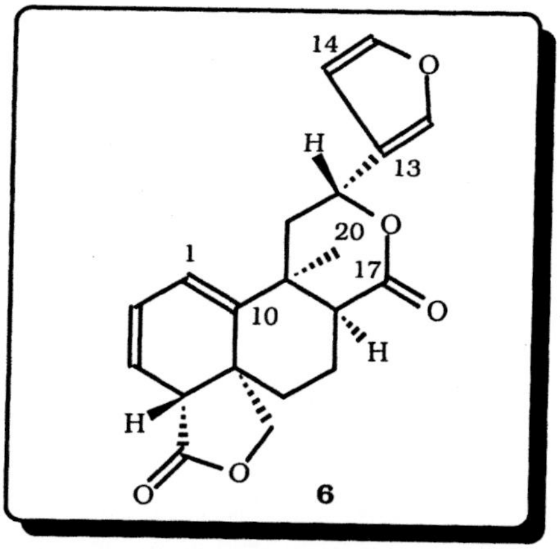

6

Almost all of the diterpenoids from *Teucrium* species have *neo*-clerodane structure as proven by the use of X-ray diffraction, ORD, CD and by chemical reactions. The stereochemistry at C-12 is usually β which is deduced from the ^{1}H NMR spectrum. However, when there is a hydroxyl or an acetyl group substituent at C-12, problems may arise. If ring-C is present, its configuration is usually 12S, eventhough 12R configuration is possible in some compounds. The presence of a furan ring in the molecule followed from the IR bands at 1490 and 870 cm[-1], substantiated by positive Erlich test [17] and ^{1}H NMR spectrum. The two α-protons and one β-proton of the furan ring usually resonate at *ca* δ 7.40-7.45 (2H, d) and δ 6.30-6.40 (1H, dd), respectively. In the MS spectrum, furan ring is recognized from the ion fragments at *m/z* 94(a) and 81(b) [18].

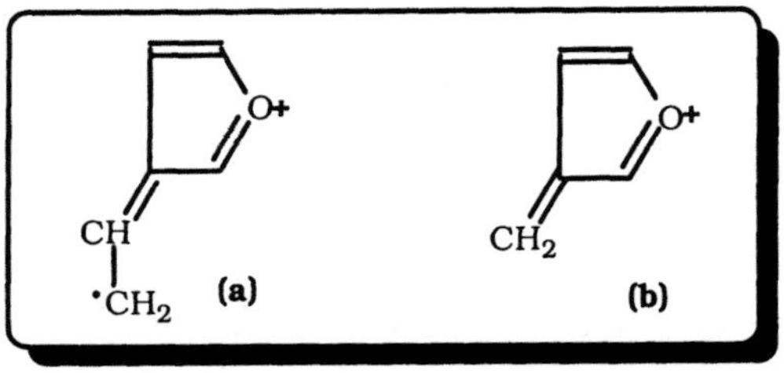

Spin-decoupling experiments showed that H-12 is the X part of an A_2X type of signal while H_2-11 A_2 part which appeared as a doublet (J=8.5 Hz) at *ca* δ 2.50-2.55, [19,20] as followed in the structure determination of teucvin (**7**). Although some chemical reactions and spectral methods were used in order to establish the structure and the absolute configuration of **7**, X-ray crystallography was also performed [19]. After treating **7** with Na_2CO_3 and refluxing in MeOH, crystalline product $C_{20}H_{24}O_6$ (**8**) was obtained in a good yield. The latter was a keto compound and reduced to a hydroxy ester with $NaBH_4$ treatment to yield $C_{20}H_{26}O_6$ (**9**). This was converted to its bromoacetate derivative $C_{22}H_{27}O_7Br$(**10**), the X-ray of the latter was recorded as seen in **11**.

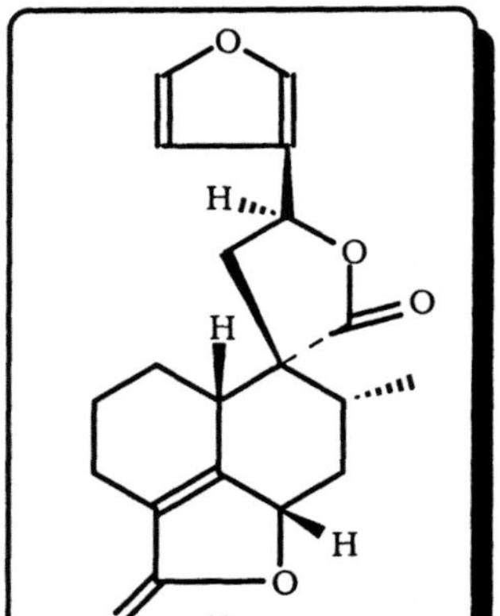

Two new compounds montanins A (**12**) and B (**13**) were isolated from *T.montanum* L. [20], the structures and stereochemistry of **12** and **13** were proved by correlation with those of teucvin (**7**). Both compounds had the furan ring at C-12 as β which followed from H-12 signal at δ 5.38 (1H, t, J=8 Hz) for **12** and δ 5.28 (1H, t, J=8 Hz) for **13**. Auto-oxidation of **12** occured within five days to yield teucvin (**7**), when kept in $CHCl_3$ [20].

From *T. flavum* [21] two new diterpenoids teucvidin (**14**) and its 3β-hydroxy derivative teuflidin (**15**) were isolated. Both compounds are the C-6 and C-10 epimers of teucvin (**7**). In compounds **14** and **15**, the presence of a triplet at δ 5.43 (J= 8.5 Hz, H-12α) and two double doublet signals at δ 2.75 (1H, dd, J= 8.5 Hz and 14.5 Hz), 1.90 (1H, dd, J= 8.5

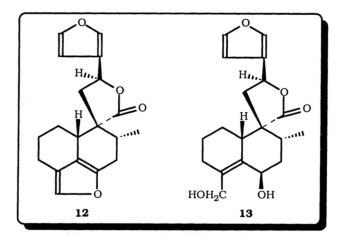

and 14.5 Hz) for C-11 protons were indicative of the β-position of the furan ring.

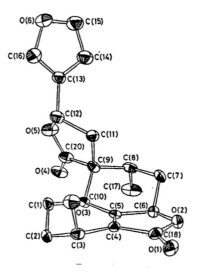

14 R=H
15 R=OH

In order to ensure the stereochemistry of these molecules at C-6 and C-10, they were subjected to X-ray analysis (Fig. **1**) which showed that ring A deviates from the preferred half-chair conformation of cyclohexane, is slightly distorted from perfect chair. This is possibly due to constraint imposed by the γ-lactone ring. The sign of circular dichroism curve associated with the α, β unsaturated lactone has been taken as evidence for the absolute stereochemistry of this series. Teuflidin (**15**) showed the same sign and the magnitude of CD curve as teucvidin (**14**) and, hence, it was assigned to the same absolute stereochemistry.

Fig. (1).

A new compound teuflin (**16**) which is C-10 epimer of teucvidin(**14**) was isolated from the same plant material [22]. Since the co-occurence of

C-10 epimers are extremely rare, the molecule was subjected to X-ray analysis (Fig. 1). Teuflin (16) showed a strong negative Cotton effect at 218 nm, suggesting the same absolute stereochemistry at C-6 as observed in teucvidin (14).

T.viscidum var. *miquelianum* [18,19,22] yielded teucvin (7), teucvidin (14) and teuflin (16), spectral data were checked in detail. In teuflin (16), the ^{1}H NMR signal at δ 2.68 as a multiplet and ^{13}C NMR signal at δ 43.0 (C-10) were shown to be the same as those of teucvin (7). These data supported that both compounds have the same configuration at C-10. On the other hand, the hydrogen at C-6 (δ 5.72, m) showed an unusual paramagnetic shift compared to the C-6 hydrogens of 7, 14 and 15. This shift can be attributed to the anisotropic effect of the lactone carbonyl at C-20 on ring B which has a fixed boat like conformation. Thus, C-6 hydrogen must have the configuration opposite to that of teucvin, supported with the similarity of the ^{13}C chemical shift at C-6 of 16 and 14. Table 1 shows the spectral data of teucvin (7), teucvidin (14), teuflidin (15) and teuflin (16). When this table is studied, the ^{1}H NMR signal of H-10α seems at lowfield compared to H-10β signal.

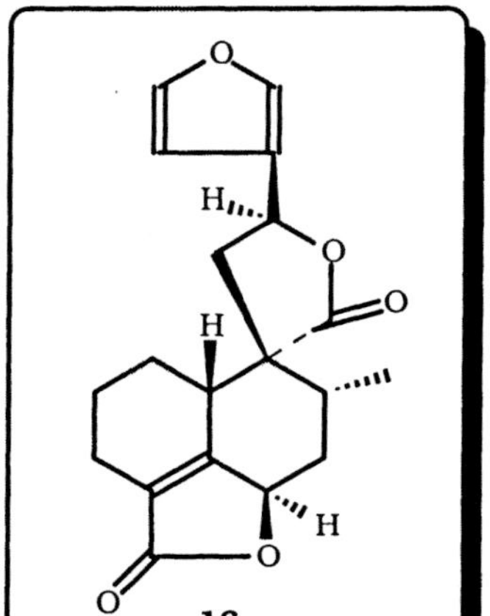

16

The stereochemistry followed from CD spectrum by the observation negative Cotton effect at *ca* 230 nm for teucvidin (14) and teuflidin (15) which assigned H-6β position [23].

The stereochemistry at C-12 is also very important. This chiral centre of *neo*-clerodane diterpenoids, C-12 was assumed to have (S) configuration. This view was maintained after X-ray analysis and, chemical correlations with the compounds of known absolute stereochemistry. Hence, *Teucrium* diterpenoids were invariably found with C-12S configuration. However, recent studies, using ^{1}H NOE difference techniques [24,25], have shown that S configuration at C-12 is by no means a common stereochemical property. During the investigation of *T. scorodonia* [26] and *T. lanigerum* [27], the authors found teupolin I (**17**) previously isolated from *T. polium* [28] and claimed having C-12R configuration on the basis of physical data such as mp, $(\alpha)_D$, and low frequency ^{1}H NMR. In order to solve this problem, a detailed highfield ^{1}H NOE study on teupolin I (**17**) and related diterpenoids (**18-30**) were undertaken [29]. Stereochemical considerations showed that *Teucrium* diterpenoids with the C-12R configuration have their Me-17 group and H-12 which are nearly parallel steric disposition on the same side of the lactone ring.

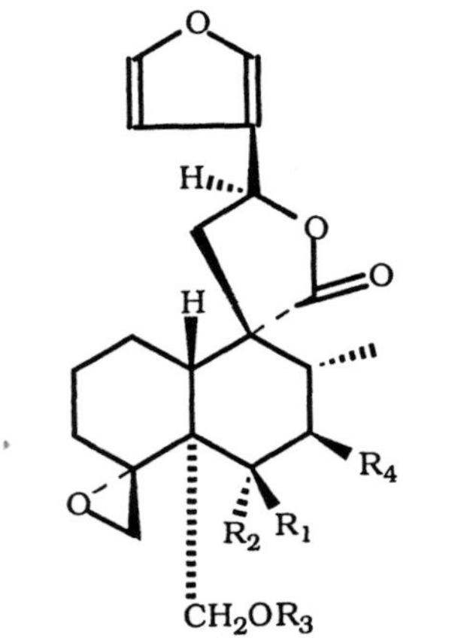

	R_1	R_2	R_3	R_4
17	H	OH	Ac	H
18		O		Ac	OH
19	H	OAc	H	H
20		O		Ac	H

The resulting spatial proximity can be easily monitored in selective NOE experiments by irradiating either one of the two pertinent resonances and observing the net enhancement of the other signal [30]. Performing these experiments on compounds **17-30**, significant enhancements were detected only in **21** and **30** which suggested that these two compounds have C-12R stereochemistry while the remaining diterpenoids belonged to C-12S series.

    According to lit. 24, the lack of enhancements in the NOE experiments is conclusive evidence for the C-12S stereochemistry. In fact, sizeable(3-5%)enhancements of the H-1β(equatorial) were observed in C-12S series compounds, **17-19** and **24-30** upon selective low-power irradiation of the proton signal due to H-12. In order to decide the stereochemistry of C-12,

NOE experiment should be performed at 400, 300 and 200 MHz instruments in $CDCl_3$ or $CDCl_3$-DMSOd$_6$ solutions of the samples using frequency cycling technique [31].

Table 1. ^1H and ^{13}C NMR Data of 7, 14, 15, 16

Position	7 ^1H	7 ^{13}C	15 ^1H	15 ^{13}C	14 ^1H	14 ^{13}C	16 ^1H	16 ^{13}C
1		21.7		17.7		21.4		18.7
2		24.7		29.8		23.5		23.7
3		19.7	4.67m	58.7		20.1		23.3
4	-	126.4	-	128.7	-	127.6	-	123.7
5	-	161.6	-	165.7	-	162.2	-	166.2
6	4.74t (6.5)	78.3	5.03q (7.13)	76.2	5.0dd (6.11)	76.0	5.72m	76.6
7		35.3	2.38ddd (2,7,13)	35.6	2.31ddd (6,11,13)	35.8		31.8
8		35.9	2.25ddd (24.7)	36.8	2.18ddq (4,3,7)	35.9		35.8
9	-	53.5	-	52.1	-	52.2	-	51.0
10	2.68m	42.0	3.27t(8)	38.6	3.27m	38.8	2.68m	43.0
11	2.54dd (8.5,15)	40.8	2.75q (7,5,14,5)	38.8	2.58dd (7,14)	39.0	2.67dd (13.7,7.3)	42.8
12	5.43t (8.5)	71.8	5.42t	72.2	5.36t (7)	71.8	5.38dd (9.9,7.3)	71.6
13	-	124.9	-	125.3	-	125.2	-	124.3
14	6.37m	107.9	6.37m	107.9	6.36dd (1,1)	107.9	6.41m	107.8
15	7.43m	144.2	7.45m	144.2	7.44d (1)	144.2	7.46m	144.2
16	7.43	139.5	7.45m	139.5	7.44 d (1)	139.5	7.46m	139.8
17	1.05d (6.5)	17.0	1.36 (7)	14.3	1.36d (7)	14.3	1.22d (7.1)	17.6
18	-	175.6	-	177.5		177.6		175.9
19	-	-	-	-		-		
20	-	173.0	-	172.0		172.5		173.5

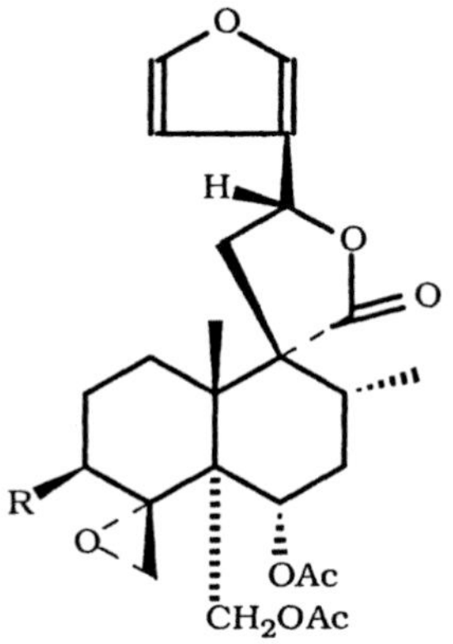

21 R=H
30 R=OH

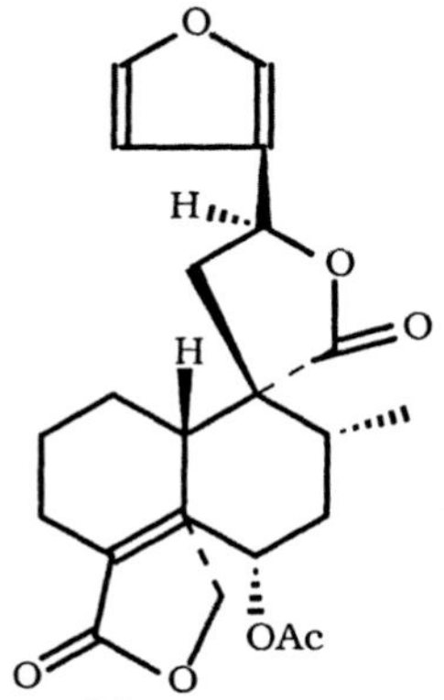

22

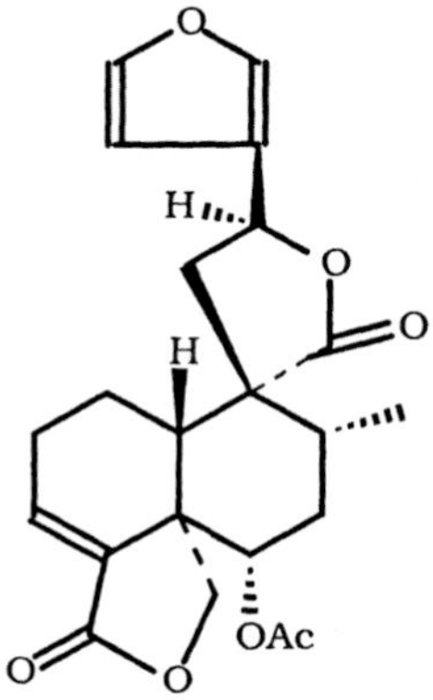

23

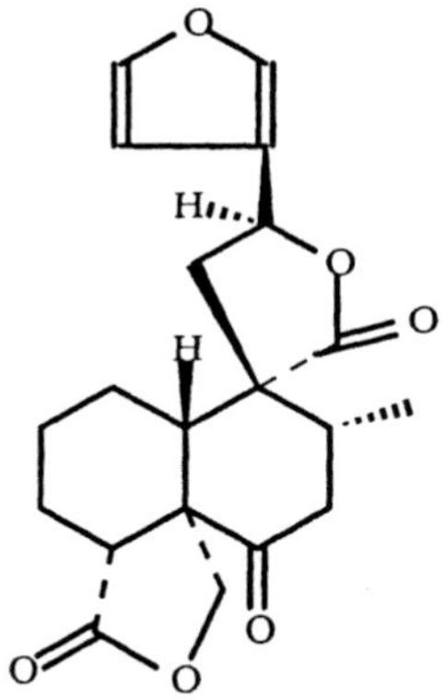

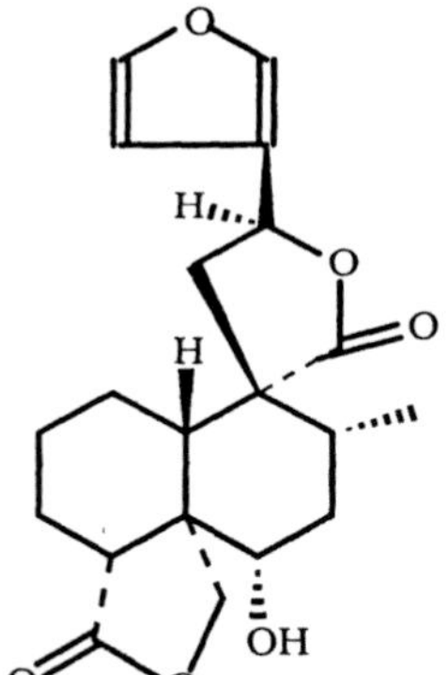

25

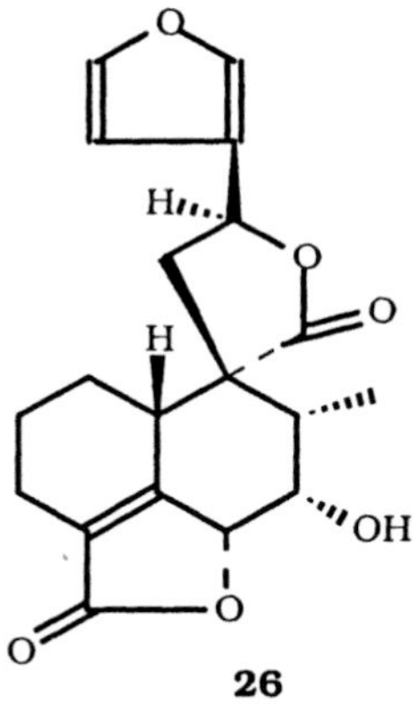

26

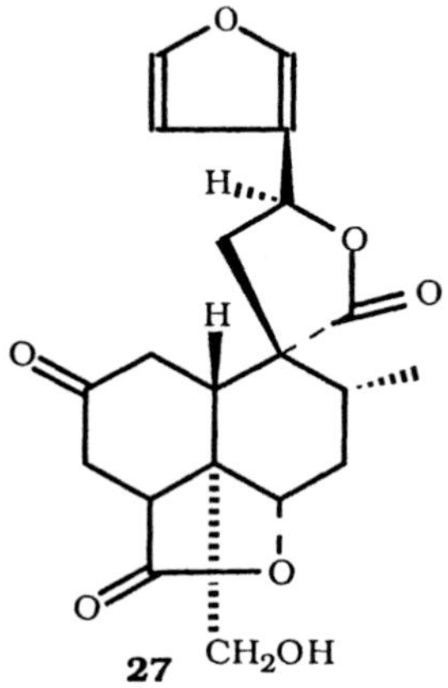

27 CH₂OH

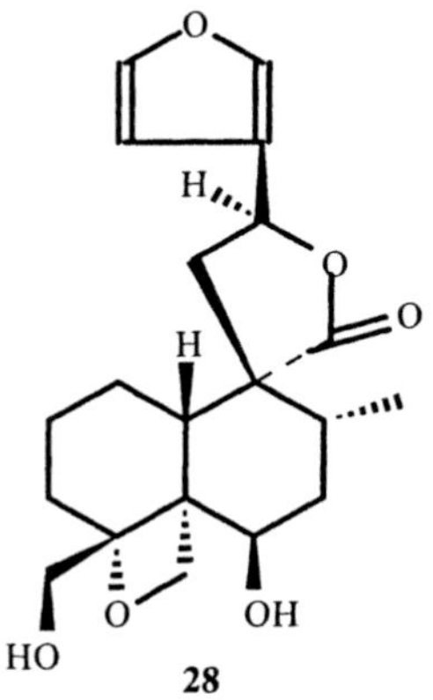

28

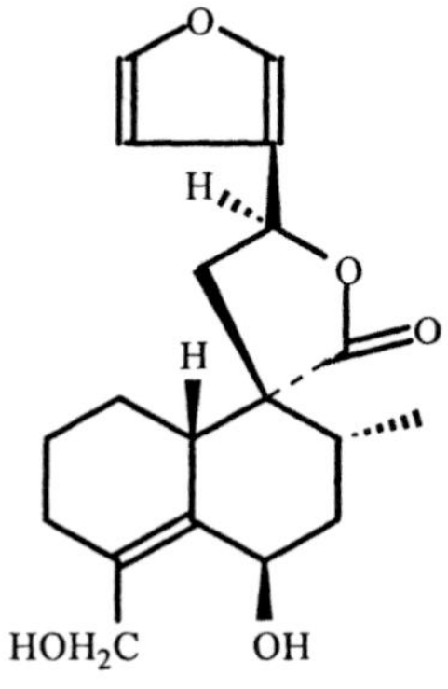

29

12-*Epi*-teucvidin (**31**) was isolated from *T. kotschyanum* [32]. The ^{1}H NMR spectrum of **31** was almost identical with that of teucvidin (**14**), with small differences in the chemical shifts corresponding to C-12 methine, C-11 methylene and C-17 methyl protons (Table 2).In principle, this could be attributed to an opposite stereochemistry at C-12 center in both diterpenoids. However, the relative stereochemistry of **31** was achieved by NOE experiments. Irradiation of C-17 methyl protons of **31** δ (1.28) produced a 2% NOE enhancement of the H-12 signal (δ 5.41), thus establishing that C-12 methine proton and C-17 methyl protons were on the same side of the plane defined by the C_{20}-C_{12} lactone ring.

When the C-17 protons of teucvidin (**14**) were irradiated, no NOE was observed in the signal of C-12 proton.

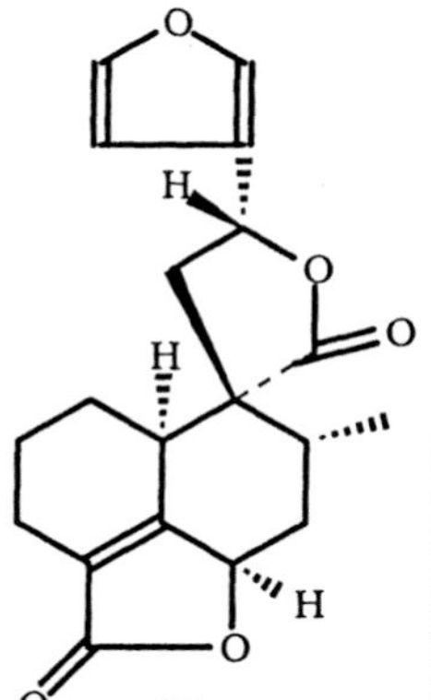

31

In both cases, strong and identical NOE enhancements were also produced in the signals of C-6α and C-10α protons. These results clearly established that **14** and **31** possessed an identical substituted decalin moiety with ring B in a distorted boat conformation in which C-17 methyl group is pseudoaxially oriented. Thus, the structural difference between these two compounds was the stereochemistry at the C-12 centre. As the C-12S configuration of **14** was well established, **31** is therefore C-12R epimer of teucvidin (**14**). There is quite a number of neo-clerodane diterpenoids with 4, 18-oxirane ring. Among them, teucjaponins A(**32**) and B(**33**) which were isolated from *T. japonicum* Houtt [33], have the signals of their C-18 protons at δ 2.26 (d, *J*=6 Hz) and 3.78 (dd, *J*=6 and 2 Hz) for **32** and at δ 2.24 (d, *J*=6 Hz), 3.56 (dd, *J*=6 and 2 Hz) for **33**. From *T. lanigerum* [27] a group of *neo*-clerodanes were isolated, all of them (**34-47**) possess a 4, 18-oxirane ring with regular structural pattern of the molecule 10β, 5α- and C-12S structure and having similar chemical shifts.The ^{13}C NMR chemical shifts of C-4 and C-18 were also quite similar in all of compounds.

Table 2.   NMR Data of Compounds 14 and 31

Position	14		31	
	^1H	^{13}C	^1H	^{13}C
1		20.0		19.6
2		23.4		22.9
3		21.4		21.4
4	-	127.8	-	126.7
5	-	162.2	-	162.9
6	5.00brt	76.1	5.01brdd	75.9
7		35.7		35.8
8		38.7		33.0
9	-	52.1	-	54.1
10	3.27brt	35.8	3.27brt	33.2
11	1, 92dd, 2.59dd	39.0	2.20dd;2.26dd	36.5
12	5.36t	71.9	5.41dd	71.3
13	-	125.5	-	123.6
14	6.36dd	107.9	6.42dd	107.9
15	7.44m	144.2	7.46t	144.3
16	7.44m	139.5	7.48m	139.9
17	1.36d	14.3	1.28d	14.8
18	-	172.6	-	172.6
20	-	177.6	-	176.8

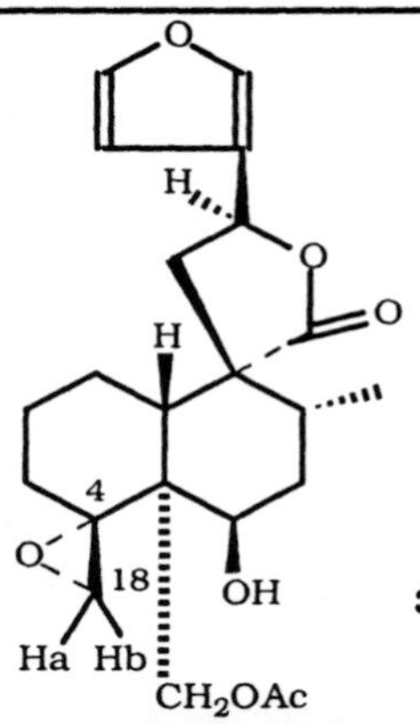

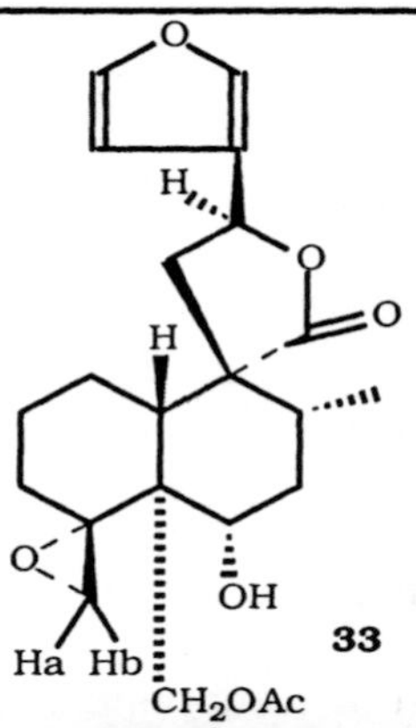

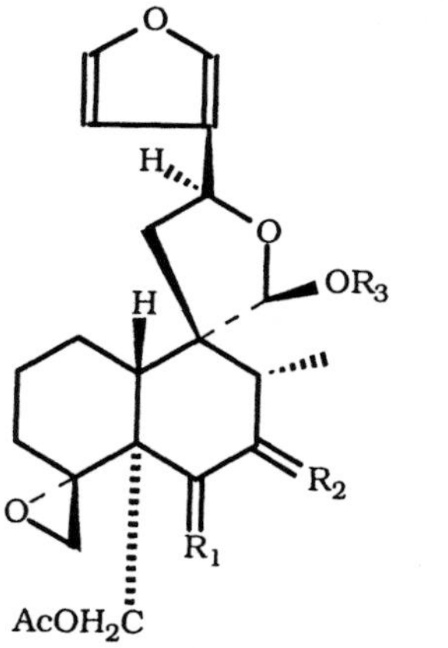

	R_1	R_2	R_3
34	O	αOH,βH	Ac
35	O	αOH,βH	H
36	O	αOAc,βH	Ac
40	αOH,βH	O	Ac
41	αOAc,βH	O	Ac
43	αOAc,βH	αOH,βH	Ac
44	αOH,βH	αOAc,βH	Ac
45	αOAc,βH	αOAc,βH	Ac
46	αOH,βH	αOH,βH	Ac
47	βOH,αH	αOH,βH	Ac

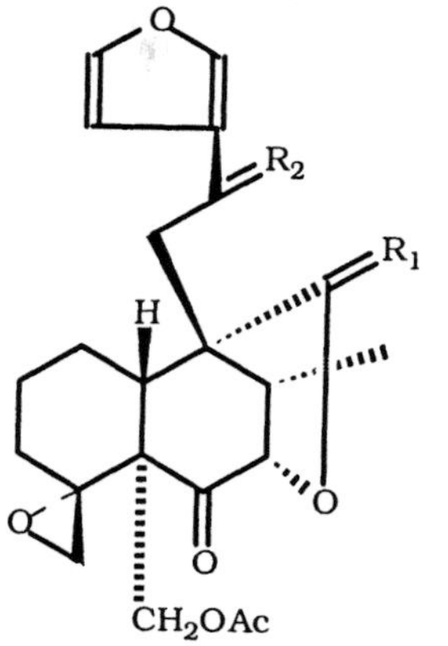

	R_1	R_2
37	O	O
38	O	H,OAc(S)
39	H,OH	H,OAc(S)

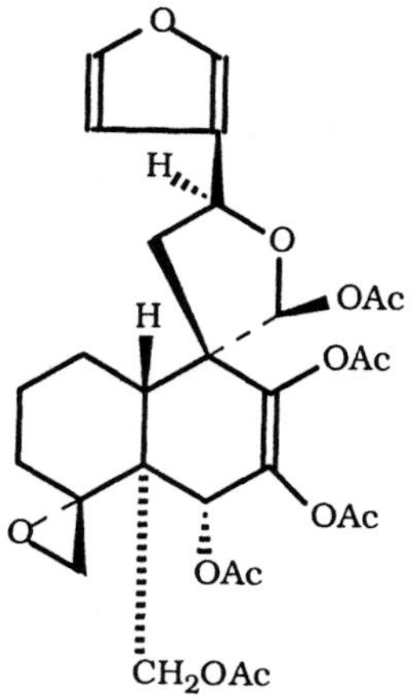

42

In case of the presence of an acetyl group at C-3 together with 4, 18-oxirane ring in a neo-clerodane diterpenoid, the ^{1}H NMR spectrum clearly shows strong deshielding effect on the H-3, due to equatorial (β) orientation of the 3-OAc group, as contrary to axial disposition of the 3α-OAc epimers.

Table 3.　　**^{1}H NMR Data of 35, 37, 40-47**

H	35	37	40	41	42	43	44	45	46	47
6 α	-	-	-	-	-	-	-			3.70d
6 β			4.07d	5.05brs	5.40q	4.60brd	3.66dd	4.75dd	3.41dd	-
7 β	4.83d	4.38s	–	-		3.80dd	5.47dd	5.26dd	3.60	3.60d
H-8		2.95q		2.67q	-					
H$_A$-11		2.92d	2.38d	2.43d	2.20dd					
H$_B$-11		3.50d	2.38d	2.43d	2.50dd		2.45dd	2.53dd	2.52dd	
H-12	5.27t	-	4.93t	5.05t	5.08t	5.12dd	5.15dd	5.19dd	5.11dd	5.10t
H-14	6.30m	6.77d	6.42m	6.40m	6.40m	6.41m	6.45m	6.42m	6.46m	6.48m
H-15	7.35m	7.48t	7.40m	7.37m	7.36m	7.33m	7.42m	7.35m	7.35m	7.40m
H-16	7.35m	8.17d	7.40m	7.37m	7.36m	7.33m	7.42m	7.35m	7.35m	7.40m
Me-17	0.88d	1.10d	1.45d	1.42d	1.69d	1.34d	1.20d	1.25d	1.40d	1.35d
H$_A$-18		2.40d	2.58d	2.42d	2.32d	2.20d	2.40d	2.21d	2.42d	2.40d
H$_B$-18		3.27dd	3.27dd	3.03dd	3.08dd	2.92dd	3.20dd	2.95d	3.18dd	3.80±
H$_A$-19	4.65s	4.55d	4.23dd	4.41brd	4.39d	4.63brd	4.71brd	4.53dd	4.95brd	4.85s
H$_B$-19	4.65s	5.03d	4.32d	4.57d	4.57d	5.80d	5.54d	5.72d	5.45d	4.85s
H-20	5.47s	-	6.10s	6.24s	6.38s	6.35s	6.48s	6.40s	6.38s	6.40s
OAc	2.05s	2.05s	2.06s	2.08s	2.05s	2.08s	2.10s	2.13s	2.09s	2.11s
	-	-	1.99s	2.00s	2.03s	2.00s	2.03s	2.10s	1.99s	2.01s
	-	-	-	1.92s	2.02s	1.95s	1.90s	1.92s	-	-
	-	-	-	-	2.00s	-	-	1.90s	-	-

Tables 3 and 4 show the ^{1}H and ^{13}C NMR chemical shifts of compounds **34-47**, respectively.

Table 4.　　**^{13}C NMR of 37, 38, 40, 41, 46, 47**

C	37	38	40	41	46	47
1	21.7	21.5	23.2	23.0	22.6	22.7
2	24.1	24.2	25.2	24.8	25.4	24.1
3	31.7	31.6	33.0	32.0	31.3	33.4
4	62.0	62.1	65.3	64.4	66.8	62.9

(Table 4). contd.....

C	37	38	40	41	46	47
5	52.7	52.8	50.0	49.2	53.8	54.0
6	199.5	199.4	79.1	76.5	73.8	69.3
7	87.9	87.3	206.8	200.8	75.0	76.8
8	47.9	47.4	51.3	52.8	42.0	37.3
9	49.7	50.4	56.8	56.7	46.2	45.9
10	48.9	50.4	50.6	50.6	54.5	47.9
11	37.1	32.1	46.0	45.3	43.1	45.1
12	191.3	64.9	71.3	72.0	73.3	73.0
13	127.2	124.7	127.3	126.7	129.6	129.4
14	108.3	108.4	109.5	108.6	109.5	109.4
15	144.9	143.8	144.3	143.7	143.6	143.5
16	148.1	140.1	140.3	139.5	139.5	139.6
17	14.1	14.6	11.0	10.7	16.4	15.9
18	49.9	49.5	51.6	51.1	49.4	53.6
19	61.1	61.1	63.1	61.7	63.8	63.8
20	177.3	176.4	97.8	97.6	99.2	99.6
C=0	170.2	170.0	170.1	169.8	110.1	170.5
CH$_3$	20.7	20.9	20.9	21.0	21.4	21.4
C=0	-	169.7	169.6	169.6	168.8	169.2
CH$_3$	-	21.5	20.7	20.5	21.0	21.2
C=0	-	-	-	169.4	-	-
CH$_3$	-	-	-	20.5	-	-

The same effect is also observed for the 3β-hydroxy epimers as seen in compounds **48-61** [34,35,53]. Table 5 shows the ¹H NMR shifts of H-3α and H-3β in acetyl and hydroxyl derivatives of compounds **48-61**.

The spiro-oxirane ring, in many cases, is replaced by a CH$_2$OR and an α-hydroxy group. In such cases, the C-18 protons are shifted from *ca* 2.7 and 3.10 to *ca* δ 4.00 and 4.50 as observed in **62-65** [35-37] while the C-3 proton, geminal to an acetyl group, appears around δ 5.20-5.40. Table 6 gives the ¹H and ¹³C NMR data of **62-65**.

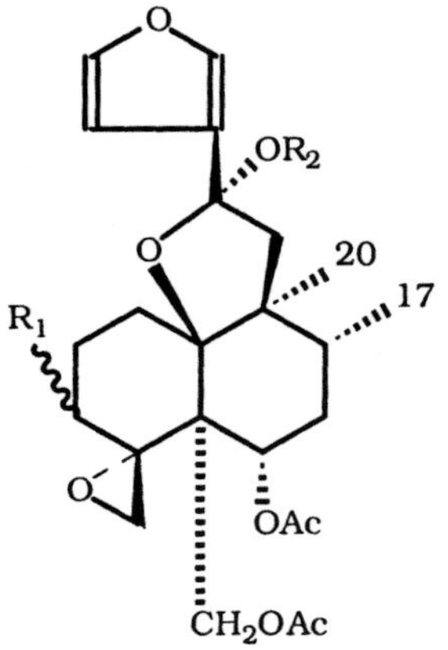

	R_1	R_2
48	αOAc	H
49	αOAc	Me
50	αOAc	Et
51	βOAc	Me
52	βOAc	H
53	βOH	Me

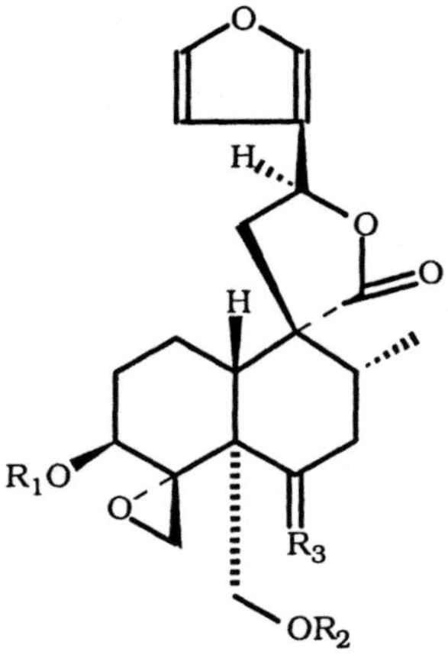

	R_1	R_2	R_3
54	Ac	Ac	αH,βOH
55	H	Ac	αH,βOH
56	H	Ac	βH,αOH

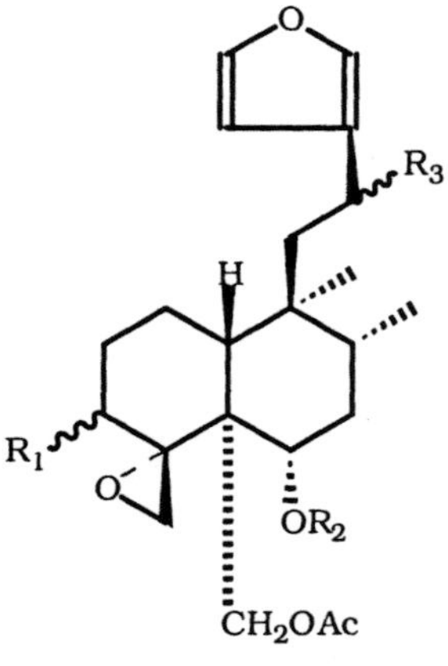

	R_1	R_2	R_3
57	αOAc	H	OH
58	αOAc	Ac	=O
59	αOAc	H	OAc

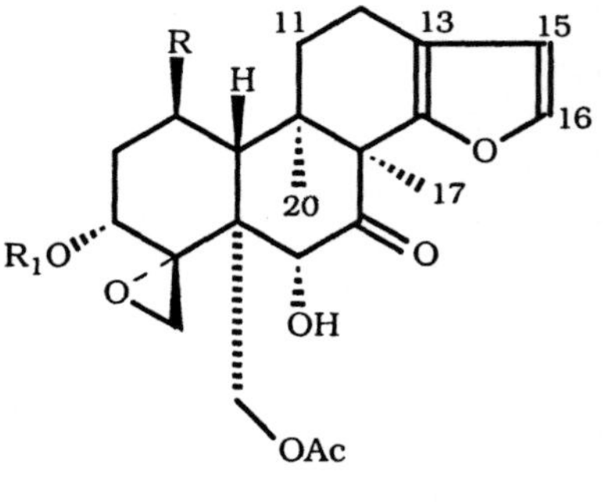

60 R=H R_1 = Ac
61 R=R$_1$=H

Table 5. ¹H NMR Data of H-3(and H-3β for 48-61

	H-3β(-OAc)	H-3β(-OH)	H-3α(-OAc)	H-3α(-OH)
48	-	-	4.62 t	-
49	-	-	4.62 brt	-
50	-	-	4.63 t	-
51	5.65 dd	-	-	-
52	5.67 dd	-	-	-
53	-	4.19 dd	-	
54	5.41 dd	-	-	
55	-	4.32 dd	-	
56	-	4.52 dd	-	
57	-	-	4.56 t	-
58	-	-	4.57 t	-
59	-	-	4.55 t	-
60	-	-	4.45 t	-
61	-	-	-	3.36

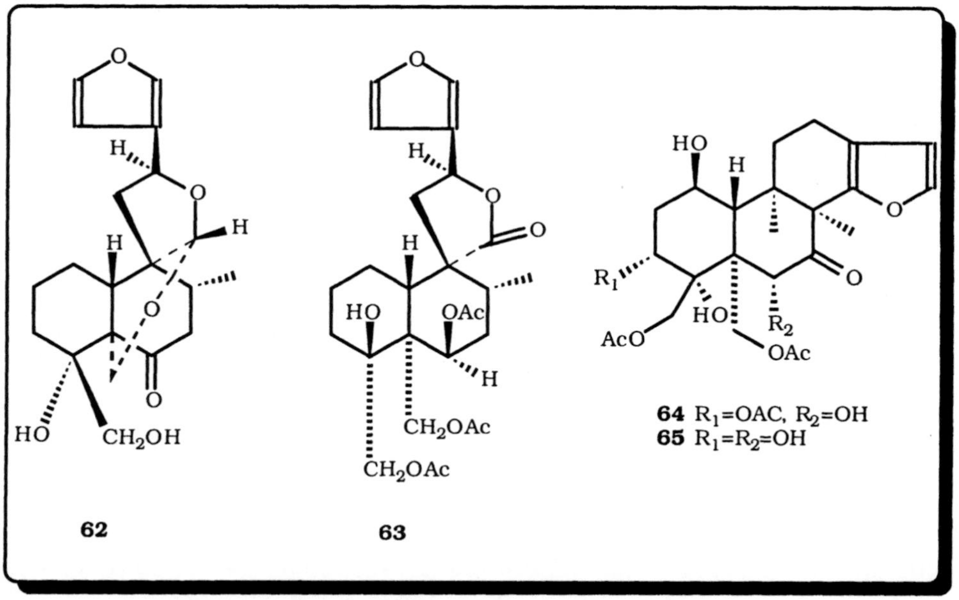

62

63

64 R₁=OAC, R₂=OH
65 R₁=R₂=OH

Table 6.   NMR Data of 62-65

Position	62	63		64		65	
	^1H	^1H	^{13}C	^1H	^{13}C	^1H	^{13}C
1	2.21,2.02	1.57	21.8	4.39	64.8	4.47	65.0
2	1.81,1.30		22.0	2.42,1.92	37.8	2.42,1.92	39.8
3	1.52,2.26		31.0	5.29	71.3	4.09	74.3
4	-	-	78.4	-	71.4	-	77.0
5	-	-	47.3	-	44.8	-	45.2
6	-	5.67	67.8	3.92	75.9	3.92	74.3
7	2.72,2.39		35.4	-	206.8	-	206.4
8	2.18		33.0	-	53.5	-	53.7
9	-		52.2	-	52.4	-	52.7
10	1.97	2.80	43.6	2.40	42.0	2.45	42.2
11	2.28,2.14	2.58,2.57	45.4	2.01,2.42	18.9	1.95,2.50	19.3
12	5.13	5.41	71.1	1.61,2.84	34.5	1.60,2.84	34.9
13	-	-	124.6	-	116.5	-	116.6
14	6.37	6.40	108.4	-	159.2	-	152.4
15	7.41	7.44	144.4	6.22	110.3	6.22	110.5
16	7.41	7.46	140.0	7.38	142.5	7.35	142.7
17	1.03		16.9	1.42	18.1	1.42	18.6
18	3.90,4.60	3.72,4.46	68.3	4.22,4.06	66.2	4.16,4.02	66.5
19	3.78,3.70	5.35,4.51	60.5	4.41,5.15	65.5	4.37,5.15	60.6
20	5.20	-	176.1	1.21	17.5	1.23	17.1
C=O	-	-	178.1	-	170.0	-	170.1
CH$_3$	-	2.10	20.2	1.99	20.2	1.96	22.6
C=O	-	-	170.1	-	169.6	-	169.8
CH$_3$	-	2.09	20.1	2.05	20.5	2.02	20.8
C=O	-	-	-	-	169.6	-	170.1
CH$_3$	-	-	-	2.08	20.8		

In some *neo*-clerodanes, the CH_2-18 group forms a lactone ring with C-6 hydroxyl group as observed in compounds **4, 7, 15, 16, 31.** In all cases, the lactone ring is an α, β-unsaturated γ-lactone and easily differentiated from their γ-spirolactone rings, and C-18 lactone carbonyl has a smaller value *ca* (δ 171.0-172.0) compared to C-20 lactone carbonyl *ca* (δ 177.0-178.0) [38,39]. The C-18 group could also form a hemiketal ring with C-19 methylene group (**66**) [40], in which case the C-18 protons were observed at δ_H 4.11 (d, J=10 Hz) and 4.20 (d, J=10Hz) and δ_C 76.2 (in pyridine-

d_5).There is a number of examples that C-18 forms a lactone with C-19 methylene group as seen in compounds teugin (**67**) [41], dihydroteugin (**68**) [42], teurins F and G (**69, 70**) [43], teusalvins A and B (**71, 72**) [44], 2-deoxy chamaedroxide (**73**) [45], bidentatin (**74**) [46], teupernins A and B

66

67

68

69

70

71 R=O
72 R=αH,βOH

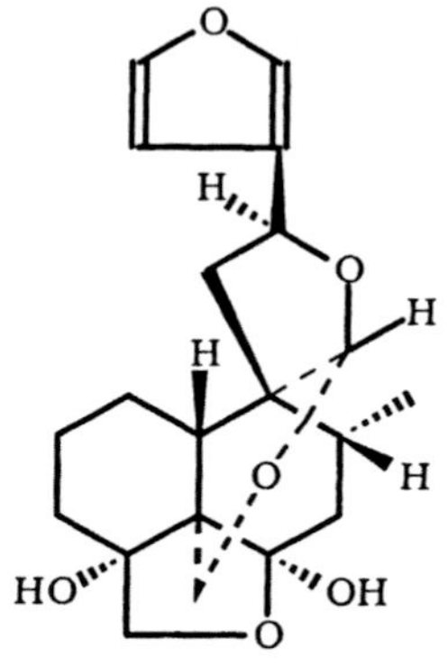

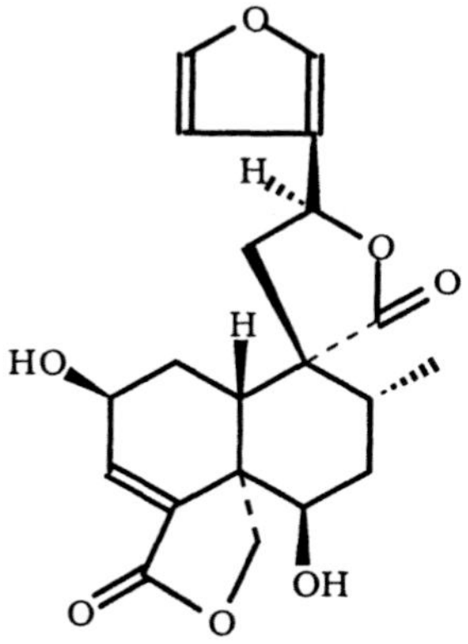

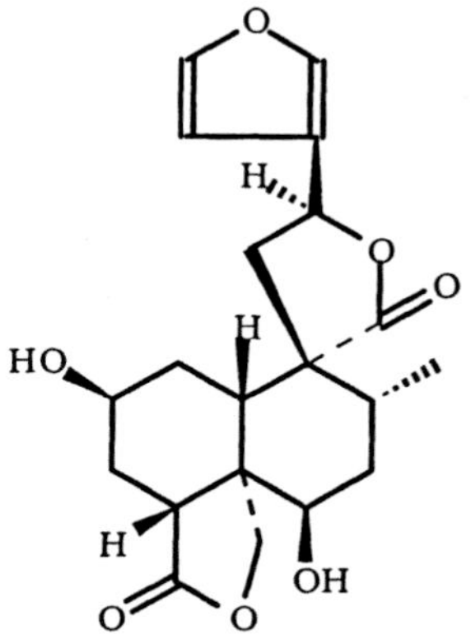

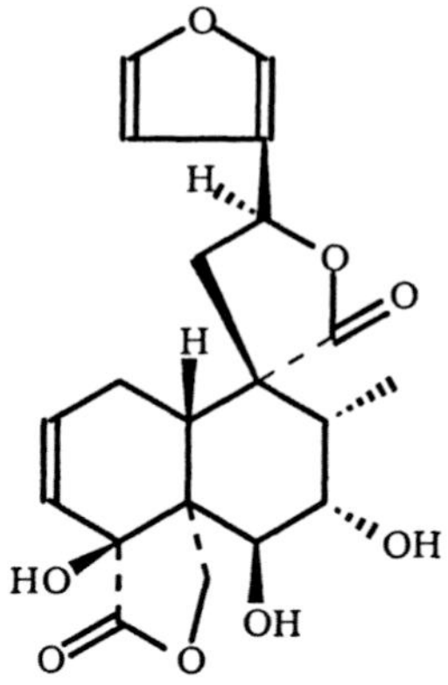

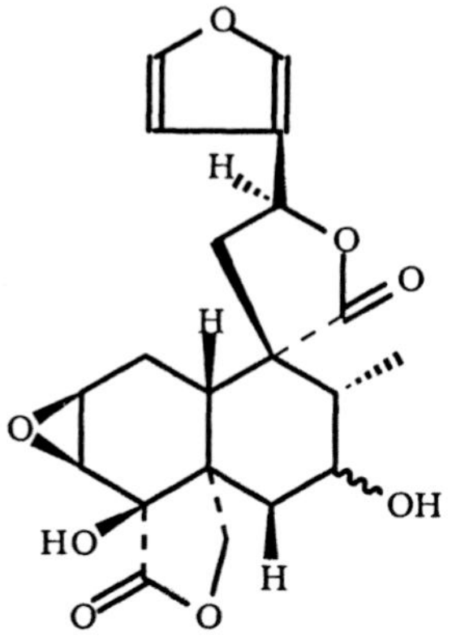

(**75-76**) [47]. In some cases, a tetrahydrofuran structure is formed as observed in **71**. Bidentatin (**74**) and teupernin (**76**) were found to be the same compounds published in the same year by two different Chinese groups. The ^{1}H and ^{13}C NMR data of **73-76** are given in Table 7.

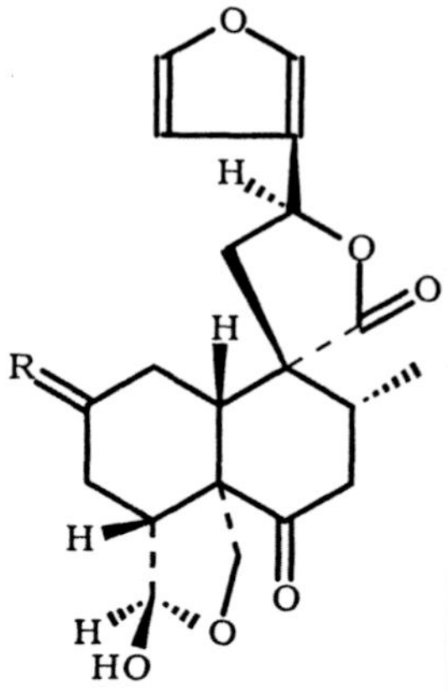

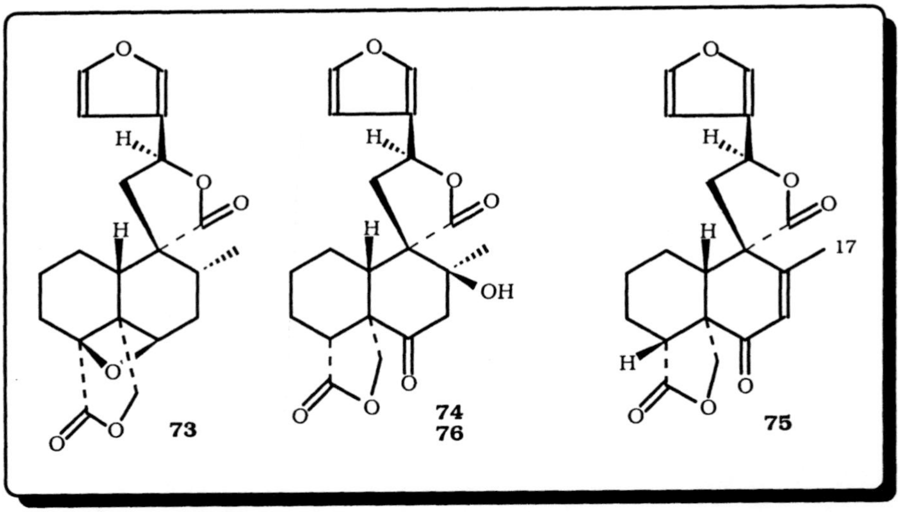

Table 7.   The NMR Data of Compounds 73-76

Position	73		74		75		76	
	^1H	^{13}C	^1H	^{13}C	^1H	^{13}C	^1H	^{13}C
1	2.05,1.63	22.5		23.2		21.3		22.7
2	1.76,1.35	24.1		24.2		23.1		25.2
3	2.05,1.63	24.3		24.1		22.9		24.8
4	3.09	42.1	2.63	41.2	3.08	41.3	3.17	38.5
5	-	56.0	-	51.3	-	54.6	-	48.4
6	-	208.5	-	197.4	-	207.6	4.10	69.7
7	4.13,2.74	48.8	6.11	127.7	2.71,4.10	36.7	2.06,2.73	37.4
8	-	76.3	-	155.5	-	75.3	-	73.8
9	-	56.5	-	54.2	-	55.4	-	57.5
10	2.88	44.7	2.31	45.9	2.85	43.2	2.62	40.3
11	3.09,2.65	39.7	2.79,2.56	41.5	2.61,3.05	47.5	2.51,3.02	39.4
12	5.69	72.5	5.57	72.9	5.65	71.4	5.55	72.3
13	-	125.9	-	124.5	-	124.7	-	126.3
14	7.93	108.9	6.40	107.7	6.66	108.2	6.62	108.9
15	7.56	145.0	7.50	144.7	7.73	144.1	7.86	144.7
16	6.69	140.9	7.47	139.7	7.62	140.2	7.69	140.6
17	1.54	26.2	1.95	20.8	1.51	25.1	1.36	26.1
18	-	180.0	-	174.9	-	176.5	-	178.3
19	5.17,4.82	69.5	4.47,4.50	69.1	4.80,5.14	68.4	4.61,4.76	70.5
20	-	177.6	-	176.8	-	177.1	-	178.5

In *Teucrium* diterpenoids C-6 position is also very important. Almost all of the *neo*-clerodane diterpenes have an oxygen function at that location, either as a hydroxyl, acetyl or a ketone, and sometimes as an ether function. In somecases, C-6 hydroxyl group forms a lactone with C-18 methylene group as in compounds **7, 14, 16**. The hydroxyl or the acetyl groups could possess α or β stereochemistry. When the literature is studied, there are some discrepancies for the chemical shifts of the C-6 proton, either α or β, it appears in the lowfield or highfield, as observed in compounds **77** and **78**.

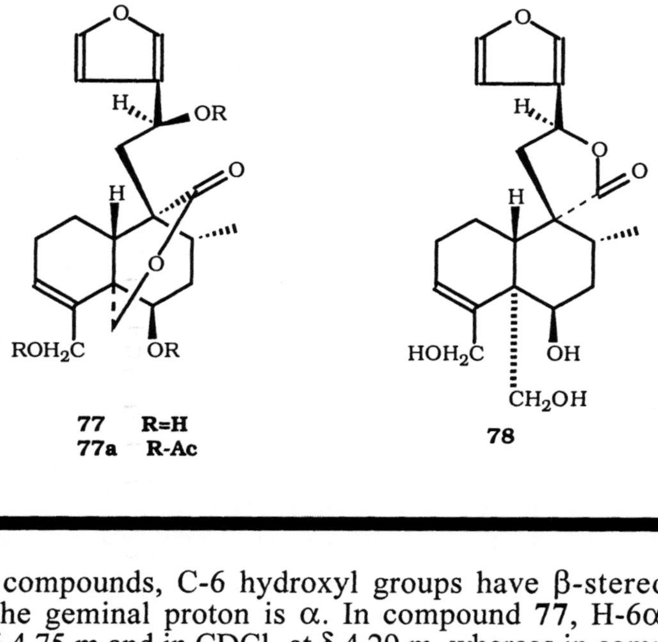

77 R=H
77a R-Ac

78

In both compounds, C-6 hydroxyl groups have β-stereochemistry therefore, the geminal proton is α. In compound **77**, H-6α in pyr.-d$_5$ appears at δ 4.75 m and in CDCl$_3$ at δ 4.29 m, whereas in compound **78**, it was observed at δ 5.09 m (in pyr.-d$_5$), and in the acetyl derivative of **77** (**77a**), it was at δ 5.17 t [48]. Another study with similar compounds (**40, 41, 43-47**) supported these observations (Table 8) [27].

Table 8.　　^{1}H NMR Data of Compounds 40, 41, 43-47 at C-6 (in CDCl$_3$)

	40	41	43	44	45	46	47
H-6α	-	-	-	-	-	-	3.70
H-6β	4.07d	5.05brs	5.40q	3.66brd	4.75dd	3.41brd	-

Usually the H-6α observed in highfield compared to its β-epimer, as an example teucjaponins A (**32**) and B (**33**) which are two epimer diterpenes at C-6 were given. While H-6α is observed at δ 4.17 as a broad doublet in **32**, H-6β appears at δ 3.66 as a dd (*J*=11 and 4 Hz) in **33**.

In the studies of Savona et al. [49,50], the chemical shifts of H-6α having a β-hydroxy group at C-6 were observed at δ 4.33 t, 4.10 t or 4.24 dt in three different compounds. The stereochemistry of these compounds was deduced by double resonance experiments while in another study, H-6α was found as a broad singlet at δ 5.16 (in pyr.-d₅). In this study, it was claimed that the broad singlet indicated the β-orientation of the hydroxyl group [51]. The stereochemistry of C-6 was investigated by studying 12S and 12R series of compounds such as 12R-*epi*-teuscordinin (**79**), montanin C (**80**), teucvin (**7**), teucrin H-2 (**81**) and teuscordinin (**82**) [52]. When the furan ring is β-substituted, H-6 appears at δ 5.05 as dd (*J*= 12.1 and 5.9 Hz) which was assigned by selective decoupling experiments, this is observed in compound **82**. A change at C-6 would necessitate a change in the conformation of ring B (from chair to boat) in order to have axial (β H-6. Irradiation at δ 1.28 (H-17) resulted 14.2 % NOE enhancement of the H-6 and 2.8 % NOE enhancement of the H-12 signals. These results confirmed

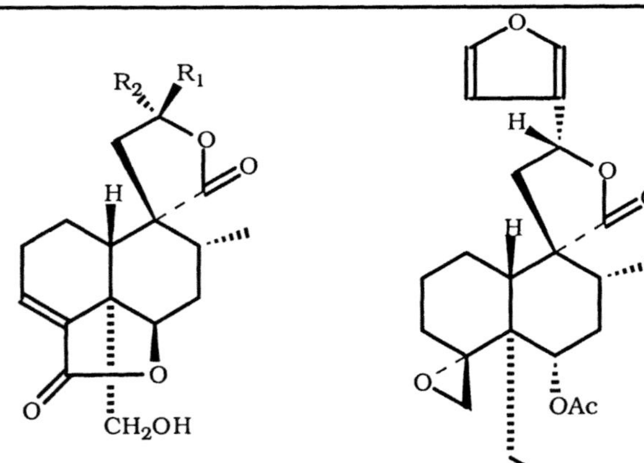

79 R₁ = H, R₂ = β-furyl
82 R₁ = β-furyl, R₂ = H   **80**

the boat conformation of ring B with H-6α and 12R configuration as shown in compound **79**. The places of the relevant groups being very close to the secondary methyl group and the small value of the NOE enhancement of H-12 (2.8 %) are also a consequence of the boat formation of ring B. At C-6, the β-substitution of the hydroxyl or acetyl groups is rather rare, however, there are some examples with the β-substituents [53]. Two epimeric compounds **83**, **84** having the only difference at C-6 showed chemical shift differences in the ¹H NMR spectra, such as at δ 4.11 brd (*J*=2.7 Hz) for H-6α (C-6β OH) in compound **83** and δ 3.94 brdd

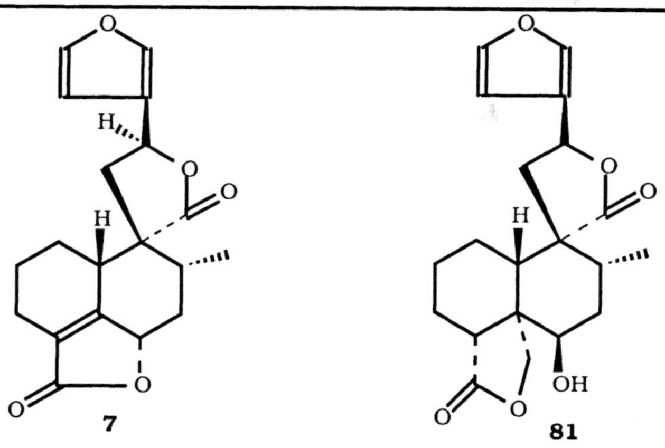

(J=11.3 and 4.1 Hz) for H-6β (C-6α OH) in compound **84**. In the ^{13}C NMR spectra of these two compounds **83** and **84** C_1-C_3, C_5, C_7, C_9, C_{11}-C_{17}, C_{19} and C_{20} resonances are almost identical. The observed differences at C_4, C_6, C_8, C_{10} and C_{18} resonances were due to the 6β-axial and 6α-equatorial configuration of the secondary hydroxyl groups.

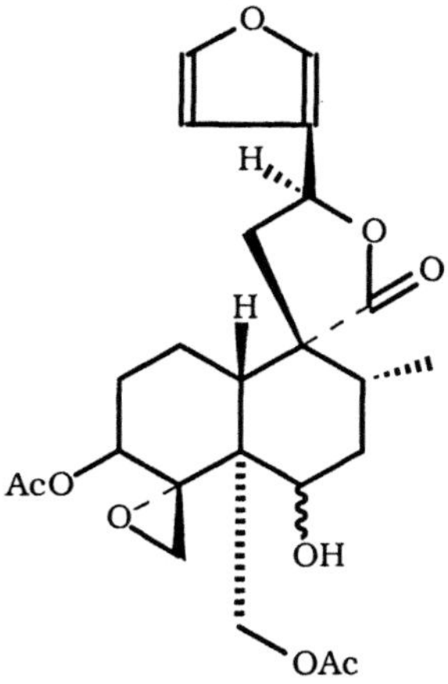

83 C-6 α-H, β-OH Teugracilin A
84 C-6 β-H, α-OH Teumicropodin

As a whole, when the C-6 substituent has an α-stereochemistry, its chemical shift would be observed in higherfield compared to its β-stereochemistry. Nevertheless, NOE experiments are always advisable, and the 12R and 12S forms of the molecule should be checked as suggested in the literature [52].

The C-19 methyl group, in most cases, appears as a hydroxymethylene, either substituted with an acetyl group or formed an epoxy ring with different centers or could form a lactone. If the stereochemistry at C-5 is α, the substituents at C-19 would always be α.

Gnaphalin (**85**) and its 19-acetyl derivative (**20**) were isolated from *T. gnaphalodes* [54] and gnaphalin has a hydroxymethylene group at C-19. The ^{1}H NMR spectrum of **85** gave an AB quartet at δ 4.36 and 5.02 (J=12 Hz), in compound **20** these signals were shifted to δ 5.05 and 5.49 (J= 12 Hz).

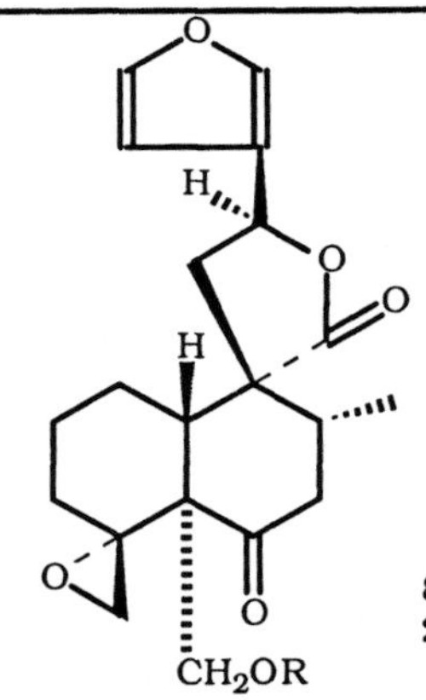

85 R=A
20 R=Ac

From *T. massiliense* [55], six neo-clerodanes were isolated, two of which have hydroxymethylene and the other four acetoxymethylene groups at C-19. In all six compounds, C-6 substituents have β-stereochemistry. The ^{1}H NMR spectra of all six compounds (**86-91**) were recorded in CDCl$_3$, so it is easy to compare their chemical shifts. The chemical shift of C-19 hydroxymethylene groups were around δ 4.00-4.30 while its acetyl derivatives at δ 4.35-4.80 (Table 9).

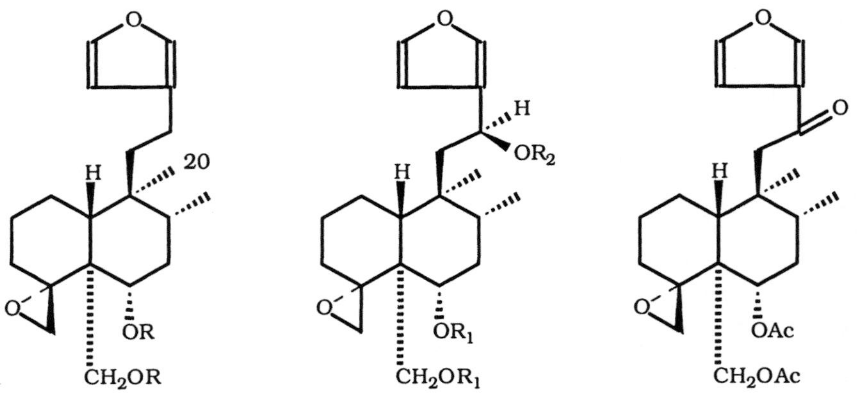

86 R=H
87 R-Ac

88 R$_1$=R$_2$=H
89 R$_1$=Ac, R$_2$=H
90 R$_1$=R$_2$=Ac

91

Table 9. ^{1}**H NMR Data of Compounds 86-91**

Position	86	87	88	89	90	91
H-6β	3.60ddd	4.7ddd	3.61ddd	4.75m	4.75ddd	4.75ddd
2H-11						2.73
H-12			4.70ddd	4.70m	5.90ddd	-
H-14	5.83tt	5.83tt	6.35m	6.38m	6.35m	6.73d
H-15	-	-	7.34m	7.37m	7.33m	7.43t
H-16	4.70d	4.73d	7.34m	7.34m	7.39m	8.00d
Me-17	0.85d	0.85d	0.79d	0.78d	0.75d	0.88d
H$_A$-18	2.42d	2.22d	2.40d	2.20d	2.18d	2.22d
H$_B$-18	3.16dd	3.00dd	3.15dd	3.00dd	2.97dd	3.05dd
H$_A$-19	4.01dd	4.35dd	4.03dd	4.40dd	4.38dd	4.40dd
H$_B$-19	4.30d	4.80d	4.27d	4.80d	4.79d	4.82d
Me-20	0.71s	0.79s	0.61s	0.70s	0.70s	0.80s
OAc	-	1.95s	-	1.93s	1.93s	1.93
		2.10s		2.08s	2.08s	2.08

When the spectra were recorded in pyridine-d$_5$, the chemical shift of hydroxymethylene group changes towards lowfield, H$_A$-19 δ 4.79 d (*J*=11 Hz) and H$_B$-19 δ 4.70 d (*J*=11 Hz) as in teupernin D (**92**) [51] or even much more towards lowfield as in case teupolin IV (**93**) [40] to δ 5.21 d (*J*=13 Hz) and 5.87 d (*J*=13 Hz) (in pyridine-d$_5$) perhaps due to the presence of a keto group at C-6.

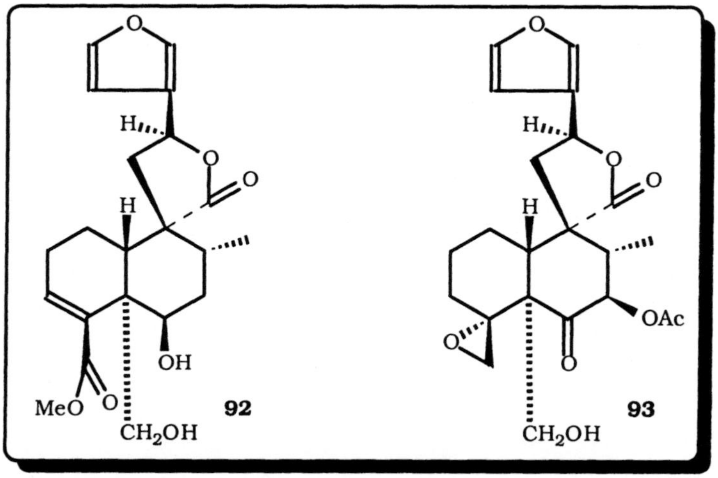

The same lowfield shift was also observed in the acetyl derivative of C-19 hydroxymethylene group with C-6 keto group.

Table 10 gives the difference between compounds **94** with a C-6 keto group and **95, 96** having C-6 acetoxyl or hydroxyl groups [53].

Table 10.   ^{1}H NMR of C-19 Protons of Compounds 94-96

	94	95	96
H_A-19	4.81 d (13.4)	4.86 d (13.5)	4.74 d (13)
H_B-19	5.50 d (13.4)	4.80 d (13.5)	4.98 d (13)

The lowfield shift effect of the keto group is induced only on H_B-19, due to their spatial closeness. The ^{13}C NMR signal of CH_2-19 whether hydroxymethylene or acetoxy methylene appears about at δ 61.0-64.0 either in $CDCl_3$ or in pyridine-d_5.

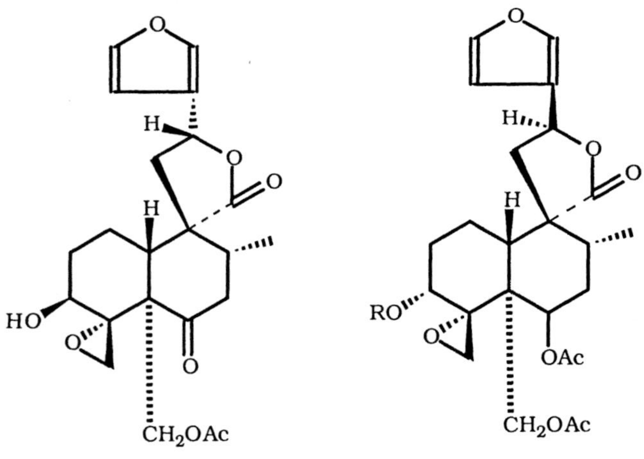

95 R=H
96 R=Ac

94

An ether bond could form between C-19 and C-20 as observed in compound **62** [36], the methylene protons of C-19 appeared at δ 3.78 d (12 Hz) H-19α and 3.70 dd (12 and 1 Hz) for H-19β. Spin-decoupling experiments confirmed these assignments, H-20 was seen at δ 5.20 as a singlet. A similar ether bond is observed in a group of compounds **97-101** isolated from *T. micropodioides* [56].

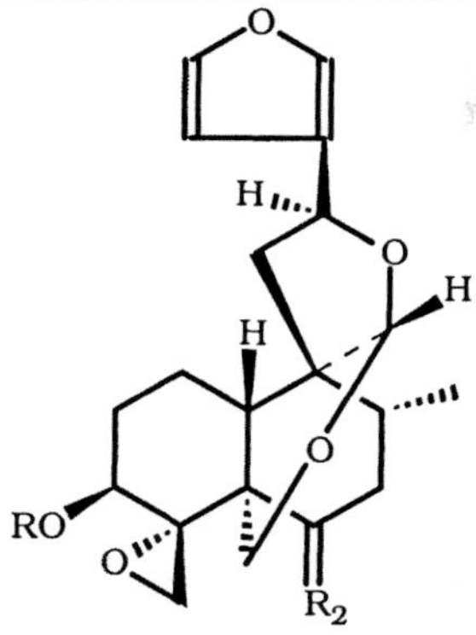

	R$_1$	R$_2$
97	Ac	αH,βOH
98	Ac	O
99	H	αH,βOH
100	Ac	αH,βOAc
101	H	O

The spectral data of 3-acetylteumicropin (**97**) was quite similar to that of **98**, the former could be converted to the latter by CrO$_3$ oxidation. By NOE experiments, irradiating Me-17 protons, enhancement was observed in the signals of C-14, C-16 and C-20 protons, but not of that in C-12. The 12S and 20S-configurations of these diterpenes were in agreement with NOE experiments. The chemical shifts of C-19 protons in **97-101** were in agreement with those of the acetal group in compound **62**. All these values are in high field compared to the chemical shift of hydroxymethylene protons. Table 11 gives the ^{1}H NMR data of C-19 protons of **97, 99-101**.

Table 11. ^{1}H NMR of 97, 99-101 at C-19 Protons

	97	99	100	101
H$_A$-19	3.34d	3.32d	3.40d	4.06d[*]
H$_B$-19	4.25d	4.16d	4.27d	4.38d

[*]The low field shift of H$_A$-19 is probably due to the presence of a keto group at C-6.

A new compound montanin H (**102**) was obtained from *T. montanum* [57]. C-19 is formed a hemiacetal ring with C-20 having AB proton doublets (δ 4.20 H$_A$-19, δ 4.60 H$_B$-19, J_{AB}=11 Hz) and a proton doublet at

δ 5.2 (*J*=4 Hz, H-20), after D$_2$O exchange, it collapsed to a singlet, showing that the hemiacetal proton had no vicinal proton.

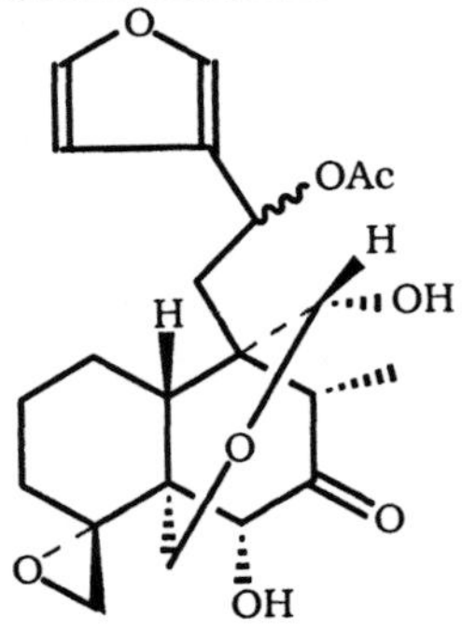

102

T.pyrenaicum [58] yielded a compound teupyrin A (**104**) again having a hemiacetal ring, the C-20 proton was observed at δ 5.01, d, *J*=3.3 Hz which collapsed to a singlet after addition of D$_2$O, indicating that no vicinal proton was present. The disclosure of the hemiacetal group revealed by an AB system due to a methylene group attached to a fully substituted carbon atom (δ 2.96 H$_A$-19, δ 4.41 H$_B$-19, J_{AB}= 12.5 Hz). The ^{13}C NMR of **103** showed C-20 at δ 94.5. Acetylation of **103** (**103a**) induced a shift on H-20 towards δ 5.99 s.

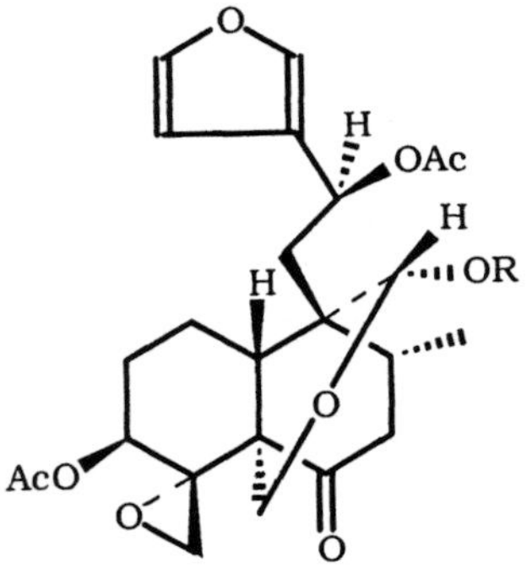

103   R=H
103a R=Ac

The formation of a six-membered lactone ring between C-19 and C-20, was encountered in a number of compounds. Such compounds were isolated from *T. chamaedrys* [59] teuchamaedrin C (**104**), from *T. botrys* [60] 15, 16-epoxy-6β, 12, 18-trihydroxy-neo-cleroda-3, 13(16), 14-trien-

20, 19-olide (**105**) from *T. gnaphalodes* [61] teugnaphalodin (**106**), from *T. lamifolium* and *T. polium* teulamifin B (**77**) [48].

In compounds **77, 104-106** and in similar other compounds C-19 protons were found quite close to those of hemiacetal group signals (Table 12). The important difference is the observation of the lactone signal at C-20 which is usually found around δ 170-175.0 and the C-19 signal around δ 62-69.

Table 12. **^{1}H NMR Data of 77, 104-106 at C-19 Protons**

	104	105	106	77*
H$_A$-19	4.62d(11.5)	4.20d(12.2)	4.92d(12.1)	4.41d(12)
H$_B$-19	4.02dd(11.5, 1)	4.38d(12.2)	5.17d(12.1)	4.71d(12)

*in pyridine-d$_5$

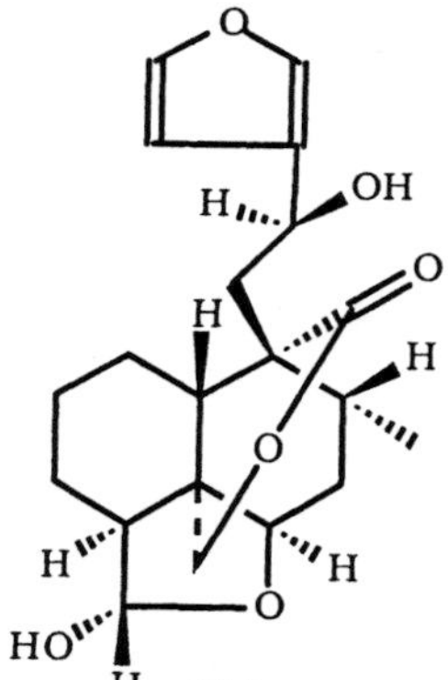

104

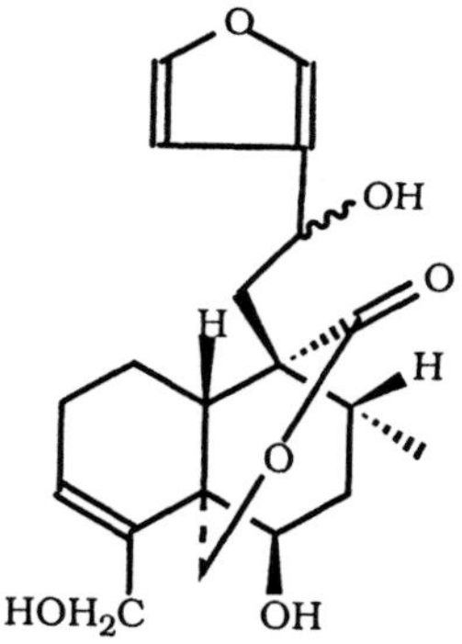

105

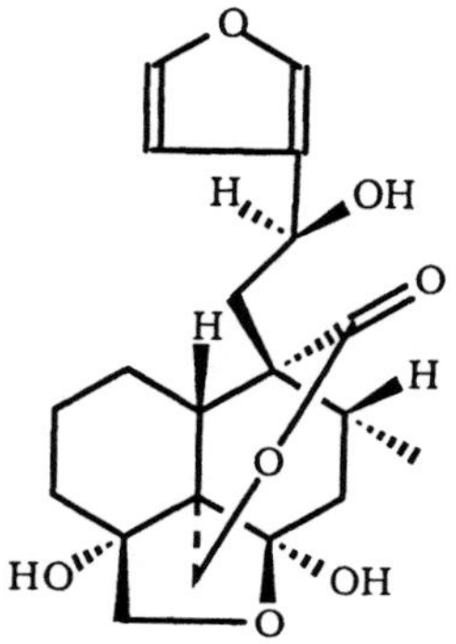

106

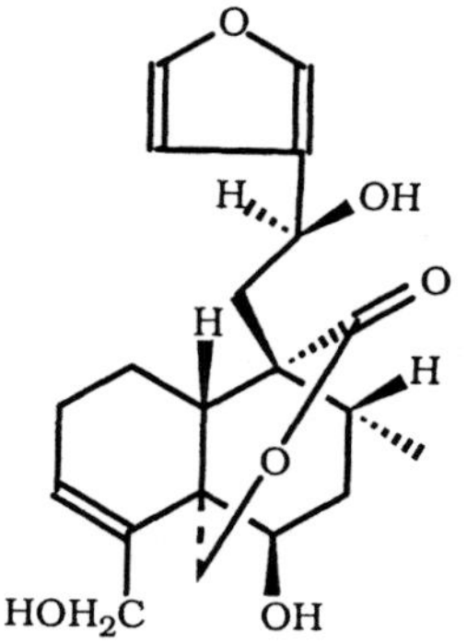

77

C_{19} Methyl group could also form five membered ether (tetrahydrofuran ring) between C-4 and C-19 [62], as in teuscorodin (**107**), or a five membered lactone in the same position [23], as in teucrin H-2 (**81**), a hemiacetal ring either between C-3 and C-19, as in teuspestalin A (**108**) [50], or between C-19 and C-7, as in teuvincentin B (**109**) [63]. The latter type hemiacetal containing compounds were isolated from *T. alyssifolium* as alysines D and E (**110, 111**) [64].

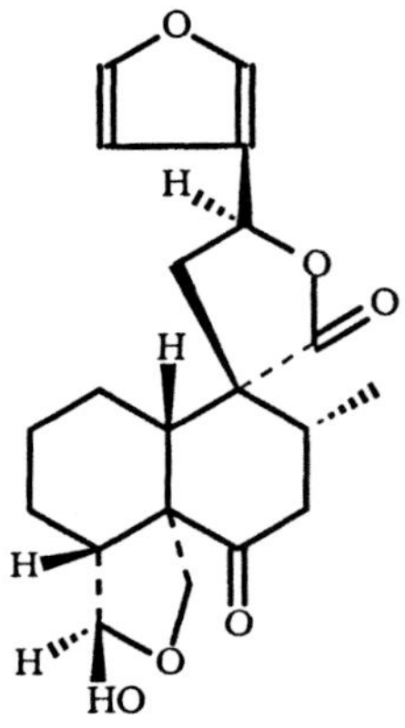

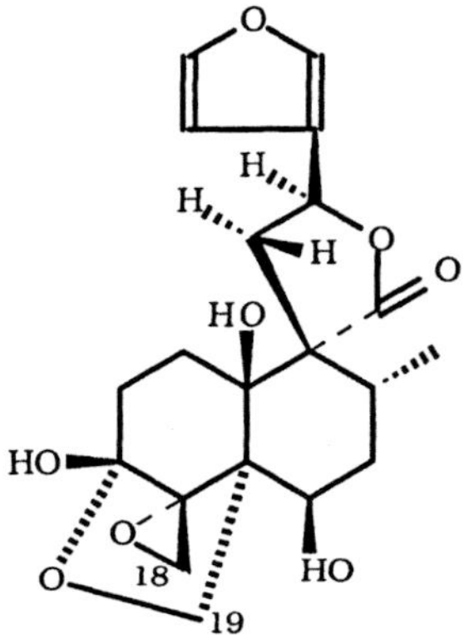

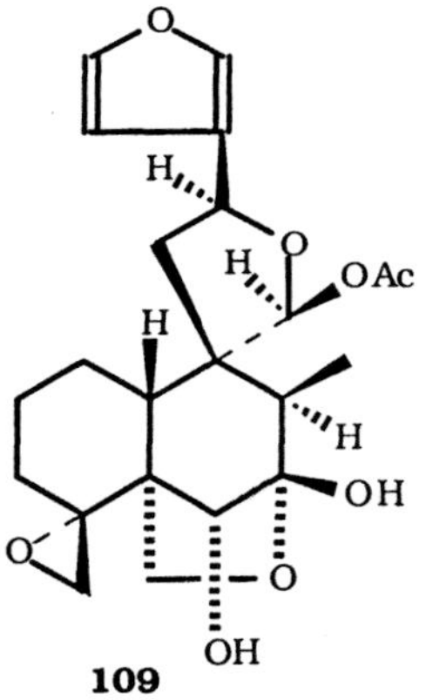

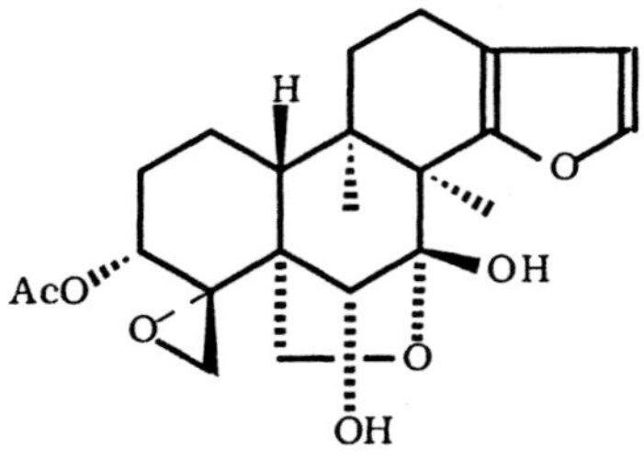

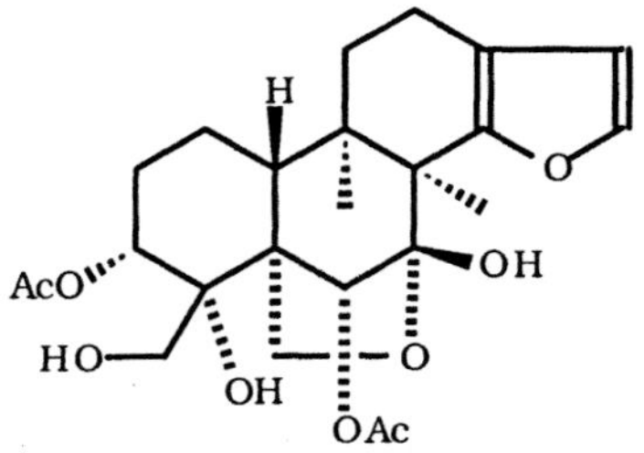

In one example, in teusalvin F (**112**), a six-membered ether ring was formed between C-19 and C-2 position.

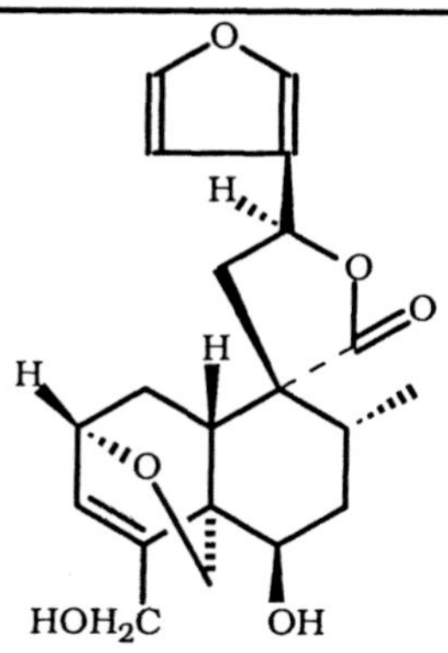

112

The structure was established by spectral methods and by X-ray analysis [44].

Although quite rare, there are a few examples with a hydroxyl group or an ether function at C-10. Teupestalin A (**108**) is one example isolated from *T. pestalozzae* [50], the structure of the compound was established by NMR spectral data including spin-decoupling and NOE experiments.

In **108**, the position of the tertiary hydroxyl group at C-10 was decided from the ^{13}C NMR spectrum. C-10 hydroxyl showed paramagnetic shift at C-1 ($\Delta\delta$ + 7.9 ppm), C-5 ($\Delta\delta$ + 3.8) and at C-9 ($\Delta\delta$ + 5.2) which are β carbons with respect to closely related *neo*-clerodane derivatives lacking the tertiary alcohol function, such as teuflavin (**113**) obtained from *T. flavum* subsp. *glaucum* [65].

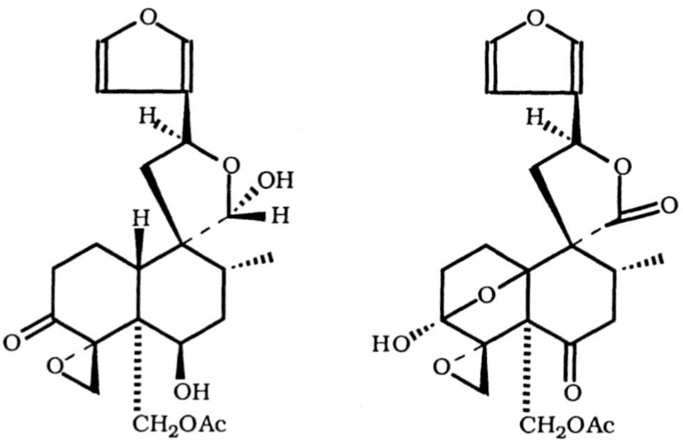

113   **114**

The β-configuration assigned to this tertiary hydroxyl group in teupestalin A (**108**) was supported by the very deshielded resonance of one of the C-11 methylene protons [H-11α (pro-S), δ 3.47 dd, J= 1.5 and 3.7 Hz] which is close to the 10β-hydroxyl group, and by comparison of the ^{13}C NMR spectral data of **113** and **114** (Table 13). Since the observed differences in the chemical shifts of the C-8 [Δδ = δ(1)- δ(3), -6.2 ppm] and C-11 (Δδ-7.7) γ-carbons were only compatible with a γ-*gauche* arrangement [66], whereas for a 10α-hydroxyl substituents, smaller shielding effects are expected on these carbon atoms, at least on C-11 (γ-*trans* carbon) [67].

Table 13. ^{13}C NMR Comparison of Compounds 108, 113, 114

C	108	113	114
1	30.6	22.7	28.2
2	32.7	42.5	31.2
3	101.5	206.5	105.2
4	67.9	68.8	66.9
5	51.4	47.6	53.4
6	67.6	66.2	205.9
7	35.3	37.5	42.6
8	29.1	35.9	39.4
9	59.3	54.1	61.5
10	77.9	45.4	87.2
11	38.5	46.2	37.4
12	72.3	70.2	73.2
13	125.5	127.1	124.9
14	108.3	110.0	107.9
15	144.1	142.2	144.4
16	139.7	139.9	139.3
17	16.6	17.5	17.3
18	50.7	56.5	51.8
19	69.5	63.0	62.0
20	177.8	100.0	175.3
C=0	-	170.2	169.9
Me	-	20.7	20.6

Teupestalin B (114) was another new compound which also isolated from *T. pestalozzae* [50], the oxygen function at C-10 was formed a hemiacetal with C-3. The ^{13}C NMR signals of C-3 at δ 105.2 and C-10 at δ 87.2 indicated their involvement to the hemiacetal group. The β-configuration of the oxygen bridge between C-3 and C-10 was supported by the upfield resonance of the C-11 (δ 37.4, γ-gauche carbon) and the deshielded positions of the C-11α proton (pro-S) (δ 3.20 dd). Moreover, NOE experiments clearly established that H-12 and Me-17 protons of compound 114 are on the opposite side of the plane, defined by C-20, C-12 lactone ring (Table 14).

Table 14.   NOE Experiments of 108-114

Compd.	Irr. δH	Observed NOE enhancement (%)							
		H-7α	H-7β	H-8β	H-11α*	H-11β	H-12	H-14	H-16
108	1.03(Me-17)	1.9	3.7	9.1	-0.4	4.2	0	3.6	1.4
114	1.06(Me-17)	5.6	6.1	6.9	0	5.9	0	3.0	1.0

*H-11α and H-11β are pro-S and pro-R protons respectively.
H-11β and Me-17 are onthe same side of the plane defined by the δ-lactone ring.

From *Teucrium oliverianum* [34,68] the two C-10 hydroxylated neo-clerodane diterpenes were isolated, teucrolivins B (115) and A (116). Both possess a C-10β hydroxyl group appeared at δ 81.53 and 81.73, respectively, the β-configuration of them was supported by NOE experiments, which showed that C-20 methyl group and the H-1α proton must be close (3.7 % NOE enhancement in H-1α when the Me-20 protons were irradiated). The NOE experiments summarized in Table 15 clearly revealed that the relative stereochemistry of the substituents of decalin ring of teucrolivin B is that depicted in its formula 115.

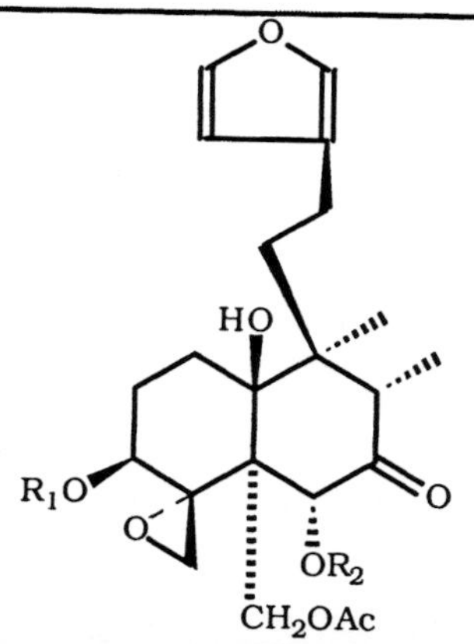

115 R_1=Ac,R_2=H
116 R_1=R_2 = H

Table 15. **NOE Experiment of Compounds 115 and 116**

Irradiated protons	115				116	
	H-3α	H-6β	H-8β	Me-20	H-6β	Me-20
Obser. NOE (%)						
H-1α	4.2	0	0	3.7	0	-
H-6β	0	-	6.0	0	-	0
H-8β	0	8.1	-	0	6.3	0
H-11α	-	-	-	3.6	1.1	1.1
Me-17	0	0	1.7	0.8	0	0.8
H$_A$-18	0	2.3	0	0	11	0
H$_B$-18	0	12.1	0	0	7.9	0
H$_A$-19	0	0	0	5.5	0	2.9
H$_B$-19	2.2	0	0	3.5	0	2.7

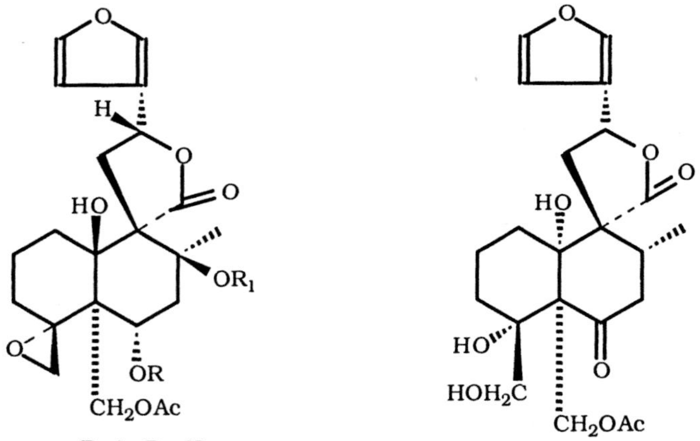

117 R=Ac,R$_1$=H
118 R=R$_1$=H

119

From *T. sandrasicum* [69] two C-10 β-hydroxylated and one α-hydroxylated neo-clerodane diterpenes were isolated, sandrasin A (117), 6-deacetylsandrasin A (118) and sandrasin B (119). All three compounds had C-12R configuration. This configuration was deduced by the observation of a NOE between Me-17 and H-12 which was extensively studied previously [25,29]. The C-10 hydroxyl group induced paramagnetic shifts on C-1, C-5 and C-9 with respect to those of closely related neo-clerodane diterpenes either lacking or with an α-hydroxyl group at C-10 [50,67]. Table 16 shows the ^{1}H and ^{13}C NMR data of compounds **117-119**.

Table 16. ^{1}H and ^{13}C NMR Data of 117-119

	117		118		119	
	^{1}H	^{13}C	^{1}H	^{13}C	^{1}H	^{13}C
1	1.18, 2.20	31.8	1.18, 2.16	30.6		25.4
2	1.80, 2.05	18.8	1.78.2.00	18.9		14.9
3	1.75, 1.90	29.4	1.75.1.85	29.3		21.0
4	-	62.9	-	63.9	-	74.8
5	-	49.8	-	51.6	-	51.9
6	5.42	67.3	4.36	64.9	-	197.0
7	1.72, 2.75	38.2	1.86, 2.52	39.3	2.28, 3.55	37.7
8	-	75.8	-	76.0	1.58	31.2
9	-	59.3	-	59.6	-	53.8
10	-	81.9	-	81.5	-	84.0
11	2.52, 3.12	34.8	2.52, 3.16	34.9	1.85, 2.78	33.0
12	5.38	72.2	5.39	72.1	5.41	71.2
13	-	124.9	-	125.0	-	119.2
14	6.33	108.0	6.39	108.0	6.40	107.3
15	7.41	144.2	7.42	144.1	7.39	143.2
16	7.47	139.6	7.49	139.6	7.47	139.4
17	1.20	26.0	1.25	26.1	1.00	15.3
18	2.40, 3.02	51.6	2.62, 3.32	51.6	3.70, 3.86	69.8
19	4.40, 5.20	63.6	4.54, 5.08	63.8	3.78, 3.98	59.2
20	-	174.6	-	174.8	-	174.2
CO	-	170.6	-	170.8	-	-
CH$_3$	1.92	21.2	2.03	21.1	-	-
CO	-	170.6	-	-	-	-
CH$_3$	2.03	21.2	-	-	-	-

The α-configuration of the hydroxyl group at C-10 was decided from the NOE correlation between C-10 α-OH and C-19 protons and from the ^{13}C NMR spectrum of C-1 at δ 25.4 indicating no paramagnetic shifts on C-1 and C-9.

From *T. oliverianum* [68,70] two neo-clerodanes with a β-oxygen function at C-10 were isolated. These compounds were named teucrolivin A (**120**) and teucrolivin F (**121**).

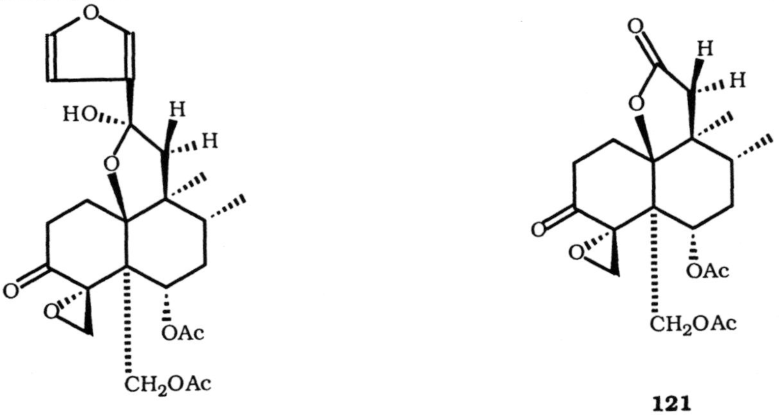

120

121

In teucrolivin A (**120**) two methyls of acetoxy groups were observed in addition two one tertiary (δ 1.08, 3H, s) and one secondary (δ 0.80, 3H, d) methyl groups (Me-20 and Me-17, respectively). A β-substituted furan ring with its proton signals at [δ_H 7.48, 1H, dd, (H-16), 7.41, 1H, t (H-15) and 6.41, 1H, dd (H-14)], a 4α, 18-oxirane ring protons are at δ 2.61 d and 3.17 d J_{gem}=5.4 Hz; an equatorial acetoxy group attached to the C-6α position coupled with H-7α and β protons, identical to quite a number of α-substituted acetyl group carrying compounds (**41, 43, 45, 80, 87, 89** etc.). In addition, teucrolivin A possessed a hemiacetal group (hemiacetal carbon at δ_C 100.77, C-12) which must involve the C-10 carbon (δ_C 90.09). The relative configuration of all the asymmetric centres of **120** was established by NOE experiments. This date clearly showed that C-6, C-8 and C-18 protons are on the same side of the plane, since irradiation at δ 5.26 (H-6β) caused NOE enhancement in the signals of H-8β and H_B-18 protons and vise versa. Also from the NOE data it was decided that Me-20 is *cis*-oriented with respect to the H-1α, H-7α, Me-17 and C-19 methylene protons with enhancements 5.7, 0.9, 1.7, 7.5%, respectively. The ¹H and ¹³C NMR spectra of teucrolivin F (**121**) were identical with those of teucrolivin A (**120**) except for the absence of the signals corresponding to the β-substituted furan ring and the downfield resonance of the C-12 carbon, which forms a 12, 10β-γ-lactone in teucrolivin F (**121**) [δ_{C-12} 173.09; δ_{C-10} 91.59; ν 1790 cm⁻¹ (γ-lactone) and no hydroxyl absorptions] instead of the 12R, 10β-hemiacetal of **120** (δ_{C-12} 100.77, δ_{C-10} 90.09). The ¹H and ¹³C NMR spectra of **120** and **121** are given in Tables 17 and 18, respectively.

Table 17. ^{1}H NMR Data of 120-121

H	120		121	
1α	2.22ddd	(5.8, 13.9, 12.0)		
1β	2.49ddd	(13.9, 1.0, 7.1)		
2α	2.60ddd	(5.8, 1.0, 17.7)		
2β	2.85ddd	(12.0, 7.1, 17.7)		
6β	5.26brdd	(10.9, 6.5)	5.20brdd	(12.1, 4.5)
7α	1.38ddd	(10.9, 13.4, 11.5)	1.47q	(12.6, 12.6)
7β	1.41ddd	(6.5, 13.4, 4.8)	1.59dt	(12.6, 4.5, 6.8)
8(	1.87ddq	(11.5, 4.8, 6.8)	1.86ddq	(6.8, 4.5, 12, 5)
11α	2.36d	(13.8)	2.61d	(17.6)
11β	2.46d	(13.8)	-	-
14	6.41dd	(1.8, 0.9)	-	-
15	7.41t	(1.8, 1.8)	-	-
16	7.48dd	(0.9, 1.8)		
Me-17	0.80d	(7)	0.98d	(7)
H$_A$-18	2.61d	(5.4)	2.67d	(5.2)
H$_B$-18	3.17d	(5.4)	3.17d	(5.2)
H$_A$-19	4.36brd	(11.8)	4.35brd	(12)
H$_B$-19	4.79d	(11.8)	4.89d	(12)
Me-20	1.08s	-	1.19s	-
C=0	-	-	-	-
CH$_3$	2.03s	-	2.04	-
C=0	-	-	-	-
CH$_3$	2.00	-	2.02	-

Table 18. ^{13}C NMR Data of 120-121

C	120	121
1	28.17	28.34
2	36.13	25.33
3	205.40	202.64
4	62.96	62.42
5	48.72	48.25
6	68.96	68.10
7	32.86	32.23
8	33.73	35.10
9	49.15	45.53
10	90.09	91.59
11	50.59	42.35
12	100.77	173.09
13	131.93	-
14	108.15	-
15	143.91	-
16	138.46	-
17	15.89	16.11
18	52.47	52.62
19	61.84	61.77
20	15.01	14.45
C=0	170.15	169.80
CH$_3$	21.19	21.08
C=0	169.96	169.62
CH$_3$	20.46	20.48

Rearranged Diterpenoids

The *neo*-clerodane diterpenoids were isolated from the aerial parts of *Teucrium* species, however, from the roots rearranged abietane diterpenes were obtained. The first study with the roots of *T. polium* subsp. *vincentinum* was carried out by the Spanish-Italian group [71]. In this study four new compounds teuvincenones A, B, C and D (**122-125**) have been isolated. Two of them, teuvincenones C (**124**) and D (**125**) possess a novel hydrocarbon skeleton which contains a cyclopropane ring constituted by C-3, C-4 and C-18 carbons of the 17(15-16)-*abeo*-abietane framework.

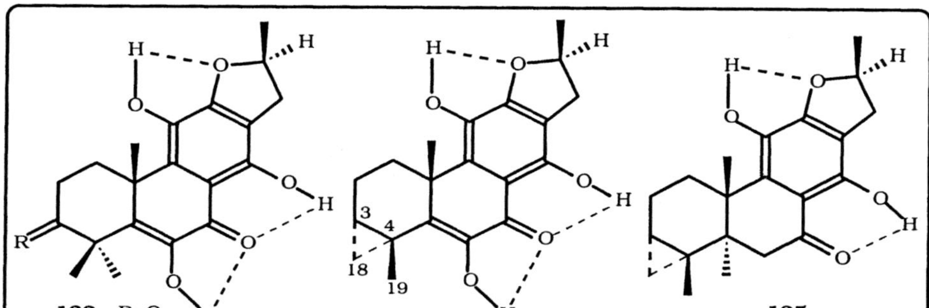

122 R=O
123 R=H,H
124
125

The structures of these compounds were established by spectral data. UV spectra of compounds **122-125** revealed of a phenolic system conjugated with a diosphenol moiety by the addition of base, $AlCl_3$ and $AlCl_3$-HCl which showed characteristic band shifts of this cross-conjugated chromophore. These results were identical with those reported for coleon U (**126**), lanugons R (**127**) and S (**128**) [72]. The spectral data of compounds **122-125** were given in Tables 19 and 20. The X-ray analyses of teuvincenones A (**122**) and C (**124**) confirmed their structures.

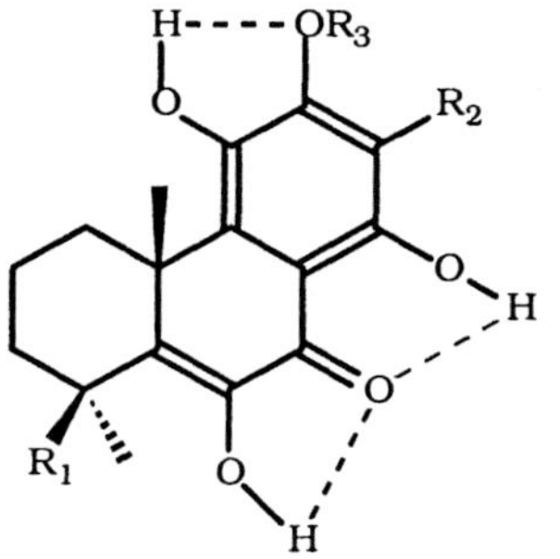

126 R=Me, R_2=CHMe, R_3=H
127 R_1=CH$_2$OOCH, R_2=CH$_2$CH=CH$_2$, R_3=H
128 R_1=Me, R_2=CHMe$_2$, R_3=Me

Table 19. ^{1}H NMR Data of Compounds 122-125, 129, 130, 132-138

Proton	122	123	124	125	129	130	132	133	134	135	136	137	138
H-1 α	1.86		0.82	0.89	2.43	2.43	2.53		1.93	1.91	0.86	0.89	1.10
H-1β	3.34	3.06	2.80	3.05	4.17	4.19	4.29	3.40	3.42	3.08	3.26	3.05	3.30
H-2α	2.75		1.88	1.80			-		2.78	2.80	1.68	1.75	1.75
H-2β	2.73		2.38	2.18			-		2.76	2.81	2.10	2.15	2.20
H-3α	-		-	-			-				1.00	1.10	1.10
H-3β	-		0.99	0.72							0.88	0.75	0.90
H-5α	-	-	-	1.90				2.39	-		1.78	1.78	1.75
H-6α	-	-	-	2.67	6.53	6.53	6.59	2.6	-	-	2.64	2.65	-
H-6β	-	-	-	2.72			-	2.6	-	-	2.74	2.70	-
H-7α	-	-	-	-					-	-	-	-	-
H-7β	-	-	-	-					-	-	-	-	-
H-15α	3.43	3.40	3.40	3.35	3.42	3.07	6.62	3.38	6.62	6.53	3.35	3.32	3.30
H-15β	2.91	2.88	2.83	2.90	2.87	-	2.85	-	-	-		3.05	3.00
H-16α	5.18	5.14	5.14	5.11	5.17	4.33	-	5.14	-	-	-	5.13	5.12
H-17	1.55	1.51	1.53	1.50	1.54	1.33	2.48	1.52	2.48	2.50	-	3.8	3.88
Me-18	1.54	1.54	-	-	2.01	1.99	2.25	1.18	1.56	1.53	0.83	0.90	1.42
Me-19	1.45	1.43	1.44	1.06	2.22	2.21	2.03	1.17	1.53	1.49	0.86	0.92	1.44
Me-20	1.58	1.66	1.65	1.38	1.63	1.63	1.69	1.44	1.60	1.58	1.35	1.34	1.65
OH-6	6.94	6.94	6.57	-	-	-	-	-	6.96	7.07	-	-	6.92
OH-11	4.91	4.73	4.72	4.64	5.11	-	5.66	4.73	5.24	-	4.65	4.75	4.95
OH-14	12.44	12.60	12.60	13.26	13.34	13.58	13.74	13.19	12.82	-	13.74	13.42	12.59

The same group observed in a similar compound teuvincenone E (**129**) isolated from the roots of *T. fruticans* [73].

The structure of (**129**) has been established by spectral data as well as by comparison with closely related compounds such as (16S)- and (16R)-plectrinone A (**130, 131**). The UV spectrum showed characteristic band shifts of a cross conjugated chromophore by an α,β,γ,δ-diunsaturated ketone moieties at 430 (sh), 370 (sh), 298, 254(sh) nm. The same group was found and studied on teuvincenones F and G (**132, 133**) together with compound **129** from different collection of *T. fruticans* and on four

Table 20. ^{13}C NMR Data of Compounds 122,124, 129, 130, 132, 134-138

C	122	124	129	130	132	134	135	136	37	138
1	27.1	25.9	45.4	45.8	46.0	27.6	28.8	36.4	27.5	27.8
2	33.1	20.3	197.3	196.5	197.5	33.2	32.5	16.2	19.0	19.0
3	214.5	18.2	131.0	131.1	136.4	214.4	212.1	41.1	40.5	41.0
4	48.7	17.9	146.2	146.1	146.2	48.8	48.5	37.2	36.4	37.0
5	139.9	141.2	160.6	161.2	161.3	140.1	141.0	49.9	49.2	49.5
6	140.0	142.4	123.9	123.8	123.9	141.2	144.3	29.7	32.9	140.1
7	182.9	183.1	188.9	189.7	190.4	184.3	179.6	185.6	185.3	182.0
8	107.1	107.9	108.8	107.4	108.9	107.4	129.4	107.6	107.5	108.0
9	135.2	137.9	134.2	136.2	131.4	131.5	157.2	139.5	138.5	138.4
10	40.6	39.8	42.6	42.5	42.6	40.6	41.9	35.4	36.0	39.6
11	131.2	132.8	136.2	136.5	127.5	128.4	178.3	131.1	131.1	131.0
12	153.8	153.2	154.6	152.0	152.2	151.2	161.6	155.6	154.0	153.3
13	111.5	111.2	111.8	112.0	117.0	117.3	127.6	110.6	110.5	110.6
14	154.8	155.9	154.7	157.0	155.5	155.4	173.4	154.8	154.7	154.6
15	34.4	34.2	34.3	31.4	101.5	101.5	104.7	34.3	35.4	35.4
16	83.5	82.2	83.4	69.6	148.8	149.1	144.8	83.2	82.2	35.4
17	21.9	21.9	21.9	22.5	14.0	14.0	14.2	24.5	65.0	65.1
18	24.4	17.6	11.9	11.6	12.0	21.1	22.2	33.2	33.0	27.7
19	21.2	19.0	17.4	17.1	17.5	20.5	20.9	21.6	21.5	21.0
20	20.1	23.0	24.9	24.5	25.4	24.3	24.5	16.2	17.2	25.2

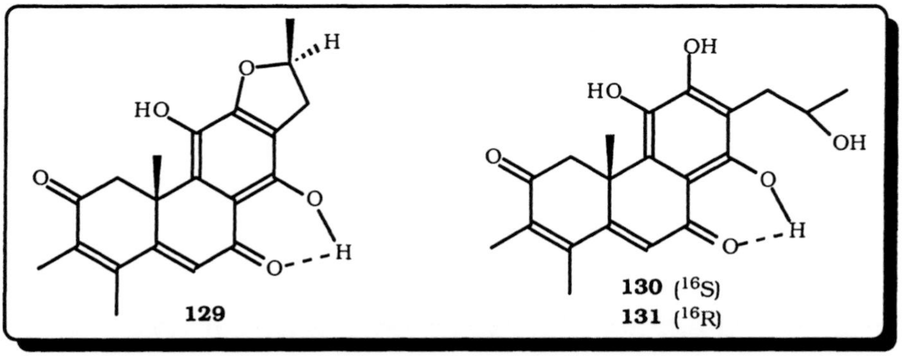

129

130 (^{16}S)
131 (^{16}R)

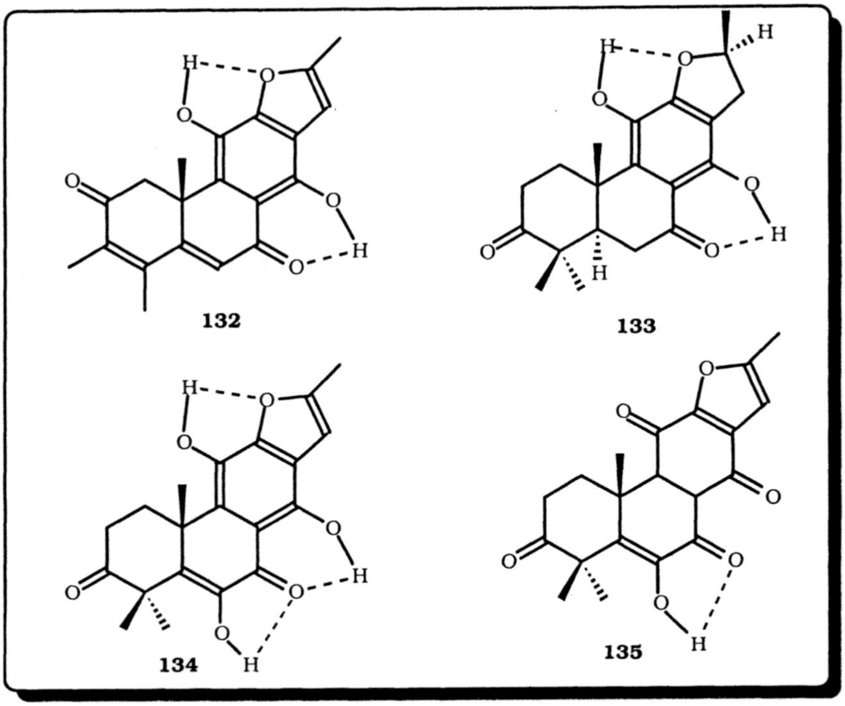

rearranged abietane diterpenoids (**132-135**) isolated from *T. polium* subsp. *expansum* [74], two of which were known (**132, 133**) and two were new compounds, teuvincenones H and I (**134, 135**).

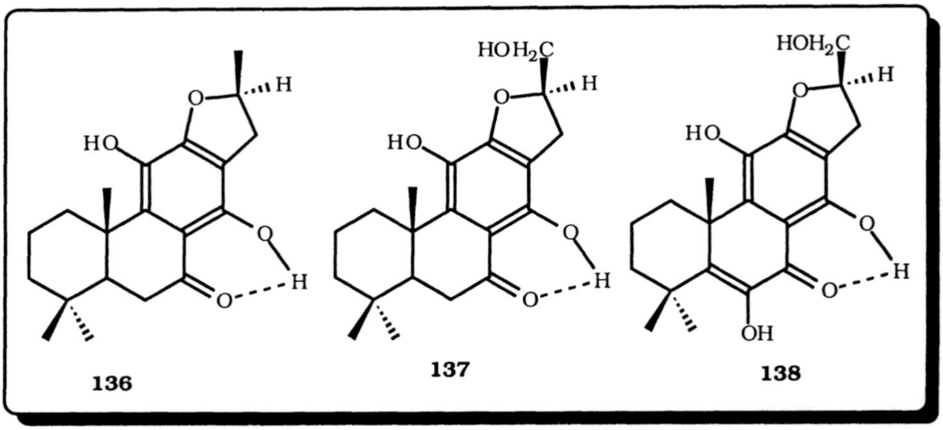

In a more recent study with the whole plant of *T. divaricatum* subsp. *villosum* [75] we have isolated one known diterpenoid teuvincenone B [71] and three new rearranged abietane diterpenoids villosins A (**136**) B

(137) and C (138), all of them were 17 (15→16)-*abeo*-abietane derivatives which were similar to teuvincenones [71-74].

Their structures were established by using the spectral data. The ^{1}H and ^{13}C NMR data of all these compounds are given in Tables 19 and 20.

In two different studies with the aerial parts of *Teucrium alyssifolium* [35,64], we have isolated six new rearranged neo-clerodane diterpenes alysines D and E (110-111) and 3-deacetylalysine B (140). The structures of all these compounds were established by spectral data including 2D NMR techniques. X-ray analysis were performed for compounds 60 (alysine A) and 139 (alysine B) in order to be sure of their stereochemistry. All six compounds had typical substitutions of *neo*-clerodane diterpenes, only instead of a monosubstituted furan, they all have α, β-disubstituted furan ring, which was the first isolation of such type of compounds.

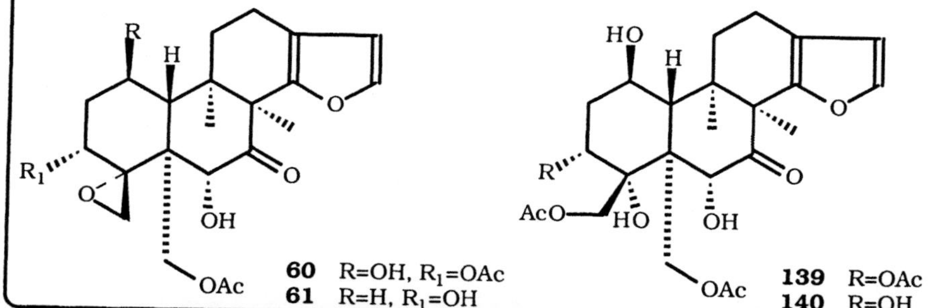

60 R=OH, R₁=OAc
61 R=H, R₁=OH
139 R=OAc
140 R=OH

In addition, the presence of a keto group together with the substitution both at C-3 and C-6 are fairly rare encountered [70]. Table 21 gives the ^{1}H and ^{13}C NMR data of 60, 139, 61, 110, 111, 140.

Sesquiterpenoids

In an early study with *Teucrium polium* [76] a sesquiterpene β-eudesmol (141) was isolated together with a group of amino acids.

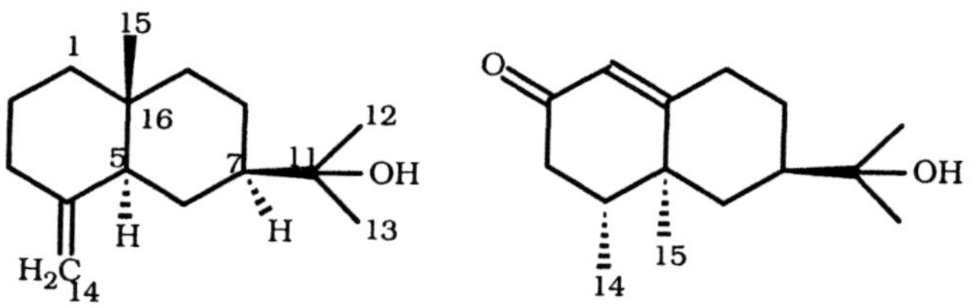

141   142

Table 21. ^{1}H and ^{13}C NMR Data of Compounds 60, 61, 110, 111, 139, 140

	64		139		140		61		110		111	
	^{1}H	^{13}C	^{1}H	^{13}C	^{1}H	^{13}C	^{1}H	^{13}C	^{1}H	^{13}C	^{1}H	^{13}C
1α	4.45	65.0	4.39	64.9	4.47	65.0		18.6	1.50	18.6	1.68	18.6
2α	2.28	40.6	2.42	37.9	2.42	39.8		31.6	1.52	32.9	1.58	32.9
3	4.59	74.6	5.29	71.3	4.09	74.3	3.36	74.0	4.45	75.1	51.8	75.3
4	-	63.3	-	71.4	-	76.9	-	66.2	-	62.3	-	75.1
5	-	45.6	-	44.8	-	45.2	-	44.1	-	42.1	-	41.8
6	3.80	75.4	3.92	75.9	3.92	74.3	3.70	74.0	2.95	75.4	4.30	75.4
7	-	206.7	-	206.8	-	206.4	-	207.4	-	106.6	-	106.6
8	-	53.9	-	53.5	-	53.7	-	50.9	-	51.6	-	51.6
9	-	49.2	-	52.4	-	52.7	-	49.5	-	48.1	-	48.1
10	2.31	45.7	2.40	42.0	2.45	42.2	2.30	41.9	2.33	42.1	2.35	41.8
11α	2.55	19.3	2.01	18.9	1.95	19.3	-	17.2	1.40	18.6	1.44	18.6
12α	1.61	33.6	1.61	34.5	1.60	34.9	-	32.4	1.38	29.1	1.40	29.6
13	-	116.3	-	116.5	-	116.6	-	115.4	-	115.7	-	115.7
14	-	151.9	-	151.2	-	152.4	-	151.8	-	154.1	-	154.2
15	6.27	110.4	6.22	110.3	6.22	110.5	6.22	110.4	6.10	109.5	6.14	109.1
16	7.37	142.5	7.38	142.5	7.35	142.7	7.25	142.6	7.31	141.9	7.31	141.9
17	1.34	19.2	1.42	18.1	1.42	18.6	1.37	18.4	1.34	17.1	1.34	17.2
18	3.08	49.2	4.22	66.2	4.16	66.5	2.58	50.9	2.78	49.9	4.11	66.4
19	4.27	-	4.41	65.5	4.37	60.6	4.49	62.2	4.18	66.9	4.24	66.4
20	1.14	17.8	1.21	17.5	1.23	17.7	1.02	18.9	1.21	20.2	1.15	20.2
CO	-	170.1	-	169.9	-	170.1	-	169.8	-	169.6	-	170.6
CH$_3$	2.02	20.7	1.99	20.2	1.96	20.6	2.02	20.7	2.10	21.2	2.01	21.2
CO	-	169.7	-	169.6	-	169.8	-	-	-	-	-	171.2
CH$_3$	2.06	20.9	2.05	20.5	2.02	20.8	-	-	-	-	2.13	20.2
CO	-	-	-	169.6	-	170.1	-	-	-	-	-	-
CH$_3$	-	-	2.08	20.8	-	-	-	-	-	-	-	-

In later years from a Spanish species of *Teucrium, T. carolipaui* C. Vicioso *ex* Pau [77]. which grows only in small areas of south-east Spain, a new sesquiterpene 11-hydroxy-valenc-1(10)-en-2-one (142) was isolated and its structure was determined by NMR techniques (Table 22). Three new guaiane sesquiterpenoids, teucladiol (143), teuclatriol (144) and 10-epiteuclatriol (145) were isolated from the aerial parts of *T. leucocladum* [78] and their structures were established spectroscopically (Table 22).

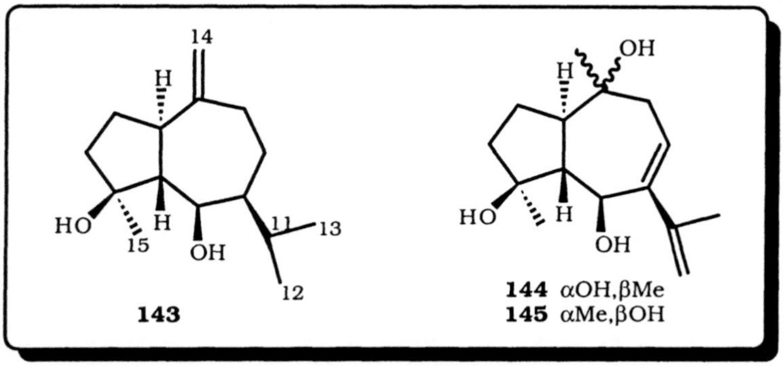

143

144 αOH,βMe
145 αMe,βOH

Table 22.   NMR Data of Compounds 142-147

	142		143		144		145		146		147	
	^{1}H	^{13}C	^{1}H	^{13}C	^{1}H	^{13}C	^{1}H	^{13}C	^{1}H	^{13}C	^{1}H	^{13}C
1	5.76	124.2		42.8		52.1		51.6	3.42	79.3	3.30	79.9
2	-	199.9		27.3		23.3		23.6		31.1		30.8
3	2.18 2.36	42.0		40.6		41.2		41.0	2.31	35.3		34.4
4	2.0	40.5	-	80.8		81.1		81.2	-	146.1		146.9
5	-	39.1	1.88	59.5		55.4	2.22	53.9	2.10	50.3		52.1
6	2.0 0.98	39.6	4.09	72.6	4.15	71.4	4.11	71.3	4.12	69.9	4.32	68.6
7	1.72	43.8		48.5		45.6		45.7	1.75	44.5		50.0
8	1.18	27.7		24.0		20.5		19.9		22.3		21.2
9	2.47	33.0	2.142.56	35.5		48.1		47.1		32.2		29.8
10	-	171.6		152.5		75.5		73.0		41.7		39.8
11	-	72.0		29.0		29.6		29.7	2.0	25.2		28.9
12	1.19	27.2	1.03	21.5	1.03	21.1	1.02	21.2	0.92	22.2	0.93	20.4

(Table 22). contd.....

	142		143		144		145		146		147	
	^1H	^{13}C	^1H	^{13}C	^1H	^{13}C	^1H	^{13}C	^1H	^{13}C	^1H	^{13}C
13	1.20	26.8	0.97	21.6	0.98	21.5	0.97	21.3	1.10	25.0	0.96	20.8
14	0.97	14.9	4.73	108.1	1.26	22.2	1.20	29.9	0.77	12.1	0.91	13.1
15	1.08	16.9	1.31	23.2	1.28	23.1	1.29	23.3	5.0	108.1	5.01	108.6

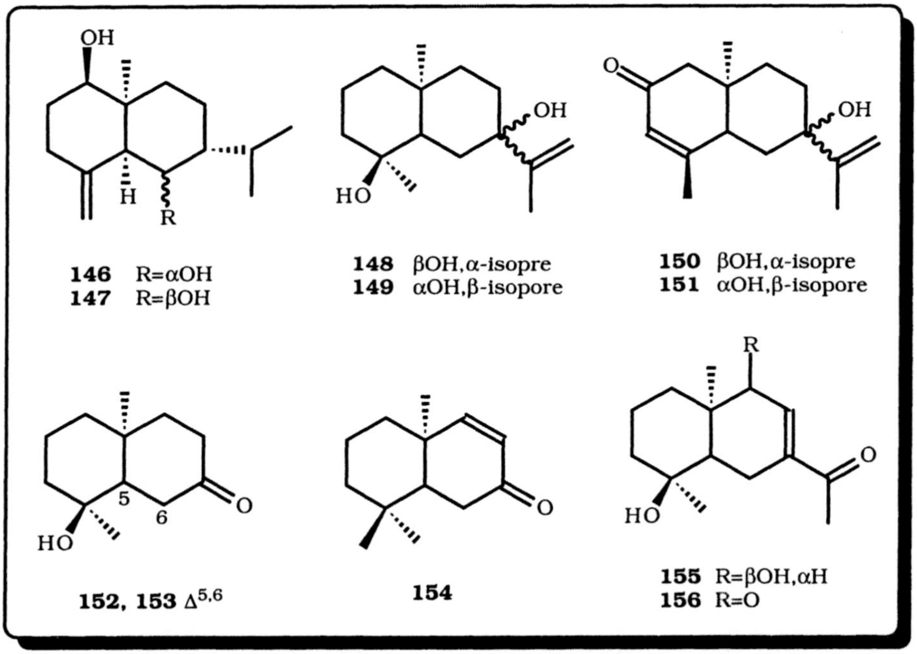

146 R=αOH
147 R=βOH

148 βOH,α-isopre
149 αOH,β-isopore

150 βOH,α-isopre
151 αOH,β-isopore

152, 153 Δ^{5,6}

154

155 R=βOH,αH
156 R=O

The GC/MS analysis of the volatile oil of *T. polium* [79] showed that sesquiterpenoids comprise the major part of the oil (78.6 %), certain sesquiterpene alcohols such as β-eudesmol and α-cadinol.Two new sesquiterpene diols 7-*epi*-eudesm-4(15)-ene-1β, 6α-diol (146) and 7-epi-eudesm-4(15)-ene-1β, 6β-diol (147) were isolated and their structures were elucidated by ^1H and ^{13}C NMR spectral analyses.

Teucrium heterophyllum [80] has yielded quite a number of sesquiterpenes, teucdiol A (148), B (149) and teucrenone (150), 7-*epi*-teucrenone B (151) and nor-sesquiterpenes teuhetone (152), teuhetones A (153) and B (154), 9β-hydroxytephyllone(155) and 9-oxo-tephyllone (156). The structures were assigned by using the NMR techniques.

Table 23. ^{13}C NMR Data of 151-153

C	151	152	153
1	54.0	37.9	40.6
2	199.1	20.2	19.4
3	126.8	38.2	40.8
4	162.5	71.8	72.3
5	44.9	54.0	175.1
6	33.0	43.5	122.4
7	74.7	212.1	200.5
8	31.5	42.9	42.2
9	37.6	40.7	33.9
10	37.4	34.5	26.1
11	145.9	-	-
12	114.1	-	-
13	18.6	-	-
14	16.9	17.3	24.5
15	22.0	22.0	29.6

Triterpenoids and Steroids

Triterpenic and steroidal compounds were also isolated from *Teucrium* species. Common triterpenic acids such as ursolic acid, oleanolic acid and small amounts of micromeric acid were isolated from *T.fruticans* [81] while α-amyrin, lupeol, sitosterol and stigmasterol were present in *T. divaricatum* subsp. *villosum* [75]. Two new steroidal compounds 3β-hydroxy-stigmast-24(24^1), 25-dien-24^2-al (**157**) and 3β-hydroxy-24α-ethylcholesta-5, 25-dien-7-one (**158**) were isolated together with two known compounds 24α-ethylcholesta-5, 25-dien-3β-ol (**159**) and sitosterol from the aerial parts of *T.chamaedrys* subsp. *chamaedrys* [82]. The plant also yielded α-amyrin and ursolic acid. The structures of the new steroidal compounds were established by NMR techniques and by comparison of similar known compounds (Table 24).

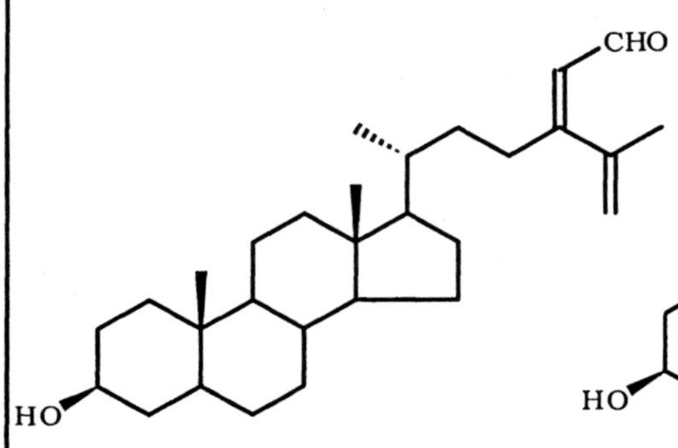

Three 24-ethylcholestane derivatives were isolated from two species of *Teucrium*, *T. abutiloides* and *T. betonicum* [83]. All of this compounds have 24β-configuration and identified as (24S)-24-ethyl- cholesta-5, 22(E), 25-trien-3β-ol (**160**), (24S)-24-ethylcholesta-5, 25-dien-3β-ol(clerosterol) (**161**) and (24R)-24-ethylcholesta-5, 22-(E)-dien-3β-ol (poriferasterol) (**162**). The structures of these compounds were decided by spectral data as well as by comparison with the well established similar steroidal structures (Table 24).

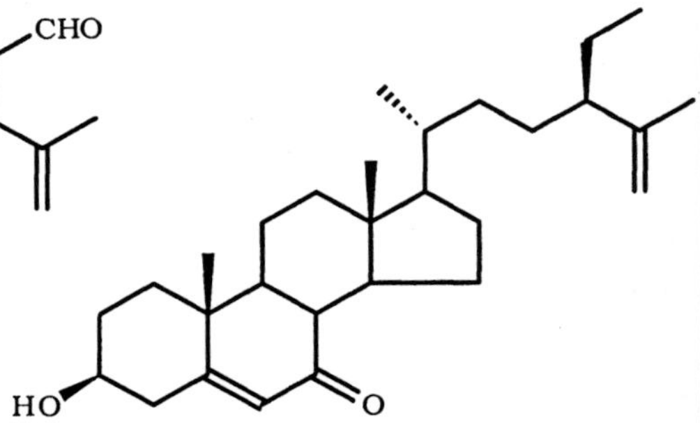

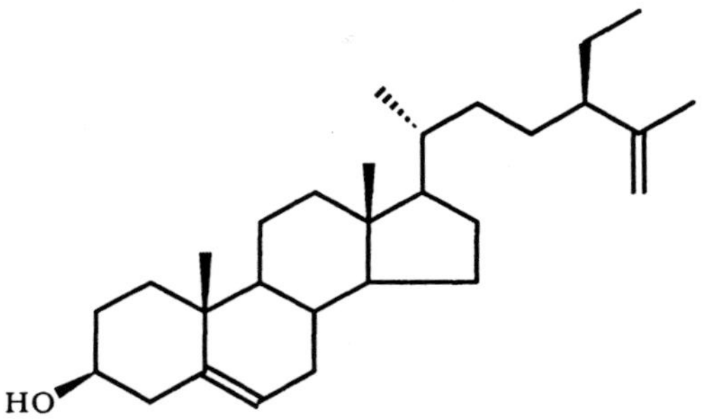

Flavonoids and Aromatic Compounds

From the aerial parts of *Teucrium* species flavonoids have been also isolated, although they are not abundant. Luteolin was obtained from *T. scorodonia* [26], salvigenin from *T. flavum* [65], salvigenin, circimaritin from *T. massiliense* [55], luteolin, luteolin 7-0-glucoside and 6-

methoxygenkwanin from *T. polium* [84], cirsiliol, apigenin from *T. polium* subsp. *vincentinum* [63], cirsiliol also from *T. montanum* [57].

Table 24. NMR Data of compounds 157-162

	157		158		159		160		161	162
	^1H	^{13}C	^1H	^{13}C	^1H	^{13}C	^1H	^{13}C	^1H	^1H
1		35.2		36.3		37.3		37.0		
2		24.5		31.2		31.6		27.8		
3	3.70	67.4	3.64	70.5	3.60	71.9	3.52	74.0	4.59	4.60
4		39.7		41.6		42.4		38.1		
5		36.4	5.69	165.4		140.8		139.7		
6		28.2		126.1	5.70	124.8	5.35	122.6	5.36	5.36
7		33.3		201.0		33.8		31.9		
8		33.7		45.4		31.9		31.9		
9		49.4		49.5		50.2		50.9		
10		35.2		38.7		39.8		39.6		
11		20.7		21.2		21.1		21.0		
12		40.1		38.7		39.8		39.6		
13		43.2		43.1		42.3		42.3		
14		56.2		54.6		56.9		56.8		
15		25.9		26.3		24.3		24.3		
16		26.8		26.5		28.3		28.7		
17		55.7		54.6		56.1		55.9		
18	0.70	12.5	0.65	11.9	0.68	11.8	0.68	12.0	0.66	0.69
19	0.78	18.4	0.80	17.3	0.80	19.5	1.00	19.3	1.00	1.02
20		25.1		35.4		35.7		40.2		
21	0.90	18.4	0.90	18.8	0.90	18.8	1.00	21.4	0.89	1.03
22		33.3		33.6		33.5	5.24	137.2		
23		29.4		29.4		29.4	5.16	130.1		
24		148.0		49.9		49.6		52.0		
25		146.2		146.6		147.6		148.6		
26	4.64	111.6	4.64	111.4	4.66	111.4	4.69	109.5	4.63	0.84
	4.70		4.70		4.72					
27	1.65	17.3	1.63	17.3	4.72	17.9	1.64	20.2	4.71	0.79
28		109.8		29.6		26.5		25.7	1.55	
29		204.6	0.85	12.1	0.85	12.1	0.82	12.1	0.79	0.81

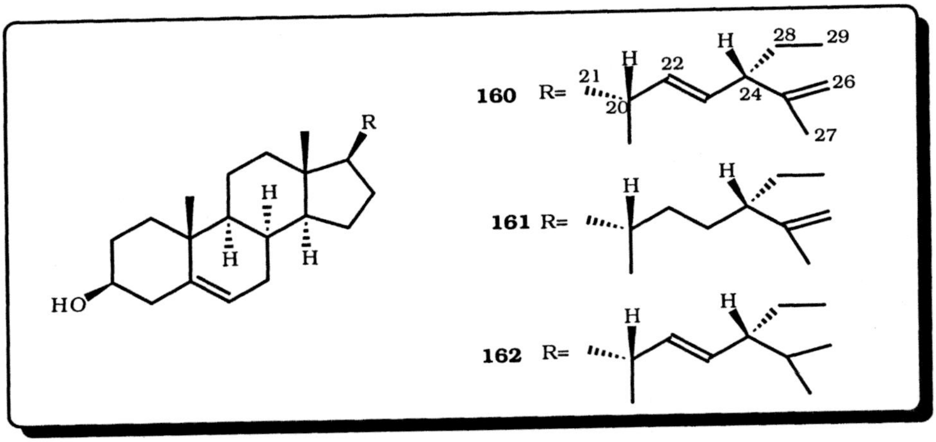

Cirsiliol, apigenin, pectolinarigenin, naringenin from *T. chamaedrys* subsp. *chamaedrys*[82], cirsiliol, eupatilin and apigenin from *T. divaricatum* subsp. *villosum* [75], salvigenin, circimaritin, 3'-O-methyleupatonin from *T. heterophyllum* [80]. In a recent study with *T. alyssifolium* [85] the known flavones, cirsiliol, cirsilineol and luteolin-6, 7, 3', 4'-tetramethyl ether, were obtained in addition to a new flavanone, alysifolinone (**163**) which was elucidated by UV shift reagants and NMR techniques.

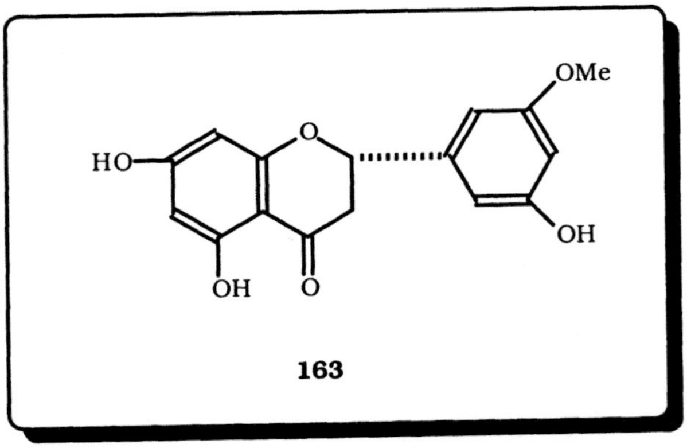

163

A new caffeic glycoside ester, poliumoside (**164**) has been isolated from *T. belion* [86] and its structure was determined as [β-(3',4'-dihydroxyphenyl)-ethyl]-(3,6-O-α-L-dirhamnopyranosyl)-(4-O-caffeoyl)-β-D-glucopyranoside.

From *T. chamaedrys* [87], a new phenylpropanoid glycoside, teucrioside(**165**) was isolated. Its structure has been elucidated as 3, 4-dihydroxy-β-phenylethoxy-O-α-L-xylopyranosyl-(1-2)α-L-rhamnopyranosyl-(1-3)-4-caffeoyl-β-D-glucopyranoside on the basis of 1D and 2D NMR experiments.

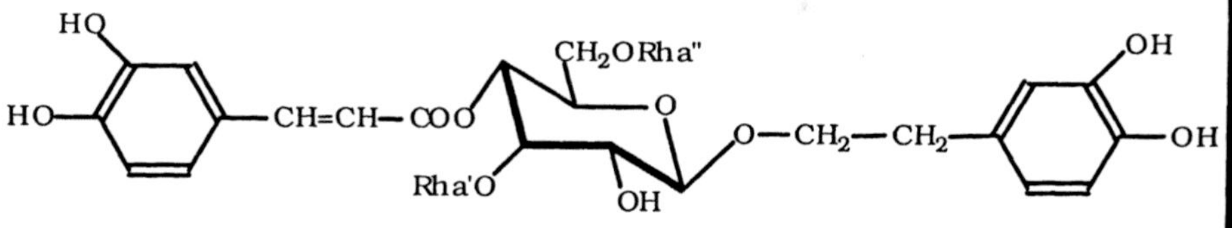

164

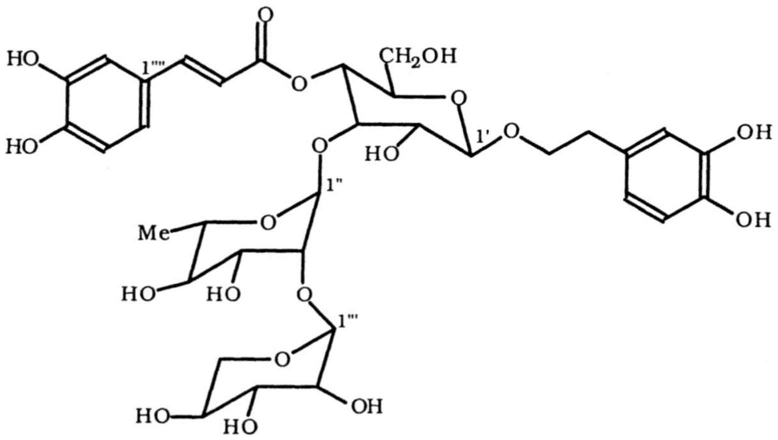

165

Volatile Compounds

The essential oil of the *Teucrium* species was scarcely studied. Twelve major and about thirty minor compounds were identified in *T. polium* [88,89,90], and in later studies using a polar (carbowax 20M) and a non-polar column (SE 30) columns GS-MS analyses were performed and many compounds were found. Individual compounds were identified by comparing their MS and retention times to those of authentic samples.

The quantitative composition of the essential oils from six *Teucrium* species collected from Iberian peninsula have been studied and sixty two constituents have been identified. Five populations of *T. scorodonia* subsp. *scorodonia*, three populations of *T. scorodonia* subsp. *baeticum*, two samples of *T. oxylepis* subsp. *marianum* and several samples *of T. asiaticum, T. salviastrum* and *T. oxylepis* subsp. *oxylepis* were studied and no significant differences were found in the terpenoid (monoterpenes and diterpenes) patterns. The GC columns were packed with 5 % silicon OV1 and 10 % carbowax 20M [91].

BIOLOGICAL ACTIVITIES

Teucrium species are used in folk medicine as diuretic, diaphoretic, carminative, astringent, stimulant and tonic, antiseptic, antipyretic and antiinflammatory agents. *T. cubense* [92] for example is used against stomach aches and as an amoebacide. Teucjaponin A (**32**) obtained from *T. japonicum* [33] was shown to be antifeedant for the larve of *Prodenia litura*. Its threshold concentration for inhibitory activity was about 400 ppm in the leaf disk test. Generally *neo*-clerodane diterpenes posses insect antifeedant activity and they are used as antifungal, antitumor and antibacterial agents [56,63]. The antitumor activity of *T. bicolor* [52] was reported. The antifeedant activity of neoclerodanes is mentioned in a

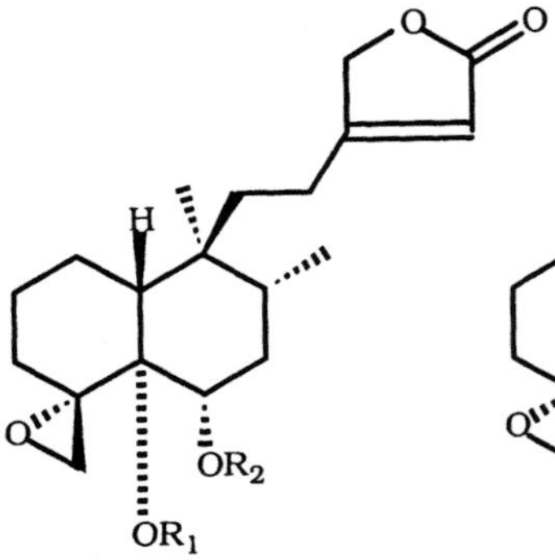

166 R₁=R₂=Ac
167 R₁=R₂=H

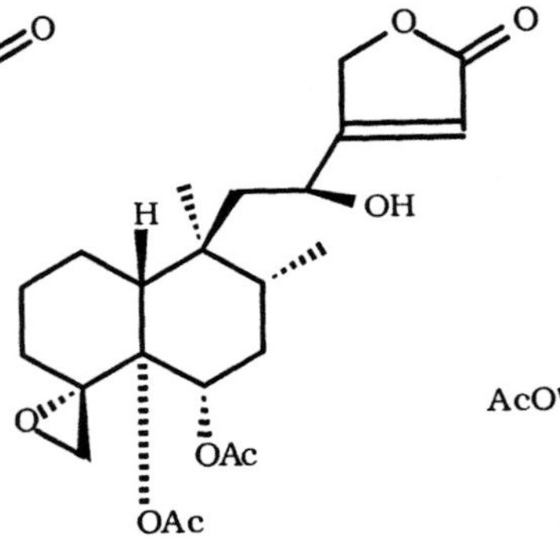

168

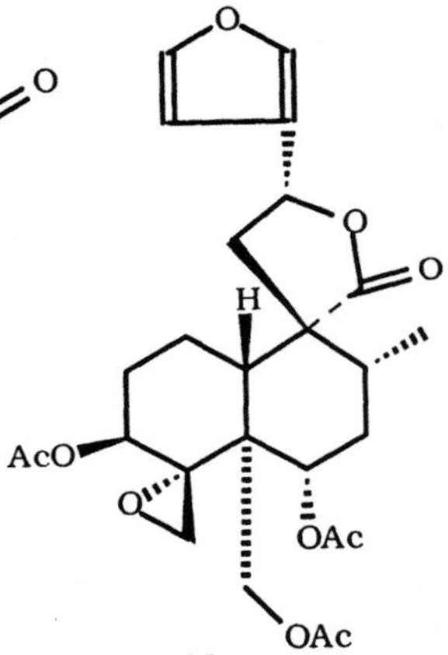

169

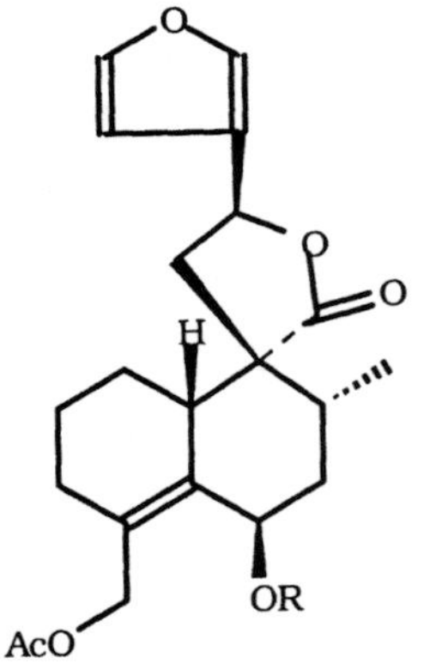

170 R= 2-acetylglucose

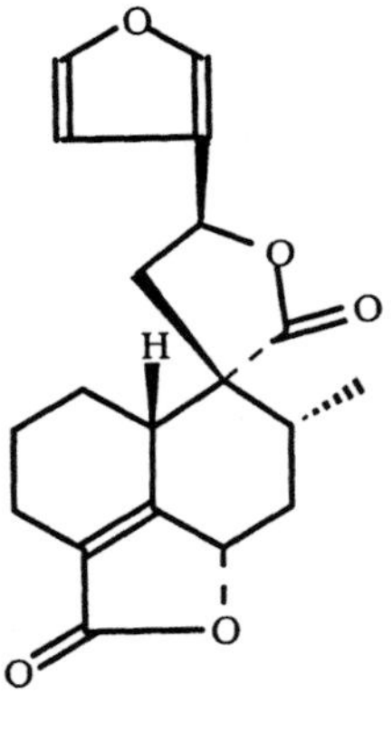

171

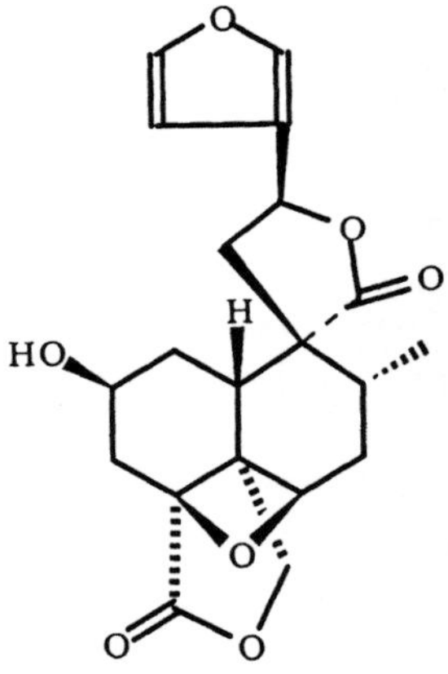

172

study indicating their economical importance against *Lepidopterous* pests [93]. In this study Simmonds et al used two insect species, *Spodoptera littoralis* and *Heliothis armigera* and used nineteen *neo*-clerodane diterpenoids isolated from various *Teucrium* species and found significant difference between control (C) disc only had sucrose and the treatment (T) disc had sucrose and a 100 ml aliquot of one of the test solutions. The duration of the bioassay varied between species but was never longer than 18 hr, so that never more than 50 % of any disc was eaten. The discs were then reweighted and the antifeedant index ((C-T)/(C+T)(% calculated on the amounts eaten. This index identifies both phogostimulants (-ve values) and antifeedants (+ve values). Table 25 gives the results of these tests. The following compounds were used for this test: 19-acetylgnaphlin(**20**), teucrin A (**26**), teucjaponin A (**32**), teucjaponin B (**33**), eriocephalin (**34**), isoeriocephalin (**40**), dihydroteugin (**68**), montanin C (**80**), ajugarin I (**166**), deacetylajugarin II (**167**), 6,19-diacetylteumassilin (**168**), teupyreinin (**169**), teuflavoside (**170**), 12-*epi*-teucvin (**171**), chamaedroxide (**172**), teucroxide (**173**), 12-ketoteugnaphalodin (**174**), 6-acetylteucjaponin (**175**), 7-O-acetyleriocephalin (**176**).

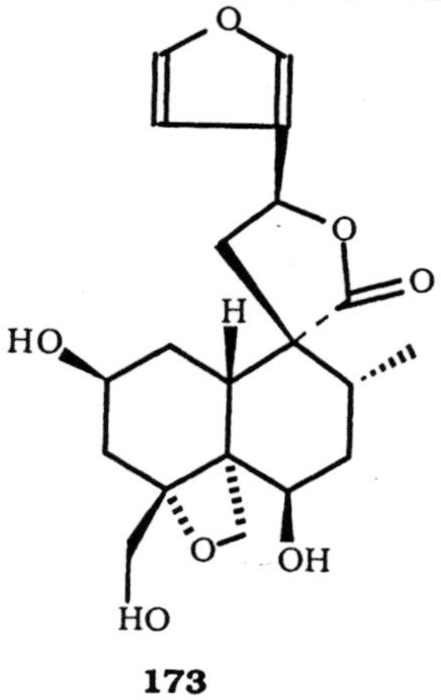

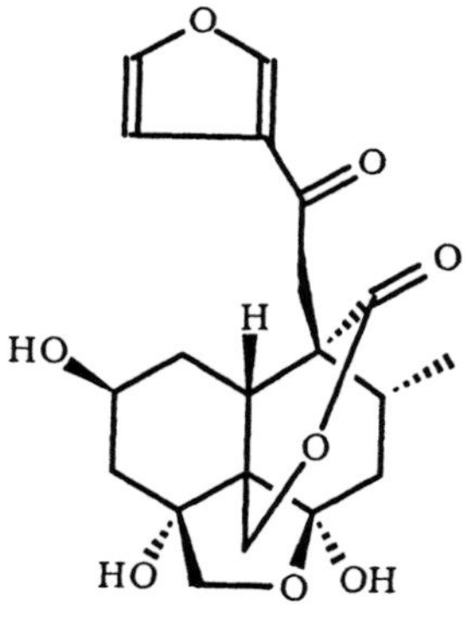

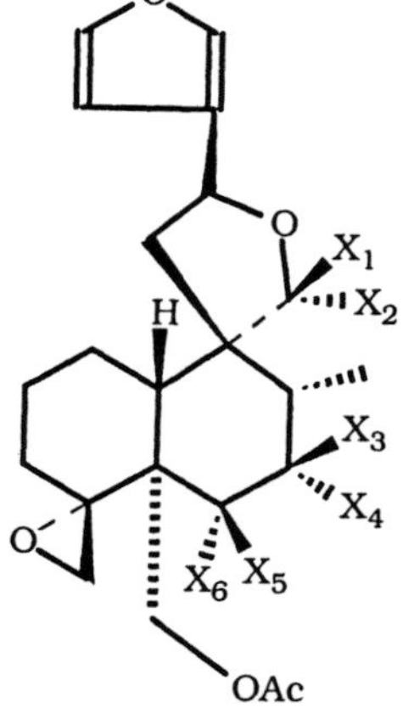

175 X_1, X_2=O, X_3=X_4=X_6=-H, X_5=OAc
176 X_1=OAc, X_2=X_3=H, X_4=OAc, X_5, X_6=O

Antifeedant activity was also studied both in natural neo-clerodanes
and their semisynthetic derivatives against *Spodoptera littoralis* using the
method given in literature 93.

Teucrium polium [79] is reputed for its medicinal value in folk
medicine. In Egypt, it is used as appetizer, expectorant and hypoglycemic.
It is also used in the treatment of stomach ache and to promote wound
healing. The plant also posseses antispasmolitic activity.

In order to evaluate the analgesic and central nervous system
depressant activity and to find out acute toxicity, four *Teucrium* species

Table 25. Antifeedant Index ((C-T)/(C+T)(% of Test Compounds [Mean+/-(S.E.M.)]
in Dual Choice Test with Glass Fibre Discs [Control (C) Versus
Treatment(T)(N=20)]

Conc. ppm Compound	Antifeedant index			
	S. littoralis		*H. armigera*	
	100	10	100	10
20	23.7(6.98)	21.9(18.04)	11.7(14.87)	19.5(7.09)
26	9.8(11.87)	8.7(15.92)	11.9(5.95)	15.6(6.52)
32	12.9(7.67)	7.5 (6.98)	3.9(14.98)	3.4(14.67)
33	48.9(5.98)*	43.0(7.64)*	29.8(14.95)	23.2(6.58)
34	48.9(5.98)*	40.9(9.78)*	23.9(14.56)	24.0(7.22)
40	38.9(16.98)	33.1(10.59)	24.8(16.98)	20.9(7.98)
68	-9.9(13.76)	-18.6(17.54)	-0.9(11.87) -	5.8(5.00)
80	6.5(7.87)	7.5(9.97)	4.9(12.98)	8.9(15.98)
166	43.1(7.31)*	34.5(6.95)*	39.6(9.65)*	23.9(9.86)
167	29.6(8.96)*	17.9(9.65)	23.9(9.62)	2.9(5.65)
168	32.2(22.16)	21.7(15.87)	63.2(5.41)*	16.9(7.78)
169	14.8(6.98)	12.6(12.25)	8.8(8.76)	6.8(3.71)
170	-9.8(23.87)	-2.6(16.85)	2.8(11.98)	-2.9(6.98)
171	49.9(3.87)*	45.4(9619*	29.6(8.93)	29.2(8.55)
172	31.8(11.87)	25.6(13.51)	18.9(7.65)	15.7(13.65)
173	14.8(6.94)	16.6(7.42)	9.8(15.56)	19.6(6.46)
174	9.6(11.41)	9.4(10.62)	8.6(10.41)	2.4(3.75)
175	10.4(13.21)	7.4(10.41)	4.6(10.21)	3.0(18.41)
176	19.9(12.87)	18.5(10.61)	16.9(12.98)	11.5(5.68)

*Significant difference between the amount eaten of the control and treatment discs (Wilcoxon's matched pairs test
$p < 0.05$).

were studied, i.e. *T. cartaginenses* Linn. (TC), *T. flavum* Linn. (TF), *T. pumillum* Linn. (TP) and *T. buxifolium* Linn [95] (TB). For the test albino mice weighting 25-30 g and female Wistar rats (150-160 g) were used. The extracts were administered intraperitonally (ip) with a constant volume of 1 ml/kg and in most studies the dose tested was usually 1/10 LD_{50} TC in MeOH, TC in $CHCl_3$ and TF in MeOH (200 mg/kg). TF in $CHCl_3$ (138 mg/kg), TP in MeOH (50 mg/kg), TF in $CHCl_3$ (132 mg/kg), TB in MeOH (26 mg/kg) and TB in $CHCl_3$ (158 mg/kg). Acute toxicity tests were done with male albino mice, anticonvulsive action in albino mice, in motility tests chemical stimules and thermal stimules albino mice (n= 6 per group) were used, in mechanical stimules female Wistor rats (n= 10 per group) were used. All compounds showed a limited toxicity. Most of the extracts exhibited central nervous system depressant action. They did not show a significant activity, however, some of them showed significant effect on mechanical and chemical stimules thus suggesting the induction of a peripheral analgesic effect. As a result of this study methanol and dichloromethane extracts of the leaves and stems of *T. cartaginenses* and *T. buxifolium* possess central depressor and modest peripheral analgesic activities.

REFERENCES

[1]   Davis, P.H., Flora of Turkey and East Aegean Island, **1988**, *7*, 53-75.
[2]   Fujita, E., Nagao, Y. and Node, M., *Heterocycles*, **1976**, *5*, 793-795.
[3]   Piozzi, F., *Heterocycles*, **1981**, *15*, 1489-1503.
[4]   Piozzi, F., Savona, G. and Rodriguez, B., *Heterocycles*, **1987**, *25*, 807-811.
[5]   Piozzi, F., *Heterocycles*, **1994**, *37*, 603-625.
[6]   Hanson, J.R., *Natural Product Reports*, **1984**, *1*, 171-180.
[7]   Hanson, J.R., *Natural Product Reports*, **1984**, *1*, 339-348.
[8]   Hanson, J.R., *Natural Product Reports*, **1986**, *3*, 307-322.
[9]   Hanson, J.R., *Natural Product Reports*, **1987**, *4*, 399-413.
[10]   Hanson, J.R., *Natural Product Reports*, **1988**, *5*, 211-227.
[11]   Hanson, J.R., *Natural Product Reports*, **1989**, *6*, 347-366.
[12]   Hanson, J.R., *Natural Product Reports*, **1990**, *7*, 149-164.
[13]   Hanson, J.R., *Natural Product Reports*, **1991**, *8*, 1-16.
[14]   Rowe, J.W., "The Common and Systematic Nomenclature of Cyclic Diterpenes", **1969**, Third revision.
[15]   Rogers, D., Ünal, G.G., Williams, D.J., Ley, S.V., Sim, G.A., G.A., Joshi, B.S.and Ravindranath, *J.C.S. Chem.Comm.*, **1979**, 97-99.
[16]   Esquivel, B., Cardenas, J., Ramamoorthy, T.P. and Rodriguez-Hahn, L., *Phytochemistry*, **1986**, *25*, 2381-2384.
[17]   Reichstein, T., *Helv. Chim. Acta*, **1932**, *15*, 1110-1112.
[18]   Fujita, E., Uchida, I. and Fujita, T., *J.C.S. Perkin Trans 1*, **1974**, 1547-1555.
[19]   Fujita, E., Uchida, I., Fujita, T., Masaki, N. and Osaki, K., *J.C.S. Chem.Comm.*, **1973**, 793-794.
[20]   Malakov, P.Y., Papanov, G.Y. and Mollov, N.M., *Tetrahedron Letters*, **1978**, 2025-2026.

[21] Savona, G., Patesnostro, M.P., Piozzi, F., Hanson, J.R., Hitchcock, P.B. and Thomas, S.A., *J.C.S. Perkin Trans I*, **1978**, 1080-1083.

[22] Node, M., Sai, M. and Fujita, E., *Phytochemistry*, **1981**, *2*, 757-760.

[23] Gacs-Baitz, E., Radics, L., Oganessian, G.B. and Mnatsakanian, V.A., *Phytochemistry*, **1978**, *17*, 1967-1973.

[24] Fayos, J., Fernandez-Gadea, F., Pascuel, C., Perales, A., Piozzi, F., Rico, M., Rodriguez, B. and Savona, G., *J. Org. Chem.*, **1984**, *49*, 1789-1790.

[25] Pascual, C., Fernandez, P., Garcia-Alvarez, M.C., Marco, J.L., Fernandez-Gadea, F., de la Torre, M.C., Hueso-Rodriguez, J.A., Rodriguez, B., Bruno, M., Paternostro, M., Piozzi, F. and Savona, G., *Phytochemistry*, **1986**, *25*, 715-718.

[26] Marco, J.L., Rodriguez, B., Savona, G. and Piozzi, F., *Phytochemistry*, **1982**, *21*, 2567-2569.

[27] Fernandez-Gadea, F., Rodriguez, B., Savona, G. and Piozzi, F., *Phytochemistry*, **1984**, *23*, 1113-1118.

[28] Malakov, P.Y., Paponov, G.Y. and Mollov, N.M., *Z. Naturforsch. Teil B.*, **1979**, *34*, 1570-1573.

[29] Gacs-Baitz, E., Papanov, G.Y., Malakov, P.Y. and Szilagyi, L., *Phytochemistry*, **1987**, *26*, 2110-2112.

[30] Sanders, J.K.M. and Mersh, J.D., Nuclear Magnetic Double Resonance: The Use of Difference Spectroscopy. *Prog. Nucl. Magn. Reson. Spectrosc.*, Vol. *15* **1983**,, Pergamon Press Oxford, p. 353.

[31] Kinns, M. and Sanders, J.K.M., *J.Magn.Reson.*, **1984**, *56*, 618-621.

[32] Simeos, F., Rodriguez, B., Bruno, M., Piozzi, F., Savona, G. and Arnold, N.A., *Phytochemistry*, **1989**, *28*, 2763-2768.

[33] Miyase, T., Kawasaki, H., Noro, T., Ueno, A., Fukushima, S. and Takemoto, T., *Chem. Pharm. Bull.* **1981**, *29*, 3561-3564.

[34] Al-Yahya, M.A., Muhammad, I., Mirza, H.H., El-Feraly, F.S. and McPhail, A.T., *Journal Natural Products*, **1993**, *56*, 830-842.

[35] Topçu, G., Eris, C., **Ulubelen**, A., Krawiec, M. and Watson, W.H., *Tetrahedron*, **1995**, *51*, 11793-11800.

[36] Malakov, P.Y., Papanov, G.Y. and Ziesche, J. *Phytochemistry*, **1982**, *21*, 2597-2598.

[37] Papanov, G.Y. and Malakov, P.Y., *Phytochemistry*, **1983**, *22*, 2787- 2789.

[38] Savona, G., Piozzi, F., Rodriguez, B., Pascual, C. and Servettaz, O., *Phytochemistry*, **1986**, *25*, 2857-2859.

[39] Xie, N., Min, Z.D., Zhao, S.-X., Lu, Y., Zheng, Q.-T., Wang, C., Mizuno, M., Iinuma, M. and Tanaka, T., *Chem. Pharm. Bull.*, **1992**, *40*, 2193-2195.

[40] Malakov, P.Y. and Papanov, G.Y., *Phytochemistry*, **1983**, *22*, 2791- 2793.

[41] Bruno, M., Savona, G., Pascual, C. and Rodriguez, B., *Phytochemistry*, **1981**, *20*, 2259-2261.

[42] Savona, G., Garcia-Alvarez, M.C. and Rodriguez, B., *Phytochemistry*, **1982**, *21*, 721-723.

[43] Rodriguez, M.-C., Barluenga, J., Savona, G., Piozzi, F., Servettaz, O. and Rodriguez, B., *Phytochemistry*, **1984**, *23*, 1465-1469.

[44] de la Torre, M.C., Pascual, C., Rodriguez, B., Piozzi, F., Savona, G. and Perales, A. *Phytochemistry*, **1986**, *25*, 1397-1403.

[45] Bruno, M., Piozzi, F., Savona, G., Rodriguez, B., de la Torre, M.C. and Servettaz, O., *Phytochemistry*, **1987**, *26*, 2859-2861.

[46] Handong, S., Xingliang, C., Tianen, W., Lutai, P., Zhonguen, L. and Duyuan, C., *Phytochemistry*, **1991**, *3*, 1721-1723.

[47] Ning, X., Zhi-Da, M., Shou-Xun, Z., Bing, W., Qi-Tai, Z. and Pei, Z., *Phytochemistry*, **1991**, *30*, 1963-1966.
[48] Malakov, P.Y., Boneva, I.M., Papanov, G.Y. and Spassov, S.L., *Phytochemistry*, **1988**, *27*, 1141-1143.
[49] Savona, G., Piozzi, F., Servettaz, O., Fernandez-Gadea, F. and Rodriguez, B., *Phytochemistry*, **1984**, *23*, 611-613.
[50] de la Torre, M.C., Rodriguez, B., Bruno, M., Savona, G., Piozzi, F., Perales, A., Torres, M.R. and Servettaz, O., *Phytochemistry*, **1990**, *29*, 2229-2233.
[51] Di-An, S. and Guang-Yi, L., *Phytochemistry*, **1993**, *33*, 716-717.
[52] Labbe, C., Polanco, M.I. and Castillo, M., *J. Nat.Prod.*, **1989**, *52*, 871-874.
[53] Bruno, M., Dominguez, G., Lourenco, A., Piozzi, F., Rodriguez, B., Savona, G., de la Torre, M.C. and Arnold, N.A., *Phytochemistry*, **1991**, *30*, 3693-3697.
[54] Savona, G., Paternostro, M., Piozzi, F. and Rodriguez, B., *Tetrahedron Letters*, **1979**, 379-382.
[55] Savona, G., Bruno, M., Piozzi, F., Servettaz, O. and Rodriguez, B., *Phytochemistry*, **1984**, 23, 849-852.
[56] de la Torre, M.C., Rodriguez, B., Bruno, M., Savona, G., Piozzi, F. and Servettaz, O., *Phytochemistry*, **1988**, *27*, 213-216.
[57] Malakov, P.Y., Papanov, G.Y. and Boneva, I.M., *Phytochemistry*, **1992**, *31*, 4029-4030.
[58] Fernandez, P., Rodriguez, B., Villages, J-A., Perales, A., Savona, G., Piozzi, F. and Bruno, M., *Phytochemistry*, **1986**, *25*, 1405-1409.
[59] Malakov, P.Y. and Papanov, G.Y., *Phytochemistry*, **1985**, *24* 301-303.
[60] de la Torre, M.C., Fernandez-Gadea, F., Michavila, A., Rodriguez, B., Piozzi, F. and Savona, G., *Phytochemistry*, **1986**, *25*, 2385-2387.
[61] de la Torre, M.C., Rodriguez, B., Savona, G. and Piozzi, F., *Phytochemistry*, **1986**, *25*, 171-173.
[62] Marco, J.L., Rodriguez, B., Pascual, C., Savona, G. and Piozzi, F., *Phytochemistry*, **1983**, *22*, 727-731.
[63] Carreiras, M.C., Rodriguez, B., Piozzi, F., Savona, G., Torres, M.R. and Perales, A., *Phytochemistry*, **1989**, *28*, 1453-1461.
[64] Topçu, G., Eris, C. and Ulubelen, A., J. Nat.Prod. **1997**, *60*, 1054-1047.
[65] Savona, G., Piozzi, F., Servattaz, O., Rodriguez, B., Fernandez-Gadea, F. and Martin-Lomas, M., *Phytochemistry*, **1984**, *23*, 843-848.
[66] Savona, G., Bruno, M., Paternostro, M., Marco, J.L. and Rodriguez, B., *Phytochemistry*, **1982**, *21*, 2563-2566.
[67] Atta-ur-Rahman and Ahmad, S., *Phytochemistry*, **1988**, *27*, 1882-1884.
[68] Bruno, M., Omar, A.A., Perales, A., Piozzi, F., Rodriguez, B., Savona, G. and de la Torre, M.C., *Phytochemistry*, **1991**, *30*, 275-282.
[69] Topçu, G., Eris, C., Che.C.-T. and Ulubelen, A., *Phytochemistry*, **1996**, *42*, 775-778.
[70] de la Torre, M.C., Bruno, M., Piozzi, F., Savona, G., Rodriguez, B. and Omar, A.A., *Phytochemistry*, **1991**, *30*, 1603-1606.
[71] Carreiras, M.C., Rodriguez, B., de la Torre, M.C., Perales, A., Torres, M.R., Savona, G. and Piozzi, F., *Tetrahedron*, **1990**, *46*, 847-860.
[72] Miyase, T., Rüedi, P., Eugster, C.H., *Helv. Chim. Acta*, **1977**, *60*, 2770-2779.
[73] Bruno, M., de la Torre, M.C., Savona, G., Piozzi, F. and Rodriguez, B., *Phytochemistry*, **1990**, *29*, 2710-2712.
[74] Cuadrado, M.J.S., Bruno, M., de la Torre, M.C., Piozzi, F., Savona, G. and Rodriguez, B., *Phytochemistry*, **1992**, *31*, 1679-1701.

[75] Ulubelen, A., Topçu, G. and Ölçal, S., *Phytochemistry*, **1994**, *37*, 1371-1375.
[76] Ghiglione, C., Lemordant, D. and Gast, M., Plantes Medicinales et *Phytotherapie*, **1976**, *X*, 221-232.
[77] Savona, G., Piozzi, F., de la Torre, M.C., Servettaz, O. and Rodriguez, B., *Phytochemistry*, **1987**, *26*, 571-572.
[78] Bruno, M., de la Torre, M.C., Rodriguez, B. and Omar, A.A., *Phytochemistry*, **1993**, *34*, 245-247.
[79] Kamel, A., *J. Nat. Prod.*, **1995**, *58*, 428-431.
[80] Fraga, B.M., Hernandez, M.G., Mestres, T., Terrero, D. and Arteaga, J.M., *Phytochemistry*, **1995**, *39*, 617-619.
[81] Passannanti, S., Paternostro, M. and Piozzi, F., *Phytochemistry*, **1983**, *22*, 1044-1045.
[82] Ulubelen, A., Topçu, G. and Kaya, Ü., *Phytochemistry*, **1994**, *36*, 171-173.
[83] Gaspar, H., Brito Palma, F.M.S., de la Torre, M.C. and Rodriguez, B., *Phytochemistry*, **1996**, *43*, 613-615.
[84] Rizk, A.M., Hammouda, F.M., Rimpler, H. and Kamel, A., *Planta Medica*, **1986**, *52*, 87-88.
[85] Topçu, G., Eris, C., Kurucu, S. and Ulubelen, A., *Tr. J. of Chemistry*, **1996**, *20*, 265-267.
[86] Andary, C., Wylde, R., Heitz, A., Rascol, J.P., Roussel, J.L. and Laffite, C., *Phytochemistry*, **1985**, *24*, 362-364.
[87] Gross, G.-A., Lahloub, M.F., Anklin, C., Schulten, H.-R. and Sticher, O., *Phytochemistry*, **1988**, *27*, 1459-1463.
[88] Wassel, G.M. and Ahmed, S.S., Pharmazie, **1974**, *29*, 351.
[89] Hassan, M.M.A., Muhtadi, F.J. and Al-Badr, A.A., *J. Pharm. Sci.* **1979**, *68*, 800-805.
[90] Vokou, D. and Bessiere, J.M., *J. Nat. Prod.*, **1981**, *48*, 498-499.
[91] Valesco-Negueruela, A. and Perez-Alonso, M.J., *Phytochemistry*, **1990**, *21*, 1165-1169.
[92] Dominguez, X.A., Merijanian, A. and Gonzales, B.I., *Phytochemistry*, **1974**, *13*, 754-755.
[93] Simmonds, M.S.J., Blaney, W.M., Ley, S.V., Savona, G., Bruno, M. and Rodriguez, B., *Phytochemistry*, **1989**, *28*, 1069-1071.
[94] Rodriguez, B., de la Torre, M.C., Perales, A., Malakov, P.Y., Papanov, G.Y., Simmonds, M., S.J. and Blaney, W.M., *Tetrahedron*, **1994**, *50*, 5451-5468.
[95] Bello, R., Barrachina, M.D., Martinez-Cuesta, M.A. and Esplugues, J., *Phytotheraphy Research*, **1995**, *9*, 277-280.

Atta-ur-Rahman (Ed.) *Studies in Natural Products Chemistry, Vol. 23*

ETHNOBOTANICAL USES OF CELASTRACEAE. BIOACTIVE METABOLITES

A.G. GONZÁLEZ[a]*, I. L. BAZZOCCHI[a], L. MOUJIR[b] and I. A. JIMÉNEZ[a]

*aInstituto Universitario de Bio-Orgánica Antonio González, Astrofísico F. Sánchez, 2, 38206, La Laguna, Tenerife, Canary Islands, Spain.
bDepartamento de Microbiología y Biología Celular, Universidad de La Laguna, 38206 La Laguna*

Abstract: Numerous ethnobotanical studies aimed at identifying new bioactive compounds have become recognized tools in the search for new compounds with biological application. The use of ethnobotanical data can give enormous value, by indicating plants with specific folk-medicinal uses which might be likely sources of biologically active chemicals.

The Celastraceae family, commonly known as the bittersweet family, consist of 55 genera and 850 species distributed pantropically, but with a fair number of species in temperate regions. This family contains several species which are said to be useful in folk medicine and several research laboratories intensively researching this family, having been inspired by its broad and varied botanical distribution, the interesting chemical nature of its secondary metabolites, the complexity of the biogenetic processes which produce them, and most of all by the different types of pharmacological actions displayed through preparations of its constituents.

In the course of these research programmes, many characteristic bioactive compounds have been reported. Polyester sesquiterpenes and pyridine-sesquiterpene alkaloids with insect antifeedant or insecticidal properties have been isolated from some species and recently sesquiterpene pyridine alkaloids with immunosuppressive or antitumoral activities have also been described. Diterpene triepoxides with potent antileukemic and immunosupressive activities and triterpenoid quinonemethides, named as "celastroloids", with antibiotic and cytostatic activities, have been isolated from species of the Celastraceae family.

1. INTRODUCTION

Since time immemorial, man has primarily depended on herbs in the treatment of diseases. It is estimated that less than 1% of the plants in developing countries are utilized as medicine, although one third of all world pharmaceuticals are of plant origin [1]. In recent years, however, there has been a reawakening of scientific interest in the fundamental role which plants play in many cultures, even for medicinal purposes.

Ethnobotany is the study of how people of a particular culture and region make use of indigenous plants. Ethnobotanists explore how plants are used for things such as food, shelter, medicine, clothing, hunting, and religious ceremonies.

Ethnomedicine refers to the use of plants by humans as medicines, but it should probably be called ethnobotanical medicine. Traditional medicine is the total sum of all non-mainstream medical practices, usually excluding so-called "Western" medicine. Generally speaking, plants used in various forms of traditional medicine have different potential medicinal values, according to the degree of authentication of the medical practice involved. For example, traditional Chinese, Ayurvedic, Unani and Unani-Tibb medical systems and naturopathy are all based on a theory, formal education and a written documented history. The systems are periodically revised on the basis of experience and current thought. Information on plants used in these systems is probably more reliable than that from other systems, such as traditional healers, *curanderos,* shamans, "witch doctors" and herbalists. In these practices, the apprentice system is practised, whereby information is passed on from person to person, father to son, etc. There is usually no formal educational component to these systems and frequently the information is considered highly secretive and is not documented in writing.

Remedies in traditional medicine consist of formulas prepared from various natural substances, animal and vegetable. The vegetable remedies account for about 90%, thus the name herbalists (Mganga). Great importance is attributed to the proper preparation of all herbal drugs, which are considered ineffective unless pretreated in a prescribed manner. These herbal remedies are either swallowed, rubbed into scarifications, poured into wounds, boiled and inhaled as fumes, splashed onto eyes, smoked in pipes or sniffed as snuff.

In spite of the scepticism of Western-trained physicians and scientists concerning the value of information resulting from any of these systems, all together they are currently serving the primary health-care needs of most of the world's population and this source must not be ignored in any programme of rational drug development starting with plant materials [2].

It is thought that about 80% of the 5200 million people in the world live in under developed countries. The World Health Organisation (WHO) estimates that about 80% of these people rely almost exclusively on traditional medicine for their primary health-care needs. Since medicinal plants are the "backbone" of traditional medicine, this means that more than 3300 million people in the under developed countries utilize medicinal plants on a regular basis. Thus, for this segment of the world population, who are generally unable to pay the cost of "Western" drugs, there is a need to study these plants for safety and efficacy and to develop Galenical products that are standardized and stable. On the other hand, from a developed country's point of view, plants are chemical factories that produce a vast array of unusual chemical structures that display a variety of biological activities.

A renewed interest in the systematic study of indigenous medicines and associated medicinal plants came about in the 1970s. This was a worldwide phenomenon encouraged by the action of the WHO, which was stimulated by the impact of the "Chinese experience" on the Western medical world at that time. The 1990's has seen a growing shift of interest once more; plants are reemerging as a significant source of new pharmaceuticals. Industries are now interested in exploring parts of the world where plant medicine remains the predominant form of dealing with illness. South America, for example, has an extraordinary diversity of plant species and has been regarded as a treasure trove of medicinal plants. The jungles and rain forests of South America contain an incredibly diverse number of plant species, many still unexplored, many unique and potentially useful as medicinal sources.

Nevertheless, the intensification of ethnobotanical field research is urgently needed because of acculturation in most areas where people of primitive societies are living and because of the increasing devastation and destruction of rain forests in the tropics. From both points of view, an incredible chemical laboratory still practically untouched by scientific investigation for new therapeutic agents is awaiting urgent attention.

Many researchers in different parts of the world are actively involved in recording the available information on traditional medicine. To contribute to this vast task we present in this review a compilation of the ethnomedical uses of the species of the Celastraceae family, recorded from the NAPRALERT data base up to September 1997.

This review also covers the literature included in Chemical Abstracts up to September 1997 and contains comprehensive discussions on biological activities of the metabolites isolated from species of Celastraceae.

2. TAXONOMIC CONSIDERATIONS

The Celastraceae family, commonly known as the bittersweet family, consists of 55 genera and 850 species distributed pantropically, but with a fair number of species in temperate or temperate-cold climates. *Maytenus* (225 species), *Euonymus* (200 species), *Cassine* (40 species), and *Celastrus* (30 species) are the largest genera [3, 4]. The family Hippocrateaceae consists of only two genera, *Salacia* (200 species) and *Hippocratea* (100 species) with species widespread in tropical regions [4]. The taxonomic position of the family Hippocrateaceae is presently being debated. Some taxonomists would combine Hippocrateaceae with the Celastraceae whilst others maintain that it should be considered a family distinct from Celastraceae. Robson argues that the Hippocrateaceae are not a natural group, but derived in two separate lines from the Celastraceae [5]. Heywood has submerged all species of Hippocrateaceae into the family Celastraceae [3]. According to Hegnauer the amalgamation of Hippocrateaceae into Celastraceae is justified by the presence of some common chemotaxonomic markers such as dulcitol, polyisoprene and triterpenoid quinonemethides [6]. A report by Ravelo *et al.* has pointed out some chemotaxonomic relationships between the families Celastraceae and Lamiaceae [7].

The Celastraceae are a family of trees and shrubs, many of which are climbing (e.g. *Salacia*) or twining (e.g. *Hippocratea*) in habit. The classification at the generic level has undergone changes due to hybridization [8], e.g. genus *Maytenus* now includes the species formerly placed in *Gymnosporia*. Similarly, some *Rhacoma* species are now included in the genus *Crossopetalum*, and this has resulted in many species having several synonyms. Two species endemic to the Macaronesia, *Maytenus canariensis* (Loes) Kunk et Sund and *Maytenus umbellata* (R Br) Mabb found in Madeira have at some time been considered as belonging to the genus *Catha* Forsk, among others, but present opinion places them in *Maytenus* Feuill, leaving *Catha* Forsk as a one-species genus, namely, *Catha edulis* (Vahl) Forsk. ex Endl. There is chemotaxonomic evidence connecting these species [9]. González *et al.* argued that the differences and similarities noted in the chemical content of the dihydro-β-agarofuran sesquiterpenes of four *Maytenus* species (*M. chubutensis*, *M. disticha*, *M. magellanica* and *M. boaria*) from Chile are in line with the taxonomic characterization of these species [10].

3. ETHNOMEDICAL USES OF CELASTRACEAE

The family Celastraceae contains several species with claims that they are useful in medicine and folk agriculture, especially in Asia and Latin America but also in other continents.

The ethnomedical uses of Celastraceae species, recorded from the NAPRALERT data base up to September 1997, are listed in Table 1. The species are arranged alphabetically and it is included the botanical names, local names, plant parts, popular uses and form of remedy are included.

Table 1.   Ethnomedical Uses of Celastraceae Species

Species	Local name	Part used	Form of remedy	Medicinal uses	Ref.
Bhesa paniculata Arn	Balam	leaves	pounded in water and the juice is drunk	headache	11
		bark	decoction is drunk	diarrhoea, to treat vomiting	12
Cassine glauca Kuntze	Karvi	leaves	made into pills, and one pill is given twice daily, for 45 days	asthma	13

(Table 1). contd.....

Species	Local name	Part used	Form of remedy	Medicinal uses	Ref.
Cassine transvaalensis (Burtt Davy) Codd	Mulumanamana= Mukuvhazwihi	bark	infusion is drunk twice per day maceration of the powder is drunk[a]	laxative, stomachache,dysmeno rrhea kidney and bladder complaints, diuretic, kidney calculi, venereal diseases	14
		bark and root	decoction is drunk	laxative	
		root	maceration of the powder is drunk	stomachaches in males, laxative	
			infusion is drunk	diarrhoea, cough	
Catha edulis (Vahl) Forsk.ex Endl.	Flower of paradise = Khat = Abyssinian tea	leaves	chewed decoction is drunk infusion is drunk	stimulant-narcotic, stimulant aphrodisiac, euphoriant stimulant, astringent epilepsy, coughs, hysteria, asthma, stimulant	15 16 17 15
		leaves and twigs	chewed	to overcome fatigue, to aid digestion, anaphrodisiac, dysentery, blenorrhea, biliousness	15
		leaves and branches	decoction is drunk	stimulant, urinary diseases,astringent	18
			infusion is drunk	coughs, asthma	
		leaves and root	decoction is drunk	influenza	15
		stem	chewed	stimulant, stimulant-narcotic aphrodisiac	16, 17
		root	decoction is drunk chewed	stomach troubles stimulant-narcotic, aphrodisiac	15, 17 16
		root and bark	decoction is drunk	gonorrhoea, general body illness	15
		entire plant	oral	stimulant, malaria, stomach troubles, coughs	19

(Table 1). contd.....

Species	Local name	Part used	Form of remedy	Medicinal uses	Ref.
Celastrus angulatus Maxim.		rootbark	hot water/oral	antipyretic, arthritis	20
*Celastrus glaucophyllus*Re hd. et Wils.		rootbark	decoction is drunk	for antiwandering arthritis,antipyretic	21
Celastrus paniculatus Willd.	Jotismati = Malkangni = Lakhan	leaves	hot water/oral made into a paste and applied	emmenagogue for cuts and wounds	22, 23
			decoction is drunk	common fever	24
		leaves and seed	decoction is drunk	abortifacient	25
		stembark	decoction is drunk twice a day for 3 days	abortifacient	26
		bark juice	oral[a]	for indigestion	27
		fruit	the paste is mixed with warmed mustard oil and applied	rheumatism, to treat gout	28
		seed	decoction is drunk	mental disorders, aphrodisiac	29
			made into a paste and applied[a]	for burns	30
			taken with ghee and sugar	tonic, help in survival of children	31
			roasted in butter oil and made into pills[a]	rheumatism	23
		seed oil	oral	tranquilizer, to treat psychiatric cases, tuberculosis	32 28
			internal and external remedy	local rheumatism, gout, paralysis to increase the intellect and memory power	31
			external	for body pain, rheumatism,scabies	28

(Table 1). contd.....

Species	Local name	Part used	Form of remedy	Medicinal uses	Ref.
Celastrus scandens L.	Bittersweet= StaffvineWaxwork	bark	oral	scrofula, syphilis, diuretic,diaphoretic, to treat obstruction of the menses	33
		rootbark	external	cancer and tumors	34
			hot water/oral	to induce menstrual flow,alterative, diuretic	34 35
Elaeodendrom buchanani Loes.	Eukanda = Mhakumo = Sunwa	rootbark	decoction/external	wounds, syphilis, diarrhoea	36
Elaeodendrom capense	Murmanyama	root	decoction is drunk	sterility in males	37
Elaeodendrom glaucum (Rottb.) Pers.	Muskaka	part not specified	external[a]	to treat poison arrow wounds	38
Elaeodendrom schweinfurthianum Loes.	Mnenekanda	leaves	chewed	diarrhoea	39
		rootbark	decoction is drunk	to treat menstrual disorders	39
		root	decoction/oral decoction mixed with milk is drunk	syphilis for coughing blood	39
Euonymus alatus (Thunb.) Makino	Komayum	twigs	hot water /oral	emmenagogue	40
		wood	infusion is drunk	stomachache, vermicide	41
		stem	hot water/oral	abortifacient, emmenagogue	42
Euonymus atropurpureus Jacq.	Sitterash = Burning bush = Wahoo	bark	hot water/oral	tonic, hydrogogue cathartic,cholagogue, diuretic	43
		stembark	hot water/oral	tonic, laxative, diuretic	44
		rootbark	tincture	cholagogue	45
Euonymus europaeus L.	Beresklet = Bonnet de Pretre	bark	decoction/external	for scab	46

(Table 1). contd.....

Species	Local name	Part used	Form of remedy	Medicinal uses	Ref.
			hot water/oral	laxative, tonic, alterative, diuretic, expectorant	35
		fruit	oral	emetic, purgative	45
		seed	decoction/oral	mouthwash	47
			powder	antiparasitic	48
Euonymus verrucosus Scop.	Sanguanella	leaves	decoction is drunk	tonic	49
		seed	oral	antiparasitic for domestic animals	49
		seed flour	spread and massaged on children's scalps with vinegar	to treat head lice	49
Gymnosporia montana (Roth) Benth	Baikal = Vikankata	leaves	powder is mixed with mustard oil and rubbed on the back	for rickets	50
			decoction is drunk	gastrointestinal disorders ulcers, dysentery, toothache to treat jaundice, hepatoprotective	51
Kokoona zeylanica Thwaites	"kokum soap"	bark	powdered and snuff made into a paste with water and used as a toilet soap	headache to improve the complexion of the skin, leech repellent	52
Maytenus acuminata Loes	Umzungulwa Umnama = Inqayi	root	decoction is drunk twice a day	stomach upset	53
Maytenus aquifolium Martius	Cangarosa	leaves	infusion is drunk	ulcers	54
Maytenus arbustifolia		stembark	infusion is drunk	fevers, malaria	55
Maytenus blepharodes (Pitt.) A. Gentry		branches	not specified	antitumor	56
Maytenus boaria Mol.	Maiten = Maghtun	aerial part	infusion/external infusion is drunk	skin eruptions, purgative febrifuge	57

(Table 1). contd.....

Species	Local name	Part used	Form of remedy	Medicinal uses	Ref.
Maytenus buxifolia Griseb.	Granny bush = Gripe-bush = Smoke-bush	leaves	decoction is drunk	fever, extended menstruation	58
		bark, leaves and twigs	hot water/drunk 1 quart/day for 5 days	abortifacient up to 3 months pregnant	59
Maytenus canariensis (Loes.) Kunk. et Sund.	Peralillo=Arbol negro	fruit juice	external	papillomas	60
		leaves	hot water/oral	cutaneous abscesses, antirheumatic	60
			chewed	to ward off fatigue	61
Maytenus chuchuasca Raymond-Hamet et Colas	Chuchuhuanso = Curicaspe = Xuxua	leaves	hot water/oral	aphrodisiac	62
		bark	external infusion	skin cancer rheumatism	63
Maytenus ebenifolia Reiss	Chuchuhuasi	bark	decoction is added to the ayahuasca beverage and drunk	hallucinogen	64
		stembark	decoction is drunk	antiinflammatory, antitumor rheumatism	65
			tincture/oral		
Maytenus emarginata (Willd.) Hou	Amti = Henkel	fruit	oral	to purify the blood	31
		root	paste is applied externally	sprains and pulled muscles	66
		entire plant	decoction is drunk	cancer	67
Maytenus heterophyla (Eckl. Zeyh.) N. Robs.	Omuseka= Ndegamau	leaves	not specified	dysmenorrhoea	
		root	decoction is drunk	antihelmintic, epilepsy, hernia, syphilis	68
Maytenus ilicifolia Mart.	Cangorosa = Yuyra = Rapo Ju	leaves	oral	sialogogue, antiasthmatic,antiseptic, vulnerary	69
				emmenagogue	70

(Table 1). contd.....

Species	Local name	Part used	Form of remedy	Medicinal uses	Ref.
			internal	digestion, antispasmodic, asthma, contraceptive, astrigent	71
			external	antiseptic, vulnerary	
			decoction is drunk	diarrhoea, respiratory and urinary tract infections	72
		stembark	hot water/oral[a]	abortifacient, emmenagogue	70
		flowers	decoction is drunk	antiinflammatory	73
		rootbark	decoction/oral	fertility-regulation	74
		entire plant	decoction is drunk	long-term contraceptiveantiasthmatic, sialogogue,	75
			hot water/oral	antiseptic, vulnerary, antitumor, fertility regulation	76
			infusion is drunk	menstrual inducer	
Maytenus krokovii A. C. Smith	Chuchuasi= Chucchu huashu	stembark	infusion/external	skin cancer	77
Maytenus laevis Reiss.	Chuchuasha = Curi-caspi	rootbark	soaked in aguardiente	rheumatism, aphrodisiac	78
Maytenus obscura	Muthuthi	leaves	decoction mixed with soup and drunk	cancer	36
Maytenus peduncularis (Sonder) Loes.	Mukwatule	root	decoction is drunk 4 times a day	backache (males only)	14
Maytenus phyllanthoides Beth.	Mangle rojo	bark	decoction is drunk / decoction/external	blood circulation, stomachache / wounds	79
Maytenus putterlic-kioides (Loes.)Exell.& Mendoça	Muthunthi = Shomafisi = Mtulavuha	leaves	hot water/oral	hook worm infestations	36

(Table 1). contd.....

Species	Local name	Part used	Form of remedy	Medicinal uses	Ref.
		root	decoction is drunk[a]	irregular menstruation, hernia, swollen testicles, aphrodisiac, dysmenorrhoea	68
			cooked with soup	laziness, internal body injuries	
Maytenus senegalensis (Lam.) Exell.	Bademi = Tshiphandwa = Mnyabuliko	leaves	powdered are taken in milk juice female sterility is drunk[a]	vermifuge	80 81
		bark	crushed bark is mixed with water and drunk	cancer	82
		stembark and root bark	decoction/oral	antimalarial	83
		root	hot water/oral hot water/external decoction is drunk or 5 days decoction[a]	antiseptic, gonorrhoea, laxative, dysentery, diuretic, cicatrizant, wounds cough menorrhagia	84 80 14 85
		stembark and rootbark	decoction/oral	antimalarial	83
Maytenus vitis-idaea Griseb.		leaves	decoction /oral	diarrhoea, respiratory and urinary tract infections	72
Mystroxylon aethiopium Eckl. et Zeyh.	Fana zava	entire plant	hot water/oral[a]	oxytocic	86
Pristimera grahamii A. C. Smith		leaves root	hot water/oral hot water/oral	chronic cough, bodyaches, joint pain, urinary disorders,skin diseases antibiotic	87
Shaefferia cuneifolia A. Gray	Capul	root	hot water/oral	venereal diseases	88

(Table 1). contd.....

Species	Local name	Part used	Form of remedy	Medicinal uses	Ref.
Simerestis welwitschii N. Halle		entire plant	hot water/oral	oxytocic	89
Tripterygium wilfordii Hook	Chi Hsueh Teng = Lei-Gong-Teng = Thunder god vine	root	dry roots are immersed in spirits and the filtered is drunk not specified	rheumatoid arthritis, ankilosing spondylitis leprosy	90, 91 92
		rootwood	decoction is drunk	to relieve joint pain autoimmune diseases	93 94
		fruit	hot water/oral	cancer	95

([a]Effect are described for a multi-components).

Catha edulis Forsk. (Khat) is an evergreen shrub which grows along the eastern coast of Africa and the Arabian peninsula. The leaves and stem tips of Khat are chewed for their stimulant effects and this has become a kind of social institution. The chewing usually takes place in parties with special patterns, where friends gather after work. The consumers get a feeling of well-being, mental alertness and excitement; the after effects are usually insomnia, numbness and lack of concentration. The excessive use of Khat may create considerable problems of social, health and economic nature. A multidisciplinary historical overview of the use, abuse and sociology of *C. edulis* was reviewed by Krikorian [15].

An important step towards identification of *C. edulis* constituents was the discovery of (+)-norpseudoephedrine (cathine) in dried leaves by Wolfes in 1930 and during several decades this alkaloid was thought to be the main active principle of *C. edulis* [96]. However, as already pointed out by Brucke (1941), (+)-norpseudoephedrine is CNS stimulant of high potency and the amount of this substance that was present in a portion of *C. edulis* was insufficient to account for the symptoms observed after consumption of the material [97]. Subsequently, in the course of a reinvestigation of the constituents of *C. edulis* leaf, the United Nations Narcotics Laboratory identified this compound as α-aminopropiophenone and the name (-)-cathinone was proposed for the new alkaloid [98]. Since then, several researches on *C. edulis* constituents and its pharmacological effects have been recorded [99, 100].

Tripterygium wilfordii Hook is a perennial twining vine which is cultivated in many parts of southern China such as Zhejiang, Anhui, Jiangxi, Fujian, and Guangdong provinces and also in Taiwan. The herb is commonly known in China as Lei Gong Teng (thunder god vine) or Mang Cao (rank grass).

The powdered roots of. *T. wilfordii* have been used in China for centuries as an insecticide. In 1930 the "thunder god wine" suddenly came into prominence in Chekiang province, China, as the result of a dispute regarding the damage to valuable and highly productive valley lands caused by rain water from the nearby hills following the harvest of *Tripterygium* roots, which left the soil loose, hence flooding the valley. The land owners of the valley had requested that the cultivation of *T. wilfordii* be forbidden. This quickly provoked a strong opposition from vegetable growers who complained that they could not grow their crops without using the powdered roots of *T. wilfordii* to kill noxious insects. Entomologists and other experts were sent to investigate the validity of this claim. As a result of these investigations, Chou and Mei published the first paper on the root constituents of *T. wilfordii* in 1936 [101]. They were able to isolate the insecticidal principle (suspected to be an alkaloid) and dulcitol along with an insecticidally inert pigment named tripterine. Tripterine was later shown to be identical to the celastrol obtained by Gisvold from another Celastraceae species, *Celastrus scandens* [102]. Tests made in 1935 demonstrated that *Bombyx mori* and *Malacosoma americana* larvae were deterred from feeding on leaves dusted with the processed plant [103]. Later, the alkaloid wilfordine and the triterpene celastrol were isolated from this plant and were considered responsible for its insecticidal activity [104-107].

In 1972, Kupchan *et al.* isolated the novel diterpenoid triepoxides, triptolide, tripdiolide, and triptonide from the roots of *T. wilfordii* with strong antileukemic activity [108]. Most recently, crude extracts and refined extracts (a so-called multi-glycoside extract, or GTW) have been used increasingly in clinical treatment of rheumatoid arthritis, ankylosing spondylitis, and a variety of dermatological disorders [90, 91, 109].

In several areas of the subandean rain forest in the Amazonian river basin, mainly in Peru, Ecuador and Colombia, a plant known as "chuchuhuasha" or "chuchuhaso" is used by several tribes: Pastazas, Ticunas, Boras, Huitotos, Pebas, Yahuas, Cocamas, Arguarunas and Jívaros. The powdered red root bark is used as an alcoholic infusion, generally in "aguardiente", as a general tonic, for the treatment of rheumatism and even as an aphrodisiac. For topical use it is employed as an antitumoral agent in skin cancer and also for the treatment of sores [78].

A large ethnobotanical literature has flourished on *chuchuhuasha* and many different plants have received this name, giving rise to certain confusion. It is usually agreed that the plants used belong to the genus *Maytenus* [110]. *Maytenus* species considered to be used as *chuchuhuasha* are *M. chuchuhuasha*, *M. krukovii*, *M. colasii* (*Salacia colasii*), and *M. laevis*. The name *chuchuhuasha* is attributed in Peru to *Heisteria pallida*. Extracts of *Maytenus ilicifolia* have been used in Guaraní folk medicine for birth control purposes [75, 111]. The practice of drinking infusions of *chuchuhuasha* is still common in the above regions. *M. laevis* has been reported in the traditional medicine of Colombia as an anti-arthritic. Extracts of trunk bark demonstrated the anti-inflammatory activity of this species [112].

The leaves of this species have also been of interest in tests for radioprotection in a comparative study with the increase in phagocitosis of the reticular endothelial system [113]. More recently, *Maytenus* species have attracted a certain interest because of the particular properties, such as antitumoral and contraceptive ones of the extracts. For example, antitumor activity of a number of phenoldienone triterpenes isolated from *Maytenus* sp. has been shown [114] and it was demonstrated that they inhibit the synthesis of proteins and of uridine incorporation into DNA [115].

A very high antitumor activity was earlier found in some extracts of African *Maytenus* species (*M. ovata* Loes, *M. buchananii*). This promoted a thorough investigation for the active principle, maytansine, which was shown to be present in minute amounts [116]. This product is structurally related to ansamycins and it is uncommon for a secondary plant metabolite to be related to metabolites from fungi.

4.   BIOLOGICAL ACTIVE CONSTITUENTS OF CELASTRACEAE

The Celastraceae family took on a new lease of life in the seventies when the maytansinoids, compounds with exceptional antitumoral properties, were discovered [116]. Nonetheless, the maytansinoids have not been made into a useful drug form as they cause serious gastro-intestinal damage when applied to rats [117].

For some time now several research laboratories have been intensively researching this family, inspired by its broad and varied botanical distribution, the interesting chemical nature of its secondary metabolites, the complexity of the biogenetic processes which produce them, and most of all by the different types of pharmacological action displayed by preparations of its constituents.

In the course of these research programmes, many characteristic bioactive compounds have been reported. Polyester sesquiterpenes and pyridine-sesquiterpene alkaloids with insect antifeedant or insecticidal properties have been isolated from some species and recently sesquiterpene pyridine alkaloids with immunosuppressive or antitumoral activities have also been described. Diterpene triepoxides with potent antileukemic and immunosuppressive activities and triterpenoid quinonemethides, named as "celastroloids", with antibiotic and cytostatic activities, have been isolated from species of Celastraceae family.

The next pages are an update on the state of the active constituents of species of the Celastraceae family detailing, the results of biological activities carried out since the last publication on the subject [6, 118].

4.1. Sesquiterpenes

The Celastraceae family is a rich source of sesquiterpene esters based on the dihydro-β-agarofuran [5,11-epoxy-5β,10α-eudesm-4(14)-ene] skeleton, and they are considered to be chemotaxonomic indicators of the family [6]. X-ray data and a conformational study by molecular mechanics showed that in every instance with only slight variations, the *trans*-fused A and B rings formed a chair-chair decalin system, slightly distorted by the presence of the 1,3-diaxial bond responsible for the tetrahydrofuran C ring, practically perpendicular to the plain formed by carbons C-5, C-7, C-8 and C-10, Fig. (1).

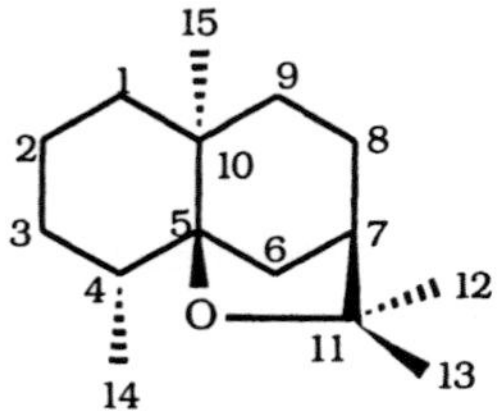

Fig. (1). Basic dihydro-β-agarofuran skeleton

These sesquiterpene polyesters are found in a variety of oxygenated forms, being esterified by either aliphatic (acetic, 2-methylbutyric, hydroxyisobutyric, etc.) or aromatic (benzoic, furoic, cinnamic, epoxycinnamic, etc.) acids. Despite their sesquiterpene core, when the ester function is due to nicotinic acid or its derivatives, they are termed sesquiterpene pyridine alkaloids [119].

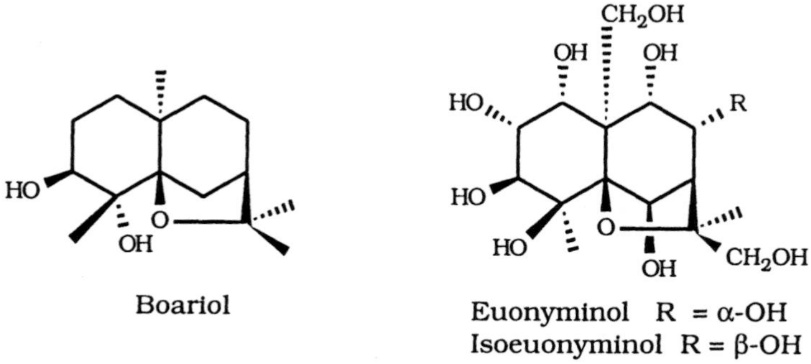

Boariol

Euonyminol R = α-OH
Isoeuonyminol R = β-OH

Fig. (2). Examples of polihydroxy dihydro-β-agarroduran sesquiterpenes skeletons

The basic polyhydroxy skeleton varies according to the position, number and configuration of the ester residues in the dihydro-β-agarofuran sesquiterpene. The complexity and increasing numbers of these sesquiterpenes makes it difficult to arrange them systematically. They can, however, be treated as derivatives of a basic polyhydroxy skeleton and therefore organized in simpler series. Accordingly, more than forty series of dihydro-β-agarofuran type sesquiterpenes have been proposed ranging from a skeleton with two hydroxyl groups (boariol) [120] to one with nine hydroxyl groups (the euonyminol and isoueonyminol series) [121], Fig. (2).

In the structural elucidation of these sesquiterpenes, the difficulty of determining the linking sites of the respective ester groups when more than three kinds of acids are involved as esters in the molecule, necessitates the use of X-ray crystallographic methods, or selective hydrolyses. Nevertheless, determination of the ester sites in such compounds has been unsuccessful in many cases. However, this problem can be solved conveniently using 2D-NMR, including [1]H-^{13}C long-range correlation (HMBC). The absolute configuration of such compounds has been resolved in many cases by circular dichroism studies using the benzoate chirality method, which is an extension of the CD exciton chirality method [122].

Since Smith's work with X-rays [123] and Lotler's corrections [124], all recent publications have agreed in ascribing an *S* configuration to

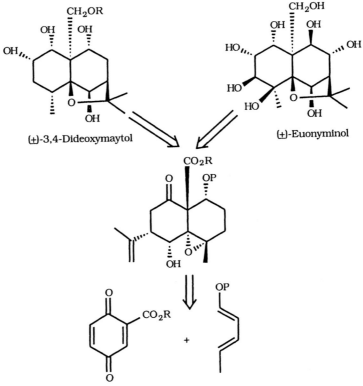

Fig. (3). Total synthesis of 3,4 dideoxymaytol and euonyminol

position C-4 in this sort of sesquiterpenes. However, boariol isolated from *Maytenus boaria* Mol. by González and coworkers [120], does not agree with the classic model of the sesquiterpenes previously described, and is in fact the simplest of all the compounds recorded from the Celastraceae; the presence in the molecule of a secondary and a tertiary hydroxyl group, the latter at C-4 but with the opposite configuration to the usual α-hydroxyl at this position and the absence of substituents at C-1, another notable feature of this structure, casts doubts on the biogenetic theory for the dihydro-β-agarofuran sesquiterpenes of this family, which presumes that such substituents are present in nature.

In spite of the comprehensive efforts leading to the structural elucidation, however, only a few synthetic routes to these complex sesquiterpenoids have been reported. Recently, White *et al.* [125] devised the total syntheses of (±)-euonyminol and (±)-3,4-dideoxymaytol using a strategy based on a "cascade cyclization" to establish the dihydroagarofuran framework of these structures, Fig. (3).

The maytolins are an interesting instance of new sesquiterpenes from the Celastraceae characterized by the presence of a tetrahydro-oxepine nucleus. It would seem that these new types of skeleton are only biosynthesized by species of the *Mortonia* genus, which consists of just four species, endemic to Mexico and the southern United States. The structures proposed for mortonins A and B, Fig. (4), are the first recorded examples of a natural product in which ring B of the eudesmane skeleton undergoes oxidative cleavage to the the γ-lactone [126].

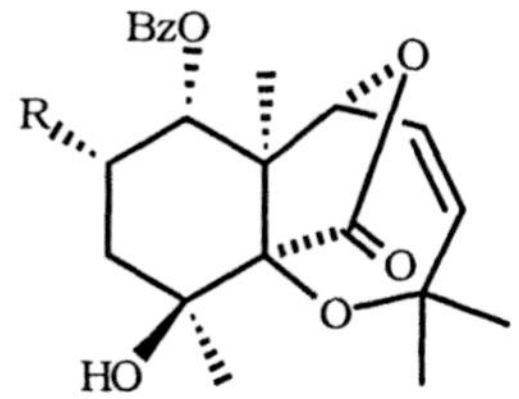

64 Mortonin A R = H
65 Mortonin B R = Ac

Fig. (4). Sesquiterpenes from the genus *Mortonia*

The interest generated by polyester sesquiterpenes from the Celastraceae has increased in line with the complexity of the substances isolated and the possibility of their being applied to combat insect plagues instead of synthetic insecticides. These compounds have also attracted a great deal of interest on account of their antitumor-promotion activity.

4.1.1. Antifeedant and Insecticidal Activities

The Chinese bittersweet *Celastrus angulatus* Max., is widely distributed in the People's Republic of China and has been used as a traditional insecticidal plant to protect other plants from insect damage. The plant is finely ground, suspended in water and sprayed over the crop to be protected [127].

The insect antifeedant activity of root bark powders and ethanol extracts of *C. angulatus* against the cruciferous leaf beetle, *Colaphellus bowringi* was recorded in 1936 and since then a body of work on the activity of the extract against several insect species has been carried out [128]. This finding prompted several scientific groups to initiate an extensive research on this species to extract the active components.

The first scientific paper dealing with a non-alkaloid sesquiterpene polyol ester showing insect antifeedant activity was published by Wakabayashi *et al.* [128], who reported that celangulin (1), isolated from the ethyl ether extract of Chinese bittersweet root bark, exhibited activity against *Spodoptera frugiperda* (Noctuidae, Lepidoptera) larvae. Thus, addition of the compound in the diet at a concentration of 5 and 10 ppm, the average body weight of larvae was reduced to 61% and 37%, respectively.

Since celangulin was reported to present insect antifeedant activity, a number of sesquiterpenes from *C. angulatus* were described as insect antifeedant. Thus, celangulins II (2), III (3), and IV (4), were isolated by activity-guided fractionation and they showed strong narcotic or insecticidal action against the armyworm, *Mythimna separata*, with values of KD_{50} (the dose required to knock down 50% of the population) of 46 and 260 µg/g for 2 and 4, respectively, and a LD_{50} (the dose required to kill 50% of the population) for 3 of 110 µg/g [129], Fig. (5).

Liu *et al.* [130] reported four new sesquiterpene alkaloids (5-8), 5 and 6 being the first naturally occurring compounds having at C-15 an isobutyryl and a 2-methylbutyryl groups, respectively. These compounds exhibited strong antifeedant activity when tested against *Pieris rapae*, *Ostrina furnacolis* and *Tribolium castaneum*. It was observed that the insects were paralysed for many hours after ingesting a small dose of the test sample, the insects then recovered, fed and became paralysed again and as a result, the insects gradually starved to death. Angulatin A (9), also isolated from *C. angulatus*, had strong insecticidal and antifeedant effects against *Heliothis armigera*, *Aphis gossypii*, *P. rapae* and *Brevicoryne brassicae* [131]. An insect antifeedant screening showed that bittersweet seed oil also possesses biological activity. From this a new sesquiterpene, angulatueoid G (10), was reported to show an insect antifeedant effect against *Aulacophora femoralis* (73.2% antifeedant rate) and *Piutella xylostella* (87.7% antifeedant rate) at a concentration of 100 ppm. This compound, with two benzoyl ester groups at C-8β and C-9β in the structure, was the first example of sesquiterpenes belonging to this family [132], Fig. (**5**).

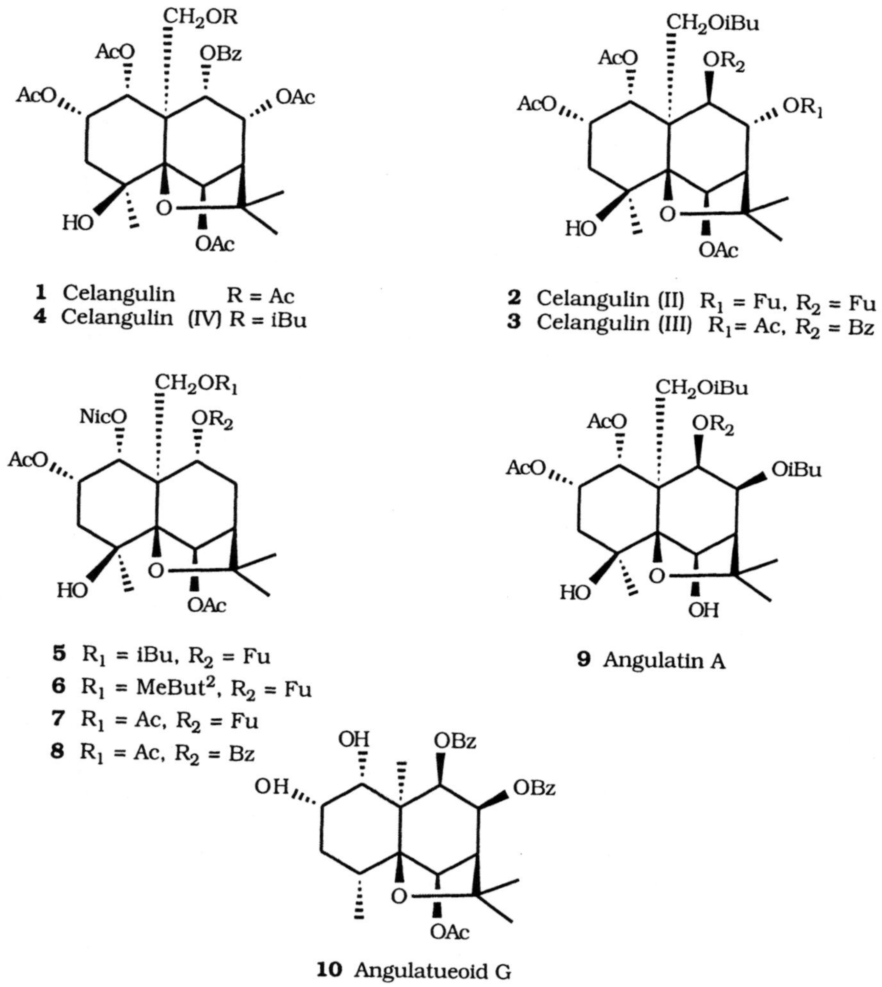
CH2OR
AcO
OBz
AcO
OAc
HO
O
OAc
1 Celangulin R = Ac
4 Celangulin (IV) R = iBu
CH2OiBu
AcO
OR2
AcO
OR1
HO
O
OAc
2 Celangulin (II) R1 = Fu, R2 = Fu
3 Celangulin (III) R1= Ac, R2 = Bz
CH2OR1
NicO
OR2
AcO
HO
O
OAc
5 R1 = iBu, R2 = Fu
6 R1 = MeBut2, R2 = Fu
7 R1 = Ac, R2 = Fu
8 R1 = Ac, R2 = Bz
CH2OiBu
AcO
OR2
AcO
OiBu
HO
O
OH
9 Angulatin A
OH
OBz
OH
OBz
O
OAc
10 Angulatueoid G

Fig. (5). Insect antifeedant and insecticidal sesquitepenes from *Celastrus angulatus*

In the course of a search for novel insect antifeedant and insecticidal compounds, Tu *et al.* isolated compound **11** (50% death rate) from *Celastrus gemmatus* Loes and compound **12** (49% antifeedant rate) from *Celastrus rosthornianus* Loes, which showed effects against the larvae of *P. rapae* [133, 134]. *Celastrus flagellaris* have also yielded sesquiterpenes with insect antifeedant activity and a preliminary test of the Me_2CO solutions of compounds **13** and **14** (500 mg/l) against *M. separata*

exhibited 93.0% and 78.6% antifeedant activity, respectively [135], Fig.
(6).

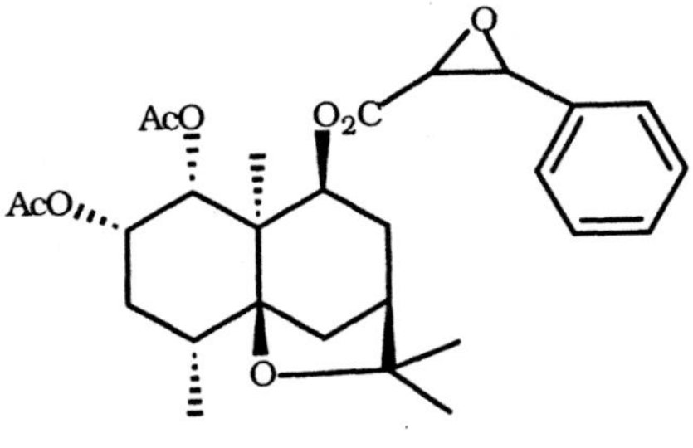

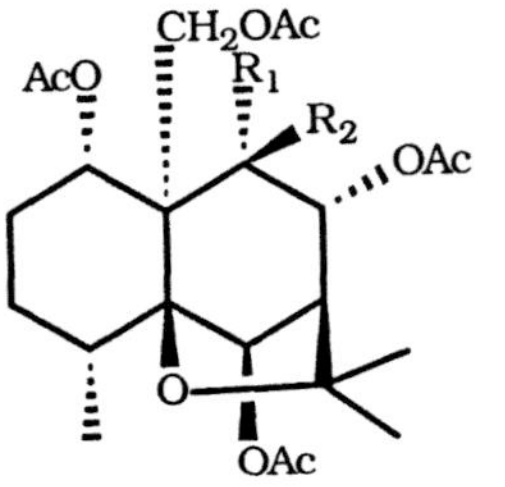

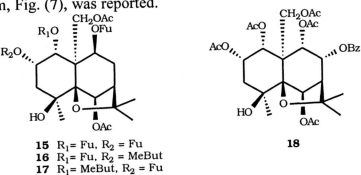

13 R_1= OBz, R_2 = H
14 R_1= H, R_2 = OBz

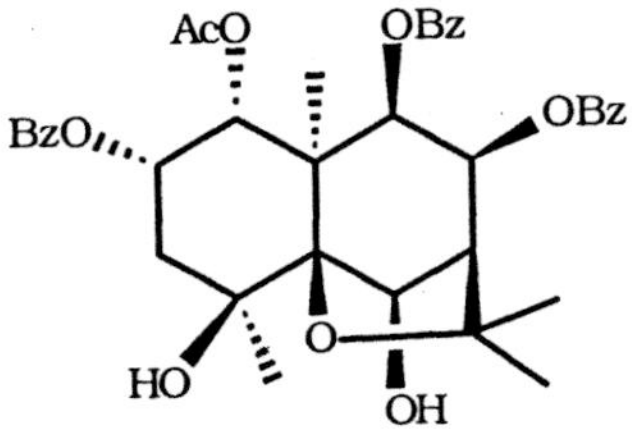

Fig. (6). Insecticidal and antifeedant sesquiterpenes from some *Celastrus* species

Although the use of *Euonymus bungeanus* Max. as a natural insecticide
has not been reported, a preliminary insecticidal test [136] indicated that
sesquiterpene polyol esters 15 and 16, isolated from the seed oil of this
species, exhibited an insect antifeedant effect against *P. rapae* with
antifeedant rates (%) of 42 and 67, respectively at a concentration of 200
ppm, and compounds 16 and 17 exhibited an insecticidal effect against *O.
furnacolis* with values of corrected death rates (%) of 70 and 50,
respectively at 300 ppm. Recently, [137] a new sesquiterpene (18) from
Euonymus fortunei with an antifeedant effect of 58.4% against *P. rapae* at
1000 ppm, Fig. (7), was reported.

15 R_1= Fu, R_2 = Fu
16 R_1= Fu, R_2 = MeBut
17 R_1= MeBut, R_2 = Fu

18

Fig. (7). Insecticidal and antifeedant sesquiterpenes from *Euonymus bungeanus* and *E.
fortunei*

The tree *Elaeodendron buchananii* Loes (Eukanda, Mhakumo, Sunwa, in local dialects) is one of the best known poisonous plants in East Africa. Ingestion of its leaves has resulted in many deaths to domestic stock although, interestingly, wild animals (e.g. giraffes) are apparently unaffected [138]. The root bark of the plant is widely used in local medicine as a remedy for wounds, syphilis and diarrhoea [36]. Tsanuo *et al.* in their search for anti-insect phytochemicals, described the isolation and structure elucidation of a novel sesquiterpene, mutangin (19), from the unripe fruits of *E. buchananii*, with moderate insect antifeedant activity against the lepidopteran, *Chilo partellus*. The following % deterrence were obtained: 64.9 ±5.0 (100 μg/ disc), 54.8 ± 2.8 (50 μg/disc), 9.2 ± 7.0 (25 μg/disc), and 3.1 ±1.8 (12.5 μg/disc) [139], Fig. (8).

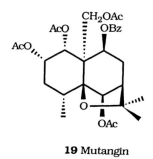

19 Mutangin

Fig. (8). Insect antifeedant sesquiterpenes from *Elaeodendron buchananii*

Maytenus canariensis (Loes) Kunk et Sund, known locally as "peralillo", is an endemic species that grows in the forest areas of the Canary Islands. Shepherds are known to chew the leaves of this species to ward off fatigue [61], in a similar way that *Catha edulis* is used in northeast Africa.

González and coworkers [140-142] reported the isolation of new sesquiterpenes from *M. canariensis* which were assayed for antifeedant and insecticidal activities against the Egyptian cottonleaf worm *Spodoptera littoralis* (Boisduval). In a structure-activity relationship study of the antifeedant and insecticidal activities of sesquiterpenes with a dihydro-β-agarofuran skeleton, forty one sesquiterpenes, isolated from seven species of Celastraceae, and fourteen related synthetic compounds, were evaluated for their activity against *S. littoralis* [143]. Thirty eight of the sesquiterpenes assayed showed antifeedant activity, and two of them demonstrated insecticidal activity. Based on this study the following conclusions were made: the most active compounds were those with isoalatol (20-23) or 4β-hydroxyalatol (24-26) skeletons (Table 2); comparing the activities of compounds with the same skeleton the activity increases with the number of acetate esters and decreases with the number of benzoate esters; products with aromatic esters at C-1 and C-9, whatever their stereochemistry, were only weakly active and the introduction of ester groups at C-2 had no effect on activity. The insecticidal sesquiterpenes (27 and 28) showed an LD_{100} of 0.1 and 1 $\mu g/cm^2$, respectively; the structural characteristic common to these compounds was a carbonyl group at C-8, which indicated that this group is a structural requirement for this activity, Fig. (9). No antifeedant nor insecticidal activity was observed in the racemic and/or optically active tetrahydrofuran products with substituents in C-2 or C-3 and 2-methyl butanoid derivatives. These results allowed the authors to conclude that the compact tricyclic system is a prerequisite for antifeedant or insecticidal activity in this type of sesquiterpene.

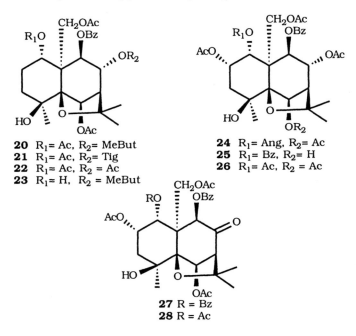

20 R_1= Ac, R_2= MeBut
21 R_1= Ac, R_2= Tig
22 R_1= Ac, R_2 = Ac
23 R_1= H, R_2 = MeBut

24 R_1= Ang, R_2= Ac
25 R_1= Bz, R_2= H
26 R_1= Ac, R_2 = Ac

27 R = Bz
28 R = Ac

Fig. (9). Antifeedant and inseciticidal sesquiterpenes from *Maytenus canariensis*

4.1.2. Antitumor-promotion Activity

Investigations on Epstein-Barr virus (EBV) activation in human lymphoblastoid cells and latently infected Raji cells, revealed a considerable overlap between the EBV activating compounds and the tumor promoters of mouse skin carcinogenesis. In a search for possible antitumor promoters, Takaishi *et al.* [144] carried out a screening of thirty-seven dihydroagarofuran sesquiterpenes isolated from *Trypterigium wilfordii* Hook fil. var. *regelii* Makino and *Euonymus sieboldianus* Blume, using their possible inhibitory effects on the Epstein-Barr virus early antigen (EBV-EA) activation, which is induced by 12-O-tetra-decanoylphorbol-13-acetate (TPA) in Raji cells. The sesquiterpenes, triptofordin F-2 (**29**) and triptogelin A-1 (**30**) and the derivative **31**, were shown to have strong inhibitory activity. The active triptogelin A-1 also exhibited remarkable inhibitory effects on mouse skin tumor promotion in an *in vivo* two-stage carcinogenesis test and it is suggested that it is a valuable new type of compound for chemoprevention and chemotherapy against papilloma and carcinoma [145].

Table 2.   Antifeeedant Activity of Sesquiterpenes from *M. Canariensis* Against *Spodoptera littoralis*

Compound	Dose ($\mu g/cm^2$)	$FR^a_{50} \pm SEM$ [b]
20	10	0
	1	0
	0.1	0.23 ± 0.07
	0.01	0.69 ± 0.08
21	10	0.06 ± 0.03
	1	0.20 ± 0.06
	0.1	0.43 ± 0.06
22	10	0.03 ± 0.01
	1	0.20 ± 0.06
	0.1	0.64 ± 0.07
23	10	0.08 ± 0.03
	1	0.23 ± 0.02
	0.1	0.57 ± 0.09
24	10	0.10 ± 0.03
	1	0.18 ± 0.02
	0.1	1.10 ± 0.02
25	10	0.10 ± 0.06
	1	0.19 ± 0.08
	0.1	1.02 ± 0.07
26	10	0.05 ± 0.02
	1	0.24 ± 0.03
	0.1	0.40 ± 0.08
TTA[c]	10	0.37 ± 0.09

[a] FR_{50}: The ratio when control disk has been 50 % consumed.

[b] Standar error mean.

[c] Triphenyl tin acetate was used as control.

The inhibition of EBV-EA activation was also assayed [146] for three new dihydro-β-agarofuran sesquiterpene polyol esters, celafolins B-2 (**32**), C-1 (**33**) and D-2 (**34**) and four known related compounds (**35-38**), isolated from the seeds of *Celastrus stephanotiifolius* Makino, Fig. (**10**).

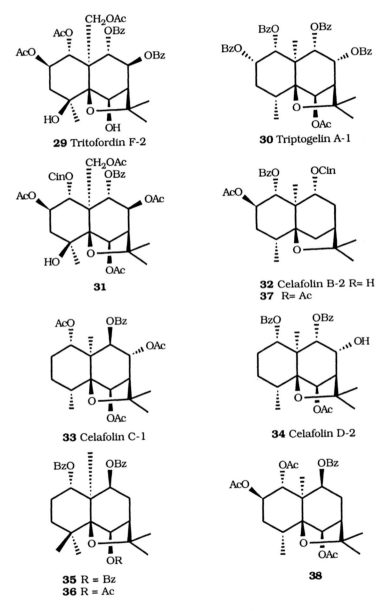

29 Tritofordin F-2

30 Triptogelin A-1

31

32 Celafolin B-2 R= H
37   R= Ac

33 Celafolin C-1

34 Celafolin D-2

35 R = Bz
36 R = Ac

38

Fig. (10). Antitumor promoters sesquiterpenes from Celastraceae species

Compounds **32** and **34** showed stronger inhibitory activities than did other compounds. Their inhibitory effects on the activation of the early antigen and the viabilities of Raji cells are shown in Table 3.

Table 3.   Inhibitory Effects of Dihydroagarofuran Sesquiterpenes on TPA-Induced EBV-EA Activation

Compound	% to control[a] (% viability)[b]		
	concentration (mol ratio/32 pmol TPA)		
	1×10^3	5×10^2	1×10^2
triptofordin F-2 (**29**)	20.7 (80)	41.5 (> 80)	63.3 (> 80)
triptogelin A-1 (**30**)	38.0 (80)	41.6 (> 80)	80.3 (> 80)
31	38.8 (80)	70.2 (> 80)	90.4 (> 80)
celafolins B-2 (**32**)	0 (70)	41.6 (> 80)	86.4 (> 80)
celafolins C-1 (**33**)	15.3 (70)	52.8 (> 80)	90.6 (> 80)
celafolins D-2 (**34**)	0 (60)	39.3 (> 80)	81.1 (> 80)
35	53.6 (70)	84.6 (> 80)	100 (> 80)
36	26.2 (70)	47.9 (> 80)	92.1 (> 80)
37	25.6 (70)	63.1 (> 80)	89.4 (> 80)
38	34.2 (70)	65.9 (> 80)	92.9 (> 80)

[a]Values represent relative percentages to the positive control values (TPA 32 pmol = 100%).

[b] Values in parentheses are viability percentages of Raji cells.

4.2. Macrocyclic Sesquiterpene Alkaloids

The presence of alkaloids in plants of the Celastraceae family has been recognized for more than 50 years [119]. Since the first publication appeared in 1934 on the presence of alkaloids in the plants of this family [147], isolation of alkaloidal components from these plants and their structural studies have been performed by a number of investigators.

This type of polyester sesquiterpenes, is a family of macrolactones that contain a pyridine dicarboxylic acid, such as, evoninic, isoevoninic, wilfordic, isowilfordic, hydroxywilfordic, cassinic, edulinic or cathaic Fig. (**11**), which bridge the positions C_3-C_{13} of the highly functionalized sesquiterpenoid cores: evoninol, euonyminol and isoeounyminol, Fig. (**12**).

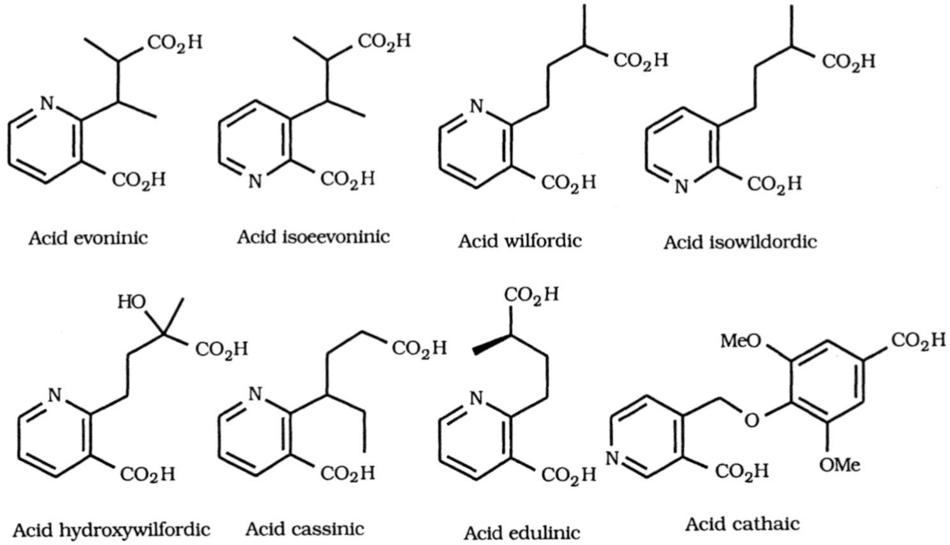

Acid evoninic   Acid isoeevoninic   Acid wilfordic   Acid isowildordic

Acid hydroxywilfordic   Acid cassinic   Acid edulinic   Acid cathaic

Fig. (11). Dicarboxylic acids that esterified the macrocyclic sesquiterpnes alkaloids

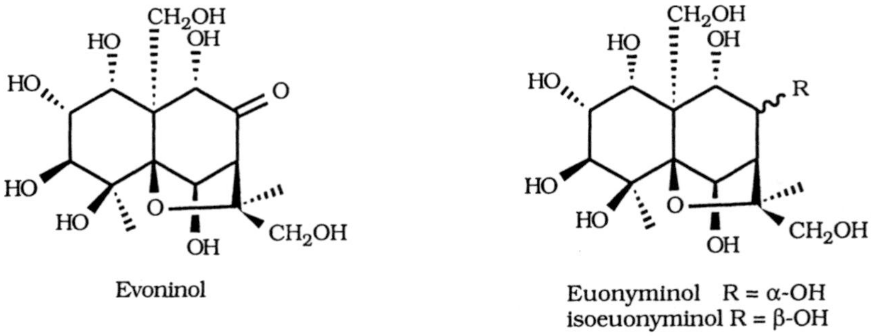

Evoninol

Euonyminol   R = α-OH
isoeuonyminol R = β-OH

Fig. (12). Basic sesquiterpene cores of the macrocyclic sesquiterpenes alkaloids

Research on the alkaloids of species of the genus *Euonymus*, have clarified certain aspects on the structure of these compounds. Evonine (**48**), one of the principal alkaloids obtained from *Euonymus europeaus* L, was reported first by Doebel and Reichstein [148] and later by Pailer and Libiseller [149], and was formulated as $C_{31}H_{39}NO_{14}$ and $C_{36}H_{43-45}NO_{17}$ by the respective groups. The latter group established the structure of evoninic acid, an alkaline hydrolysis product of evonine, whose structure was established as (2*S*, 3*S*)-2-methyl-3-methyl-3(β-carboxy-α-pyridyl)-propionic acid and further revealed that evonine consisted of a C_{15}-polyhydroxy compound, esterified by acetic acid (5 moles) and evoninic acid. The configuration of the methyl groups of the nicotinate derivative moiety of the evoninate-type compounds were determined as 7'S,8'S (evoninic acid: 2S, 3S) by an X-ray crystallographic analysis of bromoacetylneoevonine monohydrate [150].

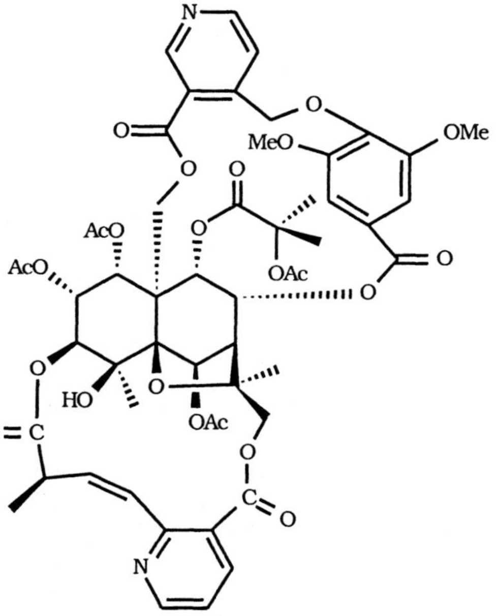

39 Cathedulin K-19

Fig. (13). Macrocyclic sesquiterpene alkaloids from *Catha edulis*

The extraction of the leaves of *Catha edulis*, has yielded over 20 cathedulins of which cathedulin K-19 (**39**) is among the most structurally complex, Fig. (**13**). Cathedulin K-19 is unique among this family of macrolactones in containing a diester bridge derived from (Z)-4-(3-carboxy-2-pyridyl)2-methyl-3-enoic acid (edulinic), which bridges C$_3$-C$_{13}$ of the highly functionalized sesquiterpenoid core of euonyminol. Although the stereochemical designation made to cathedulin K-19 by the Crombie's group [151] was complete in all other respects, the stereogenicity was left undefined. Kim and White [152] resolved the absolute configuration of edulinic acid, and hence of the C-9' centre of cathedulin K-19, which was found to be (S).

Two new evoninate-type alkaloids have been described from *Peritassa compta* Miers, peritassines A (**40**) and B (**41**), Fig. (**14**). These structures are distinguished by the macrocyclic unit which consists of an evoninic acid isomer in which the pyridine ring of the dibasic acid is substituted at 4'-5', instead of the more usual substitution of evoninic acid at 2'-3' [153].

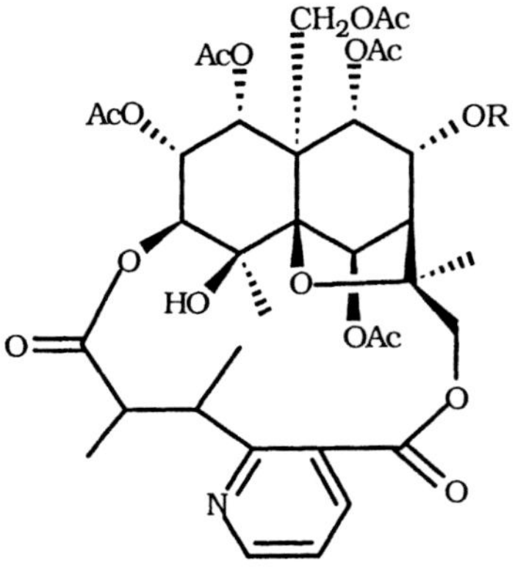

40 Peritassine A R=Ac
41 Peritassine B R=Bz

Fig. (14). Evoninate type alkaloids from *Peritassa compta*

Recently, was reported from the acetone extract of the root bark of *Tripterygium wilfordii*, two novel sesquiterpene alkaloids, triptonine A (**42**) and triptonine B (**43**) [154], Fig. (**15**) which were related to the previously isolated wilfortrine (**47**) [155]. The structures of these compounds were established as 18-O-benzoylwilfortrine and 18-O-(3-furoyl) wilfortrine, being this the first report of the isolation of 18-O-esterified alkaloids as naturally occurring compounds.

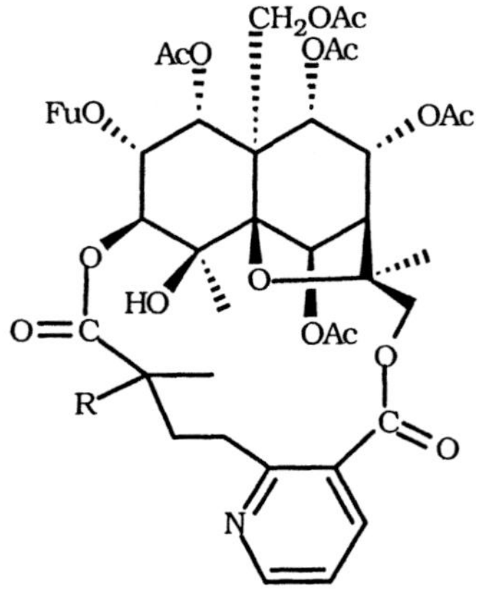

42 Triptonine A R=Bz
43 Triptonine B R=Fu

Fig. (15). Two novel sesquiterpenes alkaloids from *Tripterigium wilfordii*

4.2.1. Antifeedant and Insecticidal Activities

Insecticidal activity of *T. wilfordii* has been recorded since 1931 [103], and efforts to extract the active components were made by Acree and Haller [14] and by Beroza [105-107], resulting in the isolation of wilfordine (44), wilforine (45), wilforgine (46) and wilfortrine (47), Fig. (16) as some of the insecticidal alkaloids of this plant, being active against the European corn borer larvae. However, the structure of wilfordine (44) was not elucidated until 1978 by Yamada *et al.* [155], while the structures of wilforine, (45) wilforgine (46) and wilfortrine (47) postulated by Smith [119] on the basis of Beroza's earlier research [107], were established by Wu *et al.* [156, 157].

Delle Monache *et al.* [158] described the stepwise and bioassay methods for the isolation of the potent antifeedant wilforine (45) from *Maytenus rigida*. The activity was greater with the cruciferivorous *Pieris rapae* and the graminivorous *Locusta migratoria*, and less with the more polyphagous feeders, and negligible with *Schistocerca gregaria*.

Although the use of *Euonymus nanus* M. Bieb. as a natural insecticide has not been reported, a preliminary insecticidal screening showed that the lipophylic extract of the fruits exhibited moderate toxicity against the L_4 larvae of *Pieris brassicae* and *Hyphantria cunea*, and a marginal effect against *Oncopeltus fasciatus*. Evonine (48), Fig. (16), along with other sesquiterpene compounds, was isolated from the extract, and proved to be the most active compound when it was tested for larvicidal activity on the large white cabbage butterfly, *P. brassicae*. At a dosage of 200 mg/specimen, evonine induced a 74% of mortality [159].

Orthosphenin (49), Fig. (16), the first example of an evoninol nucleus with an oxo group at C-8 and residual cassinic acid, was isolated from the hexane extract of the roots of *Orthosphenia mexicana* Standley and was found to have a lethal dosis (LD_{100}) of 0.01 μg/cm² against the fith-instar larvae of the Egyptian cotton leafworm *Spodoptera littoralis* [143].

4.2.2. Antitumor Activity

Maytenus emarginata (Willd) Hou is known as "Lan Yu Lo Shinh" to indicate that it is a shrub indigenous to Lan Yu island southeast of Taiwan. The methanolic extract of the stem and branches of this species have been found to display significant *in vitro* cytotoxicity against KB cell cultures and *in vivo* antitumor activity against P-388 lymphocytic leukemia [160]. Kuo and coworkers studied the active chloroform fraction guided by *in vitro* KB (nasopharynx carcinoma) cell culture assay and isolated emarginatine-A (50) [161] and emarginatine-B (51) [162], which represented the two first examples of an euonymine-type polyester sesquiterpene alkaloid bearing a 5-carboxy-N-methylpyridonyl substituent at the C-2 position and the first observation of significant cytotoxicity among this type of alkaloids. Further investigation of the plant had lead to the isolation of emarginatine-C (**52**) , D (**53**), E (**54**), F (**55**) and G (**56**) and emarginatinine (**57**) [163, 164], which were assayed for their cytotoxic activity against KB cells, Fig. (**17**). Emarginatine F and G, were also evaluated on a battery of cancer cell lines and the results showed that emarginatine F demonstrated strong cytotoxicity in the HCT-8, human colon (1.29 µg/ml), P-388 (0.69 µg/ml), RPMI-7951, human melanoma (< 0.1 µg/ml), and TE-671, human medulloblastoma (0.21 µg/ml) cell lines and marginal cytotoxicity in the A-549, human lung carcinoma (5.50 µg/ml) cell line, while emarginatine G was inactive.

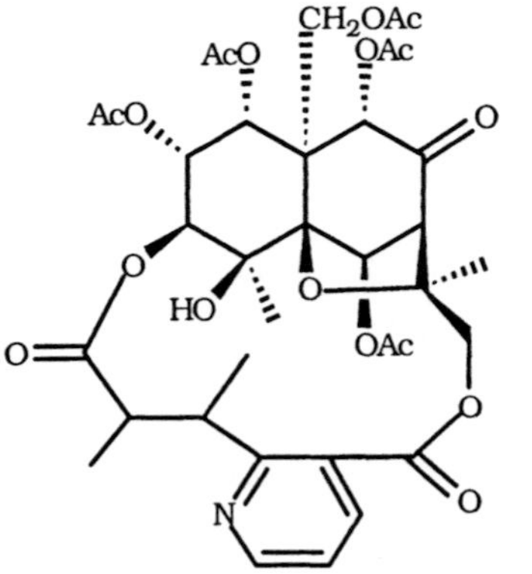

48 Evonine

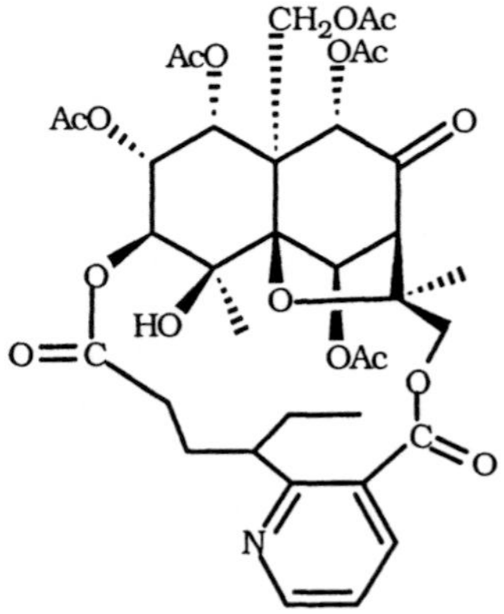

49 Orthosphenin

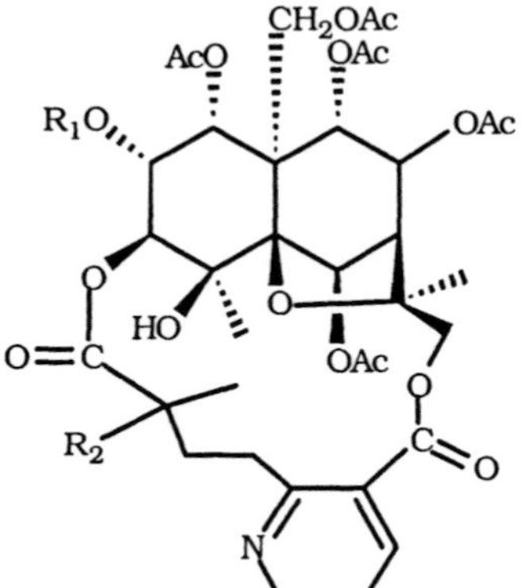

	R$_1$	R$_2$
44 Wilfordine	Bz	OH
45 Wilforine	Bz	H
46 Wilforgine	Fu	H
47 Wilfortrine	Fu	OH

Fig. (16). Insecticidal sesquiterpenes alkaloids from Celastraceae species

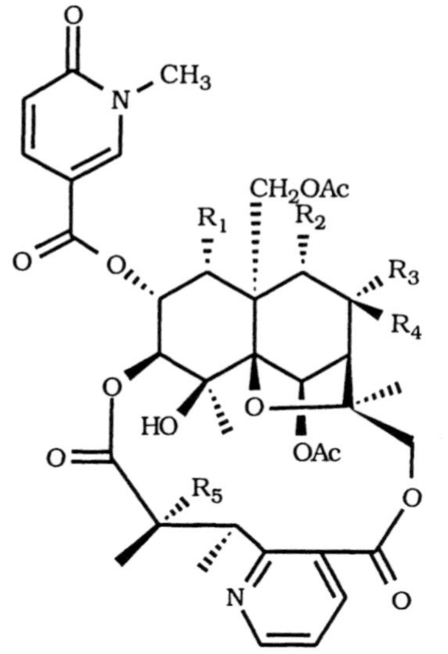

	R_1	R_2	R_3	R_4	R_5
50 Emarginatine A	OAc	OAc	OAc	H	H
51 Emarginatine B	OAc	OBz	H	OAc	H
52 Emarginatine C	OAc	OH	OAc	H	H
53 Emarginatine D	OH	OAc	OAc	H	H
54 Emarginatine E	OH	OH	H	OAc	H
55 Emarginatine F	OBz	OAc	H	OH	H
56 Emarginatine G	OMeBut	OAc	OAc	H	H
57 Emarginatinine	OAc	OAc	OAc	H	OH

Fig. (17). Cytotoxic sesquiterpenes alkaloids from *Maytenus emarginate*

The structure-activity relationship studies showed that cytotoxicity was affected noticeably by the type of functional group substitution at the C-1 and C-9 positions, and by the configuration of the proton at C-8 (α or β). Since, emarginatine B (**51**) showed very good activity with an ED_{50} value of 0.4 µg/ml and contains two ester moieties, an acetate and a benzoate, at C-1 and C-9, respectively. However, emarginatine E (**54**), with a hydroxyl group at both of these carbons, was approximately five-fold less active. In addition, the configuration at C-8 may be important to the cytotoxicity, since both of the H-8α-epimers (emarginatine B and E) were more active than the H-8β-epimers (emarginatine A, C and D). The alatamine-type polyester sesquiterpene, emarginatinine (**57**) was slightly more active than its evoninate-type analogue, emarginatine A (**50**).

The reddish brown bark, commonly known as "cancerina" of *Hippocratea excelsa* H.B.K., is valuable for treating skin ailments and gastric ulcers and for its pesticidal properties. The sesquiterpene evoninoate alkaloids hippocrateine I (**58**) and II (**59**), Fig. (**18**), were isolated from the root and stem barks of this species. Hippocrateine I showed slight activity (LC_{50} 212 μg/ml) in the brine shrimp lethality test, a preliminary bioassay used to detect antitumor agents, and in the 9PS (human nasopharyngeal carcinoma) cytotoxicity test (ED_{50} 1.85x10^{-1} μg/ml) but was inactive in HT-29 (human colon adenocarcinoma), A-549, and MCF-7 cell culture systems (ED_{50} >10 μg/ml) [165].

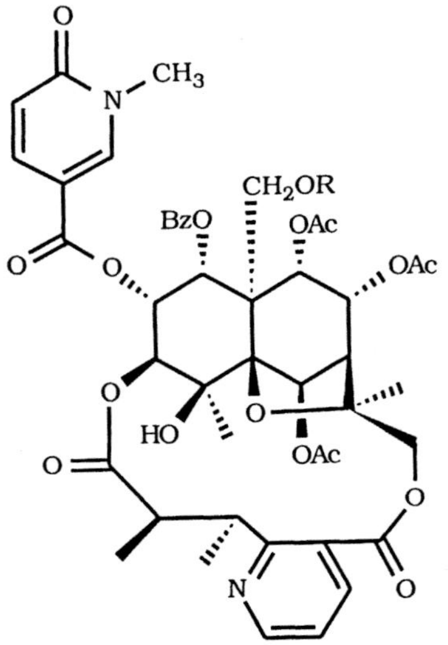

58 Hippocrateine I   R = Ac
59 Hippocrateine II R = MeBut

Fig. (18). Macrocyclic sesqiterpenes alkaloids from *Hippocratea excelsa*

4.2.3. Other Biological Activities

Immunomodulators are drugs that directly modify a specific immune function or have a net positive or negative effect on the activity of the immune system. The potential uses of immunodulators in clinical medicine include the reconstitution of immune deficiency (e.g. the treatment of AIDS) and the suppression of normal or excessive immune function (e.g. the treatment of graft rejection or autoimmune disease).

The crude extract of the xylem of *T. wilfordii* has been used to treat rheumatoid arthritis, chronic nephritis and various skin disorders, including psoriasis, systemic lupus erythematosis, allergic angitis, lepra reaction, as well as a number of other immunological disorders. Clinical observations suggest promising results in a large number of cases. It is worthwhile mentioning that two pyridine-type alkaloids, wilfortrine (**47**) and euonine (**60**), Fig. (**19**), isolated from this plant, were shown to inhibit the hemolysin reaction and significantly decreased the clearance rate of charcoal particles and the weights of spleen and thymus in mice at a dose of 800 mg/kg. In addition, wilfortrine (160 mg/kg) had a depressant effect on the graft-versus-host reaction (GVHR), while euonine (800 mg/kg) showed marked suppressive effects on the dinitrochlorobenzene (DNCB) induced a delayed hypersensitivity reaction on the skin of mice [166].

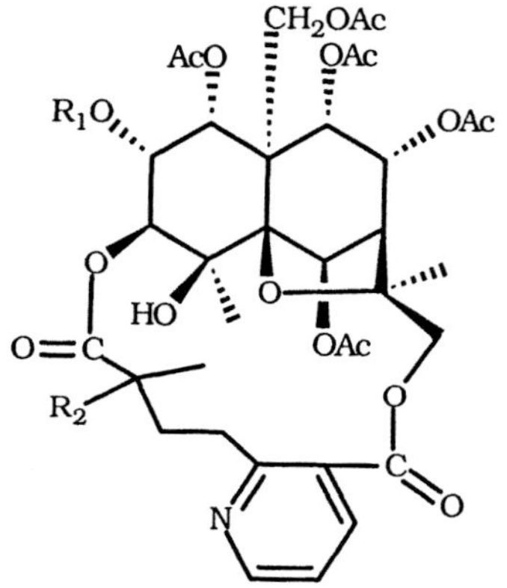

47 Wilfortine R_1=Fu R_2=OH
60 Euonine R=H R_2=H

Fig. (19). Immunosuppressive sesquiterpenes alkaloids

An ethanolic extract of the stem bark of *Maytenus krukovii* A. C. Smith demonstrated significant inhibitory activity in a protein kinase C (PKC) inhibition bioassay (IC_{50} 10 µg/ml). The phytochemical study of this extract gave mayteine (**61**) and 6-benzoyl-6-deacetylmayteine (**62**), Fig. (**20**); both compounds showed toxicity in the brine shrimp lethality assay, with LC_{50} <10 µg/ml, but neither of them demonstrated significant PKC inhibitory activity, this was concentrated in non-alkaloidal fractions [167].

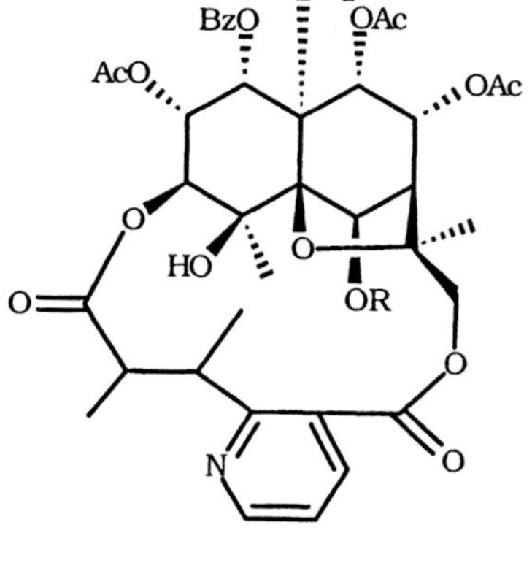

61 Mayteine R = Ac
62   R = Bz

Fig. (20). Active macrocyclic sesquiterpenes from *Maytenus krukovii*

4.3 Diterpenes

In general, very little has been written about diterpenes from Celastraceae as these structures are not often found. However, the novelty of their structures and their biological activities have stimulated several groups to research on the diterpenoid components from species of this family.

Abietatriene type diterpenes have been the general rule in the Celastraceae. Sugiol (**63**), maytenone (**64**), dispermol (**65**), dispermone (**66**) and maytenoquinone (**67**), among others, have been isolated from species of this family, Fig. (**21**). On the other hand, the chemical study of the minor constituents of *Rzedowskia tolantonguensis* [168] yielded a series of new diterpenes (**68-69**) and norditerpenes (**70-74**) with an isopimarane skeleton, Fig. (**22**), described for the first time in the Celastraceae.

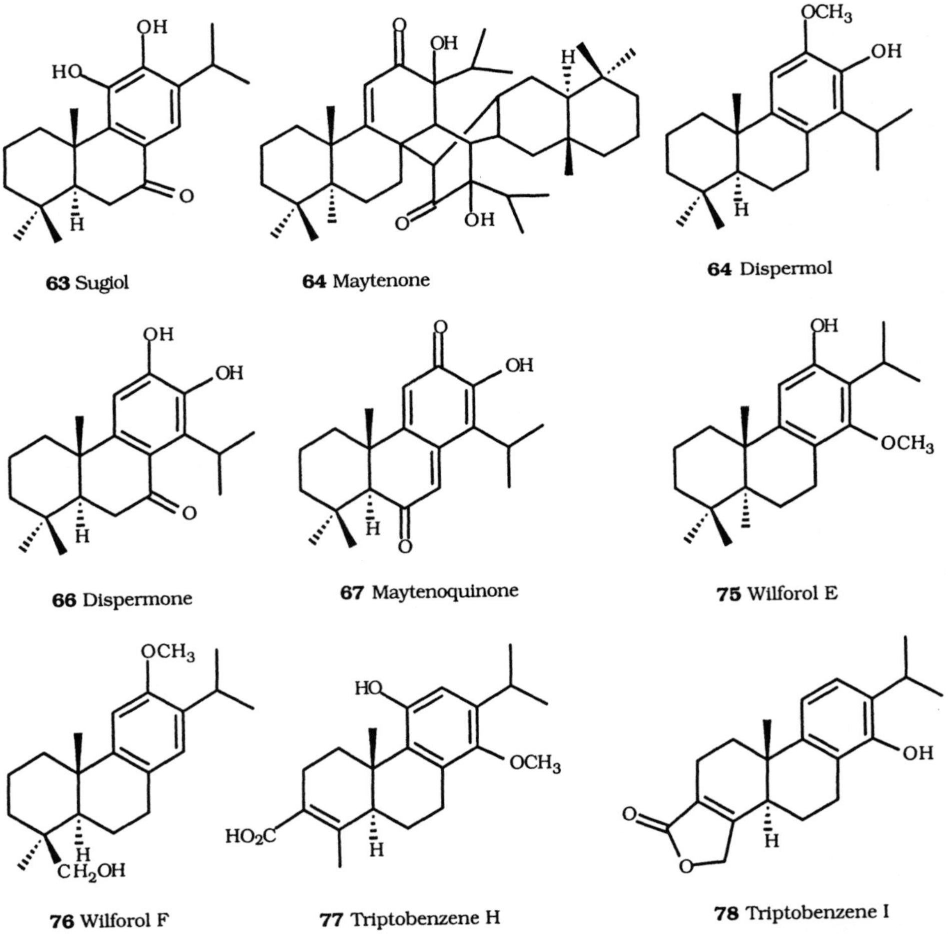

63 Sugiol   **64** Maytenone   **64** Dispermol

66 Dispermone   **67** Maytenoquinone   **75** Wilforol E

76 Wilforol F   **77** Triptobenzene H   **78** Triptobenzene I

Fig. (21). Some abietane diterpenoids from Celastraceae species

The most interesting type of diterpenoids, from a biological point of view, are a variety of novel and highly oxygenated diterpenes of the abietane and dehydroabietane family, which *T. wilfordii* plants elaborate. Among these, triptolide (**79**), and tripdiolide (**80**), with strong antitumor activities [108], were isolated and these findings stimulated a more detailed study of other novel diterpenoids isolated from Celastraceae species. As a result, several new diterpenoids have been isolated and as an example the recent isolation of the abietane diterpenoids, wilforol E (**75**) and F (**76**) [169], and triptobenzene H (**77**) and I (**78**) [170], Fig. (**21**) could be cited.

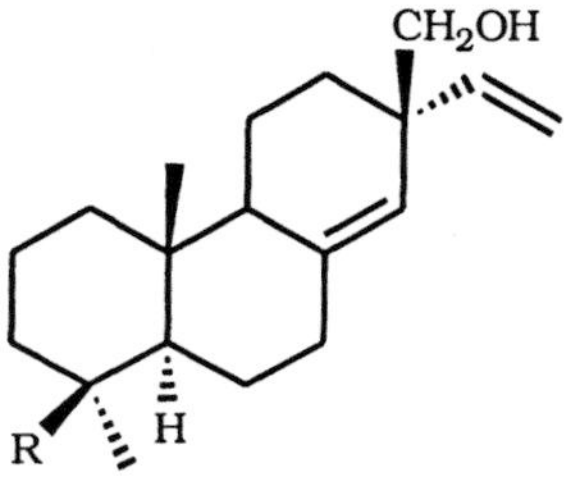

68   R = CO$_2$H
69   R = CH$_2$OH

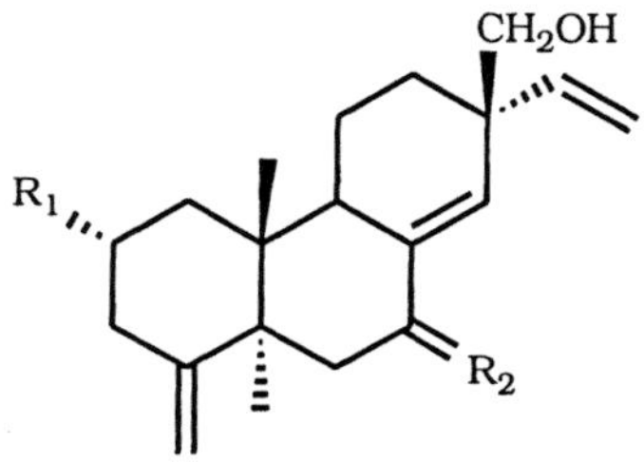

70   R$_1$=H R$_2$=O
71   R$_1$=H R$_2$=β–OH
72   R$_1$=H R$_2$=α–OH
73   R$_1$= α–OH R$_2$=O
74   R$_1$= α–OH R$_2$=β–OH

Fig. (22). Isopimaradiene diterpenes from *Rzedowskia tolantonguensis*

4.3.1. Cytotoxic and Antitumor Activities

In 1972, Kupchan *et al.* reported that the alcoholic extract of *T. wilfordii*, showed significant activity *in vivo* against the L-1210 and P-388 leukemias in the mouse, and *in vitro* against the KB cells derived from the human carcinoma of the nasopharynx. Bioassay-directed fractionation of the active extract against the cell lines mentioned above, led to the isolation of triptolide (**79**), tripdiolide (**80**), and triptonide (**81**) Fig. (**23**), which appeared to be the first reported natural products containing the 18(4-3)-*abeo*-abietane skeleton and the first recognized diterpenoid triepoxide. The configuration of compounds **79** and **80** were confirmed by direct X-ray analysis and the structure of **81** by oxidation of triptolide (**79**) with CrO$_3$-pyridine complex in dichloromethane, whereupon triptonide (**81**) was obtained in excellent yield. Triptolide and tripdiolide were found to show significant antileukemic activity against the L-1210 and P-388 leukemias at the 0.1 mg/kg level, and cytotoxicity (ED$_{50}$ 10^{-3}-10^{-4} µg/ml) against KB cell culture [108].

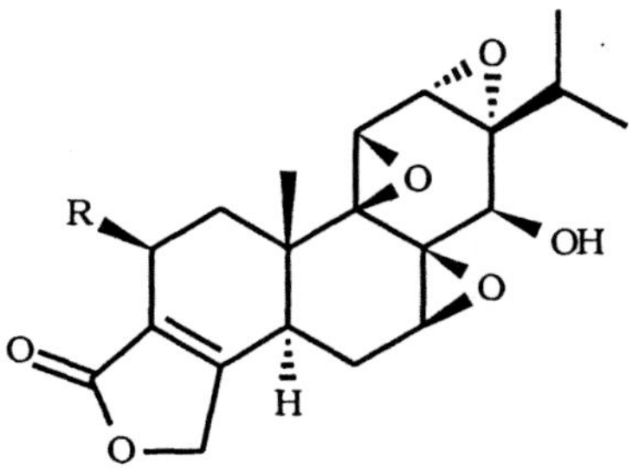

78 Triptolide   R = H
80 Tripdiolide   R = OH

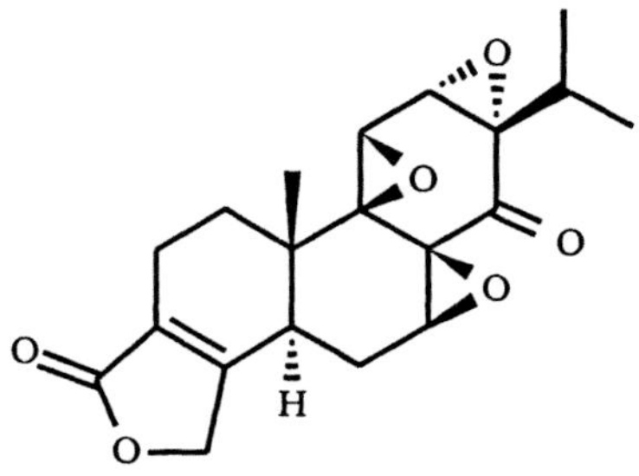

81 Triptonide

Fig. (23). Diterpenoids triepoxides from *Tripterigium wilfordii*

The potent antileukemics **79** and **80** contain a characteristic hydrogen-bonded 9,11-epoxy-14β-hydroxy system. As triptonide (**81**) showed no antileukemic activity in doses up to 0.4 mg/kg, Kupchan *et al.* [171] hypothesized that this system was necessary for the antileukemic activity of the triptolides. Subsequent testing of the minor variants 14-epitriptolide (**82**), the main product obtained from the reduction of **81**, with α-oriented 14-hydroxyl, and the thiol adducts **83** and **84**, obtained by selective alkylation of **79** and **80**, respectively (lacking 9,11-epoxide) Fig. (**24**), revealed that these triptolides derivatives, did not show any

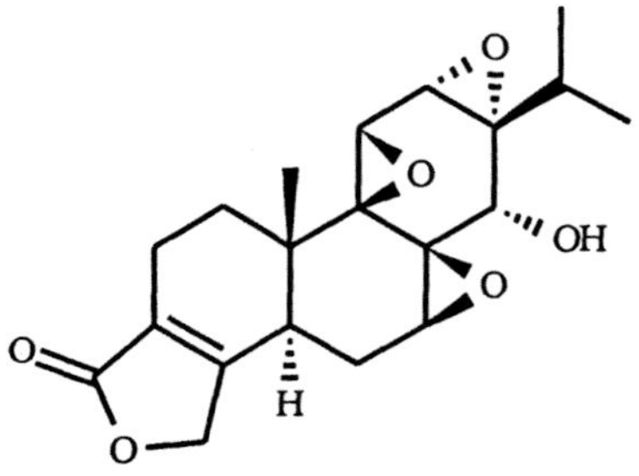

82 14-*Epi*-triptolide

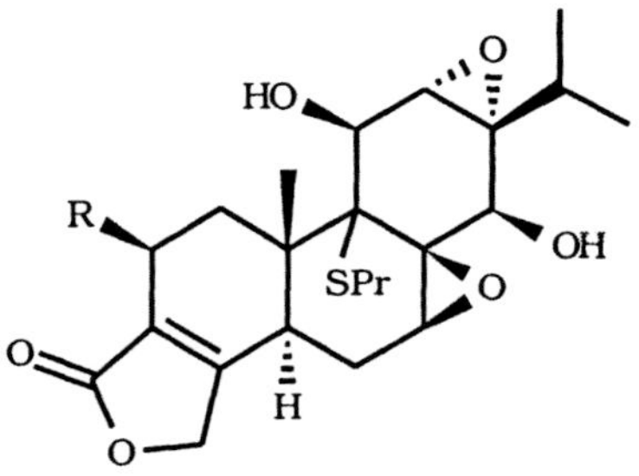

83   R = H
84   R = OH

Fig. (24). Derivatives of triptolides

antileukemic activity at doses up 0.4 mg/kg. The importance of the 9,11-epoxy-14β-hydroxy system for the biological activity of the triptolides was also indicated by their relative cytotoxicities against KB cells in culture. The median effective dose, ED_{50} values for the series were: **79**, 0.0017; **80**, 0.0042; **81**, 0.021; **82**, 0.076, and **83** and **84** >1 µg/ml. The results were in agreement with the hypothesis that the hydroxyl-assisted attack by nucleophiles on the 9,11-epoxy may mimic the mechanism by which the antileukemic triptolides exert their biological activity.

In 1991, the effects of triptolide (79) on the colony formation of breast cancer (MCFG-7 and BT-20), stomach cancer (MKN-45, MKN-7, and KATO-III), and promyelocytic leukemia (HL-60) cell lines, were reported [172]. The magnitudes of the inhibitory effect of triptolide on both breast and stomach cancer cell lines were similar to that of the leukemia cell line HL-60 (IC_{50} 0.504-1.22 µg/l). The results obtained suggested that triptolide might have a potential therapeutic effect on some types of solid tumors, e.g., breast and stomach cancers.

The isolation of these antileukemic compounds encouraged Kutney and coworkers [173, 174] to develop a plant cell culture line of *T. wilfordii* that would hopefully produce these compounds in higher yield and under laboratory controlled conditions. In this line, this group have published that the stable cell line, coded as TRP4a, was shown to produce the cytotoxic diterpene triptolide (79) and tripdiolide (80) in yields that were 3 and 16 times greater, respectively, than those observed in the parent plant source.

Recently, additional novel diterpene analogs of triptolide and tripdiolide, were obtained, via a combination of synthesis and enzyme catalyzed biotransformations, by Kutney and coworkers [175]. Pharmacological screening of triptolide, tripdiolide and, butenolide (85) and isotriptophenolide (86) Fig. (25), which were obtained by synthesis from dehydroabietic acid, in the murine P388 lymphocytic leukemia system and in six human cancer cell lines grown *in vitro* have been assayed. The more significant activity was found particularly in the triepoxides, triptolide (79) and tripdiolide (80), with values ED_{50} 0.01 and 0.04 µg/ml respectively, while butenolide (85) and isotriptophenolide (86) exhibited lower potency. Although the structure-activity relationship for this family of diterpenes is still incomplete, there is a strong suggestion that the presence of epoxide functionality in rings B and/or C is important for promoting antineoplastic activity.

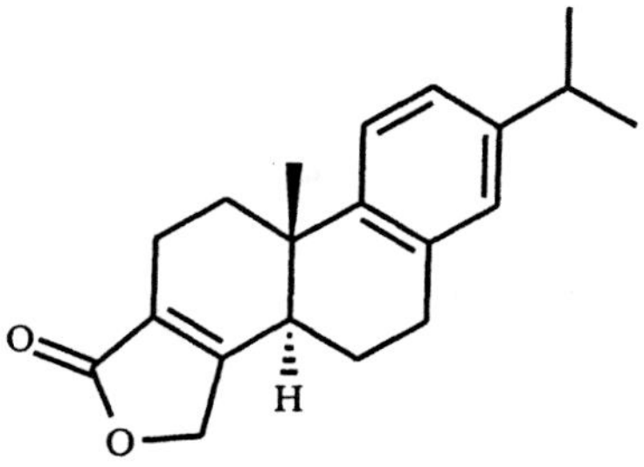

85 Butenolide 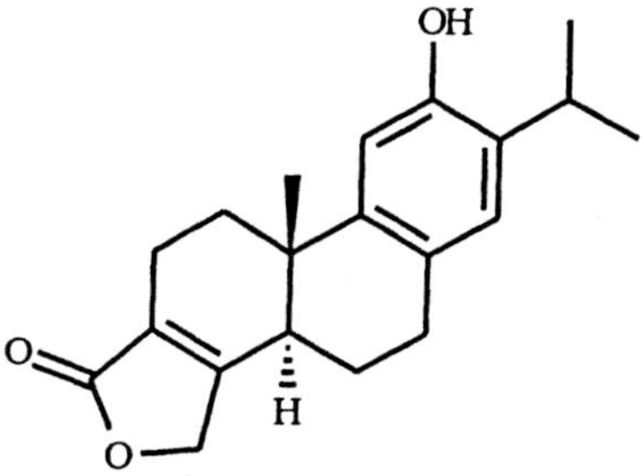   **86** Isotriptophenolide

Fig. (25). Novel diterpenes analogs with the actives triptolide and tripdiolide

Previous efforts by both van Tamelen [176] and Berchtold [177] on the total synthesis of triptolide (**79**) encountered some difficulties. In 1997, Yang *et al.* [178] reported an efficient method for the total synthesis of several triepoxide analogs of triptolide using *in situ* generated dioxiranes and basic hydrogen peroxide. Among these, compounds **88** and **89**, obtained by epoxidation of the iso-propyl dienone (**87**) Fig. (**26**) were found to be cytotoxic to human leukemia HL-60 cells (IC$_{50}$ 16 µM for both) and human lymphoma Jurkat cells (IC$_{50}$ 31 µM and 36 µM, respectively). These results suggested that the electrophilic α,β-enone groups, present in both compounds, may be responsible for their similar biological activity.

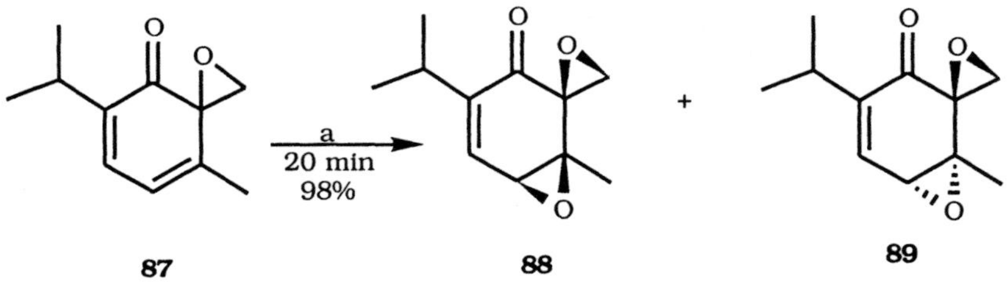

Reagents and conditions: (a) CF$_3$COCH$_3$, Oxone/NaHCO$_3$, CH$_3$CN/H$_2$O, 0-1°C, pH 7-7.5

Fig. (26). Synthesis of triepoxide analogos of Triptolide

Recently, it was reported that triptolide (**79**) was not mutagenic toward the *Salmonella typhimurium* strain TM677, either in the presence or absence of a metabolic activating system. Relatively potent but non-specific cytotoxicity was observed with a panel of cultured mammalian cell lines, and modest antitumor activity was observed when an i.p. dose of 25µg was administered three times weekly to athymic mice carrying human breast tumors. However, treatment regimens involving higher doses of triptolide (e.g. 50 µg/mouse three times weekly) were lethal [179].

4.3.2. Antiviral Activity

In the course of the search for novel potent anti-AIDS agents, the ethanolic extract of the roots of *T. wilfordii* was found to show significant anti-HIV activity. Bioassay-directed fractionation of the active extract has led to the isolation and characterization of a new anti-HIV principle, a kaurane-type diterpene lactone, tripterifordin (**90**), Fig (**27**). This compound inhibited HIV replication in H9 lymphocyte cells with an EC$_{50}$ of 1 µg/ml (6 (M) but it did not inhibit uninfected H9 cell growth at 15 µM [180]. From the same species a new kaurene type diterpene, neotripterifordin (**91**) was also isolated, which showed potent anti-HIV replication activity in H9 lymphocyte cells with an EC$_{50}$ of 25 nM and a therapeutic index (TI) of 125 [181]. Recently Corey and Liu [182] reported the enantioselective total synthesis of neotripterifordin (**91**), which was of special significance in view of its remarkable anti-HIV activity and its scarcity from natural the source.

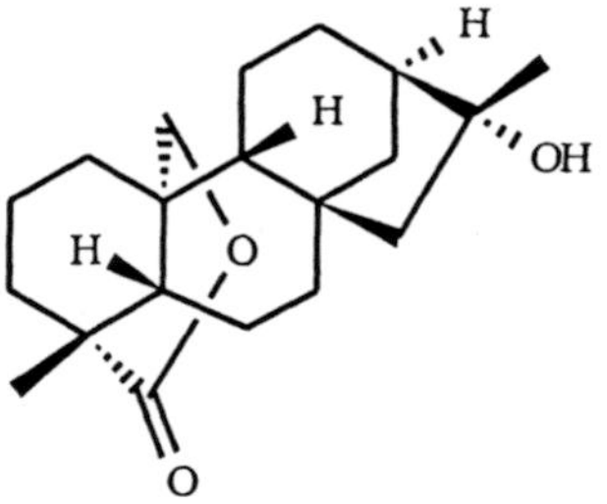

90 Tripterifordin 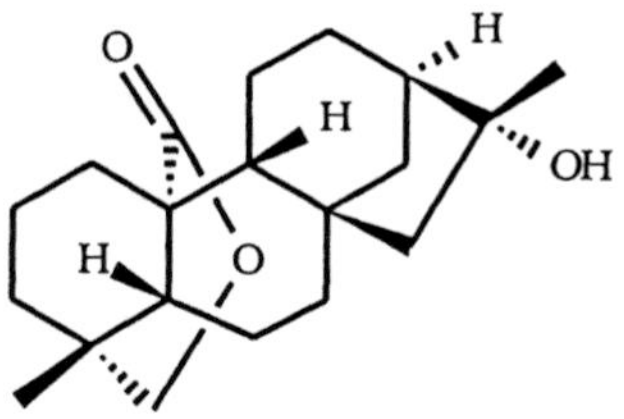   **91** Neotripterifordin

Fig. (27). Anti-HIV kaurene type diterpenoids from *Tripterigium wilfordii*

4.3.3 Immunomodulator Activity

In the past 20 years hundreds of Chinese traditional and herbal medicines with immunomodulating functions have been screened, and their active components studied. Through these studies, six diterpene lactone epoxide compounds, with antiinflammatory and immunosuppressive actions, were isolated from *T. wilfordii*. Zheng *et al.* [183] assayed the half-effective dose (ED_{50}), the therapeutic index (TI) and a certain safety factor (CSF) using croton oil induced ear swelling and hemolysin-antibody formation mouse models and found that triptolide (**79**), tripdiolide (**80**), triptonide (**81**), tripchlorolide (**92**), and triptolidenol (**93**) possessed both antiinflammatory and immunosuppressive activities, while triptriolide (**94**) showed antiinflammatory action only, Fig. (**28**). The results of the TI and CSF of both activities clearly demonstrated that the antiinflammatory and immunosuppressive compounds in *T. wilfordii* are pluralistic and the differences in these compounds can be used as a base for the evaluation and selection of them.

The immunomodulating activities of triptolide and its mechanisms of action have been further studied in China. In this line, the effects of **79** on mixed lymphocyte culture (MLC), suppressor T cell (Ts) activity, delayed type hypersensitivity (DHT) reaction, interleukin-2 (IL-2) secretion activity and Th/Ts ratio were evaluated.

In vivo, triptolide (**79**) was found to inhibit skin allograft rejection in a dose and time-dependent manner, and at 0.1 mg/kg/day it significantly prolonged the graft survival time when given for 9 days after, but not before, transplantation [184]. The DTH reaction, sensitized to dinitroflurobenzene (DNFB) monitored by the increase in ear weight after challenge with antigen, was suppressed by triptolide at a dose of 0.12-0.50 mg/kg and at 0.25-0.50 mg/kg also exhibited a prominent suppression of the enhanced DTH reaction induced by cyclophosphamide (250 mg/kg, i.p.). The Th/Ts ratio of mouse thymus cells was reduced after administering triptolide at a dose of 0,25 mg/kg for 5 consecutive days [185]. However, the Ts cells induced by suproptimal immunization were suppressed by triptolide at 0.5 and 0.75 mg/kg [186]. Therefore, T cells and their subpopulation may be the target cells of triptolide.

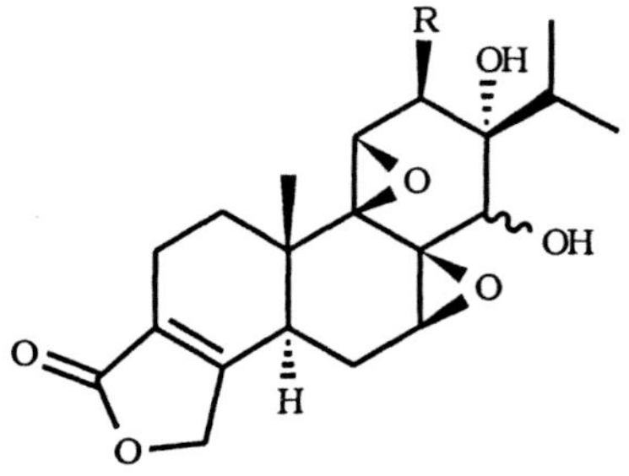

92 Tripchlorolide R = Cl
94 Triptriolide   R = OH

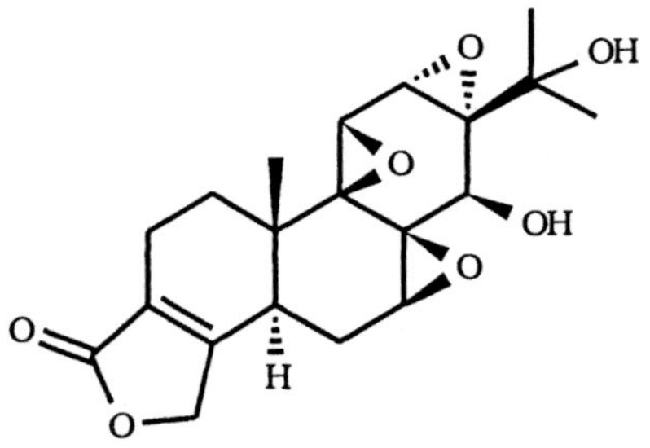

93 Triptolidenol

Fig. (28). Antiinflammatory and immunesuppressive diterpenoids from *Tripterigium wilfordii*

In vitro, triptolide (0.5-10 μg/ml) suppressed the one-way mixed lymphocyte reaction (MLR); the lymphocytes induced in the first MLR with triptolide of 5 and 10 μg/ml suppressed the second mixed lymphocyte culture (MLC) after irradiation with a 3000 rad ^{60}Co source. This suggested that triptolide can induce Ts cells [185]. Yang *et al* [184] also found that triptolide markedly suppressed cytotoxic T-lymphocyte (CTL) induction and MLC at these concentrations significantly when triptolide was added to the culture 36 h after initial incubation. Triptolide (0.4 ng/ml) significantly suppressed lymphocyte proliferation and interleukin-2 (IL-2) receptor expression stimulated by PHA in a concentration dependent fashion, but showed no effect on IL-2 production [187]. These data suggested that the immunosuppression of this compound was mediated by inhibiting IL-2 receptor expression and interfered with signal transduction of IL-2, but not through suppressing IL-2 production.

The above mentioned data suggest that triptolide has a suppressive effect on cellular immune function. The suppressive mechanism may be related to the suppression of Th cells and IL-2 secretion activity and the induction of Ts cells. Triptolide is one of the main immunosuppressive components of *T. wilfordii* and, although its mechanism of action remains to be further elucidated, represents a novel class of drugs with potential a clinical utility [188].

Interleukin-1 (IL-1) has various biological activities and is thought to be a biologically active substance necessary for the maintenance of homeostasis of the human body. However, if the controlling function for IL-1 production becomes abnormal, it could cause various diseases. In rheumatoid arthritis, it was reported that there are strong relationships between production of IL-1 by synovium and the degree of inflammation of the arthricular synovial membrane. Takaishi and coworkers [189, 190] found that the extract of the *T. wilfordii* var. *regelii*, which was used to treat rheumatoid arthritis and ankylosing spondylitis in some chinese clinics [91], showed inhibitory effects of IL-1 release. The bioassay-guided fractionation led to the isolation of seven new quinoid-type diterpenes designated triptoquinone A-G (95-101) Fig. (29), and their structures were determined by spectroscopic data, chemical evidence and X-ray analysis. Compounds 95-100 showed extremely potent inhibitory activities against interleukin-1α and β releases for lipopolysaccharide-stimulated human peripheral mononuclear cells, compared with the reference compound (prednisolone) (Table 4). A significant inhibitory activity of 95 for adjuvant-induced arthritis was confirmed in rat.

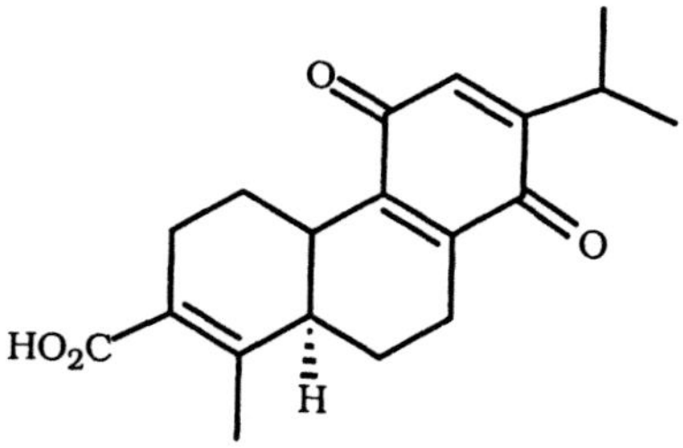

95 Triptoquinone A

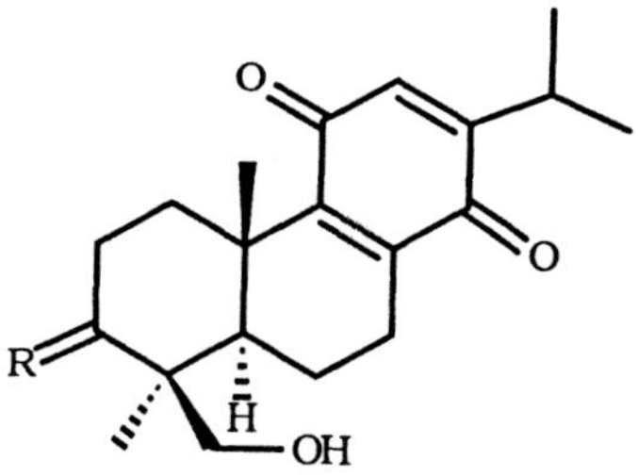

96 Triptoquinone B R = O
97 Triptoquinone C R = β-OH

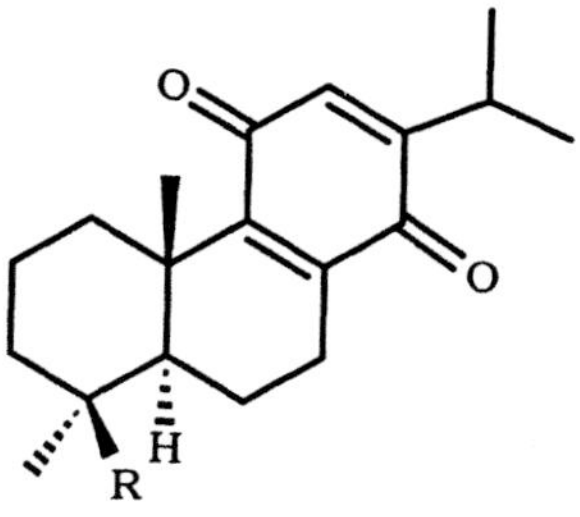

98   Triptoquinone D R = CH₂OH
99   Triptoquinone E R = CHO
100 Triptoquinone F R = CO₂H

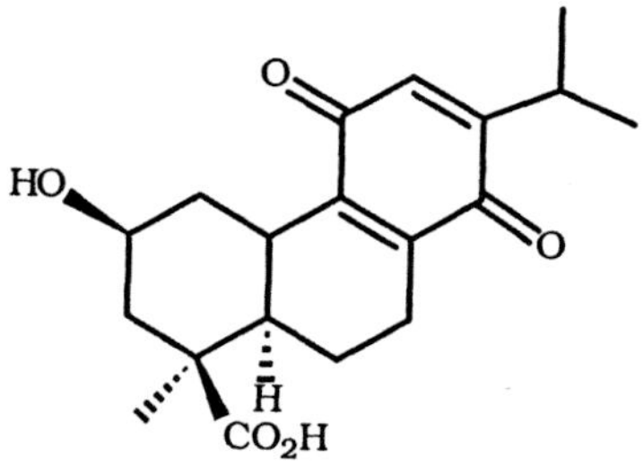

101 Triptoquinone G

Fig. (29). Interleukin-1 inhibitors from *Tripterigium wilfordii*

Table 4.   The Activities of Triptoquinones A-G as Inhibitors of   IL-1α and IL-1β
             Release

Compounds	Inhibition (%)	
	IL-1α	IL-1β
triptoquinone A (95)	63	34
triptoquinone B (96)	92	64
triptoquinone C (97)	35	11
triptoquinone D (98)	79	31
triptoquinone E (99)	74	99
triptoquinone F (100)	22	46
Prednisolone	87	76

Concentration (95-100: 1 x 10^{-5} g/ml, prednisolone: 3 x 10^{-7} g/ml).

In the course of a screening programme for anti-allergic compounds from Chinese higher plants, it was found that the ethanolic extract of the roots of T. wilfordii showed potent leukotriene D_4 (LTD$_4$) antagonistic activity in the bioassay using the smooth tracheal muscles of guinea-pig. Bioassay guided separation of the extract furnished four active compounds, two of them, triptinin A (102) and B (103), were found to be novel 19(4-3)-abeo-abietane diterpenes, while the other two compounds were identified as triptoquinone A (95) [189] and the related 11-hydroxy-14-methoxy-19 (4-3)-abeo-abieta-3,8,11,13-tetraen-19-oic acid (77) [191], Fig. (30). These compounds appeared to be competitive antagonists of LDT$_4$ with values of K_D x10^{-5} (constant dissociation between receptor and antagonist) of 4.5, 20.2, 12.4 and 3.4, for compounds 77, 95, 102 and 103 at a concentration of 50 µg/ml. These compounds are the first diterpenoid compounds to be reported to have LDT$_4$ antagonistic activity and these results explain one of the reasons why T. wilfordii is prescribed in Chinese medicines as an anti-asthmatic agent [192].

4.3.3. Other Biological Activities

The hot water and alcohol extracts of the roots of T. wilfordii have shown effects on the reproductive system and this fact has stimulated a great deal of research, particularly relating to the male reproductive system, and their possible use as a birth control agent. Studies in animals and humans have provided considerable encouragement for the development of a novel non-steroidal male contraceptive agent [173]. In this context, pharmacological experiments revealed that triptolide (**79**) and tripdiolide (**80**) showed antifertility activity [193].

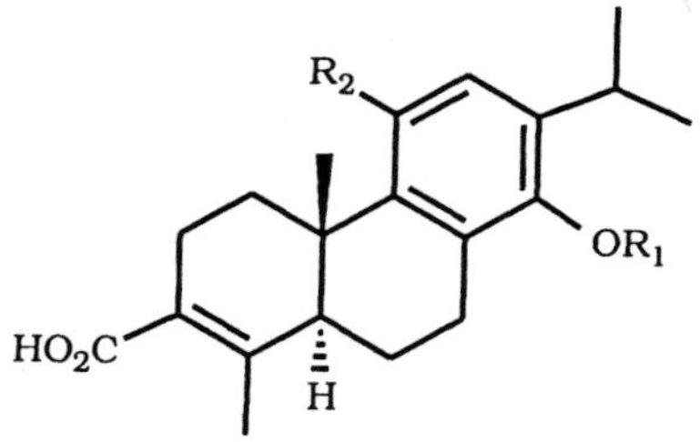

102	Triptinin A	$R_1 = CH_3$; $R_2 = H$
103	Triptinin B	$R_1 = R_2 = H$
104		$R = CH_3$; $R_2 = OH$

Fig. (30). Leukotriene D_4 antagonistic 19 (4→3)-abeo-abietanes from *Tripterigium wilfordii*

4.4. Triterpenes

The widespread reports in recent years on their useful biological activities and indeed some practical applications of triterpenes, have made these metabolites more relevant and interesting. Some plants contain large quantities of triterpenes and the physiological function of these compounds is generally believed to be a chemical defence against pathogens and herbivores. It is expected, therefore, that triterpenes could act against certain pathogens causing human and animal diseases and indicate their varied potential.

The triterpenes described for the Celastraceae almost invariably belonged to the friedo-oleane, lupane, oleane, glutinane, taraxerane, ursane, and dammarane series, Fig. (**31**). Friedo-oleane triterpenes with a hemiketal 24-hydroxy-3-oxo function are exclusive to the Celastraceae. Among these, salaspermic acid (**156**), isolated from *Salacia macrosperma* Wight [194] and orthosphenic acid (**157**), isolated from *Orthosphenia mexicana* Standley [195] are two examples of this singular group of triterpenes, Fig. (**41**). It has been proposed that these compounds may be possible intermediates in the biosynthetic pathway to the Celastraceae triterpene quinones [196].

Many oleane and ursane triterpenoids are reported to have interesting biological, pharmacological, and medicinal activities similar to those of retinoids and steroids. These include antiinflammatory activity, suppression of tumor promotion, suppression of immunoglobulin synthesis, protection of the liver against toxic injury, induction of collagen synthesis, and induction of differentiation in leukemia or teratocarcinoma cells [197].

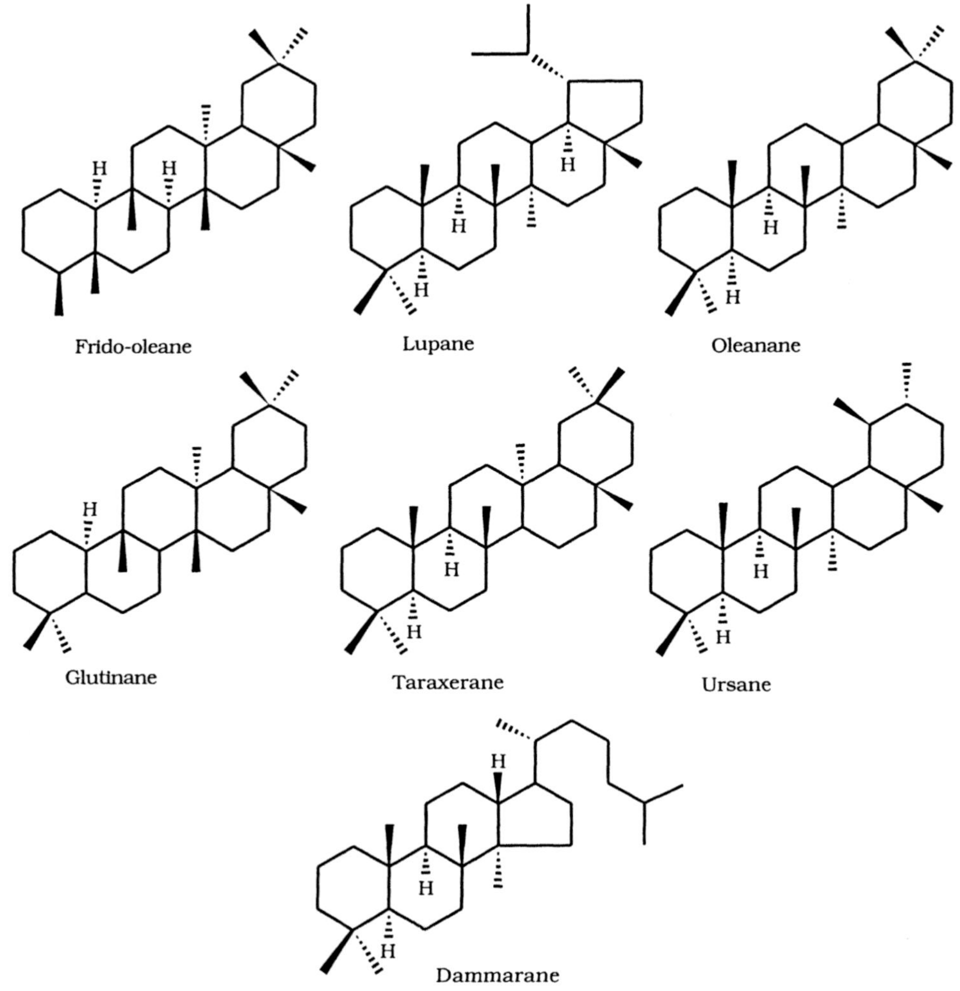

Fig. (31). Triterpenoids skeletons described from Celastraceae

Recently, the isolation of five new triterpenes from the Brazilian medicinal plant *Maytenus krukovii* (local name "chuchuhuasi"), used for the treatment of skin cancer, was reported [77]. These triterpenes, named krukovines A-E (**104-108**), have an α,β-unsaturated ketone system in the C ring and are triterpenoids based on either a 3-oxooleanane or a 3-oxoursane skeletons, Fig. (**32**). The synthesis of krukovines A and B, has also been described, which not only confirmed the proposed structures for these compounds, but also provides sufficient material for biological activities [197].

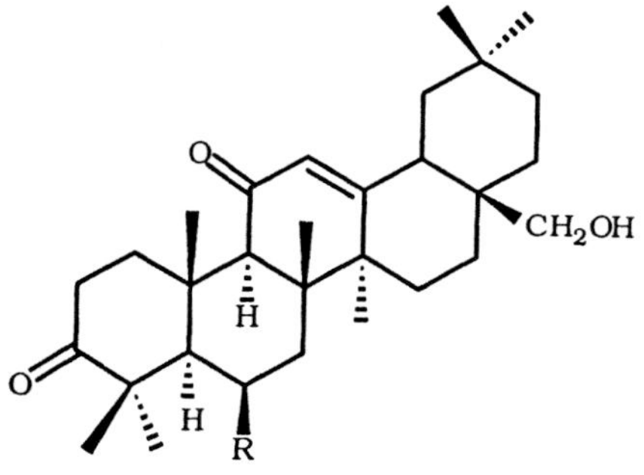

104 Krukovine A R = H
105 Krukovine C R = OH

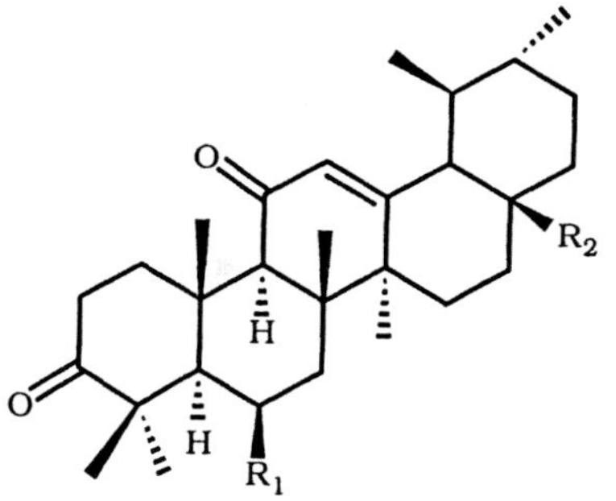

106 Krukovine B R_1 = H R_2 =CH$_2$OH
107 Krukovine D R_1 = OH R_2 =CH$_2$OH
108 Krukovine E R_1 = H R_2 =OH

Fig. (32). Krukovines from *Maytenus krukovii*

4.4.1. Cytotoxic and Antitumor Activities

The spinescent shrub *Maytenus diversifolia* (Grai) Hou is known as "Pak-Tiong (Pei-Chung)" or "Tzu-Lou-Shih" in Taiwan. The methanolic extract of the stems of this species was found to show significant inhibitory activity *in vivo* against the P-388 lymphocytic leukemia growth in BDF$_1$ mice (T/C 180%) at 50 mg/kg/day [198].

Nozaki *et al.* [199, 200] isolated, two new triterpenoids, maytenfolic acid (109) and maytenfoliol (110) from *M. diversifolia*, whose structures were determined on the basis of chemical transformations, spectral data and X-ray crystal structure analysis as 3β,22α-dihydroxyolean-12-en-30-oic acid and 28,29-dihydroxy-friedelan-3-one, respectively. Both compounds showed significant inhibitory activity against P-388 lymphocytic leukemia in BDF$_1$ mice (T/C 148 and 120 %) at the 6.25 and 10 mg/kg/day levels, respectively. Compound 110 also exhibited cytotoxicity (ED$_{50}$ 4.56 μg/ml) in the KB cell culture *in vitro*. Further investigation on the methanolic extract of *M. diversifolia* [201], which also showed potent *in vitro* cytotoxicity against A-549 lung carcinoma cells, has led to the isolation and characterization of three known triterpenes, 3-oxofriedelan-29-oic acid (polpunonic acid) (111) [202], 3-oxofriedelan-28-oic acid (112) [203], and 28,29-dihydroxyfriedelan-3-one (113) [204], as the cytotoxic principles. Compounds 111-113 demonstrated cytotoxicity against the A-549 cells with ED$_{50}$ values of 0.21, 1.18, and 0.64 μg/ml, respectively. Compound 112 was also cytotoxic against both L-1210 (ED$_{50}$ 2.95 μg/ml) and KB (ED$_{50}$ 3.70 (μg/ml) tumor cells, Fig. (33).

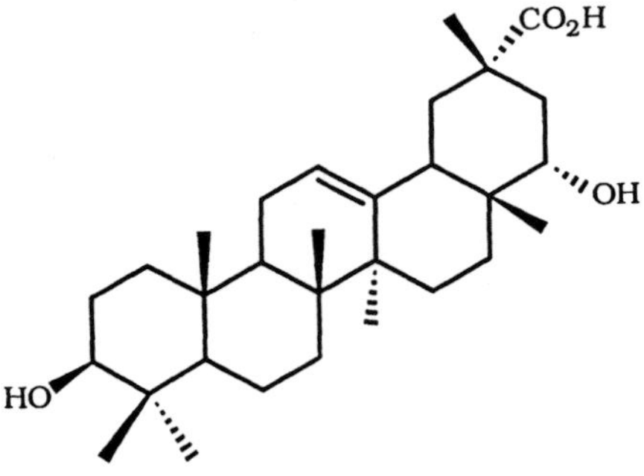

109 Maytenfolic 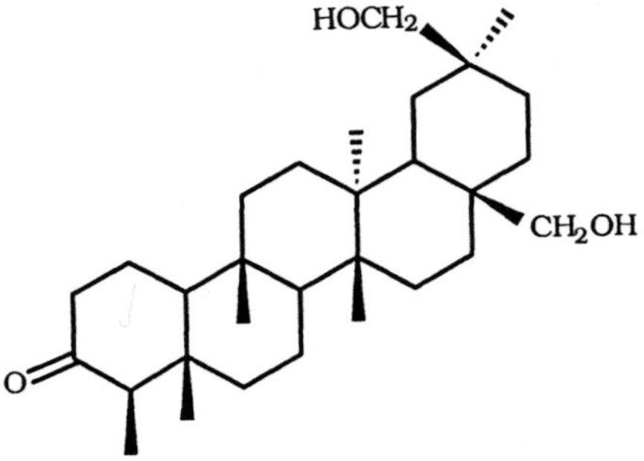   **110** Maytenfoliol

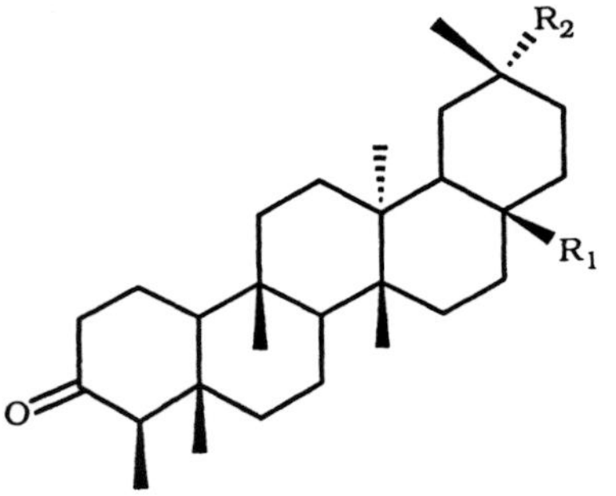

111 Polpunonic acid R$_1$=CH$_3$ R$_2$=CO$_2$H
112 3-Oxo-friedelan-28-oic acid R$_1$=CO$_2$H R$_2$=CH$_3$
113 28,29-Dihydroxyfriedelan-3-one R$_1$=R$_2$=CH$_2$OH

Fig. (33). Cytotoxic triterpenes from *Maytenus diversifolia*

The hexane extract of the dried wood and stems of *Maytenus nemerosa* demonstrated significant inhibitory activity *in vitro* against KB cells [205]. A group of lupane triterpenes (**114-116**) were isolated from this extract. Comparison of the toxicity of these metabolites, and of several analogs (**117-120**) obtained from them by synthetic transformations, has permitted correlation of cytotoxicity with certain of the structural features in these molecules. 3-Oxo-20(29)-lupen-30-al (**114**), earlier isolated from *Gymnosporia emarginata* [206], was found to be cytotoxic at a concentration of 10 μg/ml, although none of the structurally related compounds tested exhibited significant inhibition of KB cell growth at the same concentration. These results led to the conclusion that expression of the cytotoxicity by compound **114** required the presence of the carbonyl at C-3 and the α,β-unsaturated aldehyde at C-19 moieties, Fig. (**34**).

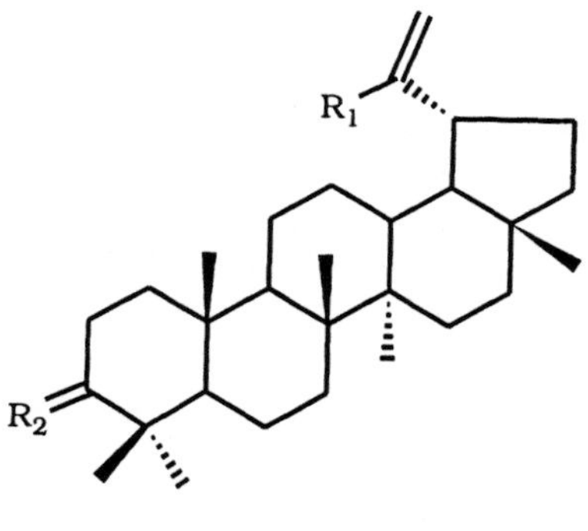

114 R_1=CHO R_2=O
115 R_1=CH$_2$OH R_2=O
116 R_1=CH$_2$OH R_2=α–H,β–OH
118 R_1=CHO R_2=O
120 R_1=CHO R_2=α–H,β–OH

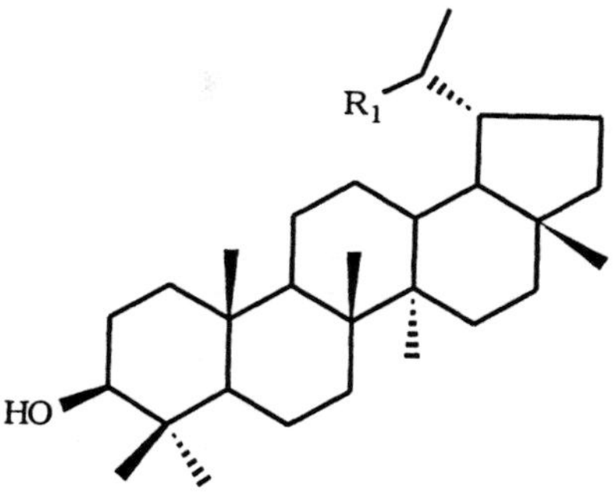

117 R=CHO
119 R=CH$_2$OH

Fig. (34). Some lupane triterpenes from *Maytenus nemerosa* and derivatives

Maytenus ilicifolia Mart. is a large shrub found in southern Brazil, Paraguay, and Argentina and is claimed to have several medicinal properties. Its reddish to orangish brown bark, known as "cangorosa", is

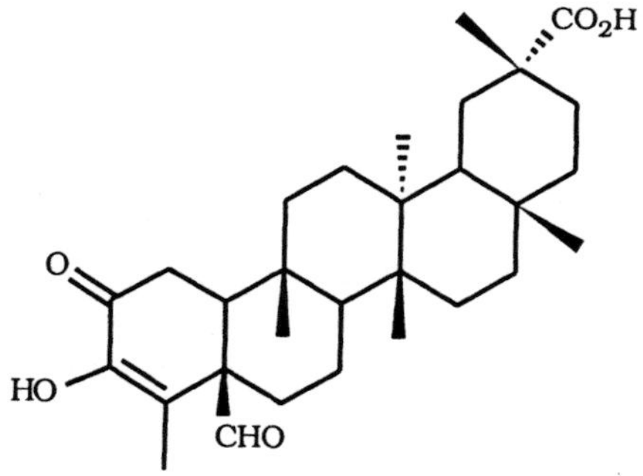

121 Cangoronine

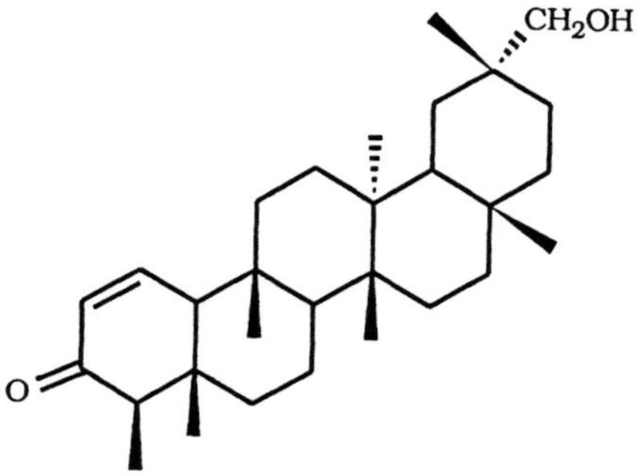

122 Ilicifoline

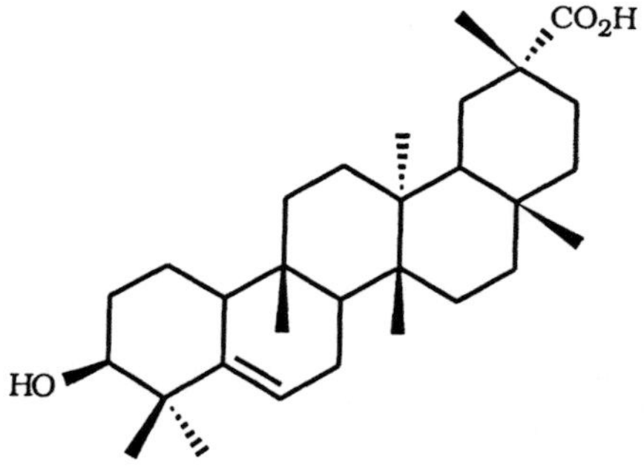

123 D:B-Friedoolean-5-en-3β,29-diol

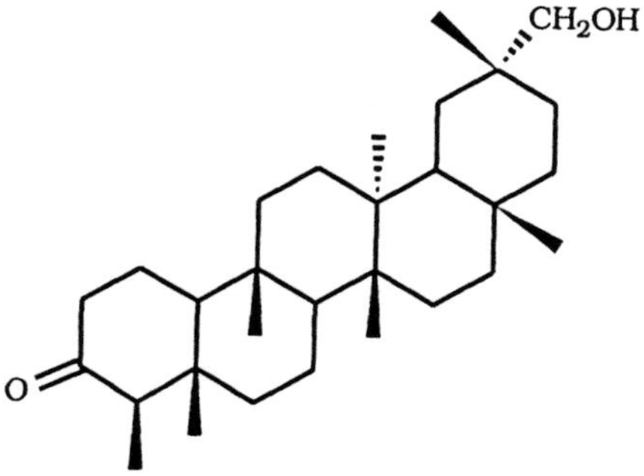

124 D:A-Friedoolean-29-ol-3-one

used by Indian tribes and rural populations in Paraguay as a fertility regulating agent.

In continuing their work on South American folk medicinal plants for the presence of anti-cancer agents, Itokawa and coworkers [207] found that the methanol extract of *M. ilicifolia* showed significant cytotoxic activity against V-79, KB and P-388 cells. Fractionation of this extract led to the isolation of two novel triterpenes, D:A-friedoolean-24-al-3-en-3-ol-2-on-29-oic acid (cangoronine, **121**) and D:A-friedoolean-1-en-29-ol-3-one (ilicifoline, **122**), whose structures were established by spectroscopic and X-ray analyses. The already known triterpenes: maytenoic acid (polpunoic acid, **111**), D:B-friedoolean-5-en-3(,29-diol (**123**) [208], D:A-friedoolean-29-ol-3-one (**124**) [200], and salaspermic acid (**156**) [194], Fig. (35) were also isolated along with some known triterpene quinone-methides. Compounds **111** and **122-124** were evaluated for cytotoxic effects against V-79, KB and P-388 cell lines, being ilicifoline the most active compound, with IC_{50} values of 38, 12 and 23 µg/ml, for the above mentioned cell lines.

Two new antitumor ursene-type triterpenoids, regelin (**125**) and regelinol (**126**) were isolated, together with the known oleanene-type regelide (wilforlide or abruslactone A) (**127**), from the roots of *Tripterygium regelii* [209], Fig. (36). The structures of the new compounds were determined by NMR spectrometric studies and X-ray analysis of **125**, as the methyl ester of 3-oxo-22α-hydroxy-urs-12-en-30-oic acid and its 23-hydroxy analog, respectively. The antitumor activities of **125-127**, against KB cell lines were ED_{50} 4.2, 10.6 and > 100 µg/ml, respectively.

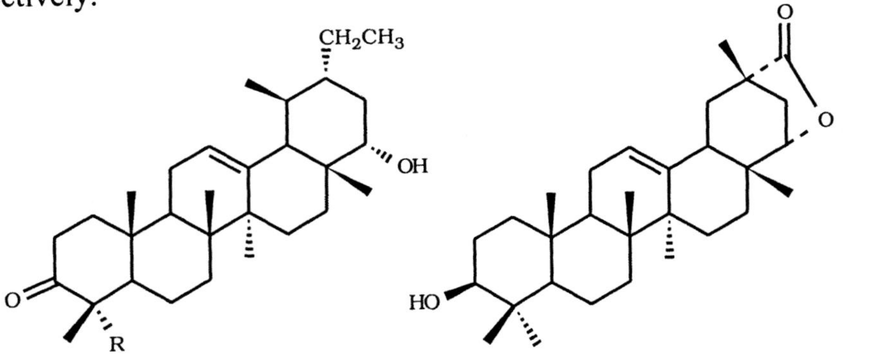

125 Regelin R=CH$_3$
126 Regelinol R=CH$_2$OH

127 Regelide = Wilforlide

Fig. (36). Antitumor ursene and oleanene type triterpenes from *Tripterigium wilfordii*

Kokoona ochracea (Elm.) Merrill is one of eight species in the genus. It is distributed in the Malay Peninsula, Borneo, and Palawan Island of the Philippines. There is no local medicinal use of this plant among the Palawan inhabitants, however, in Palawan at least, the tree finds

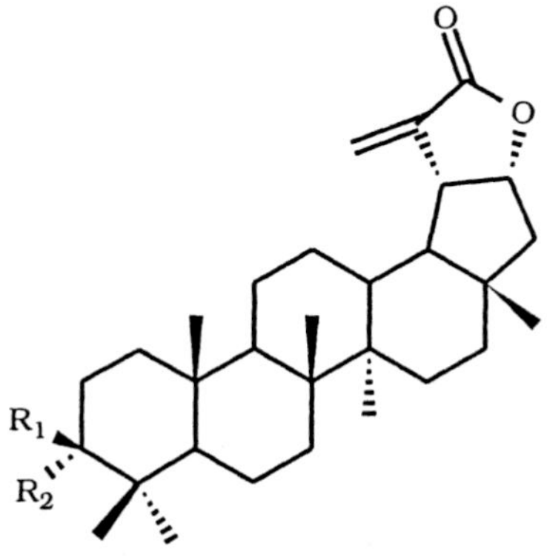

128 Ochraceolide A R₁,R₂ = O
129 Dihydroochraceolide A R₁= OH R₂ = H

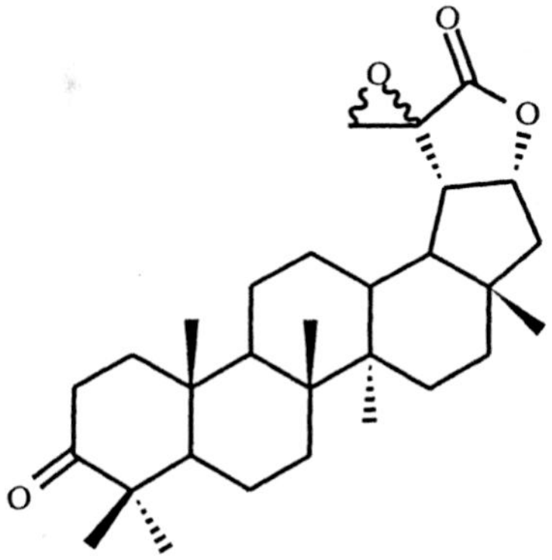

129 Ochraceolide B   20,29-α-Epoxide
130 Epi-ochraceolide B 20,29-β-Epoxide

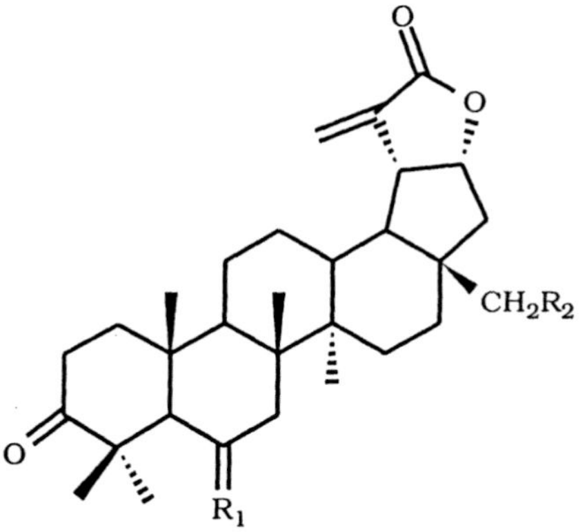

130 Ochraceolide C R₁ = O R₂ = H
132 Ochraceolide E R₁ = H R₂ = OH

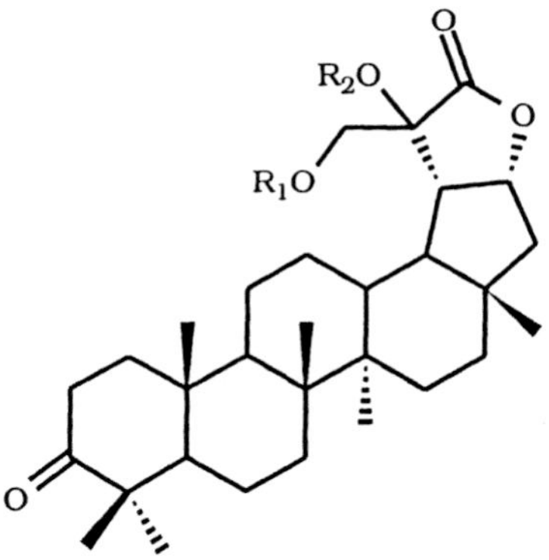

131 Ochraceolide D R₁= R₂= H
133 R₁= Ac, R₂= H
134 R₁= R₂= Ac

Fig. (37). Cytotoxic and FPTasa actives lupane lactones from *Kokoona ochracea*   and *Lophopetalum wallichii*

application as a source of timber and fire wood. The bark contains an inflammable oil, and an asphalt-like residue remains after combustion. The leaves also burn readily, producing a crackling noise like that of a firecracker. Hence, the common name "Repetik" is used in Palawan [210].

The stem bark of this species was studied by Ngassapa *et al.* [211, 212] and the petroleum ether extract and the CHCl₃ phase, were cytotoxic (ED₅₀ <20 µg/ml). Bioassay guide fractionation of the active extracts, using UISO-BCA-1 (human breast cancer) cells, led to the isolation of ochraceolides A (**128**), B (**129**), and C (**130**), which were the lupane lactones: 3-oxolup-20(29)-en-30,21α-olide, 20,29-epoxy-3-oxolupan-30,21α-olide and 3,6-dioxolup-20(29)-en-30,21α-olide. These compounds were the first example of lupanes possessing a γ-lactone located between

C-30 and C-21 of the lupane skeleton. Further investigations of this species yielded the new lupane lactones, 20,29-dihydroxy-3-oxolupan-30,21α-olide (ochraceolide D, **131**) and 28-hydroxy-3-oxolup-20(29)-en-30,21α-olide (ochraceolide E, **132**), Fig. (**37**).

Table 5. *In vitro* **Cytotoxic Activity of Ochraceolides A-D (128-132) and Derivatives from** *Kokoona ochracea*

Cell line	Compound (ED$_{50}$ μg/ml)							
	128	129	130	131	132	137	138	139
P-388	0.26	7.8	0.53	> 5	> 5	4.6	4.8	
UISO-BCA-1	10	> 20	4.5					
BC-1	7.9			> 20	> 20	> 20	> 20	10.2
UISO-COL-2	6.8	> 20	9.2					
COL-2	> 20			> 20	> 20	>20	> 20	12.1
UISO-LUC-1	12	> 20	12					
LU-1	16.1			> 20	> 20	> 20	> 20	8.9
UISO-MEL-2	8.3	> 20	8.3					
MEL-2	> 20			> 20	11.9	20.0	6.2	> 20
HT-1080	9.3	> 20	8.6	14.7	12.2	> 20	> 20	
HT	8.6							9.6
KB	> 20			> 20	> 20	16.7	16.8	> 20
KB-V	17.2							> 20
KB-3	2.7	5.2	5.5					
KB-V1	> 20	> 20	> 20					
KB-V1 (+VLB)				> 20	> 20	11.7	6.7	
KB-V1(-VLB)				> 20	> 20	14.6	18.3	
A-431	> 20			9.5	> 20	> 20	> 20	6.8
LNCaP	13.4			15.5	> 20	> 20	> 20	14.5
ZR-75-1	4.5			> 20	18.8	> 20	11.5	3.0
U-373	6.7			3.9	8.6	> 20	> 20	9.1

P388, murine lymphocytic leukemia; UISO-BCA-1 and BC-1, breast cancer; UISO-COL-2 and COL-2, colon cancer; UISO-LUC-1 and LU-1, lung cancer; UISO-MEL-2 and MEL-2, melanoma; HT-1080 and HT, fibrosarcoma; KB, oral epidermoid; KB-V, multi drug-resistant KB; KB-3, oral epidermoid carcinoma; KB-V1, durg-resistant KB3; KB-V1 (+VLB) and KB-V1 (-VLB), KB3 in the presence and absence of vinblastine, respectively; A-431, epidermoid carcinoma; LNCaP, prostate cancer; ZR-75-1, hormone-dependent breast cancer; U-373, gliobastoma.

Compounds **128-132** and the mono-and di-acetates of ochraceolide D (**133** and **134**, respectively) were evaluated for *in vitro* cytotoxic activity against P-388 murine lymphocytic leukemia cells and a panel of human cancer cell systems. As summarized in Table 5, ochraceolides A (**128**) and C (**130**) exhibited significant cytotoxic activities with cultured P-388 cells with ED_{50} values of 0.26 and 0.53 µg/ml, respectively. They exhibited less intense activity with other cancer cell lines and were not active with KB-V1 cells (ED_{50} >20 µg/ml). Ochraceolide B (**129**), on the other hand, was less active, exhibiting ED_{50} values of 7.8 and 5.2 µg/ml with P-388 and KB-3 cells, respectively. No significant activity was observed with the other cell lines. These results suggested that the presence of an α,β-unsaturated lactone may be important for the observed biological activity, since the 20,29-epoxide in **129** resulted in a reduction of *in vitro* cytotoxicity. Ochraceolide D (**131**) was significantly cytotoxic (ED_{50} 3.9 µg/ml) against human glioblastoma (U373) cells. Compounds **132-134** exhibited only a weak cytotoxic response in certain cancer cell lines. Interestingly, derivatives **133** and **134** were inactive in the U373 cell line, but exhibited a weak activity against other cell systems, including the multidrug-resistant cell line KB-V1. It was also notable that the activity of **134** was more pronounced on KB-V1 in the presence of vinblastine (+VLB), than on KB-V1 in the absence of vinblastine (-VLB) or the parent cell line, KB. These observations indicated that **134** enhances the cytotoxic effect of vinblastine.

The genus *Lophopetalum* is represented by about eighteen species of evergreen trees that grow in Cambodia, India, Laos, Malaysia, Myanmar, Thailand, and Vietnam. Only a few species of this genus have been subjected to prior phytochemical investigation, the most highly studied being *L. toxicum* from the Philippines, whose bark extract is used as arrow poison.

Both the mixed stems and stem bark, and the stems $CHCl_3$-soluble extracts of *L. wallichii* Kurz, were found to display significant inhibitory activity in a farnesyl protein transferase (FPTase) assay system. It has been suggested that inhibitors of this enzyme may be considered as potential anticancer agents for tumors in which products of the *ras* oncogene contribute to transformation. The bioassay directed fractionation of the two active extracts [213] led to the isolation of the known lupane lactones, ochraceolide A (**128**), ochraceolide B (**129**), and the new compound dihydroochraceolide A (**135**), among other known triterpenes. The structure of **135** was confirmed by reduction of **128**, Fig. (**37**) and the stereochemistry to the epoxide group of **129**, not determined when this compound was first isolated from *K. ochracea* [211], was established by preparation of both epoxide isomers, **129** and the new semisynthetic derivative, 20-*epi*-ochraceolide B (**136**) from **128**.

Compounds **128** and **129** exhibited significant inhibitory activity in the FPTasa assay (IC_{50} values of 1.0 and 0.7 µg/ml, respectively); whereas no

significant inhibition was detected for the other compounds. It is
noteworthy that reduction of the ketone carbonyl group at C-3 of
ochraceolide A led to the complete loss of activity against FPTasa assay,
suggesting that this centre is a possible binding site.

Although the $CHCl_3$-soluble extracts were not found to be significantly
cytotoxic (ED_{50} >20 µg/ml), the lupane-type triterpenoid lactones were
tested in a panel of human cancer cell lines. In this case it was reported,
that compounds **128** and **135** were modestly cytotoxic, while **129** and **136**
were not active in the panel (Table 5). On being evaluated in the National
Cancer Institute (NCI) 60-cell line human tumor panel, as well as in a
newly developed hollow fiber assay, ochraceolide A has been selected for
in vivo evaluation at the NCI in several murine xenograft systems.

Ursolic acid (**137**) is a common triterpenoid isolated from members of
the Celastraceae. Ursolic acid has been shown to possess
antiinflammatory, antiarthritic, antidiabetic, antiulcer, as well as
hypolipidemic, and antiartherosclerotic activities. The cytotoxicity of **137**
and seven related esters (**138-144**) against various human tumor cell lines,
was reported for the first time by Lee *et al.* [214], Fig. (**38**). A comparison
of their activity (Table 6) clearly indicated that ursolic acid showed
significant cytotoxicity against the growth of lymphocytic leukemia P-388
and L-1210 as well as the human carcinoma (A-549) cell. It also exhibited
marginal cytotoxicity in the KB, and the human colon (HCT-8) and
mammary (MCF-7) tumor cells. Esterification of the hydroxyl group at C-
3 and the carboxyl group at C-17, such as in derivatives **138-144**, led to
decreased cytotoxicity in the human tumor cell screens; however, in the
murine screen e.g. L-1210 and P-388 cells, the derivatives **138-144**
demonstrated cytotoxicity which was equal or slightly more potent than
the parent compound **137**. Ursolic acid also demonstrated marginal *in vivo*
antileukemic activity (T/C 125 % at 50 mg/kg) against P-388 lympocytic
leukemia.

	R_1	R_2
137	H	H
138	H	$COCH_3$
139	CH_3	H
140	CH_3	$COCH_3$
141	CH_3	$COCH=CHCH_3$
142	CH_3	$COCH_2CH_2$
143	CH_3	$COCH=C(CH_3)_2$
144	CH_3	$CO(CH_2)_{12}CH_3$

Fig. (38). Ursolic acid (137) and derivatives

Table 6. Cytotoxicity of Ursolic acid (137) and Derivitatives Against Various Tumor Cells

Compound	ED$_{50}$ (μg/ml)					
	KB	P388	L-1210	A-549	HCT-8	MCF-7
137	6,6	3.18	4.00	4.0	4.5	4.9
138	3.7	5.87	8.52	7.1	5.8	7.45
139	>4.0	4.24	4.98	6.5	7.6	>10.0
140	>4.0	3.07	3.50	>10.0	>10.0	>10.0
141	>4.0	3.17	2.10	>10.0	>10.0	5.5
142	>4.0	2.64	2.36	7.8	>10.0	>10.0
143	>4.0	2.82	2.71	6.3	>10.0	>10.0
144	>4.0	2.64	2.79	6.3	>10.0	>10.0
etoposide	0.12	2.62	1.58			
vinblastine sulfate				0.002	0.005	
adriamycin HCl				0.15	0.3	

KB, oral epidermoid; P388, lymphocytic leukemia; L-1210, lymphoid leukemia; A-549, lung carcinoma; HCT-8, human colon; MCF-7, mammary gland carcinoma.

Investigations of the ethanolic extract of the stems of *Celastrus hindsii* Benth showed potent cytotoxicity against HEPA-2B (hepatoma), HeLa (cervix carcinoma), COLO-205 (colon carcinoma) and KB (nasopharynx carcinoma) cells. This led Kuo and Kuo [215] to isolate and characterize four new triterpenes: maytenfolone-A (145) celasdin-A (146), celasdin B (147) and celasdin C (148). Their structures were determined by spectroscopic techniques and the structure of 145 was also confirmed by X-ray studies and its spectra were compared with those of maytenfolone (149) which was isolated from the leaves of *M. diversifolia* [216]. In addition, the known friedelin (150) and canophyllol (151) were also isolated. These compounds were tested for cytotoxicity in the above four cancer cell lines and in order to establish a structure-activity relationships a number of derivatives (152-154) were prepared, Fig. (39). Compound 145 showed cytotoxicity against HEPA-2B (ED$_{50}$ 2.3 μg/ml) and KB (ED$_{50}$ 3.8 μg/ml) cells, whereas the other compounds were inactive due to the ED$_{50}$ >10 μg/ml, under the reference compound, mitomycin C (ED$_{50}$ <0.04 μg/ml). These results indicated that the friedelin derived seven-membered ring lactone between C-17 and C-20 may be important if the hydroxyl is located at C-16.

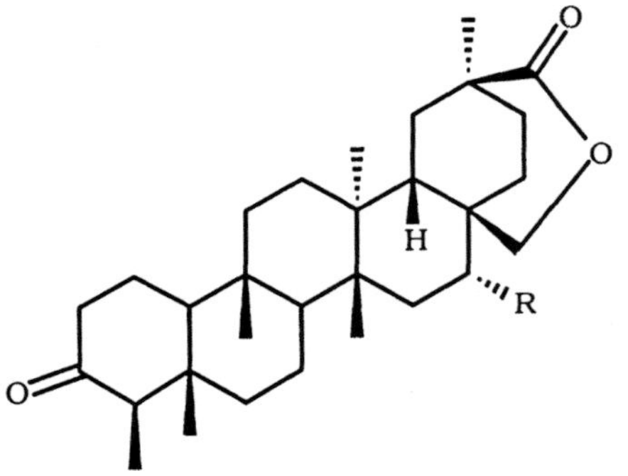

145 Maytenfolone A R = OH
149 Maytenfolone   R = H
152   R = OAc

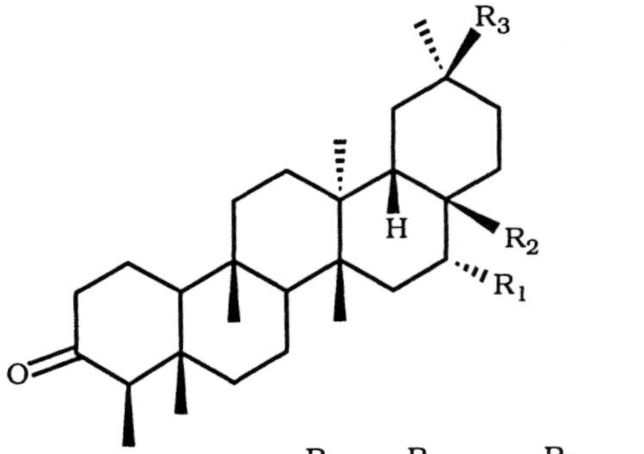

	R_1	R_2	R_3
146 Celasdin A	OH	CH_3	CH_2OH
147 Celasdin B	OH	CH_2OH	CH_3
150 Friedelin	H	CH_3	CH_3
151 Canophyllol	H	CH_2OH	CH_3
153	OAc	CH_3	CH_2OAc
154	OAc	CH_2OAc	CH_3

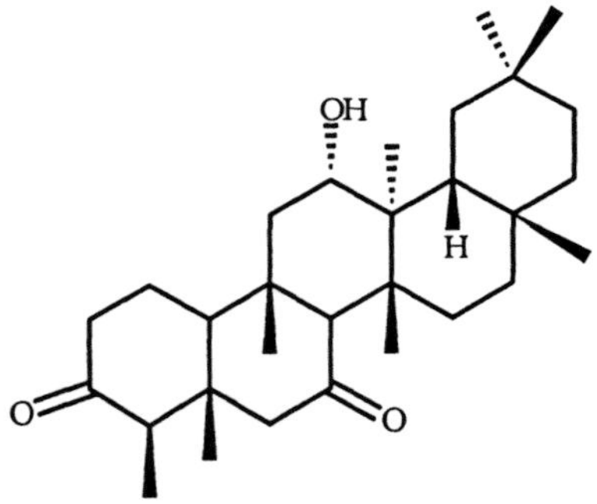

148 Celasdin C

Fig. (39). Some triterpenes and derivatives from *Celastrus hindsii* assayed for antitumor and anti-AIDS activities

Kubo and Fukuhara [217] found that the methanolic extract of the root bark of *Elaeodendron bunchananii* exhibited various biological activities, such as insecticidal, molluscicidal, antimicrobial, and cytotoxic. This extract showed cytotoxic activity (ED_{50} 100 µg/ml) against L-1210 leukemic cells in a preliminary screening. The isolation of the bioactive principle was guided by the cytotoxicity assay and led to the isolation of a new dammarane-type triterpene, 16α-hydroxy-dammara-20,24-dien-3-one (elabunin, **155**), whose structure was determined by spectroscopic analysis and CD studies, Fig. **(40)**. There are many reports of naturally occurring dammarane compounds, however, a dammarane structure with a 16-OH is a rarity. Elabunin exhibited moderate cytotoxic activity (ED_{50} 1 µg/ml) against L-1210 leukemia cells; however, lupeol which was also isolated from the active fraction, did not exhibit cytotoxicity.

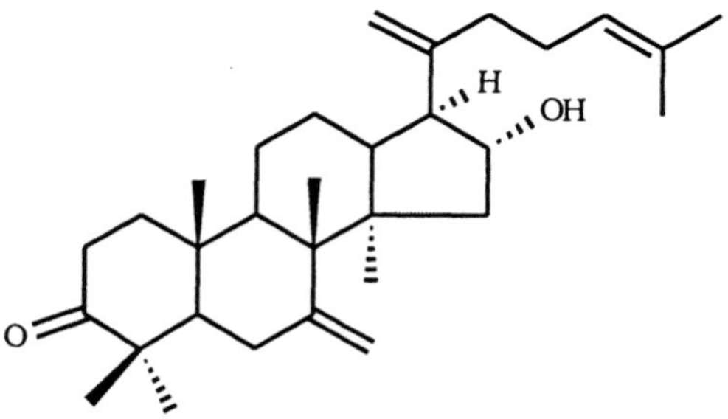

155 Elabunin

Fig. (40). Cytotoxic dammarane triterpene from *Elaeodendron buchananii*

4.4.2. Antiviral Activity

The extract of the root bark of *Tripterygium wilfordii*, after its purification and removal from the bark, furnished a fraction which is known commercially as "Lei-gong-teng duo-dai" (LGTDD), which has been used for the treatment of several diseases, such as dermatitis, rheumatoid arthritis, systemic acne rosacea, and nephritis, with good results and without hormonal side effects. LGTDD was found to show significant anti-HIV activity [218] and a bioassay-directed fractionation has led to the isolation of salaspermic acid (**156**), previously isolated from *Salacia macrosperma* [194], as the anti-HIV principle from the CHCl$_3$-soluble fraction. X-ray analysis of a crystal of the monohydrate of **156** confirmed the structure and provided details of the molecular geometry.

Salaspermic acid demonstrated an inhibitory effect against HIV-1 recombinant reverse-transcriptase-associated reverse transcriptase activity, and it also exhibited an inhibitory effect against HIV-1 reverse-transcriptase-associated DNA polymerase activity. On the other hand, **156** showed no inhibitory effect upon the HIV-reverse-transcriptase-associated RNAse H-activity, when tested at a concentration of 100 µg/ml with Poly(dT).[3H]Poly(rA) as substrate. The behaviour of **156** was also tested against HIV-2 recombinant reverse-transcriptase-associated reverse transcriptase activity, but the degree of inhibition appeared to be lower (EC$_{50}$ 100 µg/ml) with Poly(rC).(oligo(dG) as template-primer. In order to investigate structure-activity relationships, the HIV-replication-inhibitory effect of **156** and related compounds (**110-113, 151, 156-161**) isolated from various plant sources, was evaluated, Fig. (**41**). The results (Table 7) indicated that the acetal linkage in ring A and carboxyl group in ring E may be required for the anti-HIV activity.

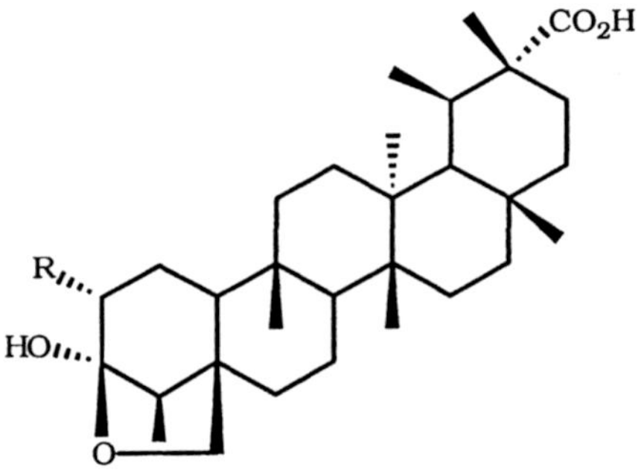

156 Salaspermic acid R = H
157 Orthosphenic acid R= OH

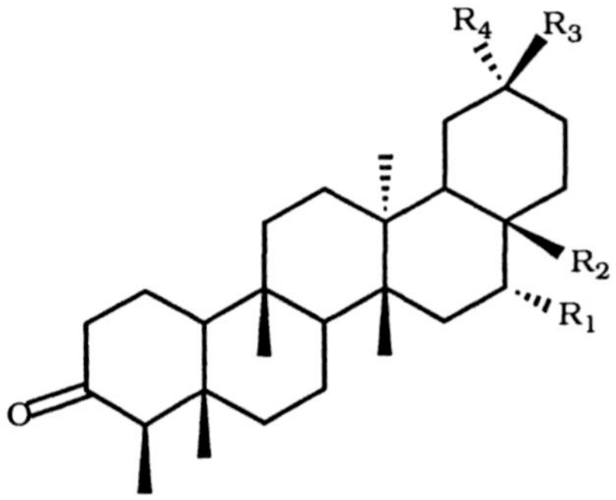

	R_1	R_2	R_3	R_4
110	H	CH_2OH	CH_3	CH_2OH
111	H	CH_3	CO_2H	CH_3
112	H	CO_2H	CH_3	CH_3
113	H	CH_2OH	CH_2OH	CH_3
151	H	CH_2OH	CH_3	CH_3
158	OH	CH_3	CH_3	CH_3
159	H	CH_3	CH_3	CO_2H
160	H	CO_2H	CH_3	CH_3

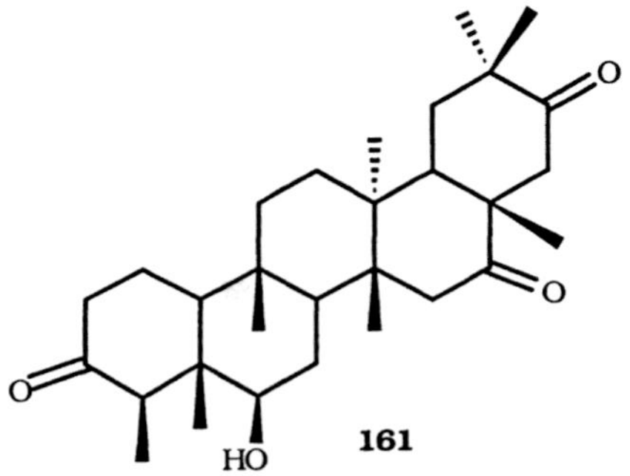

161

Fig. (41). Salaspermic acid and related compounds assayed for anti-HIV activity

Table 7.   HIV Inhibition by Salaspermic Acid (156) and Related Compounds

Compound	IC_{50}[a]		EC_{50}[b]	
	µg/ml	µM	µg/ml	µM
156	25	53	5	10
110	31	64	50	103
111	9	20	6	13
112	7	15	5	11
113	8	18	2	4
151	10	22	7	15
157	3.5	8	0.9	2
158	20	44	30	66
159	20	44	27	59
160	17	38	17	38
161	50	106	>50	>106

[a]Concentration which inhibits H9 lymphocytes cell growth by 50%.
[b]Concentration which inhibits virus replication by 50%.

Investigations of an ethanolic extract of the stems of *C. hindsii* [215], showed potent HIV replication activity in H9 lymphocytes *in vitro*. In the anti-HIV bioassay of compounds 145-151, Fig. (39), isolated from this extract, the anti-HIV potential was poor, except for 145 and 147, because the EC_{50} values exceeded 10 µg/ml. Compound 147 inhibited HIV replication in H9 lymphocyte cells with an EC_{50} of 0.8 µg/ml and exhibited toxicity at 5.5 µg/ml. This therapeutic index of 147 was better than that of 145 with an EC_{50} of 1.8 µg/ml and a lower toxicity at 7.0 µg/ml.

Betulinic acid (162), a common metabolite in the Celastraceae species, was found to be an inhibitor of HIV replication in H9 lymphocyte cells, with an EC_{50} value of 1.4 µM and inhibited uninfected H9 cell growth with an IC_{50} value of 13 µM. A comparison of the activity of 162 and eight derivatives (163-170), Fig. (42), clearly indicated that dihydrobetulinic acid (170), with EC_{50} and IC_{50} values of 0.9 and 13 µM, respectively, was as potent or slightly more potent than 162 in inhibiting HIV replication in H9 lymphocytes. The results obtained (Table 8)

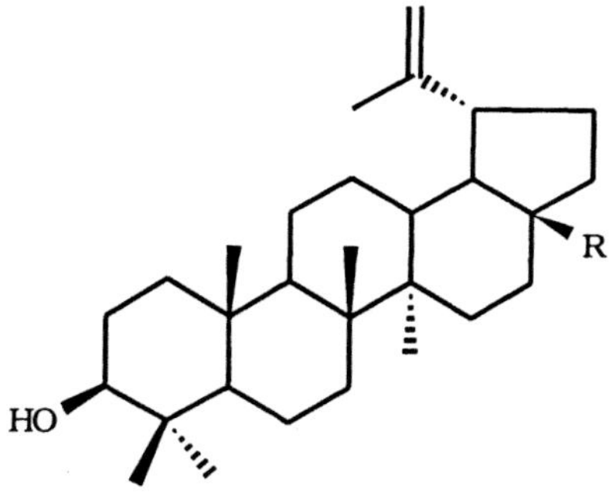

162	Betulinic acid R = CO_2H
163	Betulin R= CH_2OH
164	R = $CO_2CH_2CO_2CH_3$

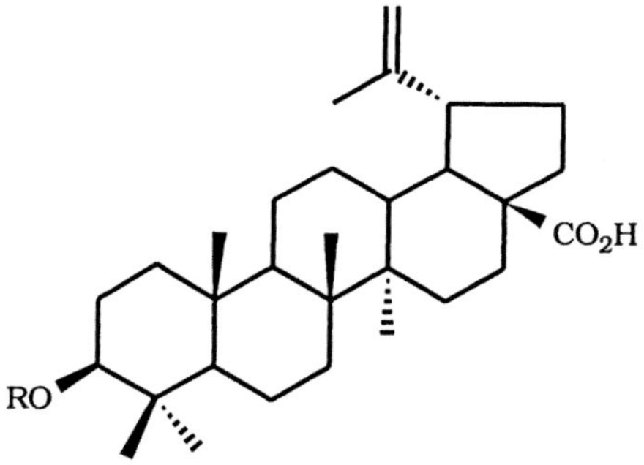

165	R = $COCH_3$
166	R = COC_6H_6
167	R = $COCH=CHCH_3$
168	R = SO_3K
169	R = $COCH_2CH_2CO_2H$

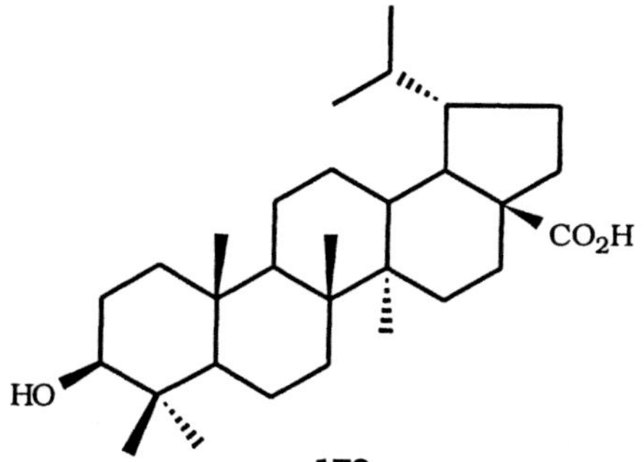

170

Fig. (42). Betulinic acid and related triterpenes assayed anti-AIDS activity implied that the free hydroxy group at C-3, the carboxylic acid group at C-17, as well as the substituents at C-19, play an important role in their activity. The inhibitory activity of these compounds against protein kinase C (PKC) was also examined, since a correlation between anti-HIV and anti-PKC activities had been suggested. However, there was no apparent correlation between anti-HIV activity and the inhibition of PKC among these compounds [219].

Table 8.   HIV Inhibitory Effects (μM) for Betulinic (162) Acid and Derivatives

Compound	EC_{50} [a]	IC_{50} [b]	T.I. [c]
162	1.4	13	9.3
163	0.9	13	14
164	23	45	1.9
165	17	26	1.5
166	25	20	0.8
167	15	15	1.0
168	19	48	2.5
169	20	35	1.7
170	4	16	4.0

[a] Concentration which inhibits virus replication by 50%.

[b] Concentration which inhibits H9 lymphocyte cell growth by 50%.

[c] Therapeutic index.

4.4.4. Other Biological Activities

β-Sitosterol, betulin, and betulinic, ursolic and oleanolic acids, common metabolites in the plant kingdom, are also present in species of the Celastraceae, and have been reported to exhibit a wide range of biological activities.

In this context, β-sitosterol has been reported [220] to possess potent anti-inflammatory and antipyretic activities and its effect was similar to hydrocortisone and oxyphenbutazone on carrageenin induced oedema and it was independent of the pituitary adrenal system; moreover, β-sitosterol had a wide margin of safety with minimal ulcerogenic activity and may be of therapeutic value.

The lupane triterpene, betulin (**163**) showed inhibitory effects on Epstein-Barr virus activation and bacteriostatic activities against *Escherichia coli*, *Salmonella typhi*, *Shigella flexneri* and *Staphylococcus aureus*. Antitumor activity was reported against human epidermoid carcinoma of the nasopharynx *in vitro* and Walker 256 (5WA16) tumor system, but not against L-1210 lymphoid leukemia in mice or Walker carcinosarcoma 256 in rats. For betulin and betulin-containing herbs antiseptic antiinflammatory, feeding deterrent and insecticidal activities have also been mentioned [221].

Besides anti-HIV activity, betulinic acid (162) has also shown antitumor activity against human melanoma by induction of apoptosis and against mouse skin two-stage carcinogenesis, leishmanicidal activity against amastigotes of *Leishmania amazonensis*, spasmogenic activity on isolated rat fundus, anti-inflammatory activity in mouse paw and ear edema tests, and antimalarial activity against asexual erythrocytic stages of the human malaria parasite *Plasmodium falciparum* [222].

Two triterpene carboxylic acids, ursolic acid (137) and oleanolic acid (171), Fig. (43) have been reported as inhibitors of 12-*O*-tetradecanoylphorbol-13-acetate (TPA) induced Epstein-Barr virus (EBV) activation in Raji cells. Both acids significantly inhibited the activation at a 1000-fold molar ratio to TPA. The dose responses of the acids were very similar to those of the antitumor promoters, retinoic acid and glycyrrhetinic acid. The results suggested that ursolic and oleanolic acids may be valuable antitumor promoters [223].

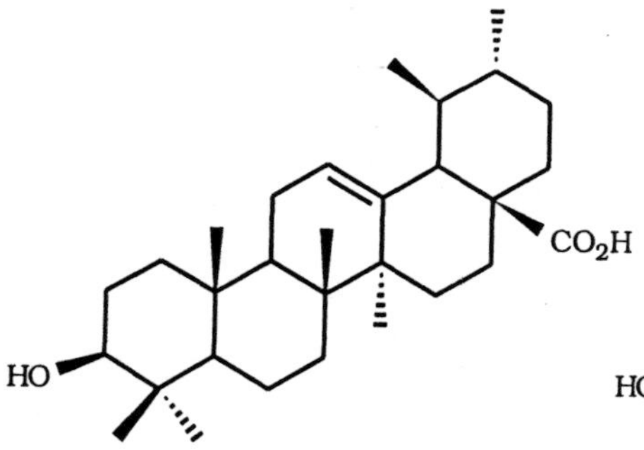

137 Ursolic acid 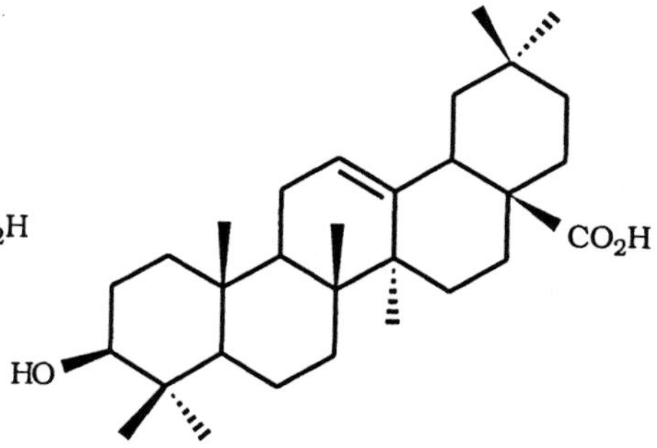   171 Oleanoic acid

Fig. (43). Antitumor promotor activity of ursolic and oleanoic acid

4.5. Triterpenoid Quinonemethides and Related Compounds

The triterpenoid quinonemethides constitute a group of unsaturated and oxygenated D:A-friedo-*nor*-oleananes, and in nature these pigments are found restricted to the Celastraceae. They are considered to be indicators of the family and Brüning and Wagner [6] coined the general name "celastroloids" for this class of compounds.

In general, celastroloids incorporate the 24-*nor*-D:A-friedo-oleanane nucleus, Fig. (**44**) and invariably contain oxygenated functionalities at C-2 and C-3. Although comparatively small in number, they constitute a group of triterpenoids with unique structural features and biological activities. These compounds have shown interesting biological properties and their antibiotic [224, 225] and cytostatic [226, 227] activities have been intensively investigated. In addition, several studies on the structure-activity relationships [228, 229] and mechanisms of action [230, 231] of this type of triterpenes have been published. Besides antimicrobial and antitumor activities celastroloids have also shown antimalarial (schizontocidal) [232], and spermicidal activities [233].

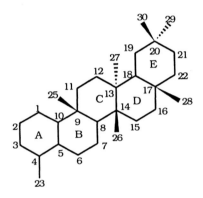

Fig. (44). Basic nucleus of celastroids

The first scientific paper dealing with celastroloids was published in 1939 by Gisvold [102] who isolated a red pigment named celastrol (**172**) from *Celastrus scandens*. In 1942 Schechter and Haller [234] reisolated from *T. wilfordii* tripterine previously reported by Chinese workers [101] and showed that it was identical to celastrol. In the early 1950's Bhatnagar and Divekar [235] were prompted to initiate chemical studies on an Indian Celastraceae species, *Pristimeria indica,* whose roots were claimed to be effective against respiratory diseases. They isolated the antibacterial principle from the root bark of *P. indica* which separated as bright orange needles and was named pristimerin (**173**). The probable relationship between pristimerin and celastrol was noted by Kulkarni and Shah [236] and later by Kamat *et al.* [237]. However, it was Nakanishi's group who established that pristimerin was the methylated derivative of celastrol [238].

The pioneering work of Bhatnagar and Divekar encouraged subsequent workers to search for bioactive compounds in the Celastraceae species and since the discovery of celastrol (**172**) and pristimerin (**173**), a variety of celastroloids have been encountered in plants of this family.

Celastroloids may be classified in several different ways depending on their structural features and a classification based on the nature of the main chromophore could be considered. According to its conjugation there are four different structures of norquinonemethide triterpenoids; those with

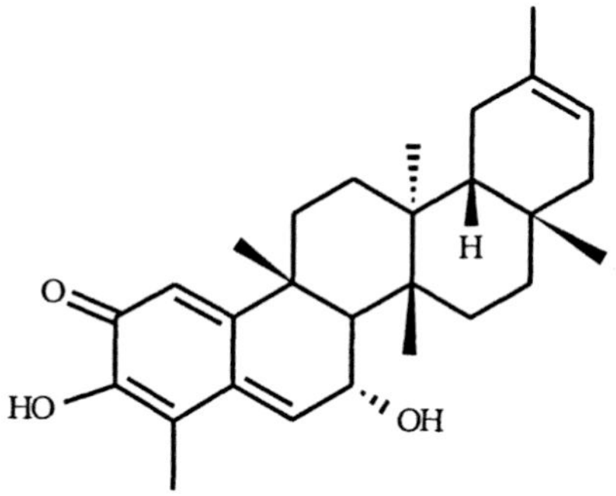

174 7α-Hydroxy-7,8-dihydro-iguesterin

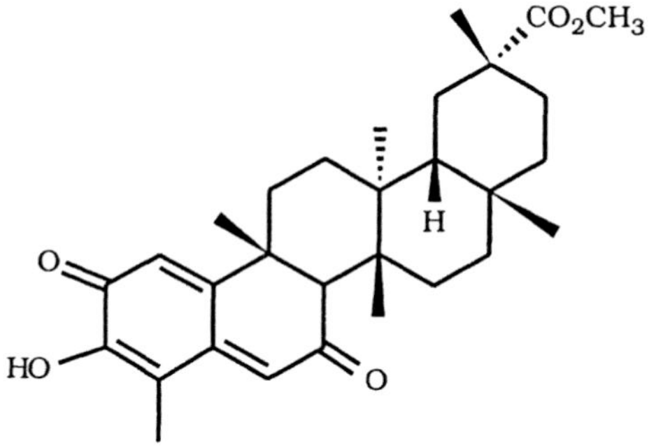

175 Dispermoquinone

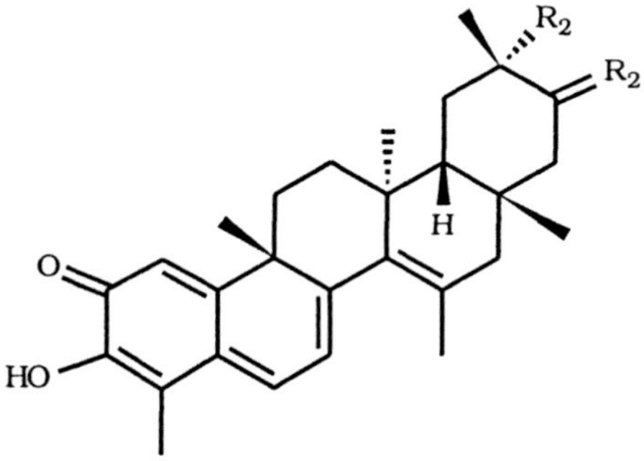

173 Pristimerin R$_1$=CO$_2$CH$_3$ R$_2$= 2H   **177** Netzahualcoyone
176 Tingenone R$_1$=H   R$_2$=O

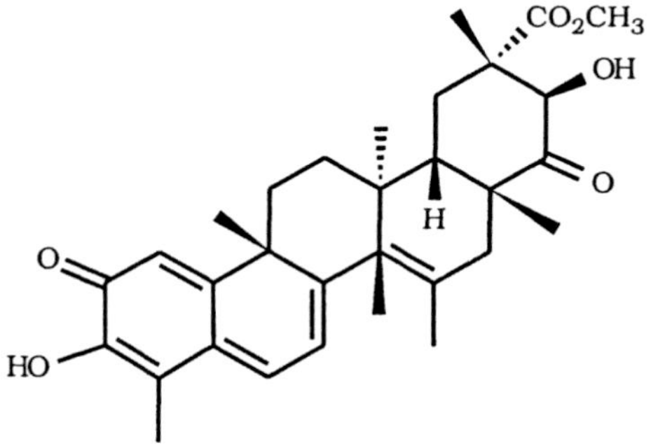

Fig. (45). Examples of quinonemethide triterpenoids with different grade of conjugation

non-extended conjugation as in 7(-hydroxy-7,8-dihydro-iguesterin (**174**) [239]; those with an extended conjugation of the quinone methide A ring with a carbonyl group as in the case of dispermoquinone (**175**) [240]; those with an additional double bond as in pristimerin (**173**) or tingenone (**176**) [241]; and the quinones with an additional dienic system as with netzahualcoyone (**177**) [196], Fig. (**45**).

The most common and basic class of celastroloids are those with a quinonemethide chromophore extended over the A and B rings of the triterpenoid skeleton. In all of them, the C-29 methyl group has undergone partial oxidation to -CH$_2$OH (eg. excelsine, **178**) [242] or complete oxidation to -CO$_2$H (e.g. celastrol, **172**) sometimes followed by methylation to produce -CO$_2$Me (e.g. pristimerin, **173**). Several examples are also known in which C-29 is lost possibly as a result of a decarboxylation process (e.g. tingenone, **176**). 29-*Nor* celastroloids with an endocyclic (e.g. iguesterin, **179**) [243] and exocyclic double bond in ring E have also been described, Fig. (**46**).

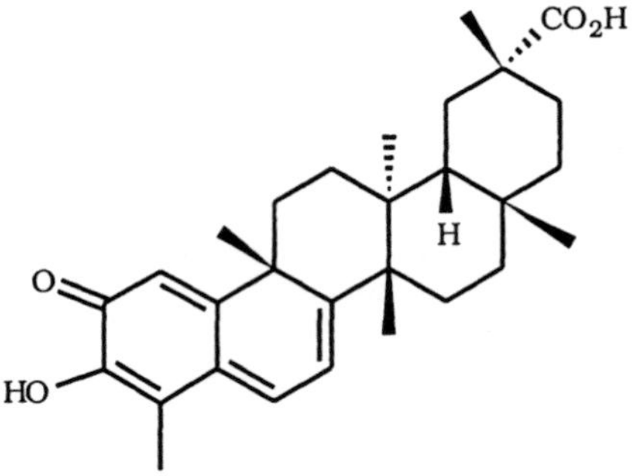

172 Celastrol

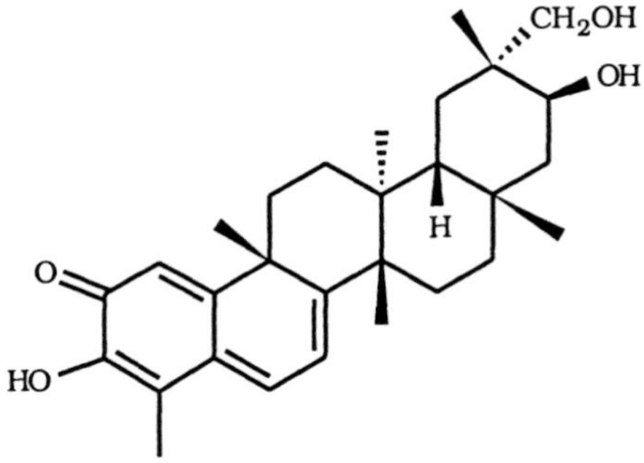

178 Exelsine

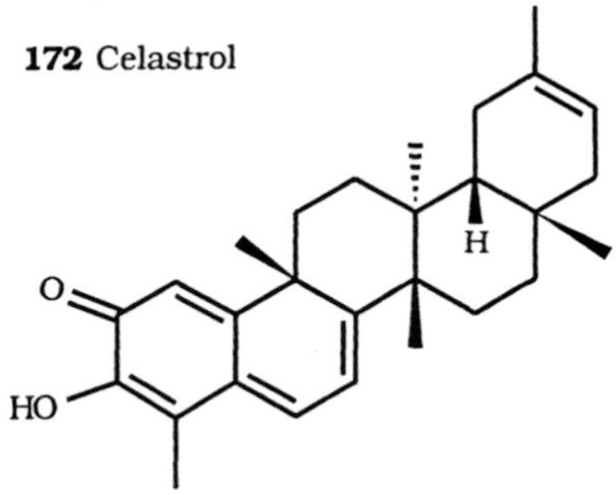

179 Iguesterin

Fig. (46). Some quinonemethide triterpenoid skeleton

The second major group of celastroloids are constituted by those compounds with a chromophore extended to the ring D of the triterpenoid

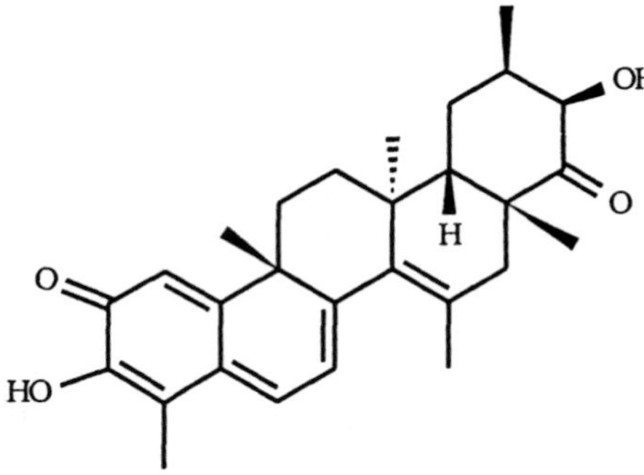

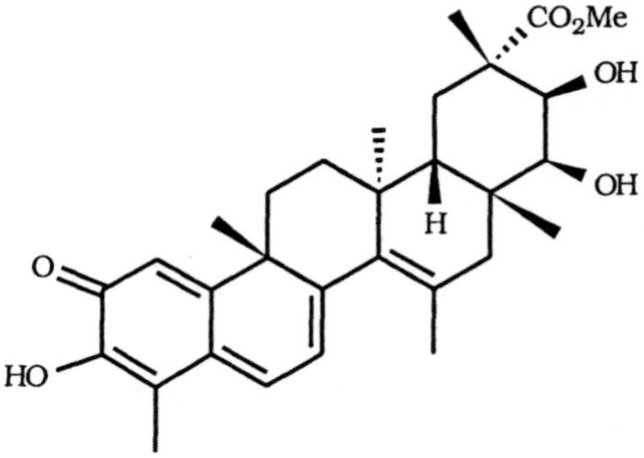

180 Balaenonol

181 Netzahualcoyondiol

Fig. (47). Examples of 14(15)-enequinonemethide triterpenoids

with an additional double bond at C-14(15) and a rearrangement of the methyl group originally present at C-14 to C-15. The majority of these compounds are oxidized at C-21, C-22 and C-29 in ring E, as exemplified by balaenonol (**180**) [244] and netzahualcoyondiol (**181**) [196], Fig. (**47**).

Phenolic and 6-oxophenolic triterpenoids have also been reported. Zeylasterone (**182**) was the first celastroloid in this class to be described [245], and contains a 6-oxo group in addition to the aromatic (phenolic) A ring. Since then a variety of phenolic triterpenoids have been isolated and characterized. Phenolic nortriterpenes with the pristimerin skeleton (e.g. 23-*nor*-6-oxopristimerol, **183**) have been isolated from *Kokoona zeylanica* [246], although phenolic compounds with the tingenone-type skeleton have been reported only as dimer constituents, Fig. (**48**). Recently, a phenol triterpenoid with an iguesterin skeleton (6-oxo-iguesterol, **205**) have been published by González *et al.* [247].

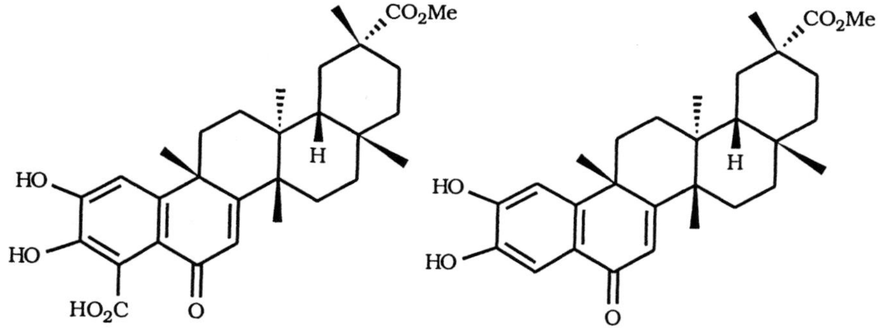

182 Zeylasterone   **183** 23-*Nor*-6-oxopristimerol

Fig. (48). Examples phenolic triterpenoids

Triterpene dimers of this class have been isolated from only five plants in the Celastraceae, *Rzedowskia tolantonguensis*, *Maytenus ilicifolia*, *M. umbellata*, *M. chuchuhuasca*, and *M. scutiodes*, and have been studied by only two groups, Itokawa *et al.* and González *et al.* Most of these compounds were found to be composed of one quinoid unit and one aromatic unit of triterpenes, derived from pristimerin, tingenone, and/or their congeners, and joined by two ether linkages between the two A rings, like cangorosins. The structures of cangorosins were reported [248] to consist of two triterpene units joined by an ether bond formed between the two A rings, however these structures were revised [249], concluding that these triterpene dimers were linked by two ether bridges between the two A rings for cangorosin B (**184**) and between the A and B rings for the cangorosin A (**185**) series, Fig. (**49**).

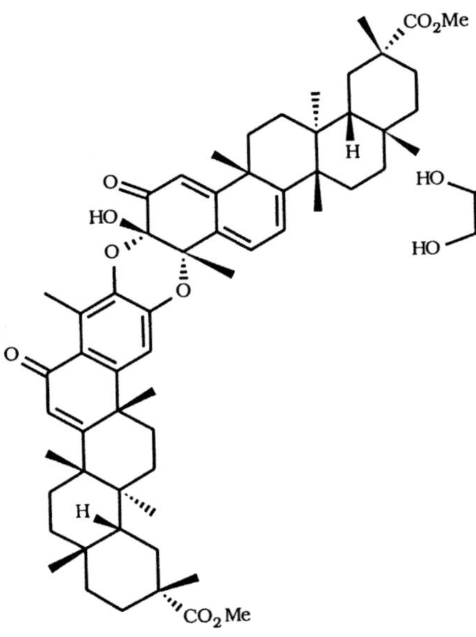

184 Cangorosin B

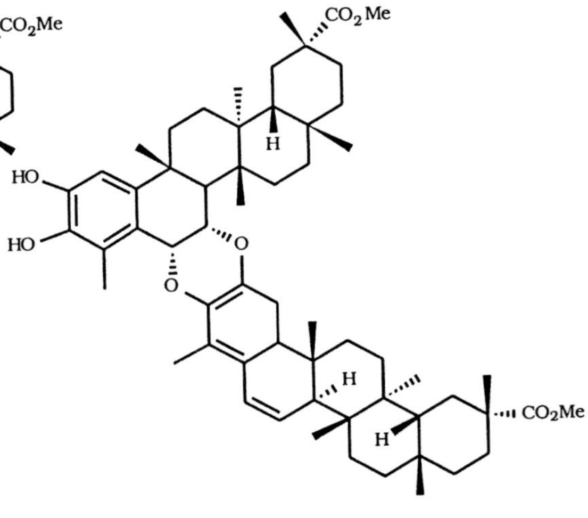

185 Cangorosin A

Fig. (49). Triterpene dimers from *Maytenus ilicifolia*

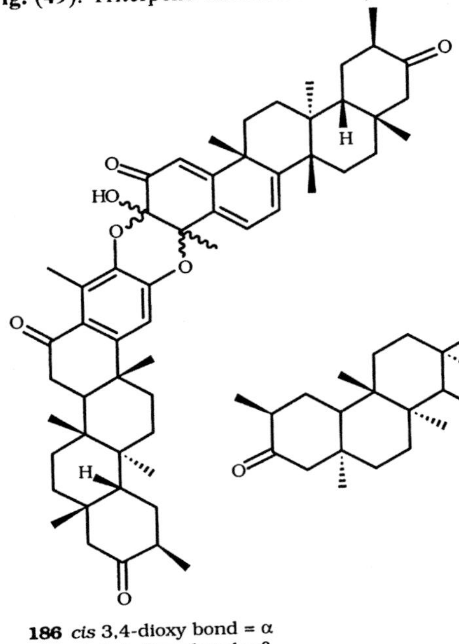

186 *cis* 3,4-dioxy bond = α
187 *cis* 3,4-dioxy bond = β
188 7,8-dihydro of **186**

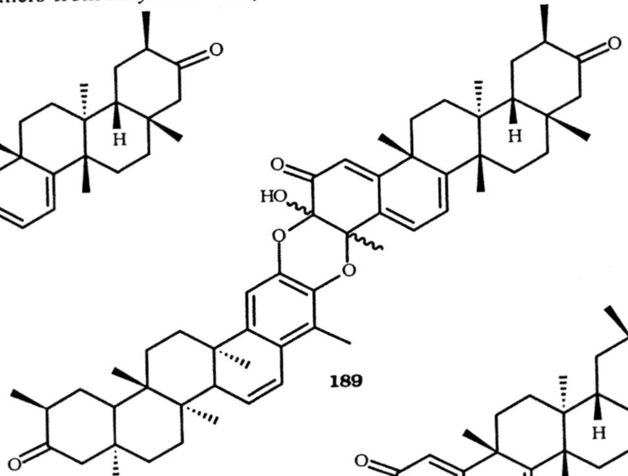

189

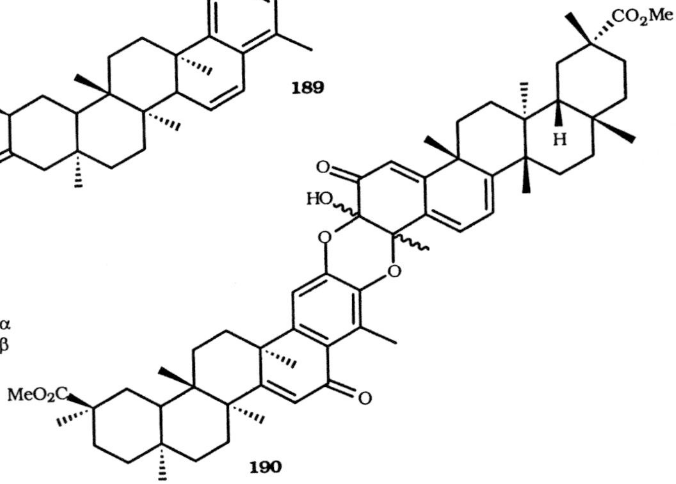

190

Fig. (50). Triterpenes dimers from *Maytenus chuchuasca*

Recently, Itokawa and coworkers [250] reported the isolation of five new triterpene dimers (186-190) from the South American plant "xuxua" (*Maytenus chuchuhuasca*), and their structures, representative regiochemical and stereochemical isomers, were determined on the basis of spectroscopic evidence including CD spectral, Fig. (50).

A review which contains comprehensive discussions on distribution, structure elucidation, chemistry, biological activity and biosynthetic aspects of celastroloids, covering the literature up to 1994, have been published by Gunatilaka [251]. For this reason, the present review only emphasises the results on the biological active celastroloids reported during the period 1994 to 1997.

4.5.1. Antimicrobial and Cytotoxic Activities

Maytenus scutioides (Lourteig and O'Donnell) is a subtropical shrub that is distributed in the Chaco (North of Argentina and South of Paraguay and Bolivia). The aerial part of this plant has been used as cardiotonic and the roots as abortive by the inhabitants of these regions. Recently, in their research for new bioactive compounds, González *et al.* [252] investigated the root bark of *M. scutioides* for cytostatic and antibiotic constituents. Thus, by bioactivity-directed fractionation, scutione (191), a new

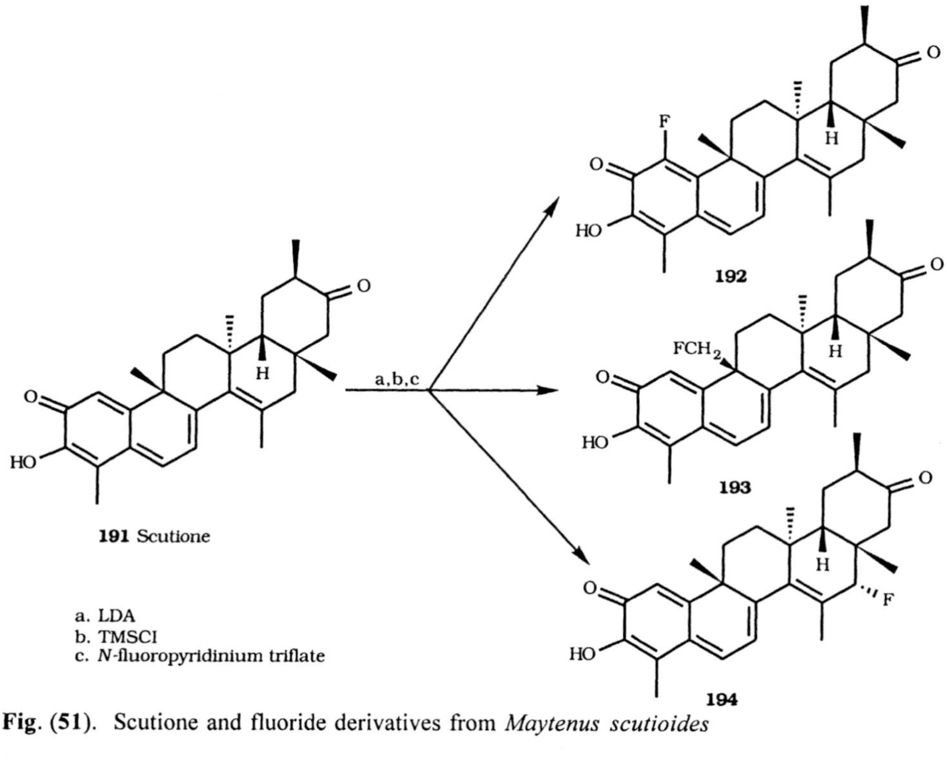

191 Scutione

a. LDA
b. TMSCI
c. *N*-fluoropyridinium triflate

192

193

194

Fig. (51). Scutione and fluoride derivatives from *Maytenus scutioides*

norquinonemethide triterpene with a netzahualcoyene type skeleton, was isolated.

Three fluoride derivatives (**192-194**) from **191** were prepared given the potential effects of the introduction of fluorine into organic compounds, Fig. (**51**). The results obtained in the antimicrobial assay showed that the effect of scutione (**191**) and its fluoride derivatives **192-194** was limited to Gram-positive bacteria (Table 9), and unexpectedly scutione was more active than the fluoride derivatives.

The antimicrobial activity of scutione (**191**) and the related triterpenequinones, netzahualcoyone (**177**) which differs from scutione in the presence of a methyl ester group on C-20, a hydroxyl group on C-21 and a carbonyl group on C-22 and tingenone (**176**), whose ring E is identical to that in scutione, but lacks the double bond in ring D, were compared. Scutione was more active that the two related compounds, indicating that the position of the ketonic group and the double bond was relevant for this activity.

Table 9. Minimal Inhibitory Concentrations (μg/ml) of Scutione (191) and Derivatives Against Gram-positive Bacteria

Bacteria	191	192	193	194	Control [a]
S. aureus	2	10-15	15	3-4	2-5
S. albus	0.5	2.8	3.6	2.7	1
S. epidermidis	0.2-0.3	0.5	1.5-2	2.7	5
S. warnieri	2				
S. saprophyticus	1-2				4
M. luteus	0.60	15	2	2.7	1
B. subtilis	1-0.16	0.2	0.7-0.9	1-1.5	2-5
B. pumilus	0.2	0.4-0.6	1.5-2	2.5-3	> 10
B. alvei	0.3				8
B. megaterium	0.2-0.4				
B. cereus	0.1-0.2				

[a] cephotaxime was used as the positive control.

Cytotoxic activity of compounds **191-194**, against three cell lines in culture, were also assayed and the results are summarized in Table 10. None of these compounds showed significant activity as compared with that of mercaptopurine (0.1 μg/ml), which was used as a positive control. Introduction of fluorine again resulted in a decrease of activity, except for **192** where a slight increase was observed.

Table 10. Cytotoxic Activity of Scutione (191) and Derivatives

Cell line	IC$_{50}$ (µg/ml)				
	191	192	193	194	Control
HeLa	4.9	2.6	24	> 40	0.1
Hep-2	5.6	9	> 40	> 40	0.6
Vero	7.2	5.2	34	> 40	

Mercaptopurine was used as the positive control.

Further investigation of the (n-hexane:Et$_2$O; 1:1) extract of *M. scutioides* yielded eight new dimer compounds (**195-202**) and their structures were determined on the basis of spectroscopic evidence and their absolute configurations by means of CD studies [253], Fig. (**52**). Their possible biogenetic route was discussed and one unnatural dimer was also synthesized by a hetero Diels-Alder reaction. All dimer compounds were found to be composed of one quinoid-type triterpene, derived from pristimerin, 7,8-dihydro-pristimerin or netzahualcoyene and one aromatic triterpene, derived from pristimerin or 6-hydroxy-pristimerin, linked together by two ether linkages between the two A rings.

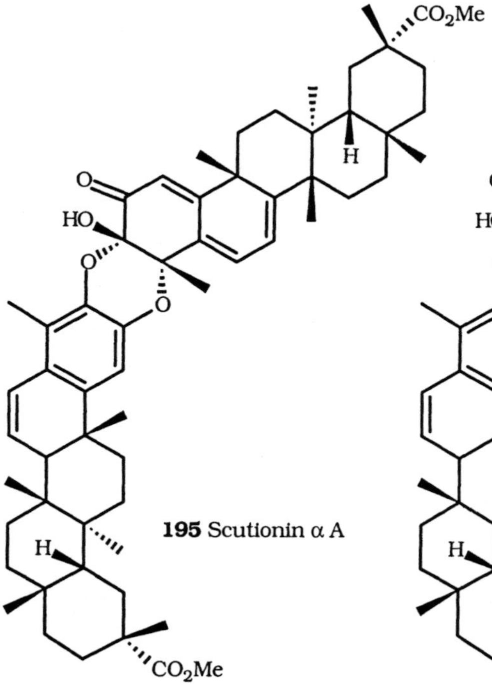

195 Scutionin α A

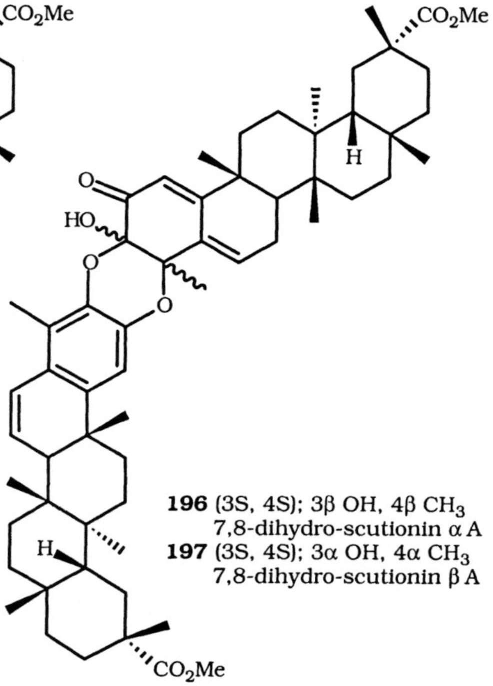

196 (3S, 4S); 3β OH, 4β CH$_3$
7,8-dihydro-scutionin α A
197 (3S, 4S); 3α OH, 4α CH$_3$
7,8-dihydro-scutionin β A

(Fig. 52). contd.....

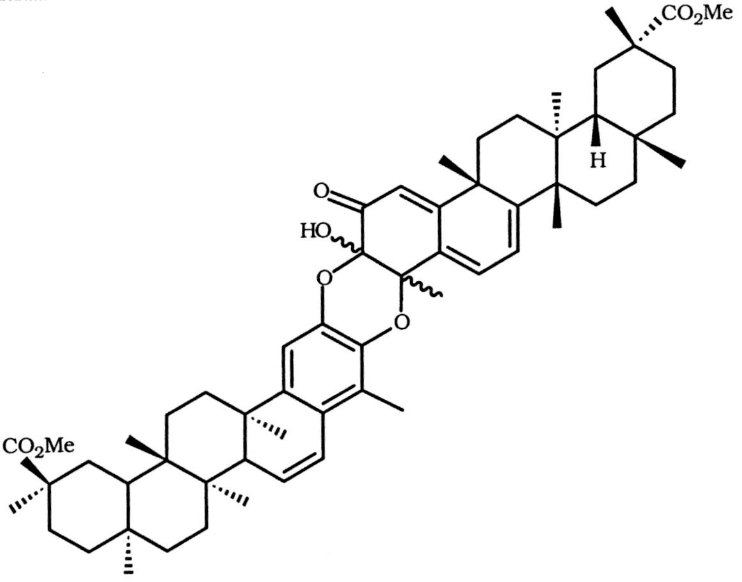

198 (3S, 4S); 3β OH, 4β CH$_3$
7,8-dihydro-scutionin α B
199 (3S, 4S); 3α OH, 4α CH$_3$
7,8-dihydro-scutionin β B

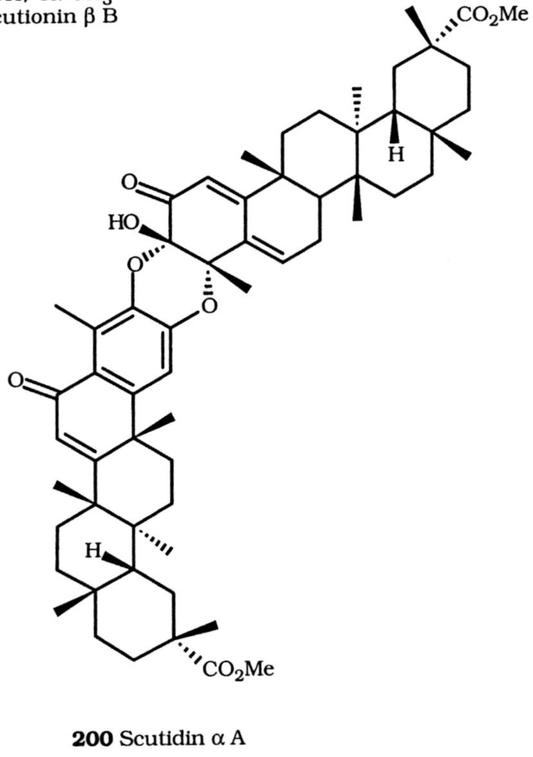

200 Scutidin α A

(Fig. 52). contd.....

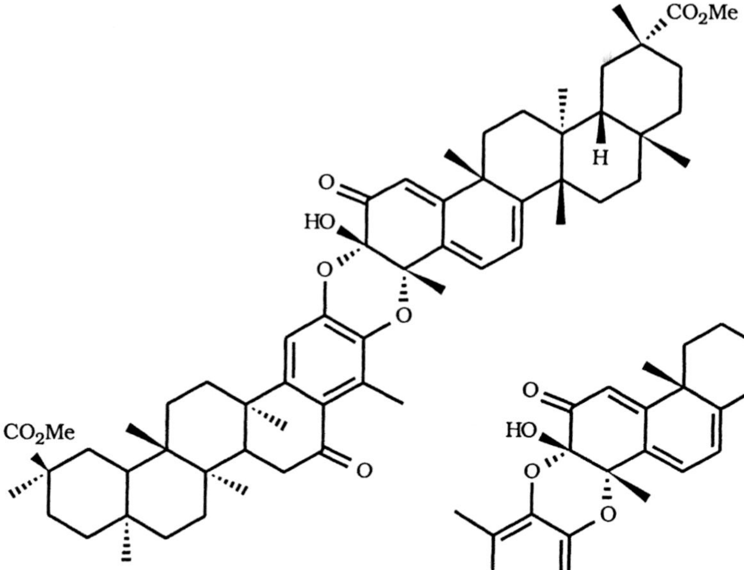

201 7,8-dihydro-scutionin α B

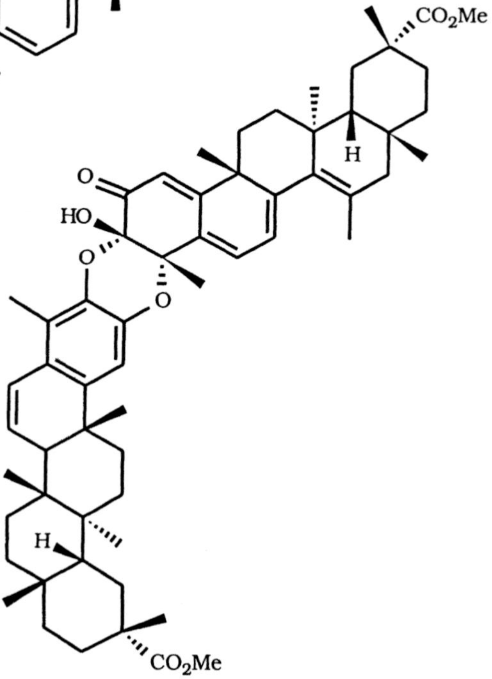

202 Netzascutionin α A

Fig. (52). Triterpene dimers from *Maytenus scutioides*

These compounds were assayed for antibiotic and cytotoxic activities and from them compound **202** was the only dimer that showed antibiotic activity against Gram positive bacteria; its MICs on *Bacillus subtilis* and *B. pumilus* were 1-2 and 20 μg/ml, respectively. None of the dimers showed cytotoxic activity against the HeLa (cervix carcinoma) and Hep-2 (larynx carcinoma) cell lines at least at 20 μg/ml. These results support the fact that the additional double bond C-14(15) plays an important role for the antibiotic activity, as **202** was the only assayed dimer with a triterpene quinoid unit derived from netzahualcoyone skeleton, instead of pristimerin.

The new phenolic nortriterpenes with the tingenone or iguesterin skeleton, 6-oxo-tingenol (203), 3-*O*-methyl-6-oxo-tingenol (204) and 6-oxo-iguesterol (205), were isolated from the root bark of *Maytenus canariensis* by González et al. [247], Fig. (53). Their structures were determined by spectroscopic studies and chemical transformation and the synthesis of 203, achieved from tingenone, was also reported.

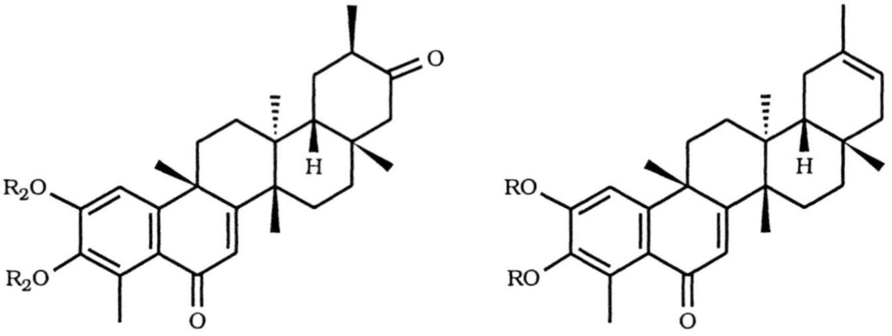

	203	204	206	207
R₁	H	H	CH₃	Ac
R₂	H	CH₃	CH₃	Ac

205   R = H
208   R = CH₃

Fig. (53). Phenolic triterpenes from *Maytenus canariensis*

The new compounds (203-205) and the derivatives 206-208 were assayed for antimicrobial activity. Compounds 203-205 showed antibiotic activity against *B. subtilis* with a MICs of 12-14, 35-39 and 25 µg/ml, respectively; 203 was also active against *Staphylococcus aureus* (MIC 40-50 µg/ml). These results suggest that the antibiotic activity may be associated with the presence of free hydroxyl groups in ring A; compound 203, with two hydroxyl groups was about three times more active than 204, with one methoxy group; while the dimethyl derivative 206 and the diacetate 207, were inactive. On the other hand, all compounds were inactive (MIC > 100 µg/ml) against the Gram-negative bacteria and the yeast *Candida albicans*.

4.5.2. Other Biological Activities

In the course of a search for bioactive metabolites from plants, Takaishi *et al.* [254] investigated the inhibitory activity for interleukin-1 secretion and anti-tumor-promoting activities of several metabolites isolated from the stem barks of *Tripterygium wilfordii* var. *regelii*. The novel triterpenes, regeol A (209), B (210) and C (211) and the known triterpenes: tingenine B (22β-hydroxy-tingenone) (212), wilforol A (213), regelide (214), celastrol (172), demethylregelin (215) isolated from this plant, and the derivatives 216 and 217 from celastrol, were tested for inhibitory activity for IL-1α and IL-1β release from lipopolysaccharide-stimulated human peripheral mononuclear cells compared to a reference compound (prednisolone) (Table 11), Fig. (54).Tingenine B (212), celastrol (172) and the derivatives 216 and 217 showed strong inhibitory activity.

The effect on EBV- EA (Epstein-Barr virus early antigen) activation of compounds 172 and 210-217, were also examined. Their inhibitory effects on the activation of the early antigen and the viabilities of Raji cells are shown in Table 12. Compounds 172, 210, 216 and 217 showed stronger activities than the other compounds.

4.6. Maytansinoids

Maytansine (218) was the first ansa macrolide shown to contain a 19-membered lactam ring with carbinolamine, epoxide, or aryl halide functions and appears to be the first member of the series reported to show significant *in vivo* tumor inhibitory activity [116].

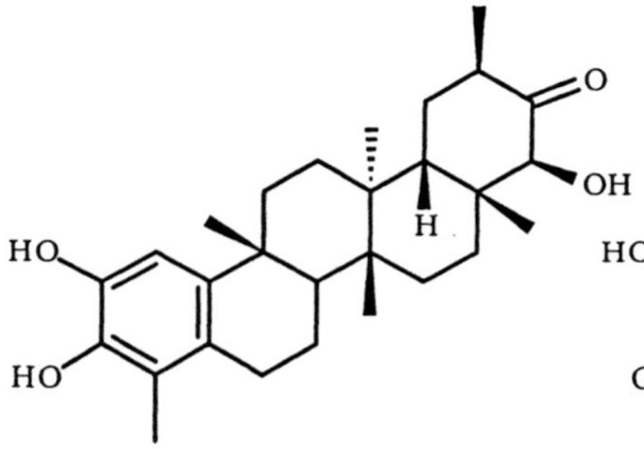

209 Regeol A

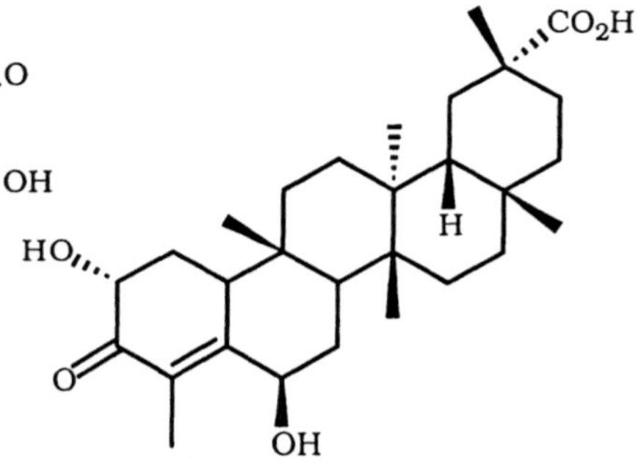

210 Regeol B

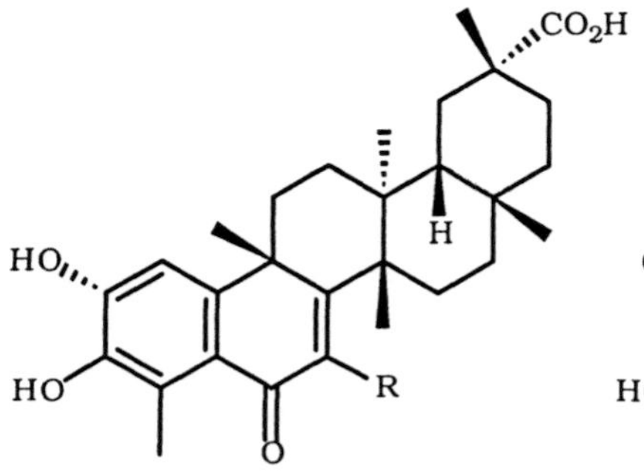

211 Regeol C   R = OH
213 Wilforol A R = H

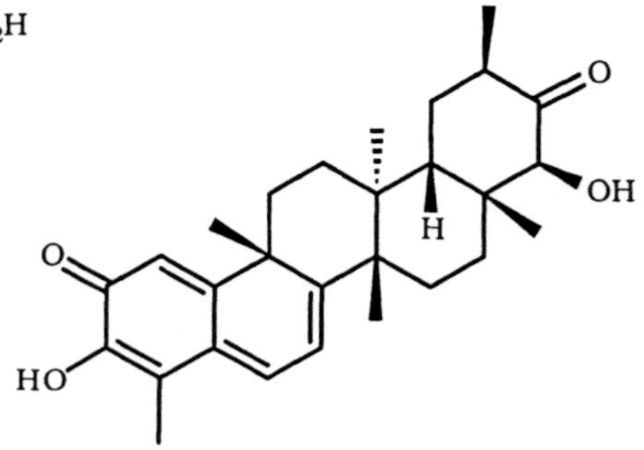

212 Tingenine B

(Fig. (54). contd.....

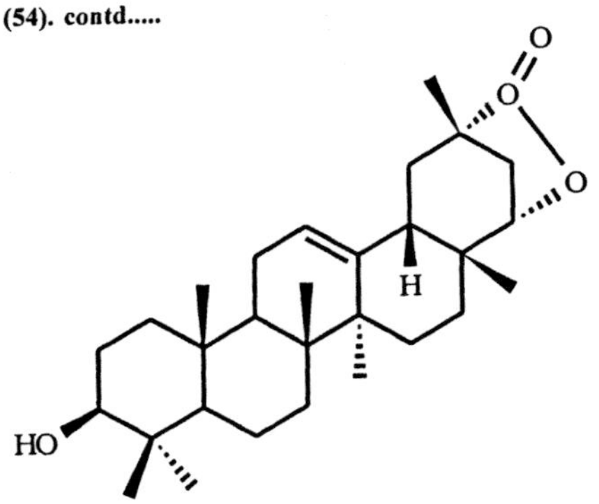

214 Regelide 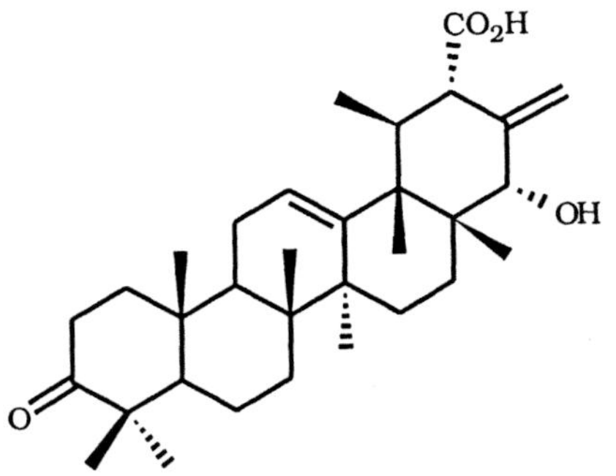   **215** Demethylregelin

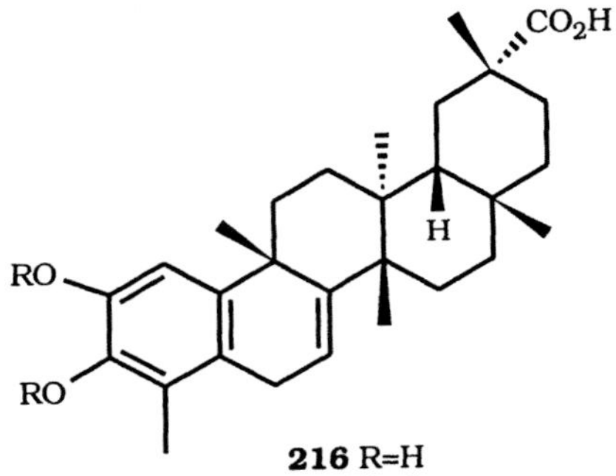

216 R=H
217 R=Ac

Fig. (54). Triterpenoid inhibitors of interleuking-1 secretion and tumor promotion from *Tripterigium wilfordii var. reglii*

Maytansine's ansa macrolide structure shows noteworthy similarities to those of the rifamycins [255], streptovaricins [256], tolypomycins [257], and geldanamycin [258]. The ansamycin antibiotics and their derivatives have aroused considerable interest as antiviral and antimicrobial agents, and as inhibitors of RNA tumor virus reverse transcriptases.

Maytansine (**218**) and other ansa macrolides, including maytanprine (**222**) and maytanbutine (**223**), were first isolated and reported by Kupchan and coworkers [116, 259, 260] as being derived from various species of the genera *Maytenus* and *Putterlickia,* both of the family Celastraceae. Shortly after, Wani *et al.* [261] reported maytansinoids, including maytanbutine (**223**), from *Colubrina texensis* Gray (Rhamnaceae); the production of ansamitocins, maytansinoids lacking the N-methyl alanyl moiety in the side chain, by fermentations of a *Nocardia* species, had also been described [262].

Table 11.   The activities of triterpenes from *T. wilfordii* as inhibitors of IL-1α and IL-1βrelease

Compound	Inhibition (%)	
	IL-1α	IL-1β
regeol A (209)	47	30
regeol B (210)	1	0
regeol C (211)	-10	2
tingenine B (212)	100	100
wilforol A (213)	-7	-2
regelide (214)	5	1
celastrol (172)	98	100
demethylregelin (215)	2	0
216	98	99
217	95	99
prednisolone	87	76

Concentration of the compounds: 1 x 10^{-6} g/ml, prednisolone: 3 x 10^{-7} g/ml.

Table 12.   Inhibition of EBV-EA activation by constituents of *T. wilfordii* and their derivatives

Compound	% to control[a] (% viability)[b]			
	concentration (mol ratio/32 pmol TPA)			
	1 x 10^3	5 x 10^2	1 x 10^2	1 x 10
regeol B (210)	0 (70)	53.7 (>80)	83.6	100
regeol C (211)	11.3 (70)	65.4 (>80)	84.5	100
tingenine B (212)	13.6 (70)	24.8 (>80)	50.3	81.8
wilforol (213)	10.5 (40)	34.7 (>80)	75.4	92.6
regelide (214)	26.4 (70)	68.9 (>80)	84.7	100
celastrol (172)	0 (50)	49.4 (>80)	78.6	95.7
demethylregelin (215)	26.2 (70)	54.8 (>80)	83.6	100
216	0 (60)	0 (>80)	42.4	75.2
217	0 (60)	0 (>80)	36.3	70.3

[a]Values represent relative percentages to the positive control values (TPA 32 pmol = 100%)

[b]Values in parentheses are viability percentages of Raji cells.

Kupchan proposed that the generic term "maytansinoids" be used for all ansa macrolide derivatives structurally related to maytansine, and "maytanside" as the term for those maytansinoids which contain the macrocyclic ring system but lack the ester moiety.

Maytansine (**218**) was an exceptionally interesting antitumor agent. It showed high inhibitory activity against several murine tumors, at the level of micrograms per kilogram body weight and over a wide dosage range. The compound has undergone extensive preclinical toxicological studies and was selected for clinical trial by the National Cancer Institute. Furthermore, biological studies have shown that maytansine is a highly active inhibitor of cell division [263, 264] and of transformation of mouse cell cultures infected with murine sarcoma virus [265].

4.5.1. Antitumor Activity

In the course of a search for tumor inhibitors from plant sources, Kupchan *et al.* carried out an exhaustive study of several species of the genus *Maytenus*. In 1972 [116], this research' group found that the alcoholic extract of *Maytenus ovatus* Loes. showed significant inhibitory activity *in vitro* against KB cells and five standard animal tumor systems and activity was noted against sarcoma 180, Lewis lung carcinoma, and L-1210 and P-388 leukemias in the mouse and Walker 256 intramuscular carcinosarcoma in the rat. Fractionation of the active extract lead to the isolation of maytansine (**218**), a novel ansa macrolide, which showed significant antileukemic activity against P-388 lymphocytic leukemia over a 50-100-fold dosage range at the mg/Kg level, and cytotoxicity (ED$_{50}$ 10^{-4}-10^{-5} μg/ml.) against KB cell culture.

In an attempt to obtain different derivatives, the methyl (**219**), ethyl (**220**) and the 3-bromopropyl (**221**) derivatives of **218** were prepared, Fig. (**55**). The reversible interrelation of maytensine and **221**, made the latter compound attractive for X-ray crystallographic analysis and led unequivocally to the structural interpretation of **221** and from this the structure of **218**, being the absolute configurations 3S, 4S, 5S, 6R, 7S, 9S, 10R, and 2'S.

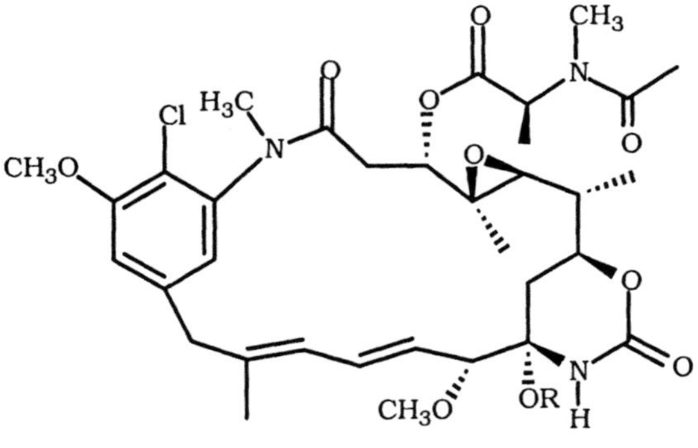

218	Maytansine	R = H
219		R = CH$_3$
220		R = CH$_3$CH$_2$
221		R = BrCH$_2$CH$_2$CH$_2$

Fig. (55). Maytansine and derivatives from *Maytenus ovatus*

In the same year Kupchan *et al.* [259] described two new and potent antileukemic ansa macrolides isolated from the alcoholic extract of stems of *Maytenus buchananii*. Thus, maytanprine (**222**) and maytanbutine (**223**) were the new principles which showed lymphocytic leukemia over a 50-100-fold dosage range at the µg/kg level, Fig. (**56**).

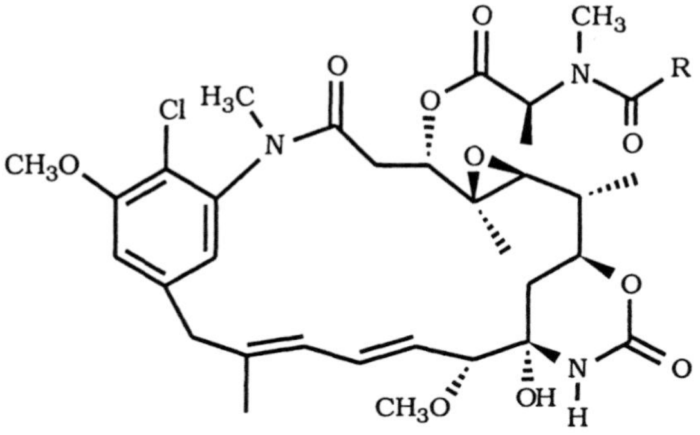

222 Maytanprine   R = CH$_2$CH$_3$
223 Maytanbutine R = CH(CH$_3$)$_2$
224 Maytanvaline R = CH$_2$CH(CH$_3$)$_2$

Fig.(56). Antileukemic maytansinoids from *Maytenus buchananii*

The potent antileukemic activity of maytansine (**218**) and related maytanside esters stimulated interest in the chemical and biological properties of related compounds. This interest was heightened by the finding that maytansine also showed significant inhibitory activity against the Lewis lung carcinoma and B-16 melanocarcinoma solid murine tumor systems, and the agent was under toxicological investigation in preparation for clinical trials.

In 1974 [266], the isolation, structural elucidation, and chemical interrelation of four new maytansinoids from the alcoholic extract of wood and bark of *M. buchananii*, were reported. Among these, maytanvaline (**224**), Fig. (**56**) was a highly active antileukemic maytanside ester, showing significant antileukemic activity against the P-388 lymphocytic leukemia over a 50-100-fold dosage range at the µg/kg level and cytotoxicity (ED$_{50}$ 10^{-5}-10^{-6} µg/ml) against KB cell culture. On the other hand, maysine (**225**), normaysine (**226**), and maysenine (**227**), were the first reported maytansides lacking antileukemic activity and showing a *ca.* 1/10,000 the cytotoxicity of maytanside esters such as maytanvaline (**224**), while **225-227**, Fig. (**57**) and maytansine ethyl ether (**220**) showed cytotoxicity at 1-10^{-2} µg/ml.

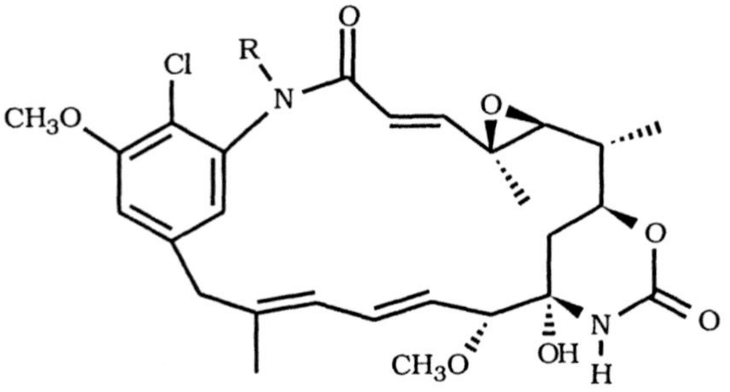

225 Maysine   R = CH₃
226 Normaysine R = H

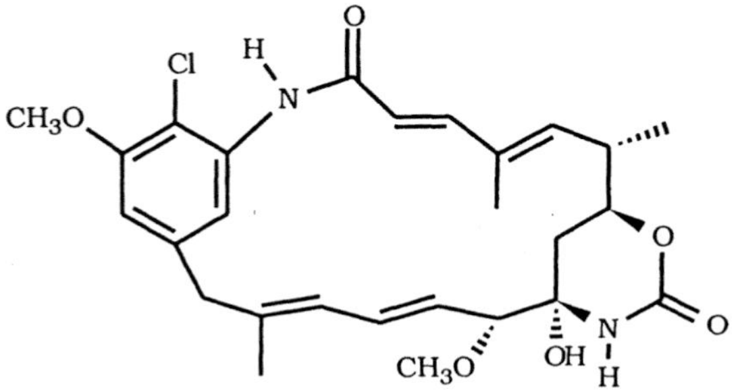

227 Maysenine

Fig. (57). Maytansides from *Maytenus buchananii*

The potent biological activity of maytanvaline (**224**) confirmed and extended the earlier observations that variations in the nature of the ester group affixed at C-3 are not accompanied by marked changes in antileukemic activity. In contrast, the strikingly diminished biological activity of the newly discovered maytansides revealed the key importance of the ester moiety for the biological activity of the antileukemic maytansinoids. The ester function in the antileukemic maytansinoids may play a key role in the formation of highly selective molecular complexes with growth-regulatory biological macromolecules. Such molecular complex formation may be crucial for the subsequent selective alkylation of specific nucleophiles by, e.g., the carbinolamide and epoxide functions. In this connection, it is noteworthy that maytansine ethyl ether (**220**), in which the reactive carbinolamide is no longer available as a potential alkylating function, showed no antileukemic activity.

Two new maytansinoids, maytanacine (**228**) and maytansinol (**229**), were isolated from the alcoholic extract of the stem of *Putterlickia verrucosa* Szyszyl. [260], the richest reported source of maytansine (**218**) and related antileukemic esters. Maytanacine (**228**), which exhibited potent antileukemic activity, was the first reported maytanside ester which does not bear an amino acid residue at C-3 and showed significant antileukemic activity against the P-388 lymphocytic leukemia over a 50-100-fold dosage range at the μg/kg level (with a T/C value of 230 at 100 μg/kg), and cytotoxicity (ED_{50} 10^{-5} μg/ml) against KB cell culture. On the other hand, maytansinol (**229**), the parent alcohol of the potent maytanside esters, lacked antileukemic activity at doses up to 70 times the optimal dose (12,5 (g/kg; T/C 258) of maytansine (**218**).

Maytansinol (**229**) was prepared by $LiAlH_4$ treatment of maytanbutine (**223**) in dry THF at -23° for 3 hr (40% yield). Similar reductive cleavage of maytanacine (**228**) also gave maytansinol (**229**) and treatment of **229** with acetic anhydride-pyridine at 53° for 18 h gave maytanacine (**228**) (53% yield).

To evaluate the effects on biological activity of variations in structure of the ester moiety, several semisynthetic esters of maytansinol were prepared. The propionate (**230**), bromoacetate (**231**), crotonate (**232**), and trifluoroacetate (**233**) esters of maytansinol (**229**) were made by procedures involving either anhydride-pyridine (**230-232**) or anhydride-acid (**233**) treatment, Fig. (**58**). The esters **228**, and **230-232** were found to show antileukemic activity comparable to those of the naturally occurring substituted alanyl esters. The trifluoroacetate ester **233** showed no antileukemic activity, possibly because of ready solvolysis *in vivo* to inactive maytansides.

228 Maytanacine   R = $COCH_3$
229 Maytansinol   R = H
230   R = $COCH_2CH_3$
231   R = $COCH_2Br$
232   R = $COCH=CHCH_3$
233   R = $COCF_3$

Fig. (**58**). Maytansinoids and derivatives from *Putterlickia verrucosa*

Unfortunately, work with these compounds has been impeded by the remoteness of source materials and their exceedingly low maytansinoid content. The ansamitocins may be converted by chemical manipulation to maytansine, but yields are not high. In order to solve this problem, Kupchan *et al.* searched methods for improving the isolation of these compounds from the active extracts of *M. serrata, M. buchananii* and *P. verrucosa*. Thus, the details of the isolation, of the species mentioned above, and the structural elucidation of the ansa macrolide principles maytansine (**218**), maytanprine (**222**), maytanbutine (**223**), maytanvaline

(224), and maytanacine (228), as well as of maytansides, maysine (225), normaysine (226), maysenine (227), and maytansinol (228), were reported. In addition, the isolation and characterization of a new antileukemic principle, the maytanside diester, from *M. serrata*, maytanbutacine (234), Fig. (59) which was found to contain two acyl ester groups, a C-3 isobutyrate ester and a C-15 acetate ester, were reported [267].

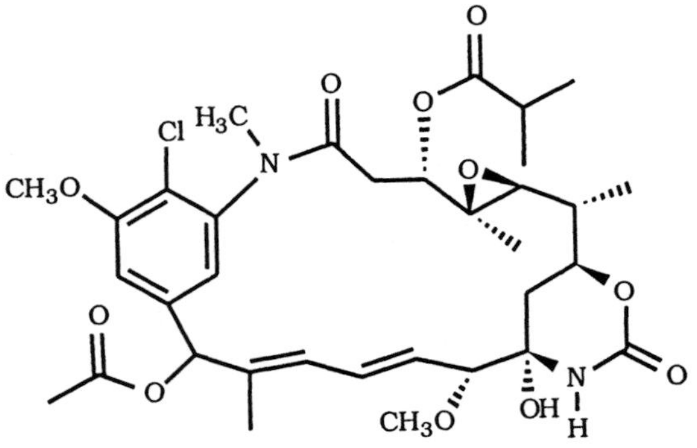

234 Maytanbutacine

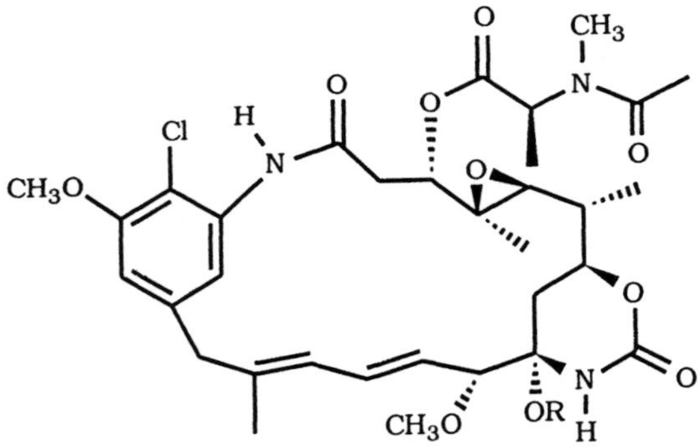

235 Nomaytansine

Fig. (59). Antileukemic ansamacrolide from *Maytenus* species

Approximately 15,000 kg of *M. buchananii* were collected in Kenya in 1976 and sent to Monsanto Research Corporation for isolation of the maytansine (218) [268] in 1980. During the course of the isolation, several fractions were obtained which did not correspond to any known maytansinoids. When two of these fractions were combined and subjected to chromatography a white solid was obtained which was crystallized from dichloromethane- hexanes to yield normaytansine, 235, Fig. (59). Normaytansine was found to have significant *in vivo* activity against the P-388 lymphocytic leukemia in mice at doses comparable to maytansine (e.g., T/C 181 at 100 μg/kg) and *in vitro* activity against the KB cell culture (ED_{50} 10^{-3} μg/ml). Thus the amide methyl at C-1 did not seem to affect the antileukemic activity of the maytansinoids.

Nettleton *et al.* [269] reported a new plant source with high maytansinoids content and a vastly improved and simplified process for the isolation of maytansine and related compounds, and also reported the investigation of a potential source of maytansinoids growing in the continental United States. Thus, a new, highly productive source of maytansine and other maytansinoids was found in the seeds of *Maytenus rothiana* (Walp.) Lobreau-Callen, a small upland tree of the western Indian Ghats near Bombay. From this source, a greatly improved and commercially feasible preparative liquid chromatography isolation procedure has been designed. Also, in an off-season collection, the wood of *Maytenus phyllanthoides* Benth, or *M. texana* (Benth) Lundell, known commonly as leatherleaf of guttapercha maytenus, from the southern Texas Gulf coast had been found to contain only a very small amount of maytansine. In a later collection, no maytansinoids were found in the fruit.

Since the highest concentration of maytansine (**218**) reported to date is from a seed fraction, the possibility arises that maytansine has some function within that part. Otherwise, it has generally been found in a wood fraction where it may well have been deposited during plant cycles and, being reasonably stable, has persisted. Antimitotic, antigibberallins, and auxin activities for the compound have been reported [270], which indicated some growth regulatory potential.

The methanolic extract of the stems of *Maytenus diversifolia* was found to show significant inhibitory activity *in vivo* against the P-388 lymphocytic leukemia growth in BDF mice (T/C 180%) at 50 mg/kg/day. Subsequent bioassay-directed fractionation in P-388 *in vivo*, led to the conclusion that the most active component of this extract was maytansine (**218**) [198]. Maytansine was isolated in a better yield of 0,0000374% of the dried plant material compared to that reported previously [267].

As an alternative source of supply of the antileukemic maytansine, Kutney and his coworkers have investigated its potential production in tissue culture of *M. buchananii* [271]. On the other hand, studies of synthetic approaches to various model compounds related to segments of the maytansine structure have been achieved by Meyers *et al.* [272] and Corey *et al.* [273]. The total synthesis of racemic maytansinol, in a highly stereocontrolled manner involving dia-stereotopic induction of all its asymmetric centers and later, the stereoselective synthesis of optically active (-)-maytansinol, among other maytansinoids, were reported [274]. Conjugates of maytansinoids derivatives with a cell binding agent such as an antibody and a hormone were prepared as cytotoxic agents to kill selected cell population in a targeted fashion. SH-containing maytansinoids was conjugated via disulfide links with antibodies anti-B4, anti-T9 and anti-A7, separately, and *in vitro* cytotoxicity of these compounds against cell line Namalwa, KB, HT-29 and SW-620 was tested. The specific affinity of the conjugate was also determined by competition binding assays to HT-29 and SW-620 cell lines [275].

Researchers from ImmunoGen, Inc. at the American Association for Cancer Research 1997 Meeting in San Diego, California, presented the results of preclinical studies of hN901-DM1, a novel tumor-targeting agent in which the drug completely eradicated human small-cell lung cancer (SCLC) tumors transplanted in mice. hN901-DM1 links the maytansinoids drug DM1, which was highly cytotoxic and was over 100 times more potent than conventional chemotherapeutic agents to hN901, a humanized antibody that binds specifically to CD56, a marker expressed on all SCLC cells. hN901-DM1 is able to harness the high potency of DM1 and deliver it specifically to the tumor rather than to healthy cells elsewhere in the body [276].

ACKNOWLEDGEMENTS

We are indebted to CYCIT Project PB96-1939 for subsidies. The authors thank Prof. Hermelo López Dorta (University of La Laguna, Spain) for critically reading this manuscript and for helpful suggestions. Prof. Mahabir Gupta and Dr. Pablo Solís (University of Panama) are thanked for providing the information of the data base NAPRALERT on Celastraceae.

REFERENCES

[1]   Loub, W.D.; Farnsworth, N.R ; Soejarto, D.D.; Quin, M.L. NAPRALERT: computer handling of natural p roduct research data. *J. Chem. Inf. Comput. Sci.*, **1985**, *25*, 99.

[2]   Chadwick, D.J.; Marsh, J. *Ethnobotany and the Search for New Drugs*, Ciba Foundation Symposium 185, John Wiley and Sons: Chichester, England, **1994**.

[3]   Heywood, V.H. *Flowering Plants of the World*, Oxford University Press: New York, **1993**.

[4]   Woodland, D.W. *Contemporary Plant Systematics*, Prentice-Hall.: New Jersey, **1991**.

[5]   Robson, N. Bot. Soc. Brot., Ser. **1965**, 2 (39), 5.

[6]   Hegnauer, R. *Chemotaxonomie der Pflanzen*, Basel: Birkhäuser, **1966**.

[7]   Ravelo, A.G.; Luis, J.G.; González, C.M.; Ferro, E.A.; Bazzocchi, I.L.; Jiménez, J.; Herrera, J.R.; Jiménez, I.A.; Aguiar, Z.E. *Rev. Latinoam. Quim.*, **1988**, *19*, 72.

[8]   Brüning, R.; Wagner, H. *Phytochemistry*, **1978**, *17*, 1821.

[9]   González, A.G.; Bazzocchi, I.L.; Ferro, E.A.; Ravelo, A.G.; Luis, J.G.; Aguilar, M.A. *Biochem. System. Ecol.*, **1986**, *14(5)*, 479.

[10]   González, A.G.; Jiménez, I.A.; Núñez, M.P.; Ravelo, A.G.; Bazzocchi, I.L.; Muñoz, O.M.; Aguilar, M.A. *J. Chem. Ecol.*, **1994**, *20(4)*, 823.

[11] Mahyar, U.W.; Burley, J.S.; Gyllenhaal, C.; Soejarto, D.D. *J. Ethnopharmacol.*, **1991**, *31*, 217.
[12] Ohashi, K.; Kojima, M.; Tanikawa, T.; Okumura, Y.; Kawazoe, K.; Tataran, N.; Shibuya, H.; Kitagawa, I. *Chem. Pharm. Bull.*, **1994**, *47(8)*, 1596.
[13] Singh, V.K.; Ali, Z.A.; Siddiovi, M.K. *Fitoterapia*, **1996**, *67(1)*, 65.
[14] Arnold H. J.; Gulumian, M. *J. Ethnopharmacol.*, **1984**, *12(1)*, 35.
[15] Der Krikorian, A. *J. Ethnopharmacol.*, **1984**, *12*, 115.
[16] Kristiansson, B.; Abdul Ghani, N.; Eriksson, M.; Garle, M.; Qirbi, A. *J. Ethnopharmacol.*, **1987**, *21(1)*, 85.
[17] Elmi, A.S. *J. Ethnopharmacol*, **1983**, *8(2)*, 163.
[18] Al-Meshal, I.A.; Ageel, A.M.; Parmar, N.S.; Tariq, M. *Fitoterapia*, **1985**, *56(3)*, 131.
[19] Kubo, I.; Kim, M.; De Boer, G. *J. Chromatogr.*, **1987**, *402(1)*, 354.
[20] Liu, J.K.; Jia, Z.J.; Wu, D.G.; Zhou, J.; Zhu, Z.Q. *Chin. Sci. Bull.*, **1989**, *34(12)*, 1041.
[21] Liu, J.; Wu, D.G., Jia, Z.; Zhou, J.; Zhu, Z. *Planta Med.*, **1991**, *57(5)*, 475.
[22] Saha, J.C.; Savini, E.C.; Kasinathan, S., *Indian J. Med. Res.*, **1961**, *49*, 130.
[23] Jain, S.P.; Puri, H.S. *J. Ethnopharmacol*, **1984**, *12*, 213.
[24] Singh, V.K.; Ali, Z.A. *Fitoterapia*, **1994**, *65(1)*, 68.
[25] Jamwal, K.S.; Anand, K.K. *Indian J. Pharm.*, **1962**, *24*, 218.
[26] Nagaraju, N.; Rao, K.N. *J. Ethnopharmacol*, **1990**, *29(2)*, 137.
[27] Manandhar, N.P. *Econ. Bot.*, **1995**, *49(4)*, 371.
[28] Singh, K.K.; Maheshwari, J.K. *Int. J. Pharmacog.*, **1994**, *32(1)*, 51.
[29] Wagner, H.; Heckel, E.; Sonnenbichler, J. *Tetrahedron*, **1975**, *31*, 1949.
[30] Jain, S.P.; Singh, S.C.; Puri, H.S. *Int. J. Pharmacog.*, **1994**, *32(1)*, 44.
[31] Sebastian, M.K.; Bhandari, M.M. *J. Ethnopharmacol.*, **1984**, *12(2)*, 223.
[32] Kakrani, H.K.; Vijaynathan-Nair, G.; Kalyani, G.A.; Satyanarayana, D. *Fitoterapia*, **1985**, *56(5)*, 293.

[33] Anon. *Lillys's Hand Book of Pharmacy and Therapeutics*, 5th Rev. Eli Lilly and Co.: Indianapolis, **1898**.
[34] Krochmal, A.; Krochmal, O. *Medicinal Plants of The United* States, Quadrangle, The New York Times Book Co.: New York, **1973**.
[35] Anon, *The Herbalist*, Hammond Book Company: Hammond Indiana, **1931**.
[36] Kokowaro, J.O. *Medicinal Plants of Esast Africa*, East African Literature Bureau: Nairobi, **1976**.
[37] Wild, H.; Gelfand, M. *Cent. Afr. J. Med.*, **1959**, *5*, 292.
[38] Bisset, N.G.; Mazars, G., *J. Ethnopharmacol.*, **1984**, *12(1)*, 1.
[39] Chhabra, S.C.; Uiso, F.C.; Mshiu, E.N. *J. Etnopharmacol.*, **1984**, *11(2)*, 157.
[40] Keys, J.D. *Chinese Herbs, Botany, Chemistry and Pharmacodynamic*, Charles E. Tottle Co.: Rutland, Vermont, USA, **1976**.
[41] Kitanaka, S.; Takido, M.; Mizoue, K.; Kakaike, S. *Chem. Pharm. Bull.*, **1996**, *44(39)*, 615.
[42] Woo, W.S.; Lee, E.B.; Shin, K.H.; Kang, S.S.; Chi, H.J. *Korean J. Pharmacog.*, **1981**, *12 (3)*, 153.
[43] Bliss, C.A.; Ramstad, E. *J. Amer. Pharm. Ass. Sci.*, **1957**, *46*, 15.
[44] Coon, N. *The Dictionary of Useful Plants*, Rodale Press, Book Div.: Emmaus, PA, **1974**.
[45] Perrot, E. *Les Plantes Medicinales, Part 1,* Paris, R.R., Presses Universitaires de France, Paris: France, **1971**.

[46] Antonone, R.; De Simone, F., Morrica, P.; Ramundo, E. *J. Ethnopharmacol.*, **1988**, *22 (3)*, 295.

[47] Leporatti, M.L.; Pavesi, A. *J. Ethnopharmacol.*, **1990**, *29(2)*, 213.

[48] De Feo, V.; Senatore, F. *J. Ethnopharmacol.*, **1993**, *39(1)*, 39.

[49] De Feo, V.; Aquino, R.; Menghini, A.; Ramundo, E.; Senatore, F. *J. Ethnopharmacol.*, **1992**, *36(2)*, 113.

[50] Dixit, R.S.; Pandey, H.C. *Int. J. Crude Drug Res.*, **1984**, *22(1)*, 47.

[51] De, S.; Ravishankar, B.; Bhavsar, G.C. *Planta Med.*, **1994**, *60*, 301.

[52] Kamal, G.M.; Gunaherath, B.; Gunatilaka, A.A.L. *J. Chem. Soc. Perkin Trans I*, **1983**, 2845.

[53] Bhat, R.B.; Jacobs, T.V. *J. Ethnopharmacol.*, **1995**, *48 (1)*, 7.

[54] Pereira, A.M.S.; Menezes, Jr.A.; Pereira, P.S.; Cerdeira, R.M.M.; Franca, S.C., Vilegas, J.H.Y.; Cordeiro, P.J.M.; Lancas, F.M. *J. Herbs. Spices Med. Plants*, **1995**, *3(2)*, 43.

[55] Gakunju, D.M.N.; Mberu, E.K.; Dossaji, S.F.; Gray, A.I.; Waigh, R.D.; Waterman, P.G.; Watkins, W.M. *Antimicrob. Agents Chemother.*, **1995**, *39(12)*, 2606.

[56] Gupta, M.P.; Monge, A.; Karikas, G.A.; López De Cerain, A.; Solis, P.N.; De León, E.; Trujillo, M.; Suárez, O.; Wilson, F.; Montenegro, G.; Noriega, Y.; Santana, A.I. *Int. J. Pharmacog.*, **1996**, *34(1)*, 19.

[57] Muñoz, O.; Galeffi, C.; Federici, E.; Garbarino, J.A.; Piovano, M.; Nicoletti, M. *Phytochemistry*, **1995**, *40(3)*, 853.

[58] Eldridge, J. *Econ. Bot.*, **1975**, *29*, 307.

[59] Morton, J.F. *J. Crude Drug Res.*, **1977**, *15*, 1.

[60] Darias, V.; Bravo, L.; Rabanal, R.; Sánchez Mateo, C.; González Luis, R.M.; Hernández Pérez, A.H. *J. Ethnopharmacol.*, **1989**, *25(1)*, 77.

[61] Pitard, J.; Proust, L. *Lees Iles Canaries. Flore de L'Archipiel*, Librarie des Sciences Naturalles, Paris, **1908**.

[62] Englert, G.; Klinga, K.; Raymond-Hamet; Schlittler, E.; Vetter, W. *Helv. Chim. Acta*, **1972**, *56*, 474.

[63] Martinod, A.; Paredes, A.; Delle Monache, F.; Marini-Bettolo, G.B. *Phytochemistry*, **1976**, *15*, 562.

[64] Luna, L.E. *J. Etnopharmacol.*, **1984**, *11(2)*, 123.

[65] Itokawa, H.; Shirota, O.; Morita, H.; Takeya, K. *Heterocycles*, **1992**, *34(5)*, 885.

[66] Sabnis, S.D.; Bedi, S.J. *Indian J. Forestry*, **1983**, *6(1)*, 65.

[67] Anon. *Probe*, **1985**, *24(4)*, 234.

[68] Hedberg, I.; Hedberg, O.; Maditi, P.J.; Mshigeni, K.E.; Mshiu, E.N.; Samuelson, G. *J. Etnopharmacol.*, **1982**, *6(1)*, 29.

[69] Bandoni, A.L.; Mendiondo, M.E.; Rondina, R.V.D.; Coussio, J.D. *Econ. Bot.*, **1976**, *30*, 181.

[70] Martinez-Crovetto, R. *Parodiana*, **1981**, *1(1)*, 97.

[71] González, A.; Ferreira, F.; Vázquez, A.; Moyna, P.; Alonso Paz, E. *J. Ethnopharmacol.*, **1993**, *39(3)*, 217.

[72] Pérez, C.; Anesini, C. *Fitoterapia*, **1994**, *65(2)*, 169.

[73] Alice, C.B.; Vargas, V.M.F.; Silva, G.A.A.B.; De Siqueira, N.C.S.; Schapoval, E.E.S.; Gleve, J.; Henriques, J.A.P.; Henriques, A.T. *J. Ethnopharmacol.*, **1991**, *35(2)*, 165.

[74] Itokawa, H.; Shirota, O.; Ichitsuka, k.; Morita, H.; Takeya, K. *J. Nat. Prod.*, **1993**, *56(9)*, 1479.

[75] Arenas, P.; Moreno-Azorero, R. *Econ. Bot.* **1977**, *31*, 298.

[76]   Ahmed, M.S.; Fong, H.H.S.; Soejarto, D.D.; Dobberstein, R.H.; Waller, D.P.; Moreno, A.R. *J. Chromatogr.*, **1981**, *213*, 340.
[77]   Shirota, O.; Tamemura, T.; Morita, H.; Takeya, K.; Itokawa, H. *J. Nat. Prod.*, **1996**, *59 (11)*, 1072.
[78]   González, J.G.; Delle Monache, G.; Delle Monache, F.; Marini-Bettolo, G.B. *J. Ethnopharmacol.*, **1982**, *5*, 73.
[79]   Dimyuga, R.E.; Agundez, J. *J. Ethnopharmacol.*, **1986**, *17*, 183.
[80]   Ayensu, E.S.*; Medicinal Plants of West Africa*, Reference Publications, Inc. Alganac.: Michigan, US, **1978**.
[81]   Haerdi, F. *Native Medicinal Plants of Olanga Distric of Tanganyika (East Africa)*, Dissertation-Univ, Verlag Fur Recht Und Gesellschaff AG, BA Sel. Basel, **1964**.
[82]   Wilson, R.T.; Marian, W.G. *Econ. Bot.*, **1979**, *33*, 29.
[83]   Gessler, M.C.; Tanner, M.; Chollet, J.; Nkunya, M.H.N.; Heinrich, M. *Phytother. Res.*, **1995**, *9*, 504.
[84]   Vasileva, B. *Plantas Medicinales de Guinea*, Conakry: Republique de Guinee, **1969**.
[85]   Chhabra, S.C.; Mahunnah, R.L.A.; Mshiu, E.N. *J. Ethnopharmacol.*, **1993**, *39(2)*, 83.
[86]   Debray, M.; Jacquemin, H.; Razafindrambao, R. *Trav. Doc. Ostrom*, **1971**, *8*, 1.
[87]   Sukumar, E.; Hamsaveni Gopal, R.; Bhima Rao, R.; Kundu, A.B.; Viswanathan, S.; Thirugnanasambandam, P.; Gopal Krishnan, C. *Fitoterapia*, **1991**, *65(5)*, 429.
[88]   Dominguez, X.A.; Franco, R.; Cano, G.; García, S.; Zamudio, A.; Amezcua, B.; Dominguez, Jr. X.A. *Phytochemistry*, **1979**, *18*, 898.
[89]   Bouquet, A.; Debray, M. *Trav. Doc. Orstom*, **1974**, *32*, 1.
[90]   Chou, W.C.; Wu, C.C.; Yang, P.C.; Lee, Y.T. *Int. J. Cardiol.*, **1995**, *49(2)*, 173.
[91]   Juling, G.; Shixiang, I.; Xichun, H.; Shixi, X.; Dada, L. *Chin. Med. J.*, **1981**, *94*, 405.
[92]   Anon. *Zhong Cao Yao Tong Xung*, **1977**, *10*, 354.
[93]   Tao, X.L.; Dong, Y.; Zhang, N.Z. *Chin J. Intern. Med.*, **1987**, *26(7)*, 445.
[94]   Wu, F.G.; Zhu, L.P.; Cui, L.X.; Wang, X.; Zhang, S.Z. *Zhonghua Weishengwuxue He*, Mianyixue Zazhi, **1993**, *13(3)*, 193. (CA 119173818B)
[95]   Takaishi, Y.; Tokura, K.; Noguchi, H.; Nakamo, K.; Murakami, K.; Tomimatsu, T. *Phytovhemistry*, **1991**, *30(5)*, 1561.
[96]   Wolfes, O. *Archive der Pharmazi*, **1930**, *268*, 81.
[97]   Brucke, F. *Archives of Experimental Pathology and Pharmakology*, **1941**, *198*, 100.
[98]   United Nations Document MNAR, **1975**, *11*.
[99]   Ripani, L.; Schiavone, S.; Garofano, L. *Forensic Sci. Int.*, **1996**, *78(1)*, 39.
[100]   Ahmed, M.B.; El-Qirbi, A.B. *J. Ethnopharmacol.*, **1993**, *39*, 213.
[101]   Chou, T. Q.; Mei, P. F. *J. Physiol.*, **1936**, *10*, 259.
[102]   Gisvold, O. *J. Amer. Pharm. Assoc.*, **1939**, *28*, 440.
[103]   Swingle, W.T.; Haller, H.L.; Siegler, E.H.; Swingle, M.C. *Science*, **1941**, *93*, 60.
[104]   Acree, Jr. F. ; Haller, H.L. *J. Am. Chem. Soc.*, **1950**, *72*, 1608.
[105]   Beroza, M. *J. Am. Chem. Soc.*, **1951**, *73*, 3656.
[106]   Beroza, M. *J. Am. Chem. Soc.*, **1952**, *74*, 1585.
[107]   Beroza, M. *J. Am. Chem. Soc.*, **1953**, *75*, 2136.

[108]   Kupchan, S.M.; Court, W.A.; Dailey, Jr. R.G.; Gilmore, C.J.; Bryan, R.F. *J. Am. Chem. Soc.,* **1972**, *94*, 7194.

[109]   Xu, W.; Zheng, J.; Lu, X. *Int. J. Dermatol.,* **1985**, *24*, 152.

[110]   Colas, R. *Les plantes amazoniennes désignées sous le nom de "Chuchuhuasha".* Thèse, Imprimerie André Lesot: Paris, **1937**.

[111]   Schwartzman, J.B.; Krimer, D.B.; Moreno Azorero, R. *Revista de la Sociedad Científica del Paraguay,* **1976**, *16*, 63.

[112]   Olarte, J.C. *Contribución al estudio de la flora colombiana,* Seminario Latinoamericano de la Química de los Productos Naturales: Bogotá, **1976**.

[113]   Flemming, K.; Flemming, C.; Graack, B. *Strahlentherapie,* **1967**, *133*, 280.

[114]   Melo, A.M.; Jardim, M.L.; De Santana, C.F.; Lacet, Y.; Lobo, J.; Gonçalves de Lima, O.; Leoncio d'Albuquerque, I. *Rev. Inst. Antibioticos (Recife),* **1974**, *14*, 9.

[115]   Angeletti, P.V.; Marini-Bettolo, G.B. *Il Farmaco, Ed. Sci.,* **1974**, *29*, 569.

[116]   Kupchan, S.M.; Komoda, Y.; Court, W.A.; Smith, R.M.; Karim, A.; Gilmore, C.J.; Haltiwanger, R.C.; Bryan, R.F. *J. Am. Chem. Soc.,* **1972**, *94(4)*, 1354.

[117]   Muguera, G.M.; Ward, J.M. *Cancer Treatm.,* **1977**, Rep. 61, 1333.

[118]   Muñoz, O.; Peñaloza, A.; González, A.G.; Ravelo, A.G.; Bazzocchi, I.L.; Alvarenga, N.L. *Studies in natural products Chemistry,* Elsevier Science Publisher: Amsterdam, **1996**, Vol. 18

[119]   Smith, R.M. *The Alkaloids;* Academic Press, Ed.; R.H.F. Manske: London, **1977**, Vol. 16.

[120]   González, A.G.; Muñoz, O.M.; Ravelo, A.G.; Crespo, A.; Bazzocchi, I.L.; Jiménez, I.A.; Solans, X.; Ruiz-Pérez, C; Rodríguez-Romero, V. *Tetrahedron Lett.,* **1992**, *33(14)*, 1921.

[121]   Shizuri, Y.; Wada, K.; Sugiura, K.; Hirata, Y. *Tetrahedron,* **1973**, *29*, 1773.

[122]   González, A.G.; Núñez, M.P.; Ravelo, A.G.; Sazatornil, J.G.; Vázquez, J.T; Bazzocchi, I.L.; Morales, E.Q.; Muñoz, O.M. *J. Chem. Perkin Trans I,* **1992**, 1437.

[123]   Smith, Jr C.R., Miller, R.W.; Weisleder, D.; Rohwedder, W.K.; Eickman, N.; Clardy, J. *J. Org. Chem.,* **1976**, *41*, 3264.

[124]   Lotler, H.; Brüning, R.; Wagner, H. *Tetrahedron Lett.* **1978**, *35*, 3243.

[125]   White, J.D.; Shin, H.; Kim, T.S.; Cutshall, N.S. *J. Am. Chem. Soc.,* **1997**, *119*, 2404.

[126]   Rodríguez-Hahn, L.; Jimenez, M.; Diaz, E.; Guerrero, C.; Ortega, S.; Romo de Vivar, A. **1977**, *33*, 657.

[127]   Jacobson, M.; Crosby, D.G. *Naturally Occurring Insecticides,* Marcel Dekker: New York, **1971**.

[128]   Wakabayashi, N.; Wu, W.J.; Waters, R.M.; Redfern, R.E.; Mills,G.D.; DeMilo, A.B.; Lusby, W.R.; Andrzejewski, D. *J. Nat. Prod.,* **1988**, *51(3)*, 537.

[129]   Wu, W.J.; Tu, Y.Q.; Liu, H.X.; Zhu, J.B. *J. Nat. Prod.,* **1992**, *55(9)*, 1294.

[130]   Liu, J.K.; Jia, Z.J.; Wu, D.G.; Zhou, J.; Wang, Q.G. *Phytochemistry,* **1990**, *29(8)*, 2503.

[131]   Wang, M.; Qin, H.; Kong, M.; Li, Y. *Phytochemistry,* **1991**, *30(12)*, 3931.

[132]   Wu, D.; Liu, J.; Cheng, C. *Phytochemistry,* **1992**, *31(12)*, 4219.

[133]   Tu, T.Q.; Wu, D.; Zhou, J.; Chen, Y. *Phytochemistry,* **1990**, *29(9)*, 2923.

[134]   Tu, Y.Q. *J. Chem. Soc. Perkin Trans I,* **1991**, 425.

[135]   Wang, M.; Chen, F. *J. Nat. Prod.,* **1997**, *60*, 602.

[136]   Tu, Y.Q.; Wu, D.G.; Zhou, J.; Chen, Y.Z.; Pan, X.F. *J. Nat. Prod.,* **1990**, *53(3)*, 603.

[137] Yuan, X.; Wu, X. *Tianran Chanwu Yanjiu Yu Kaifa*, **1994**, *6(2)*, 37.

[138] Verdcourt, B.; Trump, E.C. *Common Poisonous Plants of East Africa*, Collins, St. James's Place: London, **1969**.

[139] Tsanuo, M.K.; Hassanali, A.; Jondiko I.J.O.; Torto, B. *Phytochemistry*, **1993**, *35*, 665.

[140] González, A.G.; Jiménez, I.A.; Ravelo, A.G.; Bellés, X.; Piulachs, M.D. *Biochem. System. Ecol.*, **1992**, *20(4)*, 311.

[141] González, A.G.; Jiménez, I.A.; Ravelo, A.G.; Bazzocchi, I.L. *Tetrahedron*, **1993**, *49(30)*, 6637.

[142] González, A.G.; Jiménez, I.A.; Ravelo, A.G.; Sazatornil, J.G.; Bazzocchi, I.L.; *Tetrahedron*, **1993**, *49(3)*, 697.

[143] González, A.G.; Jiménez, I.A.; Ravelo, A.G:, Coll, J.; González, J.A.; Lloria, J. *Biochem. System.Ecol.*, **1997**, *25(6)*, 513.

[144] Takaishi, Y.; Ujita, K.; Tokuda, H.; Nishino, H.; Iwashima, A.; Fujita, T. *Cancer Lett.*, **1992**, *65*, 19.

[145] Ujita, K.; Takaishi, Y.; Tokuda, H.; Nishino, H.; Iwashima, A.; Fujita, T. *Cancer Lett.*, **1993**, 68, 129.

[146] Takaishi, Y.; Ohshima, S.; Nakano, K.; Tomimatsu, T. *J. Nat. Prod.*, **1993**, *56(6)*, 815.

[147] Orechoff, A. *Arch. Pharm.*, **1934**, *272*, 673.

[148] Doebel, K.; Reichstein, T. *Helv. Chim. Acta*, **1949**, *32*, 592

[149] Pailer, M.; Libiseller, R. *Monatsh. Chem.*, **1962**, *93*, 403.

[150] Sasaki, M.; Hirata, Y. *J. Chem. Soc.Perkin Trans 2*, **1972**, 1268.

[151] Crombie, L.; Toplis, D.; Whiting, D.A.; Rozsa, Z.; Hopmann, J.; Szendrei, K.J. *J. Chem. Soc. Perkin Tarns. I*, **1986**, 531.

[152] . Kim, T.S.; White, J.D. *Tetrahedron Lett.*, **1993**, *34(35)*, 5535.

[153] Klass, J.; Tinto, W.F. *J. Nat. Prod.*, **1993**, *56(6)*, 946.

[154] Morota, T.; Yang, C.; Ikeya, Y.; Qin, W.; Nishimura, H.; Xu, L.; Ando, M.; Miao, K.; Maruno, M.; Yang, B. *Phytochemistry*, **1995**, *39(5)*, 1219.

[155] Yamada, K.; Shizuri, Y.; Hirata,Y. *Tetrahedron*, **1978**, *34*, 1915.

[156] Wu, D.G.; Liu, L.; Chen, K.C. *Acta Bot. Yunnanica*, **1981**, *3*, 471.

[157] Wu, D.G. *Acta Bot. Yunnanica*, **1986**, *8*, 343.

[158] Monache, F.D.; Marini-Bettolo, G.B.; Bernays, E.A. *Angew. Entomol.*, **1984**, *97*, 406.

[159] Hohmann, J.; Nagy, G.; Dini, Z.; Gúnther, G.; Pelczer, Y.; Jerkouich, G.; Varjas, L. *J. Nat. Prod.*, **1995**, *58(8)*, 1192.

[160] Liu, H.C.; King, C.H.; Su, M.H.; Chen, G.L.; Wang, T.C. *J. Chin. Chem. Soc.*, **1981**, *28*, 95.

[161] Kuo, Y.; Chen, C.; Kuo, L.Y.; King, M.; Wu, T.; Lu, S.; Chen, I.; McPhail, D.R.; McPhail, A.T.; Lee, K. *Heterocycles*, **1989**, *29(8)*, 1465.

[162] Kuo, Y.; Chen, C.; Kuo, L. *J. Nat. Prod.*, **1990**, *53(2)*, 422.

[163] Kuo, Y.; Chen, C.; King, M.; Wu, T.; Lee, K. *Phytochemistry*, **1993**, *35(3)*, 803.

[164] Kuo, Y.; King, M.; Chen, G.; Chen, H.; Chen, C.; Chen, L.; Lee, K. *J. Nat. Prod.*, **1994**, *57(2)*, 263.

[165] Mata, R.; Calzada, F.; Díaz, E.; Toscano, B.A. *J. Nat. Prod.*, **1990**, *53(5)*, 1212.

[166] Zhang, L.H.; Huang, Y.; Wang, L.W.; Xiao, P.G., *Phytother. Res.*, **1995**, *9*, 315.

[167] Sekar, K.V.S.; Sneden, A.T.; Flores, F.A. *Planta Med.*, **1995**, *61*, 390.

[168] González, A.G.; Bazzocchi, I.L.; Ravelo; Luis, J.G.; Ravelo, A.G.; Fraga, B.M.; Domínguez, X.A.; Perales, A. *J. Chem. Research (S).* **1986**, 442.

[169] Morota, T.; Qin, W.Z.; Takagi, K.; Xu, L.H.; Maruno, M.; Yang, B.H. *Phytochemistry,* **1995**, *40(3),* 865.

[170] Li, K.; Duan, H.; Kawazoe, K.; Takaishi, Y. *Phytochemistry,* **1997**, *45(4),* 791.

[171] Kupchan, S.M.; Schubert , R.M. *Science,* **1974**, *185*, 791.

[172] Yisheng, W.; Isamu, A. *Zhongguo Yaoli Xuebao,* **1991**, *12(5),* 406.

[173] Kutney, J.P.; Hewitt, G.M.; Kurihara, T.; Salisbury, P.J.; Sindelar, R.D.; Stuart, K.L.; Townsley, P.M.; Chalmers, W.T.; Jacoli, G.G. *Can. J. Chem.,***1981**, *59*, 2677.

[174] Kutney, J.P.; Hewitt, G.M.; Lee, G.; Piotrowska, K.; Roberts, M.; Rettig, S.J. *Can. J. Chem.,* **1992**, *70*, 1455.

[175] Kutney, J.P.; Han, K.; Kuri-Brena, F.; Milanova, R.K.; Roberts, M. *Heterocycles,* **1997**, *44(1),* 95.

[176] Van Tamelen, E.E.; Demers, J.P.; Taylor, E.G.; Koller, K.J. *J. Am. Chem. Soc.,* **1980**, *102*, 5424.

[177] Buckanin, R.S.; Chen, S.J.; Zimmerman, D.F.; Sher, F.T.; Berchtold, G.A. *J. Org. Chem.,* **1982**, *47*, 2364.

[178] Yang, D.; Wong, M.K.; Cheung, K.K.; Chan, E.W.C.; Xie, Y. *Tetrahedron Lett.,* **1997**, *38(39),* 6865.

[179] Shamon, L.A.; Pezzuto, J.; Graves, J.M.; Mehta, R.R.; Wangcharoentrakul, S.; Sangsuwan, R.; Chaichana, S.; Tuchinda, P.; Cleason, P.; Reutrakul, V. *Cancer Lett.,* **1977**, *112(1),* 113.

[180] Chen, K.; Shi, Q.; Fujioka, T.; Zhang, D.C.; Hu, C.Q.; Jin, J.K.; Kilkuskie, R.; Lee, K.H. *J. Nat. Prod.,* **1992**, *55(1),* 88.

[181] Chen, K.; Shi, Q.; Fujioka, T.; Nakano, T.; Hu, C.Q.; Jin, J.Q.; Kilkuskie, R.E.; Lee, K.H. *Bioorg. Med. Chem.,* **1995**, *3(10),* 1345.

[182] Corey, E.J.; Liu, K. *J. Am. Chem. Soc.,* **1997**, *119*, 9929.

[183] Zheng, J.R.; Gu, K.; Xu, L.; Gao, J.; Yu, Y.; Tang, M. *Acta Acad. Med. Sin.,* **1991**, *13*, 391.

[184] Yang, S.X.; Gao, H.L.; Xie, S.S.; Zhang, W.R.; Long, Z.Z. *Int. J. Immunopharmacol.,* **1992**, *14*, 963.

[185] Pei, R.J.; Qi, L.H.; Liu, X.J. *Chin. Pharmacol. Bull.,* **1993**, *9*, 68.

[186] Pu, L.X.; Zhang, T.M. *Acta Pharmacol. Sin.,* **1990**, *11*, 76.

[187] Yang, S.X.; Xie, S.S.; Hu, G.J.; Long, Z.Z.; Tu, Y.Y. *Chin. J. Immunol.,* **1993**, *9*, 188.

[188] Lipsky, P.E.; Tao, X.L.; Cai, J. USA Patent 5, 294443 (**1994**).

[189] Takaishi, Y.; Shishido, K.; Wariishi, N.; Shibuya, M.; Goto, K., Kido, M.; Takai, M.; Ono, Y. *Tetrahedron Lett.,* **1992**, *33(47),* 7177.

[190] Shishido, K.; Nakano, K.; Wariishi, N.; Tateishi, H.; Omodani, T.; Shibuya, M.; Goto, K.; Ono, Y.; Takaishi, Y. *Phytochemistry,* **1994**, *35(3),* 731.

[191] Takaishi, Y.; Goto, K.; Takai, M.; Taniguchi, T.; Manabe, S.; Asakuni, T. Japan Patent 211035 (**1992**), C.A.. 116 28115d, **1992**.

[192] Xu, J.; Ikekawa, T.; Ohkawa, M.; Yokota, I.; Hara, N.; Fujimoto, Y. *Phytochemistry,* **1997**, *44(8),* 1511.

[193] Zhang, Z..; Ding, L.; Qian, S.; An, D. *J. Chin. Pharm. Sci.,* **1993**, *2(2),* 144.

[194] Viswanathan, N. I.; *J. Chem. Soc. Perkin Trans I,* **1979**, *2*, 349.

[195] González, A.G.; Fraga, B.M.; González, P.; González, C.M.; Ravelo, A.G.; Ferro, E.; Dominguez, X.A.; Martínez, M.A.; Perales, A.; Fayos, J. *J. Org. Chem.,* **1983**, *48*, 3759.

[196] González, A.G.; González, C.M.; Ferro, E.A.; Ravelo, A.G.; Dominguez, X.A. *J. Chem. Researchs,* **1988**, 20.

[197] Honda, T.; Finlay, H.J.; Gribble, W.G. *J. Nat. Prod.,* **1997,** *60,* 1174.

[198] Lee, K.H.; Nozaki, H.; Hall, I.H.; Kasai, R.; Hirayama, T. Suzuki, H.; Wu, R.Y. *J. Nat. Prod.,* **1982,** *45(4),* 509.

[199] Nozaki, H.; Suzuki, H.; Lee, K.H.; McPhail, A.T. *J. Chem. Soc. Chem. Commun,* **1982,** 1048.

[200] Nozaki, H.; Suzuki, H.; Hirayama, T.; Kasai, R.; Wu, R.Y.; Lee, K.H. *Phytochemistry,* **1986,** *25(2),* 479.

[201] Nozaki, H.; Matsuura, Y.; Hirono, S.; Kasai, R.; Chang, J.J.; Hsiung Lee, K.H. *J. Nat. Prod.,* **1990,** *53(4),* 1039.

[202] Monache, F.D.; Mello, J.F.; Marini-Bettolo, G.V.; Lima, O.G.; Albuquerque, I.L *Gazz. Chim. Ital.,* **1972,** *102,* 636.

[203] Govindachari, T.R.; Viswanathan, N. *Tetrahedron,* **1967,** *23,* 1901.

[204] Weeratunka, G.; Kumar, V.; Sultanbawa, M.U.S. *J. Chem. Soc. Perkin Trans I,* **1982,** 2457.

[205] Fang, S.D.; Berry, D.E.; Lynn, D.G.; Hecht, S.M.; Campbell, J.; Lynn, W.S. *Phytochemistry,* **1984,** *23(3),* 631.

[206] Wijeratne, D.B.T.; Kumar, V.; Suthanbawa, M.U.S. *J. Chem. Soc. Perkin Tran. I,* **1981,** 2724.

[207] Itokawa, H.; Shirota, O.; Ikuta, H.; Morita, H.; Takeya, K.; Iitaka, Y. *Phytochemistry,* **1991,** *30(1),* 3713.

[208] Weeratunka, G.; Kumar, V. *Phytochemistry,* **1985,** *24,* 2369

[209] Hori, H.; Pang, G.M.; Harimaya, K.; Iitaka, Y.; Inayama, S. *Chem. Pharm. Bull.,* **1987,** *35(5),* 2125.

[210] Hou, D. *Flora Malesiana,* **1962,** *6,* 258.

[211] Ngassapa, O.D.; Soejarto, D.D.; Che, C.T.; Pezzuto, J.M.; Farnsworth, N.R. *J. Nat. Prod.,* **1991,** *54(5),* 1353.

[212] Ngassapa, O.D.; Soejarto, D.D.; Pezzuto, J.M.; Farnsworth, N.R.; Che, C.T. *J. Nat. Prod.,* **1993,** *56(10),* 1676.

[213] Sturm, S.; Gil, R.R.; Chai, H.-B.; Ngassapa, O.D.; Santisuk, T.; Reutrakul, V.; Howe, A.; Moss, M.; Besterman, J.M.; Yang, S.-L.; Farthing, J.E.; Tait, R.M., Lewis, J.A.; O'Neill, M.J.; Farnsworth, N.R.; Cordell, G.A.; Pezzuto, J.M.; Kinghorn, A.D. *J. Nat. Prod.,* **1996,** *59,* 658.

[214] Lee, K.H.; Lin, Y.M.; Wu, T.S.; Zhang, D.C.; Yamagishi, T.; Hayashi, T.; Hall, I.H.; Chang, J.J.; Wu, R.Y.; Yang, T.H. *Planta Med.,* **1988,** *54,* 308.

[215] Kuo, Y.H.; Yang Kuo, L.M. *Phytochemistry,* **1997,** *44(7),* 1275.

[216] Kuo, Y.H.; Ou, J.C.; Lee, K.H.; Chen, C.F. *J. Nat. Prod.,* **1995,** *58,* 1103.

[217] Kubo, I.; Fukuhara, K. *J. Nat. Prod.,* **1990,** *53(4),* 968.

[218] Chen, K.; Shi, Q.; Kashiwada, Y.; Zhang, D.C.; Hu, C. Q.; Jin, J.Q.; Nozaki, H.; Kllkuskie, R.E.; Tramontano, E.; Chen, Y.C.; McPhail, D.R.; McPhail, A.T.; Lee, K.H. *J. Nat. Prod.,* **1992,** *55(3),* 340.

[219] Fujioka, T.; Kashiwada, Y.; Kilkuskie, R.E; Cosentino, L.M.; Ballas, L.M.; Jiang, J.B.; Janzen, W.P.; Chen, I.S.; Lee, K.H. *J. Nat. Prod.,* **1994,** *57(2),* 243.

[220] Gupta, M.B.; Nath, R.; Srivastava, N.; Shanker, K.; Kishor, K.; Bhargava, K.P. *Planta Med.,* **1980,** *39,* 157.

[221] Hayek, E.W.H.; Jordis, U.; Moche, W.; Sauter, F. *Phytochemistry,* **1989,** *28(9),* 2229.

[222] Bringmann, G.; Saeb, W.; Assi, L.A.; François, G.; Narayanan, A.S.S.; Peters, K.; Peters, E.M. *Planta Med.*, **1997**, *63*, 255.

[223] Ohigashi, H.; Takamura, H.; Koshimizu, K.; Tokuda, H.; Ito, Y. *Cancer Lett.*, **1986**, *30*, 143.

[224] Gonçalves de Lima, O.; D'alburquerque, L.; De Barros Coelho, S.S.; Medeirso maciel, G.; Martins, D.G.; Lacerda, A.L. *Rev. Inst. Antibioticos*, **1969**, *9*, 17.

[225] González, A.G.; González, C.M.; Ravelo, A.G.; Gutiérrez, A.M.; Moujir, L.; Navarro, E.; Boada, J. *Rev. Latinoam. Quim.*, **1988**, *19*, 36.

[226] Ferreira de Santana, C.; Asfora, J.J.; Cortias, C.T. *Rev. Inst. Antibioticos*, 1971, *11*, 37.

[227] Sneden, A.T. *J. Nat. Prod.*, **1981**, *44*, 503.

[228] Moujir, L.; Gutiérrez-Navarro, A.M.; González, A.G.; Ravelo, A.G.; Luis, J.G. *Biochem. System. Ecol.*, **1990**, *18*, 25.

[229] González, A.G.; Ravelo, A.G.; Bazzocchi, I.L.; Jiménez, J..; González, C.M.; Luis, J.G.; Ferro, E.A.; Gutiérrez-Navarro, A.M.; Moujir, L.; De Las heras, F.G. *Il Farmaco*, **1988**, *43*, 264.

[230] Moujir, L.; Gutiérrez-Navarro, A.M.; González, A.G.; Ravelo, A.G.; Luis, J.G. *Antimicrob. Agents Chemother.*, **1991**, *35*, 211.

[231] Campanelli, A.R.; D'Alagni, M.; Marini-Bettolo, G.B. *FEBS Lett.*, **1980**, *122*, 256.

[232] Pavanand, K.; Webster, H.K.; Yongvanitchit, K.; Kun-Anake, A.; Dechatiwongse, T.; Nutakul, W.; Bansiddhi, J. *Phytother. Res.*, **1989**, *3*, 136.

[233] Premakumara, G.A.S.; Ratnasooriya, W.D.; Balasubramaniam, S.; Dhanabalasingham, B.; Fernando, H.C.; Dias, M.N.; Karunaratne, V.; Gunatilaka, A.A.L. *Phytochemistry*, **1992**, *11*, 219.

[234] Schechter, M.S.; Haller, H.L. *J. Am. Chem. Soc.,* **1942**, *64*, 182.

[235] Bhatnagar, S.S.; Divekar, P.V. *J. Sci. Industr. Res.*, **1951**, *10B*, 56.

[236] Kulkarni, A.B.; Shah, R.C. *Nature*, **1954**, *173*, 1237.

[237] Kamat, V.N.; Fernandes, F.; Bhatnagar, S.S. *J. Sci. Industrial Res.*, **1955**, *14C*, 1.

[238] Nakanishi, K.; Kakisawa, H.; Hirata, Y. *Bull. Chem. Soc. Japan*, **1965**, *30*, 1729.

[239] González, A.G.; Alvarenga, N.L.; Rodríguez, F.; Ravelo, A.G.; Jiménez, I.A.; Bazzocchi, I.L; Gupta, M.P. *Natl. Prod. Lett.*, **1995**, *7*, 209.

[240] Martín, J.D. *Tetrahedron*, **1973**, *29*, 2997.

[241] Krishnamoorthy, V.; Ramanathan, J.D.; Seshadri, T.R. *Tetrahedron Lett.*, **1962**, 1047.

[242] Calzada, F.; Mata, R.; López, R.; Linares, E.; Bye, R.; Barreto, V.M.; Del Río, F. *Planta Med.*, **1991**, *57*, 194.

[243] González, A.; Francisco, C.G.; Freire, R.; Hernández, R.; Salazar, J.A.; Suarez, E. *Phytochemistry*, **1975**, *14*, 1067.

[244] Fernando, H.C.; Gunatilaka, A.A.L.; Tezuka, Y.; Kikuchi, T. *Tetrahedron*, **1989**, *45*, 5867.

[245] Gunaherath, G.M.K.B.; Gunatilaka, A.A.L.; Sultanbawa, M.U.S.; Wazeer, M.I.M. *Tetrahedron Lett.*, **1980**, *21*, 4749.

[246] Gamlath, C.B.; Gunatilaka, A.A.L. *Phytochemistry*, **1988**, *27*, 3221.

[247] González, A.G.; Alvarenga, N.L.; Ravelo, A.G.; Jiménez, I.A.; Bazzocchi, I.L; Canela, N.J.; Moujir, L.M. *Phytochemistry*, **1996**, *43(1)*, 129.

[248] Itokawa, H.; Shirota, O.; Morita, H.; Takeya, K.; Tomioka, N.; Itai, A. *Tetrahedron Lett.*, **1990**, *31(47)*, 6881.

[249]   Shirota, O.; Morita, H.; Takeya, K.; Itokawa, H. *J. Nat. Prod.,* **1997,** *60,* 302.
[250]   Shirota, O.; Morita, H.; Takeya, K.; Itokawa, H. *J. Nat. Prod.,* **1997,** *60,* 1100.
[251]   Gunatilaka, A.A.L. *Progress in the Chemistry of Organic Natural Products,* Springer-Verlag: New York, **1996.**
[252]   González, A.G.; Alvarenga, N.L.; Ravelo, A.G.; Bazzocchi, I.L.; Ferro, E.A.; Navarro, A.G.; Moujir, L.M. *Bioorg. Med. Chem.,* **1996,** *4(6),* 815.
[253]   González, A.G.; Alvarenga, N.L.; Estévez-Braun, A.; Ravelo, A.G.; Bazzocchi, I.L; Moujir, L.M. *Tetrahedron,* **1996,** *52(28),* 9597.
[254]   Takaishi, Y.; Wariishi, N.; Tateishi, H.; Kawazoe, K.; Nakano, K.; Ono, Y.; Tokuda, H.; Nishino, H.; Iwashima, A. *Phytochemistry,* **1997,** *45(5),* 969.
[255]   Oppolzer, W.; Prelog, V.; Sensi, P. *Experientia,* **1964,** *20,* 336.
[256]   Rinehart, K.L., Jr.; Maheshwari, M.L.; Antosz, F.J.; Mathur, H.H.; Sasaki, K.; Schacht, R.J. *J. Amer. Chem. Soc.,* **1971,** *93,* 6273.
[257]   Kishi, T.; Harada, S.; Asai, M.; Muroi, M.; Mizumo, K. *Tetrahedron Lett.,* **1969,** 97.
[258]   Sasaki, K.; Rinehart, K.L., Jr.; Slomp, G.; Grostic, M.F.; Olson, E.C. *J. Amer. Chem. Soc.,* **1970,** *92,* 7591.
[259]   Kupchan, S.M.; Komoda, Y.; Thomas, G.J.; Hintz, H.P.J. *J. C. S. Chem. Comm.,* **1972,** 1065.
[260]   Kupchan, S.M.; Branfman A.R.; Sneden, A.T.; Verma, A.K.; Dailey, R.G.; Komoda, Y.; Nagao, Y. *J. Amer. Chem. Soc.,* **1975,** *97,* 5294.
[261]   Wani, M.C.; Taylor, H.L.; Wall, M.E. *J. C. S. Chem. Comm.,* **1973,** 390.
[262]   Higashide, E.; Asia, M.; Ootsu, K.; Kozai, Y.; Hasegawa, T.; Kishi, T.; Sugino, Y.; Yoneda, M. *Nature,* **1977,** *270,* 721.
[263]   Wolpert-Defilippes, M.K.; Adamson, R.H.; Cysyk, R.L.; Johns, D.G. *Biochem. Pharmacol.,* **1975,** *24,* 751.
[264]   Remillard, S.; Rebhun, L.I.; Howie, G.A.; Kupchan, S.M. *Science,* **1975,** *189,* 1002.
[265]   O'Connor, T.E.; Aldrich, C.; Hadidi, A.; Lomax, N.; Okano, P.; Sethi, S.; Wood, H.B. *Proceeding of the 86th Annual Meeting of the American Association of Cancer Research,* **1975,** 29.
[266]   Kupchan, S.M.; Komoda, Y.; Branfman A.R.; Dailey, R.G.; Zimmerly, V.A. *J. Amer. Chem. Soc.,* **1974,** *96,* 3706.
[267]   Kupchan, S.M.; Komoda, Y.; Branfman A.R.; Sneden, A.T.; Court, W.A.; Thomas, G.J.; Hintz, H.P.; Smith, R.M.; Karim, A.; Howie, G.A.; Verma, A.K.; Nagao, Y.; Dailey, Jr. R.G.; Zimmerly, V.A., Sumner, W.C. *J. Org. Chem.,* **1977,** *42(14),* 2349.
[268]   Sneden, A.T.; Beemsterboer, G.L.; *J. Nat. Prod.,* **1980,** *43(5),* 637.
[269]   Nettleton, D.E., Jr.; Balitz, D.M.; Brown, M.; Moseley, J.E.; Myllymaki, R.W.; *J. Nat. Prod.,* **1981,** *44 (3),* 340.
[270]   Komoda, Y.; Isogai, Y. *Sci. Pap. Coll. Gen Educ., Univ. Tokio,* **1978,** *28,* 129.
[271]   Kutney, J.P.; Beale, M.H.; Salisbury, P.J.; Stuart, K.L.; Worth, B.R.; Townsley, P.M.; Chalmers, W.T.; Nilsson, K.; Jacoli, G.G. *Phytochemistry,* **1981,** *20,* 653.
[272]   Meyers, A.I.; Reider, P.; Campbell, A.L. *J. Am. Chem. Soc.,* **1980,** *102,* 6579.
[273]   Corey, E.J.; Weigel, L.O.; Chamberlin, A.R.; Cho, H.; Hua, D.H. *Ibid.,* **1980,** *102,* 6615.
[274]   Kitamura, M.; Isobe, M.; Ichikawa, Y.; Goto, T. *J. Am. Chem. Soc.,* **1984,** *106,* 3252.

[275] Chari, R.J.; Goldmacher, V.S.; Lambert, J.M.; Blattler, W.A. (Inmunogen, Inc.). Eur. Pat. Appl. EP 425,235. (Cl. A61K47/48), 02 May, 1991. US Appl. 426,247, 25 Oct 89; C.A. 116 28121 c **1992.**.

[276] Liu, C.; Bourret, L.A.; Derr, S.M.; Widdison, W.C.; Lambert, J.M.; Bláttler, W.A.; Chari, R.J (Inmunogen, Inc.). Data presented at the American Association for Cancer Research Meeting (**1997**).

Atta-ur-Rahman (Ed.) *Studies in Natural Products Chemistry, Vol. 23*
© 2000 Elsevier Science B.V. All rights reserved

ANTIOXIDANT PHENOLIC METABOLITES FROM FRUIT AND VEGETABLES AND CHANGES DURING POSTHARVEST STORAGE AND PROCESSING

F.A. TOMÁS-BARBERÁN, F. FERRERES *and* M.I. GIL

Department of Food Science and Technology, CEBAS (CSIC), P.O. Box, 4195, Murcia 30080, Spain.

ABSTRACT: Flavonoids and other phenolic metabolites have important biological activities related to their antioxidant properties and, especially, to their free-radical scavenging ability. It has been suggested that dietary phenolics might be beneficial agents for the prevention of cardiovascular diseases and cancer, a suggestion supported by epidemiological and experimental studies. In the present review, the antioxidant activity of flavonoids and other phenolic metabolites is reviewed, as well as their bioavailability, absorption and metabolism in animals and humans. In addition, the flavonoid content and that of other phenolics in different fruits and vegetables is reviewed.
The changes in flavonoids and other phenolic antioxidants during the postharvest life of fruit and vegetables are evaluated. The effect of storage, technological treatments (controlled atmospheres, UV irradiation, heat shocks, etc.), and processing (minimal processing, juice manufacturing, drying, cooking, canning, etc.) on the content of the biologically active phenolics is also reviewed.

INTRODUCTION

During recent years epidemiological studies have increasingly correlated the incidence of chronic diseases, such as cardiovascular disease and cancer, with diet. It is widely accepted that diets rich in fruit and vegetables or their derivates play a positive role in the protection against heart attacks [1], strokes [2] and cancer [3]. The beneficial effect of fruit and vegetables was first attributed to the antioxidant vitamins A, C and E which they contained [4-6]. It was claimed that they prevented the oxidation of plasma LDL and other free-radical mediated degenerative processes. However, recent studies have demonstrated the lack of effect of long-term supplementation with these vitamins in the incidence of malignant neoplasms and cardiovascular disease [7]. Other constituents of fruit and vegetables have also shown antioxidant activity, and some of them are considered even more effective than vitamins. Among such antioxidant constituents, flavonoids and other phenolic metabolites have received much attention due to their widespread occurrence in plant derived products and to their high activity.

The importance of plant flavonoids in health was first reported in 1936, when Rusznyak and Szent-Gyorgyi proposed that the intake of flavonoids decreased the capillary permeability and fragility seen in scurvy [8]. This gave rise to a claim for the vitaminic action of flavonoids (vitamin P), which were later stripped of their vitamin status around 1950. During the 70s and 80s the absorption, excretion and metabolism of flavonoids was extensively studied [9-11].

In the last decade, there has been a resurgence in the interest shown in flavonoids and other phenolics because of possible links with the prevention of cancer and cardiovascular diseases [12]. It is thought that these natural compounds may protect tissues against oxygen free radicals and lipid peroxidation, both of which play a role in several pathologies such as atherosclerosis and cancer. Atherosclerosis is an important factor in the development of cardiovascular diseases such as coronary heart disease and thrombotic stroke. It is characterised by thickening and narrowing of the arteries caused by the formation of fibrofatty and fibrous lesions that obstruct the blood flow. A major hypothesis proposes that oxidised low-density lipoprotein (LDL) particles play a key role in the development of atherosclerosis, and so the avoidance or delay of LDL oxidation by dietary antioxidants might provide a promising strategy for preventing atherosclerosis and, therefore, coronary heart disease. However, the ability of dietary antioxidants to inhibit LDL *in vivo* still remains to be established [13].

The present chapter reviews the antioxidant activity of fruit and vegetable phenolics and its relationship with health and nutrition, as well as the effect of processing and postharvest technological treatments on these antioxidant metabolites, and on the nutritional quality of fruit and vegetable derived food products.

GENERAL STRUCTURES AND BIOSYNTHESIS OF ANTIOXIDANT PHENOLICS

Phenolic compounds include a wide range of secondary metabolites that are biosynthesised from carbohydrates through the shikimate pathway [14]. This is the biosynthetic route to the aromatic amino acids, phenylalanine, tyrosine, and tryptophan, and only occurs in microorganisms and plants. In the first step, the glycolytic intermediate phosphoenol pyruvate and the pentose phosphate intermediate erythrose-4-phosphate are condensed to 3-deoxy-D-arabino-heptulosonate 7-phosphate (DAHP), a step catalysed by DAHP synthase. Intermediates of the shikimate pathway are 3-dehydroquinate, shikimate, and chorismate (Fig. 1). Phenylalanine is biosynthesised from chorismate, and from phenylalanine all the phenylpropanoids. Quinate is produced from 3-dehydroquinate and incorporated into chlorogenic and isochlorogenic acids (caffeoyl quinic acids) by combination with caffeic acid. Gallic acid is produced from shikimate.

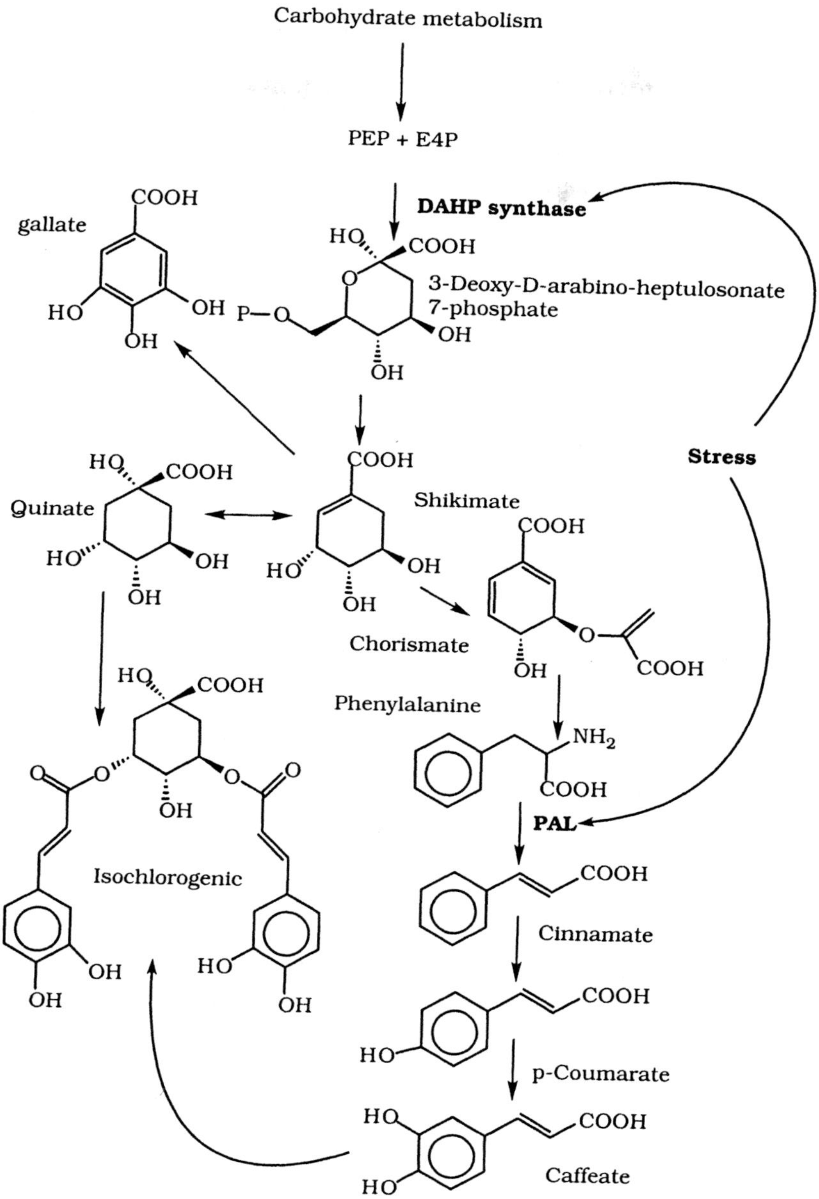

Fig. (1). Phenolic metabolism. PEP (phospho enol pyruvate); E4P (erythrose 4-phosphate); DAHP (3-Deoxy-D-arabino-heptulosonate 7-phosphate); PAL (phenylalanine ammonia lyase). Stress transcriptionally activates the key enzymes of phenolic metabolism (PAL and DAHP synthase).

The enzyme DAHP synthase regulates the carbon flow in the shikimate pathway. Different biotic and abiotic stresses, including mechanical wounding and fungal elicitation, induce the accumulation of DAHP synthase mRNA, and, therefore, of phenolic metabolites.

All phenylpropanoids are derived from cinnamic acid, which is formed from phenylalanine through the action of phenylalanine ammonia-lyase (PAL), which is considered the branch point enzyme between primary (shikimate pathway) and secondary metabolism [15].

Several simple phenylpropanoids (C_6-C_3) are produced from cinnamate via a series of hydroxylation, methylation, and dehydration reactions; these include *p*-coumaric, caffeic, ferulic and sinapic acids and simple coumarins. The free acids rarely accumulate to any great extent within plant cells but are usually conjugated to sugars, cell wall carbohydrates, or organic acids (quinic, malic, tartaric, etc.). Lignin and suberin are complex polymers formed from a mixture of simple phenylpropanoids, and will not be considered in the present review.

A large number of phenolics are derived from the C_{15} flavonoid skeleton, which is synthesised via the chalcone synthase (CHS) catalysed condensation of p-coumaroyl-coenzyme A and three molecules of malonyl-CoA. In most plant families, the initial CHS product is a tetrahydroxychalcone, which is further converted to other flavonoid classes, such as flavones, flavonols, flavanones, flavan-3-ols, isoflavones and anthocyanins (Fig. 2). Structural diversity among the phenylpropanoids arises from a variety of modifications, including

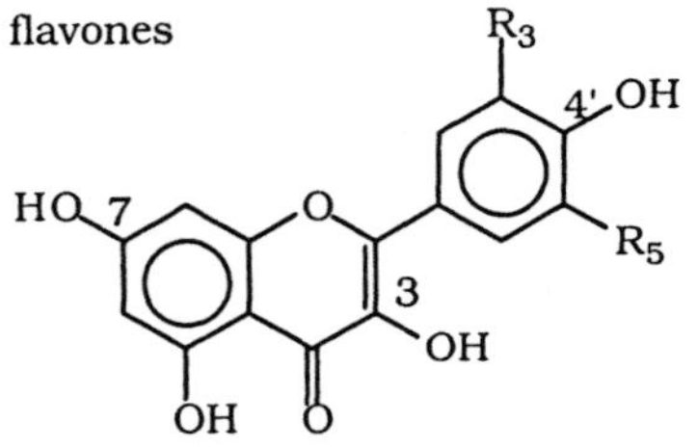

flavones		R_3	R_5
kaempferol		H	H
quercetin		OH	H
isorhamnetin		OCH_3	H
myricetin		OH	OH

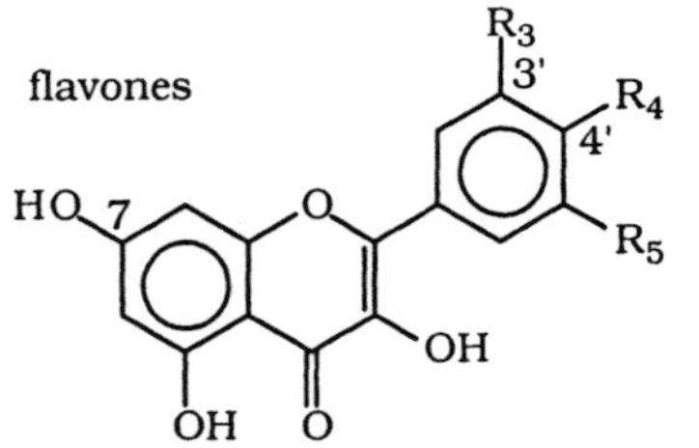

flavones		R_3	R_5
apigenin		H	OH
luteolin		OH	OH
chrysoeriol		OCH_3	OH
diosmetin		OH	OCH_3

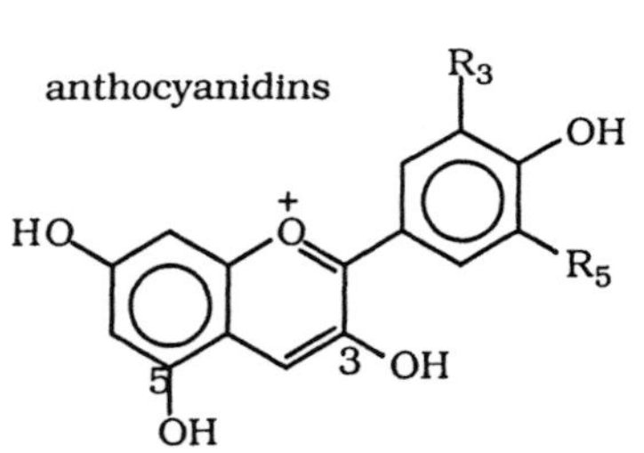

anthocyanidins		R_3	R_5
pelargonidin		H	H
cyanidin		OH	H
peonidin		OCH_3	H
delphinidin		OH	OH
petunidin		OCH_3	OH
malvidin		OCH_3	OCH_3

Fig. (2). General structures of antioxidant flavones, flavonols and anthocyanidins.

hydroxylation, glycosylation, acylation, prenylation, sulphation and methylation.

OCCURENCE OF PHENOLICS IN FRUIT AND VEGETABLES

Since phenolic metabolites may be responsible for the antioxidant properties of fruit and vegetables it is essential to know, both qualitative and quantitatively, their phenolic constituents. Although the data is incomplete, and the quantitative data is not reliable due to the very different extraction and analytical techniques used for phenolic compound analysis, the available information is summarised in this paper. As regards fruit phenolics, a comprehensive book has been published, in which qualitative and some quantitative data are provided [16]. The anthocyanins present in fruit, vegetables and grains have also been reviewed [17].

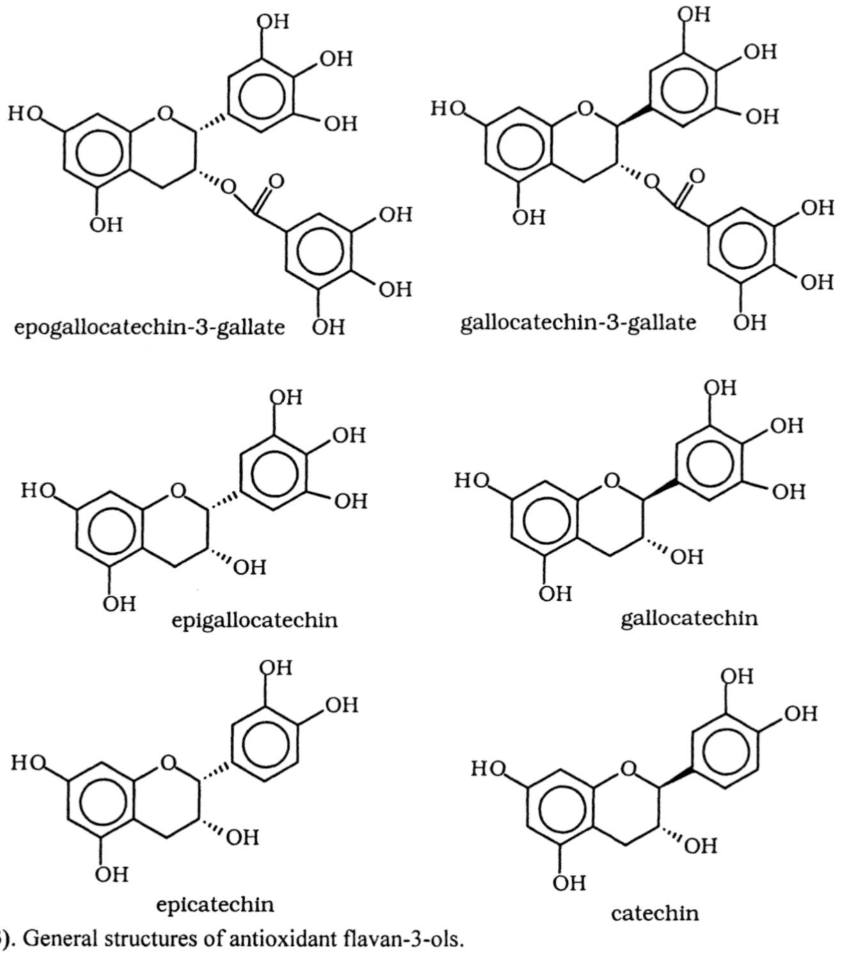

Fig. (3). General structures of antioxidant flavan-3-ols.

The principal phenolics present in fruit and vegetables can be arranged in five structural groups: planar flavonols and flavones (Fig. **2**), anthocyanin pigments (Fig. **2**), optically active flavan-3-ols (Fig. **3**), hydroxycinnamic acid derivatives (Fig. **4**) and benzoic acid derivatives (Fig. **5**). Other minor phenolics include citrus flavanones, tomato chalcones, apple dihydrochalcones, citrus coumarins, celery furocoumarins, and grape stilbenoids such as resveratrol (Fig. **5**). In these figures the general structures of fruit and vegetable phenolics can be appreciated.

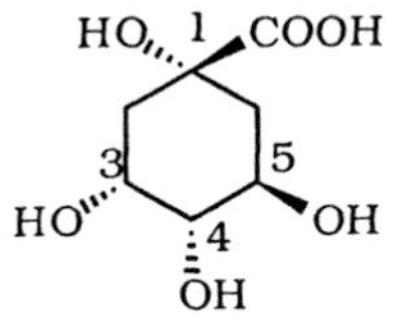

quinic acid

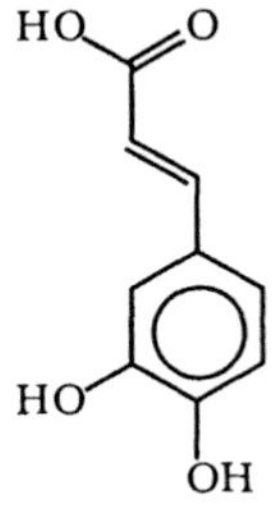

caffeic acid

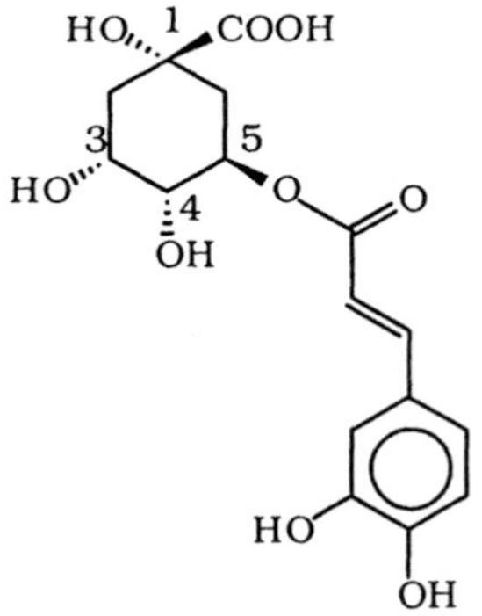

chlorogenic acid (5-O-caffeoylquinic acid)
cryptochlorogenic acid (4-O-caffeoylquinic acid)
neochlorogenicd acid (3-O-caffeoylquinic acid)
isochlorogenic acid 'a' (4,5-di-O-caffeoylquinic acid)
isochlorogenic acid 'b' (4,5-di-O-caffeoylquinic acid)
isochlorogenic acid 'c' (4,5-di-O-caffeoylquinic acid)

chlorogenic acid (5-O-caffeoylquinic acid)

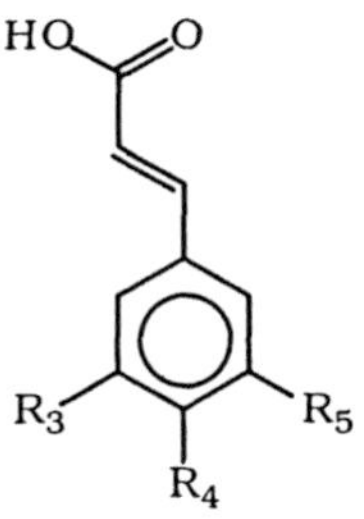

	R_3	R_4	R_5
cinnamic acid	H	H	H
p-coumaric acid	H	OH	H
caffeic acid	OH	OH	H
ferulic acid	OMe	OH	H
sinapic acid	OMe	OH	OMe

Fig. (4). General structures of antioxidant hydroxycinnamic acid derivatives.

FRUIT PHENOLICS

Apple

The main anthocyanin pigment in apple skin is cyanidin 3-galactoside [18], although cyanidin 3-glucoside, 3-arabinoside and 3-xyloside, and acylated derivatives have also been described [16]. Among the phenolic acid derivatives, chlorogenic acid (5'-caffeoylquinic) is the main constituent, although 4'-caffeoylquinic and 3'caffeoylquinic have also been reported, as well as dicaffeoylquinic derivatives. In addition, 3'-, 4', and 5'-p-coumaroyl quinic derivatives, and p-coumaroyl-, caffeoyl-, feruloyl- and sinpoyl glucose have also been reported in smaller amounts [16]. Apple contains caffeic acid in a concentration range of 52-191 mg/kg f.w. (fresh weight), but also p-coumaric acid (15-22 mg/kg) and ferulic (4-8 mg/kg) acids [19].

Several quercetin glycosides have been reported in apple peel, the principal one being quercetin 3-galactoside, although the 3-arabinoside, the 3-rhamnoside, the 3-xyloside, the 3-rutinoside and the 3-glucoside of quercetin have also been reported. Those flavonoids are present in the peel at concentrations between 1540 and 2851 mg/kg f.w. [16].

Apple flavan-3-ols include (-)-epicatechin, the main catechin derivative (31-129 mg/kg f.w.), (+)-catechin (0.5-27 mg/kg), (+)-gallocatechin and (-)-epigallocatechin [19]. Other studies have reported that these compounds in apple are in the range 34-165 mg/kg f.w., depending on the cultivar [16].

The most characteristic compounds of apple are the dihydrochalcones, analysis of which can be applied to apple-derived food characterisation (juices, jams, purées, cider, etc.). Phloretin 2'-glucoside (phloridzin) (Fig. 5) and phloretin 2'-(2''-xylosylglucoside) are the only dihydrochalcones reported in apple so far [20].

Apricot

Anthocyanin (cyanidin) pigmentation can be observed in some cultivars, although its nature has not been fully determined. 3'-Caffeoylquinic is the principal phenolic acid derivative, although 4'- and 5'-caffeoylquinic acids are also present. A concentration of 167 mg caffeic acid per kg has been reported in apricots [19]. The presence of the corresponding p-coumaroyl- and feruloyl quinic acid derivatives has also been reported [16], as have p-coumaroylglucose and feruloyl glucose.

The flavan-3-ols (-)-epicatechin (202 mg/kg) and (+)-catechin (44 mg/kg) have also been reported [19], while the flavonols, quercetin 3-rutinoside (the principal flavonol), and kaempferol 3 rutinoside, were only located in the skin at a concentration of 15 mg of flavonoids per kg of fruit [21;22].

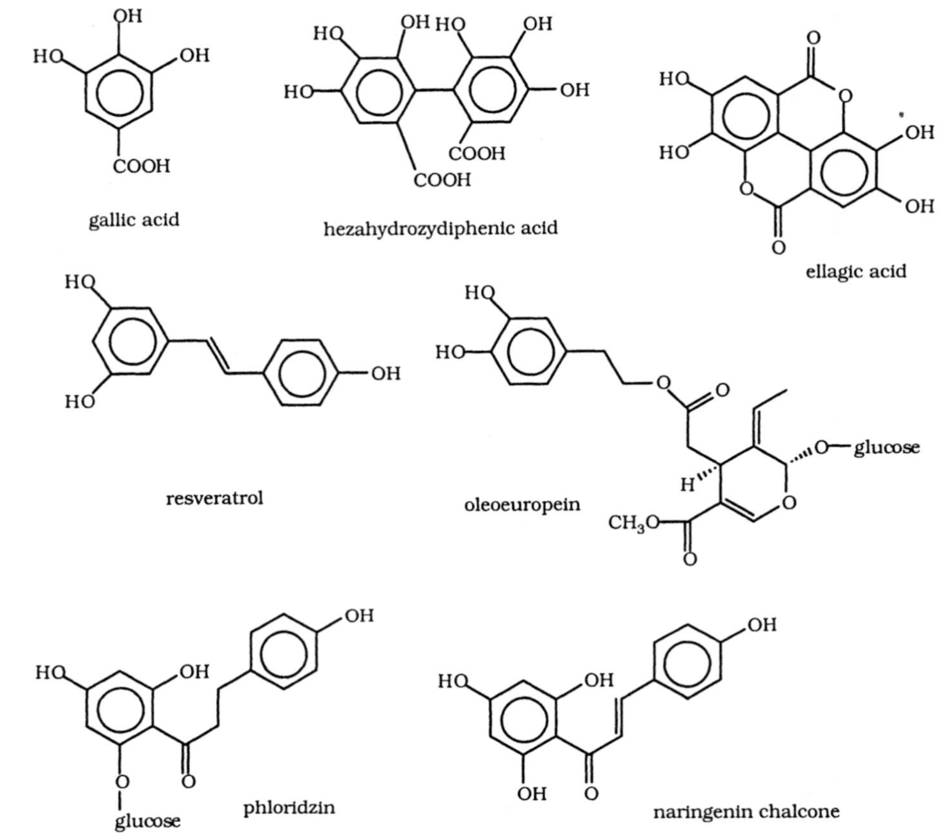

Fig. (5). Structures of miscelaneous phenolic antioxidants.

Cherry

Both *Prunus avium* and *Prunus cerasus* fruits are included in the generic name of cherry. Both contain similar phenolics, but with some minor differences. *Prunus avium* contains cyanidin 3-rutinoside as the main pigment, with cyanidin 3-glucoside, peonidin 3-rutinoside and peonidin 3-glucoside also being present [16]. *Prunus cerasus*, however, accumulates cyanidin 3-2"-glucosyl-rutinoside as the main pigment, while cyanidin 3-rutinoside, 3-sophoroside, 3-glucoside and 3-(2"-xylosylrutinoside) and peonidin 3-rutinoside are observed in smaller amounts [17]. There was a wide variation in the anthocyanin content between cultivars. Thus, pale pink cultivars contain less than 40 mg of anthocyanin per kg of peel, while in intense red skinned cultivars it can reach 300mg/kg peel. However, other intensely pigmented varieties, in which both the peel and flesh contain pigments, may present values of 3500-4500 mg anthocyanin per kg fresh weight of fruit [16].

Among the phenolic acid derivatives, 3'-caffeoylquinic is the main compound in both species, although other p-coumaroyl-, caffeoyl-, and feroulyl quinic derivatives have also been reported in minor amounts.

(-)-Epicatechin (12-48 mg/kg) and (+)-catechin (3-23 mg/kg) have also been detected.

A wide range of kaempferol and quercetin glycosides has also been reported, among them quercetin 3-rutinoside, kaempferol 3-rutinoside and quercetin 3-rutinoside 4'-diglucoside [16].

Citrus (Orange, Lemon, Grapefruit)

Within the term citrus, we can include sour orange (*Citrus aurantium*), lemon (*Citrus limon*), grapefruit (*Citrus paradisi*), mandarin, clementine and tangerine (*Citrus reticulata*) and sweet orange (*Citrus sinensis*). These fruits are characterised by the accumulation of high amounts of flavanone glycosides (1700-2800 mg naringin per kg grapefruit; 2700-6000 mg

Table 1.   Flavonoids of Citrus Fruits

Citrus fruit	Flavanones	Flavones	Methylated flavones
Sour orange	neohesperidin, naringin, hesperidin, narirutin, neoeriocitrin	apigenin 7-rutinoside	nobiletin, sinensetin, tetramethoxyflavone
Lemon	Hesperidin, eriocitrin	rutin, diosmin 6,8-di-C-glucosyl diosmetin, 6-C-glucosyldiosmetin	
Grapefruit	Naringin, narirutin, hesperidin, neohesperidin		tangeretin, heptamethoxyflavone isosinensetin
Mandarin, clementine, tangerine	narirutin, hesperidin, isosakuranetin 7-rutinoside		sinensetin, nobiletin, tangeretin, heptamethoxyflavonetetra methoxyflavone
Sweet orange	hesperidin, narirutin, eriocitrin, isosakuranetin 7-rutinoside		sinensetin, nobiletin, tangeretin, heptamethoxyflavone tetramethoxyflavone isosinensetin

Hesperetin (5,7,3'-trihydroxy-4-methoxyflavanone); naringenin (5,7,4'-trihydroxyflavanone); eriodictyol (5,7,3',4'-tetrahydroxyflavanone); isosakuranetin (5,7-dihydroxy-4-methoxyflavanone); diosmetin (5,7,3'-trihydroxy-4'-methoxyflavone); neohesperidin (hesperetin 7-neohesperidoside); naringin (naringenin 7-neohesperidoside); neoeriocitrin (eriodictyol 7-neohesperidoside); hesperidin (hesperetin 7-rutinoside); narirutin (naringenin 7-rutinoside); eriocitrin (eriodictyol 7-rutinoside); rutin (quercetin 3-rutinoside); diosmin (diosmetin 7-rutinoside); nobiletin (5,6,7,8,3',4'-hexamethoxyflavone); sinensetin (5,6,7,3',4'-pentamethoxyflavone); tetramethoxyflavone (5,6,7,4'-tetramethoxyflavone); tangeretin (5,6,7,8,4'-pentamethoxyflavone); heptamethoxyflavone (3,5,6,7,8,3',4'-heptamethoxyflavone); isosinensetin (5,7,8,3',4'-pentamethoxyflavone).

hesperidin per kg sweet orange) [16], the absence of flavan-3-ols, and very small amounts of phenolic acid derivatives. Coumarins are present in most species, as well as fully methylated flavonoid aglycones (Table 1).

In lemon, *p*-coumaric acid, caffeic, ferulic and sinapic are detected in small amounts as well as *p*-coumaryl, feruloyl and sinapoyl glucose. These phenolic acids are also present in grapefruit, although in this case ferulic acid is the main compound, while in lemon it is *p*-coumaric acid [16].

In Fig. 6 a characteristic HPLC chromatogram of lemon juice phenolics is shown. The main phenolics in lemon juice are flavanones and C-glucosylflavones. The small peaks at the beginning of the chromatogram correspond to hydroxycinnamic acid derivatives, which are minor constituents in lemon juice.

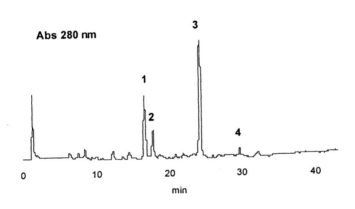

Fig. (6). HPLC chromatogram of lemon juice phenolics: (**1**) eriocitrin, (**2**) diosmetin 6,8-di-C-glucoside, (**3**) hesperidin, and (**4**) diosmetin 6-C-glucoside. HPLC conditions: RP C_{18} column (12 x 0.4 cm; particle size 5 μm). Mobile phase: acidified water (5% formic acid) (A) and methanol (B). Gradient: 0 min- 10% B, 30 min- 40% B, 40 min- 80% B. Flow rate: 1 mL/min.

In addition to the previous compounds, the glucaric and galactaric derivatives of *p*-coumaric and ferulic acids have been described in sweet orange.

Grapes

Complex anthocyanin patterns are observed in red grapes. The 3-glucosides, 3-acetylglucosides and 3-*p*-coumaroylglucosides of cyanidin, peonidin, delphinidin, petunidin and malvidin are present.

Among the phenolic acid derivatives, the most abundant and characteristic is caffeoyltartaric acid. *p*-Coumaroyl tartaric and feruloyltartaric are also present in smaller amounts. Other compounds include hydroxybenzoic acid derivatives, salycilic acid, gallic acid, cinnamic acid and p-coumaroyl and feruloyl glucose.

Grape peel and seeds are especially rich in the flavan-3-ols, catechin, epicatechin, gallocatechin, epigallocatechin, epicatechin gallate and catechin gallate (Fig. **3**). The content in the peel varies greatly (10-520 mg/kg), since it depends on the cultivar and seasonal and environmental factors [16]. In addition, the occurrence of the flavanonols astilbin (dihydroquercetin 3-rhamnoside) and engeletin (dihydrokaempferol 3-rhamnoside) has been described.

Flavonols are also present, particularly in the skin. 3-Glucosides and 3-glucuronides of kaempferol, quercetin and myricetin have been reported (8-97 mg/kg) [16].

In grapes, and derived food products (grape juice, wine), the presence of the antioxidant stilbenoid resveratrol (Fig. **5**) has also been reported [23-26].

Olive

Red pigmented olives accumulate cyanidin 3-rutinoside as the main anthocyanin, along with the 3-glucoside, the 3-caffeoylrutinoside, the 3-glucosylrutinoside and the 3-caffeoyl-glucosylrutinoside of cyanidin [16].

In addition, the phenolic acid derivatives include chlorogenic acid, caffeoylglucose and verbascoside. The flavonol rutin and the flavones luteolin 7-glucoside and apigenin 7-glucoside have also been reported.

It is important to mention oleuropein (Fig. **5**), a glycosidic ester of elenolic ester and hydroxytyrosol, which reaches 60-90 mg/g dry weight in olive leaves. It has been reported to have antioxidant properties [27].

Peach

The red-coloured cultivars contain cyanidin 3-glucoside. This is mainly located in the peel, but in some cultivars, it can also be present in the flesh, and in the tissues surrounding the stone. In some cultivars, small amounts of cyanidin 3-rutinoside are also present.

The main phenolic acid derivative is chlorogenic acid, although 3'-caffeoylquinic and 4'-caffeoylquinic acids, and p-coumaroyl- and feruloyl-quinic acid derivatives have also been reported [16]. Up to 128mg caffeic acid per kg fresh peaches have been reported [19].

Catechin is the main flavan-3-ol detected (54 mg/kg), although epicatechin (5 mg/kg), gallocatechin (4 mg/kg) and epigallocatechin (3 mg/kg) have also been found [19]. Quercetin and kaempferol derivatives (kaempferol and quercetin 3-glucoside, 3-rutinoside and 3-galactoside) were present in the fruit skin, at levels of 10 mg/kg fresh weight [21].

Pear

Red pigmented *Pyrus communis* fruits accumulate cyanidin 3-galactoside as the main pigment, although cyanidin 3-arabinoside has also been reported.

Chlorogenic acid (43-108 mg/kg f.w. as caffeic acid) [19] and arbutin (1-gucosyl-1,4-dihydroxy-benzene) are the main non-flavonoid phenolics. Catechin and, especially, epicatechin (7-12 mg/kg) have also been found in pears [19]. In the peels, a complex flavonoid pattern of quercetin and isorhamnetin 3-glycosides has been detected. The occurrence of characteristic flavonols acylated with dicarboxylic acids in pears has recently been reported [28].

Plums

Within the term plums, several *Prunus* species may be included (*Prunus domestica*, *P. salicina*, etc.) with substantial varietal differences existing in their phenolic composition.

In the red pigmented cultivars, the anthocyanin pigments are quite similar to those of cherry, and include cyanidin 3-glucoside as the main compound, along with cyanidin 3-rutinoside, peonidin 3-glucoside and peonidin 3-rutinoside.

The phenolic acids detected are characteristic, and include vanillic acid, and glucosides of *p*-hydroxybenzoic, protocatechuic, vanillic, syringic and salicylic acids. 3'-Caffeoylquinic is the principal phenolic acid derivative (63-218 mg caffeic acid per kg fresh weight) [19], although other *p*-coumaric and ferulic acid derivatives have also been reported [16].

Small amounts of catechin (8-17 mg/kg) and epicatechin (10-14 mg/kg) have also been detected [19], while higher amounts of quercetin and kaempferol 3-rutinosides have been detected, especially in the skin.

Raspberry

This fruit contains hydrolysable tannins (both gallotannins and ellagitannins), and ellagic acid isomers (Fig. 5). The pulp contains 0.43-4.64 mg ellagic acid per g d.w. (dry weight) while the achenes contain 1.37-21.65 mg/g [29]. In juice, ellagic acid content ranges from 22 to 45.5 mg/L [30].

The main pigment is cyanidin 3-sophoroside, although the 3-glucoside, the 3-rutinoside and the 3-(2"glucosyl)rutinoside of cyanidin are also detected. Pelargonidin derivatives can also be detected at trace levels in some cultivars. Total anthocyanin varies from 230-590 mg/kg f.w. [17;31].

Among the phenolic acid derivatives, 4-hydroxybenzoic gucoside is the main compound, although gallic, *p*-coumaric, caffeic and ferulic acid derivatives have also been recorded in much smaller amounts [32].

Epicatechin is the most significant flavan-3-ol present, and catechin can also be detected in trace amounts [16].

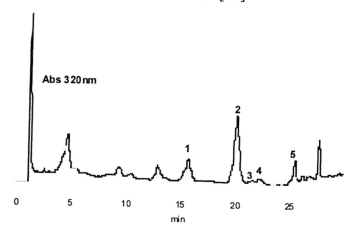

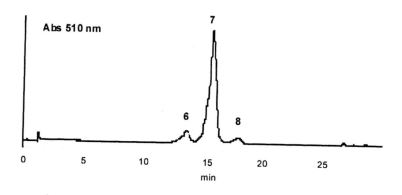

Fig. (7). HPLC chromatogram of strawberry anthocyanins and other phenols: (1) *p*-coumaroylglucoside, (2) quercetin 3-glucoside + quercetin 3-glucuronide, (3) ellagic acid isomer, (4) ellagic acid, (5) kaempferol 3-glucoside + kaempferol 3-glucuronide (6) cyanidin 3-glucoside, (7) pelargonidin 3-glucoside, and (8) pelargonidin 3-rutinoside. HPLC conditions: RP C_{18} column (12 x 0.4 cm; particle size 5 μm). Mobile phase: acidified water (2.5% formic acid) (A) and acidified methanol (2.5% formic acid) (B). Gradient: 0 min- 15% B, 15 min- 30% B, 20 min- 30% B, 25 min- 80% B. Flow rate: 1 mL/min.

The external part of the fruit contains several quercetin and kaempferol derivatives, mainly quercetin and kaempferol 3-glucosides and glucuronides. However, other less frequent glycosidic combinations have been reported in specific cultivars [32;33].

Strawberry

The pigmentation of strawberries is due to three anthocyanins: pelargonidin 3-glucoside (which accounts for more than 90% of the pigment), and smaller amounts of pelargonidin 3-rutinoside and cyanidin 3-glucoside [34].

Among the phenolic acid derivatives, the principal compound is *p*-coumaroyl glucose, although chlorogenic, *p*-coumaroylglucoside, caffeoyl glucose, feruloyl glucose, 4-hydroxybenzoic glucoside, protcatechuic and vanillic acids have also been detected [35].

The main flavan-3-ol present is catechin, although epicatechin and gallocatechin are also present in much smaller amounts.

The 3-glucuronide and 3-glucoside derivatives of kaempferol and quercetin are the main flavonoid constituents. They particularly accumulate in the external tissues of the fruit [36]. In a study of 20 cultivars the content varied from 21 to 174 mg/kg f.w.

Strawberry also contains ellagic acid isomers [29], which originate from the hydrolysable tannins (gallotannins and ellagitannins), which constitute 37% of strawberry tannins (the rest being condensed tannins) [16]. Both tannin types are located in the ripe akenes of strawberry. The ellagic acid content of cultivar Selva ranges 19.9-28.5 mg/kg f.w., and higher concentrations (33-48 mg/kg) in the external tissues than in the internal ones (8-9.4 mg/kg) were observed [36].

Fig. 7 depicts characteristic HPLC chromatograms of strawberry phenolics. The chromatogram recorded at 510 nm shows the three anthocyanin pigments present in this fruit, the predominance of pelargonidin 3-glucoside being clearly observable. In the chromatogram at 320 nm, the flavonol derivatives and the ellagic acid isomers are separated. The ellagic acid isomers are under-represented in this chromatogram, since they have maximum absorbance at 260 nm.

VEGETABLE PHENOLICS

Artichoke

Artichoke heads (*Cynara scolymus*) are a good source of antioxidant phenolic compounds since they contain large amounts of caffeic acid derivatives and flavonoids. The flavonoids include apigenin and luteolin glycosides. In bracts and receptacles, apigenin 5-glucoside and 7-

rutinoside along with luteolin 7-glucoside (cynaroside) and 7-rutinoside have been described. Depending on the stage of development of the head, naringenin glycosides can also be present [37;38]. The same compounds also occur in leaves and stems.

Caffeic acid derivatives are the main phenolics found in artichoke, and a wide range of caffeoyl quinic derivatives are biosynthesised (Table 2).

Since the chlorogenic acid content was reported to be higher in inner (younger) than in outer bracts, the blackening which is caused by the effect of polyphenol oxidase on ortho-di-phenols is much greater in the inner tissues of the head [39].

Previously published data, on a fresh weight basis, also found that chlorogenic acid was the main compound (433 mg/kg fresh weight) [40].

It has also been reported that exposure of artichoke heads to low, non-freezing temperature storage leads to increased levels of caffeoylquinic derivatives, particularly chlorogenic acid, caused by the cold-induced stimulation of PAL activity [39]. Both PAL activity and the phenolic content of artichoke heads show a peak during the early stage of cold storage, followed by a gradual return to very low levels [39].

Table 2. Caffeic Acid Derivatives Content in Artichoke. Values are mg/100 g Dry Weight [39]

Compound	Content mg/110 g (dry weight)
1-Caffeoylquinic	38
3-Caffeoylquinic	52
4-Caffeoylquinic	267
5-Caffeoylquinic (chlorogenic)	1545
1,3-di-Caffeoylquinic	61
1,4-di-Caffeoylquinic	143
4,5-di-Caffeoylquinic	225
3,5-di-Caffeoylquinic	347
1,5-di-Caffeoylquinic	837
3,4-di-Caffeoylquinic	429

Artichoke phenolics are good hepatoprotective agents against hepatotoxic substances [41].

Asparagus

Asparagus officinalis has a very low phenolic content when it is harvested in its white form. In this case, rutin, which is the main flavonoid, is restricted to the tips and is present at concentrations below 1 mg/kg f.w., although, as the colour changes, the flavonoid content in the tips may reach 10 mg/kg [42]. In green asparagus the flavonol content is much higher and can reach similar values to those found in the leaves (close to 1.4% quercetin on a dry matter basis). In this tissue, kaempferol derivatives are also detected in trace amounts. The amount of rutin in green asparagus is so high that it precipitates after the canning process to give a yellow sediment which decreases its marketability.

The pigments, cyanidin 3-glucosylrutinoside and cyanidin 3-rutinoside, have been reported in the red cultivars [43].

Carrot

In carrots, the principal phenolics are the hydroxycinnamic acid derivatives. Of these, caffeic acid derivatives are the main components (20-100 mg/kg f.w. depending on the cultivar), while ferulic acid derivatives are accumulated in intermediate quantities (10-20 mg/kg) and *p*-coumaric derivatives, 4-hydroxybenzoic acid derivatives and vanillic acid are present in trace amounts [38]. The main compound in carrots is chlorogenic acid, which has been estimated as between 23 and 121 mg/kg f.w. depending on the cultivar. However, in carrot tops, this compound can reach 375-400 mg/kg [40], while cryptochlorogenic, neochlorogenic and 5-feruloylquinic acids have also been detected.

Minimal processing of carrots induces phenylalanine ammonia lyase (PAL) activity and phenolics accumulation. In shredded carrots, chlorogenic acid, which is rapidly accumulated, represents 60% of the total phenolics. In addition traces of 3'-caffeoylquinic and 4'-caffeoylquinic acids are biosynthesised, and 3',4'-dicaffeoylquinic and 3',5'-dicaffeoyl quinic acids also accumulate [44]. *p*-Hydroxybenzoic acid derivatives are also biosynthesised but more slowly, and are related to defence against microbial attack (phytoalexin response), the degree and speed of which depends on the cultivar [45].

Under controlled atmospheres containing 30% CO_2 and 0% O_2, phenolic compounds accumulate very slowly. In such conditions, the increase in PAL activity leads to the acummulation of phenolic compounds [44].

Carrots can accumulate, in response to fungal infection, the coumarins 6-methoxymellein and scopoletin alongside several *p*-hydroxybenzoic acid derivatives [46]. In carrot leaves, apigenin 7-glucoside and luteolin 7-glucoside have been detected [47].

Celery

In celery (*Apium graveolens*), the leaves, stalks and tubers (celeriac) are consumed. The phenolic content of tubers is much smaller than that of the other organs, and is located mainly in the external tissues. The main flavonoid in the leaves is apiin (apigenin 7-apiosylglucoside, 202 mg/kg f.w.), although luteolin (48 mg/kg f.w.) and chrysoeriol (27 mg/kg f.w.) 7-apiosylglucosides are also present. In addition, trace amounts of apigenin (6 mg/kg f.w.), luteolin (11 mg/kg f.w.) and chrysoeriol 7-glucosides have been detected [48]. The flavonoid content of the tubers is much smaller (apiin 1.7 mg/kg and luteolin 7-apiosylglucoside 0.8 mg/kg f.w.).

In a recent study of the luteolin and apigenin content of commercial celery purchased in the U.K., white celery stalks and green and white celery hearts were studied [49]. The amounts detected in the white celery stalks (36-40 mg luteolin per kg f.w. and 90-104 mg apigenin per kg) were in keeping with the 22 mg luteolin and 108 mg apigenin per kg f.w. previously reported for celery purchased in The Netherlands [50]. However, significant sample-to sample variation was observed in the U.K. study, and some samples contained no flavonoid at all when purchased [49].

When the phenolic acid derivatives of celery roots were studied [38;51] those of caffeic acid (89-168 mg/kg f.w. depending on the cultivar) were the most abundant, although ferulic acid was also present in substantial amounts (34-61 mg/kg), with *p*-coumaric acid derivatives only existing at trace levels (< 0.5 mg/kg). Much greater amounts of these compounds were found in the corresponding aerial parts [51]. In a study on the characteristic phenolic compounds of celery roots which may be responsible for discolouration during cooking, the main compound was chlorogenic acid (5'-caffeoylquinic acid; 2-65 mg/kg f.w.), while 5'-feruloylquinic acid (1-8 mg/kg) and 5'-*p*-coumaroylquinic acid (1-3 mg/kg) were also present [52].

The coumarins, scopoletin and aesculetin, were present in celery roots in amounts smaller than 0.5 mg/kg f.w. [51]. The coumarin concentration was five to ten times higher in the peel of the tuber than in the edible portion. When celery tubers were coated with pectin gels and stored for 6 months at 4°C and 95% relative humidity, the furanocoumarin concentration increased considerably [53].

The furanocoumarins, psoralen, bergapten, xanthotoxin and isopimpinellin, were detected in celery leaves and stalk. The total furanocoumarin content ranged from about 12 to 50 mg/kg in Florida cultivars [54]. Treatment with fungicides increased the bergapten content 2-4 times in leaves and stalk, xanthotoxin 2-3 times in stalk, and isopimpinellin about 2-3 times in leaves, while psoralen levels remained constant. This fact is of some importance since linear furanocoumarins are the cause of photosensitive reactions in humans.

Crucifferae (Cabbage, Cauliflower, Broccoli, etc.)

The phenolic acid content of vegetables of the genus *Brassica* consists almost totally of hydroxycinnamic acid compounds, of which, unlike in other species of vegetables, sinapic acid is the predominant one [55] (Table 3).

Table 3.   Phenolic Acid Content of Brasicaceae Vegetables. Values are mg/kg Fresh Weight (Table Modified from [55]).

Brassica species	*p*-Coumaric	Caffeic	Ferulic	Sinapic
Brussels sprouts	0.5-12	34-50	10-29	102-241
Cauliflower	6-80	1-90	0.5-82	4-94
Kale	0.5-28	9-305	7-276	27-940
White cabbage	0.5-69	0.5-62	1-56	12-54
Red cabbage	9-154	6-24	11-45	35-193
Kholrabi	0-17	1-113	1-89	1-91
Savoy cabbage	13-70	4-36	2-37	15-90
Chinese cabbage	2-30	4-54	5-43	4-55
Broccoli	13-14	8-10	13-31	40-97

The glucose or quinic acid esters of caffeic, *p*-coumaric and ferulic acids and sinapylglucose have been determined in cauliflower, kale, Brussels sprouts and red, white, Savoy and Chinese cabbage [40;56]. The pattern of hydroxycinnamic acid esters in the *Brassica* species varies appreciably. Sinapylglucose (3-250 mg/kg f.w.) is predominant in the group of glucose esters alongside smaller amounts of feruloylglucose (0.2-72 mg/kg). Neochlorogenic acid (3'-caffeoylquinic) (10-140 mg/kg f.w.) is the main quinic acid derivative, while 5'- and 3'-caffeoylquinic acids are present in lower concentrations, although sinapylquinic is not detected. These reported levels vary greatly, which may be explained by seasonal, horticultural and environmental factors, or, perhaps, by the postharvest treatments to which the vegetables were submitted before analysis. All the analysed vegetables were purchased on the open market, and so details of the growth conditions and the postharvest life were unavailable.

Brassica oleracea varieties contain kaempferol and quercetin glycosides [43;57]. In addition, apigenin and luteolin have been reported in broccoli. The flavonol content of the edible parts of cabbage, cauliflower and kohlrabi is very low. Brussels sprouts, broccoli and Chinese cabbage have been reported to contain 15-60 mg/kg f.w. of flavonoids and kale 200-500 mg/kg. Cultivation under glass led to lower flavonol levels being recorded in the leaves than when plants were grown in the open field [57]. It has also been demonstrated that the flavonol content of Brussels sprouts and kale leaves increases during development. Broccoli leaves contain 3-sophoroside-7-glucoside derivatives of kaempferol, quercetin and isorhamnetin [43]. In the case of kale, the kaempferol content has been reported as 13-30 mg/kg f.w. and the quercetin content as 7-20 mg/kg, depending on the cultivar [58]. In red cabbage no kaempferol was detected and only small amounts of quercetin (2 mg/kg f.w.) were reported [58].

The pigmentation of red cabbage is caused by cyanidin derivatives which are acylated with various phenolic acids. Cyanidin 3-sophoroside-5-glucoside and its derivatives acylated with malonic acid and one or two residues of *p*-coumaric, ferulic or sinapic acids have also been detected [59;60]. The relative stability of these pigments seems to be related to their degree of acylation.

Cucurbitaceae (Cucumber, Melon, Pumpkin, Squash, etc)

The phenolic compound content of the Cucurbitaceae is very low. Only trace amounts (3mg/ kg f.w.) of neochlorogenic acid (3'-caffeoylquinic acid) were detected in one out of three zucchini (summer-squash) cultivars [40]. In an extensive study on the phenolic content of melons, watermelon, etc., only traces (<1mg/kg f.w.) of caffeic, *p*-coumaric and ferulic acid were detected after hydrolysis. Very small antioxidant activity was reported for cucumber extracts [61], consistent with the very small phenolic compound content. However, the phenolic metabolism of melons is activated under environmental stress and wounding induces the phenylalanine ammonia lyase (PAL) gene [62].

Eggplant

The pigmented cultivars of eggplant (*Solanum melongena*) accumulate delphidin glycosides in the peel. The major compound is delphinidin 3-(*p*-coumaroylrutinoside)-5-glucoside, although the 3-(*p*-coumaroylglucoside)-5-glucoside, the 3-rutinoside, the 3-rutinoside-5-glucoside, the 3-glucoside, the 3,5-diglucoside, and the 3-(caffeoyldiglucoside)-5-glucoside of delphinidin have also been detected [17].

The main phenolic acid derivative is chlorogenic acid (575-632 mg/kg f.w.), although neochlorogenic (trace) and cryptochlorogenic acids (8-11 mg/kg), 5'-feruloyl quinic acid (15-17 mg/kg), and caffeoyl, feruloyl and sinapoylglucose have also been reported [63].

The coumarins, scopoletin and scoparone, are also found in some cultivars.

Garlic (Leeks, Chive, etc.)

The principal phenolic acids in these vegetables are ferulic and *p*-coumaric acid derivatives, the green parts of chives and leeks containing more than in the white part. In leek, the white tissue contains 6-7 mg ferulic acid per kg f.w., while this compound reaches concentrations of 23-39 mg/kg in the green parts [64]. In the case of chive, *p*-coumaric acid derivatives reach 21-51 mg/kg and ferulic acid derivatives 32-76 mg/kg. In garlic a different pattern of phenolic metabolite accumulation is observed in skins and internal tissues. The external tissues contain 49-58 mg/kg *p*-coumaric acid, 27-31 mg/kg ferulic acid and 27-25 mg/kg sinapic acid, whereas the internal tissues only contain 2mg, 6-8 mg and 2 mg/kg, respectively. In addition, the internal tissues contain 12-13 mg/kg *p*-hydroxybenzoic acid [64].

As regards flavonoid compounds, the green leaves of leek and chive mainly contain kaempferol glycosides, with mono- and di-glycosides predominant in leek and di- and tri-glycosides in chive. In leek, kaempferol 3-glucoside and 3-xylosylglucoside are the main compounds, accompanied by smaller amounts of quercetin 3-glucoside. In chive, the 3-glucosides of kaempferol, quercetin and isorhamnetin have been reported [65]. In the bulbs of garlic and leek only a few mg/kg of kaempferol and quercetin can be detected. The internal leaves always contain smaller amounts of flavonoids than the external leaves. In the green portions of chives 55 mg/kg kaempferol and 9 mg/kg quercetin were detected while the amount in the white portions decreased considerably to 16 mg/kg kaempferol and quercetin was not detected at all [58]. In the case of leek, no flavonoid was found in the white portions and only 20 mg kaempferol per kg was detected in the green parts.

Lettuce (Endive, Chicory, etc.)

Lettuce contains the flavonol glycosides, quercetin 3-glucoside, 3-glucuronide and 3-malonylglucoside, and according to the light dependence of their biosynthesis, they are mainly found in the outer green leaves [66]. Up to *ca* 250mg quercetin per kg fresh weight were detected. Cultivation under glass considerably reduced the flavonoid content. In some cultivars, the quercetin content of the external leaves reaches 60 mg/kg, while the inner leaves may only contain 3 mg/kg.

In a study of the flavonoid content of American leaf lettuce and head lettuce [58], the former varieties contained 2-54 mg quercetin per kg, while the head lettuce varieties contained 1-28 mg/kg f.w. Again, the content was much higher in the outer leaves than in the inner leaves. In some of these samples, kaempferol derivatives were also detected in small amounts (up to 2 mg/kg).

In a recent paper on red lettuce ('Lollo rosso') the flavonoids, quercetin 3-glucoside, 3-glucuronide and 3-(6"-malonyl) glucoside were detected as well as the new naturally occurring flavonoid quercetin 3-(6"-malonyl) glucoside-7-glucoside [67]. In this case three different tissues were differentiated: white, green and red. White tissue contained only 43 mg flavonoids per kg f.w., while the green tissue contained 244 mg/kg and the red tissue 1384 mg/kg.

The quercetin content of lettuce cultivars purchased in the market in The Netherlands ranged from 2 to 30 mg/kg f.w. [68]. In a recent survey of the quercetin contents of lettuce cultivars purchased in the U.K., head lettuce cultivars again contained the smallest amounts (11 mg/kg). Green leaf lettuce cultivars (e.g. 'Green salad') contained 147 mg/kg, while red cultivars like 'Lollo Rosso' contained the highest amounts: 911mg / kg in external leaves and 450mg / kg in inner leaves [49].

In red lettuce cultivars, the only pigment detected was cyanidin 3-(6"-malonyl) glucoside [67;69]. This compound reaches 950mg/kg f.w. in the red tissues, but is also present in the green tissue (100 mg/kg) and midribs (20 mg/kg), where during cold storage it may reach 45 mg/kg. This pigment is degraded in minimally processed 'Lollo rosso' during storage in air.

As regards phenolic acid derivatives, lettuce, endive and chicory only contain caffeic acid derivatives (and in some cases traces of ferulic acid). The caffeic acid content of lettuce ranged from 182 to 381 mg/kg f.w. in cultivars grown in the field, and 40-108 mg/kg in greenhouse grown plants [64]. In chicory, the caffeic acid content ranges from 99 to 138 mg/kg and in endive an average of 164 mg/kg has been reported. When external and internal leaves were differentiated, the caffeic acid levels in lettuce ranged from 160 to 600 mg/kg and in endive from 160 to 870 mg/kg [38].

In freshly harvested lettuce, the caffeic acid derivatives caffeoyltartaric, dicaffeoyltartaric, caffeoyl malate and chlorogenic acids have been identified, as they were in endive and chicory [63]. In the case of lettuce, the outer leaves contained 28 mg/kg caffeoyltartaric, 160 mg/kg dicaffeoyltartaric, 15 mg/kg caffeoylmalate and 55 mg/kg chlorogenic acid, while the inner leaves contained almost half these amounts. However, after minimal processing (wounding) or ethylene exposure, the phenolic metabolism is activated and other caffeic acid derivatives such as the isochlorogenic isomers (3,5-dicaffeoylquinic, 3,4-dicaffeoylquinic and 4,5-dicaffeoylquinic acids) accumulate, especially in the white tissue (midrib) [67;70-72].

Fig. **8** shows typical HPLC chromatograms of 'Lollo rosso' phenolics, and Fig. **9** a characteristic chromatogram of 'Romaine' lettuce. The main difference in the phenolic compound patterns of both lettuce types is that

'Lollo rosso' contains large amounts of flavonoids and anthocyanins, while 'Romaine' only presents small amounts of caffeic acid derivatives. Thus, the antioxidant activity of 'Lollo rosso' should be much more important than that of 'Romaine'.

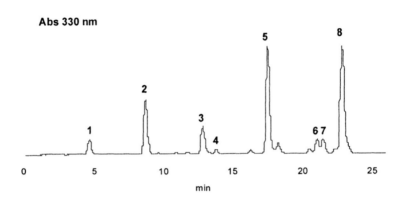

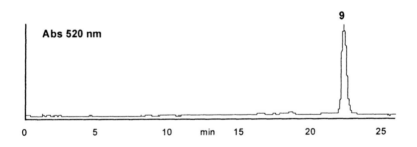

Fig. (8). HPLC chromatogram of red lettuce anthocyanins and other phenols: (**1**) caffeoyltartaric acid, (**2**) chlorogenic acid, (**3**) dicaffeoyltartaric acid, (**4**) quercetin 7-glucoside 3-(6''-malonylglucoside), (**5**) dicaffeoylquinic acid (isochlorogenic acid), (**6**) quercetin 3-glucuronido, (**7**) quercetin 3-glucosido, (**8**) quercetin 3-malonylglucoside, and (**9**) cyanidin 3-malonylglucoside. HPLC conditions: RP C_{18} column (12 x 0.4 cm; particle size 5 μm). Mobile phase: acidified water (5% formic acid) (A) and methanol (B). Gradient: 0 min- 5% B, 25 min- 40% B. Flow rate: 1 mL/min.

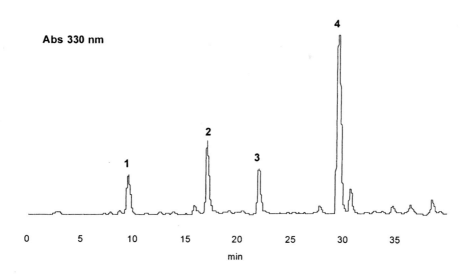

Fig. (9). HPLC chromatogram of 'Romaine' lettuce phenols: **(1)** caffeoyltartaric acid, **(2)** chlorogenic acid, **(3)** dicaffeoyltartaric acid, and **(4)** dicaffeoylquinic acid (isochlorogenic acid). HPLC conditions: RP C_{18} column (12 x 0.4 cm; particle size 5 μm). Mobile phase: acidified water (5% formic acid) (A) and methanol (B). Gradient: 0 min- 5% B, 40 min- 40% B. Flow rate: 1 mL/min.

In endive, caffeoyltartaric ranged from 21 to 31 mg/kg, dicaffeoyltartaric 163-334 mg/kg, caffeoyl malate 10-13 mg/kg, feruloyl malate 14-16 mg/kg and chlorogenic acid 36-124 mg/kg [63]. In the case of chicory smaller amounts were detected (4mg/kg caffeoyltartaric, 17mg/kg dicaffeoyltartaric, 5 mg/kg chlorogenic acid).

Kaempferol 3-glucoside and 3-glucuronide were the only flavonoids found in endive [66]. In this case, the kaempferol content of the external leaves reached 225 mg/kg, while the inner leaves only contained 5 mg/kg.

Onion

Protocatechuic acid is the main benzoic acid derivative in onions although *p*-hydroxybenzoic acid and vanillic acid may also be observed [64]. In the outer dry coloured skins protocatechuic acid may represent up to 2% of plant material. The internal pulpy tissues show lower concentrations (*ca* 20 mg/kg f.w.). Unlike the bulbs, the green leaves of onions contain almost exclusively ferulic acid and *p*-coumaric acid derivatives.

It has been reported that the epidermis of onion scales contains only glucosides of quercetin, while the dry outer skin contains quercetin mainly in its free state [73]. Quercetin 4'-glucoside and other diglucosides are biosynthesised, and the flavonoid content decreases from the outer to the inner scales.

When the quercetin contents of eight onion varieties grown in the U.S.A. were determined [74] the highest quercetin content in the edible portion was about 60 mg/kg f.w., although in some cultivars smaller amounts of kaempferol were also detected. In similar studies on the quercetin content of onions on the market in The Netherlands and U.K., the quercetin content ranged from 284 to 486 mg/kg in fresh Dutch onions [50], while their U.K. counterparts contained 185-634 mg/kg [49]. Both

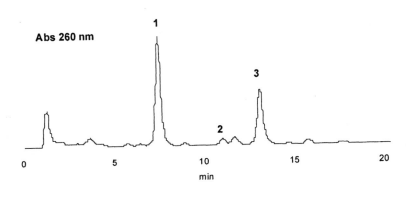

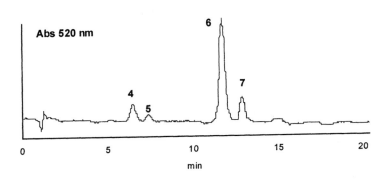

Fig. (10). HPLC chromatogram of red onion anthocyanins and other phenols: **(1)** quercetin 3,4'-diglucoside, **(2)** quercetin 4'-glucoside, **(3)** isorhamnetin 4'-glucoside, **(4)** cyanidin 3-glucoside, **(5)** cyanidin 3-arabinoside, **(6)** cyanidin 3-malonylglucoside, and **(7)** cyanidin 3-malonylarabinoside. HPLC conditions: RP C_{18} column (12 x 0.4 cm; particle size 5 μm). Mobile phase: acidified water (5% formic acid) (A) and methanol (B). Gradient: 0 min- 15% B, 15 min- 35% B, 25 min- 40% B. Flow rate: 1 mL/min.

amounts were higher than those found in the U.S.A. but no kaempferol was detected in the European cultivars.

In Spanish commercially purchased red onion the anthocyanin pigments, cyanidin 3-glucoside, cyanidin 3-arabinoside, cyanidin 3-(6"-malonyl) glucoside and cyanidin 3-(malonyl) arabinoside, were found alongside the flavonols, quercetin 3,4'-diglucoside, quercetin 7,4'-diglucoside, quercetin 3-glucoside, dihydroquercetin 3-glucoside and isorhamnetin 4'-glucoside. The anthocyanin content of this cultivar was 233 mg/kg f.w. and the flavonols content was 943 mg/kg [75]. When quercetin glycosides were quantified in other red and white onion cultivars, the contents at harvest ranged from 1187 to 1917 mg/kg f.w. although 40% of the flavonoids were lost during the curing process (drying) to which onions are submitted before retail in order to extend their storage life. After curing, the total flavonol glycosides ranged from 864 to 1390 mg/kg [76]. No significant loss of flavonoids was observed after 6 months' storage. Minimal processing to obtain shredded onion and storage in perforated films led to a slight decrease in the anthocyanin content after 7 days' storage [75].

Fig. 10 shows HPLC chromatograms of red onion phenolics. Four anthocyanins are visible in the 520 nm chromatogram and three principal flavonoids can be clearly observed in the 260 nm chromatogram. Hydroxycinnamic acid or benzoic acid derivatives cannot be observed in any significant quantity in the chromatograms.

Pepper (Sweet)

Bell peppers, both green and red, contain low amounts of phenolic acid derivatives: p-Coumaric (< 4 mg/kg), caffeic (< 10 mg/kg), ferulic (<15 mg/kg), sinapic (<5 mg/kg) and vanillic (10 mg/kg) acids [77]. Caffeoyl glucose, feruloyl glucose and sinapoyl glucose are the main hydroxycinnamic acid derivatives found, while no quinic acid derivatives have been detected [63]. Feruloyl glucose may accumulate up to 11 mg/kg f.w. and sinapoyl glucose up to 5 mg/kg in some red cultivars.

As regards flavonoids, they are mainly found in green peppers and are confined to the skin. Quercetin ranges from 12 to 18 mg per kg in whole pepper and luteolin was found to occur at 15 mg/kg [78]. However, when the pepper was divided into external (28%) and internal (72%) parts, the flavonoid content of the external tissues was much higher (43-63 mg quercetin per kg and 54 mg luteolin per kg) than that of the inner parts which contained less than 1 mg/kg of flavonoids [78].

Potato

Pelargonidin and peonidin glycosides accumulate in red skinned potato varieties. In both anthocyanidins, a rutinosyl residue is linked in the 3 position and a glucosyl residue in the 5 position. Both pigments are acylated with *p*-coumaric and ferulic acids [79].

As regards phenolic acid derivatives, potato tubers contain mainly caffeic acid derivatives, although ferulic, *p*-coumaric, sinapic and vanillic acids are also present [64]. The caffeic acid derivatives are mainly located in a thin 1-2 mm thick layer on the outer part of the tuber. The caffeic acid content ranges from 163 to 280 mg/kg f.w. depending on the cultivar. Ferulic acid (from 13 to 28 mg/kg), *p*-coumaric (from 0.5 to 4 mg/kg), sinapic (from 0.5 to 8 mg/kg) and vanillic acid (from 5 to 16 mg/kg) are present in much smaller amounts [64].

An HPLC study of the individual phenolic acid derivatives showed chlorogenic acid to be the predominant compound in potato (22-71 mg/kg). Cryptochlorogenic acid (11 mg/kg), neochlorogenic acid (7 mg/kg) and isochlorogenic acid (4,5-dicaffeoyl quinic acid) (Fig. 4) (3 mg/kg) were also present in smaller amounts [80]. It was also demonstrated that caffeic acid compounds are not distributed uniformly in the tuber, but occur in strongly decreasing concentrations from the outer to the inner section. About 50% of the compounds as a whole are located in the potato peel and adjoining tissue.

Spinach

The leaves of *Spinacia oleracea* contain *p*-coumaric (70-133 mg/kg f.w.) and ferulic acid (15-45 mg/kg) derivatives [38;64], of which *p*-coumaroyl-*meso*-tartaric acid is the main constituent (189-230 mg/kg). Feruloyl glucose (42-64 mg/kg), *p*-coumaroyl glucose (16-21 mg/kg) and *p*-coumaroyl malate (23-29 mg/kg) are also present in smaller amount [63].

Spinach leaves are also very rich in flavonoid glycosides (Fig. 11). The main compounds are the 4'-glucuronides of 5,7,4'-trihydroxy-3,6,3'-trimethoxyflavone (jaceidin), 5,3',4'-trihydroxy-3-methoxy-6:7-methylenedioxyflavone, and 5,4'-dihydroxy-3,3'-dimethoxy-6:7-methylenedioxyflavone [81]. Other studies identified new patuletin (3,5,7,3',4'-pentahydroxy-6-methoxyflavone) and spinacetin (3,5,7,4'-tetrahydroxy-6,3'-dimethoxyflavone) glycosides [82] as well as acylated derivatives [83].

The flavonoid content of fresh spinach leaves ranges from 1106 to 4875 mg/kg fresh weight depending on the harvest date and cultivars (Tomás-Barberán *et al.*, unpublished results).

Fig. 11 shows a characteristic HPLC chromatogram of spinach flavonoids. Its complexity is due to the large number of compounds that are biosynthesised by the spinach leaves. However, the analysed methanol extract does not contain many hydroxycinnamic acid derivatives (with only one peak at around 10 minutes in the HPLC chromatogram revealing their presence).

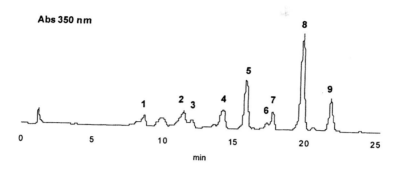

Fig. (11). HPLC chromatogram of spinach phenols: (1) patuletin 3-glucosyl(1→6)-[apiosyl(1→2)]-glucoside, (2) spinacetin 3-glucosyl(1→6)-[apiosyl(1→2)]-glucoside, (3) patuletin 3-(2''feruloylglucosyl)(1→6)-[apiosyl(1→2)]-glucoside, (4) spinacetin 3-(2''-p-coumaroylglucosyl)(1→6)-[apiosyl(1→2)]-glucoside, (5) spinacetin 3-(2''feruloyl glucosyl)(1→6)-[apiosyl(1→2)]-glucoside, (6) spinacetin 3-(2''feruloylglucosyl)(1→6)-glucoside, (7) jaceidin 4'-glucuronide, (8) 5,3',4'-trihydroxy-3-methoxy-6:7-methylenedioxyflavone 4'-glucuronide, (9) 5,4'-dihydroxy-3,3'-dimethoxy-6:7-methylenedioxy flavone 4'-glucuronide. HPLC conditions: RP C$_{18}$ column (12 x 0.4 cm; particle size 5 μm). Mobile phase: acidified water (5% formic acid) (A) and methanol (B). Gradient: 0 min- 15% B, 15 min- 35% B, 25 min- 40% B. Flow rate: 1 mL/min.

Tomato

The main phenolic acid derivative of tomato is chlorogenic acid, although 3'-caffeoylquinic, 4'-caffeoyl quinic, 5'-feruloyl quinic and 5'-p-coumaroyl quinic have also been reported. In addition, p-coumaroyl, caffeoyl, feruloyl, and sinapoyl glucose and the corresponding glucosides have been detected [84]. When the hydroxycinnamic acid derivative content of nine tomato varieties was studied, chlorogenic acid was usually found in higher amounts in unripe green fruits (22-64 mg/kg) than in ripe ones (12-31 mg/kg). The glucoside content, on the other hand, increased with ripening (caffeic acid 4-β−D-glucoside, 5-17 mg/kg in green tomato and 15-48 mg/kg in red), as did the 4'- and 3'-caffeoylquinates [84].

The flavonols quercetin 3-rutinoside and 3-glucoside and kaempferol 3-rutinoside have been detected in tomato peel [16]. The quercetin content of commercial tomatoes purchased in The Netherlands ranged from 4.6 to 11.0 mg/kg f.w. [50]. In a recent study on the quercetin content of Scottish, Spanish, Dutch and English tomatoes, similar concentrations (2.0-11.2 mg/kg) were found [49]. The lowest levels were found in Scottish, Spanish and Dutch 'long-life' tomatoes, all of which were held in cold storage for varying periods after harvest. However, cherry type tomatoes contained much higher quercetin concentrations (179-203 mg/kg in winter samples and 23-55 mg/kg in summer samples). Note that tomato also contains significant amounts (1.8-9.7 mg/kg fresh tomato weight) of naringenin chalcone (4,2',4',6'-tetrahydroxychalcone) [85], which accumulates at high levels in the peel (660-1180 mg/kg).

Fig. **12** shows a characteristic chromatogram of the phenolic compounds of tomato peel. Especially interesting is the relatively large naringenin chalcone peak. Other peaks are observed in the chromatogram, corresponding to unidentified hydroxycinnamic and benzoic acid derivatives.

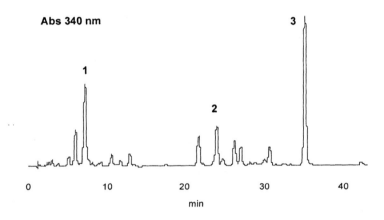

Fig. (12). HPLC chromatogram of tomato peel phenolics: (**1**) chlorogenic acid, (**2**) quercetin 3-rutinoside, and (**3**) naringenin chalcone. HPLC conditions: RP C_{18} column (12 x 0.4 cm; particle size 5 μm). Mobile phase: acidified water (5% formic acid) and methanol. Gradient: 0 min- 10% B, 30 min- 40% B, 40 min- 80% B. Flow rate: 1 mL/min.

METHODS FOR EVALUATION OF PHENOLIC ANTIOXIDANT ACTIVITY

Lipid oxidation is an important topic in food science and technology since the reaction of polyunsaturated fatty acids with oxygen leads to rancidity and quality loss. The same process is important in human health, since the polyunsaturated fatty acids from lipids present in blood plasma (low density lipoproteins, LDL) are oxidised by oxygen in a free radical mediated reaction, promoting the development of atherosclerosis. LDL enters the arterial wall from the plasma and is oxidised locally within the wall by oxidising agents derived from the cells present in atherosclerotic

Table 4.   Substrates Used to Evaluate Antioxidant Activity

Substrate	References
Linoleic, methyl linoleate and linolenic	[90]
β-Carotene/linoleate method	[188] [123]
Liposomes; soy lecithin, egg lecithin	[108] [94]
Corn oil and corn oil-in water emulsion	[108]
Lard	[122]
Human LDL	[87] [106] [130] [92]
Rabbit erythrocyte membranes	[94]
Rat liver microsomes	[94]
Rat brain homogenate	[131]
Brain phospholipids	[120]

lesions, namely macrophages, smooth muscle cells, endothelial cells and lymphocytes [86].

The Chemistry of Lipid Oxidation

In order to understand how phenolic antioxidants work, and how antioxidant activity can be measured, it is important to take into account the chemistry of lipid oxidation which can be summarised in the following simplified scheme (Fig. 13). A free radical abstracts a hydrogen atom from a polyunsaturated fatty acid. A double bond in the resulting lipid alkyl radical rearranges to form a conjugated diene and molecular oxygen then binds to form a lipid peroxyl radical. This attacks another polyunsaturated fatty acid, itself becoming a lipid hydroperoxide. In the presence of Fe^{2+} or Cu^+, this lipid hydroperoxide breaks down to form a lipid alkoxyl radical, which can fragment to form an aldehyde (hexanal, pentanal, malondialdehyde etc., depending on the lipid nature). The rate of lipid oxidation, and the antioxidant effect of phenolics can be evaluated in different lipid substrates or systems (Table 4) by measuring the different products of polyunsaturated fatty acid oxidation: conjugated dienes, the lipid hydroperoxides or the β-scission aldehydes (hexanal, malondialdehyde) etc. (Fig. 13).

Lipid Oxidation Measurement

To evaluate the oxidation process, several polyunsaturated fatty acids, or lipid rich systems (Table **4**), are submitted to oxidation. As oxidation by the oxygen present in air takes a long time, several oxidation inductors have been used in order to accelerate the process (Table **5**). Such inductors include temperature, cupric salts, or free radical generating substances. The degree of lipid oxidation can be followed by determining conjugated dienes, hydroperoxides or the aldehydes obtained after β-scission (Fig. **13**). Phenolic antioxidants prevent the formation of these lipid oxidation substances, the degree of prevention depending on their antioxidant potency.

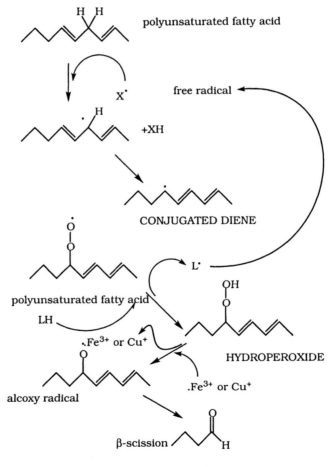

Fig. (13). Chemistry of lipid oxidation.

Conjugated diene formation can be measured by following the changes in absorbance at 234 nm [87], the wavelength maximum at which these dienes absorb. Hydroperoxide formation can be directly determined by HPLC [88;89] or indirectly by the thiocyanate method, in which hydroperoxides react with $FeCl_2$ and thiocyanate to give a red colour which can be measured at 500 nm [90].

Table 5.   **Oxidation Inductors**

Inductors	References
Temperature (40-50°C)	[108] [94]
$CuSO_4$ or cupric acetate	[87] [108]
2,2'-azobis(2,4-dimethlvaleronitrile) (AMVN)	[88]
2,2'-azobis(2-amidinopropane) hydrochloride (AAPH)	[94]
Tert-butyl hydroperoxide	[94]

During the last steps of lipid oxidation, the fatty acid chains breakdown to give aldehydes (hexanal, propanal, malondialdehyde), depending on the lipid structure. These compounds react with thiobarbituric acid to give coloured compounds the measurement of which at 535 nm can be used to follow the oxidation process in its terminal phase [91]. In addition, hexanal, which is an important decomposition product of n-6 polyunsaturated fatty acid peroxidation in rat liver samples, human red blood cell membranes, and human LDL (low density lipoproteins), can be measured by headspace gas chromatography [92]. Malondialdehyde, another important decomposition product, can also be analysed by GC (Gas Chromatography) [93], and, after reaction with urea to give 2-hydroxypyrimidine, by HPLC [94].

The antioxidant activity of a given compound, or plant or food extract, can also be measured by determining its ability to neutralise free radicals, which act as free radical scavengers. Free radicals which absorb in the visible region can be generated *in vitro*, and the free radical scavenging activity can be readily measured by following the decrease in this absorbance value since the free radical becomes discoloured when neutralised. The radical cation 2,2'-azinobis(3-ethylbenzo-thiazoline-6-sulphonate) ($ABTS^{·+}$) in aqueous phase can be generated by the interaction of ABTS, H_2O_2 and metmyoglobin [95], and the antioxidant activity of different compounds can be evaluated since they suppress the formation of $ABTS^{·+}$. Recently, another system for ABTS radical production, in this case catalysed by peroxidase, has been reported [96]. It is also possible to use the commercial free radical 2,2-diphenyl-1-picrylhydrozyl (DPPH·) [97].

Recently, an indirect method to measure the free radical scavenging activity of extracts and, thus, their antioxidant activity has been developed using HPLC coupled to a coulometric array detector system [61]. This was developed to characterise the overall antioxidant activity status of fruit and vegetables. A significant positive linear correlation was demonstrated between the total antioxidant activity determined by using the oxygen radical absorbance capacity assay and that measured using the electrochemical data obtained from the coulometric array detectors.

PHENOLICS AS ANTIOXIDANTS

The antioxidant activity of different food products has been linked to their preventive role in cardiovascular disease [68;86;98;99]. Many of them are fruit and vegetables, although cereals, herbs and spices also show important antioxidant activity. Processed fruit products also provide important antioxidant protection as has been described for orange and lemon juice, [100], grape juice [91], grape pomace [101] and red wines [102-107]. Very pronounced 'in vitro' antioxidant activity has been demonstrated in tea [99;108] and olives [27]. Many herbs and spices containing antioxidant constituents can be useful to prevent rancidity of meat and fish products [109-112]. Fibre also has antioxidant activity, mainly due to the presence of phenolic metabolites linked to the polysaccharides [87]. Thus, the antioxidant activity of barley leaves [93], pineapple peel fibre [113] and rice-hull [114] have been described.

Polyphenols are responsible for most of the antioxidant activity of the above mentioned food products. Spices have been used since ancient times to prevent the rancidity of fish and meat products, as well as to provide an appealing flavour. In recent publications, it has been demonstrated that the antioxidant activity of herbs and spices is mainly due to antioxidant phenolics and to other secondary metabolites. These plant polyphenols inhibit the oxidative rancidity of frozen cooked fish flakes [112], salted cooked ground fish [115], meat lipids [116] and lard [117]. In a recent paper, the effect of tea and green tea catechins and extracts in the prevention of oxidation in a fish meat model system was demonstrated [118]. Epigallocatechin gallate and epicatechin gallate were the most active compounds, which, in turn, were more antioxidant than epigallocatechin and, particularly, than epicatechin. The antioxidant activity of green tea phenolics and extracts and the flavonol myricetin was also tested in oil-in-water emulsions, where they showed a high activity [119]. However, some flavonoids may exhibit a pro-oxidant effect in the presence of ferric ion [119].

Antioxidant activity has been reported for different structural groups of phenolic secondary metabolites.

Benzoic Acids

The antioxidant activity of gallic acid (Fig. 5) and of its methyl, propyl and lauryl esters has been reported [120]. These compounds decreased the peroxidation of brain phospholipids, although under some conditions they acted as pro-oxidants. Gallic acid is a common constituent of gallotannins (hydrolysable tannins), and is present in the form of polymers in many plant products. Ellagic acid is a gallic acid dimeric derivative, which occurs in many plants, and is derived from ellagitannins. Its antioxidant activity has been demonstrated, in the inhibition of lipid peroxidation in rat liver microsomes [121]. This activity was higher than that of the related substances, hexahydroxydiphenic acid (Fig. 5) and ellagic acid tetraacetate. Food processing releases ellagic and gallic acids from their respective polymers, and probably increases the antioxidant activity and nutritional interest of the processed products.

Hydroxycinnamic Acid Derivatives

Caffeic, *p*-coumaric, ferulic, and sinapic acids (Fig. 4) and their esters and glycosides have all shown interesting antioxidant activity [122;123]. 3,5-Dicaffeoyl-4-succinylquinic acid and 3,5-dicaffeoylquinic acid (isochlorogenic acid) showed antioxidant activity in a β-carotene-linoleate method [123].

Five dicaffeoyl quinic acid derivatives, mono- and di-acylated with succinic acid, were isolated from an edible root used in Japan (*Arctium lappaa* L.). Their antioxidant activity increased with the number of caffeoyl residues per molecule and was in all cases higher than that of α-tocopherol[88]. The succinyl residues had no effect on this activity. In another recent experiment, the antioxidant activity of caffeic and ferulic acids, their phenylethyl esters, rosmarinic acid (a caffeic acid dimer) and chlorogenic acid was evaluated in lard [122], and their activities were compared with those of α-tocopherol and BHT. All the compounds significantly prolonged the induction time of lipid oxidation in lard. Caffeic and rosmarinic acids were the most active. When the lipid substrate was changed to corn oil, the induction time decreased and the most powerful antioxidant was rosmarinic acid, which also was the most active in scavenging free radicals in an *in vitro* assay (DPPH assay, see above). Chlorogenic acid was less effective than caffeic acid in the three models.

Ferulic acid and ester combinations may also be components of fibre. When the effect of these compounds on LDL oxidation was recently tested [87], ferulic acid, when added as free acid, showed little antioxidant effect. However, ferulic acid sugar esters had a positive effect. Both facts together indicate that affinity of the LDL particle for the bound ferulic acid is important for the antioxidant effect. It has also been reported that the antioxidants present in Japanese sake are ferulic acid sugar esters, which are cell-wall fragments of rice solubilised from insoluble dietary fibre by enzymatic hydrolysis [87].

Catechins (Flavan-3-ols)

Extensive research has been carried out into catechins, since they are the principal responsible compounds for the antioxidant activity of tea and wine. In a recent paper, the antioxidant activity of epigallocatechin (Fig. 3), epigallocatechin gallate, epicatechin gallate, epicatechin and catechin was tested in different lipid systems [108] and was compared to that of gallic acid and propylgallate. It was found that the efficacy of these compounds as antioxidant varies with the system used. Thus, in corn oil triglycerides oxidised at 50°C, epigallocatechin, epigallocatechin gallate and epicatechin gallate showed higher activity than epicatechin and catechin, which have a lower number of hydroxyls per molecule. In corn oil-in-water emulsions, all tea catechins, gallic acid and propylgallate acted as pro-oxidants at 5 and 20 µM by accelerating hydroperoxide and hexanal production. In soy lecithin liposomes oxidised at 50°C, epigallocatechin gallate and propylgallate were the best antioxidants. However, when the liposomes were oxidised at 37°C with cupric acetate, catechin and epicatechin were better than epicatechingallate, while epigallocatechingallate, epigallocatechin, propylgallate and gallic acid promoted lipid oxidation. The better antioxidant activity observed for tea catechins in liposomes than in emulsions can be explained by the greater affinity of the polar catechins towards the polar surface of the lecithin bilayers.

Flavones and Flavonols

The antioxidant activity of flavones such as luteolin and its 7-glucoside and 7-glucuronide has been demonstrated by the β-carotene/linoleic method [124]. Luteolin was more active than BHT, while the glycosides showed less activity. C-glycosylflavones also show antioxidant activity. Thus, 2"-O-glucosylvitexin from rice-hull [93] and barley leaves [125], and 6-C-glucosyldiosmin and 6,8-di-C-glucosyldiosmin from lemon peel [126] showed significant antioxidant activity. In addition, flavonoid aglycones from thyme, namely cirsilineol (5,4-fihydroxy-6,7,3'-trimethoxyflavone), genkwanin (5,4'-dihydroxy-7-methoxyflavone), and 8-methoxycirsilineol (5,4'-dihydroxy-6,7,8,3'-tetramethoxyflavone) showed greater antioxidant activity than tocopherol and in the same range as BHT [110]. Similar results were also reported for the flavonoid aglycones from oregano [109]. The aglycones, sideritoflavone (5,3',4'-trihydroxy-6,7,8-trimethoxyflavone) and cirsiliol (5,3',4'-trihydroxy-6,7-

dimethoxyflavone), showed higher antioxidant activity than cirsilineol and 8-methoxycrisilineol in a fish meat oxidation system (Tomás-Barberán, unpublished results).

Anthocyanin Pigments

Anthocyanin pigments, such as cyanidin 3-glucoside and its aglycone cyanidin, showed antioxidant activity in four different lipid oxidation sytems: the linoleic acid autoxidation system, the egg lecithin liposomes system, the rabbit erythrocyte membrane system and the rat liver microsomal system [94]. Anthocyanins from eggplant [127], grape [91] and red beans [128], also show significant antioxidant activity.

Other Phenolic Derivatives

The phenolic glucoside of olive, oleuropein (Fig. 5), which is present in olive leaves in a substantial amount (60-90 mg/g d.w.), was much more effective than BHT or vitamin E in extending the induction period of lipid oxidation [129].

The antioxidant activity of resveratrol (Fig. 5), a stilbenoid derivative, has been reported [130].

The antioxidant activity of lignans has also been established. Thus, dihydroguayaretic acid, guayacasin and isopregomisin showed to be powerful antioxidants with similar activities to that of propylgallate [131].

Structure-antioxidant Activity Relationship

The antioxidative activity of polyphenols is generally attributed to their hydroxyl groups, although this is not the only determining factor. In the case of ferulic acid, there is a single hydroxyl group *para*-substituted on an aromatic ring that is connected to a highly conjugated side chain. This *para* substitution allows the phenoxy radical of ferulic acid to be delocalised across the entire molecule and therefore be stabilised. The *ortho* substitution with the electron donor methoxy group also increases the stability of the phenoxy radical and hence increases its antioxidative efficiency [122]. The presence of a second hydroxyl group in the *ortho* or *para* position is known to increase antioxidative activity because of the additional resonance stabilisation and formation of *o*-quinone or *p*-quinone structures [132]. This knowledge was used to explain the fact that the antioxidative efficiencies of caffeic acid and its phenethyl ester are greater than those of ferulic acid and its phenethyl ester in three test model systems [122].

The antioxidant behaviour of flavonoids and their activity-structure relationships were investigated using the oxygen radical absorbance capacity assay [133]. Flavonoids including flavones, isoflavones, and flavanones acted as antioxidants against peroxyl and hydroxyl radicals, but served as pro-oxidants in the presence of Cu^{2+}. Both the antioxidant and the copper-initiated prooxidant activities of a flavonoid depend upon the number of hydroxyl substitutions. In general, the more hydroxyl substitutions, the stronger the antioxidant activity. Thus, the antioxidant activity of several flavonols against canola oil oxidation at 105°C was: myricetin (6 hydroxyls) > quercetin (5 hydroxyls) > morin (5 hydroxyls) > kaempferol (4 hydroxyls) [134]. However, this order changed to quercetin > kaempferol > myricetin > morin when the oxidation was carried out at 54°C.

Flavonoids that contain multiple hydroxyl substitutions showed antiperoxyl radical activities several times stronger than that of Trolox, an alpha-tocopherol analogue. The single hydroxyl substitution at position 5 gives rise to no activity, whereas the di-OH substitution at 3' and 4' is particularly important to the peroxyl radical absorbing activity of a flavonoid. Quercetin and cyanidin, with their 3',4'-dihydroxy substitutions and conjugation between rings A and B, show an antioxidant activity which is four times higher than that of Trolox. When the *ortho*-dihydroxy grouping is removed, as in kaempferol, or the potential electron delocalisation is reduced by removing the double bond 2-3 in ring C, as in catechin or epicatechin, the antioxidant capacity decreases by more than 50%, although these structures still remain more effective than α-tocopherol or ascorbate [135]. The O-methylation of the hydroxyl substitutions inactivates both the antioxidant and the prooxidant activities of the flavonoids.

To summarise, the structural criteria for flavonoids to effectively scavenge free radicals are [136]:

(1)   The 3-hydroxy group on the unsaturated C ring: blocking or removing the 3-OH group from participating in electron delocalisation substantially decreases antioxidant activity.
(2)   The 2,3-double bond with the 3-OH group and the 4-one in the C ring: reduction of the unsaturated bond in the C ring reduces the antioxidant activity by eliminating the possibility of electron delocalisation across the structure.
(3)   The *o*-dihydroxy grouping in the B ring: the presence of a single hydroxy group in the B ring instead of the *o*-dihydroxy structure drastically reduces antioxidant activity. An isolated OH group on the B ring makes no contribution to such activity. Whether a 3-OH group is present under these structural conditions is irrelevant, which demonstrates the contribution made by the 3-OH to electron delocalisation and radical stabilisation when the *o*-dihydroxy structure in the B ring is involved along with the 2,3-double bond and the 4-one [136].

HEALTH AND NUTRITIONAL BENEFITS OF PHENOLIC ANTIOXIDANTS

In vitro studies have pointed to the antioxidant activity of many fruit and vegetable phenolics. To ascertain whether these substances play a real role in the prevention of oxidative-originated diseases *in vivo,* other considerations should be taken into account. We review epidemiological studies on the correlation between the dietary intake of phenolics and the prevention of cardiovascular disease. In addition, we look at studies on phenolic bioavailability and metabolism, which indicate whether fruit and vegetable antioxidants that show activity *in vitro* are absorbed and may, therefore, have an antioxidant effect in the tissues and so help prevent atherosclerosis [137].

Epidemiological Studies

During the last 25-30 years several epidemiological studies have been carried out in different countries in an attempt to evaluate the effect of dietary habits on the development of coronary heart disease. These studies examined the diet of individuals in the 1960s and recorded mortality by heart attack during a 25 year follow-up. Recent studies, using modern analytical techniques, have evaluated the average flavonoid and flavone intake as it would have been around 1960 in 16 cohorts participating in the Seven Countries Study. It was found that the average flavonol and flavone intake was inversely correlated to the rate of mortality due to coronary heart disease [138]. The intake of flavonols and flavones, together with smoking and the intake of saturated fat, explained about 90% of the variance in coronary heart mortality rates in the 16 cohorts.

In addition, five prospective within-population cohort studies have been carried out, four of them on coronary heart disease and one on strokes. The four coronary heart disease studies were carried out in The Netherlands (Zutphen) [139], the USA (Health Professionals study) [140], the U.K. (the Caerphilly study) [141], and Finland [142]. In the Zutphen study, coronary heart disease was inversely associated with flavonol intake, in which a maximum intake of 42 mg/day and a minimum of 12mg/day were recorded. A clear dose-response correlation was observed. In the Health Professionals study, a modest non-significant inverse association was found (flavonoid intake between 40 mg/day and 7 mg/day). The Finnish study indicated a weak inversely associated correlation, while the Caerphilly study involving Welsh men showed that flavonoid intake increased the mortality.

The authors of the Caerphilly study offered the following possible explanation for the increase in mortality found in those individuals with a higher flavonoid intake. The main source of phenolics in the individuals of the Caerphilly study was tea. However, the subjects used milk in their tea and it is known that milk proteins form complexes with flavonoids that inhibit their absorption. The intake of tea with no added milk indeed raised plasma antioxidant capacity in volunteers whereas tea with milk did not [143]. However, the lack of absorption is an insufficient explanation since plasma concentrations of catechins and quercetin in volunteers given tea were the same whether the tea contained milk or not [12].

The only epidemiological study on stroke was that undertaken in Zutphen and this also showed that flavonoid intake reduced considerably the risk of stroke [144].

To summarise, from the available epidemiological studies, 3 out of 5 indicate the protective role of flavonoids, one shows that there is no association, and one indicates a positive association of flavonoid intake with the development of coronary heart disease. These results show that the evidence obtained is not conclusive and that more studies need to be carried out.

It is possible, however, that an evaluation of dietary flavonoid intake today, with currently used cultivars, postharvest technological treatments, storage conditions and processing, should not be compared with the intake arising from the products available during the time in which the above studies were performed. Present cultivars are grown under controlled environmental conditions in greenhouses to avoid environmental stress, using treatments to protect plants from pests and with irrigation and nutritional conditions close to the ideal situation. Plants, therefore, no longer need to produce phenolic secondary metabolites as defence mechanisms, and the phenolic content of present day fruit and vegetables is likely to be very much reduced compared with that of the cultivars available 25 years ago.

It is also essential to consider how food is ingested, and the possible effect of combinations of different products. The effect of food proteins on the absorption, bioavailability and *in vivo* antioxidant activity of fruit and vegetable phenolics may be as important as it was in the case of tea [143].

In addition, another point should be considered. What is the real availability of phenolics to be absorbed in the intestinal tract? The above mentioned epidemiological studies on the flavonoid content of fruit and vegetables, were performed on lyophilised material extracted by boiling with HCl in methanol-water (50%) [68]. These conditions are harsher than those actually occurring after the ingestion of food, when digestive enzymes will play a much more important role. Thus, the availability (*in vivo* extraction from food) of phenolics in the gastro-intestinal tract, may be very different from the numbers reported [68].

Bioavailability

For phenolic compounds to act as potential antioxidants in humans they must enter the blood stream.For this, once consumed, the compounds have to pass through the intestinal wall and resist metabolism in the liver. The fraction of molecules that successfully survive these barriers (intestine and liver) can be considered as bioavailable [138].

Oxidants, such as free radicals, are continuously generated by cells and have the potential to damage these cells. A steady-state antioxidant concentration that is sufficiently high to prevent this deleterious oxidation would be beneficial. Because food is not consumed continuously over 24 hours, the input of dietary antioxidants occurs as a series of pulses separated by many hours. Rapid elimination of these antioxidants from the body after a meal would be undesirable and it is feasible that kinetic parameters vary between foods.

There are three main questions concerning the absorption of phenolics. First, to what extent are these compounds absorbed from the gastro-intestinal tract? Which factors affect their absorption? What percentage of the phenolics present in a given product (fruit, vegetable, processed food, etc.) are available for absorption after digestion of the product?

The absorption of flavonoids from the diet was long considered to be negligible, since most of the flavonoid compounds in plants, except catechins, are bound to sugars as glycosides, and these were considered non-absorbable [9;10]. The same can be stated for other phenolic compounds (phenylpropanoids, anthocyanins, etc.). Studies with germ-free rats showed that large amounts of unchanged glycosides were excreted with faeces, whereas only small amounts of glycosides were found in the faeces of rats with normal microflora [145]. Evidently, enzymes that can split these predominant β-glycosidic bonds were not secreted into the gut or present in the intestinal wall. Bacteria in the colon were able to hydrolyse flavonoid glycosides, but at the same time degraded the released flavonoid aglycones. Since the absorption capacity of the colon is far less than that of the small intestine, it was assumed that only free flavonoids (aglycones) are absorbed in the gut, and that glycosides are not [137]. This assumption was never seriously questioned even though there was little evidence to support it.

In the 70s and 80s, balance studies with radioactively labelled flavonoids were performed to quantify the rate of absorption of different compounds including catechins, quercetin and flavanones (aglycones in all cases) in different mammals including man. Different absorption rates were detected depending on the flavonoid type. Thus, catechin was well absorbed since 47% to 58% of the administered radioactivity was excreted in urine. Quercetin was less well absorbed than catechins and only 4-13% of the administered radioactivity was recovered with urine, while around 40% was excreted with faeces. The absorption of flavanones was higher with 30% being excreted with urine [137].

The absorption of flavonoids from food was recently studied by Hollman et al [146], who studied the absorption of onion quercetin derivatives after the administration of onion. To circumvent the problem of microbial degradation, they used healthy patients who had undergone ileostomy. In these experiments, and contrary to the established knowledge, quercetin glycosides from onions were absorbed to a far greater extent than the pure aglycone (52% of the ingested amount, as opposed to 24% of the aglycone and 17% of rutin). Thus, it seems that glycosides can be absorbed in man without prior hydrolysis by microorganisms. Evidence of the direct absorption of glycosides was also found in rats, in which the oral administration of naringin and hesperidin led to the glycosides being secreted with bile [147] implying that glycosides were transported across the intestinal membranes. The epigallocatechins present in tea were well absorbed by rats [148] and humans [149]. It can be concluded, then, that contrary to the common belief that only aglycones can be absorbed, flavonol glycosides can be well absorbed in man without prior hydrolysis by microorganisms, and similar observations have been made in rats. Only a small fraction of the flavonols subsequently excreted with urine have an intact flavonoid structure.

Proteins in the diet may theoretically affect flavonoid absorption because they bind polyphenols. Circumstantial evidence for reduced absorption of tea polyphenols by complexation with milk proteins was found in humans [143]. The ingestion of tea caused a significant increase of the plasma antioxidant capacity, although not when tea was consumed with milk.

Another important point to take into consideration is that 98% of the quercetin found in human plasma was bound to proteins, and that the binding of quercetin to human albumin reached 70-80% [150].

Metabolism

The metabolism of flavonoids and other phenolics is an important aspect since a major part of the ingested compounds is excreted in urine after more or less extensive modification. Thus, a potential effect predicted from *in vitro* studies may be modulated *in vivo* because of metabolism of the parent compound after ingestion. Products can be metabolised in the liver or by the colonic flora.

In the liver two phases of biotransformation reactions can take place: phase I transformations that introduce polar groups (hydroxyls) in the molecule, and phase II reactions which include conjugations with glucuronic acid, sulphate, or glycine to yield water-soluble metabolites which are excreted in urine. O-Methylations, to inactivate catechol moieties in these molecules, are also phase II reactions. Phase I reactions are not important in natural phenolic metabolites since they generally contain several polar hydroxyl groups). Phase II reactions increase the molecular weight of the phenolics and promote their secretion into bile.

Phenolics can reach the colon in two ways: unabsorbed, by passing through the small intestine, or by secretion into the duodenum via the gall bladder after absorption. In the colon they are metabolised by the bacteria. These bacteria produce three types of reactions: hydrolysis of conjugates to release aglycones, cleavage of the heterocycle oxygen-containig ring of the flavonoids to yield phenolic acids, and degradation of these acids by oxidation processes to produce benzoic acids. The sugar moieties, glucuronic acids, and sulphates are released by glycosidases, glucuronidases and sulphatases of colonic bacteria; the released aglycones are then absorbed and enter the enterohepatic cycle. After ring cleavage, the phenolic acids produced can be absorbed or further degraded.

Catechins are extensively metabolised in humans. After administration of radioactive catechins, some 50% of the radioactivity is recovered in urine, and only 0.5-3% in the form of the catechin aglycone. Phase II transformation products are detected including sulphates, glucuronides and methyl ethers. Bacterial degradation through ring cleavage leads to valerolactones, phenylpropionic acid derivatives and benzoic acids. Nearly half of the orally administered and absorbed catechin was secreted with bile into the small intestine. Glucuronides and sulphate conjugates of catechin and catechin 3'-methyl ether, the major hepatic metabolite, were secreted with bile. These metabolites are prone to microbial degradation in the colon, and after hydrolysis of the conjugates, catechin and its phenolic acid and lactone metabolites were absorbed.

As in the case of catechins, flavonols suffered only phase II transformations in the liver including glucuronidation, sulphation and o-methylation. In the colon, the bacterial ring cleavage produced phenylacetic and phenylpropionic acids, and dehydroxylation reactions have also been reported. In contrast with catechin, quercetin in humans is metabolised to a limited extent via conjugation with sulphates, glucuronic acid or O-methylation. In human liver microsomes, fully methylated flavonoids, such as tangeretin from citrus fruits, undergo O-demethylation. Only limited data is available about anthocyanins, but it seems that these compounds are metabolised to a much more limited extent than other flavonoids. They are not transformed to phenolic acid derivatives by faecal bacteria. It has been reported that after intravenous administration to rats, 20% was excreted in an untransformed state in urine [137].

Phenolic acid derivatives (cinnamic acids) and degradation products of flavonoids (phenylpropionic and phenylacetic acids) suffer transformations by caecal bacteria. The following transformations have been observed: dehydroxylation of 3,4-dihydroxy derivatives to give 3-hydroxy compounds, demethylation of o-hydroxy-methoxyphenolic acids, reduction of the double bonds of cinnamic acids to yield the corresponding phenylpropionic acids, decarboxylation of cinnamic and phenylacetic acids (only when 4-hydroxyl is present), hydroxylation of phenylpropionic to produce phenylhydracrylic acids (only observed in man and monkeys) [151].

These compounds also suffer enzymatic transformations including conjugations with glycine, glucuronic acid, sulfate, o-methylations, and β-oxydations of phenylpropionic acids to yield benzoic acids.

EFFECT OF POSTHARVEST STORAGE AND DIFFERENT TECHNOLOGICAL TREATMENTS OF FRUIT AND VEGETABLES ON PHENOLIC ANTIOXIDANTS

The antioxidant phenolic content of fruits and vegetables, and, therefore, the antioxidant potential of the phenolic substances can be much affected by the storage conditions and technological treatments to which fresh fruit and vegetables are submitted during their postharvest life until their consumption. Cold storage is generally used to decrease the respiration and metabolism of these living products in order to extend their storage life. In some cases, controlled atmospheres, in which the air composition is modified by increasing CO_2 and/or decreasing O_2 are applied to reduce decay and maintain the quality attributes of fruits. New applications for this technological treatment have been developed for different products. The air composition is also modified by storage under polymeric films with different gas permeabilities allowing increased CO_2 and decreased O_2 levels as a result of the commodities respiration. This technique is known as Modified Atmosphere Packaging (MAP) and is especially important in the storage of minimally processed products. All these storage conditions have an effect on the phenolic metabolism and phenolic composition, and therefore on the antioxidant potential of the fruits and vegetables when they reach the consumer.

In addition, fruit and vegetables can be minimally processed (washed, pealed, cut, diced, stripped, chopped, etc.) in order to provide 'ready to eat' or 'ready to use' products (salad mixes, apple slices, carrot sticks, diced potatoes, etc.). These minimal processing treatments damage the plant surface and induce changes in phenolic metabolism and phenolic antioxidant composition.

Antioxidant phenolics can also be much affected during the manufacturing and storage of fruit juices and jams, and during the domestic processing of fruit and vegetables.

We shall now look at the effect of storage conditions, technological treatments and processing on antioxidant phenolic metabolite composition.

Storage

The cold storage of fruit and vegetables can alter phenolic metabolite composition. When the changes in caffeic acid derivatives in artichoke heads (*Cynara scolymus* L.) during postharvest storage were studied [152], the caffeic acid content increased considerably during postharvest storage of the healthy heads (2 weeks at 20°C or 1 month at 4°C). In heads injured during harvest or by postharvest manipulation (internal or external blackening), a significant decrease in caffeic acid derivatives was observed during storage (2 weeks at 20°C). This decrease was attributed to the effect of the enzyme polyphenol oxidase (PPO) on the caffeic acid derivatives, leading to the formation of brown polymers.

During the cold storage of pears (4°C up to two months) the total phenolic content increased under all the storage conditions studied [153]. Phenolic accumulation rates differed according to the storage conditions used. During the first month of storage in air, the total phenolic content slightly increased from 185 at harvest to 228 mg kg^{-1} (fresh weight). A greater increase up to 403 mg kg^{-1} was observed during the second month of storage. The most affected compounds were the flavanols, followed by flavonols and hydroxycinnamic acid derivatives.

The changes in pear phenolic metabolites during cold storage largely depend on the cultivar. The major phenolics of mature Beurre d'Anjou and Beurre Bosc pear fruit flesh at harvest were chlorogenic acid, catechin and arbutin [154]. During 160 days at -1°C the chlorogenic acid content of d'Anjou pears increased significantly, although it decreased in Bosc pears. The catechin content increased linearly while arbutin levels remained nearly constant in both cultivars. Coinciding with the completion of the cold requirement for the initiation of ripening and endogenous ethylene production, there was a slight increase in a *p*-coumaric acid derivative, and trace amounts of epicatechin and *p*-coumaroyl quinate were recorded. Ethylene production and phenolic quantity and composition during storage were related. Bruising pear fruits after 120 days of storage caused a 30% increase in chlorogenic acid and a 50% increase in catechin, but no increase in *p*-coumaric derivatives.

In the case of 'Granny Smith' apples, cold storage and rewarming induced different changes in the peel than in the pulp [155], phenolic compounds decreasing in the pulp and increasing in the peel.

In these changes, polyphenol oxidase plays an important role, its activity falling significantly during apple maturation, but remaining quite constant during cold storage [156].

Apples suffer a physiological disorder during cold storage, which is known as superficial scald, and which is characterised by browning of the skin. This disorder is thought to be due to an uncontrolled polyphenol oxidase (PPO) system oxidising the vacuolar phenolics. In scalded tissue there was a significant decline in the concentration of all the phenolics, particularly the flavonols [157]. Skin patches affected by advanced superficial scald do not contain traces of any polyphenol, and there is a clear relationship between the degree of browning and a decrease in the amount of detectable polyphenols. It was suggested that the disappearance of polyphenols from the scald affected zones was due to the reduction of quercetin aglycone to flavan-3,4-diol followed by a production of polymeric compounds [158]. When 'Granny Smith' apples were wounded with a hypodermic needle prior to cold storage for 7 months at 0°C, no superficial scald developed in the tissues adjacent to the wounded areas, while non-wounded tissue developed scald on greater than 60% of the apple surface. The increase in antioxidant hydroxycinnamic acid derivatives in the wound area suggests that it was this type of compounds which was responsible for scald prevention [159].

There are several reports in which the anthocyanin pigments of fruits and vegetables increased during cold storage. The concentration of anthocyanins of the lowbush blueberry (*Vaccinium angustifolium* Aiton) increased by 18% after 2 weeks at 1°C [160], and the total anthocyanin concentration of pomegranate arils increased during storage in air and was significantly different from the initial value after 4 weeks' storage at 10°C [161]. This increase was associated with an increase in the key enzyme of phenolic metabolism, phenylalanine ammonia lyase (PAL) [161].

Some vegetables, such as onions or potatoes, are stored at room temperature. In the case of onions, this storage starts after a curing (drying) process. When the effect of curing and storage on the flavonol composition of two onion cultivars (Red Baron and Crossbow) was studied [76], a 50% loss of quercetin 4'-monoglucoside was observed during the initial drying process (curing). However, there was little change in the phenolic content and composition during 6 months of storage.

Controlled Atmospheres

Some fruit and vegetables are stored under atmospheres enriched in CO_2 and with reduced levels of O_2. These storage conditions have a marked effect on phenolic metabolism and phenolic composition. Thus, 'Williams' pears stored in air accumulate more phenolics than fruits stored under controlled atmospheres (1% CO_2-1% O_2 and 3% CO_2-3%O_2). A controlled atmosphere, then, strongly reduces the ability of pears to synthesise phenolic compounds [153].

Carbon dioxide-enriched atmospheres (10-20% CO_2 in air) are used to extend the postharvest life of strawberries [162]. However, some adverse effects on colour, mainly a reduction in the intensity of red of the internal tissue, have been reported. Changes in strawberry anthocyanins and other polyphenols in response to carbon dioxide treatments have been studied [36]. The external and internal anthocyanin contents were significantly different in fruit stored in air compared with the initial values or with those of fruit stored under CO_2 enriched atmosphere. No differences in the external anthocyanin content between fruit stored under CO_2-enriched atmospheres and freshly picked fruit were observed. However, there was a noticeable decrease in the internal anthocyanin content, particularly at 20 and 40% CO_2. Factors such as copigmentation, pH and anthocyanin metabolism may play a significant role in the expression of colour in strawberries. The most efficient copigments are flavonols [17], which are almost exclusively located in the external tissues of strawberries. Anthocyanin stability in the external tissue could be explained by intermolecular copigmentation with flavonols and other phenolics, while the anthocyanins of internal tissues, where flavonols are present in very low concentrations, would be more susceptible to degradation. CO_2 treatments increased pH in the internal tissue and most anthocyanins appeared colourless.

The effect of different O_2 and CO_2 levels in different packages on the stability of apple anthocyanins has also been determined [163]. When the concentration of cyanidin 3-galactoside, cyanidin 3-arabinoside and an unidentified cyanidin arabinoside in the skin of 'Starkrimson' apples stored up to 30 weeks at 2°C, was determined, the three anthocyanins were severely destabilised by carbon dioxide levels above 70%.

The effect of carbon dioxide on anthocyanins, and on some enzymes related to anthocyanin biosynthesis, was investigated in the arils of pomegranates stored under 10 or 20% CO_2 for 6 weeks [161]. The total anthocyanin concentration of the arils increased during storage both in air and in air + 10% CO_2. However, after 6 weeks, the anthocyanin concentration of pomegranates stored in air + 20% CO_2 was lower than the initial concentration. Anthocyanin concentration was well correlated with the activity of phenylalanine ammonia lyase (PAL), but not with glucosyltransferase activity.

In lettuce, raised CO_2 atmospheres may cause a disorder called 'brown stain' (browning of the epidermal tissue). This is caused by the production of brown pigments generated by the oxidation of phenolic compounds in the presence of the enzyme polyphenol oxidase. It was reported that under high CO_2 atmospheres, PAL activity was induced whereas phenolic production and browning were inhibited until the lettuce was transferred from CO_2 to air [164;165]. After this transfer, tissue browning occurred, which was associated with a rapid increase in the soluble phenolics content.

It was also recently reported that the storage of lettuce under controlled atmospheres (3% CO_2 - 5% O_2 and 10% CO_2 -11% O_2) followed by 24 hr exposure to air (all at 5°C) caused a considerable increase in total phenols, and in polyphenol oxidase and peroxidase activities [166].

Minimal Processing

The preparation of minimally processed fruits and vegetables entails some degree of wounding. Cutting divides the whole fruit or vegetable into smaller segments and shortens the storage life [167]. This preparation can be accomplished by slicing, chopping, dicing or shredding. Examples include sliced apples, diced onions and shredded lettuce.

In general, minimal processing (wounding) transcriptionally induces phenolic compound biosynthesis (Fig. 1), and these compounds accumulate during the storage of processed end products. This increase has been demonstrated in wounded potato, tomato, lettuce and carrots.
During the storage of carrot sticks, an increase in soluble phenolics was reported [168]. The same increase was reported for shredded carrot during cold storage [44]. Among the hydroxycinnamic derivatives, chlorogenic acid accounted for 60% of total phenolics. The shredded carrots which showed the greater storage stability were those which accumulated chlorogenic acid fastest during the first 24 hr in oriented polypropylene films [45]. Therefore, the rate at which they accumulate chlorogenic acid may be a useful index for selecting carrot cultivars for their stability during storage under modified atmosphere.

Minimal processing also causes changes in the soluble phenolics content of lettuce [169]. After 3 days' storage at 5°C, increases in chlorogenic, isochlorogenic, caffeoyltartaric and dicaffeoyltartaric acids were detected in three minimally processed lettuce types (icerberg, butter leaf and romaine lettuce). This accumulation of phenolic compounds was especially noticeable in the midribs (white vascular tissue), which were devoid of these substances in whole lettuce leaves. However, no significant increase in phenolics was observed with processing in the green photosynthetic tissue, which was rich in these phenolics in intact leaves. Minimal processing also affects the pigment composition of red pigmented lettuce ('Lollo rosso'), and a significant decrease was observed after storage in modified atmosphere packaging compared to the values recorded after the storage in air [67].

The phenolic content of the minimally processed lettuce midribs kept under CO_2 was lower than that of tissues stored in air. The increase in phenolics in cut midribs exposed to air was caused by wounding and was associated with the development of browning symptoms [170]. Exposure to 20% CO_2 prevented such an increase in the phenolic content and reduced browning in the cut midrib tissue [171].

The stability of anthocyanin pigments in shredded red onion was evaluated during storage for 7 days at 8°C. After 1 day of storage, a slight increase in anthocyanins was observed, after which the level falls reaching the lowest levels after 7 days of storage. The stability of individual anthocyanins was very different, malonated anthocyanins being much more stable than the corresponding non-acylated pigments [75]. In addition, the arabinosides were less stable than the corresponding glucosides.

Free phenolic constituents also increased more than 2-fold in the injured peel of orange (*Citrus sinensis* Osbeck) cv. Valencia after 48 h at 30°C and 96-98% relative humidity. However, at 5°C the wound healing processes slowed down [172]. Injured fruit, held at high humidities, developed a much thicker layer of lignin type material compared to those maintained at ambient humidity levels.

The total phenolic composition of sliced whole carambola (*Averrhoa carambola* L.) fruit during storage at 4°C was higher than that of the whole fruit [173]. Total phenol levels in the whole fruit decreased after 2 weeks storage, and remained lower than initial levels throughout 6 weeks of storage. However, total phenols in sliced fruit increased after 2 and 4 weeks of storage, but a decrease was observed between 4 and 6 weeks.

Ethylene

Some fruit and vegetables can be exposed to ethylene, the plant hormone produced by other commodities as climacteric fruits, during the postharvest storage and transport. This hormone can induce changes in phenolic metabolism and affect the phenolic metabolite composition and the antioxidant potential.

The most characteristic example of this phenomenon is the development of the lettuce physiological disorder known as 'russet spotting'. It consists of numerous small brown spots on the midrib of iceberg lettuce, and is induced by exposure to part per million levels of ethylene during storage at 5 ± 2°C. In leaves suffering 'russet spotting', the increase in spotting was accompanied by a parallel increase in the amount of soluble phenolic compounds [174] and in phenolic metabolism (phenylalanine ammonia lyase activity was induced). These ethylene-induced phenolics were readily oxidised to brown substances by the enzyme polyphenol oxidase [175].

Light and UV Irradiation

Several authors have reported a stimulating effect of light on the postharvest colour development of non fully coloured berries [176]. The influence of postharvest light and temperature conditions on colour development in 'Kent' strawberries has been described at two temperatures (10 and 20 °C) [177]. Colour development for white berries was greatest in the light, and at 20°C compared to 10°C. A pronounced stimulation of anthocyanin synthesis in light was observed for white and pink berries stored at 20°C and for red berries stored at 10°C.

Similar results were found on apple skin disks of preclimacteric or climacteric apples, which were exposed to high intensity white light at different temperatures in order to find the optimum temperature for anthocyanin accumulation [178]. Pre-cooling the apple skin tissue disk for 40 h at 2°C before incubation at 25°C increased pigment accumulation.

In an attempt to develop new technological treatments to improve apple colour, ultraviolet (UV) irradiation was applied [179]. After 2 days of UV and white light exposure at 14ºC, apple skin anthocyanin content increased when apples were transferred for storage at 4ºC in the dark. It was demonstrated that, in addition to UV irradiation, cool temperatures are needed to enhance the development of red colour in apple.

Domestic Processing

Cooking affects the phenolic composition of fruit and vegetables, and, therefore, their antioxidant potential. After boiling in water or frying in oil, onions showed a 25% loss in flavonoids. However, domestic processing did not produce free quercetin or the inter-conversion of quercetin conjugates [76]. When potatoes were baked, chlorogenic acid was completely lost, while 55% remained after microwave cooking and 40% after boiling [180].

In the case of fresh spinach, boiling extracted 48% of the flavonoids, which were recovered in the cooking water. The remaining 52% was found in the cooked spinach. When frozen spinach was boiled, the percentage of flavonoids extracted in the boiling water reached 75% (Tomás-Barberán et al., unpublished results).

In the domestic processing of tomato, microwave cooking resulted in a fall in the quercetin content (35% remained) and boiling produced an even bigger reduction (18% remained) [49]. The losses that occurred during frying were less substantial (65% remained).

Thus, domestic processing considerably affects the phenolic compound content of vegetables, and their antioxidant activity can also be dramatically reduced. More research into the effect of cooking on the nutritional quality of fruit and vegetables is needed.

Juice and Jam Manufacturing

The industrial production of fruit juices entails a number of technological treatments that can affect the phenolic compounds of the processed products and their antioxidant potential. These treatments include enzymatic digestion, concentration, heat treatment, filtration, etc.

Hydroxycinnamic acids and flavan-3-ols are good substrates for the enzyme PPO (polyphenol oxidase), while the flavonols appear to be less suited as PPO substrates. Operations such as crushing, prepress enzymatic treatments, and pressing can led to PPO activity and the degradation of phenolic compounds with a resulting loss of antioxidant activity. The most significant level of oxidation occurs in apple pulp before and during pressing. Oxidation after extraction may also be substantial if a long period of time elapses before PPO inactivation. High-temperature short-time (HTST) treatment immediately after pressing protected phenolic compounds from oxidation during subsequent processing operations

[181;182]. Enzyme inactivation in apple requires heating to ca 90°C for 30s. This inactivation is also essential in pear. Pressing in the presence of agents suitable for PPO inhibition, such as sulphur dioxide or ascorbic acid, drastically increases the phenolic yields.

Commercial pectolytic enzymes used in clarification can cause hydrolysis of hydroxycinnamic acids in apple, pear and grape juice [183]. In addition, the use of gelatin as a fining agent reduces total phenolics with the polymeric phenolics being the most affected.

The application of hot water (above 57°C) in the extraction of apple juice is also used, as an alternative to conventional pressing in order to increase juice yield. The juices obtained by this technology are known as 'diffusion extracted' juices. The extraction temperature is considered very important and the phloridzin, which is concentrated in the seeds, and quercetin glycosides, which are concentrated in the skin, are extracted much more readily when the temperature is increased, since these compounds are poorly soluble in cold water. The quantitative data on the phenolic composition of 'diffusion-extracted' juices compared with juices extracted by conventional pressing show a 5 fold increase in total apple phenolics. This increase can even be 12 fold in the case of procyanidins [184].

The storage of apple juice and juice concentrates results in phenolic degradation. After storage for 9 months at 25°C, apple juice concentrates showed an approximate 36% degradation of hydroxycinnamic acids, 60% degradation of quercetin and phloretin glycosides and total loss of procyanidins. Flavan polymerisation was also detected [181].

Anthocyanin pigments are also affected by technological treatments. During the processing and storage of raspberry juice, anthocyanins are transformed into polymeric products [31]. When juices were obtained using enzymatic liquefaction with pectolytic enzymes, there was a 20% increase in the anthocyanin content. In addition, 25% of the pigments in these samples were in the form of polymers, while freshly processed juices typically contained only 10%. In addition, the occurrence of β-1-2 glucosidase activity in the pectinase preparation used for enzymatic liquefaction decreased the content of cyanidin 3-sophoroside and cyanidin 3-glucosylrutinoside in the juice since the terminal glucosyl residue was selectively removed by the enzyme. In contrast the products of this enzymatic reaction, cyanidin 3-glucoside and cyanidin 3-rutinoside, increased considerably in the treated juice. Thus, enzymatic treatment to increase juice yield and for clarification purposes can affect the phenolic composition. When raspberry juice was prepared by 'diffusion extraction', the juice had low cyanidin 3-glucoside content and a high degree of polymeric colour (30%). This could be attributed to he use of hot water for extraction. This technique extracts more anthocyanins and other phenolics that may contribute to anthocyanin polymerisation [31].

The effect of processing on strawberry juice anthocyanins has also been studied [185;186]. Clarification of juice before storage considerably decreased the anthocyanin concentration. Nitrogen atmospheres did not influence the rate of anthocyanin loss during storage or polymeric pigment formation. No polymerisation occurred in the samples during storage at -20°C, although there was substantial polymerisation in samples stored at 20°C [185].

The phenolic content of grape juice and wine is also greatly affected by the technological processes used [187]. The effect of immediate and hot pressing on the phenolic composition of grape juice (coutaric acid, caftaric acid, gallic acid, ellagic acid, catechin, epicatechin and procyanidins B_3 and B_4) was compared. Hot press extraction considerably increased the phenolic content, and this increase was mainly observed in gallic and ellagic acids, epicatechin, and procyanidins B_3 and B_4, while the other phenolics were not affected.

Other raspberry phenolics, including ellagic acid and its derivatives, and flavonols (quercetin derivatives) were also affected by the process used to obtain juice. The juices produced by high-speed centrifugation contained the highest quantity of total quercetin derivatives, while the use of pectinases decreased the concentration of flavonol derivatives [33]. 'Diffusion-extracted' juice contained even less quercetin forms, which might have resulted from the accelerated breakdown of flavonol glycosides by the combined effects of depectinisation and exposure to high temperature (63°C) for several hours. However, the same diffusion-extracted juice contained much higher concentrations of total ellagic acid forms than other juices [30]; probably because ellagic acid was released from cell walls, through the hydrolysis of ellagitannins during this slow, high temperature, extraction process. Osmotic concentration through membranes or vacuum concentration decreased the quercetin content of juice.

CONCLUSION

In vitro studies widely demonstrate the antioxidant activity of fruit and vegetable phenolics. This activity is generally higher than that of the antioxidant vitamins A, C and E. There is, however, little evidence to support this antioxidant activity *'in vivo'*. The epidemiological studies on the possible role of dietary flavonoids in cardiovascular disease prevention are conflicting. Several studies show a clear inverse association, but other evidences point to a weak association or no association at all. In the Caerphilly study, an increase in cardiovascular disease mortality was observed in those subjects with a higher flavonoid intake. However, such differences could be explained by several factors. First, the way in which flavonoids are ingested and the presence of other food constituents can be essential for flavonoid absorption and activity. The presence of proteins can cause flavonoid precipitation and decrease bioavailability. Second, the flavonoid content, as analysed in epidemiological studies, can differ substantially from the flavonoids present in the food products ingested by the individuals participating in the different studies. The phenolic content of modern fruit and vegetable cultivars may be much lower than that of the products ingested during the course of the epidemiological studies conducted 15-25 years ago. In addition, the real availability of fruit and vegetable phenolics can be very different, depending on the domestic processing habits of different countries. Third, antioxidant phenolics, other than flavonoids, could also be responsible for antioxidant activity.

Bioavailability studies are essential if we are to understand the possible role of plant phenolics in cardiovascular disease prevention. Recent studies showing that food flavonoids can be absorbed as glycosides, which goes against conventional knowledge concerning flavonoid absorption, indicate the need for such bioavailability studies.

The real availability of fruit and vegetable phenolics ready to be absorbed in the intestinal tract should also be studied. Analysis of the flavonoid and phenolic content of fruit and vegetables does not provide a real picture of the compounds which may be released from plant tissues after ingestion and which may be available for absorption. Studies concerning the real availability of food phenolics should be carried out.

The principal food products providing phenolic antioxidants to the diet of European countries are onions, apples, citrus fruits, tea, grapes and red wine. However, many other fruits and vegetables rich in phenolic antioxidants could be also included in this list. For example, berry fruits, cherry tomatoes, some red lettuce cultivars, artichokes, green asparagus, spinach, olives, broccoli and eggplant, are very rich in antioxidant phenolics.

Processing can produce changes in the phenolic metabolite composition of fruit and vegetables, and, in some cases, may increase it and the corresponding total antioxidant activity. Thus, minimal processing increases phenolic metabolism and, therefore, the accumulation of phenolic metabolites in many fruits and vegetables. It is important to demonstrate whether this increase in phenolics is also correlated with an increase in antioxidant activity. In the case of fruit juices and jams, the technological treatments can improve the bioavailability of antioxidant phenolics, especially through the transformation of insoluble phenolic polymers into soluble antioxidant monomers. This is the case of strawberry and raspberry gallotannins and ellagitannins, which can be transformed into the antioxidant compounds, gallic acid and ellagic acid, by the technological processes used for juice and jam manufacture.

REFERENCES

[1]   Verlangieri, A.J.; Kapeghian, J.C.; el-Dean, S.; Bush, M. *Med.Hypoth.* **1985**, *16*, 7-15.

[2]   Acheson, R.M.; Williams, D.R.R. *Lancet* **1983**, *1*, 1191-1193.
[3]   Doll, R. *Proc.Natl.Acad.Sci.U.S.A.* **1990**, *49*, 119-131.
[4]   Ames, B.M. *Science* **1983**, *221*, 1256-1263.
[5]   Stähelin, H.B.; Gey, K.F.; Eichholzer, M.; Lüdin, E.; Bernasconi, F.;
        Thurneysen, J.; Brubacher, G. *Am.J.Epidemiol.* **1991**, *133*, 766-775.
[6]   Ames, B.M.; Shigena, M.K.; Hagen, T.M. *Proc.Natl.Acad.Sci.U.S.A.* **1993**, *90*,
        7915-7922.
[7]   Hennekens, C.H.; Buring, J.E.; Manson, J.E.; Stampfer, M.; Rosner, B.; Cook,
        N.R.; Belanger, C.; LaMotte, F.; Gaziano, J.M.; Ridker, P.M.; Willet, W.; Peto,
        R. *N.Engl.J.Med.* **1996**, *334*, 1145-1149.
[8]   Rusznyak, S.; Szent-Gyorgyi, A. *Nature* **1936**, *138*, 27
[9]   Kühnau, J. *World Rev.Nutr.Diet.* **1976**, *24*, 117-191.
[10]   Griffiths, L.A.In *The Flavonoids: Advances in Research*; Harborne, J.B., Mabry,
        T.J., Eds.; Chapman and Hall: London, 1982; pp 681-718.
[11]   Hackett, A.M.In *Plant Flavonoids in Biology and Medicine. Biochemical,
        pharmacological, structure-activity relationships*; Cody, V., Middleton-E, J.,
        Harborne, J.B., Eds.; Alan R. Liss Inc.: New York, 1986; pp 177-199.
[12]   Katan, M.B. *Am.J.Clin.Nutr.* **1997**, *65*, 1542-1543.
[13]   Hollman, P.C.H.; Hertog, M.G.L.; Katan, M.B. *Biochem.Soc.Trans.* **1996**, *24*,
        785-789.
[14]   Kerrmann, K. *Plant Physiol.* **1995**, *107*, 7-12.
[15]   Dixon, R.A.; Paiva, N.L. *The Plant Cell* **1995**, *7*, 1085-1097.
[16]   Macheix, J.J.; Fleuriet, A.; Billot, J.*Fruit phenolics*; CRC Press: Boca Raton,
        Florida, 1990;
[17]   Mazza, G.; Miniati, E.*Anthocyanins in fruits, vegetables and grains*; CRC Press:
        Boca Raton, 1993; pp 212-214.
[18]   Harborne, J.B.*Comparative Biochemistry of the Flavonoids.*; Academic Press:
        London, 1967;
[19]   Mosel, H.D.; Herrmann, K. *Z.Lebensm.Unters.-Forsch.* **1974**, *154*, 6-11.
[20]   Tomás-Barberán, F.A.; García-Viguera, C.; Nieto, J.L.; Ferreres, F.; Tomás-
        Lorente, F. *Food Chem.* **1993**, *46*, 33-36.
[21]   Henning, W.; Herrmann, K. *Z.Lebensm.Unters.-Forsch.* **1980**, *171*, 183-188.
[22]   García-Viguera, C.; Bridle, P.; Ferreres, F.; Tomás-Barberán, F.A.
        Z.Lebensm.Unters.Forsch. **1994**, *199*, 433-436.
[23]   Lamuela-Raventós, R.M.; Waterhouse, A.L. *J.Agric.Food Chem.* **1993**, *41*, 521-
        523.
[24]   Lamuela-Raventós, R.M.; Romero-Pérez, A.; Waterhouse, A.L.; Torre-Boronat,
        M.C. *J.Agric.Food Chem.* **1995**, *43*, 281-283.
[25]   Roggero, J.P.; García-Parrilla, C. *Science des Aliments* **1995**, *15*, 411-422.
[26]   Waterhouse, A.L.; Lamuela-Raventós, R.M. *Phytochemistry* **1994**, *37*, 571-573.
[27]   Le Tutour, B.; Guedon, D. *Phytochemistry* **1992**, *31*, 1173-1178.
[28]   Wald, B.; Wray, V.; Galensa, R.; Herrmann, K. *Phytochemistry* **1989**, *28*, 663-
        664.
[29]   Maas, J.L.; Galletta, G.J.; Stoner, G.D. *HortScience* **1991**, *26*, 10-14.
[30]   Rommel, A.; Wrolstad, R.E. *J.Agric.Food Chem.* **1993**, *41*, 1951-1960.
[31]   Boyles, M.J.; Wrolstad, R.E. *J.Food Sci.* **1993**, *58*, 1135-1141.
[32]   Mosel, H.D.; Herrmann, K. *Z.Lebensm.Unters.-Forsch.* **1974**, *154*, 324-327.
[33]   Rommel, A.; Wrolstad, R.E. *J.Agric.Food Chem.* **1993**, *41*, 1941-1950.
[34]   Bakker, J.; Bridle, P.; Bellworthy, S.J. *J.Sci.Food Agric.* **1994**, *64*, 31-37.
[35]   Stöhr, H.; Herrmann, K. *Z.Lebensm.Unters.-Forsch.* **1975**, *159*, 341-348.

[36] Gil, M.I.; Holcroft, D.M.; Kader, A.A. *J.Agric.Food Chem.* **1997**, *45*, 1662-1667.

[37] El-Negoumy, S.I. *Fitoterapia* **1987**, *58*, 178-180.

[38] Herrmann, K. *Z.Lebensm.Unters.-Forsch.* **1978**, *167*, 262-273.

[39] Lattanzio, V.; Cardinalli, A.; Di Venere, D.; Linsalata, V.; Palmieri, S. *Food Chem.* **1994**, *50*, 1-7.

[40] Winter, M.; Brandl, W.; Herrmann, K. *Z.Lebensm.Unters.-Forsch.* **1987**, *184*, 11-16.

[41] Adzet, T.; Camarasa, J.; Laguna, J.C. *J.Nat.Prod.* **1987**, *50*, 612-617.

[42] Wöldecke, M.; Herrmann, K. *Z.Lebensm.Unters.-Forsch.* **1974**, *155*, 151-154.

[43] Herrmann, K. *Z.Lebensm.Unters.-Forsch.* **1977**, *165*, 151-164.

[44] Babic, I.; Amiot, M.J.; Nguyen-The, C.; Aubert, S. *J.Food Sci.* **1993**, *58*, 351-356.

[45] Babic, I.; Amiot, M.J.; Nguyen-The, C.; Aubert, S. *J.Food Sci.* **1993**, *58*, 840-841.

[46] Mercier, J.; Kuc, J. *J.Sci.Food Agric.* **1997**, *73*, 60-62.

[47] El-Moghazi, A.M.; Ross, S.A.; Halim, A.F.; Abou-Rayya, A. *Planta Med.* **1980**, *40*, 382-383.

[48] Galensa, R.; Herrmann, K. *Z.Lebensm.Unters.-Forsch.* **1979**, *169*, 170-172.

[49] Crozier, A.; Lean, M.E.J.; McDonald, M.S.; Black, C. *J.Agric.Food Chem.* **1997**, *45*, 590-595.

[50] Hertog, M. G. L. Flavonols and flavones in foods and their relation with cancer and coronary heart disease risk. 1-149. 1994. Wageningen University. Thesis.

[51] Stöhr, H.; Herrmann, K. *Z.Lebensm.Unters.-Forsch.* **1975**, *159*, 219-224.

[52] Brandl, W.; Galensa, R.; Herrmann, K. *Z.Lebensm.Unters.-Forsch.* **1983**, *177*, 325-327.

[53] Röber, M.; Pydde, E.; Knorr, D. *Lebensm.-Wiss.u.-Technol.* **1991**, *24*, 466-468.

[54] Nigg, H.N.; Strandberg, J.O.; Beier, R.C.; Petersen, H.D.; Harrison, J.M. *J.Agric.Food Chem.* **1997**, *45*, 1430-1436.

[55] Schmidtlein, H.; Herrmann, K. *Z.Lebensm.Unters.-Forsch.* **1975**, *159*, 139-148.

[56] Brandl, W.; Herrmann, K. *Z.Lebensm.Unters.-Forsch.* **1983**, *176*, 444-447.

[57] Wildanger, W.; Herrmann, K. *Z.Lebensm.Unters.-Forsch.* **1973**, *152*, 134-137.

[58] Bilyk, A.; Sapers, G.M. *J.Agric.Food Chem.* **1985**, *33*, 226-228.

[59] Tanchev, S.S.; Timberlake, C.F. *Phytochemistry* **1969**, *8*, 1825-1827.

[60] Hrazdina, G.; Iredale, H.; Mattick, L.R. *Phytochemistry* **1977**, *16*, 297-299.

[61] Guo, C.; Cao, G.; Sofic, E.; Prior, R.L. *J.Agric.Food Chem.* **1997**, *45*, 1787-1796.

[62] Diallinas, G.; Kanellis, A.K. *Plant Mol.Biol.* **1994**, *26*, 473-479.

[63] Winter, M.; Herrmann, K. *J.Agric.Food Chem.* **1986**, *34*, 616-620.

[64] Schmidtlein, H.; Herrmann, K. *Z.Lebensm.Unters.-Forsch.* **1975**, *159*, 255-263.

[65] Starke, H.; Herrmann, K. *Z.Lebensm.Unters.-Forsch.* **1976**, *161*, 25-30.

[66] Wöldecke, M.; Herrmann, K. *Z.Lebensm.Unters.-Forsch.* **1974**, *156*, 153-157.

[67] Ferreres, F.; Gil, M.I.; Castañer, M.; Tomás-Barberán, F.A. *J.Agric.Food Chem.* **1997**, *45*,

[68] Hertog, M.G.L.; Hollman, P.C.H.; Katan, M.B. *J.Agric.Food Chem.* **1992**, *40*, 2379-2383.

[69] Yamaguchi, M.A.; Kawanobu, S.; Maki, T.; Ino, I. *Phytochemistry* **1996**, *42*, 661-663.

[70] Iwagami, S.; Sawabe, Y.; Nakagawa, T. *Shoyakugaku Zasshi* **1992**, *46*, 339-347.

[71]   Ke, D.; Saltveit, M.E. *Plant Physiol.* **1988**, *88*, 1136-1140.
[72]   Tomás-Barberán, F.A.; Loaiza-Velarde, J.; Bonfanti, A.; Saltveit, M.E. *J.Amer.Soc.Hort.Sci.* **1997**, *122*, 399-404.
[73]   Starke, H.; Herrmann, K. *Z.Lebensm.Unters.-Forsch.* **1976**, *161*, 137-142.
[74]   Bilyk, A.; Cooper, P.L.; Sapers, G.M. *J.Agric.Food Chem.* **1984**, *32* , 274-276.
[75]   Ferreres, F.; Gil, M.I.; Tomás-Barberán, F.A. *Food Res.Int.* **1996**,
[76]   Price, K.R.; Bacon, J.R.; Rhodes, M.J.C. *J.Agric.Food Chem.* **1997**, *45*, 938-942.
[77]   Herrmann, K. *Z.Lebensm.Unters.-Forsch.* **1977**, *165*, 87-98.
[78]   Woldecke, M.; Herrmann, K. *Z.Lebensm.Unters.-Forsch.* **1974**, *155 (4)*, 216-219.
[79]   Sachse, J. *Z.Lebensm.Unters.-Forsch.* **1973**, *153*, 294-300.
[80]   Brandl, W.; Herrmann, K. *Z.Lebensm.Unters.-Forsch.* **1984**, *178*, 192-194.
[81]   Aritomi, M.; Kawasaki, T. *Phytochemistry* **1984** *23*, 2043-2047.
[82]   Aritomi, M.; Komori, T.; Kawasaki, T. *Phytochemistry* **1986**, *25*, 231-234.
[83]   Ferreres, F.; Castañer, M.; Tomás-Barberán, F.A. *Phytochemistry* **1997**, *45*, 1701-1705.
[84]   Lindeberg, J. *Food Chem.* **1996**, *55*, 73-94.
[85]   Krause, M.; Galensa, R. *Z.Lebensm.Unters.Forsch.* **1992**, *194*, 29-32.
[86]   Leake, D.S.In *Phytochemistry of Fruit and Vegetables*; Tomás-Barberán, F.A., Robins, R.J., Eds.; Clarendon Press: Oxford, 1997; pp 287-311.
[87]   Ohta, T.; Semboku, N.; Kuchii, A.; Egashira, Y.; Sanada, H. *J.Agric.Food Chem.* **1997**, *45*, 1644-1648.
[88]   Maruta, Y.; Kawabata, J.; Niki, R. *J.Agric.Food Chem.* **1995**, *43*, 2592-2595.
[89]   Murase, H.; Nagao, A.; Terao, J. *J.Agric.Food Chem.* **1993**, *41*, 1601-1604.
[90]   Osawa, T.; Namiki, M. *J.Agric.Food Chem.* **1985**, *33*, 777-780.
[91]   Tamura, H.; Yamagami, A. *J.Agric.Food Chem.* **1994**, *42*, 1612-1615.
[92]   Frankel, E.N.; German, J.B.; Davis, P.A. *Lipids* **1997**, *27*, 1047-1051.
[93]   Kitta, K.; Hagiwara, Y.; Shibamoto, T. *J.Agric.Food Chem.* **1992**, *40*, 1843-1845.
[94]   Tsuda, T.; Watanabe, M.; Ohshima, K.; Norinobu, S.; Choi, S.W.; Kawakishi, S.; Osawa, T. *J.Agric.Food Chem.* **1994**, *42*, 2407-2410.
[95]   Miller, N.J.; Diplock, A.T.; Rice-Evans, C.A. *J.Agric.Food Chem.* **1995**, *43*, 1794-1801.
[96]   Arnao, M.B.; Cano, A.; Hernández-Ruiz, J.; García-Cánovas, F.; Acosta, M. *Anal-Biochem* **1997**, *236*, 255-261.
[97]   Bondet, V.; Brand-Williams, W.; Berset, C. *Lebensm.-Wiss.u.-Technol.* **1997**, *30*, 609-615.
[98]   Hertog, M.G.L.; van Poppel, G.; Verhoeven, D.In *Phytochemistry of Fruit and Vegetables*; Tomás-Barberán, F.A., Robins, R.J., Eds.; Clarendon Press: Oxford, 1997; pp 313-329.
[99]   Cao, G.; Sofic, E.; Prior, R.L. *J.Agric.Food Chem.* **1996**, *44*, 3426-3431.
[100] Miyake, Y.; Yamamoto, K.; Osawa, T. *J.Agric.Food Chem.* **1997**, *45*, 3738-3742.
[101] Larrauri, J.A.; Rupérez, P.; Saura-Calixto, F. *J.Agric.Food Chem.* **1997**, *45*, 1390-1393.
[102] Kanner, J.; Frankel, E.; Granit, R.; German, B.; Kinsella, J.E. *J.Agric.Food Chem.* **1994**, *42*, 64-69.
[103] Waterhouse, A.L. *Chem.Ind.* **1995**, 338-341.
[104] Simonetti, P.; Pietta, P.; Testolin, G. *J.Agric.Food Chem.* **1997**, *45*, 1152-1155.

[105]   Soleas, G.J.; Tomlison, G.; Diamandis, E.P.; Goldberg, D.M. *J.Agric.Food Chem.* **1997**, *45*, 3995-4003.

[106]   Frankel, E.N.; Waterhouse, A.L.; Teissedre, P.L. *J.Agric.Food Chem.* **1995**, *43*, 890-894.

[107]   Hurtado, I.; Caldú, P.; Gonzalo, A.; Ramon, J.M.; Mínguez, S.; Fiol, C. *J.Agric.Food Chem.* **1997**, *45*, 1283-1289.

[108]   Huang, S.W.; Frankel, E.N. *J.Agric.Food Chem.* **1997**, *45*, 3033-3038.

[109]   Vekiari, S.A.; Oreopoulou, V.; Tzia, C.; Thomopoulos, C.D. *J.Amer.Oil Chem.Soc* **1993**, *70*, 483-487.

[110]   Miura, K.; Nakatani, N. *Agric.Biol.Chem.* **1989**, *53*, 3043-3045.

[111]   Lyckander, I.M.; Malterud, K.E. *Acta Pharm.Nord.* **1992**, *4*, 159-166.

[112]   Boyd, R.; Green, D.P.; Giesbrecht, F.B.; King, M.F. *J Sci Food Agric* **1993**, *61*, 87-93.

[113]   Larrauri, J.A.; Rupérez, P.; Saura-Calixto, F. *J.Agric.Food Chem.* **1997**, *45*, 4028-4031.

[114]   Ramarathnam, N.; Osawa, T.; Namiki, M.; Kawakishi, S. *J.Agric.Food Chem.* **1989**, *37*, 316-319.

[115]   Ramanathan, L.; Das, N.P. *J.Food Sci.* **1993**, *58*, 318-320.

[116]   Shahidi, F.; Zheng, Y.; Saleemi, Z.O. *J.Food Lipids* **1993**, *1*, 69-78.

[117]   Dziedzic, S.Z.; Hudson, B.J.F.; Barbers, G. *J.Agric.Food Chem.* **1985**, *33*, 244-246.

[118]   He, Y.; Shahidi, F. *J.Agric.Food Chem.* **1997**, *45*, 4262-4266.

[119]   Roedig-Penman, A.; Gordon, M.H. *J.Agric.Food Chem.* **1997**, *45*, 4267-4270.

[120]   Aruoma, O.I.; Murcia, A.; Butler, J.; Halliwell, B. *J.Agric.Food Chem.* **1993**, *41*, 1880-1885.

[121]   Osawa, T.; Ide, A.; Su, J.D.; Namiki, M. *J.Agric.Food Chem.* **1987**, *35*, 808-812.

[122]   Chen, J.H.; Ho, C.T. *J.Agric.Food Chem.* **1997**, *45*, 2374-2378.

[123]   Chuda, Y.; Ono, H.; Ohnishi-Kameyama, M.; Nagata, T.; Tsushida, T. *J.Agric.Food Chem.* **1996**, *44*, 2037-2039.

[124]   Igile, G.O.; Oleszek, W.; Jurzysta, M.; Burda, S.; Fafunso, M.; Fasanmade, A.A. *J.Agric.Food Chem.* **1994**, *42*, 2445-2448.

[125]   Kitta, K.; Hagiwara, Y.; Shibamoto, T. *J.Agric.Food Chem.* **1992**, *40*, 1843-1845.

[126]   Miyake, Y.; Yamamoto, K.; Morimitsu, Y.; Osawa, T. *J.Agric.Food Chem.* **1997**, *45*, 4619-4623.

[127]   Igarash, K.; Yoshida, T.; Suzuki, F. *J.Jap.Soc.Food Sci.Technol.* **1993**, *40*, 138-143.

[128]   Tsuda, T.; Ohshima, K.; Kawakishi, S.; Osawa, T. *J.Agric.Food Chem.* **1994**, *42*, 248-251.

[129]   Wiseman, S.; Mathot, J.; de Fouw, N.; Tijburg, L. *Atherosclerosis* **1996**, *120*, 15-23.

[130]   Vinson, J.A.; Jang, J.; Dabbagh, Y.A.; Serry, M.M.; Cai, S. *J.Agric.Food Chem.* **1995**, *43*, 2798-2799.

[131]   Fauré, M.; Lissi, E.; Torres, R.; Videla, L.A. *Phytochemistry* **1990**, *29*, 3773-3775.

[132]   Brand-Williams, W.; Cuvelier, M.E.; Berset, C. *Food Sci.Technol.(London)* **1995**, *28*, 25-30.

[133]   Cao, G.; Sofic, E.; Prior, R.L. *Free Radical Biol.Med.* **1997**, *22*, 749-760.

[134]   Chen, Z.; Chan, P.; Ho- KY; Fung, K.; Wang, J. *Chem-Phys-Lipids* **1996**, *79*, 157-163.

[135]   Nagai, N.; Nakai, A.; Nagata, K. *Biochem-Biophys-Res-Commun* **1995**, *208*, 1099-1105.

[136]   Rice-Evans, C.A.; Miller, N.J.; Paganga, G. *Free Radical Biol.Med.* **1996**, *20*, 933-956.

[1]   Hollman, P. C. H. Determinants of the absorption of the dietary flavonoid quercetin in man. 1-187. 1997. Landbouwuniversiteit Wageningen. Thesis.

[2]   Hertog, M.G.L.; Kromhout, D.; Aravanis, C.; Blackburn, H.; Buzina, R. *Arch.Intern.Med.* **1995**, *155*, 381-386.

[137]   Hertog, M.G.L.; Feskens, E.J.M.; Hollman, P.C.H.; Katan, M.B.; Kromhout, D. *The Lancet* **1993**, *342*, 1007-1011.

[138]   Rimm, E.B.; Katan, M.B.; Ascherio, A.; Stampfer, M.J.; Willett, W.C. *Ann.Intern.Med.* **1996**, *125*, 384-389.

[139]   Hertog, M.; Feskens, E.; Hollman, P.; Katan, M.; Kromhout, D. *Lancet.* **1993**, *342*, 1007-1011.

[140]   Knekt, P.; Järvinen, R.; Reunanen, A.; Maatela, J. *Br.Med.J.* **1996**, *312*, 478-481.

[141]   Serafini, M.; Ghiselli, A.; Ferro-Luzzi, A. *Eur,J.Clin.Nutr.* **1996**, *50*, 28-32.

[142]   Keli, S.O.; Hertog, M.G.L.; Feskens, E.J.M.; Kromhout, D. *Arch.Intern.Med.* **1996**, *156*, 637-642.

[143]   Griffiths, L.A.; Barrow, A. *Biochem J* **1972**, *130*, 1161-1162.

[144]   Hollman, P.C.; de Vries, J.H.M.; van Leeuwen, S.D.; Mengelers, M.J.B.; Katan, M.B. *Am.J.Clin.Nutr.* **1995**, *62*, 1276-1282.

[145]   Hackett, A.M.; Marsch, I.; Barrow, A.; Griffiths, L.A. *Xenobiotica* **1979**, *9*, 491-501.

[146]   Okushio, K.; Matsumoto, N.; Kohri, T.; zuki, M.; njo, F.; ara, Y. *Biol.Pharm.Bull.* **1996**, *19*, 326-329.

[147]   Lee, M.J.; Wang, Z.Y.; Li, H.; Chen, L.; Sun, Y.; Gobbo, S.; Balantine, D.A.; Yand, C.S. *Cancer Epidemiol.Biomark.Prev.* **1995**, *4*, 393-399.

[148]   Lembke, B.; Kinawi, A.; Wurm, G. *Arch.Pharm.* **1994**, *327*, 467-468.

[149]   Hollman, P.C.H.; Katan, M.B.In *Flavonoids in health and disease*; Rice-Evans, C.A., Packer, L., Eds.; Marcel Dekker Inc.: New York, 1997; pp in press

[150]   Lattanzio, V.; Van Sumere, C.F. *Food Chem.* **1987**, *24*, 37-50.

[151]   Amiot, M.J.; Tacchini, M.; Aubert, S.; Oleszek, W. *J.Agric.Food Chem.* **1995**, *43*, 1132-1137.

[152]   Blakenship, S.M.; Richardson, D.G. *J.Amer.Soc.Hort.Sci.* **1985**, *110*, 336-339.

[153]   Pérez-Ilzarbe, J.; Hernandez, T.; Estrella, I.; Vendrell, M. *Z.Lebensm.Unters.-Forsch.* **1997**, *204*, 52-55.

[154]   Coseteng, M.Y.; Lee, C.Y. *J.Food Sci.* **1987**, *52*, 985-989.

[155]   Golding, J.B.; McGlasson, W.B.; Leach, D.N.; Wyllie, S.G. *Proceedings of the ISHS Postharvest 96* , **1996**.

[156]   Piretti, M.V.; Gallerani, G.; Brodnik, U. *Postharvest Biology and Technology* **1996**, *8*, 11-18.

[157]   Abdalla, A.Y.; Gil, M.I.; Biasi, W.; Mitchan, E.J. *Postharvest Biology and Technology* **1997**,

[158]   Kalt, W.; McDonald, J.E. *J.Amer.Soc.Hort.Sci.* **1996**, *121*, 142-146.

[159]   Holcroft, D.M.; Gil, M.I.; Kader, A.A. *J.Amer.Soc.Hort.Sci.* **1997**,

[160]   Kader, A.A. *Food Technol.* **1986**, *40*, 99-100-102-104.

[161]   Lin, T.Y.; Koehler, P.E.; Shewfelt, R.L. *J.Food Sci.* **1989**, *54*, 405-407.

[162] Siriphanich, J.; Kader, A.A. *J.Amer.Soc.Hort.Sci.* **1985**, *110*, 333-335.

[163] Ke, D.; Saltveit, M.E. *J.Amer.Soc.Hort.Sci.* **1989**, *114*, 789-794.

[164] Leja, M.; Mareczek, A.; Rozek, S. *Folia Horticulturae* **1996**, *8*, 83-93.

[165] Saltveit, M.E.In *Phytochemistry of Fruit and Vegetables*; Tomás-Barberán,
       F.A., Robins, R.J., Eds.; Oxford University Press: Oford, 1997; pp 205-220.

[166] Howard, L.R.; Griffin, L.E. *J.Food Sci.* **1993**, *58*, 1065-1067,1072.

[167] Tomás-Barberán, F.A.; Loaiza-Velarde, J.; Bonfanti, A.; Saltveit, M.E.
       J.Amer.Soc.Hort.Sci. **1997**, *122*, 399-404.

[168] Ke, D.; Saltveit, M.E. *Physilogia Plantarum* **1989**, *76*, 412-418.

[169] Mateos, M.; Ke, D.; Cantwell, M.; Kader, A.A. *Postharvest Biology and
       Technology* **1993**, *3*, 225-233.

[170] Ismail, M.A.; Brown, G.E. *J.Amer.Soc.Hort.Sci.* **1975**, *100*, 249-251.

[171] Weller, A.; Sims, C.A.; Matthews, R.F.; Bates, R.P.; Brecht, J.K. *J.Food Sci.*
       1997, *62*, 256-260.

[172] Hyodo, H.; Kuroda, H.; Yang, S.F. *Plant Physiol.* **1978**, *62*, 31-35.

[173] Ke, D.; Saltveit, M.E. *Plant Physiol.* **1988**, *88*, 1136-1140.

[174] Kalt, W.; Prange, R.K.; Lidster, P.D. *Canadian J.Plant Sci.* **1993**, *73*, 541-548.

[175] Miszczak, A.; Forney, C.F.; Prange, R.K. *J.Amer.Soc.Hort.Sci.* **1995**, *120*, 650-655.

[176] Curry, E.A. *J.Hort.Sci.* **1997**, *72*, 723-729.

[177] Dong, Y.H.; Mitra, D.; Koostra, A.; Lister, C.E.; Lancaster, J.E.
       J.Amer.Soc.Hort.Sci. **1997**,

[178] Friedman, M. *J.Agric.Food Chem.* **1997**, *45*, 1523-1540.

[179] Spanos, G.A.; Wrolstad, R.E.; Heatherbell, D.A. *J.Agric.Food Chem.* **1990**, *38*,
       1572-1579.

[180] Spanos, G.A.; Wrolstad, R.E. *J.Agric.Food Chem.* **1990**, *38*, 817-824.

[181] Spanos, G.A.; Wrolstad, R.E. *J.Agric.Food Chem.* **1990**, *38*, 1565-1571.

[182] Spanos, G.A.; Wrolstad, R.E. *J.Agric.Food Chem.* **1992**, *40*, 1478-1487.

[183] Bakker, J.; Bridle, P.; Koopman, A. *J.Sci.Food Agric.* **1992**, *60*, 471-476.

[184] Bakker, J.; Bridle, P. *J.Sci.Food Agric.* **1992**, *60*, 477-481.

[185] Auw, J.M.; Blanco, V.; O'Keefe, S.F.; Sims, C.A. *Am.J.Enol.Vitic.* **1996**, *47*,
       279-286.

[186] Miller, H.E. *J.Amer.Oil Chem.Soc* **1971**, *48*, 91

Atta-ur-Rahman (Ed.) *Studies in Natural Products Chemistry, Vol. 23*

PHYTOCHEMISTRY OF THE ZINGIBERACEAE

ORASA PANCHAROEN[1], UMA PRAWAT[2] *and*
PITTAYA TUNTIWACHWUTTIKUL[1]*

1. Department of Chemistry, Silpakorn University, Nakorn Pathom 73000, Thailand.
2. Department of Chemistry, Rajabhat Institute Phuket, Phuket 83000, Thailand.

ABSTRACT: Zingiberaceae is one of the major tropical plant families. Members of the Zingiberaceae are famous for their use as spices and as medicinal herbs. Rhizomes of several species are also used as insect repellents. Many compounds with novel structures and a large number of biologically active compounds have been identified from these plants. This review focuses on the chemistry and biological activity of compounds isolated from the Zingiberaceous plants.

INTRODUCTION

The Zingiberaceae is a large family of perennial herbaceous plants. The family contains approximately 1400 species in 47 genera [1]. The species are widely distributed throughout India, tropical Asian and extend to northern Australia. Zingiberaceous plants have been used in traditional or herbal medicine throughout the tropical world. Members of the family have attracted continuous phytochemical interest due to their culinary uses and their biological and pharmaceutical activities.

This paper covers the chemistry and biological activity of compounds isolated from nine genera: *Aframomum, Alpinia, Amomum, Boesenbergia, Costus, Curcuma, Hedychium, Kaempferia and Zingiber.*

Aframomum

The essential oil of the seeds of *Aframomum mala* was evaluated by GLC, which revealed that the constituents were mainly the monoterpenes, α-pinene, β-pinene, sabinene, α-phellandrene, limonene and 1,8-cineol [2].

The composition of the essential oil of the stems of *A. giganteum* was investigated by GC-MS [3]. Forty seven terpenes were identified. Linalool and linalyl acetate were the main components and these were responsible for the distinctive smell of the stems. From the stems of *A. giganteum*, three rare naturally occurring flavonol 3-methyl ethers were isolated together with chrysophanol, physcion, 2,6-dimethoxybenzoquinone and β-sitosterol [4]. The flavonol methyl ethers were identified as kaempferol

3,7,4'-trimethyl ether (**1**), quercetin 3,7,4'-trimethyl ether (ayanin, **2**) and quercetin 3,7,3',4'-tetramethyl ether (retusine, **3**). Further investigation of the chloroform extract of the stems of the same plant yielded emodin, syringaldehyde, syringic acid and dehydrozingerone (**4**) [5].

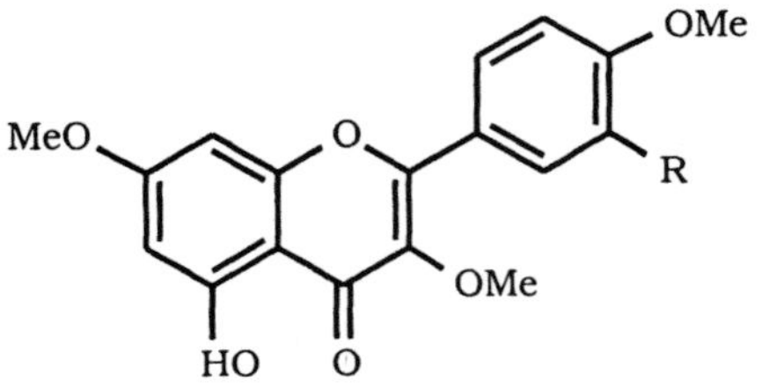

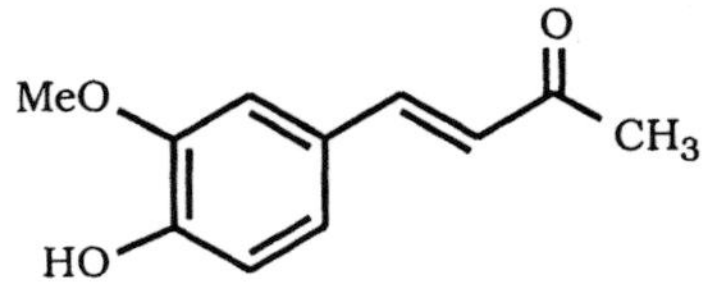

1 R = H
2 R = OH
3 R = OMe

Two 3-acetylated 3-hydroxyflavanones, 2*R*,3*R*-3-acetoxy-4',5-dihydroxy-7-methoxyflavanone (**5**) and 2*R*,3*R*-3-acetoxy-4',5,7-trihydroxyflavanone (**6**) were isolated together with 4',7-di-*O*-methylaromadendrin (**7**) and nerolidol (**8**) from the seeds of *A. pruinosum* [6].

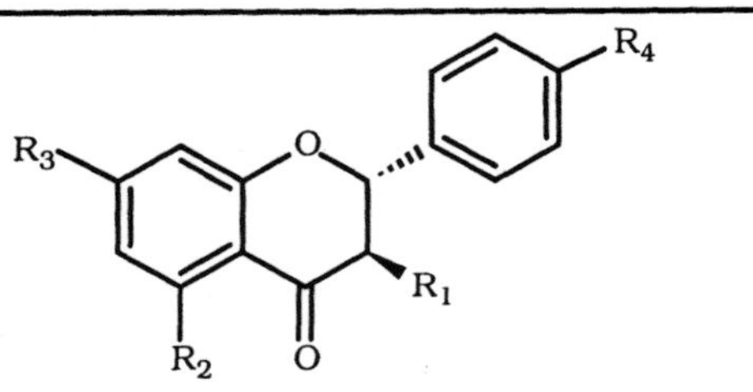

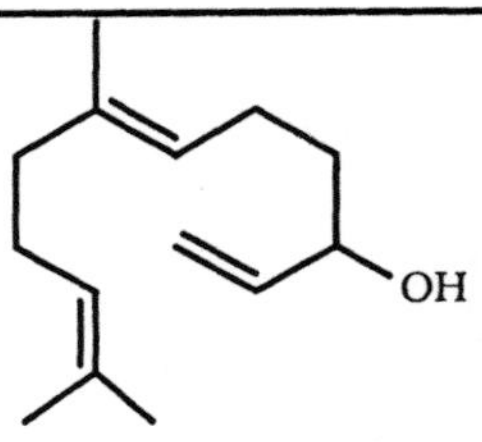

5 R$_1$ = OAc, R$_2$ = R$_4$ = OH, R$_3$ = OMe
6 R$_1$ = OAc, R$_2$ = R$_3$ = R$_4$ = OH
7 R$_1$ = R$_2$ = OH, R$_3$ = R$_4$ = OMe

From the hexane extract of the seeds of *A. daniellii*, a diterpenoid dialdehyde was isolated and its structure was assigned as (*E*)-8β,17-epoxylabd-12-ene-15,16-dial (**9**) by spectroscopic evidences and by correlation with *cis*-12-norambreinolide (**10**) [7]. Compound **9** exhibited potent anti-fungal [8, 9] and anti-hypercholesterolemic properties [10]. On further examination of the same extract, two minor compounds were isolated and identified as (*E*)-8(17),12-labdadiene-15,16-dial (**11**) and methyl (*E*)-14ξ,15-epoxy 8(17),12-labdadiene-16-oate (**12**) [11].

9

10

11

12

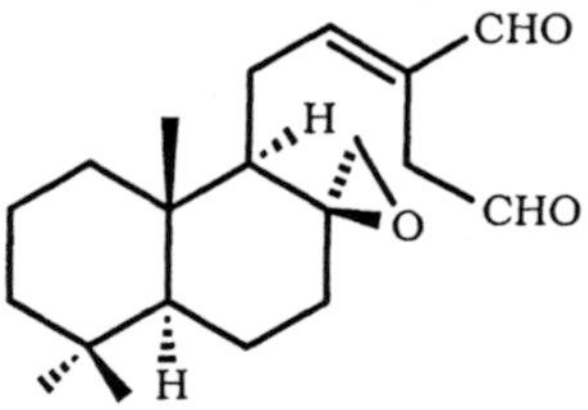

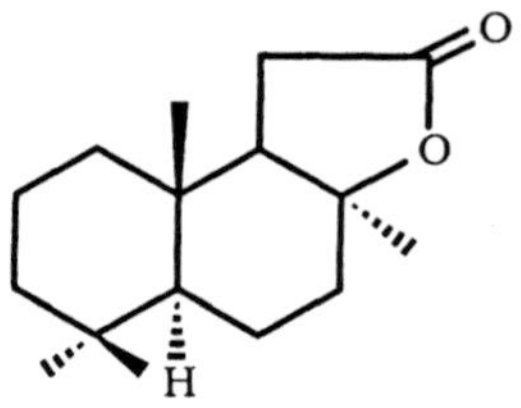

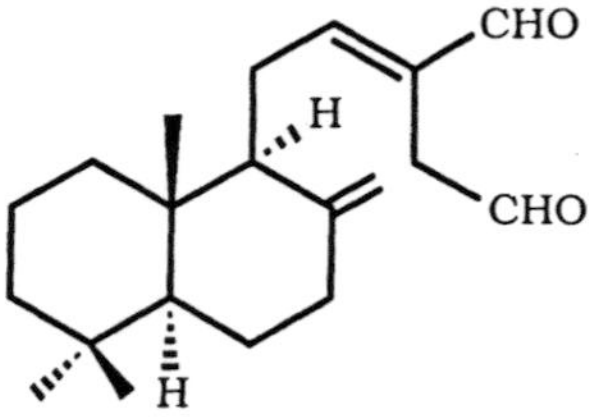

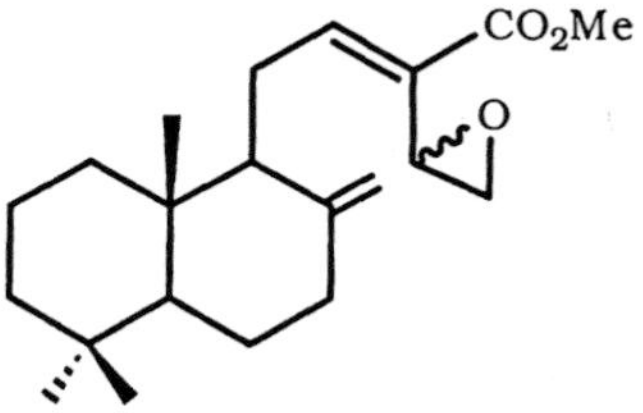

Investigation of the seeds of *A. aulacocarpos* yielded three novel diterpenoids, aulacocarpinolide (**13**), aulacocarpin A (**14**) and aulacocarpin

13

14 R = H
15 R = OH

16

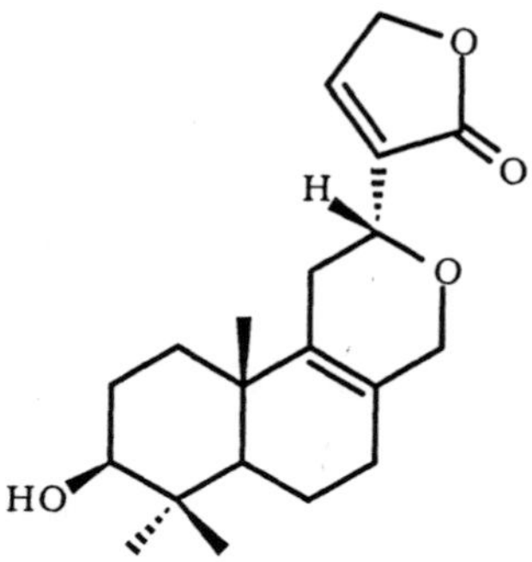

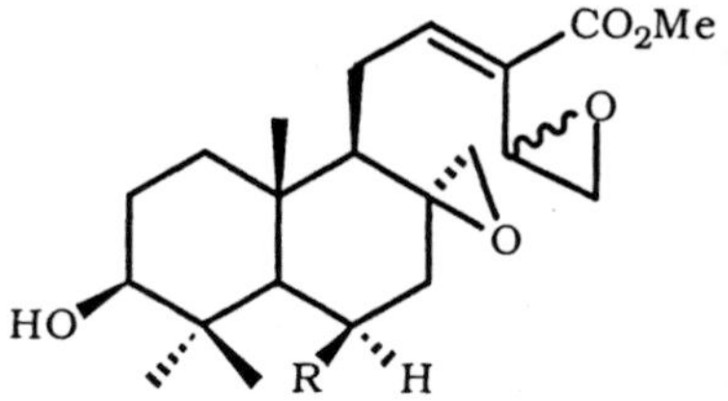

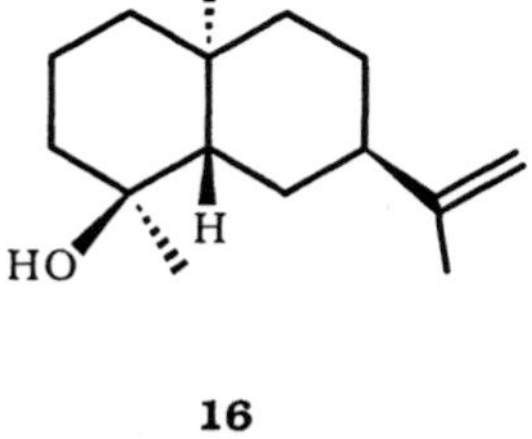

B (**15**), in addition to compound **9** [12]. Compounds **13-15** showed weak anti-microbial and cytotoxic activities. The biological activity of **9** was briefly reviewed and ten other Cameroonian *Aframomum* species were investigated as potentially better sources for **9** [13].

From the rhizomes of *A. abloviolaceum* a diterpene, (+)-intermedeol (**16**), was isolated together with compound **11** [14].

Alpinia

The rhizomes of *Alpinia galanga* are commonly used for flavouring foods in Southeast Asia. Hot water extract of the rhizomes of *A. galanga* is drunk for bronchitis [15] and also for diabetes [16] in India. The rhizomes boiled with *Curcuma* tubers in dilute vinegar and the vinegar extract is drunk for postpartum purification in Indonesia [17]. In Saudi Arabia a hot water extract of the rhizomes is used for dyspepsia, gasteralgia, chronic enteritis and anti-nausea [18] and is also used for kidney stones [19]. The rhizome macerated in vinegar (acetic acid 2%) is used externally as an anti-inflammatory agent by applying to inflamed areas [20].

Different preparations of extracts of the rhizomes of *A. galanga* have been found to have various biological activities. An ethanolic extract of the rhizomes had anti-ascariasis activity [21]. An aqueous ethanolic (50%) extract of the rhizomes showed hypotensive and hypothermic activities [22]; the extract also inhibited prostaglandin synthetase [23]. An ether extract of the commercial sample of rhizomes in Egypt possessed anti-bacterial activity [24] and the essential oil of the rhizomes had anti-bacterial and anti-fungal activities [25, 15]. A methanolic extract of the rhizomes had anti-tumor activity [26]. Anti-ulcer activity and gastric secretory inhibition were found in the ethanolic extract of the rhizomes [18]. The methanolic extract of the seeds of *A. galanga* also showed anti-ulcer activity in male rats [27]. Acute (24 days) and chronic (90 days) oral toxicity studies on the ethanolic extract of the rhizomes of *A. galanga* were carried out in mice [28].

The essential oils of the fresh and dried rhizomes of *A. galanga* from Malaysia were analyzed by capillary GC and GC-MS [29]. Forty components were identified. Apart from common monoterpenes and sesquiterpenes, methyleugenol, eugenol acetate, chavicol and chavicol acetate were present [29]. From the methanolic extract of the seeds of *A. galanga* which showed significant inhibitory activity against Shay ulcer in rats, two potent anti-ulcer constituents, 1'-acetoxychavicol acetate (**17**) and 1'-acetoxyeugenol acetate (**18**), were isolated, together with three sesquiterpenes, caryophyllene oxide (**28**), caryophyllenol-I (**29**) and caryophyllenol-II (**30**) [27]. From the hexane-soluble fraction of the alcoholic extract of the rhizomes of *A. galanga*, compounds **17** and **18** were isolated as the anti-tumor principles against Sarcoma 180 ascites in mice; **17** and **18** were also the main pungent principles of the rhizome [26].

Three other compounds with related structures **21, 22** and **25** were also obtained from the chloroform-soluble fraction of the same extract [26]. Compounds **17, 18,** 1'-hydroxychavicol acetate (**19**) and 1'-hydroxychavicol (**20**) were synthesized.The anti-tumor effect against Sarcoma 180 of compounds **17-22** and **25** was evaluated and the results showed that compounds **19-22** and **25** did not exhibit anti-tumor activity [26]. A chloroform extract of the dried rhizomes of *A. galanga* yielded five xanthine oxidase inhibitors; they were identified as compounds **17, 18,** *trans-p*-coumaryl diacetate (**23**), *trans*-coniferyl diacetate (**24**) and *p*-hydroxybenzaldehyde [30]. The xanthine oxidase inhibition effect of **17** and **23** was found to be uncompetitive [30]. Two phenolic derivatives, *p*-hydroxycinnamaldehyde (**25**) and [di-(*p*-hydroxy-*cis*-styryl)]methane (**26**), were isolated from the chloroform extract of the rhizomes of *A. galanga* and compound **26** was characterized as its diacetate (**27**) [31]. Galangin and methyl galangin were found in the roots of *A. galanga* [32]. In further examination on the constituents of the seeds of *A. galanga*, two diterpenes with new skeletons named galanal A (**31**) and galanal B (**32**) [33, 34], a new labdane diterpene named galanolactone (**33**) [34] and (*E*)-8(17),12-labdadiene-15,16-dial (**11**) [33, 34] and (*E*)-8b(17)-epoxylabd-12-ene-15,16-dial (**9**) [33, 34] were isolated from the hexane-soluble fraction of the methanolic extract of the seeds. The anti-fungal and cytotoxic activities of compounds **9, 11** and **31-33** were evaluated [34]. Compounds **9** and **11** exhibited stronger anti-fungal activities than compounds **31-33** and galanal A (**31**) and B (**32**) exhibited the strongest cytotoxic activity [34].

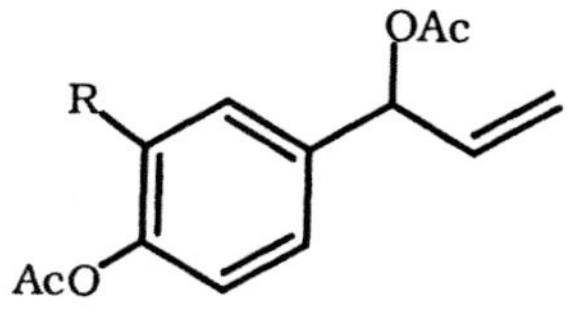

17 R = H
18 R = OMe

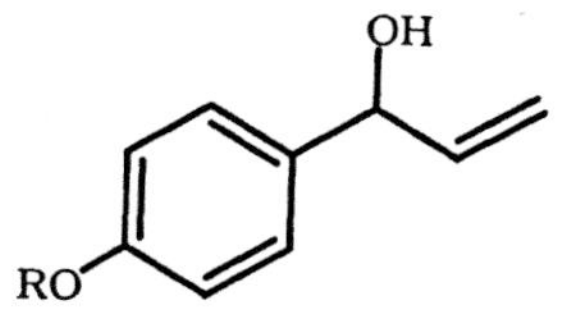

19 R = Ac
20 R = H

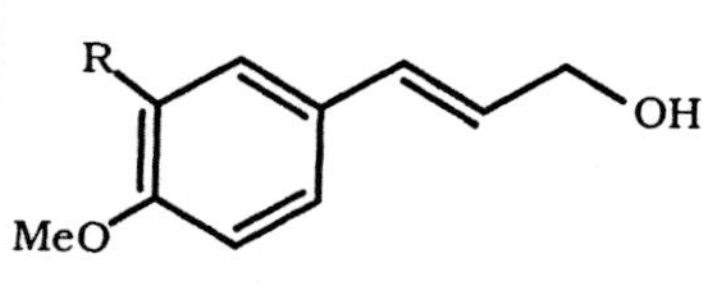

21 R = OMe
22 R = H

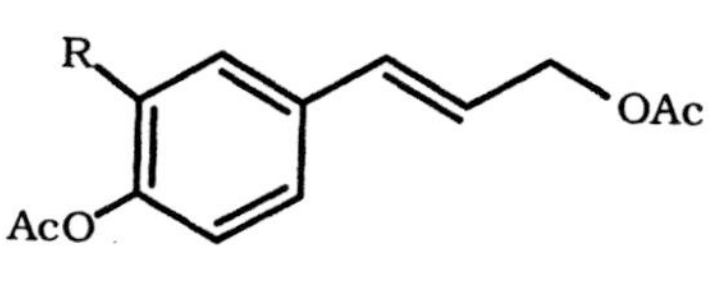

23 R = H
24 R = OMe

25

26 R = H
27 R = Ac

28

29

30

31

32

33

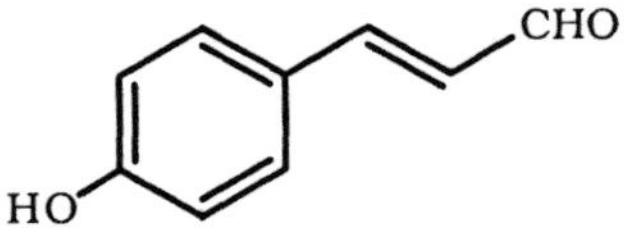

Essential oil from the seeds of *A. katsumadai*, a Chinese drug "caodoukou", was analyzed by GLC and 22 components were detected [35]. Three main constituents were isolated and identified as 1,8-cineol, α-humulene and *trans,trans*-farnesol. From the seed residue, after removal of the essential oil by steam, alpinetin and cardamonin were isolated [36]. Together with *trans,trans*-farnesol, *trans*-cinnamaldehyde and three flavonoids, alpinetin, cardamonin and pinocembrin, six diarylheptanoids **34-39** were isolated from the seeds of *A. katsumadai* and identified on the basis of chemical and spectral evidence [37].

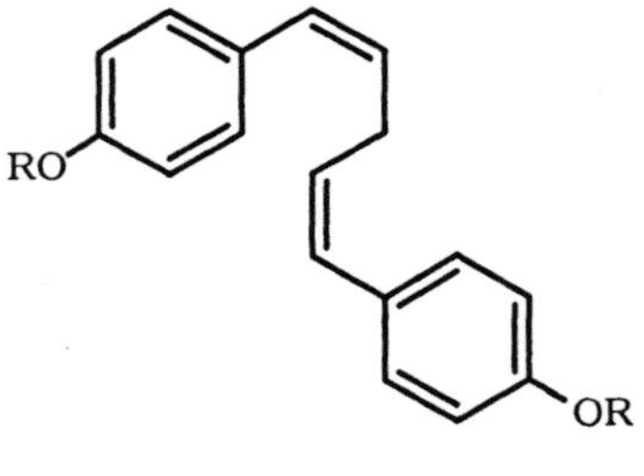

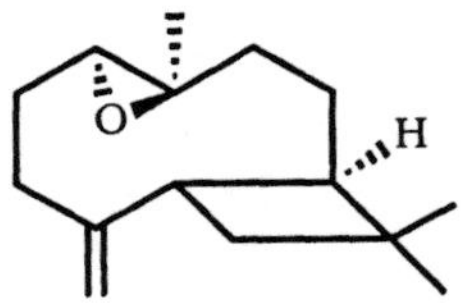

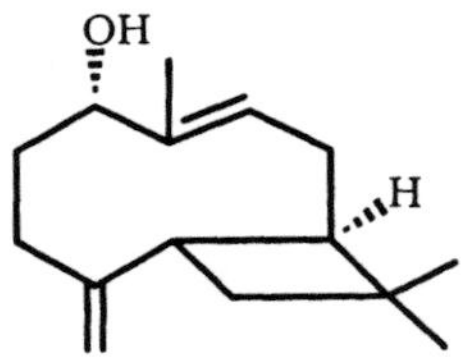

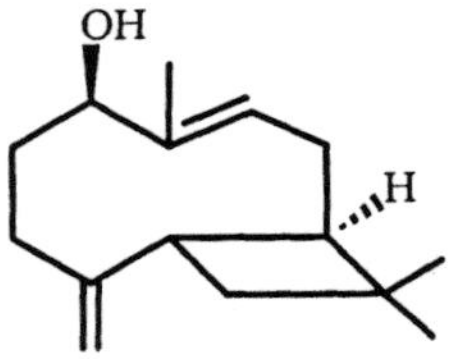

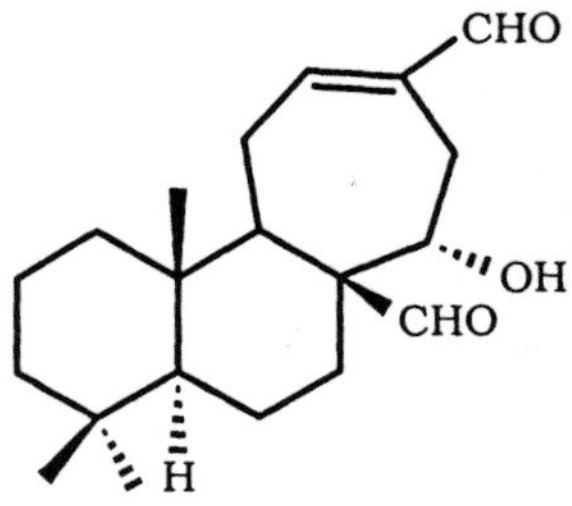

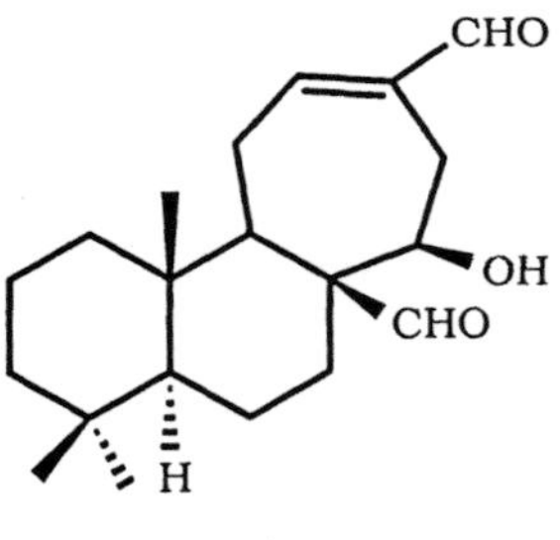

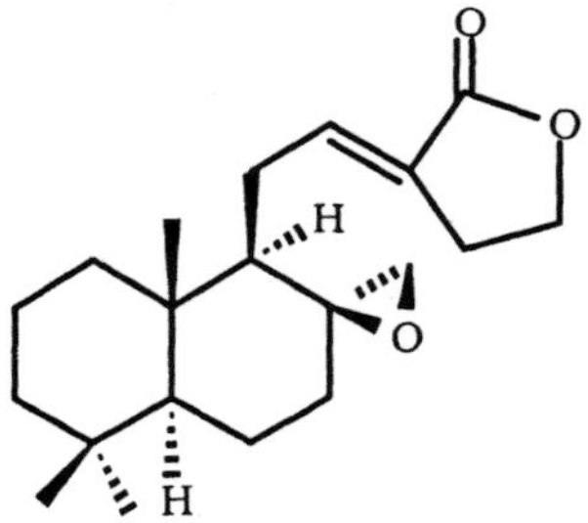

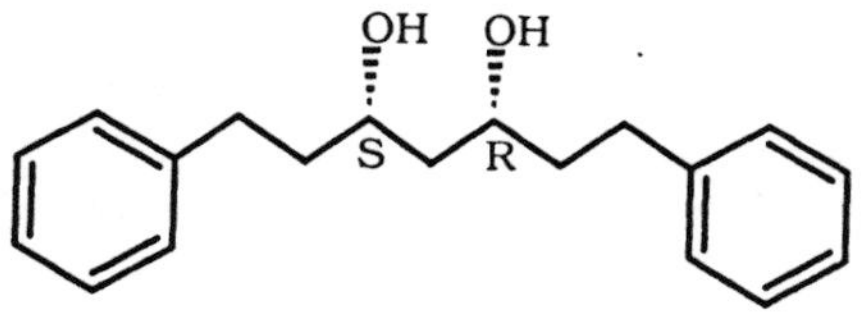

Four diarylheptanoids were isolated from the methanolic extract of the rhizomes of *A. officinarum* and identified on the basis of their spectral data and by chemical correlations as 1,7-diphenylhept-4-en-3-one (**40**), 7-(4"-hydroxy-3"-methoxyphenyl)-1-phenylhept-4-en-3-one (**41**), 1,7-diphenyl-5-hydroxy-3-heptanone (**43**) and 5-hydroxy-7-(4"-hydroxy-3"-methoxyphenyl)-1-phenyl-3-heptanone (**44**) [38]. From the chloroform-soluble fraction of the methanolic extract of the rhizomes of *A. officinarum*, three new diarylheptanoids, 5-methoxy-7-(4"-hydroxy-3"-methoxyphenyl)-1-phenyl-3-heptanone (**46**), 5-hydroxy-7-(4"-hydroxyphenyl)-1-phenyl-3-heptanone (**45**) and 7-(4"-hydroxy-3"-methoxyphenyl)-1-phenyl-3,5-heptanedione (**49**) were isolated together with **41, 43** and **44** [39]. Compound **44** was the major constituent of the chloroform-soluble fraction. Compounds **41, 43-46** and **49** showed inhibitory effect against PG synthetase and IC_{50} values of the compounds were reported [39]. In addition to compounds **43, 44** and **52**, three new diarylheptanoids, **42, 47** and **48** were isolated from the rhizomes of *A. officinarum* and the structures were elucidated on the basis of the spectroscopic data [40]. The configuration of C-5 of **43** and **52** was determined to be *R* and that of hexahydrocurcumin was *S* [40].

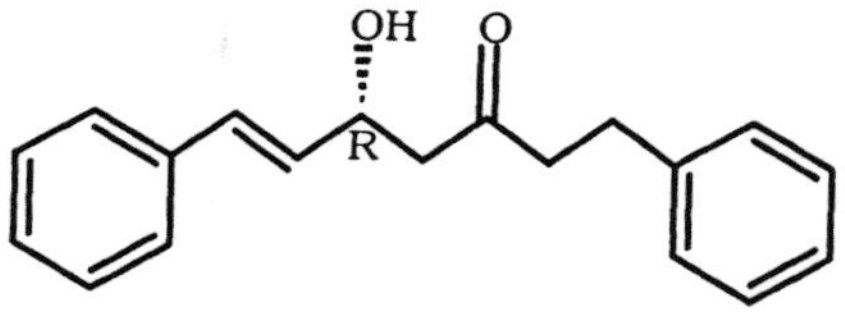

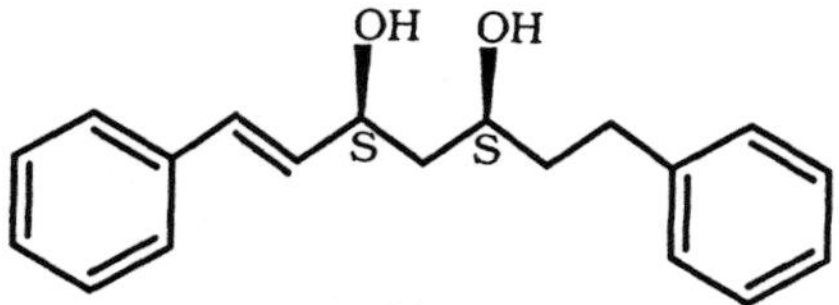

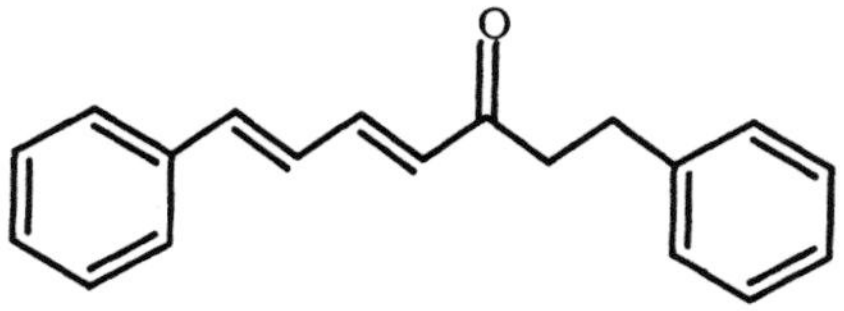

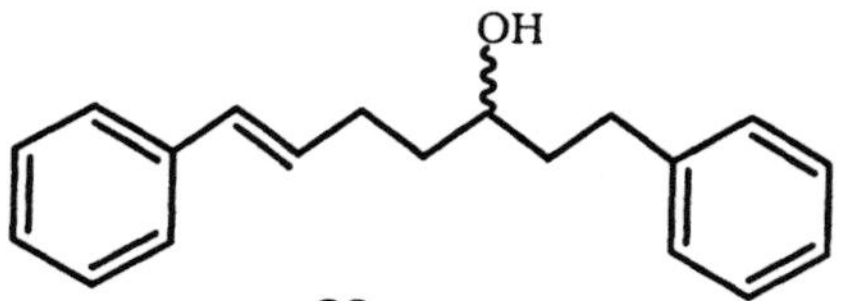

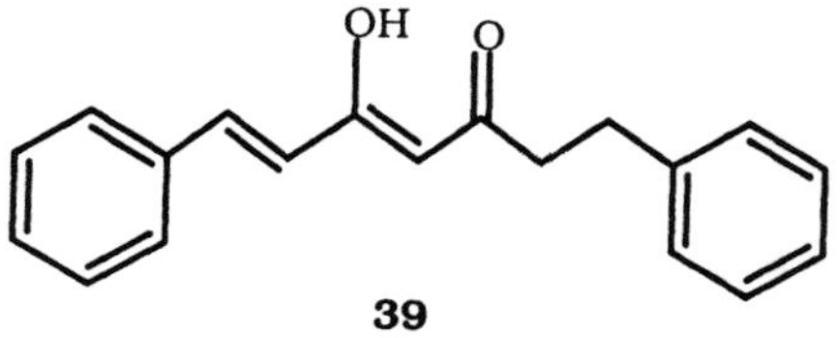

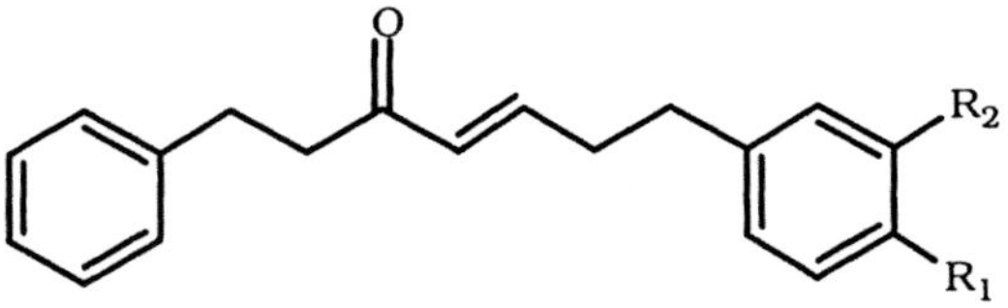

40 $R_1 = R_2 = H$
41 $R_1 = OH$, $R_2 = OMe$
42 $R_1 = OH$, $R_2 = H$

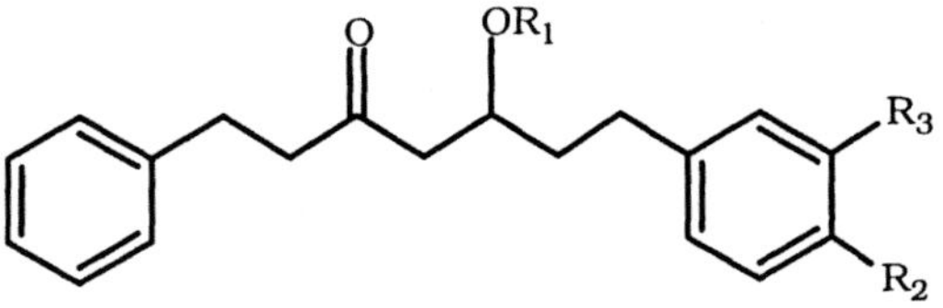

43 $R_1 = R_2 = R_3 = H$
44 $R_1 = H$, $R_2 = OH$, $R_3 = OMe$
45 $R_1 = R_3 = H$, $R_2 = OH$
46 $R_1 = Me$, $R_2 = OH$, $R_3 = OMe$
47 $R_1 = Me$, $R_2 = R_3 = H$
48 $R_1 = Me$, $R_2 = OH$, $R_3 = H$

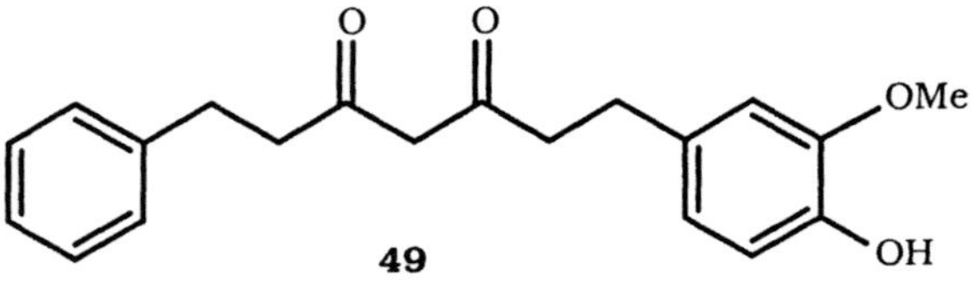

49

From the fruits of *A. oxyphylla*, two pungent priciples were isolated and identified as 1-(4'-hydroxy-3'-methoxyphenyl)-7-phenyl-3-heptanone (yakuchinone A, **50**) [41] and *trans*-1-(4'-hydroxy-3'-methoxyphenyl)-7-phenylhept-1-en-3-one (yakuchinone B, **51**) [42] on the basis of spectroscopic methods and synthesis. The pungency of **50** was found to be 125 times stronger than that of zingerone and that of **51** was about 5 times weaker than that of **50** [42]. Compounds **50** and **51** also showed inhibitory effect against PG biosynthesis with **50** being the most potent [39]. A sesquiterpene, nootkatol (**53**), possessing calcium antagonistic activity, was isolated from the seeds of *A. oxyphylla* [43].

Essential oil obtained from the fresh rhizomes of *A. conchigera* showed the presence of twelve terpenoids by GC-MS analysis, and the major component was chavicol acetate [44]. 1-Hydroxychavicol acetate [45], 4-acetoxycinnamyl alcohol and 4-acetoxycinnamyl acetate [46] were isolated from the aqueous layer obtained from the steam distillation of the rhizomes. Phenylpropanoid derivatives, chavicol acetate and eugenol acetate were present in the fruit of the plant [47]. In addition to four known diarylheptanoids **40, 41, 43** and **44**, and two known flavonoids, 3,5,7-trihydroxyflavone and 3,5,7-trihydroxy-4'-methoxyflavone, a new diarylheptanoid **54** was isolated from the rhizomes of the same plant [48].

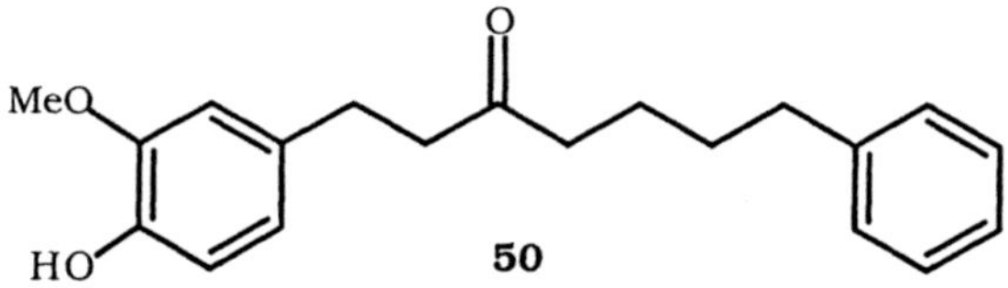

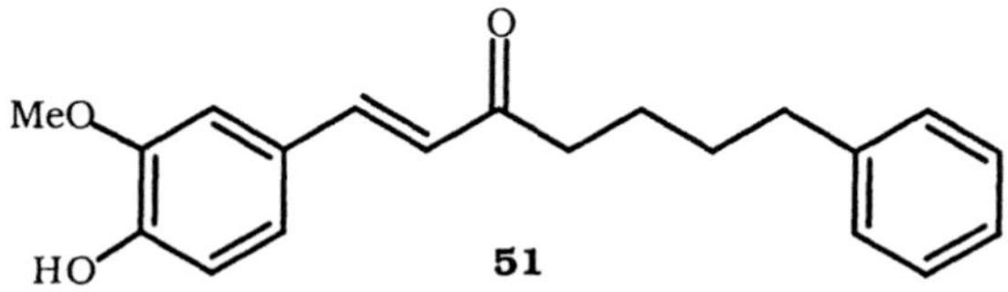

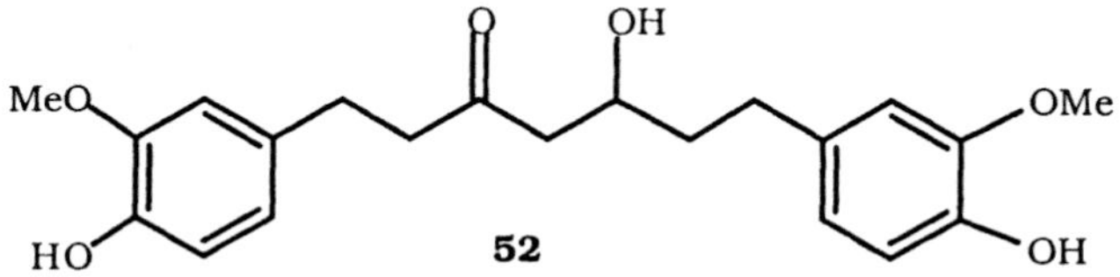

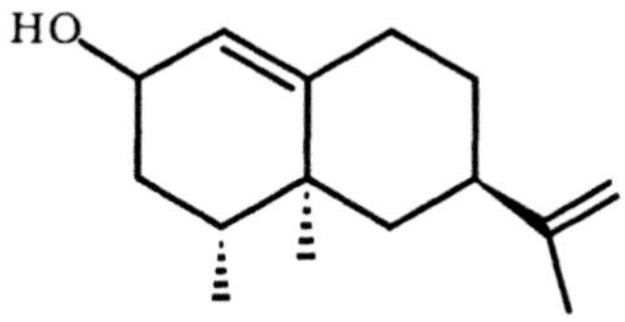

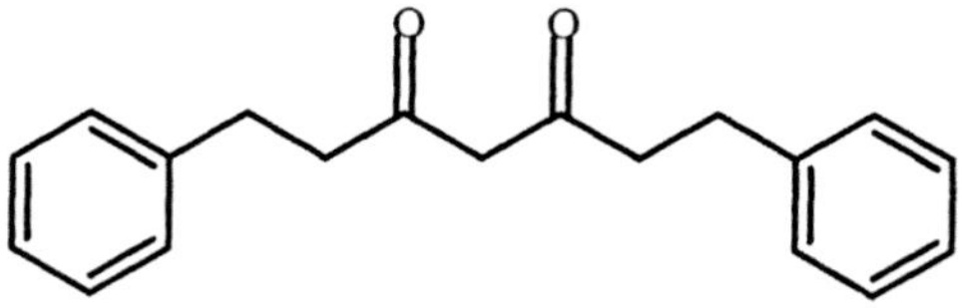

A. blepharocalyx has been used as a stomachic in South-west China [49]. Three novel diarylheptanoids, calyxin A (**55**), calyxin B (**56**) and epi-calyxin B (**57**) were isolated from ethanolic extract of the seeds of *A. blepharocalyx* and their structures were determined by 2D NMR techniques and chemical analysis [49]. These compounds showed a mild 3α-dehydrogenase (3α-HSD) inhibitory activity [49]. Further examination of the same extract yielded two additional diarylheptanoids, blepharocalyxin A (**58**) and blepharocalyxin B (**59**) [50]. Compounds **58** and **59** inhibited nitric oxide production in endotoxin-activated murine macrophages, J774.1 [50]. Six diarylheptanoids bearing a chalcone or flavanone moiety, calyxin B (60), epi-calyxin B (61), calyxin C (62), epi-calyxin C (63), calyxin D (64) and epi-calyxin D (65) were isolated from the same source [51]. Calyxin B and epi-calyxin B were previously identified as structures 56 and 57, the stereoisomers at C-3 [49]. Close examination on their absolute configurations led to the conclusion that they were stereoisomers at C-7. Calyxin B and epi-calyxin B were then assigned as structures 60 and 61, respectively [51]. These compounds, 60-65 inhibited nitric oxide production in endotoxin-activated murine macrophages, J774.1 [51].

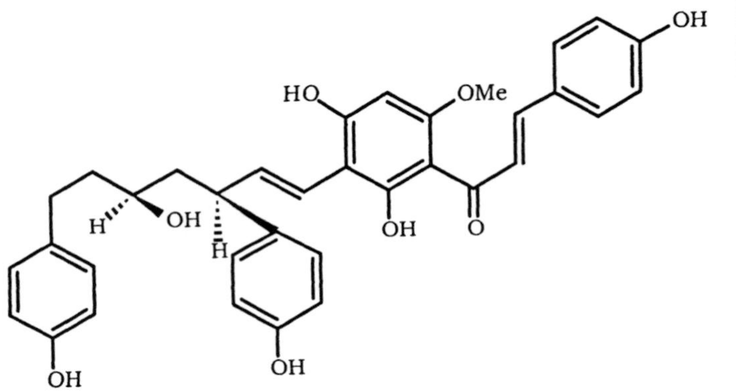

55

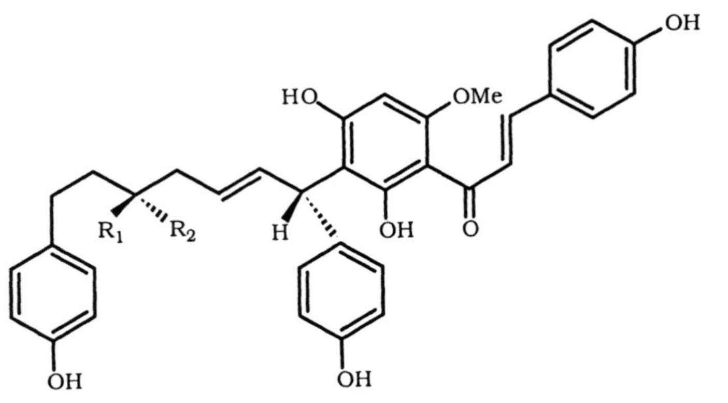

56 R_1 = H, R_2 = OH
57 R_1 = OH, R_2 = H

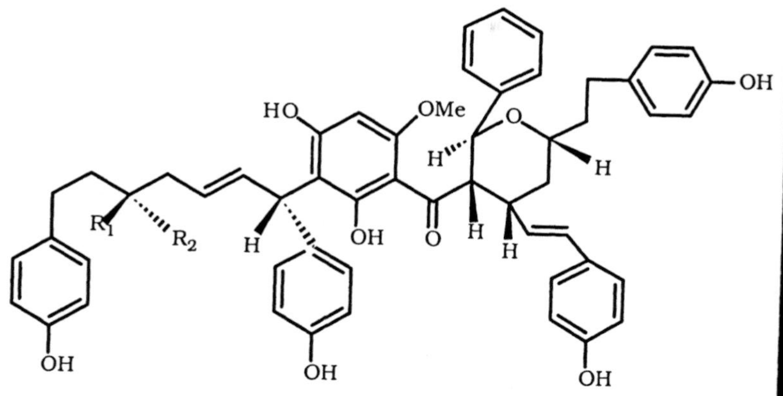

58 R₁ = H, R₂ = OH
59 R₁ = OH, R₂ = H

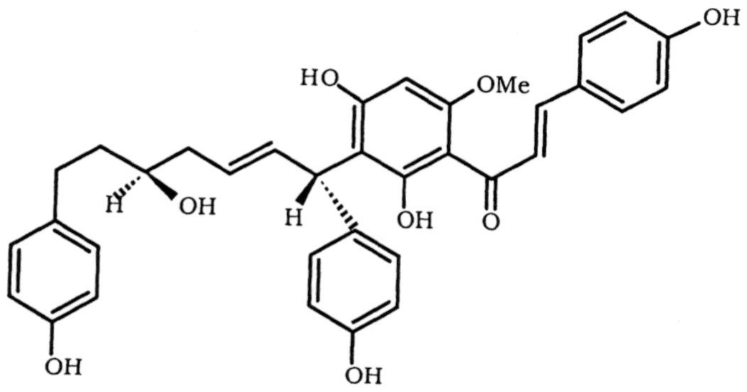

60

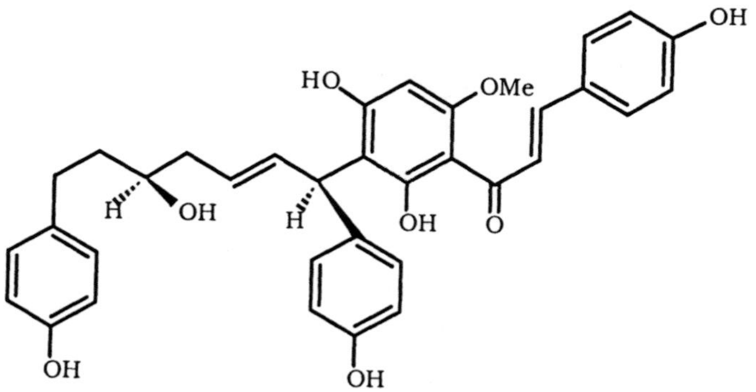

61

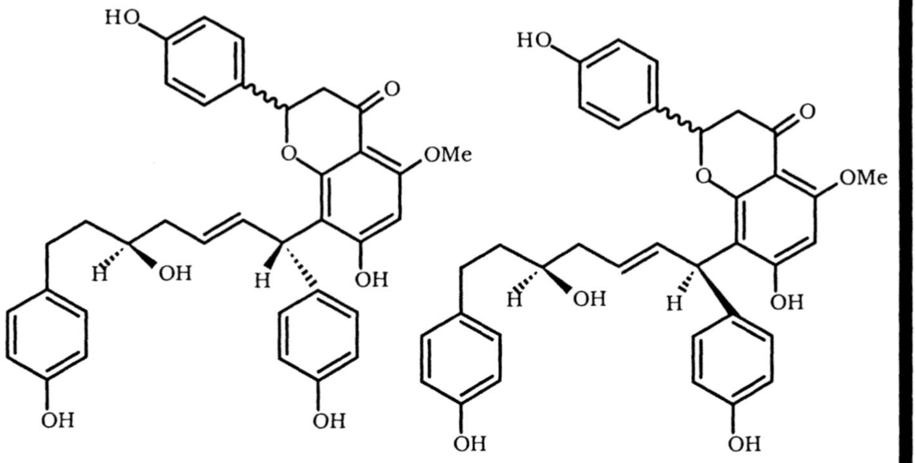

62 and 63   64 and 65

Preliminary biological evaluation showed that the methanolic extract of the rhizomes of *A. speciosa* possessed significant inhibitory activitiy against histamine and barium chloride by the Magnus method using excised guinea pig ileum [52].

From the petroleum ether extract of the rhizomes of *A. speciosa*, dihydro-5,6-dehydrokawain (66) and 5,6-dehydrokawain (67) were isolated; both compounds were also found in the rhizomes of *A. kumatake* [53]. From the leaves of *A. speciosa*, dihydro-5,6-dehydrokawain (66) and compound 68 were isolated and identified and 66 was found to be plant growth inhibitor [54]. Several derivatives related to 66 were synthesized and evaluated for their phytotoxic effects and compound 69 was found to be the most effective [54]. The petrol-soluble fraction of the aqueous methanolic extract of the fresh rhizomes of *A. speciosa* yielded methyl *trans*-cinnamate, alpinetin, cardamonin, flavokawin B (70) and dihydroflavokawin B (71) as well as 66 and 67 [52]. Two diterpenes were also isolated from the petrol-soluble fraction of the same rhizomes and elucidated as labda-8(17),12-diene-17,16-dial (11) and (*E*)-15,16-bisnorlabda-8(17),11-dien-13-one (72) [55]. Alpinetin and cardamonin were isolated from the seeds of *A. speciosa* from India [56]. Terpenoids of the essential oil obtained from the leaves of *A. speciosa* from Brazil were analyzed by GC-MS [57].

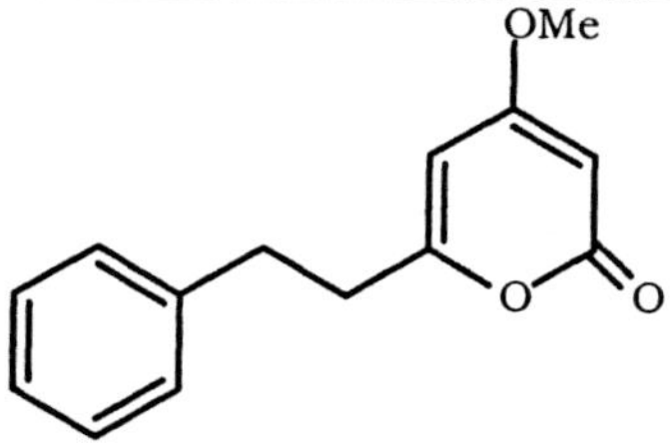

(E)-Labda-8(17),12-diene-15,16-dial and coronarin E were isolated from the chloroform extract of the rhizomes of *A. javanica* [58]. The distribution of these two labdane diterpenes within the Zingiberaceae was reported [58].

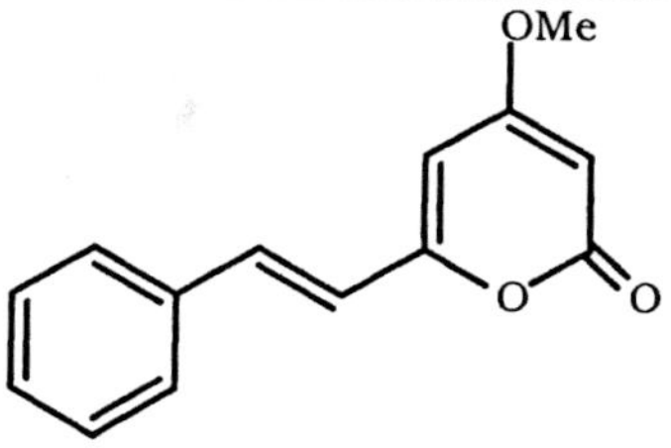

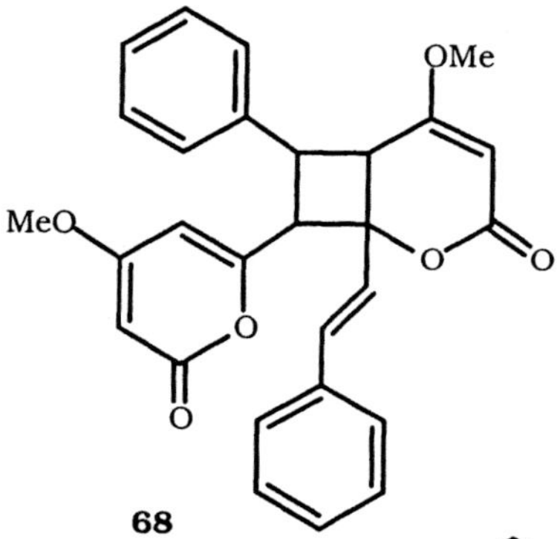

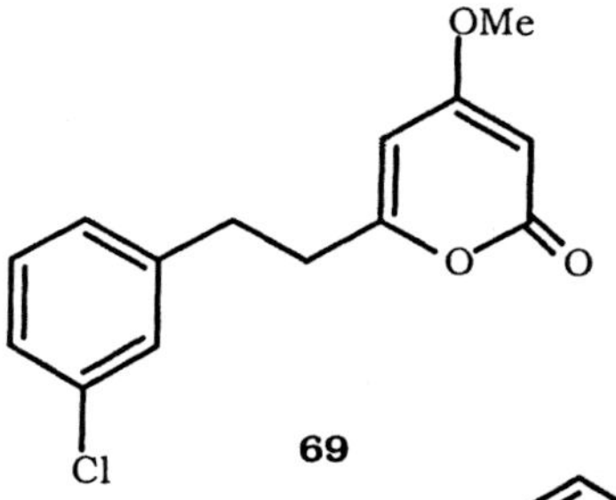

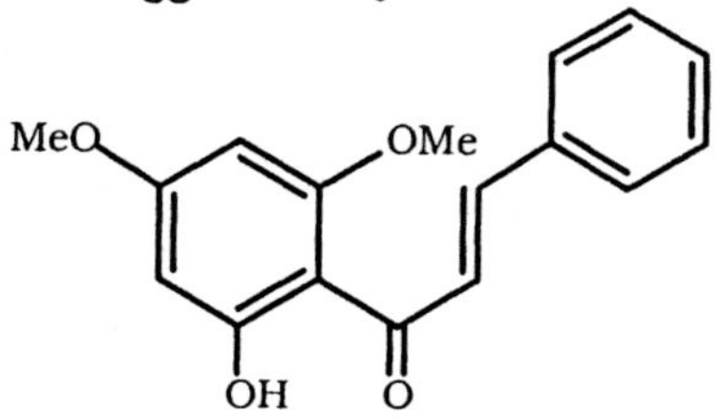

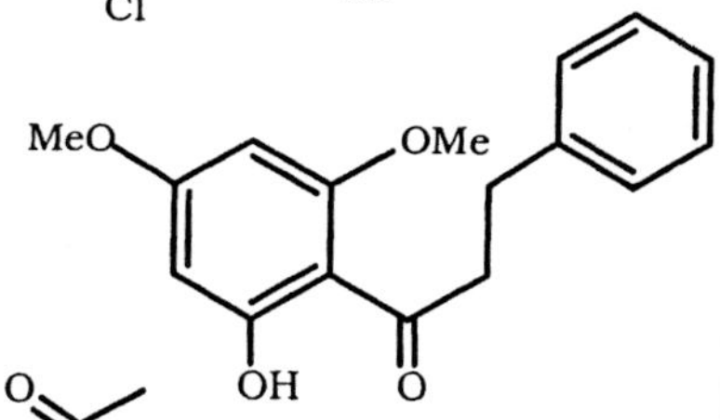

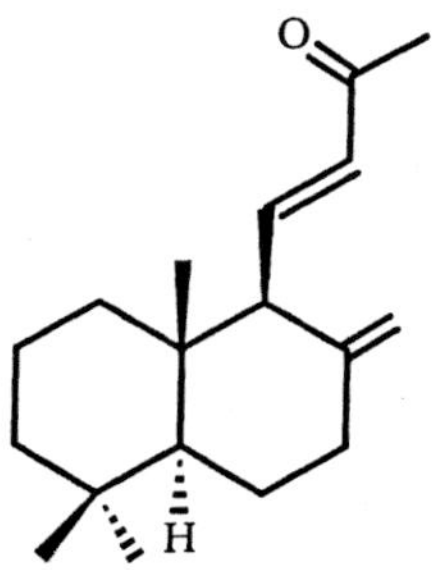

From the chloroform soluble fraction of the methanolic extract of the seeds of *A. zerumbet*, two labdane-type diterpenes, named zerumin A (**73**) and B (**74**) were isolated along with two known compounds coronarin E and (E)-15,16-bisnorlabda-8(17),11-diene-13-one [59].

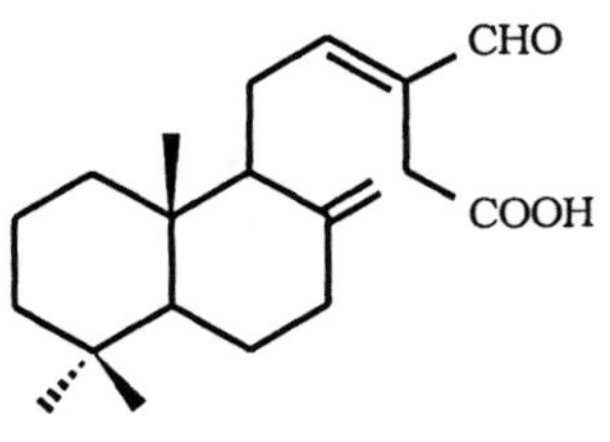

The constituents of the flowers of *A. chinensis* were investigated by GC and a number of sesquiterpenes were detected [60]. From the aerial part of *A. chinensis*, nine new labdane diterpenes (**75-83**) and two known diterpenes, (*E*)-labda-8(17),12-diene-15,16-dial and coronarin E were isolated and the structures were determined by 2D NMR spectroscopy [61]. Biogenetic relationships among these compounds were discussed [61].

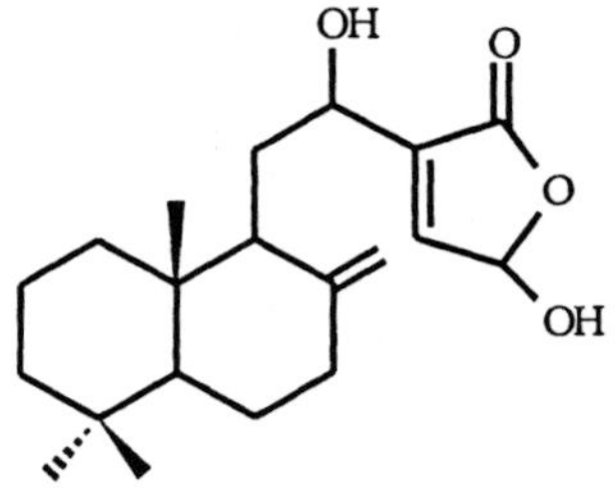

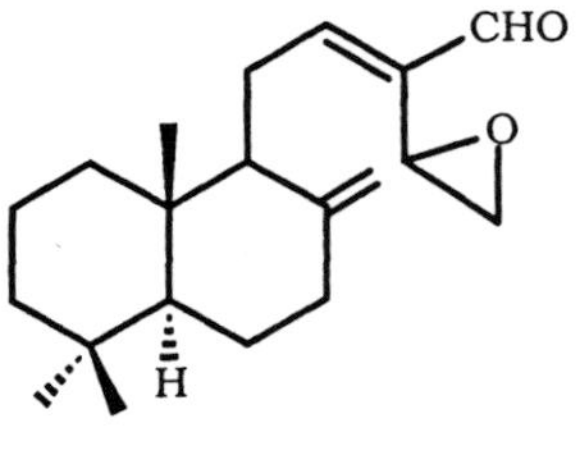

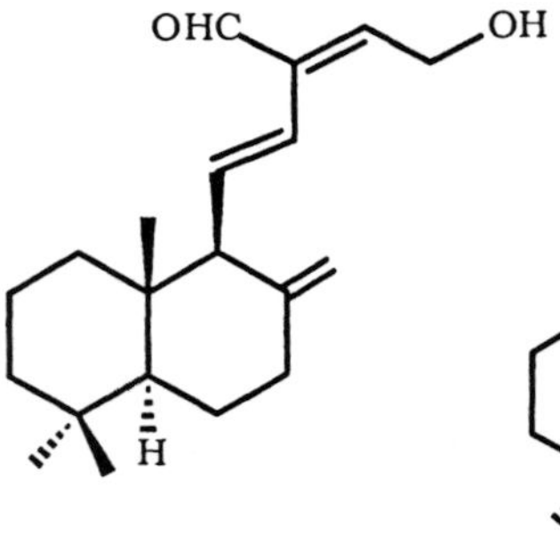

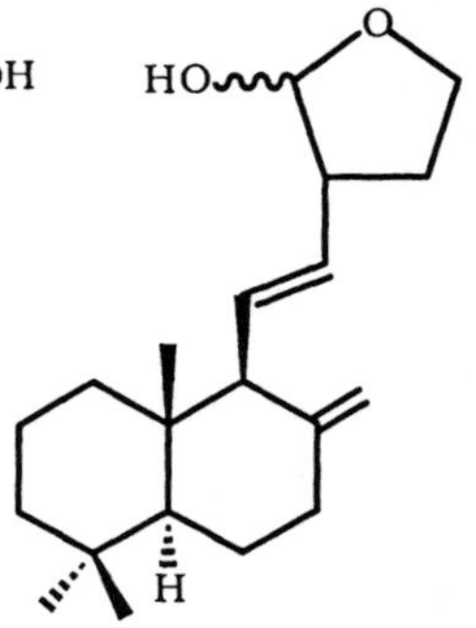

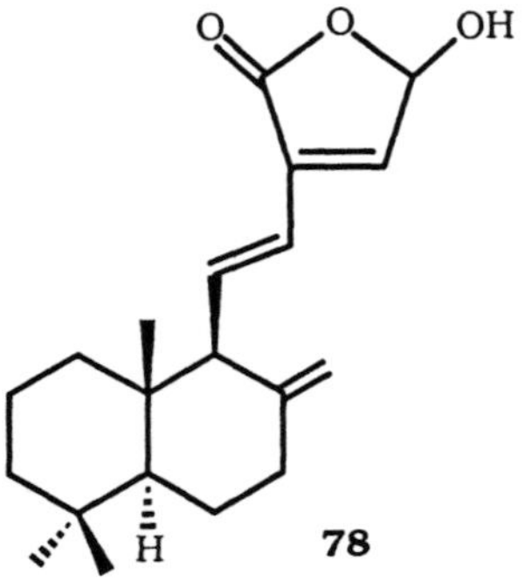

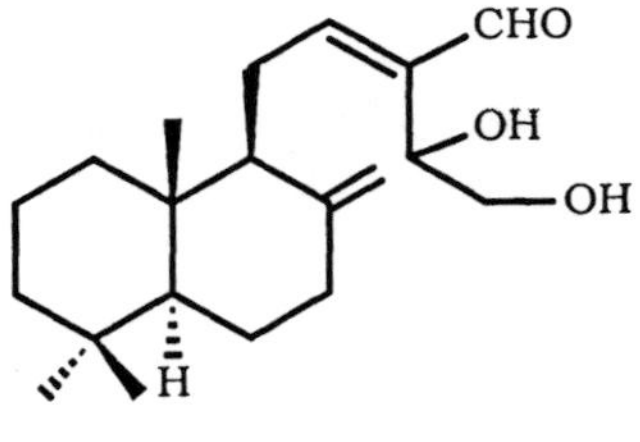

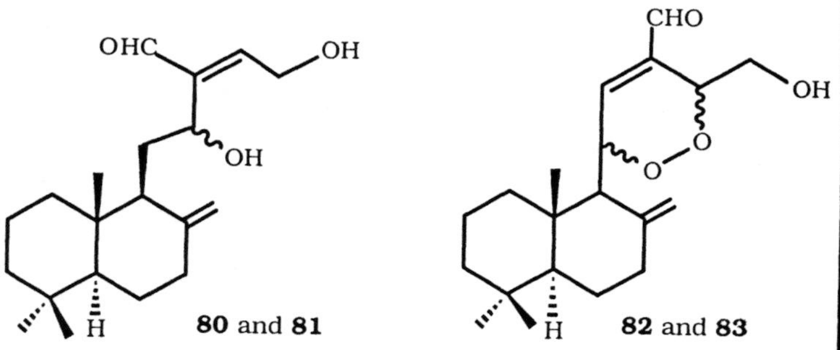

80 and 81   82 and 83

A new labdane diterpene, (E)-labda-8(17),12-diene-15-ol-16-al (**84**) was isolated from the rhizomes of *A. formosana* together with (E)-15,16-bisnorlabda-8(17),11-diene-13-one, two known sesquiterpenes, furopelargone B and humulene epoxide, and three known phenolic compounds, methyl *trans*-cinnamate, dihydro-5,6-dehydrokawain and dihydroflavokawin B [62]. The chemotaxonomic significance of these compounds was discussed briefly [62].

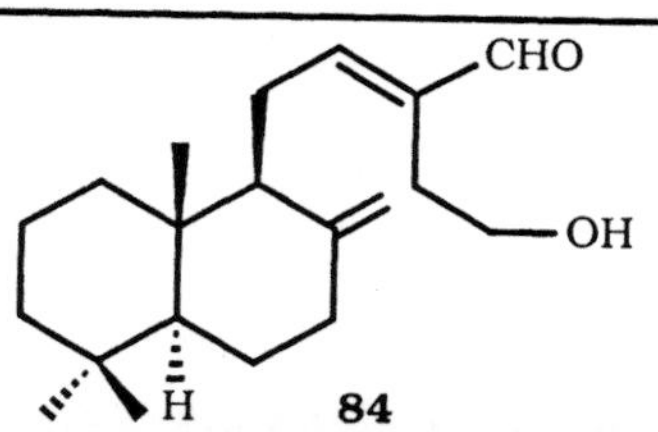

84

Three agarofurans (**85-87**) and β-eudesmol (**88**) were isolated from the petroleum ether-soluble fraction of the methanolic extract of the fresh rhizomes of *A. japonica* [63]. It is biogenetically interesting that β-eudesmol was present in the same plant with agarofurans which possess a 10-epieudesmol skeleton. From the rhizomes of *A. japonica*, three eudesmane-type sesquiterpenes, 10-*epi*-5β-hydroperoxy-β-eudesmol (**89**), 10-*epi*-5α-hydroperoxy-β-eudesmol (**90**) and 4,10-*epi*-5β-hydroxydihydroeudesmol (**91**) were isolated and their structure were determined by spectroscopic methods and chemical conversions [64]. Compounds **89-91** are considered to be biosynthesized from 10-*epi*-γ-eudesmol (**92**), which was a possible precursor of agarofuran-type

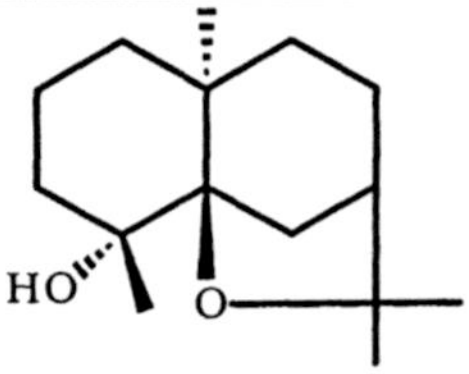

85

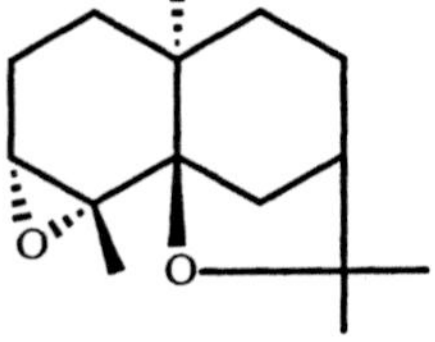

86

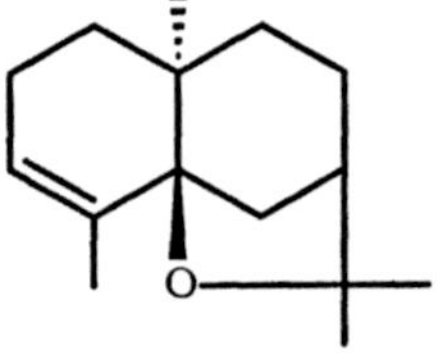

87

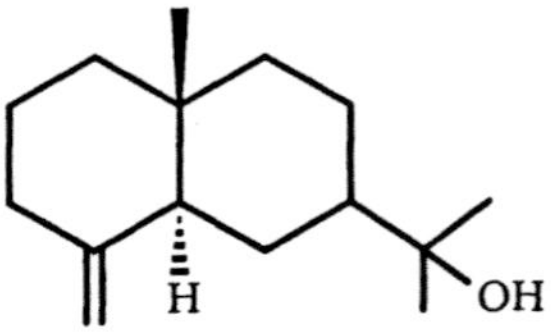

88

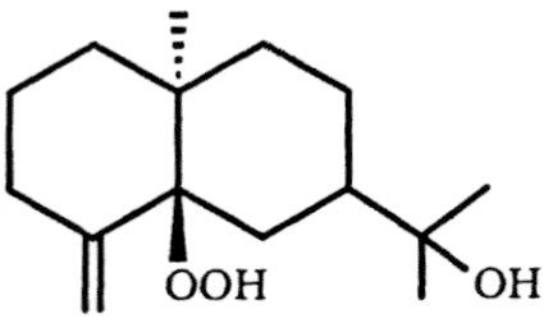

89

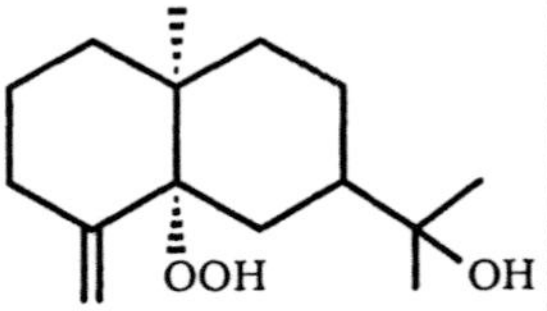

90

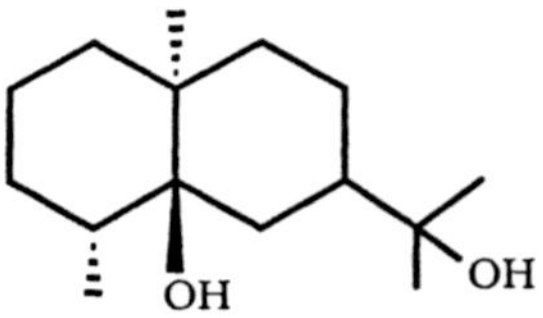

91

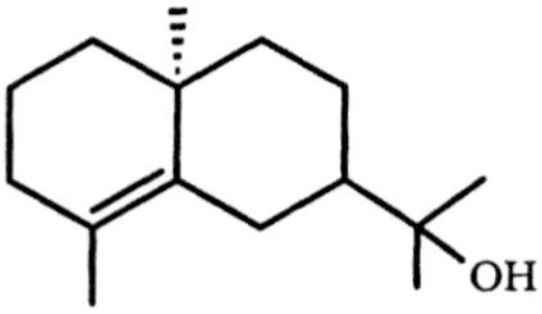

92

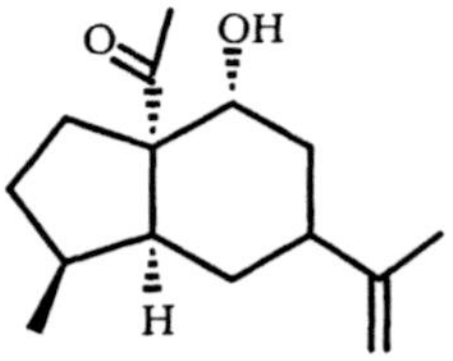

93

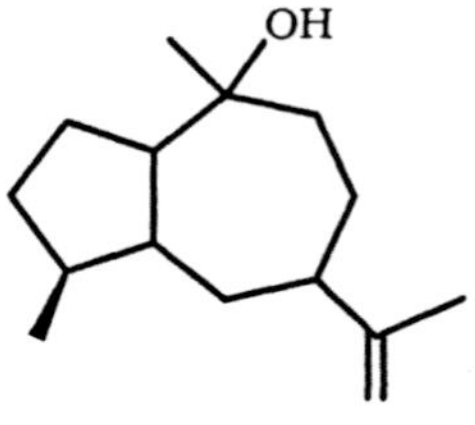

94

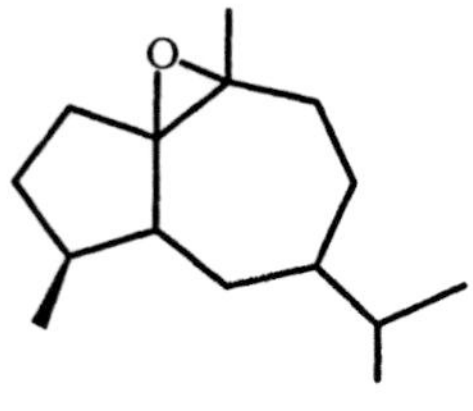

95

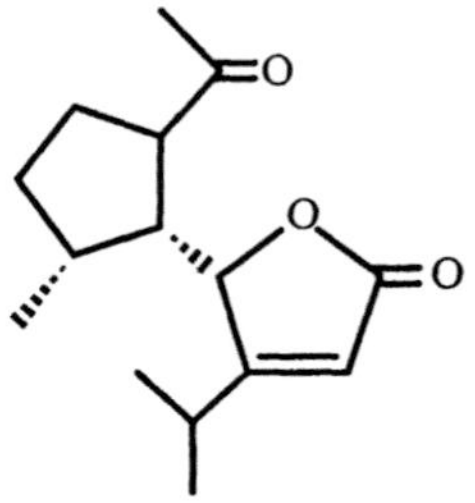

96

sesquiterpenes [64]. From the chloroform-soluble fraction of the methanolic extract of the fresh rhizomes of the same plant, sesquiterpenoid with a new skeleton, alpiniol (**93**) was isolated together with pogostol (**94**) and structure **93** was determined by X-ray analysis [65]. Biogenetically, alpiniol and pogostol may be considered to be derived from guaian-1,10-epoxide (95) through a rearrangement [65]. Further examination of the same extract yielded two new sesquiterpenoids, alpinolide (96) and hanamyol (97) [66]. The structures 96 and 97 were determined by spectral and chemical evidence. Structure 97 was also proved by X-ray analysis. Furopelargone A (98) and B (99) were also isolated from this plant[66]. Alpinolide 96 was prepared from hanalpinol (100) [66]. Oxidation of 100 with pyridinium chlorochromate in dichloromethane gave the corresponding keto compound (101), which on reaction with BF$_3$-Et$_2$O afforded alpinolide (96); the reaction pathway is shown in "Fig. (1)". Further examination of the constituents of the rhizomes of *A. japonica* resulted in the isolation and structural elucidation

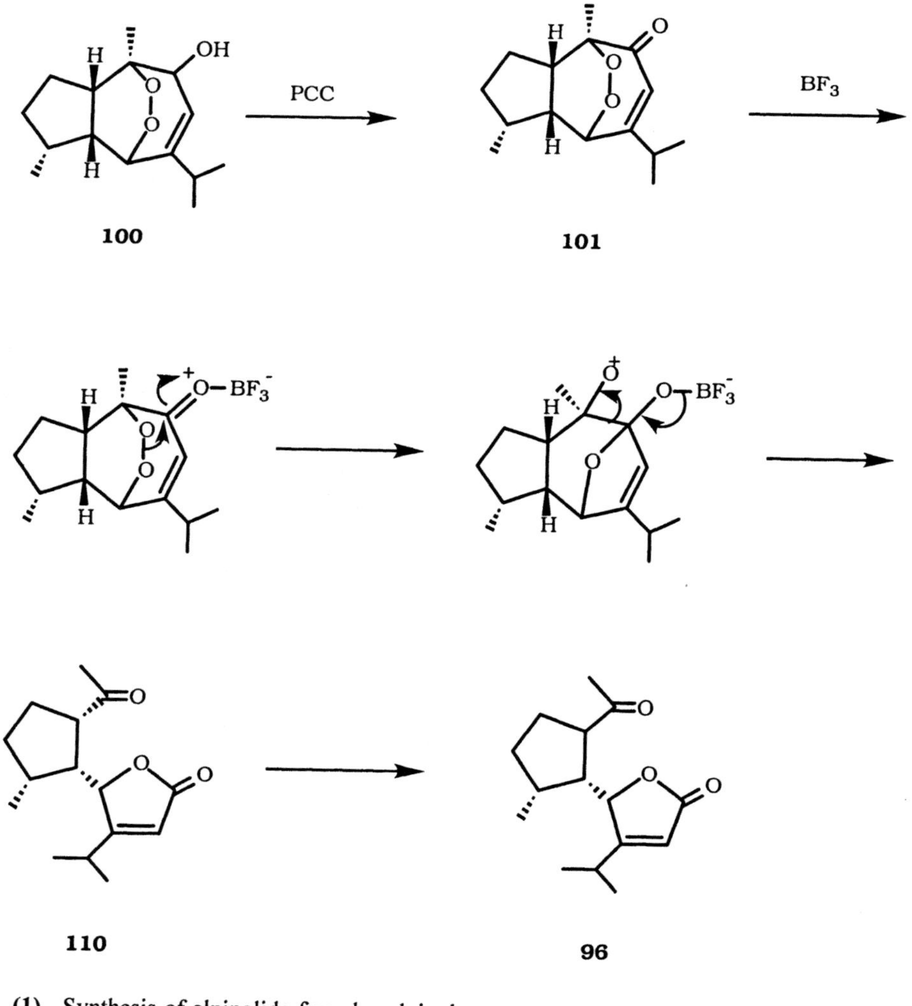

Fig. (1). Synthesis of alpinolide from hanalpinol.

97

98

99

100

101

102

103

104

105

106

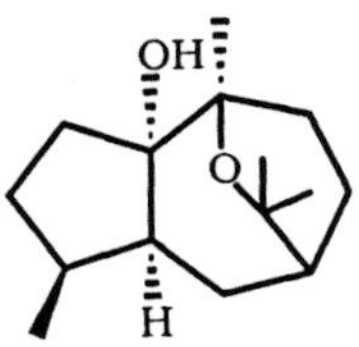

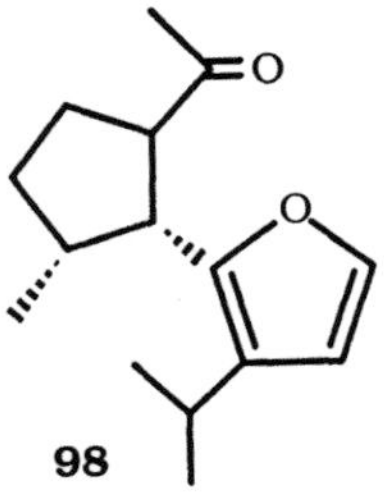

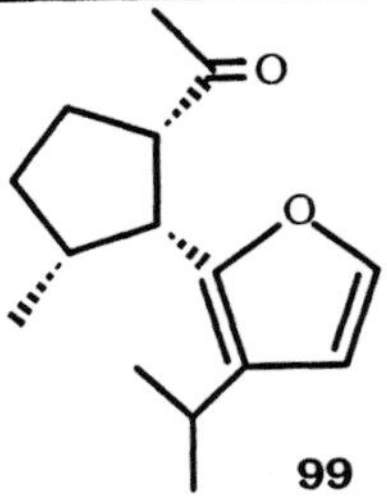

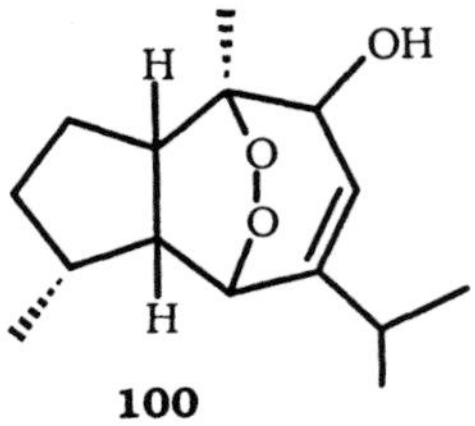

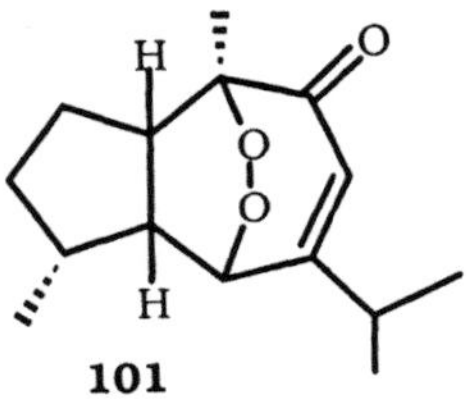

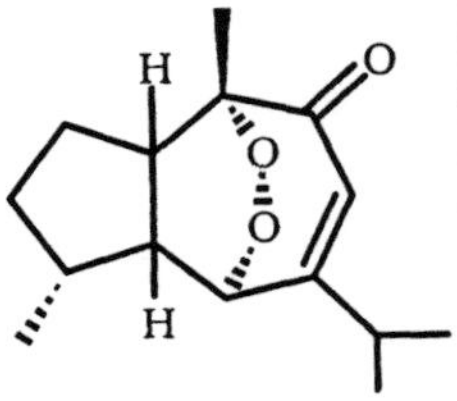

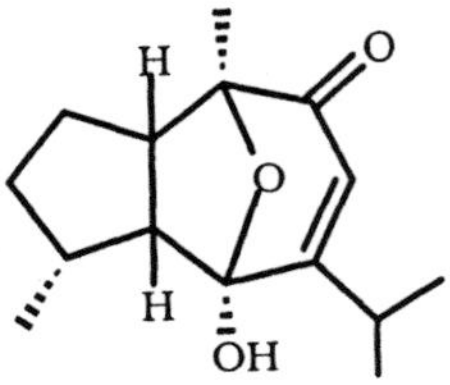

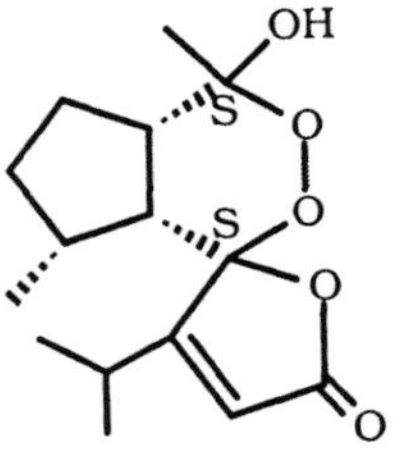

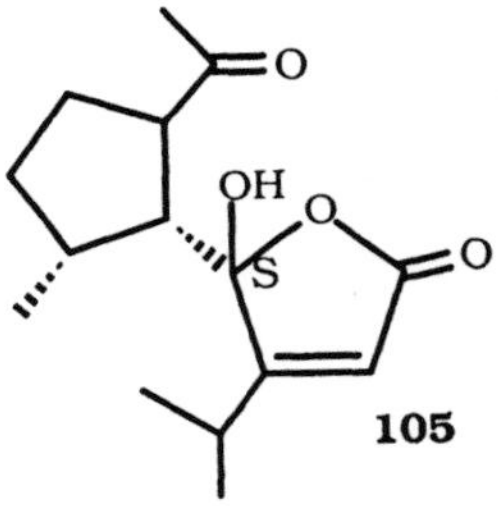

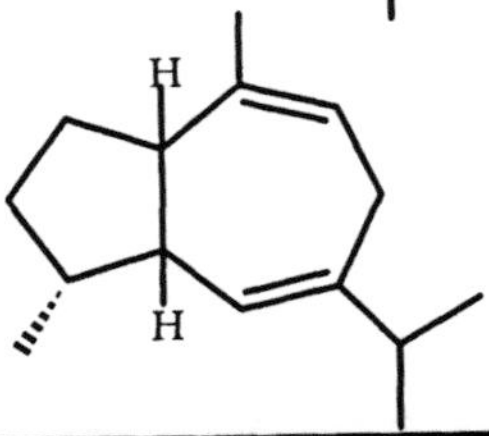

of a sesquiterpene cyclic peroxide, hanalpinol (**100**) [67]. Treatment of **100** with *p*-toluenesulfonic acid gave **99** and this was gradually converted to **98** under these conditions [67]. Three novel guaiane-type sesquiterpenes, hanalpinone (**101**), isohanalpinone (**102**) and alpinenone (**103**), and two novel secoquaiane-type sesquiterpenes, alpinolide peroxide (**104**) and 6-hydroxyalpinolide (**105**) were isolated from the rhizomes of *A. japonica* [68]. Biosynthetic relationships of the identified sesquiterpenes from *A. japonica* have been discussed [68]. The chemical conversion of alpinenone (**103**) into furopelargone (**99**) was studied and the reaction mechanism was examined using molecular mechanics calculations (MM2), molecular orbital calculations and chemical evidence [69]. Guaia-6,9-diene (**106**) was converted into the cyclic peroxide, hanalpinol (**100**) by autooxidation and photosensitized oxidation; both compounds, **106** and **100** were found in the rhizomes of *A. japonica* [70]. Diene **106** may be regarded as an intermediate precursor of hanapinol (**100**) and a biogenetic pathway for **100** was proposed, which was supported by the results of frontier electron theory calculations [70].

Sesquiterpenes isolated from *A. speciosa* and *A. japonica* and their derivatives were found to inhibit histamine- or barium chloride-induced contraction of excised guinea pig ileum when tested by the Magnus method [71]. Relationship between the chemical structures of the sesquiterpenes and their derivatives and their spasmolytic activities was discussed.

From the rhizomes of *A. intermedia*, three novel sesquiterpenes peroxides, hanalpinol peroxide (**107**), isohanalpinol (**108**) and aokumanol (**109**), a new secoguaiane-type sesquiterpene, epialpinolide (**110**) and a new elemophilane-type sesquiterpene, $\Delta^{11(12)}$-eremophilen-10β-ol (**111**) were isolated together with eight known sesquiterpenes, hanalpinol(**100**), hanalpinone (**101**), isohanalpinone (**102**), alpinenone (**103**), furopelargone B (**99**), furopelargone A (**98**), intermedeol and β-selinene [72]. The structures of **107-111** were determined by spectroscopic methods, chemical conversions and X-ray analysis. Chemical constituents of both species, *A. japonica* and *A. intermedia*, are very similar. From a chemotaxonomic viewpoint, they are closely related plants.

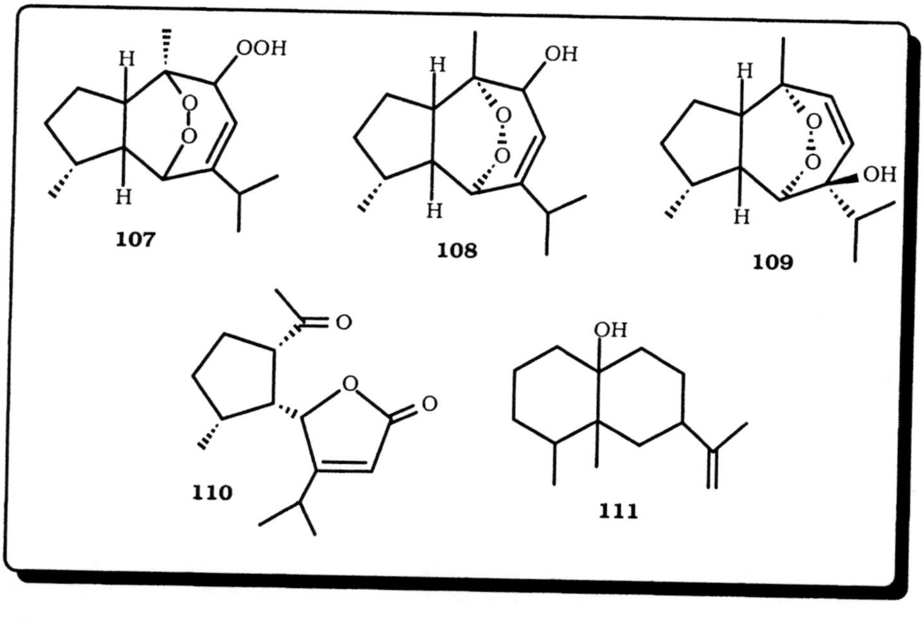

From the petroleum ether extract of the fresh roots of *A. flabellata*, a cyclohexene derivative, alflabene (**112**) was isolated and identified on the basis of its spectral properties and by synthesis [73]. The same compound was also isolated from the rhizomes of *Zingiber cassumunar*.

112

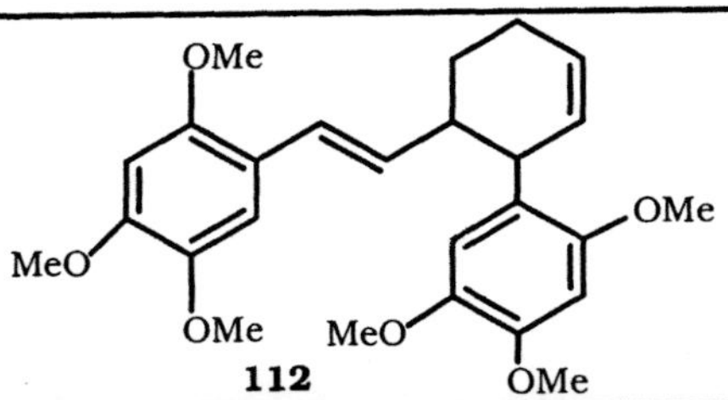

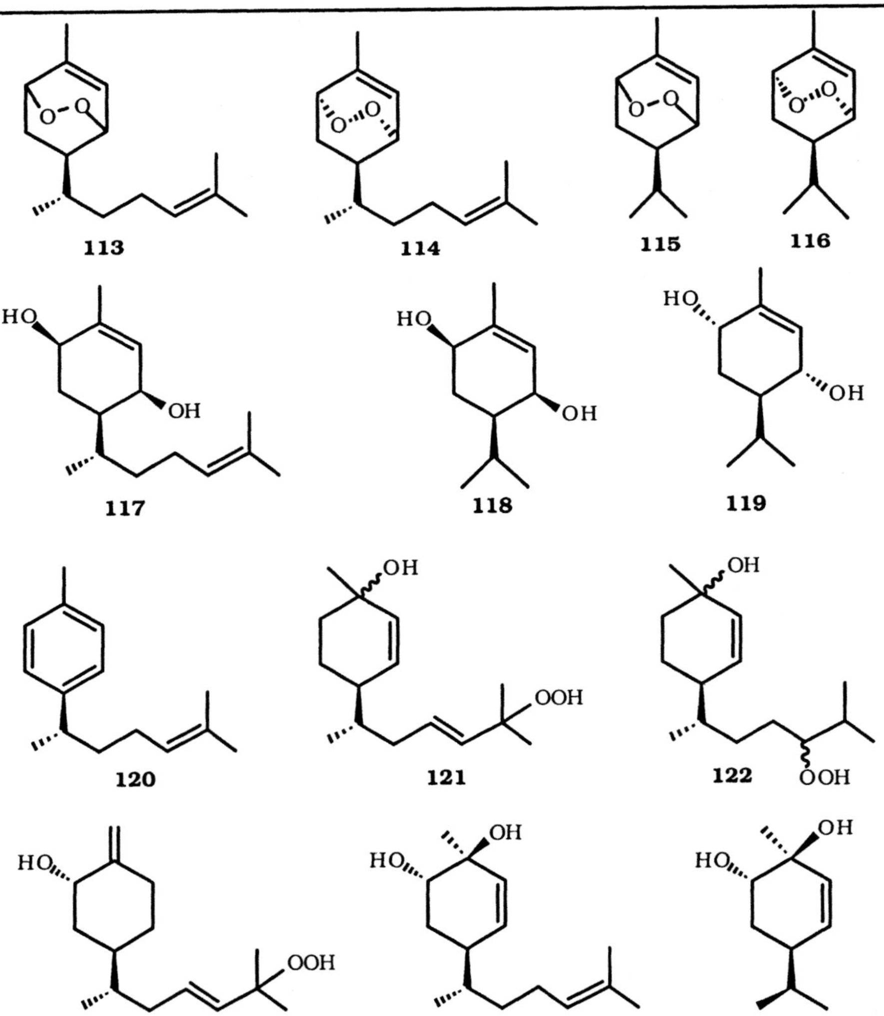

From the dichloromethane extract of the aerial part of *A. densibracteata*, numerous oxygenated bisabolane sesquiterpenes and oxygenated methane monoterpenes, compounds (**113-128**) were isolated

113 114 115 116

117 118 119

120 121 122

123 124 125

and the structures were determined by NMR spectroscopy and chemical transformations [74].

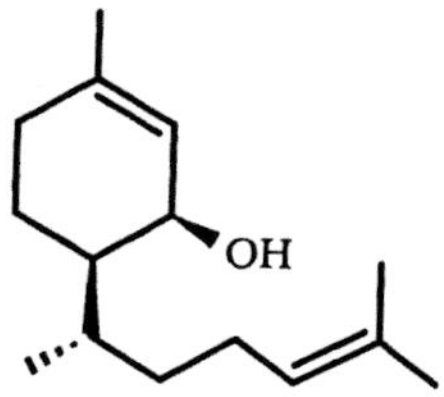

126

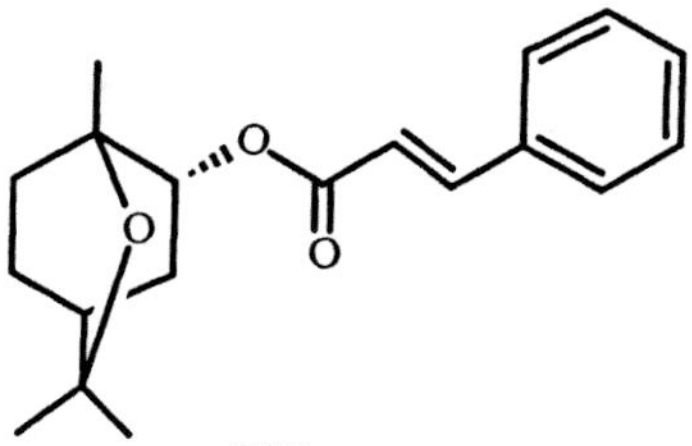

127

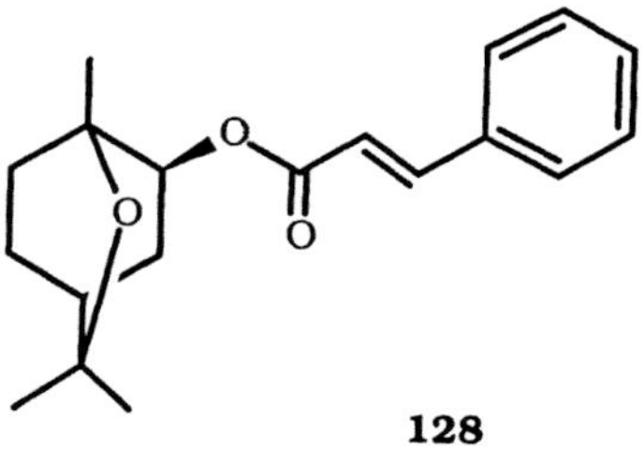

128

Amomum

Numerous species of the genus *Amomum* have been used in traditional medicine in Asia for different purposes. Dried fruits of *A. aurantiacum* and *A. xanthiodes* are used for stomach ache in China [75]. Juice of pounded fresh rhizomes of *A. coccineum* is drunk as a vermifuge in children in Indonesia [76]. Hot water extract of the rhizomes of *A. dealbatum* is used for postpartum purification in female and fruits of *A. gracile* are chewed for indigestion and nausea in Indonesia [77]. Dried seeds of *A. krervanh* are used as carminative [78] and ground dried roots of *A. ovoideum* are used as anti-asthmatic [20] in Thailand. Dried seeds of *A. xanthiodes* are used as emmenagogue in South Korea [79].

Preliminary studies indicated that extracts from various species of the genus *Amomum* possessed different activities. Aqueous ethanolic extracts of the rhizomes of *A. aromaticum* and *A. subulatum* showed hypoglycemic activity [80]. Water extract of the fruits of *A. compactum* [81] and *A. xanthiodes* [82] possessed anti-tumor activity. Ether extract of the fruits of *A. medium* had anti-inflammatory action [83]. Fresh leaf essential oil of *A. subulatum* showed anti-fungal activity [84]. Methanolic extract of the dried seeds of *A. xanthiodes* was found to have anti-hepatotoxic action [85]. Chloroform extract of the fruits of *A. costatum* inhibited cyclic nucleotide phosphodiesterase [86].

The terpenoid constituents of the oils of the fruits of *A. korarima* from Ethiopia and *A. subulatum* from India were analyzed by GC [87]. The main constituents of the oils from both species were 1,8-cineol (31.5% and 74.0%, respectively) and limonene (13.5% and 10.3%, respectively) [87]. Mono- and sesquiterpenoids of the essential oils obtained from steam distillation of the dried comminuted fruits of *A cardamomum* and of the freshly ground dried fruits of *A. globosum,* both species being from Thailand, were analyzed by GLC [88]. The oil of *A. cadamomum* was found to contain 1,8-cineol (66.8%) and β-pinene (15.5%) as the main components whereas oil from *A. globosum* fruits had camphor (38.9%) and bornyl acetate (25.2%) [88]. Gas chromatographic evaluation of the essential oils of 7 strains of the dried fruits of *A. subulatum* growing wild in Sikkim, India, showed the major component was cineol (77.89%) [89]. Essential oil from the seeds of *A aurantiacum* from China was reported to have nerolidol (75.48%) and linalool (22.29%) as the major constituents [90]. Compositions of the essential oils from the fruits of *A. aurantiacum, A. longiligare* and *A. villosum* were analyzed by GC-MS; the main components of *A. aurantiacum* fruit oil were linalool and nerolidol, those of *A. longiligare* fruit oil were borneol acetate and nerolidol and those of *A. villosum* fruit oil were borneol acetate, camphore and limonene [91]. The essential oil of *A. xanthoides* from Vietnam was analyzed by GC-MS and the main components were identified as α-pinene, camphene, δ-limonene, linalool, camphor, δ-borneol, 3-phenyl-2-butanone, nerolidol, etc. [92]. The essential oil of the fruit of *A. xanthoides* was found to contain >1% of ethyl *trans-p*-methoxycinnamate [93].

A flavonoid glycoside, leucocyanidin-3-*O*-β-D-glucoside [94], an aurone glycoside named subulin [95], cardamonin [96] and alpinetin [96] were isolated from the seeds of *A. subulatum*. From the ether extract of the seeds of *A. medium*, two simple aldehydes, 1*H*-indene-2,3-dihydro-4-carboxaldehyde (**129**) and 1*H*-indene-2,3-dihydro-5-carboxaldehyde (**130**), were isolated [97]. From the ethanolic extract of the seeds of *A. melequeta*, also known as Grains of Paradise, from equatorial Africa, the major constituents, (6)-gingerol (**135**) and (6)-paradol (**131**) with trace amounts of (6)-shogaol (**137**), (8)-gingerol (**136**) and (8)-paradol (**133**) were isolated and identified as its pungent principles [98]. Three major hydroxyphenylalkanones, **131**, (7)-paradol (**132**) and **137** with trace quantities of **133** and zingerone (**134**) were obtained from the acetone extract of Ghanian Grains of Paradise [99]. No gingerols were detected in this extract.

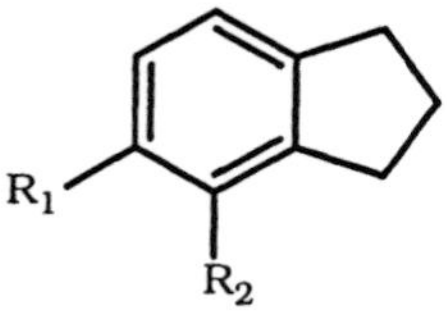

129 R_1 = CHO, R_2 = H
130 R_1 = H, R_2 = CHO

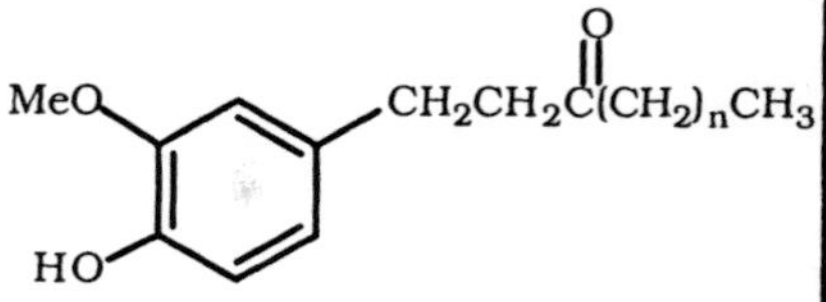

131 n = 6
132 n = 7
133 n = 8
134 n = 0

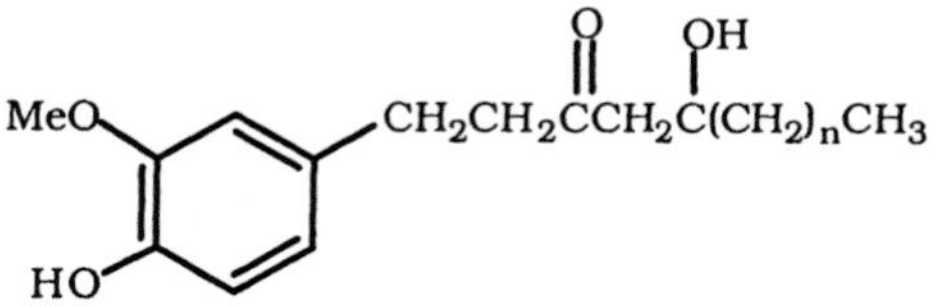

135 n = 4
136 n = 6

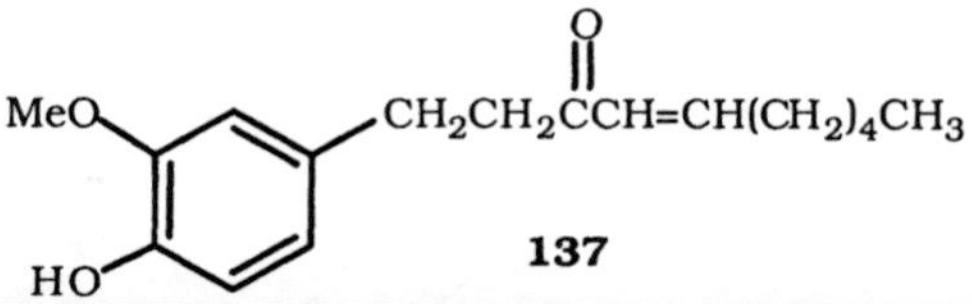

137

A novel diterpene peroxide (138) with potent activity against *Plasmodium falciparum* was isolated together with myrtenal, 4-hydroxymyrtenal, myrtenol and *trans*-pinocarveol from the hexane extract of the fruits of *A. krervanh* [100].

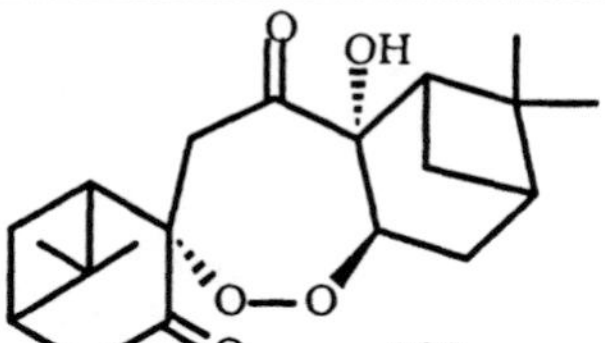

138

Boesenbergia

Fresh root of *Boesenbergia pandurata* has been eaten as a vegetable and used for food seasoning in Thailand. In Indonesia, the rhizome of *B. pandurata* is used for thrush, dry cough and sore throat by chewing with areca [101]. A methanolic extract of the fresh rhizome from Thailand was reported to have anti-tumor-promoting and cytotoxic activities [102].

Rhizome essential oil of *B. pandurata* was found to have kidney stone dissolution effect [103].

    Of all the species of the genus *Boesenbergia*, only the two varieties of *B. pandurata* (yellow and red rhizomes) have been chemically examined.

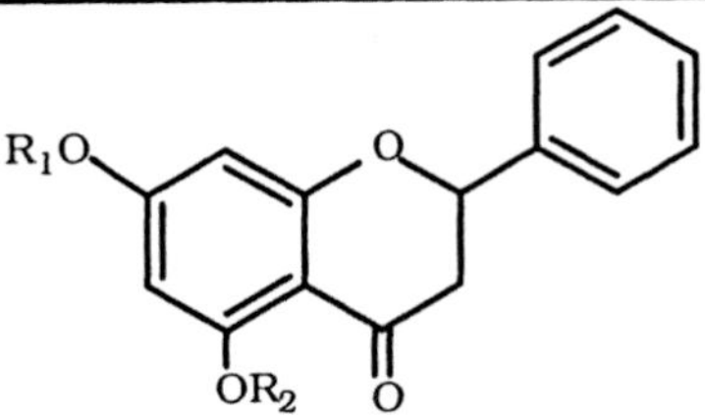

139 R_1 = Me, R_2 = H
140 R_1 = H, R_2 = Me
141 R_1 = H, R_2 = H

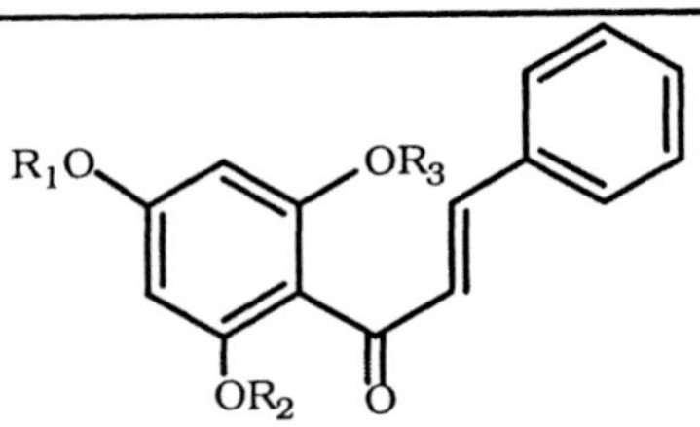

142 R_1 = Me, R_2 = R_3 = H
143 R_1 = R_2 = H, R_3 = Me

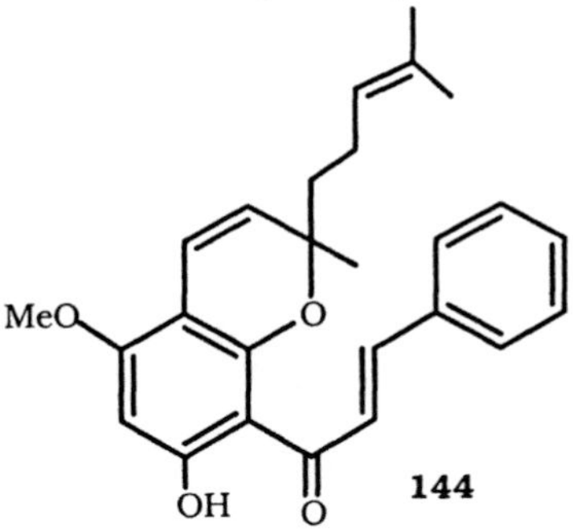

144

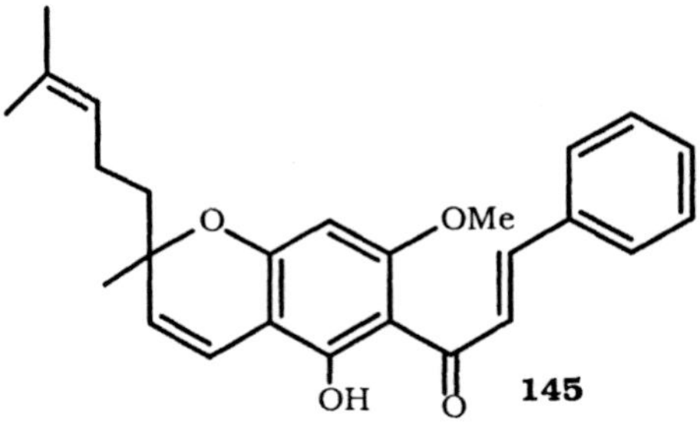

145

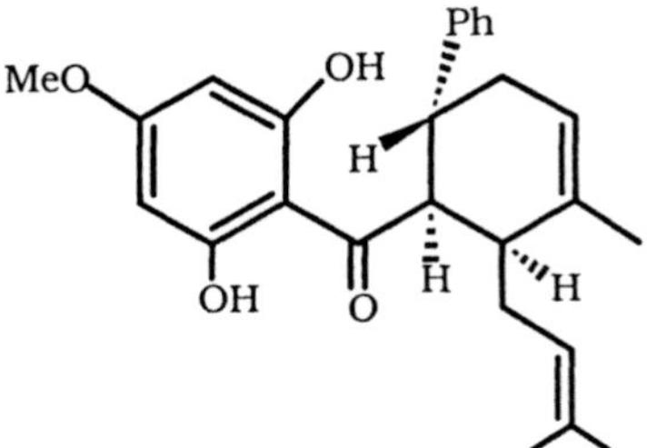

146

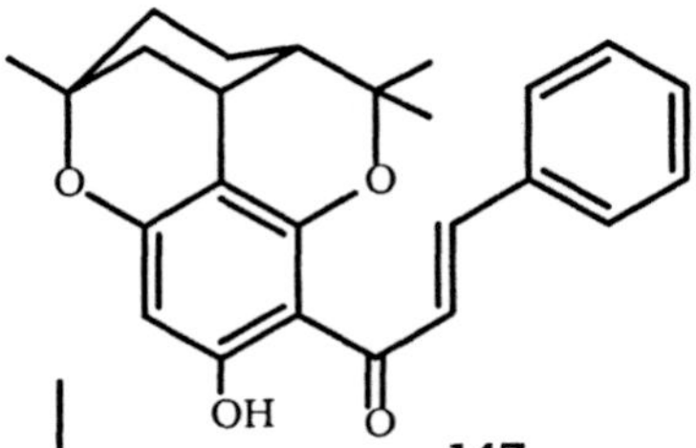

147

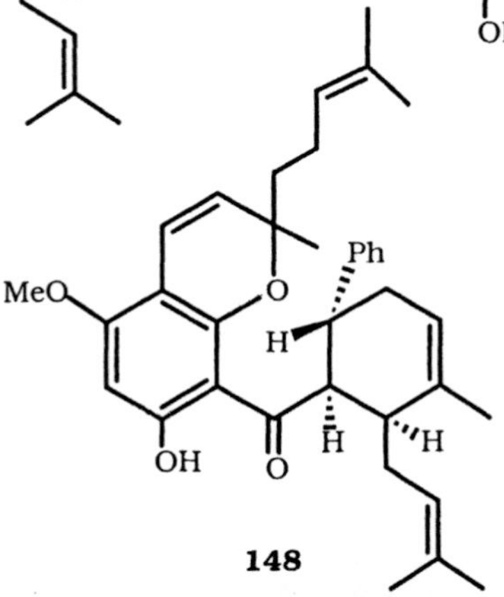

148

The terpenoids of the essential oil obtained from the rhizomes of *B. panduruta* (yellow rhizome) were examined [104]. (±)-Pinostrobin (**139**) and (±)-alpinetin (**140**) were isolated from the ether extract of the rhizomes [105]. From the chloroform extract of the fresh rhizomes compound **139**, pinocembrin (**141**) and three chalcones, 2',6'-dihydroxy-4'-methoxy- chalcone (**142**), cardamonin (**143**) and boesenbergin A (**144**) were isolated[106]. A simple synthesis of boesenbergin A has been achieved and its acid-catalyzed cyclization has been studied [106]. Another chalcone boesenbergin B (**145**) and a cyclohexane derivative, panduratin A (**146**) were later isolated from the same extract [107]. Boesenbergin B was synthesized using the procedure described for boesenbergin A and the acid catalyzed cyclization of **145** was also carried out [107]. The structure of boesenbergin B was confirmed by a single-crystal X-ray analysis [108]. From the hexane extract of the rhizomes of *B. pandurata* (red form), flavonoid **139**, two chalcone derivatives, boesenbergin A (**144**) and rubranine (**147**) and a cyclohexane derivative, panduratin A (**146**) were isolated [109]. Rubranine was synthesized in moderate yield from the reaction of citral with pinocembrin in the presence of pyridine [109]. The structure of panduratin A has been confirmed by a single-crystal X-ray structure determination [110]. The cyclohexyl derivative, panduratin A, could conceivably be derived from a Diels-Alder reaction of a related chalcone and an isoprenoid unit. Further examination of the hexane extract of *B. pandurata* (red form) has yielded panduratin B1 and panduratin B2, diastereoisomers of **148** [111]. Chromenylation of panduratin A with citral in the presence of pyridine gave panduratin B as a diastereoisomeric mixture [111].

Costus

The rhizomes of *Costus speciosus*, which possessed stimulant action on isolated uterus [112, 113], were found to be a rich source of diosgenin (**149**) [113, 114]. From the rhizomes of the plant, tigogenin, diosgenin, β-sitosterol-β-D-glucoside, prosapogenin A (**150**) and B (**151**) of dioscin, dioscin (**152**) and gracillin (**153**) were isolated [113, 115-117].From the *n*-hexane fraction of the methanolic extract of the roots of *C. speciosus*, a new sterol, 5α-stigmast-9(11)-en-3β-ol (**155**) was isolated [118]. Together with four known aliphatic hydroxyketones, 10-hydroxyhentriacontan-16-one, 13-hydroxyhentriacontan-16-one, 16-hydroxytriacontan-15-one and 3-hydroxytriacontan-11-one, 24-hydroxyhentriacontan-27-one (**156**) and 24-hydroxytriacontan-26-one (**157**) were obtained from the roots [119]. Furthermore, two aliphatic compounds were isolated and characterized as 8-hydroxytriacontan-25-one (**158**) and methyl triacontanoate [120]. From the roots of the same plant, four cycloartane derivatives (**159-162**) were

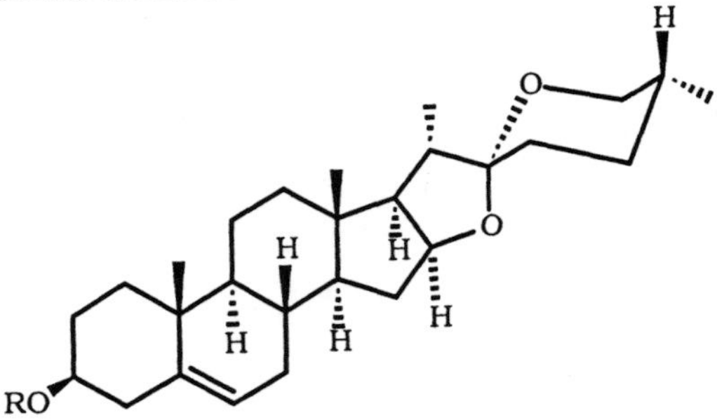

149 R = H
150 R = β-D-glc*p*(2→1)α-L-rha*p*
151 R = β-D-glc*p*(4→1)α-L-rha*p*
152 R = β-D$_2$glc*p*(4→1)α-L-rha*p*
$$\downarrow 1$$
α-L-rha*p*
153 R = β-D$_2$glc*p*(3→1)α-L-rha*p*
$$\downarrow 1$$
α-L-rha*p*
154 R = β-D$_2$glc*p*(4→1)α-L-rha*p*
$$\downarrow 1$$
β-D-epi*f*

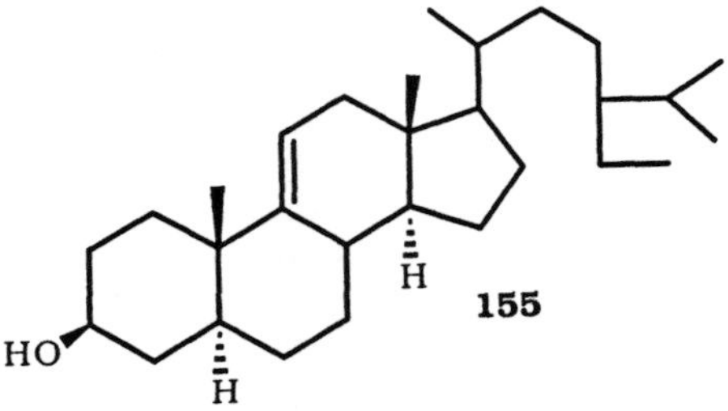

155

$$CH_3(CH_2)_zC(CH_2)_yCH(CH_2)_xCH_3$$

156 x = 22, y = 2, z = 3
157 x = 22, y = 1, z = 3
158 x = 6, y = 16, z = 4

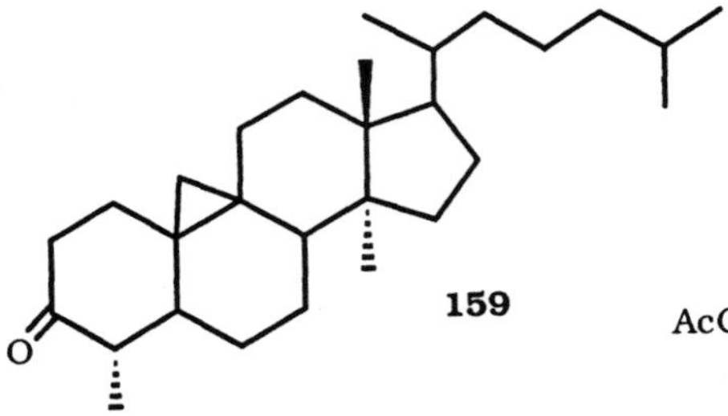

159

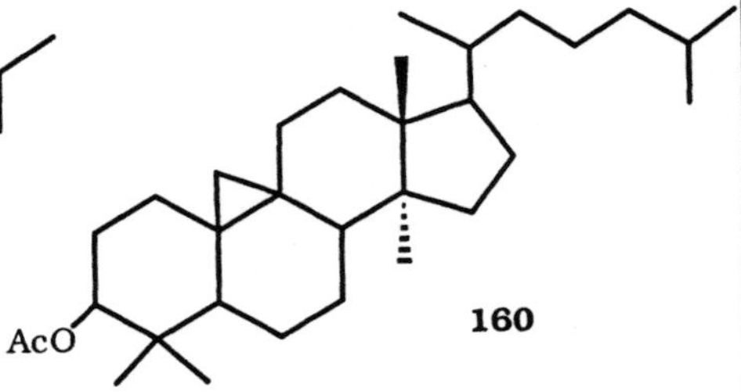

160

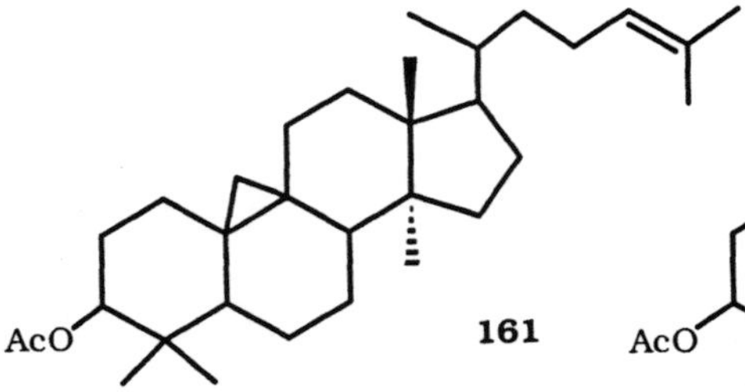

161

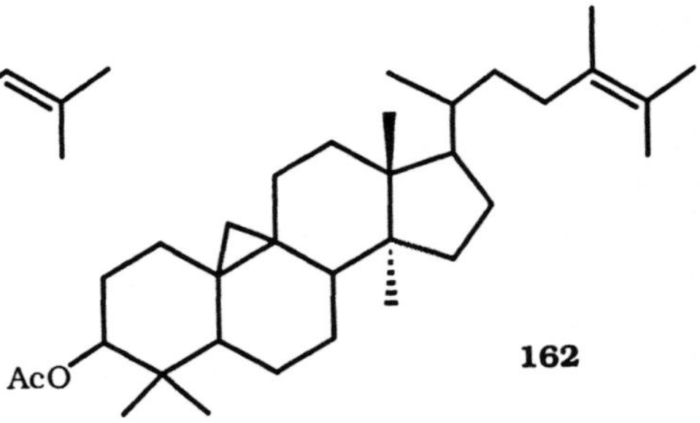

162

$$CH_3CH_2CH(CH_2)_nCO(CH_2)_{13}CH_3$$
$$CH_3$$

163 n = 11
164 n = 9

$$CH_3(CH_2)_nC(CH_2)_{12}COOH$$

165 n = 8
166 n = 12
167 n = 13

isolated and identified [121]. The mixture of compounds **160-162** was acetylated and resolved by GC-MS analysis [121]. During a large scale isolation of diosgenin from the rhizomes of *C. speciosus*, five new compounds together with five known derivatives, diosgenin, β-sitosterol, triacontanol, triacontanoic acid and 5α-stigmast-9(11)-en-3β-ol were isolated from the hexane-soluble fraction [122]. These new compounds were characterized as tetradecyl 13-methylpentadecanoate **(163)**, tetradecyl 11-methyltridecanoate **(164)**, 14-oxotricosanoic acid **(165)**, 14-oxoheptacosanoic acid **(166)** and 15-oxocotacosanoic acid **(167)**.

The mature seeds of *C. speciosus* were reported to contain an appreciable amount of diosgenin [123]. Numerous steroidal saponins were isolated from the seeds [124]. Their structures were elucidated as β-sitosterol-β-D-glucoside, prosapogenin A and B of dioscin, dioscin, gracillin, 3-*O*-[α-L-rhamnopyranosyl(1 → 2)-β-D-glucopyranosyl]-26-*O*-[β-D-glucopyranosyl]-22-α - methoxy - (25*R*) - furost-5 - en-3β, 26-diol

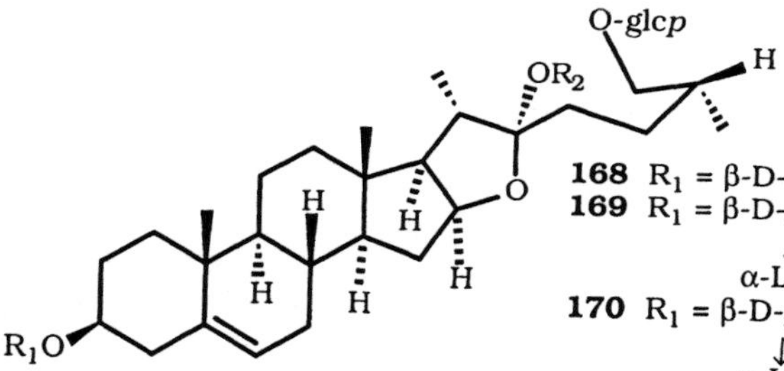

168 R_1 = β-D-glc$p$(2→1)α-L-rha$p$, R_2 = Me
169 R_1 = β-D-glc$p$(4→1)α-L-rha$p$, R_2 = Me
        ↓2
     α-L-rha$p$
170 R_1 = β-D-glc$p$(4→1)α-L-rha$p$, R_2 = H
        ↓2
     α-L-rha$p$
171 R_1 = β-D-glc$p$(4→1)α-L-rha$p$, R_2 = H
        ↓2
     α-L-rha$p$(2→1)β-D-glc$p$
172 R_1 = β-D-glc$p$(4→1)α-L-rha$p$, R_2 = Me
        ↓2
     α-L-rha$p$(2→1)β-D-glc$p$

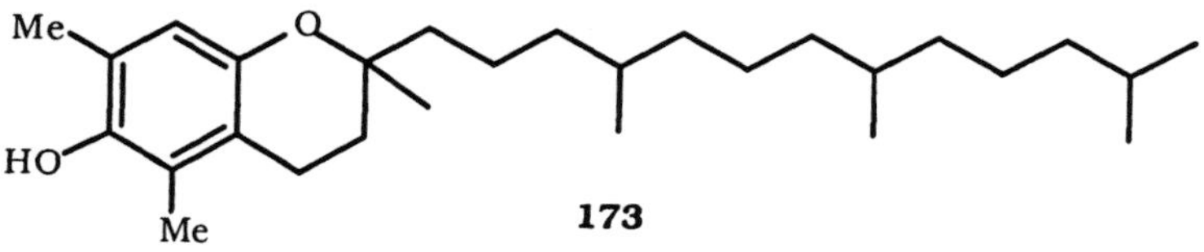

173

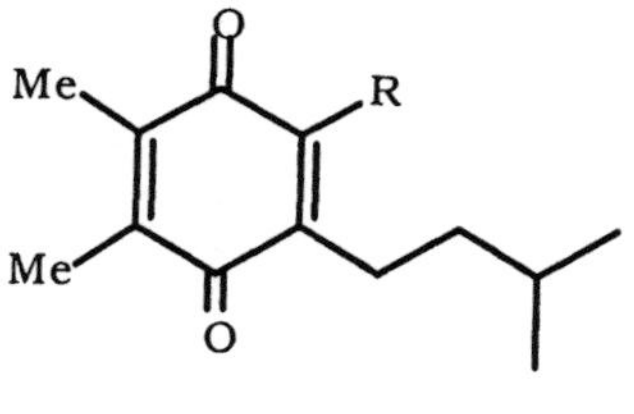

174 R = Me
175 R = H

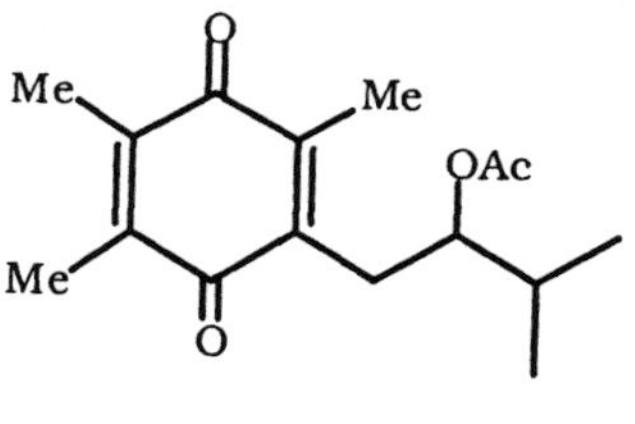

176

(168), methyl protodioscin (169) and protodioscin (170). Furthermore, two new furostanol saponins were obtained and established as 3-*O*-{β-D-glucopyranosyl(1 → 2)-α-L-rhamnopyranosyl(1 → 2)[α-L-rhamnopyranosyl (1 → 4)]-β-D-glucopyranosyl}-26-*O*-(β-D-glucopyranosyl)-(25*R*)-furost-5-en 3β,22α,26-triol (171) and its 22α-methoxy derivative (172)[125]. From the hexane fraction of the seeds, ζ₂-tocopherol (173) [126], 6-methyl dihydrophytylplastoquinone (174) and dihydrophytylplastoquinone (175) [127] and 2,3,6-trimethyl-5-(2-acetylisopentyl)-1,4-benzoquinone (176) [128] were isolated and characterized.

The rhizomes of *C. lacerus* were found to contain an appreciable amount of diosgenin [129]. The steroidal saponins of the rhizomes were identified as β-sitosterol-β-D-glucoside, prosapogenin A of dioscin, dioscin and gracillin [129].

The steroidal contents in callus cultures of *C. speciosus* were reported to be diosgenin, tigogenin, lanosterol and stigmasterol [130].

By GLC determination, various sapogenins present in the rhizomes of *C. afer* were reported to be diosgenin (0.8%), stigmasterol (1.5%) and costugenin (3.0%) [131]. The most abundant sapogenin, costugenin, is closely related in structure to samentogenin and could be an intermediate for the production of steroid hormones. A new steroidal saponin, aferoside A (154) was isolated from the roots of *C. afer* and identified by chemical transformations and spectroscopic methods [132].

From the roots and aerial parts of *C. tonkinensis*, numerous triterpenoids alcohols and ketones of the 14α-methyl-9,19-cyclocholestane type, compounds 177-181, were identified by GC-MS [133]. In addition, α-amyrin, β-amyrin, several common phytosterols and four 7-oxosterols were identified.

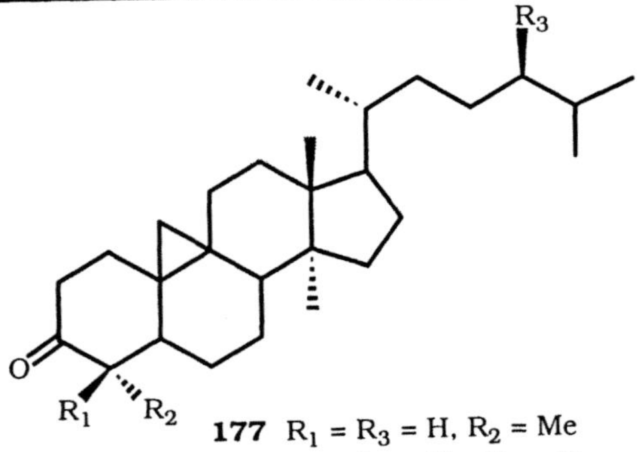

177 $R_1 = R_3 = H, R_2 = Me$
178 $R_1 = R_2 = Me, R_3 = H$
179 $R_1 = H, R_2 = R_3 = Me$
180 $R_1 = R_2 = Me, R_3 = =CH_2$
181 $R_1 = R_2 = H, R_3 = Me$

Diosgenin in the rhizomes of 3 species of *Costus, C. deistellii, C. igneus and C. lucanusianus,* was analyzed; only *C. lucanusianus* (diosgenin content = 0.61%) could be a new commercial source of diosgenin [134].

The rhizomes of *C. speciosus* was found to contain β-glucosidase which converts protogracillin to gracillin [135]. The soluble enzyme showed pH optima of 5.0 and the apparent Km for protogracillin of 0.10 mM.

An anti-fungal constituent of the rhizomes of *C. speciosus* has been identified as the methyl ester of *p*-coumaric acid [136].

Curcuma

Rhizomes of *Curcuma* sp. such as *C. aromatica, C. zedoaria, C. xanthorrhiza, C. longa* and *C. wenyujin* are used in traditional medicine in Asia, particularly, the essential oil of *C. wenyujin* which is used as a clinical remedy for uterus cancer in China [137].

Chemical constituents of the essential oils of *C. aromatica, C. zedoaria, C. xanthorrhiza, C. wenyujin* and *C. longa* have been examined by various groups and found to consist of mainly sesquiterpenoids.

From the rhizomes of *C. aromatica,* ten related sesquiterpenes, germacrone (182), (4S,5S)-germacrone 4,5-epoxide (184), curdione(188), neocurdione (189), dehydrocurdione(191), zedoarondiol (194), isozedoarondiol (195), methylzedoaron-diol (196), procurcumenol (197) and curcumenone (201) were isolated [38]. Further chemical examination of the rhizomes yielded twelve minor sesquiterpenoids, 13-hydroxygermacrone (183), (4S,5S)-12-acetoxygermacrone 4,5-epoxide (185), (4S, 5S)-13-hydroxygermacrone 4,5-epoxide (186), (4S,5S)-13-acetoxygermacrone 4,5-epoxide (187), acetoxyneocurdione (190), (4S)-13-hydroxydehydrocurdione (192), (4S)-13-acetoxydehydrocurdione(193), epiprocurcumenol (198), isoprocurcumenol (199), neoprocurcumenol (200), curcumadione (202) and isocurcumadione (203) [139]. Germacrone epoxide 184 gave methylzedoarondiol (196) when stirred in acidic methanol, and in acidic dioxane 184 gave zedoarondiol (194) and isozedoarondiol (195) [138]. The absolute stereostructure of neocurdione (189) isolated from *C. aromatica* was determined on the basis of its chemical transformation to curdione (188) and 4-epicurdione (188a) and from the X-ray crystallographic structure of the *p*-bromobenzoate of 8αH-dihydro-4-epicurdione [140].

Numerous sesquiterpenoids, germacrone (182) [141, 142], 13-hydroxygermacrone (183) [152], germacrone 4,5-epoxide (184) [142], dehydrocurdione (191) [143], zedoarondiol (194) [151], curcumenone (201) [153], curcumenol (204) [144] and isocurcumenol (205) [145], nine furanosesquiterpenoids, furanodienone (206) [146, 153], isofuranodienone (207) [153], curzerenone (208) [147, 153], epicurzerenone (209) [153],

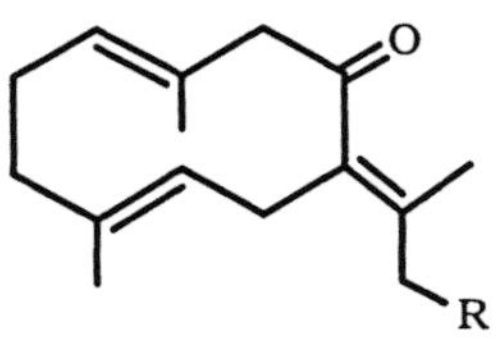

182 R = H
183 R = OH

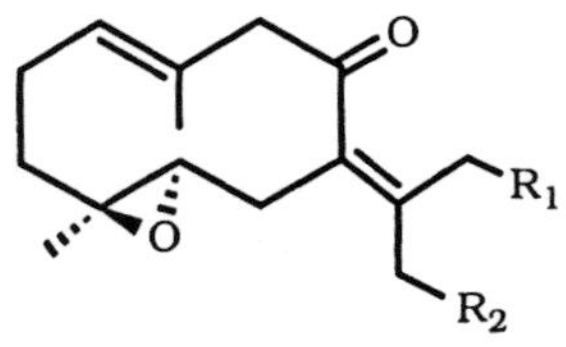

184 R₁ = R₂ = H
185 R₁ = OAc, R₂ = H
186 R₁ = H, R₂ = OH
187 R₁ = H, R₂ = OAc

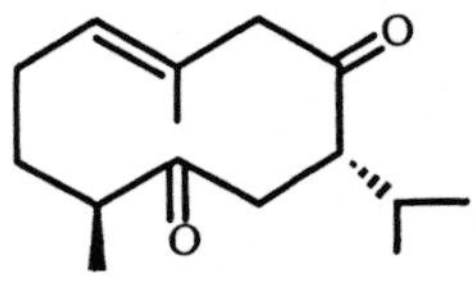

188

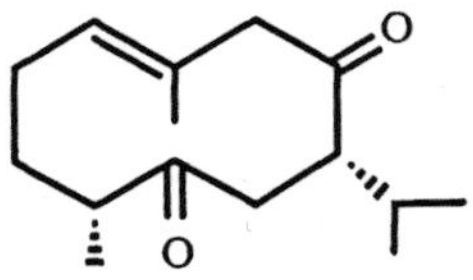

188a

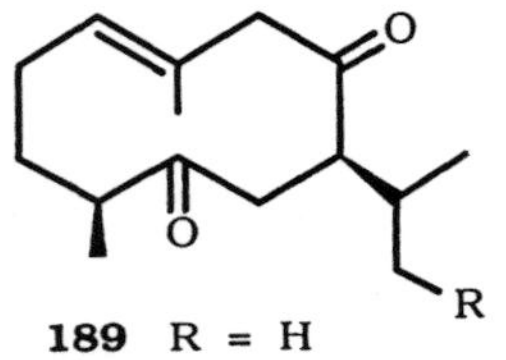

189 R = H
190 R = OAc

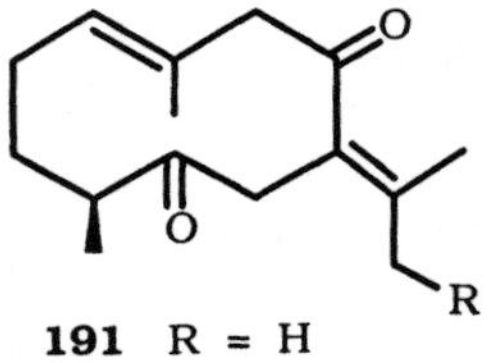

191 R = H
192 R = OH
193 R = OAc

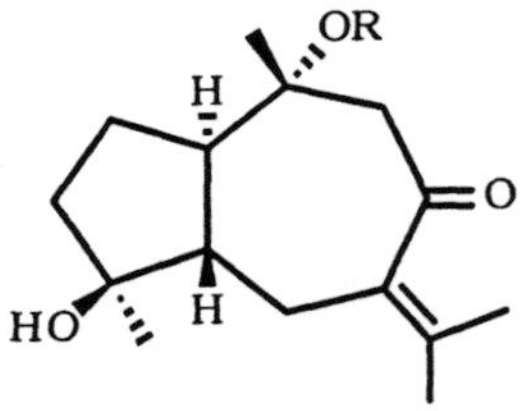

194 R = H
196 R = Me

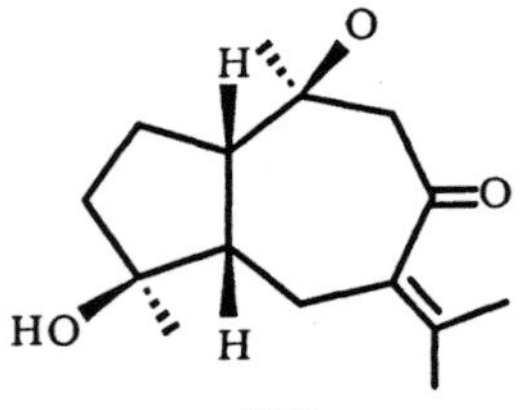

195

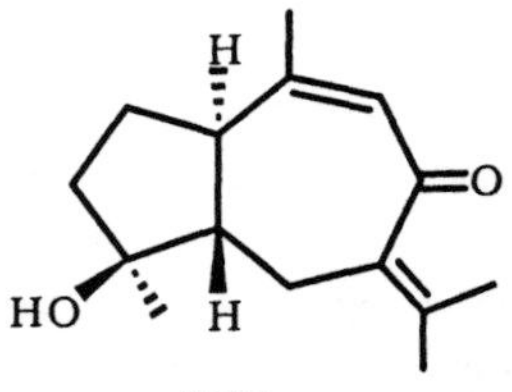

197

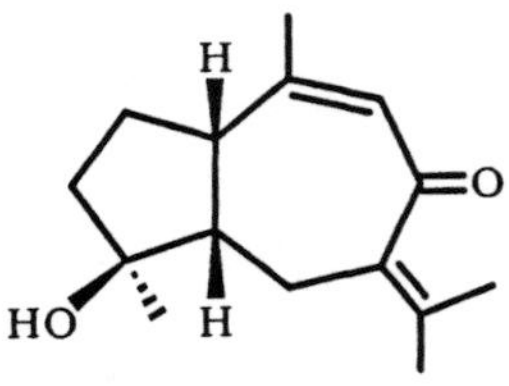

198

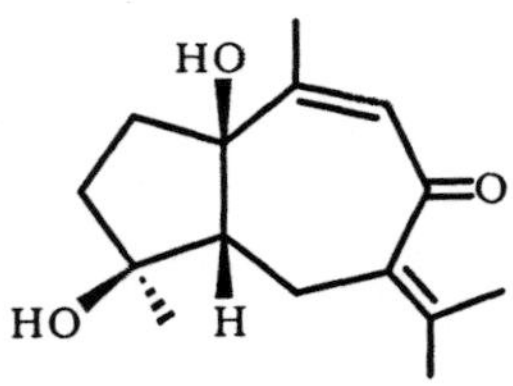

198a

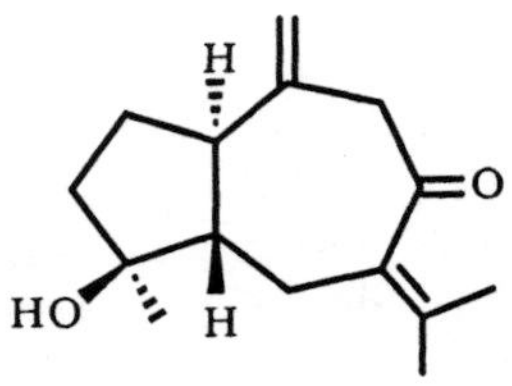

199

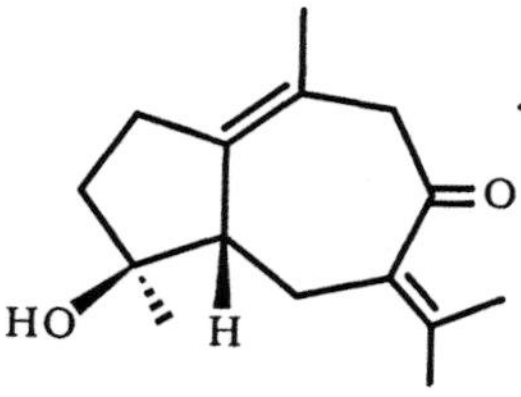

200 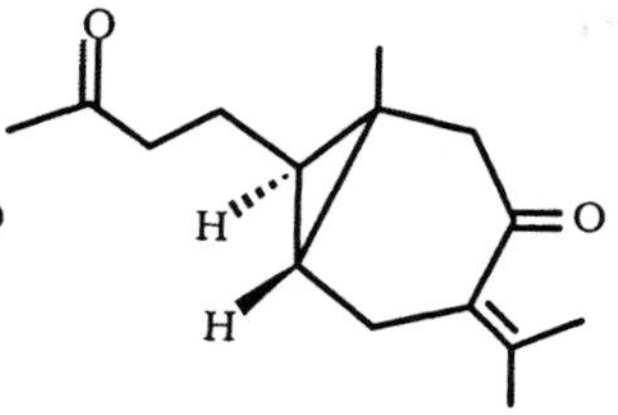　　　　　201 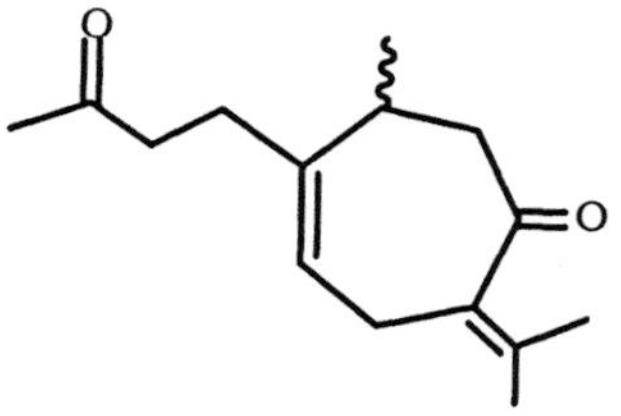　　　　　202

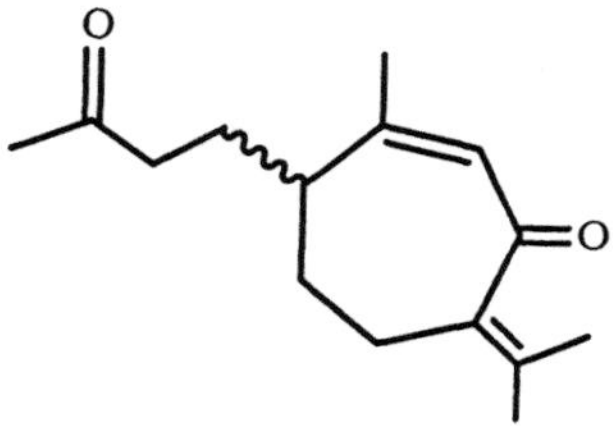

203

furanogermenone (**210**) [148], zederone (**211**) [149, 150], pyrocurzerenone (**212**) [153], zederol (**213**) [152] and curzenone (**214**) [156] and two spirolactones, curcumanolide A (**215**) [154] and curcumanolide B (**216**) [154], have been isolated from the rhizomes of *C. zedoaria*. Dehydrocurdione (**191**) was reported to be the major component of the rhizome [154]. Sesquiterpenes **182** and **184**, key intermediates in the biosynthesis of germacrone-type sesquiterpenes of *C. zedoaria*, were found to be the major constituents of the young shoot [154]. A biogenetic pathway for the formation of sesquiterpenes was proposed as shown in "Fig. (2)" [154]. The absolute stereostructure of germacrone 4,5-epoxide (**184**) isolated from *C. zedoaria* has been determined by spectroscopy, X-ray structural analysis and CD spectra [155]. The structure of zederone was revised from the previously proposed structure **211** to **211a** on the basis of NOE experiments and X-ray analysis of 12-bromozederone [156]. Furanogermenone (**210**) was found to be an anti-hepatotoxic principle of *C. zedoaria* [157]. Ethyl *p*-methoxycinnamate was the anti-fungal component of the rhizome of *C. zedoaria* [158]. *ar*-Curcumene, compounds **191, 210** and **211a**, from the rhizomes of *C. zedoaria* showed toxicity against neonate larvae of *Spodoptera littoralis* in a contact residue bioassay [159].

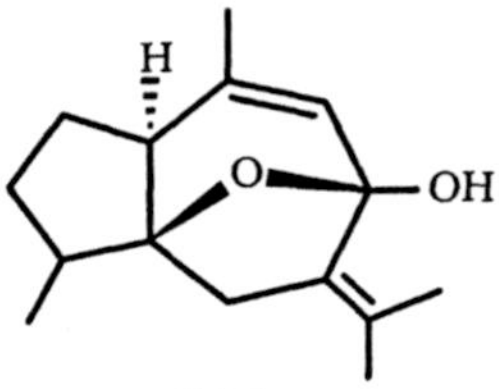

204

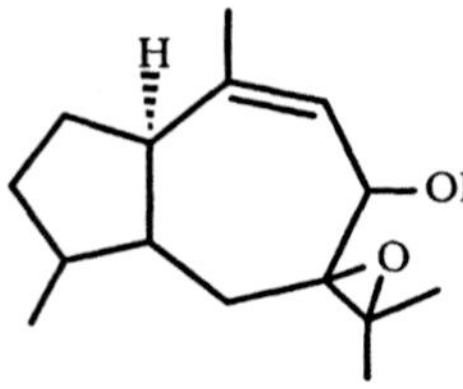

204a

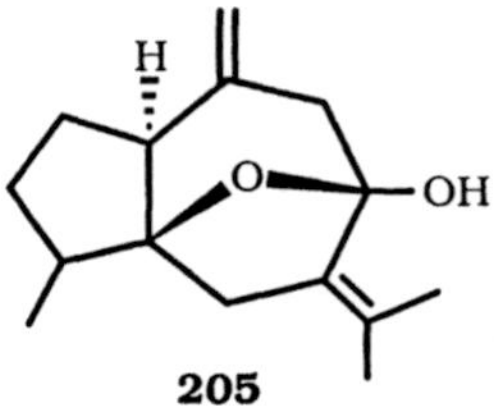

205

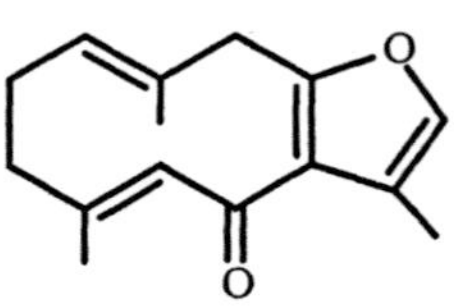

206

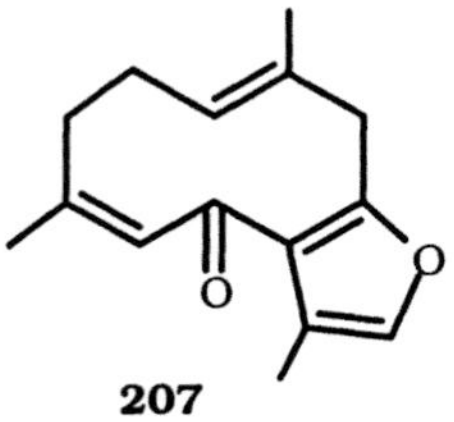

207

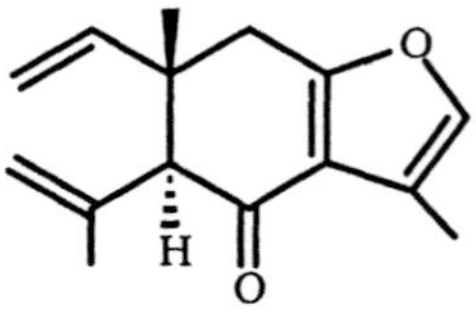

208

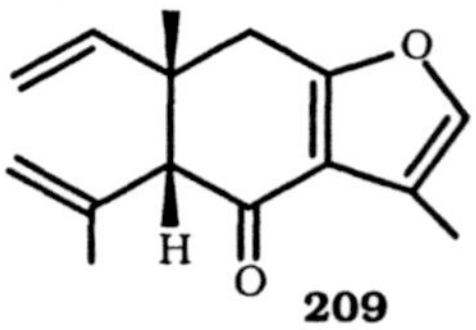

209

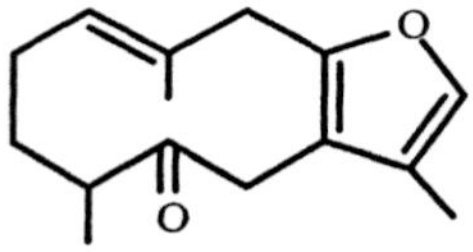

210

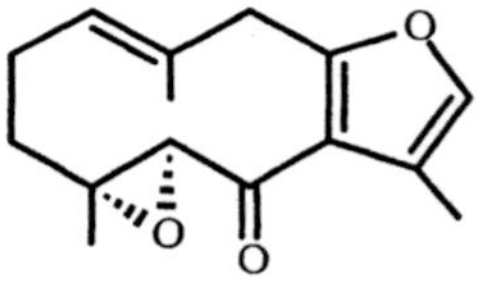

211

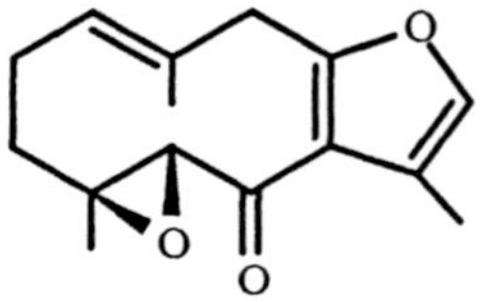

211a

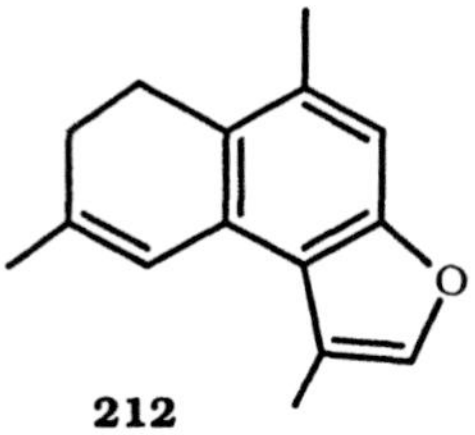

212

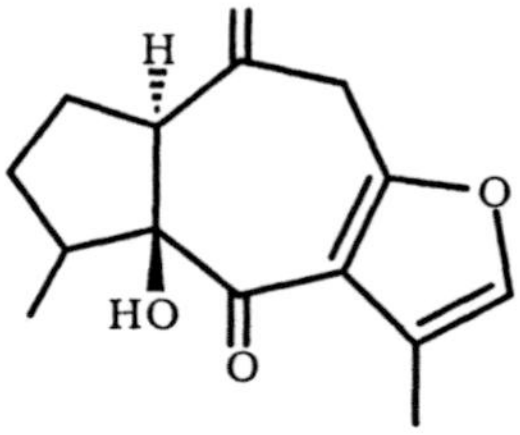

213

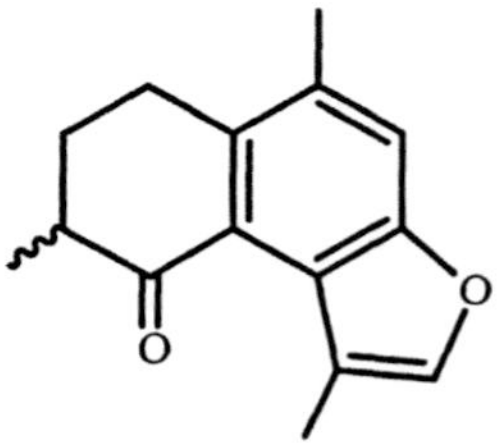

214

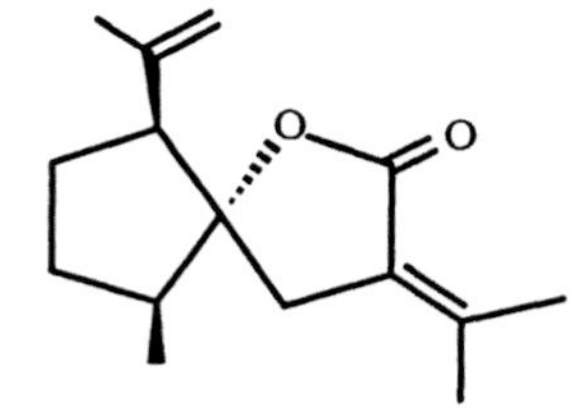

215

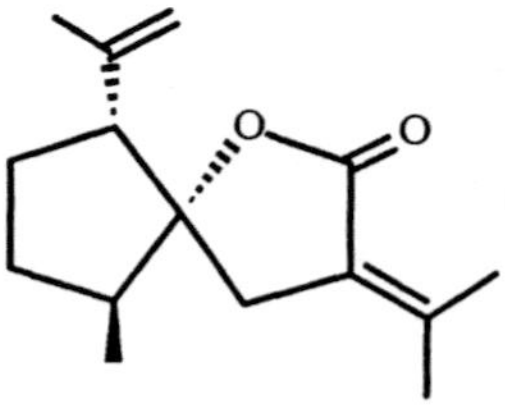

216

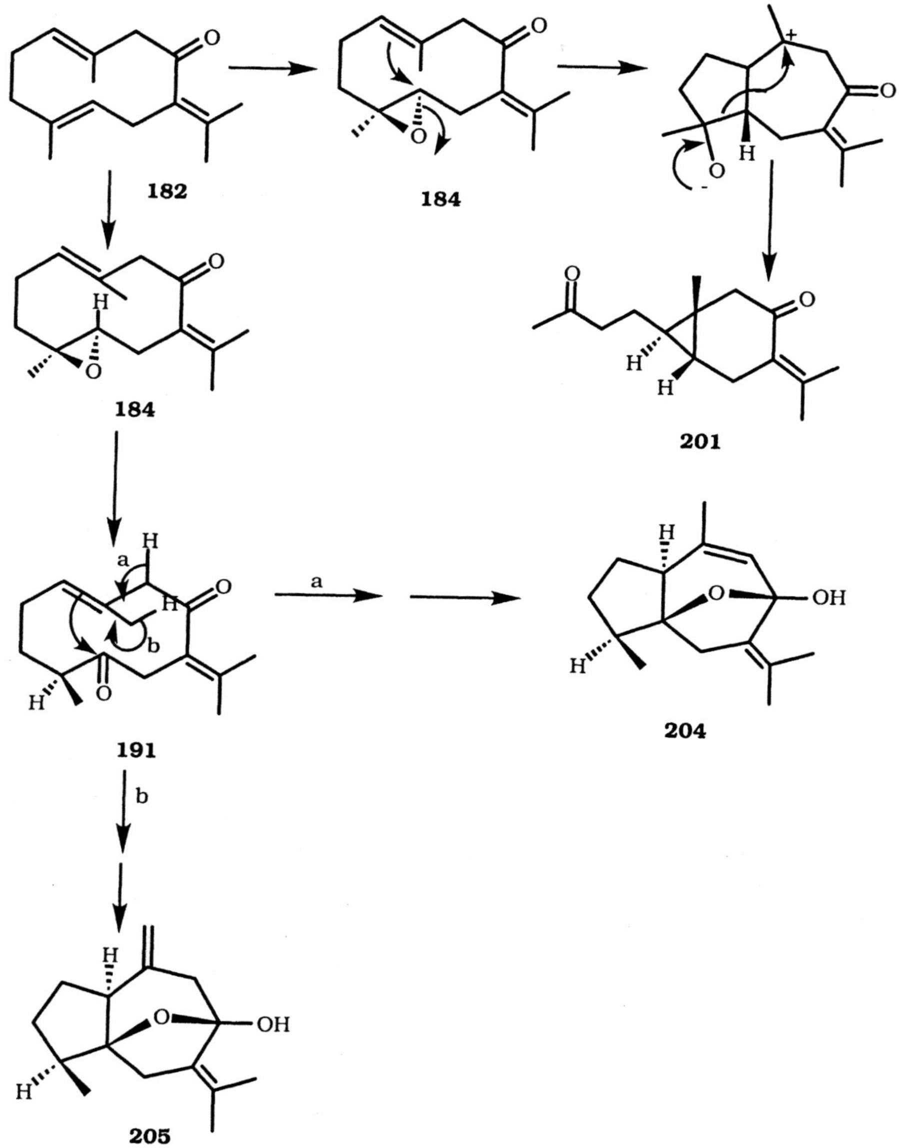

Fig. (2). Possible biogenetic pathways for the formation the sesquiterpenoids found in *C. zedoaria*.

Essential oil of *C. wenyujin* is used clinically as an anti-tumor agent in China [137]. Examination of the oil yielded eight sesquiterpenes, curdione **(188)** [160, 161], neocurdione **(189)** [162, 163], germacrone epoxide **(184)** [164, 165], (1*S*,10*S*), (4*S*,5*S*)-germacrone-1(10),4-diepoxide **(217)** [164], (-)-(1*R*,10*R*)-1,10-dihydrocurdione **(218)** [166], wenjine **(219)** [164], curcumol **(220)** [167] and curcumalactone **(221)** [161, 168]. The absolute stereostructure of curdione (**188**) was determined on the basis of its CD spectrum and the X-ray analysis of the bromobenzoate of 8αH-dihydrocurdione [160]. The absolute configuration of **221** was also elucidated on the basis of X-ray crystallographic analysis and the stereospecific transformation of **188** to **221** [161]. Compound **188**, in chloroform solution with a catalytic amount of HCl at room temperature, was stereoselectively and exclusively transformed to compound **221** as shown in "Fig (3)" [161]. The stereostructure of **221** was confirmed by X-ray diffraction study of the *p*-bromobenzoate of the lactol obtained from **221** [168]. Epoxidation of germacrone epoxide **184** with MCPBA (1 eq.) in CH$_2$Cl$_2$ at room temperature gave **217** in 86% yield [164]. Wenjine has a surprizingly stable diepoxide peroxyhemiacetal structure which is rare in nature. Photooxidation of germacrone diepoxide **217** using rose bengal or hematoporphorin as a photosensitizer gave wenjine in 25% yield [169].

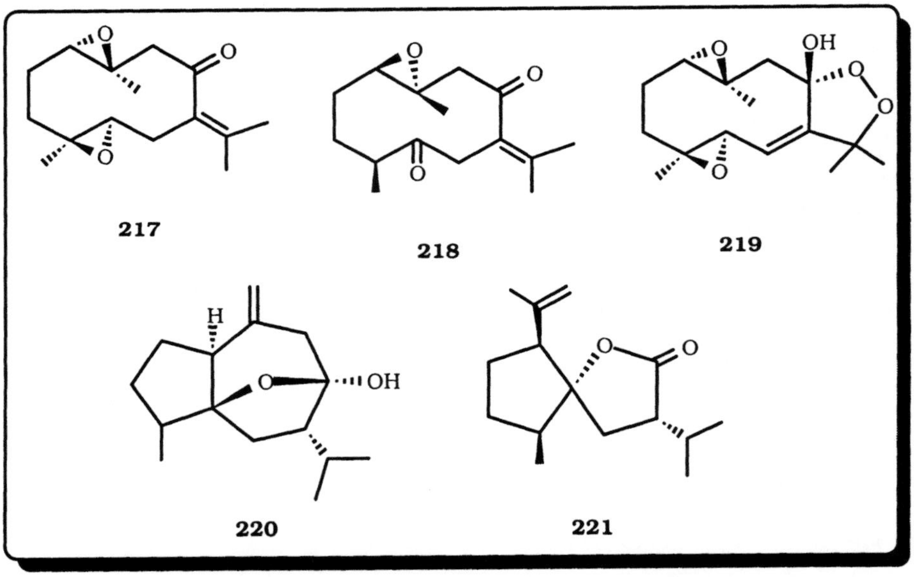

217

218

219

220   221

Four anti-tumor bisabolane sesquiterpenes, α-curcumene (**224**), *ar*-turmerone (**225**), β-atlantone (**226**) and xanthorrhizol (**227**), were isolated from the rhizomes of *C. xanthorrhiza* and **224** was reported to be the most active compound [170]. On further examination of the rhizomes, four additional bisabolanes, bisacurone (**228**), biscumol (**229**), bisacurol (**230**) and curlone (**231**) were obtained [171]. The absolute stereostructures of compounds **228-231** were determined by spectroscopy methods, chemical conversions and the CD exciton chirality method for allylic alcohols [171]. In more work by the same group, four related sesquiterpenes, bisacurone epoxide (**232**), bisacurone A (**233**), bisacurone B (**234**) and bisacurone C (**235**) were isolated from the same extract and their stereostructures were

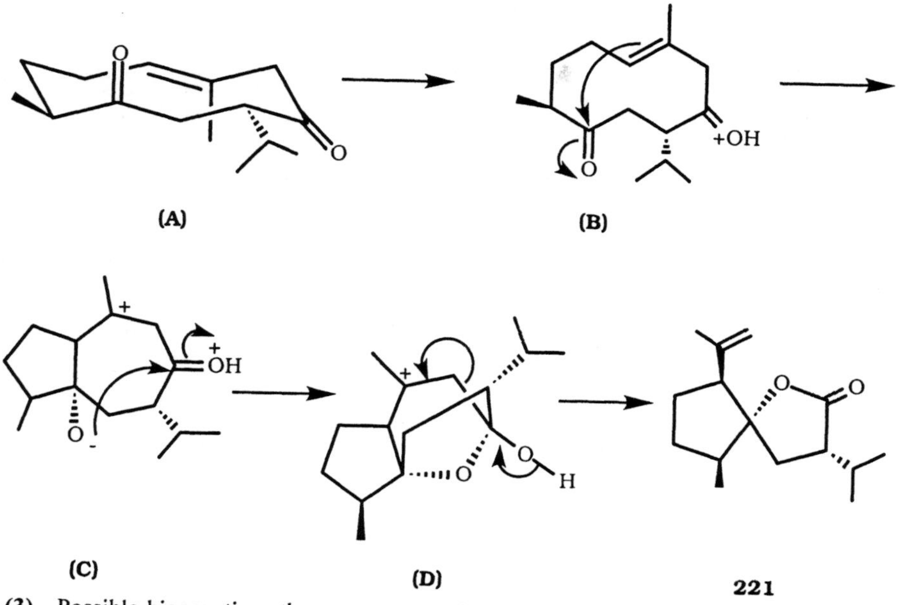

(A)

(B)

(C)

(D)

221

Fig.(3). Possible biogenetic pathway to curcumalactone (**221**) from curdione (**188**).

determined [172]. In the process of examining the absolute structure of epoxide **232**, the structure of bisacurone previously reported was revised from **228** to **228a** [36]. The chiral carbon of xanthorrhizol (**227**) was assigned to be *R* [173].

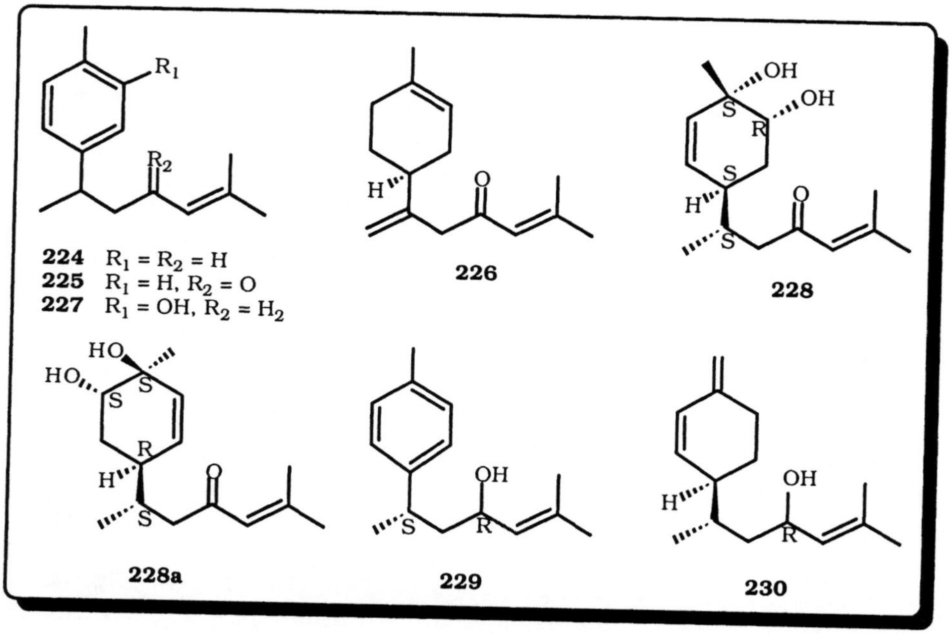

224 $R_1 = R_2 = H$
225 $R_1 = H, R_2 = O$
227 $R_1 = OH, R_2 = H_2$

226

228

228a

229

230

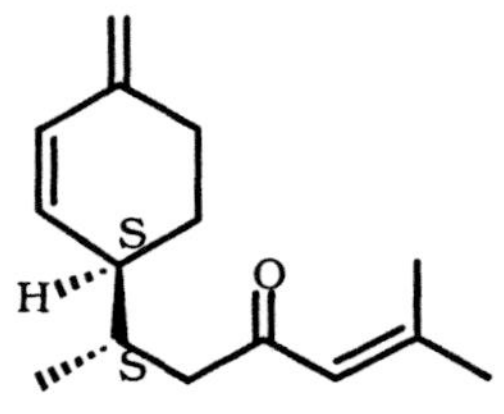

231 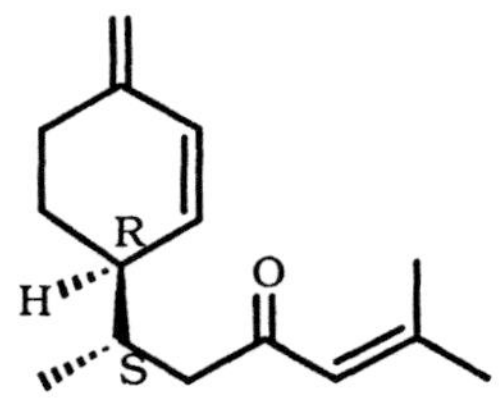   231a 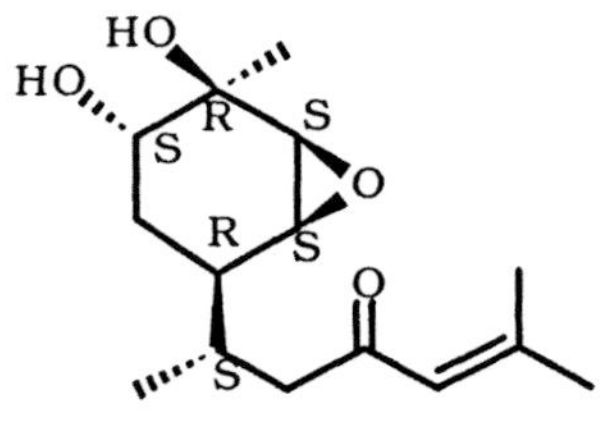   232 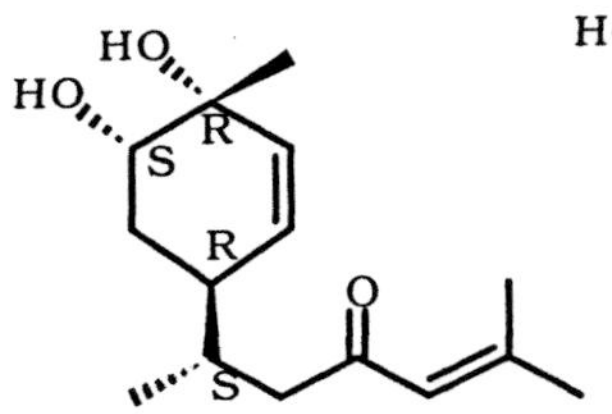   233 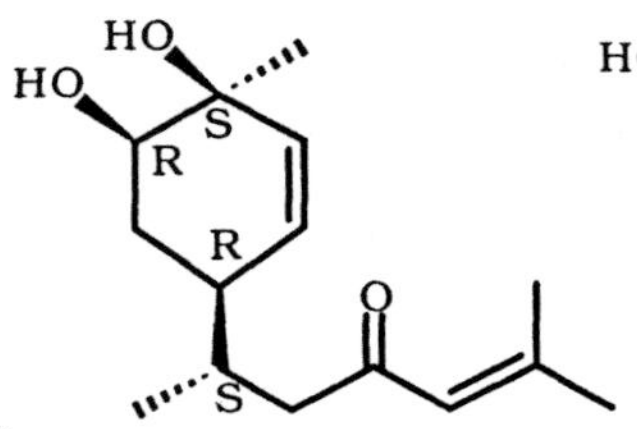   234 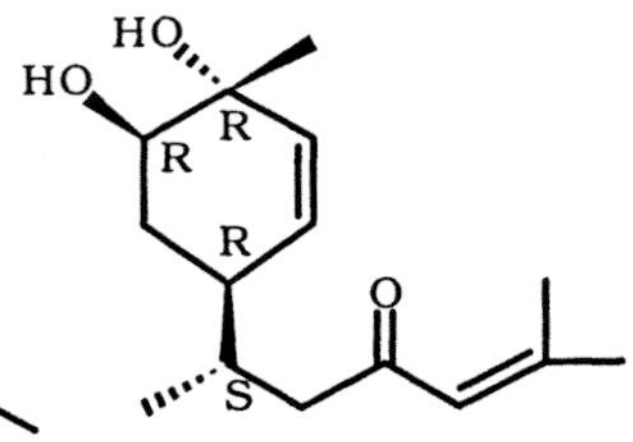   235

Several diarylheptanoids, curcumin (**237**) [174, 175], demethoxycurcumin (**238**) [174, 177], bisdemethoxycurcumin (**239**) [174, 177], curcumenoid **240** [177], dihydrocurcumin (**241**) [175], hexahydrocurcumin (**242**) [175], compound **243** [175, 176] and octahydrocurcumin (**244**) [175], have been isolated from the rhizomes of *C. xanthorrhiza*. Curcuminoids (**237-239**) have potent antioxidant activity [178]. Antioxidant activity of **240** and **243** was measured and compared with that of curcumin **237** [177]. Compound **240** showed slightly stronger activity, whereas **243** had weaker activity than that of **237** [177]. Although the high antioxidant activity of the curcuminoids was previously explained by the metal chelation of a 1,3-diketone moiety [179b, 179c], the antioxidative activity of curcumin was suggested to depend on the delocalization of the phenolic radical to the ketone function [179a]. The stronger activity of **237** and **240** than that of **243** indicated that the delocalization of the phenolic radical to the alkyl chain is essential for the antioxidant activity of the curcuminiods [177]. Three non-phenolic diarylheptanoids were isolated from the hexane extract of the rhizomes of *C. xanthorrhiza* and they were identified as *trans,trans*-1,7-diphenyl-1,3-heptadien-5-one (**249**), *trans*-1,7-diphenyl-1-hepten-5-ol (**250**) and *trans, trans*-1,7-diphenyl-1,3-heptadien-5-ol (**251**) [180]. The three diarylheptanoids exerted significant anti-inflammatory activity in the carrogeenin-induced hind paw edema in rat [180]. Two new phenolic diarylheptanoids, 5-hydroxy-7-(4-hydroxyphenyl)-1-phenyl-(1*E*)-1-heptene (**252**) and 7-(3,4-dihydroxyphenyl)-5-hydroxy-1-phenyl-(1*E*)-1-heptene (**253**), were isolated from the rhizomes of *C. xanthorrhiza* [181]. The two compounds exhibited significant hypolipidemic action by inhibiting hepatic triglyceride secretion [181]. Xanthorrhizol isolated from the rhizomes of *C. xanthorrhiza* was identified as a constituent prolonging phenobarbital-induced sleeping time and the effect was demonstrated to be due to inhibition of cytochrome P-450 activity [182]. Xanthorrhizol and furanodienone of *C. xanthorrhiza* showed potent insecticidal activity [159]. Germacrone obtained from the rhizomes of *C. xanthorrhiza* was reported to be anti-inflammatory [183] and hypothermic [184].

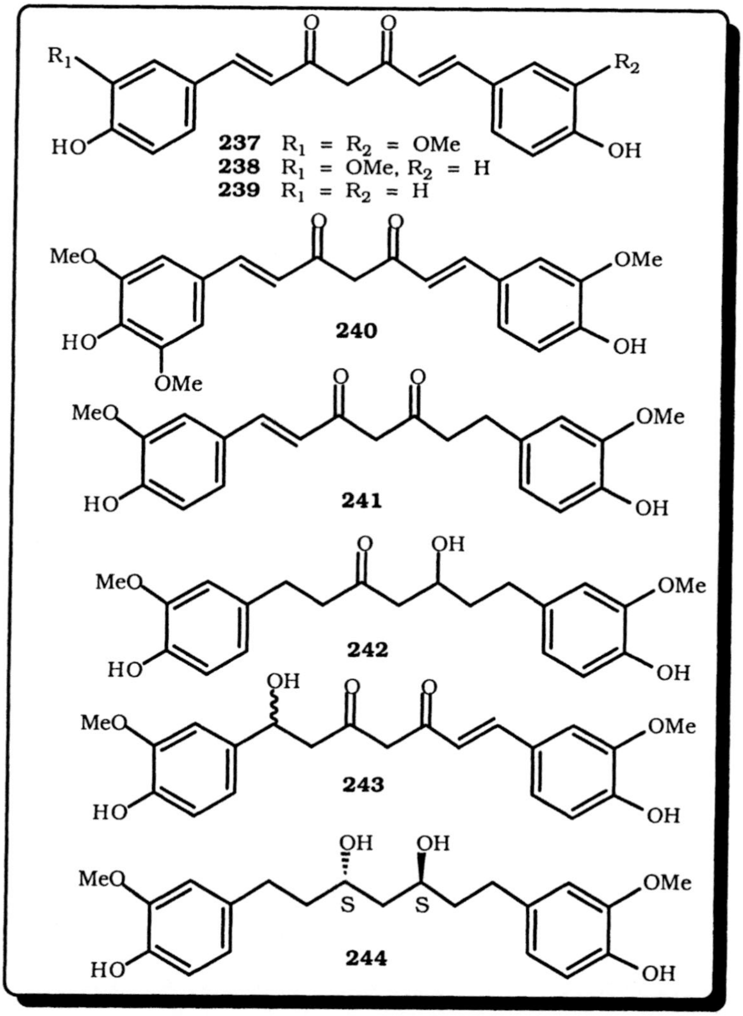

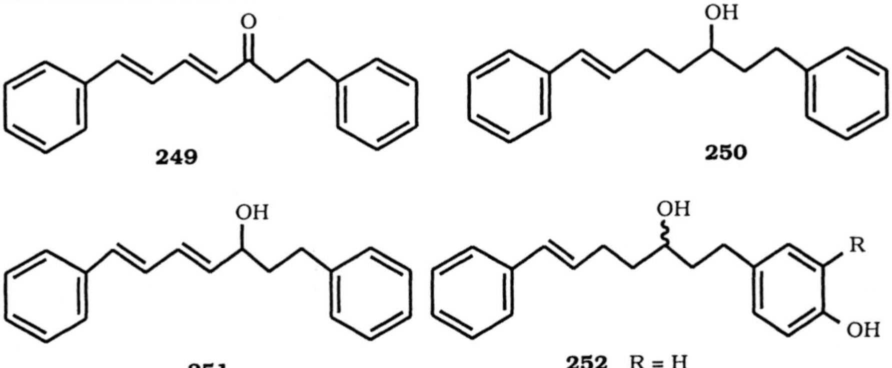

The dried rhizome of *C. longa* (*C. domestica*) is well-known under the name of turmeric. It has been used not only as a yellow coloring agent but also as food seasoning and traditional medicine in Asia.Turmeric powder was reported to increase the mucin content of gastric juice in rabbit [48], thus protecting gastric mucosa from irritants. Turmeric was shown to have gastric and duodenal anti-ulcer activities in rat [185]. The sesquiterpene constituents of the rhizomes of *C. longa* have been reported [187, 188]. Turmeric oil obtained from the cold hexane extraction of dried turmeric rhizomes was reported to contain three main constituents which were identified as *ar*-turmerone (**225**), α-turmerone (**236**) and β-turmerone (**231a**) [189]. It was reported that the anti-peptic ulcer constituent of *C. longa* was *ar*-turmerone [190].

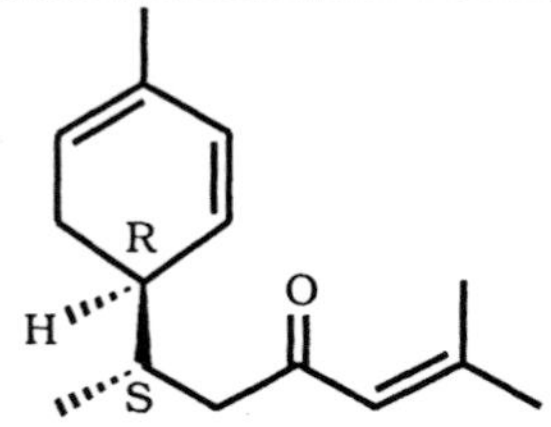

From the rhizomes of *C. longa*, several curcuminoids (**237-240**) [191], (**245, 246**) [192] and two curcumin related phenolic compounds (**247, 248**) [191] were isolated. Antioxidant activity of the two phenolic compounds **247** and **248** was determined by inhibiting the autooxidation of linoleic acid and comparing the activity with that of curcumin [191]. Compounds **247** and **248** showed stronger activity than that of curcumin. It seems to indicate that the chelation of the 1,3-diketone moiety is not so important for the activity. Curcuminoids **238-240** showed stronger anti-inflammatory action than that of curcumin and compound **239** possessed the strongest activity [191]. The anti-inflammatory activity of **247** was comparable with that of curcumin, while **248** had weaker activity [191]. Curcumin isolated from *C. longa* exhibited inhibitory effect on leukotriene B4 formation in rat [193]. Together with three known curcuminoids **237-239**, a new curcuminoid derivative, cyclocurcumin **254** was isolated from the rhizomes of *C. longa* [194]. Structure **254** was elucidated on the basis of spectral data and confirmed by synthesis from curcumin. The cyclocurcumin had nematocidal activity and the activity increased remarkably when the combinations of curcuminoids were used, suggesting a synergistic action [194]. Anti-oxidative compounds isolated from the rhizomes of *C. longa* were also identified to be curcumin and compound **239** [195]. Anti-coagulative principles isolated from *C. longa* were identified to be curcumin and compounds **238** and **239** [196].

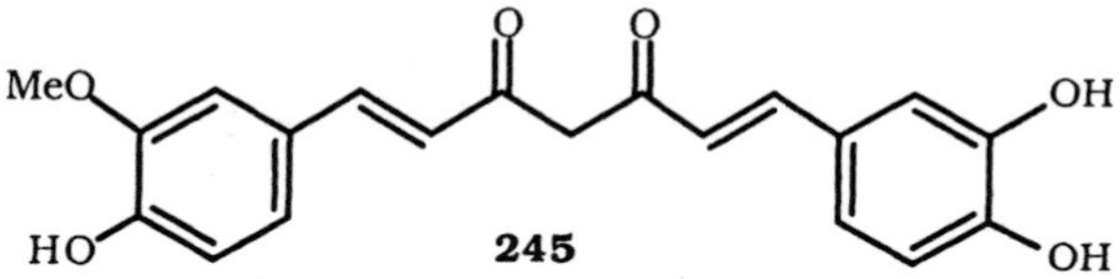

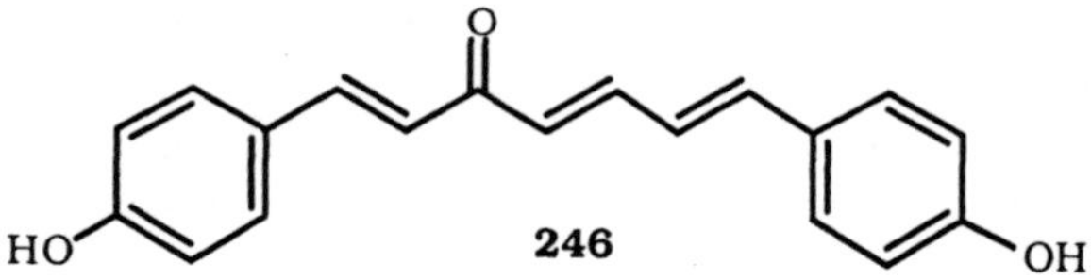

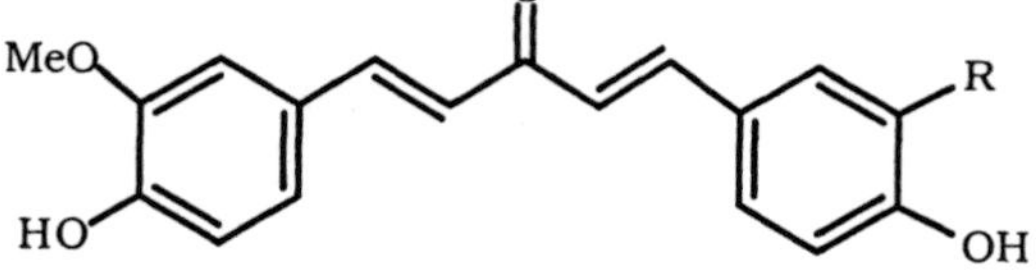

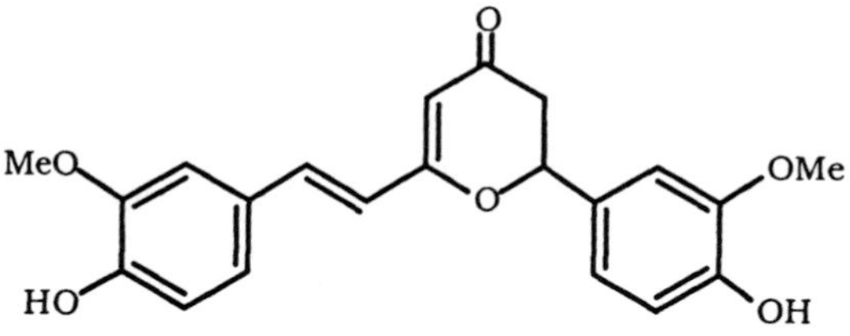

From the hot water extract of the rhizomes of *C. longa* which possesses immunological activity [197], a glycan named ukonan A was isolated [198]. The glycan showed remarkable reticuloendothelial system potentiating activity in the carbon clearance test [198]. Three acidic polysaccharides, ukonan A, ukonan B and ukonan C were isolated from the rhizomes of *C. longa* [199]. The three glycans were homogeneous on electrophoresis and gel chromatography and possessed remarkable reticuloendothelial system potentiating activity in the carbon clearance test [199]. Ukonan A, B and C have been studied further by methylation analysis, ^{13}C NMR and periodate oxidation and the molar ratios of the sugars in these polysaccharides were determined[199-201]. A neutral polysaccharide, named ukonan D, was isolated from the rhizomes of *C. longa* [202]. Ukonan D is composed of L-arabinose, D-galactose, D-glucose and D-mannose in the molar ratio of 1:1:12:0.2, in addition to small amounts of peptide moiety. It also showed remarkable RES activity [202].

Cytotoxicity of diterpenes from *Hedychium coronarium* and anti-tumor activity against sarcoma 180A of phenylpropanoids from *Alpinia galanga* and of bisabolane sesquiterpenes from *Curcuma xanthorrhiza* have been reviewed [203].

From the hexane-soluble fraction of the methanolic extract of the dried roots of *C. comosa*, a new diarylheptanoid, 1,7-diphenyl-3-acetoxy-(6*E*)-heptene (**255**), was isolated together with compounds **256, 250, 251** and **35** [204]. Compound **256** has been previously reported as a synthetic product [205].

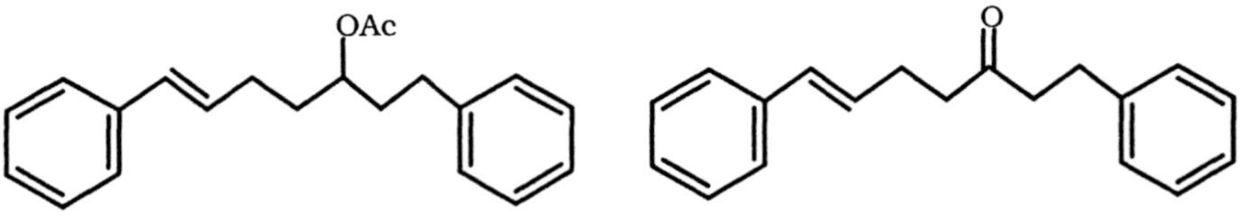

255   **256**

A sesquiterpene named oxycurcumenol (**204a**) was isolated from the steam-distilled essential oil of the fresh rhizomes of *C. heyneana* in addition to the known sesquiterpenes, germacrone (**182**), dehydrocurdione (**191**, curcumenol (**204**), isocurcumenol (**205**), curcumanolide A (**215**), B (**216**) and zerumbone [206]. From the benzene extract of the same plant, (*E*)-labda-8(17),12-dien-15,16-dial was isolated together with dehydrocurdione, curcumenol, curcumanolides A and B [206]. Epoxidation of curcumenol with MCPBA (1 eq.) in CH_2Cl_2 stereoselectively gave the β-epoxide which was identical with oxycurcumenol (**204a**) [206]. The essential oil of the rhizomes of *C. heyneana* was reported to be anthelmenthic against *Ascaris lumbricoides* Var. *suis* [206].

From the fresh rhizomes of *C. aeruginosa*, aerugidiol (**198a**) was isolated and its absolute stereostructure was determined by CD methods [207]. From the dried rhizomes of *C. aeruginosa*, two novel guaiane sesquiterpene lactones, zedoalactone A (**222**) and zedoalactone B (**223**) were isolated together with zedoarondiol (**194**) and isozedoarondiol (**195**) [208]. The structures **222** and **223** were established by ^{1}H and ^{13}C NMR spectroscopic studies and by comparison with closely related compounds.

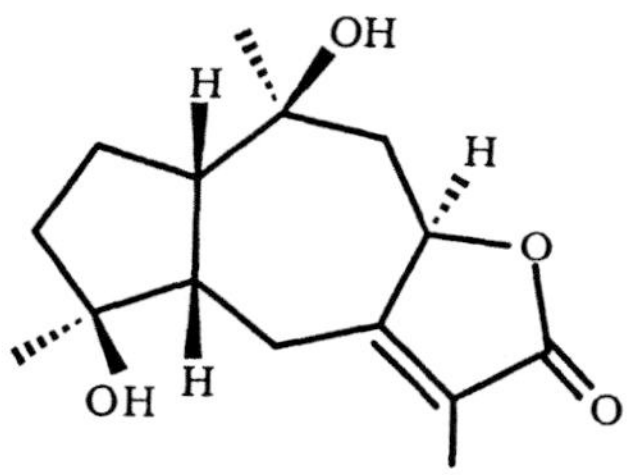

222 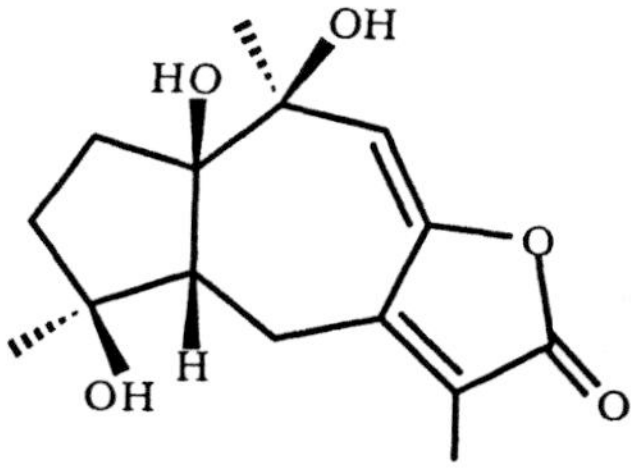   **223**

Hedychium

Several species of the genus *Hedychium* have been reported to be used in folk medicine. Hot water extract of fresh leaves of *H. coronarium* is used orally to treat hypertension or induce diuresis in Brazil [209]. Dried rhizomes powder of *H. coronarium* mixed with a cup of milk is taken daily to treat diabetes in India [210]. Decoction of the stems of *H. cylindricum* is drunk for spleen fever in Borneo [211]. Hot water extract of the rhizomes of *H. longicornutum* is drunk as vermifuge in Malaysia [212]. Fresh plant juice of *H. spicatum* in India is used for snakebite, the plant is made into a paste and applied to the bite, the juice is dropped into the nostrils, ears and navel [213]. Rhizomes of *H. spicatum* were used as an insect repellant and for stomach ailments [214] and also as an emmenagogue [215] in India.

An aqueous ethanolic (50%) extract of fresh leaves of *H. coronarium* from Brazil showed hypotensive [209] and diuretic [216] activities in rat. Leaf essential oil of *H. coronarium* had anti-bacterial activity [217]. Stems of *H. coronarium* were reported to possess anti-cercarial activity [218]. An ethanolic extract of the stems of *H. coronarium* showed anti-malarial activity against *Plasmodium falciparum* and also possessed cytotoxic activity [211]. Petroleum ether extract of the dried entire plants of *H.*

gardnerianum was active as a molluscicide [219, 220]. An aqueous ethanolic (50%) extract of the leaves of *H. gardnerianum* possessed hypoglycemic activity in rat [221]. An aqueous extract of the dried entire plants of *H. spicatum* had anti-tumor action [222]. Leaf essential oil and an ethanolic extract of the rhizomes of *H. spicatum* showed anti-bacterial and anti-fungal activities [217, 223]. Aqueous ethanolic (50%) extract of the dried roots of *H. spicatum* was active as anti-malarial against *Plasmodium berghii* [224] and also had hypotensive activity in dog [225].

Essential oil of the rhizomes of *H. acuminatum* was examined and 75 constituents were identified, most of which were known monoterpenes and sesquiterpenes [226]. Terpenoids of the essential oil from the leaves [227] and rhizomes [228] of *H. coronarium* were identified. Essential oil obtained from the rhizomes of *H. spicatum* was chemically examined [228-230]. Several sesquiterpene alcohols from the rhizomes of *H. spicatum* var. *acuminatum* were isolated and identified [231]. Tubers of *H. yunnanense* were found to contain mainly sesquiterpenoids [232]. The constituents of the leaves of several species of the genus including *H. coccineum, H. coronarium, H. cylindricum, H. ellipticum, H. gardnerianum, H. greenii, H. longicornutum, H. roxburghii, H. stenopetalum* and *H. thyrsiforme* were found to be mainly flavonoids and aromatic acids [233, 234].

β-Sitosterol, β-sitosterol-β-D-glucoside and cryptomeridiol (**257**) were isolated from *H. spicatum* [235]. From the rhizomes of *H. spicatum*, a furanoid diterpene was isolated and was deduced on the basis of spectroscopic and chemical evidences as hedychenone (**258**) [236]. From a larger quantity of rhizomes of *H. spicatum*, a second furanoditerpene, 7-hydroxyhedychenone (**259**) was obtained [237]. On the basis of spectroscopic data and chemical transformations 6-oxo-labda-7,11,14-trien-16-oic lactone (**260**) was identified in the rhizomes of *H. spicatum* [238]. Five cytotoxic labdane-type diterpenes, (*E*)-labda-8(17),12-diene-15,16-dial, coronarin A (**261**), coronarin B (**262**), coronarin C (**263**) and coronarin D (**264**) were isolated from the chloroform extract of the rhizomes of *H. coronarium* [239, 240]. Among these five diterpenes, coronarin A and B exhibited the most significant cytotoxic activity. Further examination of the same extract yielded two additional labdane-type diterpenes, coronarin E (**265**) and coronarin F (**266**) [241]. Together with two known diterpenes **264** and **265**, isocoronarin D (**267**) was isolated from the hexane extract of the rhizomes of *H. coronarium* and the ethyl ether of coronarin D which was an artefact of the extraction process was obtained from the ethanolic extract of the rhizomes [242]. In addition to (*E*)-labda-8(17),12-diene-15,16-dial, compounds **262, 264** and **267** 7-β-hydroxycoronarin B (**268**), labda-8(17),11,13-trien-15(16)-olide (**269**) and ester (**270**) were isolated from the dichloromethane extract of *H. coronarium* [243].

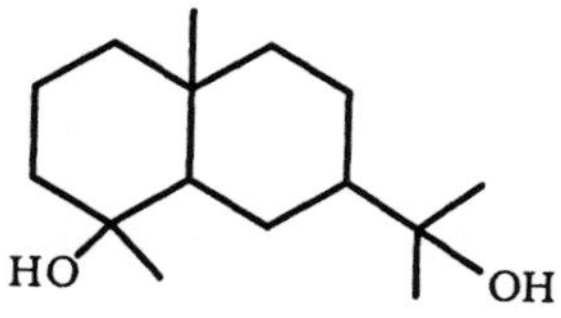

257

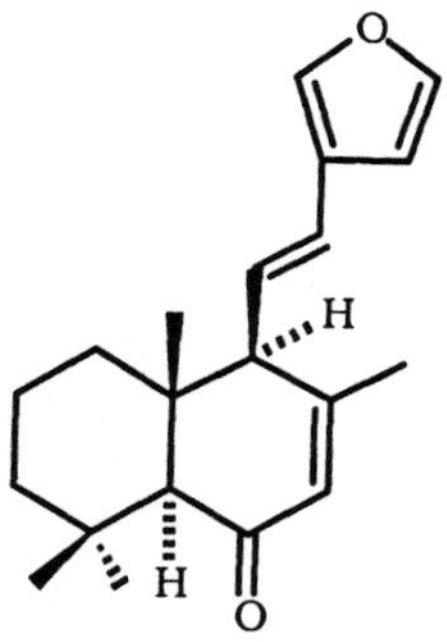

258

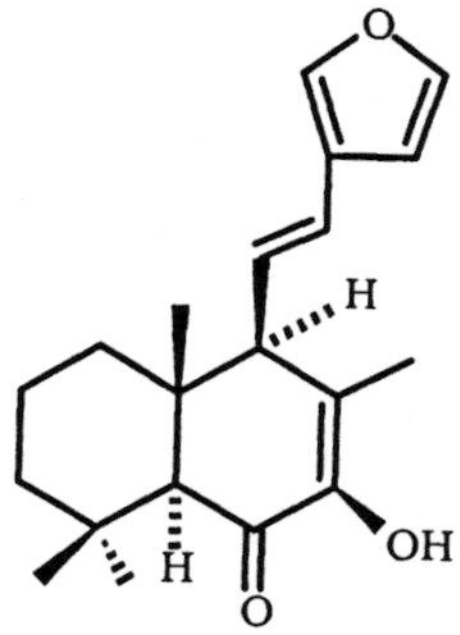

259

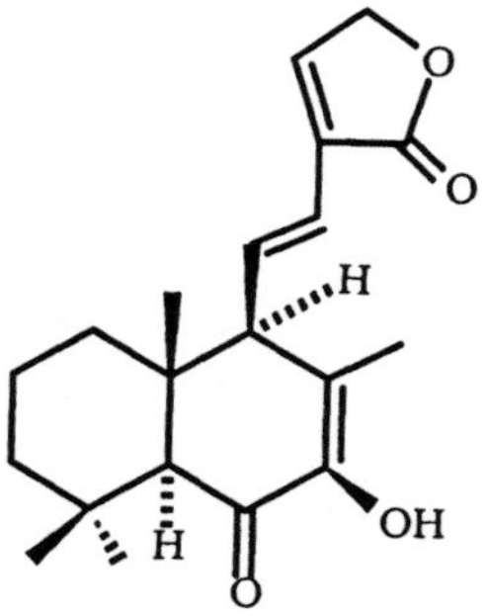

260

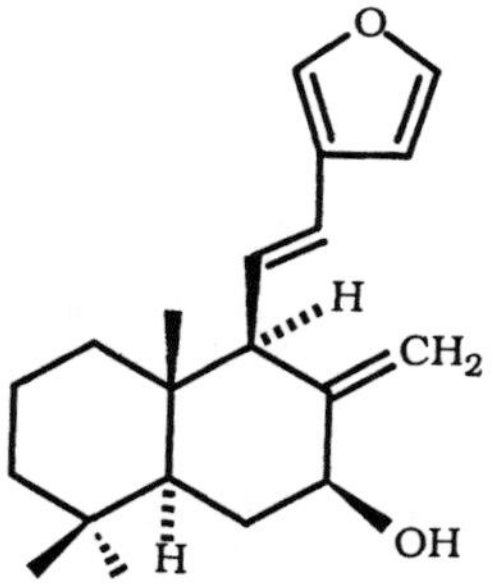

261

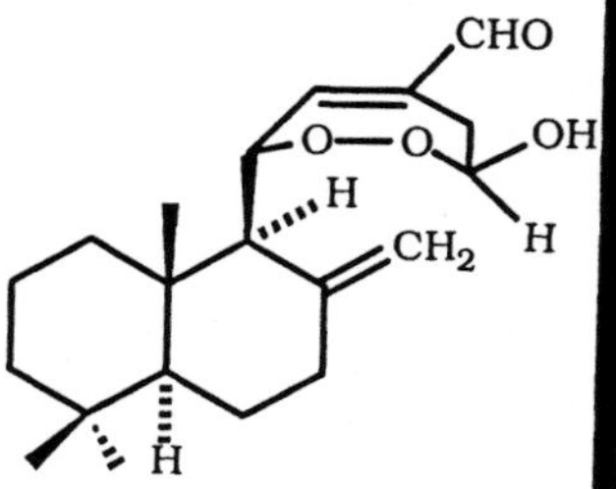

262

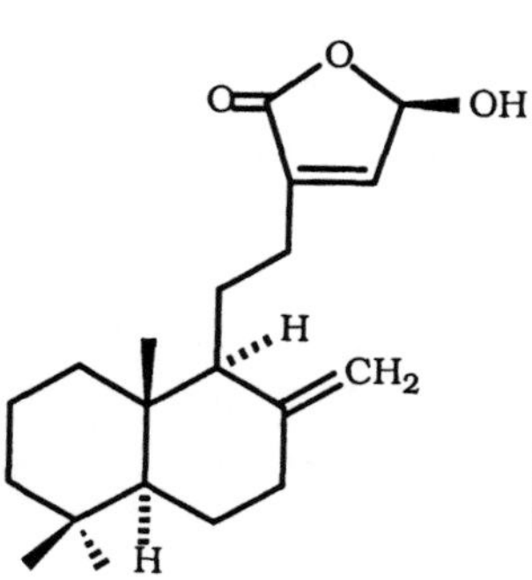

263

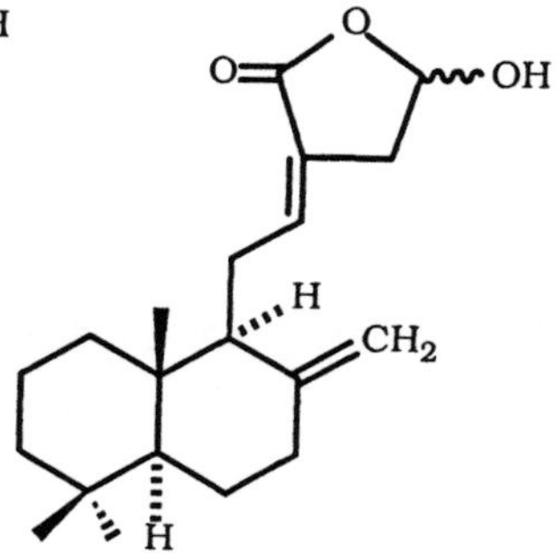

264

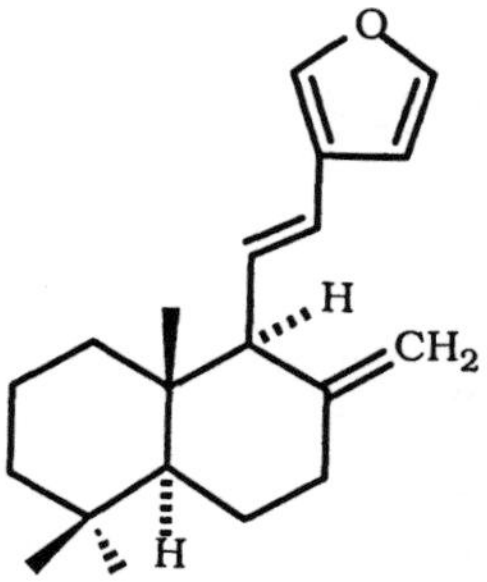

265

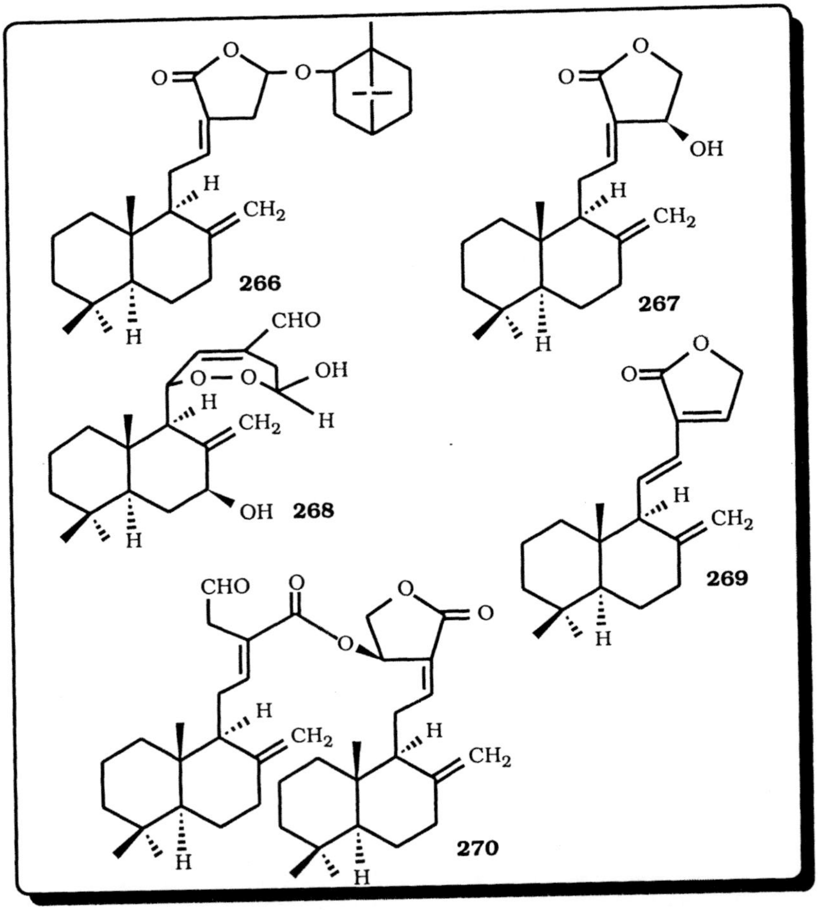

Kaempferia

Rhizome of *Kaempferia galanga* is used for abdominal pain in women by applying externally and also used as an ointment for swelling and rheumatism in Indonesia [244]. Fresh rhizome of *K. galanga* is used to treat itching and pruritus in the Philippines [245]. Root of *K. rotunda* is used for typhoid fever in India [246].

Within the genus *Kaempferia*, the preliminary biological activity investigations were carried out mainly on the extracts of the rhizome of *Kaempferia galanga*. An ethanolic extract of the rhizomes of *K. galanga* possessed anti-ascariasis activity [247]. An acetone extract of the same rhizomes inhibited monoamine oxidase [248]. Rhizomes of *K. galanga* from Sri Lanka and from Bangladesh showed nematocidal activity [249, 250]. A benzene extract of the dried rhizomes of *K. galanga* from China had colony formation inhibition in Hela cell culture [251]. A water extract of the rhizomes showed anti-tumor activity [82] and a methanolic extract of the same plant had anti-tumor promoting activity [252]. Essential oil of the rhizomes of *K. galanga* possessed glutathion-*S*-transferase induction effect [253].

From the rhizomes of *K. rotunda*, a cyclohexane oxide derivative, crotepoxide (**271**), was isolated [254]. Besides compound **271**, three additional cyclohexane diepoxides (**272-274**) and (-)-zeylenol (**275**) were obtained from the rhizomes of the same plant [255]. Crotepoxide (**271**), diepoxide (**276**), (+)-zeylenol (**277**) and two (+)-zeylenol related substances (**278, 279**) were isolated from an unnamed *Kaempferia* (Thai name; krachaikao) [256, 257].Together with **271** and **276-279**, (-)-pipoxide (**280**) was isolated from the rhizomes of *K. angustifolia* [257]. Oxygenated cyclohexanes are a small group of naturally occurring compounds. Only four plant genera, *Uvaria* (Annonaceae)[258-263], *Croton* (Euphorbiaceae)[264], *Piper* (Piperaceae) [265] and *Kaempferia* [254-257] have been found to be capable of producing cyclohexane oxide derivatives.

Benzyl benzoate and crotepoxide isolated from the rhizomes of *K. rotunda* exhibited insecticidal activity [266].

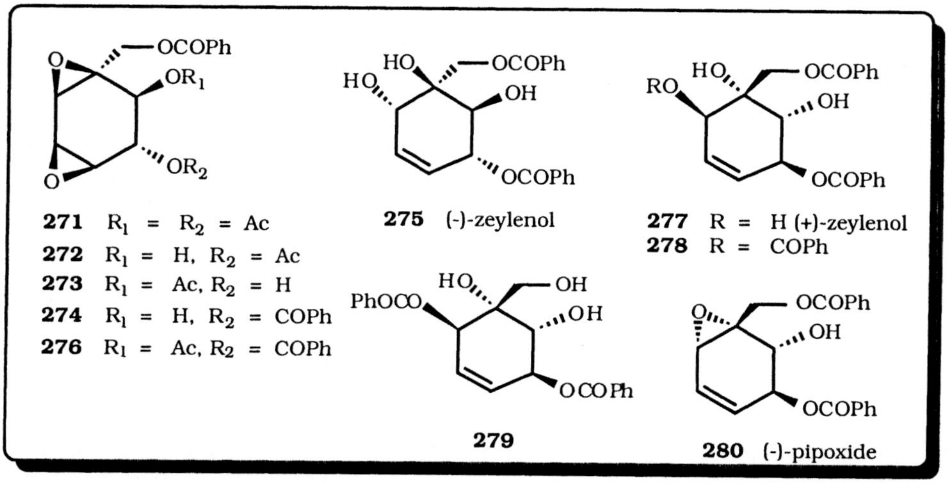

271   R_1 = R_2 = Ac
272   R_1 = H, R_2 = Ac
273   R_1 = Ac, R_2 = H
274   R_1 = H, R_2 = COPh
276   R_1 = Ac, R_2 = COPh

275   (-)-zeylenol

277   R = H (+)-zeylenol
278   R = COPh

279

280   (-)-pipoxide

Two chalcones (**281, 282**) and fourteen flavonoids (**283-296**) were isolated from the hexane extract of the rhizomes of *K. parviflora* [the plant was previously described as *Boesenbergia pandurata* (Roxb) Schlr. (black rhizome)] and compound **288** was found to be the major constituent of the extract [267, 268]. Anti-inflammatory and anti-pyretic activities of **288** were determined to be comparable to that of aspirin [269]. Several flavonoids and flavonoid glycosides were isolated from the leaves of *K. rotunda, K. pulchra* and *K. elegans* [233].

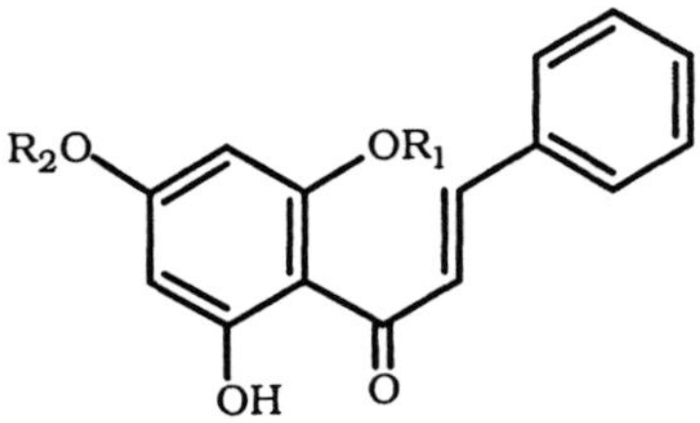

281 $R_1 = H, R_2 = Me$
282 $R_1 = R_2 = Me$

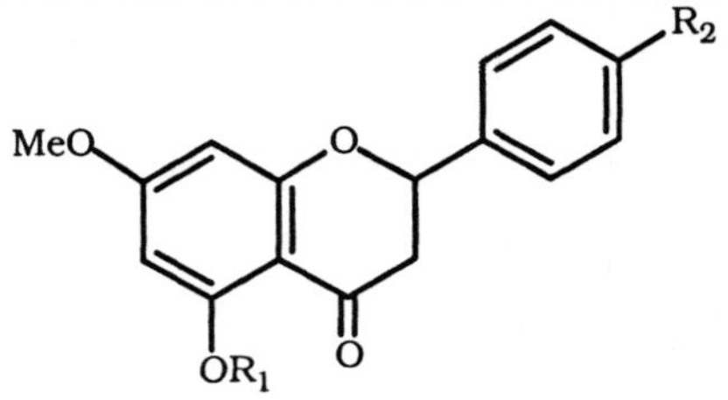

283 $R_1 = R_2 = H$
284 $R_1 = Me, R_2 = H$
285 $R_1 = H, R_2 = OMe$

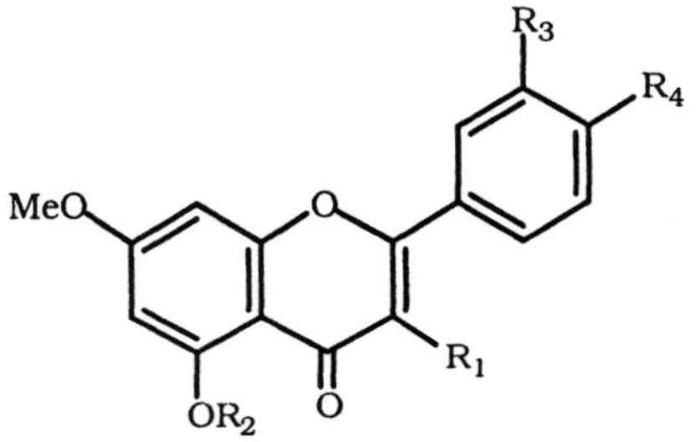

286 $R_1 = R_2 = R_3 = R_4 = H$
287 $R_1 = R_2 = R_3 = H, R_4 = OMe$
288 $R_1 = R_3 = R_4 = H, R_2 = Me$
289 $R_1 = R_3 = H, R_2 = Me, R_4 = OMe$
290 $R_1 = H, R_2 = Me, R_3 = R_4 = OMe$
291 $R_1 = OMe, R_2 = R_3 = R_4 = H$

292 $R_1 = R_4 = OMe, R_2 = R_3 = H$
293 $R_1 = OMe, R_2 = Me, R_3 = R_4 = H$
294 $R_1 = R_3 = R_4 = OMe, R_2 = H$
295 $R_1 = R_4 = OMe, R_2 = Me, R_3 = H$
296 $R_1 = R_3 = R_4 = OMe, R_2 = Me$

Six pimarane diterpenes (**297-302**) were obtained from the rhizomes of an unnamed *Kaempferia* (Thai name: jang-ngung) [270]. Rhizomes of *K. pulchra* were found to contain two pimarane diterpenes (**303, 304**) [271], which were reported to possess anti-inflammatory action [272]. Isopimaric acid, a well-known pimarane derivative, was also isolated from the chloroform extract of the rhizomes of *Kaempferia* sp.[256].

Three cinnamic acid derivatives (**305-307**) were isolated from the rhizomes of *K. galanga* [273]. Compound **306** was also found in the extract of *K. pulchra* [271]. Ethyl *p*-methoxycinnamate (**306**) was reported to inhibit monoamine oxidase [248] and to have cytotoxic activity [251]. Cinnamic acid derivatives obtained from *K. galanga* were shown to possess larvicidal activity [273]. Ethyl cinnamate as well as ethyl *p*-methoxycinnamte from the rhizomes of *K. galanga* was highly toxic against larvae of *Spodoptera littoralis* [159]. A monoterpene ketone, 3-

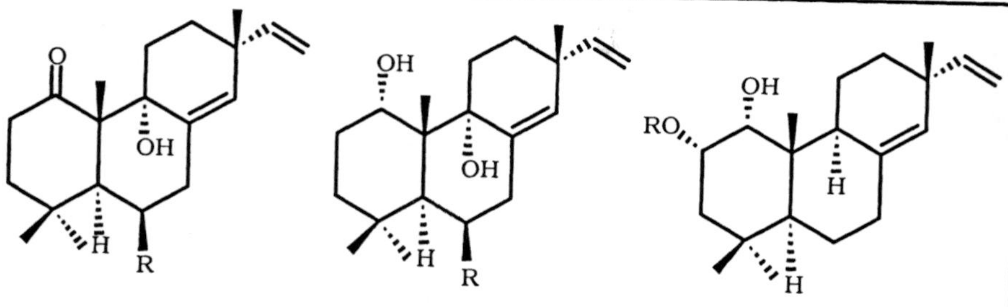

297 R = H
298 R = OH
299 R = OAc

300 R = H
301 R = OH
302 R = OAc

303 R = H
304 R = Ac

caren-5-one (**308**) was obtained from the rhizomes of *K. galanga* [274]. From the leaves of *K. galanga* and *K. rotunda*, various aromatic acid derivatives were isolated [234].

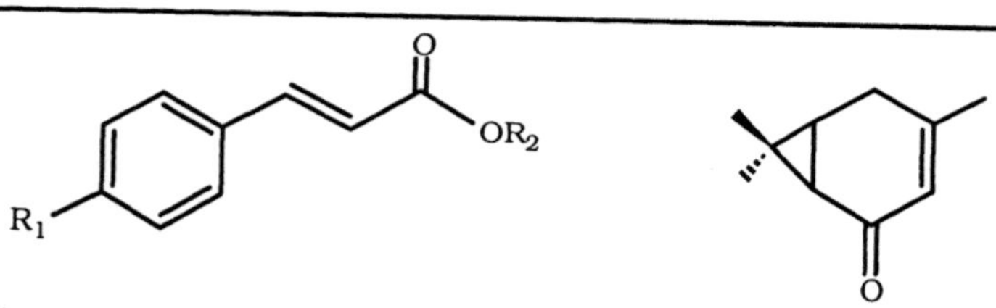

305 R_1 = H, R_2 = Et
306 R_1 = OMe, R_2 = Et
307 R_1 = OMe, R_2 = H

308

Zingiber

Oil prepared from the rhizomes of *Zingiber cassumunar* has long been used in Thai traditional medicine for relieving muscle pain and inflammation. The pulverized dried rhizomes are also given orally as an anti-asthmatic in Thailand. An ethanolic extract of the rhizomes was found to have anti-spasmodic activity [275]. A study of the anti-asthmatic activity of the rhizomes in children has been conducted [276]. Oral administration of capsules containing pulverized dried rhizomes to asthmatic children showed positive effect in relieving asthmatic symthoms with low side effects. Oil prepared from the rhizomes showed anti-inflammatory action [277]. Sauna treatment for rheumatoid arthritis patients using essential oil from the rhizomes showed significantly good effects [278].

Numerous monoterpenes were identified in the essential oil of the rhizomes of Z. *cassumunar* [279]. From the hexane extract of the rhizomes of Z. *cassumunar*, thirteen aromatic compounds 309, 310, 312, 317-324, 326 and 327 were isolated [280-282]. The structures of these compounds were determined on the basis of their spectroscopic data and by syntheses. The crystal structures of cyclohexene derivative 312 and quinone 324 were determined by X-ray diffraction analysis [280]. Evaluation of the biological activity of these compounds revealed that compound 317 possessed smooth muscle relaxant activity (guinea-pig ileum and tracheal chain) [283].The mechanism of bronchodilator activity of 317 was found to be similar to those of aminophylline and papaverine [284]. Compound 317 also exhibited uterine relaxant effect and shared a similar mechanism of action with papaverine [285]. Anti-inflammatory activity of 309, 312, 317-319, 326 and 327 was evaluated using the carragenin-induced rat paw edema model and compound 317 was reported to possess the strongest inhibitory activity on the edema formation [286]. Compounds 322 and 323 exhibited insecticidal activity towards neonate larvae of *Spodoptera littoralis* [266].

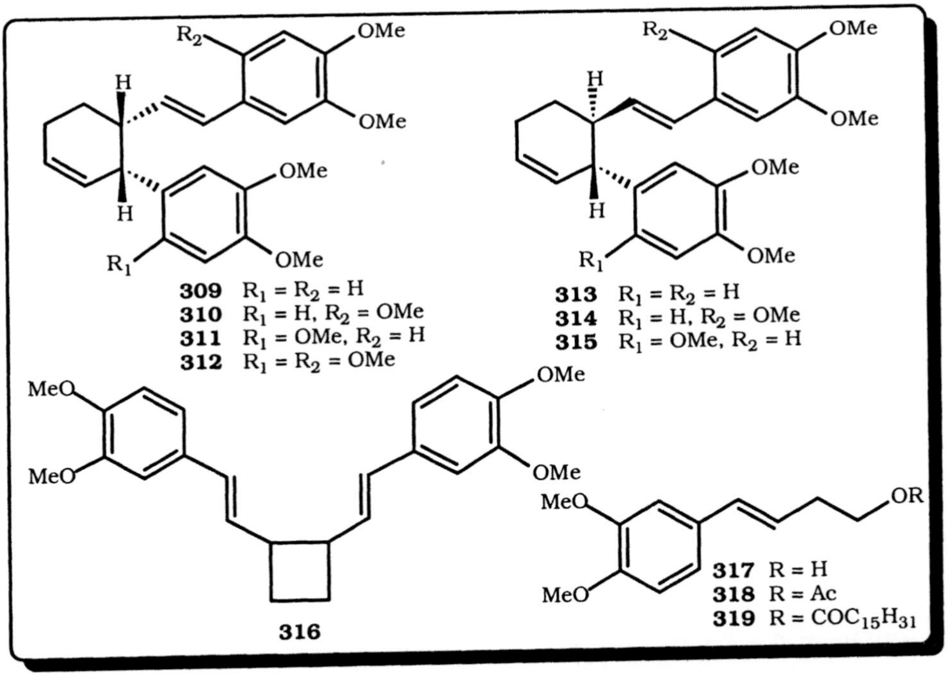

309 $R_1 = R_2 = H$
310 $R_1 = H, R_2 = OMe$
311 $R_1 = OMe, R_2 = H$
312 $R_1 = R_2 = OMe$

313 $R_1 = R_2 = H$
314 $R_1 = H, R_2 = OMe$
315 $R_1 = OMe, R_2 = H$

316

317 R = H
318 R = Ac
319 R = $COC_{15}H_{31}$

On further examination of the chemical constituents of the rhizomes of Z. *cassumunar* four aromatic compounds 311, 313, 314, and 325 , in addition to compounds 309, 310, 312, 317, 319-322 and 324, were identified[287]. Together with compounds 309 and 313, two new phenylbutanoid dimers were isolated from the fresh rhizomes of Z.

cassumunar and their structures were elucidated as **315** and **316** [288]. Three potent antioxidants, cassumunin A (**328**), cassumunin B (**329**) and cassumunin C (**330**) were isolated from the rhizomes [289]. These compounds also showed inhibitory activity against inflammation induced by the tumor promotor, TPA [289]. Recently, four new phenylbutanoid monomers were isolated from the fresh rhizomes of *Z. cassumunar* and identified as (*E*)-4-(4-hydroxy-3-methoxyphenyl)but-3-en-1-yl acetate (**331**), (*E*)-4-(4-hydroxy-3-methoxyphenyl)but-2-en-1-ol (**332**), (*E*)-4-(3,4-dimethoxyphenyl)but-3-en-1,2-diol (**333**) and (*E*)-2-methoxy-4-(3,4-dimethoxyphenyl)but-3-en-1-ol (**334**) [290].

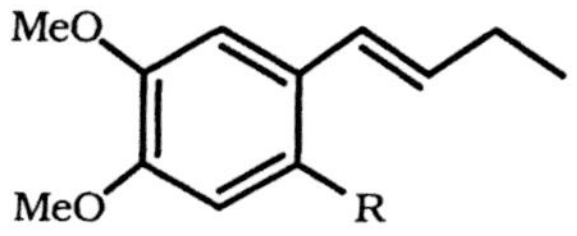

320 R = H
321 R = OMe

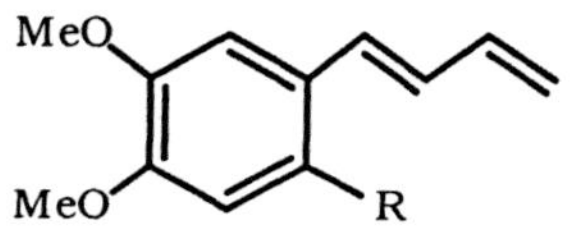

322 R = H
323 R = OMe

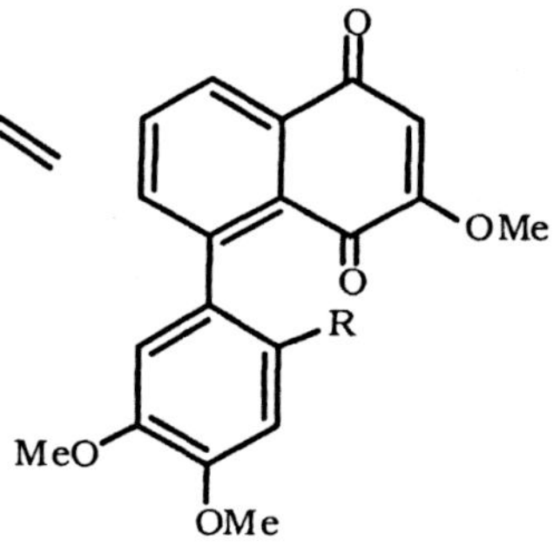

324 R = H
325 R = OMe

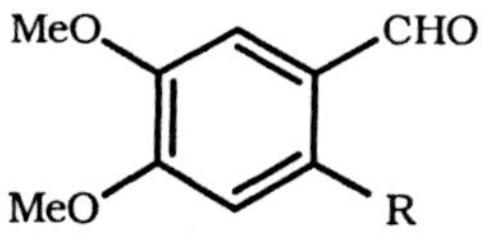

326 R = H
327 R = OMe

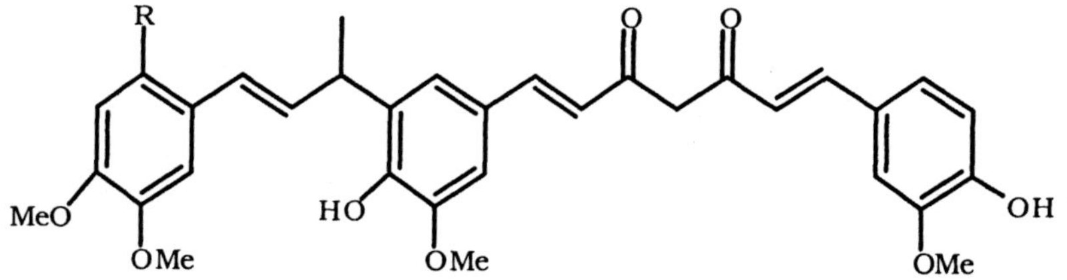

328 R = H
329 R = OMe

Ginger (*Zingiber officinale*) is one of the best known and most important spices. Ginger is sold commercially as a fresh rhizome, as preserved ginger, as a pungent extract oleoresin and as a steam distilled essential oil. This plant is widely cultivated in the tropical, subtropical and temperate regions. Constituents of the essential oil of ginger were reported to be mainly mono- and sesquiterpenes [291-297]. Changes in the volatile constituents of ginger rhizomes during storage and cultivation have been

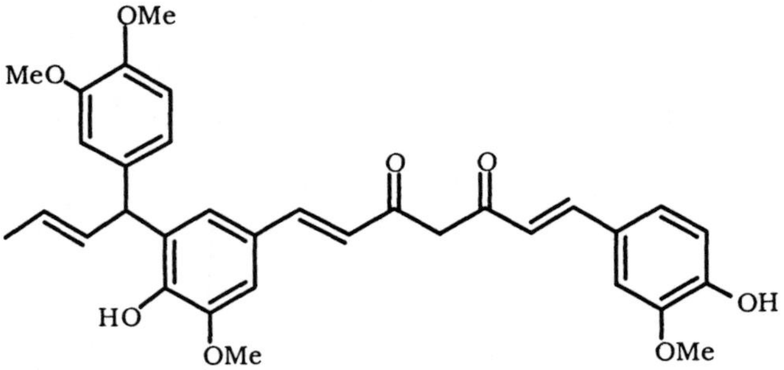

330

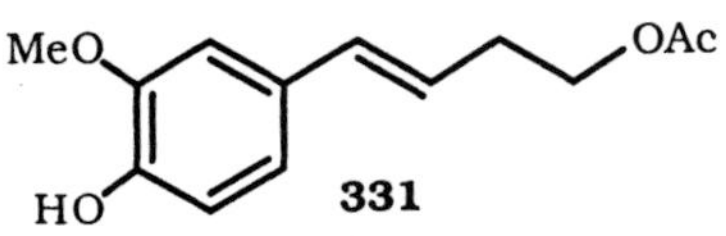

331

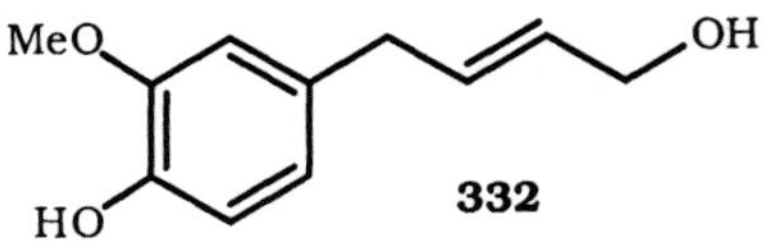

332

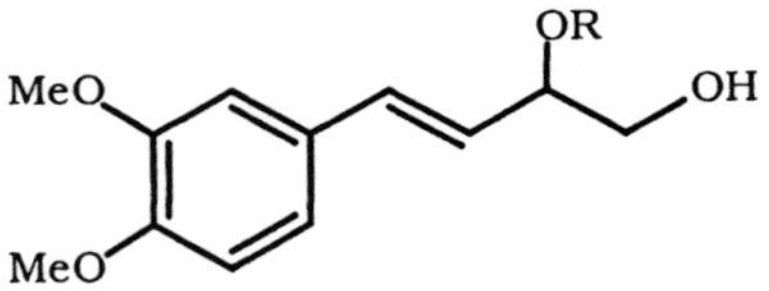

333 R = H
334 R = Me

studied [297, 298]. The major pungent constituents of ginger are gingerols
(**I**), shogaols (**II**), paradols (**III**) and related compounds including
gingerdiols (**IV**), and related compounds (**IVa**), gingerdiones (**V**),
hexahydrocurcumin (**VI**) and the methyl ethers of many of these
compounds [293, 299-302]. The absolute stereo-chemistry of C-5 in the
(+)-gingerols **Ia** (n = 4, 6 and 8) was found to be *S* [299]. C-3 and C-5 in
the gingerdiols (**IV**) were determined to have *R* and *S* configurations,
respectively [302].

The biosynthesis of [6]-gingerol was investigated by administration of
labelled precursors to whole *Z. officinale* plants [303, 305]. The
hypothetical intermediates [6]-dehydrogingerdione (**335**), [6]-gingerdione
(**337**) and [6]-dehydrogingerol (**339**), were synthesized and shown to be
incorporated into [6]-gingerol [304]. A biosynthetic pathway for [6]-
gingerol was then proposed as shown in "Fig. (4)".

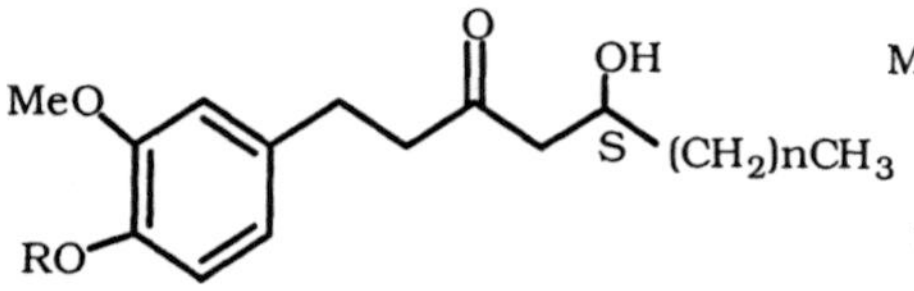

gingerols, Ia R = H
                Ib R = Me
         n = 2, 4, 6, 8, 10

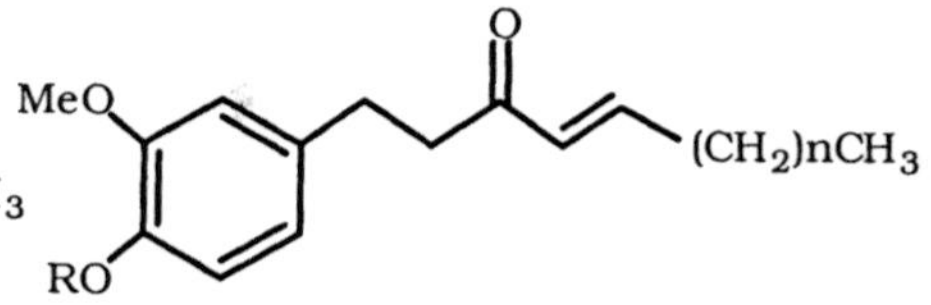

shogaols, IIa R = H
                  IIb R = Me
         n = 2, 4, 6, 8, 10

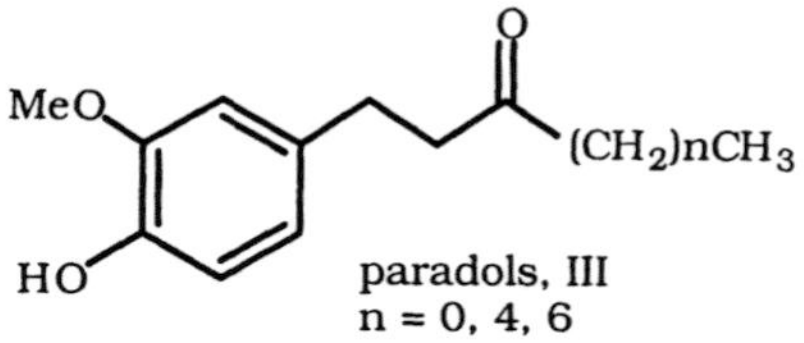

paradols, III
n = 0, 4, 6

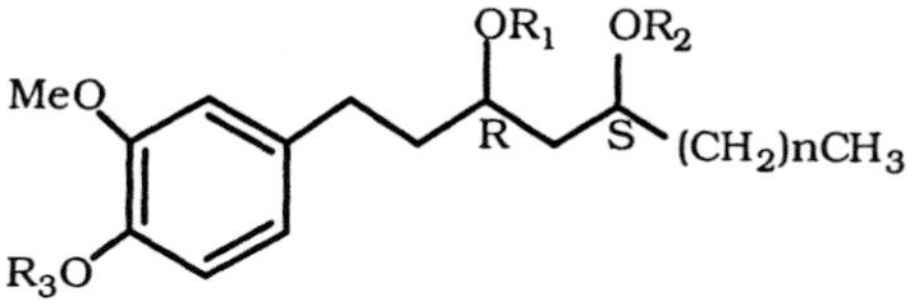

gingerdiols, IV
$R_1 = R_2 = R_3 = H$
n = 4, 6, 8
gingerdiols, IVa
n = 4
$R_1 = R_3 = H, R_2 = Ac$
$R_1 = Ac, R_2 = R_3 = H$
$R_1 = R_2 = Ac, R_3 = H$
$R_1 = R_2 = Ac, R_3 = Me$

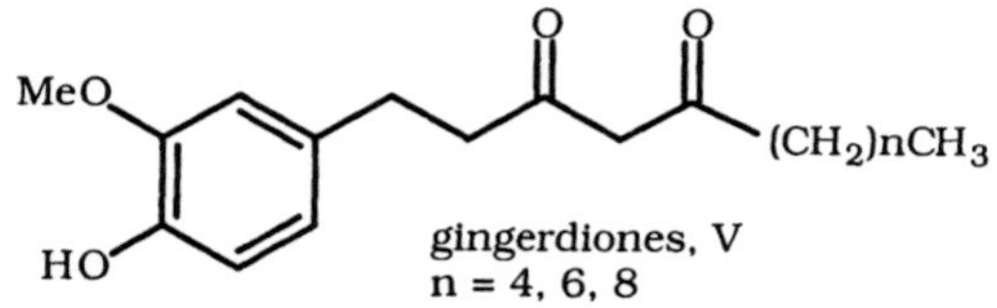

gingerdiones, V
n = 4, 6, 8

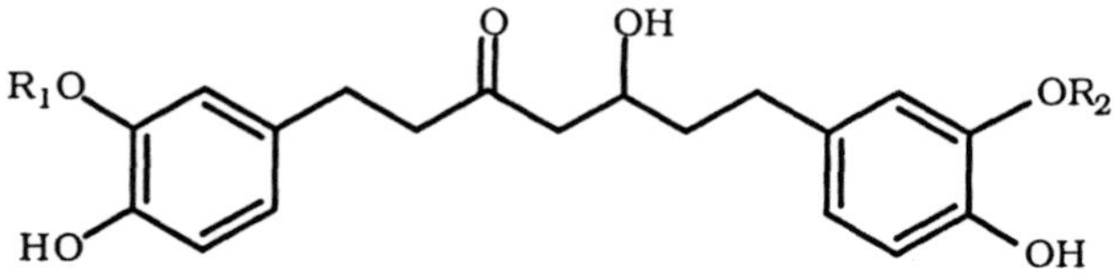

hexahydrocurcumins

VIa $R_1 = R_2 = Me$
VIb $R_1 = Me, R_2 = H$
VIc $R_1 = H, R_2 = Me$

In addition to [6]-gingerol, four gingerdione derivatives, [6]- and [10]-dehydrogingerdione (**335** and **336**) and [6]- and [10]-gingerdione (**337** and **338**) were isolated from the root of *Z. officinale* [306]. These five compounds were potent inhibitors of prostaglandin (PG) biosynthesis and compounds **335-338** were more potent inhibitors than indomethacin which is known to be one of the strongest inhibitors [306]. Isolation of **335** and

337 from ginger gives strong support to the biosynthetic scheme of [6]-gingerol as shown in "Fig. (4)".

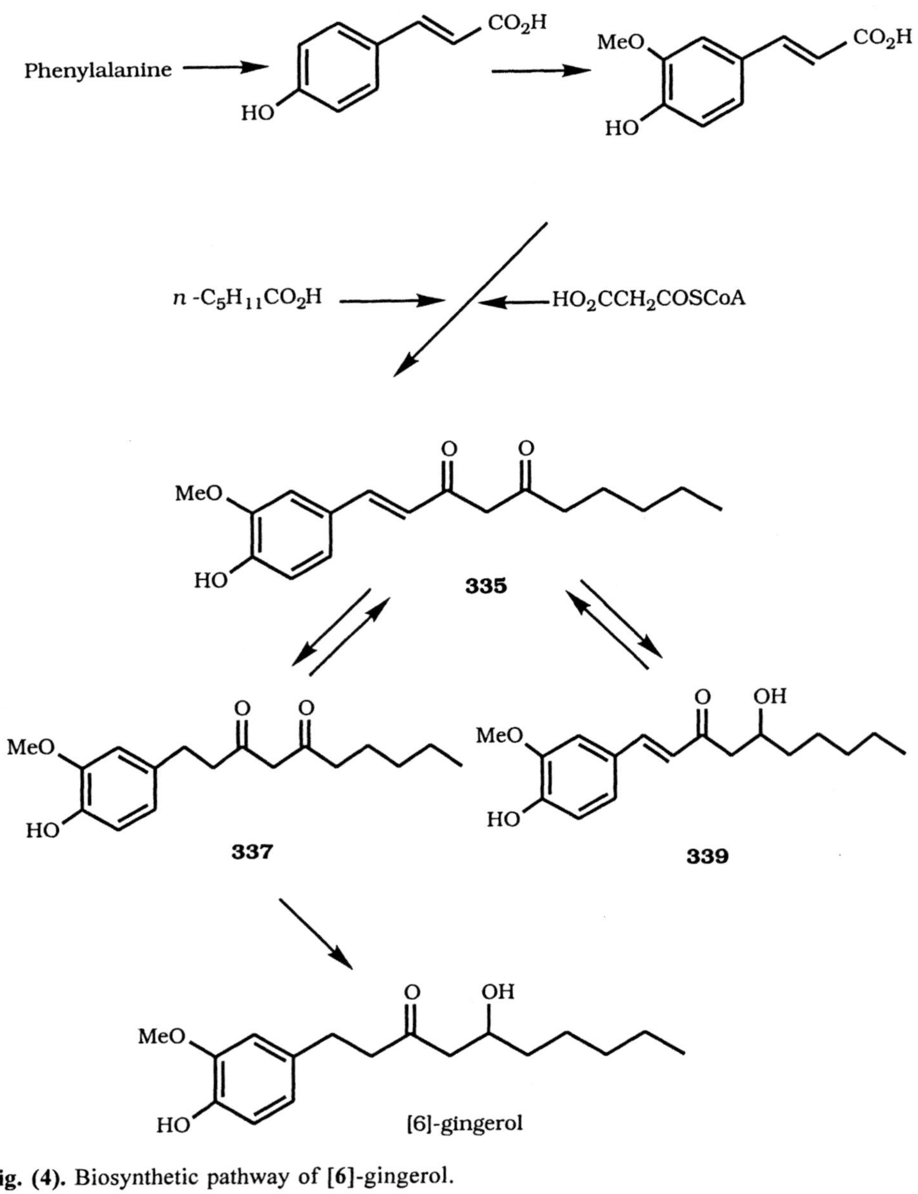

Fig. (4). Biosynthetic pathway of [6]-gingerol.

The pungent principles, [6]- and [10]-gingerol, present in the acetone extract of ginger, were found to increase bile secretion and were also mainly responsible for the cholagogic effect of ginger [307]. Zingiberene, the major constituent of the acetone extract of ginger, and [6]-gingerol

335 n = 4
336 n = 8
337 gingerdione, V n = 4
338 gingerdione, V n = 8

significantly inhibited HCl/ethanol-induced gastric lesions in rats [308]. This suggests that zingiberene and [6]-gingerol are the important constituents in the stomachic medications containing ginger [308]. The pungent principles of ginger, [6]-shogaol and [6]-, [8]- and [10]-gingerol were reported to to enhance gastrointestinal motility [309]. Gingerol and shogaol also exhibited potent molluscicidal activity on *Biomphalara glabrata* [310]. A series of gingerols, shogaols and numerous diarylheptanoids and related analoges have been tested for their anti-hepatotoxic actions using CCl₄- and GalN-induced cytotoxicity in primary cultured rat hepatocytes [311]. Gingerols, shogaols and diarylheptanoids exert anti-hepatotoxic actions with the length of the linear chain in the gingerols and shogaols and the presence of hydroxyls on the phenyl ring in diarylheptanoids being important for the activity [311]. Gingerols and diarylheptanoids were reported to be potent inhibitors against the prostaglandin biosynthesizing enzyme (PG synthetase) and the compounds were also active against arachidonate 5-lipooxygenase, an enzyme of leukotriene (LT) biosynthesis [312].

From the rhizomes of *Z. officinale*, four diarylheptanoids were isolated and their structures were established by spectral methods and chemical transformations to be 1,7-*bis*(4-hydroxy-3-methoxyphenyl)hept-4-en-3-one (gingerenone A, **340**), 7-(3,5-dimethoxy-4-hydroxyphenyl)-1-(4-hydroxy-3-methoxyphenyl)hept-4-en-4-one (gingerenone B, **342**), 1-(3,5-dimethoxy-4-hydroxyphenyl)-7-(4-hydroxy-3-methoxy-phenyl)hept-4-en-3-one (isoginge-renone B, **343**) and 1-(4-hydroxy-3-methoxyphenyl)-7-(4-hydroxyphenyl)hept-4-en-3-one (gingerenone C, **344**) [313]. Gingerenone A exhibited a moderate *in vitro* anti-coccidium activity and a strong anti-fungal effect to *Pyricularia oryzae* [313]. Together with gingerenone A and hexahydrocurcumin (**345**), two diarylheptanoids, *meso*-3,5-diacetoxy-1,7-*bis*(4-hydroxy-3-methoxyphenyl)heptane (**352**) and 3,5-diacetoxy-1-(4-hydroxy-3,5-dimethoxyphenyl)-7-(4-hydroxy-3-methoxyphenyl)heptane (**349**) were isolated from a dichloromethane extract of the rhizomes of ginger [314]. The stereochemistry of C-5 in **345** was reported to be *S* configuration [314]. From the same extract, eight additional diarylheptanoids were isolated and identified as 7-(3,4-

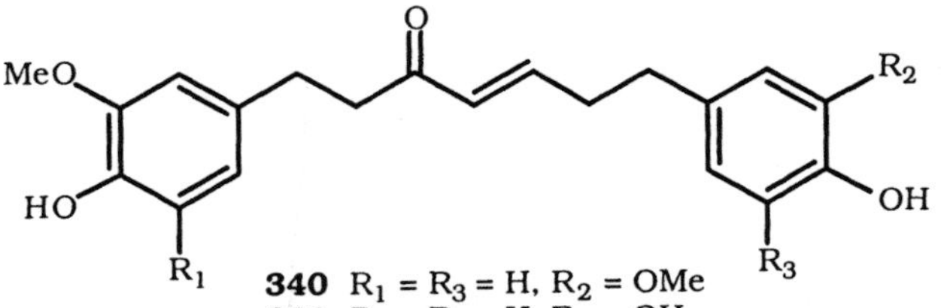

340 R$_1$ = R$_3$ = H, R$_2$ = OMe
341 R$_1$ = R$_3$ = H, R$_2$ = OH
342 R$_1$ = H, R$_2$ = R$_3$ = OMe
343 R$_1$ = R$_2$ = OMe, R$_3$ = H
344 R$_1$ = R$_2$ = R$_3$ = H

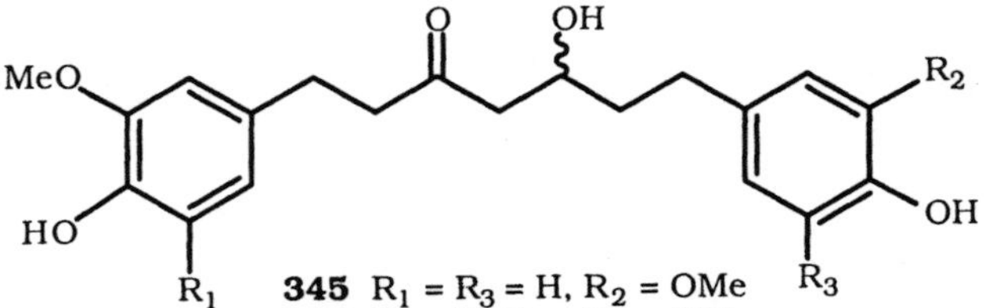

345 R$_1$ = R$_3$ = H, R$_2$ = OMe
346 R$_1$ = H, R$_2$ = R$_3$ = OMe
347 R$_1$ = R$_2$ = OMe, R$_3$ = H
348 R$_1$ = R$_2$ = R$_3$ = H

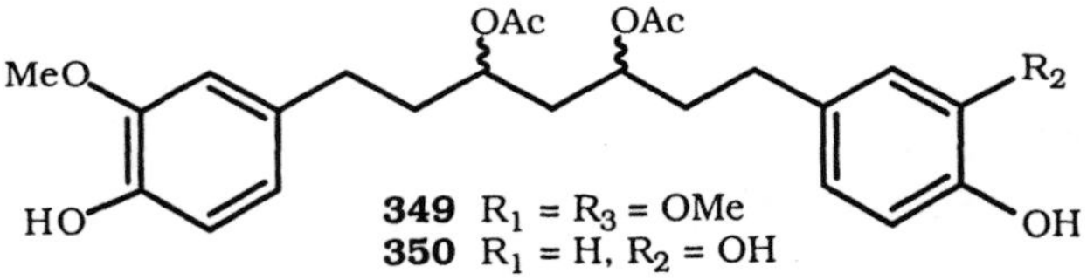

349 R$_1$ = R$_3$ = OMe
350 R$_1$ = H, R$_2$ = OH

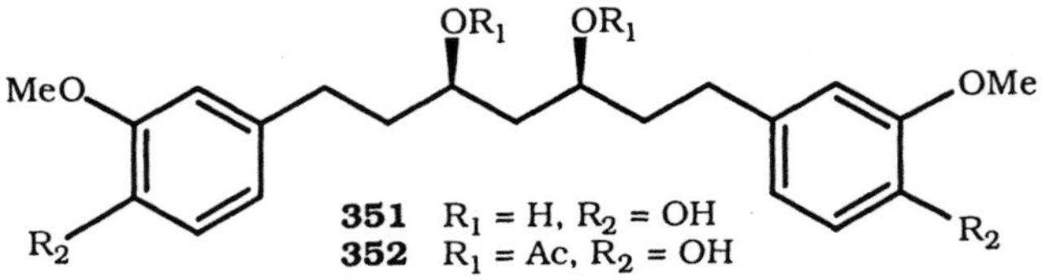

351 R$_1$ = H, R$_2$ = OH
352 R$_1$ = Ac, R$_2$ = OH

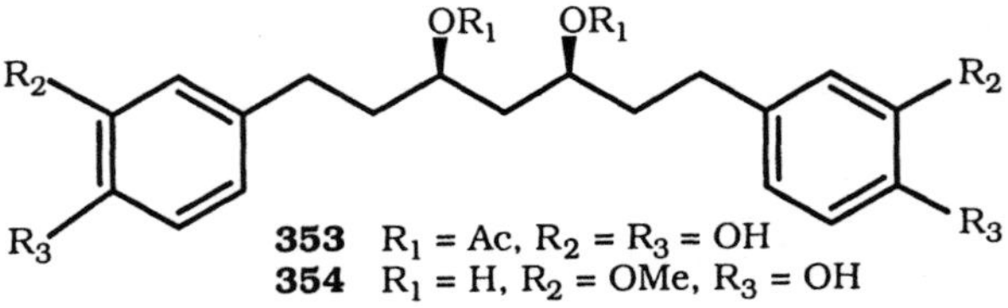

353 R$_1$ = Ac, R$_2$ = R$_3$ = OH
354 R$_1$ = H, R$_2$ = OMe, R$_3$ = OH

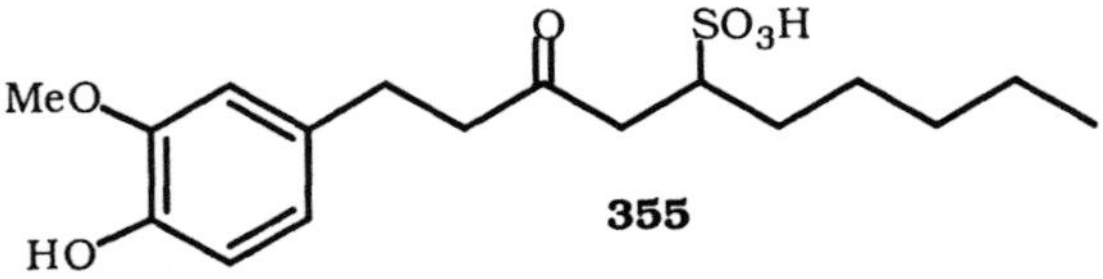

355

dihydroxyphenyl)-1-(4-hydroxy-3-methoxyphenyl)hept-4-en-3-one (**341**), 5-hydroxy-7-(4- hydroxy - 3, 5-dimethoxyphenyl)- 1 -(4-hydroxy- 3 -methoxyphenyl)- 3 - heptanone (**346**), 5-hydroxy-1-(4-hydroxy-3,5-dimethoxyphenyl)-7-(4-hydroxy-3-methoxyphenyl)-3-heptanone (**347**), 5-hydroxy-7-(4-hydroxyphenyl)-1-(4-hydroxy-3-methoxyphenyl)-3-heptanone (**348**), 3,5-diacetoxy-7-(3,5-dihydroxyphenyl)-1-(4-hydroxy-3-methoxyphenyl)-heptane (**350**), (3*R*,5*S*)-3,5-dihydroxy-1,7-*bis*(4-hydroxy-3-methoxy-phenyl)heptane (**351**), (3*S*,5*S*)-3,5-diacetoxy-1,7-*bis* (3,4-dihydroxy-phenyl)heptane (**353**) and (3*S*,5*S*)-3,5-dihydroxy-1,7-*bis*(4-hydroxy-3-methoxyphenyl)heptane (**354**) [315]. Isolation and structure elucidation of (*E*)-8β,17-epoxylabd-12-ene-15,16-dial was reported [316] and the compound was later found to possess an inhibitory effect on the cholesterol biosynthesis [317]. From the water soluble fraction of the methanolic extract of dried rhizomes of ginger, an anti-ulcer principle named 6-gingesulfonic acid (**355**) was isolated together with three new monoacyldigalactosylglycerols, gingerglycolipids A (**356**), B (**357**) and C (**358**) [318]. 6-Gingesulfonic acid showed more potent anti-ulcer

359 R_1 = Ac, R_2 = H
360 R_1 = R_2 = H
361 R_1 = H, R_2 = Me

362 R = Ac
363 R = H

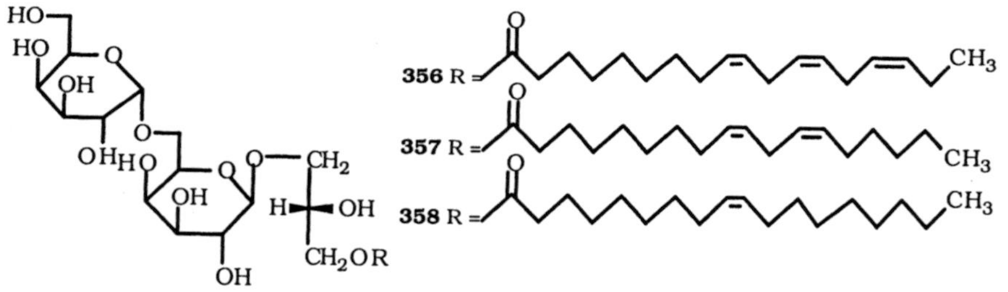

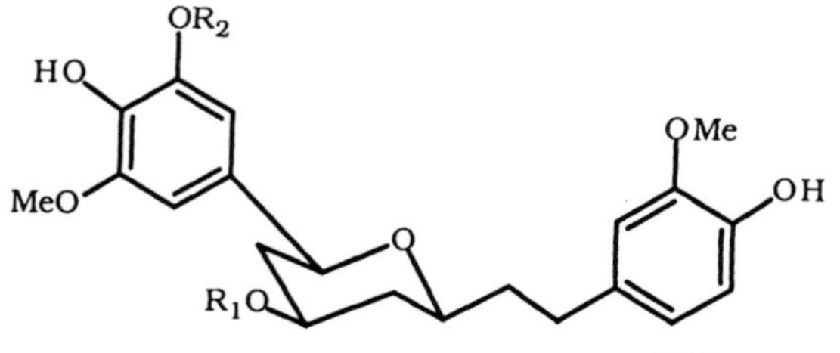

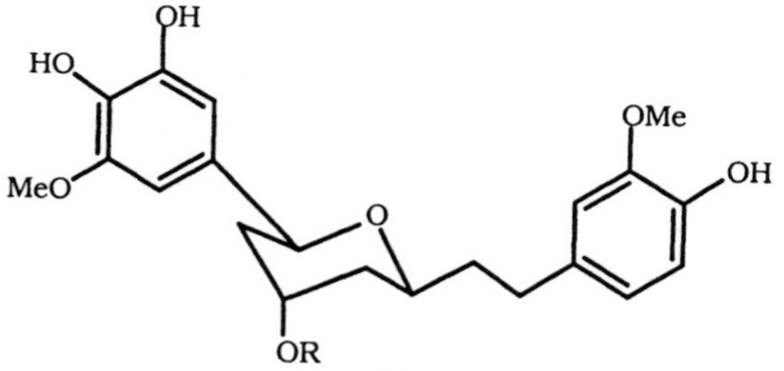

activity than [6]-gingerol and [6]-shogaol. Five cyclic diarylheptanoids
(**359-363**) were isolated from the dichloromethane extract of the rhizomes
of ginger and their structures were elucidated by spectroscopic and
chemical methods [319]. They were oxygenated at C-1, C-3 and C-5 on
the heptane chain and cyclized between C-1 and C-5 through oxygen.

Several monoterpenes [320, 321], humulene (**364**) [320] and zerumbone
(**365**) [322] have been reported from the essential oil from the rhizomes of
Zingiber zerumbet. The characterization and elucidation of structure **365**
of zerumbone was reported later [323]. The crystal structure of
zerumbone isolated from the rhizomes of *Z. zerumbet* was determined by
single-crystal X-ray diffraction [324]. The stereochemistry of the double
bonds of humulene was established by crystallographic study of its silver
nitrate adduct [325, 326]. The essential oil from the rhizomes of *Z.*

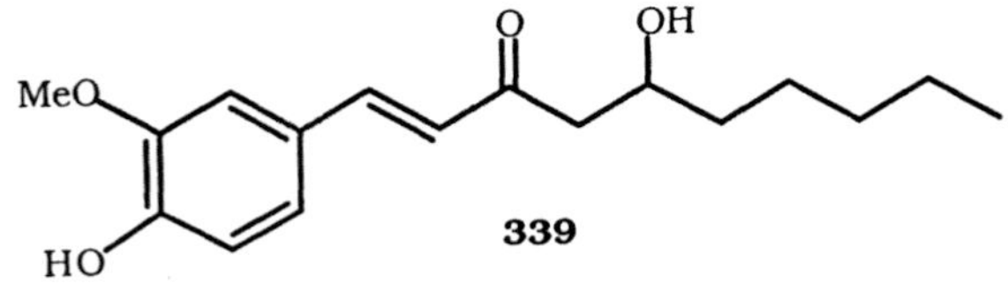

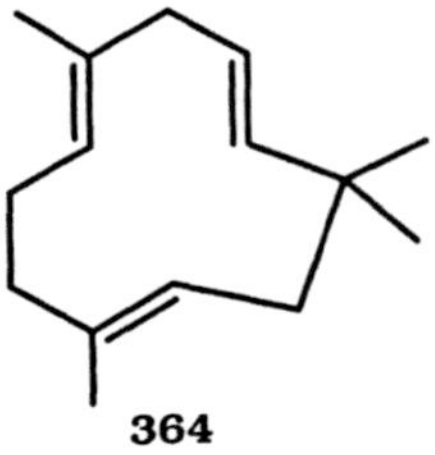

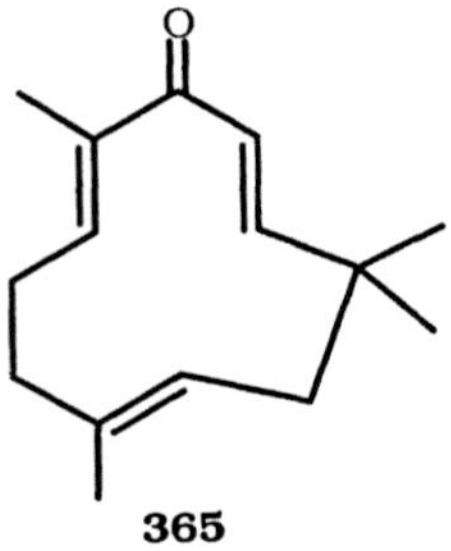

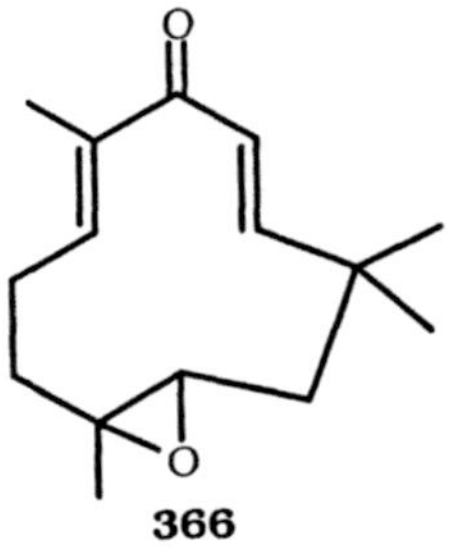

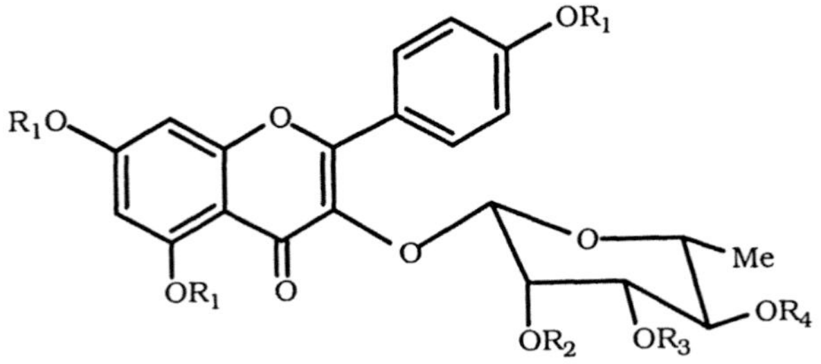

367 $R_1 = R_2 = H$, $R_3 = R_4 = Ac$
368 $R_1 = R_3 = R_4 = H$, $R_2 = Ac$
369 $R_1 = R_2 = R_4 = H$, $R_3 = Ac$
370 $R_1 = R_2 = R_3 = H$, $R_4 = Ac$
371 $R_1 = R_2 = R_3 = R_4 = H$

zerumbet was reinvestigated by GLC and TLC and the isolation of several new humulene sesquiterpenoids was described [327]. Six cytotoxic compounds were isolated from the rhizomes of *Z. zerumbet* and elucidated as zerumbone, zerumbone epoxide (**366**), curcuminoids **337-339** and 3",4"-*O*-diacetylafzelin (**367**) by spectroscopic methods [328]. From the acetone extract of the fresh rhizomes of *Z. zerumbet* from Japan, kaempferol-3-*O*-α-L-rhamnopyranoside (**371**), kaempferol-3-*O*-(2-*O*-acetyl-α-L-rhamnopyra-noside) (**368**), kaempferol-3-*O*-(3-*O*-acetyl-α-L-rhamnopyranoside) (**369**) and kaempferol-3-*O*-(4-*O*-acetyl-α-L-rhamnopyranoside) (**370**), were isolated and their structures were elucidated on the basis of spectroscopic methods [329].

CONCLUSIONS

This survey shows that plants in these nine genera of the Zingiberaceae are capable of producing a wide range of chemical structures. These can be divided into 7 different types.

1   Terpenoids. Mono- and sesquiterpenoids are the common constituents of all nine genera. Labdane-type diterpenes are found mainly in *Aframomum, Alpinia* and *Hedychium*. Only plants in the genus *Kaempferia* are able to produce pimarane diterpenes. Steroidal compounds and steroidal saponins are found only in plants of the genus *Costus*.

2   Diarylheptanoids. Yellow coloring materials, diarylheptanoids, are the common constituents of plants in three genera, *Alpinia, Curcuma* and *Zingiber*.

3   Arylalkanoids. The main pungent principles, arylalkanoids, are concentrated in two genera, *Amomum* and *Zingiber*.

4   Phenylpropanoids. Plants of the *Alpinia* and *Kaempferia* genera are capable of synthesizing phenylpropanoids.

5   Phenylbutanoids. Phenylbutanoids are rare in nature, found only in the genus *Zingiber*.

6   Cyclohexane oxides. Only plants of the genus *Kaempferia* can produce compounds of this class.

7   Flavonoids. Flavonoids and flavonoid related derivatives are the constituents of plants of the *Aframomum, Alpinia, Boesenbergia, Kaempferia* and *Zingiber* genera.

This survey has also demonstrated that compounds isolated from a number of plants of this family possess a broad spectrum of biological activities such as cytotoxic, anti-fungal, anti-bacterial, anti-malarial, anthelmintic, anti-hepatotoxic, anti-oxidative, anti-ulcer, anti-inflammatory and hypothermic, hypolipimenic, hypotensive and spasmolytic activities. Insecticidal, molluscicidal, nematocidal and larvicidal activities are also important properties of a number of compounds isolated from the Zingiberaceous plants.

ACKNOWLEDGMENTS

We are grateful to Prof. W. C. Taylor, Sydney University, Australia, for reading the manuscript.

REFERENCES

[1]   Holtum, R. E., The Gardens' Bull. Singapore, **1950**, *13*, 1.
[2]   Eglinton, G. and Hamilton, R. J., *Phytochemistry*, **1965**, *4*, 197.
[3]   De Bernardi, M., Mellerio, G., Colombo, M. P., Vidari, G. and Vita-Finzi, P., *Planta Med.*, **1981**, *41*, 359.
[4]   Vidari, G., Vita-Finzi, P. and De Bernardi, M., *Phytochemistry*, **1971**, *10*, 3335.
[5]   De Bernardi, M., Vidari, G. and Vita-Finzi, P., *Phytochemistry*, **1976**, *15*, 1785.
[6]   Ayafor, F. and Connolly, J. D., *J. Chem. Soc. Perkin Trans. I*, **1981**, 2563.
[7]   Kimbu, S. F., Njimi, T. K., Sondengam, B. L., Akinniyi, J. A. and Connolly, J. D., *J. Chem Soc. Perkin Trans. I*, **1979**, 1303.
[8]   Kim, T. H. and Isoe, S., *J. Chem. Soc. Chem. Commun.*, **1983**, 730.
[9]   Morita, H. and Itokawa, H., *Planta Med.*, **1988**, *54*, 117.
[10] Tanabe, M., Chen, Y. D., Saito, K. and Kano, Y., *Chem. Pharm. Bull.*, **1993**, *41*, 710.
[11] Kimbu, S. F., Ngadjui, B., Sondengam, L. B., Connolly, J. D. and Fakunle, C. O., *J. Nat. Prod.*, **1987**, *50*, 230.
[12] Ayafor. J. F., Tchuendem, M. H. K., Nyasse, B., Tillequin, F. and Anke, H., *J. Nat. Prod.*, **1994**, *57*, 917.
[13] Ayafor, J. F., Tchuendem, M. H. K., Nyasse, B., Tillequin, F. and Anke, H., *Pure and Appl. Chem.*, **1994**, *66*, 2327.
[14] Abreu, P. M. and Noronha, R. G., *Flavor and Fragance J.*, **1997**, *12*, 79.
[15] Chopra, I. C., Khajuria, B. N. and Chopra, C. L., *Antibiot. Chemother.*, **1957**, *7*, 378; *Chem. Abstr.*, **1958**, *52*, 13014a.
[16] Jain, S. R. and Sharma, S. N., *Planta Med.*, **1967**, *15*, 439.
[17] Hirschhorn, H. H., *J. Ethnopharmacol.*, **1983**, *7*, 123.
[18] Al-Yahya, M. A., Rafatullah, S., Massa, J. S., Ageel, A. M., Al-Said, M. S. and Tariq, M., *Phytother. Res.*, **1990**, *4*, 112.
[19] Ahsan, S. K., Shah, A. H., Tanira, M. O. M., Ahamh, M. S., Tariq, M. and Ageel, A. M., *Fitoterapia*, **1990**, *61*, 435.
[20] Panthong, A., Kanjanapothi, D. and Taylor, W. C., *J. Ethnopharmacol.*, **1986**, *18*, 213.
[21] Kaleysa Raj, R., *Indian J. Physiol. Pharmacol.*, **1975**, *19*, 47.
[22] Bhakuni, O. S., Dhar, M. L., Dhar, M. M., Dhawan, B. N. and Mehrotra, B. N., *Indian J. Exp. Biol.*, **1969**, *7*, 250.
[23] Tseng, C. F., Iwakami, S., Mikajiri, A., Shibuya, M., Hanaoka, F., Ebizuka, Y., Padmawinata, K. and Sankawa, U., *Chem. Pharm. Bull.*, **1992**, *40*, 396.
[24] Ross, S. A., Megalla. S. E., Bishay, D. W. and Awad, A, H., *Fitoterapia*, **1980**, *51*, 303.

[25] Janssen, A. M. and Scheffer, J. J. C., *Planta Med.*, **1985**, *6*, 507.
[26] Itokawa, H., Morita, H., Sumitomo, T., Totsuka, N. and Takeya, K., *Planta Med.*, **1987**, *53*, 32.
[27] Mitsui, S., Kobayashi, S., Naghori, H. and Ogiso, A., *Chem. Pharm. Bull.*, **1976**, *24*, 2377.
[28] Qureshi, S., Shah, A. H. and Ageel, A. M., *Planta Med.*, **1992**, *58*, 124.
[29] De Pooter, H. L., Omar, M. N., Coolsaet, B. A. and Schamp, N. M., *Phytochemistry*, **1985**, *24*, 93.
[30] Noro, T., Sekiya, T., Katoh, M., Oda, Y., Miyase, T., Kuroyanagi, M., Ueno, A. and Fukushima, S., *Chem. Pharm. Bull.*, **1988**, *36*, 244.
[31] Barik, B. R., Kundu, A. B. and Dey, A. K., *Phytochemistry*, **1987**, *26*, 2126.
[32] Ramachandran Nair, A. G. and Gunasegaran, R., *Indian J. Chem. Ser. B.*, **1982**, *21*, 979.
[33] Morita, H. and Itokawa, H., *Chem. Lett.*, **1986**, 1205.
[34] Morita, H. and Itokawa, H., *Planta Med.*, **1988**, *54*, 117.
[35] Saiki, Y., Ishikawa, Y., Uchida, M. and Fukushima, S. *Phytochemistry*, **1978**, *17*, 808.
[36] Kimura, Y., Takahashi, S. and Yashida, I., *Yakugaku Zasshi*, **1968**, *88*, 239; *Chem. Abstr.*, **1968**, *69*, 35863k.
[37] Kuroyanagi, M., Noro, T., Fukushima, S., Aiyama, R.,, Ikuta, A., Itokawa, H. and Morita, M., *Chem. Pharm. Bull.*, **1983**, *31*, 1544.
[38] Itokawa, H., Morita, M. and Mihashi, S., *Chem. Pharm. Bull.*, **1981**, *29*, 2383.
[39] Kiuchi, F., Shibuya, M. and Sankawa, U., *Chem. Pharm. Bull.*, **1982**, *30*, 2279.
[40] Itokawa, H., Morita, H., Midorikawa, I., Aiyama, R. and Morita, M., *Chem. Pharm. Bull.*, **1985**, *35*, 4889.
[41] Itokawa, H., Aiyama, R. and Ikuta, A., *Phytochemistry*, **1981**, *20*, 769.
[42] Itokawa, H., Aiyama, R. and Ikuta, A., *Phytochemistry*, **1982**, *21*, 241.
[43] Shoji, N., Umeyama, A., Asakawa, Y., Takemoto, T., Nomoto, K. and Ohizumi, Y., *J. Pharm. Sci.*, **1984**, *73*, 843.
[44] Pooter, H. L. D., Omar, M. N., Coolsaet, B. A. and Schamp, N. M., *Phytochemistry*, **1985**, *24*, 93.
[45] Jassen, A. M. and Scheffer, J. J. C., *Planta Med.*, **1985**, *51*, 507.
[46] Noro, T., Sekiya, T., Katoh, M., Oda, Y., Miyase, T., Kuroyangi, M., Ueno, A. and Fukushima, S., *Chem. Pharm. Bull.*, **1988**, *36*, 244.
[47] Yu, J., Fang, H., Chen, Y. and Yao, Z., *Zhongyao Tongbao*, **1988**, *13*, 354; *Chem. Abstr.*, **1988**, *109*, 134876t.
[48] Athamaprasangsa, S., Buntrarongroj, U., Dampawan, P., Ongkavoranan, N., Rukachaisirikul, V., Sethijinda, S., Sornnarintra, M., Sriwub, P. and Taylor, W. C., *Phytochemistry*, **1994**, *37*, 817.
[49] Kadota, S., Hui, D., Basnet, P., Prasain, J. K., Xu, G.-J. and Namba, T., *Chem. Pharm. Bull.*, **1994**, *42*, 2647.
[50] Kadota, S., Prasain, J. K., Li, J. X., Basnet, P., Dong, H., Tani, T. and Namba, T., *Tetrahedron Lett.*, **1996**, *37*, 7283.
[51] Prasain, J. K., Tezuka, Y., Li, J. X., Tanaka, K., Basnet, P., Dong, H., Namba, T. and Kadota, S., *Tetrahedron*, **1997**, *53*, 7833.
[52] Itokawa, H., Morita, M. and Mihashi, S., *Phytochemistry*, **1981**, *20*, 2503.
[53] Kimura, Y., Takido, M., Nakano, K. and Takishita, M., *Yakugaku Zasshi*, **1966**, *86*, 1184; *Chem. Abstr.*, **1967**, *67*, 21775e.
[54] Fujita, T., Nishimura, H., Kaburagi, K. and Mizutani, J., *Phytochemistry*, **1994**, *36*, 23.

[55] Itokawa, H., Morita, M. and Mihashi, S., *Chem. Pharm. Bull.*, **1980**, *28*, 3452.
[56] Krishna, B. M. and Chaganty, R. B., *Phytochemistry*, **1973**, *12*, 238.
[57] Luz, A. I. R., Zoghbi, M. G. B., Ramos, L. S., Maia, J. G. S. and Silva, M. L., *J. Nat. Prod.*, **1984**, *47*, 907.
[58] Sirat, H. M., Masri, D. and Rahman, A. A., *Phytochemistry*, **1994**, *36*, 699.
[59] Xu, H.-X, Dong, H. and Sim, K.-Y., *Phytochemistry*, **1996**, *42*, 149.
[60] Dung, N. X., Chinh, T. D., Rang, D. D. and Leclerq, P. A., *J. Essent. Oil Res.*, **1994**, *6*, 637.
[61] Sy, L.-K. and Brown, G. D., *J. Nat. Prod.*, **1997**, *60*, 904.
[62] Itokawa, H., Yoshimoto, S. and Morita, H., *Phytochemistry*, **1988**, *27*, 435.
[63] Itokawa, H., Watanabe, K., Mihashi, S. and Iitaka, Y., *Chem. Pharm. Bull.*, **1980**, *28*, 681.
[64] Itokawa, H., Morita, H. and Watanabe, K., *Chem. Pharm. Bull.*, **1987**, *35*, 1460.
[65] Itokawa, H., Morita, H., Watanabe, K. and Iitaka, Y., *Chem. Lett.*, **1984**, 451.
[66] Itokawa, H., Morita, H., Watanabe, K., Takase, A. and Iitaka, Y., *Chem. Lett.*, **1984**, 1687.
[67] Itokawa, H., Watanabe, K., Morita, H., Mihashi, S. and Iitaka, Y., *Chem. Pharm. Bull.*, **1985**, *33*, 2023.
[68] Itokawa, H., Morita, H., Osawa, K., Watanabe, K. and Iitaka, Y., *Chem. Pharm. Bull.*, **1987**, *35*, 2849.
[69] Morita, H., Simizu, K., Takizawa, H., Aiyama, R. and Itokawa, H., *Chem. Pharm. Bull.*, **1988**, *36*, 3156.
[70] Morita, H., Tomioka, N., Iitaka, Y. and Itokawa, H., *Chem. Pharm. Bull.*, **1988**, *36*, 2984.
[71] Morita, M., Nakanishi, H., Morita, H., Mihashi, S. and Itokawa, H., *Chem. Pharm. Bull.*, **1996**, *44*, 1603.
[72] Itokawa, H., Morita, H., Kobayashi, T., Watanabe, K. and Iitaka, Y., *Chem. Pharm. Bull.*, **1987**, *35*, 2860.
[73] Mori, I. and Nakachi, Y., *Tetrahedron Lett.*, **1978**, *26*, 2297.
[74] Sy, L.-K. and Brown, G. D., *Phytochemistry*, **1997**, *45*, 537.
[75] Pei, S.-J., *J. Ethnopharmacol.*, **1985**, *13*, 121.
[76] Mahyar, U. W., Burley, J. S., Gyllenhaal, C. and Soejarto, D. D., *J. Ethnopharmacol.*, **1991**, *31*, 217.
[77] Hirschhorn, H. H., *J. Ethnopharmacol.*, **1983**, *7*, 123.
[78] Sriyabhaya, T., *J. Pharm. Ass. Siam*, **1948**, *II*, 56.
[79] Woo, W. S., Lee, E. B., Shin, K. H., Kang, S. S. and Chi, H. J., *Korean J. Pharmacog.*, **1981**, *12*, 153.
[80] Dhar, M. L., Dhar, M. M., Dhawan, B. N., Mehrotra, B. N. and Ray, C., *Indian J. Exp. Biol.*, **1968**, *6*, 232.
[81] Itokawa, H., Hirayama, F., Tsuruoka, S., Mizuno, K., Takeya, K. and Nitta, A., *Shoyakugaku Zasshi*, **1990**, *44*, 58.
[82] Itokawa, H., *Yakugaku Zasshi*, **1988**, *108*, 824.
[83] Han, B. H., Chi, H. J., Han, Y. N. and Ryu, K. S., *Korean J. Pharmacog.*, **1972**, *4*, 205.
[84] Mishra, A. K. and Dubey, N. K., *Econ. Bot.*, **1990**, *44*, 530.
[85] Hong, N. D., Kim, J. W., Kim. B. W. and Shon, J. G., *Korean J. Pharmacog.*, **1982**, *13*, 33.
[86] Nikaido, T., Ohmoto, T., Noguchi, H., Saitoh, H. and Sankawa, U., *Planta Med.*, **1981**, *43*, 18.
[87] Lawrence, B. M., *Phytochemistry*, **1970**, *9*, 665.

[88] Lawrence, B. M., Hogg, J. W. and Stuart, J. T. and Pichitakul, N., *Phytochemistry*, **1972**, *11*, 1534.

[89] Gupta, P. N., Naqvi, A., Misra, L. N., Sen, T. and Nigam, M. C., *Parfuem. Kosmet.*, **1984**, *65*, 528; *Chem. Abstr.*, **1985**, *102*, 31918j.

[90] Chen, X., Lin, J. and Zhang, J., *Yunnan Zhiwu Yanjiu*, **1989**, *11*, 316; *Chem. Abstr.*, **1990**, *112*, 124896x.

[91] Fang, H., Yu, J., Chen., Y., Chen, R. and Sun, S., *Zhongcaopao*, **13**, 5; *Chem. Abstr.*, **1982**, *97*, 188360x.

[92] Nguyen, X. D., Dao, L. P., Leclercq, P. A., Nguyen, T. T., *Tap Chi Duoc Hoc*, **1990**, 17; *Chem. Abstr.*, **1991**, *115*, 239284a.

[93] Le, T. C., *Tap San Hoa-Hoc*, **1975**, *13*, 30; *Chem. Abstr.*, **1978**, *88*, 65857w.

[94] Lakshmi, V. and Chauhan, J. S., *J. Indian Chem. Soc.*, **1976**, *53*, 633.

[95] Lakshmi, V. and Chauhan, J. S., *Indian J. Chem. Ser. B*, **1977**, *15*, 814.

[96] Rao, C. B., Rao, T. N. and Suryaprakasam, S., *Planta Med.*, **1976**, *29*, 391.

[97] Chatterjee, A., Sen, R. and Ganguly, D., *Phytochemistry*, **1978**, *17*, 328.

[98] Connell, D. W., *Aust. J. Chem.*, **1970**, *23*, 369.

[99] Tackie, A. N., Dwuma-Badu, D., Ayim, J. S. K., Dabra, J. T., Knapp, J. E., Slatkin, D. J. and Schiff, P. L., Jr., *Phytochemistry*, **1975**, *14*, 853.

[100] Kamchonwongpaisan, S., Nilanonta, C., Tarnchompoo, B., Thebtaranonth, C., Thebtaranonth, Y., Yuthavong, Y., Kongsaeree, P. and Clardy, J., *Tetrahedron Lett.*, **1995**, *36*, 1821.

[101] Hirschhorn, H. H., *J. Ethnopharmacol.*, **1983**, *7*, 123.

[102] Murakani, A., Kondo, A., Nakamura, Y., Ohigashi, H. and Koshimizu, K., *Biosci. Biotech. Biochem.*, **1993**, *57*, 1971.

[103] Sasmito, E., *Maj. Farm. Indones.*, **1994**, *5*, 10.

[104] Lawrence, B. W., Hogg, J. W., Terhune, S. J. and Pichitakul, N., *Appl. Sci. Res. Crop (Thailand)* Report No. 2, **1971**, 1.

[105] Mongkolsuk, S. and Dean, F. M., *J. Chem. Soc.*, **1964**, 4654.

[106] Jaipetch, T., Kanghae, S., Pancharoen, O., Patrick, V. A., Reutrakul, V., Tuntiwachwuttikul, P. and White, A. H., *Aust. J. Chem.*, **1982**, *35*, 351.

[107] Mahidol, C., Tuntiwachwuttikul, P., Reutrakul, V. and Taylor, W. C., *Aust. J. Chem.*, **1984**, *37*, 1739.

[108] Mahidol, C., Tuntiwachwuttikul, P., Pakawatchai, C., Patrick, V. A., Reutrakul, V., Skelton, B. W. and White, A. H., *J. Sci. Soc. Thailand*, **1988**, *14*, 301.

[109] Tuntiwachwuttikul, P., Pancharoen, O., Reutrakul, V. and Byrne, L. T., *Aust. J. Chem.*, **1984**, *37*, 449.

[110] Pancharoen, O., Reutrakul, V., Skelton, B. W., Taylor, W. C., Tuntiwachwuttikul, P. and White, A. H., *Aust. J. Chem.*, **1984**, *37*, 2589.

[111] Pancharoen, O., Picker, K., Reutrakul, V., Taylor, W. C. and Tuntiwachwuttikul, P., *Aust. J. Chem.*, **1987**, *40*, 455.

[112] Tewari, P. Prasad, D. N., Chaturvedi, C. and Das, P. K., *J. Res. Ind. Med.*, **1967**, *1*, 196.

[113] Dasgupta, B. and Pandy, V. B., *Experientia*, **1970**, *26*, 475.

[114] Sarin, Y. K., Bedi, K. L. and Atal, C. K., *Curr. Sci.*, **1974**, *43*, 569.

[115] Sarin, Y. K., Kapahi, B. K., Kapur, S. K. and Atal, C. K., *Curr. Sci.*, **1976**, *45*, 688.

[116] Tschesche, R. and Pandy, V. B., *Phytochemistry*, **1978**, *17*, 1781.

[117] Gupta, M. M., Shukla, Y. N. and Lal, R. N., *Planta Med.*, **1983**, *48*, 64.

[118] Gupta, M. M., Lal, R. N. and Shukla, Y. N., *Phytochemistry*, **1981**, *20*, 2557.

[119] Gupta, M. M., Lal, R. N. and Shukla, Y. N., *Phytochemistry*, **1981**, *20*, 2553.

[120] Gupta, M. M., Lal, R. N. and Shukla, Y. N., *Phytochemistry*, **1982**, *21*, 230.
[121] Gupta, M. M., Singh, S. B. and Shukla, Y. N., *Planta Med.*, **1988**, *54*, 268.
[122] Gupta, M. M., Verma, R. K. and Akhila, A., *Phytochemistry*, **1986**, *25*, 1899.
[123] Singh, S. B., Gupta, M. M., Lal, R. N. and Thakur, R. S., *Planta Med.*, **1980**, *38*, 185.
[124] Singh, S. B. and Thakur, R. S., *J. Nat. Prod.*, **1982**, *45*, 667.
[125] Singh, S. B. and Thakur, R. S., *Phytochemistry*, **1982**, *21*, 911.
[126] Mahmood, U., Shukla, Y. N. and Thakur, R. S., *Chem. Ind. (London)*, **1985**, 56.
[127] Mahmood, U., Shukla, Y. N. and Thakur, R. S., *Phytochemistry*, **1984**, *23*, 1725.
[128] Mahmood, U., Shukla, Y. N. and Thakur, R. S., *Planta Med.*, **1989**, *55*, 204.
[129] Prawat, U., Tuntiwachwuttikul, P. and Taylor, W. C., *J. Sci. Soc. Thailand*, **1989**, *15*, 139.
[130] Rathore, A. K. and Khanna, P., *Planta Med.*, **1979**, *35*, 289.
[131] Iwu, M., *Planta Med.*, **1981**, *43*, 413.
[132] Lin, R.-C., Hanquet, B. and Lacaille-Debois, M.-A., *Phytochemistry*, **1996**, *43*, 665.
[133] Bohme, F., Schmidt, J., Sung, T. V. and Adam, G., *Phytochemistry*, **1997**, *45*, 1041.
[134] Lambert, N., Baccou, J. C. and Sauvaire, Y., *Planta Med.*, **1988**, *54*, 366.
[135] Inoue, K., Shimomura, K., Kobayashi, S., Sankawa, U. and Ebizuka, Y., *Phytochemistry*, **1996**, *41*, 725.
[136] Bandara, B. M. R., Hewage, C., Karunaratne, V. and Adikaram, N. K. B., *Planta Med.*, **1988**, *54*, 477.
[137] Xu, H.-X., *Zhong Cao Yao Tong Xun*, **1979**, *10*, 433.
[138] Kuroyanagi, M., Ueno, A., Ujiie, K. and Sato, S., *Chem. Pharm. Bull.*, **1987**, *35*, 53.
[139] Kuroyanagi, M., Ueno, A., Koyama, K. and Natori, S., *Chem. Pharm. Bull.*, **1990**, *38*, 55.
[140] Ohkura, T., Gao, J.-F., Harimaya, K., Hikichi, M., Iitaka, Y., Kawamata, J., Kuroyanagi, M., Fukushima, S. and Inayama, S., *Chem. Pharm. Bull.*, **1986**, *34*, 4435.
[141] Hikino, H., Konno, C., Nagashima, T., Kohama, T. and Takemoto, T., *Tetrahedron Lett.*, **1971**, 337.
[142] Matthes, H. W., Luu, B. and Ourisson, G., *Phytochemistry*, **1980**, *19*, 2643.
[143] Hikino, H., Konno, C. and Takamoto, T., *Chem. Pharm. Bull.*, **1972**, *20*, 987.
[144] Hikino, H., Sakurai, Y., Numake, S. and Takemoto, T., *Chem. Pharm. Bull.*, **1968**, *16*, 39.
[145] Hikino, H., Agatsuma, K. and Takemoto, T., *Chem. Pharm. Bull.*, **1969**, *17*, 959.
[146] Hikino, H., Konno, C., Takemoto, T., Tori, K., Ohtsuru, M. and Horibe, I., *J. Chem. Soc. Chem. Commun.*, **1969**, 662.
[147] Hikino, H., Agatsuma. K. and Takemoto, T., *Tetrahedron Lett.*, **1968**, 2855.
[148] Shibuya, H., Yamamoto, Y., Miura, I. and Kitagawa, I., *Heterocycles*, **1982**, *17*, 215.
[149] Hikino, H., Takahashi, H., Sakurai, Y., Takemoto, T. and Bhacca, N. S., *Chem. Pharm. Bull.*, **1966**, *14*, 550.
[150] Hikino, H., Tori, K., Horibe, I. and Kuriyama, K., *J. Chem. Soc. (C)*, **1971**, 688.

[151] Kouno, I. and Kawano, N., *Phytochemistry*, **1985**, *24*, 1845.

[152] Shiobara, Y., Asakawa, Y., Kodama. M. and Takemoto, T., *Phytochemistry*, **1986**, *25*, 1351.

[153] Hikino, H., Komo, C., Agatsuma, K. and Takemoto, T., *J. Chem. Soc. Perkin Trans I*, **1975**, 478.

[154] Shiobara, Y., Asakawa, Y., Kodama, M., Yasuda, K. and Takemoto, T., *Phytochemistry*, **1985**, *24*, 2629.

[155] Yoshihara, M., Shibuya, H., Kitano, E., Yanagi, K. and Kitagawa, I., *Chem. Pharm. Bull.*, **1984**, *32*, 2059.

[156] Shibuya, H., Hamamoto, Y., Cai, Y. and Kitagawa, I., *Chem. Pharm. Bull.*, **1987**, *35*, 924.

[157] Yamahara, J., Matsuda, H., Sawada, T., Kushida, H., Shibuya, H.and Kitagawa, I., *Yakugaku Zasshi*, **1982**, *102*, 306; *Chem. Abstr.*, **1982**, *97*, 719u.

[158] Gupta, S. K. and Banerjee, A. B., *Lloydia*, **1976**, *39*, 218.

[159] Pandji, C., Grimm, C., Wray, V., Witte, L. and Proksch, P., *Phytochemistry*, **1993**, *34*, 415.

[160] Inayama, S., Gao, J.-F., Harimaya, K., Iitaka, Y., Guo, Y.-T. and Kawamata, T., *Chem. Pharm. Bull.*, **1985**, *33*, 1323.

[161] Inayama, S., Gao, J.-F., Hariyama, K., Hikichi, M., Iitaka, Y., Guo, Y.-T. and Kawamata, T., *Chem. Pharm. Bull.*, **1985**, *33*, 2179.

[162] Ohkura, T., Gao, J.-F., Hariyama, K., Hikichi, M., Kawamata, T. and Iitaka, Y., Wu, X.-Y., Nishishita, T. and Inayama, S., *Shoyakugaku Zasshi*, **1986**, *40*, 352.

[163] Ohkura, T., Gao, J.-F., Harimaya, K., Hikichi, M., Iitaka, Y., Kawamata, T., Kuroyanagi, M., Fukushiga and Inayama, S., *Chem. Pharm. Bull.*, **1986**, *34*, 4435.

[164] Gao, J.-F., Xie, J.-H., Iitaka, Y. and Inayama, S., *Chem. Pharm. Bull.*, **1989**, *37*, 233.

[165] Hariyama, K., Gao, J.-F., Ohkura, T., Kawamata, T., Iitaka, Y., Guo, Y.-T. and Inayama, S., *Chem. Pharm. Bull.*, **1991**, *39*, 843.

[166] Gao, J.-F., Xie, J.-H., Iitaka, Y. and Inayama, S., *Shoyakugaku Zasshi*, **1988**, *42*, 347.

[167] Inayama, S., Gao, J.-F., Harimaya, K., Kawamata, T., Iitaka, Y. and Guo, Y.-T., *Chem. Pharm. Bull.*, **1984**, *32*, 3783.

[168] Gao, J.-F., Ohkura, T., Harimaya, K., Hikichi, M., Kawamata, T., Ying, W.-X., Iitaka, Y. and Inayama, S., *Chem. Pharm. Bull.*, **1986**, *34*, 5122.

[169] Gao, J.-F., Xie, J.-H., Harimaya, K., Kawamata, T., Iitaka, Y. and Inayama, S., *Chem. Pharm. Bull.*, **1991**, *39*, 854.

[170] Itokawa, H., Hirayama, F., Funakoshi, K. and Takeya, K., *Chem. Pharm. Bull.*, **1985**, *33*, 3488.

[171] Uehara, S.-I., Yasuda, I., Takeya, K. and Itokawa, H., *Chem. Pharm. Bull.*, **1989**, *37*, 237.

[172] Uehara, S.-I., Yasuda, I., Takeya, K., Itokawa, H. and Iitaka, Y., *Chem. Pharm. Bull.*, **1990**, *38*, 261.

[173] John, T. K. and Rao, G. S. K., *Indian J. Chem. Ser. B*, **1985**, *24*, 35.

[174] Kuroyanagi, M. and Natori, S., *Yakugaku Zasshi*, **1970**, *90*, 1467; *Chem. Abstr.*, **1971**, *74*, 61612a

[175] Uehara, S.-I., Yasuda, I., Akiyama, K., Morita, H., Takaya, K. and Itokawa, H., *Chem. Pharm. Bull.*, **1987**, *35*, 3298.

[176] Uehara, S.-I., Yasuda, I., Akiyama, K., Morita, H., Takeya, K. and Itokawa, H., *Chem. Pharm. Bull.*, **1987**, *35*, 3298.

[177] Masuda, T., Isobe, J., Jitoe, A. and Nakatani, N., *Phytochemistry*, **1992**, *31*, 3645.

[178] Toda, S., Miyase, T., Arichi, H., Tanizawa, H. and Takino, Y., *Chem. Pharm. Bull.*, **1985**, *33*, 1725.

[179] a) Cuvelier, M.-E., Richard, H. and Berset, C., *Biosci. Biotech. Biochem.*, **1992**, *56*, 324. b) Larson, R., *Phytochemistry*, **1988**, *27*, 969. c) Osawa, T. and Namikki, M., *J. Agric. Food Chem.*, **1985**, *33*, 780.

[180] Claesen, P., Panthong, A., Tuchinda, P., Reutrakul, V., Kanjanapothi, D., Taylor, W. C. and Santisuk, T., *Planta Med.*, **1993**, *59*, 451.

[181] Suksamrarn, A., Eiamong, S., Piyachaturawat, P. and Charoenpiboonsin, J., *Phytochemistry*, **1994**, *36*, 1505.

[182] Yamazaki, M., Maebayashi, Y., Iwase, N. and Kaneko, T., *Chem. Pharm. Bull.*, **1988**, *36*, 2070.

[183] Ozaki, Y., *Chem. Pharm. Bull.*, **1990**, *38*, 1045.

[184] Yamazaki, M., Maebayashi, Y., Iwase, N. and Kaneko, T., *Chem. Pharm. Bull.*, **1988**, *36*, 2075.

[185] Ammon, H. P. T. and Wahl, M. A., *Planta Med.*, **1991**, *57*, 1.

[186] Rafatullan, S., Tariq, M., Al-Yahya, M. A., Mossa, J. S. and Agoal, A. M., *J. Ethnopharmacol.*, **1990**, *29*, 25.

[187] Ohshiro, M., Kuroyamagi, M. and Ueno, K., *Phytochemistry*, **1990**, *29*, 2201.

[188] Kiso, Y., Suzuki, Y., Oshima, Y. and Hikino, H., *Phytochemistry*, **1983**, *22*, 596.

[189] Golding, B. T. and Pombo-Villar, E., *J. Chem. Soc. Perkin Trans I*, **1992**, 1519.

[190] Nutakul, W., *Bull. Med. Sci. Thailand*, **1994**, *36*, 21.

[191] Masuda, T., Jitoe, A., Isobe, J., Nakatani, N. and Yonemori, S., *Phytochemistry*, **1993**, *32*, 1557.

[192] Nakayama, R., Tamura, Y., Yamanaka, H., Kikuzaki, H. and Nakatani, N., *Phytochemistry*, **1993**, *33*, 501.

[193] Ammon, H. P. T.,, Anazodo, M. I., Safayhi, H., Dhawan, B. N. and Srimal, R. C., *Planta Med.*, **1992**, *58*, 226.

[194] Kiuchi, F., Goto, Y., Sugimoto, N., Akao, N., Kondo, K. and Tsuda, Y., *Chem. Pharm. Bull.*, **1993**, *41*, 1640.

[195] Toda, S., Miyase, T., Arichi, H., Tanizawa, H. and Takino, Y., *Chem. Pharm.Bull.*, **1985**, *33*, 1725.

[196] Kosuge, T., Ishida, H. and Yamazaki, H., *Chem. Pharm. Bull.*, **1985**, *33*, 1499.

[197] Kinoshita, G., Nakamura, F. and Maruyama, T., *Shoyakugaku Zasshi*, **1986**, *40*, 325.

[198] Tomoda, M., Gonda, R., Shimizu, N., Kanari, M. and Kimura, M., *Phytochemistry*, **1990**, *29*, 1083.

[199] Gonda, R., Tomoda, M., Shimizu,N. and Kanari, M., *Chem. Pharm. Bull.*, **1990**, *38*, 482.

[200] Gonda, R., Tomoda, M., Takeda, K., Ohara, N. and Shimizzu, N., *Chem. Pharm. Bull.*, **1992**, *40*, 990.

[201] Gonda, R. and Tomoda, M., *Chem. Pharm. Bull.*, **1991**, *39*, 441.

[202] Gonda, R., Takeda, K., Shimizu, N. and Tomoda, M., *Chem. Pharm. Bull.*, **1992**, *40*, 185.

[203] Itokawa, H. and Takeya, K., *Heterocycles*, **1993**, *35*, 1467.

[204] Jurgens, T. M., Frazier, E. G., Schaeffer, J. M., Jones, T. E., Zink, D. L., Borris, R. P., Nanakorn, W., Beck, H. T. and Balick, M. J., *J. Nat. Prod.*, **1994**, *57*, 230.

[205] Muchowski, J. M., Greenhouse, R. J., Ackrell, J., Li, T. T. and Pfister, J. R., Eur. Pat. Appl. EP71,399, 09 Feb.1983.

[206] Firman, K., Kinoshita, T., Itai, A. and Sankawa, U., Phytochemistry, 1988, 27, 3887.

[207] Masuda, T., Jitoe, A. and Nakatani, N., Chem. Lett., 1991, 1625.

[208] Takano, I., Yasuda, I., Takeya, K. and Itokawa, H., Phytochemistry, 1995, 40, 1197.

[209] De A. Ribeiro, R., Fiuza De Melo, M. M. R., De Barros, F., Gomes, C. and Trolin, G., J. Ethnopharmacol. , 1986, 15, 261.

[210] Bhandary, M. J., Chandrashekar, K. R. and Kaveriappa, K. M., J. Ethnopharmacol., 1995, 47, 149.

[211] Leaman, D. J., Arnason, J. T., Yusul., R., Sangat-Roemantyo, H., Soedjito, H., Angerhofer, C. K. and Pezzuto, J. M., J. Ethnopharmacol., 1995, 49, 1.

[212] Nitta, A., Yakugaku Zasshi, 1984, 104, 256.

[213] Jain, S. P. and Puri, H. S., J. Ethnopharmacol., 1984, 12, 213.

[214] Sharma, S. C., Tandon, J. S., Uprety, H., Shukla, Y. N. and Dhar, M. M., Phytochemistry, 1975, 14, 1059.

[215] Saha, J. C., Savini, E. C. and Kasinathan, S., Indian J. Med. Res., 1961, 49, 130.

[216] De A. Ribeiro, R., Barros, F., Margarida, M., Melo, R. F., Muniz, C., Chiela, S., Wanderley, M. G., Gomes, C. and Trolin, G., J. Ethnopharmacol., 988, 24, 19.

[217] Sinha, A. K., Mehra, M. S., Sinha, G. K. and Pathak, R. C., Indian Perfum., 1979, 20, 25.

[218] Warren, K. S. and Peters, P. A., Nature (London), 1968, 217, 647.

[219] Saleh, M. M., Shabana, M. and Torki, M. A., Planta Med., 1982, 45, 138.

[220] Saleh, M. M., Kamal, R. M. and Abdulla, W. A., Planta Med., 1982, 45, 166.

[221] Abraham, Z., Bhakuni, S. D., Garg, H. S., Goel, A. K., Mehrotra, B. N. and Patnaik, G. K., Indian J. Exp. Biol., 1986, 24, 48.

[222] Kosuge, T., Yokota, M., Sugiyama, K., Yamamoto, T., Ni, M. Y. and Yan, S. C., Yakugaku Zasshi, 1985, 105, 791.

[223] Ray, P. G. and Majumdar, S. K., Econ. Bot., 1976, 30, 317.

[224] Misra, P., Pal, N. L., Guru, P. Y., Katiyar, J. C. and Tandon, J. S., Int. J. Pharmacog, 1991, 29, 19.

[225] Dhar, M. L., Dhar, M. N., Dhawan, B. N., Mehrotra, B. N., Srimal, R. C. and Tandon, J. S., Indian. J. Exp. Biol, 1973, 11, 43.

[226] Weyerstahl, P., Marschall, H., Schneider, S. and Scubba, G. C., Flavour Fragrance, J., 1995, 10, 179.

[227] Hagga, M. Y. and El-Shamy, A. M., Egypt J. Pharm, Sci., 1977, 18, 465.

[228] Dixit, V. K., Varma, K. C. and Vashisht., V. N., Indian J. Pharmacy., 1977, 39, 58.

[229] Nigam, M. C., Siddiqui, M. S., Misra, L. N. and Sen. T., Parfuem Kosmet, 1979, 60, 245; Chem. Abstr., 1979, 91, 181259p.

[230] Sinha, A. K. and Mehra, M. S., Indian Perfum., 1977, 22, 129.

[231] Bottini, A. T., Garfagnoli, D. J., Delgado, L. S., Dev. V., Duong, S. T., Kelley, C. G., Keyer, R., Raffel, R., Joshi, P. and Mathela, C. S., J. Nat. Prod., 1987, 50, 732.

[232] Zhao, K. Q., Hao, X. J., Chen, Y. Z. and Zou, C., Yunnan Zhiwu Yanjiu, 1995, 17, 201.

[233] William, C. A. and Harborne, J. B., Biochem. Syst. Ecol., 1977, 5, 221.

[234] Merp., P.S., Daniel, M. and Sabnis, S. D., *Curr. Sci*, **1986**, *55*, 835.

[235] Sharma, S. C., Shukla, Y. N. and Tandon, J. S., *Phytochemistry*, **1975**, *14*, 578.

[236] Sharma, S. C., Tandon, J. S., Uprety, H., Shukla, Y. N. and Dhar, M. M., *Phytochemistry*, **1975**, *14*, 1056.

[237] Sharma, S. C., Tandon, J. S. and Dhar, M. M., *Phytochemistry*, **1976**, *15*, 827.

[238] Sharma, S. C. and Tandon, J. S., *Indian J. Chem.*, **1983**, *22B*, 93.

[239] Itokawa, H., Morita, H., Katou, J., Takeya, K. Cavalheiro, A. J., De Oliveira, R. C. B, Ishige, M. and Motidome, M., *Planta Med.*, **1988**, *54*, 311.

[240] Itokawa, H., Takeya, K., Watanabe, K., Morita, H., Ichihara, Y., Totsuka, N., Izumi, H., Satake, M., Yasuda, I, Sankuwa, U., Motidome, M., Flores, F. A. *J. Pharmacobio. Dyn.*, **1992**, *15*, S-2.

[241] Itokawa, H., Morit, H., Takeya, K. and Motidome, M., *Chem. Pharm. Bull.*, **1988**, *36*, 2682.

[242] Singh, S., Gray, A. I., Skelton, B. W., Waterman, P. G. and White, A. H., *Aust. J. Chem.*, **1991**, *44*, 178.

[243] Nakatani, B., Kikuzaki, H., Yamaji, H., Yoshio, K., Kitora, C., Okada, K. and Padolina, W. G., *Phytochemistry*, **1994**, *37*, 1383.

[244] Hirschhorn, H. H., *J. Ethnopharmacol.*, **1983**, *7*, 123.

[245] Madulid, D. A., Gaerlan, E. J. M., Romero, E. M. and Agoo, E. M. G., *Acta Manilana*, **1989**, *38*, 25.

[246] Singh, V. K. and Ali, Z. A., *Fitoterapia*, **1994**, *65*, 68.

[247] Kaleysa Raj, R., *Indian J. Physiol. Pharmacol.*, **1975**, *19*, 47.

[248] Noro, T., Miyase, T., Kuroyanagi, M., Ueno, A. and Fukushima, S., *Chem. Pharm. Bull.*, **1983**, *31*, 2708.

[249] Kiuchi, F., Hioki, M., Nakamura, N., Miyashita, N., Tsuda, Y. and Kondo, K., *Shoyakugaku Zasshi*, **1989**, *43*, 288.

[250] Ali, M. A., Mikage, M., Kiuchi, F., Tsuda, Y. and Kondo, K., *Shoyakugaku Zasshi*, **1991**, *45*, 206.

[251] Kosuge, T., Yokota, M., Sukiyama, K., Saito, M., Iwata, Y., Nakura, M. and Yamamoto, T., *Chem. Pharm. Bull.*, **1985**, *33*, 5565.

[252] Murakami, A., Kondo, A., Nakamura, Y., Ohigashi, H. and Koshimizu, K., *Biosci. Biotech. Biochem.*, **1993**, *57*, 1971.

[253] Lam, L. K.T. and Zheng, B. L., *J. Agr. Food Chem.*, **1991**, *39*, 660.

[254] Pai, B. R., Rao, N. N. and Wariya, N. S., *Indian J. Chem.*, **1970**, *8*, 468.

[255] Pancharoen, O., Tuntiwachwuttikul, P. and Taylor, W. C., *Phytochemistry*, **1996**, *43*, 305.

[256] Tuntiwachwuttikul, P., Pancharoen, O., Bubb, W. A., Hambly, T. W., Taylor, W. C. and Reutrakul, V., *Aust. J. Chem.*, **1987**, *40*, 2049.

[257] Pancharoen, O., Tuntiwachwuttikul, P. and Taylor, W. C., *Phytochemistry*, **1989**, *28*, 1143.

[258] Jolad, S. D., Hoffmann, J. J., Schram, K. H., Cole, J. R., Tempesta, M. S. and Bates, R. B., *J. Org. Chem.*, **1981**, *46*, 4267.

[259] Hollands, R., Becher, D., Gaudemer, A. and Polonsky, J., *Tetrahedron*, **1968**, *24*, 1633.

[260] Holbert, G. W., Ganem, B., Borsub, L., Chantrapromma, K., Van Engen, D., Clardy, J., Sadavongvivad, C. and Thebtaranonth, Y., *Tetrahedron Lett.*, **1979**, *20*, 715.

[261] Schutte, G. R., Ganem, B., Chantrapromma, K., Kodpinid, M. and Sudsuansri, K., *Tetrahedron Lett.*, **1982**, *23*, 289.

[262] Schutte, G. R., Kodpinid, M., Thebtaranonth, C. and Thebtaranonth, Y., *Tetrahedron Lett.*, **1982**, *23*, 4303.

[263] Kodpinid, M., Sadavongvivad, C., Thebtaranonth, C. and Thebtaranonth, Y., *Tetrahedron Lett.*, **1983**, *24*, 2019.

[264] Kupchan, S. M., Hemingway, R. J. and Smith, R. M., *J. Org. Chem.*, **1969**, *34*, 3898.

[265] Singh, J., Dhar, K. L. and Atal, C. K., *Tetrahedron*, **1970**, *26*, 4403.

[266] Nugroho, B. W., Schwarz, B., Wray, V. and Proksch, P., *Phytochemistry*, **1996**, *41*, 129.

[267] Jaipetch, T., Reutrakul, V., Tuntiwachwuttikul, P. and Santisuk, T., *Phytochemistry*, **1983**, *22*, 625.

[268] Herunsalee, A., Pancharoen, O. and Tuntiwachwuttikul, P., *J. Sci. Soc. Thailand*, **1987**, *13*, 119.

[269] Panthong, A., Tassaneeyakul, W., Kanjanapothi, D., Tuntiwachwuttikul, P. and Reutrakul, V., *Planta Medica*, **1989**, *55*, 133.

[270] Prawat, U., Tuntiwachwuttikul, P., Taylor, W. C., Engelhardt, L. M., Skelton, B. W. and White, A. H., *Phytochemistry*, **1993**, *32*, 991.

[271] Tuchinda, P., Udchachon, J., Reutrakul, V., Santisuk, T., Skelton, B. W., White, A. H. and Taylor, W. C., *Phytochemistry*, **1994**, *36*, 731.

[272] Pongyrayoon, U., Sematong, T., Tuchinda, P., Claeson, P., Reutrakul, V. and Nahar, N., *Phytother. Res.*, **1996**, *10*, 534.

[273] Kiuchi, F., Nakamura, N., Tsuda, Y., Kondo, K. and Yoshimura, H., *Chem. Pharm. Bull.*, **1988**, *36*, 412.

[274] Kiuchi, F., Nakamura, N. and Tsuda, Y., *Phytochemistry*, **1987**, *26*, 3350.

[275] Dechatiwongse, T. and Yoshihira, K., *Bull. Med. Sci. Thailand*, **1973**, *15*, 1.

[276] Tuchinda, M., Srimaruta, N., Habanananda, P., Kanchanapee, P. and Dechatiwongse, T., *Siriraj Hospital Gaz., Thailand*, **1984**, *36*, 1.

[277] Wasuwat, S., Soonthron Sarathoon, P., Boonkong, P., Rojanapkodi, W., Phicha, P., Nandhasri, P., Sankamnerd, D., Chamchaang, W. and Jenbhanit, K., *The 10th Conference of Science and Technology of Thailand*, Chiang Mai University, Chiang Mai, Thailand, **1984**, A79.

[278] Sukontasan, A., *Master of Science Thesis*, Mahidol University, Bangkok, Thailand, **1986**.

[279] Casey, T. E., Dougan, J., Matthews, W. S.. and Nabney, J., *Trop. Sci.*, **1971**, *13*, 199; *Chem. Abstr.*, **1972**, *76*, 14474s.

[280] Amatayakul, T., Cannon, J. R., Dampawan, P., Dechatiwongse, T., Giles, R. G. F., Huntrakul, C., Kusamran, K., Mokkhasamit, M., Raston, C. L., Reutrakul, V. and White, A. H., *Aust. J. Chem.*, **1979**, *32*, 71.

[281] Tuntiwachwuttikul, P., Limchawfar, B., Reutrakul, V., Pancharoen, O., Kusamran, K. and Byrne, L. T., *Aust. J. Chem.*, **1980**, *33*, 913.

[282] Tuntiwachwuttikul, P., Pancharoen, O., Jaipetch, T. and Reutrakul, V., *Phytochemistry*, **1981**, *20*, 1164.

[283] Kiatiyingunsulee, N., Wangmad, M., Sawasdimonkol, K. and Mokkhasamith, M., *Bull. Med. Sci. Thailand*, **1979**, *21*, 13.

[284] Srisawatdi, P., *Master of Science Thesis*, Chiang Mai University, Chiang Mai, Thailand, **1987**.

[285] Kanjanapothi, D., Soparat, P., Panthong, A, Tuntiwachwuttikul, P. and Reutrakul, V., *Planta Med.*, **1987**, *53*, 329.

[286] Niwatananun, W., *Master of Science Thesis*, Chiang Mai University, Chiang Mai, Thailand, **1989**.

[287] Kuroyanagi, M., Fukushima, S., Yoshihira, K., Natori, S., Dechatiwongse, T., Mihashi, K., Nishi, M. and Hara, S., *Chem. Pharm. Bull.*, **1980**, *28*, 2948.
[288] Jitoe, A., Masuda, T. and Nakatani, N., *Phytochemistry*, **1993**, *32*, 357.
[289] Masuda, T., Jitoe, A. and Nakatani, N., *Chem. Lett.*, **1993**, 189.
[290] Masuda, T. and Jitoe, A., *Phytochemistry*, **1995**, *39*, 459.
[291] Herout, V., Benesova, V. and Pliva, J., *Coll. Czech. Chem. Commun.*, **1953**, *18*, 248.
[292] Nigam, M. C. Nigam, I. C., Levi, V. and Handa, K. L., *Can. J. Chem.*, **1964**, *42*, 2610.
[293] Connell, D. W., *Flavour Ind.*, **1970**, *1*, 677; *Chem. Abstr.*, **1971**, *74*, 34534v.
[294] Connell, D. W. and Sutherland, M. D., *Aust. J. Chem.*, **1966**, *19*, 283.
[295] Kami, T., Nakayama, M. and Hayashi, S., *Phytochemistry*, **1972**, *11*, 3377.
[296] Smith, R. M. and Robinson, J. M., *Phytochemistry*, **1981**, *20*, 203.
[297] Macleod, A. J. and Pieris, N. M., *Phytochemistry*, **1984**, *23*, 353.
[298] Sakamura, F., *Phytochemistry*, **1987**, *26*, 2207.
[299] Connell, D. W. and Sutherland, M. D., *Aust. J. Chem.*, **1969**, *22*, 1033.
[300] Connell, D. W. and Mclachlan, R., *J. Chromatography*, **1972**, *67*, 29.
[301] Harvey, D. J., *J. Chromatography*, **1981**, *212*, 75.
[302] Kikuzaki, H., Tsai, S.-M. and Nakatani, N., *Phytochemistry*, **1992**, *31*, 1786.
[303] Denniff, P. and Whiting, D. A., *J. Chem. Soc. Chem. Comm.*, **1976**, 711.
[304] Macleod, I. and Whiting, D. A., *J. Chem. Soc. Chem. Comm.*, **1979**, 1152.
[305] Denniff, P., Macleod, I. and Whiting, D. A., *J. Chem. Soc. Perkin I*, **1980**, 2637.
[306] Kiuchi, F., Shibuya, M. and Sankawa, U., *Chem. Pharm. Bull.*, **1982**, *30*, 754.
[307] Yamahara, J., Miki, K., Chisaka, T., Sawada, T., Fujimura, H., Tomimatsu, T., Nakano, K. and Nohara, T., *J. Ethnopharmacol.*, **1985**, *13*, 217.
[308] Yamahara, J., Mochizuki, M., Rong, H. Q., Matsuda, H. and Fujimura, H., *J. Ethnopharmacol.*, **1988**, *23*, 299.
[309] Yamahara, J., Huang, Q., Li, Y., Xu, L. and Fujimura, H., *Chem. Pharm. Bull.*, **1990**, *38*, 430.
[310] Adewunmi, C. O., Oguntimein, B. O. and Furu, P., *Planta Med.*, **1990**, *56*, 347.
[311] Hikino, H., Kiso, Y., Kato, N., Hamada, Y., Shioiri, T., Aiyama, R., Itokawa, H., Kiuchi, F. and Sankawa, U., *J. Ethnopharmacol.*, **1985**, *14*, 31.
[312] Kiuchi, F., Iwakami, S., Shibuya, M., Hanaoka, F. and Sankawa, U., *Chem. Pharm. Bull.*, **1992**, *40*, 387.
[313] Endo, K., Kanno, E. and Oshima, Y., *Phytochemistry*, **1990**, *29*, 797.
[314] Kikuzaki, H., Usuguchi, J. and Nakatani, N., *Chem. Pharm. Bull.*, **1991**, *39*, 120.
[315] Kikuzaki, H., Kobayashi, M. and Nakatani, N., *Phytochemistry*, **1991**, *30*, 3647.
[316] Kano, Y., Tanabe, M. and Yasuda, M., *Shoyakugaku Zasshi*, **1990**, *44*, 55; *Chem. Abstr.*, **1991**, *114*, 12041z.
[317] Tanabe, M., Chen, Y.-D., Saito, K. and Kano, Y., *Chem. Pharm. Bull.*, **1993**, *41*, 710.
[318] Yoshikawa, M., Hatakeyama, S., Toniguchi, K., Matuda, H. and Yamahara, J., *Chem. Pharm. Bull.*, **1992**, *40*, 2239.
[319] Kikuzaki, H. and Nakatani, N., *Phytochemistry*, **1996**, *43*, 273.
[320] Parihar, D. B. and Dutt, S., *Indian Soap J.*, **1950**, *16*, 123, 154.
[321] Balakrishnan, V. K., Razdan, R. K. and Bhattacharyya, S. C., *Perfum. Essent. Oil Rec.*, **1956**, *47*, 274.

[322] Varier, N. S., *Proc. Indian Acad. Sci.*, **1944**, *20A*, 257; *Chem. Abstr.*, **1945**, *39*, 3331[1].

[323] Dev, S., *Tetrahedron*, **1960**, *8*, 171.

[324] Hall, S. R., Nimgirawath, S., Raston, C. L., Sittatrakul, A., Thadaniti, S., Thirasasana, N. and White, A. H., *Aust. J. Chem.*, **1981**, *34*, 2243.

[325] Hartsuck, J. A. and Paul, I. C., *Chem. Ind. (London)*, **1964**, 977.

[326] McPail, A. T. and Sim, G. A., *J. Chem. Soc. B.*, **1966**, 112.

[327] Damodaran, N. P. and Dev, S., *Tetrahedron*, **1968**, *24*, 4113.

[328] Matthes, H. W. D., Luu, B. and Ourisson, G., *Phytochemistry*, **1980**, *19*, 2643.

[329] Masuda, t., Jitoe, A., Kato, S. and Nakatani, N., *Phytochemistry*, **1991**, *30*, 2391.

SUBJECT INDEX

899

902